Reinforced Concrete Design
A Practical Approach
Second Edition

Svetlana Brzev
Ph.D., P.Eng.
British Columbia
Institute of Technology

John Pao
M.Eng., P.Eng., Struct. Eng., S.E.
Bogdonov Pao Associates Ltd.

Taken from:
Reinforced Concrete Design: A Practical Approach
by Svetlana Brzev and John Pao

Pearson Learning Solutions, 501 Boylston Street, Suite 900, Boston, MA 02116
A Pearson Education Company
www.pearsoned.com

Printed in Canada

1 2 3 4 5 6 7 8 9 10 V011 17 16 15 14 13 12

0002000102710803107

BK/CM

ISBN 10: 1-256-87384-5
ISBN 13: 978-1-256-87384-6

Table of Contents

About the Authors

Svetlana Brzev, Ph.D., P.Eng., is faculty in the Department of Civil Engineering of the British Columbia Institute of Technology (BCIT) in Vancouver, Canada. Dr. Brzev received her B.Eng. and M.A.Sc. degrees in Civil/Structural Engineering from the University of Belgrade, Serbia in 1894 and 1989 respectively, and a Ph.D. degree in Earthquake Engineering from the Indian Institute of Technology, Roorkee, India in 1994. She has over 25 years of combined teaching, research, and consulting experience related to structural and seismic design and rehabilitation of reinforced concrete and masonry buildings, municipal and industrial facilities. She is registered professional engineer in British Columbia. Her keen interest in teaching and sharing professional experience motivated her to join BCIT in 2000. She has developed and taught courses related to structural and seismic design of concrete and masonry structures. Her passion for teaching was recognized in 2011 through a BCIT teaching excellence award. Dr. Brzev is a member of two CSA Technical Committees responsible for developing national standards for masonry design and construction. She has co-authored a guide on seismic design of masonry structures and more than 80 papers and reports. She is actively involved in major international initiatives related to earthquake risk reduction, and has co-authored many publications on seismic safety of concrete and masonry buildings that have been translated in several languages. She is a member of the American Concrete Institute and the Masonry Society, and served as a Director and Vice-President of the Earthquake Engineering Research Institute.

John Pao, M.Eng., P.Eng., Struct.Eng., S.E., is the President of Bogdonov Pao Associates Ltd. of Vancouver, Canada, and BPA Group of Companies with offices in Seattle and Los Angeles. Mr. Pao received his B.A.Sc. and M.Eng. degrees in Civil Engineering from the University of British Columbia in 1980 and 1984 respectively. Mr. Pao has more than 30 years of experience related to structural design of high-rise residential and office buildings, shopping centers, sport arenas, parking garages, and institutional buildings. Mr. Pao's design projects involve a variety of structural materials, including reinforced and post-tensioned concrete, structural steel, wood, and masonry. His creative and innovative design approaches have won accolades from clients and colleagues from Canada and the United States. He is a licensed engineer in over twenty jurisdictions in the United States and Canada. His passions for structural design and engineering education have inspired him to give back to the profession and co-author this textbook. Mr. Pao has taught reinforced concrete design courses at various institutions since the mid-1980s. Along with a few colleagues, he pioneered the Certificate in Structural Engineering Program, a unique continuing education program for practicing structural engineers in British Columbia. He has served as the chair of the Organizing Committee for the program since its inception in 2001. The program has been very successful and has attracted students from across Canada and abroad. Mr. Pao's knowledge and teaching experience related to design of reinforced concrete structures are reflected in this textbook for benefit of engineering students and practising engineers.

Note to the Reader

While the authors have tried to be as accurate as possible, they cannot be held responsible for the designs of others that might be based on the material presented in this book. The material included in this book is intended for the use of design professionals who are competent to evaluate the significance and limitations of its contents and recommendations and able to accept responsibility for its application. The authors disclaim any and all responsibility for the applications of the stated principles and for the accuracy of any of the material included in the book. Software provided with the book was checked for accuracy, however there are no warranties or guarantees regarding its suitability or results for any specific application.

Preface

The authors are pleased to announce the second edition of the *Reinforced Concrete Design: A Practical Approach*, the only Canadian textbook that covers the design of reinforced concrete structural members in accordance with the National Building Code of Canada 2010 and CSA Standard A23.3-04 Design of Concrete Structures and its 2005, 2007, and 2009 amendments. This edition contains a new chapter on the design of two-way slabs and numerous revisions of the original manuscript. Design of two-way slabs is a challenging topic for engineering students and young engineers. The authors have made an effort to give a practical design perspective to this topic, and have focused on analysis and design approaches that are widely used in structural engineering practice. The book covers the key topics contained in the curriculum of undergraduate reinforced concrete design courses, and it is intended to be a useful learning resource for the students and a practical reference for design engineers. Since its original release in 2005 the book has been well received by readers from Canadian colleges, universities, and design offices. The authors have been commended for a simple and practical approach to the subject by students and course instructors who have adopted the book.

APPROACH

The authors believe that the structural designer should thoroughly understand the fundamental concepts before using them in a design. However, a good structural design should strike a balance between sound design concepts and feasibility from a practical field implementation perspective. Therefore, this book highlights the important aspects of reinforced concrete design from the point of view of all important stakeholders, including the owner, architect, contractor, rebar installer, and supplier, and discusses their role in the design and construction of a particular structure.

The material in this book is presented in the logical sequence in which a structural design would be performed in practice. The topics are covered at different levels of complexity. The book takes a non–calculus-based, practical approach to the analysis and design of reinforced concrete members rather than a rigorous theoretical approach. Modern analysis and design procedures consistent with design practice have been used. The book contains many numerical examples solved in a step-by-step format. Metric (SI) units have been used throughout the book.

The book is a collaborative effort between an academic and a practising engineer and reflects their unique perspectives on the subject. It is recommended for college and university students taking reinforced concrete design courses in the areas of civil engineering and architecture; however, it may also be used by practising professionals interested in reviewing familiar topics or gaining insight into new topics. The authors have attempted to help the reader understand the design of reinforced concrete buildings and their components in an efficient and accurate manner. The practical recommendations related to the design and construction of reinforced concrete components offered in the book are largely based on the authors' own experiences. It is the authors' hope that this book will provide practical insight into reinforced concrete design early on in a designer's professional life.

CONTENTS

This book covers the design of the main reinforced concrete structural members in accordance with the limit states design method and is based on the CSA Standard A23.3-04 *Design of Concrete Structures*. The load provisions are consistent with the *National Building Code of Canada 2010* (NBC 2010). The book is restricted to the design of cast-in-place concrete structures.

Chapter 1 presents an overview of design and construction aspects as related to reinforced concrete structures and introduces the main structural systems and components. Chapter 2 provides an overview of concrete and steel material properties used in the design-related chapters of the book.

The behaviour of reinforced concrete beams (including rectangular and T-beam sections) and one-way slabs in flexure is explained in Chapter 3, whereas the design of these structures is covered in Chapter 5. Serviceability criteria (covered in Chapter 4) and strength requirements need to be considered while designing flexural members in Chapter 5. Chapters 6 and 7 discuss the design of reinforced concrete beams and one-way slabs for shear and torsion, respectively. The main focus of these chapters is on simply supported reinforced concrete beams and slabs.

Chapter 8 discusses the behaviour and design of reinforced concrete columns subjected to axial load and flexure; the main focus is short columns, although an overview of the design procedures for slender columns is also provided. The critical issues related to bond and anchorage of reinforced concrete structures are discussed in Chapter 9.

Chapters 10 and 11 focus on continuous concrete construction. Chapter 10 introduces key concepts related to the behaviour of continuous beams and slabs as well as the analysis methods that are used in the design of these structures. Chapter 11 builds on the background of Chapter 10 and exposes the reader to the design and detail of continuous structures. An overview of various floor systems employing continuous concrete construction is presented in this chapter, along with a case study that demonstrates the design of a floor system.

Relevant approaches and considerations related to the analysis and design of two-way slabs have been covered in Chapter 12. Chapter 13 provides an overview of various types of reinforced concrete walls, including bearing walls, basement walls, and shear walls. Finally, Chapter 14 covers the design of shallow foundation systems, including spread and strip footings, eccentrically loaded footings, and combined footings.

Appendix A includes design aids, while Appendix B includes notation for commonly used symbols. A list of key references is included at the end of the book. This list, while not exhaustive, is useful as initial supplementary reading for those interested in pursuing the subject further. A number of references included in the book were published by the American Concrete Institute (ACI). The students can join ACI through a free student membership program and access these publications online at www.concrete.org.

FEATURES

This text includes the following features:

- **Learning Objectives** provide an idea of what the reader will be able to accomplish after reading a particular chapter.
- **Key boxes** summarize important concepts, recommendations, and equations within a chapter.
- **Did You Know? boxes** contain interesting facts about reinforced concrete design and construction.
- **Examples** provide practical applications of theoretical material.
- **Learning from Examples boxes** point out the important implications of one or more examples.
- **Checklists** outline design procedures in a tabular form for easy reference.
- A **Summary** at the end of each chapter provides a review of the key topics covered, organized by the learning objectives.
- **Problems** at the end of each chapter provide an opportunity for readers to practise the concepts learned in the chapter.

SUPPLEMENTS

The book includes a few useful supplements for the readers, including the column design software BPA COLUMN and two spreadsheets related to foundation design and column load take down. A few Power Point presentations showcasing reinforced concrete structures under construction and in completed form are also included. These supplements are posted on the book web site.

Instructors may access an additional web site, which contains electronic version of the Instructor's Solution Manual with complete solutions to the end-of-chapter problems, and Power Point presentations including illustrations from the book.

REVISIONS IN THIS EDITION

This edition includes new content and several revisions to the original edition of the text. Comprehensive revisions were made to Chapter 4 to reflect the changes contained in the 2009 amendment to CSA A23.3-04. Chapters 6 and 7 have been revised to correct an oversight related to the transverse reinforcement spacing requirements (Cl.11.3.8.3) in the previous edition of the book. Chapter 8 includes a new example on slender columns and a few problems. Several errors and omissions (both text and illustrations) have also been corrected. More than 300 pages of the original book have been revised in this edition.

ACKNOWLEDGMENTS

The authors would like to acknowledge several colleagues and friends who have helped with the development of this book by providing invaluable input. The authors acknowledge valuable input by the reviewers of the second edition of the book: Scott Alexander (AECOM), Sreekanta Das (University of Windsor), Bryan Gallagher (Stantec), Reza Kianoush (Ryerson University), Jonathan Klop (Quantum Engineering), Solomon Tesfamariam (University of British Columbia), and Andrew Vizer (Cement Association of Canada). The authors are thankful for information and resources provided by Richard McGrath of the Cement Association Canada. The assistance provided by Pamela Barbosa and Adam Lubell of Read Jones Christoffersen Ltd, and Dmitry Itskovich of Axiom Builders is gratefully acknowledged. The authors gratefully acknowledge the contributions made by the following colleagues who have reviewed the first edition of the book: Martin Bollo, Bryan Folz, and Patrick Stewart, all of British Columbia Institute of Technology (BCIT), and Professor C.V.R. Murty of the Indian Institute of Technology, Chennai, India. Suggestions and review comments by Graham Finch and Neal Damgaard, graduates of the BCIT Civil Engineering Program, were invaluable in providing a student perspective on this work.

The authors gratefully acknowledge Greg Beaveridge for providing overall technical editing and review of the book and the Instructor's Manual. The authors and publisher would also like to thank the following reviewers for their valuable feedback regarding the first edition of this book: Yanglin Gong (Lakehead University), Amgad Hussein (Memorial University of Newfoundland), Nino Sirianni (St. Clair College), Jacqueline Vera (Southern Alberta Institute of Technology), and Lionel Wolpert (Humber College).

The authors are especially grateful for outstanding contribution made by graphic artists Denise and Bernice Pao, twin sisters who prepared illustrations both for the first and the second edition of the book. Their patience and dedication are highly appreciated. The second edition of the book was edited by Isaak Olian, and his contribution is gratefully acknowledged.

The authors greatly appreciate the support of Bogdonov Pao Associates Ltd. in the development of this book. The authors appreciate assistance provided by Weihan Timmie. Special thanks are due to Lisa Xue for assisting with developing solutions for several examples in the second edition of the book, and preparing the instructors manual. The assistance provided by Emma Kirkham and Anmar Saman is gratefully acknowledged. The authors acknowledge Catharine Martin for typing the solutions for the first edition of the Instructor's Manual.

The authors are grateful to Andreas Felber of EDI for developing the column design software, and to Garry Kirkham for developing the foundation design spreadsheet. The authors acknowledge Liliana Dapcevic for assisting with developing solutions for several examples in the book.

The following individuals assisted the review and/or collection of the material for the presentations posted on the Web site: Meiric Preece, Martin Bollo, A.P. Sukumar, Paul Lam, Rasko P. Ojdrovic, John H. Thomsen, Vanja Alendar, Predrag Stefanovic, Bill McEwen, Dave Nedelec, Kosta Marcakis, Maria Ofelia Moroni, Matej Fischinger, Dusan Radojevic, Sasa Popovic, Sudhir Jain, Keith Griffiths, Mark Van Bockhoven, V.R. Kulkarni, Luis Gonzalo Mejia, Mohammed Farsi, Maria Bostenaru, and Lilly Wang. Special thanks are due to Marjorie Greene, Martin Bollo, and Bill McEwen for reviewing the draft presentations, and to Marija Brzev for her assistance in developing the presentations.

The authors are grateful to Leslie Carson and Brad Keist of Pearson Learning Solutions for their support during the development of the current edition. The authors are indebted to the following staff of Pearson Education Canada for their support related to the first edition: Angela Kurmey, Kelly Torrance, Judith Scott, and Marisa D'Andrea.

Svetlana Brzev developed a portion of the manuscript during the Professional Development (PD) Leave granted by the BCIT in the fall of 2003. The support provided by the BCIT Department of Civil Engineering, School of Construction and the Environment, and the PD Leave Committee is gratefully acknowledged.

John Pao wishes to acknowledge the late Dr. Noel Nathan, Professor in the Department of Civil Engineering at the University of British Columbia, Vancouver. Many years of wisdom and practical advice contained in Dr. Nathan's reinforced concrete design lectures inspired many engineering students. This book reflects some of Dr. Nathan's teaching philosophy and approach.

The authors are grateful to Dr. Reza Joghataie (Sharif University of Technology, Tehran) and Jacqueline Gaudet of BCIT for the suggested revisions to the first printing of the book.

Last, but not least, the development of this book would not have been possible without the continuous support and encouragement of the authors' families.

Svetlana Brzev, PhD, PEng
John Pao, MEng, PEng, Struct.Eng

CREDITS

The authors are grateful to the Cement Association of Canada, Portland Cement Association, American Concrete Institute, the American Society of Civil Engineers, International Code Council, Inc., and Pearson Education Inc. for their permission to reproduce the copyright material presented in this book.

The authors gratefully acknowledge the following individuals and organizations who have kindly given permission to reproduce photographs in the book and the companion Web sites: Ron Hopen, Gustavo Parra-Montesinos, Getty Images, Nebojsa Ojdrovic, Stavroula J. Pantazopoulou, Graham Finch, Greater Vancouver Regional District, Bogue Babicki Associates Inc., Buckland and Taylor Ltd., Sportsco International, L.P., Strait Crossing Bridge Ltd., Canada Lands Company CLC Ltd., Masonry Institute of BC, Paul Thurston, Marita Luk, Thomas Abbuhl, Weiler Smith Bowers Consulting Engineers, Rapid Transit Project 2000 Ltd., Lafarge North America Inc., Ledcor Construction Ltd., Giancarlo Garofalo Architect, Montreal Olympic Park, Read Jones Christoffersen Ltd., Canadian Society for Civil Engineering, Michael Sherman Photography, Portland Cement Association, Simpson Gumpertz & Heger Inc., Predrag Stefanovic, DVBU-DSFB-SNO, Crom Corporation, Instituto del Cemento y Hormigón de Chile, Judith Siess, Aga Khan Award for Architecture, Earthquake Engineering Research Institute, Associated Cement

Companies Ltd. India, Alpa Sheth, Tandon Consultants Pvt Ltd., Vanja Alendar, Matej Fischinger, Anna F. Lang, PERI GmbH, Grand Hyatt Shanghai, Chitr Lilavivat, SKYCITY Auckland Ltd., and the Milwaukee Art Museum.

With the permission of Canadian Standards Association (CSA), material is reproduced from CSA Standard, A23.1-04/A23.2-04, Concrete Materials and Methods of Concrete Construction/Methods of Test and Standard Practices for Concrete, and A23.3-04, Design of Concrete Structures, which is copyrighted by Canadian Standards Association, Toronto, Ontario. While the use of this material has been authorized, CSA shall not be responsible for the manner in which the information is presented, nor for any interpretations thereof. CSA Standards are available by contacting CSA.

1 Concrete Basics

LEARNING OUTCOMES

After reading this chapter, you should be able to

- explain the basic concepts of reinforced and prestressed concrete
- describe the main components of reinforced concrete buildings
- explain the three structural systems typical of reinforced concrete building construction in Canada
- review the concepts of gravity and lateral load paths
- outline the four key considerations related to the design of reinforced concrete structures
- describe the five major steps in the design process
- describe the four basic methods of concrete construction
- review the types of loads considered in the design of reinforced concrete structures
- outline the four key categories of limit states considered in the design of reinforced concrete structures

1.1 INTRODUCTION

Concrete is one of the most versatile construction materials, offering potentially unlimited opportunities for developing diverse forms of construction. Concrete is what is known as a universal material, as its ingredients, namely cement, sand, aggregates, and water, are available all over the globe. Furthermore, concrete structures can be built with all different levels of technology, ranging from the simplest hand tools to computerized equipment. Concrete also has the excellent characteristics of fire resistance and durability and requires substantially less maintenance than other materials. Its mechanical properties can be enhanced by the use of steel reinforcement. Reinforced concrete construction makes use of the high compressive strength of concrete and the high tensile strength of steel reinforcement.

This chapter presents an overview of the basic aspects of concrete design and construction. The basic concepts of reinforced and prestressed concrete are explained in Section 1.2. The main components of a concrete building and the common structural systems found in the design practice are outlined in Section 1.3. The key considerations related to the design of reinforced concrete structures are discussed in Section 1.4. Construction methods and processes related to reinforced concrete structures are outlined in Section 1.5. Canadian design codes and standards related to reinforced concrete design are outlined in Section 1.6, while design loads are discussed in Section 1.7. The limit states design method used to design reinforced concrete structures in Canada is explained in general terms in Section 1.8. The accuracy of design calculations is discussed in Section 1.9. Finally, the use of computer-aided design tools is discussed in Section 1.10.

1

1.2 | CONCRETE AS A BUILDING MATERIAL

1.2.1 A Historical Overview

The earliest concrete applications date back to the ancient Greeks and Romans. The word *concrete* derives from the Latin verb *concrescere*, "to grow together" or "to harden." To make concrete, the Romans used a special type of volcanic sand called pozzuolana, first found near Pozzuoli in the bay of Naples, Italy. This substance was found to react with lime and water to solidify into a rock-like mass. In the period from 300 BC to 476 AD, this early form of concrete was used in bridges, aqueducts, and several renowned structures such as the Colosseum and the Pantheon in Rome. However, these early developments in concrete construction were discontinued for many centuries thereafter, mainly due to the lack of availability of cementitious materials such as pozzuolana in other parts of the world.

The invention of Portland cement in 1824 by Joseph Aspdin, an English stonemason, represented a major breakthrough in the development of modern concrete materials. The name "Portland" came from the high-quality building stones quarried at Portland, England.

The concept of reinforced concrete was first introduced in 1855 by J. Lambot of France, who made a rowboat out of a wire framework coated on both sides with cement mortar and exhibited it at the Paris exposition in 1855. Also in France, François Coignet proposed the use of concrete with steel reinforcement for the construction of roofs, pipes, and shells in 1861. However, the French gardener Joseph Monier is often given credit for being the inventor of reinforced concrete. In 1867, Monier patented his invention of flowerpots made of wire mesh plastered on both sides with cement mortar. He subsequently patented the use of reinforced concrete for reservoirs, pipes, slabs, and shells.

Nevertheless, reinforced concrete construction only became established in terms of practical applications after two German construction firms acquired Monier's patent in 1884. The first textbook on reinforced concrete, called *The Monier System*, by Gustav Wayss and Mathias Koenen, was published in Germany in 1887. For the first time this book proposed that steel reinforcement should be provided in concrete structures where tensile stresses can occur. The French contractor François Henebrique introduced stirrups in concrete beams and suggested the use of T-beams at the end of the 19th century. Henebrique made a major contribution to the application of reinforced concrete technology in Europe.

The development of modern concrete technology in North America began in the second half of the 19th century. Ernest L. Ransome, who patented the use of twisted square steel rods for an improved bond in reinforced concrete, was the creator of several reinforced concrete buildings in the United States between 1880 and 1900. The first concrete high-rise, the 16-storey Ingalls Building in Cincinnati, was built in 1902 using the Ransome system. By that time, the basic theoretical principles of reinforced concrete design were generally well understood. Since then, reinforced concrete has been successfully used in all types of designs, from simple building frames to complex industrial buildings, bridges, and skyscrapers.

The development of prestressed concrete began in the 1930s with the pioneering work of French engineer Eugene Freyssinet, who was the first to use pretensioned high-tensile steel bars as reinforcement. Freyssinet is called "the father of prestressing."

The development of modern concrete construction in Canada started at the beginning of the 20th century. The first reinforced concrete bridge in Canada was built in Massey, Ontario, in 1906. Since then, concrete has been used for the construction of many bridges, buildings, dams, and other significant structures. Most notably, concrete was used for the construction of Toronto's CN Tower in 1976. This 550 m tall broadcast and observation tower is the world's tallest free-standing structure. Another more recent accomplishment with this material is the 13 km long Confederation Bridge linking Prince Edward Island and New Brunswick completed in 1997.

There are many notable concrete buildings in Canada; in fact, a number of Canada's historical landmark buildings are reinforced concrete structures. For example, a portion of the Fairmont Empress Hotel in Victoria, British Columbia, a national heritage building, was built around 1910 using reinforced concrete columns and composite concrete and masonry floors. A few other reinforced concrete buildings were built at the same period in other Canadian cities (Vancouver, Toronto, etc.).

Selected facts and details related to the design and construction of notable reinforced concrete structures in Canada and abroad will be referred to in this book and the companion Web site.

DID YOU KNOW?

The Scotia Tower is a 68-storey concrete high-rise built in Toronto in 1987 (see Figure 1.1). The main structural feature of this high-rise is the exterior tube that provides the strength and rigidity in the building and carries 40% of its weight (including the contents). The use of self-elevating jump forms enabled rapid construction (less than 3 days per floor). The construction was performed using an off-site custom-made concrete pump that was able to pump concrete to a height of 275 m for the first time in Canada. At the time of its construction, the Scotia Tower was the world's tallest reinforced concrete tube structure (RSIO, 1988).

Figure 1.1 Scotia Tower under construction, showing self-elevating jump forms at the exterior.

(Portland Cement Association)

1.2.2 Concrete

Concrete is artificial stone made from two main components: cement paste and aggregates. Aggregates usually consist of natural sand and gravel or crushed stone. The paste hardens as a result of the chemical reaction between cement and water and glues the aggregates into a rock-like mass. The concrete-manufacturing process is quite complex and includes a number of steps, such as proportioning, batching, mixing, placing, compacting, finishing, and curing. The versatility of concrete construction, in terms of forms and shapes, is due to the fact that fresh concrete has no form of its own until it hardens. Concrete has a high compressive strength; however, its tensile strength is low (only about 10% of its compressive strength). For this reason, concrete is rarely used without the addition of steel reinforcement.

1.2.3 Reinforced Concrete

Reinforced concrete structures utilize the best qualities of concrete and steel — concrete's high compressive strength and steel's high tensile strength. The concept of reinforced concrete will be explained in an example.

Consider a simply supported concrete beam subjected to the uniform load shown in Figure 1.2a. The beam deforms under the load and flexural (bending) stresses develop

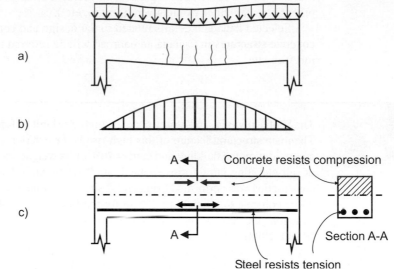

Figure 1.2 Tension steel reinforcement in a reinforced concrete beam subjected to flexure: a) flexural cracks develop under the load; b) bending moment diagram; c) distribution of internal stresses in a reinforced concrete beam.

throughout the span. The top portion of the beam is subjected to compression, whereas the bottom is under tension. The concrete has a limited ability to carry tension so it cracks once its tensile strength has been reached. The cracks develop in the region of maximum bending moments (M), in this case around the beam midspan (see the bending moment diagram shown in Figure 1.2b). A plain concrete beam will fail in the midspan region shortly after the cracks have developed. To prevent such behaviour, steel reinforcement is placed inside the beam near the bottom to resist the tensile stresses, as illustrated in Figure 1.2c. Steel reinforcement that resists tensile stresses is often called *tension reinforcement* or tension steel. The top portion of the beam section resists compression stresses, as illustrated in Figure 1.2c (see section A-A).

Consider a simply supported beam subjected to the uniform load shown in Figure 1.3a. When the beam deflects, shear forces develop in the beam in addition to bending moments. The maximum shear forces (V) and the corresponding shear stresses develop in the support region (see shear-force diagram in Figure 1.3b). Diagonal cracks develop in the beam in the region of high shear stresses, as illustrated in Figure 1.3c. To prevent the diagonal tension cracks, U-shaped reinforcing bars called stirrups are placed vertically along the beam length in the regions where shear cracks are expected to develop.

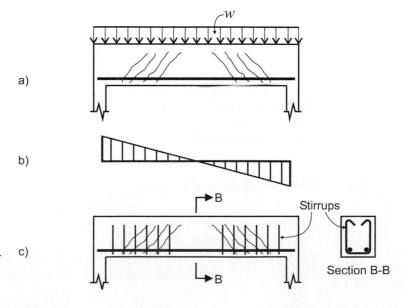

Figure 1.3 Shear reinforcement in concrete beams: a) beam cracks under the distributed load; b) shear force diagram; c) stirrup reinforcement resisting shear effects.

The main idea behind reinforced concrete is to provide steel reinforcement at locations where tensile stresses exist that the concrete cannot resist. Due to its high strength, only a relatively small amount of steel is needed to significantly increase load-carrying capacity of concrete. In general, the area of steel reinforcement accounts for only a small fraction of the overall cross-sectional area of a concrete member, typically on the order of 0.2% to 2% for slabs and beams and 1% to 4% for columns.

It is very important to note that reinforcement in concrete structures is effective *only* when it is appropriately used, strategically placed, and in proper quantity.

In reinforced concrete structures, steel and concrete act together, thus enabling effective load transfer between them. This is an essential feature of reinforced concrete called *bond*.

1.2.4 Prestressed Concrete

Prestressed concrete is a special type of reinforced concrete in which internal compression stresses are introduced to reduce potential tensile stresses in the concrete resulting from external loads. High-strength steel tendons are embedded within the concrete and subjected to a tensile stress imposed by special equipment (jacks). A few different types of tendons are used in prestressed concrete construction, such as wires, cables, bars, rods, and strands. The two main methods of prestressed concrete construction are

- *pretensioning:* when the tendons are tensioned before the concrete has hardened;
- *posttensioning:* when the tendons are tensioned after the concrete has hardened.

The concept of prestressed concrete can be explained with an example. Consider a concrete beam subjected to the external load shown in Figure 1.4. The beam deforms and flexural stresses develop. The top part of the beam is under compression, whereas the bottom part is under tension, as illustrated in section A-A in Figure 1.4a. Concrete is rather weak under tension, and hence it cracks at rather small loads. However, when an axial compressive force (P) is externally applied at the beam ends, compression stresses develop in the beam, as shown in Figure 1.4b. Note that the axial load is applied eccentrically, resulting in larger compression stresses at the bottom of the beam. The effect of prestressing is therefore equivalent to an externally applied axial force. The combined effects of the external load

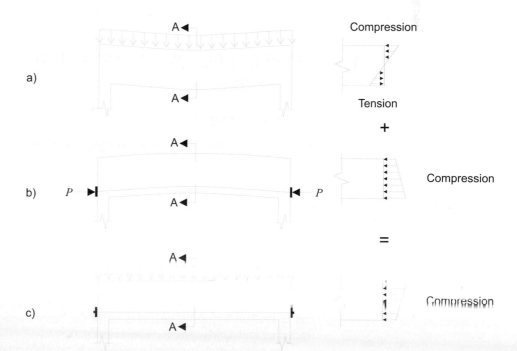

Figure 1.4 A prestressed concrete beam: a) stress distribution for a beam without reinforcement subjected to a uniform load; b) stresses due to the external compressive force (*P*); c) combined stresses.

(see Figure 1.4a) and prestressing (see Figure 1.4b) result in the compressive stress distribution shown in Figure 1.4c. In some cases, minor tensile stresses might still be present in the beam. The major feature and benefit of prestressing is its ability to reduce or entirely eliminate development of tensile stresses and cracking in concrete sections under service loads. Prestressed concrete members require smaller cross-sectional dimensions than reinforced concrete members with similar spans and loads. As a result, prestressed concrete structures are lighter than reinforced concrete structures designed to similar requirements.

KEY CONCEPTS

Concrete is artificial stone made from two main components: cement paste and aggregates. Aggregates usually consist of natural sand and gravel or crushed stone. The paste hardens as a result of the chemical reaction between cement and water and glues the aggregates into a rock-like mass.

Reinforced concrete structures utilize the best qualities of concrete and steel — concrete's high compressive strength and steel's high tensile strength. The main idea behind reinforced concrete is to provide steel reinforcement at locations where tensile stresses exist that the concrete cannot resist. Due to its strength, only a relatively small amount of steel is needed to reinforce concrete. Steel's ability to resist tension is around 10 times greater than concrete's ability to resist compression. It is very important to note that reinforcement in concrete structures is effective *only* if it is appropriately used, strategically placed, and in proper quantity.

Prestressed concrete is a special type of reinforced concrete in which internal compression stresses are introduced to reduce potential tensile stresses in the concrete resulting from external loads. High-strength steel tendons are embedded within the concrete and subjected to a tensile stress imposed by special equipment (jacks). The two main methods of prestressed concrete construction are

- *pretensioning:* when the tendons are tensioned before the concrete has hardened;
- *posttensioning:* when the tendons are tensioned after the concrete has hardened.

1.3 REINFORCED CONCRETE BUILDINGS: STRUCTURAL COMPONENTS AND SYSTEMS

1.3.1 Structural Components

Reinforced concrete buildings consist of several structural components (or members). The basic components of a reinforced concrete building are (see Figure 1.5)

- floor and roof systems
- beams
- columns
- walls
- foundations

These structural components can be classified into *horizontal* components (floors, roofs, and beams) and *vertical* components (columns and walls). According to another classification, the part of the building above ground is called the *superstructure,* while the part below ground (including foundations, basement, and other underground structures) is called the *substructure.*

The role of each structural component is briefly explained below.

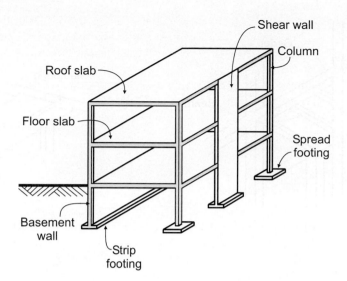

Figure 1.5 Components of a reinforced concrete building.

The *floor and roof systems* are the main horizontal structural components in a building. They carry gravity loads and transfer them to the vertical components (columns and/or walls), and also act as horizontal diaphragms by transferring the lateral load to the vertical components of a structure. The most common floor and roof systems are listed below (see Figure 1.6):

*   *Slab-beam-and-girder:* The slabs are supported by beams, which are in turn supported by girders (see Figure 1.6a). A *girder* is a large beam that carries loads from the beams framing into it. Beams around the outside edges of the floor are called *spandrel beams*.
*   *Slab band*: This is a uniform slab with a thickened slab portion along the column lines parallel to the longer spans (see Figure 1.6b).
*   *Flat slab:* This is a system without beams, where a slab is supported by round or square columns (see Figure 1.6c). In this system, the design may also require a flared cone-shaped cap on the top of the column, called the capital, and a thickened slab above it, called the drop panel.
*   *Flat plate:* This is similar to the flat slab, except that there are no drop panels or capitals, as shown in Figure 1.6d. Columns are typically of circular or square shape.
*   *Slab with beams:* The beams frame into columns and support floor or roof slabs, as illustrated in Figure 1.6e. They provide moment interaction with the columns (this interaction is essential for the frame to resist lateral loads).
*   *Joist floor (pan joist):* This system consists of a series of closely spaced joists (similar to small beams), spanning in one or two directions, topped by a reinforced concrete slab cast integrally with the joists, and beams spanning between the columns perpendicular to the joists (see Figure 1.6f).
*   *Waffle slab:* This is a two-way reinforced concrete joist floor. Waffles are hollow spaces between the joists.

The design of reinforced concrete floor systems will be covered in Chapters 11 and 12.

Slab on grade is a very common form of slab construction that is placed directly on the ground. It is also called "floor on ground." It is possible to confuse this term with the term "floor system." The basic difference is that a slab on grade is supported by the earth beneath it, whereas a floor system is supported only by columns at a few distinct locations. The design of slabs on grade will be covered in Chapter 14.

Beams transmit the loads from the floors to the vertical supports (columns). Beams are usually cast monolithically with the slab and are subjected to bending and shear. The design of reinforced concrete beams will be introduced in Chapter 7.

Columns are vertical components that support a structural floor system. Columns are usually subjected to combined axial load and bending. Column design will be covered in Chapter 8.

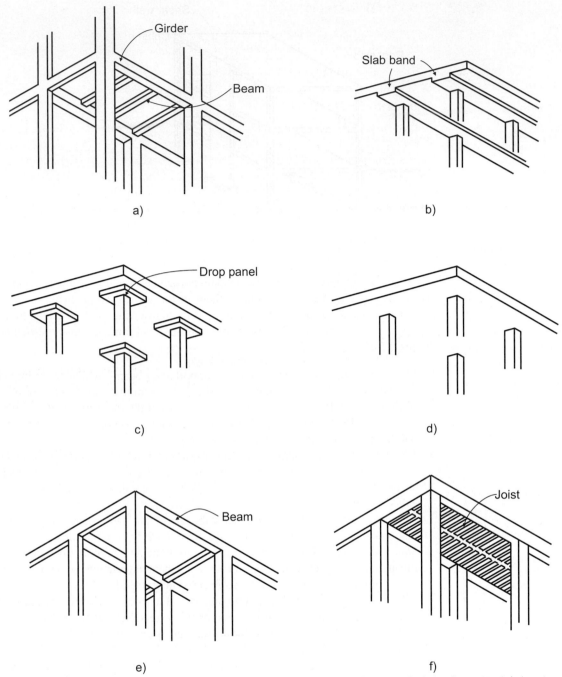

Figure 1.6 Floor systems in reinforced concrete buildings: a) slab-beam-and-girder floor; b) slab bands; c) flat slab; d) flat plate; e) slab with beams; f) joist floor.

Walls provide the vertical enclosure for a building. *Bearing walls* carry gravity loads only, whereas *shear walls* have a major role in carrying lateral loads due to wind and earthquakes. Concrete walls built in the basements of buildings are subjected to lateral soil pressure in addition to gravity loads — such walls are called *basement walls*. The design of reinforced concrete walls will be covered in Chapter 13.

Foundations transmit the weight of the superstructure to the supporting soil. There are several types of foundations. *Spread footings* transfer the load from the columns to the soil. Walls are supported by *strip footings*. Other types of foundations include combined footings, which support more than one column; piles which may be driven into dense soil strata beneath; and raft foundations, where several columns rest upon a raft or a mat distributing the column or wall loads over a uniform soil bearing area. The design of foundations will be discussed in Chapter 14.

1.3.2 Structural Systems

The structural (or framing) system is the skeleton of a building, and it supports the rest of the structure. Structural systems characteristic of reinforced concrete buildings are (see Figure 1.7)

- moment-resisting frames
- bearing-wall systems
- frame/shear-wall hybrid systems

A *moment-resisting frame* (or moment frame) consists of columns and beams that act as a three-dimensional (3-D) space frame system, as shown in Figure 1.7a. Both gravity and lateral forces are resisted by bending in beams and columns, while strong rigid joints between columns and beams have a special role in providing stability in moment frames. The frames are often infilled with masonry partitions (usually hollow concrete blocks or hollow clay tiles); such a system is called a *concrete frame with masonry infills*. The moment-frame system has been used often for office buildings in Canada. The predominant use is in the five- to ten-storey range.

A *bearing-wall system* consists of reinforced concrete bearing walls located along exterior wall lines and at interior locations as required (see Figure 1.7b). These bearing walls are also used to resist lateral forces, in which case they are called shear walls. *Shear walls* are designed to resist lateral forces from floor structures and transmit them to the ground. Ideally, these shear walls are continuous structures, extending from the foundation to the roof of the building. This system is usually characterized by a rectangular plan, with a centrally located elevator and stair core and uniformly distributed walls. Many residential and office buildings in Canadian cities utilize the bearing-wall system.

A *frame/shear-wall hybrid system* utilizes a complete 3-D space frame to support gravity loads and shear walls to resist lateral loads (see Figure 1.7c). The main lateral load-resisting system consists of reinforced concrete shear walls forming the elevator core (central core formed by the elevators and stairs in the building), and additional walls located elsewhere in the building as required. The role of the concrete frame is to transfer gravity loads only, so it is often called the *gravity frame*. The columns typically support concrete flat slab structures or two-way slabs with beams. The interaction of the frame and shear walls is essential for limiting lateral deformations due to wind and earthquake loads. These buildings are generally characterized by a symmetrical plan of square, circular, or hexagonal shape with a centrally located elevator core. This system is commonly found in modern office and residential high-rise buildings in Canada.

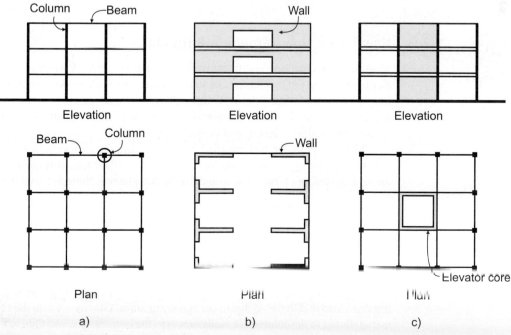

Figure 1.7 Schematic elevations and plans for structural systems in reinforced concrete: a) moment-resisting frame; b) bearing-wall system; c) frame/shear-wall hybrid system.

KEY **CONCEPTS**

Reinforced concrete buildings consist of several structural components (or members). The basic components of a reinforced concrete building are as follows:

- The *floor and roof systems* carry gravity loads and transfer them to the vertical components (columns and/or walls). They are the main horizontal structural components in a building and also act as horizontal diaphragms by transferring the lateral load to the vertical components of a structure.
- *Beams* transmit the loads from the floors to the vertical components (columns). They are usually cast monolithically with the slab and are subjected to bending and shear.
- *Columns* support a structural floor system. They are vertical components that are usually subjected to combined axial compression and bending.
- *Walls* provide the vertical enclosure for a building. *Load-bearing walls* carry gravity loads only, whereas *shear walls* have a major role in carrying lateral loads due to wind and earthquakes.
- *Foundations* transmit the weight of the superstructure to the supporting soil. There are several types of foundations. *Spread footings* transfer the load from the columns to the soil. Walls are supported by *strip footings*.

The structural system (or framing system) is the skeleton of a building and supports the rest of the structure. Structural systems characteristic of reinforced concrete buildings are as follows:

- A *moment-resisting frame* (or moment frame) consists of columns and beams that act as a 3-D space frame system. Both gravity and lateral forces are resisted by bending in beams and columns mobilized by strong rigid joints between columns and beams.
- A *bearing-wall system* consists of reinforced concrete bearing walls located along exterior wall lines and at interior locations as required. These bearing walls are also used to resist lateral forces, in which case they are called shear walls.
- A *frame/shear-wall hybrid system* utilizes a complete 3-D space frame to support gravity loads and shear walls to resist lateral loads. In modern buildings of this type, the main lateral load-resisting system consists of reinforced concrete shear walls forming the elevator core and additional walls located elsewhere in the building as required.

1.3.3   How Loads Flow Through a Building

Multiple elements are used to transmit and resist external loads within a building. These elements define the mechanism of load transfer in a building known as the *load path*. The load path extends from the roof through each structural element to the foundation. An understanding of the critical importance of a complete load path is essential for everyone involved in building design and construction.

The load path can be identified by considering the elements in the building that contribute to resisting the load and by observing how they transmit the load to the next element. Depending on the type of load to be transferred, there are two basic load paths:

- gravity load path
- lateral load path

Both the gravity and lateral load paths utilize a combination of horizontal and vertical structural components, as explained below.

Gravity load path   Gravity load is the vertical load acting on a building structure, including dead load and live load due to occupancy or snow. Gravity load on the floor and roof slabs is transferred to the columns or walls, down to the foundations, and then to the supporting soil beneath. Figure 1.8 shows an isometric view of a concrete structure and a gravity load path.

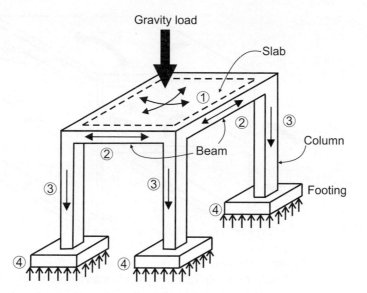

Figure 1.8 An isometric view of a concrete structure showing a gravity load path.

The vertical gravity load acts on a slab (1), which transfers the load to the beams (2), which in turn transfer the load to the columns (3) and then down to the foundations (4).

The gravity load path depends on the type of floor slab, that is, whether a slab is a one-way or a two-way system. In the *one-way system* in Figure 1.9a, the effect of external loads is transferred primarily in one direction, shown with an arrow. The slab-beam-and-girder floor (discussed in Section 1.3.1) is an example of a one-way system. The gravity load acting on this system is transferred from the slab (1) to the beams (2) and then to the girders (3). Finally, the girders transfer the load to the columns (4).

The load path in a *two-way system* is not as clearly defined. The slab transfers gravity load in two perpendicular directions; however, the amount carried in each direction depends on the ratio of span lengths in the two directions, the type of end supports, and other factors. For example, in the slab with beams system shown in Figure 1.9b, the load is transferred from the slab (1) to the beams aligned in the two directions (2) and then to the columns (3).

Lateral load path   The lateral load path is the way lateral loads (mainly due to wind and earthquakes) are transferred through a building. The primary elements of a lateral load path are as follows:

- vertical components: shear walls and frames;
- horizontal components: roof, floors, and foundations.

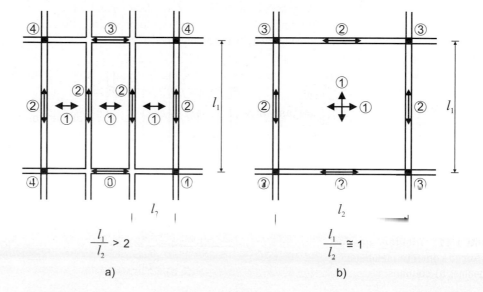

Figure 1.9 Gravity load path in a floor slab: a) one-way system; b) two-way system.

$$\frac{l_1}{l_2} > 2$$

a)

$$\frac{l_1}{l_2} \cong 1$$

b)

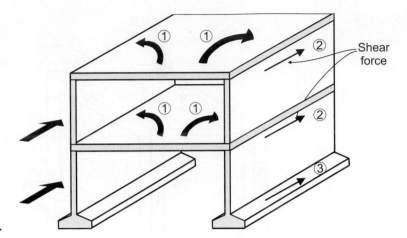

Figure 1.10 Lateral load path.

Figure 1.10 shows a reinforced concrete structure and the elements constituting the lateral load path: roof and floor systems (1) transfer the load to the walls (2), which in turn transfer the load to the foundations (3).

Roof and floor systems (also called diaphragms) take horizontal forces from the storeys at or above their level and transfer them to walls or frames in the storey immediately below.

Shear walls and *frames* are the primary lateral–load resisting elements; however, these members also carry gravity loads. Shear walls receive lateral forces from diaphragms and transmit them to the foundations.

Foundations form the final link in the load path by collecting the lateral forces from all storeys and transmitting them to the ground.

Tributary area   The *tributary area* is related to the load path, and is used to determine the loads that beams, girders, columns, and walls carry. The reader is expected to be familiar with the concept of tributary area from other design courses, as it also applies to design of timber and steel structures; however, a brief overview is presented in this section. The tributary area for a beam or a girder supporting a portion of the floor is the area enclosing the member and bounded by the lines located approximately halfway between the lines of support (columns or walls), as shown in Figure 1.11. For example, a tributary area for the reinforced concrete beam AB that is a part of the one-way floor system is shown hatched in Figure 1.11a. A typical column has a tributary area bounded by the lines located halfway from the line of support in both directions (shown hatched in Figure 1.11b). In the case of uniformly loaded floors, tributary areas are approximately bounded by the lines of zero shear, that is, the lines corresponding to zero shear forces in the slabs, beams, or girders supported by the element for which the tributary area is determined. Zero-shear locations are generally determined by the analysis. For buildings with a fairly regular column spacing, the zero-shear locations may be approximated to be halfway between the lines of support.

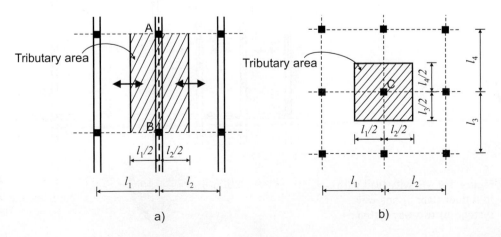

Figure 1.11 Tributary area for reinforced concrete members: a) beams; b) columns.

KEY CONCEPTS

Within every building, multiple elements are used to transmit and resist external loads. These elements define the mechanism of load transfer in a building known as the *load path*. The load path can be identified by considering the elements in the building that contribute to resisting the load and by observing how they transmit the load to the next element. Depending on the type of load to be transferred, there are two basic types of load path:

- *Gravity load path:* The gravity load acting on the slab is transferred by the floor and roof systems to the columns or walls, down to the foundations, and then to the ground. The gravity load path depends on the type of floor system. In a *one-way system,* the external loads are transferred primarily in one direction. The gravity load acting on this system is transferred from the slab to the beams and then to the girders. Finally, the girders transfer the load to the columns. In a *two-way floor system,* the load path is not as clearly defined. The slab transfers the gravity load in two directions; however, the amount carried in each direction depends on the type of end supports and other factors.
- *Lateral load path:* This is the way lateral loads (mainly due to wind and earthquakes) are transferred through a building. The primary elements of a lateral load path are vertical components (shear walls and frames) and horizontal components (roof, floors, and foundations).

Example 1.1

A partial floor plan of a reinforced concrete building is shown in the figure below. The roof is subjected to a total uniform area load (*w*) of 15.0 kPa (including the floor self-weight). *Determine the uniform load on the typical interior beam B1 and the typical interior girder G1 in the figure.*

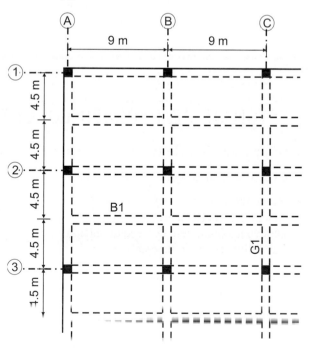

SOLUTION:

a) Load on a typical interior beam B1

It can be observed from the floor plan that the beams are spaced at a distance of 4.5 m, equal to the tributary width (*b*) as shown hatched in the sketch below. Therefore, the uniform load acting on a typical beam B1 is equal to the product of the area load (w) and the tributary width (b); that is,

$$w_{B1} = w \times b = 15 \text{ kPa} \times 4.5 \text{ m} = 67.5 \text{ kN/m}$$

b) Load on a typical interior girder G1

In the building under consideration, the girders are provided at 9 m on centre spacing, which is equal to the tributary width (s) as shown cross-hatched on the sketch below. Therefore, the uniform load acting on a typical girder G1 is equal to the product of the area load (w) and the tributary width (s); that is,

$$w_{G1} = w \times s = 15 \text{ kPa} \times 9.0 \text{ m} = 135 \text{ kN/m}$$

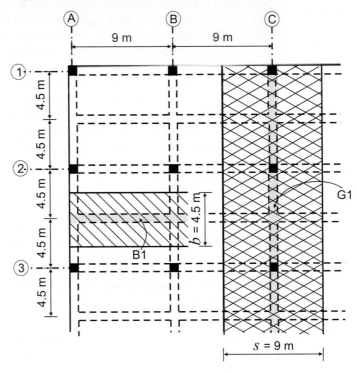

1.4 DESIGN OF REINFORCED CONCRETE STRUCTURES

1.4.1 Design Considerations

The key structural design considerations are

- structural safety and serviceability
- functional requirements
- economics (minimum cost requirements)
- durability and environmental impact

Structural safety and serviceability Building structures must be designed to be safe from collapse during construction and throughout their useful life. Structural failures might lead to human and economic losses and must be prevented.

In Canada, the design of concrete structures is performed according to the limit states design method, which will be explained in Section 1.8. In general, each structure should be capable of carrying the expected loads without excessive deflections, damage or collapse. However, when the structure is overloaded in extreme circumstances (such as unexpectedly large loads, major earthquakes, or unusually high wind), it should be designed to fail in a "desirable way" by giving enough warning to the occupants or users before collapse. One of the main objectives of this book is to provide guidance on how adeqate structural safety can be provided in reinforced concrete buildings.

The structure must also behave in a satisfactory manner during its service life: deflections must remain within acceptable limits and premature or excessive cracking must be prevented.

Functional requirements Functional requirements are driven by architectural and structural engineering considerations. In general, a building should enclose space and contents, provide security to occupants, and facilitate work in the enclosed space. In addition to being safe from collapse due to imposed loads, a building also needs to meet fire-resistance requirements. Architectural requirements often pose constraints related to the dimensions and location of walls, columns, and other elements. In some cases, a building needs to present aesthetically pleasing exposed surfaces. The intended building function often has consequences with regard to structural design; for example, a structural system used for the design of a museum would be different from that for a residential building.

Economics In today's world, there is a demand for speed and efficiency, and structural engineers are challenged to design buildings that are cost-effective and fast to build. A good reinforced concrete structure must achieve a proper balance between cost, speed of construction, and quality. The balance point varies in different parts of the world, where the cost of labour and material varies, as do the skill and knowledge levels of the construction workers.

A good reinforced concrete design takes into account the material and labour costs, the construction schedule (which has financial implications related to the speed of construction), and the quality and durability of the finished building.

Durability and environmental impact The designer needs to ensure that the concrete structure will remain functional throughout its expected useful life, which is considered to be at least 50 years. Satisfactory durability is to be ensured by giving proper consideration to freeze-thaw cycles (where required), potential exposure to aggressive environments, and adequate concrete cover. In general, properly designed and constructed concrete structures are durable and require less maintenance than structures built with other materials. Environmental impact is a function of durability and the use of recycled materials. Reinforced concrete can be made of recycled flyash and recycled steel (see Chapter 2 for information on concrete admixtures).

KEY CONCEPTS

The four key structural design considerations are as follows:

1. *Structural safety and serviceability:* Building structures must be designed to be safe from collapse during construction and throughout their useful life.
2. *Functional requirements:* These requirements are driven by architectural and structural engineering considerations. In general, a building should enclose space and contents, provide security to its occupants, and facilitate work in the enclosed space.
3. *Economics:* A good design must achieve a proper balance between cost, speed of construction, and quality.
4. *Durability:* The designer needs to ensure that a concrete structure will remain functional throughout its expected useful life, which is considered to be at least 50 years, and meet environmental impact requirements.

1.4.2 Design Process

It is very important for novice designers to thoroughly understand the design process, including the steps involved and the time allocated to and spent on performing each step. Major steps in the design process are

- schematic (conceptual) design
- detailed (final) design
- development of contract documents
- coordination
- services during construction

Schematic (conceptual) design   This step involves the identification of project constraints, including cost, building shape, and architectural form, and functional constraints, including column spacing, materials, and serviceability limits. Conceptual design is the most important part of the entire design process. At this stage, the structural engineer has the pivotal role of developing a practical structural concept that strikes the proper balance between the external constraints and the project objectives. To develop a good structural concept, the designer needs to have not only a sound background in reinforced concrete design but also a strong appreciation for the architectural aspects of the project, constructability issues, and the owner's overall design objectives. The final goal is to develop a structural concept that is simple to build, aesthetically pleasing, functionally effective, and affordable to the owner. Once all the design issues have been identified, the designer should be able to make a schematic drawing of the structural system and decide on the general arrangement of structural elements. Further on, the structural designer can estimate gravity and lateral loads and develop trial sizes of key structural members. This process may require a few iterations before the optimal solution is found. Next, the preliminary concrete outlines to be used for both architectural and structural drawings are developed. Finally, a preliminary construction budget is determined. Depending on the complexity of the project, this phase could take 10% to 20% of the total time on the project.

Detailed (final) design   This step involves the detailed analysis, evaluation, and sizing of members and more refined calculation of gravity and lateral loads. At this stage, the designer needs to ensure the safety and serviceability of the structure by carefully following the requirements of pertinent building codes. However, the designer also needs to keep in mind that several external factors may have an adverse effect on the performance and safety of reinforced concrete structures. Whenever possible, the designer should take advantage of available opportunities to increase the structural capacity, that is, to provide a reserve capacity in the structure. In general, reserve capacity may be required to account for construction errors in the field, errors in load estimates, load increases due to design modifications by the owner or architect, change of building use leading to load increase, variations in soil capacities, etc.

The best design solutions involve good judgment based on experience and knowledge, consideration of the economy and construction issues, repetition, simplicity in rebar placements, reduction of potential field errors, etc. A detailed design takes approximately 20% of the total time spent on the project.

Development of contract documents   The main focus in this step is to transfer the concept and details from the mind of the designer to those involved in the construction. The contract documents include drawings and specifications. *Drawings* are a graphical representation of the design, whereas *specifications* are written descriptions of materials and construction procedures. A well-designed building will not perform in a satisfactory manner when poorly constructed. Therefore, preparation of clear and correct contract documents is an essential part of the design and may take up to 50% of the total time spent on the project.

Coordination   The designer needs to keep in mind that structural design is only one component of the overall building design. Other disciplines include architectural, electrical, mechanical, geotechnical, and civil engineering. The structural designer needs to coordinate the design with other disciplines. This stage might be rather time-consuming and may take up to 10% of the total time spent on the project.

Services during construction   Structural designers are routinely involved in the review of shop drawings, concrete mix designs, and laboratory testing reports. The designer must make regular visits to the construction site to observe the construction, answer questions, and clarify contract documents. The main objective of these visits is to verify that the work is progressing in the manner intended by the design. The involvement of a designer at this stage varies with design and construction complexity; however, it may take about 10% to 20% of the total time spent on the project.

KEY CONCEPTS

The five major steps in the design process are as follows:

1. *Schematic (conceptual) design:* This includes the identification of project constraints, including cost, building shape, and architectural form, and functional constraints, including column spacing, materials, and serviceability criteria. This step also involves selection of the structural system, development of the general arrangement of structural elements, and trial sizes of key structural members based on estimated loads.
2. *Detailed (final) design:* Here, detailed analysis, evaluation, and redesign of preliminary member sizes and more refined calculation of gravity and lateral loads must be performed. The safety and serviceability of the structure need to be ensured by carefully following the requirements of pertinent building codes.
3. *Development of contract documents (drawings and specifications):* The main focus is to transfer the concept and details from the mind of the designer to those involved in the construction. A well-designed building is worthless if it is poorly constructed.
4. *Coordination:* The designer must coordinate with other disciplines, including architectural, electrical, mechanical, geotechnical, and civil engineering.
5. *Services during construction:* These include the review of shop drawings, concrete mix designs, and laboratory testing reports. The designer must make regular visits to the construction site to observe the construction, answer questions, and clarify contract documents.

1.5 CONSTRUCTION OF REINFORCED CONCRETE STRUCTURES

1.5.1 Construction Process

Once the design has been completed, a team of experienced and knowledgeable construction workers and supervisors is required to transform the structural design into the finished construction. Concrete construction is a complex process that involves several activities, from batching and mixing the ingredients to pouring fresh concrete into forms and curing the new construction. There are several factors that might influence the quality of the final product in this process, that is, the built structure. The critical activities associated with the construction of cast-in-place reinforced concrete structures are described below.

Development of structural drawings   Structural design should be based on the sound application of fundamental principles of reinforced concrete design and a knowledgeable use of building codes and standards. However, structural design is ineffective if it is not possible to ensure its proper implementation. The structural engineer has a critical role in communicating the design information to personnel involved in the construction process. In the world of structural engineering, structural drawings and specifications are a critical means of communication between the structural engineer and the contractor. Consequently, the importance of an accurate set of structural drawings that communicate clear and concise information cannot be underestimated.

In reality, even a well-designed structure may result in a variety of problems, both financial and legal, when structural drawings do not properly communicate the design intent. A poor set of structural drawings may expose the engineer to various problems, such as extra costs, delays in the construction schedule, disputes in the field, construction errors, unhappy contractor/owner, and even legal disputes.

Construction estimating and tendering   Once the structural drawings have been completed, the contractor produces an estimate of the concrete and reinforcing steel quantities, as well as the amount of formwork and shoring. This estimate is based on structural drawings and specifications that are issued for construction. Based on these drawings, the estimator should be able to estimate the quantity of materials and also the extent of complexity in the construction procedures, any unusual complications associated with the erection, formwork, and shoring, etc. The drawings must be clear and concise and without ambiguities. When drawings are ambiguous, there could be large variations in price between competitive bids submitted by different contractors in the bidding process. In general, a variation in the bid price of over 10% is considerable, as it leaves the owner confused about the proper market price for the project. This is an unfavourable situation for the owner, who ultimately pays for the construction. On the one hand, the owner may feel that (s)he is overpaying if the highest bid is accepted. On the other hand, a bid below fair market value may lead to problems during construction, usually in the form of numerous requirements for extra payments by the contractor in trying to recover losses caused by the low bid.

Detailing of reinforcement   The reinforcing contractor interprets the information on structural drawings and produces reinforcing steel detail sheets. These sheets are primarily used by the shop to cut and bend the reinforcing bars required for the project. Each bar has a unique code that is referred to by the person responsible for rebar placement in the field.

Placing of reinforcement   Once the reinforcement has been detailed and cut in the shop, it is delivered to the site in bundles. In some cases, rebars are cut and bent at the construction site. Placement of the reinforcing steel is carried out by rebar placers. Their role is to interpret structural drawings along with the detail sheets and then place the reinforcing steel as precisely as possible (within the acceptable tolerances specified by CSA standards). Rebar clearances must be properly shown on the structural drawings in order for the rebar placer to accurately place bar supports to ensure proper bar position.

Supplying the concrete   The concrete ready-mix company will supply concrete based on the concrete strengths specified on the structural drawings and the mix designs that meet the specific climate and workability requirements of the site. Special concrete placement procedures, such as placement by pumps or wheelbarrows; site batching; or cylinder testing by an independent testing agency, must be carried out in accordance with the structural drawings and specifications.

Installation of concrete formwork and shoring   Formwork and shoring support the weight of the wet concrete and other construction loads. Formwork is a temporary structure made of wood, metal, or plastic, and it is constructed to form the final shape of a concrete member. The concrete formwork contractor must hire an engineer who is responsible for designing and producing drawings showing the formwork and shoring supports for concrete beams, slabs, walls, columns, and foundations. The formwork must be built precisely in accordance with the structural drawings and must allow for the proper placement of reinforcing steel and wet concrete without any major deviations in shape and stability.

Placing and curing the concrete   The proper amount of concrete must be placed such that the slab thickness, beam sizes, and wall and column dimensions are in accordance with the structural drawings. In some cases, special procedures and construction sequences are required to minimize the development of cracks in the concrete due to structural or architectural restraints or temperature changes. For example, pour strips are often used in large floor structures to separate different floor sections to be placed in the same pour. In some cases, a continuous gap of minimum 300 mm width is left between two adjacent floor sections. A delay period (usually 28 days) is specified to allow for shrinkage to take place between these pours. Subsequently, pour strips between floor sections are filled with concrete. In some cases, allowance for various types of joints (such as control joints and expansion joints) needs to be made during construction. For more information on joints, refer to Sections 12.12.4 and 13.8.

The curing process is critical for fresh concrete to gain the required design strength. Improperly cured concrete tends to exhibit extensive cracking and creep-induced deflections. Structural drawings need to specify the curing procedures in accordance with the design requirements. It is commonly required to provide curing for fresh concrete over a 28-day period.

Reshoring and special construction procedures   By and large, construction projects require an accelerated concrete placement schedule. In some cases, concrete formwork may need to be stripped (removed) as quickly as 3 days after the placement. Green concrete (concrete that has not attained design strength) must develop sufficient strength to support its self-weight plus the construction load. However, concrete stripped shortly after casting may develop larger creep deflections than concrete stripped after proper curing. Hence, a proper reshoring sequence must be specified by the engineer responsible for construction to prevent sustained loads from being prematurely applied to green concrete.

Procedures that are relevant to the overall design objective need to be outlined on the structural drawings. Such special procedures are generally related to the mitigation of the effects of shrinkage and cracking, as well as creep and long-term deflections. These procedures usually bear certain cost implications and must be clearly communicated on the structural drawings before construction starts. When the structural engineer requests a special procedure after construction has begun that was not specified on the drawings, there is a chance that the contractor will charge extra to carry it out.

DID YOU KNOW?

In 2003, the Canadian construction industry had a gross output of $123 billion; this accounts for 12% of the national GDP (CCA, 2003). Major areas of construction activity in Canada are as follows:

- *Buildings* made up over 50% of total construction activity in 2003, including 35% residential buildings and over 15% other types of buildings (such as commercial, institutional, etc.).
- *Engineering projects* made up over 30% of the total, including industrial-based construction, roadbuilding, bridges, and others.
- *Repair and renovation of existing structures* made up over 15% of the total gross output.

1.5.2   Construction Methods

The four basic methods of concrete construction are

- cast in place construction
- precast construction

- tilt-up construction
- concrete masonry construction

Cast-in-place construction   In cast-in-place concrete construction, each element is built at the construction site. Fresh concrete, obtained by mixing the ingredients (cement, sand, and aggregates) and adding water, is a shapeless slurry without any strength and has to be supported and shaped by the formwork. Before the concrete is poured into the formwork, reinforcing bars need to be placed in the position prescribed by the design. Once the concrete has hardened and developed sufficient strength, the formwork is removed (*stripped*) and the curing continues for a few weeks by keeping the new construction moist until it gains full strength (for more details refer to Section 2.2.7). Concrete members under construction are often supported by a temporary structure called *falsework* or *scaffolding*. The cast-in-place construction process can be accelerated where required by the project schedule. The speed of curing can be increased by using chemical admixtures. The formwork system can be designed to increase the speed of the forming cycle. Also, much of the reinforcing steel can be prefabricated.

Formwork constitutes a significant portion of the overall cost in cast-in-place concrete construction. There are a number of proven systems available in the industry to facilitate concrete construction. Experienced concrete designers generally have a good knowledge of practical formwork systems and are able to develop designs that are easily constructible.

Cast-in-place construction is weather dependent. Special measures are often taken to maintain the required construction schedule under extreme temperatures.

Precast construction   Concrete is cast into permanent, reusable forms at an industrial plant. Fully cured structural units are then transported to the job site, where these units are hoisted into place and connected in a manner similar to structural steel shapes. Precast construction has certain advantages over cast-in-place construction: it offers improved quality control as well as uniform and fast construction on site since the elements are manufactured away from the construction site. In general, precast concrete elements are usually prestressed and hence lighter weight than those that are cast in place. However, precasting also has some disadvantages: it involves the shipping cost of precast members from the plant to the construction site and it requires the use of special equipment (large cranes) to hoist and place precast members in position. Precast construction can be very cost effective if there is much repetition on a project. In Canada, precast concrete elements are commonly used for bridge construction. The design of precast concrete structures is a specialized subject beyond the scope of this book. For detailed coverage of precast concrete construction, the reader is referred to CPCI (1996).

Tilt-up construction   Tilt-up wall construction is a form of precast construction. Tilt-up wall panels are cast horizontally on the floor slab. Once the panels have attained sufficient strength, a mobile crane sets them on footings. The panels are temporarily braced while the floor and roof structures are erected. Finally, the panels are connected to other structural members by welding embedded steel plates. Tilt-up construction is mainly used in low-rise warehouse or office buildings. The design of tilt-up walls is beyond the scope of this book. For more details, the reader is referred to PCA (1994).

Concrete masonry construction   Concrete masonry is used for warehouse, industrial, and institutional construction in Canada. Masonry load-bearing walls are assembled by laying masonry units in mortar and providing vertical and horizontal steel reinforcement. The principles of masonry design are beyond the scope of this book. For more details, the reader is referred to Drysdale, Hamid, and Baker (1999).

KEY CONCEPTS

Concrete construction is a complex process that involves many steps, from batching and mixing the ingredients to pouring fresh concrete into forms and curing the new construction. The four basic methods of concrete construction are as follows:

1. *Cast-in-place construction:* Each element is built at the construction site. Fresh concrete, obtained by mixing the ingredients (cement, sand, and aggregates) and adding water, is a shapeless slurry without any strength that has to be supported and shaped by the formwork.
2. *Precast construction:* The concrete is cast into permanent, reusable forms at an industrial plant, then transported in the form of fully cured structural units to the job site, where these units are hoisted into place and connected in a manner similar to structural steel shapes. Precast concrete elements are mainly used for bridge construction in Canada.
3. *Tilt-up construction:* Precast wall panels are cast horizontally on the floor slab. Once the panels have attained sufficient strength, a mobile crane sets them on footings. Tilt-up construction is mainly used in low-rise warehouse or office buildings.
4. *Concrete masonry construction:* Masonry load-bearing walls are assembled by laying masonry units in mortar and providing vertical and horizontal steel reinforcement. Concrete masonry construction is mainly used for warehouse, industrial, and institutional construction.

1.6 CANADIAN DESIGN CODES AND STANDARDS FOR CONCRETE STRUCTURES

The design and construction of buildings in Canada is regulated by the *National Building Code of Canada* (NBC), prepared in the form of a recommended model code to permit adoption by provincial and territorial governments. The main objective of the NBC is to protect the health and safety of the general public. The code is divided into nine parts, and it spells out the minimum requirements for fire protection, structural design, heating, ventilation and air-conditioning, plumbing, construction site safety, and housing and small building construction. The NBC pertains to the design and construction of new buildings and also to renovations of existing ones. The first edition of the NBC was published in 1941, and the most recent previous edition was published in 2005. The current edition of the NBC, published in 2010, is referred to as NBC 2010 in this book. The NBC is published by the National Research Council of Canada's Institute for Research in Construction.

NBC refers to material standards regulating design in various building materials (steel, concrete, timber, etc.). These material standards are published by the Canadian Standards Association (CSA). The design of concrete structures in Canada is performed according to the CSA Standard A23.3-04 *Design of Concrete Structures*. The current edition of the standard, published in 2004, is referred to as A23.3 in this book, whereas the previous edition was published in 1994 and is referred to as A23.3-94. This book is largely based on the provisions of CSA A23.3-04 standard, and the reader should preferably have a copy as a companion to this book.

1.7 LOADS

1.7.1 Types of Loads

One of the main NBC 2010 structural design requirements states that a building structure should be designed to have sufficient structural capacity and structural integrity to safely and effectively resist all loads and effects of loads that may reasonably be expected

(NBC 2010 Cl.4.1.1.3(1)). The term *load* means the forces, pressures, and imposed deformations applied to the building structure. In general, building codes offer guidelines on how to determine the magnitude of various loads. Part 4 of NBC 2010 prescribes types of loads to be considered in structural design. The reader should be familiar with design loads from previous design courses; however, a brief overview of loads will be presented in this section.

The following types of loads need to be considered in the design of buildings (NBC 2010 Cl.4.1.2.1):

- dead load (D);
- live load due to use and occupancy (L);
- snow load (S);
- wind load (W);
- earthquake load (E);
- effects of temperature change (T),
- permanent load due to lateral earth pressure (including groundwater) (H);
- permanent effect caused by prestress (P).

Loads can be classified by the predominant direction in which they act. Loads are commonly classified into gravity loads and lateral loads. *Gravity loads* act in the vertical direction (mainly downward), such as dead load, live load, and snow load. *Lateral loads* act mainly in the horizontal direction, such as wind and earthquakes. (It should be noted, however, that wind and earthquake loads can also act in the vertical direction.) This load classification can be presented in diagram form, as shown in Figure 1.12.

Loads can also be classified as permanent, variable, or rare. A *permanent load* is a load that does not vary once it has been applied to the structure. Examples of permanent loads are dead load, load due to lateral earth pressure, and the effect of prestress. A *variable load* is a load that frequently changes in magnitude, direction, or location. Live load and snow load are variable loads. A *rare load* can be defined as a load that occurs infrequently and for a short time only. Earthquake ground motion is an example of a rare load. An earthquake of an intensity that most buildings in Canada are designed for has a probability of occurrence once in approximately 500 years.

Static loads, such as dead load, snow, and live load, are applied slowly to the structure. Loads occurring over a short period of time are called *dynamic loads* (for example, wind and earthquake loads).

NBC 2010 contains an important new feature related to loads — it introduces a way to differentiate loads based on building occupancy. Specified loads are multiplied by an appropriate *importance factor* (I). One of the inspirations for this approach was the Eastern Ice Storm of 1998. After the event happened, it became evident that many of the buildings used as places of refuge, such as schools and community centres, were potentially at or near their design limit due to a large amount of ice and snow on their roofs. It was concluded that structures used as areas of refuge, which by definition accommodate many people during or after a disaster, should be designed for higher loads than other buildings. It should be noted that the importance factor existed in previous editions of NBC, but it was related to earthquake load only.

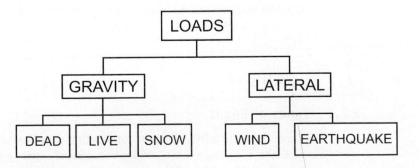

Figure 1.12 Load classification.

The following four importance categories have been established based on building use:

- *Low importance category* buildings are those where failure is not likely to lead to loss of human life (for example, barns and warehouses).
- *High importance category* buildings are those typically used as emergency shelters in a catastrophic event (for example, elementary, middle, and secondary schools and community centres). Also included in this category are structures that house hazardous substances.
- *Post-disaster buildings* must remain fully functional in a disaster, and therefore warrant their own importance category. These are essential services buildings such as hospitals, emergency treatment facilities, etc.
- *Normal importance category* buildings are all other buildings.

It should be noted that the loads discussed in this section are *specified loads*, that is, actual or nominal loads. However, in the limit states design discussed in Section 1.8, specified loads are multiplied by the load factor to take into account safety considerations — such loads are called *factored loads*.

1.7.2 Dead Load

Dead load (*D*) is a permanent load due to the weight of building components. According to NBC 2010 Cl.4.1.4.1, the specified dead load for a structural member consists of

- self-weight (weight of the member itself)
- superimposed dead load

The *self-weight* of a building member always needs to be considered in the design. The size and weight of reinforced concrete elements may be significant in relation to the total load on the structure. The self-weight of a structural member is calculated based on a consideration of the unit weight of the material involved and its volume. The unit weight of normal-weight reinforced concrete (γ_w) can be taken as 24 kN/m^3; this value will be used for calculations in this book. Note that some designers take the unit weight of reinforced concrete as 23.5 kN/m^3, that is, the exact conversion of 150 pcf (pound-force per cubic foot) in Imperial or U.S. Customary units. Unit weights of commonly used building materials are listed in Appendix A (Table A.11).

In many cases, determining the self-weight is an iterative process. The designer needs to start the design by taking a certain member size and weight, perform design calculations, and then verify whether the original sizes are suitable for the design.

The *superimposed dead load* (SDL) is a dead load that is typically added, that is, superimposed, onto the primary structure after it is complete. SDL is a dead load of everything permanent that is not part of the primary structural system, such as materials of construction, partitions, and permanent equipment. Some equipment items, such as generators that may be removed and replaced, are treated as live load instead of SDL. Such items may impose some dynamic loads onto the structure due to their operation.

Example 1.2

A reinforced concrete beam of rectangular cross-section (350 mm width and 700 mm overall depth) is shown in the figure below. The beam is made of normal-density concrete with unit weight (γ_w) of 24 kN/m^3.

Determine the beam self-weight (in kN/m) to be used for the dead load calculation.

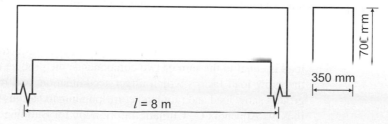

SOLUTION: The beam self-weight should be determined as a product of its cross-sectional area and unit weight; that is,

$$w = b \times h \times \gamma_w$$
$$= 0.35 \text{ m} \times 0.7 \text{ m} \times 24 \text{ kN/m}^3 = 5.88 \text{ kN/m} \cong 6.0 \text{ kN/m}$$

A sketch of the beam under this self-weight is shown below.

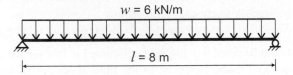

$$w = 6 \text{ kN/m}$$

$$l = 8 \text{ m}$$

1.7.3 Live Load

Live load (L) includes those forces that may or may not be present and acting upon a structure at any time. NBC 2010 defines live load as a variable load due to the intended use and occupancy of a building. Occupancy loads include personnel, furniture, stored materials, and other similar items. In the previous editions of NBC (for example, NBC 1995), snow load was also considered a live load. Earthquake and wind loads are considered to be special forms of live load; however, they are treated separately due to their dynamic nature.

According to NBC 2010, live load should be considered in the design as a *uniformly distributed load* (Cl.4.1.5.3) or a *concentrated (point) load* (Cl.4.1.5.10), whichever produces the most critical effects.

The values of specified uniformly distributed live load due to use and occupancy are provided in Table 4.1.5.3 of NBC 2010. NBC 2010 also requires that the structure accommodate minimum point loads applied over an area of 750 mm by 750 mm (Cl.4.1.5.10). This corresponds to point loads from mechanical equipment and heavy load in localized areas, such as filing cabinets, vehicle tires, etc. When a member supports a large tributary area of the floor and/or roof (for example, a column in the lower storey of a multistorey building), it is highly unlikely that the entire effective area would be loaded uniformly with the full specified live load. Therefore, NBC 2010 Cl.4.1.5.9 permits a reduction in the specified live load intensity depending on the tributary area.

It should be noted that a structure initially designed to carry the load derived from one occupancy type should be carefully checked before being subjected to loads from other occupancy types. For example, a structure designed as a residential apartment building would be inadequate in terms of live load if the building were converted into an office building. In this case, the load-bearing capacity of the existing building would need to be evaluated and some structural members might need to be retrofitted.

1.7.4 Snow Load

Snow load (S) is a variable environmental load acting mainly on the roof areas. Considering the severe winter conditions throughout Canada, due consideration of snow load (including ice and associated rain) and its effect on building structures is of high importance. Snow loads on roofs vary depending on geographical location (climate), site exposure, and shape and type of roof. Calculation of snow load in Canada is to be performed according to the NBC 2010 provisions outlined in Cl.4.1.6. The specified snow load is equal to the sum of two components: the product of the ground snow load (S_s) and the snow load factors which affect accumulation, and the associated rain load (S_r). The ground snow load and rain load corresponding to various locations in Canada are specified in NBC Appendix C "Climatic Information for Building Design in Canada."

There are a couple of important changes related to the snow load provisions in NBC 2010 as compared to NBC 1995:

1. Ground snow load and rain load are determined as loads corresponding to a 1-in-50-year return period. In NBC 1995, these loads were based on a 1-in-30-year return period. As a result, design according to NBC 2010 results in an approximately 10% increase in snow load as compared to NBC 1995.
2. Snow load is treated separately from occupancy and use live load; this allows for a more rational treatment of snow load. It is now possible to differentiate snow loads based on building type by assigning different importance factors, as discussed in Section 1.7.1.

1.7.5 Wind Load

Wind load (W) is a variable dynamic load that acts on wall and roof surfaces within a building. The magnitude of wind load depends on the orientation, area, and shape of the surface; wind velocity; and air density. Wind speed increases with height and depends on the roughness of the terrain. NBC 2010 prescribes two basic approaches for wind design: the simple procedure (static approach) and the dynamic procedure.

The *simple procedure* (Cl.4.1.7.1) represents wind loads by an equivalent static load. This procedure is appropriate for most building design applications where the structure is relatively rigid, including low- and medium-rise buildings. This procedure is also used for claddings and windows in high-rise buildings.

The specified external pressure or suction due to wind on part or all of the surface of a building (p) is calculated as the product of the referential velocity pressure (q) and several wind load modification factors. The referential velocity pressure corresponding to various locations in Canada is specified in NBC 2010 Appendix C "Climatic Information for Building Design in Canada."

The *dynamic procedure* (Cl.4.1.7.2) treats wind load as a dynamic load. The magnitude and distribution can be determined either using special wind tunnel tests or other experimental methods, or according to the "detailed procedure" — a dynamic approach to the action of wind gusts. These procedures should be used for buildings susceptible to vibration, such as tall and slender buildings.

The three important changes related to the NBC 2010 wind load provisions are as follows:

1. The referential wind velocity pressure is determined as a load corresponding to a 1-in-50 year return period. In NBC 1995, this load was based on a 1-in-30-year return period. As a result, design according to NBC 2010 results in an approximately 10% increase in wind load as compared to NBC 1995. However, this difference is offset by the reduced basic wind load factor from 1.5 to 1.4 (see Table 4.1.3.2 of NBC 2010).
2. Wind load can be differentiated according to building type by assigning different importance factors, as discussed in Section 1.7.1.
3. Cladding design is to be performed using 1-in-50-year loads (instead of 1-in-10-year loads); cladding design is a life safety issue.

A few other revisions related to wind design are outlined in NBC 2010 Cl.4.1.7.

1.7.6 Earthquake Load

Earthquake load (E), also known as *seismic load,* is a dynamic load that occurs rather rarely. An earthquake is a vibration of the earth's crust caused by slippage along a fault zone below the surface of the earth. When earthquake ground shaking occurs, a building gets thrown from side to side and/or up and down. That is, while the ground is violently moving from side to side, the building tends to stand at rest, similar to a passenger standing on a bus that accelerates quickly. Once the building starts moving, it tends to continue in the same direction, but by this time the ground is moving back in the opposite direction (as if the bus driver first accelerated quickly, then suddenly stopped). The forces developed in a building while the ground shakes are called *inertial forces* (or *seismic forces*). The design of buildings to resist earthquakes is called *earthquake-resistant design* or *seismic design*.

NBC 2010 prescribes seismic design criteria for all new buildings in Canada. Seismic design according to NBC 2010 has the following intents:

* to protect the life and safety of building occupants and the general public as the building responds to strong ground shaking;
* to limit damage due to low to moderate levels of ground shaking;
* to ensure that post-disaster buildings continue to be occupied and functional following strong ground shaking despite minimal damage.

The design of buildings to resist earthquakes is an advanced topic beyond the scope of this book. However, the main aspects of seismic design according to the NBC 2010 seismic provisions (Cl.4.1.8) are outlined in this section.

General requirements There are several general requirements stipulated in NBC 2010 Cl.4.1.8.3; however, the two key requirements are as follows:

1. The structure must have a clearly defined load path (refer to Section 1.3.3 for a discussion on load path).
2. The structure must have a clearly defined seismic force resisting system (SFRS) capable of resisting 100% of the earthquake loads and effects (the main types of structural systems that act as SFRSs are discussed in Section 1.3.2).

Methods of analysis The basic seismic design procedure prescribed by NBC 2010 Cl.4.1.8.7 is the *dynamic analysis* procedure. This method takes into account the dynamic nature of earthquakes. A specific design earthquake is applied to a mathematical model of the building. The response of the structure is calculated and used to determine the forces in each member as a function of time.

Alternatively, the *equivalent static force* procedure can be used in special cases (where the seismic hazard is low or in the case of regular structures). The effect of an earthquake is represented as an equivalent lateral force calculated as a fraction of the building weight. The magnitude of the force depends on the dynamic properties of a building (whether it is a rigid low-rise or a slender high-rise), type of structural system (bearing-wall, moment-resisting frame, etc.), soil properties (soft soil, rock, etc.), and building importance (an apartment building or a hospital of post-disaster importance). The seismic force calculated in this way is distributed over the building height according to an inverse triangular distribution (maximum force at the top decreasing toward the base of the building).

It should be noted that, according to the NBC 1995 seismic provisions, the equivalent static procedure used to be the default method of analysis, whereas the dynamic procedure was prescribed only for seismic analysis of complex structures. There are several reasons for the transition toward the dynamic analysis procedure. Dynamic analysis simulates the effects of earthquakes on a structure much better than the equivalent static force procedure. Also, the dynamic procedure has become straightforward with modern developments in computer software technology.

Seismic hazard It is a complicated process to attempt to predict *seismic hazard,* that is, probable shaking at a site from a given earthquake. The amount of shaking is influenced by the size of the earthquake, the distance from the source of fault slippage, the type of soil at the site, etc. According to NBC 2010, seismic hazard corresponding to specific geographical locations in Canada is presented in the form of *uniform hazard spectra* (UHS). UHS consist of spectral acceleration values expressed as a fraction of the acceleration of gravity (g) corresponding to the different dynamic properties (vibration periods) of a structure. UHS spectral accelerations are a direct representation of structural response. Spectral acceleration values for various locations in Canada are listed in Appendix C of NBC 2010.

UHS are introduced for the first time in NBC 2005. In NBC 1995, Canada was divided into seven seismic zones (zones 0 to 6), depending on the expected seismic hazard and intensity of ground shaking. Each seismic zone was characterized with zonal acceleration (Z_a) and zonal velocity (Z_v) factors.

Earthquake return period The design earthquake most building structures in Canada need to accommodate, according to NBC 2010, corresponds to a ground motion with 2% probability of exceedance in 50 years. NBC 1995 used a design earthquake with 10% probability of exceedance in 50 years (corresponding to a return period of 500 years). The design earthquake according to NBC 2010 is considered more realistic in terms of the probability of failure or collapse that might be expected in an earthquake.

1.7.7 Practical Considerations Related to Load Calculations

In performing structural design, a designer should consider realistic loads, in terms of both their magnitude and their distribution. The following three important considerations need to be kept in mind while performing load calculations:

1. *Critically evaluate the code-prescribed load values.* In general, building codes recommend the minimum load values required to ensure safe design. What if your client's "office" were really more like a library due to huge amounts of file storage? In consultation with the client, you may choose to perform a design using a higher live load than required by the building code.
2. *Use conservative load values in the design.* Experienced designers routinely apply an average area load instead of taking into account the specific location of every partition, pipe, mechanical unit, or ceramic tile. The exact location of some items is often not known until after the structure is designed — or even after it is built! For this reason, the loads used in the design need to be somewhat conservative. However, every project is different, and the designer should always make sure that the assumptions made are appropriate.
3. *Clearly indicate the load values on the structural drawings.* The design notes and the finished structural drawings should always provide a clear summary of the loads used. This recommendation is very important, keeping in mind chances for future building renovations and expansions.

KEY CONCEPTS

The term *load* means the forces, pressures, and imposed deformations applied to the building structure. In general, building codes offer guidelines on how to determine the magnitude of various loads. Part 4 of NBC 2010 prescribes loads to be considered in structural design.

The following five types of loads are commonly considered in the design of concrete structures:

1. *Dead load (D):* a permanent load due to the weight of building components. According to NBC 2010 Cl.4.1.4.1, the specified dead load for a structural member consists of *self-weight* (weight of the member itself) and superimposed dead load (the weight of all materials of construction incorporated into the building to be supported permanently by the member).
2. *Live load (L):* those forces that may or may not be present and acting upon a structure at any time. NBC 2010 defines live load as a variable load due to the intended use and occupancy of a building. Occupancy loads include personnel, furniture, stored materials, and other similar items.

3. *Snow load (S):* a variable environmental load acting mainly on roof structures. Considering the severe winter conditions throughout Canada, due consideration of snow load (including ice and associated rain) and its effect on building structures is of high importance.

4. *Wind load (W):* a variable dynamic load that acts on wall and roof surfaces within a building structure. The magnitude of wind load depends on the orientation, area, and shape of the surface; wind velocity; and air density.

5. *Earthquake load (E)* or *seismic load:* a dynamic load that occurs rather rarely. An earthquake is a vibration of the earth's crust caused by slippage along a fault zone below the surface of the earth. NBC 2010 prescribes seismic design criteria for all new buildings in Canada.

In performing structural design, a designer should consider realistic loads, in terms of both the magnitude and the distribution. The following three considerations need to be kept in mind while performing load calculations:

1. The code-prescribed load values need to be critically evaluated.
2. Conservative load values should be used in the design.
3. The load values should be clearly indicated on the construction drawings.

1.8   THE LIMIT STATES DESIGN METHOD

1.8.1   Limit States

All building structures should be designed to be *safe* from collapse and *serviceable* during their useful life. *Limit states* refer to those conditions under which the structure becomes unfit for its intended use. The following four categories of limit states are considered in the design of concrete structures:

1. *Strength* or *ultimate limit states* concern safety and include such failures as loss of load-carrying capacity, overturning, sliding, and fracture.
2. *Serviceability limit states* restrict the intended use and occupancy of the building and include deflection, vibration, permanent deformation, and local structural damage such as cracking, and fatigue limit states, which represent failure under repeated loading.
3. *Fire resistance* is the property of a material or a structure to withstand fire or to provide protection from fire.
4. *Durability* is related to the long-term satisfactory performance of concrete structures.

It should be noted that these four limit states have been referred to in CSA A23.3 Cl.8.1; however, NBC 2010 refers only to ultimate limit states and serviceability limit states.

    The primary aim of the limit states design is to prevent failure of a structural member or the entire structure.

1.8.2   Ultimate Limit States

The ultimate limit states (ULS) method, also known as "strength design" or load and resistance factor design (LRFD), is used to design concrete structures in Canada. According to the ULS method, loads are factored to calculate an ultimate load, which is then applied to an elastic model of the structure to calculate the internal forces. The load-resisting capacities of structural members under axial loads, flexure, and shear are calculated assuming some inelastic behaviour.

    Uncertainties in the loads are considered using load factors and load combinations. The load factors are based on a statistical interpretation of measured conditions and thus reflect maximum variations in the loads from their mean estimate.

The main objective of ULS design according to NBC 2010 Cl.4.1.3.2(1) is that a building and its structural components should be designed to have sufficient strength and stability so that the factored resistance is greater than or equal to the effect of factored loads; that is,

Factored resistance of a structure ≥ Factored load effect

The *factored load effect*, according to NBC 2010 Cl.4.1.3.2(2), is determined according to the companion action load format, as follows:

Factored load effect = permanent load × load factor + principal load × load factor
+ companion load × load factor

The dead load (DL) is the *permanent load*. The *principal load* is a specified variable load or rare load that dominates in a given load combination. The *companion load* is a specified variable load that accompanies the principal load in a given load combination.

It should be noted that permanent load, principal load, and companion load are *specified loads* (see Section 1.7.1). *Factored loads* used in the ULS method are determined as the product of a specified load and the corresponding load factor.

The companion action load format uses a clear set of load combinations with direct physical meaning. Table 4.1.3.2 of NBC 2010 prescribes seven load cases that deal with typical load combinations.

The *factored resistance* of a structure can be determined as

Factored resistance = ϕR

where R is the calculated resistance of a member, connection, or structure based on the specified material properties, and ϕ is the resistance factor applied to the resistance or specified material property, which takes into account the variability of material properties and dimensions, quality of work, type of failure, and uncertainty in the prediction of resistance.

Example 1.3

A partial roof plan of a reinforced concrete building is shown in the figure below. The roof is subjected to a dead load (*D*) of 6 kPa (including its self-weight), a live load (*L*) of 1.0 kPa, and a snow load (*S*) of 3.0 kPa.

Determine the factored axial compression load for a typical interior column C1 supporting the roof according to NBC 2010 requirements.

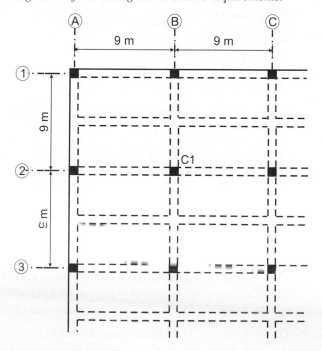

SOLUTION: **1. Determine the tributary area**

In this example, the column spacing is equal to 9 m. Therefore, a typical interior column carries load over the tributary area as shown on the sketch below:

$$A = 9 \text{ m} \times 9 \text{ m} = 81 \text{ m}^2$$

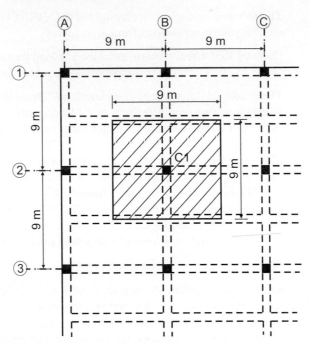

2. Determine the factored roof load

The factored roof load should be determined according to Cl.4.1.3.2 of NBC 2010. In this example, the following loads need to be considered:

$D = 6$ kPa

$L = 1$ kPa

$S = 3$ kPa

By inspection, load cases 1, 2, and 3 of Table 4.1.3.2 of NBC 2010 need to be considered in this design.

Case 1:

$w_{f1} = 1.4D = 1.4 \times 6 \text{ kPa} = 8.4 \text{ kPa}$

Case 2:

$w_{f2} = 1.25D + 1.5L + 0.5S = 1.25 \times 6 \text{ kPa} + 1.5 \times 1.0 \text{ kPa} + 0.5 \times 3.0 \text{ kPa}$

$\quad = 10.5$ kPa

Case 3:

$w_{f3} = 1.25D + 1.5S + 0.5L = 1.25 \times 6 \text{ kPa} + 1.5 \times 3.0 \text{ kPa} + 0.5 \times 1.0 \text{ kPa}$

$\quad = 12.5$ kPa

It can be concluded that case 3 governs as it results in the largest factored load; that is,

$w_f = 12.5$ kPa

3. Determine the factored axial load on the column C1

The factored axial compression load on the column C1 can be determined as a product of the design factored load and the tributary area; that is,

$P_f = w_f \times A = 12.5 \text{ kPa} \times 81 \text{ m}^2 \cong 1010 \text{ kN}$

1.8.3 Serviceability Limit States

Serviceability limit states (SLSs) restrict the intended use and occupancy of the building and include deflection, vibration, permanent deformation, and local structural damage such as cracking. SLSs also include fatigue limit states, which represent failure under repeated loading. The most common SLS that needs to be considered in the design of all buildings relates to deflections. It may happen that a structural component satisfies the ULS criteria, yet it has excessive deflections (sagging) and is therefore considered unacceptable. According to NBC 2010 Cl.4.1.3.5, the following considerations should be taken into account to limit the problems resulting from excessive deflections:

- intended use of the building or a structural member;
- prevention of damage to structural and nonstructural members in a building due to excessive deflections;
- creep, shrinkage, and temperature effects.

Concrete structures are susceptible to cracking, so cracking control is one of the main serviceability design requirements. The design of reinforced concrete structures for serviceability is discussed in Chapter 4.

1.8.4 Fire Resistance

Fire resistance is the ability of a material or a structure to withstand fire or provide protection from fire. The expected fire-resistance performance of a specific building is usually expressed through a *fire-resistance rating,* which can be defined as the time in hours during which a building element or an assembly maintains the ability to confine a fire, continues to perform a given structural function, or both. Section 3 of NBC 2010 specifies the required fire-resistance ratings for various building assemblies. In general, the required fire-resistance rating ranges from 1 to 4 hours.

Reinforced concrete is considered to be one of the most highly fire resistant building materials. In general, reinforced concrete construction does not require additional fireproofing (unlike steel construction). The fire resistance of concrete structures is influenced by the following major factors:

- *Type of concrete:* Temperatures greater than 95°C sustained for a prolonged period of time can have a significant effect on concrete (CAC, 2002). At high temperatures, the cement paste shrinks due to dehydration, while the aggregate expands. Some aggregates can produce volume-stable concrete at high temperatures, whereas others undergo abrupt volume changes at high temperatures, causing distress in concrete. Low-density aggregates generally exhibit better fire performance than natural stone aggregate concretes. Appendix D of NBC 2010 specifies seven different types of concrete as related to fire-resistance requirements. For more details on the fire resistance of different types of concrete, refer to Section 2.5.
- *Member dimensions:* The fire resistance of structural members is proportional to their thickness — the larger the cross-sectional dimensions are, the longer the member takes to heat up and lose strength in case of fire. Appendix D of NBC 2010 specifies the thickness requirements for various structural members (beams, slabs, columns, etc.).
- *Reinforcing steel:* The fire-resistance rating depends on the type of steel and the level of stress in the steel. In general, steel loses strength at high temperatures; however, hot rolled reinforcing steel used for deformed bars loses its strength less quickly than the cold drawn steel used for prestressing tendons.
- *Concrete cover:* This protects the reinforcing steel from fire. The concrete cover requirements for reinforced concrete beams and slabs to retain their fire resistance are prescribed by NBC 2010 Appendix D (refer to Chapter 5 for more details on concrete cover).

The designer must refer to the fire-resistance requirements when selecting member sizes. In some cases, a smaller member thickness may be satisfactory based on the ULS and SLS criteria, but the fire-rating requirements prescribe a larger value.

1.8.5 Durability

Durability is related to the long-term satisfactory performance of concrete structures and their ability to satisfy their intended use and exposure conditions during the specified or traditionally expected service life. CSA A23.3 Cl.8.1 identifies durability as one of the four limit states to be considered in the design of concrete structures. Durability requirements for concrete structures are specified by CSA A23.1–04 "Concrete Materials and Methods of Concrete Construction." In particular, CSA A23.1 Cl.4.1.1 outlines the specific durability requirements for concrete structures subjected to weathering, sulphate attack, corrosive environment, or any other process of deterioration. For more details related to the durability of concrete, refer to Section 2.4.

KEY CONCEPTS

All building structures should be designed to be *safe* from collapse and *serviceable* during their useful life. *Limit states* refer to those conditions under which the structure becomes unfit for its intended use. The following four categories of limit states are considered in the design of concrete structures:

- *Strength or ultimate limit states* concern safety and include loss of load-carrying capacity, overturning, sliding, and fracture.
- *Serviceability limit states* restrict the intended use and occupancy of the building and include deflection, vibration, permanent deformation, and local structural damage such as cracking. SLSs also include fatigue limit states, which represent failure under repeated loading.
- *Fire resistance* is the property of a material or a structure to withstand fire or to provide protection from fire.
- *Durability* relates to the long-term satisfactory performance of concrete structures.

The primary aim of the limit states design is to prevent failure of a structural member or the entire structure.

1.9 ACCURACY IN DESIGN AND CONSTRUCTION

1.9.1 Research Studies Versus Real-Life Design Applications

The design of reinforced concrete structures is based on the concepts and principles developed as a result of laboratory-based research and observations of real-life structural behaviour. After being evaluated through extensive research studies, these concepts are often adopted as provisions of building codes and standards. Structural engineers are faced with the challenging task of applying these code provisions in practice, where a sound design solution presents a balance between economy, ease of construction, prevention of construction errors, and the client's requirements.

The concepts of reinforced concrete design resulting from research studies performed in an ideal laboratory environment must be applied with due consideration for the approximate nature of design and practical limitations of the construction process. In cast-in-place concrete construction, several factors may influence the load-carrying capacity of structural components in the building, including batching process (accuracy of measuring ingredients); duration of the concrete truck travel and waiting time before the actual pour; weather conditions (including air temperature and moisture content during pour); method of placement (pumped, conveyed, or placed with a wheelbarrow); thoroughness of vibration; curing process; proper representation of cylinder tests; variation in reinforcement placement as compared to design drawings; and a construction schedule that can affect quality control and the curing process. All of these factors are subject to human error. The nature of concrete construction is unique, because it relies on the collaborative effort of a large number of different tradespeople. Errors made by any of these tradespeople may adversely affect the load-carrying capacity of a structure. Further on, deviations in the curing process from the minimum required duration of 28 days can also influence the capacity of structural components. The designer can influence this process to a certain extent by paying special attention to quality control during each phase of construction. However, the designer makes the largest contribution to the satisfactory final product, that is, the finished building, when performing design by applying appropriate design concepts and procedures within reasonable bounds of accuracy. In design practice, a design within 5% of the theoretical solution is adequate provided that it adheres to the required principles. The designer should exercise judgment in placing emphasis on practical aspects of construction in favour of too many significant digits in the solution.

1.9.2 Accuracy of Calculations

It is far more important for the designer to understand how a structure behaves under loading, how to arrange structural members, and where to place steel reinforcement to resist tension, than to calculate the reinforcement area to an unnecessary level of accuracy. Most mistakes in structural design arise from the following sources:

- failure to fully understand the behaviour of the structure being designed,
- errors due to unit conversion,
- errors in looking up or writing down numbers.

Failure to understand the behaviour of the structure   This is a very serious mistake. In order to be able to perform his/her work in a competent manner, the designer must have a thorough understanding of the broad principles of statics and structural behaviour. Without this background, any effort to perform structural design may be useless and dangerous. If the analysis is performed using computer software applications, the designer needs to be able to perform a quick approximate check of the results using hand calculations, rather than blindly believing that the output of computer calculations is correct.

Errors due to unit conversion   Many students make conversion errors while performing numerical calculations, leading to completely unrealistic results — either extremely large or extremely small. The following guidelines can be used to ensure that this kind of error does not occur in practice:

- Be knowledgeable about units and their relationships (refer to Section A.3 to review units).
- Always indicate the units within the computation, as shown in the examples in this book.
  Always indicate the units in the final result of a calculation.
- Use common sense while checking the results of your calculations, especially when you get very large or very small values. Relate your results to examples from real life. For example, a small car weighs 1 t or 10 kN, and one large step while walking is approximately equal to 1 m.

Errors in looking up or writing down numbers   These errors are typically a result of the designer's attempt to perform the calculations in a rush while trying to meet project deadlines. However, the designer must find time to properly backcheck the calculations and ensure that all loads and dimensions have been recorded correctly.

It is recommended to record the values in numerical computations with three significant figures. In general, more than three figures can lead to a false sense of accuracy, as structural loads (the input to the calculations) are rarely known to more than two significant figures. In rounding to three significant figures, round to the nearest number. For example, the number 1543 should be rounded to 1540.

If computations are performed with three significant figures, the final answer will have two significant figures.

The guidelines summarized in Table 1.1 may be used as a reference in rounding the results of design calculations.

To minimize errors in arithmetic and logic, the designer should prepare neat and systematic calculations in a consistent format. Properly formatted calculations can be more easily reviewed by someone else and even by the designer, especially when referring to them a few years after the original design has been done. In many cases, owners retain the same designer to design an expansion or renovation of the building, which may occur 5 or 10 years after the original design has been done.

Table 1.1   Recommendations related to the rounding of design calculations

Type of calculation	Record/round to the nearest	Type of element designed
Loading	0.1 kPa	Slabs
	1 kN/m	Beams, girders
	0.1 kN	Point loads
	5 kN	Column and footing loads
Dimensions	10 mm	Span length and location of load
	5 mm	Effective beam and slab depth
Computations	10 kN·m	Bending moments
	10 mm^2	Reinforcement area
Design selection	5 mm	Slab thickness
	25 mm	Beam depth and width
	25 mm	Column cross-sectional dimensions
	25 mm	Bar spacings in slabs and walls

1.10 USE OF COMPUTER-AIDED DESIGN TOOLS

With the wide use of personal computers in a design office environment, various computer-aided design tools can assist structural designers in performing design in a more efficient way and in making better design decisions. Many designers develop their own computer-aided design tools in addition to numerous commercially available software packages for structural analysis and design. Computer spreadsheets are among the most popular design tools and can be developed by designers themselves and then routinely used to increase productivity in a design office.

The use of self-developed computer spreadsheets in design applications carries the following benefits:

- The development of a design spreadsheet requires that a designer go through the entire design process and realize the intricacies of the design.
- In the process of spreadsheet development, problems need to be solved to allow for the maximum flexibility of the spreadsheet application; this leads to a broader understanding of the design problem by a designer developing the spreadsheet.
- The development of spreadsheets is very helpful in avoiding the 'black-box' syndrome associated with using third-party software. An example of the black-box syndrome is "I cannot explain this result, but it must be right because it is an output of a computer program." This is a true downside of computer-aided design. The black-box syndrome is especially dangerous if the designer does not completely understand structural design — such a design may have disastrous consequences. Structural designers should refrain from using third-party analysis and design software unless there is an independent means of verifying the results.
- Finally, the development of spreadsheets allows for sensitivity analysis of design solutions combined with a thorough understanding of the background behind the black-box. Computer-aided design tools are most effective when they offer flexibility. For example, some design parameters can be varied within a range of values, thus enabling a designer to make more informed design decisions that could benefit the overall project objective.

The authors of this book advocate the use of spreadsheets in the design of reinforced concrete structures. Two sample spreadsheets are posted on the companion Web site, one of them is related to the column design and load take down discussed in Chapter 8 and another is related to spread-footing design, in Chapter 14.

KEY CONCEPTS

Most mistakes in structural design arise from the following three sources:

1. failure to fully understand the behaviour of the structure being designed,
2. errors due to unit conversions,
3. errors in looking up or writing down numbers.

It is far more important for a designer to understand how a structure behaves under loading, how to arrange structural members, and where to place steel reinforcement to resist tension than to calculate the reinforcement area to an unnecessary level of accuracy.

SUMMARY AND REVIEW—CONCRETE BASICS

Concrete is one of the most versatile construction materials; it provides potentially unlimited opportunities for developing diverse forms of construction. Concrete is what is known as a universal material, as its ingredients, namely cement, sand, aggregates, and water, are available all over the globe. Concrete also has the excellent characteristics of fire resistance and durability, and requires substantially less maintenance than other materials. Competent design and skilful construction of concrete structures are essential for their satisfactory long-term performance.

Basic concepts of reinforced and prestressed concrete

Concrete is artificial stone made from two main components: cement paste and aggregates. Aggregates usually consist of natural sand and gravel or crushed stone. The paste hardens as a result of the chemical reaction between cement and water and glues the aggregates into a rock-like mass.

Reinforced concrete structures utilize the best qualities of concrete and steel — concrete's high compressive strength and steel's high tensile strength. The main idea behind reinforced concrete is to provide steel reinforcement at locations where tensile stresses exist that the concrete cannot resist. Due to its strength, only a relatively small amount of steel is needed to reinforce concrete. Steel's ability to resist tension is around 10 times greater than concrete's ability to resist compression. It is very important to note that reinforcement in concrete structures is effective *only* if it is appropriately used, strategically placed, and in proper quantity.

Prestressed concrete is a special type of reinforced concrete in which internal compression stresses are introduced to reduce potential tensile stresses in the concrete resulting from external loads. High-strength steel tendons are embedded within the concrete and subjected to a tensile stress imposed by special equipment (jacks). The two main methods of prestressed concrete construction are as follows:

* *pretensioning:* when the tendons are tensioned before the concrete has hardened;
* *posttensioning:* when the tendons are tensioned after the concrete has hardened.

Main components of reinforced concrete buildings

Reinforced concrete buildings consist of several structural components (or members). The basic components of a reinforced concrete building are as follows:

* The *floor and roof systems* are the main horizontal structural components in a building. They carry gravity load and transfer it to the vertical components (columns and/or walls) and also act as horizontal diaphragms by transferring the lateral load to the vertical components of a structure.
* *Beams* transmit loads from the floors to the vertical components (columns). Beams are usually cast monolithically with the slab and are subjected to bending and shear.
* *Columns* are the vertical components that support a structural floor system. They are usually subjected to combined axial compression and bending.
* *Walls* provide the vertical enclosure for a building. *Load-bearing walls* carry gravity loads only, whereas *shear walls* have a major role in carrying lateral loads due to winds and earthquakes.
* *Foundations* transmit the weight of the superstructure to the supporting soil. There are several types of foundations. *Spread footings* transfer the load from the columns to the soil. Walls are supported by *strip footings*.

Structural systems characteristic of reinforced concrete buildings

The structural system (or framing system) represents the skeleton of a building and supports the rest of the structure. Structural systems characteristic of reinforced concrete buildings are as follows:

* The *moment-resisting frame* (or moment frame) consists of columns and beams that act as a 3-D space frame system. Both gravity and lateral forces are resisted by bending in beams and columns mobilized by strong rigid joints between them.
* The *bearing-wall system* consists of reinforced concrete bearing walls located along exterior wall lines and at interior locations as required. These

bearing walls are also used to resist lateral forces, in which case they are called shear walls.

- The *frame/shear-wall hybrid system* utilizes a complete 3-D space frame to support gravity loads and shear walls to resist lateral loads. In modern buildings of this type, the main lateral load-resisting system consists of reinforced concrete shear walls forming the elevator core (central core formed by the elevators and stairs in the building) and additional walls located elsewhere in the building as required.

The concept of load path

Within every building, multiple elements are used to transmit and resist external loads. These elements define the mechanism of load transfer in a building known as the *load path*. Depending on the type of load to be transferred, there are two basic types of load path:

- *Gravity load path:* The gravity load acting on the slab is transferred by the floor and roof systems to the columns or walls, down to the foundations, and then to the ground. The gravity load path depends on the type of floor system. In a *one-way system,* external loads are transferred primarily in one direction. The gravity load is transferred from the slab to the beams and then to the girders. Finally, the girders transfer the load to the columns. In a *two-way floor system,* the gravity load path is not as clearly defined. The slab transfers load in two directions; however, the amount carried in each direction depends on the type of end supports, among other factors.
- *Lateral load path:* This is the way lateral loads (mainly due to wind and earthquakes) are transferred through a building. The primary elements of a lateral load path are vertical components (shear walls and frames) and horizontal components (roof, floors, and foundations).

Key considerations related to the design of reinforced concrete structures

The main objective of structural design is to ensure structural safety and serviceability on the one hand and economy on the other. The four key structural design considerations are as follows:

1. *Structural safety and serviceability:* Building structures must be designed to be safe from collapse and serviceable throughout their useful life.
2. *Functional requirements:* These requirements are driven by architectural and structural engineering considerations. In general, a building should enclose space and contents, provide security to occupants, and facilitate work in the enclosed space.
3. *Economics:* A good design must achieve a proper balance between cost, speed of construction, and quality.
4. *Durability:* the designer needs to ensure that the concrete structure will remain functional throughout its expected useful life, which is considered to be at least 50 years, and meet environmental impact requirements.

The design process

The five major steps in the design process are as follows:

1. *Schematic (conceptual) design:* This involves the identification of project constraints, including cost, building shape, and architectural form, and functional constraints, including column spacing, materials, and serviceability criteria. This step also involves selection of the structural system, development of the general arrangement of structural elements, and trial sizes of key structural members based on estimated loads.
2. *Detailed (final) design:* This involves the detailed analysis, evaluation, and redesign of preliminary member sizes and more refined calculation of gravity and lateral loads. The safety and serviceability of the structure need to be ensured by carefully following the requirements of pertinent building codes.
3. *Development of contract documents (drawings and specifications):* The main focus is to transfer the concept and details from the mind of the designer to those involved in the construction. A well-designed building is worthless if it is poorly constructed.

4. *Coordination:* The designer has to cooperate with other disciplines, including architectural, electrical, mechanical, civil, and geotechnical engineering.
5. *Services during construction:* These include review of shop drawings, concrete mix designs, and laboratory testing reports. The designer must make regular visits to the site to observe the construction, answer questions, and clarify contract documents.

Methods of concrete construction

Concrete construction is a complex process that involves many steps, from batching and mixing the ingredients to pouring fresh concrete into forms and curing the new construction. The four basic methods of concrete construction are as follows:

1. *Cast-in-place construction:* Each element is built at the construction site. Fresh concrete, obtained by mixing the ingredients (cement, sand, and aggregates) and adding water, is a shapeless slurry without any strength, so it has to be supported and shaped by the formwork.
2. *Precast construction:* The concrete is cast into permanent, reusable forms at an industrial plant, then transported in the form of fully cured structural units to the job site, where these units are hoisted into place and connected in a manner similar to structural steel shapes. Precast concrete elements are mainly used for bridge construction in Canada.
3. *Tilt-up construction:* Precast wall panels are cast horizontally on the floor slab. Once the panels have attained sufficient strength, a mobile crane sets them on footings. Tilt-up construction is mainly used in low-rise warehouse or office buildings.
4. *Concrete masonry construction:* Masonry load-bearing walls are assembled by laying masonry units in mortar and providing vertical and horizontal steel reinforcement. Masonry is used for warehouse, industrial, and institutional construction.

Loads considered in the design of reinforced concrete structures

The term *load* means the forces, pressures, and imposed deformations applied to the building structure. In general, building codes offer guidelines on how to determine the magnitude of various loads. Part 4 of NBC 2010 prescribes loads to be considered in structural design.

The following five types of loads are commonly considered in the design of concrete structures:

1. *Dead load* (D) is a permanent load due to the weight of building components; it consists of *self-weight* (weight of the member itself) and *superimposed dead load* (weight of all construction materials supported permanently by the member).
2. *Live load* (L) is a variable load due to the intended use and occupancy of a building. Occupancy loads include personnel, furniture, stored materials, and other similar items.
3. *Snow load* (S) is a variable environmental load acting mainly on the roof structures. Considering the severe winter conditions throughout Canada, due consideration of snow load (including ice and associated rain) and its effect on building structures is of high importance.
4. *Wind load* (W) is a variable dynamic load that acts on wall and roof surfaces within a building structure. The magnitude of wind load depends on the orientation, area, and shape of the surface; wind velocity; and air density.
5. *Earthquake load* (E) or *seismic load* is a dynamic load that occurs rarely. An earthquake is a vibration of the earth's crust caused by slippage along a fault zone below the surface of the earth. NBC 2010 prescribes seismic design criteria for all new buildings in Canada.

In performing the design, a designer should consider realistic loads in terms of both the magnitude and the distribution. The following three considerations need to be kept in mind while performing load calculations:

1. The code-prescribed load values need to be critically evaluated.
2. Conservative load values should be used in the design.
3. The load values should be clearly indicated on the construction drawings.

Limit states

All building structures should be designed to be *safe* from collapse and *serviceable* during their useful life. *Limit states* refer to those conditions under which the structure becomes unfit for its intended use. The following four categories of limit states are considered in the design of concrete structures:

- *Strength or ultimate limit states* concern safety and include such failures as loss of load-carrying capacity, overturning, sliding, and fracture.
- *Serviceability limit states* restrict the intended use and occupancy of the building and include deflection, vibration, permanent deformation, and local structural damage such as cracking.
- *Fire resistance* represents the property of a material or a structure to withstand fire or to provide protection from fire.
- *Durability* relates to the long-term satisfactory performance of concrete structures.

The primary aim of the limit states design is to prevent failure, that is, the attainment of a limit state.

PROBLEMS

1.1. Determine the specified uniformly distributed live load according to NBC 2010 for the following design applications:
  a) the second floor in an office building
  b) a floor in a parking garage for passenger cars
  c) the residential area in a 20-storey apartment building
  d) a college library (at maximum load)

1.2. Consider a typical classroom in your building.
  Determine the average occupancy live load and compare the obtained value with the NBC 2010 prescribed value for this type of occupancy.

1.3. A simply supported reinforced concrete beam of rectangular cross-section (350 mm width and 700 mm overall depth) spans 8 m. The beam supports a specified live load of 15 kN/m in addition to its own weight. Use the unit weight of 24 kN/m³ for reinforced concrete.
  Determine the maximum bending moment in the beam
  a) due to the total specified load, M_s
  b) due to the factored load, M_f

LL =15 kN/m

700 mm

350 mm

l = 8 m

1.4. A partial floor plan of a reinforced concrete building is shown in the figure below. The floor structure is subjected to a specified dead load (D) of 10.0 kPa (including the floor self-weight) and a specified live load (L) of 5.0 kPa.
  a) Determine the uniform factored load on the typical beam B1 and girder G1.
  b) Determine the factored axial compression load on the column C1.

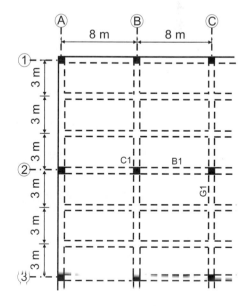

1.5. Find the self-weight (expressed in kilonewtons per metre) for the reinforced concrete T-beam in the figure that follows. Use the unit weight of 24 kN/m³ for reinforced concrete.

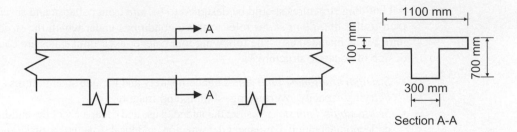

Section A-A

1.6. A section through a floor slab in a parking garage for passenger cars is shown in the figure below. The slab is supported by T-beams spaced at 4.5 m on centre. The slab is subjected to a superimposed dead load of 0.3 kPa due to lighting fixtures and water-proof coatings. The slab is made of normal-density concrete with unit weight 24 kN/m^3.

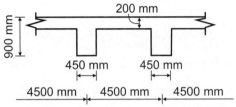

a) Determine the specified uniformly distributed live load on the slab according to NBC 2010 requirements.

b) Determine the self-weight for a typical T-beam expressed in kilonewtons per metre.

c) Determine the total factored load for a typical T-beam considering all possible loads and load combinations according to NBC 2010.

1.7. The column C1 discussed in Problem 1.4 is subjected to the following specified axial loads due to dead load (*D*), live load (*L*), and snow load (*S*):

$P_D = 1000$ kN
$P_L = 800$ kN
$P_S = 500$ kN

Determine the maximum factored load for which this column should be designed based on the NBC 2010 requirements.

2 Materials

LEARNING OUTCOMES

After reading this chapter, you should be able to

- identify the main concrete-producing materials and describe their characteristics as related to concrete construction
- describe the most important mechanical properties of concrete
- discuss the factors influencing creep and shrinkage effects in concrete structures
- list the basic exposure classes for concrete structures and discuss the durability requirements
- identify the types of reinforcement in concrete structures and outline the related design requirements and mechanical properties

2.1 INTRODUCTION

Concrete is artificial stone made from two main components: cement paste and aggregates. It is a composite material consisting of natural (but processed) ingredients, such as water, aggregate, and air, and manufactured ones, such as cement and admixtures. The concrete manufacturing process is quite complex and has a number of steps, including proportioning, batching, mixing, placing, compacting, finishing, and curing. In building structures, concrete is most often used in the form of reinforced concrete, which is a composite material consisting of concrete and steel reinforcement.

Concrete is a highly versatile building material, but the design of safe and durable concrete structures is closely related to the quality of concrete in finished structures. The authors share the view of Neville (2003) that "a good knowledge of concrete technology is as important as a good knowledge of structural analysis and design." There are several stakeholders in the design and construction of concrete structures. The structural designer and the specifier determine the required structural qualities of concrete, while the materials engineer, contractor, and supplier control the actual quality of concrete in the finished structure. All of these stakeholders must be thoroughly conversant in the properties of concrete and must understand the structural aspects of what is being built (Neville, 2003).

It is expected that the reader has already gained a background in concrete technology. This chapter provides an overview of concrete-producing materials and the mechanical characteristics of concrete and steel relevant to the structural design of concrete structures covered in subsequent chapters. Section 2.2 gives an overview of concrete materials and production. The relevant mechanical properties of hardened concrete used in structural design are discussed in Section 2.3. Section 2.4 covers the durability of concrete structures, classes of exposure, and related durability requirements. Fire-resistance requirements for concrete structures are outlined in Section 2.5. Types and properties of steel reinforcement in concrete structures are discussed in Section 2.6.

In Canada, concrete is manufactured according to the pertinent CSA standards. For the reader's convenience, materials-related standards referred to in this chapter are listed below.

Concrete-related standards:

CSA A23.1-04 Concrete Materials and Methods of Concrete Construction
CSA A23.2-04 Methods of Test for Concrete
CSA A23.3-04 Design of Concrete Structures
CSA A23.4-00 Precast Concrete — Materials and Construction
CSA A266.4 Guidelines for the Use of Admixtures

Steel-related standards:

G30.18-M1992(R2002) Billet Steel Bars for Concrete Reinforcement
ASTM A185-02 Standard Specification for Steel Welded Wire Reinforcement, Plain, for Concrete
ASTM A496-02 Standard Specification for Steel Wire, Deformed, for Concrete Reinforcement
ASTM A497/497M-02 Standard Specification for Steel Welded Wire Reinforcement, Deformed, for Concrete

2.2 CONCRETE MATERIALS AND PRODUCTION

Concrete is a composite material made from a binder (cement paste) and a filler (aggregate). The paste usually contains Portland cement, water, and air. Aggregates usually consist of natural sand and gravel or crushed stone. The paste hardens as a result of the chemical reaction between cement and water and glues the aggregates into a rock-like mass. In addition to these basic ingredients, supplementary cementing materials, chemical admixtures, and/or fibre reinforcement are often added to the concrete mixture. It is important to note that the paste accounts for 25% to 40% of the total volume of concrete, while the aggregates account for the remaining 60% to 75% (CAC, 2002). A typical mix of concrete ingredients is shown in Figure 2.1. A typical concrete mix is expressed as the ratio by mass of cement to fine aggregate to coarse aggregate, in that order (for example, roughly 1:2:4).

Concrete is a composite material with several components, so designers need to know the properties that are required for satisfactory concrete construction. This section presents a brief overview of concrete materials and their role in concrete construction. For more details on this subject, the reader is referred to CSA A23.1-04 and CAC (2002).

2.2.1 Portland Cement and Supplementary Materials

Portland cement is a mixture made by heating limestone, clay, and other raw materials in a rotary kiln at very high temperatures. Portland cement is hydraulic cement, which can set when combined with water and hardens in a chemical process called hydration. CSA A23.1 Cl.3.1.2 specifies five types of Portland cement that meet different physical and chemical requirements for specific purposes:

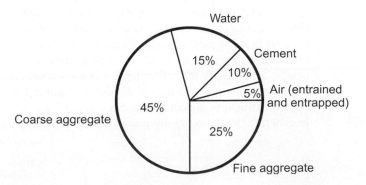

Figure 2.1 Concrete ingredients.

Type 10 (normal Portland cement) is most commonly used. It is a general-purpose cement used where cement or concrete is not subjected to specific exposures such as sulphate attack from soil or water.

Type 20 (moderate Portland cement) is used where precaution against moderate sulphate attack is important, such as in drainage structures.

Type 30 (high-early-strength Portland cement) provides high strength at an early age (usually a week or less). It is used when forms need to be removed as soon as possible or when the structure must be quickly put into service.

Type 40 (low-heat-of-hydration Portland cement) is not commonly available, but is used where the rate and amount of heat generated as a result of the hydration process must be minimized. This is often a requirement in the construction of massive concrete structures such as gravity dams.

Type 50 (sulphate-resistant Portland cement) should be used only for concrete exposed to high sulphate content in soil or groundwater.

In addition to Portland cement, supplementary cementitious materials are often used in concrete construction. These materials are mixed with Portland cement for various reasons, either related to economy or to improve the mechanical properties of hardened concrete. Most supplementary cementitious materials are by-products of other processes and therefore their use is desirable from environmental and energy-saving points of view. Fly ash, silica fume (microsilica), and granulated slag are among the most commonly used supplementary cementing materials. Fly ash, a by-product of coal-burning power plants, usually replaces a part of Portland cement when used in concrete. Fly ash enhances the workability of concrete, especially in mixes with low cement content. Silica fume is a by-product of the manufacture of silicon metals, and it is mainly used in high-strength concrete construction. Granulated slag (also known as ground-granulated blast-furnace slag) can be used as a partial replacement for cement to improve the sulphate resistance of hardened concrete. It should be noted that fly ash and silica fume are pozzolans, that is, materials that react with one of the products of the hydration of Portland cement to form compounds with cementitious properties.

To acknowledge the frequent use of supplementary cementitious materials, the term *cementitious* has replaced the term *cement* in technical publications. In this case, cementitious refers to Portland cement and possible supplementary materials and is used even when no supplementary materials are added to the mixture.

2.2.2 Water and Water-Cement Ratio

Water has an important role in concrete construction and throughout the service life of a concrete structure. Water is used to wash aggregates for a concrete mixture, and a certain amount of water is added to the concrete mixture as prescribed by the mix design. Water is also used to cure the concrete and to wash out mixers and other equipment.

In general, water constitutes from 14% to 21% of the total volume of fresh concrete (CAC, 2002). Water used in concrete mix (also called *mix water*) serves a dual purpose: it combines with cement in a chemical process called hydration and also provides the workability required for proper placing and compaction. The quality of water used for the mix is rarely a problem. In general, drinking water is suitable. Water containing harmful ingredients, contamination, salt, oil, sugar, silt, or chemicals has a negative effect on the strength and setting properties of cement and should not be used.

As discussed above, water and cement make up the cement paste that bonds the aggregates together in the hardened concrete. The amount of water used in the mix significantly influences the quality and especially the strength of concrete and should be carefully controlled. The *water-cement (w/c) ratio* denotes the ratio of water and cement in a concrete mix (by mass). The w/c ratio is used as a basis for concrete mix proportioning. A minimum amount of water is necessary to ensure complete hydration (w/c ratio of approximately 0.4), but beyond that, the higher the w/c ratio, the weaker the concrete, with all other factors remaining constant. Figure 2.2 shows a decrease in the 28-day compressive strength of a concrete cylinder with increasing w/c ratio. Some advantages of a lower w/c ratio are increased

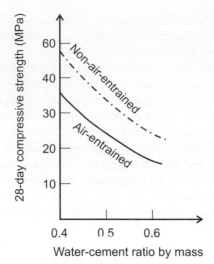

Figure 2.2 Variation of 28-day compressive strength with w/c ratio (adapted from CAC, 2002 with the permission of the Cement Association of Canada).

compressive and flexural strength, increased watertightness, increased resistance to weathering, better bond between successive layers and between concrete and reinforcement, and reduced shrinkage-cracking tendencies.

Typical values for the w/c ratio required to achieve a workable consistency in the field range from 0.4 to 0.6. The maximum w/c ratios for various exposure conditions are specified in Table 2 of CSA A23.1. It should be noted that some technical publications use the term *water-cementitious material ratio* (w/cm ratio) instead of w/c ratio (see the discussion on cementitious materials in Section 2.2.1).

2.2.3   Air

Air content in concrete varies from 0% (non-air-entrained concrete) to 8% (air-entrained concrete). There are two different sources of air in concrete, namely *entrapped air* and *entrained air*. *Entrapped air* voids occur in all concrete and can be reduced by proper placement and consolidation of fresh concrete. However, intentionally *entrained air* bubbles are produced by adding an air-entraining agent that stabilizes bubbles formed during the mixing process. These air bubbles are well distributed and completely separated. Entrained air bubbles are extremely small in size and range from 10 μm to 1000 μm in diameter, whereas the size of entrapped voids is 1000 μm or larger.

The main benefits of intentionally entrained air are improved concrete resistance to freeze-thaw cycles and workability, decreased bleeding, increased resistance to scaling from de-icing chemicals, and improved resistance to sulphate attack. However, as entrained air is increased there is a corresponding decrease in compressive strength. Table 4 of CSA A23.1 outlines two air-content categories (categories 1 and 2), characterized by different air-content requirements depending on the size of coarse aggregate. In general, higher air content (on the order of 5% to 8%) is required for concrete subjected to severe exposure conditions. These air-content categories are specified for concrete exposure classes (to be discussed in Section 2.4).

2.2.4   Aggregates

Aggregates are a nearly inert filler material constituting from 60% to 75% of the total volume of concrete (CAC, 2002). Although aggregates do not participate in the chemical reactions that cause cement paste to harden, they significantly affect the quality of concrete. Aggregates should consist of particles with adequate strength and resistance to exposure conditions and should not contain materials that will cause concrete to deteriorate.

Aggregates can be *fine* or *coarse*. *Fine aggregates* consist of natural or manufactured sand with most particle sizes less than 5 mm. *Coarse aggregates* consist of gravels or crushed stone with particle sizes predominantly larger than 5 mm, and a maximum size generally between 10 mm and 40 mm. The most commonly used maximum aggregate size is 20 mm.

Aggregates must be clean, hard, strong, durable, and free of coatings of clay and other fine materials that could affect the hydration and bond of the cement paste. Soft, porous aggregates can result in weak concrete with low wear resistance, while the use of hard aggregates results in rather high-strength concrete and a high resistance to abrasion (but less workability).

The key aggregate properties are as follows:

- *Grading* is the particle-size distribution of the aggregate. Aggregates should be well graded to improve packing efficiency and minimize the amount of cement paste needed. This makes the concrete more workable.
- *Particle shape* and *surface texture* influence the properties of freshly mixed concrete more than the properties of hardened concrete. The bond between cement paste and a given aggregate generally increases as particles change from smooth and rounded to rough and angular. Rough-textured, angular, elongated particles require more water to produce workable concrete than do smooth, rounded, compact aggregates.
- *The density of aggregate* determines the density of concrete. The most commonly used aggregates, such as sand, gravel, crushed stone, and air-cooled blast-furnace slag, produce wet or freshly mixed normal-density concrete (density from 2200 kg/m^3 to 2400 kg/m^3).

Other aggregate properties are absorption and surface moisture, resistance to freezing and thawing, wetting and drying, abrasion and skid resistance, strength and shrinkage, resistance to corrosive substances, and alkali reactivity.

2.2.5 Admixtures

Admixtures are the ingredients in concrete other than cementitious materials, water, and aggregates that are added to the mix before or during mixing. The major reasons for using admixtures are

- to reduce the cost of concrete construction;
- to achieve certain properties of concrete more effectively than by other means;
- to ensure the quality of concrete during the stages of mixing, transporting, placing, and curing in adverse weather conditions;
- to overcome certain emergencies during concreting operations.

A brief overview of most important admixtures is given below. For more details, the reader is referred to CSA A266.4 and CAC (2002).

Air-entraining admixtures purposely cause the formation of billions of tiny air bubbles in the concrete. Air entrainment must be used in concrete that will be exposed to freezing and thawing cycles and/or de-icing chemicals and can also be used to improve workability. The tiny bubbles provide interconnected pathways such that the water near a concrete surface can escape as it expands due to freezing temperatures. Without air entrainment, freeze-thaw cycles will cause the concrete to spall (degenerate). Similarly, air entrainment reduces surface scaling caused by de-icing chemicals and salts.

Water-reducing strength-increasing admixtures reduce the quantity of water required to produce concrete of a certain slump, reduce w/c ratio (and thus increase strength), reduce cement and water content, or increase slump.

Set-retarding admixtures retard the rate of concrete setting. In some cases, these admixtures are used to offset the accelerating effect of hot weather, to delay the initial set of concrete or grout when difficult or unusual placement conditions occur (such as in large piers and foundations), or to delay the set for special finishing processes.

Accelerating admixtures accelerate the strength development of concrete at an early age. Calcium chloride ($CaCl_2$) is a commonly used accelerating admixture; however, its amount is limited to 2%. Note that $CaCl_2$ has some negative effects: it causes an increase in drying shrinkage, potential corrosion of reinforcement, discoloration, and scaling.

Superplasticizers (high-range water reducers) are added to concrete with a low to normal slump and w/c ratio to make high-slump flowing concrete for special applications. Alternatively, these admixtures are used as high-range water reducers to get high-strength concrete. Superplasticizers are used in applications such as thin section placements, areas of closely spaced and congested reinforcing steel, and underwater placements.

2.2.6 Concrete Mix Design and Fresh Concrete

Proportioning and mixing concrete components is one of the most important steps in the concrete manufacturing process, a step that strongly influences the quality of hardened concrete. The objective of concrete mix design is to determine the most economical and practical combination of readily available materials to produce concrete that will satisfy the performance requirements under particular conditions of use. To produce satisfactory fresh concrete, the consistency of the mix needs to be such that the concrete can be placed and compacted by means available at the site; also, the mix should be cohesive enough to be transported and placed without segregation. The hardened concrete needs to be durable, and it should have adequate strength (according to the design specifications) and a uniform appearance.

Before a concrete mix can be proportioned, mix characteristics are selected based on the intended use of the concrete, the exposure conditions, the size and shape of members, and the physical properties of the concrete (such as strength) required for the structure (ACI, 1998).

The *mix proportions* need to meet the following major requirements:

- *compressive strength* (based on w/c ratio);
- *durability* (including the requirements for air entrainment related to a particular exposure class, compressive strength, and cement type);
- *slump* (based on minimum workability requirements for method of placement); and
- *maximum aggregate size* (limited by section dimensions and reinforcement spacing).

The concrete mix proportion is usually expressed as the ratio, by volume, or more accurately by mass, of cement to fine aggregate to coarse aggregate, in that order (for example, 1:2:4), together with a w/c ratio. The first step in proportioning a concrete mix is the selection of the appropriate w/c ratio or minimum 28-day compressive strength. As discussed in Section 2.2.2, the lower the w/c ratio, the stronger the concrete, all other factors remaining constant. A higher w/c ratio reduces strength and adversely affects other concrete properties as well.

Economy is an important consideration in designing a concrete mix. It would seem that by keeping the cement content high, enough water could be used to provide a satisfactory workability while keeping the w/c ratio suitably low. However, cement is the most expensive basic ingredient of concrete. Since larger aggregate sizes have relatively less surface area for the cement paste to coat, and since less water means less cement, one way of achieving economy is to use the largest practical maximum aggregate size and the stiffest practical mix.

The key characteristics of *fresh concrete* are as follows:

1. *Workability* is a measure of how easy or difficult it is to place, consolidate, and finish concrete.
2. *Consistency* is the ability of freshly mixed concrete to flow (measured with a standard slump test, which is an indicator of workability).
3. *Plasticity* determines the concrete's ease of moulding. Freshly mixed concrete should be plastic or semifluid and generally capable of being moulded by hand.

The quality of the cement paste strongly influences the mechanical properties of hardened concrete. In properly produced concrete, each aggregate particle is completely coated and all of the spaces between aggregate particles are completely filled with paste.

2.2.7 Hardened Concrete

Freshly mixed concrete should be plastic and generally capable of being moulded by hand. Concrete sets in about 2 hours after mixing. The hardening of concrete is caused by a chemical reaction between the cement and water called *hydration*. Concrete gains strength as it cures; this strength continues to increase as long as any unhydrated cementing materials are present, provided that the concrete remains moist or has a relative humidity above about 80% and the temperature remains favourable. Note that concrete does not harden or cure by drying: it needs moisture to hydrate and harden. Figure 2.3 illustrates an increase in concrete strength with age (note the significant difference in strength between the moist-cured specimen and the air-cured specimen). The rate of strength gain slows with time but most of its final strength will have been

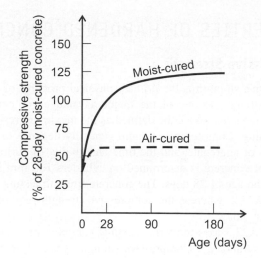

Figure 2.3 Variation of concrete compressive strength with age and curing condition (adapted from CAC, 2002 with the permission of the Cement Association of Canada).

attained within 28 days. Normal concrete under proper curing gains approximately 75% of its design strength by the end of 14 days. The 14-day strength is usually adequate to carry dead load and construction loads (depending on mix design), so the formwork can be removed at about that time. However, a contractor may wish to alter the concrete mix to more quickly attain strength in order to reduce the need for shoring or to improve construction sequencing. This can be achieved by varying the mix proportions, using admixtures, or using a stronger mix than required. The properties of the proposed mix should always be verified by the structural designer because the design relies on concrete with specified characteristics.

KEY CONCEPTS

Proportioning and mixing concrete components is one of the most important steps in the concrete manufacturing process. The concrete mix proportion is usually expressed as the ratio by mass of cement to fine aggregate to coarse aggregate, in that order (for example, 1:2:4), together with a w/c ratio. The mix proportions need to meet the following major requirements:

- *compressive strength* (based on w/c ratio);
- *durability* (including the requirements for air entrainment related to a particular exposure class, compressive strength, and cement type);
- *slump* (based on minimum workability requirements for method of placement);
- *maximum aggregate size* (limited by section dimensions and reinforcement spacing).

The key characteristics of *fresh concrete* are as follows:

- *Workability* is a measure of how easy or difficult it is to place, consolidate, and finish concrete.
- *Consistency* is the ability of freshly mixed concrete to flow (measured with a standard slump test).
- *Plasticity* is the ease of moulding the concrete. Freshly mixed concrete should be plastic or semifluid and generally capable of being moulded by hand.

Hardened concrete needs to be durable and of adequate strength (according to the design specifications) and should have a uniform appearance. Concrete will set in about 2 hours after mixing and will continue to gain strength as it cures. Concrete gains most of its final strength within 28 days. Normal concrete under proper curing gains approximately 75% of its design strength in 14 days.

2.3 PROPERTIES OF HARDENED CONCRETE

2.3.1 Compressive Strength

Compressive strength is the primary physical property of hardened concrete. High compressive strength is one of the major advantages of concrete as a building material. Compressive strength can be defined as the maximum resistance of concrete cylinders to axial loading. Compressive strength generally depends on the type of concrete mix, the properties of aggregate, and the time and quality of curing. In Canada, the compressive strength of concrete is determined on cylinders 100 mm in diameter and 200 mm high, tested at the age of 28 days. The concrete strength testing procedure has been prescribed by CSA A23.2, whereas the compressive strength requirements are prescribed by CSA A23.1 Cl.4.3.5.

CSA A23.3 uses the term *specified compressive strength of concrete* and the corresponding symbol f_c'. Compressive strength is usually expressed in megapascals (MPa). General-use concrete has a compressive strength between 20 MPa and 40 MPa, whereas high-strength concrete is characterized with an f_c' of over 70 MPa. Compressive strengths as high as 200 MPa have been used in building applications.

Consider a typical stress-strain curve for concrete obtained from tests on cylinders loaded in axial compression, as shown in Figure 2.4. The initial portion of the curve, up to about 40% of the maximum stress, can be considered linear for most practical purposes. However, with further load increase (at approximately 70% of the maximum stress), the material loses a large portion of its stiffness, which is reflected by the curvilinear shape of the stress-strain diagram in that range. Before the ultimate load has been reached, cracks of an hourglass shape become distinctly visible, and most concrete cylinders fail suddenly shortly thereafter. A typical concrete cylinder after failure is shown in Figure 2.5.

The ultimate stress at the peak of the stress-strain diagram is equal to a characteristic compressive strength (f_c') corresponding to a strain of approximately 0.002. A further increase in strain is accompanied by a decrease in stress until the failure. For the usual range of concrete strengths, the strain at failure is on the order of 0.0035 to 0.0045. It should be noted that higher-strength concretes experience a brittle failure characterized by smaller failure strains. For design purposes, CSA A23.3 Cl.10.1.3 limits the maximum strain at the extreme concrete compression fibre to 0.0035.

It should be noted that the strength of concrete cast in a structure tends to be somewhat lower as compared to the strength of test cylinders made from the same concrete. Also, the concrete near the top of deep members tends to be weaker than the concrete lower down. For more details, see Bartlett and MacGregor (1999).

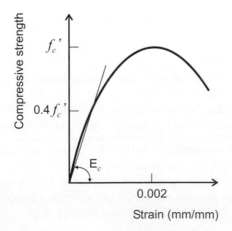

Figure 2.4 Typical stress-strain curve for concrete under compression.

Figure 2.5 A concrete cylinder after compression testing to failure.

(Svetlana Brzev)

Other characteristic concrete strengths, such as flexural, tensile, torsional, and shear strength, are related to the compressive strength, but the correlation varies with concrete ingredients and environmental factors.

DID YOU KNOW?

A concrete with 70 MPa compressive strength was used in the construction of the Scotia Tower in Toronto, featured in Section 1.2. The concrete in the columns was able to reach 12 MPa compressive strength in 12 hours, thus allowing a 3-day pouring cycle. The concrete reached its full strength of 70 MPa in 91 days. In total, around 70 000 m^3 of concrete were used in the construction of this skyscraper (RSIO, 1988).

2.3.2 Tensile Strength

The tensile strength of concrete is relatively low (about 8% to 15% of the compressive strength) and it is often assumed to be equal to zero in structural calculations. The actual value is strongly affected by the compressive strength of concrete, the type of test used to determine the tensile strength, and other factors (Raphael, 1984). Tensile strength can be determined either by the flexure test (CSA Test Method A23.2-8C), in which a plain concrete beam is loaded to failure under third-point loading, or by the splitting test (CSA Test Method A23.2-13C), in which a standard concrete cylinder of the same type as used for the compression test is loaded in compression on its side until splitting failure has occurred. CSA A23.3 uses the *modulus of rupture* (f_r) as an indicator of concrete tensile strength. The modulus of rupture is the theoretical maximum tensile stress reached in the extreme fibre of a flexural member. Cracks develop when the tensile stress in concrete reaches the modulus of rupture and the concrete is unable to resist any tension beyond that point. According to CSA A23.3 Cl.8.6.4, f_r can be determined as

A23.3 Eq. 8.3 $\qquad f_r = 0.6\,\lambda\,\sqrt{f_c'}\ \text{(MPa)}$ [2.1]

where

f_c' = the specified compressive strength of concrete (MPa)

λ = the factor to account for the density of concrete (A23.3 Cl.8.6.5); λ can assume the following values:

  $\lambda = 1.0$ for normal-density concrete;
  $\lambda = 0.85$ for structural semi–low-density concrete, in which all the fine aggregate is natural sand;
  $\lambda = 0.75$ for structural low-density concrete, in which none of the fine aggregate is natural sand.

It should be noted that some other concrete properties (such as shear strength and bond strength) are strongly influenced by the tensile strength.

2.3.3 Shear Strength

The shear strength of concrete is generally not used in design. Shear failure in concrete members usually occurs due to concrete reaching its tensile resistance. It is rather difficult to determine the concrete shear strength experimentally — shear strength is usually determined from tests on specimens subjected to combined stresses. In general, shear strength is on the order of 20% of the compressive strength (CAC, 2002). Shear stresses in concrete structures are generally limited to rather low values in order to prevent the occurrence of brittle shear failure. For more details related to concrete shear resistance, the reader is referred to Chapter 6.

2.3.4 Modulus of Elasticity

The modulus of elasticity is an important property of concrete used in deflection calculations and member stiffness calculations in analyses of concrete structures. The modulus of elasticity varies considerably as a function of concrete strength, concrete density, and, especially for higher-strength concretes, the type of coarse aggregate. In general, the modulus of elasticity for materials that demonstrate elastic behaviour is equal to the slope of the stress-strain curve, that is, the ratio of normal stress and strain. However, concrete is characterized by a nonlinear stress-strain curve, so the modulus of elasticity can be defined as either a tangent or a secant to the stress-strain curve in the range of "nearly" elastic strain.

A23.3 Cl.8.6.2.1 defines the modulus of elasticity (E_c) as the average secant modulus equal to the slope of the straight line that connects the origin of the stress-strain diagram to a stress of $0.4\,f_c'$ (see Figure 2.4). A23.3 Cl.8.6.2.3 prescribes the following simple E_c equation suitable for normal-density concretes with compressive strength range from 20 MPa to 40 MPa:

A23.3 Eq. 8.2 $\qquad E_c = 4500\sqrt{f_c'}\ \text{(MPa)}$ [2.2]

Note that f_c' is expressed in megapascals. In the general case, for concrete with mass density (γ_c) between 1500 kg/m^3 and 2500 kg/m^3, CSA A23.3 Cl.8.6.2.2 prescribes the equation

A23.3 Eq. 8.1 $\qquad E_c = \left(3300\sqrt{f_c'} + 6900\right)\left(\dfrac{\gamma_c}{2300}\right)^{1.5}\ \text{(MPa)}$ [2.3]

where γ_c is the concrete mass density in kilograms per cubic metre (refer to Section 2.3.8 for more details).

The actual modulus of elasticity of concrete generally varies from 80% to 120% of the values obtained by using the equation prescribed by A23.3 Cl.8.6.2.2 and 8.6.2.3.

KEY CONCEPTS

Compressive strength can be defined as the maximum resistance of a concrete cylinder to axial compression. Compressive strength is the primary physical property of hardened concrete and generally depends on the type of concrete mix, the properties of the aggregate, and the time and quality of curing. In Canada, concrete compressive strength is determined on cylinders 100 mm in diameter and 200 mm high, tested at the age of 28 d. CSA A23.3 uses the term *specified compressive strength of concrete* and the corresponding symbol f_c'. General-use concrete has a compressive strength of between 20 MPa and 40 MPa, whereas high-strength concrete is characterized by an f_c' value of over 70 MPa.

The *tensile strength* of concrete is relatively low (about 8% to 15% of the compressive strength) and it is strongly affected by the compressive strength, the type of test used to determine the tensile strength, etc. CSA A23.3 uses the *modulus of rupture* (f_r) as an indicator of concrete tensile strength. The modulus of rupture is the theoretical maximum tensile stress reached in the extreme fibre of a flexural member. When the tensile stress in concrete reaches the modulus of rupture, cracks develop and concrete is unable to resist any tension beyond that point.

The *modulus of elasticity* (E_c) is used in the calculation of deflections and member stiffnesses in the analyses of concrete structures. CSA A23.3 Cl.8.6.2.1 defines the modulus of elasticity as an average secant modulus equal to the slope of the straight line that connects the origin of the stress-strain diagram to the stress equal to $0.4 f_c'$. The modulus of elasticity may vary considerably as a function of concrete strength, concrete density, and, especially for higher-strength concretes, the type of coarse aggregate.

Example 2.1

A simply supported unreinforced concrete test beam spanning 1.4 m is shown in the figure below. The beam is of rectangular cross-section (100 mm width by 150 mm overall depth). The beam failed when subjected to two point loads of 3 kN each, as shown in the figure. The beam was made of normal-density concrete.

Determine the modulus of rupture (f_r), the specified compressive strength (f_c'), and the modulus of elasticity (E_c) for the concrete the beam was made of. Neglect the beam self-weight in the calculations.

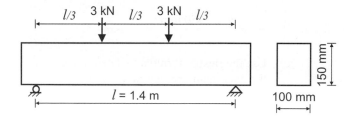

SOLUTION: The beam under consideration was subjected to flexure and failed when the maximum tensile stress due to flexure reached the modulus of rupture.

1. **Determine the bending moment diagram for the beam**

   The bending moment diagram for the beam subjected to the given load is shown below. It can be determined from either statics or using the beam diagrams and formulas included in Appendix A.

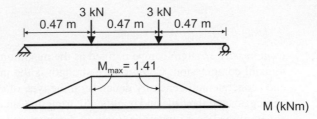

It can be observed from the diagram that the maximum bending moment occurs at the central one-third of the beam span; that is,

$$M_{max} \cong 1.4 \text{ kNm}$$

2. **Determine the modulus of rupture (f_r)**

The modulus of rupture is determined by using the flexure formula covered in mechanics of materials textbooks (also reviewed in Section 4.2 of this book):

$$f = \frac{M \cdot y_t}{I_g}$$

where f is the flexural stress, M is the bending moment, y_t is the distance from the centroid of the section to the extreme tension fibre, and I_g is the moment of inertia of the gross cross-section of a concrete beam around the axis of bending.

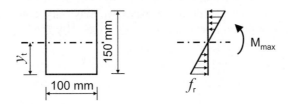

The following tasks need to be performed to find f_r:

a) Determine the moment of inertia for the beam section:

$$I_g = \frac{bh^3}{12}$$

$$= \frac{(100 \text{ mm})(150 \text{ mm})^3}{12} \cong 28 \times 10^6 \text{ mm}^4$$

Note that expressing I as units $\times 10^6$ will greatly ease calculations with large numbers.

b) Determine y_t:

$$y_t = \frac{h}{2} = \frac{150 \text{ mm}}{2} = 75 \text{ mm}$$

c) Use the flexure formula to determine the tensile stress in the section subjected to the maximum bending moment by taking $f_t = f_r$ and $M = M_{max}$:

$$f_r = \frac{M_{max} \times y_t}{I_g}$$

$$= \frac{(1.4 \times 10^6 \text{ N} \cdot \text{mm}) \times 75 \text{ mm}}{28 \times 10^6 \text{ mm}^4} = 3.75 \text{ MPa}$$

3. **Determine the specified compressive strength of concrete (f_c')**

In this case, f_c' can be determined based on f_r:

A23.3 Eq. 8.3

$$f_r = 0.6\lambda \sqrt{f_c'} \qquad \qquad \text{[2.1]}$$

In this case, we can take $\lambda = 1.0$ (normal-density concrete).

Hence,

$$f_c' = \left(\frac{f_r}{0.6\lambda}\right)^2$$

$$= \left(\frac{3.75 \text{ MPa}}{0.6 \times 1.0}\right)^2 = 39.0 \text{ MPa}$$

4. **Determine the modulus of elasticity of concrete (E_c)**

   Finally, E_c can be determined as follows:

A23.3 Eq. 8.2

$$E_c = 4500\sqrt{f_c'} \qquad\qquad\qquad\qquad [2.2]$$

$$= 4500\sqrt{39 \text{ MPa}} \cong 28\,100 \text{ MPa}$$

Failure of an unreinforced concrete beam very similar to the one used in the above example is shown in Figure 2.6. The beam failed by breaking into three pieces when the tensile stress reached the modulus of rupture. The failure occurred suddenly and without any warning, accompanied by a loud noise.

2.3.5 Creep

Creep is inelastic time-dependent deformation of concrete under sustained stress. When a load is applied to a concrete member, it causes an instantaneous elastic strain, as illustrated in Figure 2.7. However, if the load is sustained, the strain continues to increase with time. This time-dependent strain is called creep. Creep is a unique property of concrete (for example, steel does not creep).

Creep is related to the hydration process in cement paste that starts in fresh concrete when it is placed and continues for many years thereafter. Let us imagine that concrete consists of a pile of pebbles stuck together by a glue that takes 5 years to harden completely. Any load applied to this assemblage between the stages of the soft glue and the completely hardened glue will cause the glue to deform and change shape. The glue deforms gradually and its incremental deformation is not immediately noticeable. As a consequence, the overall shape of the assemblage changes as the glue deforms over time.

In concrete structures, cement acts as a glue and aggregates act as pebbles. It is important to note that of all concrete ingredients, cement is the one that creeps under sustained loads. The hardening process starts shortly after the concrete is placed into the forms, but

Figure 2.6 Failure of an unreinforced concrete test beam in the laboratory.

(Svetlana Brzev)

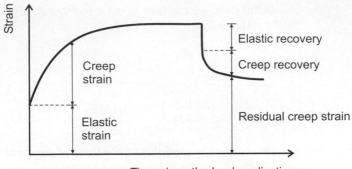

Figure 2.7 Variation of creep strain with time.

the hydration process within the cement paste continues for several years. Concrete is expected to creep until the hydration process in the cement paste is complete. When the concrete is placed, the rate of creep is initially high. At that stage, the cement paste is still semiplastic. The rate of creep gradually decreases with time under a constant sustained load. The process continues until hydration is complete, usually over a period of 5 years or so. However, some long-term experiments have shown that creep strain continues to increase for as long as 30 years.

It should be noted that the creep process is partially reversible. When the load is removed, an immediate (but partial) *elastic recovery* occurs, followed by a *creep recovery* that occurs over a period of time. However, there is always a substantial residual deformation *(residual creep strain);* see Figure 2.7. Creep is closely related to shrinkage, and both phenomena are related to hydrated cement paste. In general, a concrete resistant to shrinkage also shows a low tendency to creep.

Factors that affect the magnitude of creep strain are as follows:

- *W/c ratio:* The higher the w/c ratio, the larger are the creep effects. (Note that a higher w/c ratio indicates a larger amount of water in the cement paste.)
- *Age of loading:* Creep effects are more pronounced in members at an early age and can be reduced by delaying loading until the concrete is more mature. (Elastic strains are also reduced when load application is delayed.)
- *Relative air humidity:* For a given concrete, creep is higher when the relative humidity of the ambient air is lower (for example, in desert areas).
- *Stress level:* At service stresses, both the elastic and creep deformations are essentially proportional to the applied stress; hence the higher the stress, the higher the creep.
- *Reinforcement in concrete structures:* Creep effects are much more pronounced in reinforced concrete members than in those made of plain concrete. Creep causes an increase in stresses developed in the compression steel in reinforced concrete columns and other members; in some cases, these compression stresses may double due to the creep effect. The creep effect on tension steel stresses is not significant.

Creep does not induce stresses in concrete members, but it causes deflections that increase with time. In reinforced concrete structures, total deflections (including initial and long-term components) could be on the order of 2.5 to 3 times the initial deflections. Several illustrative examples of possible creep effects in building structures are presented by Neville (2003). Columns in buildings (and piers in bridges) are subject to shortening due to creep. The magnitude of creep-induced shortening may be significant, on the order of 100 mm to 150 mm for columns in concrete high-rises (Neville, 2003). Even if all columns shorten by the same amount, problems may occur in the cladding panels (typically either concrete panels or tiles or bricks), which generally do not carry any sustained load other than self-weight and are therefore not subject to creep. If these panels are rigidly connected to the structure, they often bulge or crack and even fail in some cases. Figure 2.8 illustrates creep-induced cladding-panel damage in a concrete high-rise building. The creep in concrete columns can negatively affect vertical attachments in the building (for example, water and gas pipes and guide rails for elevators), which are usually rigidly connected to the structure.

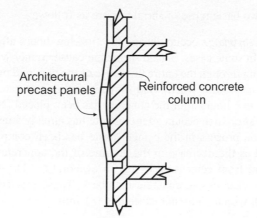

Figure 2.8 Cladding-panel damage in a concrete high-rise building (adapted from Neville, 2003 with the permission of the American Concrete Institute).

In some cases, differential creep may occur in columns. For example, exterior columns are subjected to high humidity, whereas interior columns are exposed to dry air, which causes a different amount of creep in these columns. Due to their smaller tributary area, exterior columns and especially corner columns are subjected to lower normal stresses than interior columns. As a result of the differential shortening in exterior and interior columns, beams and slabs in a floor system may be subject to differential settlement (similarly as in the case of differential soil settlement), with a possible significant impact on the internal force distribution in these members. A difference in level within a floor could be significant, on the order of 40 mm in a 50-storey building, as reported by Neville (2003). Effects of creep on long-term deflections in concrete structures will be discussed in Section 4.4.3.

Creep is a complex phenomenon and there is no single universally valid expression for creep prediction. This is mainly due to the fact that "we are dealing with a composite material, consisting both of manufactured and of natural ingredients, and partly processed in a semi-industrial manner" (Neville, 2003). However, as in the case of shrinkage, there are several empirical methods for creep prediction. CSA A23.3 does not recommend a specific method for creep prediction, but the reader is referred to the ACI Committee 209 report (ACI 209R-92) for an overview of various creep-prediction methods.

2.3.6 Shrinkage

Small changes in the volume of concrete occur during and after hardening. This decrease in volume (excluding that caused by externally applied forces and temperature changes) is called *shrinkage*. Unlike creep, shrinkage is not related to applied stress, but it is a time-dependent phenomenon. The rate of shrinkage reduces with time — older concretes are more resistant to stress and consequently undergo less shrinkage. Figure 2.9 illustrates the variation of shrinkage strain with time. Shrinkage is usually expressed as a linear strain (units mm/mm) and is denoted by the symbol ε_{sh}. Shrinkage strain in reinforced concrete is on the order of 0.0002 to 0.0003 mm/mm. For example, shrinkage deformation in a 30 m long slab is less than 10 mm. For more details on shrinkage effects, refer to Section 4.7.2.

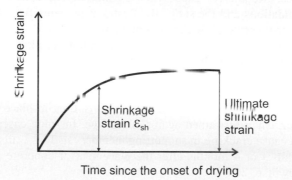

Figure 2.9 Variation of shrinkage strain with time.

The two basic types of shrinkage are as follows:

- *Plastic shrinkage* occurs during the first few hours after placing fresh concrete in the forms. In some cases, plastic shrinkage causes rather wide and deep cracks, sometimes extending through the entire thickness of the concrete member (typical for slabs). These cracks may form when wind or heat causes concrete to lose water rapidly, usually 30 min to 6 hours after the concrete has been placed (Newman, 2001).
- *Drying shrinkage* occurs after concrete has already attained its final set and most of the hydration process in the cement paste has been completed. Drying shrinkage can be defined as the decrease in the volume of the concrete member caused by the loss of moisture from cement paste due to evaporation. Drying-shrinkage cracks develop at random locations in concrete members. These cracks are usually very small (barely visible), with a width not exceeding 0.1 mm.

It should be noted that there is another type of shrinkage in concrete structures called *autogenous shrinkage* (CAC, 2002), which is caused by the loss of water used up in hydration. This type of shrinkage occurs even when moisture movement to or from concrete is not possible. Except in massive structures (for example, dams), it is hard to distinguish autogenous shrinkage from the drying shrinkage of hardened concrete.

The major concern with shrinkage in concrete structures is its potential to cause cracking and deformation. Therefore, it is worthwhile for a designer to be familiar with the key factors influencing the magnitude of *drying shrinkage* (Nawy, 2003):

- *Water/cement (w/c) ratio:* The higher the w/c ratio, the higher the shrinkage effects; water content is the main factor affecting shrinkage.
- *Curing:* Concrete that is not properly cured will almost certainly crack as a result of shrinkage.
- *Aggregates:* Concrete with a high aggregate content is less vulnerable to shrinkage.
- *Size of the concrete member:* The magnitude of shrinkage decreases with the volume of the concrete member; however, the duration of shrinkage is longer for larger members since more time is needed for drying to reach the internal regions.
- *Ambient conditions:* The rate of shrinkage is lower at higher humidity; humidity is a significant factor influencing the magnitude of shrinkage effects. Environmental temperature also influences shrinkage (shrinkage becomes stabilized at low temperatures).
- *Amount of reinforcement:* Reinforced concrete shrinks less than plain concrete; the relative difference is a function of the amount of reinforcement.
- *Admixtures:* Some admixtures, such as accelerating admixtures (calcium chloride), increase shrinkage.
- *Type of cement:* Rapid-hardening cement shrinks somewhat more than other types of cement, while shrinkage-compensating cements minimize or entirely eliminate shrinkage cracking.

In general, concrete tends to crack when its shrinkage is restrained (Newman, 2001) — this phenomenon is also called *restrained shrinkage.* In practice, there is always some form of external restraint, such as concrete or masonry walls, columns, elevator or stair cores, irregularly shaped floors, continuity of construction, or friction between the foundations and the subgrade. Restraint prevents free shrinkage and exposes concrete to tensile stresses, which in turn cause cracks to develop. Steel reinforcing bars placed closer to one side of the beam may cause shrinkage (if possible, symmetrical reinforcement should be used as it minimizes shrinkage). Shrinkage cracks are especially common at re-entrant corners in slabs on grade, but they can also occur over reinforcing bars and embedded conduits placed close to the surface, around anchor bolts in footings, or at the top of foundation walls. Cast-in-place beams and floor slabs may also crack when restrained against shrinkage. In some cases, stresses from slab shrinkage can cause cracking in supporting members such as walls. Restraint cracks usually form within 3 or 4 months after the placement of concrete and remain dormant thereafter (Newman, 2001).

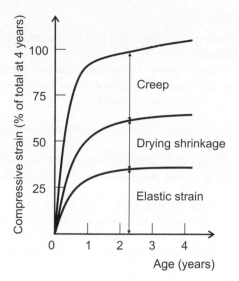

Figure 2.10 Variation of strain components with age for a column in a 76-storey concrete building.

(Reproduced from CAC, 2002 with the permission of the Cement Association of Canada)

Restrained shrinkage in reinforced concrete construction cannot be prevented. However, the effect of shrinkage can be mitigated by controlling shrinkage cracking; this can be accomplished by using smaller reinforcing bars at closer spacing placed at proper locations. As a result, the cracks caused by the restrained shrinkage are distributed into many small cracks that are not usually visible with the naked eye. A23.3 Cl.7.8.2 prescribes shrinkage and temperature reinforcement for crack control in reinforced concrete slabs (refer to Section 4.7 for more details on cracks in concrete structures). Alternatively, the effects of restrained shrinkage can be controlled when restraint is prevented by providing control joints that temporarily disconnect the slab from the restraint (for example, in slabs on grade), as discussed in Section 14.12.

The significance of shrinkage effects should not be underestimated. In 1989, a survey by the U.S. National Ready Mixed Concrete Association ranked uncontrolled shrinkage cracking as the number one problem its members faced (ACI/ASCC, 1998).

Due to the number of interacting variables involved, shrinkage strain can only be approximated. There are several empirical methods for shrinkage prediction — for more details refer to the ACI Committee 209 report (ACI 209R-92). Note that CSA A23.3 does not recommend a specific method for shrinkage prediction.

The magnitude of the combined creep and shrinkage strains can be significant as compared to the elastic strains due to gravity and lateral loads. Figure 2.10 shows superimposed creep and shrinkage and elastic strains as a function of time for a column in a concrete high-rise (CAC, 2002). The building under consideration was 76 storeys high, the column cross-sectional dimensions were 400 mm by 1200 mm, and the vertical reinforcement ratio was 2%. It can be noted from Figure 2.10 that the combined creep and shrinkage strains exceed the magnitude of the elastic strains. After 4 years, elastic strains account for only 37% of the total strain in the column — the remaining strain is due to creep and drying shrinkage.

KEY CONCEPTS

Creep is inelastic time-dependent deformation of concrete under sustained stress. Factors that tend to increase creep effects are high w/c ratio, loading at an early age, exposing concrete to drying conditions, presence of compression reinforcement, etc. In general, creep decreases with the age of concrete, but the effects are not entirely reversible (there is a residual creep strain after the load is removed). Creep does not cause stresses in concrete members, but it causes deflections that increase with time.

In reinforced concrete structures, total deflections could be on the order of 2.5 to 3 times the initial deflections.

Shrinkage is the decrease in volume of concrete during and after hardening. Unlike creep, shrinkage is not related to applied stress, but it is a time-dependent phenomenon.

There are two basic types of shrinkage:

- *Plastic shrinkage* occurs during the first few hours after placing fresh concrete in the forms.
- *Drying shrinkage* occurs after concrete has already attained its final set and a good portion of the hydration process in the cement paste has been completed.

The effects of shrinkage are aggravated when an external restraint (such as steel reinforcing bars, formwork, or continuity of construction) prevents free shrinkage and subjects concrete to tensile stresses, which in turn cause cracking.

Proper consideration of creep and shrinkage effects is a significant component of design in reinforced concrete.

2.3.7   Temperature Effects

Concrete, like most other materials, expands with increasing temperatures and contracts with decreasing temperatures. Thermal contraction causes tensile stresses, which are often combined with shrinkage stresses. The amount of thermal contraction and expansion varies with the type and amount of aggregate, w/c ratio, temperature range, concrete age, etc. Thermal contraction can cause cracks in concrete, especially in statically indeterminate structures. Temperature effects in concrete structures can be controlled by providing movement joints (to be discussed in Section 13.8). The coefficient of thermal expansion for concrete (α_t) used in design is on the order of $10 \times 10^{-6}/^\circ$C. It should be noted that the coefficient of thermal expansion for reinforcing steel is of a similar value, that is, $12 \times 10^{-6}/^\circ$C. This is an important characteristic of reinforced concrete as a composite material.

2.3.8   Mass Density

Concrete density varies depending on the amount and relative density of the aggregate and the amount of air (entrapped or purposely entrained), water, and cementitious material (mainly influenced by the maximum-size aggregate). Mass density is denoted as γ_c per CSA A23.3.

CSA A23.1 Cl.4.3.4 identifies the following three types of concrete as a function of mass density:

- *normal-density concrete:* with air-dry density between 2150 kg/m^3 and 2500 kg/m^3;
- *semi-low-density concrete:* with a specified 28-day compressive strength of 20 MPa or higher and an air-dry mass density between 1850 kg/m^3 and 2150 kg/m^3;
- *low-density concrete:* with a specified 28-day compressive strength of 20 MPa or higher and an air-dry mass density not exceeding 1850 kg/m^3.

In practice, concrete density can be as low as 240 kg/m^3 (insulating concrete) or as high as 6000 kg/m^3 (concrete used for radiation shielding). However, normal-density concrete is used in most applications, characterized by mass densities of 2300 kg/m^3 and 2450 kg/m^3 for plain and reinforced concrete, respectively.

Note that the *unit weight* of concrete (γ_w) used for dead load calculations is determined by multiplying the mass density by a gravitational constant ($g = 9.81$ m/s^2). For example, normal-density reinforced concrete with a mass density of 2450 kg/m^3 has a unit weight of 24 kN/m^3 (obtained as the product 9.81 m/s^2 × 2450 kg/m^3).

2.3.9 Poisson's Ratio

Poisson's ratio is the ratio between transverse and longitudinal strain in a concrete specimen subjected to axial load. Poisson's ratio is generally in the range of 0.15 to 0.20 for concrete, and the most common value used in the design of concrete structures is 0.2. Poisson's ratio for steel is equal to 0.3.

However, caution must be used when performing computer analyses to ensure that Poisson's ratio is compatible with values used for shear modulus of elasticity for concrete, or else serious errors may occur.

KEY CONCEPTS

Temperature effects in concrete structures can be controlled by providing movement joints. The coefficient of thermal expansion for concrete (α_t) used in design is on the order of 10×10^{-6}/°C.

The concrete mass density (γ_c) varies depending on the amount and relative density of the aggregate and the amount of air (entrapped or purposely entrained), water, and cementitious material (mainly influenced by the maximum-size aggregate). CSA A23.1 Cl.4.3.4 identifies the following three types of concrete as a function of mass density:

- *normal-density concrete:* with air-dry density between 2150 kg/m^3 and 2500 kg/m^3;
- *semi-low-density concrete:* with a specified 28-d compressive strength of 20 MPa or higher and an air-dry mass density between 1850 kg/m^3 and 2150 kg/m^3;
- *low-density concrete:* with a specified 28-d compressive strength of 20 MPa or higher and an air-dry mass density not exceeding 1850 kg/m^3.

The mass densities can be taken as 2300 kg/m^3 and 2450 kg/m^3 for plain and reinforced normal-density concrete, respectively.

The *unit weight* (γ_w) of normal-density reinforced concrete is usually taken as 24 kN/m^3.

Poisson's ratio is the ratio between the transverse and longitudinal strain in a concrete specimen subjected to axial load. The most common value of Poisson's ratio used in the design of concrete structures is 0.2.

2.4 DURABILITY OF CONCRETE

Satisfactory durability is expected if concrete meets the requirements related to strength, w/c ratio, air entrainment, and cement type chosen to satisfy the intended use and exposure conditions of a particular concrete structure. In addition to the above, CSA A23.1 Cl.4.1.1.1.1 states that "although minimum requirements for concrete durability are specified, it should be stressed that a durable concrete also depends upon the use of high-quality materials, an effective quality control program, and good quality of work in manufacturing, placing, finishing and curing the concrete." CSA A23.1 Cl.4.1.1 outlines specific durability requirements for concrete structures subjected to weathering, sulphate attack, corrosive environment, or any other process of deterioration.

CSA A23.1 specifies three basic exposure classes for concrete. Class C relates to concrete subjected to de-icing chemicals (chlorides), Class F relates to concrete subjected to freezing and thawing, and Class N relates to normal concrete (not exposed to either chlorides or freezing and thawing). Definitions of various exposure classes are given in Table 2.1 (adapted from Table 1 of CSA A23.1).

Table 2.1 Definitions of C, F, and N exposure classes

Class	Description
C-XL	Structurally reinforced concrete exposed to chlorides or other severe environments with or without freezing and thawing conditions, with higher durability performance expectations than the C-1, A-1, or S-1 classes.
C-1	Structurally reinforced concrete exposed to chlorides with or without freezing and thawing conditions. Examples: bridge decks, parking decks and ramps, portions of marine structures located within tidal and splash zones, concrete exposed to seawater spray and salt-water pools.
C-2	Nonstructurally reinforced (i.e., plain) concrete exposed to chlorides and freezing and thawing. Examples: garage floors, porches, steps, pavements, sidewalks, curbs, and gutters.
C-3	Continuously submerged concrete exposed to chlorides but not to freezing and thawing. Examples: underwater portions of marine structures.
C-4	Nonstructurally reinforced concrete exposed to chlorides but not to freezing and thawing. Examples: underground parking slabs on grade.
F-1	Concrete exposed to freezing and thawing in a saturated condition but not to chlorides. Examples: pool decks, patios, tennis courts, freshwater pools, and freshwater control structures.
F-2	Concrete in an unsaturated condition exposed to freezing and thawing but not to chlorides. Examples: exterior walls and columns.
N	Concrete not exposed to chlorides or to freezing and thawing. Examples: footings and interior slabs, walls and columns.

(*Source:* Table 1 of CSA A23.1-04, reproduced with the permission of the Canadian Standards Association)

Requirements for various exposure classes are specified in Table 2.2 (adapted from Table 2 of CSA A23.1). Note that the main requirements include maximum w/c ratio, minimum specified compressive strength (f_c'), and air-content category. Air content is one of the most important durability requirements. In general, 5% to 8% air (usually produced by using air-entraining admixtures) is strongly recommended for concrete subjected to severe exposure conditions.

CSA A23.1 also specifies the requirements for concrete in contact with sulphates. Sulphates may occur in the soil, in groundwater, or in industrial wastes. Table 1 of CSA A23.1 identifies three different exposure classes (S-1, S-2, and S-3) depending on the extent of sulphate exposure. CSA A23.1-04 introduces requirements for concrete exposed to manure and/or sileage gasses and liquids. Table 1 of CSA A23.1 identifies 4 different exposure classes (A-1 to A-4) depending on the extent of exposure.

Table 2.2 Requirements for C, F, N, and A classes of exposure

Class of exposure	Maximum w/c ratio	Minimum specified 28-day compressive strength f_c' (MPa)	Air-content category
C-XL	0.37	50 within 56 days	1 or 2
C-1 or A-1	0.4	35	1 (for concrete exposed to freezing and thawing) and 2 otherwise
C-2 or A-2	0.45	32	1
C-3 or A-3	0.50	30	2
C-4 or A-4	0.55	25	2
F-1	0.50	30	1
F-2	0.55	25	2
N	For structural design	For structural design	None

(*Source:* Table 2 of CSA A23.1-04, reproduced with the permission of the Canadian Standards Association)

Durability is related to the long-term satisfactory performance of concrete structures and the ability to satisfy their intended use and exposure conditions during the specified or traditionally expected service life. A satisfactory durability is expected if concrete meets the requirements related to strength, w/c ratio, air entrainment, and cement type chosen to satisfy the intended use and exposure conditions of a particular concrete structure.

CSA A23.1 specifies five basic exposure classes for concrete:

- Class C relates to concrete subjected to de-icing chemicals (chlorides).
- Class F relates to concrete subjected to freezing and thawing.
- Class N relates to normal concrete, not exposed to chlorides or freezing and thawing.
- Class S relates to concrete subjected to sulphates.
- Class A relates to concrete subjected to manure and/or sileage gasses.

The main requirements of the exposure classes are maximum w/c ratio, minimum specified compressive strength (f_c'), and air-content category. Air content is one of the most important durability requirements.

2.5 FIRE-RESISTANCE REQUIREMENTS

Reinforced concrete is one of the most highly fire-resistant building materials. However, the properties of concrete and reinforcing steel change significantly at high temperatures. The most significant effects include reduction in strength and modulus of elasticity and increased creep. The fire resistance of concrete structures is influenced by several factors, including type of concrete, member dimensions, reinforcing steel, concrete cover, structural restraints, and continuity; whether the construction is protected or not; and column eccentricities. In general, temperatures greater than 95°C sustained for a prolonged period of time can have a significant effect on concrete (CAC, 2002). At high temperatures, the cement paste shrinks due to dehydration while the aggregate expands. Some aggregates can produce a volume-stable concrete at high temperatures whereas others undergo abrupt volume changes causing distress in concrete. Low-density aggregates generally exhibit better fire performance than natural stone aggregate concretes. Appendix D of NBC 2010 specifies seven different types of concrete as related to fire-resistance requirements. Out of these, there are two basic types of normal-density concrete (Type S and Type N) and five types of low-density concrete. The main difference between these is the type of aggregate used. Type S concrete is composed of coarse aggregate such as granite, quartzite, siliceous gravel, or other dense materials containing at least 30% quartz, chert, or flint. Type N concrete is composed of coarse aggregate such as cinders, broken brick, blast-furnace slag, limestone, calcareous gravel, trap rock, sandstone, or similar dense material containing not more than 30% quartz, chert, or flint. Structural low-density concrete, as defined by CSA A23.3, generally corresponds to Type L40S as described by NBC 2010. Type N and Type L40S concrete is superior in terms of fire performance to Type S concrete. As these various types of concrete demonstrate different fire performance characteristics, the designer should specify the type of aggregate to be used to satisfy the prescribed fire requirements. For more details on the fire resistance of concrete structures (including example calculations), the reader is referred to CAC (2005).

In general, steel has inferior fire resistance to concrete. Steel reinforcement loses strength at high temperatures. Hot-rolled reinforcing steel used for deformed bars loses its strength less quickly than the cold-drawn steel used for prestressing tendons. Therefore, the fire-resistance rating of a concrete structure depends on the type of reinforcing steel and the magnitude of tensile stresses developed in the reinforcement.

Concrete is considered to be one of the most highly fire-resistant building materials. However, the properties of concrete and reinforcing steel change significantly at high temperatures. The most significant effects include reduction in strength and modulus of elasticity and increased creep.

Appendix D of NBC 2010 specifies seven different types of concrete as related to fire-resistance requirements. Of these, there are two basic types of normal-density concrete (Type S and Type N) and five types of low-density concrete. The main difference between these is the type of aggregate used. As various types of concrete demonstrate different fire performance, the designer should specify the type of aggregate to be used in order to satisfy the prescribed fire requirements.

2.6 | REINFORCEMENT

In reinforced concrete structures, reinforcement has a major role in sustaining tensile stresses due to flexure, shear, and other loads. The presence of reinforcement is one of the key reasons for the success of reinforced concrete as a material: concrete is strong in compression but weak in tension, so plain concrete has very limited use in design applications. Steel is the main material used for reinforcement due to its favourable mechanical characteristics, such as high tensile strength and the ability to deform by a significant amount before failure. Steel reinforcement has been widely used since the inception of reinforced concrete as a building material. However, steel is susceptible to corrosion, which has been a major threat to the durability of concrete structures. For more details on corrosion in reinforced concrete structures, refer to Section 4.7.3.

Recently, substantial research has been conducted in an attempt to identify alternative materials potentially suitable for reinforcement in concrete structures, such as fibre-reinforced composites (FRCs). FRCs are made out of glass or carbon fibres impregnated with a resin matrix. These innovative materials have excellent corrosion resistance and other favourable characteristics. However, due to rather high cost and inadequate technical expertise on the design of concrete structures reinforced with composite reinforcement, the application of FRCs is currently limited to special structures. For more details on FRC reinforcement, the reader is referred to ISIS (2001). Alternatively, epoxy coating can be applied to provide improved corrosion protection in steel reinforcement to be used in bridge decks, parking garages, or marine structures. Care must be taken during unloading, handling, and storage of epoxy-coated bars at the job site to prevent damage to the coating. This book is focused on the design of concrete structures reinforced with steel. Note that steel reinforcing bars are often denoted as *rebars* in the design practice.

2.6.1 Types of Reinforcement

CSA A23.3 Cl.3.1 outlines the types of reinforcement for concrete structures. In general, several types of deformed reinforcement are used in reinforced concrete structures, while prestressing tendons are used in prestressed concrete structures. According to CSA A23.3 Cl.3.1.3, *deformed reinforcement* includes the following types:

- deformed bars (most common), conforming to CSA Standard G30.18;
- welded wire fabric (fabricated using smooth or deformed wires), conforming to ASTM Standards A185 or A497/497M;
- deformed wire reinforcement, conforming to ASTM Standard A496.

Although deformed bars are prescribed for most design applications, plain reinforcing bars may be used for spirals; if smaller than 10 mm in diameter, these bars may also be used for stirrups and ties (CSA A23.3 Cl.3.1.2). In the past, plain reinforcing bars were used for most applications, but their use in Canada has been abandoned since the 1960s due to their inferior bond characteristics to deformed reinforcement.

According to CSA A23.3 Cl.3.1.4, *prestressing tendons* need to conform to ASTM A416/416M, ASTM A421/421M, and ASTM A722A/722M. Prestressing steel includes high-strength smooth wire, seven-wire strands, and high-strength alloy bars (such as Dywidag threaded rods).

The primary design requirements related to steel reinforcement are

- minimum specified yield strength (f_y) (specified by grade),
- minimum ultimate tensile strength (f_u),
- minimum elongation (this requirement is related to ductility),
- bendability (ability to bend and form hooks),
- weldability (ability to be welded).

These requirements are prescribed by the pertinent standards mentioned above (note that different standards apply to different types of reinforcement). For more information about steel reinforcement, the reader is also referred to RSIC (2004).

The mass density of reinforcing steel is equal to 7850 kg/m^3 and the unit weight is equal to 77 kN/m^3. The coefficient of thermal expansion is 12×10^{-6}/°C (similar to concrete).

2.6.2 Mechanical Properties of Steel

Consider the typical stress-strain diagram for nonprestressed steel reinforcement in Figure 2.11. Several distinct stages of behaviour characteristic of a steel bar subjected to axial loading until failure are as follows:

- *Linear-elastic behaviour:* At a low load level, steel shows elastic behaviour resulting in a linear stress-strain relationship.
- *Yielding:* The behaviour of steel changes when the stress in steel reaches the yield stress (f_y). The strain continues to increase while the stress remains constant and equal to f_y, so the stress-strain diagram looks like a straight horizontal line. This behaviour is called plastic flow or simply "yielding."
- *Strain hardening:* At a strain approximately equal to 1%, plastic flow stops and the steel enters the strain-hardening range. Strain hardening is characterized by a linear-elastic behaviour wherein the stress increases with the increase in strain. This behaviour ceases at the point where the stress reaches its maximum value, termed the *ultimate strength* (f_u).
- *Failure:* A further load increase results in a decrease in stress until failure takes place. The corresponding stress is called the breaking stress or *rupture stress*.

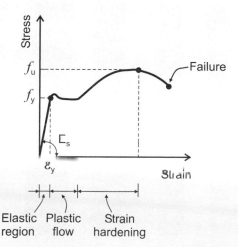

Figure 2.11 Stress-strain diagram for reinforcing steel.

The key mechanical properties of steel used in modern design are as follows:

- *The specified yield strength* (f_y) depends on the type of steel. In general, f_y is in the range from 400 MPa to 500 MPa for nonprestressed steel reinforcement.
- The modulus of elasticity (Young's modulus) (E_s) is equal to the slope on the stress-strain curve in the elastic range, as shown in Figure 2.11. E_s is a constant taken as 200 000 MPa for nonprestressed reinforcement (CSA A23.3 Cl.8.5.4.1).

In some cases, it is required to determine the steel yield strain (ε_y) corresponding to the yield stress (f_y) (see Figure 2.11). At that point, the behaviour of steel is still elastic and Hooke's law applies, so ε_y can be obtained as

$$\varepsilon_y = \frac{f_y}{E_s}$$
[2.4]

CSA A23.3 specifies certain limitations related to the mechanical properties of steel reinforcement:

- The specified yield strength (f_y) used for the design purpose should not exceed 500 MPa, except for prestressing tendons (CSA A23.3 Cl.8.5.1); this requirement is meant to limit the crack size in steel with f_y of over 300 MPa.
- For compression reinforcement with f_y greater than 400 MPa, the f_y value assumed in design calculations shall not exceed the stress corresponding to a strain of 0.35%. This requirement is meant to ensure that the compression steel in concrete members (if existent) reaches the yielding point. According to CSA A23.3, the maximum strain in the concrete compression zone is equal to 0.35% (or 0.0035). If the steel yield strength is rather high, the compressive steel might remain elastic. This requirement should be satisfied for steel with a yield strength of 500 MPa or less.

2.6.3 Deformed Bars

Deformed bars are the main type of reinforcement used in structural concrete. Deformed bars are round steel bars with lugs (also called deformations) rolled into the surface of the bar during manufacture (see Figure 2.12). These deformations provide additional bond or anchorage between the concrete and the steel.

Reinforcing steel is classified into different grades on the basis of the minimum specified yield strength (f_y) and the chemical composition. CSA Standard G30.18 specifies

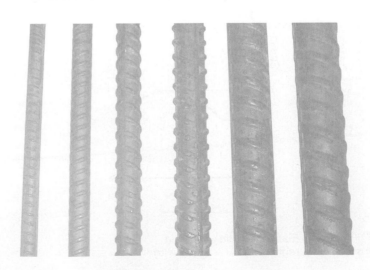

Figure 2.12 Sample deformed reinforcing bars of different sizes available in Canada.

(Svetlana Brzev)

5 grades of steel: 300R, 400R, 500R, 400W, and 500W. The steel grade must be specified by the following two characteristics:

- a *number* that indicates a minimum specified f_y value in MPa, and
- a *letter* (R or W) that is related to the weldability of a particular steel grade; the letter R denotes the *regular* steel grade while the letter W denotes the *weldable* steel grade.

It should be noted that W grades are intended for special applications where improved weldability or bendability is required. Also, W-grade steel has less variation in its range of minimum and maximum yield strength (f_y) values; the minimum f_y corresponds to the 95th-percentile strength, whereas the maximum f_y corresponds to the 5th-percentile strength. W steel grade is more favourable for certain seismic applications, where an upper limit in the maximum yield strength for the reinforcing steel has to be specified by the design. If a stress higher than the specified yield strength develops during an earthquake, it might lead to unsatisfactory seismic performance of the building.

The designer should specify a steel grade on the structural drawings. When the grade specification is omitted from the drawings, it is likely that 400R steel will be supplied (this is the basic steel grade).

Deformed bars are sized to have specific cross-sectional areas and are designated by a number based on the rounded-off nominal diameter expressed in millimetres (for example, 25M). The letter M stands for "metric," indicating metric bars. The standard metric-sized bars are listed in Table A.1. Note that 45M and 55M bars are rarely used due to their heavy mass and large development lengths. 10M and 15M bars are normally used for stirrups and ties. Rebars come in 12 m lengths for 10M bars and 18 m lengths for the other bar sizes. Sample deformed reinforcing bars of different sizes are shown in Figure 2.12.

Imperial-sized bars are designated by numbers representing the number of eighths of an inch in their nominal diameter; for example, a #8 bar has a 1 inch diameter (and a cross-sectional area of 0.79 in²), while a #4 bar has a 1/2 inch diameter. In Canada, Imperial bar designations can be found on older construction drawings (developed before the 1970s).

All deformed reinforcing bars are required to be marked with identifying symbols rolled into one side of each bar. The symbols consist of a letter or symbol to show the producer's mill, a number to show bar size, a marking to indicate type of steel (R or W), and a number to show the grade; alternatively, a continuous line through five or more spaces can be used to denote Grade 400 steel. Figure 2.13 illustrates a marking system for deformed reinforcing bars.

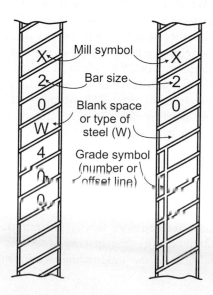

Figure 2.13 Identification requirements for deformed steel bars.

DID YOU KNOW?

Approximately 15 000 t of reinforcing steel were used in the construction of the *Rogers Centre* (formerly known as the SkyDome), the home of the Toronto Blue Jays. The bars were of weldable grade, and the sizes varied from 10M to 55M. The epoxy-coated reinforcement was used mainly in high-traffic areas, as well as in places where wintertime use of salt called for a durable component (RSIO, 1989).

The Rogers Centre in Toronto had the only completely retractable stadium roof in the world at the time of construction. The roof is close to 100 m high — tall enough to build a 30-storey building inside! Construction was completed in 1989 at a cost of around $600 million.

Figure 2.14 A view of the Rogers Centre, Toronto.
(Sportsco International, L.P.)

2.6.4 Welded Wire Fabric

Welded wire fabric (WWF) can be used as a primary reinforcement for structural slabs and shear walls and is often used as temperature and shrinkage reinforcement in concrete slabs. WWF consists of a square or rectangular mesh of wires welded at all intersections. In general, cold-drawn wire is used for WWF. Cold-drawn wire can be smooth, designated with a letter W, or deformed, designated with a letter D. The letter M (denoting metric reinforcement) is used as a prefix. For example, a typical designation for a smooth wire is MW25.8, where M stands for metric, W stands for smooth wire, and 25.8 denotes a wire with a cross-sectional area of 25.8 mm^2. The yield strength for WWFs ranges from 450 MPa to 515 MPa, depending on the use and the size and on whether the wire is deformed or smooth. The spacing and sizes of wires are identified by a style designation. An example of a style designation for WWF is

$$\text{WWF}102 \times 152 - \text{MW}18.7 \times 25.8$$

where WWF denotes welded wire fabric, 102 mm denotes the spacing of the longitudinal wires, 152 mm denotes the spacing of the transverse wires, 18.7 mm^2 denotes the cross-sectional area of the longitudinal wires, and 25.8 mm^2 is the area of the transverse wires.

KEY CONCEPTS

Reinforcement has a major role in reinforced concrete structures in sustaining tensile stresses due to flexure, shear, and other loads.

According to CSA A23.3 Cl.3.1.3, deformed reinforcement includes the following types:

- *deformed bars* (most common), conforming to CSA Standard G30.18;
- *WWF* (fabricated using smooth or deformed wires), conforming to ASTM A185 or ASTM A497/497M;
- *deformed wire reinforcement*, conforming to ASTM A496.

The primary design requirements for each reinforcement type are:

- minimum specified yield strength (f_y) (specified by grade);
- minimum ultimate tensile strength (f_u);
- minimum elongation (related to ductility);
- bendability (ability to bend and form hooks);
- weldability (ability to be welded).

The key mechanical properties of steel used in structural concrete are as follows:

- The specified yield strength (f_y) ranges from 400 MPa to 500 MPa for nonpre-stressed steel reinforcement.
- The modulus of elasticity (E_s) is equal to the slope on the stress-strain curve in the elastic range; E_s is constant and can be taken as 200 000 MPa for nonprestressed reinforcement.

Other steel properties used in reinforced concrete design are

- mass density of 7850 kg/m³,
- unit weight of 77 kN/m³,
- coefficient of thermal expansion of 12×10^{-6}/°C.

SUMMARY AND REVIEW — MATERIALS

Concrete is artificial stone made from the following basic components:

- cement paste: Portland cement, water, and air (25% to 40% of the total volume of fresh concrete);
- aggregates: natural sand and gravel or crushed stone (60% to 75% of the total volume of concrete).

The paste hardens as a result of the chemical reaction between cement and water and glues the aggregates into a rock-like mass. In building structures, concrete is most often used in the form of reinforced concrete, which is a composite material consisting of concrete and steel reinforcement.

The main concrete-producing materials

Portland cement is a mixture of several compounds made by heating limestone and clay together at very high temperatures. There are five basic types of Portland cement complying with different CSA A23.1 requirements for specific purposes; out of these, Type 10 (normal Portland cement) is most frequently used if concrete is not subjected to specific exposures.

Water serves a dual purpose in a concrete mix: it combines with cement in a process called hydration and provides the workability required for proper placing and compaction. The *water-cement ratio (w/c)*, that is, the ratio of water and cement in a concrete mix, is used as a basis for concrete mix proportioning. Typical values for the w/c ratio required to achieve a workable consistency in the field are in the range from 0.4 to 0.6.

The *air content* in concrete varies from 0% to 8%. There are two different sources of air in concrete, namely *entrapped air* and *entrained air*. *Entrapped air* voids occur in all concretes and are reduced by proper placement and consolidation of fresh concrete. *Entrained air* is produced by adding an air-entraining agent that stabilizes bubbles formed during the mixing process, with the main objective of improving concrete resistance to freezing and thawing. Higher air content (5% to 8%) is required for concretes subjected to severe exposure conditions.

Aggregates are a nearly inert filler material that significantly influences the quality of concrete. Aggregates must be clean, hard, strong, durable, and free of coatings of clay and other fine materials that may affect hydration and bond of the cement paste. Aggregates are classified into *fine* and *coarse*. *Fine aggregates* consist of natural or manufactured sand with most particle sizes less than 5 mm. *Coarse aggregates* consist of one or a combination of gravels or crushed concrete with size between 10 mm and 40 mm. The most commonly used maximum aggregate size is 20 mm.

Mechanical properties of concrete

Compressive strength can be defined as the maximum resistance of a concrete cylinder to axial compression. Compressive strength is the primary physical property of hardened concrete and generally depends on the type of concrete mix, the properties of aggregate, the time and quality of curing, etc. In Canada, the compressive strength of concrete is determined on cylinders 100 mm in diameter and 200 mm high, tested at the age of 28 days. CSA A23.3 uses the term *specified compressive strength of concrete* (f_c'). General-use concrete has a compressive strength between 20 and 40 MPa, whereas high-strength concrete is characterized by an f_c' value of over 70 MPa.

The *tensile strength* of concrete is relatively low (about 8% to 15% of the compressive strength) and is strongly affected by the compressive strength of concrete, the type of test used to determine the tensile strength, etc. CSA A23.3 uses the *modulus of rupture* (f_r) as an indicator of concrete tensile strength. The modulus of rupture is the theoretical maximum tensile stress reached in the extreme fibre of a flexural member. Cracks develop when the tensile stress in concrete reaches the modulus of rupture, and concrete is unable to transmit any tension beyond that point. The modulus of rupture can be determined according to CSA A23.3 Cl.8.6.4 as follows:

A23.3 Eq. 8.3

$$f_r = 0.6 \, \lambda \, \sqrt{f_c'} \text{ (MPa)}$$

The *modulus of elasticity* (E_c) is used in deflection calculations and member stiffness calculations in the analyses of concrete structures. CSA A23.3 Cl.8.6.2.1 defines the modulus of elasticity as the average secant modulus equal to the slope of the straight line that connects the origin of the stress-strain diagram with the stress of $0.4\,f_c'$. A23.3 Cl.8.6.2.3 prescribes the following simple E_c equation that is applicable to normal-density concrete with compressive strength between 20 MPa and 40 MPa:

A23.3 Eq. 8.2

$$E_c = 4500 \, \sqrt{f_c'} \text{ (MPa)}$$

Creep and shrinkage effects in concrete structures

Creep is the inelastic time-dependent deformation of concrete under sustained stress. Factors that tend to increase creep effects include high w/c ratio, loading at an early age, exposing concrete to drying conditions, and the presence of compression reinforcement. In general, creep decreases with the age of the concrete, but the effects are not reversible (there is a residual creep strain after the load is removed). Creep does not induce stresses in concrete members, but it causes deflections that increase with time. In reinforced concrete structures, the total deflections could be on the order of 2.5 to 3 times the initial deflections.

Shrinkage is the decrease in volume of concrete during and after hardening. Unlike creep, shrinkage is not related to the applied stress, but it is a time-dependent phenomenon. There are two basic types of shrinkage:

- *Plastic shrinkage* occurs during the first few hours after placing fresh concrete in the forms.
- *Drying shrinkage* occurs after concrete has already attained its final set and a good portion of the hydration process in the cement paste has been completed.

The effects of shrinkage are aggravated when an external restraint (such as steel reinforcement, formwork, or continuity of construction) prevents free shrinkage and exposes concrete to tensile stresses, which cause crack development.

Concrete exposure classes and durability requirements

Durability is the long-term satisfactory performance of concrete structures and the ability to satisfy the intended use and exposure conditions during the specified or traditionally expected service life.

CSA A23.1 specifies the following three basic exposure classes for concrete:

- Class C relates to concrete subjected to de-icing chemicals (chlorides).
- Class F relates to concrete subjected to freezing and thawing.
- Class N relates to normal concrete not exposed to chlorides or freezing and thawing.

The main requirements related to the exposure classes are maximum w/c ratio, minimum specified compressive strength f_c', and air-content category. Air content is one of the most important durability requirements.

Types of reinforcement in concrete structures and their mechanical properties

Reinforcement has a major role in sustaining tensile stresses due to flexure, shear, and other loads in reinforced concrete structures.

According to A23.3 Cl.3.1.3, *deformed reinforcement* includes the following types:

- deformed bars (most common);
- welded wire fabric (fabricated using smooth or deformed wires);
- deformed wire reinforcement.

The primary reinforcement *design requirements* are

- minimum specified yield strength (f_y);
- minimum ultimate tensile strength (f_u);
- minimum elongation (related to ductility);
- bendability (ability to bend and form hooks);
- weldability (ability to be welded).

The key *mechanical properties* of steel reinforcement are as follows:

- The specified yield strength (f_y) ranges from 400 MPa to 500 MPa for nonprestressed steel reinforcement.
- The modulus of elasticity (E_s) is equal to the slope of the stress-strain curve in the elastic range. E_s is constant and can be taken as 200 000 MPa for nonprestressed reinforcement.

Other steel properties used in reinforced concrete design are

- mass density of 7850 kg/m³,
- unit weight of 77 kN/m³,
- coefficient of thermal expansion of $12 \times 10^{-6}/°C$.

PROBLEMS

2.1. a) How is the concrete mix proportion expressed? Give a typical example.

b) What are the four main requirements the concrete mix proportion needs to meet? Explain.

2.2. What are the two most important mechanical properties of hardened concrete? List the factors influencing their magnitude. Hint: One of these properties is considered a major positive feature of concrete, while the other is considered its weakness.

2.3. A simply supported reinforced concrete beam has been subjected to a uniform load of constant magnitude for 10 years.

a) By how much (approximately) are the beam deflections expected to increase as compared to the initial deflections at the time of load application?

b) What causes the increase in deflections in reinforced concrete structures over time? Explain.

c) If the load is completely removed after 10 years, what will happen to the deflections? Will they increase or decrease? Explain.

2.4. A standard concrete cylinder of height 200 mm and diameter 100 mm carries a maximum load of 300 kN before failure. The cylinder is cast using normal-density concrete.

Find the specified compressive strength and the modulus of elasticity of this concrete.

2.5. Develop a table to determine the E_c values for concrete of specified compressive strength ranging from 20 MPa to 40 MPa (in 5 MPa increments). Consider normal-density concrete only.

2.6. Consider a simply supported unreinforced concrete test beam of square cross-section (dimension 150 mm) spanning over 900 mm. The beam was loaded by a point load at the midspan and failed when the load reached 7.5 kN. The beam was constructed using normal-density concrete. Ignore the effects of the beam self-weight.

a) Determine the modulus of rupture (f_r) for this beam.

b) Compare the result obtained in part a) with the f_r value prescribed by CSA A23.3. Use an f_c' value of 25 MPa.

2.7. A bundle of 3-25M reinforcing bars is subjected to a tensile force of 300 kN. The modulus of elasticity for steel (E_s) is equal to 200 000 MPa and the yield stress (f_y) is 400 MPa. Use $\phi_s = 1.0$.

a) Calculate the stress and strain in the bars.

b) In theory, what is the maximum force the bars can carry?

2.8. A beam section has to be reinforced using 20M, 25M, or 30M size bars.

Suggest suitable bar combinations using an even number of bars if the required area of steel is

a) 1650 mm²

b) 2320 mm²

c) 3300 mm²

d) 4700 mm²

e) 5450 mm²

2.9. A 40 m wide by 70 m long floor plan of a reinforced concrete building is shown in the figure that follows. The floor structure consists of a 250 mm thick reinforced concrete slab on beams (not shown on the plan). The shear walls are expected to restrain the slab from movement horizontally in the east-west (E-W) direction. The stair shafts are expected to restrain horizontal movement in the slab in both directions.

a) If the entire floor was placed in one pour, estimate the magnitude of the total long-term shrinkage deformation in the E-W direction. Express the result in millimetres.

b) If the shrinkage-induced free movement in the slab has been prevented in the E-W direction, what measures would you propose to mitigate the effects of shrinkage cracking? Explain.

c) If an average acceptable crack width is 0.33 mm, estimate the amount of shrinkage-induced cracks expected to develop in the N-S direction. Express the result in number of cracks per metre.

d) Consider that 5% of all cracks have a width that is 5 times larger than the average value specified in part c). Is this performance acceptable to you as the designer? Explain. (Keep in mind that crack widths in concrete structures are generally not uniform and that larger crack widths may be expected in the slab at the locations of higher stresses.)

e) Consider the problem discussed in part d). If this were a floor in a parking structure, in which way would this amount of cracking affect the durability of the slab structure? Explain.

f) Suggest a way or ways in which the concrete placement sequence can be modified to reduce the amount of cracking in the slab in the E-W direction.

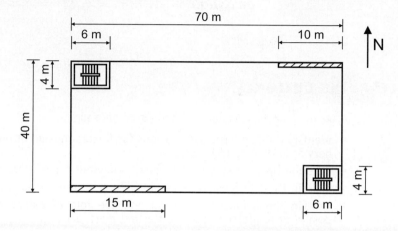

3

Flexure: Behaviour of Beams and One-Way Slabs

LEARNING OUTCOMES

After reading this chapter, you should be able to

- identify the five basic assumptions for flexural design of reinforced concrete members according to CSA A23.3
- describe the four stages of flexural behaviour for reinforced concrete members
- explain the two flexural modes of failure for reinforced concrete flexural members
- explain the balanced condition and the role of balanced reinforcement in the design of reinforced concrete flexural members
- analyze rectangular beams for flexure
- analyze one-way slabs for flexure
- analyze T- and L-beams for flexure
- analyze rectangular beams with tension and compression reinforcement for flexure

3.1 INTRODUCTION

Flexure (bending) is associated with lateral deformation of a member under a transversely applied load. Consider a reinforced concrete beam subjected to the uniform load shown in Figure 3.1a. In cast-in-place concrete construction, beams act as a monolithic unit with the supporting columns. However, for design purposes the beam can be modelled as a simply supported member, as shown in Figure 3.1b. A bending moment diagram for this beam is shown in Figure 3.1c. The top portion of the beam is subjected to compression whereas the bottom is under tension. Concrete has a limited ability to carry tension and cracks once its tensile strength has been reached in the region of maximum bending moments (in this case at the beam midspan). To increase the bending resistance of a cracked beam, steel reinforcement (often called *tension reinforcement* or tension steel) is placed inside the beam near the bottom to resist the tensile stresses.

The main objective of this chapter is to explain the behaviour and resistance of reinforced concrete beams and one-way slabs subjected to flexure. Section 3.2 outlines the types of flexural members discussed in this chapter. General assumptions related to flexure in reinforced concrete beams and slabs are discussed in Section 3.3. The behaviour of reinforced concrete beams and one-way slabs under flexure and the corresponding failure modes are explained in Section 3.4. The moment resistance of rectangular beams with tension steel only is discussed in Section 3.5. The flexural resistance of one-way slabs is discussed in Section 3.6, and the resistance of T-beams is discussed in Section 3.7. The moment resistance of rectangular beams with tension and compression reinforcement is outlined in Section 3.8.

This chapter focuses on the *analysis* of reinforced concrete members for flexure. Analysis involves computing the factored moment resistance of a member when the cross-sectional dimensions, the amount of steel reinforcement, and the material properties are known. The objective of the analysis is to verify whether a section of a concrete member

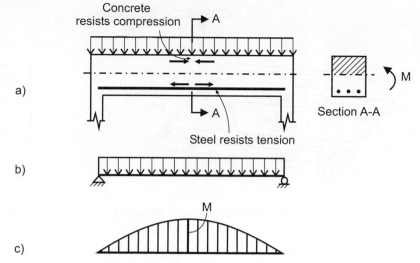

Figure 3.1 Concept of flexure in reinforced concrete members: a) actual beam, showing the distribution of internal forces; b) beam model; c) bending moment diagram.

has adequate flexural resistance to sustain the effects of factored loads without failure. Another related problem is *design,* where the cross-sectional dimensions of a structural member and the area of reinforcement need to be determined for given loads and material properties. The design of beams and one-way slabs for flexure is discussed in Chapter 5. It should be noted that flexure also causes the development of shear stresses in beams and other members; shear design of reinforced concrete beams will be discussed in detail in Chapter 6.

DID YOU KNOW?

Positive bending causes tension at the bottom and compression at the top of a flexural member.

The positive bending moment is shown on the bending moment diagram above the longitudinal axis of the member according to the North American convention followed in this book.

3.2 TYPES OF FLEXURAL MEMBERS

An isometric view of a concrete floor structure subjected to gravity load is shown in Figure 3.2. The structure consists of a slab supported by beams, which are in turn supported by columns. In this case, beams are provided in one direction only. Consequently, the slab transfers the applied load in the direction perpendicular to the beams. This type of slab is called a *one-way slab.* One-way slabs are flexural members that behave essentially like wide beams. In cast-in-place concrete construction, beams are usually cast monolithically with the slab. The portion of the slab cast monolithically with a beam contributes to the beam moment resistance; this additional capacity is taken into consideration in the beam design. The term *T-beam* is used in this case because the slab and the beam form a T section for positive bending. In design practice, beams cast monolithically with the slab may be designed either as *T-beams* or as *rectangular beams* for positive bending. If the slab contribution is ignored, the beam is considered to act as a rectangular beam. The analysis of rectangular beams is simpler than that of T-beams, and it is a good starting point for understanding the flexural resistance of concrete members.

Reinforced concrete beams and slabs can be classified into *simple* and *continuous* structures. Continuous structures are statically indeterminate. Simple (or simply supported) structures span across two supports, as shown in Figure 3.3a, whereas continuous structures span

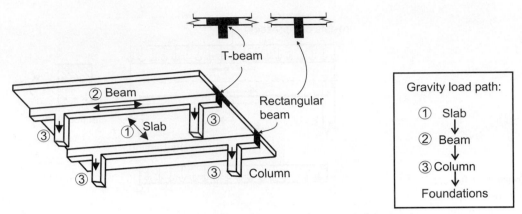

Figure 3.2 An isometric view of a reinforced concrete beam and slab structure.

across three (or more) supports (see Figure 3.3c). When subjected to a gravity load such as the uniform load in Figure 3.3a, only positive bending moments develop in simple structures (see Figure 3.3b), while both positive and negative bending moments develop in continuous structures subjected to the same load, as shown in Figure 3.3d. The deformed shape of the continuous beam subjected to a uniform load is shown in Figure 3.3e. The points where the curvature of the deflected shape changes from the sagging to the hogging shape are called *inflection points* (or points of contraflexure) and are denoted by IP on the diagram in Figure 3.3e. The same points correspond to the locations of zero moments on the bending moment diagram in Figure 3.3d. In this book, the basic concepts of reinforced concrete design

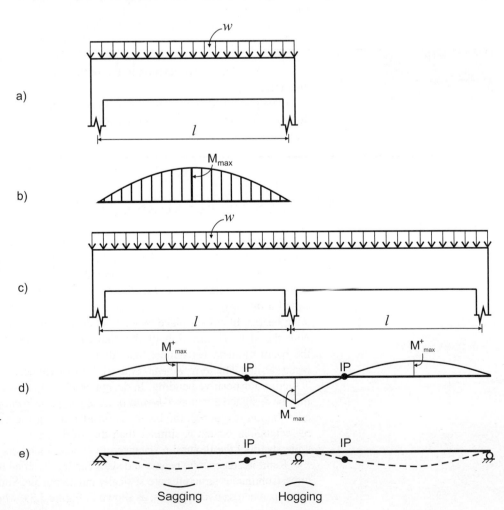

Figure 3.3 Simple versus continuous beams: a) simple beam; b) bending moment diagram for a simple beam; c) continuous beam; d) bending moment distribution in a continuous beam; e) deflected shape for a continuous beam.

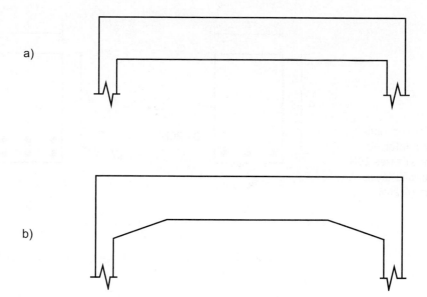

Figure 3.4 Prismatic versus nonprismatic members: a) prismatic beam; b) nonprismatic (haunched) beam.

will be initially explained with examples of simply supported structures. However, continuous structures are more common in cast-in-place concrete construction, and the analysis and design of these structures will be discussed in Chapters 10 and 11.

In general, reinforced concrete members are characterized by regular cross-sectional dimensions. Members with constant cross-sectional dimensions within one span are called *prismatic members* (see Figure 3.4a). In some cases, the cross-sectional properties of a beam within a span are varied. Beams and slabs with variable cross-sectional dimensions along their length are called *nonprismatic members*. Haunched beams are nonprismatic members commonly found in design practice. These beams are characterized by larger cross-sectional dimensions in the support regions, which could also taper toward the midspan (see Figure 3.4b). Haunched beams are sometimes considered to be a more effective design solution for longer-span continuous structures.

3.3 GENERAL ASSUMPTIONS RELATED TO FLEXURE IN REINFORCED CONCRETE BEAMS AND SLABS

3.3.1 Notation

Consider the beam cross-section with reinforcement in Figure 3.5. The beam section in Figure 3.5a is subjected to positive bending (note the tension steel in the bottom part of the section), whereas the section in Figure 3.5b is subjected to negative bending

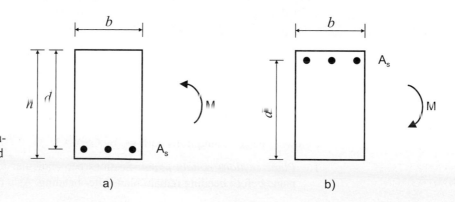

Figure 3.5 A typical rectangular cross-section of a reinforced concrete beam; a) positive bending; b) negative bending.

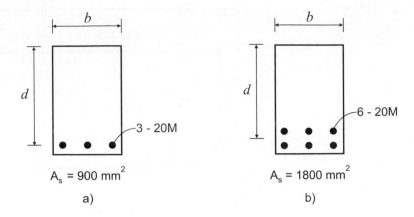

Figure 3.6 Tension reinforcement in a reinforced concrete beam: a) three 20M bars (3-20M) in one layer; b) six 20M bars (6-20M) in two layers.

(note the tension reinforcement in the top part of the section). The notation used in the figure is as follows:

b = width;

d = effective depth, equal to the distance from the *extreme compression fibre* to the centroid of the tension steel; note from Figure 3.5 that the extreme compression fibre may be located at either the top or the bottom of the section, depending on whether the bending moment is positive or negative;

h = overall depth;

A_s = total cross-sectional area of steel reinforcement resisting tension, usually called longitudinal steel or *tension steel*.

Consider the beam cross-section reinforced with three 20M bars shown in Figure 3.6a. Each 20M bar has an area of 300 mm² according to Table A.1, so the total area of tension steel (A_s) is equal to 900 mm².

It should be noted that the tension reinforcement is specified by the number of bars and their size. For example, the three 20M bars in Figure 3.6a are labelled as 3-20M, while the six 20M bars used for the section in Figure 3.6b are labelled as 6-20M. This labelling scheme is used in design practice as well as throughout this book.

In some cases, reinforcement must be placed in two or more layers, as shown in Figure 3.6b. This is usually done to avoid congestion and facilitate concrete placement. The effective depth (d) is the distance from the compression face of the beam to the centroid of the tension steel, as shown in Figure 3.6 (note the difference in d for one- and two-layer reinforcement arrangements). The effective depth can be calculated when the concrete cover is specified (to be discussed in Chapter 5). In this chapter, the effective depth will be specified in all examples and problems.

The amount of tension reinforcement is an important parameter in the design of reinforced concrete structures. The area of tension reinforcement relative to the cross-sectional area is commonly used in design and it is called the *reinforcement ratio* (ρ). For a beam of rectangular cross-section such as the one in Figure 3.5, the reinforcement ratio (ρ) is defined as

$$\rho = \frac{A_s}{b\,d} \tag{3.1}$$

3.3.2 Limit States Design Assumptions

The ultimate limit states design of reinforced concrete flexural members is based on the following basic assumptions outlined by CSA A23.3:

1. *Plane sections remain plane:* Sections perpendicular to the axis of bending that are plane before bending remain plane after bending. As a result, the strain varies linearly

throughout the member depth and is equal to zero at the neutral axis. This assumption is also known as Bernoulli's hypothesis (Cl.10.1.2).

2. *Strain compatibility:* The strain in the reinforcement is equal to the strain in the concrete at the reinforcement location; that is, strain compatibility exists between concrete and steel. This assumption implies a perfect bond between the steel and the concrete.

3. *Stress-strain relationship:* The stresses in the concrete and steel reinforcement can be computed from the stress-strain diagrams for concrete and steel (Cl.10.1.6); however, Cl.10.1.7 proposes the use of an equivalent rectangular stress block instead of the actual stress distribution for concrete for limit states "strength" design.

4. *Concrete tensile strength:* This strength is approximately equal to one tenth of the compressive strength, but it is neglected in flexural strength calculations (Cl.10.1.5).

5. *Maximum concrete compressive strain:* According to Cl.10.1.3, the maximum compressive strain in concrete (ε_{cmax}) is equal to 0.0035. It should be noted that this values does not necessarily represent the actual maximal value—numerous experiments have shown that the maximum strain in concrete generally ranges from 0.003 to 0.005.

KEY ASSUMPTIONS

The limit states design of reinforced concrete flexural members is based on the following basic assumptions outlined by CSA A23.3:

1. Plane sections remain plane (linear strain distribution across the section). This is Bernoulli's hypothesis (Cl.10.1.2).
2. Strains in the concrete and steel at the reinforcement location are equal.
3. Equivalent rectangular stress block is used instead of the actual stress distribution for concrete for strength limit states (Cl.10.1.7).
4. Concrete tensile strength is neglected in flexural strength calculations (Cl.10.1.5).
5. Maximum concrete compressive strain (ε_{cmax}) is equal to 0.0035 (Cl.10.1.3).

3.3.3   Factored Material Strength

Stress-strain diagrams for concrete and steel are presented in Figure 3.7 for reference (refer to Chapter 2 for more details). Note that two different stress-strain curves are presented for concrete and steel, respectively. The actual stress-strain diagram is shown with a solid line, while the dashed line shows the same diagram with the stress values multiplied by the material resistance factors (ϕ_s) and (ϕ_c) for steel and concrete, respectively. The material resistance factors are intended to account for

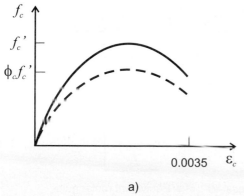
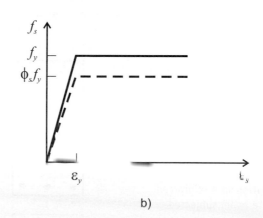

a)   b)

Figure 3.7   Stress-strain diagrams: a) concrete; b) steel.

variations in actual cross-sectional dimensions and material strengths, as well as uncertainties in the strength equations. CSA A23.3 Cl.8.4.2 and Cl.8.4.3 prescribe the following values for material resistance factors to be used in the design of reinforced concrete structures:

Resistance factor for concrete is

$$\phi_c = 0.65 \tag{3.2}$$

and the resistance factor for reinforcing steel is

$$\phi_s = 0.85 \tag{3.3}$$

The reader should note that a ϕ_c value of 0.6 was used in the previous (1994) edition of CSA A23.3.

Consequently, the design of reinforced concrete structures according to CSA A23.3 is performed using the factored specified strengths $\phi_s f_y$ and $\phi_c f_c'$ for steel reinforcement and concrete, respectively.

3.3.4 Equivalent Rectangular Stress Distribution in Concrete

The actual stress distribution in concrete as presented in Figure 3.7a is nonlinear with regard to strain (a parabolic function), while the stress distribution for steel can be approximated with an elastoplastic diagram (see Figure 3.7b). The use of the actual nonlinear stress-strain curve for concrete is considered impractical for design applications. For that reason, CSA A23.3 Cl.10.1.7 permits the use of an equivalent rectangular stress block to replace the actual stress distribution, as shown in Figure 3.8. The equivalent rectangular stress block and the actual stress distribution are characterized by an approximately equal compression force (C_r). This force acts at a distance $a/2$ away from the compressed face of the beam cross-section, hence the term *equivalent* (see Figure 3.8). The use of an equivalent rectangular stress block greatly simplifies the calculation of moment resistance for flexural members.

The rectangular stress block has a depth (a) given by

$$a = \beta_1 c \tag{3.4}$$

where c is the distance from the extreme compression fibre to the neutral axis. The parameters α_1 and β_1 are defined by CSA A23.3 Cl.10.1.7 as follows:

| A23.3 Eq. 10.1 | $\alpha_1 = 0.85 - 0.0015 f_c' \geq 0.67$ | [3.5] |

| A23.3 Eq. 10.2 | $\beta_1 = 0.97 - 0.0025 f_c' \geq 0.67$ | [3.6] |

α_1 and β_1 are purely mathematical parameters based on the requirement of equal compression stress resultants corresponding to the actual and equivalent rectangular stress distribution. In most common applications, where concrete strength varies between 25 MPa

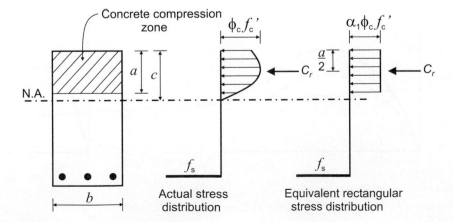

Figure 3.8 Actual versus equivalent rectangular stress distribution in a reinforced concrete beam subjected to flexure.

and 40 MPa, the calculations can be performed using the approximate values

$$\alpha_1 \cong 0.8$$ [3.7]

$$\beta_1 \cong 0.9$$

These values need not be reduced further unless the design specifies concrete strengths of 45 MPa or higher. The α_1 and β_1 values given by Eqn 3.7 will be used in this book. Note that the depth of the concrete compression zone (a) determined using the equivalent rectangular stress block (hatched area shown in Figure 3.8) is different from the neutral axis depth (c) corresponding to the actual compression zone.

3.4 BEHAVIOUR OF REINFORCED CONCRETE BEAMS IN FLEXURE

3.4.1 Unreinforced Beams

Consider a simply supported beam subjected to two point loads (P) shown in Figure 3.9a. Let us observe the behaviour of the beam as the loads increase gradually from zero to a level sufficient to cause failure. The external loads (P) cause internal bending moments that in turn cause flexural stresses in the beam. The bending moment diagram for this beam is shown in Figure 3.9b. Let us first examine the flexural behaviour of a plain concrete beam without reinforcing — the case of an *unreinforced concrete beam*. In this case, concrete behaves like a homogeneous material. Initially, the beam demonstrates elastic behaviour, characterized by a linear stress-strain relationship according to Hooke's law. If the load is removed before the failure, the beam will restore to its original unloaded position. Internal bending moments develop in the beam due to applied loads that cause bending stresses. The stress distribution across a beam cross-section at any point along the span is essentially linear, increasing from zero at the neutral axis to the maximum value at the top and bottom fibres of the section (see Figure 3.9c). The neutral axis (denoted as N.A. in the figures) is a longitudinal axis of zero strain along the member subjected to flexure. At low strains, the concrete resists both tension and compression. However, at a higher load level, tensile stress in the concrete bottom fibre reaches the modulus

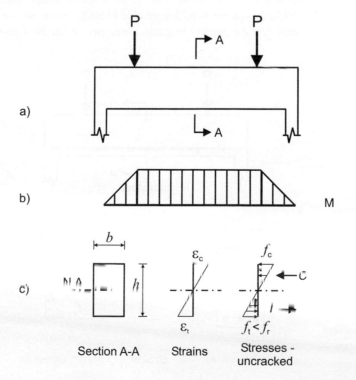

Figure 3.9 Flexural behaviour of an unreinforced concrete beam: a) beam elevation; b) bending moment diagram; c) stress and strain distribution in section A-A.

of rupture (f_r) (see Section 2.3.2 for more information). At that stage, concrete cracks and can no longer resist tension. The beam fails suddenly due to the rapid crack propagation through the critical cross-section.

3.4.2 Reinforced Beams

As the next step, let us now consider the same beam reinforced with three steel bars at the bottom, as shown in Figure 3.10. Note that the tension reinforcement has been provided in the tension zone of the beam (below the neutral axis) as shown on section A-A in Figure 3.10. The beam is subjected to the same load pattern as the unreinforced concrete beam discussed above. Concrete is characterized by a specified compressive strength (f_c') and a tensile strength expressed by the modulus of rupture (f_r), while steel is characterized by the yield strength (f_y). A properly reinforced concrete beam undergoes the following four distinct stages of flexural behaviour:

1.   elastic uncracked
2.   elastic cracked
3.   yielding
4.   failure

The behaviour of a reinforced concrete beam in each of these four stages is discussed in the following text.

Elastic uncracked behaviour   Initially, at *very small loads,* the beam remains uncracked, as shown in Figure 3.10. The stress and strain distribution for concrete is linear, increasing from zero at the neutral axis to the maximum at the top and bottom fibres of the section. At this stage, the beam demonstrates *elastic uncracked* behaviour. Both concrete and steel resist tension; however, only the part of the concrete section above the neutral axis resists compression. It should be noted that the strains in steel and concrete at the reinforcement location are equal; however, the stress in steel (f_s) is significantly larger than the stress in concrete (f_c), as shown on the stress diagram in Figure 3.10.

Elastic cracked behaviour (service load level)   The beam cracks when the flexural stresses at the bottom fibre reach the concrete tensile strength (f_r), as shown in Figure 3.11. Cracked concrete is no longer able to resist tensile stresses. From this stage onward, tensile stresses at the cracked locations are carried by the steel reinforcement only. It should be noted

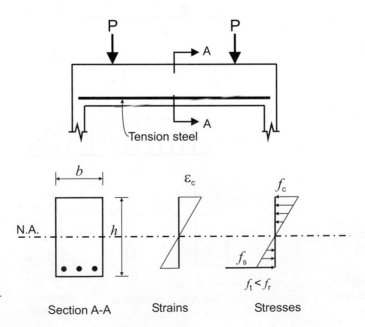

Figure 3.10 Flexural behaviour of a reinforced concrete beam — elastic uncracked phase.

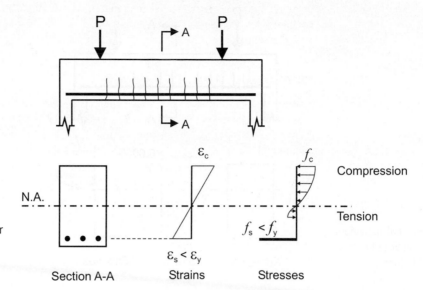

Figure 3.11 Flexural behaviour of a reinforced concrete beam — elastic cracked phase (service load).

that the stress in the steel at this load level is generally well below the yield stress; that is, $f_s \ll f_y$. The concrete continues to resist compression. As a result of the cracking, the neutral axis moves upward and the area of the concrete compression zone decreases. Maximum compressive stresses up to $f_c'/2$ develop in concrete. The stress distribution in the concrete is still close to linear at this stage. This behaviour is known as *elastic cracked behaviour*. Note that most reinforced concrete beams are proportioned such that cracks develop at the service loads; however, these cracks, often called "hairline cracks," are barely visible.

Yielding As the load continues to increase, the stress in the steel reinforcement eventually reaches the yield point ($f_s = f_y$), as shown in Figure 3.12. The compressive stresses in concrete are still elastic. Once yielding has occurred, beam deformations increase rapidly while the corresponding increase in the bending moment values are rather insignificant. The beam is still capable of carrying the applied loads, however this stage is considered the beginning of failure.

Failure With any further load increase, very large deformations develop in the beam and the beam ultimately fails, as shown in Figure 3.13. It should be noted that the tensile strain in the steel is much larger than the yield strain ($\varepsilon_s \gg \varepsilon_y$), while the stress in the steel is equal to the yield stress ($f_s = f_y$). The compressive strain in the concrete (ε_c) at the top fibre has reached or exceeded the maximum value of 0.0035, and the corresponding stress

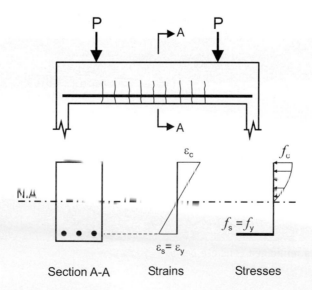

Figure 3.12 Flexural behaviour of a reinforced concrete beam — onset of yielding.

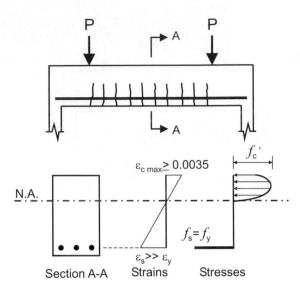

Figure 3.13 Flexural behaviour of a reinforced concrete beam — steel-controlled failure.

is equal to the compressive strength (f_c'). The failure of an actual reinforced concrete beam subjected to a similar load pattern as the above-discussed beam is illustrated in Figure 3.14. Note the flexural cracks in the bottom part of the beam (especially around the midspan) and the crushed concrete in the top region at the beam midspan.

The stages in the behaviour of a reinforced concrete member subjected to flexure are presented in Figure 3.15, which depicts the relationship between the applied load (P) and the corresponding displacement at the midspan (Δ). There are four distinct points on the diagram corresponding to the different stages of flexural behaviour. Initially, at low load levels, the $P-\Delta$ relationship is linear elastic and the beam is said to be in the *elastic uncracked* stage. Once cracking takes place (point C), the slope of the diagram changes (although it still remains linear) as the beam becomes "softer" due to the cracking; this is the *elastic cracked* stage. With further load increase, the service load level is reached (point S on the diagram), but the stress in the steel is below the yield level. Once the load has reached the point where steel reinforcement begins to *yield* (point Y), there is a significant increase in the displacement corresponding to a small amount of load increase. Finally, as the load increase continues, the beam fails after experiencing an excessive amount of deformation (point F on the diagram).

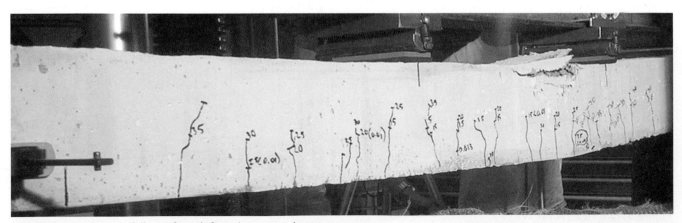

Figure 3.14   Flexural failure of a reinforced concrete beam.

(Svetlana Brzev)

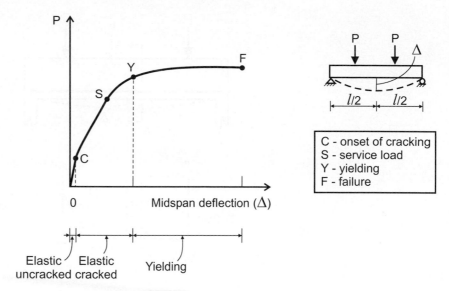

Figure 3.15 Force-deflection diagram for a reinforced concrete beam subjected to flexure.

It should be noted that reinforced concrete members subjected to flexure can fail in different ways, also called *failure modes* (to be discussed in Section 3.4.3). The mode of failure in Figure 3.13 is called *steel-controlled failure,* as the steel yields before the concrete reaches the maximum strain, at which the crushing takes place. The failure is preceded by significant visible deformation and cracking. This is a desirable failure mode for reinforced concrete members because it gives visible warning before the actual failure takes place.

KEY CONCEPTS

The following four stages of flexural behaviour are characteristic of properly reinforced concrete beams and slabs:

1. elastic uncracked
2. elastic cracked
3. yielding
4. failure

Most reinforced concrete flexural members demonstrate elastic cracked behaviour under service loads.

3.4.3 Failure Modes Characteristic of Reinforced Concrete Flexural Members

The behaviour of reinforced concrete beams subjected to flexure was explained in the previous section in an example of a beam failing in the steel-controlled mode. Reinforced concrete flexural members can fail in one of the following two modes:

- steel-controlled (initiated by yielding of tension reinforcement);
- concrete-controlled (initiated by crushing of concrete in compression).

The threshold between the steel-controlled and concrete-controlled failure modes is called the *balanced condition*; it is characterized by the amount of tension reinforcement that causes the simultaneous yielding of steel and crushing of concrete at the ultimate. The mode of failure for a reinforced concrete flexural member is strongly influenced by the amount of tension reinforcement and the mechanical properties of steel and concrete, as explained below.

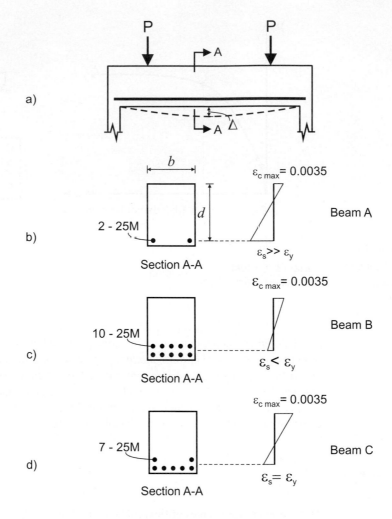

Figure 3.16 Beam cross-sections and strain distributions: a) the loading scheme; b) Beam A; c) Beam B; d) Beam C.

The key features of these failure modes are explained below.

Consider the three reinforced concrete beams with identical cross-sectional dimensions and material properties in Figure 3.16. Beam A is reinforced with two 25M bars (2-25M), Beam B with ten 25M bars (10-25M), and Beam C with seven 25M bars (7-25M). All three beams have been subjected to the two point loads (*P*), as shown in Figure 3.16a. The load increases from zero to the level that causes failure in each beam. Let us study the behaviour of these beams using the relationship between the gradually increasing load (*P*) and the corresponding midspan deflection (Δ) at section A-A.

Steel-controlled failure   Beam A, characterized by the least amount of reinforcement of the three beams, fails in the steel-controlled mode (also called "tension failure" in some references). This mode of failure occurs when the reinforcement yields before the concrete crushes and is characterized by the following strain distribution:

- The steel yields, which means the tensile strain in the steel is greater than or equal to the yield strain ($\varepsilon_s \geq \varepsilon_y$), while the stress in the steel is equal to the yield stress ($f_s = f_y$).
- The maximum compressive strain in concrete (ε_{cmax}) reaches the value of 0.0035 ($\varepsilon_{cmax} = 0.0035$).

The strain distribution in Beam A at the ultimate is shown in Figure 3.16b.

This behaviour is said to be *ductile,* as such a beam is able to sustain large deformations before failure, as shown with a solid curve in Figure 3.17. If a beam or a slab in a building fails in a ductile manner, the occupants have warning of the impending failure and have an opportunity to leave the building before the collapse, thus avoiding potential casualties. Steel-controlled failure is therefore the desirable mode of failure

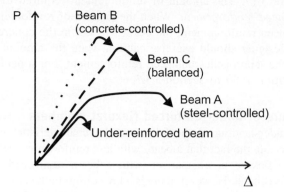

Figure 3.17 Force-deflection curves for reinforced concrete beams failing in different failure modes.

for reinforced concrete flexural members. Beams with an appropriate amount of reinforcement that would fail in a steel-controlled failure mode are said to be *properly reinforced beams*.

Concrete-controlled failure   Beam B, characterized by the maximum amount of reinforcement of the three beams, fails in the concrete-controlled mode (also called "compression failure" in some references). In this mode of failure, the concrete reaches the maximum compressive strain before the steel yields, and the strain distribution is as follows:

- The steel remains elastic ($\varepsilon_s < \varepsilon_y$ and $f_s < f_y$).
- The maximum compressive strain in concrete (ε_{cmax}) reaches the value of 0.0035 ($\varepsilon_{cmax} = 0.0035$).

The strain distribution in Beam B at the ultimate is shown in Figure 3.16c. The beam exhibits elastic behaviour and fails as a result of the sudden crushing of concrete compression zone shortly after the maximum load (P) has been reached. The beam fails without experiencing significant deformations, which is characteristic of brittle behaviour, as illustrated by the dotted curve in Figure 3.17. Beams of this type fail suddenly in a brittle manner without warning to the occupants of the building; consequently, the occupants might not have enough time to escape from the building before it collapses. This is clearly an undesirable situation. Beams that would fail in the concrete-controlled mode are said to be *overreinforced*.

Balanced condition   Beam C is reinforced with an "intermediate" amount of reinforcement, as shown in Figure 3.16d. This amount of reinforcement corresponds to the *balanced condition*, which is associated with the simultaneous crushing of concrete and yielding of steel reinforcement (see Figure 3.16d) and the following strain distribution:

- The steel has just yielded; that is, the tensile strain in steel is equal to the yield strain ($\varepsilon_s = \varepsilon_y$), whereas the corresponding stress in steel is equal to the yield stress ($f_s = f_y$).
- The maximum compressive strain in concrete (ε_{cmax}) reaches the value of 0.0035 ($\varepsilon_{cmax} = 0.0035$).

This case marks the balance point between a ductile (steel-controlled) failure and a brittle (concrete-controlled) failure, hence the name *balanced condition*. The beam shows elastic behaviour until the maximum load (P) has been reached, and a sudden brittle failure occurs shortly thereafter. The balanced condition is sometimes referred to as a special failure mode called "balanced failure"; however, in reality it is a special case of concrete-controlled failure. This behaviour is shown with a dashed curve in

Figure 3.17. The amount of reinforcement required to create this condition is called *balanced reinforcement*. When the amount of reinforcement is equal to or greater than balanced reinforcement, the beam will fail in the concrete-controlled mode. Therefore, the designer should exercise caution when the amount of reinforcement in a flexural member approaches balanced reinforcement. For a detailed discussion of the balanced condition, refer to Section 3.5.3.

Failure of underreinforced flexural members   By now, the reader has developed an understanding of the failure modes characteristic of flexural members and should appreciate the fact that a beam with less reinforcement will perform better when subjected to flexure in comparison with a beam reinforced with a rather large amount of reinforcement. However, there is a lower limit to the amount of reinforcement. Let us consider that Beam A was reinforced with only two 10M bars. When the load that causes cracking has been reached, reinforcing bars will rupture as the ultimate steel tensile strength is simultaneously attained. This is followed by the crushing of the concrete and the complete collapse of the beam. This behaviour is similar to that characteristic of an unreinforced concrete beam described in Section 3.4.1, and is shown with a thin solid curve in Figure 3.17. This is a brittle and hence undesirable failure mode. Beams with a very small amount of reinforcement are called *underreinforced*. To prevent the design of underreinforced beams, CSA A23.3 prescribes the minimum amount of reinforcement in flexural members. This will be explained in Chapter 5.

KEY CONCEPTS

There are two basic failure modes characteristic of reinforced concrete flexural members:

* steel-controlled
* concrete-controlled

These failure modes are related to the behaviour of steel and concrete, which are the main ingredients of reinforced concrete. Steel is a ductile material with the ability to deform extensively before failure. It is considered good practice to design reinforced concrete flexural members to fail in the *steel-controlled mode* initiated when the steel reinforcement yields and characterized by ductile behaviour. Flexural members failing in the steel-controlled mode usually have a moderate amount of reinforcement.

Concrete is a material characterized by brittle behaviour. It is therefore *not* recommended to design flexural members to fail in the *concrete-controlled mode* initiated by the crushing of concrete. This mode of failure is typically brittle and occurs suddenly and without warning. Flexural members failing in the concrete-controlled mode are characterized by a rather large amount of reinforcement.

The *balanced condition* represents the threshold between the steel-controlled and the concrete-controlled failure modes. This condition is associated with the simultaneous crushing of concrete and yielding of steel reinforcement, and it represents a special case of concrete-controlled failure. The designer should exercise caution when the required reinforcement ratio approaches balanced reinforcement.

Flexural members characterized by a very small amount of reinforcement (also called underreinforced members) demonstrate brittle behaviour and fail suddenly and without warning. The behaviour of such members under flexure is similar to the behaviour of unreinforced concrete members. CSA A23.3 prescribes the minimum amount of reinforcement in flexural members.

3.5 MOMENT RESISTANCE OF RECTANGULAR BEAMS WITH TENSION STEEL ONLY

3.5.1 Properly Reinforced Beams (Steel-Controlled Failure)

Figure 3.18a shows a simply supported rectangular beam subjected to point loads (*P*). The beam has a rectangular cross-section with width *b* and effective depth *d,* as illustrated in Figure 3.18b. The beam is considered to be *properly reinforced;* that is, it fails in the steel-controlled mode (as discussed in Section 3.4.3). The following design assumptions are made in this case:

- The steel yields; that is, the tensile strain in the steel is greater than or equal to the yield strain ($\varepsilon_s \geq \varepsilon_y$), while the stress in the steel is equal to the yield stress ($f_s = f_y$).
- The maximum compressive strain in concrete (ε_{cmax}) reaches the value of 0.0035 ($\varepsilon_{cmax} = 0.0035$).

Let us consider the equilibrium of forces acting on the cross-section in Figure 3.18c. Concrete resists compression whereas steel reinforcement resists tension. The internal bending moment developed at any location within a beam is resisted by a force couple T_r and C_r separated by a lever arm, as illustrated in Figure 3.18c. The resultant concrete compression force (C_r) is equal to the product of the area of the compressive zone of depth *a* and width *b* and the uniform stress of magnitude $\alpha_1 \phi_c f_c$ (see Figure 3.18d); that is,

$$C_r = \alpha_1 \phi_c f_c' a\, b \qquad\qquad [3.8]$$

The tension force in the reinforcement (T_r) is equal to the product of the factored stress in steel ($\phi_s f_y$) and the reinforcement area (A_s). This is based on the expression for normal stress presented in mechanics of materials textbooks:

$$T_r = \phi_s f_y A_s \qquad\qquad [3.9]$$

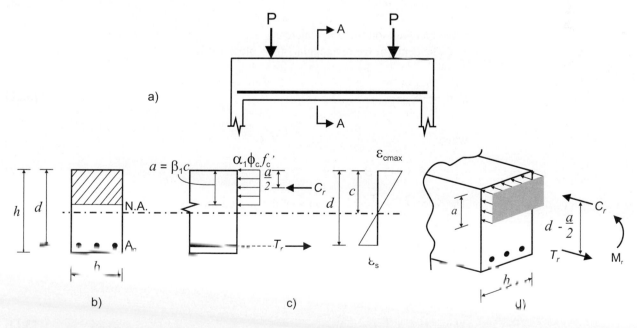

Figure 3.18   Reinforced concrete beam at the ultimate: a) beam elevation; b) typical cross-section; c) equivalent rectangular stress distribution; d) isometric view of stress distribution.

As the beam is not subjected to any external axial loads, the equilibrium of forces in the horizontal direction gives

$$C_r = T_r \qquad\qquad\qquad [3.10]$$

The equation of 3.10 can be rewritten using the T_r and C_r values from Eqns 3.8 and 3.9, respectively:

$$\alpha_1 \phi_c f_c' \, a \, b = \phi_s f_y A_s \qquad\qquad\qquad [3.11]$$

The depth of the rectangular stress block (a) can be determined from (3.11) as

$$a = \frac{\phi_s f_y A_s}{\alpha_1 \phi_c f_c' b} \qquad\qquad\qquad [3.12]$$

Once the depth of the compression zone (a) has been determined, the *factored moment resistance* of the section (M_r) can be obtained from either of the forces T_r or C_r by using the expression: moment = force × lever arm, a notion introduced in statics courses

KEY EQUATIONS

The equation of equilibrium is

$$C_r = T_r \qquad\qquad\qquad [3.10]$$

where

$$T_r = \phi_s f_y A_s \qquad\qquad\qquad [3.9]$$

is the tension force in the reinforcement and

$$C_r = \alpha_1 \phi_c f_c' \, a \, b \qquad\qquad\qquad [3.8]$$

is the compression force in concrete.

The depth of the rectangular stress block (a) is

$$a = \frac{\phi_s f_y A_s}{\alpha_1 \phi_c f_c' \, b} \qquad\qquad\qquad [3.12]$$

where

$$\alpha_1 \cong 0.8 \quad \text{and} \quad \beta_1 \cong 0.9 \qquad\qquad\qquad [3.7]$$

The *factored moment resistance* of the section (M_r) is

$$M_r = T_r \left(d - \frac{a}{2} \right) \qquad\qquad\qquad [3.13]$$

or

$$M_r = \phi_s f_y A_s \left(d - \frac{a}{2} \right) \qquad\qquad\qquad [3.14]$$

(see Figure 3.18d), while the lever arm length is $(d - a/2)$. M_r can then be determined as follows:

$$M_r = T_r\left(d - \frac{a}{2}\right) \qquad \text{[3.13]}$$

If we substitute the expression for T_r from Eqn 3.9 into Eqn 3.13, M_r can be expressed in expanded form as

$$M_r = \phi_s f_y A_s\left(d - \frac{a}{2}\right) \qquad \text{[3.14]}$$

The above equation determines the *factored moment resistance* (M_r) for a beam or a slab cross-section. The word "resistance" implies capacity and is calculated using the material resistance factors (ϕ_s) and (ϕ_c) discussed in Section 3.3.3.

Example 3.1

A typical cross-section of a reinforced concrete beam is shown in the figure below. The beam is reinforced with two 25M bars (2-25M) in the tension zone. The beam is properly reinforced. Concrete and steel material properties are given below.
Find the factored moment resistance for the beam section.

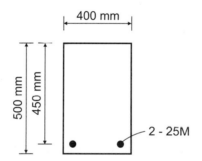

Given: $f_c = 25$ MPa
$f_y = 400$ MPa
$\phi_c = 0.65$
$\phi_s = 0.85$

SOLUTION: 1.   **Calculate the area of tension steel (A_s) and the effective depth (d)**
Tension steel: 2-25M bars

The area of one bar is 500 mm^2 (see Table A.1). The total area of tension steel (A_s) is given by

$$A_s = 2 \times 500 \text{ mm}^2 = 1000 \text{ mm}^2$$

The effective depth (d) is given (see the sketch above); that is

$$d = 450 \text{ mm}$$

2.   **Calculate the depth of the equivalent rectangular stress block (a)**
First, let us calculate the tension force in the reinforcement (T_r). As the beam is properly reinforced, the tension steel yields; that is,

$$f_s = f_y$$

Therefore,

$$T_r = \phi_s f_y A_s \tag{3.9}$$

$$= 0.85 \times 400 \text{ MPa} \times 1000 \text{ mm}^2 = 340 \times 10^3 \text{ N} = 340 \text{ kN}$$

Let us use the equation of equilibrium

$$C_r = T_r \tag{3.10}$$

Since the compression force in concrete (C_r) is given by

$$C_r = \alpha_1 \phi_c f_c' \, a \, b \tag{3.8}$$

where

$$\alpha_1 \cong 0.8$$

the depth of the rectangular stress block (a) can be calculated from the equation of equilibrium as

$$a = \frac{T_r}{\alpha_1 \phi_c f_c' \, b} = \frac{340 \times 10^3 \text{ N}}{0.8 \times 0.65 \times 25 \text{ MPa} \times 400 \text{ mm}} = 65.4 \text{ mm} \cong 65 \text{ mm}$$

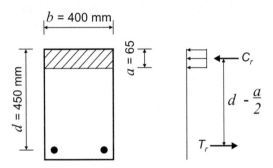

3. **Calculate the factored moment resistance (M_r)**

The moment resistance can be calculated from the equation

$$M_r = T_r \left(d - \frac{a}{2} \right) \tag{3.13}$$

$$= (340 \times 10^3 \text{ N}) \left(450 \text{ mm} - \frac{65 \text{ mm}}{2} \right) = 141.9 \times 10^6 \text{ N} \cdot \text{mm}$$

The factored moment resistance is

$$M_r = 142 \text{ kN} \cdot \text{m}$$

3.5.2   Overreinforced Beams (Concrete-Controlled Failure)

Reinforced concrete beams characterized by a large amount of reinforcement generally fail in the concrete-controlled mode; these beams are also called *overreinforced beams*. The concrete-controlled mode is an undesirable mode of failure and it is expected that the reader will not design overreinforced beams; however, an overview of the analysis procedure is presented in this section.

When the concrete crushes before the steel yields, the following stress and strain distribution take place:

- The maximum compressive strain in the concrete (ε_{cmax}) reaches the value of 0.0035 ($\varepsilon_{cmax} = 0.0035$).
- The tensile strain in the steel is less than the yield strain ($\varepsilon_s < \varepsilon_y$), while the stress in the steel remains in the elastic range ($f_s < f_y$), so the tension reinforcement does not yield.

The equation of equilibrium remains the same as in the case of properly reinforced beams discussed in Section 3.5.1; that is,

$$C_r = T_r \qquad\qquad [3.15]$$

When the T_r and C_r are substituted from Eqns 3.8 and 3.9 into Eqn 3.15, it follows that

$$(\alpha_1 \phi_c f_c')\,(a\,b) = \phi_s f_s A_s \qquad\qquad [3.16]$$

As the tension steel remains in the elastic stress range, Hooke's law applies and the stress in the steel is proportional to the strain:

$$f_s = E_s \varepsilon_s \qquad\qquad [3.17]$$

The steel strain (ε_s) can be determined using the similarity of triangles from the strain distribution diagram in Figure 3.18c:

$$\frac{\varepsilon_{cmax}}{c} = \frac{\varepsilon_{cmax} + \varepsilon_s}{d} \qquad\qquad [3.18]$$

where $\varepsilon_{cmax} = 0.0035$ and c can be determined from Eqn 3.4 as

$$c = \frac{a}{\beta_1}$$

The ε_s value can be substituted in Eqn 3.17 to determine f_s. Subsequently, f_s can be substituted in Eqn 3.16. The quadratic equation 3.19 obtained in this way can be solved for a:

$$\frac{(\alpha_1 \phi_c f_c')b}{\varepsilon_{cmax}\phi_s E_s A_s}\, a^2 + a - d\beta_1 = 0 \qquad\qquad [3.19]$$

Once the a value has been obtained from Eqn 3.19, the factored moment resistance (M_r) can be obtained as follows:

$$M_r = T_r \left(d - \frac{a}{2} \right) \qquad\qquad [3.13]$$

Although it is possible to gain additional moment resistance by increasing the amount of reinforcement, the design of overreinforced beams should be discouraged for the following reasons:

It is not economical to place a large amount of reinforcement in a beam.
- Excessively large amounts of reinforcement will lead to the concrete controlled failure mode, which is brittle and does not provide any warning to building occupants.

A more appropriate solution in this case would be to redesign a beam using a deeper or wider section.

KEY EQUATIONS

The steel strain (ε_s) can be determined from the strain distribution diagram as

$$\frac{\varepsilon_{cmax}}{c} = \frac{\varepsilon_{cmax} + \varepsilon_s}{d}$$ [3.18]

where $\varepsilon_{cmax} = 0.0035$ and c can be determined as

$$c = \frac{a}{\beta_1}$$

If

$$\varepsilon_s < \varepsilon_y = \frac{f_y}{E_s}$$

then the beam should be treated as overreinforced. The depth of the rectangular stress block (a) is determined from the quadratic equation

$$\frac{(\alpha_1 \phi_c f_c') b}{\varepsilon_{cmax} \phi_s E_s A_s} a^2 + a - d\beta_1 = 0$$ [3.19]

where

$$\alpha_1 \cong 0.8 \quad \text{and} \quad \beta_1 \cong 0.9$$ [3.7]

The *factored moment resistance* of the section (M_r) is given by

$$M_r = T_r \left(d - \frac{a}{2} \right)$$ [3.13]

Example 3.2

Consider the beam section from Example 3.1 reinforced with ten 25M bars (10-25M) in the tension zone. The concrete and steel material properties are given below.

Verify whether the beam is overreinforced and find the factored moment resistance.

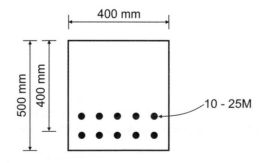

Given: $f_c' = 25$ MPa
$f_y = 400$ MPa
$\phi_c = 0.65$
$\phi_s = 0.85$
$E_s = 200\ 000$ MPa

SOLUTION:

1. **Calculate the area of tension steel (A_s) and the effective depth (d)**

   Tension steel: 10-25M bars

   The area of one 25M bar is 500 mm^2 (see Table A.1). The total area of tension steel (A_s) is

   $$A_s = 10 \times 500 \text{ mm}^2 = 5000 \text{ mm}^2$$

   The effective depth (d) is given as

   $$d = 400 \text{ mm}$$

   Note that this beam is reinforced with 10 bars distributed in 2 layers. The effective depth (d) is the distance from the top of the beam to the *centroid* of the reinforcement.

2. **Check whether the tension reinforcement has yielded**

   Assume that the tension steel has yielded.

   Also, consider the following values (this is OK since $f_c' = 25$ MPa):

   $$\alpha_1 \cong 0.8 \quad \text{and} \quad \beta_1 \cong 0.9 \tag{3.7}$$

   The tension force in the reinforcement (T_r) is given by

   $$T_r = \phi_s f_y A_s \tag{3.9}$$
   $$= 0.85 \times 400 \text{ MPa} \times 5000 \text{ mm}^2 = 1700 \times 10^3 \text{ N} = 1700 \text{ kN}$$

   Let us use the equation of equilibrium

   $$C_r = T_r \tag{3.10}$$

   Since the compression force in concrete (C_r) is given by

   $$C_r = \alpha_1 \phi_c f_c' \, a \, b \tag{3.8}$$

   the depth of the rectangular stress block (a) can be calculated from the equation 3.10 as

   $$a = \frac{T_r}{\alpha_1 \phi_c f_c' \, b} = \frac{1700 \times 10^3 \text{ N}}{0.8 \times 0.65 \times 25 \text{ MPa} \times 400 \text{ mm}} = 327 \text{ mm}$$

   The neutral axis depth can be then determined as

   $$c = \frac{a}{\beta_1} = \frac{327 \text{ mm}}{0.9} = 363 \text{ mm}$$

   The strain in the steel reinforcement can be calculated from the strain distribution diagram using the similarity of triangles (see the sketch below):

   $$\frac{\varepsilon_{cmax}}{c} = \frac{\varepsilon_{cmax} + \varepsilon_s}{d} \tag{3.18}$$

   or

   $$\frac{0.0035}{363 \text{ mm}} = \frac{0.0035 + \varepsilon_s}{400 \text{ mm}}$$

   The steel strain can be calculated from the above equation as

   $$\varepsilon_s = 0.00036$$

   Calculate the yield strain:

   $$\varepsilon_y = \frac{f_y}{E_s} = \frac{400 \text{ MPa}}{200\,000 \text{ MPa}} = 0.002$$

Since

$$\varepsilon_s = 0.00036 \ll 0.002$$

the tension steel has not yielded, and the beam should be analyzed as an overreinforced beam.

The strain distribution for this beam section is shown on the sketch below.

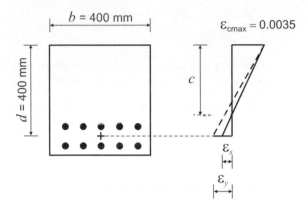

3. **Calculate the depth of the equivalent rectangular stress block (a)**

    The depth of the rectangular stress block (a) can be determined from the quadratic equation

    $$\frac{(\alpha_1 \phi_c f_c') b}{\varepsilon_{cmax} \phi_s E_s A_s} a^2 + a - d\beta_1 = 0 \qquad [3.19]$$

    $$\left(\frac{(0.8 \times 0.65 \times 25 \text{ MPa}) \times 400 \text{ mm}}{0.0035 \times 0.85 \times 200\,000 \text{ MPa} \times 5000 \text{ mm}^2} \right) a^2 + a - (400 \text{ mm} \times 0.9) = 0$$

    Therefore,

    $$(1.75 \times 10^{-3}) a^2 + a - 360 = 0$$

    and

    $$a = \frac{-1 \pm \sqrt{1^2 + 4 \times 360 \times 1.75 \times 10^{-3}}}{2 \times 1.75 \times 10^{-3}} = 250 \text{ mm}$$

    The neutral axis depth can be then calculated as

    $$c = \frac{a}{\beta_1} = \frac{250 \text{ mm}}{0.9} = 278 \text{ mm} \cong 280 \text{ mm}$$

4. **Calculate the factored moment resistance (M_r)**

    First, let us calculate the steel strain using the procedure from step 2 above.

    The steel strain (ε_s) can be determined from the strain distribution diagram as

    $$\frac{\varepsilon_{cmax}}{c} = \frac{\varepsilon_{cmax} + \varepsilon_s}{d} \qquad [3.18]$$

    or

    $$\frac{0.0035}{280 \text{ mm}} = \frac{0.0035 + \varepsilon_s}{400 \text{ mm}}$$

From the above equation it follows that

$$\varepsilon_s = 0.0015$$

The stress in the steel can be calculated from Hooke's law as

$$f_s = E_s \varepsilon_s = 200\,000 \text{ MPa} \times 0.0015 = 300 \text{ MPa}$$

As a result, the tension force in the reinforcement is given by

$$T_r = \phi_s f_s A_s = 0.85 \times 300 \text{ MPa} \times 5000 \text{ mm}^2 = 1275 \times 10^3 \text{ N} = 1275 \text{ kN}$$

The moment resistance can be calculated from the equation

$$M_r = T_r \left(d - \frac{a}{2} \right) \tag{3.13}$$

$$= (1275 \times 10^3 \text{ N}) \left(400 \text{ mm} - \frac{250 \text{ mm}}{2} \right) = 350.6 \times 10^6 \text{ N} \cdot \text{mm}$$

The factored moment resistance for this beam section is

$$M_r = 350 \text{ kN} \cdot \text{m}$$

3.5.3 Balanced Condition

The balanced condition is characterized by the simultaneous crushing of concrete and yielding of the tension reinforcement. The strain in the concrete reaches the maximum value $\varepsilon_{cmax} = 0.0035$ while the strain in the steel reaches the yield strain ($\varepsilon_s = \varepsilon_y$). Consequently, the stress in the steel is equal to the yield stress; that is, $f_s = f_y$. The balanced condition is illustrated in Figure 3.19.

The following proportion can be obtained from the strain diagram using the similarity of triangles (Eqn 3.18):

$$\frac{0.0035}{c} = \frac{0.0035 + \varepsilon_y}{d}$$

The c/d ratio for the balanced strain condition can be determined as

$$\frac{c}{d} = \frac{0.0035}{0.0035 + \varepsilon_y} \tag{3.20}$$

Equation 3.20 can be used to determine the neutral axis depth (c) at the balanced condition.

The yield strain (ε_y) can be obtained from Hooke's law as (see Section 2.6.2)

$$\varepsilon_y = \frac{f_y}{E_s} \tag{2.4}$$

where E_s is the modulus of elasticity for steel ($E_s = 200\,000$ MPa).

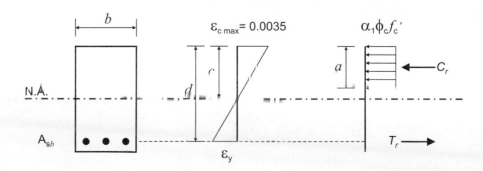

Figure 3.19 Balanced condition.

The depth of the equivalent rectangular stress block (a) can be determined as

$$a = \beta_1 c \tag{3.4}$$

The area of tension reinforcement corresponding to the balanced condition is called the *balanced reinforcement* (A_{sb}), and it can be found from the equation of equilibrium of internal forces presented in Section 3.5.1:

$$C_r = T_r \tag{3.10}$$

where

$$T_r = \phi_s f_y A_{sb} \tag{3.9}$$

is the tension force in the reinforcement and

$$C_r = \alpha_1 \phi_c f_c' \, a \, b \tag{3.8}$$

is the compression force in the concrete.
From Eqn 3.10 it follows that

$$A_{sb} = \frac{C_r}{\phi_s f_y} = \frac{\alpha_1 \phi_c f_c' \, a \, b}{\phi_s f_y} \tag{3.21}$$

The corresponding reinforcement ratio, called the *balanced reinforcement ratio* (ρ_b), can be determined as

$$\rho_b = \frac{A_{sb}}{b \, d} \tag{3.22}$$

The above equation can be used to calculate the balanced reinforcement ratio (ρ_b). It should be noted that this ratio does not depend on the cross-sectional dimensions of a particular beam; the ρ_b value is constant for all beams with the same concrete and steel material properties (f_c' and f_y).

KEY EQUATIONS

The *c/d* ratio corresponding to the *balanced condition* is

$$\frac{c}{d} = \frac{0.0035}{0.0035 + \varepsilon_y} \tag{3.20}$$

The depth of the equivalent rectangular stress block (a) is

$$a = \beta_1 c \tag{3.4}$$

The balanced reinforcement area (A_{sb}) is

$$A_{sb} = \frac{C_r}{\phi_s f_y} = \frac{\alpha_1 \phi_c f_c' \, a \, b}{\phi_s f_y} \tag{3.21}$$

The balanced reinforcement ratio (ρ_b) is

$$\rho_b = \frac{A_{sb}}{b \, d} \tag{3.22}$$

The factored moment resistance (M_r) for the balanced condition is

$$M_r = \phi_s f_y A_{sb} \left(d - \frac{a}{2} \right) \tag{3.14}$$

The *factored moment resistance* (M_r) for a rectangular beam section in the balanced condition can be determined using the equation

$$M_r = \phi_s f_y A_{sb} \left(d - \frac{a}{2} \right)$$ [3.14]

As discussed in Section 3.4.3, the balanced condition represents the threshold between overreinforced beams failing in the concrete-controlled mode and properly reinforced beams failing in the steel-controlled mode. For that reason, structural designers use the balanced reinforcement ratio (ρ_b) to predict the potential failure mode for a beam with reinforcement ratio (ρ). The three possible scenarios are summarized below.

1. *The balanced condition* ($\rho = \rho_b$) is characterized by the concrete crushing and the steel yielding simultaneously. The amount of steel required to create this condition is considered to be large.
2. *Concrete-controlled failure* ($\rho > \rho_b$) is a brittle failure characterized by the concrete crushing. The amount of steel required to create this condition is larger than for the balanced condition, and such a beam is called overreinforced.
3. *Steel-controlled failure* ($\rho < \rho_b$) is characterized by the yielding of steel reinforcement. The amount of reinforcement required to create this condition is less than in balanced reinforcement, and such a beam is said to be *properly reinforced*.

It should be noted that, given the variability in the actual material strengths and stress-strain relationship for concrete and steel, the actual value of the balanced reinforcement ratio may be lower than the value calculated using Eqn 3.22. The designer should exercise caution when the required reinforcement ratio approaches the balanced ratio. A recommended upper bound for the reinforcement ratio to ensure steel-controlled failure is around 75% of the ρ_b value; that is, $\rho \leq 0.75\rho_b$.

KEY CONCEPTS

The balanced condition, characterized by the simultaneous crushing of concrete and yielding of steel reinforcement, represents the threshold between overreinforced beams failing in the concrete-controlled mode and properly reinforced beams failing in the steel-controlled mode. Structural designers use the balanced reinforcement ratio (ρ_b) to predict the potential failure mode for a beam with reinforcement ratio (ρ). The three possible scenarios are summarized below.

• If $\rho = \rho_b$, the amount of reinforcement corresponds to the *balanced condition*. The amount of steel required to create this condition is considered to be large— it is the upper bound of the amount of reinforcement permitted in the flexural design.
• If $\rho > \rho_b$, the amount of reinforcement corresponds to the *concrete-controlled failure mode*, characterized by an excessively high reinforcement ratio (overreinforced beams).
• If $\rho < \rho_b$, the amount of reinforcement corresponds to the *steel-controlled failure mode*, characterized by a moderate amount of reinforcement (properly reinforced beams).

It is desirable to use the "proper" amount of steel reinforcement in the design to ensure the steel-controlled mode of failure. The reinforcement ratio should preferably be less than 75% of the ρ_b value, that is, $\rho \leq 0.75\rho_b$.

Example 3.3

Consider the beam cross-section discussed in Example 3.1. The concrete and steel material properties are given below. The effective depth *(d)* is equal to 400 mm.

Find the following:

- *the balanced area of reinforcement,*
- *the balanced reinforcement ratio,*
- *the factored moment resistance corresponding to the balanced condition.*

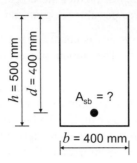

Given: $f_c' = 25$ MPa
$f_y = 400$ MPa
$\phi_c = 0.65$
$\phi_s = 0.85$
$E_s = 200\ 000$ MPa

SOLUTION: **1. Calculate the depth of the equivalent compression stress block (*a*)**
The effective depth *(d)* is given by

$$d = 400 \text{ mm}$$

The *c/d* ratio corresponding to the balanced condition can be obtained as

$$\frac{c}{d} = \frac{0.0035}{0.0035 + \varepsilon_y} \qquad \text{[3.20]}$$

The yield strain can be calculated from Hooke's law as

$$\varepsilon_y = \frac{f_y}{E_s} = \frac{400 \text{ MPa}}{200\ 000 \text{ MPa}} = 0.002$$

Subsequently, the neutral axis depth can be calculated from Eqn 3.20 as

$$c = \frac{0.0035}{0.0035 + 0.002}\, d$$

$$= 0.64\, d = 0.64 \times 400 \text{ mm} \cong 255 \text{ mm}$$

Also, consider the following values (this is OK since $f_c' = 25$ MPa):

$$\alpha_1 \cong 0.8 \text{ and } \beta_1 \cong 0.9 \qquad \text{[3.7]}$$

The depth of the equivalent rectangular stress block (*a*) can be calculated as

$$a = \beta_1 c \qquad \text{[3.4]}$$

$$= 0.9 \times 255 \text{ mm} = 230 \text{ mm}$$

2. **Calculate the balanced reinforcement (area and ratio)**

The compression force in concrete (C_r) is given by

$$C_r = \alpha_1 \phi_c f_c' \, a \, b \qquad \text{[3.8]}$$

$$= 0.8 \times 0.65 \times 25 \text{ MPa} \times 230 \text{ mm} \times 400 \text{ mm} = 1196 \times 10^3 \text{ N}$$

The area of reinforcement can be calculated from the equation of equilibrium

$$C_r = T_r \qquad \text{[3.10]}$$

where

$$T_r = \phi_s f_y A_{sb} \qquad \text{[3.9]}$$

is the tension force in the reinforcement.

As a result, A_{sb} can be calculated as

$$A_{sb} = \frac{C_r}{\phi_s f_y} \qquad \text{[3.21]}$$

$$= \frac{1196 \times 10^3 \text{ N}}{0.85 \times 400 \text{ MPa}} = 3518 \text{ mm}^2 \cong 3500 \text{ mm}^2$$

7-25M bars can be used in this case (one 25M bar has an area of 500 mm^2; see Table A.1):

$$A_s = 7 \times 500 \text{ mm}^2 = 3500 \text{ mm}^2$$

The balanced reinforcement ratio can be calculated as

$$\rho_b = \frac{A_{sb}}{b \, d} = \frac{3500 \text{ mm}^2}{400 \text{ mm} \times 400 \text{ mm}} = 0.022$$

Therefore, $\rho_b = 2.2\%$.

3. **Calculate the factored moment resistance (M_r)**

The moment resistance can be calculated from the equation

$$M_r = \phi_s f_y A_{sb} \left(d - \frac{a}{2} \right) \qquad \text{[3.14]}$$

$$= 0.85 \times 400 \text{ MPa} \times 3500 \text{ mm}^2 \left(400 \text{ mm} - \frac{230 \text{ mm}}{2} \right) = 339 \times 10^6 \text{ N} \cdot \text{mm}$$

The factored moment resistance corresponding to the balanced condition is

$$M_r = 339 \text{ kN} \cdot \text{m}$$

4. **Provide a design summary**

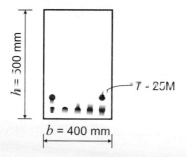

Learning from Examples

At this point, it is important to compare the results of the analyses performed in Examples 3.1 to 3.3. A rectangular beam section with the width of 400 mm and overall depth of 500 mm and material properties: $f_c' = 25$ MPa and $f_y = 400$ MPa has been considered in all three examples; the only difference is in the amount of reinforcement provided in the section. The relevant results of the three examples are summarized in Table 3.1 and Figure 3.20.

Table 3.1 Summary of the results — Examples 3.1 to 3.3

Example (type of beam)	Reinforcement	Area (mm²)	Neutral axis depth (mm)	Factored moment resistance (kN·m)
3.1 (properly reinforced)	2-25M	1000	73	142
3.2 (overreinforced)	10-25M	5000	280	350
3.3 (balanced)	7-25M	3500	255	339

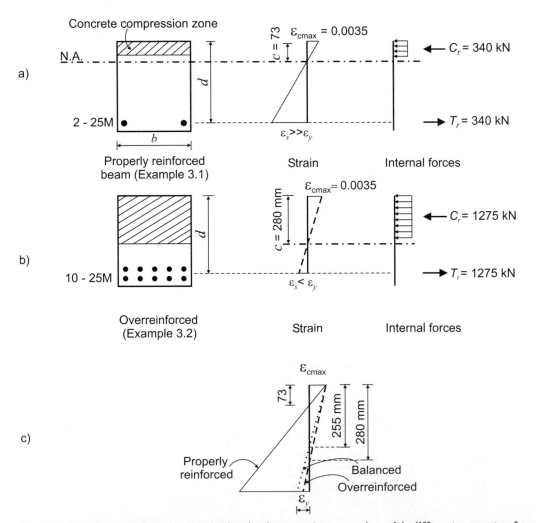

Figure 3.20 Strain and stress distribution in the same beam section with different amounts of reinforcement: a) properly reinforced beam; b) overreinforced beam; c) a combined strain distribution diagram for the three beam sections.

The following conclusions can be drawn by comparing the results of these examples:

1. The increase in the amount of reinforcement between the properly reinforced beam and the overreinforced beam is fivefold (from 1000 mm² to 5000 mm²). However, there is not a proportional increase in the factored moment resistance (the factored moment resistance increases from 142 kN·m to 350 kN·m). It is not economical to design overreinforced beams, as the steel strength is not fully utilized before failure.

2. The amount of reinforcement provided in reinforced concrete beams should preferably not exceed the balanced reinforcement. Overreinforced beams pose a risk to building occupants in the case of failure that occurs suddenly and without warning. The reader should be discouraged from designing overreinforced beams.

3. It is very important for the reader to gain some sense of the importance of strain and stress distribution within a concrete flexural member. A summary of the strain distributions in the three beam sections is shown in Figure 3.20c. The strain distribution in the properly reinforced beam is shown with a solid line, the distribution in the overreinforced beam is shown with a dashed line, and the balanced strain distribution is shown with a dotted line. The reader should acknowledge the possibility of a wide variation in neutral axis depth and the corresponding strain and stress distributions depending on the amount of reinforcement placed in a section.

It can be concluded that the designer should be able to control the behaviour of a reinforced concrete flexural member by choosing the proper amount and distribution of reinforcement.

3.6 FLEXURAL RESISTANCE OF ONE-WAY SLABS

3.6.1 One-Way and Two-Way Slabs

A one-way slab is a reinforced concrete floor panel that transfers load to the supporting beams in one direction only, as shown on the partial floor plan in Figure 3.21. The slab is supported by beams and girders that are in turn supported by the columns at points A, C, D, and F. A slab panel can be defined as a portion of the slab between the two adjacent beams and girders (for example, see panel ABED in Figure 3.21). Consider that the span (L_y) in the north-south direction is significantly larger than the span (L_x) in the east-west direction such that $L_y/L_x > 2$. In that case, the slab panel ABED transfers most of the floor load (over 90%) to the beams AD and BE. It is said that this slab spans in the "short" direction (L_x) and is considered a *one-way slab*.

It should be noted that, if the same floor system were characterized with a span ratio $L_y/L_x < 2$, this slab would not transfer load primarily in one direction; that is, the supporting beams and girders on all four sides would carry a portion of the floor load. Such slabs are called *two-way slabs*. The design of two-way slabs will be discussed in Chapter 12. However, a basic discussion of the concept is presented below.

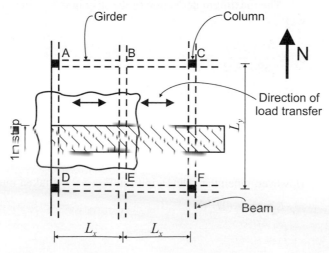

Figure 3.21 A partial floor plan of a one-way slab and beam floor system.

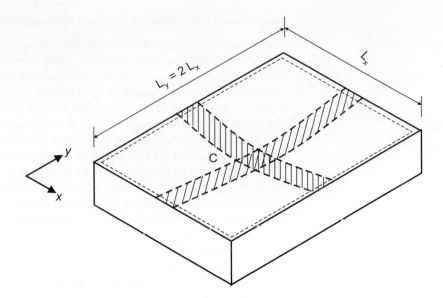

Figure 3.22 A slab with an L_y/L_x ratio of 2.0.

Consider a slab panel supported by walls along the edges, as shown in Figure 3.22. The slab is subjected to a uniformly distributed load (w). The span ratio for this slab is equal to 2.0; that is,

$$\frac{L_y}{L_x} = 2 \qquad [3.23]$$

The total load acting on the slab is shared by the slab strips aligned in the x and y directions. The slab strips must maintain compatibility in deflections caused by the applied loads; this will be used to estimate the portion of load carried in each direction.

Consider the strips in the x and y directions in the midspan region of the slab. Each strip is treated as a simple beam subjected to a uniform load. At this stage, the portion of the load carried by each strip is not known, but the sum of the load components in the x and y directions must be equal to the total load; that is,

$$w = w_x + w_y$$

The maximum deflection of the strip in the x direction can be determined by treating the strip as a simply supported beam of span L_x subjected to a uniform load w_x:

$$\delta_x = \frac{5 \cdot w_x \cdot L_x^{\,4}}{384 \cdot E \cdot I} \qquad [3.24]$$

where E is the modulus of elasticity and I is the moment of inertia for the beam section (these values are considered to be equal in both directions).

The maximum deflection of the strip in the y direction can be determined in a similar way, as

$$\delta_y = \frac{5 \cdot w_y \cdot L_y^{\,4}}{384 \cdot E \cdot I} \qquad [3.25]$$

The maximum deflection at the centre of the slab must be the same in both directions, so

$$\delta_x = \delta_y \qquad [3.26]$$

When δ_x and δ_y from Eqns 3.24 and 3.25 are substituted into Eqn 3.26, it follows that

$$w_x \cdot L_x^{\,4} = w_y \cdot L_y^{\,4} \qquad [3.27]$$

However, when the span ratio from Eqn 3.23 is taken into account, it follows that

$$w_x \cdot L_x^{\,4} = w_y (2L_x)^4$$

or

$$w_x = 16w_y$$

It follows that the load carried in the short (x) direction is 16 times larger than that carried in the long (y) direction, or approximately 94% (16/17) of the total load. It is generally considered that when the span ratio (L_y/L_x) for a slab supported by beams is 2.0 or larger, the slab can be considered to behave like a one-way slab. One-way slab is a common application of concrete slab construction and will be covered in detail in this book.

3.6.2 Moment Resistance of One-Way Slabs

For design purposes, one-way slabs are treated as if they are composed of a series of beams, also called *slab strips,* placed side by side, as illustrated in Figure 3.23. Each slab strip is considered as a rectangular beam with a width (b) of 1 m and a depth (h) equal to the slab thickness (see Figure 3.23). Note that the slab dimensions will be expressed in millimetres in this text, and so $b = 1000$ mm. As the slab strips are of unit width, the uniformly distributed load acting on a slab (generally specified in kPa or kN/m^2) can be expressed in kN/m.

The primary tension reinforcement in one-way slabs always runs parallel to the short direction, as shown in Figure 3.23. Depending on the direction of bending moment, reinforcement is placed either at the bottom of the slab (for positive bending) or at the top (for negative bending). Examples and problems presented in this chapter are related to slabs subjected to positive bending. In reality, slabs are statically indeterminate continuous structures spanning across several supports. Continuous slabs are subjected to both positive and negative bending (to be discussed in detail in Chapters 10 and 11).

In the one-way slab design, the area of reinforcement for a 1000 mm wide strip is denoted as A_s, as shown in Figure 3.24 (note that the same symbol has been used earlier in this chapter to denote the total area of tension reinforcement in a beam). A_s can be determined as follows

$$A_s = A_b \frac{1000}{s}$$

[3.28]

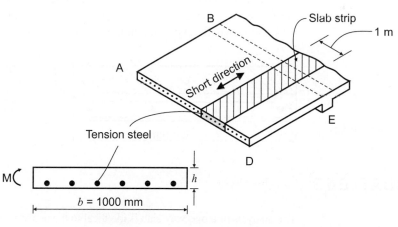

Figure 3.23 An isometric view of a one-way slab system.

Typical slab strip cross section

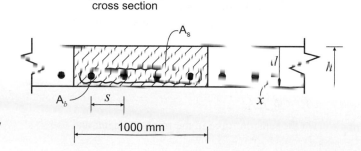

Figure 3.24 Cross-section of a typical unit strip of a one-way slab.

where

A_s = the area of tension reinforcement for a 1000 mm wide strip (usually expressed in square millimetres per metre)

A_b = the area of one reinforcing bar (see Table A.1)

s = the centre-to-centre bar spacing (mm)

Slab reinforcement is usually denoted by bar size and spacing; for example, 15M@300 denotes 15M bars at 300 mm spacing. Since the area (A_b) of a 15M bar is 200 mm^2 (see Table A.1), the corresponding reinforcement area (A_s) for a 1000 mm wide slab strip can be calculated from Eqn 3.28 as

$$A_s = 200 \text{ mm}^2 \times \frac{1000 \text{ mm}}{300 \text{ mm}} = 667 \text{ mm}^2/\text{m}$$

Note that the reinforcement area is expressed per metre width (mm^2/m). Alternatively, the required bar spacing (s) for a given reinforcement area can be determined from Eqn 3.28 as

$$s \le A_b \frac{1000}{A_s} \qquad\qquad\qquad [3.29]$$

Note that the less than or equal to sign has been used in the above expression because the spacing (s) corresponding to the required area of reinforcement (A_s) is the maximum acceptable for a specific design.

The factored moment resistance (M_r) of a one-way slab is determined in the same manner as the flexural resistance of properly reinforced rectangular beams, as discussed in Section 3.5.1. The design assumptions are:

- The tensile strain in the steel is greater than or equal to the yield strain ($\varepsilon_s \ge \varepsilon_y$) and the stress in the steel is equal to the yield stress ($f_s = f_y$).
- The maximum compressive strain in the concrete (ε_{cmax}) reaches the value of 0.0035 ($\varepsilon_{cmax} = 0.0035$).

The analysis of one-way slabs is performed for a typical slab strip of a width (b) equal to 1000 mm and a depth (h) equal to the slab thickness, as shown in Figure 3.25. The resulting factored moment resistance (M_r) per unit slab width is determined from Eqns 3.13 or 3.14 and is usually expressed in kilonewton metres per metre (kN·m/m).

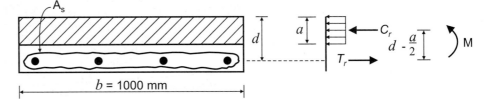

Figure 3.25 Moment resistance for a unit slab strip.

The analysis of a one-way slab is identical to the analysis of a properly reinforced rectangular beam, where b = 1000 mm is the width of a unit strip and h is the overall slab thickness.

The key equations are the same as in Section 3.5.1 for rectangular beams with proper reinforcement. These equations are repeated below for the reader's convenience.

The equation of equilibrium is

$$C_r = T_r \qquad\qquad\qquad [3.10]$$

where

$$T_r = \phi_s f_y A_s \qquad\qquad\qquad [3.9]$$

is the tension force in the reinforcement, and

$$C_r = \alpha_1 \phi_c f_c' \, a \, b \qquad \text{[3.8]}$$

is the compression force in concrete.

The depth of the rectangular stress block (a) is given by

$$a = \frac{\phi_s f_y A_s}{\alpha_1 \phi_c f_c' b} \qquad \text{[3.12]}$$

where

$$\alpha_1 \cong 0.8 \quad \text{and} \quad \beta_1 \cong 0.9 \qquad \text{[3.7]}$$

The *factored moment resistance* of the section (M_r) is given by

$$M_r = T_r \left(d - \frac{a}{2} \right) \quad (\text{kN} \cdot \text{m/m}) \qquad \text{[3.13]}$$

or

$$M_r = \phi_s f_y A_s \left(d - \frac{a}{2} \right) \quad (\text{kN} \cdot \text{m/m}) \qquad \text{[3.14]}$$

The area of reinforcement for a 1000 mm wide strip (usually expressed in square millimetres per metre) is

$$A_s = A_b \frac{1000}{s} \qquad \text{[3.28]}$$

The required bar spacing (s) for a given reinforcement area is given by

$$s \le A_b \frac{1000}{A_s} \qquad \text{[3.29]}$$

The analysis of one-way slabs will be illustrated with Example 3.4.

Example 3.4

A typical cross-section of a one-way slab is shown in the figure below. The slab is 200 mm thick and is reinforced with 15M bars at 300 mm spacing (15M@300) at the bottom face. For analysis purposes, use an effective depth of 170 mm. The slab is properly reinforced. The concrete and steel material properties are given below.

Find the factored moment resistance for the slab.

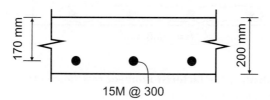

15M @ 300

Given: $f_c' = 25$ MPa

$f_y = 400$ MPa

$\phi_c = 0.65$

$\phi_s = 0.85$

SOLUTION: 1. **Calculate the area of tension steel (A_s) and the effective depth (d)**
The analysis will be carried out based on a 1 m wide slab strip, so

$$b = 1000 \text{ mm}$$

The slab reinforcement is 15M@300 mm. The area of one 15M bar is 200 mm² (see Table A.1); that is

$$A_b = 200 \text{ mm}^2$$

and

$$s = 300 \text{ mm}$$

As a result, the total area of tension steel (A_s) per metre of slab width is given by

$$A_s = A_b \frac{1000}{s} \tag{3.28}$$

$$= 200 \text{ mm}^2 \times \frac{1000}{300 \text{ mm}} = 667 \text{ mm}^2/\text{m}$$

The effective depth (d) is given as

$$d = 170 \text{ mm}$$

2. **Calculate the depth of the equivalent rectangular stress block (a)**

First, calculate the tension force in the reinforcement (T_r). Because the slab is properly reinforced, the tension steel yields; that is,

$$f_s = f_y$$

Therefore,

$$T_r = \phi_s f_y A_s \tag{3.9}$$

$$= 0.85 \times 400 \text{ MPa} \times 667 \text{ mm}^2/\text{m} \cong 227 \times 10^3 \text{ N/m}$$

Use the equation of equilibrium (see the sketch below)

$$C_r = T_r \tag{3.10}$$

Since the compression force in concrete (C_r) is given by

$$C_r = \alpha_1 \phi_c f_c' a b \tag{3.8}$$

and

$$\alpha_1 \cong 0.8$$

the depth of the rectangular stress block (a) can be calculated from Eqn 3.10 as

$$a = \frac{T_r}{\alpha_1 \phi_c f_c' b} = \frac{227 \times 10^3 \text{ N}}{0.8 \times 0.65 \times 25 \text{ MPa} \times 1000 \text{ mm}} = 17 \text{ mm}$$

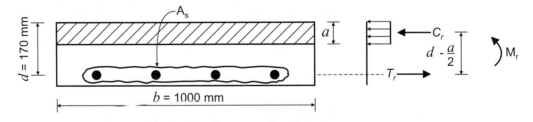

3. **Calculate the factored moment resistance (M_r)**

The moment resistance can be calculated from the following equation:

$$M_r = T_r \left(d - \frac{a}{2} \right) \tag{3.13}$$

$$= (227 \times 10^3 \text{ N}) \left(170 \text{ mm} - \frac{17 \text{ mm}}{2} \right)$$

$$\cong 37.0 \text{ kN} \cdot \text{m/m}$$

The factored moment resistance for a one-way slab determined *per metre width* is

$$M_r = 37.0 \text{ kN} \cdot \text{m/m}$$

3.7 T-BEAMS

3.7.1 Background

In a typical cast-in-place concrete slab and beam floor system, the slab supports gravity loads and transfers the loads to the beams, which are in turn supported by the columns. Figure 3.26 shows a partial floor plan of a one-way slab that transfers the gravity load to the beams AC and BD. In cast-in-place concrete floor construction, rectangular beams that are part of the floor structure are usually cast monolithically with the slab. The continuity between the slab and the beam permits the slab to be considered part of the beam, acting as a top flange, as shown with a hatched pattern on section 1-1 in Figure 3.26. In general, flanges in interior beams extend on both sides of a web; such beams are called *T-beams*, like the beam BD shown in Figure 3.26. Beams with a flange on one side only (usually located around the perimeter of a floor structure) are referred to as *L-beams* (also called *spandrel beams* or *edge beams*), like the beam AC shown in Figure 3.26.

The main components of T- and L-beams are (see section 1-1 in Figure 3.26)

- web
- flange
- overhangs

It should be stressed that a T-beam carries floor or roof load in proportion to its tributary area, as discussed in Section 1.3.3. Consider the beam BE that is a part of the floor system in Figure 3.27. This T-beam carries a portion of the floor load over a distance L_x (equal to the spacing between the adjacent parallel beams); this distance is called the *tributary width*. The tributary area for the beam BE is shown with a hatched pattern in Figure 3.27.

Figure 3.26 Partial floor plan of a slab and beam floor system and a typical section showing T- and L-beams.

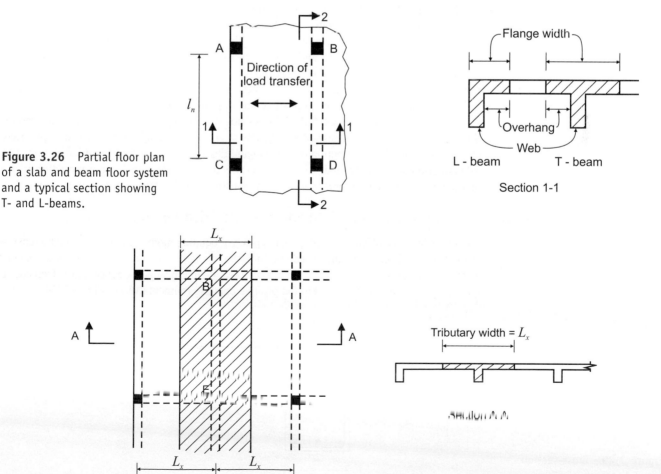

Figure 3.27 Tributary area for a T-beam.

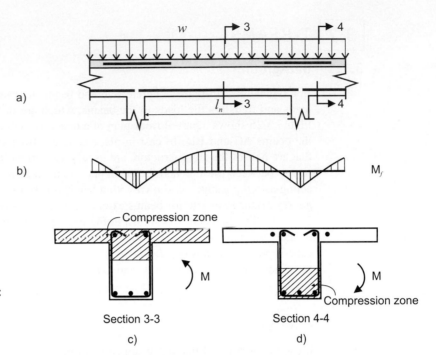

Figure 3.28 Continuous T-beam: a) elevation; b) bending moment diagram; c) section at midspan; d) section at the support.

In a typical building, cast-in-place concrete beams are continuous structures spanning across several supports. An elevation showing a typical interior span of a continuous T-beam is shown in Figure 3.28a (note the same beam BD is part of the floor system in Figure 3.26). The bending moment diagram due to a uniformly distributed load for this beam is shown in Figure 3.28b. The zone of positive bending moment is around the beam midspan, and a typical beam section at the midspan is shown in Figure 3.28c. In this case, the flange and (possibly) a portion of the web are under compression. However, the beam web in the negative moment zone (usually near the supports) is subjected to compression, as illustrated in Figure 3.28d. It should be stressed that the moment resistance of a beam section may be significantly different for sections in positive and negative moment zones.

It should be noted that CSA A23.3 does not specifically require that beams cast monolithically with the slab be designed as T-beams in positive bending. Instead, the designer may choose to ignore the contribution of the slab and design a beam as a rectangular section. The T-beam approach usually leads to more cost-effective designs with a slight reduction in the amount of tension reinforcement; this will be illustrated with a few examples in Section 3.7.2.

3.7.2 Flexural Resistance of T-Beams for Positive Bending

Flexural resistance of T-beams subjected to positive bending moments characterized by tension at the bottom of the beam and compression at the top can be determined in the same manner as the flexural resistance of properly reinforced rectangular beams in Section 3.5.1. Consider a typical T-beam section subjected to the positive bending moment (M_f) in Figure 3.29.

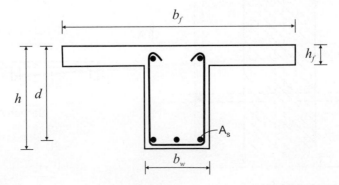

Figure 3.29 Cross-section of a T-beam.

The following notation will be used:

b_f = effective flange width,
b_w = web width,
b = width of the compression zone (varies depending on the neutral axis location),
h_f = flange (slab) thickness,
d = effective depth of T-beam,
h = overall depth of T-beam.

It should be noted that the effective flange width is the width of the slab that serves as a flange in T-beams. The effective flange width is affected by the type of loading, structural system (simple/continuous beam), the spacing of the beams, and the relative stiffness of the slabs and the beams. CSA A23.3 provides rules for estimating the effective flange width for design purposes (to be discussed in Section 5.8).

The procedure for calculating the factored moment resistance of a T-beam subjected to positive bending depends on whether the neutral axis is located within the web or within the flange, as discussed below.

Factored moment resistance for a T-beam with the neutral axis within the flange

This scenario, illustrated in Figure 3.30, is very common for T-beams. In general, the effective flange width in T-beams is rather large and, as a result, a very small depth of compressive stress block (a) is required to generate the compression force (C_r) in the flange to balance the tension force in the reinforcement (T_r). Consequently, the depth of the compression zone $a \leq h_f$. The amount of tension steel in these T-beams is usually moderate.

In this case, the analysis of a T-beam is identical to that of a rectangular beam of width equal to the effective flange width (b_f), as illustrated in Figure 3.31. This model is valid because the concrete below the neutral axis is considered to be cracked and does not contribute to the flexural resistance of the section.

The analysis of a T-beam with the neutral axis in the flange will be illustrated with Example 3.5. Subsequently, the same beam section will be considered in Example 3.6 as a rectangular section, and the results will be compared.

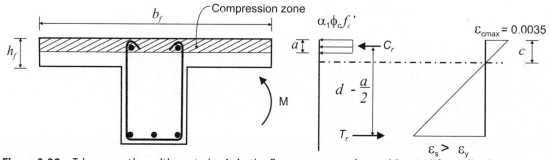

Figure 3.30 T-beam section with neutral axis in the flange: cross-section and internal force distribution.

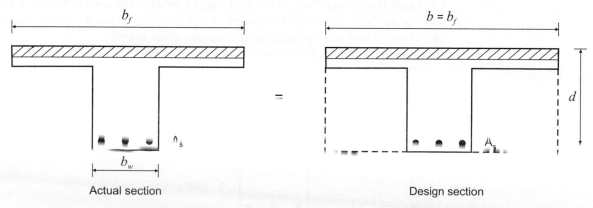

Actual section

Design section

Figure 3.31 T-beam with neutral axis in the flange acts as a rectangular section.

KEY EQUATIONS

The key equations for a T-beam section with neutral axis in the flange are the same as those stated in Section 3.5.1 related to properly reinforced rectangular beams. These equations are repeated below for the reader's convenience.

The equation of equilibrium is

$$C_r = T_r \tag{3.10}$$

where

$$T_r = \phi_s f_y A_s \tag{3.9}$$

is the tension force in the reinforcement, and

$$C_r = \alpha_1 \phi_c f_c' \, a \, b \tag{3.8}$$

is the compression force in concrete.

The depth of the rectangular stress block (a) is given by

$$a = \frac{\phi_s f_y A_s}{\alpha_1 \phi_c f_c' b} \tag{3.12}$$

where the section width is equal to the flange width, that is,

$$b = b_f$$

and

$$\alpha_1 \cong 0.8 \quad \text{and} \quad \beta_1 \cong 0.9 \tag{3.7}$$

The neutral axis is within the flange provided that

$$a \leq h_f$$

The *factored moment resistance* of the section (M_r) is given by

$$M_r = T\left(d - \frac{a}{2}\right) \tag{3.13}$$

or

$$M_r = \phi_s f_y A_s \left(d - \frac{a}{2}\right) \tag{3.14}$$

Example 3.5

A typical cross-section of a reinforced concrete T-beam is shown in the figure below. The beam is reinforced with three 20M bars (3-20M) in the tension zone. For the purposes of analysis, use an effective depth of 650 mm. The beam is properly reinforced. The concrete and steel material properties are given below.

Find the factored moment resistance for the beam section.

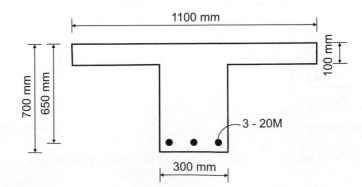

Given: $f_c' = 25$ MPa
$f_y = 400$ MPa
$\phi_c = 0.65$
$\phi_s = 0.85$

SOLUTION: **1.   Calculate the area of tension steel (A_s) and the effective depth (d)**
Tension steel: 3-20M bars

The area of one 20M bar is 300 mm^2 (see Table A.1). The total area of tension steel (A_s) is

$$A_s = 3 \times 300 \text{ mm}^2 = 900 \text{ mm}^2$$

The effective depth (d) is given as

$$d = 650 \text{ mm}$$

2.   Calculate the depth of the equivalent rectangular stress block (a) and confirm that the neutral axis is within the flange

First, calculate the tension force in the reinforcement (T_r). Since the beam is properly reinforced, the tension steel yields; that is,

$$f_s = f_y$$

Therefore,

$$T_r = \phi_s f_y A_s \tag{3.9}$$
$$= 0.85 \times 400 \text{ MPa} \times 900 \text{ mm}^2 = 306 \times 10^3 \text{ N} = 306 \text{ kN}$$

Use the equation of equilibrium

$$C_r = T_r \tag{3.10}$$

Since the compression force in concrete (C_r) is given by

$$C_r = \alpha_1 \phi_c f_c' \, a \, b \tag{3.8}$$

where

$$b = b_f = 1100 \text{ mm}$$
$$\alpha_1 \cong 0.8$$

the depth of the rectangular stress block (a) can be calculated from Eqn 3.10 as

$$a = \frac{T_r}{\alpha_1 \phi_c f_c' \, b} = \frac{306 \times 10^3 \text{ N}}{0.8 \times 0.65 \times 25 \text{ MPa} \times 1100 \text{ mm}} = 21.4 \text{ mm} \cong 21 \text{ mm}$$

The distribution of internal forces in the beam is shown in the sketch below.

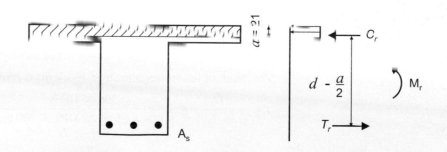

Finally, confirm that the neutral axis is within the flange. Because

$$a = 21 \text{ mm} < h_f = 100 \text{ mm}$$

it can be concluded that the neutral axis is within the flange.

3. **Calculate the factored moment resistance (M_r)**

    The moment resistance can be calculated from the following equation:

    $$M_r = T_r \left(d - \frac{a}{2} \right) \qquad\qquad [3.13]$$

    $$= (306 \times 10^3 \text{ N}) \left(650 \text{ mm} - \frac{21 \text{ mm}}{2} \right) \cong 196 \times 10^6 \text{ N} \cdot \text{mm}$$

    The factored moment resistance is

    $$M_r = 196 \text{ kN} \cdot \text{m}$$

Example 3.6

Consider the same beam as discussed in Example 3.5. In this example, use a rectangular cross-section of width 300 mm and overall depth of 700 mm (ignore the slab contribution). For analysis purposes, use an effective depth of 650 mm. The beam is properly reinforced. The concrete and steel material properties are given below.

Find the factored moment resistance for the beam section.

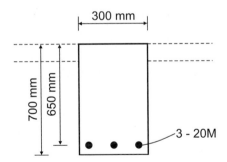

Given: $f_c' = 25 \text{ MPa}$
        $f_y = 400 \text{ MPa}$
        $\phi_c = 0.65$
        $\phi_s = 0.85$

SOLUTION: 1. **Calculate the area of tension steel (A_s) and the effective depth (d) — same as Example 3.5**

    $$A_s = 900 \text{ mm}^2$$
    $$d = 650 \text{ mm}$$

2. **Calculate the depth of the equivalent rectangular stress block (a)**

    $T_r = 306 \text{ kN}$ (same as Example 3.5).

    The depth of the rectangular stress block (a) is (note that $b = 300$ mm)

    $$a = \frac{T_r}{\alpha_1 \phi_c f_c' b} = \frac{306 \times 10^3 \text{ N}}{0.8 \times 0.65 \times 25 \text{ MPa} \times 300 \text{ mm}} \cong 78 \text{ mm}$$

3. **Calculate the factored moment resistance (M_r)**

The moment resistance can be calculated from the following equation:

$$M_r = T_r \left(d - \frac{a}{2} \right) \tag{3.13}$$

$$= (306 \times 10^3 \text{ N}) \left(650 \text{ mm} - \frac{78 \text{ mm}}{2} \right) \cong 187 \times 10^6 \text{ N} \cdot \text{mm}$$

The factored moment resistance is

$$M_r = 187 \text{ kN} \cdot \text{m}$$

The distribution of the internal forces in the beam section is shown below.

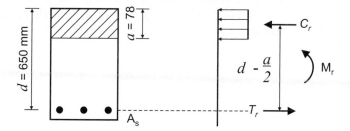

Learning from Examples

At this point, it is important to compare the results of the analyses performed in Examples 3.5 and 3.6. The same beam cross-section and material properties were used in both examples. First, the T-beam section was analyzed in Example 3.5, and then the same beam was analyzed considering a rectangular section in Example 3.6. The following conclusions can be drawn by comparing the results of these two examples:

1. There is a rather small (approximately 5%) difference in the factored moment resistance (M_r) values obtained for a rectangular beam (187 kN·m) and a T-beam section (196 kN·m). In a practical design situation, a T-beam section will be considered if the analysis is carried out by using a computer program. On the other hand, if the analysis is performed manually, a rectangular section may be considered adequate. The analysis of a rectangular beam section is faster and might be preferred by a designer performing a manual calculation.

2. T-beams with neutral axis in the flange are generally lightly reinforced (characterized by a rather small reinforcement ratio). In this case, the reinforcement ratio is

$$\rho = \frac{A_s}{b\,d} = \frac{900 \text{ mm}^2}{300 \text{ mm} \times 650 \text{ mm}} = 0.005 = 0.5\,\%$$

corresponding to a rather small depth of the equivalent stress block; in this case, the a/d ratio is less than 4%.

It should be noted that the above observations are valid only for T-beams with neutral axis in the flange. Different conclusions will be drawn for T-beams with neutral axis in the web, as discussed below.

Factored moment resistance for a T-beam with the neutral axis in the web

This scenario is less common and it is characteristic of heavily reinforced beams with rather small web width (b_w). When such a T-beam section is subjected to flexure, the compression force generated in the flange is not sufficient to balance the tension force in the reinforcement, hence an additional concrete compression zone is provided by the web area. Consequently, the neutral axis moves down in the web.

In this case, more complex calculations are required to account for the nonrectangular shape of the concrete compression zone; however, the analysis can be simplified if it is considered that the tension steel yields, that is, $f_s = f_y$.

Consider the T-beam section shown in Figure 3.32a. In this case, the depth of the equivalent stress block falls in the web ($a > h_f$), as shown in Figure 3.32a. The concrete compression zone consists of the flange area and a portion of the web area and is denoted by A_c (see Figure 3.32a). The compression force in the concrete (C_r) acts at the centroid (CG) of this zone at a distance $\overline{a}'$ from the top of the section, as shown in Figure 3.32a.

To find A_c, use the equation of equilibrium of internal forces in steel and concrete as follows:

$$C_r = T_r \tag{3.10}$$

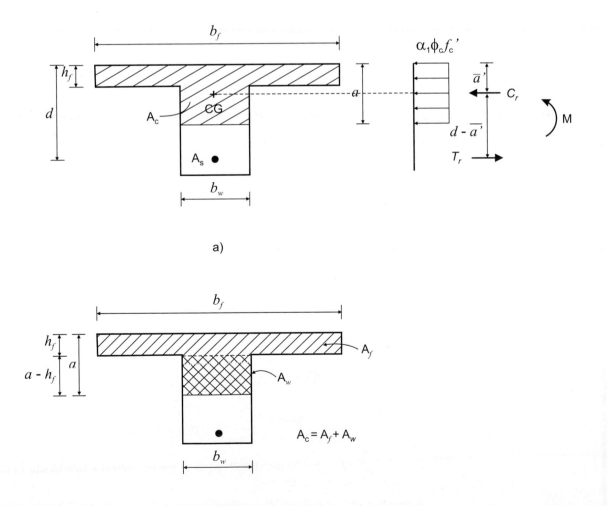

a)

b)

Figure 3.32 T-beam section with neutral axis in the web: a) cross-section and the internal force distribution; b) concrete compression zone.

where

$$T_r = \phi_s f_y A_s \tag{3.9}$$

is the tension force in the reinforcement,

$$C_r = \alpha_1 \phi_c f_c' A_c \tag{3.8}$$

is the compression force in the concrete acting over an area A_c, and

$$\alpha_1 \cong 0.8 \quad \text{and} \quad \beta_1 \cong 0.9 \tag{3.7}$$

Next, the A_c value can be found by substituting T_r and C_r from Eqns 3.9 and 3.8 into Eqn 3.10 as follows:

$$\alpha_1 \phi_c f_c' A_c = \phi_s f_y A_s$$

Hence,

$$A_c = \frac{\phi_s f_y A_s}{\alpha_1 \phi_c f_c'} \tag{3.30}$$

However, A_c can also be expressed as the sum of the flange area (A_f) and the web area under compression (A_w) as follows (see Figure 3.32b):

$$A_c = A_f + A_w = A_f + b_w (a - h_f) \tag{3.31}$$

where

$$A_f = b_f \times h_f$$

is the flange area, and

$$A_w = b_w (a - h_f)$$

is the web area under compression.

Consequently, a can be obtained from Eqn 3.31 as

$$a = h_f + \frac{A_c - A_f}{b_w} \tag{3.32}$$

Next, the location of the centroid (CG) of the concrete compression zone (A_c) (denoted as $\overline{a'}$ in this discussion) can be determined as

$$\overline{a'} = \frac{A_f \dfrac{h_f}{2} + A_w \left(h_f + \dfrac{a - h_f}{2} \right)}{A_c} \tag{3.33}$$

Finally, the factored moment resistance can be determined in the same manner as for the case of the rectangular beam, as shown on the stress distribution diagram in Figure 3.32a:

$$M_r = T_r (d - \overline{a'}) = \phi_s f_y A_s (d - \overline{a'}) \tag{3.34}$$

It should be noted that some alternative procedures can be used to determine the factored moment resistance for T-beams with neutral axis in the web. One of the procedures is the steel beam analogy, in which the actual T-beam section is represented by two idealized rectangular sections. The procedure presented in this section has been selected because it is commonly used in design practice.

The analysis of a T-beam section with neutral axis in the web will be illustrated with Example 3.7. Subsequently, the same beam will be analyzed by considering a rectangular section, and the results will be compared.

KEY EQUATIONS

Key equations related to moment resistance of a T-beam section with neutral axis in the web are summarized below.

The area of the concrete compression zone (A_c) is

$$A_c = \frac{\phi_s f_y A_s}{\alpha_1 \phi_c f_c'} \qquad \text{[3.30]}$$

or

$$A_c = A_f + A_w = A_f + b_w(a - h_f) \qquad \text{[3.31]}$$

where

$$A_f = b_f \times h_f$$

is the flange area, and

$$A_w = b_w(a - h_f)$$

is the web area under compression.

The depth of the compression stress block (a) is

$$a = h_f + \frac{A_c - A_f}{b_w} \qquad \text{[3.32]}$$

The location of the centroid of the concrete compression zone (distance $\overline{a'}$) is

$$\overline{a'} = \frac{A_f \dfrac{h_f}{2} + A_w\left(h_f + \dfrac{a - h_f}{2}\right)}{A_c} \qquad \text{[3.33]}$$

The factored moment resistance (M_r) for a T-beam section is

$$M_r = T_r(d - \overline{a'}) = \phi_s f_y A_s(d - \overline{a'}) \qquad \text{[3.34]}$$

Example 3.7

A typical cross-section of a reinforced concrete T-beam is shown in the figure below. The beam is reinforced with eight 30M bars (8-30M) in the tension zone. For analysis purposes, use an effective depth of 600 mm. The beam is properly reinforced. The concrete and steel material properties are given below.

Find the factored moment resistance for the beam section.

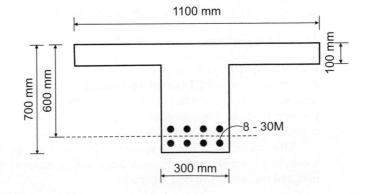

Given: $f_c' = 25$ MPa

$f_y = 400$ MPa

$\phi_c = 0.65$

$\phi_s = 0.85$

SOLUTION:

1. **Calculate the area of tension steel (A_s) and the effective depth (d)**

   Tension steel: 8-30M bars

   The area of one 30M bar is 700 mm² (see Table A.1). The total area of tension steel (A_s) is

   $$A_s = 8 \times 700 \text{ mm}^2 = 5600 \text{ mm}^2$$

   The effective depth (d) is given as

   $$d = 600 \text{ mm}$$

2. **Calculate the depth of the equivalent rectangular stress block (a) and confirm that the neutral axis is in the web**

   First, calculate the tension force in the reinforcement (T_r). Since the beam is properly reinforced, the tension steel yields; that is,

   $$f_s = f_y$$

   Therefore,

   $$T_r = \phi_s f_y A_s \qquad\qquad\qquad [3.9]$$
   $$= 0.85 \times 400 \text{ MPa} \times 5600 \text{ mm}^2 = 1904 \times 10^3 \text{ N} = 1904 \text{ kN}$$

   Use the equation of equilibrium

   $$C_r = T_r \qquad\qquad\qquad [3.10]$$

   Since the compression force in concrete (C_r) is

   $$C_r = \alpha_1 \phi_c f_c' \, a \, b \qquad\qquad\qquad [3.8]$$

   where

   $$\alpha_1 \cong 0.8$$
   $$b = b_f = 1100 \text{ mm}$$

   the depth of the rectangular stress block (a) can be calculated from Eqn 3.10 as

   $$a = \frac{T_r}{\alpha_1 \phi_c f_c' b} = \frac{1904 \times 10^3 \text{ N}}{0.8 \times 0.65 \times 25 \text{ MPa} \times 1100 \text{ mm}} = 133 \text{ mm}$$

   Finally, confirm that the neutral axis is in the web. Because

   $$a = 133 \text{ mm} > h_f = 100 \text{ mm}$$

   it can be concluded that the neutral axis is in the web.

3. **Determine the factored moment resistance (M_r)**

   In order to determine M_r, the following will have to be calculated:

   a) The area of the concrete compression zone (A_c):

   $$A_c = \frac{\phi_s f_y A_s}{\alpha_1 \phi_c f_c'} \qquad\qquad\qquad [3.30]$$

   $$= \frac{0.85 \times 400 \text{ MPa} \times 5600 \text{ mm}^2}{0.8 \times 0.65 \times 25 \text{ MPa}} = 146\,461 \text{ mm}^2$$

b) The flange area (A_f):

$$A_f = b_f \times h_f$$
$$= 1100 \text{ mm} \times 100 \text{ mm} = 110\,000 \text{ mm}^2$$

c) The depth of the compression stress block (a):

$$a = h_f + \frac{A_c - A_f}{b_w} \qquad\qquad [3.32]$$

$$= 100 \text{ mm} + \frac{146\,461 \text{ mm}^2 - 110\,000 \text{ mm}^2}{300 \text{ mm}} = 222 \text{ mm}$$

d) The web area under compression (A_w):

$$A_w = b_w (a - h_f)$$
$$= 300 \text{ mm} (222 \text{ mm} - 100 \text{ mm}) = 36\,600 \text{ mm}^2$$

e) The location of the centroid of the concrete compression zone (distance $\overline{a}'$):

$$\overline{a}' = \frac{A_f \dfrac{h_f}{2} + A_w \left(h_f + \dfrac{a - h_f}{2} \right)}{A_c} \qquad\qquad [3.33]$$

$$= \frac{(110\,000 \text{ mm}^2) \times \left(\dfrac{100 \text{ mm}}{2} \right) + (36\,600 \text{ mm}^2) \left(100 \text{ mm} + \dfrac{221 \text{ mm} - 100 \text{ mm}}{2} \right)}{146\,461 \text{ mm}^2}$$

$$= 78 \text{ mm}$$

The concrete compression zone is shown below.

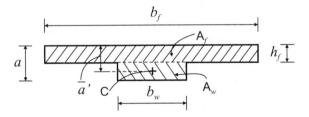

f) The factored moment resistance (M_r):

$$M_r = \phi_s f_y A_s (d - \overline{a}') \qquad\qquad [3.34]$$
$$= 0.85 \times 400 \text{ MPa} \times 5600 \text{ mm}^2 (600 \text{ mm} - 78 \text{ mm})$$
$$= 994 \times 10^6 \text{ N} \cdot \text{mm}$$

The factored moment resistance is

$$M_r = 994 \text{ kN} \cdot \text{m}$$

The distribution of the internal forces in the beam section is shown below.

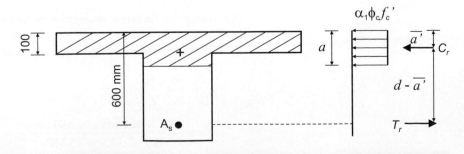

Example 3.8

Consider the same beam as discussed in Example 3.7. In this example, use a rectangular cross-section of width 300 mm and overall depth 700 mm (neglect the slab contribution). For the purposes of analysis, use an effective depth of 600 mm. The beam is properly reinforced. The concrete and steel material properties are given below.

Find the factored moment resistance for the beam section.

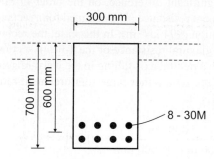

Given: $f_c' = 25$ MPa
$f_y = 400$ MPa
$\phi_c = 0.65$
$\phi_s = 0.85$

SOLUTION:

1. **Calculate the area of tension steel (A_s) and the effective depth (d) — same as Example 3.7**

   $A_s = 5600$ mm^2
   $d = 600$ mm

2. **Calculate the depth of the equivalent rectangular stress block (a)**

   $T_r = 1904$ kN (same as Example 3.7).

   The depth of the rectangular stress block (a) is (note that $b = 300$ mm)

   $$a = \frac{T_r}{\alpha_1 \phi_c f_c' \, b} = \frac{1904 \times 10^3 \text{ N}}{0.8 \times 0.65 \times 25 \text{ MPa} \times 300 \text{ mm}} = 488 \text{ mm}$$

3. **Calculate the factored moment resistance (M_r)**

   The moment resistance can be calculated from the following equation:

   $$M_r = T_r \left(d - \frac{a}{2} \right) \qquad\qquad [3.13]$$

   $$= (1904 \times 10^3 \text{ kN}) \left(600 \text{ mm} - \frac{488 \text{ mm}}{2} \right) \cong 678 \times 10^6 \text{ N} \cdot \text{mm}$$

   The factored moment resistance is

   $M_r = 678$ kN $\cdot$ m

   The distribution of the internal forces in the beam section is shown below.

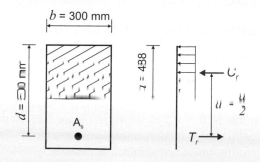

Learning from Examples

At this point, it is important to compare the results of the analyses performed in Examples 3.7 and 3.8. The same beam was analyzed as a T-beam section in the first example and as a rectangular section in the second one. The following conclusions can be drawn by comparing the results of these examples:

1. There is a significant difference, on the order of 30%, between the values of the factored moment resistance (M_r) obtained for a rectangular section (678 kN·m) and a T-beam section (994 kN·m). In this case, the rectangular section analysis under-estimates the flexural resistance of the T-beam section.

2. T-beams with a neutral axis falling in the web are generally heavily reinforced, that is, characterized by a rather large reinforcement ratio. In this case, the reinforcement ratio is

$$\rho = \frac{A_s}{bd} = \frac{5600 \text{ mm}^2}{300 \text{ mm} \times 600 \text{ mm}} = 0.031 = 3.1\%$$

3. For the material properties given, the balanced reinforcement ratio ρ_b is equal to 2.2%, as determined in Example 3.3; hence, $\rho > \rho_b$ (see Section 3.5.3). Had this been a rectangular beam section, it would have failed in the concrete-controlled failure mode, which is generally not desirable.

4. However, the T-beam section discussed in Example 3.7 is characterized by a wide flange (provided by the slab) capable of resisting the concrete compression forces to counterbalance the large steel tensile forces developed due to a significant amount of tension steel. Therefore, if the reinforcement ratio is calculated considering a T-beam section, by using the flange width $b = b_f = 1100$ mm, then

$$\rho = \frac{A_s}{bd} = \frac{5600 \text{ mm}^2}{1100 \text{ mm} \times 600 \text{ mm}} = 0.0085 = 0.85\%$$

Hence $\rho < \rho_b$, indicating a ductile steel-controlled failure mode. Therefore, it is clear that the presence of a wide flange ensures that a proper steel-controlled failure mode will occur since the reinforcement ratio is well below the balanced ratio (ρ_b) of 2.2%.

5. It should be stressed that, for lightly reinforced beams with $\rho < 0.005$, there is no significant difference in the moment resistance values obtained for T-beam sections versus rectangular sections.

    However, in heavily reinforced beams where $0.5\rho_b < \rho < \rho_b$, there is a significant increase in the moment resistance values when a T-beam section is used in the design in lieu of a rectangular section.

3.7.3   Flexural Resistance of T-Beams in Negative Bending

T-beam sections subjected to negative bending moments are also found in continuous concrete construction. Negative bending usually occurs in areas in the vicinity of the supports, as illustrated on section 4-4 in Figure 3.28d. In that case, the compression zone is always in the web. The analysis is performed using a rectangular section with width equal to the web width; that is, $b = b_w$, as shown in Figure 3.33. In this case, the analysis procedure for a beam with tension reinforcement only, presented in Section 3.5.1, can be used. The tension steel reinforcement is distributed in the slab.

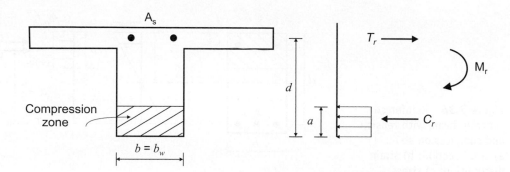

Figure 3.33 T-beam section subjected to negative bending.

Due to the continuity of concrete construction, rectangular beams that are part of the floor structure are usually placed integrally with the slab in a monolithic pour. As a result, a portion of the slab works as the top flange of the beam. In general, flanges in interior beams extend on both sides of a web; such beams are called *T-beams*, whereas beams with a flange on one side only (usually located around the perimeter of a floor structure) are referred to as *L-beams* (also known as *spandrel beams* or *edge beams*).

There are two possible scenarios related to the flexural resistance of T-beams subjected to positive bending, depending on the neutral axis location:

- *T-beams with a neutral axis in the flange* are beams with a moderate amount of reinforcement or a rather large effective flange width. The flexural resistance can be determined by treating a T-beam as a rectangular beam with a width equal to the effective flange width ($b = b_f$).
- *T-beams with a neutral axis in the web* are heavily reinforced beams with a rather small web width. More complex calculations are required to account for the nonrectangular shape of the compression block; however, the analysis can be simplified if it is considered that the tension steel yields.

3.8 RECTANGULAR BEAMS WITH TENSION AND COMPRESSION REINFORCEMENT

3.8.1 Background

It is often required in a practical design situation to increase the factored moment resistance (M_r) of a beam with tension steel only. Two solutions can be used in this case.

1. *Increase the beam cross-sectional dimensions.* The moment resistance (M_r) can be increased by increasing either the beam width or its depth. It should be noted that the increase in moment resistance is more effectively achieved by increasing the effective depth (d) than the width (b) of a beam cross-section (due to the fact that the moment resistance is proportional to the square of the beam effective depth). However, architectural or engineering considerations very often limit the beam depth.

2. *Provide additional reinforcement to resist the concrete compression stresses.* This solution allows for an increase in the amount of tension reinforcement, as well as a provision of additional reinforcement in the concrete compression zone with a *compression steel* or *compression reinforcement*. A beam with tension and compression steel is shown in Figure 3.34. Beams reinforced with tension and compression steel are also called *doubly reinforced beams*, whereas beams with tension steel only are called *singly reinforced beams*.

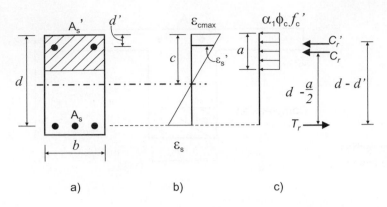

Figure 3.34 Reinforced concrete beam with tension and compression steel: a) cross-section; b) strain distribution; c) stress distribution.

The benefits of providing compression reinforcement are as follows.

1. *There is an increased amount of reinforcement and moment resistance while ensuring the steel-controlled failure mode.* When beam cross-sectional dimensions are limited due to architectural constraints, the moment resistance can be increased beyond that provided by the balanced amount of reinforcement. This can be accomplished by providing additional tension and compression steel. The compression force developed in the compression steel couples with the tension force developed in the additional tension steel to provide additional moment resistance in a section. It is important to note that doubly reinforced beams fail in the ductile steel-controlled failure mode.

2. *Long-term deflections are reduced.* This is one of the major reasons for adding compression steel to a flexural member. Research studies have shown that long-term deflections in beams with tension steel only can be greatly reduced by adding compression reinforcement. As much as a 50% reduction in long-term deflection under sustained load has been observed when equal areas of compression and tension steel are provided in the section. (For more details see MacGregor and Bartlett (2000).) Compression steel is effective in reducing compressive stresses in concrete and reducing the rate of creep and long-term deflection caused by sustained loads. CSA A23.3 accounts for the effect of compression steel in reducing long-term deflections (this will be discussed in more detail in Chapter 4).

3. *Effects of moment envelope.* A moment envelope is a series of bending moment diagrams for various load cases and combinations drawn on top of one another. In most cases, the tension zones of positive and negative moments overlap; as a result, tension reinforcement extends into the compression zone.

3.8.2 Flexural Resistance of Doubly Reinforced Rectangular Beams

The moment resistance of doubly reinforced beam sections can be determined in a similar manner to the resistance of beams with tension steel only, as presented in Section 3.5.1. A few additional terms will be introduced into the equations to account for the effect of compression reinforcement, as explained in this section.

Consider the beam with tension and compression reinforcement in Figure 3.34a. The beam is reinforced with the tension steel reinforcement and the corresponding reinforcement ratio is

$$\rho = \frac{A_s}{b\,d} \qquad\qquad [3.1]$$

where

A_s = total area of tension steel,
b = beam width,
d = effective depth.

The area of compression steel corresponds to the reinforcement ratio

$$\rho' = \frac{A_s'}{b\,d} \qquad\qquad [3.35]$$

where A_s' is the total area of compression steel.

The distance from the top compression fibre to the centroid of the compression steel is denoted as d'.

The following assumptions are made in the analysis of doubly reinforced sections:

1. The tension steel has yielded; that is, $f_s = f_y$.
2. The compression steel has either yielded or remained elastic:
   - If $\varepsilon_s' \geq \varepsilon_y$, then the compression steel has yielded and $f_s' = f_y$.
   - If $\varepsilon_s' < \varepsilon_y$, then the compression steel has not yielded; that is, $f_s' = E_s \varepsilon_s' < f_y$.

Note that ε_y can be determined from Hooke's law as $\varepsilon_y = f_y/E_s$ (see Section 2.6.2).

The strain distribution over the beam section is shown in Figure 3.34b. The strain in the compression steel ε_s' can be determined from the similarity of triangles:

$$\frac{\varepsilon_s'}{c - d'} = \frac{\varepsilon_{cmax}}{c}$$

The strain in the compression steel (ε_s') can be expressed from the above equation as

$$\varepsilon_s' = \varepsilon_{cmax}\left(1 - \frac{d'}{c}\right) \tag{3.36}$$

Note that $\varepsilon_{cmax} = 0.0035$, as discussed earlier in this chapter.

Let us consider the equilibrium of forces acting on the cross-section in Figure 3.34c. In this case, the compression is resisted by the concrete (force C_r) and the compression reinforcement (force C_r'). C_r' is equal to the product of the factored compression stress in steel ($\phi_s f_s'$) and the reinforcement area (A_s') (similar to the T_r calculation presented in Section 3.5.1), as follows:

$$C_r' = \phi_s f_s' A_s'$$

The resultant concrete compression force (C_r) is equal to the product of the area of the compressive zone of depth (a) and width (b) and the uniform stress of magnitude $\alpha_1 \phi_c f_c'$. Note that the area of compression steel (A_s') is deducted from the concrete area as it displaces the corresponding area of concrete as

$$C_r = (\alpha_1 \phi_c f_c')(a b - A_s') \cong \alpha_1 \phi_c f_c' a b$$

It is assumed that the tension steel has yielded; that is, $f_s = f_y$. The tension force in the steel reinforcement (T_r) is equal to the product of the factored stress in steel ($\phi_s f_y$) and the reinforcement area (A_s) (as discussed in Section 3.5.1):

$$T_r = \phi_s f_y A_s$$

The equilibrium of internal forces acting on a beam cross-section in the horizontal direction gives

$$C_r + C_r' = T_r \tag{3.37}$$

The depth of the compression stress block (a) can be determined from Eqn 3.37 as

$$a = \frac{T_r - C_r'}{\alpha_1 \phi_c f_c' b} \tag{3.38}$$

An iterative procedure can be used to determine the factored moment resistance of a doubly reinforced beam section with the following steps:

1. *Calculate the depth of the equivalent rectangular stress block (a).* Assume that the compression steel has yielded; that is, $f_s' = f_y$. Then, calculate the depth of the compression stress block (a) from (3.38) and the neutral axis depth (c) from Eqn 3.4 as $c = a/\beta_1$.

2. *Calculate the strain (ε_s') and check whether the compression steel has yielded.* First, calculate the ε_s' value from Eqn 3.36. Then, check whether the compression steel has yielded or remained elastic.

3. *Estimate the f_s' value and repeat steps 1 and 2.*

   a) If the compression steel has not yielded ($\varepsilon_s' < \varepsilon_y$), then a trial value of the steel stress (f_s') needs to be estimated, where $f_s' < f_y$. Steps 1 and 2 need to be repeated until the difference in f_s' values obtained in two successive trials is very small (less than 5%). The factored moment resistance can be determined from Eqn 3.39.

   b) If the compression steel has yielded, that is, $\varepsilon_s' = \varepsilon_y$ and $f_s' = f_y$, go to the next step using the a value determined in step 1.

4. *Calculate the factored moment resistance (M_r).* The factored moment resistance (M_r) can be determined by summing up compression forces around the centroid of the tension steel as follows:

$$M_r = C_r'\,(d - d') + C_r\left(d - \frac{a}{2}\right) = (\phi_s f_s' A_s')(d - d') + (\alpha_1 \phi_c f_c'\, a\, b)\left(d - \frac{a}{2}\right)$$

[3.39]

The analysis of doubly reinforced beams will be demonstrated with an example.

KEY CONCEPTS

Beams reinforced with tension and compression steel are often called *doubly reinforced beams*. The benefits of compression reinforcement in the beams are as follows:

1. increased amount of reinforcement and moment resistance while ensuring the steel-controlled failure mode;
2. reduced long-term deflection in doubly reinforced beams.

The factored moment resistance (M_r) for beams with tension and compression reinforcement can be determined from the equilibrium of internal forces in the section. The following assumptions are made in the analysis:

1. The tension steel has yielded; that is, $f_s = f_y$.
2. The compression steel has either yielded or remained elastic:

   - If $\varepsilon_s' \geq \varepsilon_y$, then the compression steel has yielded and $f_s' = f_y$
   - If $\varepsilon_s' < \varepsilon_y$, then the compression steel has not yielded; that is, $f_s' = E_s \varepsilon_s' < f_y$

The key equations used in the iterative procedure to calculate the M_r value are summarized below.

The strain in the compression steel (ε_s') is

$$\varepsilon_s' = \varepsilon_{cmax}\left(1 - \frac{d'}{c}\right)$$

[3.36]

where

$$\varepsilon_{cmax} = 0.0035$$

The equation of equilibrium is

$$C_r + C_r' = T_r$$

[3.37]

where

$$C_r' = \phi_s f_s' A_s'$$

is the force in the compression steel,

$$C_r = (\alpha_1 \phi_c f_c')(a\,b - A_s') \cong \alpha_1 \phi_c f_c'\, a\, b$$

is the resultant concrete compression force, and

$$T_r = \phi_s f_y A_s$$

is the force in the tension steel.

The depth of the compression stress block (a) is

$$a = \frac{T_r - C_r'}{\alpha_1 \phi_c f_c' b} \qquad [3.38]$$

The neutral axis depth (c) is

$$c = \frac{a}{\beta_1}$$

The factored moment resistance (M_r) is

$$M_r = C_r'(d - d') + C_r\left(d - \frac{a}{2}\right) = (\phi_s f_s' A_s')(d - d') + (\alpha_1 \phi_c f_c' \, a \, b)\left(d - \frac{a}{2}\right) \qquad [3.39]$$

Example 3.9

A typical cross-section of a reinforced concrete beam is shown in the figure below. The beam is reinforced with four 25M bars (4-25M) in the tension zone and two 25M bars (2-25M) in the compression zone. For analysis purposes, use an effective depth (d) of 450 mm and d' of 50 mm. The concrete and steel material properties are given below.

Find the factored moment resistance for the beam section.

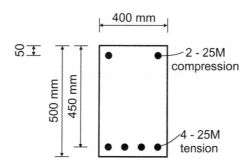

Given: $f_c' = 25$ MPa
$f_y = 400$ MPa
$\phi_c = 0.65$
$\phi_s = 0.85$
$b = 400$ mm
$h = 500$ mm

SOLUTION: 1.   Calculate the tension steel area (A_s), the compression steel area (A_s'), and the effective depth (d)

a)   Find the area of tension steel (4-25M bars).

The area of one 25M bar is 500 mm^2 (see Table A.1). The total area of tension steel (A_s) is

$$A_s = 4 \times 500 \text{ mm}^2 = 2000 \text{ mm}^2$$

b)   The area of compression steel (2-25M bars) is

$$A_s' = 2 \times 500 \text{ mm}^2 = 1000 \text{ mm}^2$$

c)   The effective depth (d) is given as

$$d = 450 \text{ mm}$$

2.   **Determine the depth of the equivalent rectangular stress block (a)**
To determine the depth of the compression stress block, compute the following:

a)   The tension force in the reinforcement (T_r):
Assume that the tension steel has yielded; that is,

$$f_s = f_y$$

Therefore,

$$T_r = \phi_s f_y A_s$$
$$= 0.85 \times 400 \text{ MPa} \times 2000 \text{ mm}^2 = 680 \text{ kN}$$

b)   The force in the compression steel (C_r'):
Assume that the compression steel has yielded; that is, $f_s' = f_y$ (the validity of this assumption will be checked in the next step). Then,

$$C_r' = \phi_s f_s' A_s' = 0.85 \times 400 \text{ MPa} \times 1000 \text{ mm}^2 = 340 \text{ kN}$$

c)   The depth of the compression stress block (a):

$$a = \frac{T_r - C_r'}{\alpha_1 \phi_c f_c' b} \qquad\qquad\qquad \text{[3.38]}$$

$$= \frac{680 \times 10^3 \text{ N} - 340 \times 10^3 \text{ N}}{0.8 \times 0.65 \times 25 \text{ MPa} \times 400 \text{ mm}} = 65 \text{ mm}$$

(note that $\alpha_1 \cong 0.8$).

3.   **Calculate the strain (ε_s') and check whether the compression steel has yielded**

a)   First, calculate the neutral axis depth:

$$c = \frac{a}{\beta_1} = \frac{65 \text{ mm}}{0.9} = 72 \text{ mm}$$

b)   Next, calculate the strain in the compression steel (ε_s'):

$$\varepsilon_s' = \varepsilon_{cmax}\left(1 - \frac{d'}{c}\right) \qquad\qquad\qquad \text{[3.36]}$$

$$= 0.0035\left(1 - \frac{50 \text{ mm}}{72 \text{ mm}}\right) = 0.0011$$

c)   Finally, calculate the yield strain as follows:

$$\varepsilon_y = \frac{f_y}{E_s} = \frac{400 \text{ MPa}}{200\,000 \text{ MPa}} = 0.002$$

Since

$$\varepsilon_s' = 0.0011 < 0.002$$

the compression steel has not yielded.

4.   **Estimate the f_s' value and repeat steps 2 and 3**

a)   Estimate f_s':
Assume that the compression steel is at 75% of yield strength (f_y); that is,

$$f_s' = 0.75 \times 400 \text{ MPa} = 300 \text{ MPa}$$

where

$f_y = 400$ MPa

b) Recalculate the force in the compression steel:

$C_r' = \phi_s f_s' A_s' = 0.85 \times 300$ MPa $\times 1000$ mm$^2 = 255$ kN

c) Recalculate the depth of the compression stress block:

$$a = \frac{T_r - C_r'}{\alpha_1 \phi_c f_c' b}$$ [3.38]

$$= \frac{680 \times 10^3 \text{ N} - 255 \times 10^3 \text{ N}}{0.8 \times 0.65 \times 25 \text{ MPa} \times 400 \text{ mm}} = 82 \text{ mm}$$

d) Recalculate the neutral axis depth as

$$c = \frac{a}{\beta_1} = \frac{82 \text{ mm}}{0.9} = 91 \text{ mm}$$

e) Check the strain in the compression steel:

$$\varepsilon_s' = \varepsilon_{cmax}\left(1 - \frac{d'}{c}\right) = 0.0035\left(1 - \frac{50 \text{ mm}}{91 \text{ mm}}\right) = 0.00158$$

Since

$\varepsilon_s' = 0.00158 < 0.002$

the assumption that the compression steel has not yielded is confirmed.

f) Next, calculate the stress in the compression steel:

$f_s' = E_s \varepsilon_s' = 200\,000$ MPa $\times 0.00158 = 316$ MPa

This f_s' value is within 5% of the initial estimate of $f_s' = 300$ MPa. The calculation can be repeated using $f_s' = 316$ MPa. The value $f_s' = 316$ MPa is within 5% of the actual value and should be considered acceptable. The average value of the last two iterations can be used to determine the factored moment resistance (M_r) as follows:

$$f_s' = \frac{300 \text{ MPa} + 316 \text{ MPa}}{2} = 308 \text{ MPa}$$

5. **Calculate the factored moment resistance (M_r)**
   a) Calculate the force in the compression steel using $f_s' = 308$ MPa:

   $C_r' = \phi_s f_s' A_s' = 0.85 \times 308$ MPa $\times 1000$ mm$^2 = 262$ kN

   b) Calculate the compression force in the concrete:

   $C_r = \alpha_1 \phi_c f_c' a b = 0.8 \times 0.65 \times 25$ MPa $\times 82$ mm $\times 400$ mm $= 426.4$ kN

   c) Calculate the factored moment resistance:

   $$M_r = C_r'(d - d') + C_r\left(d - \frac{a}{2}\right)$$ [3.39]

   $$= (262 \times 10^3 \text{ N})(450 \text{ mm} - 50 \text{ mm})$$

   $$+ (426.4 \times 10^3 \text{ N})\left(450 \text{ mm} - \frac{82 \text{ mm}}{2}\right)$$

   $$= 279 \times 10^6 \text{ N} \cdot \text{mm} = 279 \text{ kN} \cdot \text{m}$$

d) Verify whether the equation of equilibrium has been satisfied.
    The equation of equilibrium is

$$C_r + C_r' = T_r \qquad\qquad [3.37]$$

where

$$C_r + C_r' = 426.4 \text{ kN} + 262 \text{ kN} \cong 688 \text{ kN}$$

and

$$T_r = 680 \text{ kN}$$

Since

$$688 \text{ kN} \cong 680 \text{ kN}$$

the equation of equilibrium is satisfied.
The results of this analysis are summarized in the sketch below.

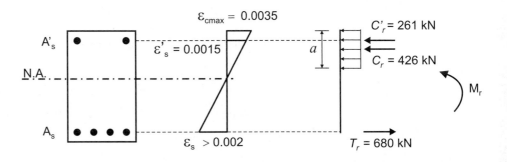

SUMMARY AND REVIEW — FLEXURE: BEHAVIOUR OF BEAMS AND ONE-WAY SLABS

Flexure (bending) is associated with the lateral deformation of a member under a transversely applied load. In a reinforced concrete flexural member with positive bending, the top portion is under compression whereas the bottom is under tension. Concrete has a limited ability to resist tension, and it cracks once its tensile strength has been reached in the region of maximum bending moments (in this case at the beam midspan). To prevent cracking, steel reinforcement (often called tension reinforcement or tension steel) is placed inside the beam near the bottom to resist the tensile stresses.

The following types of reinforced concrete flexural members are discussed in this chapter:

*   rectangular beams (singly and doubly reinforced),
*   T- and L-beams
*   one-way slabs

Basic assumptions of flexural design of concrete members according to CSA A23.3.

The ultimate limit states design of reinforced concrete flexural members is based on the following five basic assumptions outlined by CSA A23.3:

1. Plane sections remain plane (linear strain distribution across the section) (Cl.10.1.2).
2. Strains in concrete and steel at the reinforcement location are equal.
3. Equivalent rectangular stress block is used instead of the actual stress distribution for concrete (Cl.10.1.7).
4. Concrete tensile strength is neglected in flexural strength calculations (Cl.10.1.5).
5. Maximum concrete compressive strain (ε_{cmax}) is equal to 0.0035 (Cl.10.1.3).

Stages of flexural behaviour for reinforced concrete members

Four stages of flexural behaviour are characteristic of properly reinforced concrete beams and slabs:

1. elastic uncracked
2. elastic cracked
3. yielding
4. failure

Most reinforced concrete flexural members demonstrate elastic cracked behaviour under service loads.

The flexural modes of failure for concrete flexural members

There are two basic failure modes characteristic of reinforced concrete flexural members:

- steel-controlled
- concrete-controlled

These failure modes are related to the behaviour of steel and concrete, which are the main ingredients of reinforced concrete.

It is considered good practice to design reinforced concrete flexural members to fail in the *steel-controlled mode* initiated when the steel reinforcement yields and character-ized by ductile behaviour. Flexural members failing in the steel-controlled mode are characterized by a moderate amount of reinforcement.

Concrete is characterized by a brittle behaviour, and so is the *concrete-controlled fail-ure* mode. It is therefore *not* recommended to design flexural members to fail in the concrete-controlled mode initiated by the crushing of concrete and occurring suddenly and without warning. Flexural members that fail in the concrete-controlled mode are charac-terized by a rather large amount of reinforcement.

Flexural members characterized by a very small amount of reinforcement (underreinforced members) demonstrate brittle behaviour and fail suddenly and without warning. The behaviour of such members under flexure is similar to the behaviour of unreinforced concrete members.

Role of the balanced condition and balanced reinforcement

The balanced condition, characterized by the simultaneous crushing of concrete and yielding of the steel reinforcement, represents the threshold between overreinforced beams failing in the concrete-controlled mode and properly reinforced beams failing in the steel-controlled mode. The area of tension steel corresponding to the balanced condition is called *balanced reinforcement*.

Structural designers use the balanced reinforcement ratio (ρ_b) to predict the potential failure mode for a beam with the reinforcement ratio ρ. The three possible scenarios are summarized below.

- If $\rho = \rho_b$, the amount of reinforcement corresponds to the *balanced condition*.
- If $\rho > \rho_b$, the amount of reinforcement corresponds to the *concrete-controlled failure mode*, characterized by an excessively high reinforcement ratio (overreinforced beam).
- If $\rho < \rho_b$, the amount of reinforcement corresponds to the *steel-controlled failure mode*, characterized by a moderate amount of reinforcement (properly reinforced beam).

It is desirable to use the "proper" amount of steel reinforcement in the design to ensure the steel-controlled mode of failure. The reinforcement ratio should preferably be less than 75% of the ρ_b value; that is, $\rho \le 0.75\rho_b$.

One-way slabs

One-way slabs are reinforced concrete floor panels that transfer load in one direction. For design purposes, one-way slabs are treated as a series of unit (1 m wide) beams placed side by side (spanning in the short direction). The amount and spacing of reinforcement deter-mined for one strip applies to the entire slab.

One-way slabs can be analyzed as properly reinforced rectangular beams, where $b = 1000$ mm is the unit width and h is the overall slab thickness.

The main tension reinforcement in one-way slabs runs in the short direction. The ten-sion reinforcement can be placed either at the bottom of the slab (for positive bending) or at the top (for negative bending).

T-beams

Due to the continuity of concrete construction, rectangular beams that are part of the floor structure are usually cast with the slab in a monolithic pour. As a result, a portion of the slab works as the top flange of the beam. In general, flanges in interior beams extend on both sides of a web; such beams are called *T-beams*, whereas beams with a flange on one side only (usually located around the perimeter of a floor structure) are referred to as *L-beams* (also known as *spandrel beams* or *edge beams*).

There are two possible scenarios related to the flexural resistance of T-beams subjected to positive bending, depending on the neutral axis location:

- *T-beams with a neutral axis in the flange* are beams with a moderate amount of reinforcement or rather large effective flange width. The flexural resistance can be determined by treating a T-beam as a rectangular beam with a width equal to the effective flange width ($b = b_f$).
- *T-beams with a neutral axis in the web* are heavily reinforced beams with a rather small web width. More complex calculations are required to account for the nonrectangular shape of the compression block; however, the analysis can be simplified if it is considered that the tension steel yields.

Rectangular beams with tension and compression reinforcement

Beams reinforced with tension and compression steel are often called doubly reinforced beams. The benefits of the compression reinforcement in the beams are as follows:

1. increased amount of reinforcement and moment resistance while ensuring the steel-controlled failure mode;
2. reduced long-term deflection in doubly reinforced beams.

The factored moment resistance (M_r) for beams with tension and compression reinforcement can be determined from the equilibrium of internal forces in the section. The following assumptions are taken in the analysis:

1. The tension steel has yielded.
2. The compression steel has either yielded or remained elastic.

PROBLEMS

3.1. According to CSA A23.3, what are the five basic limit states design assumptions related to reinforced concrete flexural members? Explain.

3.2. a) What are the four stages of flexural behaviour characteristic of reinforced concrete beams?
   b) What stage are the members expected to be at under service loads?
   c) What stage is considered for the ultimate limit states design according to CSA A23.3? Explain.

3.3. a) What are the two basic failure modes characteristic of reinforced concrete flexural members?
   b) Which mode is preferred in the design of flexural members and why?

3.4. a) Calculate the balanced reinforcement ratio for a rectangular beam section with $f_c' = 30$ MPa and $f_y = 400$ MPa.
   b) When does the balanced condition occur in reinforced concrete flexural members?
   c) How is the balanced reinforcement ratio useful in the design of flexural members? Explain.

3.5. A typical cross-section of a reinforced concrete beam is shown in the figure that follows. The concrete and steel material properties are given below. Use an effective depth of 650 mm.
   a) Determine the depth of the compression zone (*a*), the neutral axis depth (*c*), and the strain distribution at the ultimate condition. Draw the beam cross-section showing *a, c,* and the strain values at the critical locations.
   b) Is the beam properly reinforced/overreinforced/balanced? What are the implications of the amount of reinforcement with regard to the mode of failure in the beam?

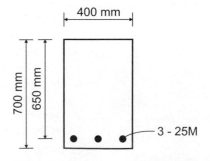

c) Calculate the factored moment resistance for the beam section.

Given:

$$f_c' = 30 \text{ MPa}$$
$$f_y = 400 \text{ MPa}$$
$$\phi_c = 0.65$$
$$\phi_s = 0.85$$
$$E_s = 200\,000 \text{ MPa}$$

3.6. A typical cross-section of a properly reinforced concrete beam is shown in the figure below. The material properties are summarized below. Use an effective depth of 650 mm.
a) Determine the factored moment resistance for the following cases:

i) $f_c' = 20 \text{ MPa}$

ii) $f_c' = 25 \text{ MPa}$

iii) $f_c' = 30 \text{ MPa}$

iv) $f_c' = 35 \text{ MPa}$

b) Comment on the variation in concrete strength on the moment resistance of the beam.

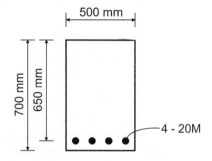

Given:

$$f_y = 400 \text{ MPa}$$
$$\phi_c = 0.65$$
$$\phi_s = 0.85$$
$$E_s = 200\,000 \text{ MPa}$$

3.7. Consider the beam section from Problem 3.6. Suppose that you have designed the beam with $f_c' = 35 \text{ MPa}$. However, due to embedded items the reinforcing is raised by 60 mm such that the effective depth is reduced to 590 mm (see the figure that follows).
a) Find the factored moment resistance considering the new reinforcement layout.
b) Comment on the effect of the change in reinforcement location on the moment resistance (refer to the solution in Problem 3.6a)iv)

Given:

$$f_y = 400 \text{ MPa}$$
$$\phi_c = 0.65$$
$$\phi_s = 0.85$$

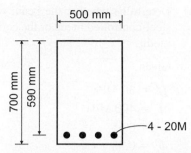

3.8. Refer to the beam section from Problem 3.7, designed using $f_c' = 35 \text{ MPa}$. The same contractor reported that, due to some problems related to batching, the actual 28-day concrete strength is only 31.5 MPa.
a) Find the factored moment resistance for the new strength.
b) Refer to the solution to Problem 3.7. What do you believe has a larger influence on the magnitude of the factored moment resistance: a 10% decrease in the concrete strength (as discussed in this problem) or placement of reinforcement resulting in a 10% decrease in the effective depth?

Given:

$$f_y = 400 \text{ MPa}$$
$$\phi_c = 0.65$$
$$\phi_s = 0.85$$

3.9. A typical section of a precast concrete beam with 100 mm by 100 mm cuts at the corners is shown in the figure below. Use an effective depth of 590 mm. The material properties are summarized below.

Find the factored moment resistance for the beam section.

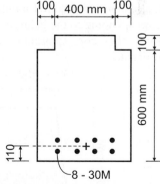

Given:

$$f_c' = 30 \text{ MPa}$$
$$f_y = 400 \text{ MPa}$$
$$\phi_c = 0.65$$
$$\phi_s = 0.85$$

3.10. Refer to the beam section from Problem 3.9. The amount of tension steel has been increased to 8-35M bars.
a) Find the factored moment resistance for the beam section.

b) Determine whether the beam would fail in the steel-controlled mode or the concrete-controlled mode.

Given:

$f_c' = 30$ MPa
$f_y = 400$ MPa
$\phi_c = 0.65$
$\phi_s = 0.85$
$E_s = 200\,000$ MPa

3.11. Refer to the beam section from Problem 3.9. The tension steel consists of 8-30M bars. The contractor has misplaced the reinforcement such that the effective depth is reduced to 440 mm.
a) Find the factored moment resistance (M_r) for the section.
b) Determine whether the beam would fail in the steel-controlled mode or the concrete-controlled mode.
c) Compare the M_r values obtained in Problems 3.9, 3.10, and 3.11 and analyze the effect of the amount of steel and the neutral axis depth on M_r value.

Given:

$f_c' = 30$ MPa
$f_y = 400$ MPa
$\phi_c = 0.65$
$\phi_s = 0.85$
$E_s = 200\,000$ MPa

3.12. A typical cross-section of a reinforced concrete beam is shown in the figure below. Use an effective depth of 600 mm.
a) Determine the required number of 35M rebars corresponding to the balanced condition.
b) Find the corresponding factored moment resistance.

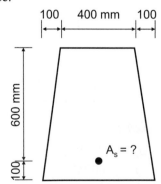

Given:

$f_c' = 25$ MPa
$f_y = 400$ MPa
$\phi_c = 0.65$
$\phi_s = 0.85$
$E_s = 200\,000$ MPa

3.13. The architect has specified a 200 mm thick concrete slab cast with the patterns shown as a feature ceiling. The slab is reinforced with 15M bars at 250 mm spacing (15M@250). Use an effective depth of 170 mm. The concrete and steel material properties are given below.
a) Find the factored moment resistance per metre width of the slab.
b) Draw the slab cross-section showing the concrete compression zone.

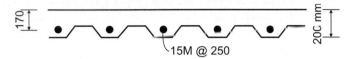

Given:

$f_c' = 25$ MPa
$f_y = 400$ MPa
$\phi_c = 0.65$
$\phi_s = 0.85$
$E_s = 200\,000$ MPa

3.14. A 150 mm thick concrete slab is reinforced with 15M bars at 300 mm spacing (15M@300). The effective depth is 120 mm. The concrete and steel material properties are given below.
a) Find the factored moment resistance per metre width of the slab.
b) If the slab is simply supported in one direction over a 5 m span, find the safe specified live load this slab can carry in addition to its self-weight. Refer to NBC 2010 for the load combinations.

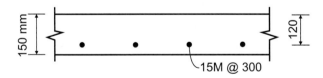

Given:

$f_c' = 30$ MPa
$f_y = 400$ MPa
$\phi_c = 0.65$
$\phi_s = 0.85$
$\gamma_w = 24$ kN/m^3

3.15. A typical cross-section of a reinforced concrete T-beam is shown in the figure that follows. The beam is reinforced with 2-30M bars. Use an effective depth of 530 mm. The material properties are summarized below.

Find the factored moment resistance for the beam section.

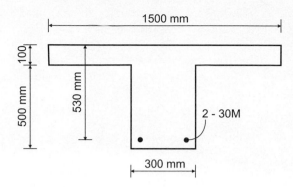

Given:

$f_c' = 30$ MPa
$f_y = 400$ MPa
$\phi_c = 0.65$
$\phi_s = 0.85$

3.16. A typical-cross section of a reinforced concrete T-beam is shown in the figure below. The beam is reinforced with 7-35M bars. The material properties are summarized below.

Find the factored moment resistance for the beam section.

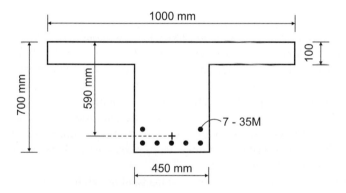

Given:

$f_c' = 30$ MPa
$f_y = 400$ MPa
$\phi_c = 0.65$
$\phi_s = 0.85$

3.17. A cross-section of an inverted T-beam is shown in the figure that follows. The beam is reinforced with 8-30M tension rebars at the bottom. Assume a 50-mm clear cover to the reinforcing. The material properties are summarized below.

a) Determine whether the beam is expected to fail in the steel-controlled mode or the concrete-controlled mode upon initial bending

b) To prevent the concrete-controlled mode of failure, compression steel needs to be provided in addition to the existing 8-30M tension steel. Find the amount of compression steel required to ensure steel-controlled failure. Use 25M bars for the compression steel and a 40 mm clear cover to the bars.

c) Find the factored moment resistance for the section with tension and compression steel. Use the amount of compression steel determined in part b) of this problem.

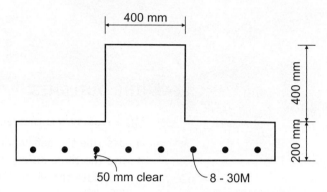

Given:

$f_c' = 25$ MPa
$f_y = 400$ MPa
$\phi_c = 0.65$
$\phi_s = 0.85$
$E_s = 200\,000$ MPa

3.18. A typical cross-section of a doubly reinforced concrete beam is shown in the figure below. The beam is reinforced with 6-25M bars at the bottom and 2-20M bars at the top. The material properties are summarized below.

a) Find the reinforcement ratios for tension and compression steel.

b) Determine the neutral axis location and the strain distribution for the beam section at the ultimate condition.

c) Determine whether the beam is expected to fail in the steel-controlled or the concrete-controlled mode.

d) Find the factored moment resistance for the beam section.

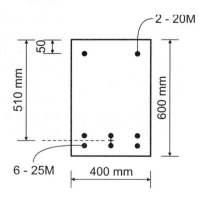

Given:

$f_c' = 30$ MPa
$f_y = 400$ MPa
$\phi_c = 0.65$
$\phi_s = 0.85$

Serviceability

4

LEARNING OUTCOMES

After reading this chapter, you should be able to

- describe the behaviour of reinforced concrete flexural members under service loads
- use the three different moment of inertia properties for deflection calculations
- calculate the immediate and long-term deflections in reinforced concrete flexural members
- apply the two CSA A23.3 approaches to deflection control
- identify the causes of cracking in reinforced concrete flexural members and apply the CSA A23.3 cracking control requirements

4.1 INTRODUCTION

Every building structure is expected to be able to support service loads without experiencing excessive cracking, deflections, or vibrations. Serviceability limit states represent the limit for maintaining a satisfactory structural performance under normal service load conditions.

Cracking in concrete structures occurs when the concrete tensile resistance has been reached. Excessive cracking can lead to poor appearance, reduced durability and serviceability in extreme cases, and loss of strength that could in turn cause the failure of a reinforced concrete structure. Visible cracks and sagging of concrete members could lead to the perception that the entire building is unsafe. Buildings with excessive cracking problems require repair to maintain structural competence.

Any flexural member deflects when subjected to external loads, although in most cases these deflections are too small to be visible to the naked eye. Accurate estimation of deflections in reinforced concrete flexural members is perhaps the most complex task in reinforced concrete design. This is due to the composite nature of reinforced concrete as well as the fact that a flexural member changes its properties once it cracks. Deflections in reinforced concrete structures due to sustained loads increase over a period of time as a result of creep. These long-term deflections can be quite significant, on the order of two to three times the initial deflections developed after the load has been initially applied.

Serviceability limit states, in particular cracking and deflections, need to be paid the same attention in design as ultimate limit states. A structural member may have sufficient strength to support the imposed load; however, it may be unacceptable if it experiences excessive deflections (sagging). Such deflections might result in excessive floor or roof slopes, which may cause operational or roof drainage problems or possible damage to architectural components.

This chapter is focused mainly on describing the basic concepts of the behaviour of reinforced concrete flexural members under service loads, explaining the methods used to estimate the extent of deflections and cracking in these members, and verifying whether these values are within the limits prescribed by CSA A23.3. The behaviour of reinforced concrete flexural members under service loads is explained in Section 4.2. The properties of reinforced concrete flexural members for the serviceability considerations are presented in Section 4.3. Immediate and long-term deflections are discussed in Section 4.4. The deflection control requirements per CSA A23.3 are outlined in Section 4.5. Deflection calculation procedures for simply supported and continuous flexural members are presented in Section 4.6, along with the design examples. Causes of cracking in reinforced concrete structures are discussed in Section 4.7, while the CSA A23.3 cracking control requirements are outlined in Section 4.8.

DID YOU KNOW?

The cost of repairing deteriorated concrete structures is extremely high. For example, it has been estimated that the rehabilitation costs of Canadian reinforced concrete parking structures are on the order of $4 to $6 billion, while the repair cost for all concrete structures in the U.S. has been estimated to be as high as $1 to $3 trillion (Bedard, 1992). Many older bridges in Canada have deteriorated over time due to corrosion and other effects, as shown in Figure 4.1.

Figure 4.1 A severely deteriorated reinforced concrete bridge girder in Ontario; the deterioration is due to corrosion and the freeze and thaw effect.

(Graham Finch)

4.2 BEHAVIOUR OF REINFORCED CONCRETE FLEXURAL MEMBERS UNDER SERVICE LOADS

Consider a simple span reinforced concrete beam subjected to a uniformly distributed load, shown in Figure 4.2a. The bending moment distribution for this beam is shown in Figure 4.2b. Initially, at a low load level, the entire beam remains uncracked and it demonstrates elastic uncracked behaviour (as discussed in Section 3.4.2). The beam cracks when the flexural tensile stresses at the bottom fibre of the beam reach the modulus of rupture (f_r) (see section B-B in Figure 4.2c). The internal moment developed in the beam at the onset of cracking is called the *cracking moment* (M_{cr}). Concrete in the cracked condition is no longer able to resist tensile stresses. As a result of the cracking, the neutral axis moves upward and the area of concrete above the neutral axis is subjected to compression while the reinforcing steel sustains tensile stresses below the neutral axis, as shown on section C-C in Figure 4.2c. As the load increases, cracks develop over a large region of the beam, which is called the cracked region in Figure 4.2a; this region corresponds to bending moments of magnitude exceeding the cracking moment; that is, $M > M_{cr}$. The remaining portion of the beam, where the bending moments are less than the cracking moment ($M < M_{cr}$) is uncracked.

It should be noted that the reinforced concrete structures are expected to develop cracks under service loads; however, the extent and magnitude of cracking should be controlled, as discussed later in this chapter. This type of behaviour is called *elastic cracked behaviour* (refer to Section 3.4.2). The stress and strain in concrete and steel are elastic at this stage, as illustrated by the stress-strain diagrams in Figure 4.3.

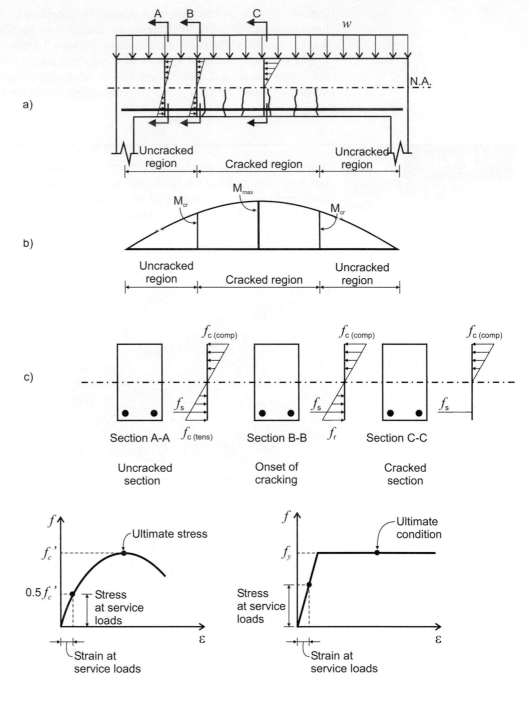

Figure 4.2 A reinforced concrete beam under service loads: a) beam elevation; b) bending moment distribution; c) beam sections and corresponding stresses.

Figure 4.3 Stress-strain diagrams: a) concrete; b) steel.

It can be observed from the above discussion that the beam cross-sectional properties vary along the span depending on the load magnitude and the corresponding bending moment and the amount of cracking. The concept of variable cross-sectional properties along the member length is unique to reinforced concrete structures and will be discussed in detail in Section 4.3.

The cracking moment (M_{cr}) can be calculated based on the *flexure formula*, which states that the flexural stress (f) developed in an elastic homogeneous beam due to the bending moment (M) can be determined as

$$f = \frac{M \cdot y_t}{I_g}$$

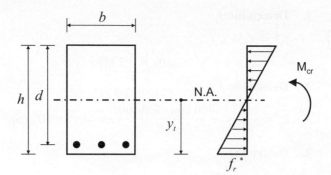

Figure 4.4 Concrete section at the onset of cracking.

For more details on the flexure formula, the reader is referred to textbooks on mechanics of materials. At the onset of cracking, the concrete tensile stress reaches the modulus of rupture (f_r) (refer to Section 2.3.2 for more information on the modulus of rupture). At this stage, the internal moment developed in the beam section just before cracking is equal to M_{cr}, as shown in Figure 4.4. As a result, the cracking moment can be obtained from the above equation by setting $M = M_{cr}$ and $f = f_r^*$ as follows:

$$M_{cr} = \frac{f_r^* \cdot I_g}{y_t} \qquad\qquad [4.1]$$

where

$$f_r^* = 0.5 f_r = 0.3\lambda \sqrt{f_c'} \text{ (MPa)}$$

Note that f_r^* is a reduced modulus of rupture for deflection estimations in flexural members (A23.3 Cl.9.8.2.3). This revision was introduced in 2009 as an attempt to provide a more consistent estimation of deflections across a broad range of reinforcement ratios, and to more adequately account for the effects of shrinkage and restraint cracking in determining deflections. Also,

$$y_t = \frac{h}{2}$$

and I_g is the moment of inertia of the gross cross-section of a concrete beam around the axis of bending (the effect of reinforcement is ignored); therefore, for a rectangular section of width b and depth h, it follows that

$$I_g = \frac{bh^3}{12} \qquad\qquad [4.2]$$

Example 4.1

Consider the section of a concrete beam in the sketch below. The beam properties are summarized below.

Determine the cracking moment for this section.

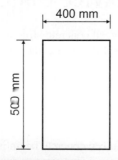

Given: $f_c' = 25$ MPa
$\lambda = 1$
$b = 400$ mm
$h = 500$ mm

SOLUTION: **1. Determine f_r**

A23.3 Eq. 8.3

$$f_r^* = 0.3\lambda \sqrt{f_c'}$$ [2.1]

$$= 0.3 \times 1 \times \sqrt{25 \text{ MPa}} = 1.5 \text{ MPa}$$

2. Determine I_g

$$I_g = \frac{bh^3}{12} = \frac{(400 \text{ mm})(500 \text{ mm})^3}{12} = 4.17 \times 10^9 \text{ mm}^4$$ [4.2]

3. Determine y_t

$$y_t = \frac{h}{2} = \frac{500 \text{ mm}}{2} = 250 \text{ mm}$$

Finally,

$$M_{cr} = \frac{f_r^* I_g}{y_t}$$ [4.1]

$$= \frac{(1.5 \text{ MPa})(4.17 \times 10^9 \text{ mm}^4)}{250 \text{ mm}} = 25 \times 10^6 \text{ N·mm} = 25 \text{ kN·m}$$

KEY CONCEPTS

Reinforced concrete flexural members crack when the tensile stresses reach the modulus of rupture (f_r). The internal moment developed in the member at the onset of cracking is called the cracking moment (M_{cr}). The cracking moment can be obtained as

$$M_{cr} = \frac{f_r^* I_g}{y_t} \quad \text{where } f_r^* = 0.5f_r \text{ for deflection calculations}$$ [4.1]

There are two distinct regions within a reinforced concrete flexural member:

- the *cracked region,* where the bending moments exceed the cracking moment, that is, $M \geq M_{cr}$;
- the *uncracked region,* where the bending moments are less than the cracking moment, that is, $M < M_{cr}$.

4.3 | PROPERTIES OF REINFORCED CONCRETE FLEXURAL MEMBERS UNDER SERVICE LOADS

One of the unique features of reinforced concrete structures is that their flexural properties change when sufficient load is applied to cause cracking in certain regions of the structural member. The two critical and mutually related properties, namely flexural stiffness and moment of inertia, will be discussed in this section.

4.3.1 Flexural Stiffness

Consider the simply supported reinforced concrete beam in Figure 4.5a with the corresponding bending moment diagram in Figure 4.5b. Let us study the relationship between the bending moment at the midspan (M) and the corresponding deflection (Δ) illustrated on the diagram in Figure 4.5c. Initially, at a low load level, the beam is uncracked and demonstrates

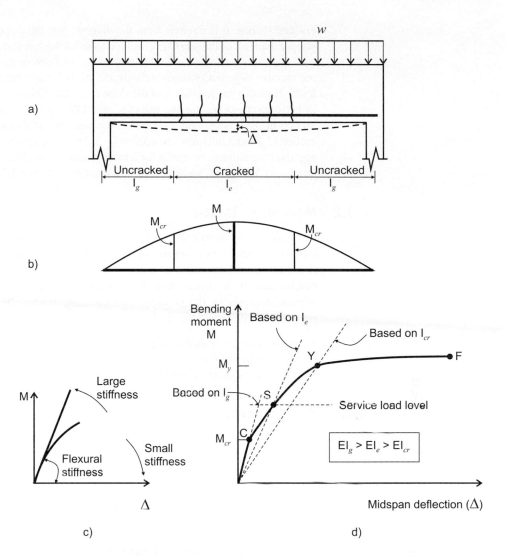

Figure 4.5 Deflection behaviour for a reinforced concrete beam: a) beam elevation; b) bending moment diagram; c) flexural stiffness; d) bending moment deflection at the midspan.

elastic uncracked behaviour, as discussed earlier in this chapter. At this stage, the beam cross-sectional properties can be represented by a gross transformed moment of inertia (I_g). Once the bending moment has reached the cracking moment (M_{cr}), the beam cracks (point C on the diagram); this behaviour is known as *elastic cracked behaviour*. From that point onward, the moment of inertia of the beam section at the midspan decreases with increasing loads. It should be noted that, at the service load level, the bending moments in the midspan region of the beam usually exceed the cracking moment (point S on the diagram). With further load increase, the tension reinforcement in the beam starts to yield (point Y on the diagram). At that stage, the beam section can be considered fully cracked and its properties can be represented by the moment of inertia of the cracked transformed section (I_{cr}). With further load increase, the beam finally fails (point F on the diagram).

Now observe the slope of a radial line connecting the origin of the coordinate system (point O on the diagram) with any point of interest on this diagram. The slope is equal to the angle between a radial line and the horizontal axis on the diagram. In the structural sense, the slope reflects the magnitude of the *flexural stiffness* of the member (see Figure 4.5d). Flexural stiffness (also known as *flexural rigidity* in technical literature) is a resistance to deformation of a flexural member. The larger the flexural stiffness, the smaller are the flexural stresses and deflections developed in the member. Flexural stiffness is equal to $E_c I$, that is, the product of the modulus of elasticity (E_c for reinforced concrete structures) and the moment of inertia (I). In general, I is a variable in cracked reinforced concrete flexural members. Consider the two extreme cases represented on the diagram in Figure 4.5c, that is, the line OC corresponding to the uncracked section and the line OY corresponding to the fully cracked section. It is obvious from the diagram that the slope of the line OC is significantly steeper than that of the line OY. This indicates that the cracked section has a much smaller stiffness than the uncracked section. Therefore, the stiffness of the cracked member decreases and the member experiences larger deflections; this behaviour becomes more pronounced at larger loads. Note that the magnitude of the slope of the line OS corresponding to the service load falls in between the slopes for the lines OC and OY. Therefore, the flexural stiffness for a beam section at the service load level falls in between the stiffness values of an uncracked (gross) section (I_g) and a fully cracked section (I_{cr}). The effective moment of inertia (I_e) represents an approximate moment of inertia for a beam under service loads and is used in deflection calculations according to CSA A23.3, as discussed in Section 4.3.5.

4.3.2 Moment of Inertia

The moment of inertia (I) is a key cross-sectional property used in the deflection calculations of reinforced concrete flexural members. It depends on several factors, including the shape and dimensions of a cross-section, the extent of cracking, and the amount of reinforcement. As discussed in the previous section, the moment of inertia value for the section depends on the bending moment magnitude. The following moment of inertia properties will be used in deflection calculations:

* gross moment of inertia (I_g) (the moment of inertia of an uncracked section),
* moment of inertia of the cracked transformed section (I_{cr}),
* effective moment of inertia (I_e).

4.3.3 Gross Moment of Inertia

The moment of inertia of an uncracked concrete section is called the *gross moment of inertia* (I_g). The gross moment of inertia for a flexural member of a rectangular section of width b and overall depth h can be determined from Eqn 4.2. It should be noted that Eqn 4.2 disregards the effect of reinforcing steel. The effect of reinforcement can be accounted for by considering the transformed section properties. The concept of a transformed section is used to design composite structures. (Refer to mechanics of materials textbooks for more information.) A composite section consisting of two different materials (for example, steel and concrete) can be treated as an equivalent *transformed section* that consists of one material only. In the case of a reinforced concrete section, the actual area of steel reinforcement is replaced with an equivalent area of concrete.

The concept of transformed sections will be explained on an example of an uncracked reinforced concrete beam section (see Figure 4.6). The beam is reinforced with the tension steel of area A_s. The stress and strain distributions for the beam section subjected to a positive bending moment are shown in Figure 4.6. Suppose that the loads are small enough such that there are no cracks in the tension zone of the section.

The stresses in steel and concrete are expected to be in the elastic range, so Hooke's law is valid (refer to textbooks on mechanics of materials). Therefore, the strain in the steel reinforcement can be determined as

$$\varepsilon_s = \frac{f_s}{E_s} \qquad\qquad \text{[4.3]}$$

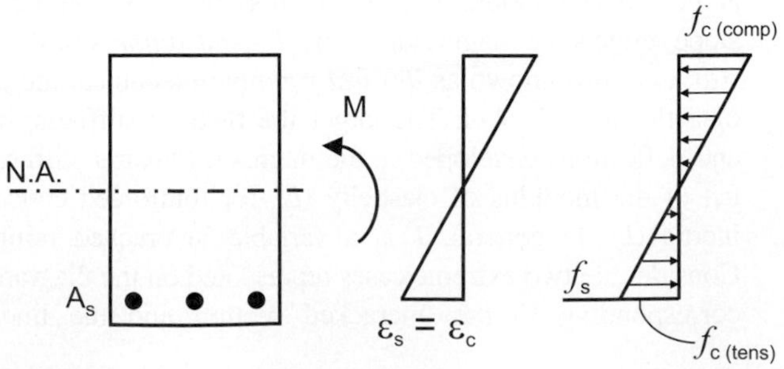

Figure 4.6 Stress and strain distribution in an uncracked beam section.

The strain in the concrete at the reinforcement location can be expressed as

$$\varepsilon_c = \frac{f_{c(tens)}}{E_c} \qquad\qquad [4.4]$$

Based on the strain compatibility between steel and concrete, the strains in the steel and concrete at the reinforcement location are equal; that is,

$$\varepsilon_s = \varepsilon_c \qquad\qquad [4.5]$$

or

$$\frac{f_s}{E_s} = \frac{f_{c(tens)}}{E_c}$$

The above equation can be expressed in terms of f_s as follows:

$$f_s = \left(\frac{E_s}{E_c}\right) f_{c(tens)}$$

or

$$f_s = n f_{c(tens)} \qquad\qquad [4.6]$$

where

$$n = \frac{E_s}{E_c} \qquad\qquad [4.7]$$

The *modular ratio* (n) is the ratio between the moduli of elasticity of steel and concrete.

As discussed in Section 2.6.2, the modulus of elasticity of steel (E_s) is constant and can be taken as 200000 MPa for nonprestressed reinforcement (A23.3 Cl.8.5.4.1). As a result, the modular ratio typically varies from 8 to 10 for a normal range of f_c' values (25 MPa to 40 MPa).

In order to form a transformed section, the area of steel is replaced with an equivalent area of concrete. This equivalent area can be determined by expressing the tension force in steel reinforcement as follows:

$$T = f_s A_s = (n f_{c(tens)}) A_s = f_{c(tens)}(n A_s)$$

Based on this numerical transformation, the force in steel reinforcement of area A_s and stress f_s is equal to the force in concrete corresponding to area $n A_s$ and stress $f_{c(tens)}$. The transformed section is shown in Figure 4.7. To account for the displaced steel area A_s, the equivalent steel area is $(n-1)A_s$ (this is true only in the case of an uncracked cross-section).

The area of the transformed uncracked section (A_{tr}) is

$$A_{tr} = A_c + (n-1)A_s$$

In general, it is not necessary to use transformed section properties to calculate deflections in an uncracked reinforced concrete beam; gross cross-sectional properties represent a satisfactory approximation for this purpose. However, in the case of cracked reinforced concrete members, it is essential to use the transformed section properties for the deflection calculations, as discussed in the following sections.

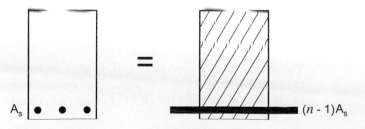

Figure 4.7 Transformed section: uncracked beam section.

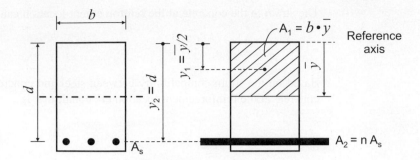

Figure 4.8 Transformed section for a cracked beam.

4.3.4 Cracked Moment of Inertia

The moment of inertia of a cracked section (or cracked moment of inertia) is denoted as I_{cr}. The calculation of I_{cr} is more complex due to the fact that concrete is effective in the compression zone only, while only steel reinforcement is effective in the tension zone. The concept of a transformed section is used to calculate I_{cr} by taking into account the steel and concrete properties, as illustrated in Figure 4.8.

The transformed section comprises the following two areas (shown hatched in Figure 4.8): the area (A_1) above the neutral axis (N.A.) equal to the compression zone of the concrete section and the area (A_2) equal to the "transformed" steel area (nA_s), that is, $A_2 = nA_s$.

The area of the transformed cracked section (A_{tr}) is

$$A_{tr} = A_1 + A_2 = b\bar{y} + nA_s$$

The neutral axis depth ($\bar{y}$) can be determined from the equation of the centroid of a transformed section as follows:

$$\bar{y} = \frac{A_1 y_1 + A_2 y_2}{A_1 + A_2} = \frac{(b\bar{y})\dfrac{\bar{y}}{2} + (nA_s)d}{b\bar{y} + nA_s} \qquad [4.8]$$

The above equation can be transformed into the quadratic equation

$$\left(\frac{b}{2}\right)\bar{y}^2 + (nA_s)\bar{y} - nA_s d = 0$$

The solution of this equation expressed in terms of A_s is

$$\bar{y} = \frac{-nA_s + \sqrt{(nA_s)^2 + 2bd(nA_s)}}{b}$$

This solution can be more conveniently expressed in terms of ρ as

$$\bar{y} = d\left(\sqrt{(n\rho)^2 + 2n\rho} - n\rho\right) \qquad [4.9]$$

where

$$\rho = \frac{A_s}{bd} \qquad [3.1]$$

The moment of inertia of the cracked section (I_{cr}) around the neutral axis can be determined based on the parallel axis theorem (covered in mechanics of materials texts) as

$$I_{cr} = \frac{b\bar{y}^3}{3} + nA_s(d - \bar{y})^2 \qquad [4.10]$$

The above equation applies to sections with the tension steel only. In the case of doubly reinforced sections (with both tension and compression steel), an additional term needs to be added to account for the effect of compression steel.

Example 4.2

Consider the typical cross-section of a reinforced concrete beam in the sketch below. The beam is reinforced with 4-20M longitudinal bars and its properties are summarized below.

Determine the moment of inertia for the uncracked transformed section.

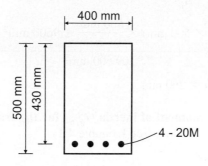

Given: $f_c' = 25$ MPa
$f_y = 400$ MPa
$b = 400$ mm
$h = 500$ mm

SOLUTION: **1. Compute the modular ratio (n)**

$E_s = 200\,000$ MPa

| A23.3 Eq. 8.2 |

$E_c = 4500\sqrt{f_c'}$ **[2.2]**

$\quad = 4500\sqrt{25 \text{ MPa}} = 22\,500$ MPa

$n = \dfrac{E_s}{E_c}$ **[4.7]**

$\quad = \dfrac{200\,000 \text{ MPa}}{22\,500 \text{ MPa}} = 8.89 \cong 9$

2. Calculate the transformed section area (A_{tr})

The area of a 20M bar is 300 mm² (refer to Table A.1), so

$A_s = 4 \times 300 \text{ mm}^2 = 1200 \text{ mm}^2$

Convert the area of steel reinforcement (A_s) to the equivalent area of concrete $(n - 1)A_s$:

$(n - 1)A_s = 8 \times 1200 \text{ mm}^2 = 9600 \text{ mm}^2$

Compute the area of the transformed uncracked section (A_{tr}):

$A_{tr} = bh + (n - 1)A_s$

$\quad = 400 \text{ mm} \times 500 \text{ mm} + 9600 \text{ mm}^2 = 209\,600 \text{ mm}^2$

The transformed section is shown in the sketch below.

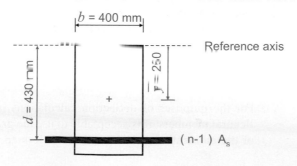

3. **Determine the location of the centroid of the transformed section ($\bar{y}$)**
    The $\bar{y}$ value will be determined with regard to the top compression fibre:

$$\bar{y} = \frac{(bh) \times \dfrac{h}{2} + (n-1)A_s \times d}{A_{tr}}$$

$$= \frac{(400 \text{ mm} \times 500 \text{ mm}) \times \dfrac{500 \text{ mm}}{2} + 9600 \text{ mm}^2 \times 430 \text{ mm}}{209\ 600 \text{ mm}^2}$$

$$= 258.2 \text{ mm} \cong 260 \text{ mm}$$

4. **Compute the moment of inertia ($I_{g, tr}$) for the transformed section**
    Note that $y_t = 250$ mm (see Example 4.1).

$$I_{g, tr} = \frac{bh^3}{12} + bh\,(\bar{y} - y_t)^2 + (n-1)A_s(d - \bar{y})^2$$

$$= \frac{400 \text{ mm} \times (500 \text{ mm})^3}{12} + 400 \text{ mm} \times 500 \text{ mm} \times (260 \text{ mm} - 250 \text{ mm})^2$$
$$+ 9600 \text{ mm}^2 \times (430 \text{ mm} - 260 \text{ mm})^2$$

$$= 4.5 \times 10^9 \text{ mm}^4$$

Learning from Examples

At this point, it is important to compare the moment of inertia values obtained in Examples 4.1 and 4.2. A beam section with the same cross-sectional dimensions (width 400 mm and overall depth 500 mm) and the same material properties ($f_c' = 25$ MPa and $f_y = 400$ MPa) has been considered in both examples.

In Example 4.1, gross moment of inertia (I_g) for a beam cross-section was calculated without considering the effect of reinforcement. The same beam cross-section was considered in Example 4.2, except that the moment of inertia for the transformed section ($I_{g,tr}$) was calculated, taking into consideration the effect of longitudinal reinforcement (4-20M bars). The following observations can be made with regard to these two examples:

1. The amount of computational effort required to calculate $I_{g,tr}$ is significant compared to the quick calculation performed to determine the I_g value; however, the difference in the moment of inertia values is not significant (on the order of 8%) (see Figure 4.9).

2. Due to the presence of reinforcement, the location of the centroid of the transformed section ($\bar{y}$) is lower than that of the gross section (y_t); however, the difference is only 10 mm (on the order of 4%) in this case:

$$\bar{y} - y_t = 260 \text{ mm} - 250 \text{ mm} = 10 \text{ mm}.$$

For the purposes of deflection calculations in uncracked reinforced concrete flexural members, it is acceptable to use the gross moment of inertia (I_g) instead of the moment of inertia for the transformed section ($I_{g,tr}$). This will result in

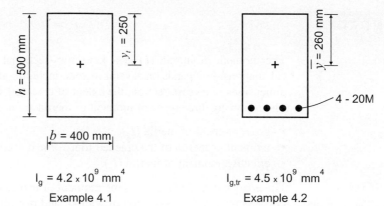

Figure 4.9 Moment of inertia values for uncracked section: Examples 4.1 and 4.2.

$I_g = 4.2 \times 10^9 \text{ mm}^4$

Example 4.1

$I_{g,tr} = 4.5 \times 10^9 \text{ mm}^4$

Example 4.2

a slight increase in the magnitude of deflections, but the difference is not significant.

4.3.5 Effective Moment of Inertia

As discussed in the previous section, I_g represents the moment of inertia of the section before cracking and I_{cr} represents the moment of inertia of the fully cracked section. Some portions of the beam are cracked under service loads, while the remaining portions remain uncracked. As a result, the actual flexural stiffness (and the corresponding moment of inertia) is larger than that corresponding to a fully cracked member, and less than the stiffness corresponding to the uncracked member. To account for the variations in the moment of inertia values along the beam length, CSA A23.3 Cl.9.8.2.3 prescribes the use of the *effective moment of inertia* (I_e), which is a reasonable approximation of the actual moment of inertia. I_e can be determined based on the equation

A23.3 Eq. 9.1
$$I_e = I_{cr} + (I_g - I_{cr})\left(\frac{M_{cr}}{M_a}\right)^3 \leq I_g$$
[4.11]

where

I_g = the moment of inertia of a gross concrete section
I_{cr} = the moment of inertia of the transformed cracked section
M_{cr} = the cracking moment
M_a = the maximum bending moment in the member at the load stage at which the deflection is computed, or at any previous stage. Depending on the type of deflections that need to be determined, M_a can take the following values:

$M_a = M_D$ if the deflections are to be determined due to dead load (DL) only.
$M_a = M_L$ if the deflections are to be determined due to live load (LL) only.
$M_a = M_{D+L}$ if the deflections are to be determined due to total dead and live load.

In any case, bending moments used for deflection calculations are due to specified service loads, that is, unfactored loads.

The above empirical equation, developed by Dan Branson in 1960s, is based on extensive experimental research and was adopted by the ACI Code and also by CSA A23.3 in the 1970s.

Based on Eqn 4.11, the I_e value falls in between the I_g and I_{cr} values; that is,

$$I_{cr} < I_e < I_g$$

The following recommendations can be made regarding I_e calculations:

• If the maximum bending moment acting on the beam (M_a) is significantly larger than M_{cr}, that is, $M_a/M_{cr} \geq 2$, then I_e is very close to I_{cr}. In such cases, it is recommended to use I_{cr} for deflection calculations.
• If the maximum bending moment (M_a) is less than M_{cr}, then the entire member is expected to remain uncracked, so I_g can be used instead of I_e for deflection calculations.

KEY CONCEPTS

The moment of inertia (I) is a key cross-sectional property used in the deflection calculations. It depends on several factors, including the load magnitude, the shape and dimensions of a cross-section, the extent of cracking, and the amount of reinforcement. The following three types of moment of inertia are used in deflection calculations:

- gross moment of inertia (I_g),
- moment of inertia of the cracked transformed section (I_{cr}),
- effective moment of inertia (I_e).

The *gross moment of inertia* is the moment of inertia of an uncracked concrete section. For a rectangular cross-section of width b and overall depth h, I_g can be determined as

$$I_g = \frac{bh^3}{12} \tag{4.2}$$

The *moment of inertia of a fully cracked section* (I_{cr}) with tension steel only can be determined based on the transformed section properties as

$$I_{cr} = \frac{b\bar{y}^3}{3} + nA_s(d - \bar{y})^2 \tag{4.10}$$

To account for the variations in the moment of inertia values along the beam length, CSA A23.3 Cl.9.8.2.3 prescribes the use of the *effective moment of inertia* (I_e), which can be determined based on the following equation:

| A23.3 Eq. 9.1 |

$$I_e = I_{cr} + (I_g - I_{cr})\left(\frac{M_{cr}}{M_a}\right)^3 \tag{4.11}$$

It is very important to note that bending moment M_a used for the deflection calculations is due to specified service loads, that is, unfactored loads.

Some portions of the member crack at service loads; however, the remaining portions are uncracked. Therefore, I_e falls in between I_g and I_{cr}, that is,

$$I_{cr} < I_e < I_g$$

4.4 DEFLECTIONS IN REINFORCED CONCRETE FLEXURAL MEMBERS

4.4.1 Background

Deflections in reinforced concrete members are estimated based on the assumption of the elastic behaviour at service loads, which is appropriate in most cases. Depending on the load duration, deflections can be classified as immediate or long-term.

Immediate (or short-term) *deflections* occur immediately on the application of the load. These deflections can be calculated from the properties of uncracked or cracked members or some combination of these, as discussed in Section 4.3. Variable cross-sectional properties, continuity of concrete structures, and construction-related factors can add to the complexity of deflection calculations.

In addition to immediate deflections, reinforced concrete members subjected to sustained loads such as dead load cause additional *long-term deflections*. These deflections are mainly caused by creep and shrinkage. Creep causes the deflections in concrete structures to increase under constant sustained load for a period of several years. Total deflections (after several years of service) may be on the order of two or three times the immediate deflections and can cause functional problems for the occupants and damage to the building components.

Deflection calculations must take into account both immediate and long-term deflections.

4.4.2 Immediate Deflections

As discussed in Section 4.2, some portions of a reinforced concrete flexural member develop cracks at service load levels while the other portions remain uncracked. Consequently, the properties of concrete sections, in particular the moment of inertia and the stiffness, cannot be easily determined from the basic concepts of mechanics of materials. This adds to the complexity of deflection calculations.

Immediate deflections are also influenced by construction-related factors. Concrete structures loaded at an early age will experience larger initial deflections than the structures that are properly cured and protected from early stresses until the concrete reaches its design strength. Hence, the reshoring sequence during construction can significantly influence the magnitude of deflections. Deflections in concrete members also depend on other factors, such as the time delay between batching and placement of concrete, the method of concrete placement, the actual w/c ratio during placement, the method of curing, and the temperature gradient during the curing period.

Given the above discussion, it is clear that the accurate prediction of deflections in reinforced concrete members can be rather complex. CSA A23.3 recognizes this complexity and offers a simplified deflection calculation procedure in lieu of a detailed procedure. Therefore, Cl.9.8.2.2 of CSA A23.3 permits the computation of immediate deflections by using simple formulas for elastic deflections. The effective moment of inertia (I_e) is used in this procedure to account for the effects of cracking and amount of reinforcement. For most design applications, the simplified procedure should offer a reasonable estimate of the expected deflections in a reinforced concrete beam or slab.

For members subjected to uniformly distributed loads, the immediate deflection (Δ) can be determined based on the equation

$$\Delta = k\left(\frac{5}{48}\right)\frac{Ml^2}{E_c I_e}$$ [4.12]

where

k = the coefficient that takes into account different end-support conditions
M = the bending moment at the midspan
l = the span length
E_c = the modulus of elasticity of concrete (see Section 2.3.4)
I_e = the effective moment of inertia.

The above general equation can be applied to various types of structural members subjected to uniform loads by varying k value (see Figure 4.10). Note that, for cantilevered members (Case 1), deflections due to rotation at the supports must also be included in the calculations.

It should be noted that Eqn 4.12 gives the same deflection values as formulas for elastic deflection calculations expressed in terms of uniform loads. For example, consider a simply supported beam with span l subjected to the uniformly distributed load w in Figure 4.11. The maximum elastic deflection (Δ) for this beam can be calculated based on the standard formula found in mechanics of materials textbooks:

$$\Delta = \frac{5}{384} \times \frac{wl^4}{EI}$$ [4.13]

The bending moment at the midspan of this beam is

$$M = \frac{wl^2}{8}$$

Hence, w can be expressed from the above equation as

$$w = \frac{8M}{l^2}$$

Case	Structural System	k
1	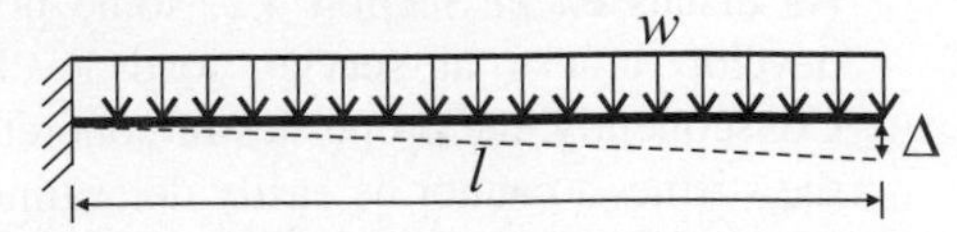 w l Δ	2.4*
2	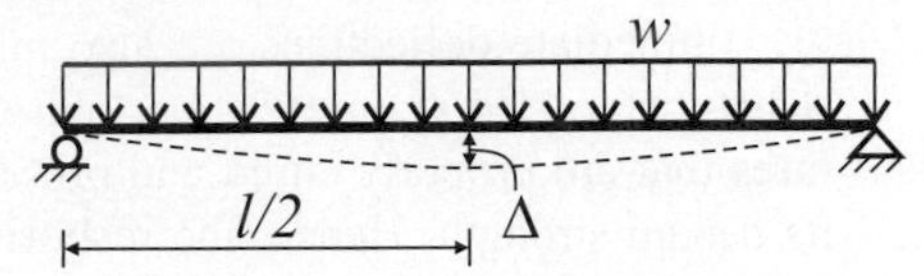 w $l/2$ Δ	1.0
3a	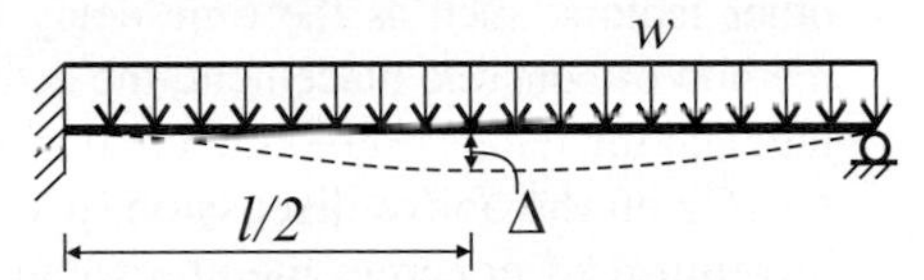 w $l/2$ Δ	0.8
3b	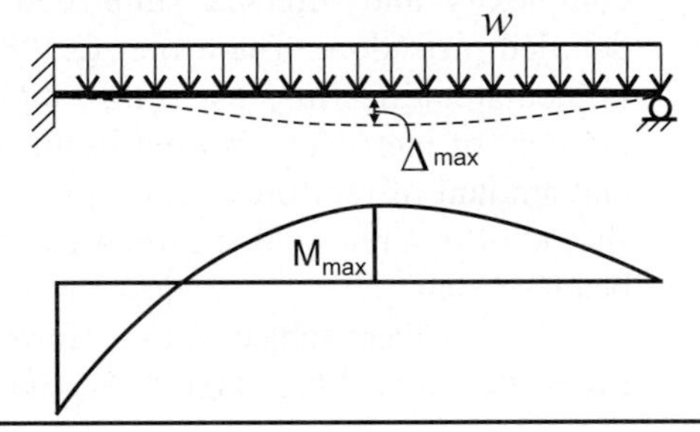 w Δ_{max} M_{max}	0.74
4	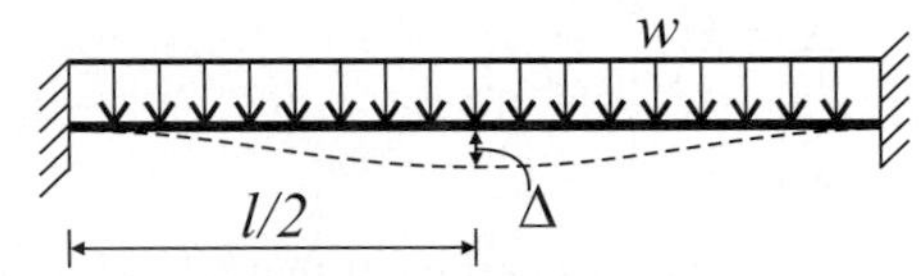 w $l/2$ Δ	0.6

Figure 4.10 k values for various structural systems.

Note: * - Deflections due to end support rotation must also be included

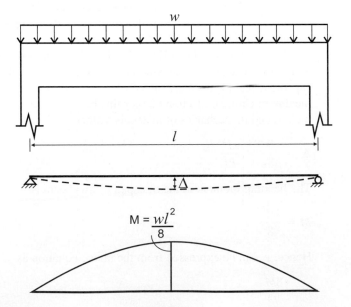

Figure 4.11 Deflections in a simply supported beam.

Finally, w can be substituted into Eqn 4.13 as follows:

$$\Delta = \frac{5}{384} \times \frac{8M}{l^2} \times \frac{l^4}{EI} = \frac{5}{48} \times \frac{Ml^2}{EI}$$

The transformed form of the deflection formula would be the same as that given in Eqn 4.12 (provided that $k = 1$).

Note that the k value for a simply supported beam is equal to 1 based on case 2 in Figure 4.10.

Deflection formulas for other types of loads are provided in Appendix A.

The effective moment of inertia (I_e) should be calculated based on the load level under consideration. The immediate deflections can be computed by the following procedure.

1. The dead load deflection (Δ_D) is calculated based on the I_e value corresponding to the dead load bending moment (M_D).
2. Deflection due to the combined dead and live loads (Δ_{D+L}) is calculated based on the I_e value corresponding to the total bending moment (M_{D+L}).
3. The live load deflection (Δ_L) is calculated as

$$\Delta_L = \Delta_{D+L} - \Delta_D \qquad \text{[4.14]}$$

Note that the use of M_{D+L} to calculate all deflection terms tends to overestimate the Δ_D value and underestimate the Δ_L value.

4.4.3 Long-Term Deflections

Long-term deflections in reinforced concrete structures may be caused by superimposed creep, shrinkage, and temperature strains.

Creep is the inelastic, time-dependent deformation of concrete under sustained stress, as discussed in Section 2.3.5. Creep can cause a significant increase in deflections in reinforced concrete members. The increase in the magnitude of initial deflections due to creep effects may be two- to threefold, so creep effects must be taken into consideration in deflection calculations. The detailed treatment of creep deformations can be extremely complex and is beyond the scope of this book; for more information, refer to Branson (1977). Alternatively, CSA A23.3 prescribes a simple method to account for long-term effects. According to Cl.9.8.2.5, long-term deflections in a reinforced concrete flexural member are obtained by multiplying the immediate deflection caused by the sustained load by the factor ζ_s, given by

| A23.3 Eq. 9.5 | $\zeta_s = 1 + \dfrac{s}{1 + 50\rho'}$ | [4.15] |

where

  s = the time-dependent factor for creep deflection under sustained loads; s varies between 0 and 2 depending on the load duration (see Table 4.1)

  ρ' = the compression reinforcement ratio; the presence of compression steel will substantially reduce long-term creep deflection, in some cases by as much as 20% or 30%.

Therefore, the total long-term deflection (Δ_t) in a reinforced concrete flexural member can be obtained when the immediate deflection (Δ_i) is multiplied by the factor ζ_s; that is,

$$\Delta_t = \zeta_s \Delta_i$$

It should be noted that immediate deflection needs to account for the combined effects of dead and live loads; that is,

$$\Delta_i = \Delta_D + \Delta_L$$

Table 4.1 s factor values

s	Duration of sustained loads
2.0	5 years or more
1.4	12 months
1.2	6 months
1.0	3 months
0	Less than 3 months

(*Source:* CSA A23.3 Cl.9.8.2.5, reproduced with the permission of the Canadian Standards Association)

Dead load is generally present throughout the service life of a structure; however, live loads are not necessarily present at all times. Therefore, only a part of the live load in addition to the more permanent dead load needs to be considered as the sustained load. As a result, the total long-term deflection can be determined as

$$\Delta_t = \zeta_s(\Delta_D + \Delta_{LS}) + (\Delta_L - \Delta_{LS}) \qquad [4.16]$$

where

Δ_D = the immediate deflection due to the total dead load
Δ_L = the immediate deflection due to the total live load
Δ_{LS} = the immediate deflection due to the sustained portion of the live load.

If the dead load and the sustained live load are applied at different times, it may be desirable to use different ζ_s values for the dead load and sustained live load. In that case, the total long-term deflection can be obtained as

$$\Delta_t = \zeta_{sD}\Delta_D + \zeta_{sL}\Delta_{LS} + (\Delta_L - \Delta_{LS})$$

where ζ_{sD} depends on the dead load duration and ζ_{sL} depends on the live load duration.

KEY CONCEPTS

Deflections in reinforced concrete members are estimated based on the assumption of elastic behaviour at service loads, which is appropriate in most cases. Depending on the load duration, deflections can be classified into immediate and long-term.

Immediate (or short-term) *deflections* occur immediately when the load is applied. According to A23.3 Cl.9.8.2.2, immediate deflections can be computed by using the simple formulas for elastic deflections. To account for the effects of cracking and reinforcement on member stiffness, the effective moment of inertia (I_e) is used in deflection calculations.

For members subjected to uniformly distributed loads, the immediate deflection (Δ) can be determined based on the equation

$$\Delta = k\left(\frac{5}{48}\right)\frac{Ml^2}{E_cI_e} \qquad [4.12]$$

Long-term deflections are mainly caused by creep and shrinkage. Creep causes an increase in deflections in concrete structures under constant sustained load for a period of several years. The total long-term deflection (Δ_t) in a reinforced concrete flexural member can be obtained when the immediate deflection (Δ_i) is multiplied by the factor ζ_s (Cl.9.8.2.5):

$$\Delta_t = \zeta_s\Delta_i$$

where

A23.3 Eq. 9.5 $\qquad \zeta_s = 1 + \dfrac{s}{1 + 50\rho'} \qquad\qquad [4.15]$

Shrinkage is a decrease in the volume of concrete (other than that caused by externally applied forces and temperature changes) that depends on the ambient weather conditions during and after construction. As a result, concrete members cast and maintained in a damp, cool atmosphere can be expected to experience smaller deflections than those cast in a hot and dry atmosphere. Also, since the amount of drying shrinkage and creep is affected by the volume/surface ratio, large concrete beams will have lower deflections than thin slabs (the difference may be as high as 40% to 50%). It should be noted, however, that an unequal amount of tension and compression steel will cause shrinkage deflections. The detailed treatment of shrinkage effects is beyond the scope of this book.

4.5   CSA A23.3 DEFLECTION CONTROL REQUIREMENTS

4.5.1   Background

Building structures are expected to support service loads without experiencing excessive deflections that might cause cracking of partitions or architectural elements, sagging of floors and roofs, or drooping of overhang canopies. A23.3 Cl.9.8.1 prescribes that reinforced concrete flexural members have adequate stiffness to limit deflections or any deformations that may adversely affect the strength or serviceability of the structure. CSA A23.3 prescribes two approaches to deflection control:

- indirect approach
- detailed deflection calculations

The *indirect approach* consists of setting suitable upper limits on the member span/depth ratio, as discussed below. This approach is simple and satisfactory for applications where spans, loads, load distributions, and member sizes and proportions are within the prescribed ranges.

    Detailed deflection calculations are performed to estimate immediate and long-term deflections in reinforced concrete members and compare these predicted values with the CSA A23.3 prescribed allowable values summarized in Section 4.5.3. Deflection calculation procedures will be discussed in Section 4.6.

4.5.2   Indirect Approach

Detailed deflection calculations can be very tedious and time-consuming. However, detailed calculations may not be required if the flexural member has sufficient depth and stiffness such that the deflections are not a concern. CSA A23.3 Cl.9.8.2.1 (Table 9.2) prescribes the span/depth (l_n/h) ratio limits for beams and slabs that will result in members with sufficient stiffness that explicit deflection calculations are not deemed necessary; l_n denotes the clear span of a beam or a slab and h its overall depth (see Figure 4.12). These values apply to most concrete flexural members subjected to uniformly distributed loads. Exceptions include members supporting partitions and other elements that are likely to be damaged by large deflections. Caution is required for concentrated loads or structures where the ratio of dead to live loads is significantly different from the above assumptions.

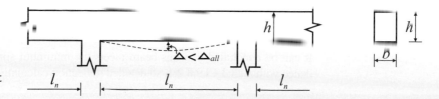

Figure 4.12 CSA A23.3 indirect approach to deflection control.

The table considers that Grade 400 reinforcement is used; for f_y values other than 400 MPa, the values in the table should be multiplied by

$$(0.4 + f_y/670)$$

CSA A23.3 Table 9.2 is reproduced in this book (Table A.3). The minimum thicknesses proposed in this table are frequently used in selecting the overall depths of beams or slabs.

An application of the CSA A23.3 indirect approach to deflection control is illustrated by Example 4.3.

Example 4.3

Consider a simply supported reinforced concrete beam with a span of 6 m and an overall depth of 500 mm, as shown in the figure below.
Check whether detailed deflection calculations are required according to the CSA A23.3 deflection control requirements.

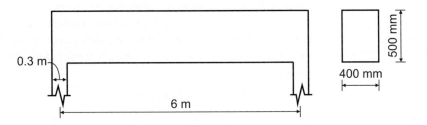

Given: $h = 500$ mm

SOLUTION: First, determine the clear span for this beam as

$$l_n = 6 \text{ m} - \left(\frac{0.3 \text{ m}}{2} + \frac{0.3 \text{ m}}{2}\right) = 5.7 \text{ m}$$

Second, determine the span/depth ratio:

$$\frac{l_n}{h} = \frac{5.7 \text{ m}}{0.5 \text{ m}} = 11.4$$

Finally, refer to Table A.3 (CSA A23.3 Table 9.2), which prescribes the span/depth ratios. For simply supported beams, it follows that deflection calculations are not required provided that

$$h \geq \frac{l_n}{16} \quad \text{or} \quad \frac{l_n}{h} \leq 16$$

For the beam considered in this example,

$$\frac{l_n}{h} = 11.4 < 16$$

It can be concluded that this beam meets the minimum span/depth requirements in accordance with A23.3 Cl.9.8.2.1, so detailed deflection calculations are not required.

4.5.3 Allowable Deflections

The allowable deflection limits prescribed by building codes are based on scientific principles and observations of previous cases of structures experiencing large deflections. In general, deflections in a structural member become visible to the naked eye when they exceed the magnitude of approximately 0.4% of its span (MacGregor and Bartlett, 2000). This section discusses the allowable deflection limits prescribed by CSA A23.3 to prevent problems caused by excessive deflections.

In general, deflection problems can be divided into four categories (ACI Committee 435, 2004):

1. *Sensory problems* include floor vibrations and the appearance of droopy members. For example, building occupants may become annoyed or alarmed by the structural frame vibrations under conditions of normal use.
2. *Serviceability problems* are generally related to the function of a building structure or a portion thereof; examples are roofs that do not drain properly and floors that are not properly levelled (in gymnasia or bowling alleys). For example, the ponding of water on roofs occurs due to deflections caused by the weight of rain water — in extreme cases, roof failure due to ponding (called ponding failure) takes place; long-span roofs in regions with light roof design loads are particularly susceptible to ponding.
3. *Damage of nonstructural elements,* such as ceilings and partitions, can occur. For example, windows and brittle masonry and glass partitions can be damaged if a structural member above them deflects. These problems can be controlled by limiting the deflections that occur after the installation of nonstructural elements or by designing the nonstructural elements to accommodate the required amount of movement. It should be noted that damage to brittle partitions can occur under deflections as small as 0.1% of the span.
4. *Structural problems* include structural damage caused by excessively large deflections.

A23.3 Cl.9.8.5.3 (Table 9.3) prescribes deflection limits for various concrete structural members, expressed as a fraction of the span (l) for the structural member under consideration. It should be noted that some deflection limits apply to only a fraction of the total deflection (see Cases 3 and 4 in A23.3 Table 9.3). For example, masonry walls are stressed and cracked only by the portion of the total deflection that occurs after the walls are constructed. This portion is called *incremental deflection*. The designer is responsible for applying an appropriate deflection limit for a structural member considering its function and the characteristics of nonstructural elements, such as partitions, ceilings, or façades.

4.6 DEFLECTION CALCULATION PROCEDURES

4.6.1 Background

Due to the composite nature of reinforced concrete and the fact that the member cross-sectional properties change after cracking, deflection calculation procedures for reinforced concrete structures are more complex than those of other building materials. In fact, accurate estimation of deflections in a reinforced concrete flexural member is perhaps the most complex task in reinforced concrete design. Once the flexural member cracks, its stiffness in the cracked regions decreases and the member becomes nonprismatic. An iterative computer-aided calculation procedure is required to determine the relationship between the load and the deflection in cracked reinforced concrete flexural members.

This section presents two procedures for estimating deflections in reinforced concrete flexural members:

- the CSA A23.3 approximate procedure
- the computer-aided iterative procedure

These procedures are based on the assumption of the elastic behaviour of a flexural member.

Approximate procedure   This procedure determines elastic deflections in reinforced concrete members using the effective moment of inertia (I_e) prescribed by CSA A23.3 Cl.9.8.2.3. Deflections can be easily determined by hand calculations. This procedure is approximate and can be used when the following conditions are met:

- The member is prismatic; that is, it has constant cross-sectional properties over its length. This procedure does not apply to nonprismatic members such as beams and slabs with haunches.
- Span lengths do not vary too greatly between adjacent spans (maximum difference within 20% to 30%).
- The member is subjected to uniformly distributed loads.

As a result, the use of approximate procedure is not suitable for nonprismatic members, continuous members with large variations in span lengths or flexural stiffness between adjacent spans, or members subjected to large point loads. In these cases, computer-aided analysis using cracked section properties may be more appropriate. In any case, the designer must exercise caution in using this procedure and ensure that (s)he has a complete understanding of its suitability for a particular design application.

Computer-aided iterative procedure   This procedure uses a computer-aided structural analysis to determine the deflections in reinforced concrete members. The member is subdivided into several segments that can be characterized by different section properties. Cracked section properties are used where the bending moment exceeds the cracking moment, and gross section properties are used elsewhere along the span. With the wide availability of computer structural analysis programs in design offices, this procedure can be used for design applications for which the approximate procedure is not deemed suitable.

As part of the calculation procedure, it is necessary to determine the bending moment distribution and the corresponding bending moment diagram for the flexural member under consideration. In the case of the approximate procedure, a bending moment diagram can be developed either by performing the structural analysis of a member either by using hand calculations or a computer program. Beam diagrams in various handbooks and manuals can also be used. Some of these diagrams are included in Appendix A for the reader's benefit.

Regardless of the calculation procedure used, the designer should keep in mind a few general thoughts related to deflection calculations, as summarized below:

- It is important to use good engineering judgment when performing deflection calculations.
- Due to the large number of construction-related parameters influencing the quality of reinforced concrete structures, it can be difficult to accurately predict the magnitude of deflections. The designer should be aware that the calculated deflections represent an estimate of the deflection the structure is likely to experience. This is true even when advanced computer-aided tools and procedures are used in the design process.

- When the designer needs to gain better insight into the probable maximum range of deflection magnitudes, (s)he should perform a sensitivity analysis, that is, vary the design parameters within a probable range of values and observe how these variations affect the magnitude of the deflections.

KEY CONCEPTS

CSA A23.3 Cl.9.8.1 prescribes that reinforced concrete flexural members have adequate stiffness to limit deflections or any deformations that might adversely affect the strength or serviceability of the structure.

Two approaches can be taken to ensure that deflections are within acceptable limits:

- indirect approach
- detailed deflection calculations

The indirect approach is based on A23.3 Cl.9.8.2.1 and it consists of setting suitable upper limits on the member span/depth ratio (see Table A.3). This approach is simple and satisfactory for applications where spans, loads, load distributions, member sizes and proportions are in the usual ranges.

Detailed deflection calculations are performed to estimate immediate and long-term deflections in reinforced concrete members and compare these predicted values with the CSA A23.3 prescribed allowable values (Table 9.3). Two deflection calculation procedures are used to estimate deflections in reinforced concrete flexural members:

- the CSA A23.3 approximate procedure
- the computer-aided iterative procedure

These procedures are based on the assumption of the elastic behaviour of a flexural member, however the effect of cracking is taken into account.

The approximate procedure determines elastic deflections in reinforced concrete members using the effective moment of inertia (I_e) prescribed by CSA A23.3 Cl.9.8.2.3. Deflections can be easily determined by hand calculations.

The computer-aided iterative procedure uses computer structural analysis software to determine the elastic deflections in reinforced concrete members. The member is subdivided into several segments that can be characterized by different section properties; cracked section properties are used where the bending moment exceeds the cracking moment, and gross section properties are used elsewhere along the span.

4.6.2 Deflections in Simply Supported Flexural Members

As discussed in the previous section, the CSA A23.3 approximate procedure is suitable for deflection calculations in regular reinforced concrete structures. Deflections can be calculated by following the standard procedure for elastic deflection calculations in flexural members which uses the effective moment of inertia (I_e).

In simply supported flexural members, the I_e value used for deflection calculations is determined at the location of maximum bending moment.

Checklist 4.1 outlines the steps used in the CSA A23.3 approximate deflection calculation procedure. An application of the approximate procedure is illustrated by Example 4.4.

Checklist 4.1 Deflection Calculations According to the CSA A23.3 Approximate Procedure

Step	Description	Code Clause
1	Calculate the gross moment of inertia (I_g). For a rectangular section, $$I_g = \frac{bh^3}{12} \qquad \text{[4.2]}$$	
2	Calculate the moment of inertia of the cracked section (I_{cr}). For a rectangular section with tension steel only, $$I_{cr} = \frac{b\bar{y}^3}{3} + nA_s(d - \bar{y})^2 \qquad \text{[4.10]}$$	
3	Calculate the effective moment of inertia (I_e). $$\boxed{\text{A23.3 Eq. 9.1}} \qquad I_e = I_{cr} + (I_g - I_{cr})\left(\frac{M_{cr}}{M_a}\right)^3 \qquad \text{[4.11]}$$ where $$I_{cr} < I_e < I_g$$ For continuous members, also use Eqns 4.18 and 4.19 (Cl.9.8.2.4) (see Section 4.6.3).	9.8.2.3
4	Calculate the maximum immediate deflection (Δ) due to the total load. For beams subjected to uniformly distributed loads, $$\Delta = k\left(\frac{5}{48}\right)\frac{Ml^2}{E_c I_e} \qquad \text{[4.12]}$$ where $$\Delta = \Delta_D + \Delta_L$$ Refer to Figure 4.10 to select an appropriate k value. *An important note: always use specified (unfactored) loads and corresponding bending moments for deflection calculations!*	
5	Calculate the long-term deflections. $$\Delta_t = \zeta_s(\Delta_D + \Delta_{LS}) + (\Delta_L - \Delta_{LS}) \qquad \text{[4.16]}$$ and $$\boxed{\text{A23.3 Eq. 9.5}} \qquad \zeta_s = 1 + \frac{s}{1 + 50\rho'} \qquad \text{[4.15]}$$ Refer to Table 4.1 for the s values.	9.8.2.5
6	Check whether the deflections are within the limits prescribed by CSA A23.3.	9.8.5.3 (Table 9.3)

Example 4.4

A simply supported reinforced concrete beam of rectangular cross-section is shown in the figure below. The beam supports a uniform dead load (DL) of 16 kN/m and a uniform live load (LL) of 10 kN/m. The beam is reinforced with 4-20M bars at the bottom. Use an effective depth of 430 mm. The beam dimensions and material properties are given below.

Estimate the maximum immediate deflection for this beam due to the total service loads using the CSA A23.3 approximate procedure.

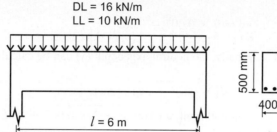

Given: $f_c' = 25$ MPa
$f_y = 400$ MPa
$b = 400$ mm
$h = 500$ mm

SOLUTION:
1. **Compute the moment of inertia of the gross section (I_g)**

$$I_g = \frac{bh^3}{12}$$ [4.2]

$$= \frac{(400 \text{ mm})(500 \text{ mm})^3}{12} = 4.17 \times 10^9 \text{ mm}^4 \cong 4.2 \times 10^9 \text{ mm}^4$$

Note that the moment of inertia for the gross section (I_g) will be used in this example — there is no need to calculate the moment of inertia for the transformed section ($I_{g,tr}$) (as discussed in Section 4.3.4, Example 4.2).

2. **Compute the moment of inertia of the cracked section (I_{cr})**
   a) Compute the modular ratio (n).

   $$E_s = 200\,000 \text{ MPa}$$

A23.3 Eq. 8.2
   $$E_c = 4500\sqrt{f_c'}$$ [2.2]

   $$= 4500\sqrt{25 \text{ MPa}} = 22\,500 \text{ MPa}$$

   $$n = \frac{E_s}{E_c}$$ [4.7]

   $$= \frac{200\,000 \text{ MPa}}{22\,500 \text{ MPa}} = 8.89 \cong 9$$

   b) Compute the neutral axis depth (y).
   The area of a 20M bar is equal to 300 mm² (refer to Table A.1), so

   $$A_s = 4 \times 300 \text{ mm}^2 = 1200 \text{ mm}^2$$

Subsequently, the reinforcement ratio (ρ) can be calculated as

$$\rho = \frac{A_s}{bd}$$ **[3.1]**

$$= \frac{1200 \text{ mm}^2}{400 \text{ mm} \times 430 \text{ mm}} = 0.007$$

and

$$n\rho = 9 \times 0.007 = 0.063$$

Finally, the neutral axis depth ($\bar{y}$) can be calculated as

$$\bar{y} = d\left(\sqrt{(n\rho)^2 + 2n\rho} - n\rho\right)$$ **[4.9]**

$$= 430 \text{ mm}\left(\sqrt{(0.063)^2 + 2 \times 0.063} - 0.063\right) = 128 \text{ mm}$$

The transformed cracked section is shown in the sketch below.

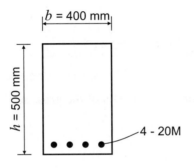

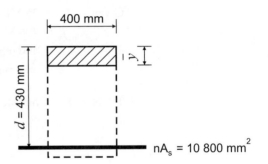

Actual cross section   Transformed cracked section

c)   Compute I_{cr}.

$$I_{cr} = \frac{b\bar{y}^3}{3} + nA_s(d - \bar{y})^2$$ **[4.10]**

$$= \frac{400 \text{ mm} (128 \text{ mm})^3}{3} + 9(1200 \text{ mm}^2)(430 \text{ mm} - 128 \text{ mm})^2$$

$$= 1.3 \times 10^9 \text{ mm}^4$$

3.   **Compute the effective moment of inertia (I_e)**
   a)   Calculate the cracking moment (M_{cr}).
        The cracking moment for this section was calculated in Example 4.1 as

$$M_{cr} = 25 \text{ kN} \cdot \text{m}$$

   b)   Calculate the maximum bending moment due to the service load (M_a).
        M_a is to be determined due to the total specified (service) load (w_s) as follows:

$$w_s = \text{DL} + \text{LL}$$

$$= 16 \text{ kN/m} + 10 \text{ kN/m} = 26 \text{ kN/m}$$

For a simply supported beam subjected to a uniform load, the maximum bending moment develops at the midspan; therefore,

$$M_a = \frac{w_s l^2}{8}$$

$$= \frac{26 \text{ kN/m} \times (6 \text{ m})^2}{8} = 117 \text{ kN·m}$$

c) Calculate I_e.

A23.3 Eq. 9.1

$$I_e = I_{cr} + (I_g - I_{cr})\left(\frac{M_{cr}}{M_a}\right)^3 \qquad\qquad\qquad [4.11]$$

$$= 1.3 \times 10^9 \text{ mm}^4 + (4.2 \times 10^9 \text{ mm}^4 - 1.3 \times 10^9 \text{ mm}^4)\left(\frac{25 \text{ kN·m}}{117 \text{ kN·m}}\right)^3$$

$$= 1.33 \times 10^9 \text{ mm}^4$$

4. **Calculate the maximum immediate deflection (Δ)**

The deflection will be calculated according to A23.3 Cl.9.8.2.3 using the equation presented in Section 4.4.2 with $k = 1.0$ for simply supported beams:

$$\Delta = k\left(\frac{5}{48}\right)\frac{M_a l^2}{E_c I_e} \qquad\qquad\qquad [4.12]$$

$$= (1.0)\left(\frac{5}{48}\right)\frac{(117 \times 10^6 \text{ N·mm})(6000 \text{ mm})^2}{(22\ 500 \text{ MPa})(1.33 \times 10^9 \text{ mm}^4)} \cong 15 \text{ mm}$$

Subsequently, the same beam was analyzed using the computer-aided iterative procedure (see Figure 4.13a). The beam was subdivided into 20 segments and analyzed by means of computer structural analysis software. The analysis procedure is described below:

1. Initially, the section properties for each segment were considered to be uncracked; that is, I_g has been used (see Figure 4.13b).
2. The beam model subjected to the total service load (w_s) was analyzed. The bending moment diagram was obtained as the output (see Figure 4.13c).
3. Subsequently, the beam was divided into uncracked and cracked portion (see Figure 4.13d). Segments in the region where the bending moment exceeds the cracking moment (M_{cr}) (based on the bending moment diagram) are considered to be cracked, and the cracked moment of inertia (I_{cr}) is used instead of the previously used gross moment of inertia (I_g). The properties of other segments remained the same as in the initial analysis; that is, they were uncracked.
4. Finally, the beam was analyzed again, and a midspan deflection of 16 mm was obtained as the output, as shown in Figure 4.13e.

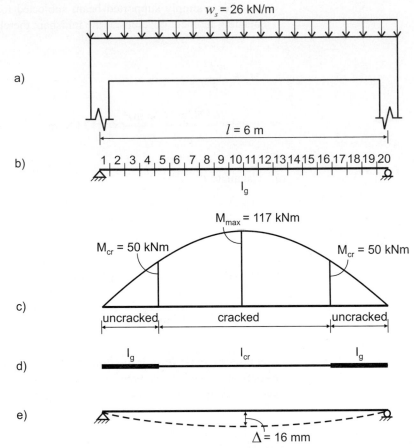

Figure 4.13 Deflections calculated using the computer-aided procedure: a) actual beam; b) structural model; c) bending moment diagram; d) moments of inertia for cracked and uncracked regions; e) deflection diagram.

Learning from Examples

The simply supported beam discussed in Example 4.4 was analyzed using a simple *approximate procedure* prescribed by CSA A23.3 with a maximum deflection of 15 mm.

Subsequently, the same beam was analyzed using the more comprehensive *computer-aided procedure,* resulting in a maximum deflection of 16 mm.

The difference in the magnitudes of maximum deflections obtained using these two procedures is only 1 mm. For the 6 m span beam discussed in this example the difference is insignificant, so using the approximate procedure is considered suitable for this application.

4.6.3   Deflections in Continuous Flexural Members

Most cast-in-place reinforced concrete beams and slabs are designed and constructed as continuous structures that span across several supports. This is a unique feature of reinforced concrete — in general, multi-span structural members of other materials (such as timber and steel) are designed as a series of simply supported structures spanning across two adjacent supports. When subjected to gravity loads, continuous structures develop positive and negative moments; usually, positive and negative bending moments develop at the midspan and the support regions, respectively. This behaviour adds to the complexity of deflection calculations since both the cracked and uncracked section properties need to be considered in the regions of positive and negative bending moments.

Deflection magnitudes in continuous flexural members are influenced by

- the relative magnitude of positive and negative bending moments within each span;
- the extent of the cracked regions, where positive and negative bending moments exceed the cracking moment;

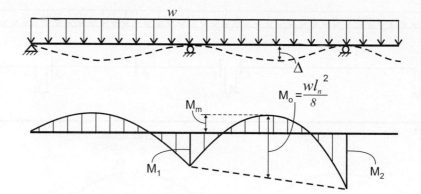

Figure 4.14 Parameters for deflection calculations in continuous beams and slabs.

- the presence of nonprismatic sections in cracked reinforced concrete members;
- the effect of cracked stiffness within one span on the effective cracked stiffnesses within adjacent spans of a continuous member.

Detailed deflection calculations in continuous members require the use of an iterative analysis procedure that simulates behaviour of cracked reinforced concrete members. However, CSA A23.3 permits the use of an *approximate procedure* similar to the one used for simply supported members. The conditions under which the approximate procedure can be used are summarized in Section 4.6.1.

Similarly to the case of simply supported members, the deflections can be calculated according to the standard procedure for elastic deflection calculations in flexural members by using the effective moment of inertia. The deflections can be calculated based on the following equation

$$\Delta = k\left(\frac{5}{48}\right)\frac{Ml^2}{E_c I_e} \qquad [4.12]$$

where M is equal to the maximum positive moment at midspan (M_m), and k is equal to

$$k = 1.2 - 0.2\frac{M_o}{M_m} \qquad [4.17]$$

where

$M_m = M_o - (M_1 + M_2)/2$ (the net moment at the midspan)

$M_o = \dfrac{wl_n^2}{8}$ the maximum bending moment for a simply supported beam based on a clear span (l_n).

To account for cracking in the positive and negative moment regions, CSA A23.3 Cl.9.8.2.4 prescribes the use of a weighted average effective moment of inertia ($I_{e,avg}$) for each span. The value of $I_{e,avg}$ for each span should be determined based on the following equations:

- For spans with two continuous ends (interior spans),

A23.3 Eq. 9.3

$$I_{e,avg} = 0.7\,I_{em} + 0.15\,(I_{e1} + I_{e2}) \qquad [4.18]$$

- For spans with one continuous end (end spans),

A23.3 Eq. 9.4

$$I_{e,avg} = 0.85\,I_{em} + 0.15\,I_{ec} \qquad [4.19]$$

where

I_{em} = the effective moment of inertia at the midspan

I_{e1} = the value of I_e at one end of an interior beam span

I_{e2} = the value of I_e at the other end of an interior beam span

I_{ec} = the value of I_e at the continuous end span.

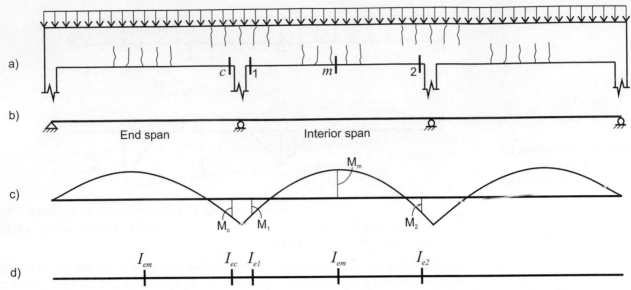

Figure 4.15 Effective moment of inertia in continuous flexural members: a) actual beam; b) structural model; c) bending moment diagram; d) moment of inertia values.

It should be noted that I_{em}, I_{e1}, and I_{e2} should be determined from Eqn 4.11 using the appropriate cross-sectional properties and bending moments, as illustrated in Figure 4.15.

The use of $I_{e,avg}$ simplifies the deflection calculations in continuous members while offering a reasonable approximation for the magnitude of deflections.

Example 4.5 illustrates the application of the CSA A23.3 approximate procedure for estimating deflections in continuous beams (refer to Checklist 4.1 for an outline of the procedure). Subsequently, the same beam is analyzed using the computer-aided iterative procedure and the results are compared.

It should be noted that the deflection calculation procedure for continuous reinforced concrete slabs is the same as the procedure for beam deflections illustrated by Example 4.5. Therefore, a separate example illustrating slab deflection calculations is not provided in this chapter.

Example 4.5

A four-span continuous (statically indeterminate) reinforced concrete beam is shown in the figure that follows. The beam supports a uniform dead load (DL) of 15 kN/m and a uniform live load (LL) of 15 kN/m. The beam is reinforced with 4-20M bars. The top reinforcement is continuous over the supports, whereas the bottom reinforcement is continuous in the midspan region. Use an effective depth of 340 mm. The beam dimensions and material properties are given below.

First, estimate the maximum immediate deflection for this beam due to the total service loads using the CSA A23.3 approximate procedure.

Then, estimate the maximum long-term deflection for this beam after 6 years of service life according to the CSA A23.3 procedure. Consider that only the dead load remains sustained for the purpose of long-term deflection calculations.

Given: $f_c' = 25$ MPa (normal-density concrete)
$f_y = 400$ MPa
$b = 600$ mm
$h = 400$ mm
$d = 340$ mm

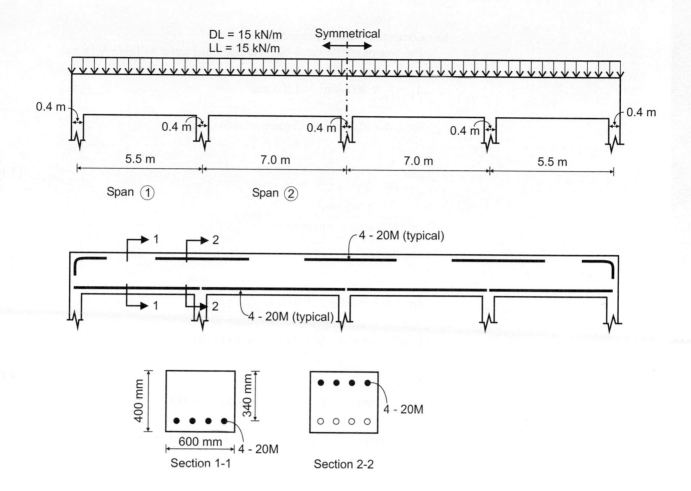

SOLUTION: In this example, Steps 1 to 4 focus on the calculations of immediate deflections, whereas Step 5 focuses on long-term deflections.

1. **Compute the moment of inertia of the gross section (I_g)**

$$I_g = \frac{bh^3}{12} \qquad\qquad \textbf{[4.2]}$$

$$= \frac{600 \text{ mm} \times (400 \text{ mm})^3}{12} = 3.2 \times 10^9 \text{ mm}^4$$

Note that the moment of inertia for the gross section (I_g) will be used in this example — there is no need to calculate the moment of inertia for the transformed section ($I_{g,tr}$) (as discussed in Section 4.3.4, Example 4.2).

2. **Compute the moment of inertia of the cracked section (I_{cr})**
   a) Compute the modular ratio (n).

$$E_s = 200\,000 \text{ MPa}$$

A23.3 Eq. 8.2

$$\bar{E}_c = 4500 \sqrt{f_c'} \qquad\qquad \textbf{[2.2]}$$

$$= 4500 \sqrt{25 \text{ MPa}} = 22\,500 \text{ MPa}$$

$$n = \frac{E_s}{E_c} \qquad\qquad \textbf{[4.7]}$$

$$= \frac{200\,000 \text{ MPa}}{22\,500 \text{ MPa}} = 8.89 \cong 9$$

b) Compute the neutral axis depth ($\bar{y}$).
The area of a 20M bar is equal to 300 mm^2 (refer to Table A.1), so

$$A_s = 4 \times 300 \text{ mm}^2 = 1200 \text{ mm}^2$$

Next, calculate the reinforcement ratio (ρ) as

$$\rho = \frac{A_s}{bd} \qquad\qquad\qquad\qquad \textbf{[3.1]}$$

$$= \frac{1200 \text{ mm}^2}{600 \text{ mm} \times 340 \text{ mm}} = 0.0059$$

and

$$n\rho = 9 \times 0.0059 = 0.053$$

Finally, calculate the neutral axis depth ($\bar{y}$) as

$$\bar{y} = d\left(\sqrt{(n\rho)^2 + 2n\rho} - n\rho \right) \qquad\qquad \textbf{[4.9]}$$

$$= 340 \text{ mm} \left(\sqrt{0.053^2 + 2 \times 0.053} - 0.053 \right) = 94 \text{ mm}$$

The transformed cracked section is shown on the sketch below.

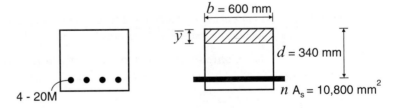

Actual cross section   Transformed cracked section

c) Compute I_{cr}:

$$I_{cr} = \frac{b\bar{y}^3}{3} + nA_s(d - \bar{y})^2 \qquad\qquad \textbf{[4.10]}$$

$$= \frac{600 \text{ mm} (94 \text{ mm})^3}{3} + 9(1200 \text{ mm}^2)(340 \text{ mm} - 94 \text{ mm})^2$$

$$= 820 \times 10^6 \; mm^4$$

It should be noted that, in this example, the same I_{cr} value can be used for positive and negative moment regions. This is due to the fact that the same reinforcement (4-20M bars) is used at the top and bottom. However, this might not be true in a general case; that is, different top and bottom reinforcement can be used in the design of continuous flexural members.

3. Compute the effective moment of inertia (I_e)

a) Calculate the cracking moment (M_{cr}).

i) Determine f_r^*:

A23.3 Eq. 8.3

$$f_r^* = 0.3\lambda \sqrt{f_c'}$$ [2.1]

$$= 0.3(1)\sqrt{25 \text{ MPa}} = 1.5 \text{ MPa}$$

ii) Determine y_t:

$$y_t = \frac{h}{2} = \frac{400 \text{ mm}}{2} = 200 \text{ mm}$$

iii) Calculate M_{cr}:

$$M_{cr} = \frac{f_r I_g}{y_t}$$ [4.1]

$$= \frac{(1.5 \text{ MPa}) \times (3.2 \times 10^9 \text{ mm}^4)}{200 \text{ mm}} = 24.0 \text{ kN} \cdot \text{m}$$

b) Determine the maximum bending moment (M_a) due to the total specified (service) load (w_s):

$$w_s = DL + LL$$

$$= 10 \text{ kN/m} + 20 \text{ kN/m} = 30 \text{ kN/m}$$

The maximum bending moment (M_a) values at the critical sections have been obtained as a result of a linear elastic analysis. The analysis can be performed using a computer structural analysis software (as was done in this example). Alternatively, the designer can use the appropriate beam diagram from Appendix A.

The following assumptions have been made for the analysis:

1. Each span was divided into 20 segments with $I_g = 3.2 \times 10^9 \text{ mm}^4$ and $E_c \cong 22\,500 \text{ MPa}$.
2. The supports are assumed to be pinned and do not influence the moment distribution in the beam.
3. The beam shows elastic behaviour; that is, the load-deflection relationship is linear.

The bending moment diagram that follows is obtained as the result of elastic analysis. The diagram illustrates the moment distribution within each span as well as the variation in bending moment distribution within different spans.

It should be noted that this beam is symmetric with regard to the vertical axis and is subjected to a uniform load throughout its length. Therefore, a consideration of only two spans (1 and 2) is justified

c) Calculate the effective moment of inertia according to A23.3 Cl.9.8.2.3 and 9.8.2.4.

A23.3 Eq. 9.1

$$I_e = I_{cr} + (I_g - I_{cr})\left(\frac{M_{cr}}{M_a}\right)^3$$ [4.11]

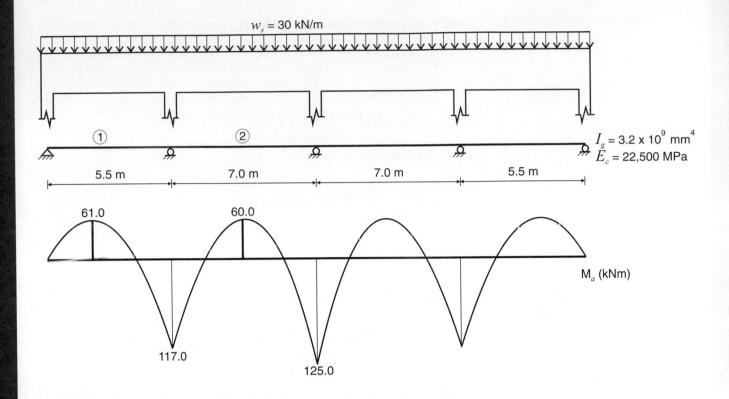

Span 1

i) The effective moment of inertia at the midspan ($M_a = 61$ kN · m) is

$$I_{em} = \left(0.82 + (3.2 - 0.82)\left(\frac{24}{61}\right)^3\right) \times 10^9 \text{ mm}^4 = 0.97 \times 10^9 \text{ mm}^4$$

ii) The effective moment of inertia at the support ($M_a = 117$ kN · m) is

$$I_{ec} = \left(0.82 + (3.2 - 0.82)\left(\frac{24}{117}\right)^3\right) \times 10^9 \text{ mm}^4 = 0.84 \times 10^9 \text{ mm}^4$$

iii) The average effective moment of inertia for the span with one continuous end (A23.3 Cl.9.8.2.4) is

A23.3 Eq. 9.4

$$I_{e,avg} = 0.85I_{em} + 0.15_{ec}$$ **[4.19]**

$$= (0.85 \times 0.97 + 0.15 \times 0.84) \times 10^9 \text{ mm}^4 = 0.95 \times 10^9 \text{ mm}^4$$

Span 2

i) The effective moment of inertia at the midspan ($M_a = 60$ kN · m) is

$$I_{em} = \left(0.82 + (3.2 - 0.82)\left(\frac{24}{60}\right)^3\right) \times 10^9 \text{ mm}^4 = 0.97 \times 10^9 \text{ mm}^4$$

ii) The effective moment of inertia at the left support ($M_a = 117$ kN · m) is

$$I_{e1} = \left(0.82 + (3.2 - 0.82)\left(\frac{24}{117}\right)^3\right) \times 10^9 \text{ mm}^4 = 0.84 \times 10^9 \text{ mm}^4$$

iii) The effective moment of inertia at the right support ($M_a = 125$ kN · m) is

$$I_{e2} = \left(0.82 + (3.2 - 0.82)\left(\frac{24}{125}\right)^3\right) \times 10^9 \text{ mm}^4 = 0.84 \times 10^9 \text{ mm}^4$$

iv) The average effective moment of inertia for a span with two continuous ends (A23.3 Cl.9.8.2.4) is

A23.3 Eq. 9.3

$$I_{e,avg} = 0.7I_{em} + 0.15(I_{e1} + I_{e2})$$ [4.18]

$$= (0.7 \times 0.97 + 0.15(0.84 + 0.84)) \times 10^9 \text{ mm}^4 = 0.93 \times 10^9 \text{ mm}^4$$

4. **Calculate the maximum immediate deflection due to the total service load (Δ_{D+L})**

The deflections will be calculated using the general equation 4.12 and the average effective moment of inertia ($I_{e,avg}$) values due to the total service load (w_s):

$$\Delta = k\left(\frac{5}{48}\right)\frac{M_m l^2}{E_c I_e}$$ [4.12]

where

$$k = 1.2 - 0.2\frac{M_o}{M_m}$$ [4.17]

M_m = the net moment at the midspan obtained from the elastic analysis performed in Step 3

M_o = the maximum bending moment for a simply supported beam subjected to the total service load w_s based on a clear span l_n

$I_e = I_{e,avg}$

Span 1

$$l_n = 5.5 \text{ m} - 0.4 \text{ m} = 5.1 \text{ m}$$

$$M_o = \frac{w_s l_n^2}{8} = \frac{30 \text{ kN/m} \times (5.1 \text{ m})^2}{8} = 97.5 \text{ kN} \cdot \text{m}$$

$M_m = 61 \text{ kN} \cdot \text{m}$ (see bending moment diagram developed in Step 3)

$$k = 1.2 - 0.2 \times \frac{97.5 \text{ kN} \cdot \text{m}}{61 \text{ kN} \cdot \text{m}} = 0.88$$

$$\Delta_{D+L} = 0.88\left(\frac{5}{48}\right)\frac{61 \times 10^6 \text{ N} \cdot \text{mm} \times (5100 \text{ mm})^2}{22\,500 \text{ MPa} \times (0.93 \times 10^9 \text{ mm}^4)} = 7 \text{ mm}$$

Span 2

$$l_n = 7 \text{ m} - 0.4 \text{ m} = 6.6 \text{ m}$$

$$M_a = \frac{30 \text{ kN/m} \times (6.6 \text{ m})^2}{8} = 163.3 \text{ kN} \cdot \text{m}$$

$M_m = 60 \text{ kN} \cdot \text{m}$ (see bending moment diagram developed in Step 3)

$$k = 1.2 - 0.2 \times \frac{163.3 \text{ kN} \cdot \text{m}}{60 \text{ kN} \cdot \text{m}} = 0.66$$

$$\Delta_{D+L} = 0.66\left(\frac{5}{48}\right)\frac{60 \times 10^6 \text{ N} \cdot \text{mm} \times (6600 \text{ mm})^2}{22\,500 \text{ MPa} \times (0.93 \times 10^9 \text{ mm}^4)} = 8.6 \text{ mm} \cong 9 \text{ mm}$$

It can be concluded that the maximum immediate deflections in the beam are on the order of 3 mm to 5 mm (see the sketch below).

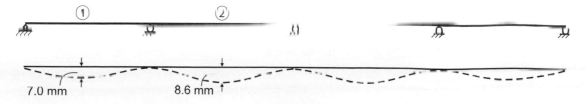

7.0 mm   8.6 mm

5. **Calculate the long-term deflections**
   a) Determine the maximum bending moment distribution due to the service dead load. In this example, the live load is treated as a transient load that does not cause any long-term deflections. Therefore, only the dead load will be considered as the sustained load influencing long-term deflections.

   The bending moment distribution due to the uniform dead load (DL = 15 kN/m) was determined using the elastic analysis. The bending moment diagram for the beam is shown below. The bending moment (M_a) values will be used for dead the load deflection calculations at various sections.

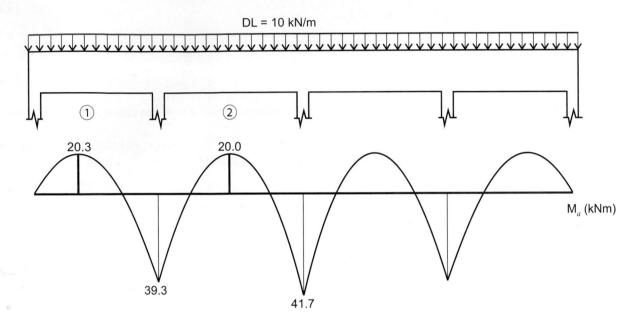

   b) Calculate the maximum immediate deflection due to dead and live load. The deflections will be calculated using the following general equation:

$$\Delta = k\left(\frac{5}{48}\right)\frac{M_m l^2}{E_c I_e}$$ [4.12]

   where

$$k = 1.2 - 0.2\frac{M_a}{M_m}$$ [4.17]

   M_m = the net moment at the midspan obtained from the elastic analysis in Step 3
   M_o = the maximum bending moment for a simply supported beam subjected to the service dead load (*DL*) based on a clear span l_n.

   Based on the bending moment diagram shown above, it can be concluded that the entire beam remains uncracked in both the midspan and support regions since the bending moments are less than M_{cr} of 24 kN·m. Hence, the gross moment of inertia can be used in the deflection calculation instead of the effective moment of inertia; that is,

$$I_e = I_g = 3.2 \times 10^9 \text{ mm}^4$$

Span 1

$$l_n = 5.5 \text{ m} - 0.4 \text{ m} = 5.1 \text{ m}$$

$$M_o = \frac{\text{DL} \times l_n^2}{8} = \frac{15 \text{ kN/m} \times (5.1 \text{ m})^2}{8} = 49 \text{ kN} \cdot \text{m}$$

$M_m = 30.5 \text{ kN} \cdot \text{m}$ (see bending moment diagram above)

$$k = 1.2 - 0.2 \times \frac{49 \text{ kN} \cdot \text{m}}{30.5 \text{ kN} \cdot \text{m}} = 0.88$$

$$\Delta_D = 0.88 \left(\frac{5}{48} \right) \frac{30.5 \times 10^6 \text{ N} \cdot \text{mm} \times (5100 \text{ mm})^2}{22\ 500 \text{ MPa} \times (3.2 \times 10^9 \text{ mm}^4)} = 1.0 \text{ mm}$$

Based on the immediate deflection calculations performed in Step 4, it follows that

$$\Delta_{D+L} = 7 \text{ mm}$$

Based on the procedure outlined in Section 4.4.2, the live load deflection can be determined as

$$\Delta_L = \Delta_{D+L} - \Delta_D \hspace{3cm} \textbf{[4.14]}$$
$$= 7 \text{ mm} - 1 \text{ mm} = 6 \text{ mm}$$

Span 2

$$l_n = 7 \text{ m} - 0.4 \text{ m} = 6.6 \text{ m}$$

$$M_o = \frac{15 \text{ kN/m} \times (6.6 \text{ m})^2}{8} = 81.7 \text{ kN} \cdot \text{m}$$

$M_m = 30 \text{ kN} \cdot \text{m}$ (see bending moment diagram above)

$$k = 1.2 - 0.2 \times \frac{49 \text{ kN} \cdot \text{m}}{30 \text{ kN} \cdot \text{m}} \cong 0.87$$

$$\Delta_D = 0.87 \left(\frac{5}{48} \right) \frac{(30 \times 10^6 \text{ N} \cdot \text{mm}) \times (6600 \text{ mm})^2}{22\ 500 \text{ MPa} \times 3.2 \times 10^9 \text{ mm}^4} = 1.7 \text{ mm}$$

Based on the immediate deflection calculations performed in Step 4, it follows that

$$\Delta_{D+L} = 8.6 \text{ mm}$$

Based on the procedure outlined in Section 4.4.2, the live load deflection can be determined as

$$\Delta_L = \Delta_{D+L} - \Delta_D \hspace{3cm} \textbf{[4.14]}$$
$$= 8.6 \text{ mm} - 1.7 \text{ mm} = 6.9 \text{ mm}$$

c) Calculate the long-term deflections.
  The long-term deflection calculations will be performed following the procedure outlined in Section 4.4.3. According to Cl.9.8.2.5, long-term deflections for the flexural member can be obtained by multiplying the immediate deflection caused by the sustained load by the factor ζ_s, given by

A23.3 Eq. 9.5 $\hspace{2cm} \zeta_s = 1 + \dfrac{s}{1 + 50\rho'} \hspace{4cm}$ **[4.17]**

$$= 1 + \frac{2}{1 + 50 \times 0} - 3$$

where

$s = 2$ for a load duration of 6 years (A23.3 Cl.9.8.2.5) (see Table 4.1)
$\rho' = 0$ because there is no compression reinforcement.

As a result, the total long-term deflection can be determined as

$$\Delta_t = \zeta_s(\Delta_D + \Delta_{LS}) + (\Delta_L - \Delta_{LS}) \qquad \textbf{[4.16]}$$

Because $\Delta_{LS} = 0$ (live load is not sustained in this example), the above equation can be restated as

$$\Delta_t = \zeta_s\Delta_D + \Delta_L$$

Span 1

$$\Delta_t = 3 \times 1.0 \text{ mm} + 6.0 \text{ mm} = 9.0 \text{ mm}$$

Span 2

$$\Delta_t = 3 \times 1.7 \text{ mm} + 6.9 \text{ mm} = 12.0 \text{ mm}$$

The beam deflections are summarized on the sketch below.

9 mm   12 mm

Subsequently, the same beam was analyzed using the computer-aided iterative procedure (see Figure 4.16a). The beam was subdivided into 20 segments and analyzed using a computer structural analysis program in order to more closely model the behaviour of the cracked beam. The results are summarized in Figure 4.16, and the analysis procedure is described below.

1. Initially, the section properties for each segment were assumed to be uncracked; that is, $I_g = 3.2 \times 10^9 \text{ mm}^4$ and $E_c \cong 22\,500 \text{ MPa}$ were used in the analysis (see Figure 4.16b).
2. The beam analysis was performed for a total (unfactored) service load (w_s) of 30 kN/m. The bending moment diagram was obtained as the output (see Figure 4.16c).
3. Subsequently, the beam was divided into uncracked and cracked portions. Segments in the regions of positive and negative bending moments where $M > M_{cr}$ were identified from the bending moment diagram. Those segments were considered to be cracked. Consequently, the gross moment of inertia (I_g) used in the previous analysis was substituted by the cracked moment of inertia (I_{cr}) along the cracked segments. The remaining segments were located in the uncracked portion of the beam, so the gross moment of inertia (I_g) was used (as in the previous analysis), as shown in Figure 4.16d.
4. Finally, the analysis was repeated using the modified section properties as described above. The resulting bending moment diagram is shown in Figure 4.16e.
5. The deflection diagram due to the total service load is shown in Figure 4.16f. The maximum midspan deflection of 11 mm at span 2 was obtained as a result of this analysis.

It should be noted that this is an iterative process. The analysis may need to be repeated a few more times until the moment distribution is consistent with the cross-sectional properties. For example, cracked section properties should be used only within the regions of positive and negative bending moments where $M > M_{cr}$, whereas the uncracked (gross) section properties need to be used in the remaining portion of the beam.

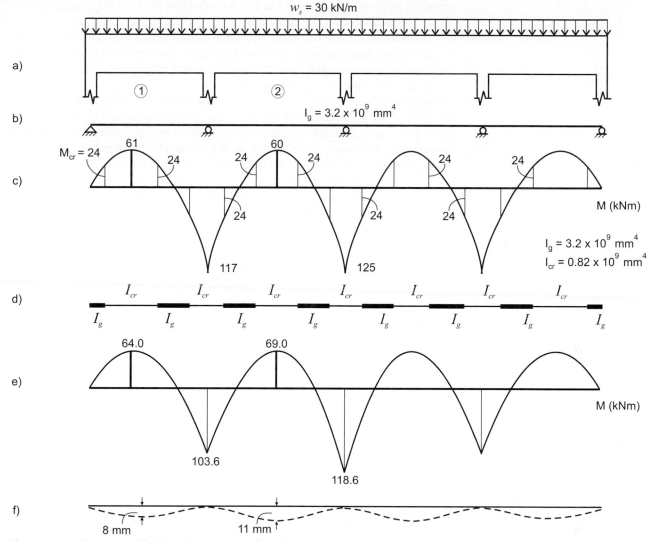

Figure 4.16 Computer-aided iterative analysis of the four-span continuous beam: a) actual beam; b) structural model; c) bending moment diagram for uncracked beam; d) moment of inertia distribution for cracked and uncracked regions; e) bending moment diagram for cracked beam; f) deflection diagram — immediate deflections due to the total service load.

Learning from Examples

1. The continuous beam discussed in Example 4.5 was analyzed using the CSA A23.3 approximate procedure. Maximum immediate deflections of 4 mm and 5 mm were obtained in spans 1 and 2, respectively (step 4). The beam considered in this example satisfies the criteria for the application of the CSA A23.3 approximate procedure.

    The computer-aided iterative analysis procedure was also performed to determine the maximum deflections. Maximum immediate deflections of 8 mm and 11 mm were obtained in spans 1 and 2, respectively (see Figure 4.16f). Although more tedious, this procedure should give more realistic estimates of the actual anticipated deflections. It should be noted that deflection values obtained using the iterative procedure are sensitive with regard to the selection of cracked and uncracked regions within a flexural member, and the computer software used to perform the analysis. The reader should be aware of this sensitivity, but not be overly concerned keeping in mind a broad variation in design and construction factors.

    The results obtained using these two procedures compare favourably, resulting in similar deflection values. Considering the low magnitude of the service loads resulting in small deflections, it can be concluded that the CSA A23.3 approximate procedure is suitable for this application.

      For more complex continuous structures such as beams with nonprismatic sections or where large variations in flexural stiffness exist between adjacent spans, the iterative analysis procedure should be used because it more accurately simulates the cracked section properties of each span.

2. Long-term deflections in the beam under consideration are significantly larger than the immediate deflections. For example, the maximum immediate deflection in span 1 is 7 mm, whereas the total long-term deflection is 9 mm; this reflects an increase in deflections by 50% due to long-term effects. Similarly, maximum immediate deflection in span 2 is 9 mm, whereas the long-term deflection is 12 mm; this shows that the deflections are increased by 60% due to long-term effects.

3. Another important observation can be made with regard to the variation in the moment of inertia values. For example, the moment of inertia of the cracked section amounts to only about 26% of the gross moment of inertia; that is,

$$\frac{I_{cr}}{I_g} = \frac{0.82 \times 10^9 \text{ mm}^4}{3.2 \times 10^9 \text{ mm}^4} = 0.26 = 26\%$$

The average effective moment of inertia $I_{e,avg}$ is less than the gross moment of inertia I_g. For span 1, it follows that

$$\frac{I_{e,avg}}{I_g} = \frac{2.2 \times 10^9 \text{ mm}^4}{3.2 \times 10^9 \text{ mm}^4} = 0.69 \cong 70\%$$

The similar ratio can be obtained for span 2:

$$\frac{I_{e,avg}}{I_g} = \frac{2.1 \times 10^9 \text{ mm}^4}{3.2 \times 10^9 \text{ mm}^4} = 0.66 \cong 66\%$$

Comparing these ratios, it is clear that the largest difference in values is in the gross and cracked moment of inertia values (around 74%), whereas the difference between the average effective and gross moments of inertia is on the order of 30% to 35%. This is because the beam under consideration did not experience significant cracking under the given load.

4. The beam discussed in this example has been lightly reinforced and relatively lightly loaded. Consequently, the ratio $I_{e,avg}/I_g$ is rather large (on the order of 66% to 70%) as compared to the I_{cr}/I_g ratio, which is on the order of 26%. However, heavily reinforced concrete beams subjected to higher loads are more extensively cracked, thereby resulting in the smaller $I_{e,avg}/I_g$ ratio. In the extreme case of an entirely cracked beam, the $I_{e,avg}/I_g$ ratio would approach the I_{cr}/I_g ratio.

Example 4.6

The beam in Example 4.5 is expected to support architectural components likely to be damaged by large deflections. These architectural components will be installed after the beam formwork has been removed.

Check whether the deflections in span 1 of the beam are within the limits prescribed by CSA A23.3.

SOLUTION: CSA A23.3 Cl.9.8.5.3 prescribes the allowable deflections for reinforced concrete members. According to the criteria for roof or floor construction that supports architectural components likely to be damaged by large deflections (see Table 9.3 of CSA A23.3), the maximum allowable deflections are $l_n/480$. The deflections considered in this example occur after the attachment of nonstructural elements.

From Example 4.5 we know that for span 1

$l_n = 5100$ mm

The corresponding allowable deflection (Δ_{all}) can be determined as

$$\Delta_{all} = \frac{l_n}{480} = \frac{5100 \text{ mm}}{480} = 10.6 \text{ mm}$$

Next, determine the magnitude of deflection that could develop after the formwork is released. Immediately after the release of the formwork, the structure would experience dead load deflection (under the assumption that no other significant load is expected).

According to the calculations performed in Example 4.5, the immediate dead load deflection is

$$\Delta_D = 1 \text{ mm}$$

and the total (immediate plus long-term) deflection due to dead load only is

$$\Delta_{tD} = \zeta_s \Delta_D = 3 \times 1 \text{ mm} = 3 \text{ mm}$$

where $\zeta_s = 3$ (see Step 5 of Example 4.5).

Therefore, the long-term deflection due to sustained dead load (Δ_{DS}) is

$$\Delta_{DS} = \zeta_s \Delta_D - \Delta_D = 3 \text{ mm} - 1 \text{ mm} = 2 \text{ mm}$$

The live-load deflection ($\Delta_L = 6.0$ mm) should be added at this point, thus resulting in the total long-term deflection after the installation of the architectural components:

$$\Delta = \Delta_{DS} + \Delta_L = 2 \text{ mm} + 6.0 \text{ mm} = 8.0 \text{ mm}$$

Since

$$\Delta = 8.0 \text{ mm} < \Delta_{all} = 10.6 \text{ mm}$$

it can be concluded that the deflections are well within the limits prescribed by CSA A23.3.

4.7 | CAUSES OF CRACKING IN REINFORCED CONCRETE STRUCTURES

Cracks are the most frequent and recognizable sign of concrete problems. Cracks develop in reinforced concrete members as a result of volume changes (creep, shrinkage, and temperature effects) and external loads. As discussed in Section 2.3.2, the concrete tensile strength is rather limited and can be easily exceeded under service loads, thus resulting in the development of cracks. Some cracks develop even before the concrete has hardened, while others develop at a later stage during the service life of a concrete structure. Minor cracks are part of normal concrete behaviour, whereas larger cracks indicate a problem that might adversely affect the structural performance. Cracks that do not increase in size and length over time are called *dormant cracks,* while cracks that open and close as a result of external loads are known as *active cracks.*

The main causes of cracking in concrete structures are

* plastic settlement
* shrinkage
* corrosion of steel reinforcement
* weathering
* structural distress
* poor construction practices

It is important to note that cracking caused by one factor could contribute to further deterioration and cause other kinds of cracks. For example, shrinkage cracks facilitate the corrosion of reinforcing steel, which could eventually rust and cause further cracking and spalling of concrete. Another example includes the corrosion cracks, which may cause loss of strength in a concrete member due to rebar delamination, eventually leading to the spalling of concrete.

The types of cracks commonly found in reinforced concrete structures will be discussed in this section from the perspective of their causes and patterns of appearance.

4.7.1　Plastic Settlement

Concrete has a tendency to continue to consolidate even after the initial placement, vibration, and finishing have been completed. During this period, the plastic concrete may be locally restrained by reinforcing steel, a previous pour, or formwork. This local restraint may result in voids and/or cracks adjacent to the restraining element. Plastic settlement cracks caused by steel reinforcement increase with increasing bar size and slump and decreasing cover. Similar cracks are caused by lateral shifting of the formwork during placement or by premature stripping of the forms. Formwork-related cracks are easy to spot, as they run parallel to the forms.

4.7.2　Shrinkage

Shrinkage is a decrease in the volume of concrete (other than that caused by externally applied forces and temperature changes) during and after hardening. In general, concrete tends to crack where its shrinkage is restrained. This phenomenon is discussed in more detail in Section 2.3.6. There are three main types of shrinkage cracks: plastic shrinkage cracks, drying shrinkage cracks, and restraint cracks.

Plastic shrinkage cracks　Plastic shrinkage occurs during the first few hours (usually between 30 minutes and 6 hours) after the fresh concrete has been placed in the forms and wind or heat causes the concrete to lose water rapidly. Plastic shrinkage usually occurs before the start of curing. Plastic shrinkage cracks are usually short and run in all directions. In some cases, plastic shrinkage causes wide and deep cracks, sometimes extending through the entire thickness of the concrete member (usually slab). The cracks can be spaced from a few centimetres to as much as a few metres apart (ACI Committee 224, 2004).

Drying shrinkage cracks　Drying shrinkage is a decrease in the volume of a concrete member caused by the loss of moisture from cement paste due to evaporation. It occurs after the concrete has already attained its final set and a good portion of the hydration process in the cement paste has been completed. Drying shrinkage cracks develop at random locations in concrete members, for example in slabs on grade with inadequate joint spacing. These cracks are usually very small (barely visible), with a width not exceeding 0.1 mm.

Restraint cracks　Cast-in-place beams and slabs may develop cracks when restrained against shrinkage. If a concrete floor is free to shrink without restraints, there will be no measurable structural consequences. However, columns and walls that are part of the structural system in a building usually prevent free shrinkage in the floors. As a result, cracks develop in localized areas where the restraining stresses exceed the concrete tensile resistance. These cracks usually run perpendicular to the length of the member and extend through its thickness; they are evenly spaced and are of uniform width. They usually form within 3 or 4 months after the concrete placement and remain dormant thereafter.

In addition to the horizontal restraints that might cause cracking in concrete structures, other permanent restraints (such as heavy masonry walls) may prevent the concrete floors from deflecting in the vertical direction. If such wall restraints are not properly detailed, they may cause cracks to develop at the top surface of the slab.

4.7.3 Corrosion

Corrosion of steel reinforcement is a common problem associated with reinforced concrete structures. A proper concrete cover provides corrosion protection to the steel reinforcement. Initially, when steel is exposed to moisture in fresh concrete, a tightly adhering iron oxide layer forms around the reinforcement bars. However, due to the variability of the thickness of the protective layer and mill scale and concrete imperfections and cracking, the same reinforcing steel rebars can have areas with different electric potentials. Eventually, these areas become electric (galvanic) cells that act as anodes and cathodes (as in an electric battery), and water contained in concrete pores and chemicals dissolved in the water act as the electrolyte. Corrosion develops when the concrete pores contain a sufficient amount of electrolyte, that is, moisture and chlorides, thereby establishing an electrochemical circuit. Chlorides are one of the most common corrosion triggers. De-icing salts used in cold climate conditions are the most common source of chlorides that affect concrete structures. Other sources of chlorides are unwashed beach sand (containing sea salt, which is essentially sodium chloride) used in concrete construction and saltwater spray affecting waterfront buildings. Concrete cover over the reinforcement acts as a barrier that delays the start of corrosion. Some reports indicate that the speed of corrosion can be increased by over 80% when the cover thickness is reduced by half. The thinner and more permeable the cover and the higher the temperature and the level of humidity, the earlier is the start of corrosion (Newman, 2001).

It should be noted that both water and oxygen are required to initiate corrosion. For example, one would expect that concrete piles placed in salted sea water would easily corrode; however, that is not the case. The immersed portions of the piles rarely corrode, although both water and chlorides (sea salt) are present, due to the lack of oxygen; however, piles do corrode in the splash zone due to their exposure to oxygen.

Corrosion-induced cracks develop due to iron oxides accumulated on the reinforcement bars. Iron oxide (known as rust) is the main product of corrosion. This accumulated layer of rust can occupy up to six times the volume of the original steel. As a result, the concrete is pushed apart and cracks develop. As the corrosion progresses, more and more rust accumulates and the cracks continue to grow. In reinforced concrete beams, these internal radial cracks form around the bars and appear on the concrete surface as longitudinal cracks parallel to the bars; these cracks form below the bars for tension steel and above the bars for compression steel (see Figure 4.17).

When the cracks become wide enough, a portion of the concrete beneath the bars spalls off, leaving the bars completely exposed. Once the bars lose their bond with the concrete due to spalling, the section behaves more like a plain concrete section. As a result of corrosion, a structural member might lose a significant portion of its strength and might even collapse. Corrosion-induced concrete spalling in a reinforced concrete bridge structure is shown in Figure 4.18.

Corrosion cracks may widen if water gets into the cracks and freezes. The expanded ice within the cracks causes them to widen and leads to further deterioration.

In slabs, spalling proceeds in a similar fashion. Deterioration starts with cracks along the rebars. As the rust products accumulate, two additional cracks form near each bar. At some point, the concrete spalls in two wedge-like pieces. If the bars are spaced close together, the concrete may spall as one whole sheet. The top bars in the beams are hidden within the slab and the corrosion may not be apparent until the slab is literally lifted off the beams due to the expanded rust products.

Figure 4.17 Corrosion cracks in a reinforced concrete beam: a) radial cracks formed around the bars; b) corrosion-induced spalling of concrete cover.

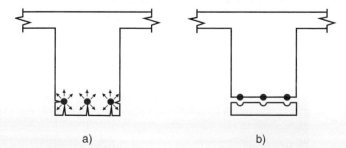

a)   b)

Figure 4.18 The concrete cover in this bridge girder in Ontario has spalled off as a result of the corrosion.

(Graham Finch)

4.7.4   Weathering

In cold climates, freezing and thawing are the most common source of weather-related cracking in concrete structures. Concrete may be damaged by freezing of water in the paste, the aggregate, or both. Concrete is best protected against freezing and thawing through the use of the lowest practical w/c ratio and total water content, durable aggregate, adequate air entrainment, and adequate curing. Weather-related cracking can also be caused by volume changes in concrete as a result of alternate wetting and drying and heating and cooling of concrete structures. Thermal expansion and contraction cause cracking when these movements are not accommodated by properly designed and spaced expansion and contraction joints (see Sections 12.12.4 and 13.8 for more details on joints in concrete structures). Typical examples of thermal cracks are cracks at re-entrant corners in doors and windows, and at dapped ends of precast beams. In general, properly anchored diagonal reinforcement should be provided at these locations to control the size and to limit the propagation of these cracks.

4.7.5   Structural Distress

External loads are a common source of cracking and distress in concrete structures. Cracks caused by external loads occur when tensile stresses in the concrete exceed its tensile resistance, that is, its modulus of rupture. Structural distress cracks are often caused by improper design and detailing of concrete members. In general, experienced designers should be able to identify the source of cracking. For example, diagonal cracks developed in concrete walls may be caused by foundation settlement. Flexural cracks generally run perpendicular to the tension reinforcement near the midspan at the bottom surface of beams and elevated slabs (these cracks are usually closely spaced); see Figure 4.19. In two-way slabs, the top surface above the column can show either the common flexural cracks radiating from the column or circumferential cracks indicating a dangerous punching shear failure.

4.7.6   Poor Construction Practices

Some contractors add water to concrete in order to improve the workability of the concrete mix during construction. The addition of water results in reduced strength, increased settlement, and increased ultimate drying shrinkage in concrete structures. Lack of curing also results in an increased degree of cracking within a concrete structure. Early termination of curing leads to increased shrinkage at the stage of low concrete strength. Other causes of cracking related to poor construction practices are inadequate form work supports and improper consolidation.

Figure 4.19 Flexural cracks in a simply supported reinforced concrete beam.

(Svetlana Brzev)

4.7.7 Crack Width

Cracks that develop in properly designed concrete members at service condition (also called *hairline cracks*) are very fine and barely visible. A hairline crack is typically defined as being less than 0.003 mm wide to 0.008 mm wide.

In general, cracks less than 0.1 mm wide can be noticed only when drying out and only in strong light. Cracks of width ranging from 0.1 mm to 0.5 mm are visible to the naked eye. Cracks of 0.5 mm width or larger are visible as a distinct interruption in the concrete surface. Crack width can be measured using various devices, the simplest being a crack gauge.

It is noteworthy that CSA A23.3 prescribes maximum crack widths of 0.33 mm and 0.4 mm for the exterior and interior exposures, respectively. These requirements apply to regular reinforced concrete structures; however, more stringent requirements may apply to special structures such as water reservoirs.

4.8 | CSA A23.3 CRACKING CONTROL REQUIREMENTS

Excessive cracking of concrete structures can lead to poor appearance, lack of serviceability, or even catastrophic failure. In most cases, cracking can be controlled by proper design and construction. Therefore, the main objective of the cracking control provisions is to ensure that the cracks in concrete members remain fine and well distributed and are of a limited width. Adherence to code requirements in combination with a common-sense approach to mitigation of restraint problems should result in a design that is both structurally sound and visually pleasing, without any unsightly cracks.

According to CSA A23.3, satisfactory crack control can be accomplished by ensuring that the crack control parameter z (related to crack width) is within acceptable limits and that skin reinforcement is provided where required, as outlined in this section.

| A23.3 Cl.10.6.1 |

Crack control parameter (z) Experimental studies related to cracking in reinforced concrete flexural members have shown that cracks develop in a random fashion and that the crack width is widely variable. The most critical factors that influence crack widths are the magnitude of tensile stress in steel, thickness of concrete cover, the number of bars in the tension zone, and the effective tension area of concrete surrounding the tension steel. The existing methods for crack width prediction are mainly based on the statistical analysis of experimental data. The crack control provisions of CSA A23.3 are based on the Gergely–Lutz equation for estimating the probable crack width. The equation was developed in 1973 based on the statistical analysis of a large sample of experimental test data. A23.3 Cl.10.6.1 introduces the crack control factor (z), which is proportional to the probable crack width. The z factor can be determined based on the equation

| A23.3 Eq. 10.6 | $z = f_s \sqrt[3]{d_c A}$ (N/mm)   [4.20]

where (note the units indicated in brackets)

d_c = the distance from the extreme tension fibre to the centre of the longitudinal bar located closest thereto (mm)

A = the effective tension area of concrete surrounding the flexural tension reinforcement (mm²)

f_s = the stress in steel at the maximum service (unfactored) load computed assuming an elastic stress distribution and cracked section properties. In lieu of detailed computations, A23.3 Cl.10.6.1 recommends that f_s be conservatively estimated as

$$f_s = 0.6 f_y \quad \text{(MPa).} \tag{4.21}$$

A typical beam cross-section is shown in Figure 4.20a. The total effective tension area (A_e) extends from the extreme tension fibre to the centroid of the flexural tension reinforcement

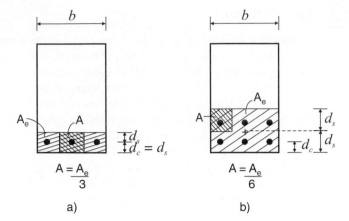

Figure 4.20 Effective tension area (A_e) for a reinforced concrete beam: a) one layer of reinforcement; b) two layers of reinforcement.

and an equal distance past that centroid. The distance from the extreme tension fibre to the centroid of the flexural reinforcement is denoted as d_s (note the difference between d_s and d_c values as shown in Figure 4.20b). The effective tension area per bar (A) for a beam with one layer of reinforcement (shown cross-hatched in Figure 4.20a) can be obtained by dividing the total effective tension area for all bars (A_e) (shown hatched) by the number of bars (N):

$$A = \frac{A_e}{N}$$ [4.22]

where the total effective tension area (A_e) is

$$A_e = b(2d_s)$$

It should be noted that, in the case of a beam with one layer of reinforcement, $d_s = d_c$. In the particular example of the beam in Figure 4.20a, there are three reinforcement bars ($N = 3$). However, beams typically contain more than one layer of reinforcement. For example, the beam in Figure 4.20b has two layers of reinforcement. In that case, the effective tension area (A) can still be determined using Eqn 4.22; however, note that in this case $d_s \neq d_c$ and $N = 6$.

When the tension steel reinforcement consists of two or more different bar sizes, N should be calculated as the total reinforcement area divided by the area of the largest bar size used.

In calculating d_c and A, the effective clear concrete cover need not be taken greater than 50 mm.

According to A23.3 Cl.10.6.1, the upper limit for z is

- 30 000 N/mm for interior exposure (corresponding to a crack width of 0.4 mm);
- 25 000 N/mm for exterior exposure (corresponding to a crack width of 0.33 mm).

Since the number of bars affects the magnitude of the effective tension area (A), it is obvious that a larger number of smaller diameter bars is preferred to ensure that the crack width and the magnitude of z are within the acceptable limits.

According to Cl.10.6.1, the z factor limits apply to normal conditions and may not be sufficient for structures subjected to very aggressive exposure or designed to be watertight.

It should be also noted that detailed f_s computations are usually required in the design of watertight structures (water reservoirs) or structures exposed to aggressive environments (chemical factories or offshore structures), where a high level of cracking control is essential. The stress in steel at the service level can be computed from the elastic analysis of the beam in Figure 4.21. The bending moment (M_a) develops in the beam due to service loads. Consider section A-A at the crack location. An elastic stress distribution is assumed for both concrete and steel. The neutral axis depth of the cracked section is denoted as $\bar{y}$ (as discussed in Section 4.3.4). To resist the effects of flexure, an internal force couple forms in the section, consisting of the compressive force in concrete (C_c)

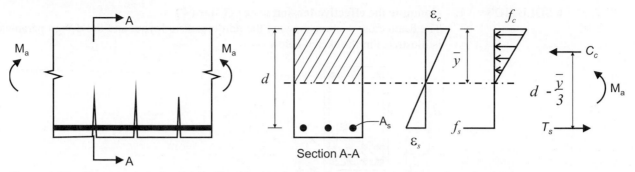

Figure 4.21   Stress and strain distribution in a cracked reinforced concrete beam at service load.

and the tensile force in steel (T_s) with the lever arm ($d - \bar{y}/3$). The force in tension steel can be expressed as

$$T_s = f_s A_s$$

Based on the equilibrium of the external and internal bending moments, it follows that

$$M_a = T_s\left(d - \frac{\bar{y}}{3}\right) = f_s A_s\left(d - \frac{\bar{y}}{3}\right)$$

The stress in steel (f_s) can be determined based on the above equation as

$$f_s = \frac{M_a}{A_s\left(d - \dfrac{\bar{y}}{3}\right)} \qquad [4.23]$$

The calculation of the z factor will be illustrated by Example 4.7.

Example 4.7

A typical cross-section of a reinforced concrete beam is shown in the figure below. The beam is subjected to the maximum bending moment of 250 kN · m at the service (unfactored) load level. Assume interior exposure. The clear cover to the stirrups is 30 mm, and a 35 mm clear bar spacing has been specified. The beam properties are summarized below.

Check whether this beam section satisfies the CSA A23.3 cracking control requirements.

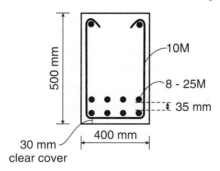

Given:

$f_u' = 25$ MPa (maximum aggregate size 20 mm)
$f_y = 400$ MPa
$E_s = 200\,000$ MPa
$b = 400$ mm
$h = 500$ mm

SOLUTION: 1. **Compute the effective tension area per bar (A)**

The beam cross-section showing the reinforcement layout and the design parameters is illustrated in the sketch below.

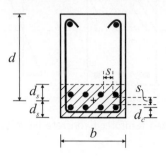

a) Calculate the distance d_c:

A 30 mm clear cover to the stirrups has been specified.
For 25M tension reinforcement, $d_b = 25$ mm (see Table A.1).
For 10M stirrups, $d_s = 10$ mm (see Table A.1).
Clear cover to the stirrups is 30 mm (given), therefore

$$d_c = \text{cover} + d_s + \frac{d_b}{2}$$

$$= 30 \text{ mm} + 10 \text{ mm} + \frac{25 \text{ mm}}{2} = 52.5 \text{ mm}$$

b) Find the distance d_s:

The clear spacing between the bar layers is $s = 35$ mm (given).

$$d_s = d_c + \frac{d_b}{2} + \frac{s}{2}$$

$$= 52.5 \text{ mm} + \frac{25 \text{ mm}}{2} + \frac{35 \text{ mm}}{2} = 82.5 \text{ mm}$$

c) Find the total effective tension area for all bars (A_e):

$$A_e = b(2d_s) = (400 \text{ mm})(2 \times 82.5 \text{ mm}) = 66\,000 \text{ mm}^2$$

d) Finally, calculate the effective tension area per bar (A):

There are eight bars in total, so $N = 8$:

$$A = \frac{A_e}{N} = \frac{66\,000 \text{ mm}^2}{8} = 8250 \text{ mm}^2$$

2. **Determine the stress in the steel reinforcement (f_s)**

The stress in the steel reinforcement can be computed in one of two ways:

a) An approximate estimate permitted by CSA A23.3 in lieu of detailed calculations (Cl.10.6.1) can be made:

$$f_s = 0.6f_y \qquad\qquad\qquad \textbf{[4.21]}$$

$$= 0.6 \times 400 \text{ MPa} = 240 \text{ MPa}$$

b) Alternatively, a more accurate f_s estimate can be made based on Eqn 4.23.
  i) The n value can be determined as follows:

A23.3 Eq. 8.2

$$E_c = 4500\sqrt{f_c'} \qquad\qquad\qquad \textbf{[2.2]}$$

$$= 4500\sqrt{25 \text{ MPa}} = 22\,500 \text{ MPa}$$

$$n = \frac{E_s}{E_c} = \frac{200\,000 \text{ MPa}}{22\,500 \text{ MPa}} = 8.9 \cong 9$$

ii) The ρ value can be determined as follows:

$$d = h - d_s = 500 \text{ mm} - 82.5 \text{ mm} = 417.5 \text{ mm}$$

$$A_s = 8 \times 500 \text{ mm}^2 = 4000 \text{ mm}^2$$

$$\rho = \frac{A_s}{bd} \qquad\qquad\qquad\qquad\qquad \textbf{[3.1]}$$

$$= \frac{4000 \text{ mm}^2}{400 \text{ mm} \times 417.5 \text{ mm}} = 0.024$$

iii) Finally, the $\bar{y}$ value can be determined as

$$\bar{y} = d\left(\sqrt{(n\rho)^2 + 2n\rho} - n\rho\right) \qquad\qquad \textbf{[4.9]}$$

$$= (416.5 \text{ mm})\left(\sqrt{(8.9 \times 0.024)^2 + 2 \times 8.9 \times 0.024} - 8.9 \times 0.024\right)$$

$$= 197 \text{ mm}$$

iv) The f_s value can be calculated as

$$f_s = \frac{M_a}{A_s\left(d - \dfrac{\bar{y}}{3}\right)} \qquad\qquad\qquad \textbf{[4.23]}$$

$$= \frac{250 \times 10^6 \text{ N·mm}}{(4000 \text{ mm}^2)\left(416.5 - \dfrac{197}{3}\right)} = 178.1 \text{ MPa} \cong 178 \text{ MPa}$$

It should be noted that the f_s value (178 MPa) obtained using the more accurate calculation procedure is smaller than the approximate value of 240 MPa, but the f_s value of 240 MPa will be used in further calculations.

3. **Determine the value of z factor**

A23.3 Eq. 10.6
$$z = f_s \sqrt[3]{d_c A} \qquad\qquad\qquad\qquad\qquad \textbf{[4.20]}$$

$$= 240 \text{ MPa} \sqrt[3]{(52.5 \text{ mm})(8250 \text{ mm}^2)} = 18159 \text{ N/mm}$$

Since

$$z = 18159 \text{ N/mm} < 30\,000 \text{ N/mm}$$

the CSA A23.3 cracking control requirements for interior exposure are satisfied.

4.8.1 Skin Reinforcement — Beams and Slabs

A23.3 Cl.10.6.2
In general, tension reinforcement should provide adequate tension control in beams. However, in deep beams with overall depth (h) exceeding 750 mm there is a chance for wider cracks to form on the side faces in the zone between the neutral axis and the main tension reinforcement (see Figure 4.22a). For that reason, CSA A23.3 Cl.10.6.2 prescribes the use of skin reinforcement. The area of skin reinforcement (A_{sk}) provided at each exposed face of a beam (as illustrated in Figure 4.22b) is

$$A_{sk} = \rho_{sk} A_{cs} \quad (\text{mm}^2) \qquad\qquad\qquad\qquad \textbf{[4.24]}$$

where

A_{cs} = the area of a concrete strip along an exposed face
ρ_{sk} = 0.008 for interior exposure and 0.01 for exterior exposure.

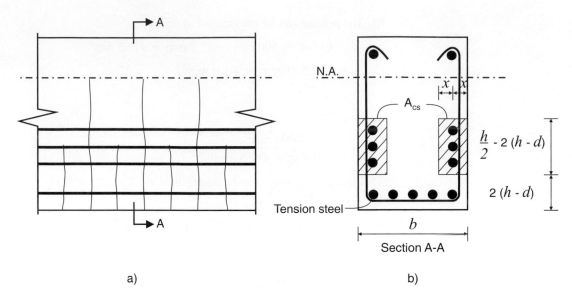

Figure 4.22   Skin reinforcement in a reinforced concrete deep beam: a) elevation showing cracks and skin reinforcement; b) beam cross-section.

The area of a concrete strip can be calculated as (refer to Figure 4.22b)

$$A_{cs} = [0.5h - 2(h - d)](2x)$$

where x is the distance from the side face to the centre of the skin reinforcement. Note that

$$x \leq \frac{b}{4}$$

The maximum spacing (s) of the skin reinforcement is given by

$$s \leq 200 \text{ mm}$$

The skin reinforcement concept can also be applied to slabs. To prevent the development of wide shrinkage and temperature cracks, A23.3 Cl.7.8.2 prescribes that skin reinforcement in excess of the minimum amount specified by A23.3 Cl.7.8.1 needs to be provided (the minimum slab reinforcement requirements will be discussed in Chapter 5). The area of skin reinforcement (A_{sk}) provided at an exposed slab face (as shown in Figure 4.23) can be determined from Eqn 4.24. In this case, A_{cs} denotes the area of a concrete strip along each exposed slab face. For the 1000 mm long slab strip in Figure 4.23, it follows that

$$A_{cs} = 1000(2x) \quad (\text{mm}^2)$$

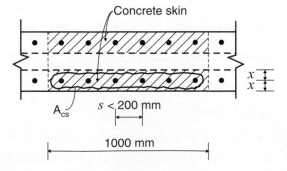

Figure 4.23   Skin reinforcement in a reinforced concrete slab.

KEY CONCEPTS

Cracks develop in reinforced concrete members as a result of volume changes (creep, shrinkage, and temperature effects) and external loads. The main causes of cracking in concrete structures are

- plastic settlement
- shrinkage
- corrosion of steel reinforcement
- weathering
- structural distress
- poor construction practices

The cracks developed in properly designed concrete members at service loads (also called *hairline cracks*) are very fine and barely visible. A hairline crack is typically defined as being less than 0.003 mm wide to 0.008 mm wide. CSA A23.3 prescribes maximum crack widths of 0.33 mm and 0.4 mm for exterior and interior exposures, respectively.

The main objective of the CSA A23.3 cracking control provisions is to ensure that the cracks in concrete members remain fine and well distributed and are of a limited width. According to CSA A23.3, satisfactory crack control can be accomplished by ensuring that the *crack control parameter z* (related to crack width) is within acceptable limits and that skin reinforcement is provided where required. The z factor can be determined based on the following equation (A23.3 Cl.10.6.1):

A23.3 Eq. 10.6

$$z = f_s \sqrt[3]{d_c A} \quad \text{(N/mm)} \qquad \qquad [4.20]$$

CSA A23.3 Cl.10.6.2 prescribes the use of *skin reinforcement* to prevent the development of wider cracks on the side faces of deep beams and the exposed slab faces.

SUMMARY AND REVIEW — SERVICEABILITY

Each building structure is expected to be able to support service loads without experiencing excessive cracking, deflections, or vibrations. Excessive cracking can lead to poor appearance, reduced durability and serviceability, and loss of strength that could in turn cause the failure of a reinforced concrete structure. Visible cracks and sagging of concrete members can lead to the perception that the entire building is unsafe. Serviceability limit states represent the limit for maintaining satisfactory structural performance under normal service load conditions.

Behaviour of reinforced concrete flexural members at service loads

Reinforced concrete flexural members crack when the tensile stresses reach a reduced modulus of rupture (f_r^*). The internal moment developed in the member at the onset of cracking is called the cracking moment (M_{cr}). The cracking moment can be obtained as

$$M_{cr} = \frac{f_r^* \cdot I_g}{y_t} \qquad \qquad [4.1]$$

There are two distinct regions within a reinforced concrete flexural member:

- the *cracked region*, where the bending moment exceeds the cracking moment; that is, $M \geq M_{cr}$;
- the *unmarked region*, where the bending moment is less than the cracking moment; that is, $M < M_{cr}$.

Moment of inertia properties for deflection calculations

The moment of inertia (I) is a key cross-sectional property used in deflection calculations. It depends on several factors, including the load magnitude, the shape and dimensions of a cross-section, the extent of cracking, and the amount of reinforcement. Three types of moments of inertia are used in deflection calculations:

- gross moment of inertia (I_g),
- moment of inertia of the cracked transformed section (I_{cr}),
- effective moment of inertia (I_e).

The *gross moment of inertia* is the moment of inertia of an uncracked concrete section. For a rectangular cross-section of width b and overall depth h, I_g can be determined as

$$I_g = \frac{bh^3}{12}$$ [4.2]

The *moment of inertia of a cracked section* (I_{cr}) with tension steel only can be determined based on the transformed section properties as

$$I_{cr} = \frac{b\bar{y}^3}{3} + nA_s(d - \bar{y})^2$$ [4.10]

To account for the variations in the moment of inertia values along the beam length, CSA A23.3 Cl.9.8.2.3 prescribes the use of the *effective moment of inertia* (I_e), which can be determined based on the equation

A23.3 Eq. 9.1 $$I_e = I_{cr} + (I_g - I_{cr})\left(\frac{M_{cr}}{M_a}\right)^3$$ [4.11]

Some portions of the member are cracked at service loads; however, the remaining part is uncracked. Therefore, the I_e value falls in between I_g and I_{cr}; that is,

$$I_{cr} < I_e < I_g$$

Immediate and long-term deflections

Deflections in reinforced concrete members are estimated based on the assumption of elastic behaviour at service loads, which is appropriate in most cases. Depending on the load duration, deflections can be classified as immediate and long-term.

Immediate (or *short-term*) *deflections* occur at once upon the load being applied. According to A23.3 Cl.9.8.2.2, immediate deflections can be computed by the usual methods or formulas for elastic deflections. To account for the effects of cracking and reinforcement on member stiffness, the effective moment of inertia (I_e) is used in deflection calculations.

For members subjected to uniformly distributed loads, the immediate deflection (Δ) can be determined based on the equation

$$\Delta = k\left(\frac{5}{48}\right)\frac{Ml^2}{E_c I_e}$$ [4.12]

Long-term deflections are mainly caused by creep and shrinkage. Creep causes an increase in deflections in concrete structures under constant sustained load for a period of several years. The total long-term deflection (Δ_t) in a reinforced concrete flexural member can be obtained when the immediate deflection (Δ_i) is multiplied by the factor ζ_s (A23.3 Cl.9.8.2.5):

$$\Delta_t = \zeta_s \Delta_i$$

where

A23.3 Eq. 9.5 $$\zeta_s = 1 + \frac{s}{1 + 50\rho'}$$ [4.15]

Deflection control according to CSA A23.3

CSA A23.3 Cl.9.8.1 prescribes that reinforced concrete flexural members have adequate stiffness to limit deflections or any deformations that might adversely affect the strength or serviceability of the structure.

Two approaches can be taken to ensure that deflections are within acceptable limits:

- indirect approach
- detailed deflection calculations

The *indirect approach* is based on A23.3 Cl.9.8.2.1 and consists of setting suitable upper limits on the member span/depth ratio (see Table A.3). This approach is simple and satisfactory for applications where spans, loads, load distributions, and member sizes and proportions are in the usual ranges.

Detailed deflection calculations are performed to estimate immediate and long-term deflections in reinforced concrete members and compare these predicted values with the CSA A23.3 prescribed allowable values (Table 9.3). Two calculation procedures are used to estimate the deflections in reinforced concrete flexural members:

- CSA A23.3 approximate procedure
- computer-aided iterative procedure

These procedures are based on the assumption of the elastic behaviour of a flexural member.

The *approximate procedure* determines elastic deflections in reinforced concrete members using the effective moment of inertia (I_e) prescribed by CSA A23.3 Cl.9.8.2.3. Deflections can be easily determined by hand calculations.

The *computer-aided iterative procedure* uses a computer structural analysis software to determine elastic deflections in reinforced concrete members. The member is subdivided into several segments that can be characterized by different section properties; cracked section properties are used where the bending moment exceeds the cracking moment, while gross section properties are used elsewhere along the span.

Causes of cracking in reinforced concrete flexural members and the CSA A23.3 cracking control requirements

Cracks develop in reinforced concrete members as a result of volume changes (creep, shrinkage, and temperature effects) and external loads. The main causes of cracking in concrete structures are

- plastic settlement
- shrinkage
- corrosion of steel reinforcement
- weathering
- structural distress
- poor construction practices

The cracks in properly designed concrete members at service loads (also called *hairline* cracks) are very fine and barely visible. A hairline crack width typically ranges from less than 0.003 mm to 0.008 mm. CSA A23.3 prescribes maximum crack widths of 0.33 and 0.4 mm for exterior and interior exposures, respectively.

The main objective of the CSA A23.3 cracking control provisions is to ensure that the cracks in concrete members remain fine and well distributed and are of limited width. According to CSA A23.3, satisfactory crack control can be accomplished by ensuring that the *crack control parameter* (z) (related to crack width) is within the acceptable limits and that skin reinforcement is provided where required. The z factor can be determined based on the following equation (A23.3 Cl.10.6.1)

$$z = f_s \sqrt[3]{d_c A} \quad (N/mm) \qquad [4.20]$$

CSA A23.3 Cl.10.6.2 prescribes the use of *skin reinforcement* to prevent the development of wider cracks on the side faces of deep beams and the exposed slab faces.

PROBLEMS

4.1. A typical reinforced concrete beam of rectangular cross-section is shown in the figure below. The beam is reinforced with 6-25M bars in the tension zone. Use an effective depth of 430 mm. The clear cover to the stirrups is equal to 30 mm and the maximum aggregate size is 20 mm. Find the following properties:
a) cracking moment (M_{cr})
b) gross moment of inertia (I_g) for the uncracked section (ignore the effect of reinforcement)
c) moment of inertia for the uncracked section using the transformed section properties ($I_{g,tr}$)
d) moment of inertia for the cracked section (I_{cr})

Given:

$f_c' = 30$ MPa (normal-density concrete)
$f_y = 400$ MPa
$E_s = 200\,000$ MPa

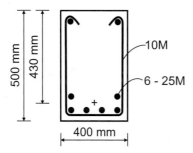

4.2. A typical cross-section of a 200 mm thick reinforced concrete slab is shown in the figure below. The slab is reinforced with 15M bars at 300 mm spacing on centre. Assume the effective depth of 170 mm. The material properties are summarized below.
Find the following properties for the unit strip (1 m width):
a) gross moment of inertia (I_g) for the uncracked section (ignore the effect of reinforcement)
b) cracking moment (M_{cr})
c) moment of inertia for the uncracked section using the transformed section properties ($I_{g,tr}$)
d) moment of inertia for the cracked section (I_{cr})

Given:

$f_c' = 25$ MPa (normal-density concrete)
$f_y = 400$ MPa
$E_s = 200\,000$ MPa

4.3. A typical cross-section of a reinforced concrete T-beam is shown in the figure below. The beam is reinforced with 6-25M bars in the tension zone. Use an effective depth of 600 mm. The material properties are summarized below.
a) Find the gross moment of inertia (I_g) for the uncracked section, ignoring the effect of reinforcement.
b) Find the moment of inertia for the uncracked section using the transformed section properties ($I_{g,tr}$).
c) Compare the results of the calculations performed in parts a) and b). Express the difference in the moment of inertia values as a percentage.
d) Do you believe that the difference in moment of inertia values obtained in part c) is significant from a practical design perspective? Explain.

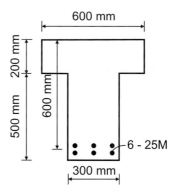

Given:

$f_c' = 35$ MPa (normal-density concrete)
$f_y = 400$ MPa
$E_s = 200\,000$ MPa

4.4. Refer to the beam section from Problem 4.3.
a) Find the cracking moment (M_{cr}) for this section.
b) Calculate the moment of inertia for the cracked section (I_{cr}).
c) Find the effective moment of inertia (I_e) values corresponding to the following specified bending moments:

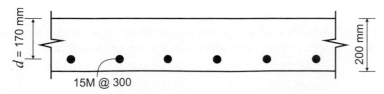

M (kN · m)
50
100
200
300

d)   Present the results of the calculations performed in part c) in the form of a diagram with bending moments (M) on the horizontal axis and I_e values on the vertical axis. Comment on the pattern of variation in the I_e values as a function of the bending moment (M).

4.5.   A continuous reinforced concrete beam ABCD is shown in the figure below. The beam is characterized with a cracking moment (M_{cr}) of 50 kN · m and a gross

moment of inertia $I_g = 10 \times 10^{10}$ mm⁴. The moment of inertia values for the cracked sections are as follows:

*   positive moment region: $I_{cr}{}^+ = 2 \times 10^{10}$ mm⁴
*   negative moment region: $I_{cr}{}^- = 4 \times 10^{10}$ mm⁴

The material properties are summarized below. Determine the effective moment of inertia (I_e) values along the beam length.

Given:

$f_c' = 30$ MPa (normal-density concrete)
$f_y = 400$ MPa
$E_s = 200\,000$ MPa

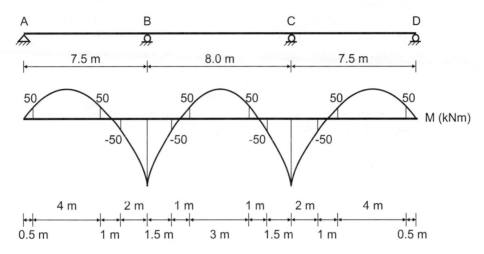

4.6.   A simply supported reinforced concrete beam of rectangular cross-section is shown in the figure below. The beam is reinforced with 4-25M bars at the bottom. The beam supports a uniform dead load (DL) of 7.5 kN/m (including the beam self-weight) and a uniform live load (LL) of 7.5 kN/m. Ignore the effects of the support width.

Find the following design parameters:
a)   maximum immediate deflection due to the total service load
b)   maximum immediate deflection due to the dead load only
c)   maximum deflection due to the live load only
d)   maximum long-term deflection due to the total load after 6 years of service life
Use the CSA A23.3 approximate procedure to determine the deflections.

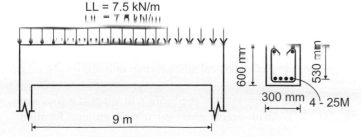

Given:

$f_c' = 30$ MPa (normal-density concrete)
$f_y = 400$ MPa
$E_s = 200\,000$ MPa

4.7.   Refer to the simply supported beam from Problem 4.6.
a)   Assume that the beam supports architectural elements that are likely to be damaged due to large deflections. Are the beam deflections within the limits prescribed by CSA A23.3?
b)   Suppose that the architectural elements described in part a) are installed 1 year after the construction. Find the maximum long-term deflection for this beam after the architectural elements have been installed.
c)   Comment on the effect of the installation schedule for architectural components on the magnitude of long-term deflections.

4.8.   Refer to the simply supported beam from Problem 4.6. After performing the detailed deflection calculations, the designer has started wondering whether those calculations were indeed necessary as per the CSA A23.3 requirements.
Find out whether the deflection calculations were required in this case and explain your answer.

4.9. Refer to the simply supported beam from Problem 4.6.
  a) Apply the computer-aided iterative procedure to find the maximum immediate deflection due to the total service load. Divide the beam into 20 segments.
  b) Compare the deflections with the values obtained in part a) of Problem 4.6. Do you believe that using the computer-aided iterative procedure to estimate deflections is justified in this application?

4.10. A four-span continuous reinforced concrete beam ABCDE of rectangular cross-section is shown in the figure below. The beam supports a uniform dead load (DL) of 7.5 kN/m (including the beam self-weight) and a uniform live load (LL) of 7.5 kN/m. The beam is reinforced with 4-25M bars at the bottom in the positive moment regions and 4-25M bars at the top in the negative moment regions.

Assume that the columns (200 mm by 200 mm cross-sectional dimensions) act as pinned supports. Find the following design parameters:
  a) maximum immediate deflection due to the total service load
  b) maximum immediate deflection due to the dead load only
  c) maximum deflection due to the live load only
  d) maximum long-term deflection due to the total load after 6 years of service life

Use the CSA A23.3 approximate procedure to determine the deflections.

Given:

$f_c' = 30$ MPa (normal-density concrete)
$f_y = 400$ MPa
$E_s = 200\,000$ MPa

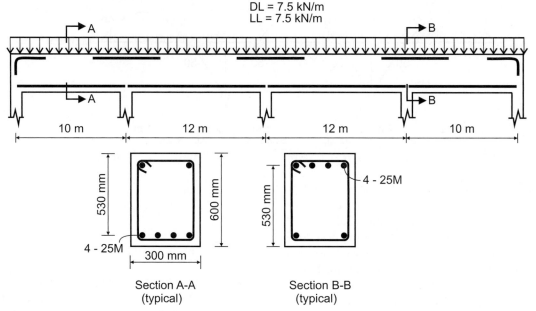

Section A-A
(typical)

Section B-B
(typical)

4.11. Refer to the continuous beam from Problem 4.10.
  a) Apply the computer-aided iterative procedure to find the maximum immediate deflection due to the total service load. Divide each beam span into 20 segments.
  b) Compare the deflections with the values obtained in part a) of Problem 4.10. Do you believe that using the computer-aided iterative procedure to estimate deflections for this application is justified?

4.12. The deflection estimates for a continuous reinforced concrete beam with a clear span of 10 m are as follows:

  - maximum deflection due to dead load is $\Delta_D = 10$ mm;
  - maximum deflection due to live load is $\Delta_L = 3$ mm.

Assume that the architectural elements are likely to be damaged by large deflections.
  a) Do you believe that the above estimated deflections will cause damage to the architectural elements?
  b) As a designer, what modifications could you make in the construction procedure to mitigate this problem? Explain.
  c) Suppose that it is important for this beam to have small deflections during its service life. What provisions in the construction procedure would you specify to ensure controlled deflections after 5 years of service? Assume in the deflection calculations that only the dead load is sustained.
  d) Suppose that the beam will support a roof slab with drains provided at each column location.

What provisions would you make to ensure that the beam has a minimum positive slope of 2% in order to maintain proper drainage? Assume in the deflection calculations that both dead and live loads are sustained in this case.

4.13. A typical cross-section of a reinforced concrete T-beam is shown in the figure below. Assume an effective depth of 630 mm. The beam is subjected to a maximum bending moment of 100 kN · m at the service load level. Assume exterior exposure. The beam properties are summarized below.

Check whether this beam section satisfies the CSA A23.3 cracking control requirements.

Given:

f_c = 25 MPa (normal-density concrete)
f_y = 400 MPa
E_s = 200 000 MPa

4.14. A typical cross-section of a 300 mm thick reinforced concrete slab is shown in the figure below. Assume the effective depth of 250 mm. The slab is subjected to the maximum bending moment of 35 kN · m per metre width at the service load level. Assume interior exposure. The slab properties are summarized below.

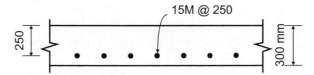

Check whether this slab section satisfies the CSA A23.3 cracking control requirements.

Given:

f_c' = 25 MPa (normal-density concrete)
f_y = 400 MPa
E_s = 200 000 MPa

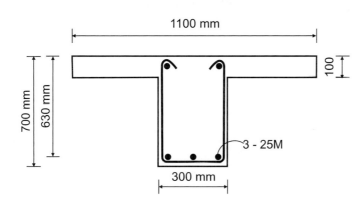

5

Flexure: Design of Beams and One-Way Slabs

LEARNING OUTCOMES

After reading this chapter, you should be able to

- outline the key considerations used in the design of reinforced concrete flexural members
- use the practical guidelines related to the design of reinforced concrete flexural members
- apply the two different design procedures for beams and slabs
- design rectangular beams with tension steel only for flexure according to CSA A23.3
- design one-way slabs for flexure according to CSA A23.3
- design T- and L-beams for flexure according to CSA A23.3
- design rectangular beams with tension and compression reinforcement for flexure according to CSA A23.3

5.1 INTRODUCTION

The objective of flexural design is to find cross-sectional dimensions and the amount and distribution of reinforcement in flexural members subjected to given design loads. In general, flexural design consists of several steps. First, the specified loads are assigned, and the concrete and steel material properties are selected. Subsequently, the geometry of the flexural member, including the cross-sectional dimensions and span, is chosen based on the understanding of the overall design requirements. Next, the appropriate amount of steel reinforcement is calculated based on the previously defined dimensions. The design process is iterative — a few cycles, including the calculations of beam dimensions and the corresponding amount of reinforcement, may be required before the designer is satisfied with the solution.

This chapter builds on the background provided in Chapters 3 and 4. The focus of Chapter 3 was mainly on the behaviour of reinforced concrete members subjected to flexure and the analysis of beam and slab sections with given cross-sectional dimensions and material properties. The criteria related to the serviceability of reinforced concrete flexural members, including cracking and deflections, were explained in Chapter 4. This chapter covers the design of simple beams and one-way slabs; the design of continuous beams and slabs will be discussed in Chapters 10 and 11.

This chapter is mainly focused on providing guidance on the design of beams and one-way slabs. Section 5.2 discusses general design requirements. Detailing requirements are outlined in Section 5.3. Practical guidelines related to design and construction are discussed in Section 5.4. Procedures for the flexural design of beams and slabs with tension steel only are explained in Section 5.5. The design of rectangular beams with tension steel only is discussed in Section 5.6, while the design of one-way slabs is discussed in Section 5.7. The design of T-beams is outlined in Section 5.8, whereas the design of rectangular beams with tension and compression reinforcement is discussed in Section 5.9.

In many cases, external factors control the selection of member dimensions. These factors include architectural considerations, formwork systems, contractor's experience, economy, and repetition and speed of construction. Experienced designers are able to develop a satisfactory design solution by optimizing and prioritizing these considerations. Students and novice designers need to gain a good grasp of the fundamentals in order to perform designs in an effective manner.

5.2 GENERAL DESIGN REQUIREMENTS

The selection of the cross-sectional dimensions and the corresponding reinforcement based on flexural requirements will be discussed in this chapter. However, the following important considerations must also be taken into account in the design of beams and one-way slabs:

- detailing of reinforcement placed in the beam or slab section, including the selection of adequate concrete cover to ensure protection from corrosion and environmental exposure and adequate bar spacing to enable satisfactory construction (to be discussed in Section 5.3);
- serviceability considerations (calculations of member deflections and the crack control parameter), which were discussed in Chapter 4;
- design for shear and torsion (to be discussed in Chapters 6 and 7);
- bar cutoffs and anchorage requirements (to be discussed in Chapter 9).

5.3 DETAILING REQUIREMENTS

5.3.1 Concrete Cover

The concrete cover can be defined as the amount of concrete provided between the surface of a concrete member and the reinforcing bars. According to CSA A23.1 Cl.6.6.6.1, the cover should be measured from the concrete surface to the nearest deformation of the reinforcement (including ties, stirrups, and tension steel). The cover is critical in ensuring a safe design suitable for the intended conditions. An adequate concrete cover needs to be provided for the following reasons:

- to protect the reinforcement against corrosion
- to ensure adequate bond between the reinforcement and the surrounding concrete
- to protect the reinforcement from the loss of strength in case of a fire
- to protect the reinforcement in case of concrete abrasion and wear (for example, floors in parking garages)

When the concrete cover provided in a reinforced concrete structure is inadequate, corrosion of reinforcing bars may take place only a few years after the construction is completed, as illustrated in the example of a slab in an underground parking garage in Figure 5.1. Such performance is unacceptable from the perspective of safety and durability.

According to CSA A23.1 Cl.6.6.6.2.1, the specified cover for the reinforcement shall be based on the consideration of life expectancy, exposure conditions, protective systems, maintenance, and the consequences of corrosion. Basic cover requirements according to CSA A23.1 Cl.6.6.6.3 are summarized in Table A.2 (same as A23.1 Table 17). It should be noted that Table A.2 prescribes different cover values depending on the concrete exposure class. The three main categories of exposure classes are as follows:

- not exposed — concrete that will be continually dry within the conditioned space (mainly interior members) (class N);

Figure 5.1 An example of a reinforced concrete slab with inadequate cover (note the reinforcing steel visible from the underside of the slab).

(John Pao)

- exposed to freezing and thawing, or manure and/or sileage gasses (classes F-1, F-2, S-1, and S-2);
- exposed to chlorides (classes C-XL, C-1, C-2, A-1, A-2, and A-3).

For more information on exposure classes, the reader is referred to Section 2.4.

It should be stressed that the concrete cover values specified in Table A.2 denote *cover to ties or stirrups* (for *beams*) and *cover to tension reinforcement* (for *slabs*).

The position of reinforcing bars on the job site might vary from the specifications of the construction drawings. CSA A23.1 takes this into account by allowing certain tolerances (Cl.6.6.8). According to the same clause, the tolerance for concrete cover is ±12 mm; however, the cover shall in no case be reduced by more than one third of the CSA A23.1 prescribed value. Therefore, because the positioning of the reinforcement is not exact, it may be advisable to increase the specified cover to ensure adequate protection (Cl.6.6.6.2.1 Note 4).

As previously discussed, one of the reasons for providing concrete cover is to prevent loss of strength (and potential collapse) in case of fire. Fire-resistance requirements for concrete structures are discussed in Sections 1.8.4 and 2.5. Concrete cover requirements related to the fire resistance of concrete structures are included in Appendix D of NBC 2010. Note that the concrete cover values based on the fire-resistance requirements may be larger than the values listed in Table A.2. When choosing the concrete cover for a structural member, the designer must ensure that both the fire-resistance and exposure-related requirements are met.

5.3.2 Bar Spacing Requirements

Bar spacing requirements, including minimum and maximum limits, are prescribed by CSA A23.1 and CSA A23.3. Bar spacing must be chosen to avoid reinforcement congestion and to enable the proper placement of concrete. In general, bar spacing requirements in one-way slabs can be met without problems, because these slabs do not require a heavy concentration of reinforcement, so congestion is generally not an issue (refer to Section 5.7 for bar spacing requirements related to slabs). However, in negative moment regions of the slabs, the area above beams can become crowded especially where beams pass over columns. Flat slab and flat plate floor systems have very heavy concentration of reinforcing above columns.

Concrete beams often need to be reinforced with a significant amount of tension steel. If the beam reinforcement is spaced too closely, the concrete will not flow smoothly between the bars, and air voids are likely to form. Therefore, the beam width needs to be adequate to accommodate bar spacing and concrete cover requirements.

Bar spacing requirements for beams specified by CSA A23.1 Cl.6.6.5.2 (also reproduced in Annex A of CSA A23.3) are shown in Figure 5.2. In general, the clear distance (*s*) between the parallel bars should not be less than

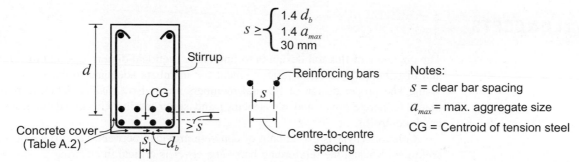

Figure 5.2 CSA A23.3 bar spacing requirements for beams.

1. 1.4 times the bar diameter ($1.4 d_b$),
2. 1.4 times the maximum size of the coarse aggregate ($1.4 a_{max}$), or
3. 30 mm.

When the beam width is inadequate to meet these spacing requirements, there are several solutions:

- Increase the beam width.
- Place the reinforcement in two or more layers.
- Bundle groups of parallel bars in contact.

A simple way of resolving the reinforcement congestion problem is to increase the beam width (when that option is available).

Reinforcement often needs to be placed in two or more layers, especially in heavily reinforced beams. In that case, bars in the upper layers must be placed directly above those in the bottom layer (CSA A23.1 Cl.6.6.5.3), as shown in Figure 5.2.

Note that the clear distance between the layers should be at least equal to the distance (s) provided for bars placed in one layer!

The designer needs to be aware that the actual position of reinforcement at the job site may differ from the construction drawings due to several factors, including the presence of electrical conduits or boxes, plumbing, or inaccurate bar placement. CSA A23.1 Cl.6.6.8 prescribes allowable tolerances related to bar spacing.

Bundling of reinforcing bars can also be used as an alternative solution to reduce reinforcement congestion. According to CSA A23.3 Cl.7.4.2, a *bundle* is a group of up to four parallel bars bundled in contact that are considered to act as a unit. A cross-section of a beam reinforced with bundled bars is presented in Figure 5.3a. Possible arrangements of bundled bars, with two-, three-, and four-bar bundles, are shown in Figure 5.3b. Bundles can be made of bar sizes 35M or smaller (CSA A23.3 Cl.7.4.2.2). Also, bundles must be enclosed by stirrups or ties (CSA A23.3 Cl.7.4.2.1). In determining the concrete cover and bar spacing requirements, a bundle shall be treated as a single bar with a diameter derived from the equivalent total area of the bars forming the bundle (CSA A23.3 Cl.7.4.2.4), as illustrated in Figure 5.3c.

Bundled bars are rarely used in building applications due to the large volume of concrete required to develop the structural properties of the reinforcing steel. However, bundles are often used in bridge construction (and other types of heavy construction) where reinforcement congestion problems are encountered.

Figure 5.3 Bundled bar requirements for beams: a) beam cross-section showing bundles; b) arrangements for bundled bars; c) equivalent bar properties.

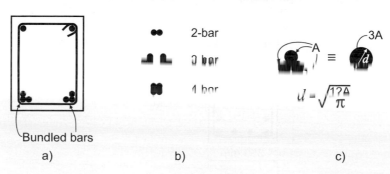

a)   b)   c)

KEY CONCEPTS

The objective of flexural design is to find cross-sectional dimensions and the amount and distribution of tension reinforcement for members adequate to carry the design loads. The proper design of flexural members must also consider detailing requirements (concrete cover and bar spacing) and serviceability checks (deflections and crack control).

Concrete cover is the amount of concrete provided between the surface of a concrete member and the reinforcing bars. The cover is critical in ensuring a safe design suitable for the intended conditions. Adequate concrete cover needs to be provided for the following reasons:

- to protect the reinforcement against corrosion
- to ensure adequate bond between the reinforcement and the surrounding concrete
- to protect the reinforcement from loss of strength in case of fire
- to protect reinforcement in case of concrete abrasion and wear (for example, floors in parking garages)

Basic cover requirements specified by CSA A23.1 Cl.6.6.6.2.3 are summarized in Table A.2.

Bar spacing requirements, including minimum and maximum limits, are related to the chances of reinforcement congestion and the ability to place the concrete. Bar spacing requirements for beams are prescribed by CSA A23.1 Cl.6.6.5.2. When the beam width is inadequate to meet these spacing requirements, there are several possible solutions: the reinforcement can be placed in two or more layers, the beam width can be increased, or groups of parallel bars in contact can be bundled.

5.3.3 Computation of the Effective Beam Depth Based on the Detailing Requirements

At this point, the reader should be familiar with the concept of effective depth in beams and slabs. The effective depth (d) was defined in Section 3.3 as the distance from the compression face of the beam to the centroid of the tension steel, as shown in Figure 3.5. The following example illustrates the calculation of the effective depth based on the given concrete cover and reinforcing bar sizes.

Example 5.1

A reinforced concrete beam with a width of 350 mm and a depth of 600 mm is shown in the figure below. The beam is reinforced with 6-30M bars at the bottom and 10M stirrups. The maximum aggregate diameter is 20 mm. The beam is located at the exterior of an office building.

Determine the effective depth to be used in a design based on the CSA A23.3 detailing requirements.

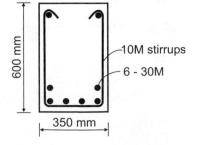

SOLUTION:

1. Determine the concrete cover

Because the beam is located at the exterior of the building, it can be considered as exposure class F-1 or F-2 according to the CSA A23.1 requirements. Therefore, the concrete cover can be determined from Table A.2 as

cover = 40 mm

2. Determine the minimum bar spacing

Given:

- 30M tension reinforcement; that is, d_b = 30 mm (see Table A.1);
- 10M stirrups; that is, d_s = 10 mm (see Table A.1);
- maximum aggregate size of 20 mm; that is, a_{max} = 20 mm.

According to CSA A23.1 Cl.6.6.5.2, the minimum spacing between reinforcing bars is equal to the largest of

1. $1.4 \times d_b = 1.4 \times 30$ mm = 42 mm
2. $1.4 \times a_{max} = 1.4 \times 20$ mm = 28 mm
3. 30 mm

Hence, the minimum spacing between reinforcing bars is s = 42 mm.

3. Determine the number of bars that can fit in one layer

The beam width is given as

b = 350 mm

It is estimated that four bars can fit in the bottom layer. Check whether this assumption is correct. The minimum required beam width (b_{min}) is

$$b_{min} = 4d_b + 3s + 2d_s + 2 \times \text{cover}$$
$$= 4 \times 30 \text{ mm} + 3 \times 42 \text{ mm} + 2 \times 10 \text{ mm} + 2 \times 40 \text{ mm} = 346 \text{ mm}$$

Since

$$b_{min} = 346 \text{ mm} < 350 \text{ mm}$$

four bars can be placed in the bottom layer. The remaining two bars must be placed in the upper layer.

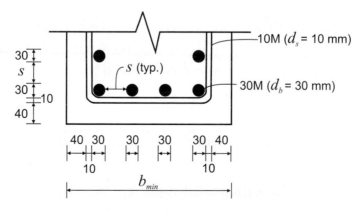

4. Determine the effective depth (d)

a) Calculate the distance from the top of the beam compression face to the centroid of the bottom layer (d_1) with four bars in total:

h = 600 mm is the overall beam depth (given)

$$d_1 = h - \text{cover} - d_s - \frac{d_b}{2}$$

$$= 600 \text{ mm} - 40 \text{ mm} - 10 \text{ mm} - \frac{30 \text{ mm}}{2} = 535 \text{ mm}$$

b) Calculate the distance from the top of the beam compression face to the centroid of the top layer (d_2) with two bars in total:

$$d_2 = d_1 - s - 2\left(\frac{d_b}{2}\right)$$

$$= 535 \text{ mm} - 42 \text{ mm} - 2 \times \left(\frac{30 \text{ mm}}{2}\right) = 463 \text{ mm}$$

c) Finally, calculate the effective depth as the distance from the top of the beam to the centroid of the reinforcement (point C on the sketch below):

$$d = \frac{4 \times d_1 + 2 \times d_2}{6}$$

$$= \frac{4 \times 535 \text{ mm} + 2 \times 463 \text{ mm}}{6} = 511 \text{ mm}$$

The final bar arrangement and the effective depth are shown in the sketch below.

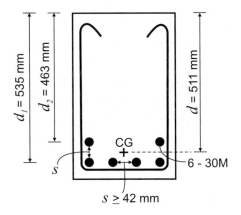

5.4 PRACTICAL GUIDELINES FOR THE DESIGN AND CONSTRUCTION OF BEAMS AND ONE-WAY SLABS

5.4.1 Design Guidelines

The design of concrete structures is a creative process and "cookbook solutions" are generally not available, but some practical guidelines on the selection of member properties are outlined below.

Overall beam/slab depth (h)   There is no unique procedure for estimating the overall beam or slab depth (h) for a design. However, a simple guideline, often followed by designers, is to select a beam depth such that the designer can be confident that the deflection criteria prescribed by the CSA A23.3 indirect approach to deflection control (Cl.9.8.2.1) can be satisfied without the need for detailed deflection calculations. Table A.3 prescribes the values of maximum span/depth ratios for beams and slabs with different support conditions to be used in the design.

Alternatively, the overall beam depth (h) can be estimated as roughly 8% to 10% of the span length; this corresponds to a traditional rule of thumb of 1 inch (25 mm) of beam depth per foot (300 mm) of span length (a good rule of thumb borrowed from Imperial units).

Note that a smaller amount of flexural reinforcement is required when deeper beams are used; however, shallower and wider beams are more efficient with regard to shear and torsion resistance, they tend to be easier to form, and they result in reduced storey height and overall building height.

Beam width (b) In general, the beam width (b) should not be less than 250 mm and preferably not less than 300 mm. Note that the arrangement of reinforcing bars is one of the key factors determining the beam width. It is very important for the designer to follow the minimum bar spacing requirements discussed in Section 5.3.2. In the case of beams with several bar layers, a space large enough for the concrete compaction vibrator to pass through should be provided between the layers. In general, a beam should be designed to be wider than the supporting columns; this facilitates the forming of the beam–column joint.

Effective beam/slab depth (d) The effective beam depth (d) can be estimated once the overall depth (h) is known. It is generally satisfactory to estimate the effective depth of a beam that is not exposed to earth or weather (for example, an interior beam) using the following recommendations:

- For beams with one layer of reinforcement,

  $d \cong h - 70$ mm

- For beams with two layers of reinforcement,

  $d \cong h - 110$ mm

For reinforced concrete slabs that span in one direction, usually only one layer of flexural reinforcement is used and shear reinforcement is not required. Bar sizes and concrete cover values are significantly smaller in slabs than in beams. As a result, the effective slab depth (d) can be estimated as

$d \cong h - 30$ mm

When the required cover is greater than 20 mm, then the effective depth can be estimated from the equation

$d \cong h - (\text{cover} + 10 \text{ mm})$

If the beam or slab is exposed to weather, the d value determined using the above equations should be reduced by 10 mm. In the case of more severe exposure conditions, d should be estimated based on the minimum cover specified in CSA A23.1, as discussed in Section 5.3.

The estimated d value should err on the lower side, thus resulting in a conservative estimate. Normal construction practices often result in smaller effective depths than indicated on construction drawings, thus resulting in the reduction of the flexural capacity of a member.

Cross-sectional depth/width (h/b) ratio For rectangular beams, cross-sections with a depth/width ratio between 1.5 and 2 are most often used, with the upper bound value of 2 being the most common; that is,

$b = \dfrac{h}{2}$

Nominal cross-section sizes It is good practice to use nominal sizes in selecting beam and slab dimensions. This reduces the chance of making errors in design calculations (this was also discussed in Section 1.10.2). Moreover, the accuracy of formwork measurement at the construction site has a practical limit. Note that member dimensions within ±10 mm are subject to some measurement errors in the field.

The following simple exercise can be used to develop a better sense of the accuracy of measurement at the construction site. Take a ruler with millimetre measurements and try to draw a line with a carpenter's pencil with an accuracy to the nearest millimetre. You will find that it is difficult to accomplish this precision because the thickness of the lead in a carpenter's pencil is 2 mm or larger. It is therefore much easier to draw a line to the nearest

5 mm, and you can be even more comfortable if asked to draw a line with a precision to the nearest 10 mm.

In structural engineering terms, there is no practical quantifiable difference in the flexural resistance for two beams with the same reinforcement and within ±10 mm difference in size. Hence, the following recommendations related to rounding slab and beam dimensions can be followed:

• The slab thickness should be chosen in increments of 10 mm.
• Beam dimensions should be rounded to the nearest 50 mm.
• Girder dimensions should be rounded to the nearest 100 mm.

An important consideration, apart from architectural requirements, is "ease" of construction and the use of common formwork lumber sizes. Labour cost issues, re-use of formwork, site accessibility for placing concrete and the type of building influence the selection of nominal cross-section dimensions.

Reinforcement ratio (ρ)   As discussed in Section 3.3.1, the reinforcement ratio (ρ) is used as an indicator of the relative amount of steel reinforcement in a cross-section. For a beam of rectangular cross-section, ρ is defined as

$$\rho = \frac{A_s}{b\,d} \qquad\qquad [3.1]$$

It is often necessary to select a trial reinforcement ratio (ρ) for the design. This choice is affected by ductility requirements and economic and construction considerations. The use of a reinforcement ratio (ρ) in the range of $0.35\rho_b$ to $0.40\rho_b$ results in ductile behaviour. Note that ρ_b denotes the balanced reinforcement ratio explained in Section 3.5.3. For Grade 400 reinforcement, ρ_b is equal to 0.022 for $f_c' = 25$ MPa, 0.027 for $f_c' = 30$ MPa, and 0.034 for $f_c' = 40$ MPa (see Table A.4).

For common steel and concrete material properties, flexural members should preferably be designed such that the required ρ value ranges from 0.0035 (0.35%) to 0.01 (1%) and remains less than $0.5\rho_b$. A reinforcement ratio greater than $0.5\rho_b$ can be considered when external constraints restrict member dimensions.

Reinforcing bar sizes   The selection of optimal rebar sizes for beam and slab design is not a simple task. In general, using bar sizes that are too large is not recommended. Larger bar sizes may result in larger flexural cracks and require a larger length to develop their full strength. On the other hand, there is a greater labour cost related to the placement of a larger number of smaller bars than the placement of a smaller number of larger bars. For example, the unit costs related to the placement of 2-10M bars are larger than those for the placement of 1-15M bar, even though the total steel area is equal in these two cases.

Guidelines related to bar sizes to be used in the design of beams and slabs are summarized in Table 5.1.

Table 5.1   Recommended reinforcing bar sizes for beams and slabs

Slabs			Beams		
overall thickness	tension steel	shrinkage and temperature steel	overall depth	tension steel	shear reinforcement (stirrups)
<150 mm	15M	10M	<300 mm	15M or 20M	10M
150 to 250 mm	15M or 20M	15M	300 mm to 600 mm	20M or 25M	10M or 15M
> 250 mm	20M or 25M	15M or 20M	600 mm to 1200 mm	25M to 35M	15M or 20M

Concrete and steel material strengths The value of the specified concrete compressive strength (f_c') is usually prescribed in the initial phase of the design. If durability is not a problem, reinforced concrete beams and floor slabs are generally constructed of concrete with $f_c' = 25$ MPa, 30 MPa, or 35 MPa, with 30 MPa strength being the most common. Due consideration needs to be given to the exposure requirements, which may be critical in prescribing the f_c' value. Refer to Section 2.4 for more details on exposure classes and durability requirements for concrete structures.

The most important mechanical property of reinforcing steel used in the design of concrete structures is its yield strength (f_y). As discussed in Section 2.6, the basic reinforcing steel grade in Canada is 400R, characterized by an f_y value of 400 MPa.

5.4.2 Construction Considerations and Practices

Beam and slab construction practice In one-way slab and beam construction, beam reinforcement is usually placed first. The bottom beam reinforcement and the stirrups for each span are assembled in the form of a cage. On larger projects, these cages are prefabricated on site or in a fabrication shop and lifted by crane into the beam forms. The top reinforcement and any hooked rebars are then added in place.

Subsequently, the bottom slab reinforcement is placed. After the beam reinforcement is completed, the top slab reinforcement is placed on top of the beam reinforcement at each span. At the same time, any hooked rebars at the perimeter and around the openings are also placed.

In every building there are mechanical and electrical systems which require openings through the beam or slab or ducting within the slab. These openings are usually accommodated with a minimal impact on the construction process provided that there is proper coordination and communication during the design phase.

Bar supports Flexural rebars must be properly supported and tied so that displacement does not occur during concrete placement. Proper supports to the reinforcing bars in the form of *bolsters* (or *chairs*) are required to prevent the sagging of rebars. Bar supports are used to firmly hold the bars at the required clearance from the forms before and during the placing of concrete. Bolsters must be sufficiently strong and properly spaced to provide support under construction conditions. They may consist of steel wire, plastic, or precast concrete blocks, but factory-made wire bar supports are most common. Unless structural drawings or specifications show otherwise, bar supports will be furnished according to the standards developed by the Reinforcing Steel Institute of Canada (RSIC) or the Concrete Reinforcing Steel Institute (CRSI). For more information, the reader is referred to RSIC (2004), CRSI (1997), and CRSI (2001).

Bar placement in beams In many cases, beam reinforcement is preassembled into a cage usually formed by the beam stirrups plus a longitudinal bar in each corner. The top bars are placed after all the longitudinal bars are in place. Even when not required by the design, at least two top rebars need to be provided in the beams to support the stirrups and the slab top reinforcing in the perpendicular direction (these bars are called *stirrup support bars*). The top bars must be securely tied to the stirrups and the column vertical reinforcement. Bar separators are often used to support the top layer of reinforcement. An example of beam reinforcement placed in the forms is shown in Figure 5.4.

Bolsters are placed in the form before the cage. In general, beam bolsters are spaced at approximately 1.5 m (5 feet) on centre.

Bar placement in slabs The designer needs to be aware of realistic rebar placement tolerance. In particular, a good understanding of the problems related to rebar placement in the field is required. A few important recommendations related to bar placement are summarized below.

- The spacing of reinforcing bars should be specified such that it reduces the chances of numerical errors in the field. For example, 250 mm spacing is preferred over 245 mm

Figure 5.4 Placement of beam reinforcement at the construction site.

(John Pao)

spacing. In the design sense, there is no real difference between 245 mm and 250 mm, but the use of rounded dimensions (like 250 mm spacing) permits the workers to use mental math without making errors. In this example, the rebar placer knows that four bars need to be placed per metre length.

- Bar spacing should be uniform wherever possible. For example, the designer should refrain from specifying 240 mm bar spacing in one area and 260 mm spacing in the adjacent area whenever possible. Instead, an experienced designer would specify 250 mm spacing for both areas to minimize the chances of field errors. Therefore, it is recommended to achieve uniformity in bar spacing rather than to adhere precisely to the design requirements.

In practice, rebar placers often lay out rebars by marking the spacing with coloured spray paint on the formwork using a tape measure and mental math. It should be noted that the width of each marking is on the order of 20 mm to 30 mm and that a high level of precision in marking the bar spacing is not a reality at the construction site.

Another issue involves the use of U.S. customary or imperial units. It is not uncommon for bar placers to use nominal dimensions in inches "soft" converted.

KEY CONCEPTS

The following practical guidelines regarding the selection of beam and slab properties can be offered based on practical design experience:

Overall beam/slab depth (h): A simple guideline is to select the overall beam depth such that a detailed deflection calculation is not required, provided that the requirements of A23.3 Cl.9.8.2.1 (summarized in Table A.3) are satisfied. Table A.3 prescribes maximum span/depth ratios for beams and slabs with different support conditions to be used in the design. Alternatively, the overall beam depth (h) can be taken as 8% to 10% of the span length (this corresponds to a traditional rule of thumb of a 1 inch deep beam per foot of span length).

Beam width (b): This should be not less than 250 mm and preferably not less than 300 mm. The arrangement of reinforcing bars is one of the key factors determining the beam width. A beam can be designed to be wider than the column; this considerably facilitates forming the beam–column joint.

Effective beam/slab depth (d): This can be estimated once the overall depth (h) is known. The d value should not be overestimated because normal construction practice tends to result in smaller effective depths than indicated on construction drawings.

Cross-sectional depth/width (h/b) ratio: For rectangular beams, cross-sections with depth/width ratio between 1.5 and 2 are most often used, with the upper bound value being the most common; that is,

$$b = \frac{h}{2}$$

Nominal cross-section sizes: The use of nominal sizes reduces the chances of making errors in design calculations. Moreover, the accuracy of formwork measurement at the construction site is limited. The following recommendations related to rounding of slab and beam dimensions can be followed:

- The slab thickness should be chosen in increments of 10 mm.
- Beam dimensions should be rounded to the nearest 50 mm.
- Girder dimensions should be rounded to the nearest 100 mm.

Reinforcement ratio (ρ): The selection of the reinforcement ratio is affected by economic considerations, ductility requirements, and construction considerations. It is considered good practice to design beams and slabs using a small amount of reinforcement. For common concrete and steel material properties, flexural members should preferably be designed such that the required ρ value ranges from 0.0035 (0.35%) to 0.01 (1%) and remains less than $0.5\rho_b$ (provided there are no major architectural constraints).

Reinforcing bar sizes: The selection of optimal bar sizes for beam and slab design is not simple. In general, the use of bar sizes that are too large is not recommended, as larger development lengths are required to ensure proper bar anchorage. However, there is a larger labour cost related to the placement of a large number of smaller bars than the placement of a small number of larger bars. Guidelines for bar sizes to be used in the design of beams and slabs are summarized in Table 5.1.

Concrete and steel strengths: Reinforced concrete beams and floor slabs are generally constructed of concrete with f_c' of 25 MPa, 30 MPa, or 35 MPa, with a strength of 30 MPa being the most common. The basic reinforcing steel grade in Canada is 400R, characterized by an f_y value of 400 MPa.

5.5 DESIGN PROCEDURES FOR RECTANGULAR BEAMS AND SLABS WITH TENSION STEEL ONLY

A key design requirement for concrete members subjected to flexure according to CSA A23.3 Cl.8.1.3 is the ultimate limit states (or strength) requirement; that is,

Factored Resistance $\geq$ Effects of Factored Loads

or

$$M_r \geq M_f \qquad\qquad [5.1]$$

The above requirement says that the factored moment resistance (M_r) must be greater than or equal to the factored bending moment (M_f) acting at any section along a concrete flexural member. In design practice, it is often required to determine cross-sectional dimensions and reinforcement for a flexural member subjected to given loads. In that case, it is possible to calculate the factored bending moment at a critical section. As an initial estimate of moment resistance, it is usually taken that

$$M_r = M_f$$

Two procedures related to the design of beams and one-way slabs will be outlined in this section:

- direct procedure
- iterative procedure

These design procedures have been developed with the assumption of *properly reinforced beams* that fail in the desirable steel-controlled mode (as discussed in Section 3.5.1).

5.5.1 Direct Procedure

The objective of this procedure is to determine the area of tension reinforcement (A_s) directly from the moment resistance (M_r) of a beam section, as explained below.

Consider the cross-section of a reinforced concrete beam in Figure 5.5. The moment resistance (M_r) for the beam section can be determined from the force couple T_r and C_r with a lever arm ($d - a/2$), as discussed in Section 3.5.1.

The moment resistance of a beam section can be determined as

$$M_r = \phi_s f_y A_s \left(d - \frac{a}{2} \right) \qquad\qquad [3.14]$$

and the depth of the rectangular stress block (a) can be determined as

$$a = \frac{\phi_s f_y A_s}{\alpha_1 \phi_c f_c' b} \qquad\qquad [3.12]$$

When a is substituted from Eqn 3.12 into Eqn 3.14, the latter equation can be rewritten as

$$M_r = \phi_s f_y A_s \left(d - \frac{\phi_s f_y A_s}{2\alpha_1 \phi_c f_c' b} \right) \qquad\qquad [5.2]$$

Equation 5.2 can be rearranged in the form of a quadratic equation with variable A_s as follows:

$$\left(\frac{(\phi_s f_y)^2}{2\alpha_1 \phi_c f_c' b} \right) (A_s)^2 - (\phi_s f_y d) A_s + M_r = 0$$

This quadratic equation can be solved for A_s as

$$A_s = \frac{-B \pm \sqrt{B^2 - 4 \times A \times C}}{2A}$$

where

$$A = \frac{(\phi_s f_y)^2}{2\alpha_1 \phi_c f_c' b}$$

$$B = -\phi_s f_y d$$

$$C = M_r$$

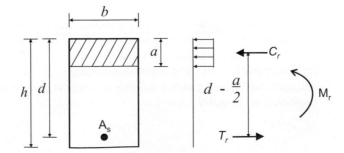

Figure 5.5 Cross-section of a reinforced concrete beam showing the distribution of internal forces.

The equation for A_s can be presented in the form

$$A_s = \frac{\alpha_1 \phi_c f_c' b}{\phi_s f_y} \left(d \pm \sqrt{d^2 - \frac{2M_r}{\alpha_1 \phi_c f_c' b}} \right) \qquad [5.3]$$

This equation can be further simplified using the following assumptions:

1. The solution containing the + term can be discarded as it results in a larger A_s value.
2. Because Grade 400 steel is used in Canada, it can be taken that $f_y = 400$ MPa.
3. Substitute the values for α_1, ϕ_c, and ϕ_s as follows:

$\alpha_1 \cong 0.8$ (valid for f_c' values ranging from 25 MPa to 45 MPa)

$\phi_c = 0.65 \qquad [3.2]$

$\phi_s = 0.85 \qquad [3.3]$

As a result, the final form of Eqn 5.3 can be obtained as

$$A_s = 0.0015 f_c' b \left(d - \sqrt{d^2 - \frac{3.85 M_r}{f_c' b}} \right) \quad [\text{mm}^2] \qquad [5.4]$$

The following units should be used in the above equation:

M_r [N·mm]
b [mm]
d [mm]
f_c' [MPa]

Eqn 5.4 can be easily programmed by the reader, provided that (s)he has access to a programmable calculator or a computer.

Example 5.2

A typical cross-section of a reinforced concrete beam is shown in the figure below. The factored bending moment (M_f) acting at a critical section is equal to 800 kN·m. The concrete and steel material properties are given below.

Use the direct procedure to find the economical amount of tension reinforcement for the beam section such that the beam is capable of carrying the given loads. It is not required to check the detailing requirements (concrete cover and bar spacing).

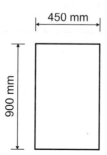

450 mm

900 mm

Given: $f_c' = 25$ MPa
$f_y = 400$ MPa
$\phi_c = 0.65$
$\phi_s = 0.85$
$M_f = 0.8$

SOLUTION: **1. Estimate the effective beam depth**

First, let us determine the effective depth (d) based on the guideline provided in Section 5.4.1 (use one layer of reinforcement):

$d = h - 70$ mm

$= 900$ mm $- 70$ mm $= 830$ mm

2. **Calculate the required area of tension reinforcement**

   a) Take $M_r = M_f = 800$ kN $\cdot$ m.

   b) Calculate the required area of tension reinforcement from Eqn 5.4 as

   $$A_s = 0.0015 f_c' \, b \left(d - \sqrt{d^2 - \frac{3.85 \, M_r}{f_c' \, b}} \right) \qquad [5.4]$$

   $$= 0.0015 \times (25 \text{ MPa}) \times (450 \text{ mm})$$

   $$\times \left(830 \text{ mm} - \sqrt{(830 \text{ mm})^2 - \frac{3.85 \, (800 \times 10^6 \text{ N} \cdot \text{mm})}{25 \text{ MPa} \times 450 \text{ mm}}} \right)$$

   $$= 3134 \text{ mm}^2$$

   c) Select the amount of reinforcement in terms of the number and size of the bars. Based on Table A.1, 5-30M bars are a suitable choice (the area of one 30M bar is 700 mm^2):

   $$A_s = 5 \times 700 \text{ mm}^2 = 3500 \text{ mm}^2$$

   d) Check whether the provided area of reinforcement is greater than or equal to the required amount of reinforcement:

   $$A_s = 3500 \text{ mm}^2 > 3134 \text{ mm}^2 : \text{okay}$$

3. **Confirm that the strength requirement is satisfied**

   a) First, calculate the depth of the compression stress block (a) as

   $$a = \frac{\phi_s f_y A_s}{\alpha_1 \phi_c f_c' \, b} \qquad [3.12]$$

   $$= \frac{0.85 \times 400 \text{ MPa} \times 3500 \text{ mm}^2}{0.8 \times 0.65 \times 25 \text{ MPa} \times 450 \text{ mm}} \cong 203 \text{ mm}$$

   b) Then, calculate the moment resistance as

   $$M_r = \phi_s f_y A_s \left(d - \frac{a}{2} \right) \qquad [3.14]$$

   $$= 0.85(400 \text{ MPa})(3500 \text{ mm}^2) \left(830 \text{ mm} - \frac{203 \text{ mm}}{2} \right)$$

   $$= 867 \times 10^6 \text{ N} \cdot \text{mm} = 867 \text{ kN} \cdot \text{m}$$

   c) Finally, check the strength requirement per A23.3 Cl.8.1.3:

   $$M_r \geq M_f \qquad [5.1]$$

   Since

   $$M_f = 800 \text{ kN} \cdot \text{m}$$

   and

   $$M_r = 867 \text{ kN} \cdot \text{m} > 800 \text{ kN} \cdot \text{m} : \text{okay}$$

   the design is adequate. The section properties are summarized in the sketch below.

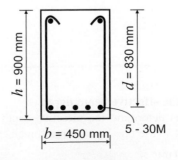

5.5.2 Iterative Procedure

The iterative procedure can be used to find the required area of tension reinforcement by assuming the depth of the compression stress block and by recalculating the reinforcement area until the required accuracy has been achieved. The factored moment resistance is then computed using the most accurate value of the reinforcement area obtained in an iterative way.

The moment resistance of a beam section can be determined as

$$M_r = \phi_s f_y A_s \left(d - \frac{a}{2}\right) \qquad \qquad [3.14]$$

The required area of reinforcement (A_s) can be determined from the above equation by setting $M_r = M_f$:

$$A_s = \frac{M_f}{\phi_s f_y \left(d - \frac{a}{2}\right)} \qquad \qquad [5.5]$$

For preliminary design purposes, the depth of the compression stress block (a) used in Eqn 5.5 can be taken as

$a = 0.3d$

for rectangular beams, and

$a = 0.2d$

for slabs.

The A_s value determined from Eqn 5.5 can be used to determine a more accurate a value based on Eqn 3.12:

$$a = \frac{\phi_s f_y A_s}{\alpha_1 \phi_c f_c' b} \qquad \qquad [3.12]$$

This a value can then be substituted in Eqn 5.5 to obtain a better A_s estimate. The process can be repeated until the desired accuracy has been achieved. The key steps performed in the iterative procedure are summarized below:

1. Estimate the depth of the compression stress block (a).
2. Calculate the reinforcement area (A_s) from Eqn 5.5.
3. Use the A_s value obtained in the previous step to find a (Eqn 3.12).
4. Continue until the difference between the A_s values obtained in the two successive cycles is within 5%.

KEY CONCEPTS

Two procedures are used to design beams and one-way slabs:

- The *direct procedure* determines the area of tension reinforcement (A_s) from the moment resistance (M_r) of a beam section.
- The *iterative procedure* determines the required area of tension reinforcement by estimating the depth of the compression stress block and then recalculating the reinforcement area until the required accuracy has been achieved. The factored moment resistance is computed using the most accurate value of the reinforcement area.

These design procedures have been developed with the assumption of *properly reinforced beams* that fail in the desirable steel-controlled mode.

Example 5.3

A cross-section of the beam designed in Example 5.2 is shown in the figure below.
Use the iterative procedure to find the economical amount of tension reinforcement for the beam section such that the beam is capable of carrying the given loads. It is not necessary to check the detailing requirements (concrete cover and bar spacing).

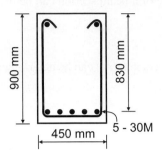

Given: $f_c' = 25$ MPa
$f_y = 400$ MPa
$\phi_c = 0.65$
$\phi_s = 0.85$
$\alpha_1 \cong 0.8$

SOLUTION: 1.   **Make the initial estimate for the compression stress block depth (a)**
The effective depth is $d = 830$ mm (same as Example 5.2). Take the depth of the compression stress block (a) as

$$a = 0.3d = 0.3(830 \text{ mm}) = 249 \text{ mm}$$

Use

$$a \cong 250 \text{ mm}$$

Remember that the numbers used to estimate the a value are approximate and therefore round figures should be used for ease of calculation.

2.   **Calculate the required area of tension reinforcement**
a)   Take $M_r = M_f = 800$ kN·m.
b)   Calculate the required area of tension reinforcement as

$$A_s = \frac{M_f}{\phi_s f_y \left(d - \dfrac{a}{2} \right)} \tag{5.5}$$

$$= \frac{800 \times 10^6 \text{ N·mm}}{0.85 \times 400 \text{ MPa} \left(830 \text{ mm} - \dfrac{250 \text{ mm}}{2} \right)} = 3338 \text{ mm}^2$$

c)   Check the accuracy of the initial estimate for a:

$$a = \frac{\phi_s f_y A_s}{\alpha_1 \phi_c f_c' b} \tag{3.12}$$

$$= \frac{0.85 \times 400 \text{ MPa} \times 3338 \text{ mm}^2}{0.8 \times 0.65 \times 25 \text{ MPa} \times 450 \text{ mm}} = 194 \text{ mm} \cong 200 \text{ mm}$$

Note that the initial estimate was $a = 250$ mm. Therefore, it is necessary to continue the calculation process and obtain a new A_s value.

d)   Recalculate A_s using the improved estimate for a:

$$a = 200 \text{ mm}$$

$$A_s = \frac{M_r}{\phi_s f_y \left(d - \dfrac{a}{2} \right)} = \frac{800 \times 10^6 \text{ N} \cdot \text{mm}}{0.85 \times 400 \text{ MPa} \left(830 \text{ mm} - \dfrac{200 \text{ mm}}{2} \right)} = 3223 \text{ mm}^2$$

This A_s value is within 5% of the initially estimated value of 3338 mm^2 and should be acceptable.

e)   Select the amount of reinforcement in terms of the number and size of the bars: based on Table A.1, 5-30M bars are a suitable choice, since the area of one 30M bar is 700 mm^2. The area of tension steel can be calculated as

$$A_s = 5 \times 700 \text{ mm}^2 = 3500 \text{ mm}^2$$

f)   Check whether the provided area of reinforcement is greater than or equal to the required amount of reinforcement.

$$A_s = 3500 \text{ mm}^2 > 3223 \text{ mm}^2 : \text{okay}$$

The use of the iterative procedure has resulted in a less than 3% difference in the reinforcement area as compared to the value obtained using the direct procedure in Example 5.2.

3.   **Confirm that the strength requirement is satisfied**
Since the amount of reinforcement is the same as in Example 5.2, the strength requirement check is identical. It can therefore be concluded that this design is adequate.

Learning from Examples

1.   The reader needs to recognize that the actual area of steel reinforcement is governed by predetermined rebar sizes specified in whole numbers. Both examples (5.2 and 5.3) ultimately lead to the same solution (5-30M bars); this is due to the fact that between 3000 mm^2 and 4000 mm^2, only two practical rebar configurations are available. Table A.8 lists several practical rebar configurations with area ranging from 1000 mm^2 to 7000 mm^2 (note that each configuration considers only bars of the same size). The right-hand column lists areas for configurations that include an even number of bars; it is easier to stagger or alternate bars in a configuration that consists of an even number of bars (to be discussed in Chapter 11). Above the value of 7000 mm^2, the reinforcement area increases in increments of 700 mm^2, since 30M bars would be the smallest bar sizes used in such configurations.

2.   It is interesting to note that the initial guess of $a = 0.3d$ in Example 5.3 would have resulted in the same solution (5-30M bars) without performing any iterations. However, a high level of precision in determining the area of reinforcement (A_s) is not necessary, provided that the designer always rounds up the area to the next practical rebar configuration.

5.6 DESIGN OF RECTANGULAR BEAMS WITH TENSION STEEL ONLY

5.6.1 CSA A23.3 Flexural Design Provisions for Rectangular Beams with Tension Steel Only

The key CSA A23.3 design requirements for flexural design of rectangular beams with tension steel only, related to minimum and maximum limits for the amount of tension reinforcement to be provided in a beam section, are discussed below.

> A23.3 Cl.10.5.1

Minimum tension reinforcement Reinforcement in concrete members becomes effective after cracking has occurred at the tension face. In general, well-designed flexural members are characterized by a moderate proportion of tension reinforcement. However, if a member is very lightly reinforced or *underreinforced* (as discussed in Section 3.4.3), the moment resistance of the cracked member could be less than the bending moment at the onset of cracking. As a result, the reinforcement will yield immediately after concrete cracks in the tension zone; this is followed by the sudden and brittle crushing failure of the concrete. This is clearly an undesirable scenario, and for that reason CSA A23.3 Cl.10.5.1 specifies two alternative requirements related to the minimum amount of reinforcement in flexural members. Either requirement should be satisfied at every section of a flexural member where tension reinforcement is required by analysis.

> A23.3 Cl.10.5.1.1

1. *The cracking moment criterion.* When the cracking moment (M_{cr}) exceeds the factored moment resistance of a beam section (M_r), a sudden and brittle failure will occur with little or no warning shortly after the cracking. (For more details on M_{cr}, refer to Section 4.2.) In very lightly reinforced beams, once the cracks have developed in the concrete tension zone, the steel tensile capacity is insufficient to carry the cracking moment. As a result, sudden failure of the beam can be expected. For this reason, CSA A23.3 Cl.10.5.1.1 requires that a sufficient amount of reinforcement be provided to ensure that the design moment capacity exceeds the cracking moment by at least 20%; that is,

> A23.3 Eq. 10.3

$$M_r \geq 1.2\,M_{cr} \qquad\qquad\qquad \textbf{[5.6]}$$

> A23.3 Cl.10.5.1.2

2. *Minimum area of tension reinforcement.* In lieu of the above requirement, CSA A23.3 Cl.10.5.1.2 prescribes that the minimum reinforcement requirement for beams be satisfied when the actual steel area exceeds A_{smin}; that is,

$$A_s > A_{smin}$$

where

> A23.3 Eq. 10.4

$$A_{smin} = \frac{0.2\sqrt{f_c'}}{f_y}\,b_t\,h \qquad\qquad\qquad \textbf{[5.7]}$$

where

  b_t = the width of the tension zone; for a rectangular section, $b_t = b$ (see Figure 5.6a); for a T-beam section under positive bending, b_t is equal to the web width (see Figure 5.6b);

  h = the overall beam depth; this is one of the rare flexural design provisions that requires the use of h (d is used for most calculations).

The purpose of this requirement is to ensure the minimum amount of reinforcement in the tension zone to make up for the loss of tensile strength caused by the concrete cracking.

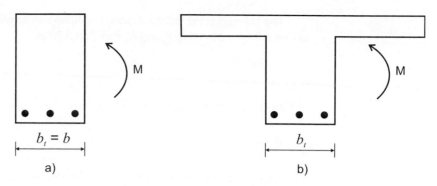

Figure 5.6 Width of the tension zone (b_t): a) rectangular section; b) T-section.

It should be noted that Cl.10.5.1.3 of CSA A23.3 permits the above two requirements to be waived when the factored moment resistance is greater than

$$M_r \geq 1.33 \, M_f \qquad\qquad [5.8]$$

The latter requirement offers a simple alternative to the CSA A23.3 minimum reinforcement requirements without checking the cracking moment criterion (Cl.10.5.1.1) or the minimum amount of tension reinforcement (Cl.10.5.1.2).

A23.3 Cl.10.5.2

Maximum amount of tension reinforcement = balanced reinforcement   CSA A23.3 Cl.10.5.2 states that "the tension reinforcement shall not be assumed to reach yield unless"

A23.3 Eq. 10.5

$$\frac{c}{d} \leq \frac{700}{700 + f_y} \qquad\qquad [5.9]$$

According to the same clause, "for flexural members without axial loads, the area of tension reinforcement should be limited such that the above condition is satisfied." This requirement implies that beams and slabs contain a "proper" amount of reinforcement and fail in the desirable steel-controlled mode. It should be noted that if the c/d ratio exceeds the limit specified by Eqn 5.9 the beam would be overreinforced and fail in the concrete-controlled mode. (Refer to Section 3.5.3 for more information on eam failure modes and corresponding amounts of reinforcement.)

The upper limit for the c/d ratio corresponds to the balanced strain condition. In fact, Eqn 5.9 has been derived from the balanced strain condition discussed in Section 3.5.3. The balanced strain condition occurs when the maximum compressive strain in concrete ($\varepsilon_{cmax} = 0.0035$) develops simultaneously with the yield strain in tension reinforcement ($\varepsilon_s = \varepsilon_y$), as shown in Figure 5.7. The following proportion can be obtained from similar triangles:

$$\frac{c}{d} = \frac{0.0035}{0.0035 + \varepsilon_y} \qquad\qquad [3.20]$$

The yield strain (ε_y) can be obtained from Hooke's law as (see Section 2.6.2)

$$\varepsilon_y = \frac{f_y}{E_s} \qquad\qquad [2.4]$$

where E_s is the modulus of elasticity for steel ($E_s = 200\,000$ MPa).

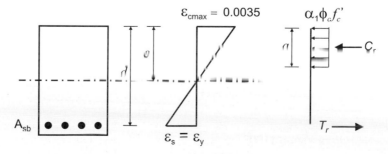

Figure 5.7 Balanced condition.

The c/d ratio at the balanced strain condition can be obtained in the same form as stated in Eqn 5.9 by substituting ε_y and E_s into Eqn 3.20 as

$$\frac{c}{d} = \frac{700}{700 + f_y} \qquad \text{[5.10]}$$

In Canada, Grade 400 steel is most frequently used, that is, $f_y = 400$ MPa, and so

$$\frac{c}{d} = 0.64$$

The above equation can be further approximated and presented in an easily remembered form as

$$c \cong \frac{2}{3}d \qquad \text{[5.11]}$$

The area of tension reinforcement corresponding to the balanced condition, also called the balanced reinforcement (A_{sb}), can be calculated from Eqn 3.21 as

$$A_{sb} = \frac{\alpha_1 \phi_c f_c' \, a \, b}{\phi_s f_y} \qquad \text{[3.21]}$$

The depth of the compression stress block (a) can be determined as

$$a = \beta_1 c \qquad \text{[3.4]}$$

where the neutral axis depth (c) can be determined from Eqn 5.10 as

$$c = \frac{700}{700 + f_y} d$$

Equation 3.21 can then be rewritten using the above equations as

$$A_{sb} = \left(\frac{\alpha_1 \beta_1 \phi_c f_c'}{\phi_s f_y} \right) \left(\frac{700}{700 + f_y} \right) b \, d \qquad \text{[5.12]}$$

Note that the following approximate values for α_1 and β_1 can be used when the concrete strength varies between 25 MPa and 40 MPa:

$$\alpha_1 \cong 0.8, \quad \beta_1 \cong 0.9 \qquad \text{[5.13]}$$

The corresponding reinforcement ratio, called the *balanced reinforcement ratio* (ρ_b), can be obtained as

$$\rho_b = \frac{A_{sb}}{bd} \qquad \text{[3.22]}$$

or

$$\rho_b = \left(\frac{\alpha_1 \beta_1 \phi_c f_c'}{\phi_s f_y} \right) \left(\frac{700}{700 + f_y} \right) \qquad \text{[5.14]}$$

It should be noted that ρ_b is constant for all beams and slabs with the same concrete and steel material properties (f_c' and f_y).

In Canada, Grade 400 steel is mainly used for steel reinforcement ($f_y = 400$ MPa), so the following approximate ρ_b value can be obtained by substituting the values for α_1, β_1, ϕ_c, and ϕ_s into Eqn 5.14:

$$\rho_b \cong \frac{f_c'}{1100} \qquad \text{[5.15]}$$

Note that ρ_b values for f_c' values from 25 MPa to 40 MPa and Grade 400 steel are given in Table A.4.

As discussed in Section 3.4.3, the balanced condition represents the threshold between overreinforced beams failing in the concrete-controlled mode and properly reinforced beams failing in the steel-controlled mode. For that reason, structural designers routinely compare the value of the reinforcement ratio (ρ) provided in a beam with the balanced reinforcement ratio (ρ_b) to predict the potential failure mode and to confirm the adequacy of the design. The three possible scenarios, outlined in Section 3.5.3, are reviewed below.

1. When $\rho = \rho_b$, the amount of reinforcement corresponds to the *balanced condition*. The amount of steel for this condition is considered to be large.
2. When $\rho > \rho_b$, the amount of reinforcement corresponds to overreinforced beams failing in the *concrete-controlled failure mode*, a brittle failure characterized by the crushing of concrete in the compression zone.
3. When $\rho < \rho_b$, the amount of reinforcement corresponds to properly reinforced beams failing in the *steel-controlled failure mode*, characterized by the yielding of the steel reinforcement.

To ensure steel-controlled failure in reinforced concrete flexural members, it is recommended to use values of the reinforcement ratio up to 75% of ρ_b, that is, $\rho \leq 0.75\rho_b$.

Example 5.4

A reinforced concrete beam needs to be designed using the following properties of steel reinforcement and concrete: $f_y = 400$ MPa and $f_c' = 25$ MPa.
Determine the balanced reinforcement ratio (ρ_b).

SOLUTION: Since the concrete and steel material properties are given, the balanced reinforcement ratio ρ_b can be obtained as

$$\rho_b = \left(\frac{\alpha_1 \beta_1 \phi_c f_c'}{\phi_s f_y} \right) \left(\frac{700}{700 + f_y} \right)$$ [5.14]

$$= \left(\frac{0.8 \times 0.9 \times 0.65 \times 25 \text{ MPa}}{0.85 \times 400 \text{ MPa}} \right) \left(\frac{700}{700 + 400 \text{ MPa}} \right) = 0.022$$

Therefore,

$$\rho_b = 0.022 = 2.2\%$$

Use the approximate formula (Eqn 5.15):

$$\rho_b = \frac{f_c'}{1100} = \frac{25 \text{ MPa}}{1100} = 0.0227$$

A slightly larger ρ_b value is obtained in this way; however, the difference between the accurate and the approximate value is not significant.

Note that the ρ_b value can also be obtained from Table A.4; the table gives the value of 0.022 (same as the above calculation).

CSA A23.3 prescribes the following requirements relating to the design of rectangular beams reinforced with tension steel only:

- The *strength requirement* (A23.3 Cl.8.1.3) is the key requirement related to the design of concrete members.
- The *maximum amount of reinforcement* (A23.3 Cl.10.5.2) represents the upper limit for the amount of tension reinforcement for beams without axial loads and is equal to the balanced reinforcement.
- The *minimum amount of reinforcement* (A23.3 Cl.10.5.1) is one of two alternative requirements to prevent the design of underreinforced beams that fail in a sudden and brittle manner.

5.6.2 Design of Rectangular Beams with Tension Steel Only: Summary and a Design Example

The main design objective is to determine the required beam dimensions and reinforcement such that a beam is able to carry given design loads. In the case of a beam subjected to the factored bending moment (M_f), the following strength requirement needs to be satisfied according to CSA A23.3 Cl.8.1.3:

$$M_f \leq M_r \tag{5.1}$$

It is common practice to design reinforced concrete beams using the design procedures explained in Section 5.5. In addition to the strength requirement, a sound design takes into account other requirements, including detailing requirements (concrete cover and bar spacing), discussed in Section 5.3, and serviceability criteria (cracking control factor and deflections), discussed in Chapter 4. These requirements will be addressed in examples and problems presented in this chapter. Other considerations, including shear and torsion effects and anchorage requirements, will be discussed in Chapters 6, 7, and 9.

This section is focused on assisting the reader in his/her effort to determine the required cross-sectional dimensions and tension reinforcement in the beam subjected to given loads, as shown in Figure 5.8. This is a more challenging task than that performed in Chapter 3, where the beam dimensions and reinforcement were given and it was required to determine the factored moment resistance. As mentioned earlier in this chapter, "cookbook" design solutions are generally not available. However, general steps related to beam design are outlined in Checklist 5.1. Although the steps have been presented in a certain sequence, it is not necessary to follow the same sequence in all design situations. It should also be stressed that design is an iterative process and that one or more design steps may need to be repeated, as shown in Example 5.5.

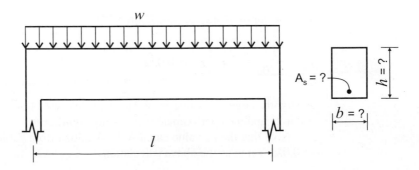

Figure 5.8 Beam design problem.

Checklist 5.1 Design of Rectangular Beams with Tension Steel Only for Flexure

Step	**Given:** - Specified loads acting on the beam - Concrete and steel material properties (f_c' and f_y) **Description**	**Code Clause**
1	Calculate the factored bending moment (M_f) at certain critical sections. Use the specified loads and the NBC 2010 load combinations to calculate factored loads acting on the beam. Perform a structural analysis or use the beam load diagrams from Appendix A to calculate the factored bending moment for the design.	
2	Estimate the beam dimensions b and h. Use the guidelines outlined in Section 5.4.1.	
3	Estimate the effective depth (d). Beams with one layer of reinforcement: $d \cong h - 70 \text{ mm}$ Beams with two layers of reinforcement: $d \cong h - 110 \text{ mm}$	
4	Estimate the required area of tension reinforcement (A_s). Use either the direct procedure or the iterative procedure (see Section 5.5).	
5	Confirm that the maximum tension reinforcement requirement is satisfied (ensure a properly reinforced beam section). $\rho \leq \rho_b$ where $\rho = \dfrac{A_s}{bd}$ **[3.1]** or $\boxed{\text{A23.3 Eq. 10.5}}$ $\dfrac{c}{d} \leq \dfrac{700}{700 + f_y}$ **[5.9]** Note that the balanced reinforcement corresponds to $c \cong \dfrac{2}{3}d$ **[5.11]** or $\rho_b \cong \dfrac{f_c'}{1100}$ **[5.15]** (provided that Grade 400 steel is used); alternatively, use Table A.4 to find ρ_b.	10.5.2
6	Determine the actual effective depth (d). This is based on the detailing requirements and actual reinforcement size (see Section 5.3).	

(Continued)

Checklist 5.1   Continued

Step	Given: - Specified loads acting on the beam - Concrete and steel material properties (f_c' and f_y) Description		Code Clause
7	Confirm that the minimum reinforcement requirement is satisfied.		
	$\boxed{\text{A23.3 Eq. 10.3}}$ $M_r \geq 1.2 M_{cr}$	[5.6]	10.5.1.1
	or		
	$\boxed{\text{A23.3 Eq. 10.4}}$ $A_s > A_{smin} = \dfrac{0.2\sqrt{f_c'}}{f_y} b_t h$	[5.7]	10.5.1.2
	Note that $b_t - b$ for rectangular sections.		
8	Calculate M_r.		
	$M_r = \phi_s f_y A_s \left(d - \dfrac{a}{2} \right)$	[3.14]	
9	Confirm that the strength requirement is satisfied.		
	$M_r \geq M_f$	[5.1]	8.1.3
10	Check the crack control parameter (z) (see Section 4.8).		10.6.1
	$z \leq 30\ 000$ N/mm $\rightarrow$ interior exposure; $z \leq 25\ 000$ N/mm $\rightarrow$ exterior exposure.		
11	Provide a design summary. Summarize the design with a sketch showing a) the beam width b) the total beam depth c) the tension reinforcement size and number of bars d) the clear concrete cover and effective depth e) the stirrup size (if available)		

Example 5.5

Consider the reinforced concrete beam of rectangular cross-section in the figure below. The beam spans 8 m (centre-to-centre between the supports) and it supports a uniform dead load (DL) of 36 kN/m (including self-weight) and a uniform live load (LL) of 36 kN/m. There are no constraints related to the selection of the beam cross-sectional dimensions. The beam is located in the interior of a building. Use 15M bars for stirrups and 30M bars for tension reinforcement and 20 mm maximum aggregate size for the concrete. The concrete and steel material properties are given below.

Design the beam of rectangular cross-section that is capable of carrying the given loads according to the CSA A23.3 requirements.

Given: $f_c' = 25$ MPa
$f_y = 400$ MPa
$\phi_c = 0.65$
$\phi_s = 0.85$
$\alpha_1 \cong 0.8$

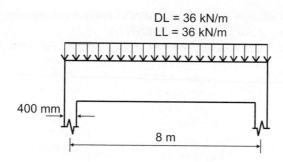

SOLUTION: **1.   Calculate the factored bending moment**

The factored load (w_f) is determined according to NBC 2010 Cl.4.1.3.2 (see Section 1.8):

$$w_f = 1.25DL + 1.5LL = 1.25(36 \text{ kN/m}) + 1.5(36 \text{ kN/m})$$

$$= 99 \text{ kN/m} \cong 100 \text{ kN/m}$$

This is a simply supported beam and the maximum bending moment at the midspan can be calculated from the equations of equilibrium or by referring to the beam load diagrams included in Appendix A. (Experienced designers memorize the formula for the maximum bending moment of a simply supported beam under uniform load!)

$$M_f = \frac{w_f \times l^2}{8} = \frac{100 \text{ kN/m} \times (8 \text{ m})^2}{8} = 800 \text{ kN} \cdot \text{m}$$

2.   Estimate the beam dimensions b and h

a)   Estimate the beam depth (h).

The beam cross-sectional dimensions can be estimated by using the practical guidelines outlined in Section 5.4.1. Table A.3 (based on A23.3 Cl.9.8.2.1) prescribes that a simply supported beam with a depth of $l_n/16$ or larger (where l_n is the clear span for the beam under consideration) satisfies the CSA A23.3 deflection requirements, and detailed deflection calculations are not required. In general, this is considered to be a good initial estimate for the overall depth. Based on design experience, overall beam depths in the range from $l_n/10$ to $l_n/16$ should lead to an economical design. Choose an intermediate value of $l_n/12$; therefore,

$$l_n = 8000 \text{ mm} - 2\left(\frac{400 \text{ mm}}{2}\right) = 7600 \text{ mm} \quad \text{(clear span)}$$

and

$$h \cong \frac{l_n}{12} = \frac{7600 \text{ mm}}{12} = 633 \text{ mm}$$

Round the overall depth value to

$$h = 700 \text{ mm}$$

b)   Estimate the beam width (b).

The beam width can be estimated considering the beam depth/width ratio (h/b) in the range from 1.5 to 2.0 (see Section 5.4.1). Use the ratio of 2.0 in this case; that is,

$$b \cong 0.5 \times h = 0.5 \times 700 \text{ mm} = 350 \text{ mm}$$

3.   Estimate the effective beam depth

Estimate the effective depth (d) based on the guidelines provided in Section 5.4.1 (use one layer of reinforcement):

$$d = h - 70 \text{ mm}$$

$$= 700 \text{ mm} - 70 \text{ mm} = 630 \text{ mm}$$

4. **Calculate the required area of tension reinforcement**

   a) Take $M_r = M_f = 800$ kN·m.

   b) Calculate the required area of tension reinforcement using the direct procedure as

   $$A_s = 0.0015 f_c' b \left(d - \sqrt{d^2 - \frac{3.85\, M_r}{f_c'\, b}} \right) \qquad \textbf{[5.4]}$$

   $$= 0.0015 \times 25 \text{ MPa}$$

   $$\times 350 \text{ mm} \left(630 \text{ mm} - \sqrt{(630 \text{ mm})^2 - \frac{3.85(800 \times 10^6 \text{ N·mm})}{25 \text{ MPa} \times 350 \text{ mm}}} \right)$$

   $$= 5488 \text{ mm}^2$$

   c) Select the amount of reinforcement in terms of the number and size of the bars. Refer to Table A.1; 8-30M bars are a suitable choice because the area of one 30M bar is 700 mm², and

   $$A_s = 8 \times 700 \text{ mm}^2 = 5600 \text{ mm}^2$$

   Check whether the provided area of reinforcement is greater than or equal to the required amount of reinforcement:

   $$A_s = 5600 \text{ mm}^2 > 5488 \text{ mm}^2$$

   The provided area of reinforcement is slightly larger than the required area, which is okay.

5. **Confirm that the maximum tension reinforcement requirement is satisfied (A23.3 Cl.10.5.2)**

   Check the reinforcement ratio:

   $$\rho = \frac{A_s}{bd} \qquad \textbf{[3.1]}$$

   $$= \frac{5600 \text{ mm}^2}{350 \text{ mm} \times 630 \text{ mm}} = 0.0254$$

   This ratio can be compared with the balanced reinforcement ratio (ρ_b) for $f_c' = 25$ MPa given in Table A.4. Since $\rho_b = 0.022$,

   $$\rho = 0.0254 > 0.022$$

   Therefore,

   $$\rho > \rho_b$$

   It follows that this beam is overreinforced and would fail in the brittle concrete-controlled mode of failure, which should be avoided. Also, a 15M stirrup requires at least 100 mm bend diameter, which is difficult to place in a 350 mm wide beam; 8-30M bars will require more space, if placed in one layer, than available within 350 mm beam width. Therefore, the beam cross-sectional dimensions need to be changed; that is, the beam should be deeper or wider (or both). Calculation Steps 2 to 5 need to be repeated. These steps will be denoted as 2R, 3R, 4R, and 5R.

2R. **Estimate the beam dimensions b and h**

   As the next trial, assume a deeper beam; note that increasing the member depth is the most efficient way of increasing the moment resistance. Therefore, consider a beam section that is 450 mm wide by 900 mm deep, that is,

   $$h = 900 \text{ mm}$$

   and

   $$b \cong 0.5 \times h = 0.5 \times 900 \text{ mm} = 450 \text{ mm}$$

3R. Estimate the effective beam depth

Determine the effective depth (d) using the same guideline as in Step 3:

$$d = h - 70 \text{ mm}$$

$$= 900 \text{ mm} - 70 \text{ mm} = 830 \text{ mm}$$

4R. Calculate the required area of tension reinforcement

a) Take $M_r = M_f = 800$ kN·m (same as Step 3).

b) Calculate the required area of tension reinforcement as

$$A_s = 0.0015 f_c' b \left(d - \sqrt{d^2 - \frac{3.85 M_r}{f_c' b}} \right) \qquad \text{[5.4]}$$

$$= 0.0015 \times 25 \text{ MPa}$$

$$\times 450 \text{ mm} \left(830 \text{ mm} - \sqrt{(830 \text{ mm})^2 - \frac{3.85(800 \times 10^6 \text{ N·mm})}{25 \text{ MPa} \times 450 \text{ mm}}} \right)$$

$$= 3134 \text{ mm}^2$$

c) Select the amount of reinforcement in terms of the number and size of the bars. Refer to Table A.1; 5-30M bars are a suitable choice since the area of one 30M bar is 700 mm², and

$$A_s = 5 \times 700 \text{ mm}^2 = 3500 \text{ mm}^2$$

Check whether the provided area of reinforcement is greater than or equal to the required amount of reinforcement:

$$A_s = 3500 \text{ mm}^2 > 3134 \text{ mm}^2$$

Therefore, the provided area of reinforcement is slightly greater than the required area; this is okay.

5R. Confirm that the maximum tension reinforcement requirement is satisfied (A23.3 Cl.10.5.2)

Check the steel ratio:

$$\rho = \frac{A_s}{bd} = \frac{3500 \text{ mm}^2}{450 \text{ mm} \times 830 \text{ mm}} = 0.0094$$

This ratio can be compared with the balanced reinforcement ratio (ρ_b) for $f_c' = 25$ MPa given in Table A.4; that is, $\rho_b = 0.022$. Hence,

$$\rho = 0.0094 < 0.022$$

or

$$\rho < \rho_b$$

The above check indicates that a beam would fail in the steel-controlled mode. The chosen beam dimensions and the reinforcement are acceptable and the design process should continue.

6 Determine the actual effective depth (d)

a) Determine the concrete cover.
Since the beam is of interior exposure, it must be considered in exposure class N according to the CSA A23.1 requirements. The concrete cover can be determined from Table A.2 as

$$\text{cover} = 30 \text{ mm}$$

b) Calculate the minimum required bar spacing (s_{min}).
   - 30M tension reinforcement (d_b = 30 mm) (see Table A.1),
   - 15M stirrups (d_s = 15 mm) (see Table A.1),
   - maximum aggregate size of 20 mm (a_{max} = 20 mm).

According to A23.1 Cl.6.6.5.2, the minimum spacing (s_{min}) between the reinforcement bars should be equal to the greatest of

1. $1.4 \times d_b = 1.4 \times 30$ mm = 42 mm
2. $1.4 \times a_{max} = 1.4 \times 20$ mm = 28 mm
3. 30 mm

Hence, the minimum spacing between reinforcement bars is s_{min} = 42 mm.

c) Determine the number of bars that can fit in one layer.
   Check whether that five bars can fit in the bottom layer. The minimum required beam width (b_{min}) is

$$b_{min} = 5d_b + 4s + 2d_s + 2 \times \text{cover}$$
$$= 5 \times 30 \text{ mm} + 4 \times 42 \text{ mm} + 2 \times 15 \text{ mm} + 2 \times 30 \text{ mm} = 408 \text{ mm}$$

Since

$$b_{min} = 408 \text{ mm} < b = 450 \text{ mm}$$

the maximum number of bars per layer is five. In this case, all of the bars can fit in one layer.

d) Determine the effective depth (d).
   The effective depth is the distance from the compression face of the beam to the centroid of the tension steel and can be determined as follows:

$$d = h - \text{cover} - d_s - \frac{d_b}{2}$$
$$= 900 \text{ mm} - 30 \text{ mm} - 15 \text{ mm} - \frac{30 \text{ mm}}{2} = 840 \text{ mm}$$

7. **Confirm that the minimum reinforcement requirement is satisfied (A23.3 Cl.10.5.1.2)**

   The minimum required reinforcement area (A_{smin}) can be determined based on A23.3 Cl.10.5.1.2 as

   | A23.3 Eq.10.4 |

   $$A_{smin} = \frac{0.2\sqrt{f_c'}}{f_y} b_t h \qquad [5.7]$$
   $$= \frac{0.2\sqrt{25 \text{ MPa}}}{400 \text{ MPa}} \times 450 \text{ mm} \times 900 \text{ mm} = 1012 \text{ mm}^2$$

   where

   $$b_t = b = 450 \text{ mm}$$

   is the width of the tension zone.

   Next, confirm that $A_s > A_{smin}$

   $$A_s = 3500 \text{ mm}^2 > A_{smin} = 1012 \text{ mm}^2 : \text{okay}$$

8. **Calculate M_r**
   a) First, calculate the depth of the compression stress block as

   $$a = \frac{\phi_s f_y A_s}{\alpha_1 \phi_c f_c' b} \qquad [3.12]$$

   $$= \frac{0.85 \times 400 \text{ MPa} \times 3500 \text{ mm}^2}{0.8 \times 0.65 \times 25 \text{ MPa} \times 450 \text{ mm}} = 203 \text{ mm}$$

   b) Then, calculate the moment resistance as

   $$M_r = \phi_s f_y A_s \left(d - \frac{a}{2} \right) \qquad [3.14]$$

   $$= 0.85 (400 \text{ MPa})(3500 \text{ mm}^2) \left(840 \text{ mm} - \frac{203 \text{ mm}}{2} \right)$$

   $$= 879 \times 10^6 \text{ N} \cdot \text{mm} = 879 \text{ kN} \cdot \text{m}$$

9. **Confirm that the strength requirement is satisfied**
   The strength requirement per A23.3 Cl.8.1.3 must be satisfied in the design; that is,

   $$M_r \geq M_f \qquad [5.1]$$

   Since

   $$M_f = 800 \text{ kN} \cdot \text{m}$$

   and

   $$M_r = 879 \text{ kN} \cdot \text{m} > 800 \text{ kN} \cdot \text{m}$$

   the strength requirement is satisfied.

10. **Check the crack control parameter (z) (A23.3 Cl.10.6.1)**
    a) Compute the effective tension area per bar (A).
       i) Calculate the distance between the centroid of the tensile reinforcement and tensile face of the concrete section (d_s) (see the sketch that follows):

       $$d_s = d_c = h - d$$

       $$= 900 \text{ mm} - 840 \text{ mm} = 60 \text{ mm}$$

       ii) Then, calculate the total effective tension area for all bars (A_e):

       $$A_e = b(2d_s)$$

       $$= (450 \text{ mm})(2 \times 60 \text{ mm}) = 54\,000 \text{ mm}^2$$

       iii) Finally, calculate the effective tension area per bar (A):
       For five bars in total, $N = 5$.
       Hence,

       $$A = \frac{A_e}{N} \qquad [1.77]$$

       $$= \frac{54\,000 \text{ mm}^2}{5} = 10\,800 \text{ mm}^2$$

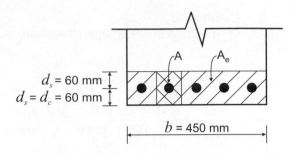

$d_s = 60$ mm

$d_s = d_c = 60$ mm

$b = 450$ mm

b) Determine the stress in steel reinforcement (f_s) under the service load level. The stress in steel reinforcement can be computed based on an approximate estimate permitted by CSA A23.3 Cl.10.6.1 in lieu of the detailed calculations, as follows:

$$f_s = 0.6f_y \qquad\qquad\qquad \textbf{[4.21]}$$

$$= 0.6 \times 400 \text{ MPa}$$

$$= 240 \text{ MPa}$$

c) Determine the value of z:

A23.3 Eq.10.6

$$z = f_s \sqrt[3]{d_c A} \qquad\qquad\qquad \textbf{[4.20]}$$

$$= 240 \text{ MPa} \sqrt[3]{(60 \text{ mm})(10\ 800 \text{ mm}^2)} = 20\ 768 \text{ N/mm}$$

Since

$$z = 20\ 768 \text{ N/mm} < 30\ 000 \text{ N/mm (okay)}$$

the CSA A23.3 cracking control requirements for interior exposure are satisfied.

11. Provide a design summary
Finally, a design summary showing the beam cross-section and the reinforcement is presented below.

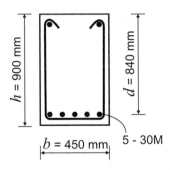

$h = 900$ mm

$d = 840$ mm

$b = 450$ mm   5 - 30M

5.7 | DESIGN OF ONE-WAY SLABS

5.7.1 CSA A23.3 Flexural Design Provisions for One-Way Slabs

The CSA A23.3 flexural design requirements for one-way slabs are very similar to the provisions for beam design discussed in Section 5.6.1. The *strength requirement* and the *maximum tension reinforcement requirement* for slabs are identical to the corresponding

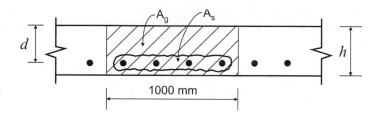

Figure 5.9 Slab reinforcement for a strip of unit width.

requirements for beams. The *minimum tension reinforcement requirement* for slabs is different from the one for beams and will be discussed in this section, along with a few other slab-specific requirements.

A23.3 Cl.7.8.1

Minimum slab reinforcement Minimum reinforcement in slabs is intended to provide cracking control due to shrinkage and temperature effects and to preserve slab integrity after cracking. The minimum reinforcement specified for slabs of uniform thickness is

$$A_{smin} = 0.002A_g \qquad\qquad [5.16]$$

where A_g is the gross cross-sectional area of a slab. For a slab strip of width $b = 1000$ mm and depth h, $A_g = 1000\,h$, as shown in Figure 5.9.

Note that the minimum slab reinforcement needs to be provided in both directions!

A23.3 Cl.7.4.1.2

Maximum tension bar spacing The maximum permitted bar spacing (s_{max}) is equal to the lesser of $3h$ and 500 mm, where h denotes the slab thickness.

A23.3 Cl.7.8.1 and 7.8.3

Shrinkage and temperature reinforcement As discussed in Section 2.3.6, concrete shrinks during the curing process. Columns, walls, and beams tend to restrain slab expansion caused by shrinkage; as a result, shrinkage cracks develop in reinforced concrete slabs. However, if steel reinforcement is placed at close spacing, small hairline cracks develop in a relatively distributed manner; this is considered as a desirable cracking pattern. CSA A23.3 prescribes the minimum amount and spacing of reinforcement in each slab direction to ensure uniform crack distribution due to shrinkage and temperature variations during the service life of the structure.

- The minimum amount of temperature reinforcement is the same as for tension steel ($A_{smin} = 0.002A_g$).
- The maximum permitted spacing (s_{max}) is equal to the lesser of $5h$ and 500 mm, where h denotes the slab thickness.

Reinforcement spacing requirements for one-way slabs, including tension steel and temperature and shrinkage reinforcement, are summarized in Figure 5.10.

A23.3 Cl.8.1.2

Fire-resistance requirements Fire resistance is one of the four limit states to be considered in the design of concrete structures, as discussed in Section 2.5. Detailed fire-resistance requirements for various structural members are specified in Appendix D of NBC 2010. In particular, minimum concrete cover and thickness requirements apply to slabs, depending on the required fire-resistance rating, which is expressed in hours. For example, a minimum slab thickness of 130 mm is required for a 2 hour fire rating according to Appendix D of NBC 2010 (for Type S concrete).

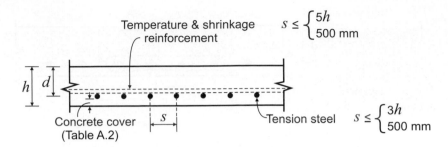

Figure 5.10 Reinforcement spacing requirements for slabs.

Similarly, minimum concrete cover requirements depending on the fire rating are also specified (as discussed in Section 5.3.1). The designer must ensure that the prescribed fire-resistance requirements have been met. In some cases, the fire-resistance requirements can govern over the strength requirements when choosing the dimensions of concrete structural members. As an example, one-way joist floor construction was abandoned in the 1970s due to more stringent fire-resistance requirements prescribing a larger slab thickness than is economical for joist floor construction. As a result, an alternative floor system, called slab band system, that meets the fire-resistance requirements is used in some parts of Canada. (For more details on floor systems, refer to Chapter 11.)

KEY REQUIREMENTS

The CSA A23.3 flexural design provisions for one-way slabs are very similar to the provisions for beams with tension steel only. However, there are some additional slab-specific requirements:

- minimum slab reinforcement (Cl.7.8.1)
- minimum tension bar spacing (Cl.7.4.1.2)
- shrinkage and temperature reinforcement (Cl.7.8.1 and 7.8.3)

Fire-resistance requirements should also be considered when designing the slab thickness (CSA A23.3 Cl.8.1.2 and Appendix D of NBC 2010).

5.7.2   Design of One-Way Slabs: Summary and a Design Example

The analysis of one-way slabs subjected to flexure was discussed in detail in Section 3.6. For design purposes, one-way slabs are treated as wide beams of unit width (b) equal to 1000 mm and depth (h) equal to the slab thickness. The general steps related to the design of one-way slabs are outlined in Checklist 5.2, followed by a design example. Although the steps have been presented in sequence, it is not necessary to follow the same sequence in all design situations.

Checklist 5.2 Design of One-Way Slabs for Flexure

Step	Given: - Specified loads acting on the slab - Unit width $b = 1000$ mm - Concrete and steel material properties (f_c' and f_y) Description	Code Clause
1	Calculate the factored bending moment (M_f) at a critical section. Use the specified loads and the NBC 2010 load combinations to calculate the factored loads acting on the slab. Perform a structural analysis or use the beam load diagrams given in Appendix A to calculate the factored bending moment to be used in the design.	
2	Estimate the slab thickness (h). Use the guidelines in Section 5.4.1.	
3	Estimate the effective depth (d). $d \cong h - 30$ mm If the required cover is greater than 20 mm, then the effective depth can be estimated from the equation $d \cong h - (\text{cover} + 10 \text{ mm})$	
4	Estimate the required area of tension reinforcement (A_s). Use either the direct procedure or the iterative procedure (see Section 5.5).	
5	Confirm that the maximum tension reinforcement requirement is satisfied. Ensure a properly reinforced beam section: $\rho \le \rho_b$ where $\rho = \dfrac{A_s}{bd}$ [3.1] Alternatively, check whether the following condition is satisfied A23.3 Eq. 10.5 $\dfrac{c}{d} \le \dfrac{700}{700 + f_y}$ [5.9] Note that the balanced condition corresponds to $c \cong \dfrac{2}{3} d$ [5.11] or $\rho_b \cong \dfrac{f_c}{1100}$ [5.15] (provided that Grade 400 steel is used); alternatively, use Table A.4 to find ρ_b.	10.5.2
6	Determine the actual effective depth (d). This is based on the detailing requirements and actual reinforcement sizes (see Section 5.3).	

(Continued)

Checklist 5.2 Continued

Step	Given: - Specified loads acting on the slab - Unit width $b = 1000$ mm - Concrete and steel material properties (f_c' and f_y) **Description**	Code Clause
7	Confirm that the minimum reinforcement requirement is satisfied, that is, $A_s > A_{smin} = 0.002\, A_g$ **[5.16]** where A_g is the gross cross-sectional area of a unit slab strip ($A_g = 1000h$).	7.8.1
8	Determine the required bar spacing. The required bar spacing can be determined when the reinforcing bar size (area A_b) has been assumed: $s \le A_b \dfrac{1000}{A_s}$ **[3.29]** Confirm that the maximum bar spacing requirement is satisfied, that is, $s \le s_{max}$ where s_{max} is the lesser of $3h$ and 500 mm.	7.4.1.2
9	Calculate M_r (assume properly reinforced slab). $M_r = \phi_s f_y A_s \left(d - \dfrac{a}{2} \right)$ **[3.14]**	
10	Confirm that the strength requirement is satisfied. $M_r \ge M_f$ **[5.1]**	8.1.3
11	Check the crack control parameter (z) (see Section 4.8). $z \le 30\,000$ N/mm $\rightarrow$ interior exposure; $z \le 25\,000$ N/mm $\rightarrow$ exterior exposure.	10.6.1
12	Design the shrinkage and temperature reinforcement. • The minimum area $A_{smin} = 0.002\, A_g$ **[5.16]** • s_{max} is the lesser of $5h$ and 500 mm. Note: Shrinkage and temperature reinforcement should be placed in the direction perpendicular to the main tension steel!	7.8.1 7.8.3
13	Provide a design summary. Summarize the design with a sketch showing the following information: a) total slab thickness b) tension reinforcement size and spacing c) shrinkage and temperature reinforcement size and spacing d) clear concrete cover	

Example 5.6

Consider the reinforced concrete slab shown in the figure below. The slab spans 6 m (centre-to-centre between the supports) and it supports a uniform dead load (DL) of 6 kPa (including self-weight) and a uniform live load (LL) of 5 kPa. There are no constraints related to the selection of the slab cross-sectional dimensions. Consider a 25 mm clear cover to the tension steel. Use 15M bars for tension reinforcement and 20 mm maximum aggregate size for the concrete. The concrete and steel material properties are given below.

Design a slab that is adequate to carry the given loads according to the CSA A23.3 requirements. Determine the slab thickness such that a detailed deflection calculation is not required.

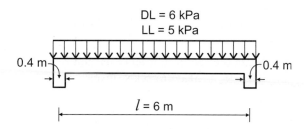

Given: $f_c' = 25$ MPa
$f_y = 400$ MPa
$\phi_c = 0.65$
$\phi_s = 0.85$
$\alpha_1 \cong 0.8$

SOLUTION:

1. **Calculate the factored bending moment**

   The factored load (w_f) is determined according to Cl.4.1.3.2 of NBC 2010:

   $$w_f = 1.25DL + 1.5LL = 1.25(6 \text{ kPa}) + 1.5(5 \text{ kPa}) = 15 \text{ kPa}$$

   However, for design purposes, one-way slabs are considered as rectangular beams of width

   $$b = 1 \text{ m}$$

   Therefore, the factored uniform load per metre of slab width is

   $$w_f = 15 \text{ kPa} \times 1 \text{ m} = 15 \text{ kN/m}$$

   This is a simply supported slab and the maximum bending moment at the midspan can be calculated from the equations of equilibrium or by referring to the beam load diagrams in Appendix A:

   $$M_f = \frac{w_f l^2}{8} = \frac{15 \text{ kN/m} (6 \text{ m})^2}{8} = 67.5 \text{ kN} \cdot \text{m/m}$$

2. **Estimate the slab thickness (h)**

   The slab thickness can be estimated by using the practical guidelines offered in Section 5.4.1. Table A.3 (based on A23.3 Cl.9.8.2.1) prescribes that a simply supported slab with a depth of $l_n/20$ or larger (where l_n is the clear span for the slab under consideration) satisfies the CSA A23.3 deflection requirements such that detailed deflection calculations are not required. In general, this is considered to be a good initial estimate for the slab thickness. Therefore,

   $$l_n = 6000 \text{ mm} - 2\left(\frac{400 \text{ mm}}{2}\right) = 5600 \text{ mm} \quad \text{(clear span)}$$

and

$$h \geq \frac{l_n}{20} = \frac{5600 \text{ mm}}{20} = 280 \text{ mm} \cong 300 \text{ mm}$$

3. **Estimate the effective slab depth (d)**
   Note that 15M tension reinforcement is specified for this design, that is, $d_b = 15$ mm (see Table A.1), and

   cover = 25 mm (given)

   Hence,

   $$d = h - \text{cover} - \frac{d_b}{2}$$

   $$= 300 \text{ mm} - 25 \text{ mm} - \frac{15 \text{ mm}}{2} = 267 \text{ mm} \cong 270 \text{ mm}$$

4. **Calculate the required area of tension reinforcement**
   a) Take $M_r = M_f = 67.5$ kN·m/m
   b) Calculate the required area of tension reinforcement using the direct procedure:

   $$A_s = 0.0015 f_c' b \left(d - \sqrt{d^2 - \frac{3.85 M_r}{f_c' b}} \right) \qquad [5.4]$$

   $$= 0.0015 \times 25 \text{ MPa}$$

   $$\times 1000 \text{ mm} \left(270 \text{ mm} - \sqrt{(270 \text{ mm})^2 - \frac{3.85 (67.5 \times 10^6 \text{ N·mm})}{25 \text{ MPa} \times 1000 \text{ mm}}} \right)$$

   $$= 750 \text{ mm}^2/\text{m}$$

   A typical slab section is shown on the sketch below.

   $d = 270$ mm   •A_s $h = 300$ mm

   $b = 1000$ mm

   c) Determine the required bar spacing.
   Refer to Table A.1; the area of one 15M bar is 200 mm^2; that is,

   $$A_b = 200 \text{ mm}^2$$

   The required bar spacing can be determined as

   $$s \leq A_b \frac{1000}{A_s} \qquad [3.29]$$

   $$= (200 \text{ mm}^2) \frac{1000}{750 \text{ mm}^2/\text{m}} = 267 \text{ mm} \cong 250 \text{ mm}$$

   Note that the required spacing of 267 mm has been rounded down to 250 mm!
   d) Check whether the provided area of reinforcement is greater than or equal to the required amount of reinforcement:

   $$A_s = A_b \frac{1000}{s} \qquad [3.28]$$

   $$= (200 \text{ mm}^2) \frac{1000}{250 \text{ mm}} = 800 \text{ mm}^2/\text{m}$$

   $$A_s = 800 \text{ mm}^2/\text{m} > 750 \text{ mm}^2/\text{m}$$

The provided area of reinforcement is slightly larger than the required area, which is okay. Therefore, the selected tension reinforcement consists of 15M bars at 250 mm spacing, that is, 15M@250.

5. **Confirm that the maximum tension reinforcement requirement is satisfied (A23.3 Cl.10.5.2)**
Check the reinforcement ratio

$$\rho = \frac{A_s}{bd} \tag{3.1}$$

$$= \frac{800 \text{ mm}^2}{1000 \text{ mm} \times 270 \text{ mm}} = 0.003$$

This ratio can be compared with the balanced reinforcement ratio (ρ_b) for $f_c' = 25$ MPa given in Table A.4. Since $\rho_b = 0.022$,

$$\rho = 0.003 < 0.022$$

or

$$\rho < \rho_b$$

The above check indicates that a beam would fail in the steel-controlled mode. The chosen slab thickness and the reinforcement are acceptable and the design process should proceed.

6. **Determine the actual effective depth**
In this case, the actual effective depth (d) is the same as estimated in Step 3; that is,

$$d = 270 \text{ mm}$$

7. **Confirm that the minimum reinforcement requirement is satisfied (A23.3 Cl.7.8.1)**
a) Calculate the gross cross-sectional area for the unit strip as

$$A_g = b \times h$$
$$= 1000 \text{ mm} \times 300 \text{ mm} = 300\,000 \text{ mm}^2$$

b) Determine the minimum reinforcement area:

$$A_{smin} = 0.002 \, A_g \tag{5.16}$$
$$= 0.002 \, (300\,000 \text{ mm}^2) = 600 \text{ mm}^2/\text{m}$$

c) Check whether the provided reinforcement area (A_s) is adequate:

$$A_s = 800 \text{ mm}^2/\text{m} > 600 \text{ mm}^2/\text{m} : \text{okay}$$

8. **Confirm that the maximum bar spacing requirement is satisfied (A23.3 Cl.7.4.1.2)**
a) Calculate the maximum bar spacing (s_{max}) as the lesser of

$$3 \times h = 3 \times 300 \text{ mm} = 900 \text{ mm} \text{ and } 500 \text{ mm}$$

The smaller value governs, so

$$s_{max} = 500 \text{ mm}$$

b) Compare the actual bar spacing with the maximum bar spacing:

$$s = 250 \text{ mm} < 500 \text{ mm} : \text{okay}$$

9. **Calculate M_r**
   a) First, calculate the depth of the compression stress block as

$$a = \frac{\phi_s f_y A_s}{\alpha_1 \phi_c f_c' b} \tag{3.12}$$

$$= \frac{0.85 \times 400 \text{ MPa} \times 800 \text{ mm}^2}{0.8 \times 0.65 \times 25 \text{ MPa} \times 1000 \text{ mm}} = 21 \text{ mm}$$

   b) Then, calculate the moment resistance as

$$M_r = \phi_s f_y A_s \left(d - \frac{a}{2} \right) \tag{3.14}$$

$$= 0.85 \,(400 \text{ MPa})(800 \text{ mm}^2/\text{m}) \left(270 \text{ mm} - \frac{21 \text{ mm}}{2} \right)$$

$$= 70.6 \times 10^6 \text{ N} \cdot \text{mm/m} \cong 71 \text{ kN} \cdot \text{m/m}$$

10. **Confirm that the strength requirement is satisfied (A23.3 Cl.8.1.3)**
    The strength requirement per A23.3 Cl.8.1.3 must be satisfied in the design; that is,

$$M_r \geq M_f \tag{5.1}$$

Because

$$M_f = 67.5 \text{ kN} \cdot \text{m/m}$$

and

$$M_r = 71 \text{ kN} \cdot \text{m/m} > 67.5 \text{ kN} \cdot \text{m/m}$$

the strength requirement is satisfied (note that the M_f and M_r values are quite close).

11. **Check the crack control parameter (z) (A23.3 Cl.10.6.1)**
    a) Compute the effective tension area per bar (A).
       i) Calculate the distance (d_s) from the centroid of the tension reinforcement to the tension face of the concrete section:

$$d_s = d_c = h - d$$

$$= 300 \text{ mm} - 270 \text{ mm} = 30 \text{ mm}$$

       ii) Then, calculate the effective tension area per bar (A).
          In this case, the A value can be determined directly for one bar with a spacing s, as follows (see the sketch that follows):

$$A = s(2d_s)$$

$$= (250 \text{ mm})(2 \times 30 \text{ mm})$$

$$= 15\,000 \text{ mm}^2$$

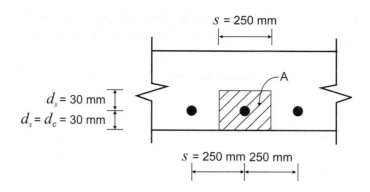

b) Determine the stress in the steel reinforcement (f_s) under the service load level. An approximate estimate for f_s permitted by A23.3 Cl.10.6.1 will be used in lieu of the detailed calculations:

$$f_s = 0.6f_y \qquad \text{[4.21]}$$
$$= 0.6 \times 400 \text{ MPa} = 240 \text{ MPa}$$

c) Determine the value of z.

A23.3 Eq. 10.6

$$z = f_s \sqrt[3]{d_c A} \qquad \text{[4.20]}$$
$$= 240 \text{ MPa} \sqrt[3]{(30 \text{ mm})(15\,000 \text{ mm}^2)} = 18\,391 \text{ N/mm}$$

Since

$$z = 18\,391 \text{ N/mm} < 30\,000 \text{ N/mm}$$

the CSA A23.3 cracking control requirements for interior exposure are satisfied.

12. **Design the shrinkage and temperature reinforcement (A23.3 Cl.7.8.1 and 7.8.3)**
   a) The minimum area of shrinkage and temperature reinforcement is the same as for the tension steel determined in Step 7:

   $$A_{smin} = 0.002A_g \qquad \text{[5.16]}$$
   $$= 0.002(300 \text{ mm} \times 1000 \text{ mm}) = 600 \text{ mm}^2/\text{m}$$

   b) The maximum bar spacing (s_{max}) is the lesser of

   $$5 \times h = 5 \times 300 \text{ mm} = 1500 \text{ mm}$$

   and

   $$500 \text{ mm}$$

   The smaller value governs, so

   $$s_{max} = 500 \text{ mm}$$

   c) Determine the required bar spacing.
   Refer to Table A.1; the area of one 15M bar is 200 mm², that is,

   $$A_b = 200 \text{ mm}^2$$

   The required spacing can be determined as

   $$s = A_b \frac{1000}{A_s} \qquad \text{[5.29]}$$

   $$= (200 \text{ mm}^2)\frac{1000}{600 \text{ mm}^2/\text{m}} = 333 \text{ mm} \cong 300 \text{ mm}$$

   Note that the bar spacing of 333 mm has been rounded down to 300 mm!

d) Check whether the provided area of reinforcement is greater than or equal to the required amount of reinforcement:

$$A_s = A_b \frac{1000}{s} \qquad\qquad \textbf{[3.28]}$$

$$= (200 \text{ mm}^2) \frac{1000}{300 \text{ mm}} = 667 \text{ mm}^2/\text{m}$$

$$A_s = 667 \text{ mm}^2/\text{m} > 600 \text{ mm}^2/\text{m} : \text{okay}$$

The provided area of reinforcement is slightly larger than the required area, which is okay. Therefore, the selected tension reinforcement consists of 15M bars at 300 mm spacing, that is, 15M@300.

13. Provide a design summary

Finally, a design summary showing the slab cross-section and the reinforcement is presented below.

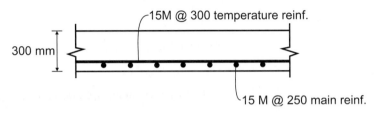

5.8 DESIGN OF T-BEAMS

5.8.1 CSA A23.3 Flexural Design Provisions for T-Beams

The CSA A23.3 flexural design provisions for beams with tension steel only also apply to T-beams (see Section 5.6.1). However, a few additional provisions specific to T-beams, mainly contained in CSA A23.3 Cl.10.3 and 10.5.3, are discussed in this section.

A23.3 Cl.10.3.3 and 10.3.4

Effective flange width (b_f)   Extensive tests have been conducted to determine the slab width that acts as a flange in a T-beam; this width is termed the *effective flange width* (b_f). The size of the effective flange width is affected by the type of loading (uniform or concentrated), the structural system (simple/continuous), the beam spacing, and the relative stiffness of the slabs and the beams. CSA A23.3 provides rules for estimating the effective flange width for design purposes.

The CSA A23.3 provisions for the effective flange width in T- and L-beams are summarized in Table 5.2, and the notation related to the effective flange width in T- and L-beams is summarized below (see also Figure 5.11):

b_T = the overhanging flange width (T-beams)
b_L = the overhanging flange width (L-beams)
h_f = the flange thickness
l_n = the clear span of a beam (for example, note the span l_n shown in Figures 3.26 and 3.28), and
l_w = the clear distance between two adjacent webs (see Figure 5.11); in the case of a beam with different l_w values on the sides of the web, the smaller value should be used to calculate b_T or b_L.

Table 5.2 Overhanging flange width requirements for T- and L-beams per CSA A23.3

T-beams (b_T) Cl.10.3.3	**L-beams** (b_L) Cl.10.3.4
$b_T \leq$ the smallest of	$b_L \leq$ the smallest of
a) $\dfrac{l_n}{5}$ for a simple beam,	a) $\dfrac{l_n}{12}$,
$\dfrac{l_n}{10}$ for a continuous beam;	b) $6\,h_f$, and
b) $12h_f$; and	c) $\dfrac{l_w}{2}$
c) $\dfrac{l_w}{2}$	

Once the overhanging flange width has been calculated based on CSA A23.3 provisions, the effective flange width (b_f) can be determined as follows:

$$b_f = b_w + 2b_T \quad \text{(T-beams)} \tag{5.17}$$

$$b_f = b_w + b_L \quad \text{(L-beams)}$$

A23.3 Cl.10.5.3 | **Reinforcement in T-beam flanges**   T-beams subjected to negative bending, where the web is under compression, should be designed as rectangular beams with the width $b = b_w$. In this case, the tension steel is placed at the top of the section, however a small portion of the top reinforcement should be distributed in the slab to ensure compatibility between the beam and the slab. This reinforcement, usually provided close to the web side faces, is effective in reducing congestion at the column or support locations. According to A23.3 Cl.10.5.3.1, part of the flexural tension reinforcement shall be distributed in the slab over the overhang width (b'), as illustrated with a hatched pattern in Figure 5.12.

According to A23.3 Cl.10.5.3.1, the b' value should be less than the smaller of the following:

a) $\dfrac{l_n}{20}$ and

b) b_T or b_L determined from A23.3 Cl.10.3.3 or 10.3.4.

The minimum required reinforcement area (A_{sf}) provided within the overhang (b'), shown in Figure 5.12, is

$$A_{sf} \geq 0.004\, b'\, h_f$$

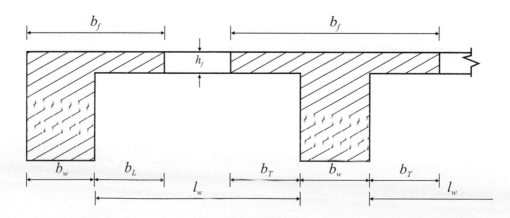

Figure 5.11 Effective flange width for T- and L-beams.

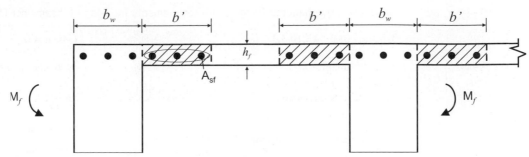

Figure 5.12 Reinforcement in flanges of T-beams (negative moment zones).

KEY REQUIREMENTS

The CSA A23.3 flexural design provisions specific to T-beams are summarized below.

The *effective flange width* is the width of the slab that serves as a flange in T-beams. The effective flange width is affected by the type of loading (uniform, concentrated), the structural system (simple/continuous), the spacing of the beams, and the relative stiffness of the slabs and the beams. Cl.10.3.3 and 10.3.4 of CSA A23.3 provide rules for estimating the effective flange width for design purposes.

Reinforcement in T-beam flanges needs to be provided to ensure an adequate beam-to-slab connection in T-beams subjected to negative bending; this reinforcement consists of a small portion of the top reinforcement distributed in the slab close to the beam web side faces (CSA A23.3 Cl.10.5.3.1).

5.8.2   Design of T-Beams: Summary and a Design Example

The design of T-beams is based on the moment resistance equations presented in detail in Section 3.7. The general steps related to the design of T-beams are outlined in Checklist 5.3, followed by a design example. Although the steps have been presented in a certain sequence, it is not necessary to follow the same sequence in all design situations.

It is usually considered in T-beam design that the neutral axis is within the flange; this corresponds to T-beams with a moderate amount of reinforcement, which are more common in design practice. A typical T-beam section with the neutral axis in the flange is shown in Figure 5.13.

It should also be noted that the slab thickness is usually a given parameter in T-beam design. The slab thickness criterion related to the effective slab width often governs (see Table 5.2).

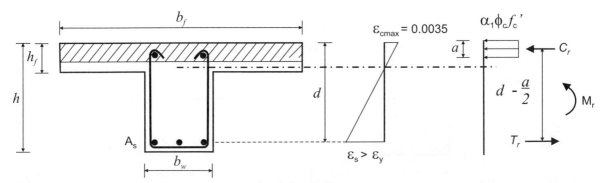

Figure 5.13 A T-beam section with the neutral axis in the flange.

Checklist 5.3   Design of T-Beams for Flexure

	Given: - Specified loads acting on the beam - Concrete and steel material properties (f_c' and f_y)	
Step	**Description**	**Code Clause**
1	Calculate the factored bending moment (M_f) at certain critical sections. Use the specified loads and the NBC 2010 load combinations to calculate the factored loads acting on the beam. Perform a structural analysis or use the beam load diagrams given in Appendix A to calculate the factored bending moment to be used in the design.	
2	Estimate the beam web width (b_w) and overall depth (h). Use the guidelines related to rectangular beam sections presented in Section 5.4.1.	
3	Estimate the effective depth (d). Beams with one layer of reinforcement: $$d \cong h - 70 \text{ mm}$$ Beams with two layers of reinforcement: $$d \cong h - 110 \text{ mm}$$	
4	Calculate the effective flange width (b_f). Refer to Table 5.2.	10.3.3 and 10.3.4
5	Estimate the required area of tension reinforcement (A_s). Use either the direct procedure or the iterative procedure (see Section 5.5).	
6	Confirm that the maximum tension reinforcement requirement is satisfied (ensure a properly reinforced beam section). $$\rho \le \rho_b$$ where $$\rho = \frac{A_s}{bd} \qquad [3.1]$$ Alternatively, check whether the following condition is satisfied $$\boxed{\text{A23.3 Eq. 10.5}} \qquad \frac{c}{d} \le \frac{700}{700 + f_y} \qquad [5.9]$$ It is recommended use the above equation to confirm that the section is properly reinforced. Note that the alternative check ($\rho \le \rho_b$) can be used only for sections with rectangular-shaped compression zone.	10.5.2
7	Determine the actual effective depth (d). This is based on the detailing requirements and actual reinforcement sizes (see Section 5.3).	

(Continued)

Checklist 5.3   Continued

Step	Description		Code Clause
	Given: - **Specified loads acting on the beam** - **Concrete and steel material properties (f_c' and f_y)**		
8	Confirm that the minimum reinforcement requirement is satisfied. $M_r \geq 1.2\, M_{cr}$ **[5.6]** or $\boxed{\text{A23.3 Eq. 10.4}}$ $A_s > A_{smin} = \dfrac{0.2\sqrt{f_c'}}{f_y}\, b_t\, h$ **[5.7]** Note that $b_t - b_w$ for T-beams in positive bending.		10.5.1.1 10.5.1.2
9	Confirm that the neutral axis is in the flange. The depth of the compression stress block can be calculated as $a = \dfrac{\phi_s f_y A_s}{\alpha_1 \phi_c f_c'\, b}$ **[3.12]** where $b = b_f$ The neutral axis is located in the flange when $a < h_f$		
10	Calculate M_r. $M_r = \phi_s f_y A_s \left(d - \dfrac{a}{2} \right)$ **[3.14]**		
11	Confirm that the strength requirement is satisfied. $M_r \geq M_f$ **[5.1]**		8.1.3
12	Check the crack control parameter (z) (see Section 4.8). $z \leq 30\ 000\ \text{N/mm} \rightarrow$ interior exposure; $z \leq 25\ 000\ \text{N/mm} \rightarrow$ exterior exposure.		10.6.1
13	Provide a design summary. Summarize the design with a sketch showing the following information: a) beam width (web and flange) b) total beam depth c) tension reinforcement (size and number of bars) d) clear concrete cover e) stirrup size (if available)		

Example 5.7

Consider a beam with the same span, loading, and material properties as discussed in Example 5.5. Consider that the beam has been cast integrally with a 200 mm thick slab, as shown in the figure below. The clear distance between beams is 4 m. The beam has been supported by 400 mm square columns at each end. Use 15M bars for stirrups and 30M bars for tension reinforcement, and 20 mm maximum aggregate size for the concrete.

Design a T-beam capable of carrying the given loads according to the CSA A23.3 requirements.

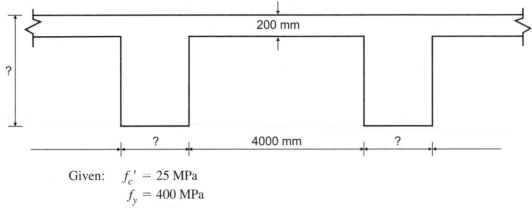

Given: $f_c' = 25$ MPa
$f_y = 400$ MPa
$\phi_c = 0.65$
$\phi_s = 0.85$
$\lambda = 1$ (normal-density concrete)
$\alpha_1 \cong 0.8$

SOLUTION:

1. Calculate the factored bending moment

The factored bending moment is the same as calculated in Example 5.5; that is,

$$M_f = 800 \text{ kN} \cdot \text{m}$$

2. Estimate the web width (b_w) and the overall depth (h)

Cross-sectional dimensions for T-beams can be estimated by using the guidelines outlined in Section 5.4.1. The web width and overall depth can be estimated in the same manner as for rectangular beams. In this case, use the dimensions from Example 5.5; that is,

$$h = 900 \text{ mm}$$

and

$$b_w = 450 \text{ mm}$$

Note that the slab thickness is given as

$$h_f = 200 \text{ mm}$$

3. Estimate the effective beam depth

Estimate the effective depth (d) based on the guidelines provided in Section 5.4.1 (use one layer of reinforcement):

$$d = h - 70 \text{ mm}$$
$$= 900 \text{ mm} - 70 \text{ mm} = 830 \text{ mm}$$

4.　**Calculate the effective flange width (b_f) (A23.3 Cl.10.3.3 and 10.3.4)**
 i) Calculate the clear span (l_n).
 The clear span is the distance between the beam supports; that is,

$$l_n = l - \left(\frac{b_s}{2} + \frac{b_s}{2} \right)$$

$$= 8000 \text{ mm} - \left(\frac{400 \text{ mm}}{2} + \frac{400 \text{ mm}}{2} \right) = 7600 \text{ mm}$$

 where $b_s = 400$ mm is the column width.
 ii) Calculate the clear distance between adjacent webs (l_w):

$$l_w = 4000 \text{ mm (given)}$$

 iii) Calculate the overhanging flange width (b_T) as the smallest of (refer to Table 5.2)

 a)　$\dfrac{l_n}{5} = \dfrac{7600 \text{ mm}}{5} \cong 1500$ mm (simply supported beam),

 b)　$12\,h_f = 12 \times 200$ mm $= 2400$ mm, and

 c)　$\dfrac{l_w}{2} = \dfrac{4000 \text{ mm}}{2} = 2000$ mm.

 The smallest of the three values governs; that is, $b_T = 1500$ mm.
 iv) Finally, the effective flange width (b_f) can be calculated as

$$b_f = b_w + 2b_T \qquad\qquad \text{[5.17]}$$

$$= 450 \text{ mm} + 2 \times 1500 \text{ mm} = 3450 \text{ mm}$$

 Therefore, the effective flange width is $b_f = 3450$ mm.
 The T-beam section used in this design is shown on the sketch below.

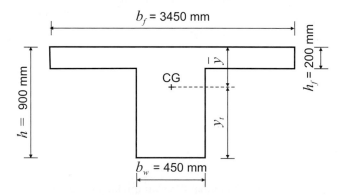

5.　**Calculate the required area of tension reinforcement**
 a)　Take $M_r = M_f = 800$ kN·m.
 b)　Estimate the width of the concrete compression zone.
 Since the T-beam is under positive bending, it is reasonable to consider that the neutral axis is located in the flange. In that case, the width of the concrete compression zone is equal to the effective flange width; that is,

$$b = b_f = 3450 \text{ mm}$$

c) Calculate the required area of tension reinforcement using the direct procedure:

$$A_s = 0.0015 f_c' b \left(d - \sqrt{d^2 - \frac{3.85 \, M_r}{f_c' b}} \right) \qquad \text{[5.4]}$$

$$= 0.0015 \times 25 \text{ MPa}$$

$$\times 3450 \text{ mm} \left(830 \text{ mm} - \sqrt{(830 \text{ mm})^2 - \frac{3.85 \, (800 \times 10^6 \text{ N} \cdot \text{mm})}{25 \text{ MPa} \times 3450 \text{ mm}}} \right)$$

$$= 2820 \text{ mm}^2$$

d) Select the amount of reinforcement in terms of the number and size of the bars. Refer to Table A.1; 4-30M bars are a suitable choice since the area of one 30M bar is 700 mm^2, and so

$$A_s = 4 \times 700 \text{ mm}^2 = 2800 \text{ mm}^2$$

Check whether the provided area of reinforcement is greater than or equal to the required amount of reinforcement:

$$A_s = 2800 \text{ mm}^2 < 2820 \text{ mm}^2$$

The provided area of reinforcement is slightly less than the required area; however, the difference is less than 1%. In general, a design that is within 5% of the actual requirement is considered acceptable. Therefore, use 4-30M bars.

6. **Confirm that the maximum tension reinforcement requirement is satisfied (A23.3 Cl.10.5.2)**

$$\rho = \frac{A_s}{bd} = \frac{2800 \text{ mm}^2}{3450 \text{ mm} \times 830 \text{ mm}} = 0.001 \qquad \text{[3.1]}$$

Since the reinforcement ratio is very small, there is no need to check whether the maximum tension reinforcement requirement has been satisfied. However, if such check is performed, it is recommended to use the following strain-based check

$$\frac{c}{d} \leq \frac{700}{700 + f_y} \qquad \text{[5.9]}$$

7. **Determine the actual effective depth (d)**
   a) Determine the concrete cover.
      Since the beam is of interior exposure, it can be considered as exposure class N according to the CSA A23.1 requirements. The concrete cover can be determined from Table A.2 as

      cover = 30 mm

   b) Determine the minimum required bar spacing (s_{min}):

      - 30M tension reinforcement ($d_b = 30$ mm) (see Table A.1),
      - 15M stirrups ($d_s = 15$ mm) (see Table A.1),
      - maximum aggregate size of 20 mm ($d_{max} = 20$ mm).

According to A23.1 Cl.6.6.5.2, the minimum bar spacing (s_{min}) should be the greatest of

1. $1.4 \times d_b = 1.4 \times 30$ mm $= 42$ mm,
2. $1.4 \times a_{max} = 1.4 \times 20$ mm $= 28$ mm, and
3. 30 mm.

Hence, the minimum spacing between reinforcement bars is $s_{min} = 42$ mm.

c)   Determine the number of bars that can fit in one layer.
     Estimate that all four bars can fit in the bottom layer. The minimum required beam width (b_{min}) is

$$b_{min} = 4d_b + 3s + 2d_s + 2 \times \text{cover}$$

$$= 4 \times 30 \text{ mm} + 3 \times 42 \text{ mm} + 2 \times 15 \text{ mm} + 2 \times 30 \text{ mm} = 336 \text{ mm}$$

Since

$$b_{min} = 336 \text{ mm} < b_w = 450 \text{ mm}$$

all four bars can fit in one layer.

d)   Determine the actual effective depth (d).
     The effective depth is the distance from the compression (top) face of the beam to the centroid of the tension steel and can be determined as follows:

$$d = h - \text{cover} - d_s - \frac{d_b}{2}$$

$$= 900 \text{ mm} - 30 \text{ mm} - 15 \text{ mm} - \frac{30 \text{ mm}}{2} = 840 \text{ mm}$$

8. **Confirm that the minimum reinforcement requirement is satisfied (A23.3 Cl.10.5.1.1 and 10.5.1.2)**

    There are two alternative CSA A23.3 minimum reinforcement requirements, and both of them will be checked in this example. However, it should be noted that only one of these requirements needs to be checked in a practical design situation.

    a)   Check the cracking moment criterion (CSA Cl.10.5.1.1):

| A23.3 Eq. 10.3 |

$$M_r \geq 1.2 M_{cr} \qquad \text{[5.6]}$$

    i)   Find the centroid of the gross concrete section ($\bar{y}$) as

$$\bar{y} = \frac{(b_f - b_w) \times \dfrac{h_f^2}{2} + b_w \dfrac{h^2}{2}}{(b_f - b_w) h_f + b_w \times h}$$

$$= \frac{3000 \text{ mm} \times \dfrac{(200 \text{ mm})^2}{2} + 450 \text{ mm} \times \dfrac{(900 \text{ mm})^2}{2}}{3000 \text{ mm} \times 200 \text{ mm} + 450 \text{ mm} \times 900 \text{ mm}} = 241 \text{ mm}$$

    ii)   Find the moment of inertia of the gross concrete section (I_g) as

$$I_g = (b_f - b_w) \times \frac{h_f^3}{12} + (b_f - b_w) \times h_f \times \left(\bar{y} - \frac{h_f}{2}\right)^2 + b_w \times \frac{h^3}{12}$$

$$+ b_w \times h \times \left(\frac{h}{2} - \bar{y}\right)^2$$

$$= (3450 \text{ mm} - 450 \text{ mm}) \times \frac{(200 \text{ mm})^3}{12} + (3450 \text{ mm} - 450 \text{ mm})$$

$$\times 200 \text{ mm} \times \left(241 \text{ mm} - \frac{200 \text{ mm}}{2} \right)^2 + 450 \text{ mm} \times \frac{(900 \text{ mm})^3}{12}$$

$$+ 450 \text{ mm} \times 900 \text{ mm} \times \left(\frac{900 \text{ mm}}{2} - 241 \text{ mm} \right)^2$$

$$= 5.9 \times 10^{10} \text{ mm}^4$$

iii) Calculate the cracking moment as

$$M_{cr} = \frac{f_r I_g}{y_t}$$ [4.1]

where

A23.3 Eq. 8.3

$$f_r = 0.6 \lambda \sqrt{f_c'} = 0.6 \times 1.0 \sqrt{25 \text{ MPa}} = 3.0 \text{ MPa}$$ [2.1]
$$\lambda = 1.0 \text{ (given)} \text{ and}$$
$$y_t = h - \bar{y} = 900 \text{ mm} - 241 \text{ mm} = 659 \text{ mm}$$

(Note that y_t denotes the distance from the centroid of the section to the extreme tension fibre.)
    Therefore,

$$M_{cr} = \frac{(3.0 \text{ MPa}) \times (5.9 \times 10^{10} \text{ mm}^4)}{659 \text{ mm}}$$

$$= 269 \times 10^6 \text{ N} \cdot \text{mm} = 269 \text{ kN} \cdot \text{m}$$

iv) Check the minimum reinforcement requirement (Eqn 5.6):

$$1.2 M_{cr} = 1.2 \times 269 \text{ kN} \cdot \text{m} = 323 \text{ kN} \cdot \text{m}$$

For the purposes of this check, take that

$$M_r = M_f = 800 \text{ kN} \cdot \text{m}$$

Since

$$800 \text{ kN} \cdot \text{m} > 323 \text{ kN} \cdot \text{m}$$

the minimum reinforcement requirement is satisfied.

b) Alternatively, find the minimum area of tension reinforcement (A23.3 Cl.10.5.1.2).
    i) Calculate the minimum reinforcement area (A_{smin}) as

A23.3 Eq. 10.4

$$A_{smin} = \frac{0.2 \sqrt{f_c'}}{f_y} b_t h$$ [5.7]

$$= \frac{0.2 \sqrt{25 \text{ MPa}}}{400 \text{ MPa}} \times 450 \text{ mm} \times 900 \text{ mm} = 1012 \text{ mm}^2$$

Note that b_t denotes the width of the tension zone. In the case of positive bending, the width of the tension zone is equal to the beam web width:

$$b_t = b_w = 450 \text{ mm}$$

ii) Confirm that $A_s > A_{smin}$.

$$A_s = 2800 \text{ mm}^2 > A_{smin} = 1012 \text{ mm}^2$$

Therefore, the second minimum reinforcement requirement is satisfied as well.

9. **Confirm that the neutral axis is in the flange**

The depth of the compression stress block can be calculated as

$$a = \frac{\phi_s f_y A_s}{\alpha_1 \phi_c f_c' b} \qquad\qquad \textbf{[3.12]}$$

$$= \frac{0.85 \times 400 \text{ MPa} \times 2800 \text{ mm}^2}{0.8 \times 0.65 \times 25 \text{ MPa} \times 3450 \text{ mm}} = 21 \text{ mm}$$

Since

$$a = 21 \text{ mm} < h_f = 200 \text{ mm}$$

the neutral axis is located in the flange.

A sketch showing the T-beam section and the internal force distribution is shown below.

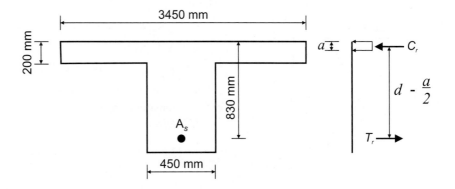

10. **Calculate M_r**

The factored moment resistance can be calculated as

$$M_r = \phi_s f_y A_s \left(d - \frac{a}{2} \right) \qquad\qquad \textbf{[3.14]}$$

$$= 0.85 \,(400 \text{ MPa})\,(2800 \text{ mm}^2) \left(840 \text{ mm} - \frac{21 \text{ mm}}{2} \right)$$

$$= 790 \times 10^6 \text{ N} \cdot \text{mm} = 790 \text{ kN} \cdot \text{m}$$

11. **Confirm that the strength requirement is satisfied (A23.3 Cl.8.1.3)**

The strength requirement per A23.3 Cl.8.1.3 must be satisfied in the design; that is,

$$M_r \geq M_f \qquad\qquad \textbf{[5.1]}$$

Since

$$M_f = 800 \text{ kN} \cdot \text{m}$$

and

$$M_r = 790 \text{ kN} \cdot \text{m} < 800 \text{ kN} \cdot \text{m}$$

the strength requirement is not formally satisfied; however, the difference (of less than 2%) is not significant enough to repeat the calculations.

12. **Check the crack control parameter (z) (A23.3 Cl.10.6.1)**

   a) Compute the effective tension area per bar (A).

       i) Find the distance (d_s) from the centroid of the tension reinforcement to the tension face of the concrete section:

$$d_s = d_c = h - d$$

$$= 900 \text{ mm} - 840 \text{ mm} = 60 \text{ mm}$$

ii) Find the total effective tension area for all bars (A_e).

$$A_e = b_w\,(2d_s)$$

$$= (450\ \text{mm})\,(2 \times 60\ \text{mm})$$

$$= 54\ 000\ \text{mm}^2$$

iii) Find the effective tension area per bar (A) (see the sketch below). For four bars in total, that is, $N = 4$,

$$A = \frac{A_e}{N}$$

$$= \frac{54\ 000\ \text{mm}^2}{4} = 13\ 500\ \text{mm}^2$$

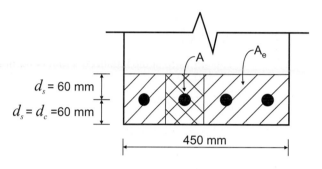

$d_s = 60$ mm

$d_s = d_c = 60$ mm

450 mm

b) Determine the stress in the steel reinforcement (f_s) under the service load. An approximate f_s estimate permitted by CSA A23.3 Cl.10.6.1 will be used in lieu of the detailed calculations, as follows:

$$f_s = 0.6 f_y \qquad \qquad \textbf{[4.21]}$$

$$= 0.6 \times 400\ \text{MPa} = 240\ \text{MPa}$$

c) Determine the value of z:

A23.3 Eq. 10.6

$$z = f_s\sqrt[3]{d_c\,A} \qquad \qquad \textbf{[4.20]}$$

$$= 240\ \text{MPa}\sqrt[3]{(60\ \text{mm})(13\ 500\ \text{mm}^2)} - 22\ 370\ \text{N/mm}$$

Since

$$z = 22\ 370\ \text{N/mm} < 30\ 000\ \text{N/mm}$$

the CSA A23.3 cracking control requirements for interior exposure are satisfied.

13. Provide a design summary

Finally, a design summary showing the beam cross-section and the reinforcement is presented below.

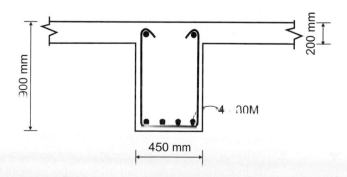

300 mm

200 mm

4 - 00M

450 mm

5.9 DESIGN OF RECTANGULAR BEAMS WITH TENSION AND COMPRESSION REINFORCEMENT

5.9.1 CSA A23.3 Flexural Design Provisions for Beams with Tension and Compression Reinforcement

The CSA A23.3 flexural design provisions for beams with tension steel only also apply to beams with tension and compression steel (see Section 5.6.1). However, additional two provisions characteristic of beams with tension and compression steel will be discussed below.

A23.3 Cl.10.5.2   **Maximum amount of reinforcement**   The provision related to the upper limit for the amount of reinforcement in beams with tension and compression steel is the same as in beams with tension steel only, discussed in Section 5.6.1. The maximum reinforcement requirement is met if the following condition is satisfied:

A23.3 Eq. 10.5

$$\frac{c}{d} \le \frac{700}{700 + f_y} \qquad [5.9]$$

The c/d ratio at the balanced strain condition is

$$\frac{c}{d} = \frac{700}{700 + f_y} \qquad [5.10]$$

As discussed in Section 3.8, the depth of the compression stress block (a) for beams with tension and compression steel is

$$a = \frac{T_r - C_r'}{\alpha_1 \phi_c f_c' b} = \frac{\phi_s f_y A_s - \phi_s f_y A_s'}{\alpha_1 \phi_c f_c' b} \qquad [3.38]$$

and the neutral axis depth is

$$c = \frac{a}{\beta_1} \qquad [3.4]$$

where

$$\rho = \frac{A_s}{b\,d} \qquad [3.1]$$

$$\rho' = \frac{A_s'}{b\,d} \qquad [3.35]$$

Equation 5.10 can then be rewritten as

$$\rho - \rho' = \left(\frac{\alpha_1 \beta_1 \phi_c f_c'}{\phi_s f_y}\right)\left(\frac{700}{700 + f_y}\right) \qquad [5.18]$$

or

$$\rho - \rho' = \rho_b$$

The above equation represents the reinforcement requirements related to the balanced strain condition in beams with tension and compression steel. The value $\rho - \rho'$ is also called the *effective reinforcement ratio* and it can be used to predict the mode of failure in beams with tension and compression steel as follows:

1. When $\rho - \rho' = \rho_b$, the amount of reinforcement corresponds to the *balanced condition*. The amount of steel required to create this condition is quite large from a design perspective.
2. When $\rho - \rho' > \rho_b$, the amount of reinforcement corresponds to overreinforced beams failing in the *concrete-controlled failure mode* (a brittle failure mode characterized by the concrete crushing before the steel yields).

3. When $\rho - \rho' < \rho_b$, the amount of reinforcement corresponds to properly reinforced beams failing in the *steel-controlled failure mode,* characterized by the yielding of steel reinforcement.

A23.3 Cl.7.6.5.1 and 7.6.5.2

Ties for compression steel   Compression steel in beams will act similarly to other compression members (such as columns) and will tend to buckle under compressive forces. To prevent buckling, CSA A23.3 requires that the compression bars be tied into a beam in a manner similar to columns. The compression reinforcement must be enclosed by ties or stirrups.

The tie diameter should be at least 30% of the largest longitudinal bar for 30M or smaller bars, and at least 10M size for larger bars (Cl.7.6.5.1).

The spacing of ties shall not exceed the smallest of (see A23.3 Cl.7.6.5.2)

- 16 times the diameter of the smallest longitudinal bar,
- 48 tie diameters,
- the least dimension of the beam, and
- 300 mm in beams containing bundled bars.

KEY REQUIREMENTS

The CSA A23.3 flexural design provisions for beams with tension and compression reinforcement are similar to the provisions for beams with tension steel only. In addition compression reinforcement needs to be tied in order to prevent buckling (Cl.7.6.5.1 and 7.6.5.2).

The provision related to the maximum amount of reinforcement is the same as in the case of beams with tension steel only (A23.3 Cl.10.5.2.). However, the effective reinforcement ratio $(\rho - \rho')$ is used in this case because it takes into account both the tension and the compression steel.

5.9.2   Design of Rectangular Beams with Tension and Compression Reinforcement: Summary and a Design Example

The design of beams with tension and compression steel is based on the moment resistance equations presented in Section 3.8. The design is based on the assumption that both the tension and the compression steel have yielded. These assumptions need to be checked in the course of the design. A typical cross-section of a rectangular beam with tension and compression reinforcement is shown in Figure 5.14.

The general design steps are outlined in Checklist 5.4, followed by a design example. Although these steps are presented in a certain sequence, it is not necessary to follow the same sequence in all design situations.

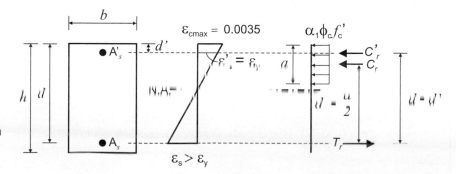

Figure 5.14   Rectangular beam with tension and compression reinforcement.

Checklist 5.4   Design of Rectangular Beams with Tension and Compression Reinforcement for Flexure

	Given: **- Specified loads acting on the beam** **- Concrete and steel material properties (f_c') and f_y)**	
Step	**Description**	**Code Clause**
1	Calculate the factored bending moment (M_f) at certain critical sections Use the specified loads and the NBC 2010 load combinations to calculate the factored loads acting on the beam. Perform a structural analysis or use the beam load diagrams given in Appendix A to calculate the factored bending moment to be used in the design.	
2	Estimate the beam dimensions b and h. Use the guidelines presented in Section 5.4.1.	
3	Estimate the effective depth (d). Beams with one layer of reinforcement: $d \cong h - 70$ mm Beams with two layers of reinforcement: $d \cong h - 110$ mm	

4	a) Estimate the required area of tension reinforcement (A_s).
	    Use either the direct procedure or the iterative procedure (see Section 5.5).
	b) Estimate the required area of compression reinforcement (A_s').
	    As a rule of thumb, the area of compression steel can be estimated to be less than 50% of the tension steel area; that is,
	
	$A_s' \le 0.5 A_s$	

5	Confirm that the maximum reinforcement requirement is satisfied (ensure a properly reinforced beam section).
	For beams with tension and compression steel, there are two alternative maximum reinforcement requirements:
	
	a) $\rho - \rho' < \rho_b$
	
	   where
	
	   $\rho = \dfrac{A_s}{bd}$ and **[3.1]**
	
	   $\rho' = \dfrac{A_s'}{bd}$ **[3.35]**
	
	   or
	
	b) $\boxed{\text{A23.3 Eq. 10.5}}$ $\dfrac{c}{d} \le \dfrac{700}{700 + f_y}$ **[5.9]**
	
	   Note that the balanced condition corresponds to
	
	   $c \cong \dfrac{2}{3}d$ **[5.11]**
	
	   or
	
	   $\rho_b \cong \dfrac{f_c'}{1100}$ **[5.15]**
	
	   (provided that Grade 400 steel is used); alternatively, use Table A.4 to find ρ_b.
		10.5.2

(Continued)

Checklist 5.4 Continued

Step	Given: - Specified loads acting on the beam - Concrete and steel material properties (f_c') and f_y) Description		Code Clause
6	Confirm that the minimum reinforcement requirement is satisfied. $M_r \geq 1.2 M_{cr}$ **[5.6]** or $A_s > A_{smin} = \dfrac{0.2\sqrt{f_c'}}{f_y} b_t h$ **[5.7]** Note that $b_t = b$ for rectangular sections.		10.5.1.1 10.5.1.2
7	Calculate M_r from Eqn 3.39. $M_r = C_r'\,(d-d') + C_r\left(d - \dfrac{a}{2}\right)$ **[3.39]** where $C_r' = \phi_s f_s' A_s'$ is the force in the compression steel, $C_r = (\alpha_1 \phi_c f_c')(ab - A_s') \cong \alpha_1 \phi_c f_c'\, a b$ is the resultant compression force in concrete, and $T_r = \phi_s f_y A_s$ is the force in the tension steel. The depth of the compression stress block (a) is $a = \dfrac{T_r - C_r'}{\alpha_1 \phi_c f_c'\, b}$ **[3.38]**		
8	Confirm that the strength requirement is satisfied. $M_r \geq M_f$ **[5.1]**		8.1.3
9	Determine the actual effective depth (d). This is based on the detailing requirements and actual reinforcement sizes (see Section 5.3).		
10	Check the crack control parameter (z) (see Section 4.8). $z \leq 30\ 000$ N/mm $\rightarrow$ interior exposure; $z \leq 25\ 000$ N/mm $\rightarrow$ exterior exposure.		10.6.1
11	Provide a design summary. Summarize the design with a sketch showing the following information: a) beam width b) total beam depth c) tension and compression reinforcement (size and number of bars) d) clear concrete cover e) stirrup size (if available)		

Example 5.8

Consider a beam with the same span, loading, and material properties as in Example 5.5. Due to architectural constraints, the overall beam depth is restricted to 650 mm and the width is restricted to 400 mm. The beam is located in the interior of a building. Use 15M bars for stirrups, 30M bars for flexural reinforcement, and 20 mm maximum aggregate size for the concrete. The concrete and steel material properties are given below.

Design an economical amount of reinforcement such that the beam is capable of carrying the given loads according to the CSA A23.3 requirements.

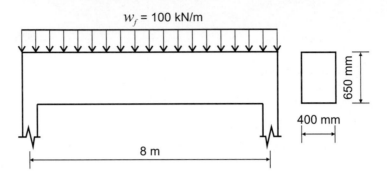

Given:
$f_c' = 25$ MPa

$f_y = 400$ MPa

$\phi_c = 0.65$

$\phi_s = 0.85$

$\alpha_1 \cong 0.8$

SOLUTION:

1. Calculate the factored bending moment

The factored bending moment is the same as that calculated in Example 5.5; that is,

$M_f = 800$ kN · m

2. Estimate the beam dimensions b and h

In this case, the beam dimensions are given as

$b = 400$ mm

$h = 650$ mm

3. Estimate the effective beam depth

Estimate the effective depth (d) based on the guidelines provided in Section 5.4.1 (use one layer of reinforcement):

$d = h - 70$ mm

$= 650$ mm $- 70$ mm $= 580$ mm

4. Calculate the required area of tension reinforcement

a) Set $M_r = M_f = 800$ kN · m

b) Calculate the required area of tension reinforcement using the direct procedure:

$$A_s = 0.0015 f_c' \, b \left(d - \sqrt{d^2 - \frac{3.85\, M_r}{f_c' b}} \right) \qquad \text{[5.4]}$$

$$= 0.0015 \times 25 \text{ MPa} \times 400 \text{ mm}$$

$$\left(580 \text{ mm} - \sqrt{(580 \text{ mm})^2 - \frac{3.85\,(800 \times 10^6 \text{ N} \cdot \text{mm})}{25 \text{ MPa} \times 400 \text{ mm}}} \right)$$

$$= 6172 \text{ mm}^2$$

c)   Select the amount of reinforcement in terms of the number and size of the bars. Refer to Table A.1; 9-30M bars are a suitable choice, since the area of one 30M bar is 700 mm², and

$$A_s = 9 \times 700 \text{ mm}^2 = 6300 \text{ mm}^2$$

Check whether the provided area of reinforcement is greater than or equal to the required amount of reinforcement:

$$A_s = 6300 \text{ mm}^2 > 6172 \text{ mm}^2$$

The provided area of reinforcement is slightly greater than the required area, which is okay.

5.   **Confirm that the maximum reinforcement requirement is satisfied (A23.3 Cl.10.5.2)**
   a)   Check the tension reinforcement ratio:

$$\rho = \frac{A_s}{b\,d}$$ [3.1]

$$= \frac{6300 \text{ mm}^2}{400 \text{ mm} \times 580 \text{ mm}} = 0.0272$$

   b)   Check whether $\rho \leq \rho_b$ (properly reinforced beam).
   This ratio can be compared with the balanced reinforcement ratio (ρ_b) for $f_c' = 25$ MPa given in Table A.4, that is, $\rho_b = 0.022$; therefore,

$$\rho = 0.0272 > 0.022$$

or

$$\rho > \rho_b$$

It follows that this beam is overreinforced and would fail in the brittle concrete-controlled mode of failure, which should be avoided. In this case, the beam cross-sectional dimensions are restricted and therefore the solution is to add compression steel and design the beam as a doubly reinforced section.
   c)   Provide 4-30M bars at the compression zone (top of the beam). The area of compression steel (A_s') is

$$A_s' = 4 \times 700 \text{ mm}^2 = 2800 \text{ mm}^2$$

Assume the distance from the top of the beam to the centroid of compression steel (d') as

$$d' = 70 \text{ mm}$$

   d)   Calculate the reinforcement ratio for the compression steel as

$$\rho' = \frac{A_s'}{b\,d}$$ [3.35]

$$= \frac{2800 \text{ mm}^2}{400 \text{ mm} \times 580 \text{ mm}} = 0.012$$

   e)   Determine the beam failure mode (steel controlled or concrete controlled) Since

$$\rho - \rho' = 0.0272 - 0.012 = 0.0152$$

and

$$0.0152 < \rho_b = 0.022$$

then

$$\rho - \rho' < \rho_b$$

The above check indicates that the doubly reinforced beam would fail in the ductile steel–controlled mode, which is okay.

6. **Confirm that the minimum reinforcement requirement is satisfied (A23.3 Cl.10.5.1.2)**
   a) Calculate the minimum area of tension reinforcement (A_{smin}) as

A23.3 Eq. 10.4

$$A_{smin} = \frac{0.2\sqrt{f_c'}}{f_y} b_t h \qquad [5.7]$$

$$= \frac{0.2 \sqrt{25 \text{ MPa}}}{400 \text{ MPa}} \times 400 \text{ mm} \times 650 \text{ mm} = 650 \text{ mm}^2$$

   where $b_t = b = 400$ mm.
   b) Confirm that $A_s > A_{smin}$:

$$A_s = 6300 \text{ mm}^2 > A_{smin} = 650 \text{ mm}^2 : \text{okay}$$

   In this example, the above check may be redundant given that the beam tension reinforcement without compression reinforcement exceeds ρ_b.

7. **Calculate M_r**
   a) Consider that both the tension and compression steel have yielded; that is,

$$f_s = f_y$$

   and

$$f_s' = f_y$$

   b) Calculate the depth of the compression stress block (a).
      i) Calculate the force in the tension steel:

$$T_r = \phi_s f_y A_s = 0.85 \times 400 \text{ MPa} \times 6300 \text{ mm}^2 = 2142 \text{ kN}$$

      ii) Calculate the resultant force in the compression steel:

$$C_r' = \phi_s f_y A_s' = 0.85 \times 400 \text{ MPa} \times 2800 \text{ mm}^2 = 952 \text{ kN}$$

      iii) Calculate the depth of the compression stress block (a):

$$a = \frac{T_r - C_r'}{\alpha_1 \phi_c f_c' b} \qquad [3.38]$$

$$= \frac{2142 \times 10^3 \text{ N} - 952 \times 10^3 \text{ N}}{0.8 \times 0.65 \times 25 \text{ MPa} \times 400 \text{ mm}} = 229 \text{ mm}$$

   c) Check whether the compression steel has yielded.
      i) Calculate the neutral axis depth from Eqn 3.4:

$$c = \frac{a}{\beta_1} = \frac{229 \text{ mm}}{0.9} = 254 \text{ mm}$$

      where $\beta_1 \cong 0.9$ from Eqn 3.7.
      ii) Calculate the strain in the compression steel from Eqn 3.36 as:

$$\varepsilon_s' = \varepsilon_{cmax}\left(1 - \frac{d'}{c}\right) = 0.0035\left(1 - \frac{70 \text{ mm}}{254 \text{ mm}}\right) = 0.0025$$

Calculate the yield strain from Eqn 2.4:

$$\varepsilon_y = \frac{f_y}{E_s} = \frac{400 \text{ MPa}}{200\,000 \text{ MPa}} = 0.002$$

Since

$$\varepsilon_s' = 0.0025 > 0.002$$

the compression steel has yielded.

d) Calculate the factored moment resistance (M_r):

$$M_r = C_r' (d - d') + C_r \left(d - \frac{a}{2} \right)$$ [3.39]

where

$$C_r = \alpha_1 \phi_c f_c' \, a \, b$$
$$= 0.8 \times 0.65 \times 25 \text{ MPa} \times 229 \text{ mm} \times 400 \text{ mm} = 1191 \text{ kN}$$
$$C_r' = 952 \text{ kN}$$

Therefore,

$$M_r = (952 \times 10^3 \text{ N}) (580 \text{ mm} - 70 \text{ mm})$$
$$+ (1191 \times 10^3 \text{ N}) \left(580 \text{ mm} - \frac{229 \text{ mm}}{2} \right)$$
$$= 1040 \text{ kN} \cdot \text{m}$$

A beam cross-section showing the distribution of internal forces is presented on the sketch below.

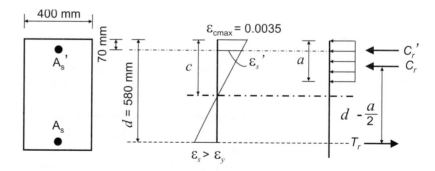

8. **Confirm that the strength requirement is satisfied (A23.3 Cl.8.1.3)**

The strength requirement per A23.3 Cl.8.1.3 must be satisfied in the design; that is,

$$M_r \geq M_f$$ [5.1]

Since

$$M_f = 800 \text{ kN} \cdot \text{m}$$

and

$$M_r = 1040 \text{ kN} \cdot \text{m} > 800 \text{ kN} \cdot \text{m}$$

The strength requirement is satisfied; however, the moment resistance is greater than the required value by approximately 30%. The objective of the design is to come up with an economical solution. Therefore, let us redesign this beam assuming 8-30M bars as the tension steel and 3-30M bars as the compression steel. Steps 7 and 8 need to be repeated; these steps will be denoted as 7R and 8R.

7R. Calculate M_r

a)   Calculate the area of tension and compression steel.

   i) Tension steel: 8-30M bars:

$$A_s = 8 \times 700 \text{ mm}^2 = 5600 \text{ mm}^2$$

   ii) Compression steel: 3-30M bars:

$$A_s' = 3 \times 700 \text{ mm}^2 = 2100 \text{ mm}^2$$

   iii) Note that $\rho - \rho'$ is the same as in the previous design since there is one bar fewer at the bottom and top each, so the difference between the tension and compression steel areas remains the same.

b)   Consider that both the tension and compression steel have yielded; that is,

$$f_s = f_y$$

and

$$f_s' = f_y$$

c)   Calculate the depth of the compression stress block (a):

   i) Calculate the force in the tension steel:

$$T_r = \phi_s f_y A_s = 0.85 \times 400 \text{ MPa} \times 5600 \text{ mm}^2 = 1904 \text{ kN}$$

   ii) Calculate the resultant force in the compression steel:

$$C_r' = \phi_s f_y A_s' = 0.85 \times 400 \text{ MPa} \times 2100 \text{ mm}^2 = 714 \text{ kN}$$

   iii) Calculate the depth of the compression stress block (a).

$$a = \frac{T_r - C_r'}{\alpha_1 \phi_c f_c' b} \qquad\qquad\qquad \textbf{[3.38]}$$

$$= \frac{1904 \times 10^3 \text{ N} - 714 \times 10^3 \text{ N}}{0.8 \times 0.65 \times 25 \text{ MPa} \times 400 \text{ mm}} = 229 \text{ mm}$$

   Note that the a value has remained unchanged as compared to the previous design. This is logical since the differential force between the tension steel and the compression steel ($T_r - C_r'$) has not changed.

d)   Check whether the compression steel has yielded.
   Given that the a value remained the same as in the previous design, it can be concluded that the compression steel has yielded.

e)   Calculate the factored moment resistance (M_r).

$$M_r = C_r' (d - d') + C_r \left(d - \frac{a}{2} \right) \qquad\qquad \textbf{[3.39]}$$

where

$C_r = 1190 \text{ kN}$ (same as in the previous design),

$C_r' = 714 \text{ kN}$

Therefore,

$$M_r = (714 \times 10^3 \text{ N})(580 \text{ mm} - 70 \text{ mm})$$

$$+ (1190 \times 10^3 \text{ N}) \left(580 \text{ mm} - \frac{229 \text{ mm}}{2} \right)$$

$$= 918 \text{ kN} \cdot \text{m}$$

8R. Confirm that the strength requirement is satisfied (A23.3 Cl.8.1.3)

The strength requirement per A23.3 Cl.8.1.3 must be satisfied in the design; that is,

$$M_r \geq M_f \qquad\qquad\qquad\qquad [5.1]$$

Since

$$M_f = 800 \text{ kN} \cdot \text{m}$$

and

$$M_r = 918 \text{ kN} \cdot \text{m} > 800 \text{ kN} \cdot \text{m}$$

the strength requirement is satisfied.

9. **Determine the actual effective depth**

a) Determine the concrete cover.

The clear cover is the same as determined in Example 5.5:

$$\text{cover} = 30 \text{ mm}$$

b) Determine minimum bar spacing (s_{min}):
    - 30M tension reinforcement: $d_b = 30$ mm (see Table A.1);
    - 15M stirrups: $d_s = 15$ mm (see Table A.1);
    - maximum aggregate size of 20 mm: $a_{max} = 20$ mm (given).

According to A23.1 Cl.6.6.5.2, the minimum spacing between reinforcement bars (s_{min}) should be the greatest of

1. $1.4 \times d_b = 1.4 \times 30 \text{ mm} = 42 \text{ mm}$,
2. $1.4 \times a_{max} = 1.4 \times 20 \text{ mm} = 28 \text{ mm}$,
3. 30 mm.

Hence, the minimum required bar spacing is $s_{min} = 42$ mm.

c) Calculate the number of bars that can fit in one layer.

It is assumed that four bars can fit in the bottom layer. Check whether this assumption is correct. The minimum required beam width (b_{min}) is

$$b_{min} = 4d_b + 3s + 2d_s + 2 \times \text{cover}$$
$$= 4 \times 30 \text{ mm} + 3 \times 42 \text{ mm} + 2 \times 15 \text{ mm} + 2 \times 30 \text{ mm}$$
$$= 336 \text{ mm}$$

Since

$$b_{min} = 336 \text{ mm} < b = 400 \text{ mm}$$

Four bars can fit in the bottom layer. Since we need 8-30M bars, the bars will be placed in two layers.

e) Determine the effective depth (d).

i) Calculate the distance from the compression face of the beam to the centroid of the bottom layer (d_1):

$$d_1 = h - \text{cover} - d_s - \frac{d_b}{2}$$

$$= 650 \text{ mm} - 30 \text{ mm} - 15 \text{ mm} - \frac{30 \text{ mm}}{2} = 590 \text{ mm}$$

ii) Calculate the distance from the top of the beam to centroid of the top layer (d_2):

$$d_2 = d_1 - s_{min} - 2\left(\frac{d_b}{2}\right)$$

$$= 590 \text{ mm} - 42 \text{ mm} - 2\left(\frac{30 \text{ mm}}{2}\right) = 518 \text{ mm}$$

iii) Determine the effective depth from the top of the beam to the centroid of the reinforcement as

$$d = \frac{4 \times d_1 + 4 \times d_2}{8}$$

$$= \frac{4 \times 590 \text{ mm} + 4 \times 518 \text{ mm}}{8} = 554 \text{ mm}$$

Since the actual effective depth (554 mm) is less than the initial estimate of 580 mm, the actual M_r value is equal to 868 kN·m; this is larger than the M_f value of 800 kN·m, so the design is okay.

iv) Calculate the distance from the compression face of the beam to the centroid of the compression reinforcement (d'):

$$d' = \text{cover} + d_s + \frac{d_b}{2}$$

$$= 30 \text{ mm} + 15 \text{ mm} + \frac{30 \text{ mm}}{2} = 60 \text{ mm}$$

10. Check the crack control parameter (z) (A23.3 Cl.10.6.1)

a)   Compute the effective tension area per bar (A).

i)   Compute the distance (d_c) from the centroid of the tension reinforcement to the tension face of the concrete section.
First, find the distance d_c from the extreme tension fibre to the centre of the longitudinal bar located closest thereto as

$$d_c = \text{cover} + d_s + \frac{d_b}{2} = 30 \text{ mm} + 15 \text{ mm} + \frac{30 \text{ mm}}{2}$$

$$= 60 \text{ mm}$$

Then, find d_s as

$$d_s = h - d$$

$$= 650 \text{ mm} - 554 \text{ mm} = 96 \text{ mm}$$

ii)   The total effective tension area for all bars (A_e) (see the sketch below) is:

$$A_e = b\,(2d_s)$$

$$= (400 \text{ mm})(2 \times 96 \text{ mm}) = 76\,800 \text{ mm}^2$$

iii)   Calculate the effective tension area per bar (A).
For eight bars in total, $N = 8$:

$$A = \frac{A_e}{N}$$

$$= \frac{76\,800 \text{ mm}^2}{8} = 9600 \text{ mm}^2$$

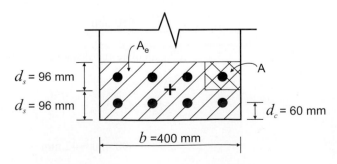

b) Determine the stress in the steel reinforcement (f_s) under the service load level. An approximate f_s estimate permitted by A23.3 Cl.10.6.1 will be used in lieu of the detailed calculations:

$$f_s = 0.6 f_y \qquad\qquad \textbf{[4.21]}$$
$$= 0.6 \times 400 \text{ MPa} = 240 \text{ MPa}$$

c) Determine the value of z:

A23.3 Eq. 10.6

$$z = f_s \sqrt[3]{d_c A} \qquad\qquad \textbf{[4.20]}$$
$$= 240 \text{ MPa} \sqrt[3]{(60 \text{ mm})(9600 \text{ mm}^2)} = 19\,969 \text{ N/mm}$$

Since

$$z = 19\,969 \text{ N/mm} < 30\,000 \text{ N/mm}$$

the CSA A23.3 cracking control requirements for interior exposure are satisfied.

11. Provide a design summary

Finally, a design summary showing the beam cross-section and the reinforcement is presented below.

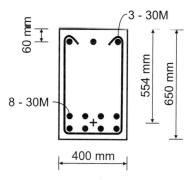

Learning from Examples

In this example, two alternative doubly reinforced beam designs were developed: 9-30M bars as tension steel plus 4-30M bars as compression steel, and 8-30M bars as tension steel plus 3-30M bars as compression steel. However, the most economical design solution would have 7-30M bars as tension steel plus 2-30M bars as compression steel (the reader can confirm this by performing design calculations). It is interesting to note that, in a sense, this arrangement can be considered equivalent to the original trial reinforcement of 9-30M (tension steel only; see Step 4). In the design solution involving doubly reinforced beams, 2-30M bars are moved to the top of the beam to resist compression.

In design practice, when an overreinforced condition is encountered, the designer may choose to move a small portion of the required tension steel (say 25%) to the compression zone; this can be done as a starting point to carry out a design check. In most cases, this rule of thumb will lead to a satisfactory solution. However, the designer must always keep in mind that the compression steel must be properly tied to prevent buckling.

SUMMARY AND REVIEW — FLEXURE: DESIGN OF BEAMS AND ONE-WAY SLABS

The objective of flexural design is to find cross-sectional dimensions and the amount and distribution of reinforcement in flexural members subjected to given loads. In general, flexural design consists of several steps. First, the specified loads are assigned and the concrete and steel material properties are selected. Subsequently, the beam geometry (cross-sectional dimensions and span) is chosen based on the understanding of the overall design objectives. Next, the appropriate amount of steel reinforcement is calculated based on the previously defined beam dimensions. The design process is iterative, and so the calculations of beam dimensions and the corresponding amount of reinforcement may need to be repeated a few times before the designer is satisfied with the solution.

Key considerations related to the design of reinforced concrete flexural members

The proper design of flexural members must take into account detailing requirements (concrete cover and bar spacing) and serviceability considerations (deflections and crack control).

Concrete cover is the amount of concrete provided between the surface of a concrete member and the reinforcing bars. The cover is critical in ensuring a safe design suitable for the intended conditions. An adequate concrete cover needs to be provided for the following reasons:

- to protect the reinforcement against corrosion;
- to ensure adequate bond between the reinforcement and the surrounding concrete;
- to protect the reinforcement from the loss of strength in case of a fire;
- to protect reinforcement from concrete abrasion and wear (for example, floors in parking garages).

Basic cover requirements specified by CSA A23.1 Cl.6.6.6.2.3 are summarized in Table A.2.

Bar spacing requirements, including minimum and maximum limits, are prescribed to minimize reinforcement congestion and to facilitate concrete placement. Bar spacing requirements for beams are specified by CSA A23.1 Cl.6.6.5.2. When the beam width is inadequate to meet these spacing requirements, there are several possible solutions: the reinforcement can be placed in two or more layers, the beam width can be increased, or groups of parallel bars in contact can be bundled.

Serviceability considerations (deflections and crack control factor) were discussed in Chapter 4.

Practical guidelines related to the design of reinforced concrete beams and slabs

The following practical guidelines related to the selection of beam and slab properties can be offered based on practical design experience:

- *Overall beam/slab depth (h):* A simple guideline is to select the overall beam depth so that a detailed deflection calculation is not required, according to the indirect approach to deflection control outlined in CSA A23.3 Cl.9.8.2.1 (see Table A.3).
- *Beam width (b):* This should not be less than 250 mm and preferably not less than 300 mm. The arrangement of reinforcing bars is one of the key factors determining the beam width. The beam can be designed to be wider than the column; this considerably facilitates the forming of the beam–column joint.
- *Effective beam/slab depth (d):* This can be estimated once the overall depth (h) is known. The d value should not be overestimated since normal construction practice tends to result in smaller effective depths than indicated on construction drawings.
- *Cross-sectional depth/width (h/b) ratio:* For rectangular beams, cross-sections with a depth/width ratio between 1.5 and 2 are most often used with the upper bound value being the most common; that is, $b = h/2$.
- *Nominal cross-section sizes:* The use of nominal section sizes reduces the chances of making errors in design calculations. The following recommendations related to rounding slab and beam dimensions can be followed:

- The slab thickness should be chosen in increments of 10 mm.
- Beam dimensions should be rounded to the nearest 50 mm.
- Girder dimensions should be rounded to the nearest 100 mm.

- *Reinforcement ratio* (ρ): It is considered good practice to design beams and slabs with an amount of reinforcement that ensures failure in the ductile steel-controlled mode. Flexural members should preferably be designed such that the required ρ value ranges from 0.0035 (0.35%) to 0.01 (1%) and remains less than $0.5\rho_b$ (provided there are no major architectural constraints).
- *Reinforcing bar sizes:* In general, the use of oversized bars is not recommended because larger development lengths are required to ensure proper bar anchorage. However, there is a larger labour cost related to the placement of a larger number of pieces of smaller bars than the placement of a smaller number of larger bars. Guidelines related to bar sizes to be used in the design of beams and slabs are summarized in Table 5.1.
- *Concrete and steel strengths:* Reinforced concrete beams and floor slabs are generally constructed of concrete with f_c' of 25 MPa, 30 MPa, or 35 MPa, with 30 MPa strength being the most common. The basic reinforcing steel grade in Canada is 400R, characterized by an f_y value of 400 MPa.

Procedures for flexural design of beams and slabs

The following two procedures are used to design beams and one-way slabs:

- *Direct procedure:* This determines the area of tension reinforcement (A_s) directly from the moment resistance (M_r) of a beam section.
- *Iterative procedure:* This determines the required area of tension reinforcement by assuming the depth of the compression stress block and by recalculating the reinforcement area until the required accuracy has been achieved. The factored moment resistance is computed using the most accurate value of the reinforcement area.

These design procedures have been developed with the assumption of *properly reinforced beams* that fail in the steel-controlled mode.

Flexural design of beams with tension steel only

CSA A23.3 prescribes the following requirements for the design of rectangular beams reinforced with tension steel only:

- The *strength requirement* (Cl.8.1.3) is a key requirement related to the design of concrete members.
- The *maximum amount of reinforcement* (Cl.10.5.2) represents the upper limit for the amount of tension reinforcement for beams without axial loads and is equal to the balanced reinforcement.
- The *minimum amount of reinforcement* (Cl.10.5.1) is one of two alternative requirements to prevent the design of underreinforced beams that fail in a sudden and brittle manner.

General steps related to beam design are presented in Checklist 5.1.

Flexural design of one-way slabs

The CSA A23.3 flexural design provisions for one-way slabs are very similar to the provisions for beams with tension steel only. However, there are some additional slab-specific requirements:

- minimum slab reinforcement (Cl.7.8.1)
- minimum tension bar spacing (Cl.7.4.1.2)
- shrinkage and temperature reinforcement (Cl.7.8.1 and 7.8.3)

Fire resistance requirements should also be considered when designing the slab thickness (CSA A23.3 Cl.8.1.2 and Appendix D of NBC 2010).

General steps related to slab design are presented in Checklist 5.2.

Flexural design of T- and L-beams

The CSA A23.3 flexural design provisions specific to T-beams are summarized below.

The *effective flange width* (the width of the slab that serves as a flange in T-beams) is affected by the type of loading (uniform/concentrated), structural system (simple/continuous), the spacing of the beams, and the relative stiffness of the slabs and the beams. CSA A23.3 Cl.10.3.3 and 10.3.4 provide rules for estimating the effective flange width for design purposes.

Reinforcement in T-beam flanges needs to be provided to ensure an adequate beam-to-slab connection in T-beams subjected to negative bending; a small portion of the reinforcement is distributed in the slab close to the beam web side faces (CSA A23.3 Cl.10.5.3.1).

General steps related to T-beam design are presented in Checklist 5.3.

Flexural design of rectangular beams with tension and compression reinforcement

The CSA A23.3 flexural design provisions for beams with tension and compression reinforcement are similar to the provisions for beams with tension steel only. The two additional requirements specific to doubly reinforced beams are as follows:

- The *compression reinforcement* needs to be tied to prevent buckling (Cl.7.6.5.1 and 7.6.5.2).
- The provision related to the *maximum amount of reinforcement* is the same as in the case of beams with tension steel only (Cl.10.5.2); however, the effective reinforcement ratio $(\rho - \rho')$ is used in this case because it takes into account both the tension and the compression steel.

General steps related to the design of beams with tension and compression reinforcement are presented in Checklist 5.4.

PROBLEMS

5.1. A typical cross-section of a reinforced concrete beam is shown in the figure below. Consider an effective depth of 830 mm. The concrete and steel material properties are given below.
   a) Calculate the factored moment resistance for the beam section.
   b) Check whether the amount of tension reinforcement meets the CSA A23.3 requirements.

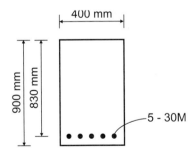

Given:

$$f_c' = 30 \text{ MPa}$$
$$f_y = 400 \text{ MPa}$$
$$\phi_c = 0.65$$
$$\phi_s = 0.85$$

b) Determine the clear spacing between the reinforcing bars in the bottom layer and check whether the CSA A23.1 bar spacing requirements are satisfied.
   c) Determine the effective depth to be used in the design based on the CSA A23.3 detailing requirements.
   d) Due to the additional loading that had to be considered in the design, an additional 30M bar needs to be placed in the upper reinforcement layer, resulting in a total of 6-30M bars. Determine the effective depth for the revised design.
   e) Calculate the difference in the effective depth determined in parts c) and d) and express the answer as a percentage. Is this difference significant from a design perspective?

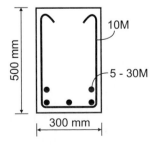

5.2. A reinforced concrete beam with a width of 300 mm and a depth of 500 mm is shown in the figure that follows. The beam is reinforced with 5-30M bars at the bottom and 10M stirrups. The maximum aggregate size is of 20 mm diameter. The beam is located in the interior of a residential building.
   a) Determine the required clear cover according to the CSA A23.1 requirements.

5.3. A typical cross-section of a reinforced concrete beam is shown in the figure that follows. Consider an effective depth of 330 mm. The concrete and steel material properties are given below.
   a) Calculate the factored moment resistance for the beam section.
   b) If the contractor has accidentally misplaced the reinforcing by 30 mm, so that the effective depth

is reduced to 300 mm, find the factored moment resistance for the beam section considering the new reinforcement layout.

c) If the actual 28-day concrete strength for the beam designed in part a) is only 27 MPa, find the factored moment resistance for the beam section.

d) Due to construction-related problems, the effective depth has been reduced to 300 mm (as in part b)) and the 28-day concrete strength has reached only 27 MPa (as in part c)). Find the corresponding moment resistance for the beam section.

e) What conclusions can you draw from parts a) to d) with regard to the effects of construction errors on reinforcing bar placement and concrete strengths, respectively?

f) Once you learned about the above construction errors, you reviewed the design calculations and realized that there is a certain reserve in moment resistance of the beam section. Which of the construction errors discussed in parts b) to d) can be tolerated when

i) a 5% reduction in the moment resistance value is acceptable?

ii) a 10% reduction in the moment resistance value is acceptable?

g) In spite of the specified tolerances, errors in the placement of reinforcement on the order of $\pm 10\%$ are common in Canada. How would you attempt to compensate for these construction errors in the design of reinforced concrete beams and slabs?

Given:

$$f_c' = 30 \text{ MPa}$$
$$f_y = 400 \text{ MPa}$$
$$\phi_c = 0.65$$
$$\phi_s = 0.85$$

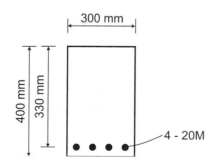

300 mm

400 mm

330 mm

4 - 20M

5.4 Consider a simply supported reinforced concrete beam with a span of 6 m. The beam has a rectangular cross-section with a width of 400 mm and an overall depth of 600 mm. The beam supports its own weight plus a superimposed uniform dead load (DL) of 5 kN/m and a uniform live load (LL) of 35 kN/m. The beam is located in the interior of a building.

The maximum aggregate size is 20 mm. The beam is reinforced with 10M stirrups. The concrete and steel material properties are given below.

a) Find the minimum required number of 25M tension rebars required to support the given load. Confirm that the design meets the pertinent CSA A23.3 requirements.

b) Provide a design summary.

Given:

$$f_c' = 30 \text{ MPa}$$
$$f_y = 400 \text{ MPa}$$
$$E_s = 200\ 000 \text{ MPa}$$
$$\phi_c = 0.65$$
$$\phi_s = 0.85$$
$$\gamma_w = 24 \text{ kN/m}^3$$

5.5. A reinforced concrete cantilevered beam with a span of 5 m extends from the wall, as shown in the figure below. The beam has a rectangular cross-section and supports a uniform dead load (DL) of 15 kN/m (excluding the self-weight) and a uniform live load (LL) of 25 kN/m. The beam width is restricted to 400 mm. Use 10M stirrups and 25M bars for tension steel. The maximum aggregate size is 20 mm. The beam is located in the interior of a building. The concrete and steel material properties are given below.

a) Find the minimum required beam depth, such that the CSA A23.3 requirements are satisfied, without a detailed deflection calculation.

b) Design the required tension reinforcement for the beam such that it can carry the design loads. Use the beam depth determined in part a). Confirm that the design meets the pertinent CSA A23.3 requirements.

c) Provide a design summary.

Given:

$$f_c' = 30 \text{ MPa}$$
$$f_y = 400 \text{ MPa}$$
$$\phi_c = 0.65$$
$$\phi_s = 0.85$$
$$\gamma_w = 24 \text{ kN/m}^3$$

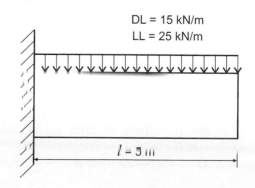

DL = 15 kN/m
LL = 25 kN/m

$l = 5$ m

5.6. A simply supported reinforced concrete beam with a span of 8 m is shown in the figure below. The beam has a rectangular cross-section and supports a uniform dead load (DL) of 20 kN/m (excluding the self-weight). There are two 50 kN point loads due to the live load located at the third points along the length. The beam width is restricted to 400 mm. Use 10M stirrups and 25M bars for tension steel. The maximum aggregate size is 20 mm. The beam is located in the interior of a building. The concrete and steel material properties are given below.

a) Find the minimum required beam depth, such that the CSA A23.3 requirements are satisfied without a detailed deflection calculation.

b) Design the required tension reinforcement for the beam such that it can carry the design loads. Use the beam depth determined in part a). Confirm that the design meets the pertinent CSA A23.3 requirements.

c) Provide a design summary.

Given:

$$f_c' = 25 \text{ MPa}$$
$$f_y = 400 \text{ MPa}$$
$$\phi_c = 0.65$$
$$\phi_s = 0.85$$
$$\gamma_w = 24 \text{ kN/m}^3$$

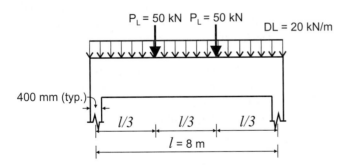

5.7. A simply supported reinforced concrete beam of rectangular cross-section is shown in the figure that follows. The beam supports a uniform dead load (DL) of 40 kN/m (including the beam self-weight) and a uniform live load (LL) of 40 kN/m. The beam width is restricted to 400 mm. The maximum aggregate size is 20 mm. Use 10M stirrups and 25M bars for tension steel. The beam is located in the interior of a building. The concrete and steel material properties are given below.

a) Find the minimum required overall beam depth and the amount of tension reinforcement such that the beam can carry the design loads while the reinforcement ratio at the midspan is $\rho \le 0.75\rho_b$.

b) Provide a design summary.

Given:

$$f_c' = 25 \text{ MPa}$$
$$f_y = 400 \text{ MPa}$$
$$\phi_c = 0.65$$
$$\phi_s = 0.85$$

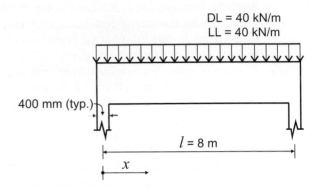

5.8. Refer to the beam designed in Problem 5.7. Use the same beam cross-sectional dimensions as determined in Problem 5.7 and 25M bars for tension steel.

Find the amount of tension steel (A_s) required at the following distances from the left support such that the beam can carry the design loads (see the sketch in Problem 5.7):

a) $x = 2.67$ m (1/3 span)

b) $x = 2$ m (1/4 span)

c) $x = 1$ m (1/8 span)

d) $x = 0.2$ m

Confirm that the design meets the pertinent CSA A23.3 requirements.

5.9. Consider a beam with the same loading and span as discussed in Problem 5.7. The beam is part of the floor system shown in the figure that follows. Use 10M stirrups and 30M bars for tension steel. The concrete and steel material properties are given below.

a) Consider the beam as a T-beam. Find the minimum required overall depth such that the beam can carry the design loads while the reinforcement ratio at the midspan is $\rho \le 0.75\rho_b$. Confirm that the design meets the pertinent CSA A23.3 requirements. Provide a design summary.

b) Compare the beam designs from Problems 5.7 and 5.9 in terms of the amount of reinforcement and the cross-sectional dimensions. Which design is more economical?

Given:

$$f_c' = 25 \text{ MPa}$$
$$f_y = 400 \text{ MPa}$$
$$\phi_c = 0.65$$
$$\phi_s = 0.85$$

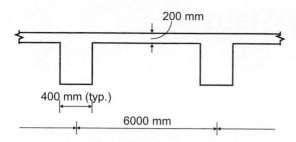

200 mm

400 mm (typ.)

6000 mm

5.10. Refer to the beam discussed in Problem 5.7. Due to architectural constraints, the maximum beam depth is restricted to 75% of the depth determined in Problem 5.7. Use 10M stirrups and 30M bars for tension steel.

a) Design a beam with tension and compression reinforcement that is capable of carrying the given design loads. Confirm that the design meets the pertinent CSA A23.3 requirements. Provide a design summary.

b) Compare the beam designs from Problems 5.7 and 5.10 in terms of the amount of reinforcement and the cross-sectional dimensions.

5.11. A simply supported reinforced concrete slab with a projection is shown in the figure that follows. The slab is supported at points A and B. The slab supports a uniform dead load (DL) of 6 kPa (including the slab self-weight) and a uniform live load (LL) of 5 kPa. The slab is located in the interior of a building. Use 15M bars for the slab reinforcement. The concrete and steel material properties are given.

a) Find the minimum required slab thickness, such that the CSA A23.3 requirements are satisfied without a detailed deflection calculation (consider that the slab thickness is constant throughout its span).

b) Plot the factored bending moment diagram for the slab.

c) Design the required tension reinforcement (including the area and spacing) for the slab at the location of the maximum negative factored bending moment. Use the slab thickness determined in part a). Confirm that the design meets the pertinent CSA A23.3 requirements.

d) Design the required tension reinforcement (including the area and spacing) for the slab at the location of the maximum positive factored bending moment. Use the slab thickness

determined in part a). Confirm that the design meets the pertinent CSA A23.3 requirements.

e) Design the temperature and shrinkage reinforcement for the slab.

f) Sketch a vertical section through the slab and the corresponding reinforcement layout. Indicate the locations of the top and bottom slab reinforcement.

Given:

$$f_c' = 25 \text{ MPa}$$
$$f_y = 400 \text{ MPa}$$
$$\phi_c = 0.65$$
$$\phi_s = 0.85$$

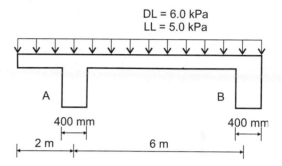

DL = 6.0 kPa
LL = 5.0 kPa

A   B

400 mm   400 mm

2 m   6 m

5.12. A typical span of a continuous reinforced concrete slab of variable thickness and the corresponding factored bending moment diagram are shown in the figure below. The slab is located in the interior of a building.

a) Determine the required amount of tension steel at the sections where $M_f = 10 \text{ kN} \cdot \text{m/m}$ and $M_f = 20 \text{ kN} \cdot \text{m/m}$, as shown in the figure below. Use 15M bars. Confirm that the design meets CSA A23.3 requirements.

b) Refer to the solution obtained in part a). If this were a problem from a design practice, would you design the reinforcement for the negative moment region of the slab based on the larger or the smaller required area determined in part a)? Explain.

Given:

$$f_c' = 25 \text{ MPa}$$
$$f_y = 400 \text{ MPa}$$
$$\phi_c = 0.65$$
$$\phi_s = 0.85$$

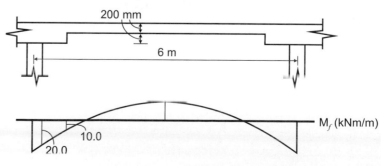

200 mm

6 m

10.0

20.0

M_f (kNm/m)

6

Shear Design of Beams and One-Way Slabs

LEARNING OUTCOMES

After reading this chapter, you should be able to

- identify the three failure modes of reinforced concrete beams subjected to flexure and shear
- explain the role and types of shear reinforcement in reinforced concrete beams
- outline the key factors influencing the shear resistance of concrete beams and one-way slabs
- design reinforced concrete beams and one-way slabs for shear according to the CSA A23.3 simplified method

6.1 INTRODUCTION

A shear force is an internal force that acts parallel to a cross-section of a structural member subjected to flexure. Flexure and shear effects are related — a shear force can be considered as an internal force caused by flexure. As a result, the shear design of concrete members needs to be performed in addition to the flexural design in virtually all design applications. Various mechanisms of the behaviour of reinforced concrete beams subjected to shear effects and the corresponding design approaches and procedures will be discussed in this chapter.

The behaviours of reinforced concrete beams in shear and flexure are substantially different. The behaviour of uncracked concrete beams subjected to shear is discussed in Section 6.2. Section 6.3 outlines various shear failure modes and discusses the shear resistance of beams without shear reinforcement. Section 6.4 explores the behaviour of concrete beams with shear reinforcement. The shear design provisions of CSA A23.3 are discussed in Section 6.5. CSA A23.3 proposes two basic methods for shear design: the *simplified method* and the *general method*. The main emphasis in this chapter is on the *simplified method* of shear design, both in terms of the theoretical background and the design examples. Other shear design methods recommended by CSA A23.3 will be discussed at a conceptual level only. Shear design considerations are outlined in Section 6.6. The shear design of reinforced concrete beams according to the CSA A23.3 simplified method is presented in Section 6.7, whereas the shear design of one-way slabs is discussed in Section 6.8. Detailing of shear reinforcement is discussed in Section 6.9. The *shear friction* concept, related to interface shear transfer, is applicable in situations involving the possibility of sliding shear failure, as discussed in Section 6.10.

The August 1955 shear failure of reinforced concrete beams in the warehouse at the Wilkins Air Force Depot in Shelby, Ohio, brought into question the traditional shear design procedures followed by the ACI and other concrete codes. Continuous roof beams supporting precast roof panels were reinforced with inadequate shear reinforcement, as shown in Figure 6.1. This failure triggered extensive studies of shear behaviour that have resulted in the development of more advanced shear design procedures used in modern codes (Anderson, 1957).

Figure 6.1 The 1955 shear failure of reinforced concrete beams at the Wilkins Air Force Depot, Shelby, Ohio.

(Judith Siess)

6.2 BEHAVIOUR OF UNCRACKED CONCRETE BEAMS

Prior to cracking, a reinforced concrete beam subjected to combined flexure and shear effects behaves like a homogeneous elastic beam. The behaviour of homogeneous beams under flexure and shear is well established and is covered in detail in textbooks on the mechanics of materials. Once cracking takes place due to increased load levels, the mechanism of internal load transfer in a concrete beam becomes substantially different. Although only the behaviour of cracked concrete beams is relevant to design practice, a brief overview of the behaviour of uncracked beams will be provided as background.

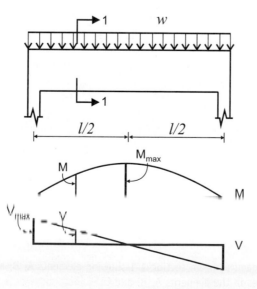

Figure 6.2 Reinforced concrete beam under uniform load and the corresponding bending moment (M) and shear force (V) diagrams.

Consider a simply supported reinforced concrete beam of rectangular cross-section subjected to a uniformly distributed load (w), as illustrated in Figure 6.2. Internal bending moments and shear forces develop in the beam as a result of the applied load. The maximum bending moment (M_{max}) occurs at the beam midspan, whereas the maximum shear forces (V_{max}) develop at the supports, where the bending moment is equal to zero. The reader is expected to have a background related to bending moment and shear force distribution in beams from statics courses.

Let us now review the stresses developed at section 1-1 of the same beam due to the internal bending moment (M) and shear force (V) (see Figure 6.3a). The bending moment (M) causes a flexural stress (f), which acts in a perpendicular direction to the beam's longitudinal

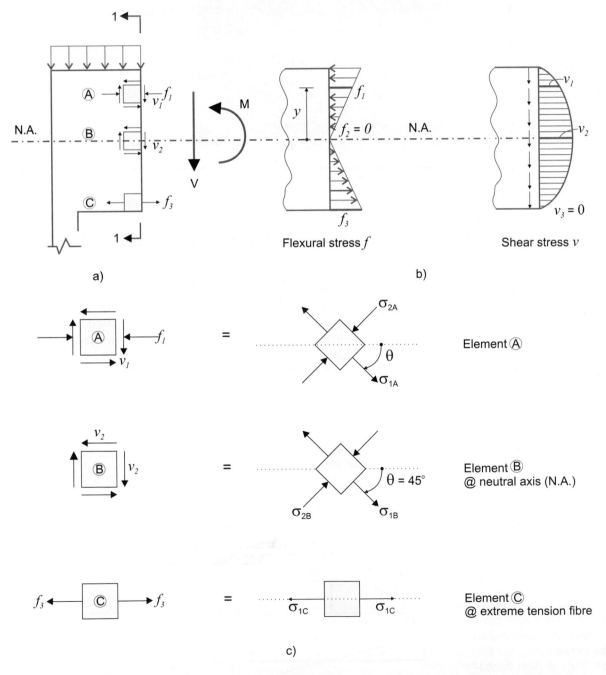

Figure 6.3 Stresses in a concrete beam: a) beam section 1-1 showing elements A, B, and C; b) flexural and shear stress distribution; c) principal stresses in the elements A, B, and C.

axis. The flexural stress varies linearly along the section depth from zero at the neutral axis to the extreme values at the top and bottom fibres of the section (see Figure 6.3b). Depending on the direction of the bending moment, the flexural stresses will act either in tension or in compression. The shear stress (v) develops in a beam subjected to flexure, acting in a direction parallel to the beam's longitudinal axis. In theory, the variation of shear stress along the section depth is curvilinear, with a maximum value at the neutral axis and zero value at the top and bottom fibres (see Figure 6.3b).

Let us consider three small beam elements, A, B, and C, located at the section 1-1 (shown as shaded squares in Figure 6.3a). Element A is located above the neutral axis and is subjected to combined shear and flexural compressive stresses. Element B is located at the neutral axis and is subjected to shear stress only (this stress state is also known as *pure shear*). Element C is located at the bottom fibre and is subjected to flexural tensile stress only.

The extreme (maximum/minimum) normal stresses acting on an element are known as *principal stresses*. For the beam under consideration, there are two principal stresses acting on each element. The principal stresses can be determined using the stress transformation equations based on the stress equilibrium of a small element extracted from a structure. For example, the principal stresses for element B can be calculated from

$$\sigma_1 = f/2 + \sqrt{(f/2)^2 + v^2} \qquad\qquad\qquad [6.1]$$

$$\sigma_2 = f/2 - \sqrt{(f/2)^2 + v^2} \qquad\qquad\qquad [6.2]$$

where σ_1 and σ_2 denote the maximum and the minimum principal stress, respectively. In general, the maximum principal stress is tensile stress (positive value) and the minimum stress is compressive stress (negative value), although other alternatives are possible; that is, both stresses can be either tensile or compressive. Principal stresses corresponding to the elements A, B, and C are shown in Figure 6.3c, along with the component shear and flexural stresses acting on each element.

It can be noted from Figure 6.3c that the principal stresses σ_1 and σ_2 act on perpendicular planes. In general, the direction of the principal stresses is defined by an angle (θ) formed by the horizontal axis and the direction of the maximum principal stress (σ_1) (see element A in Figure 6.3c). The θ value can be determined based on the shear and flexural component stresses acting on an element. Observe the directions of the principal stresses in elements A, B, and C. It can be proved that θ is equal to 45° for element B subjected to pure shear, whereas θ is equal to 0° for element C subjected to flexural tensile stresses only.

If the values and directions of the principal stresses for several elements at various beam locations are known, then it is possible to draw curves tangential to the principal stress directions at all these elements. These curves are called *stress trajectories*. Stress trajectories for the principal tensile and compressive stresses in the beam under consideration are illustrated in Figure 6.4a. (Note that beam elements A, B, and C are shown as shaded squares on the diagram.) Due to the fact that principal stresses act on perpendicular planes, the tangents to the trajectories at the intersecting points always form a 90° angle.

Let us discuss the tensile stress trajectories shown with full lines in Figure 6.4a. The beam cracks when the concrete tensile stress at a certain location reaches the modulus of rupture, as discussed in Section 2.3.2. Figure 6.4b shows cracks in the concrete beam overlaid on the stress trajectories diagram. The cracks open in the direction of the compression stress trajectories, that is, at a 90° angle to the tensile stress trajectories. Note that the cracks open in the lower part of the beam (below the neutral axis) due to the maximum flexural tensile stresses developed in that region. (Observe the direction of the bending moment (M) in Figure 6.3a and the corresponding flexural stresses in Figure 6.3b.)

It should be stressed that cracks developed due to shear effects are actually caused by tensile stresses. These inclined cracks (shown in Figure 6.4b) are often called *diagonal tension cracks* in the technical literature. In the case of a homogeneous beam, diagonal tension cracks

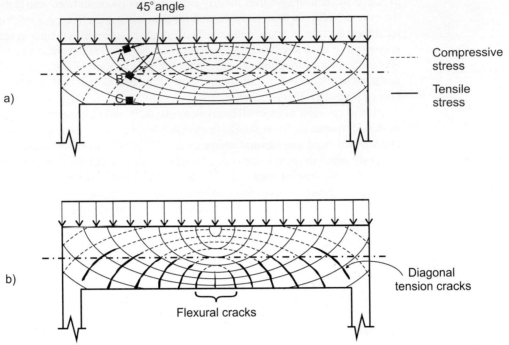

Figure 6.4 Stress trajectories in a homogeneous concrete beam: a) tensile and compressive stress trajectories; b) typical cracks coincide with the compressive stress trajectories in the beam tension zone.

occur at a 45° angle to the horizontal axis (refer to element B in Figure 6.4a). The inclination of diagonal tension cracks in reinforced concrete beams depends on several parameters; however, it can be considered that these cracks develop at an angle approximately equal to 45°.

In the case of an unreinforced concrete beam, extensive diagonal tension cracks would cause a sudden and brittle failure (see Figure 6.4b). In a reinforced concrete beam subjected to uniform load, the first cracks to open are usually vertical cracks at the midspan caused by flexure. Due to the increase in shear forces from the midspan toward the supports, diagonal tension cracks usually open at an angle of approximately 45° to the horizontal axis of the beam (as discussed above). Shear reinforcement is usually provided to prevent the development of excessive diagonal tension cracks and brittle shear failure in reinforced concrete members. (The design of shear reinforcement will be discussed in Section 6.4.)

6.3 | BEHAVIOUR OF REINFORCED CONCRETE BEAMS WITHOUT SHEAR REINFORCEMENT

After a reinforced concrete beam cracks due to the combined effects of flexural and shear stresses, a complex mechanism of shear load transfer develops. Reinforced concrete beams subjected to flexure and shear can fail in different ways, depending on several parameters. The behaviour of concrete beams reinforced with tension steel only is discussed in this section.

6.3.1 Failure Modes

Consider a simply supported reinforced concrete beam, as shown in Figure 6.5a. The beam is subjected to two point loads (P) located a distance a away from the face of the support. The distance (a) between the support and the point of load application is termed the *shear span*. The beam has rectangular cross-section with an effective depth d. The increasing applied load

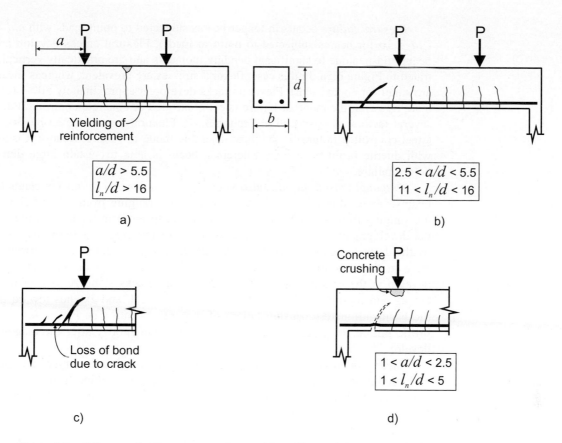

Figure 6.5 Failure modes for a concrete beam without shear reinforcement: a) flexural failure; b) diagonal tension failure; c) crack pattern characteristic of diagonal tension failure; d) shear compression failure (adapted from ASCE-ACI, 1973; the material is reproduced by permission of the American Society of Civil Engineers).

causes flexural and shear stresses in the beam, which in turn cause cracking and ultimately failure. The cracking patterns and modes of failure characteristic of concrete beams are strongly affected by the relative magnitudes of the flexural and shear stresses. Studies have shown that the *shear span/depth ratio* (a/d) is one of the key factors influencing the mode of failure. The value of the a/d ratio is directly proportional to the corresponding flexural/shear stress ratio at a certain beam location. The majority of beams with dimensions commonly found in design practice are characterized with an a/d ratio greater than 2.5 (so-called *normal* or *long beams*). It should be noted that the shear span for beams subjected to a uniform load is equal to the *clear span* (l_n), that is, the clear distance between the supports; the clear span/depth ratio l_n/d should be used instead of the a/d ratio in the discussion of failure modes in uniformly loaded beams. For detailed coverage of beam failure modes, the reader is referred to Bresler and MacGregor (1967) and ASCE-ACI (1973).

In general, the following three modes of failure or their combinations may occur depending on the beam's a/d ratio: *flexural failure, diagonal tension failure,* and *shear compression failure.* It should be noted that diagonal tension and shear compression failure are shear-controlled modes of failure. Also, note that "normal beams" ($a/d > 2.5$ or $l_n/d > 11$) fail either in flexure (flexural failure) or in shear (diagonal tension failure). The mode of failure depends on several factors such as the amount of longitudinal reinforcement, the yield strength of reinforcing steel, and the concrete tensile strength. A larger a/d value indicates that there is a predominant flexural stress in the beam (as opposed to shear stress), and therefore there is a greater chance that flexural failure will occur (as opposed to shear failure). In general, flexural failure occurs in beams with an a/d value greater than 5.5 (Nawy, 2003).

Flexural failure occurs in longer beams subjected to point loads with $a/d > 5.5$ (or $l_n/d > 16$ for beams subjected to uniform loads). Flexural cracks develop around the beam midspan due to significant bending moments and are practically vertical, as illustrated in Figure 6.5a. In this case, flexural stresses are prevalent, whereas shear stresses are small (or nonexistent). Flexural cracks develop at approximately 50% of the failure load; however, the mode of failure (steel-controlled or concrete-controlled) depends on several factors, as discussed in Section 3.4.3. Flexural failure caused by steel yielding (steel-controlled failure) is the most desirable mode of failure because it is associated with ductile beam behaviour, wherein a beam is able to sustain large deformations before failure.

Diagonal tension failure (also known as *shear tension failure*) is characteristic of beams subjected to point loads with an a/d value ranging from 2.5 to 5.5, or l_n/d values ranging from 11 to 16 for beams subjected to uniform load. Cracking starts with the development of a few very fine vertical flexural cracks at the midspan, followed by the destruction of the bond between the reinforcing steel and the surrounding concrete at the support (see Figure 6.5b). Subsequently, two or three inclined cracks develop in the area close to the support. As the cracks stabilize, one of the inclined cracks widens into a principal *diagonal tension* crack and extends toward the top of the beam, ultimately causing failure of the entire beam (see Figure 6.5c). This is a brittle failure mode, and it occurs without warning shortly after the diagonal cracks develop.

Shear compression failure occurs mainly in shorter beams subjected to point loads with an a/d ratio ranging from 1 to 2.5 (or an l_n/d ratio ranging from 1 to 5 for beams with uniform load). This mode is somewhat similar to the diagonal tension mode of failure. Initially, a few fine flexural cracks develop at the midspan. The propagation of these cracks stops as destruction of the bond occurs between the longitudinal bars and the surrounding concrete at the supports. Thereafter, a steep inclined crack suddenly develops and extends toward the neutral axis. Subsequently, the concrete crushes in the zone above the crack, as illustrated in Figure 6.5d. The failure occurs when the inclined crack joins the crushed concrete zone. This mode is considered to be less brittle than the diagonal tension failure mode due to the stress redistribution in the concrete compression zone.

An experimental study of a simply supported reinforced concrete beam is illustrated in Figure 6.6. The beam was subjected to two point loads at the third points, and the a/d ratio was around 4.0. The beam was reinforced with longitudinal reinforcement only. The failure occurred in a shear-controlled mode shortly after the principal diagonal tension crack had formed near the left support.

It is important to emphasize that the inclined cracks characteristic of shear-controlled modes of failure actually develop in the vicinity of or as an extension of a previously opened (mainly vertical) flexural crack, as illustrated in Figure 6.7a. These cracks are known as *flexure-shear cracks*.

Modes of failure characteristic of very short and deep beams (characterized by $a/d < 1$ or $l_n/d < 1$) are different from the above-discussed modes characteristic of shallow beams. In deep beams, the internal load transfer occurs in the form of a *tied-arch mechanism*, in which the concrete forms compression struts, whereas longitudinal steel reinforcement forms a tension tie. The failure of a tied-arch mechanism can occur in five different modes, as discussed by ASCE-ACI (1973).

In this section, the discussion of failure modes has focused on concrete beams of rectangular cross-section. It should be noted that flanged beams (I-shaped beams) fail in a slightly different manner, characterized by *web-shear cracks* developed in the vicinity of the neutral axis at a 45° angle, as illustrated in Figure 6.7b. These cracks lead to *web-crushing failure,* which is characteristic of I-shaped beams. Alternatively, the tied-arch mechanism develops, leading to *arch-rib failure*. For a detailed discussion of the shear behaviour of I-beams, the reader is referred to Bresler and MacGregor (1967).

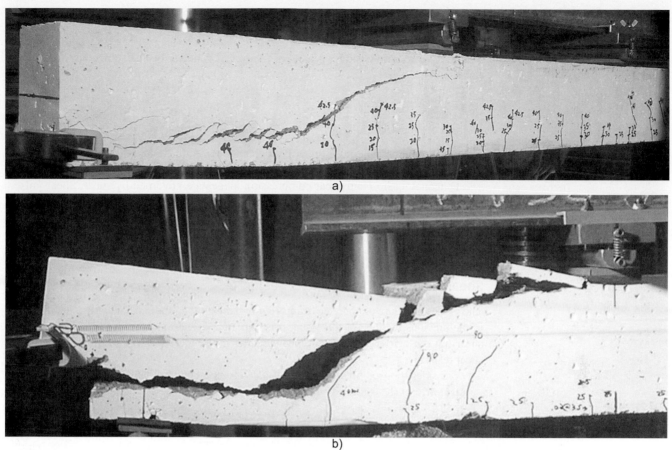

a)

b)

Figure 6.6   Shear failure of a reinforced concrete beam without shear reinforcement: a) principal diagonal tension crack before the failure; b) detail of the left support after the failure.

(Svetlana Brzev)

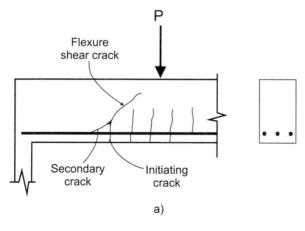

a)

Figure 6.7   Types of inclined cracks: a) flexure-shear crack; b) web-shear crack (adapted from ASCE-ACI, 1973; the material is reproduced by permission of the American Society of Civil Engineers).

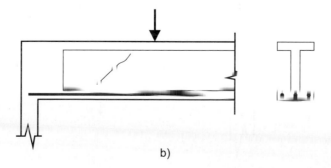

b)

The three main failure modes typical of reinforced concrete beams subjected to flexure and shear are

- flexural failure
- diagonal tension failure
- shear compression failure

Flexural (steel-controlled) failure is a ductile mode of failure and consequently the most desirable of all three modes.

Diagonal tension and shear compression modes are governed by shear. Shear failure modes in concrete beams are influenced by the tensile and compressive strength of concrete. Shear failures are generally sudden and brittle and should be avoided in design applications.

6.3.2   Shear Resistance of Cracked Beams

The modes of failure of concrete beams reinforced with longitudinal bars only were discussed in the previous section. At this point, it is important to gain insight into the mechanisms of shear transfer across the cracked concrete section and the key factors influencing concrete shear resistance.

Consider the free-body diagram of a reinforced concrete beam subjected to bending moment M and shear force V in Figure 6.8a. An inclined shear crack has developed in the

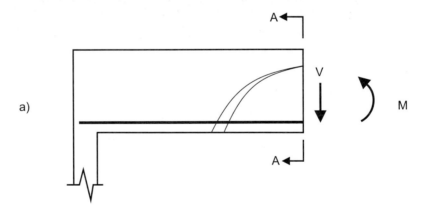

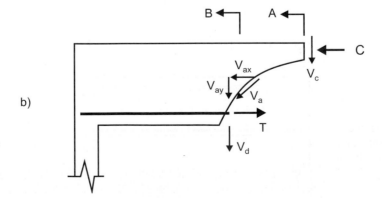

Figure 6.8 A reinforced concrete beam subjected to flexure and shear: a) a section at the inclined crack; b) forces acting at the cracked section.

beam at the section under consideration. The bending moment is resisted internally by a force couple consisting of the tensile force (T) developed in the steel reinforcement and the compression force (C) developed in the concrete section above the crack, as illustrated in Figure 6.8b. Once an inclined crack develops across the beam depth, the shear is resisted by the following internal forces:

- V_{ay} is the vertical component of the interface shear force (V_a), which is the tangential shear force transmitted across a crack by the interlocking of aggregate particles (also known as aggregate interlock). According to ASCE-ACI (1973), interface shear transfer accounts for 33% to 50% of the total force on the beam.
- V_d is the dowel-shear force developed by the longitudinal reinforcement bars crossing the crack. According to ASCE-ACI (1973), dowel shear is on the order of 15% to 25% of the total shear force.
- V_c is the concrete shear force developed in the concrete compression zone and accounts for the remaining strength (25% to 50%).

Before the cracking takes place, the shear force is entirely carried by the concrete, as discussed in previous sections. Immediately after the inclined cracks develop, the concrete shear force (V_c) accounts for up to 40% of the total shear force (ASCE-ACI, 1973), whereas the remaining portion is resisted by the dowel force (V_d) and the interface shear force (V_a). However, as the crack widens, the interface-shear transfer mechanism becomes less effective and V_a drops. Subsequently, the dowel-shear mechanism causes a splitting crack to develop in the concrete along the reinforcement, resulting in the complete loss of the dowel force (V_d). Finally, the entire shear force is carried by the concrete compression zone in the area above the crack, causing the crushing of this region. Beyond the development of the inclined cracking, it is difficult to predict which mechanism of force transfer breaks down to cause failure in concrete beams without shear reinforcement. It should be noted that the above outlined mechanism of shear load transfer applies to normal beams characterized by $a/d > 2.5$ and that the mechanism of shear load transfer in deep beams is substantially different.

It should also be noted that the shear design procedure according to the CSA A23.3 simplified method accounts for only the V_c component in determining the shear resistance of a concrete beam. Therefore, the contribution of interface-shear and dowel mechanisms in transferring shear across the cracked beam section are neglected, which is somewhat conservative. Although the shear design procedure according to CSA A23.3 will be covered later in this chapter, the factors influencing the concrete shear resistance (V_c) will be discussed here.

Considering the brittle and sudden nature of shear failure occurring shortly after the inclined cracks develop, it is reasonable to consider that the shear resistance of a cracked concrete beam (V_c) without shear reinforcement is equal to the shear force that causes the first inclined crack. This is the common approach adopted by concrete design codes.

The concrete shear resistance depends on the following factors: concrete tensile strength, longitudinal reinforcement ratio, shear span/depth (a/d) ratio, beam size, and axial forces in the beam (MacGregor and Bartlett, 2000).

The *concrete tensile strength* influences the inclined cracking load. As discussed in Section 6.2, diagonal tension cracking occurs when the principal tensile stress reaches the tensile strength. However, in the more common case of beams with flexure-shear cracks, the cracking occurs at the principal tensile stress level approximately equal to one third of the concrete tensile strength. Recall from Section 2.3.2 that the modulus of rupture for concrete (f_r), according to CSA A23.3, is

$$f_r = 0.6 \lambda \sqrt{f_c'} \qquad\qquad [2.1]$$

thus resulting in an inclined cracking shear stress of $0.2\sqrt{f_c'}$. If this stress value is taken as the *average shear stress*, that is, the ratio of shear force acting on the section and the area of the cross-section, on a rectangular section of width b and effective depth d, then the corresponding shear force is

$$V_c = 0.2\sqrt{f_c'}\, bd \qquad\qquad [6.3]$$

The *longitudinal reinforcement ratio,* which indicates the amount of longitudinal steel, is another factor influencing concrete shear resistance. Based on the statistical correlation of results for tests performed on beams with various ratios of steel reinforcement reported by MacGregor and Bartlett (2000), the concrete shear resistance can be determined by the equation

$$V_c = 0.167\sqrt{f_c'}\, bd \qquad\qquad [6.4]$$

Equation 6.4 tends to overestimate the shear resistance for beams with smaller reinforcement ratios (less than approximately 0.75%).

According to Eqns 6.3 and 6.4, the concrete shear resistance depends on the concrete compressive strength (f_c') and the beam cross-sectional dimensions: width (b) and effective depth (d). The reader will be interested to note that the equation for concrete shear resistance (V_c) used in CSA A23.3 is very similar to Eqns 6.3 and 6.4.

Axial tensile forces tend to decrease concrete shear resistance, while axial compressive forces tend to increase it.

It is important to note that inclined cracking of beams influences the design of flexural steel reinforcement. Consider the free-body diagram of a beam subjected to bending moment (M) and shear force (V) acting at section A-A (see Figure 6.8a). It can be determined from the equilibrium of the beam segment between sections A-A and B-B in Figure 6.8b that the tensile force (T) in the longitudinal reinforcement at section B-B is dependent on the bending moment (M) at section A-A. This must be considered in designing the anchorage for longitudinal beam reinforcement, which will be covered in detail in Sections 9.6 and 9.7.

KEY CONCEPTS

The concrete shear resistance (V_c) is usually taken to be equal to the shear force that causes the first inclined crack.

The main factors influencing the concrete shear resistance are the concrete tensile strength, the longitudinal reinforcement ratio, and the axial forces in the beam.

6.4 BEHAVIOUR OF REINFORCED CONCRETE BEAMS WITH SHEAR REINFORCEMENT

6.4.1 Types of Shear Reinforcement

Based on the previous discussion on the behaviour of reinforced concrete beams without shear reinforcement, it is clear that the shear resistance of such beams is limited to the shear resistance of concrete. Because the shear load often exceeds the shear resistance of concrete, it is necessary to provide *shear reinforcement* capable of carrying the excess shear beyond the concrete shear resistance. Furthermore, one of the main objectives of providing shear reinforcement is to prevent the sudden and brittle shear mode of failure in reinforced concrete beams.

By observing the directions of stress trajectories and the inclined cracks in Figure 6.4b, we can see that shear reinforcement is most effective when provided in the direction perpendicular to the cracks. However, it is very difficult to install inclined shear reinforcement correctly in the field. Therefore, in most design applications, shear reinforcement (also known as web reinforcement and transverse reinforcement) is provided in the vertical direction.

The most common type of shear reinforcement is in the form of U-shaped *vertical stirrups* anchored to the longitudinal reinforcement at both ends, as illustrated in Figure 6.9a. The two beam cross-sections show examples of two-legged and four-legged stirrups (the vertical side of a stirrup is called a *leg*). Other types of shear reinforcement, rarely used in design practice, are: *inclined stirrups* at an angle of 45° or more to the longitudinal

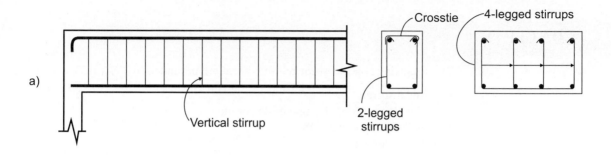

a)

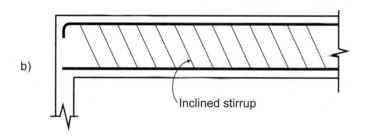

b)

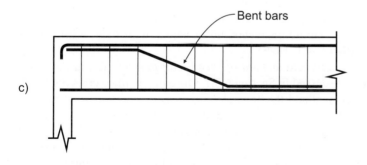

c)

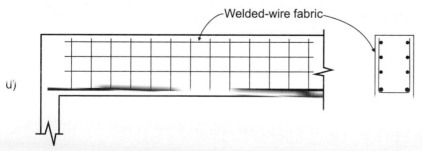

d)

Figure 6.9   Types of shear reinforcement: a) vertical stirrups; b) inclined stirrups; c) bent bars; d) welded wire fabric.

beam axis (Figure 6.9b); *bent-up longitudinal bars,* where the portion of the bar no longer needed for flexure is bent up (Figure 6.9c); *welded wire fabric,* with the wires in the direction perpendicular to the axis of the member serving as shear reinforcement (Figure 6.9d); and *spirals* (refer to Section 8.3 for a discussion on spiral reinforcement). The types of shear reinforcement and their respective characteristics are prescribed by CSA A23.3 Cl.11.2.4.

KEY CONCEPTS

The following types of shear reinforcement are permitted by CSA A23.3 (Cl.11.2.4):

- vertical stirrups
- inclined stirrups at an angle of 45° or more
- bent up longitudinal bars
- welded wire fabric
- spirals

6.4.2 Effect of Shear Reinforcement

Let us now explore the effect of shear reinforcement on the behaviour of concrete beams subjected to shear. A free-body diagram of a concrete beam reinforced with vertical stirrups (in addition to the longitudinal reinforcement) is shown in Figure 6.10. Shear reinforcement becomes effective only after the inclined cracking in the beam has begun. At that moment, tensile stresses in the shear reinforcement suddenly increase, causing the development of tensile forces in the bars. However, it should be noted that only the stirrups intersecting the cracks are effective in resisting shear loads. As illustrated in Figure 6.10, all stirrups crossing the crack will develop tensile forces, and V_s denotes the sum of all these forces in the individual stirrups. The stresses in the shear reinforcement continue to increase in proportion to the applied loads. It is desirable that shear reinforcement yield in tension prior to the beam failure in shear. Depending on the amount of shear reinforcement and other factors, there are three possible shear-controlled modes of failure characteristic of beams with shear reinforcement, as discussed below (ASCE-ACI, 1973).

Beams with a *moderate amount of shear reinforcement* fail in the *shear compression* failure mode discussed in Section 6.3.1. Once the inclined cracking has begun, both the shear reinforcement and the concrete compression zone resist shear until the shear reinforcement has yielded. From that point onward, any further increase in the shear load must be resisted by the concrete compression zone alone. Failure occurs when the concrete in the compression zone is crushed due to the combined compression and shear stresses. The failure is not sudden because the yielding shear reinforcement allows the diagonal cracks to widen, thus giving sufficient warning. This is the most desirable of all shear failure modes.

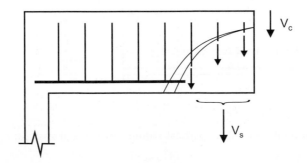

Figure 6.10 Cracked section of a reinforced concrete beam with shear reinforcement.

Beams with a *large* amount of *shear reinforcement* fail in a mode that is essentially the same as shear compression failure but with little or no warning. Due to the large amount of shear reinforcement, the concrete compression zone is crushed before the steel reinforcement has yielded.

Long beams (of a large *a/d* ratio) with a *very small* amount of *shear reinforcement* fail in the *diagonal tension* mode discussed in Section 6.3.1. In this case, the shear reinforcement yields immediately after the opening of inclined cracks, and the beam fails due to concrete crushing. There is practically no redistribution of shear forces between concrete and steel, and the failure occurs suddenly and without warning.

The reader needs to recognize that the addition of shear reinforcement by itself does not prevent the occurrence of shear-controlled failure in concrete beams. However, shear-controlled failure can be prevented when the beam is designed to have a larger shear capacity than the flexural capacity. In that case, flexural failure will be initiated before shear failure takes place. As discussed in Chapter 3, flexural failure in properly reinforced beams is characterized by steel yielding. Once the steel yields, there will be a very small increase in the internal shear forces in the beam. Hence, the designer should aim to specify a sufficient amount of shear reinforcement to ensure that the shear resistance is greater than the flexural resistance.

Adequately designed shear reinforcement can contribute to the beam shear resistance in the following two ways:

- by carrying a portion of the shear load through a sudden increase in tensile forces after the formation of inclined cracking
- by effectively arresting the growth of inclined cracking and preventing the deep penetration of the cracks into the compression zone and the occurrence of sudden beam failure

When shear reinforcement is provided in the form of closed ties, it also provides a certain level of confinement to the concrete in the compression zone, thus resulting in increased concrete shear resistance.

KEY CONCEPTS

The shear resistance of beams susceptible to shear failure can be greatly increased by providing properly designed shear reinforcement.

Shear reinforcement can effectively contribute to enhancing beam shear resistance in the following two ways:

- by carrying a portion of the shear load through a sudden increase in tensile forces after the formation of inclined cracking
- by limiting the width and spread of inclined cracks and preventing the occurrence of sudden and brittle failure

Shear reinforcement is usually designed for excess shear, generally defined as the difference between the factored shear force (V_f) acting on the beam section and the concrete shear resistance (V_c).

6.4.3 Truss Analogy for Concrete Beams Failing in Shear

At this point, the reader needs to gain insight into the relevant approaches to modelling the behaviour of cracked beams with shear reinforcement. It is very important to understand the concepts of mechanics-based models of beam behaviour because these models form the basis for shear design procedures prescribed by the codes. Traditionally, shear transfer in a cracked concrete beam with shear reinforcement has been explained by the *truss analogy,* a concept proposed independently by a Swiss engineer, W. Ritter, in

1899, and a German engineer, Mörsch, in 1902 (MacGregor and Bartlett, 2000). The truss analogy depicts a beam as a pin-connected truss in which the concrete compression zone is represented by the compression chord, the longitudinal steel reinforcement by the tension chord, and the shear reinforcement by the vertical truss members. Concrete compression struts formed between the inclined cracks are represented by the diagonal truss members. According to the basic truss model, these diagonals form a 45° angle with the horizontal beam axis. All external loads are considered to act only at the nodes. The forces in the truss can be determined from considerations of equilibrium only.

The truss model of a concrete beam is presented in Figure 6.11. Observe the similarity between the cracked beam with stirrups and longitudinal reinforcement in Figure 6.11a and the corresponding truss model in Figure 6.11b. The truss height (d_v) is equal to the lever arm of the forces T and C resisting the bending moment (refer to Section 3.5 for a discussion of the flexural resistance of reinforced concrete beams). Also, note that the diagonal truss members in Figure 6.11b form an angle (θ) with the longitudinal beam axis. In general, θ is on the

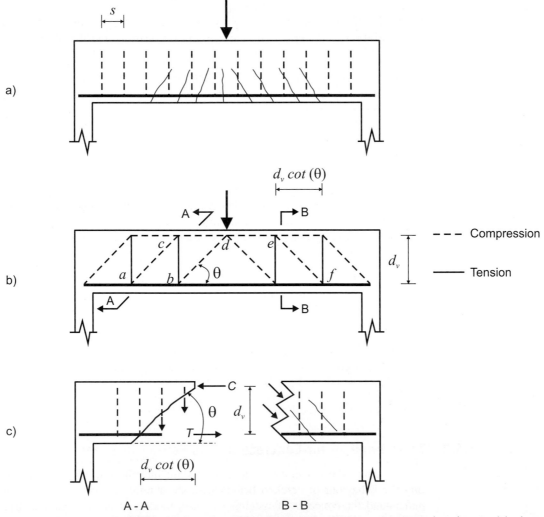

Figure 6.11 Truss model for reinforced concrete beams: a) typical crack pattern in a beam with shear reinforcement; b) truss model; c) internal forces in a cracked beam (adapted from MacGregor and Bartlett, 2000 with permission of Pearson Education Canada Inc.).

order of 45°. Also, note that the truss panel length (the distance between adjacent vertical members) is equal to $d_v \cot \theta$ (see Figure 6.11b). The model assumes that all stirrups crossing an inclined plane over a horizontal distance ($d_v \cot \theta$) are lumped into one vertical member. For example, the two stirrups crossing the inclined plane A-A in Figure 6.11c are lumped into the truss member bc in Figure 6.11b. Similarly, all diagonal concrete members (struts) crossing a vertical plane are lumped into one member. For example, the forces in the two diagonal struts crossing the vertical section B-B in Figure 6.11c are lumped together in the member *ef* in Figure 6.11b. It is also important to note that the compression chord in the truss actually represents a force in the concrete compression zone. This method expects that the stirrups have yielded in order for the truss to become statically determinate, hence the name *plastic truss model* used in some references.

The truss model correctly presents the role of shear reinforcement and the concrete compression struts that form between the inclined cracks. The CSA A23.3 simplified method is based on the truss analogy. However, the truss analogy has certain limitations, since it is not able to model the effect of interface-shear and dowel-shear forces. Also, the deformations in beams subjected to shear cannot be obtained based on the truss mechanism due to the strain incompatibility associated with the truss analogy (this becomes less important at the ultimate load level).

KEY CONCEPTS

The truss analogy offers a relatively simple model for the shear behaviour of concrete beams with shear reinforcement. A concrete beam is represented by a pin-connected truss with concrete and steel elements modelled as compression and tension truss members, respectively.

The truss analogy forms the basis for the simplified method of shear design according to CSA A23.3.

6.4.4 Shear Resistance of Beams with Shear Reinforcement

Based on the discussions of the failure modes of beams with shear reinforcement presented in Section 6.4.2, it is clear that the shear resistance of such beams is provided by concrete and steel. The factors influencing concrete shear resistance were discussed in Section 6.3.2. The discussion in this section is focused on steel shear resistance. The shear resistance of a concrete beam can be determined based on the equation

$$V_r = V_c + V_s \tag{6.5}$$

where

V_r = the factored shear resistance of a beam section
V_c = the concrete shear resistance
V_s = the steel shear resistance.

A free-body beam diagram of a cracked beam section is presented in Figure 6.12 as an idealization of the similar diagram in Figure 6.10. Note that the crack is shown as a straight line at an angle θ to the longitudinal beam axis, as shown in Figure 6.12a. The crack extends close to the top of the beam. The internal tension force in steel reinforcement (T_r) and the internal compression force in concrete (C_r) are also shown in the figure. The vertical distance between these two forces is called the *effective shear depth* and it is denoted by the symbol d_v. Note that d_v is less than the effective depth (d) at the same location; this will be discussed in Section 6.5.4.

The shear reinforcement is in the form of vertical stirrups at a spacing s, as shown in Figure 6.12a. Note that A_v denotes the total cross-sectional area of all vertical legs belonging to one stirrup (see Figure 6.12b). Because the inclined crack develops over the entire beam

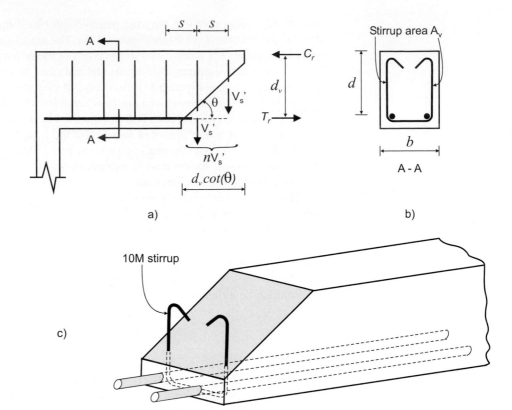

Figure 6.12 A section of a cracked reinforced concrete beam with vertical stirrups: a) forces in the stirrups; b) beam cross-section; c) isometric view of a stirrup crossing the plane of an inclined crack.

width, all vertical legs in a stirrup are considered equally effective in resisting shear. Consider the two-legged 10M stirrup in Figure 6.12c. The corresponding A_v value is 200 mm². (Note that the area of a 10M bar is 100 mm².) Note that two-legged stirrups are most common in design practice, although there are four-legged stirrups, as illustrated in Figure 6.9a.

As discussed in Section 6.4.2, only stirrups crossing the crack are effective in resisting shear at a particular section. The steel shear resistance is equal to the sum of the tensile forces developed in each stirrup (see Figure 6.12a); that is,

$$V_s = nV_s'$$ **[6.6]**

where

> n = the number of stirrups crossing an inclined crack
> V_s' = the tensile force developed in a stirrup.

Based on the discussions of shear failure modes of beams with shear reinforcement (refer to Section 6.4.2), shear reinforcement yields in tension after an inclined crack develops. The tensile force in a stirrup V_s' is

$$V_s' = \phi_s A_v f_y$$ **[6.7]**

where

> A_v = the stirrup cross-sectional area
> f_y = the yield stress
> ϕ_s = the material resistance factor for steel.

The number of stirrups crossing an inclined crack can be determined based on the geometry of the cracked section as

$$n = \frac{d_v \cot \theta}{s}$$ **[6.8]**

The final equation for V_s is obtained when Eqs 6.7 and 6.8 are substituted into Eqn 6.6:

A23.3 Eq. 11.7

$$V_s = \frac{\phi_s A_v f_y d_v \cot \theta}{s} \qquad [6.9]$$

Note that CSA A23.3 uses Eqn 6.9 to determine the steel shear resistance (V_s).

In general, inclined cracks occur at an angle θ (approximately equal to 45°) to the longitudinal axis. CSA A23.3 prescribes θ values to be used for the shear design of various reinforced concrete members (to be discussed in Section 6.5.4).

KEY CONCEPTS

The shear resistance of concrete beams with shear reinforcement is provided by concrete and steel.

According to the CSA A23.3 simplified method, the steel shear resistance of a beam section is provided only by the stirrups that cross an inclined crack plane under an angle θ to the longitudinal beam axis. Stirrup reinforcement yields in tension.

6.5 SHEAR DESIGN ACCORDING TO CSA A23.3

6.5.1 General Design Philosophy

Shear failure in concrete structures is brittle and must be avoided in design applications. The objective of a prudent design is to ensure that a beam fails in flexure before reaching its shear capacity. This can be accomplished if the shear capacity of a beam exceeds its flexural capacity. As discussed in Chapter 3, flexural failure of a properly designed reinforced concrete beam is ductile and offers warning to building occupants before collapse. This is a desirable mode of failure according to CSA A23.3.

Numerous research studies have confirmed the brittle and sudden nature of shear failure in concrete structures. Brittle shear failure in concrete members subjected to flexure can be avoided if the guidelines listed below are followed:

- *In beams:* Add shear reinforcement (usually vertical stirrups) designed to resist excess shear beyond the concrete shear capacity.
- *In slabs* (without shear reinforcement): Ensure that the shear capacity of concrete in regions with maximum shear forces is greater than the flexural capacity in regions with maximum bending moments.

6.5.2 Simplified and General Method for Shear Design

Research studies have shown that the behaviour of concrete beams subjected to shear is complex and cannot be modelled accurately in a simple manner. In most applications, it is not practical to perform the design using exhaustive engineering modelling of shear behaviour, given that the variations in quality control during construction might significantly influence the mode of shear failure and the shear resistance of a concrete member. Therefore, most designers are comfortable with a simple model for shear design as long as it represents a reasonable prediction of shear failure modes and leads to minimal risks of structural failure. CSA A23.3 prescribes two basic methods for shear design, namely the *simplified method* and the *general method*.

The simplified method is covered in CSA A23.3 Cl.11.3.6.3 and consists of a simple procedure to determine shear resistance. This method is applicable to most practical design applications, with the exception of beams subjected to significant axial tension, in which case the general method has to be used. According to the simplified method, concrete shear resistance does not depend on the state of stress and strain in a beam.

The general method for shear design (covered in CSA A23.3 Cl.11.3.6.4) is based on the *compression field theory* originally developed by Collins and Mitchell (1987). The compression field theory recognizes that a reinforced concrete beam subjected to shear can be modelled like a truss with a force distribution analogous to a stiffened thin-plate girder. When the girder is subjected to flexure and shear, the thin web plates form tension diagonals between the stiffeners to resist the shear forces. The stiffeners act in compression to resist the vertical components of the diagonal tension forces, while the flanges provide resistance to the horizontal components. According to the compression field theory, the shear forces in a reinforced concrete beam are resisted by diagonal concrete struts resisting compression and vertical shear reinforcement (stirrups) resisting tension. The stirrups resist the vertical component of the force developed in diagonal compression struts, while the truss chords, that is, the flexural couple provided by longitudinal reinforcing and the concrete compression block, resist the horizontal components of the diagonal compression force. The general method assumes that the shear resistance of both the concrete and the reinforcing steel depends on the variations in shear stress and flexural strain along the beam. It is believed that the use of this method results in a more accurate estimate of shear resistance at any section along the member length. However, such an accurate estimate of shear resistance is not necessary in most practical design applications, so detailed coverage of the general method is outside the scope of this book.

6.5.3   Major Revisions in CSA A23.3-04

CSA A23.3-04 contains significant revisions of the simplified method for shear design as compared to the previous (1994) edition. The revisions include the following items:

- concrete shear resistance (V_c) (Cl.11.3.4)
- steel shear resistance (V_s) (Cl.11.3.5.1)
- maximum factored shear resistance (Cl.11.3.3)
- maximum spacing of transverse shear reinforcement (Cl.11.3.8.1)
- required provision of shear reinforcement (Cl.11.2.8.1)

The reader who is familiar with the previous Code edition (CSA A23.3-94) should carefully study the Code changes in the next section because the revised provisions may result in different cross-sectional dimensions and amount of shear reinforcement.

6.5.4   CSA A23.3 Requirements Related to the Simplified Method for Shear Design

As discussed in Section 6.5.2, the simplified method for shear design is considered to be suitable for most practical design applications. According to CSA A23.3 Cl.11.3.6.3, this method can be used when members are not subjected to significant axial tension; that is, the tension-induced stress increase in the longitudinal reinforcement at crack locations is less than 50 MPa; otherwise the general method (CSA A23.3 Cl.11.3.6.4) should be used.

The simplified method is based on the concepts and equations presented in Section 6.4.4. Design requirements and limitations of the CSA A23.3 simplified method will be outlined in this section.

| A23.3 Cl.11.3.1 |

Strength requirement   The main design requirement for concrete members subjected to shear as stated in A23.3 Cl.11.3.1 is the *strength requirement*

| A23.3 Eq. 11.3 |

$$V_r \geq V_f \qquad\qquad\qquad [6.10]$$

The above requirement states that the factored shear resistance (V_r) must be greater than the factored shear force (V_f) acting at any section along a concrete member.

A23.3 Cl.11.3.3

Factored shear resistance According to A23.3 Cl.11.3.3, the shear resistance of a concrete member (V_r) is provided by the concrete and the steel shear reinforcement (where required):

A23.3 Eq. 11.4

$$V_r = V_c + V_s + V_p \qquad \text{[6.11]}$$

where

V_c = the concrete shear resistance
V_s = the steel shear resistance
V_p = the component of the effective prestressing force in the direction of applied shear; for nonprestressed members, $V_p = 0$.

A23.3 Cl.11.3.4

Concrete shear resistance The concrete shear resistance (V_c) should be determined based on the following equation

A23.3 Eq. 11.6

$$V_c = \phi_c \lambda \beta \sqrt{f_c'}\, b_w d_v \qquad \text{[6.12]}$$

where

λ = the factor to account for concrete density ($\lambda = 1$ for normal-density concrete)
ϕ_c = the material resistance factor for concrete
f_c' = the specified compressive strength of concrete $\left(\sqrt{f_c'} \le 8 \text{ MPa when used in the above equation}\right)$
β = the factor accounting for the shear resistance of cracked concrete, determined according to A23.3 Cl.11.3.6
d_v = the effective shear depth taken as the greater of $0.9d$ and $0.72h$
b_w = the minimum beam web width within depth d (A23.3 Cl.11.2.10.1).

In design according to the CSA A23.3 simplified method for shear design, the β value should be determined based on Table 6.1.

In cases when β is determined from Eqn 6.14, the equivalent crack spacing parameter s_{ze} can be computed from the following equation:

A23.3 Eq. 11.10

$$s_{ze} = \frac{35 s_z}{15 + a_g} \qquad \text{[6.15]}$$

where

s_z = the crack spacing parameter taken as the lesser of d_v and s
s = the maximum distance between layers of distributed longitudinal reinforcement, as shown in Figure 6.13 (each layer of distributed reinforcement shall have an area of at least $0.003\, b_w s_z$)
a_g = the specified nominal maximum size of coarse aggregate.

Table 6.1 β values according to the CSA A23.3 simplified method for shear design

Equation		Design application
$\beta = 0.18$		The section contains minimum transverse reinforcement as required by Cl.11.2.8.2.
$\beta = \dfrac{230}{1000 + d_v}$	[6.13]	The section contains no transverse reinforcement and the specified nominal maximum size of coarse aggregate is not less than 20 mm.
$\beta = \dfrac{230}{1000 + s_{ze}}$	[6.14]	The section contains no transverse reinforcement (applies to all aggregate sizes).

A23.3 Eq. 11.9

(*Source:* CSA A23.3 Cl.11.3.6.3, reproduced with the permission of the Canadian Standards Association)

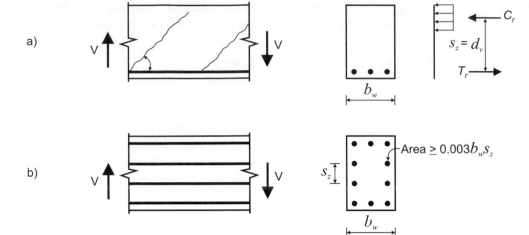

Figure 6.13 Crack spacing parameter (s_z): a) member without stirrups and with concentrated longitudinal reinforcement; b) member without stirrups but with well-distributed longitudinal reinforcement.

Note that $\beta = 0.21$ can be used for the following special member types identified by CSA A23.3 Cl.11.3.6.2:

a)   slabs or footings with an overall thickness not greater than 350 mm;
b)   footings in which the distance from the point of zero shear to the face of the column, pedestal, or wall is less than three times the effective shear depth of the footing;
c)   beams with an overall thickness not greater than 250 mm;
d)   concrete joist construction;
e)   beams cast integrally with slabs, where the depth of the beam below the slab is not greater than one-half of the web width or 350 mm.

| A23.3 Cl.11.3.5.1 |

Steel shear resistance   The shear resistance provided by the transverse steel reinforcement (V_s) perpendicular to the longitudinal axis can be determined according to Cl.11.3.5.1 as follows (see Section 6.4.4):

| A23.3 Eq. 11.7 | $$V_s = \frac{\phi_s A_v f_y d_v \cot \theta}{s} \qquad\qquad [6.9]$$

where

s = the spacing of shear reinforcement measured parallel to the longitudinal axis of the member

A_v = the area of shear reinforcement perpendicular to the axis of the member within a distance s (also discussed in Section 6.4.4)

f_y = the specified yield strength of the steel shear reinforcement

ϕ_s = the resistance factor for steel reinforcement (Cl.8.4.3)

θ = the angle of inclination of the diagonal compressive stresses to the longitudinal axis of the member.

The following θ values should be used in shear design calculations:

• $\theta = 35°$ when the CSA A23.3 simplified method (Cl.11.3.6.3) is used
• $\theta = 42°$ for special member types when $\beta = 0.21$ (Cl.11.3.6.2), as discussed earlier in this section

The above equation can be used to determine the required spacing of shear reinforcement (s):

$$s = \frac{\phi_s A_v f_y d_v \cot \theta}{V_s} \qquad\qquad [6.16]$$

The above-discussed provisions will be illustrated by Example 6.1.

Example 6.1

A typical cross-section of a reinforced concrete beam is shown in the figure below. The beam is reinforced with 10M stirrups at a 250 mm spacing on centre (10M@250). The clear cover to the stirrups is equal to 40 mm. The concrete and steel material properties are given below.

Determine the maximum shear force (V_f) this beam section is able to sustain according to the CSA A23.3 simplified method.

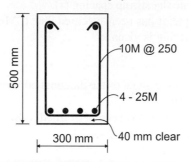

Given: $f_c' = 25$ MPa
$f_y = 400$ MPa
$\phi_c = 0.65$
$\phi_s = 0.85$
$\lambda - 1$

SOLUTION: **1.   Find the concrete shear resistance (V_c)**

a)   Calculate the effective depth (d).
Given:

- 10M stirrups; that is, $d_s = 10$ mm (see Table A.1);
- 25M tension reinforcement; that is, $d_b = 25$ mm (see Table A.1);
- cover = 40 mm;

Hence,

$$d = h - \text{cover} - d_s - \frac{d_b}{2}$$

$$= 500 \text{ mm} - 40 \text{ mm} - 10 \text{ mm} - \frac{25 \text{ mm}}{2} = 437 \text{ mm}$$

b)   Calculate the effective shear depth (d_v).
d_v is the effective shear depth taken as the greater of

$$d_v = \begin{bmatrix} 0.9d \\ 0.72h \end{bmatrix} = \begin{bmatrix} 0.9 \times 437 \text{ mm} \\ 0.72 \times 500 \text{ mm} \end{bmatrix} = \begin{bmatrix} 393 \text{ mm} \\ 360 \text{ mm} \end{bmatrix}$$

Hence, $d_v = 393$ mm $\cong 390$ mm.

c)   Calculate the factor accounting for the shear resistance of cracked concrete (β). This beam section is reinforced with 10M@250 stirrups, so the amount of shear reinforcement can be considered to be greater than the minimum (the actual check will be performed in Example 6.2).
According to A23.3 Cl.11.3.6.3,

$$\beta = 0.18 \text{ (see Table 6.1)}$$

d)   Finally, determine the V_c value:

| A23.3 Eq. 11.6 |

$$V_c = \phi_c \lambda \beta \sqrt{f_c'} \, b_w d_v \qquad \qquad [6.12]$$

$$= 0.65 \times 1.0 \times 0.18 \times \sqrt{25 \text{ MPa}} \times 300 \text{ mm} \times 390 \text{ mm}$$

$$= 68.4 \times 10^3 \text{ N} = 68.4 \text{ kN}$$

2. **Find the steel shear resistance (V_s)**
   a) Calculate the area of shear reinforcement (A_v).
      In this beam, two-legged 10M stirrups have been used. The area of a 10M bar can be obtained from Table A.1 as $A_b = 100$ mm^2. Therefore,

      $$A_v = (\text{\# of legs})(A_b)$$
      $$= 2 \times 100 \text{ mm}^2 = 200 \text{ mm}^2$$

   b) Then, calculate the stirrup spacing (s).
      In this example, it has been specified that 10M@250 stirrups have been used, so the stirrup spacing is given as

      $$s = 250 \text{ mm}$$

   c) Next, calculate the angle of the diagonal compressive stresses to the longitudinal member axis (θ).
      Because this design has been performed according to the CSA A23.3 simplified method, Cl.11.3.6.3 prescribes that

      $$\theta = 35°$$

   d) Finally, determine the V_s value:

   | A23.3 Eq. 11.7 |

   $$V_s = \frac{\phi_s A_v f_y d_v \cot\theta}{s} \qquad [6.9]$$
   $$= \frac{0.85 \times 200 \text{ mm}^2 \times 400 \text{ MPa} \times 390 \text{ mm} \times \cot(35°)}{250 \text{ mm}}$$
   $$= 151.5 \times 10^3 \text{ N} = 151.5 \text{ kN}$$

3. **Find the factored shear resistance (V_r)**
   Determine the factored shear resistance:

   | A23.3 Eq. 11.4 |

   $$V_r = V_c + V_s + V_p \qquad [6.11]$$
   $$= 68.4 \text{ kN} + 151.5 \text{ kN} + 0 = 219.9 \text{ kN} \cong 220 \text{ kN}$$

   The maximum factored shear force this beam is able to carry is equal to the factored shear resistance:

   $$V_f = V_r = 220 \text{ kN}$$

In addition to the above-stated requirements, CSA A23.3 prescribes the following *limitations,* mainly related to the amount and spacing of shear reinforcement.

| A23.3 Cl.11.3.3 |

Maximum factored shear resistance To prevent the occurrence of brittle shear compression failure, described in Section 6.3.1 (characteristic of beams heavily reinforced with shear reinforcement), CSA A23.3 places an upper limit on the factored shear resistance. This limit is intended to ensure that the stirrups will yield prior to crushing the web concrete and that diagonal cracking at the specified loads is limited. According to Cl.11.3.3, the maximum factored shear resistance ($maxV_r$) is

| A23.3 Eq. 11.5 |

$$maxV_r = 0.25\phi_c f_c' b_w d_v \qquad [6.17]$$

Note that the above requirement is relaxed as compared to the previous (1994) version of CSA A23.3. The dimensions of the beam cross-section may need to be increased when this requirement is not satisfied.

A23.3 Cl.11.3.8

Maximum spacing of transverse shear reinforcement   Shear reinforcement becomes effective only if crossed by inclined cracks, as discussed in Section 6.4.4. To ensure that at least one stirrup crosses an anticipated diagonal crack in any region along the beam length, the stirrup spacing should be generally less than or equal to the shear depth (d_v).

According to Cl.11.3.8.1, the maximum spacing (s_{max}) for transverse shear reinforcement shall not exceed the smaller of

- 600 mm
- $0.7d_v$

However, when $V_f >$ max $V_r/2$, the maximum spacing (s_{max}) should not exceed the smaller of (A23.3 Cl.11.3.8.3)

- 300 mm
- $0.35d_v$

A23.3 Cl.11.2.8.1

The required provision of shear reinforcement   Beams subjected to low shear stress corresponding to a factored shear force (V_f) less than V_c do not have to be reinforced for shear. To prevent the occurrence of brittle shear failure in members subjected to high shear stresses, A23.3 Cl.11.2.8.1 prescribes a minimum amount of transverse shear reinforcement under the following conditions:

a)   when $V_f \geq V_c$ (this applies to all types of reinforced concrete members),
b)   in regions of reinforced concrete beams with an overall thickness greater than 750 mm.

Note that the above statement applies to cast-in-place reinforced concrete structures. (There is an additional term to be used in the case of prestressed concrete structures.)

A23.3 Cl.11.2.8.2

Minimum area of shear reinforcement   Where shear reinforcement is required by Cl.11.2.8.2, the minimum required area is

A23.3 Eq. 11.1

$$A_{v,\,min} = 0.06\sqrt{f_c'}\left(\frac{b_w\,s}{f_y}\right) \qquad \textbf{[6.18]}$$

The above formula is often used to determine the maximum spacing (s) corresponding to the stirrup area (A_v):

$$s = \frac{A_v f_y}{0.06\sqrt{f_c'}\,b_w} \qquad \textbf{[6.19]}$$

Example 6.2

The beam section considered in Example 6.1 is subjected to a factored shear force (V_f) of 200 kN.
Check whether the specified shear reinforcement meets the requirements prescribed by CSA A23.3.

SOLUTION: In total, four steps need to be performed to check whether the shear design performed in Example 6.1 is adequate according to CSA A23.3.

1.   **Check the maximum factored shear resistance (A23.3 Cl.11.3.3)**
     The factored shear resistance for the beam section considered in Example 6.1 is

     $V_r = 220$ kN

     According to A23.3 Cl.11.3.3, the maximum factored shear resistance (max V_r) is

A23.3 Eq. 11.5

$$max\ V_r = 0.25\phi_c f_c'\, b_w\, d_v \qquad \textbf{[6.17]}$$

$$= 0.25 \times 0.65 \times 25\ \text{MPa} \times 300\ \text{mm} \times 390\ \text{mm}$$

$$= 475.3 \times 10^3\ \text{N} \cong 475\ \text{kN}$$

Thus,

$$V_r = 220 \text{ kN} < 475 \text{ kN} : \text{okay}$$

2. **Check the maximum spacing of shear reinforcement (A23.3 Cl.11.3.8)**

    Since $V_f < \dfrac{V_r}{2} = \dfrac{475}{2} = 238 \text{ kN}$

    It follows that

    $$s_{max} \leq \begin{bmatrix} 600 \text{ mm} \\ 0.7d_v \end{bmatrix} = \begin{bmatrix} 600 \text{ mm} \\ 0.7 \times 390 \text{ mm} \end{bmatrix} = \begin{bmatrix} 600 \text{ mm} \\ 273 \text{ mm} \end{bmatrix}$$

    Therefore,

    $$s_{max} = 273 \text{ mm}$$

    Thus,

    $$s = 250 \text{ mm} < 273 \text{ mm} : \text{okay}$$

3. **Check whether shear reinforcement is required (A23.3 Cl.11.2.8.1)**

    A minimum amount of transverse shear reinforcement is required when $V_f \geq V_c$. The V_c value was determined in Example 6.1 as

    $$V_c = 68.4 \text{ kN}$$

    and

    $$V_f = 200 \text{ kN}$$

    Hence,

    $$V_f = 200 \text{ kN} > 68.4 \text{ kN}$$

    According to A23.3 Cl.11.2.8.1, shear reinforcement is required.

4. **Check the minimum area of shear reinforcement (A23.3 Cl.11.2.8.2)**

    Where shear reinforcement is required by Cl.11.2.8.2, the minimum required area is

    A23.3 Eq. 11.1

    $$A_{v, min} = 0.06\sqrt{f_c'} \left(\frac{b_w s}{f_y} \right) \qquad \qquad \text{[6.18]}$$

    $$= 0.06\sqrt{25 \text{ MPa}} \left(\frac{300 \text{ mm} \times 250 \text{ mm}}{400 \text{ MPa}} \right) = 56.25 \text{ mm}^2$$

    The actual area of shear reinforcement determined in Example 6.1 is

    $$A_v = 200 \text{ mm}^2$$

    Since

    $$A_v = 200 \text{ mm}^2 > 56 \text{ mm}^2$$

    the above design meets all shear requirements according to CSA A23.3.

The *strength requirement* is the main CSA A23.3 requirement related to the simplified method for shear design (Cl.11.3.1).

CSA A23.3 also prescribes the procedure to determine

- the factored shear resistance (V_r) (Cl.11.3.3)
- the concrete shear resistance (V_c) (Cl.11.3.4)
- the steel shear resistance (V_s) (Cl.11.3.5.1)

The following *limitations* are set by CSA A23.3 as related to the amount and spacing of shear reinforcement:

- maximum factored shear resistance (Cl.11.3.3)
- maximum spacing of transverse shear reinforcement (Cl.11.3.8.1)
- required provision of shear reinforcement (Cl.11.2.8.1)
- minimum area of shear reinforcement (Cl.11.2.8.2)

6.6 SHEAR DESIGN CONSIDERATIONS

The best way to demonstrate the application of the CSA A23.3 shear design requirements is through design examples. However, there are a few important design considerations that need to be discussed first, such as

- shear force distribution
- critical section for shear design
- variable stirrup spacing

Shear force distribution   In general, shear force develops due to dead and live loads acting on the beam. The dead load (including beam self-weight) is considered a permanent load, whereas the live load (for example, occupancy load) is of a transient nature and may or may not be present at all times, or its distribution along the beam may be variable. The transient nature of the live load has to be accounted for in shear design by using a shear force envelope corresponding to the live load.

In a simply supported beam subjected to uniformly distributed live load, the shear force is equal to zero at the midspan, whereas the maximum shear force develops at the supports and is equal to $V_L/2$, as illustrated in Figure 6.14a. (Note that V_L denotes the uniform load resultant on the beam.) However, the shear force distribution is different when the live load acts over half of the beam span (see Figure 6.14b). In that case, the shear force at the midspan is equal to one eighth of the total live load resultant, that is, $V_L/8$.

For design purposes, it is generally considered acceptable to develop the shear force envelope by connecting the maximum shear force at the support (corresponding to the uniform load distribution) and the shear force at the midspan (corresponding to the live load acting on the half-span), as illustrated in Figure 6.14c.

A23.3 Cl.11.3.2

Critical section for shear design   In general, the maximum shear force in a concrete member occurs at the face of the support (for uniformly loaded beams and slabs or in the vicinity of point loads. When a member supported by columns or walls acts in compression, the compressive stresses in the support regions of the member cause an increase in shear strength to the member in that region. CSA A23.3 recognizes this behaviour, and Cl.11.3.2 permits the use of reduced design shear force in the support region. Consequently, sections located within a distance d_v from the face of the support can be designed for the shear force (V_f) computed at the distance d_v. A typical simple beam with a shear force diagram is illustrated in Figure 6.15.

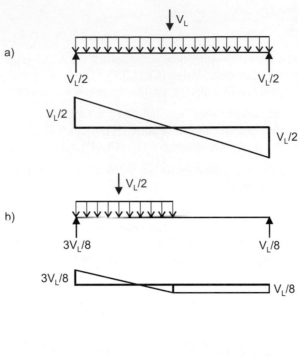

a)

h)

Figure 6.14 Shear force distribution in a concrete beam: a) a beam under uniformly distributed live load; b) partially loaded beam; c) shear force envelope.

c)

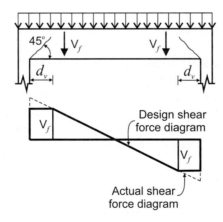

Figure 6.15   Design shear force diagram for a beam subjected to uniformly distributed load.

When the support reaction introduces tension, such as in the case of a hanging support, the critical section needs to be taken at the face of the support, as illustrated in Figure 6.16. When a load is applied to the side of the member, through brackets, ledges, or cross beams, the strut-and-tie model may be used to detail the reinforcement and the connection region (for more details on strut-and-tie model provisions, refer to CSA A23.3 Cl.11.4).

Variable stirrup spacing   In most design applications, the shear force varies along the beam length. Consequently, if designed strictly according to code equations, the stirrup size and spacing will also vary continuously along the beam. This is, however, impractical due to additional labour involved in placing stirrups accurately in the field when the spacing is variable. Ideally, the designer will arrive at a cost-effective solution by optimizing the amount of steel reinforcing and the labour time required to accurately place the stirrups. In the majority

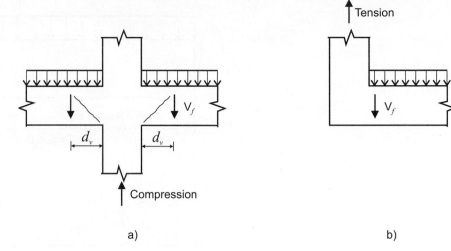

Figure 6.16 Influence of support conditions on locations of critical section for shear: a) reaction introduces compression; b) reaction introduces tension.

of cases, a reasonable rule-of-thumb involves dividing the beam into three regions of shear reinforcement, each with a constant stirrup spacing. These regions can be defined as follows:

- *Region 0,* where shear reinforcement is not required;
- *Region 1,* where the minimum shear reinforcement is required; and
- *Region 2,* where the full shear reinforcement is required.

The shear force requirements corresponding to these regions are summarized in Table 6.2.

A typical simply supported reinforced concrete beam subjected to a uniform load is shown in Figure 6.17a, and the corresponding shear envelope diagram is shown in Figure 6.17b. In order to define Regions 1, 2, and 3, the designer needs to do the following:

Table 6.2 Regions with constant stirrup spacing in a beam

	Shear force range	**Stirrup spacing**	**CSA A23.3 clause**
Region 0	$V_f \leq V_c$	Stirrups not required	11.2.8.1
Region 1	$V_c < V_f < V_{rmin}$	Spacing s_1 (maximum permitted stirrup spacing)	11.2.8.2
		$V_{rmin} = V_c + V_{smin}$	11.3.8.1
		V_{smin} corresponds to the maximum permitted stirrup spacing; that is, $s_1 = s_{max}$	
Region 2	$V_{rmin} \leq V_f < V_{fmax}$	Spacing s_2 (full shear reinforcement)	11.3.5.1
		$V_s = V_{fmax} - V_c$	
		$s_2 = \dfrac{\phi_s A_v f_y d_v \cot \theta}{V_s}$	

- First, perform linear interpolation to determine the distance x_1 from the face of the support corresponding to the force $V_f = V_{rmin}$ and the distance x_2 corresponding to the force $V_f = V_c$.
- Next, determine the stirrup spacing s_1 and s_2 corresponding to Regions 1 and 2 (see Figure 6.17c); this is illustrated in Example 6.5.

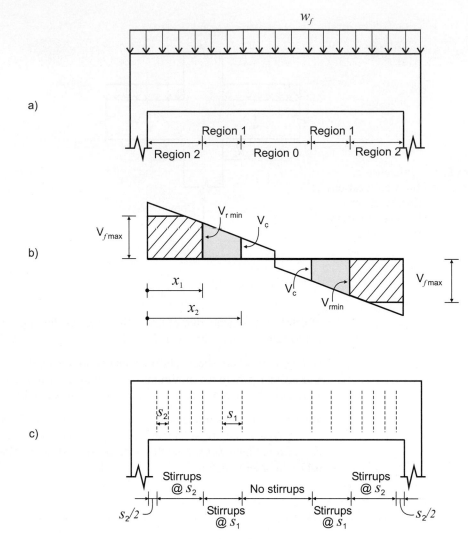

Figure 6.17 Variable stirrup spacing: a) beam elevation; b) shear envelope diagram showing Regions 0, 1, and 2; c) stirrup arrangement.

Once the stirrup spacing has been determined, a sketch showing the stirrup arrangement in the beam can be developed. In design practice, the first stirrup is usually placed at a distance $s_2/2$ away from the face of the support, where s_2 denotes the stirrup spacing in the support region (see Figure 6.17c).

6.7 SHEAR DESIGN OF REINFORCED CONCRETE BEAMS ACCORDING TO THE CSA A23.3 SIMPLIFIED METHOD: SUMMARY AND DESIGN EXAMPLES

The shear design of concrete structures was discussed in detail in the previous sections. The purpose of this section is to summarize the shear design provisions according to the CSA A23.3 simplified method. As mentioned in previous chapters, "cookbook" design solutions are generally not available. However, the general steps related to shear design are outlined in Checklist 6.1. Although the steps are presented in a certain sequence, it is not necessary to follow the same sequence in all design situations. A few design examples will be presented to illustrate the application of the CSA A23.3 simplified method for the shear design of beams.

Checklist 6.1 Shear Design According to the CSA A23.3 Simplified Method

Step	Description	Code Clause
1	Confirm that there is no significant tensile stress caused by the axial load in the beam and that the simplified method can be used.	11.3.6.3
2	Determine the factored shear force (V_f) at a distance d_v from the support (if applicable) and sketch the shear design envelope.	11.3.2
3	Determine the concrete shear resistance (V_c): $\boxed{\text{A23.3 Eq. 11.6}}$ $\qquad V_c = \phi_c \lambda \beta \sqrt{f_c'}\, b_w d_v$ [6.12]	11.3.4
4	Determine whether shear reinforcement is required and identify the region(s) in which the reinforcement is required. If $V_f < V_c$, shear reinforcement is not required. If $V_f \geq V_c$, shear reinforcement is required.	11.2.8.1
5	a) If $V_f \geq V_c$, determine the required steel shear resistance (V_s): $V_s \geq V_f - V_c$ (Set $V_f = V_r$.) b) Determine the required stirrup spacing (s): $s = \dfrac{\phi_s A_v f_y d_v \cot\theta}{V_s}$ [6.16]	11.3.5.1
6	Check whether the spacing of shear reinforcement is within the maximum limits prescribed by CSA A23.3. When $V_f \leq \dfrac{V_r \max}{2}$: $s_{max} \leq \begin{bmatrix} 600 \text{ mm} \\ 0.7 d_v \end{bmatrix}$ when $V_f > \dfrac{V_r \max}{2}$: $s_{max} \leq \begin{bmatrix} 300 \text{ mm} \\ 0.35 d_v \end{bmatrix}$ Confirm that $s \leq s_{max}$.	11.3.8.1
7	Determine the minimum required shear reinforcement area ($A_{v,\,min}$): A23.3 Eq. 11.1 $\qquad A_{v,\,min} = 0.06\sqrt{f_c'}\left(\dfrac{b_w s}{f_y}\right)$ [6.18] Confirm that $A_v \geq A_{v,\,min}$. Alternatively, calculate the maximum stirrup spacing corresponding to A_v as $s = \dfrac{A_v f_y}{0.06\sqrt{f_c'}\, b_w}$ [6.19]	11.2.8.2

Checklist 6.1   Continued

Step	Description	Code Clause
8	a) Determine V_r: A23.3 Eq. 11.4 $V_r = V_c + V_s + V_p$ **[6.11]** b) Check whether V_r is less than the maximum allowable value: A23.3 Eq. 11.5 $V_r < max\ V_r = 0.25\phi_c f_c' b_w d_v$ **[6.17]**	11.3.3
9	Sketch a beam elevation showing the distribution of the shear reinforcement.	

Example 6.3

A simply supported reinforced concrete beam of rectangular cross-section is shown in the figure below. The beam supports a uniform dead load (DL) of 24 kN/m and a uniform live load (LL) of 10 kN/m. Use 25M flexural rebars, 10M stirrups, 40 mm clear cover to the stirrups, and maximum 20 mm aggregate size.

Design the beam for shear according to the CSA A23.3 simplified method.

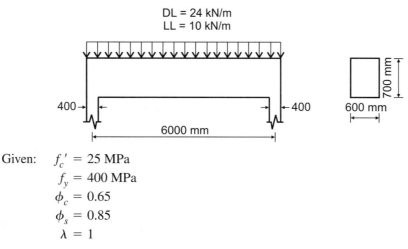

Given: $f_c' = 25$ MPa

           $f_y = 400$ MPa

           $\phi_c = 0.65$

           $\phi_s = 0.85$

           $\lambda = 1$

SOLUTION: **1. Confirm that the CSA A23.3 simplified method can be used**

       Since the beam is not subjected to tensile stress due to axial load, the shear design can be performed according to the CSA A23.3 simplified method.

     2. Calculate the design shear force ($V_{f@dv}$) at a distance d_v from the face of the support (A23.3 Cl.11.3.2)

       a) In order to determine $V_{f@dv}$, the following will have to be calculated:

         i) Effective depth (d):

           • 10M stirrups: $d_s = 10$ mm (see Table A.1)

           • 25M flexural rebars: $d_b = 25$ mm (see Table A.1)

           • cover to the stirrups: cover $= 40$ mm

$$d = h - \text{cover} - d_s - \frac{d_b}{2}$$

$$= 700 \text{ mm} - 40 \text{ mm} - 10 \text{ mm} - \frac{25 \text{ mm}}{2} = 637 \text{ mm} \cong 635 \text{ mm}$$

ii) Effective shear depth (d_v):

d_v is the effective shear depth taken as the greater of

$$d_v = \begin{bmatrix} 0.9d \\ 0.72h \end{bmatrix} = \begin{bmatrix} 0.9 \times 635 \text{ mm} \\ 0.72 \times 700 \text{ mm} \end{bmatrix} = \begin{bmatrix} 572 \text{ mm} \\ 504 \text{ mm} \end{bmatrix}$$

Hence, $d_v \cong 570$ mm.

iii) Clear span (l_n):

$$l_n = l - \frac{b_s}{2} - \frac{b_s}{2}$$

$$= 6 \text{ m} - \frac{0.4 \text{ m}}{2} - \frac{0.4 \text{ m}}{2} = 5.6 \text{ m}$$

where $b_s = 0.4$ m is the support width.

iv) Maximum shear force at the face of the support:

The factored load (w_f) is determined according to NBC 2010 Cl.4.1.3.2:

$$w_f = 1.25 \text{ DL} + 1.5 \text{ LL} = 1.25(24 \text{ kN/m}) + 1.5(10 \text{ kN/m}) = 45 \text{ kN/m}$$

and

$$V_f = \frac{w_f \times l_n}{2} = \frac{45 \text{ kN/m} \times 5.6 \text{ m}}{2} = 126 \text{ kN}$$

v) Maximum shear force at the midspan ($V_{fmidspan}$):

$$V_{fmidspan} = \frac{\left(1.5 \text{ LL}\right)\left(\dfrac{l_n}{2}\right)}{4} = \frac{(1.5 \times 10 \text{ kN/m}) \times \left(\dfrac{5.6 \text{ m}}{2}\right)}{4}$$

$$= \frac{42 \text{ kN}}{4} = 10.5 \text{ kN}$$

Now the factored shear force ($V_{f@dv}$) at a distance d_v from the support can be determined as

$$V_{f@dv} = V_{fmidspan} + \left(\frac{\dfrac{l_n}{2} - d_v}{\dfrac{l_n}{2}}\right)(V_f - V_{fmidspan})$$

$$= 10.5 \text{ kN} + \left(\frac{\dfrac{5.6 \text{ m}}{2} - 0.57 \text{ m}}{\dfrac{5.6 \text{ m}}{2}}\right)(126 \text{ kN} - 10.5 \text{ kN})$$

$$= 102.3 \text{ kN} = 103 \text{ kN}$$

b) Sketch the shear envelope diagram for the beam.

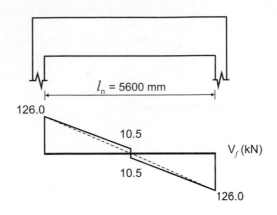

c) Sketch the shear envelope diagram for the left half of the beam.

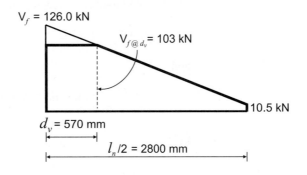

3. **Determine the concrete shear resistance (V_c) (A23.3 Cl.11.3.4)**

Assume that the amount of shear reinforcement will be less than the minimum code requirement (a conservative assumption). According to A23.3 Cl.11.3.6.3,

$$\beta = \frac{230}{1000 + d_v} = \frac{230}{1000 + 570} = 0.146 \cong 0.15 \text{ (see Table 6.1)},$$

so

| A23.3 Eq. 11.6 |

$$V_c = \phi_c \lambda \beta \sqrt{f_c'}\, b_w\, d_v \qquad\qquad\qquad\qquad\qquad\qquad \text{[6.12]}$$
$$= 0.65 \times 1.0 \times 0.15 \times \sqrt{25 \text{ MPa}} \times 600 \text{ mm} \times 570 \text{ mm}$$
$$= 166.7 \times 10^3 \text{ N} \cong 166 \text{ kN}$$

4. **Determine whether shear reinforcement is required and identify the region(s) in which the reinforcement is required (A23.3 Cl.11.2.8.1)**

If $V_f < V_c$, then shear reinforcement is not required.

$$V_{f@dv} = 103 \text{ kN}$$

and

$$103 \text{ kN} < 166 \text{ kN}$$

Therefore, $V_f < V_c$ and shear reinforcement is not required.

Example 6.4

Consider the beam discussed in Example 6.3. The beam needs to be designed for loads with larger intensity, namely a uniformly distributed dead load (DL) of 70 kN/m and a uniformly distributed live load (LL) of 30 kN/m. Use 25M longitudinal rebars, 10M stirrups, 40 mm clear cover to the stirrups, and maximum 20 mm aggregate size.

Design the beam for shear according to the CSA A23.3 simplified method.

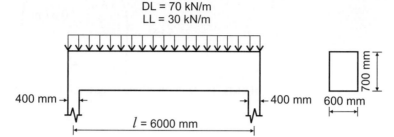

Given: $f_c' = 25$ MPa

  $f_y = 400$ MPa

  $\phi_c = 0.65$

  $\phi_s = 0.85$

  $\lambda = 1$

SOLUTION: **1. Confirm that the CSA A23.3 simplified method can be used**

Since the beam is not subjected to tensile stress due to axial load, the shear design can be performed according to the CSA A23.3 simplified method.

2. Calculate the design shear force ($V_{f@dv}$) at a distance d_v from the face of the support (A23.3 Cl.11.3.2)

a) In order to determine $V_{f@dv}$, the following will have to be calculated:

  i) Effective depth (d):

   $d = 635$ mm (from Example 6.3).

  ii) Effective shear depth (d_v):

   $d_v = 570$ mm (from Example 6.3).

  iii) Clear span (l_n):

   $l_n = 5.6$ m (from Example 6.3).

  iv) Maximum shear force at the face of the support:

   The factored load (w_f) is determined according to NBC 2010 Cl.4.1.3.2 (see Section 1.8):

   $w_f = 1.25$ DL $+ 1.5$ LL $= 1.25(70$ kN/m$) + 1.5(30$ kN/m$)$

   $= 132.5$ kN/m

   and

   $$V_f = \frac{w_f \times l_n}{2} = \frac{132.5 \text{ kN/m} \times 5.6 \text{ m}}{2} = 371 \text{ kN}$$

  v) Maximum shear force at the midspan ($V_{f\,midspan}$):

   $$V_{fmidspan} = \frac{1.5 \text{ LL}\left(\frac{l_n}{2}\right)}{4} = \frac{(1.5 \times 30 \text{ kN/m}) \times \left(\frac{5.6 \text{ m}}{2}\right)}{4}$$

   $$= \frac{126 \text{ kN}}{4} = 31.5 \text{ kN}$$

Now the factored shear force ($V_{f@dv}$) at a distance d_v from the support is

$$V_{f@dv} = V_{fmidspan} + \left(\frac{\frac{l_n}{2} - d_v}{\frac{l_n}{2}} \right)(V_f - V_{fmidspan})$$

$$= 31.5 \text{ kN} + \left(\frac{\frac{5.6 \text{ m}}{2} - 0.57 \text{ m}}{\frac{5.6 \text{ m}}{2}} \right)(371 \text{ kN} - 31.5 \text{ kN}) = 302 \text{ kN}$$

b)   Sketch the shear envelope diagram for the beam.

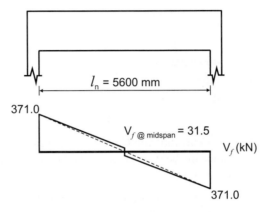

c)   Sketch the shear envelope diagram for the left half of the beam.

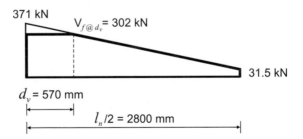

3.   **Determine the concrete shear resistance (V_c) (A23.3 Cl.11.3.4)**

$V_c = 166$ kN (from Example 6.3).

CSA A23.3 prescribes a higher V_c value for the regions where the beam contains at least the minimum shear reinforcement according to Cl.11.2.8.2. In this case,

$\beta = 0.18$ (A23.3 Cl.11.3.6.3)

Hence,

| A23.3 Eq. 11.6 |

$$V_c = \phi_c \lambda \beta \sqrt{f_c'} \, b_w d_v \qquad\qquad \text{[6.12]}$$
$$= 0.65(1)(0.18)\sqrt{25 \text{ MPa}} \, (600 \text{ mm})(570 \text{ mm}) = 200 \text{ kN}$$

To simplify the design, the lower V_c value (166 kN) may be used when shear reinforcement is intended to be provided only where it is required by CSA A23.3. On the other hand, the higher V_c value (200 kN) may be used when at least the minimum shear reinforcement per Cl.11.2.8.2 will be provided throughout the beam span. Therefore, use

$V_c = 166$ kN

4. **Determine whether shear reinforcement is required and identify the region(s) in which the reinforcement is required (A23.3 Cl.11.2.8.1)**

   If $V_f < V_c$, then shear reinforcement is not required.

   $V_{f@dv} = 302 \text{ kN}$ and $V_c = 166 \text{ kN}$

   $302 \text{ kN} > 166 \text{ kN}$

   Therefore, shear reinforcement is required. At least minimum shear reinforcement needs to be provided when $V_f > V_c$ (that is, when $V_f > 166$ kN).

   a) Sketch a shear design envelope showing the following two regions in the beam:

   - Region 1, where minimum shear reinforcement is required, $V_f > V_c$ (shaded); and
   - Region 0, where shear reinforcement is not required, $V_f < V_c$ (white).

   Perform a linear interpolation to locate the distance from the face of the left support (x), where

   $V_{f@x} = 166 \text{ kN}$

   $$V_{f@x} = V_{fmidspan} + \left(\frac{\frac{l_n}{2} - x}{\frac{l_n}{2}} \right) (V_f - V_{fmidspan})$$

   $$166 \text{ kN} = 31.5 \text{ kN} + \left(\frac{\frac{5.6 \text{ m}}{2} - x}{\frac{5.6 \text{ m}}{2}} \right) (371 \text{ kN} - 31.5 \text{ kN})$$

   $$\frac{166 \text{ kN} - 31.5 \text{ kN}}{339.5 \text{ kN}} = \frac{2.8 \text{ m} - x}{2.8 \text{ m}}$$

   $x = 1.69 \text{ m} \cong 1.7 \text{ m}$

   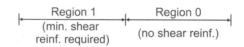

   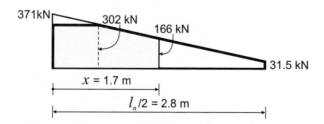

5. **Determine the required stirrup spacing based on the steel shear resistance (V_s) (A23.3 Cl.11.3.5.1)**

   a) Find the required V_s value:

   $V_s \geq V_{f@dv} - V_c$

   $= 302 \text{ kN} - 166 \text{ kN} = 136 \text{ kN}$

   $V_s = 136 \text{ kN}$

   b) Find the required stirrup spacing (s).

   i) Use two-legged 10M stirrups; the area of a 10M bar can be obtained from Table A.1 as $A_b = 100 \text{ mm}^2$; hence

   $A_v = (\text{\# of legs}) (A_b)$

   $= 2 \times 100 \text{ mm}^2 = 200 \text{ mm}^2$

ii) Calculate the s value.

Since this design has been performed according to the CSA A23.3 simplified method, Cl.11.3.6.3 prescribes that $\theta = 35°$, so $\cot\theta = 1.43$:

$$s = \frac{\phi_s A_v f_y d_v \cot\theta}{V_s} \qquad\qquad \textbf{[6.16]}$$

$$= \frac{0.85 \times 200\ mm^2 \times 400\ MPa \times 570\ mm \times 1.43}{136\ 000\ N} = 408\ mm$$

The s value obtained in this calculation needs to be compared with the CSA A23.3 prescribed limits, and the smallest of all values will be chosen for the design.

6. **Calculate the stirrup spacing according to the CSA A23.3 requirements (A23.3 Cl.11.2.8.2 and Cl.11.3.8.1)**

Determine the maximum spacing of shear reinforcement (A23.3 Cl.11.2.8.2):

$$s = \frac{A_v f_y}{0.06\sqrt{f_c'}\ b_w} \qquad\qquad \textbf{[6.19]}$$

$$= \frac{200\ mm^2 \times 400\ MPa}{0.06\sqrt{25\ MPa}\ (600\ mm)} = 444\ mm \cong 440\ mm$$

Check whether the spacing is within the maximum limits prescribed by A23.3 Cl.11.3.8.1: According to A23.3 Cl.11.3.3, the maximum permitted factored shear resistance is

$$\max V_r = 0.25\phi_c f_c' b_w d_v = 0.25 \times 0.65 \times 25\ MPa \times 600\ mm \times 570\ mm = 1390\ kN$$

Since

$$V_f = 302\ kN < V_{rmax}/2 = 1390/2 = 695\ kN$$

the following spacing requirement applies

$$s_{max} \le \begin{bmatrix} 600\ mm \\ 0.7d_v \end{bmatrix} = \begin{bmatrix} 600\ mm \\ 0.7 \times 570\ mm \end{bmatrix} = \begin{bmatrix} 600\ mm \\ 399\ mm \end{bmatrix}$$

Hence, the maximum permitted spacing s_{max} is the smallest of

- 440 mm
- 600 mm
- 399 mm

Hence, $s_{max} = 399\ mm < 408\ mm$

In this case, the maximum permitted spacing according to CSA A23.3 governs. This is due to a low factored shear force (V_f).

Therefore, use

$$s = 400\ mm$$

7. **Check whether the strength requirement is satisfied (A23.3 Cl.11.3.1)**

a) Find the shear capacity (V_s) of the shear reinforcement (A23.3 Cl.11.3.5.1).

Because this design has been performed according to the CSA A23.3 simplified method, Cl.11.3.6.3 prescribes that $\theta = 35°$, so $\cot\theta = 1.43$:

| A23.3 Eq. 11.7 |

$$V_s = \frac{\phi_s A_v f_y d_v \cot\theta}{s} \qquad\qquad \textbf{[6.9]}$$

$$= \frac{0.85 \times 200\ mm^2 \times 400\ MPa \times 570\ mm \times 1.43}{400\ mm}$$

$$= 139 \times 10^3\ N = 139\ kN$$

b) Find the factored shear resistance (V_r) (A23.3 Cl.11.3.3):

| A23.3 Eq. 11.4 |

$$V_r = V_c + V_s + V_p \qquad\qquad\qquad \textbf{[6.11]}$$
$$= 166 \text{ kN} + 139 \text{ kN} + 0 = 305 \text{ kN}$$

c) Check whether the CSA A23.3 strength requirement is satisfied; that is,

| A23.3 Eq. 11.3 |

$$V_r \geq V_f \qquad\qquad\qquad \textbf{[6.10]}$$

Since

$$V_r = 305 \text{ kN} > V_{f@d} = 302 \text{ kN}, \text{ the strength requirement is satisfied.}$$

8. **Check the maximum factored shear resistance (V_r) (A23.3 Cl.11.3.3)**
The factored shear resistance for the beam section is

$$V_r = 305 \text{ kN}$$

According to A23.3 Cl.11.3.3, the maximum permitted factored shear resistance (*max V_r*) is

| A23.3 Eq. 11.5 |

$$max\ V_r = 0.25\phi_c f_c' b_w d_v \qquad\qquad\qquad \textbf{[6.17]}$$
$$= 0.25 \times 0.65 \times 25 \text{ MPa} \times 600 \text{ mm} \times 570 \text{ mm}$$
$$= 1389 \times 10^3 \text{ N} \cong 1390 \text{ kN}$$

Hence,

$$V_r = 305 \text{ kN} < 1390 \text{ kN}, \text{ which is okay.}$$

9. **Sketch a beam elevation showing the distribution of shear reinforcement**
Note that the first stirrup is at a distance $s/2 = 200$ mm away from the face of the support (this is common practice in the design of shear reinforcement).

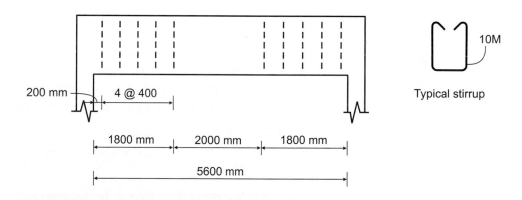

Example 6.5

Consider the beam discussed in Example 6.3. The loads are further increased and the beam needs to support a uniformly distributed dead load (DL) of 100 kN/m and a uniformly distributed live load (LL) of 60 kN/m. Use 25M longitudinal rebars, 10M stirrups, 40 mm clear cover to the stirrups, and maximum 20 mm aggregate size.

Design the beam for shear according to the CSA A23.3 simplified method.

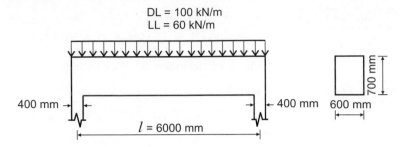

Given: $f_c' = 25$ MPa
$f_y = 400$ MPa
$\phi_c = 0.65$
$\phi_s = 0.85$
$\lambda = 1$

SOLUTION: 1. **Confirm that the CSA A23.3 simplified method can be used**
Since the beam is not subjected to tensile stress due to axial load, the shear design can be performed according to the CSA A23.3 simplified method.

2. **Calculate the design shear force ($V_{f@dv}$) at a distance d_v from the support (A23.3 Cl.11.3.2)**
a)   In order to determine $V_{f@dv}$, the following will have to be calculated:
i)   Effective depth (d):
$d = 635$ mm (from Example 6.3).
ii)   Effective shear depth (d_v):
$d_v = 570$ mm (from Example 6.3).
iii)   Clear span (l_n):
$l_n = 5.6$ m (from Example 6.3).
iv)   Maximum shear force at the face of the support:
The factored load (w_f) is determined according to NBC 2010 Cl.4.1.3.2:
$w_f = 1.25\,DL + 1.5\,LL = 1.25(100\text{ kN/m}) + 1.5(60\text{ kN/m}) = 215\text{ kN/m}$
and
$$V_f = \frac{w_f \times l_n}{2} = \frac{215\text{ kN/m} \times 5.6\text{ m}}{2} = 602\text{ kN}$$
v)   Maximum shear force at the midspan ($V_{fmidspan}$):
$$V_{fmidspan} = \frac{(1.5\,LL)\left(\dfrac{l_n}{2}\right)}{4} = \frac{(1.5 \times 60\text{ kN/m}) \times \left(\dfrac{5.6\text{ m}}{2}\right)}{4}$$
$$= \frac{252\text{ kN}}{4} = 63\text{ kN}$$

Now determine the factored shear force ($V_{f@dv}$) at a distance d_v from the support:
$$V_{f@dv} = V_{fmidspan} + \left(\frac{\dfrac{l_n}{2} - d_v}{\dfrac{l_n}{2}}\right)(V_f - V_{fmidspan})$$

$$= 63 \text{ kN} + \left(\frac{\dfrac{5.6 \text{ m}}{2} - 0.57 \text{ m}}{\dfrac{5.6 \text{ m}}{2}} \right) (602 \text{ kN} - 63 \text{ kN})$$

$$= 492 \text{ kN}$$

b) Sketch the shear envelope diagram for the beam.

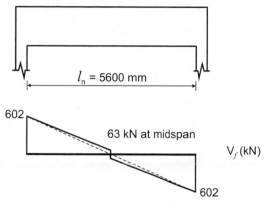

c) Sketch the shear design envelope for the left half of the beam.

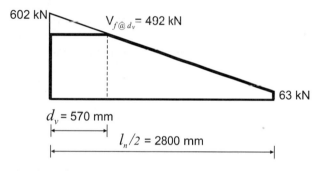

3. **Determine the concrete shear resistance (V_c) (A23.3 Cl.11.3.4)**
    Based on Step 3 in Example 6.4, there are two V_c values:
    a) $V_c = 166$ kN, where less than minimum shear reinforcement is provided;
    b) $V_c = 200$ kN, where shear reinforcement exceeds the minimum as per A23.3 Cl.11.2.8.2.
    In this example, both values will be used, as discussed in the following steps.

4. **Determine the region(s) where shear reinforcement is required (A23.3 Cl.11.2.8.1)**
    The following regions can be identified depending on the magnitude of the shear force:
    a) Region 0: Shear reinforcement is not required.
        When $V_f < V_c$, shear reinforcement is not required; that is,

        $V_f < V_c = 166$ kN

        Note that, in this case, a smaller V_c value is used from Step 3 as it relates to the region without shear reinforcement.
    b) Region 1: Minimum shear reinforcement is required.
        Region 1 is defined by the shear resistance corresponding to the code-prescribed maximum spacing of reinforcement. This was determined in Example 6.1, as follows.
        i) Steel shear resistance (see Step 7 in Example 6.4):

            $V_s = 139$ kN

        ii) Concrete shear resistance V_c (see the previous step):
            It is anticipated that a larger than minimum shear reinforcement will be provided, so we can use the larger V_c value; that is,

            $V_c = 200$ kN

iii) Factored shear resistance V_r:

| A23.3 Eq. 11.4 |

$$V_r = V_c + V_s + V_p$$ **[6.11]**

$$= 200 \text{ kN} + 139 \text{ kN} + 0 = 339 \text{ kN} \cong 340 \text{ kN}$$

iv) The upper bound shear force value for Region 1 is

$$V_f = 340 \text{ kN}$$

v) The lower bound value is

$$V_c = 166 \text{ kN}$$

Hence, Region 1 is defined within the range

$$166 \text{ kN} < V_f < 340 \text{ kN}$$

c) Region 2: Full shear reinforcement is required.
Region 2 requires the provision of shear reinforcement based on the steel shear resistance corresponding to the design shear force ($V_{f@dv}$); that is,

$$340 \text{ kN} < V_f < V_{f@dv} = 492 \text{ kN}$$

d) Sketch a design shear envelope diagram showing the following three regions in the beam:

- Region 0, where shear reinforcement is not required:

$$V_f < V_c = 166 \text{ kN (shown without a pattern on the diagram)};$$

- Region 1, where the minimum shear reinforcement is required:

$$166 \text{ kN} < V_f < 340 \text{ kN (shown shaded on the diagram)};$$

- Region 2, where the full shear reinforcement is required:

$$340 \text{ kN} < V_f < V_{f@dv} = 492 \text{ kN (shown hatched on the diagram)}.$$

e) Locate Regions 0, 1, and 2 on the shear envelope diagram.
This will be done using linear interpolation on the shear envelope diagram shown on the sketch that follows.

i) Find the distance from the left support (x_1), where

$$V_{f@x_1} = 340 \text{ kN}$$

Therefore,

$$V_{f@x_1} = V_{fmidspan} + \left(\frac{\frac{l_n}{2} - x_1}{\frac{l_n}{2}} \right) (V_f - V_{fmidspan})$$

$$340 \text{ kN} = 63 \text{ kN} + \left(\frac{\frac{5.6 \text{ m}}{2} - x_1}{\frac{5.6 \text{ m}}{2}} \right) (602 \text{ kN} - 63 \text{ kN})$$

$$\frac{340 \text{ kN} - 63 \text{ kN}}{539 \text{ kN}} = \frac{2.8 \text{ m} - x_1}{2.8 \text{ m}}$$

$$x_1 = 1.36 \text{ m} \cong 1.4 \text{ m}$$

ii) Find the distance from the left support (x_2), where

$$V_{f@x_2} = 166 \text{ kN}$$

$$V_{f@x_1} = V_{fmidspan} + \left(\frac{\frac{l_n}{2} - x_2}{\frac{l_n}{2}} \right)(V_f - V_{fmidspan})$$

$$166 \text{ kN} = 63 \text{ kN} + \left(\frac{\frac{5.6 \text{ m}}{2} - x_2}{\frac{5.6 \text{ m}}{2}} \right)(602 \text{ kN} - 63 \text{ kN})$$

$$\frac{166 \text{ kN} - 63 \text{ kN}}{539 \text{ kN}} = \frac{2.8 \text{ m} - x_2}{2.8 \text{ m}}$$

$$x_2 = 2.26 \text{ m} \cong 2.3 \text{ m}$$

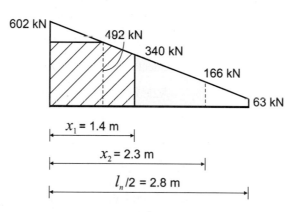

5. **Determine the required spacing of shear reinforcement in Regions 1 and 2 (A23.3 Cl.11.3.5.1)**
    a) Region 1:

    $$s_1 = 400 \text{ mm (from Example 6.4)}$$

    b) Region 2:
        In order to determine the required reinforcement spacing, the following will need to be calculated:
        i) The required V_s value:

        $$V_s \geq V_f - V_c$$
        $$= 492 \text{ kN} - 200 \text{ kN} = 292 \text{ kN}$$

        $$V_s = 292 \text{ kN}$$

        ii) The required stirrup spacing (s):
             Use two legged 10M stirrups, the area of a 10M bar can be obtained from Table A.1 as $A_b = 100 \text{ mm}^2$; hence

        $$A_v = (\# \text{ of legs})(A_b)$$

        $$= 2 \times 100 \text{ mm}^2 = 200 \text{ mm}^2$$

Find θ according to A23.3 Cl.11.3.6.3:

$\theta = 35°$, so cot $\theta = 1.43$

Therefore,

$$s_2 = \frac{\phi_s A_v f_y d_v \cot \theta}{V_s} \qquad \text{[6.16]}$$

$$= \frac{0.85 \times 200 \text{ mm}^2 \times 400 \text{ MPa} \times 570 \text{ mm} \times 1.43}{292\ 000 \text{ N}}$$

$$= 190 \text{ mm}$$

6. **Calculate the maximum stirrup spacing according to the CSA A23.3 requirements (A23.3 Cl.11.2.8.2 and Cl.11.3.8.1)**
   a) Region 1: This check was performed in Step 6 of Example 6.4.
   b) Region 2:
      i) Find the maximum permitted spacing of shear reinforcement based on the equation (A23.3 Cl.11.2.8.2)

$$s = \frac{A_v f_y}{0.06\sqrt{f_c'}\, b_w} \qquad \text{[6.19]}$$

$$= \frac{200 \text{ mm}^2 \times 400 \text{ MPa}}{0.06\sqrt{25 \text{ MPa}}\,(600 \text{ mm})} = 444 \text{ mm} \cong 440 \text{ mm}$$

   ii) Check whether the spacing is within the maximum limits prescribed by A23.3 Cl.11.3.8.1. Since $V_f < \dfrac{V_{rmax}}{2} = \dfrac{1390}{2} = 695$ kN, it follows that

$$s_{max} \leq \begin{bmatrix} 600 \text{ mm} \\ 0.7d_v \end{bmatrix} = \begin{bmatrix} 600 \text{ mm} \\ 0.7 \times 570 \text{ mm} \end{bmatrix} = \begin{bmatrix} 600 \text{ mm} \\ 399 \text{ mm} \end{bmatrix}$$

   iii) Hence, the maximum permitted spacing s_{max} is the smallest of
   - 440 mm
   - 600 mm
   - 399 mm

   Hence, $s_{max} = 399 \text{ mm} > 190 \text{ mm}$

   In this case, the spacing corresponding to the required V_s value for Region 2 determined in the previous step governs. Therefore, use

$$s_2 = 190 \text{ mm}$$

7. **Check whether the strength requirement is satisfied (A23.3 Cl.11.3.1)**
   a) Region 1: This check was performed in Step 7 of Example 6.4.
   b) Region 2:
      i) Find the shear capacity (V_s) of the shear reinforcement (A23.3 Cl.11.3.5.1). Since this design was performed according to the CSA A23.3 simplified method, Cl.11.3.6.3 prescribes that

$$\theta = 35°, \text{ so cot }(\theta) = 1.43$$

A23.3 Eq. 11.7

$$V_s = \frac{\phi_s A_v f_y d_v \cot \theta}{s_2} \qquad \text{[6.9]}$$

$$= \frac{0.85 \times 200 \text{ mm}^2 \times 400 \text{ MPa} \times 570 \text{ mm} \times 1.43}{190 \text{ mm}}$$

$$= 292 \times 10^3 \text{ N} = 292 \text{ kN}$$

ii) Find the factored shear resistance (V_r) (A23.3 Cl.11.3.3):

A23.3 Eq. 11.4

$$V_r = V_c + V_s + V_p \qquad \text{[6.11]}$$
$$= 200 \text{ kN} + 292 \text{ kN} + 0 = 492 \text{ kN}$$

iii) Check whether the CSA A23.3 strength requirement is satisfied; that is,

A23.3 Eq. 11.3

$$V_r \geq V_f \qquad \text{[6.10]}$$

Since

$V_r = 492$ kN and $V_{f@d} = 492$ kN, the strength requirement is satisfied.

8. **Check the maximum permitted factored shear resistance (V_r) (A23.3 Cl.11.3.3)**
This calculation needs to be performed only for Region 2 reinforcement. The calculation for Region 1 reinforcement was done in Example 6.4 (Step 8).
    The factored shear resistance for the beam section is

$V_r = 492$ kN

According to A23.3 Cl.11.3.3, the maximum permitted factored shear resistance ($maxV_r$) is

A23.3 Eq. 11.5

$$maxV_r = 0.25\phi_c f_c' b_w d_v \qquad \text{[6.17]}$$
$$= 0.25 \times 0.65 \times 25 \text{ MPa} \times 600 \text{ mm} \times 570 \text{ mm} = 1389 \times 10^3 \text{ N} \cong 1390 \text{ kN}$$

Hence,

$V_r = 492$ kN < 1390 kN

which is okay.

9. **Sketch a beam elevation showing the distribution of the shear reinforcement**
Note the two different stirrup spacings $s_1 = 400$ mm and $s_2 = 190$ mm. Also note that the first stirrup in the beam is located at a distance $s_2/2 \cong 100$ mm away from the face of the support (this is common practice related to the design of shear reinforcement).

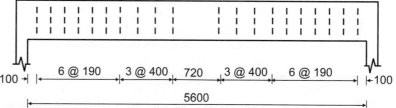

Note from the above sketch that there are only three sets of stirrups at the 400 mm spacing at each side of the beam. Therefore, a more practical alternative solution would be to use the same stirrup spacing throughout the beam, as shown in the sketch below. In that case, the smaller stirrup spacing corresponding to Region 2 should be used; that is, $s = 190$ mm. This solution would result in a simplified on-site installation and a potentially shorter installation time.

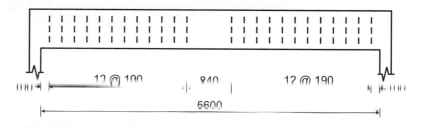

Learning from Examples

The above presented examples illustrated shear design for the same beam subjected to different load levels, as follows:

    Example 6.3: factored load $w_f = 45$ kN/m
    Example 6.4: factored load $w_f = 132.5$ kN/m
    Example 6.5: factored load $w_f = 215$ kN/m

Note that there is a more than fourfold load increase between the designs considered in Examples 6.3 and 6.5.

    As a result of the difference in applied loads, different shear reinforcement requirements were obtained for these examples, as follows:

    Example 6.3: shear reinforcement not required
    Example 6.4: 10M@ 400 mm (10 stirrup sets in total)
    Example 6.5: 10M@ 190 mm (24 stirrup sets in total)

It is interesting to note that the load increase between Examples 6.4 and 6.5 is on the order of 60%; this has resulted in a doubling of the amount of shear reinforcement (from 10 to 24 stirrup sets) between these two examples.

6.8   SHEAR DESIGN OF SIMPLE ONE-WAY SLABS

The shear design of reinforced concrete one-way slabs and beams is very similar. For design purposes, one-way slabs are treated as wide beams of a unit width (b_w) equal to 1000 mm and a depth (h) equal to the slab thickness. The general steps related to the shear design of one-way slabs according to the CSA A23.3 simplified method are outlined in Checklist 6.1.

    Flexure and deflections usually govern in slab design. (For more details on the behaviour of one-way slabs under flexure, the reader is referred to Sections 3.6 and 5.7.) Due to the low shear loads in one-way slabs, shear reinforcement is usually not required, but the shear check needs to be performed.

    The shear design of simple one-way slabs will be illustrated with Example 6.6. Note that the shear design of continuous one-way slabs will be discussed in Chapters 10 and 11.

Example 6.6

Consider the 300 mm thick reinforced concrete slab in the figure below (the same slab was designed for flexure in Example 5.6). The slab spans 6 m (centre-to-centre between the supports) and it supports a uniform dead load (DL) of 6 kPa (including self-weight) and a uniform live load (LL) of 5 kPa. There are no constraints related to the selection of the slab cross-sectional dimensions. Consider a 25 mm clear cover to the tension steel; 15M bars are used for tension reinforcement and 20 mm maximum aggregate size for the concrete. The concrete and steel material properties are given below.
Design the slab for shear according to the CSA A23.3 simplified method.

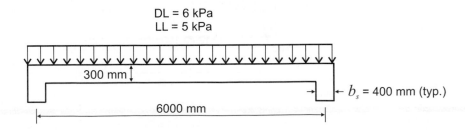

Given: $f_c' = 25$ MPa
$f_y = 400$ MPa
$\phi_c = 0.65$
$\phi_s = 0.85$
$\lambda = 1$

SOLUTION: **1. Confirm that the CSA A23.3 simplified method can be used**
Because the slab is not subjected to tensile stress due to axial load, the shear design can be performed according to the CSA A23.3 simplified method.

2. Calculate the design shear force (V_f)
a) Find the clear span (l_n):

$$l_n = l - \frac{b_s}{2} - \frac{b_s}{2}$$

$$= 6\text{ m} - \frac{0.4\text{ m}}{2} - \frac{0.4\text{ m}}{2} = 5.6\text{ m}$$

where

$b_s = 0.4$ m is the column width.

b) Calculate the factored load (w_f).
According to NBC 2010 Cl.4.1.3.2,

$w_f - 1.25\text{ DL } | \ 1.5\text{ LL} = 1.25(6\text{ kPa}) + 1.5(5\text{ kPa}) = 15\text{ kPa}$

However, for design purposes, one-way slabs are considered as rectangular beams of unit width; that is,

$b_w = 1$ m

Therefore, the factored uniform load per metre of slab width is

$w_f = 15\text{ kPa} \times 1\text{ m} = 15\text{ kN/m}$

c) Find the maximum shear force at the face of the support.
Finally, the shear force at the face of the support can be determined as

$$V_f = \frac{w_f \times l_n}{2} = \frac{15\text{ kN/m} \times 5.6\text{ m}}{2} = 42\text{ kN/m}$$

In this example, there is no need to develop the shear force envelope because shear effects in slabs are usually rather small.

d) Sketch the shear force diagram for the slab.

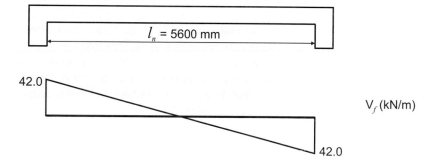

3. **Determine the concrete shear resistance (V_c) (A23.3 Cl.11.3.4)**
    a) Find the effective depth (d):
        - 15M flexural rebars: $d_b = 15$ mm (see Table A.1)
        - cover to the tension steel: cover = 25 mm (see Table A.1)

        $$d = h - \text{cover} - \frac{d_b}{2}$$

        $$= 300 \text{ mm} - 25 \text{ mm} - \frac{15 \text{ mm}}{2} = 267 \text{ mm}$$

    b) Find the effective shear depth (d_v).
        The effective shear depth is taken as the greater of

        $$d_v = \begin{bmatrix} 0.9d \\ 0.72h \end{bmatrix} = \begin{bmatrix} 0.9 \times 267 \text{ mm} \\ 0.72 \times 300 \text{ mm} \end{bmatrix} = \begin{bmatrix} 240 \text{ mm} \\ 216 \text{ mm} \end{bmatrix}$$

        Hence, $d_v = 240$ mm.
    c) Determine the β value. Since

        $$h = 300 \text{ mm} < 350 \text{ mm}$$

        according to A23.3 Cl.11.3.6.2, for slabs with an overall thickness not greater than 350 mm, $\beta = 0.21$ can be used.
    d) Finally, the V_c value can be determined as

        | A23.3 Eq. 11.6 |

        $$V_c = \phi_c \lambda \beta \sqrt{f_c'} \, b_w d_v \qquad\qquad [6.12]$$

        $$= 0.65 \times 1.0 \times 0.21 \times \sqrt{25 \text{ MPa}} \times 1000 \text{ mm} \times 240 \text{ mm}$$

        $$= 163.8 \times 10^3 \text{ N} \cong 164 \text{ kN/m}$$

4. **Determine whether shear reinforcement is required (A23.3 Cl.11.2.8.1)**
    If $V_f < V_c$, then shear reinforcement is not required:

    $$V_f = 42 \text{ kN/m}$$

    and

    $$42 \text{ kN/m} < 164 \text{ kN/m}$$

    Therefore, $V_f < V_c$ and shear reinforcement is not required.

6.9 | DETAILING OF SHEAR REINFORCEMENT

Shear reinforcement has a significant role in providing the required shear resistance in concrete beams. In most cases, shear reinforcement in the form of stirrups is used for this purpose, as discussed in Section 6.4.1. Effective detailing of shear reinforcement is of primary importance for the satisfactory performance of reinforced concrete flexural members. Inadequate detailing can lead to excessive cracking and brittle failure of concrete members subjected to shear. Various aspects of the design and detailing of shear reinforcement are discussed in this section.

| A23.3 Cl.12.13.1 | **Stirrup dimensions**   In general, shear reinforcement should be carried as close to the top and bottom face of the member as cover requirements and the proximity of other reinforcement permit (see Figure 6.18a).

Stirrup bar diameter   It is recommended to use the smallest bar sizes possible for shear reinforcement, in general 15M bars or smaller. Smaller bar sizes ensure that the stirrups crossing diagonal tension cracks can develop yield stress at the ultimate. The use of smaller bar sizes will result in closer stirrup spacing and will prevent the development of excessively wide cracks.

| A23.3 Cl.12.13.2 |
| Cl.12.13.3 |
| Cl.12.13.5 |

Anchorage of shear reinforcement   In order for the stirrups to be effective, they need to develop their full yield strength in tension over almost the full height (for example, see Figure 6.12). This is impossible to accomplish with straight bars since the required development length would be rather large, resulting in very deep beams (refer to Section 9.3

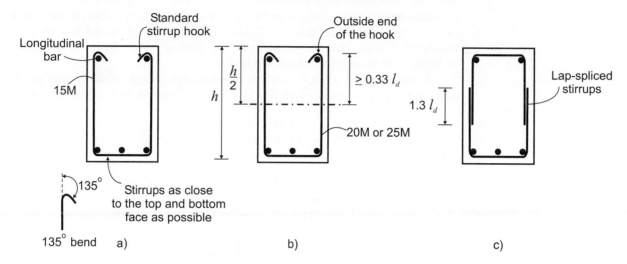

Figure 6.18   Anchorage of shear reinforcement: a) 15M and smaller stirrups; b) 20M and 25M stirrups; c) lap-spliced stirrups.

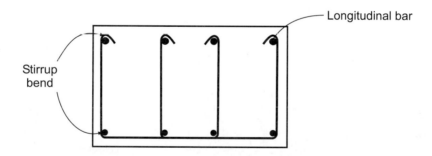

Figure 6.19   Support for longitudinal bars.

for more details on development length). This is not feasible, and therefore stirrups are normally provided with 90° or 135° hooks at their upper end.

According to A23.3 Cl.7.1.2, a *standard stirrup hook* is characterized by a 135° (or larger) bend. (For more details on stirrup hooks, refer to Section 9.5.) However, the anchorage of shear reinforcement is also influenced by the size of stirrup reinforcement; the related CSA A23.3 requirements are summarized in Figure 6.18. These requirements can be summarized as follows:

- For 15M or smaller stirrups, a standard hook should be used around each longitudinal bar, as shown in Figure 6.18a (Cl.12.13.2).
- For 20M and 25M stirrups, a standard stirrup hook should be used around each longitudinal bar; in addition, the embedment between the middepth of the member and the outside end of the hook should be greater than or equal to $0.33l_d$, where l_d is the development length of the stirrup, as shown in Figure 6.18b (Cl.12.13.2).
- In deep members, particularly if the depth varies gradually, it is sometimes feasible to use lap-spliced stirrups with a lap length of $1.3l_d$, as shown in Figure 6.18c (Cl.12.13.5).
- Each bend in a stirrup should enclose a lon
  gitudinal bar (Cl.12.13.3); this requirement is illustrated on an example of a beam with four-legged stirrups in Figure 6.19.

6.10 SHEAR FRICTION (INTERFACE SHEAR TRANSFER)

6.10.1 Background

Let us examine the shear resistance of the plain concrete member shown in Figure 6.20. The member is subjected to a shear force (V) of increasing magnitude. At a low force level, the shear resistance is provided by the concrete. However, the concrete shear resistance (V_c) is rather limited and the shear failure of concrete is brittle and sudden. (For more details on concrete shear resistance, refer to Section 6.3.2.) Therefore, concrete shear resistance will be neglected in this discussion.

When the shear force in the member becomes large enough, cracking takes place across a *shear plane* (also called *interface*), as shown in Figure 6.20. Cracks in monolithically

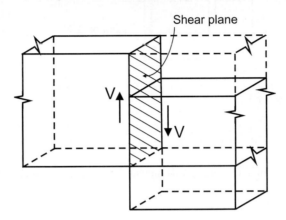

Figure 6.20 Shear force in a plain concrete member.

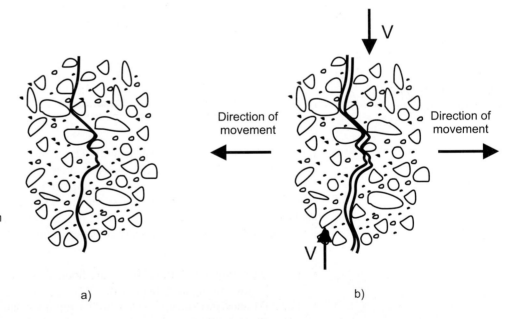

Figure 6.21 Cracking in a monolithically placed plain concrete member: a) a random crack propagates either through the cement paste or at the interface between the cement paste and the aggregate; b) aggregate interlock mechanism.

placed concrete members develop in a random fashion, either in the cement paste or at the interface between the cement paste and the aggregates, as shown in Figure 6.21a.

Once the shear force is sufficiently large to cause cracking, portions of the member on either side of the crack tend to move apart in the perpendicular direction before they slide relative to one another in the direction of the shear force, as shown in Figure 6.21b. This phenomenon, known as the *aggregate interlock* mechanism, is critical for concrete shear resistance at an interface.

When a crack develops, the reinforcing steel across the cracked plane provides the shear resistance. Tension stresses in steel are caused by the shear force across the cracked interface; this is known as the *shear friction* mechanism (see Figure 6.22). In order to provide the required shear resistance, the steel must be properly developed on either side of the crack. (Refer to Section 9.3 for more details related to anchorage requirements for reinforcing bars.)

Until now, the discussion has been focused on monolithically placed concrete members, that is, concrete members placed simultaneously (or within a short time-frame). However, in renovations or upgrades of existing buildings, it is often necessary to add a new concrete member adjacent to an older existing member. In this case, only a marginal bond exists

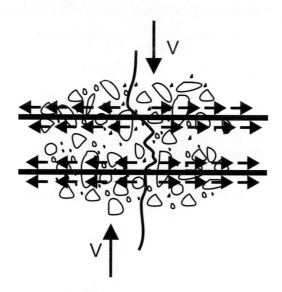

Figure 6.22 Tension stress developed in reinforcing steel provides shear friction resistance across the concrete surface.

between the new and the existing concrete; this can be considered as if a crack exists at the interface between the members. In a typical case, the existing member has a smooth formed surface, hence the aggregate interlock resistance characteristic of monolithic concrete is not available. However, some shear resistance can be supplied through the friction between the two surfaces (smooth concrete-to-concrete interfaces are characterized by a coefficient of friction of approximately 0.45). The reinforcing steel crossing the shear plane provides shear resistance by means of the shear friction mechanism. However, the shear resistance at the interface between the two smooth surfaces is significantly lower than at the interfaces in monolithic structures that can develop resistance through the aggregate interlock mechanism.

The shear resistance at the interface between the existing hardened concrete and the new concrete can be improved if the existing concrete can be intentionally roughened; this will minimize the chances of sliding between the new and the existing concrete. For example, the shear resistance across an interface can be significantly increased when the surface is randomly roughened to amplitudes of at least 5 mm in order to develop the aggregate interlock mechanism.

6.10.2 CSA A23.3 Design Requirements

The CSA A23.3 provisions related to shear resistance across interfaces are included in Cl.11.5. The factored shear resistance across an interface (v_r) can be determined as follows:

| A23.3 Eq. 11.24 |

$$v_r = \lambda \phi_c (c + \mu \sigma) + \phi_s \rho_v f_y \cos \alpha_f \ \ (MPa) \qquad \textbf{[6.20]}$$

where

$$\lambda \phi_c (c + \mu \sigma) \le 0.25 \phi_c f_c'$$

c = the cohesion stress (MPa)
μ = the coefficient of friction
σ = the effective normal stress
α_f = the angle between the shear friction reinforcement and the shear plane (see Figure 6.23)
$\rho_v = A_{vf}/A_{cv}$ is the ratio of shear friction reinforcement
A_{vf} = the area of shear friction reinforcement
A_{cv} = the area of the concrete section resisting shear transfer.

The effective normal stress (σ) can be determined as

| A23.3 Eq. 11.26 |

$$\sigma = \rho_v f_y \sin \alpha_f + \frac{N}{A_g} \qquad [6.21]$$

where

A_g = the gross area of the section
N = the unfactored permanent load perpendicular to the shear plane (positive for compression and negative for tension).

CSA A23.3 prescribes different c and μ values depending on the condition of the interface (see Table 6.3).

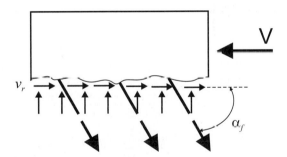

Figure 6.23 The angle (α_f) between the shear friction reinforcement and the shear plane.

Table 6.3 c and μ values according to CSA A23.3 Cl.11.5.2

	c (MPa)	μ
a) Concrete placed against hardened concrete with the surface clean but not intentionally roughened	0.25	0.6
b) Concrete placed against hardened concrete with the surface clean and intentionally roughened to a full amplitude of at least 5 mm	0.5	1.0
c) Concrete placed monolithically	1.0	1.4
d) Concrete anchored to as-rolled structural steel by headed studs or by reinforcing bars	0	0.6

(*Source:* CSA A23.3 Cl.11.5.2, reproduced with the permission of the Canadian Standards Association)

Example 6.7

A partial plan view of a 200 mm thick reinforced concrete slab with a stair opening is shown in the figure below. The design requires that a seismic shear force of 2000 kN be transferred from the slab to the shear wall below via the 5 m long section of the slab adjacent to the stair opening. The concrete will be placed monolithically in this construction.

Design the slab to resist the shear force by providing 20M vertical steel rebars across the shear plane in accordance with CSA A23.3 Cl.11.5.

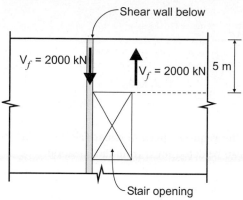

Given: $f_c' = 25$ MPa
$f_y = 400$ MPa
$\phi_c = 0.65$
$\phi_s = 0.85$
$\lambda = 1$

SOLUTION:

1. Determine c and μ

Since the concrete is placed monolithically, c and μ according to A23.3 Cl.11.5.2(c) are

$c = 1.00$ MPa
$\mu = 1.40$

2. Find the angle α_f

For reinforcement perpendicular to the plane of shear force,

$\alpha_f = 90°$, hence $\cos 90° = 0$ and $\sin 90° = 1$

3. Determine the factored shear stress (v_c) across the interface

The interface is defined by the length of the slab section as

$l = 5000$ mm

and the slab thickness is

$t = 200$ mm

Therefore, the area of the concrete section resisting the shear transfer (A_{cv}) is

$A_{cv} = l \times t = 5000 \text{ mm} \times 200 \text{ mm} = 1 \times 10^6 \text{ mm}^2$

The factored shear stress is equal to the ratio of the factored shear force acting to the area A_{cv}:

$$v_f = \frac{V_f}{A_{cv}}$$

$$= \frac{2000 \times 10^3 \text{ N}}{1 \times 10^6 \text{ mm}^2} = 2 \text{ MPa}$$

For further calculations, assume that the shear resistance is equal to the shear stress; that is,

$$v_r = v_f$$

4. **Determine the effective normal stress (σ)**

σ can be determined from the v_r equation:

A23.3 Eq. 11.24

$$v_r = \lambda \phi_c (c + \mu\sigma) + \phi_s \rho_v f_y \cos \alpha_f \qquad [6.20]$$

Since $\alpha_f = 90°$ and $\cos 90° = 0$

$$v_r = \lambda \phi_c (c + \mu\sigma)$$

Because all the parameters except σ are known, we can determine σ as

$$\sigma = \frac{\dfrac{v_r}{\lambda \phi_c} - c}{\mu}$$

$$= \frac{\dfrac{2\ \text{MPa}}{1 \times 0.65} - 1.00\ \text{MPa}}{1.40} = 1.48\ \text{MPa}$$

5. **Find the required area of shear friction reinforcement (A_{vf})**

This can be determined from the equation

A23.3 Eq. 11.26

$$\sigma = \rho_v f_y \sin \alpha_f + \frac{N}{A_g} \qquad [6.21]$$

Since $N = 0$ and $\sin \alpha_f = 1$ (due to $\alpha_f = 90°$),

$$\sigma = \rho_v f_y = 1.48\ \text{MPa}$$

Hence,

$$\rho_v = \frac{\sigma}{f_y} = \frac{1.48\ \text{MPa}}{400\ \text{MPa}} = 0.0037$$

Next, determine the required area of reinforcement from the equation

$$\rho_v = \frac{A_{vf}}{A_{cv}}$$

It follows that

$$A_{vf} = \rho_v \times A_{cv}$$
$$= 0.0037 \times 1 \times 10^6\ \text{mm}^2 = 3700\ \text{mm}^2$$

The area of one 20M bar is 300 mm² (see Table A.1). Therefore, use 13-20M bars; that is,

$$A_{vf} = 13 \times 300\ \text{mm}^2 = 3900\ \text{mm}^2 > 3700\ \text{mm}^2$$

This area is equivalent to 20M@350 mm o.c., as shown in the sketch below. Note that the reinforcing bars must have sufficient development length (l_d) on either side of the interface (see Section 9.3 for more details on reinforcement anchorage requirements).

13 - 20M @ 350 mm o.c.

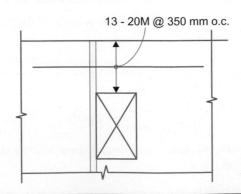

SUMMARY AND REVIEW — SHEAR DESIGN OF BEAMS AND ONE-WAY SLABS

In most cases, bending moments in reinforced concrete members are accompanied by internal forces, acting parallel to a beam cross-section, called shear forces. Shear and flexure effects are related, and a shear force can be considered as an internal force caused by flexure. The shear design of concrete members needs to be performed in most design applications.

Failure modes in reinforced concrete beams subjected to flexure and shear

There are three main failure modes typical of reinforced concrete beams subjected to flexure and shear:

* flexural
* diagonal tension
* shear compression

Flexural (steel-controlled) failure is a ductile failure mode characterized by the yielding of tension steel. It is the most desirable of all three modes.

The diagonal tension and shear compression failure modes are governed by shear. Shear failures are generally sudden and brittle and should be avoided in design applications.

Shear reinforcement in reinforced concrete beams

The shear resistance of beams susceptible to shear failure can be greatly increased by providing properly designed shear reinforcement.

Shear reinforcement can effectively contribute to enhancing the shear resistance of the beam in two ways:

* by carrying a portion of the shear load through a sudden increase in tensile forces after the formation of inclined cracking
* by limiting the width and spread of inclined cracks and preventing the occurrence of sudden and brittle failure

Shear reinforcement is usually designed for excess shear; that is, the difference between the shear force (V_f) acting on the beam section and the concrete shear resistance (V_c).

The following types of shear reinforcement are permitted by CSA A23.3 (Cl.11.2.4):

* vertical stirrups
* inclined stirrups at an angle of 45° or more
* bent-up longitudinal bars
* welded wire fabric
* spirals

Key factors influencing shear resistance of reinforced concrete beams

The shear resistance of concrete beams with shear reinforcement is provided by concrete and steel.

The concrete shear resistance (V_c) is usually taken as the shear force that causes the first inclined crack. The main factors influencing concrete shear resistance are: the concrete tensile strength, the longitudinal reinforcement ratio, and the axial forces in the beam.

The steel shear resistance (V_s) is provided by stirrups crossing an inclined crack plane at an angle θ to the longitudinal beam axis. It is assumed that stirrups resist shear by yielding in tension.

CSA A23.3 simplified method for shear design

The strength requirement is the main CSA A23.3 requirement related to the simplified method for shear design (Cl.11.3.1).

CSA A23.3 also prescribes the procedure to determine

* the factored shear resistance (V_r) (Cl.11.3.3)
* the concrete shear resistance (V_c) (Cl.11.3.4)
* the steel shear resistance (V_s) (Cl.11.3.5.1)

The following *limitations* are set by CSA A23.3 as related to the amount and spacing of shear reinforcement:

- maximum factored shear resistance (Cl.11.3.3)
- maximum spacing of transverse shear reinforcement (Cl.11.3.8.1)
- required provision of shear reinforcement (Cl.11.2.8.1)
- minimum area of shear reinforcement (Cl.11.2.8.2)

PROBLEMS

6.1. a) Which failure modes are characteristic of reinforced concrete beams subjected to flexure and shear?
   b) Which of these modes is the most desirable and why? Explain.

6.2. Can shear failure in reinforced concrete beams be avoided by design? Explain.

6.3. An unreinforced concrete beam is shown on the figure below.

Sketch approximate stress trajectories for this beam for the following load cases:
   a) uniformly distributed load (w)
   b) point load (P) at the tip of the cantilever

Use a dashed line to denote compression and a solid line to denote tension.

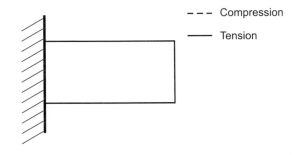

6.4. A reinforced concrete beam of 600 mm by 800 mm rectangular cross-section has been designed using four-legged 10M stirrups at 400 mm spacing and 25M longitudinal rebars. The clear cover to the stirrups is 40 mm. The maximum aggregate size is 20 mm.

Determine the factored shear resistance for this section according to the CSA A23.3 simplified method.

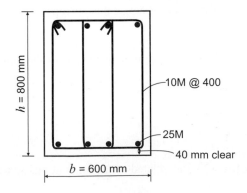

Given:

$$f_c' = 25 \text{ MPa (normal-density concrete)}$$
$$f_y = 400 \text{ MPa}$$

6.5. A typical cross-section of a reinforced concrete beam is shown in the figure that follows. The beam is subjected to a factored shear force (V_f) of 300 kN. There is a 30 mm clear cover to the stirrups. The maximum aggregate size is 20 mm. The concrete and steel material properties are given.

Determine the required spacing of 10M stirrups according to the CSA A23.3 simplified method.

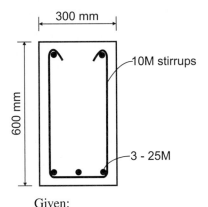

Given:

$$f_c' = 25 \text{ MPa (normal-density concrete)}$$
$$f_y = 400 \text{ MPa}$$

6.6. A simply supported reinforced concrete beam of rectangular cross-section is shown in the figure below. The beam supports a uniform dead load (DL) of 120 kN/m (including its own weight) and a uniform live load (LL) of 60 kN/m, and it is reinforced with 30M longitudinal rebars with a 40 mm clear cover to the stirrups. The maximum aggregate size is 20 mm. The concrete and steel material properties are given below.
   a) Design the required spacing of 10M stirrups for this beam according to the CSA A23.3 simplified method for shear design. Sketch a beam elevation showing the stirrup arrangement. Develop a solution using the same stirrup spacing throughout the beam (in the region where the stirrups are required).
   b) Develop an alternative solution using two different stirrup spacings (in the region where the stirrups are required).

c) Which solution would you use if you were faced with a similar problem in a design office? Explain.

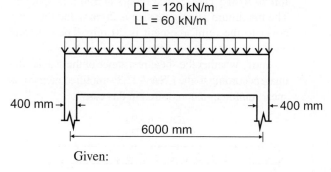

Given:

$$b = 600 \text{ mm}$$
$$h = 800 \text{ mm}$$
$$f_c' = 25 \text{ MPa (normal-density concrete)}$$
$$f_y = 400 \text{ MPa}$$

6.7. A simply supported reinforced concrete beam of rectangular cross-section is shown in the figure below. The beam depth tapers from 1200 mm to 600 mm, while the width is constant throughout the span. The beam supports a uniform dead load (DL) of 150 kN/m (including its own weight) and a uniform live load (LL) of 60 kN/m. The beam is reinforced with 25M longitudinal rebars with a 40 mm clear cover to the stirrups. The maximum aggregate size is 20 mm. The concrete and steel material properties are given below.

Determine the required spacing of 10M stirrups in the beam according to the CSA A23.3 simplified method and sketch the stirrup arrangement.

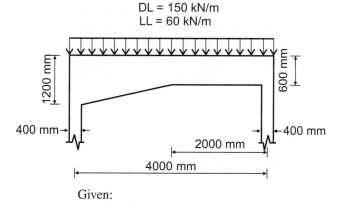

Given:

$$b = 600 \text{ mm}$$
$$f_c' = 25 \text{ MPa (normal-density concrete)}$$
$$f_y = 400 \text{ MPa}$$

6.8. A simply supported reinforced concrete beam of rectangular cross-section spans over 7500 mm; the support width is equal to 400 mm. The beam is subjected to a total factored load (w_f) (including self-weight) of 75 kN/m. The beam is reinforced with 25M longitudinal rebars (a 30 mm clear cover

to the stirrups has to be provided). The maximum aggregate size is 20 mm. Consider that the shear force at the midspan is equal to zero. The concrete and steel material properties are given below.

Determine the required spacing of 10M stirrups in the beam according to the CSA A23.3 simplified method. Sketch the stirrup arrangement.

Given:

$$b = 600 \text{ mm}$$
$$h = 800 \text{ mm}$$
$$f_c' = 25 \text{ MPa (normal-density concrete)}$$
$$f_y = 400 \text{ MPa}$$

6.9. A simply supported reinforced concrete beam of rectangular cross-section is shown in the figure that follows. The beam supports a uniform dead load (DL) of 100 kN/m (including its own weight) and a uniform live load (LL) of 40 kN/m. The beam also supports a point dead load (P_{DL}) of 500 kN and a point live load (P_{LL}) of 100 kN. The beam is reinforced with 30M longitudinal rebars, and a 40 mm clear cover to the stirrups has been provided. The maximum aggregate size is 20 mm. The concrete and steel material properties are given below.

Determine the required spacing for 15M stirrups according to the CSA A23.3 simplified method. Summarize the design on a sketch showing the stirrup arrangement.

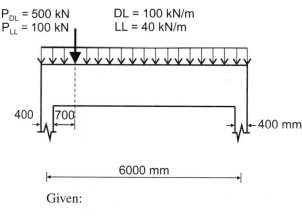

Given:

$$b = 600 \text{ mm}$$
$$h = 800 \text{ mm}$$
$$f_c' = 25 \text{ MPa (normal-density concrete)}$$
$$f_y = 400 \text{ MPa}$$

6.10. The reinforced concrete beam of rectangular cross-section in the figure below supports a point dead load (P_{DL}) of 1000 kN and a point live load (P_{LL}) of 400 kN (neglect the effect of the beam self-weight). The beam is reinforced with 30M longitudinal rebars with a 40 mm clear cover to the stirrups. The maximum aggregate size is 20 mm. The concrete and steel material properties are given below.

Design the shear reinforcement for this beam according to the CSA A23.3 simplified method. Use two-legged 15M stirrups for the design. Sketch the stirrup arrangement.

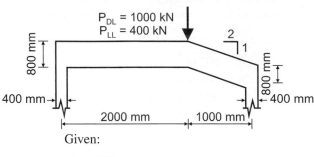

Given:

$$b = 500 \text{ mm}$$
$$f_c' = 25 \text{ MPa (normal-density concrete)}$$
$$f_y = 400 \text{ MPa}$$

6.11. A simply supported reinforced concrete beam with an overhang is shown in the figure below. The beam supports a uniform dead load (DL) of 100 kN/m (including its own weight) and a uniform live load (LL) of 40 kN/m. The beam is also subjected to an uplift point load (P_{LL}) of 500 kN, as shown in the figure. The beam is of rectangular cross-section and it is reinforced with 30M longitudinal rebars and it has a 40 mm clear cover to the stirrups. The maximum aggregate size is 20 mm. The concrete and steel material properties are given below.

Design the shear reinforcement for this beam according to the CSA A23.3 simplified method using two-legged 10M stirrups. Summarize the design on a sketch showing the stirrup arrangement.

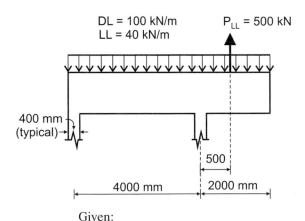

Given:

$$b = 500 \text{ mm}$$
$$h = 700 \text{ mm}$$
$$f_c' = 25 \text{ MPa (normal-density concrete)}$$
$$f_y = 400 \text{ MPa}$$

6.12. Consider the 200 mm thick reinforced concrete slab in the figure below. The slab is supported by a 200 mm thick wall at one end and a 400 mm wide

beam at the other. The slab supports a uniform dead load (DL) of 6 kPa (in addition to its own weight) and a uniform live load (LL) of 2 kPa and it is reinforced with 15M flexural rebars at 300 mm spacing. The maximum aggregate size is 20 mm and the clear cover to the reinforcement is 20 mm. The concrete and steel material properties are given below.

Verify whether the shear resistance of the slab is adequate according to the CSA A23.3 simplified method. Is shear reinforcement required in this case?

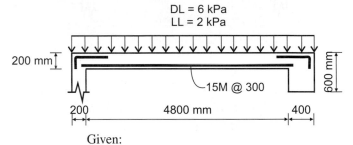

Given:

$$f_c' = 25 \text{ MPa (normal-density concrete)}$$
$$f_y = 400 \text{ MPa}$$
$$\gamma_w = 24 \text{ kN/m}^3$$

6.13. A simply supported reinforced concrete beam of rectangular cross-section is hung on the left end by a 400 mm square concrete post working in tension, as shown in the figure below. The beam supports a uniform dead load (DL) of 100 kN/m (excluding its own weight) and a uniform live load (LL) of 40 kN/m. The beam is reinforced with 25M longitudinal rebars with a 40 mm clear cover to the stirrups. The maximum aggregate size is 20 mm. The concrete and steel material properties are given below.

Design the shear reinforcement for this beam according to the CSA A23.3 simplified method. Use 10M stirrups in the design. Summarize the results of your design on a sketch showing the stirrup arrangement.

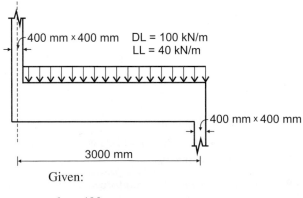

Given:

$$b = 400 \text{ mm}$$
$$h = 600 \text{ mm}$$
$$f_c' = 25 \text{ MPa (normal-density concrete)}$$
$$f_y = 400 \text{ MPa}$$
$$\gamma_w = 24 \text{ kN/m}^3$$

6.14. A 500 mm wide reinforced concrete beam is shown in the figure that follows. The beam is subjected to a factored uniform load (w_f) of 25 kN/m (including its own weight) and a factored point load (P_f) of 500 kN, as shown in the figure. The beam is reinforced with 30M longitudinal rebars, and a 40 mm clear cover to the stirrups has been provided. The maximum aggregate size is 20 mm. The concrete and steel material properties are given below.

Design the shear reinforcement for this beam according to the CSA A23.3 simplified method. Use 10M stirrups in the design. Summarize the results of your design on a sketch showing the stirrup arrangement.

Given:

$b = 500$ mm

$f_c' = 30$ MPa (normal-density concrete)

$f_y = 400$ MPa

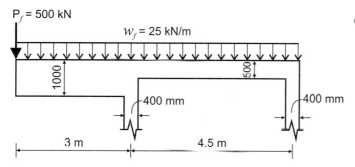

6.15. A 300 mm wide by 600 mm deep beam shown in the figure below is proposed to be cast between two existing walls A and B. One proposal is to roughen wall A only to minimum 5 mm amplitude at the interface of the new beam. Wall B was originally cast with good-quality smooth formply and will be cleaned prior to casting the new beam.

a) Determine the area of shear friction reinforcement required at the beam supports according to the CSA A23.3 requirements.

b) Do you believe that this solution is good for the right support?

c) What measures would you take during the site inspection to ensure that the aggregate interlock mechanism develops at the left support?

Given:

$f_c' = 30$ MPa (normal-density concrete)

$f_y = 400$ MPa

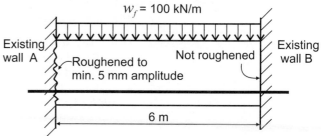

6.16. Consider the beam design discussed in Problem 6.15. Suppose you specified the use of an epoxy adhesive that would require the rebars to be drilled to a depth of 250 mm into the existing walls to develop their full strength. However, the existing concrete has inconsistencies and the contractor experienced problems while drilling 250 mm deep holes at some locations. For approximately 50% of the holes, the contractor can drill only 150 mm deep (or 100 mm shallower than the specified depth). To compensate for the inability to meet the design specifications, the contractor offers to drill the remaining holes 350 mm deep, that is, 100 mm deeper than the specified depth.

Is this solution acceptable if the design requires every rebar to develop to its full strength? Explain your reasoning.

7 Torsion

LEARNING OUTCOMES

After reading this chapter, you should be able to

- identify the main torsional failure modes
- outline the key factors influencing the behaviour and resistance of concrete beams subjected to torsion
- design reinforced concrete beams for torsion and the combined torsion, shear, and/or flexure loads according to the CSA A23.3 simplified method

7.1 INTRODUCTION

Torsion (torque) is a moment that tends to twist a member around its longitudinal axis. Reinforced concrete structures are often subjected to torsional moments, which cause shear stresses. In the past, engineers often ignored torsional effects in the design of concrete structures. Until the 1960s, code references to torsion were mainly based on the assumptions of concrete as a homogeneous isotropic material. Torsion is often combined with shear, flexure, and/or axial loads.

The fundamental principles of design for torsional effects are discussed in this chapter. CSA A23.3 proposes two basic design methods for torsion: the *simplified method* and the *general method*. The main focus in this chapter is on the *simplified method* of design for torsion, in terms of both the theoretical background and the design examples.

Torsional effects on reinforced concrete members are significantly influenced by the type of structural system (statically determinate or indeterminate), as explained in Section 7.2. The behaviour of concrete beams subjected to torsion is discussed in Section 7.3. The torsional resistance of reinforced concrete beams is explained in Section 7.4. Design for combined torsion, shear, and flexure effects is outlined in Section 7.5. The CSA A23.3 requirements for the simplified method for torsion design are summarized in Section 7.6. Recommendations for the detailing of torsional reinforcement are presented in Section 7.7. Practical considerations related to design for torsion are outlined in Section 7.8. A summary of the CSA A23.3 torsional design provisions and a design example are presented in Section 7.9.

7.2 TORSIONAL EFFECTS

Torsional moments develop in *statically determinate* structures as a result of the equilibrium between external and internal forces — this type of torsion is often called *equilibrium torsion*. In the case of equilibrium torsion, a structure may collapse when torsional resistance cannot be provided. The beam supporting an eccentrically built masonry wall in Figure 7.1a is subjected to equilibrium torsion.

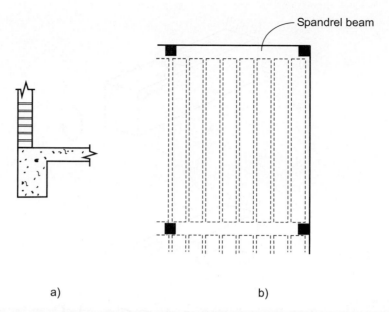

Spandrel beam

Figure 7.1 Examples of structures subjected to torsion: a) statically determinate torsion; b) statically indeterminate torsion.

a)   b)

In *statically indeterminate* structures, torsional moments are caused by the continuity of concrete construction. The torsion develops in a structural element that is restrained at its ends and has to twist in order to maintain the compatibility of deformations, hence the name *compatibility torsion* (or statically indeterminate torsion). When this effect is disregarded in the design, concrete members will develop extensive cracks; however, failure may be avoided due to the redistribution of moments to the adjacent structural members in a continuous structure. *Spandrel beams* (edge beams) in concrete floor structures are often subjected to compatibility torsion, as illustrated in Figure 7.1b. Compatibility torsion is more common in concrete structures than equilibrium torsion.

KEY CONCEPTS

Torsional effects on reinforced concrete members are significantly influenced by the type of structural system (statically determinate or indeterminate).

In *statically determinate* structures, torsional moments develop as a result of the equilibrium between external and internal forces — this type of torsion is often called *equilibrium torsion*.

Torsional moments in *statically indeterminate* structures develop in structural elements restrained at the ends that have to twist in order to maintain the compatibility of deformations, hence the name *compatibility torsion* (or statically indeterminate torsion).

7.3 BEHAVIOUR OF CONCRETE BEAMS SUBJECTED TO TORSION

Consider the cantilevered concrete beam of rectangular cross-section subjected to an external torque (T) in Figure 7.2a. The torsion causes shear stresses all around the beam section (called *torsional shear stresses* in the rest of this chapter), as shown on section A-A in Figure 7.2b. When a beam is subjected to pure torsion, the principal stresses are equal to the torsional shear stresses and they develop under a 45° angle to the longitudinal beam axis. The principal tensile stresses ultimately cause cracking in the beam. The cracking pattern is illustrated in Figure 7.2b. Note that the cracks spiral around the beam. The torsional moment that causes the cracking is called the torsional cracking resistance and is denoted as T_{cr}. A plain concrete beam will fail shortly after the cracking has begun.

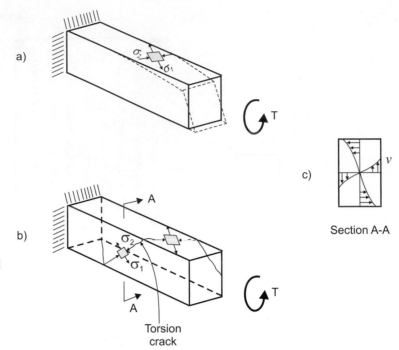

Figure 7.2 Stress distribution and cracking in a beam subjected to torsion: a) deformed shape and principal stresses; b) cracking pattern; c) distribution of torsional shear stresses within a cross-section.

As in the case of shear design, it is clear that steel reinforcement needs to be provided in order to improve the torsional resistance of a beam. To be effective in resisting torsional effects, *torsional reinforcement* needs to cross the torsion-induced cracks. The cracking pattern characteristic of the behaviour of beams subjected to torsion spirals around the beam, crossing all four faces of a beam section. Consequently, it is necessary to provide both longitudinal reinforcement, distributed all around the section, and transverse reinforcement (stirrups). Note that the addition of longitudinal steel reinforcement by itself has a very limited effect in resisting torsion and that the stirrups have a more significant role in providing torsional resistance. Given that torsional cracks spiral around the section, closed stirrups need to be provided, which is not a requirement in the case of beam shear design. The design of torsional reinforcement will be discussed in this chapter.

It is important to note that the torsional shear stresses are largest at the outer fibres of a cross-section, as illustrated in section A-A of Figure 7.2c. In fact, the outer face of the beam cross-section (analogous to a concrete tube containing the reinforcement) is essential for providing torsional resistance. This was confirmed by tests on hollow and solid beams with the same amount and distribution of reinforcement, as reported by MacGregor and Bartlett (2000). Tests have shown that the torsional resistance of hollow and solid beams is virtually identical, which indicates that the concrete core is ineffective in resisting torsional effects. Section A-A in Figure 7.2c shows that the torsional shear stresses increase from zero at the centroid of the section to the maximum value at the outer face.

It should also be noted that torsional stiffness is significantly reduced after cracking. Before cracking, the torsional stiffness is equal to the slope of the torque-twist diagram in Figure 7.3. Once the cracking takes place, the stiffness can be considered equal to the slope of the radial line through the origin and the point under consideration on the torque-twist diagram. According to the tests reported by MacGregor and Bartlett (2000), immediately after cracking the stiffness drops to 20% of its initial value. At failure, the stiffness amounts to less than 10% of the uncracked value.

The ultimate torsional resistance of beams with torsional reinforcement exceeds T_{cr}. The behaviour of cracked concrete beams can be studied by observing the torque–angle-of-twist diagram for the concrete beam in Figure 7.3. Before cracking, there is no difference in torsional behaviour between a reinforced and a plain concrete beam with the same dimensions. After cracking, the angle of twist increases at constant torque due to the redistribution of forces between the concrete (which has lost its torsional resistance) and the steel reinforcement (which becomes effective only after cracking).

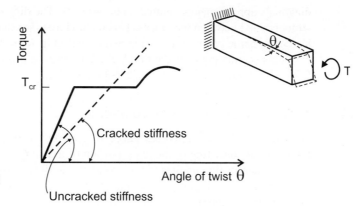

Figure 7.3 Torque–angle-of-twist diagram for a reinforced concrete beam.

Failure in reinforced concrete beams subjected to torsion can occur by either the yielding of stirrups and longitudinal reinforcement (in lightly reinforced beams) or the crushing of concrete between the inclined cracks before the reinforcement yields (in overreinforced beams). The former failure mode is ductile and therefore more desirable than the latter.

KEY CONCEPTS

Failure in reinforced concrete beams subjected to torsion can occur by either the yielding of reinforcement (in lightly reinforced beams) or the crushing of concrete (in overreinforced beams). Reinforcement yielding failure is more desirable than concrete crushing failure.

The cracking pattern characteristic of the behaviour of beams subjected to torsion spirals around the beam; therefore, both longitudinal bars distributed all around the beam cross-section and closed transverse reinforcement (stirrups) are essential for resisting the torsional effects.

Torsional shear stresses are largest at the outer fibres of a beam cross-section. Therefore, the outer face of a beam cross-section is essential in providing torsional resistance (similar to a concrete tube containing reinforcement). The concrete core is not effective in resisting torsion due to the small shear stresses developed in that zone.

7.4 TORSIONAL RESISTANCE OF REINFORCED CONCRETE BEAMS

7.4.1 Background

The two main methods related to the design of concrete structures for torsion are the *skew bending theory* and the *thin-walled tube/space truss analogy* method.

The skew bending theory considers the internal deformation of the series of transverse warped surfaces along the beam. According to this theory, the mode of failure involves bending on a skew surface resulting from the crack spiralling around the sides of the member. The skew bending theory formed the basis of the ACI Code torsion design provisions from 1971 to 1989. For more details on the skew bending theory, the reader is referred to Nawy (2003).

The *thin-walled tube/space truss analogy* method is similar to the truss analogy method for shear design presented in Section 6.1.2. The thin-walled tube/space truss analogy method has been adopted by CSA A23.3 since 1971 and it forms the basis of both the simplified method and the general method for torsional design.

The thin-walled tube/space truss analogy method includes two different mechanics-based models for the behaviour of concrete beams subjected to torsion. Before cracking, the beam is idealized as a thin-walled hollow tube. After cracking, the tube is idealized as a hollow truss consisting of closed stirrups, longitudinal bars in the corners, and compression diagonals approximately centred on the stirrups. The diagonals consist of inclined concrete struts formed between the cracks. This method assumes that the reinforcement has yielded, hence the name *plastic space truss analogy* used in some references.

This section presents the thin-walled tube/space truss analogy method as used by the CSA A23.3 simplified method of torsional design.

7.4.2 Torsional Resistance of Concrete

To determine the torsional resistance of concrete, a beam is idealized as a thin-walled hollow tube consisting of the outer portion of a concrete section with reinforcement; note that this idealization applies to both solid and hollow cross-sections. The properties of the tube are shown in Figure 7.4. Assume that the actual section is of rectangular shape with gross cross-sectional area A_c and perimeter p_c (see Figure 7.4a). The section is idealized as the tube section with wall thickness t in Figure 7.4b. Before cracking, torsional shear stresses develop in the tube as a result of an external torque (T). The accurate analysis of torsional shear stress in a tube is rather complex. However, a simple torsional shear stress formula widely used in engineering applications is

$$v_t = \frac{T}{2A_o t} \qquad\qquad [7.1]$$

where

v_t = the torsional shear stress in the thin-walled tube section
T = the torque acting on the section
A_o = the area enclosed by the centre-line of the walls of the tube (shown hatched in Figure 7.4b) with perimeter p_o
t = the tube wall thickness.

Equation 7.1 is also known as Bredt's formula. For more details on shear stresses in a thin-walled section, including the derivation of Eqn 7.1, the reader is referred to textbooks on the mechanics of materials. Bredt's formula was developed for thin-walled tubes; however, it should be noted that the walls of equivalent tubes in concrete structures tend to be quite thick, on the order of 1/6 to 1/4 of the smaller cross-sectional dimension (width or depth) of a rectangular section (MacGregor and Bartlett, 2000).

Note that the product of the torsional shear stress (v_t) and the wall thickness (t) is constant at any point around the cross-section; that is,

$$q = v_t t = \text{constant} \qquad\qquad [7.2]$$

The quantity q is called the *shear flow* (analogous to water flow through a channel). It is assumed that shear flow acts around the perimeter (p_o), as illustrated in Figure 7.4c.

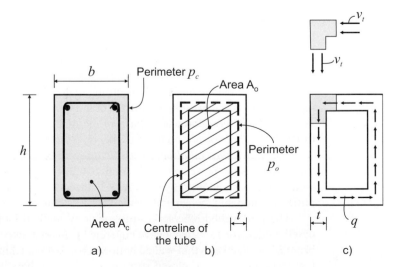

Figure 7.4 Thin-walled tube analogy for a concrete beam: a) actual section; b) equivalent tube; c) shear flow (q) and torsional shear stress (v_t).

According to CSA A23.3-94, the thickness of the equivalent tube section (t) can be taken as

$$t = \frac{0.75A_c}{p_c} \qquad [7.3]$$

where

$$A_o \cong \frac{2A_c}{3} \qquad [7.4]$$

After Eqns 7.3 and 7.4 are incorporated into Eqn 7.1, the equation for the torsional shear stress (v_t) can be restated as

$$v_t = \frac{T}{\dfrac{A_c^2}{p_c}} \qquad [7.5]$$

Torsional cracking is considered to occur when the principal tensile stress (which is equal to the torsional shear stress (v_t)) reaches the tensile strength of concrete in biaxial tension-compression, taken as $0.38\lambda\phi_c\sqrt{f_c'}$ per CSA A23.3. The torsional moment corresponding to the onset of cracking can be obtained by substituting v_t into Eqn 7.5 and setting $T = T_{cr}$:

A23.3 Eq. 11.2

$$T_{cr} = \left(\frac{A_c^2}{p_c}\right) 0.38\lambda\phi_c\sqrt{f_c'} \qquad [7.6]$$

where

A_c — the area enclosed by the outside perimeter of the section (p_c) (see Figure 7.4)
λ = a factor to account for concrete density
ϕ_c = the material resistance factor for concrete
f_c' = the specified compressive strength of concrete.

T_{cr} is defined per CSA A23.3 Cl.11.2.9.1 as the *factored torsional resistance of concrete*. Note that Eqn 7.6 is a simplified form of equation 11.2 stated in Cl.11.2.9.1 of CSA A23.3. The original code equation contains some additional terms applicable to prestressed concrete structures; however, it reduces to Eqn 7.6 when applied to reinforced (nonprestressed) concrete structures.

7.4.3 Ultimate Torsional Resistance of a Cracked Beam

Reinforcement becomes effective in beams subjected to torsion after cracking, when the beam transforms into a space truss according to the space truss analogy. Figure 7.5 shows a space truss model of a beam subjected to an external torque (T). The beam is of rectangular cross-section; however, it is idealized as a thin-walled tube section for the purposes of torsional design (as discussed above). Longitudinal bars are idealized as tension ties between the beam supports, while the stirrup legs are idealized as transverse ties spaced at a distance s. The concrete between diagonal cracks is idealized as a series of compression diagonals at an angle θ to the longitudinal beam axis. The stress in the concrete compression diagonals is denoted as (c) in Figure 7.5a.

The torque (T) causes shear stress in the tube and the corresponding shear flow (q). The relationship between the torque and the shear flow can be determined based on Eqns 7.1 and 7.2 as

$$q = \frac{T}{2A_o} \qquad [7.7]$$

where A_o is the area enclosed by the centre-line of the tube walls corresponding to the perimeter (p_o) shown in Figure 7.4b. It should be noted, however, that as the concrete cover spalls off in a cracked beam subjected to larger loads, it is considered more appropriate to

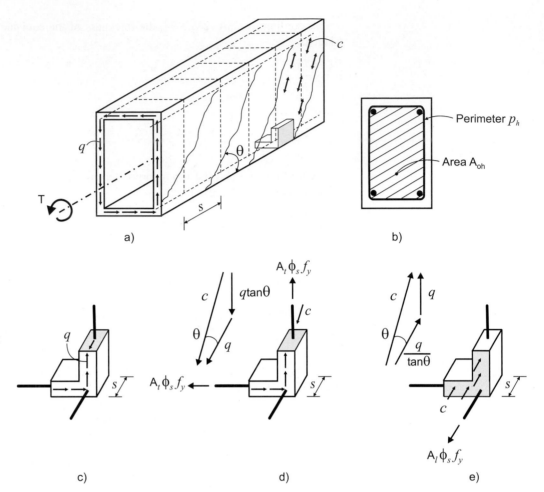

Figure 7.5 Space truss model of a concrete beam: a) truss members; b) equivalent outer tube dimensions; c) tube section showing shear flow on vertical and horizontal sections; d) equilibrium of forces: a horizontal section; e) equilibrium of forces: a vertical section.

use the outer dimensions coinciding with the centre-line of the exterior closed transverse reinforcement (stirrups) with area A_{oh} and perimeter p_h, as shown in Figure 7.5b.

Consider a portion of the corner section isolated from the tube, as illustrated in Figure 7.5c. The width of this section is equal to the stirrup spacing (s). Note the shear flow (q) acting along the centre-line of the tube walls, denoted with arrows. The shear stresses acting on the two perpendicular planes in a beam element are equal; consequently, the shear flow on the vertical beam section is equal to the shear flow on the horizontal section.

Let us consider the equilibrium of forces acting on the shaded horizontal section of the beam element in Figure 7.5d. The force developed in the stirrup is balanced by the vertical component of the concrete stress resultant. Assume that the stirrup legs have yielded and that the tensile force in one leg (bar area A_t) is equal to $A_t \phi_s f_y$. The concrete compressive stress (c) developed in the diagonal inclined at an angle θ has a horizontal component (q) and a vertical component ($q \tan \theta$), as shown on the force triangle in Figure 7.5d. The corresponding force acting along the length (s) is equal to

$$A_t \phi_s f_y = (q \tan \theta)s \qquad\qquad [7.8]$$

where

    A_t = the area of one leg of closed stirrup reinforcement
    ϕ_s = the material resistance factor for steel
    f_y = the specified yield strength of reinforcement
    q = the shear flow
    s = the spacing of the stirrups parallel to the longitudinal axis of the member.

Next, let us consider the equilibrium of forces acting on the shaded vertical section of the beam element shown in Figure 7.5e. The tensile force in the longitudinal reinforcement is balanced by the horizontal component of the concrete compressive stress. Consider that the longitudinal reinforcement has yielded and that the tensile force is equal to $A_l \phi_s f_y$. Note that (A_l) denotes the total cross-sectional area of all longitudinal reinforcement bars in a beam cross-section (and not only the area of a single bar, as shown in Figure 7.5e)! The concrete compressive stress (c) has a vertical component equal to the shear flow (q) and a horizontal component $(q/\tan \theta)$, as shown on the force triangle in Figure 7.5e. Because shear flow acts along the perimeter(P_h), the horizontal component of the concrete compression force is equal to $(q/\tan \theta) \, p_h$; that is

$$A_l \phi_s f_y = \left(\frac{q}{\tan \theta}\right) p_h \qquad \qquad [7.9]$$

where

$\quad A_l$ = the total area of the longitudinal reinforcement bars resisting torsion in a beam cross-section

$\quad p_h$ = the perimeter along the centre-line of the closed stirrup reinforcement corresponding to the area A_{oh} (see Figure 7.5b).

The *factored torsional resistance* of a beam section (T_r) can be determined by substituting (q) from Eqn 7.7 into Eqn 7.8, as follows:

| A23.3 Eq. 11.17 |

$$T_r = 2A_o \frac{A_l \phi_s f_y}{s} \cot \theta \qquad \qquad [7.10]$$

Longitudinal reinforcement on the flexural tensile side of a reinforced concrete beam should be proportioned such that the factored tension force in the reinforcement T_r is at least equal to the force F_{lt} developed in the reinforcement due to the combined effects of flexure, shear and torsion (A23.3 Cl.11.3.10.6), that is,

$$T_r \geq F_{lt} = \frac{M_f}{d_v} + \left(\sqrt{(V_f - 0.5V_s)^2 + \left(\frac{0.45 p_h T_f}{2A_o}\right)^2}\right) \cot(0) \qquad [7.11]$$

where

$\quad T_r = \phi_s f_y A_{lt}$

and A_{lt} is the area of tension steel.

The equations derived in this section form the basis for the CSA A23.3 simplified method for torsional design. The CSA A23.3 general method is based on the same concept; however, the angle θ is determined as a function of the shear stress due to combined shear and torsion and the strain in the longitudinal steel reinforcement.

KEY CONCEPTS

The thin-walled tube/space truss analogy method forms the basis of the CSA A23.3 simplified method for torsional design. The method includes the following two mechanics-based models for the behaviour of concrete beams subjected to torsion:

- Before cracking, the beam is idealized as a thin-walled hollow tube.
- After cracking, the beam is idealized as a hollow truss consisting of closed stirrups, longitudinal bars in the corners, and compression diagonals approximately centred on the stirrups.

The torsional resistance of concrete (T_{cr}), determined using the thin-walled tube analogy, is defined as the torsional moment at the onset of cracking. T_{cr} depends only on the concrete strength and is not influenced by the amount and distribution of reinforcement in the beam.

The ultimate factored torsional resistance of a concrete beam (T_r) is determined from the equilibrium of forces on a cracked beam section wherein a beam is modelled as a space truss. T_r is based on the resistance provided by steel reinforcement only.

7.5 COMBINED TORSION, SHEAR, AND FLEXURE LOADS

Torsion in reinforced concrete structures most often occurs in combination with shear, flexure, and/or axial loads. The simplified method per CSA A23.3 (Cl.11.3.10.1) accounts for the effects of combined loads by simply adding the reinforcement area required to resist torsion to that required to resist shear.

7.5.1 Combined Shear and Torsion

The distribution of torsional and shear stress components in a rectangular cross-section is shown in Figure 7.6. Torsional shear stresses developed due to the torsional moment (T_f) are denoted by v_t in Figure 7.6a, while the shear stresses due to the shear force (V_f) are denoted by v_s in Figure 7.6b. Note that the shear stresses act in the same direction throughout the section width, while the torsional shear stresses act in opposite directions at the perimeter of the section. Therefore, shear stresses caused by the shear force (V_f) will increase due to the torsional moment (T_f) at some portions of the beam section (for example, see point A in Figure 7.6) and decrease in other portions (see point B in Figure 7.6).

When torsion and shear act simultaneously, the CSA A23.3 simplified method prescribes that the shear is carried by the concrete (V_c) and the shear reinforcement (stirrups) (V_s). However, torsional resistance is provided by the reinforcement only, including the longitudinal steel reinforcement and the stirrups. (As discussed in the previous section, the torsional resistance of concrete is neglected in the case of pure torsion as well.) The assumption that no torsion is carried by concrete is somewhat conservative. According to the truss analogy, torsion causes diagonal compressive stresses in the walls of the thin-walled tube, thereby increasing the compressive stresses induced by the shear forces. These compressive stresses can cause the failure of a beam due to concrete crushing, which is a brittle and hence undesirable failure mode. In order to prevent the occurrence of concrete crushing failure in beams subjected to combined torsion and shear, A23.3 Cl.11.3.10.4 prescribes a limit to the combined torsion and shear stress depending on the type of beam section, as follows:

a) For box sections,

$$\boxed{\text{A23.3 Eq. 11.18}} \qquad \frac{V_f}{b_w d_v} + \frac{T_f p_h}{1.7 A_{oh}^2} \le 0.25\, \phi_c f_c' \qquad\qquad \textbf{[7.12]}$$

b) For other sections,

$$\boxed{\text{A23.3 Eq. 11.19}} \qquad \sqrt{\left(\frac{V_f}{b_w d_v}\right)^2 + \left(\frac{T_f p_h}{1.7 A_{oh}^2}\right)^2} \le 0.25\, \phi_c f_c' \qquad\qquad \textbf{[7.13]}$$

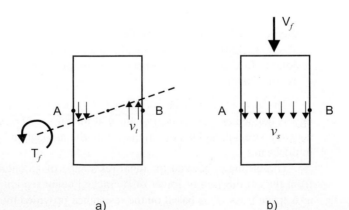

Figure 7.6 Stress distribution due to combined shear and torsional effects: a) torsional shear stresses; b) shear stresses.

where

d_v = the effective shear depth used in shear design according to the CSA A23.3 simplified method (see Section 6.5.4)

p_h = the perimeter of closed stirrups (see Figure 7.5b).

Transverse reinforcement (stirrups) needs to be designed to resist tensile stresses caused by the effects of combined shear and torsion. The approach taken by A23.3 Cl.11.3.10.1 is to first determine the areas of transverse reinforcement required to resist torsion (A_t) and shear (A_v); subsequently, these areas should be added to determine the area (A_{t+v}) and spacing (s) of the transverse reinforcement required to resist the combined torsion and shear effects.

The design of shear reinforcement was discussed in Section 6.4.4. The shear resistance provided by the steel reinforcement is equal to

A23.3 Eq. 11.7
$$V_s = \frac{\phi_s A_v f_y d_v \cot\theta}{s} \qquad [6.9]$$

where

A_v = the area of shear reinforcement perpendicular to the axis of the member within a distance s (note that A_v includes the area of all stirrup legs)

θ = the angle of inclination of the diagonal compressive stress to the longitudinal axis of the member.

Consequently, the stirrup area (A_v) and the spacing (s) need to satisfy the following equation

$$\frac{A_v}{s} = \frac{V_s}{\phi_s f_y d_v \cot\theta}$$

The area of the torsional transverse reinforcement (A_t) denotes the area of one stirrup leg only, so the area of shear reinforcement per leg ($A_v{}'$) needs to be determined; that is,

$$\frac{A_v{}'}{s} = \frac{\dfrac{A_v}{2}}{s} = \frac{V_s}{2\phi_s f_y d_v \cot\theta} \qquad [7.14]$$

Equation 7.14 considers two-legged stirrups. In general, there are stirrups with four or more legs; however, only the outer stirrup legs are considered to contribute to the torsional resistance.

The area of stirrups required to resist torsion (A_t) can be determined from Eqn 7.10 as

$$\frac{A_t}{s} = \frac{T_r}{2A_o \phi_s f_y \cot\theta} \qquad [7.15]$$

The total amount of transverse reinforcement required to resist the combined effects of torsion and shear is equal to

$$\frac{A_t}{s} + \frac{A_v{}'}{s} \qquad [7.16]$$

7.5.2 Combined Flexure and Torsion

Torsion causes an axial tensile force (N) that is resisted by the longitudinal reinforcement bars with area A_l and distributed around the beam cross-section, as discussed in Section 7.4.3. For design purposes, one-half of this force is considered to act in the top chord of the space truss; the remaining portion acts in the bottom chord, as shown in Figure 7.7a. However, flexure causes a tensile force (T) in the bottom chord of the truss and a compressive force (C) in the top chord, as shown in Figure 7.7b. Note that

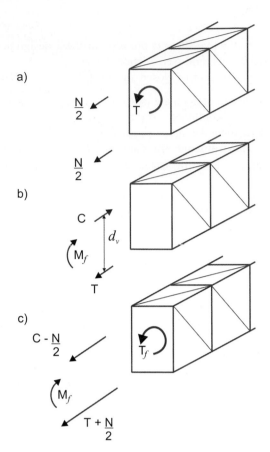

Figure 7.7 Internal forces due to combined torsion and flexure: a) torsion; b) flexure; c) torsion and flexure.

these two forces are equal; that is, $C = T$. The compressive force C is approximately given by

$$C \cong \frac{M_f}{d_v}$$

where d_v is the effective shear depth (equal to the lever arm of the couple formed by the forces T and C).

The reinforcement areas required to resist torsion and flexure effects can be determined separately and the corresponding areas added together. It is clear from Figure 7.7c that the amount of tension steel needs to be increased when there are flexural loads in addition to torsion in a beam section; however, the torsion causes a reduction in the compressive force in the flexural compression zone. CSA A23.3-04 does not include a provision to reduce the amount of compression steel, which existed in the previous (1994) edition (CSA A23.3-94 Cl.11.3.9.6).

KEY CONCEPTS

The CSA A23.3 simplified method (A23.3 Cl.11.3.10.1) accounts for the combined load effects by simply adding the reinforcement areas required to resist torsion and shear effects.

To prevent the occurrence of concrete crushing failure in beams subjected to combined torsion and shear effects, A23.3 Cl.11.3.10.4 prescribes a limit to the combined compressive stress caused by torsion and shear; if this limit is exceeded, a beam with a larger cross-section should be used.

In the case of *combined shear and torsion*, the transverse reinforcement (stirrups) resists tensile stresses caused by the combined shear and torsion effects, whereas the distributed longitudinal reinforcement resists the torsional effects only.

7.6 CSA A23.3 REQUIREMENTS FOR THE SIMPLIFIED METHOD FOR TORSIONAL DESIGN

The simplified method for torsional design is based on the concepts and equations presented in Section 7.4. The design requirements and limitations prescribed by the CSA A23.3 simplified method will be outlined in this section.

| A23.3 Cl.11.3.10.2 | **Strength requirement** According to A23.3 Cl.11.3.10.2, the main design requirement for concrete structures subjected to torsion is the strength requirement

$$T_r \geq T_f \qquad\qquad [7.17]$$

where T_r is the factored torsional resistance and T_f is the torsional moment due to factored loads.

| A23.3 Cl.11.2.9.1 | **Factored torsional resistance of concrete** The *factored torsional resistance of concrete* (T_{cr}) is taken as equal to the torsional moment at the onset of cracking (refer to Section 7.4 for the derivation of this equation):

| A23.3 Eq. 11.2 |
$$T_{cr} = \left(\frac{A_c^2}{p_c}\right) 0.38\lambda\phi_c \sqrt{f_c'} \qquad\qquad [7.6]$$

| A23.3 Cl.11.3.10.3 | **Factored torsional resistance of a beam section** The ultimate factored torsional resistance of a concrete beam (T_r) is based on the resistance provided by steel reinforcement only and it can be determined as follows (refer to Section 7.4 for the derivation):

| A23.3 Eq. 11.17 |
$$T_r = 2A_o \frac{A_t\phi_s f_y}{s} \cot\theta \qquad\qquad [7.10]$$

where

$A_o = 0.85 A_{oh}$ is the area enclosed by the centre-line of the walls in the equivalent tube (see Figure 7.5b)

A_t = the area of one leg of closed stirrup reinforcement

ϕ_s = the material resistance factor for steel

f_y = the specified yield strength of the reinforcement

s = the spacing of the stirrups measured parallel to the longitudinal axis of the member

θ = the angle of inclination of the diagonal compressive stresses to the longitudinal axis of the member.

The following θ values should be used in shear design calculations:

- $\theta = 35°$ when the CSA A23.3 simplified method is used (Cl.11.3.6.3), or
- $\theta = 42°$ for special member types (Cl.11.3.6.2).

To ensure that the member does not fail suddenly in a brittle manner after the development of torsional cracking, it is recommended that the factored torsional resistance (T_r) exceed the factored torsional resistance of concrete (T_{cr}); that is,

$$T_r \geq T_{cr}$$

Note that the above requirement is not prescribed by CSA A23.3.

CSA A23.3 also prescribes the following *limitations* related to torsion design.

| A23.3 Cl.11.2.9.1 | **Criteria related to the provision of torsional reinforcement** According to A23.3 Cl.11.2.9.1, torsional effects can be neglected if the design torsion (T_f), determined

by analysis using stiffness based on an uncracked section, is less than 25% of the concrete torsional resistance (T_{cr}); that is,

$$T_f \leq 0.25\ T_{cr}$$

The above limit on the design torsion corresponds to a nominal shear stress of $0.1\lambda\phi_c\sqrt{f_c'}$. It is considered that torsion of such a small magnitude will not cause a significant reduction in either flexural or shear strength and can therefore be neglected.

| A23.3 Cl.11.3.10.4 |

Cross-sectional dimensions   When torsion and shear act simultaneously, the CSA A23.3 simplified method prescribes a limit to the combined compressive stress caused by torsion and shear, as follows:

a)   For box sections,

| A23.3 Eq. 11.18 |

$$\frac{V_f}{b_w d_v} + \frac{T_f p_h}{1.7\ A_{oh}^2} \leq 0.25\phi_c f_c' \qquad \text{[7.12]}$$

b)   For other sections,

| A23.3 Eq. 11.19 |

$$\sqrt{\left(\frac{V_f}{b_w d_v}\right)^2 + \left(\frac{T_f p_h}{1.7 A_{oh}^2}\right)^2} \leq 0.25\ \phi_c f_c' \qquad \text{[7.13]}$$

In Eqns 7.12 and 7.13, d_v is the effective shear depth, taken as the greater of $0.9d$ and $0.72h$.
When this condition is not satisfied, a larger cross-section is required to resist the effects of combined shear and torsion.

| A23.3 Cl.11.2.8.2 |

Minimum area of transverse reinforcement   The minimum area of transverse reinforcement according to A23.3 Cl.11.2.8.2 is as follows:

| A23.3 Eq. 11.1 |

$$A_{v,\min} = 0.06\sqrt{f_c'}\left(\frac{b_w s}{f_y}\right) \qquad \text{[7.18]}$$

where $A_{v,\min}$ is the total area of shear reinforcement; in order to obtain the area per leg (A_v'), this value needs to be divided by two. In the case of pure torsion, the designer needs to perform the following check:

$$\frac{A_t}{s} \geq 0.03\sqrt{f_c'}\left(\frac{b_w}{f_y}\right) \qquad \text{[7.19]}$$

In the case of combined torsion and shear, this limit applies to both the transverse torsional reinforcement (A_t) and the shear reinforcement (A_v') as follows:

$$\frac{A_t}{s} + \frac{A_v'}{s} \geq 0.03\sqrt{f_c'}\left(\frac{b_w}{f_y}\right) \qquad \text{[7.20]}$$

Maximum spacing of transverse reinforcement   According to A23.3 Cl.11.3.8.3, when $T_f > 0.25\ T_{cr}$, the maximum spacing of transverse reinforcement resisting torsion (s_{max}) should not exceed the smaller of

- 300 mm
- $0.35\ d_v$

7.7 DETAILING OF TORSIONAL REINFORCEMENT

Torsional reinforcement has a very significant role in providing the required torsional resistance in concrete beams, as discussed in Section 7.3. Torsional cracks spiral around the beam, so both longitudinal and transverse reinforcement needs to be provided. The effective detailing of torsional reinforcement is of primary importance for the satisfactory performance of concrete members subjected to torsion. Inadequate detailing can lead to excessive cracking and brittle failure of reinforced concrete members subjected to torsion, as reported by Mitchell and Collins (1976). Various aspects of the design and detailing of torsional reinforcement are discussed in this section.

A23.3 Cl.11.2.6 **Types of torsional reinforcement** According to A23.3 Cl.11.2.6, torsional reinforcement consists of *transverse* and *longitudinal reinforcement*. There are three different types of transverse reinforcement; however, the most common type of reinforcement is in the form of *closed stirrups* perpendicular to the longitudinal axis of the member (generally vertical in the case of beams in floor and roof structures). Note that stirrups need to be provided on all four faces of a member due to the spiral distribution of torsional cracks, hence the term *closed stirrups*. (Note that, in general, stirrups used for shear reinforcement can be open at the top, as discussed earlier in this chapter.) Other types of transverse reinforcement include a *closed cage of welded wire fabric* (with wire perpendicular to the axis of the member), and *spirals*. The latter two types of transverse reinforcement are used less often in design practice.

A23.3 Cl.11.2.7 **Transverse reinforcement** The following considerations need to be taken into account when detailing transverse reinforcement in members subjected to torsion:

- *Anchorage:* Stirrups need to be anchored around longitudinal bars by means of standard 135° hooks in order to prevent the concrete cover from spalling. When the stirrups are subjected to excessive tensile stresses that are balanced by the compressive stresses in the concrete outer shell, the concrete outside the reinforcing cage is poorly anchored and there is a large chance of spalling off (Mitchell and Collins, 1976). The same tests have shown premature failure due to spalling of the concrete cover and consequent loss of anchorage when 90° hooks are used. The ACI Committee 315 (1994) offered recommendations for the anchorage of transverse torsional reinforcement (illustrated in Figure 7.8). Note that the

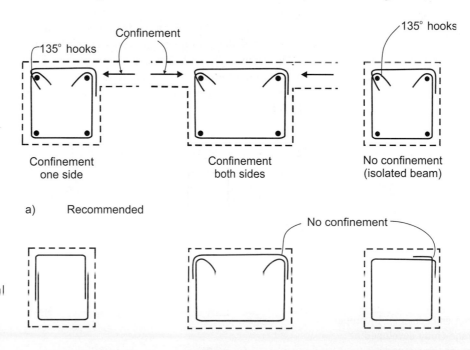

Figure 7.8 Anchorage of closed stirrups a) recommended arrangements; b) arrangements not permitted (adapted from ACI Committee 315, 1994 by permission of the American Concrete Institute).

use of 90° hooks is allowed only when the concrete around the anchorage is confined, as in the case of a slab cast continuously with the beam (see A23.3 Cl.7.1.2).

- *Spacing:* CSA A23.3-04 does not prescribe the maximum spacing of transverse torsional reinforcement. However, the 1984 version of the same code prescribed that the maximum spacing should be the smaller of $p_h/8$ and 300 mm, where p_h is the perimeter of the closed stirrups. The spacing limit of $p_h/8$ was also recommended by Mitchell and Collins (1976). Adequately spaced stirrups are of assistance in controlling crack widths — closely spaced stirrups result in a better distributed cracking pattern and smaller crack widths.

| A23.3 Cl.11.2.7 |

Longitudinal reinforcement   The following considerations need to be taken into account when detailing longitudinal reinforcement in members subjected to torsion:

- *Distribution:* One longitudinal reinforcement bar has to be placed in each corner of closed torsional transverse reinforcement (A23.3 Cl.11.2.7). It is good practice to distribute the longitudinal bars symmetrically around the section, with the bar spacing not exceeding 300 mm.
- *Anchorage:* Longitudinal reinforcement must be developed at both ends. Since torsional moments are generally largest at the ends of the beam, the longitudinal bars need to be anchored to ensure that the yield strength is developed at the face of the support. A designer needs to keep in mind that all longitudinal bars (both tension and compression steel) are subjected to torsion-induced tension. Therefore, it would be incorrect to extend the bottom bars in the continuous spandrel beam only 150 mm into the support, as prescribed by A23.3 Cl.12.11.1 (although this is adequate in the case of flexural loads only), as illustrated in Figure 7.9a. Mitchell and Collins (1976) recommend providing a full development length (l_d) for longitudinal bars beyond the face of the support, as shown in Figure 7.9b (anchorage design for flexural reinforcement in support regions will be discussed in Section 9.7). This is a common design error, resulting in premature beam failure, as reported by MacGregor and Bartlett (2000).
- *Bar diameter:* The nominal bar diameter for longitudinal reinforcement should not be less than $s/16$, where s is the stirrup spacing.

KEY CONCEPTS

The most common type of transverse torsional reinforcement is in the form of *closed stirrups*. (Note that stirrups acting as shear reinforcement need not be closed.)

The anchorage for transverse reinforcement is very important and needs to be provided in the form of 135° standard hooks (A23.3 Cl.11.2.7).

Longitudinal torsional reinforcement needs to be fully developed at both ends (A23.3 Cl.11.2.7).

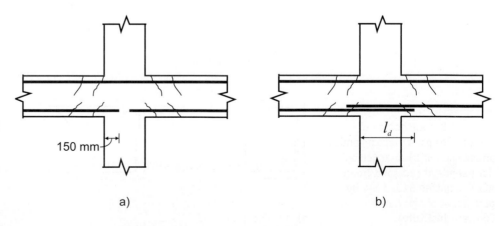

Figure 7.9 End anchorage for longitudinal steel in a reinforced concrete beam subjected to torsion: a) improper anchorage (150 mm embedment); b) proper anchorage (fully developed at the face of the support).

7.8 TORSIONAL DESIGN CONSIDERATIONS

There are a few important considerations related to torsional design that remain to be discussed:

- distribution of torsional moment
- critical section for torsional design
- shape of a beam cross-section
- torsional stiffness

Distribution of torsional moment   In the case of a statically determinate beam, the torsional moment (T_f) at any section can be calculated from statics (for example, using a free-body diagram for a portion of the beam). However, in statically indeterminate beams the ends are generally restrained and the magnitudes of torsional moments depend on the relative stiffness of the end supports (usually columns). In general, torsional moments at the supports vary from 0 to $m_t\,l$, where m_t is the distributed torque and l is the beam span. If the torsional stiffness at the beam supports is equal, the corresponding torsional moments are equal to $m_t\,l/2$. The distribution of torsional moments along the beam length is linear, with maximum values at the ends and zero value in the midspan (very similar to the shear force distribution in a beam subjected to uniform load).

Critical section for torsional design   The critical section for torsional design is located at the face of the support. It should be noted that the previous edition of the code (CSA A23.3-94) permitted the reduction in the design torsion in regions near supports (analogous to the case of shear design discussed in Section 6.6).

Shape of a beam cross-section   The shape of the cross-section is important for efficient torsional design. Based on the distribution of torsional shear stresses in a cross-section, it can be concluded that a tube section is more efficient than a rectangular section. This was also confirmed by the findings of the tests referred to in Section 7.3; tests have shown that the torsional resistance of hollow and solid beams with the same amount and distribution of reinforcement is very similar because the concrete core becomes ineffective after cracking occurs. Hollow cross-sections (also called box sections) are used in bridge design, whereas solid rectangular sections are generally used in buildings. The least favourable shape of cross-section from the aspect of torsional design is a U-shaped section characterized by a small torsional stiffness.

Torsional stiffness   The torsional stiffness of a concrete member drops significantly after cracking, as discussed in Section 7.3. The stiffness is equal to the slope of the radial line through the origin and a point under consideration on the torque-twist ($T-\theta_t$) diagram (see Figure 7.3). This relationship can be expressed as

$$T = K_t\theta_t \tag{7.21}$$

where

  T = the torque acting on a section
  K_t = the torsional stiffness
  θ_t = the angle of twist.

The torsional stiffness of a gross (uncracked) section can be determined as

$$K_t = \frac{GC_t}{l}$$

where

  G = the modulus of rigidity (or shear modulus of elasticity)
  C_t = the torsion property of a section
  l = the beam span.

It is generally acceptable to take G as

$$G = \frac{E_c}{2}$$

where

| A23.3 Eq. 8.2 | $E_c = 4500\sqrt{f_c'}$ [2.2]

is the modulus of elasticity for concrete.

C_t can be obtained from the torque-twist relationship for a thin-walled concrete tube as

$$C_t = \frac{4A_o^2 t}{p_o}$$

The values for A_o, t, and p_o can be determined based on the uncracked beam properties (see Eqns 7.3 and 7.4).

Consequently, the torsional stiffness of the uncracked beam section is

$$K_t = \frac{2E_c A_o^2 t}{p_o l} \qquad [7.22]$$

When a designer wishes to check whether torsional effects need to be considered in the design, the torsional moment due to factored loads (T_f) should be determined based on the stiffness of the uncracked section (A23.3 Cl.11.2.9.1). It should be noted that T_f might be significantly different in magnitude depending on whether cracked or uncracked section properties have been used in modelling the structural system.

When cracks develop in the beam and the beam is transformed into a space truss structure, the torsional stiffness can be determined based on the properties of an equivalent steel tube.

KEY CONCEPTS

The critical section for torsion design is located at the face of the support.

Based on the distribution of torsional shear stresses in a beam cross-section, it can be concluded that a tube section is more efficient in providing torsional resistance than a rectangular section. (This is true for other materials, such as steel, timber, etc.)

The torsional stiffness drops significantly after cracking. When a designer wishes to check whether torsional effects need to be considered in the design, the factored torsional moment (T_f) should be determined based on the stiffness of an uncracked section (A23.3 Cl.11.2.9.1) using the thin-walled tube analogy.

7.9 DESIGN FOR TORSION PER THE CSA A23.3 SIMPLIFIED METHOD: A SUMMARY AND A DESIGN EXAMPLE

The design of concrete structures for torsion was discussed in detail in the previous sections. The purpose of this section is to summarize the torsion design provisions according to the CSA A23.3 simplified method (see Checklist 7.1). A design example will be presented to illustrate the application of the CSA A23.3 simplified method for torsional design.

Checklist 7.1 Torsion Design According to the CSA A23.3 Simplified Method

Step	Description	Code Clause
1	Determine the factored torsional moment (T_f) based on the uncracked section properties.	11.2.9.1
2	In the case of combined shear and torsion, determine the factored shear force (V_f) and the shear reinforcement requirements (see Checklist 6.1).	
3	Determine the factored torsional resistance of concrete (T_{cr}) corresponding to the start of cracking: $\boxed{\text{A23.3 Eq. 11.2}} \quad T_{cr} = \left(\dfrac{A_c^2}{p_c}\right) 0.38\lambda\phi_c \sqrt{f_c'}$ **[7.6]**	11.2.9.1
4	Determine whether torsional effects should be considered in the design. If $T_f \le 0.25\,T_{cr}$, torsional effects can be ignored. If $T_f > 0.25\,T_{cr}$, torsional effects should be considered.	11.2.9.1
5	In the case of combined shear and torsion, verify whether the dimensions of the beam cross-section are okay: $\boxed{\text{A23.3 Eq. 11.19}} \quad \sqrt{\left(\dfrac{V_f}{b_w d_v}\right)^2 + \left(\dfrac{T_f p_h}{1.7\,A_{oh}^2}\right)^2} \le 0.25\phi_c f_c'$ **[7.13]** Note that the above condition applies to rectangular sections.	11.3.10.4
6	Follow the strength requirement, which states that the torsional resistance should be greater than or equal to the factored torsion; that is, $T_r \ge T_f$ **[7.17]** If the reinforcement is not given, set $T_f = T_r$.	11.3.10.2
7	If the reinforcement is given, determine the torsional resistance (T_r) based on the equation $\boxed{\text{A23.3 Eq. 11.17}} \quad T_r = 2A_o \dfrac{A_t \phi_s f_y}{s}\cot\theta$ **[7.10]**	11.3.10.3
8	Determine the required area (A_t) and spacing (s) of the transverse torsional reinforcement corresponding to T_r: $\dfrac{A_t}{s} = \dfrac{T_r}{2\,A_o \phi_s f_y \cot\theta}$ **[7.15]**	11.3.10.3
9	In the case of combined shear and torsion, determine the area of shear reinforcement *per leg*: $\dfrac{A_v'}{s} = \dfrac{V_s}{2\phi_s f_y\, d_v \cot\theta}$ **[7.14]**	11.3.5.1
10	when $T_f > 0.25\,T_{cr}$, the maximum spacing of transverse reinforcement (s_{max}) should not exceed the smaller of • 300 mm • $0.35d_v$	11.3.8.3

(Continued)

Checklist 7.1 Continued

Step	Description	Code Clause
11	Check whether the area of combined shear and torsional transverse reinforcement satisfies the minimum requirements: $$\frac{A_t}{s} + \frac{A_v{'}}{s} \geq 0.03\sqrt{f_c{'}}\left(\frac{b_w}{f_y}\right) \qquad [7.20]$$	11.2.8.2
12	Proportion longitudinal reinforcement on the flexural tensile side based on the following equation $$T_r \geq F_{lt} = \frac{M_f}{d_v} + \left(\sqrt{(V_f - 0.5V_s)^2 + \left(\frac{0.45p_hT_f}{2A_o}\right)^2}\right)\cot(\theta) \qquad [7.11]$$	11.3.10.6
13	Distribute the longitudinal reinforcement according to the code requirements.	11.2.7

Example 7.1

A partial floor plan of a reinforced concrete slab and beam system is shown in the figure below. The two-way slab is supported at the edge by the 500 mm wide by 600 mm deep spandrel beam. The spandrel beam can be considered as simply supported beam spanning across the columns with a 6 m span. The slab supports a uniform dead load (w_{DL}) of 6.0 kPa and a uniform live load (w_{LL}) of 1.9 kPa. The beam self-weight can be taken equal to 4.0 kN/m (this load should be taken in addition to the slab load). Consider 20M flexural reinforcement bars, 10M stirrups, and a 40 mm clear cover to the stirrups. A maximum aggregate size of 20 mm has been specified.

Design the beam for shear and torsion according to the CSA A23.3 simplified method. Make the following assumptions in the design:

1. *The flexural tension reinforcement has been adequately designed. Compression steel is not required.*
2. *The spandrel beam supports the portion of the floor slab shown shaded in the figure below.*
3. *The triangular distributed load transferred to the spandrel beam from the floor slab can be considered equivalent to the uniform load acting through the centroid (C) of the shaded triangle located at a distance (a) of 1 m from the beam centre-line.*
4. *The spandrel beam has sufficient torsional stiffness at its ends such that half of the total applied torsion is resisted by the beam supports.*

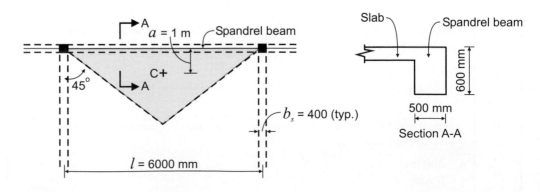

Given: $f_c' = 25$ MPa   (normal-density concrete)

$f_y = 400$ MPa

$\phi_c = 0.65$

$\phi_s = 0.85$

Spandrel beam:

self-weight $g = 4$ kN/m

$b = 500$ mm

$h = 600$ mm

SOLUTION: **1. Determine the factored torsional moment (T_f) based on the uncracked section properties**

  a) Calculate the resultant of the slab load (shaded portion only) carried by the spandrel beam.

Note that the load distribution along the spandrel beam is triangular. The resultant of the triangular load (R) is equal to

$$R = (\text{slab load}) \times (\text{tributary area})$$

The tributary area A is equal to the area of the shaded triangle shown in the figure:

$$A = \frac{\text{base} \times \text{height}}{2} = \frac{6 \text{ m} \times 3 \text{ m}}{2} = 9 \text{ m}^2$$

$$R_{DL} = w_{DL} \times A = 6.0 \text{ kPa} \times 9 \text{ m}^2 = 54 \text{ kN}$$

$$R_{LL} = w_{LL} \times A = 1.9 \text{ kPa} \times 9 \text{ m}^2 = 17 \text{ kN}$$

  b) Calculate the equivalent uniformly distributed load.

For design purposes, the triangular distributed load acting on the spandrel beam is going to be represented by an equivalent uniform load (see Design Assumption 3). The equivalent dead load (DL) and live load (LL) are determined as follows:

$$DL = \frac{R_{DL}}{l} = \frac{54 \text{ kN}}{6 \text{ m}} = 9 \text{ kN/m}$$

$$LL = \frac{R_{LL}}{l} = \frac{17 \text{ kN}}{6 \text{ m}} = 2.8 \text{ kN/m}$$

The factored uniform load (w_f) acting on the slab is determined according to NBC 2010 Cl.4.1.3.2:

$$w_f = 1.25 \text{ DL} + 1.5 \text{ LL} = 1.25(9 \text{ kN/m}) + 1.5(2.8 \text{ kN/m})$$

$$= 15.5 \text{ kN/m}$$

  c) Determine the clear span (l_n):

$$l_n = l - \frac{b_s}{2} - \frac{b_s}{2}$$

$$= 6 \text{ m} - \frac{0.4 \text{ m}}{2} - \frac{0.4 \text{ m}}{2} = 5.6 \text{ m}$$

where $b_s = 0.4$ m is the column width.

  d) Determine the distributed torsional moment caused by the eccentricity of the tributary slab load with regard to the spandrel beam axis (refer to Design Assumption 3), as shown in the sketch that follows.

$$m_{tf} = w_f a = 15.5 \text{ kN/m} \times 1 \text{ m} = 15.5 \text{ kN} \cdot \text{m/m}$$

where $a = 1$ m (given).

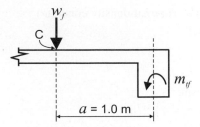

Calculate the factored torsional moment at the supports:

$$T_f = \frac{m_{tf}l_n}{2} = \frac{15.5 \text{ kN} \cdot \text{m/m} \times 5.6 \text{ m}}{2} = 43.4 \text{ kN} \cdot \text{m} \cong 43 \text{ kN} \cdot \text{m}$$

A diagram showing the distribution of the torsional moment along the beam is shown below.

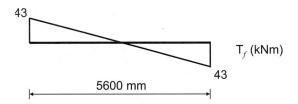

2. **Calculate the design shear force**
    a) Calculate the effective depth (d):

    - 10M stirrups: $d_s = 10$ mm (see Table A.1);
    - 20M longitudinal rebars: $d_b = 20$ mm (see Table A.1);
    - cover to the stirrups: cover = 40 mm

    $$d = h - \text{cover} - d_s - \frac{d_b}{2}$$

    $$= 600 \text{ mm} - 40 \text{ mm} - 10 \text{ mm} - \frac{20 \text{ mm}}{2} = 540 \text{ mm}$$

    b) Calculate the effective shear depth (d_v):
    The effective shear depth (d_v) is taken as the greater of

    $$d_v = \begin{bmatrix} 0.9d \\ 0.72h \end{bmatrix} = \begin{bmatrix} 0.9 \times 540 \text{ mm} \\ 0.72 \times 600 \text{ mm} \end{bmatrix} = \begin{bmatrix} 486 \text{ mm} \\ 432 \text{ mm} \end{bmatrix}$$

    Hence, $d_v = 486$ mm $\cong 490$ mm.
    c) Calculate the shear force at the face of the support.
    The total factored uniformly distributed gravity load on the spandrel beam (including the beam self-weight) is

    $$w_f' = w_f + 1.25 g = 15.5 \text{ kN/m} + 1.25 (4 \text{ kN/m}) = 20.5 \text{ kN/m}$$

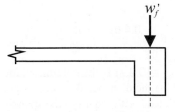

Shear force at the face of the support is equal to

$$V_f = \frac{w_f' \times l_n}{2} = \frac{20.5 \text{ kN/m} \times 5.6 \text{ m}}{2} = 57.4 \text{ kN}$$

d) Find the maximum shear force at the midspan ($V_{fmidspan}$):

$$V_{fmidspan} = \frac{(1.5 \, LL)\left(\frac{l_n}{2}\right)}{4} = \frac{(1.5 \times 2.8 \, \text{kN/m}) \times \left(\frac{5.6 \, \text{m}}{2}\right)}{4} = 2.9 \, \text{kN}$$

e) Next, the factored shear force ($V_{f@dv}$) at a distance d_v from the support can be determined as:

$$V_{f@dv} = V_{fmidspan} + \left(\frac{\frac{l_n}{2} - d_v}{\frac{l_n}{2}}\right)(V_f - V_{fmidspan})$$

$$= 2.9 \, \text{kN} + \left(\frac{\frac{5.6 \, \text{m}}{2} - 0.49 \, \text{m}}{\frac{5.6 \, \text{m}}{2}}\right)(57.4 \, \text{kN} - 2.9 \, \text{kN}) = 47.9 \, \text{kN} \cong 48 \, \text{kN}$$

For design purposes, use $V_{f@dv} = 48$ kN.

Sketch a diagram showing the distribution of the design shear force (left half of the beam).

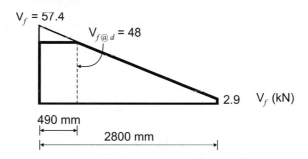

3. **Check whether shear reinforcement is required (A23.3 Cl.11.2.8.1)**

    If $V_f < V_c$, then shear reinforcement is not required.

    a) Find the V_c value (A23.3 Cl.11.3.4).

    Assume that the amount of shear reinforcement will be less than the minimum code requirement (a conservative assumption). According to A23.3 Cl.11.3.6.3,

    $$\beta = \frac{230}{1000 + d_v} = \frac{230}{1000 + 490} = 0.15 \, \text{(see Table 6.1)}$$

    Hence,

    | A23.3 Eq. 11.6 |

    $$V_c = \phi_c \lambda \beta \sqrt{f_c'} \, b_w \, d_v \qquad \qquad \text{[6.12]}$$

    $$= 0.65 \times 1.0 \times 0.15 \times \sqrt{25 \, \text{MPa}} \times 500 \, \text{mm} \times 490 \, \text{mm}$$

    $$= 119 \times 10^3 \, \text{N} = 119 \, \text{kN}$$

    b) Check whether shear reinforcement is required.

    Since

    $$V_{f@d} = 48 \, \text{kN}$$

    and

    $$48 \, \text{kN} < 119 \, \text{kN}$$

    $V_f < V_c$, so shear reinforcement is not required.

4. **Check whether torsional effects need to be considered in the design (A23.3 Cl.11.2.9.1)**

   a) Calculate the torsional resistance of the concrete:

   $$A_c = b \times h = (500 \text{ mm}) \times (600 \text{ mm}) = 300\ 000 \text{ mm}^2$$

   $$p_c = 2(b + h) = 2(500 \text{ mm} + 600 \text{ mm}) = 2200 \text{ mm}$$

   A23.3 Eq. 11.2
   $$T_{cr} = \left(\frac{A_c^2}{p_c}\right) 0.38 \lambda \phi_c \sqrt{f_c'} \qquad\qquad [7.6]$$

   $$= \left[\frac{(300\ 000 \text{ mm}^2)^2}{2200 \text{ mm}}\right] 0.38(1)(0.65)\sqrt{25 \text{ MPa}}$$

   $$= 50.5 \times 10^6 \text{ N}\cdot\text{mm} \cong 50 \text{ kN}\cdot\text{m}$$

   b) Check whether torsional effects need to be considered in the design. Torsional effects can be ignored when $T_f \le 0.25\ T_{cr}$.

   $$T_f = 43 \text{ kN}\cdot\text{m}$$

   $$0.25\ T_{cr} = 0.25(50 \text{ kN}\cdot\text{m}) = 12.5 \text{ kN}\cdot\text{m}$$

   $$43.4 \text{ kN}\cdot\text{m} > 12.5 \text{ kN}\cdot\text{m}$$

   Therefore, $T_f > 0.25\ T_{cr}$ and torsional effects have to be considered in the design.

5. **Check whether the dimensions of the beam cross-section are okay (case of combined shear and torsion) (A23.3 Cl.11.3.10.4)**

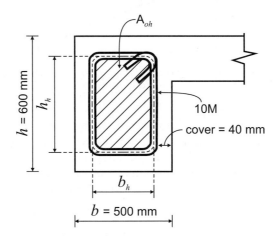

   a) Find A_{oh}.
      A_{oh} is the area enclosing the centre-line of the exterior closed stirrups (dimensions b_h and h_h) with perimeter p_h (see the sketch above).
      i) Find b_h and h_h:

   $$b_h = b - 2\left(\text{cover} + \frac{d_s}{2}\right)$$

   $$= 500 \text{ mm} - 2\left(40 \text{ mm} + \frac{10 \text{ mm}}{2}\right) = 410 \text{ mm}$$

   $$h_h = h - 2\left(\text{cover} + \frac{d_s}{2}\right)$$

   $$= 600 \text{ mm} - 2\left(40 \text{ mm} + \frac{10 \text{ mm}}{2}\right) = 510 \text{ mm}$$

ii) Find p_h:

$$p_h = 2(b_h + h_h)$$

$$= 2(410 \text{ mm} + 510 \text{ mm}) = 1840 \text{ mm}$$

iii) Find A_{oh}:

$$A_{oh} = b_h \times h_h = (410 \text{ mm})(510 \text{ mm}) = 209\ 100 \text{ mm}^2$$

b) Check the CSA A23.3 requirement:

A23.3 Eq. 11.19

$$\sqrt{\left(\frac{V_{f@dv}}{b_w\, d_v}\right)^2 + \left(\frac{T_f \times p_h}{1.7\, A_{oh}{}^2}\right)^2} \le 0.25\phi_c f_c' \qquad \text{[7.13]}$$

or

$$\sqrt{\left(\frac{48\ 000 \text{ N}}{(500 \text{ mm})\,(490 \text{ mm})}\right)^2 + \left(\frac{(43 \times 10^6 \text{ N}\cdot\text{mm})(1840 \text{ mm})}{1.7(209\ 100 \text{ mm}^2)^2}\right)^2}$$

$$\le 0.25\,(0.65)(25 \text{ MPa})$$

As $1.08 \text{ MPa} \le 4.06 \text{ MPa}$, the dimensions of the beam cross-section are okay.

6. **Determine the required torsional resistance T_r**
   The strength requirement states that $T_r \ge T_f$. For design purposes, use

   $$T_r = T_f = 43 \text{ kN}\cdot\text{m}$$

7. **Determine the required spacing of transverse reinforcement based on the torsional resistance (T_r) (A23.3 Cl.11.3.10.3)**
   CSA A23.3 prescribes the following equation to obtain the T_r value:

   A23.3 Eq. 11.17

   $$T_r = 2A_o\frac{A_t\phi_s f_y}{s}\cot\theta \qquad \text{[7.10]}$$

   The above equation can be used to determine the transverse reinforcement spacing (s) as

   $$s = 2A_o\frac{A_t\phi_s f_y}{T_r}\cot\theta$$

   a) Find A_o (the area enclosed by the centre-line of the tube walls; see Figure 7.4b):

   $$A_o = 0.85\, A_{oh} = 0.85(209\ 100 \text{ mm}^2) = 177\ 735 \text{ mm}^2$$

   b) Find A_t (area of one leg of closed stirrup reinforcement).
      In this case, 10M stirrups are used; the area of a 10M bar is 100 mm^2 (see Table A.1), hence

      $$A_t = 100 \text{ mm}^2$$

   c) Determine θ (angle of inclination of diagonal compressive stresses to the longitudinal axis of the member) according to A23.3 Cl.11.3.6.3:

      $$\theta = 35°, \text{ so } \cot\theta = 1.43$$

   d) Find the spacing s:

      $$s = 2(177\ 735 \text{ mm}^2)\frac{(100 \text{ mm}^2)(0.85)(400 \text{ MPa})}{43 \times 10^6 \text{ N}\cdot\text{mm}}(1.43)$$

      $$= 402 \text{ mm} = 400 \text{ mm}$$

8. **Check the maximum spacing requirements for transverse reinforcement (A23.3 Cl.11.2.8.2 and 11.2.8.3)**

According to A23.3 Cl.11.2.8.3:

$$S_{max} \leq \begin{bmatrix} 300 \text{ mm} \\ 0.35d_v \end{bmatrix} = \begin{bmatrix} 300 \text{ mm} \\ 0.35 \times 490 \text{ mm} \end{bmatrix} = \begin{bmatrix} 300 \text{ mm} \\ 171 \text{ mm} \end{bmatrix}$$

Hence

$$s_{max} = 171 \text{ mm}$$

The maximum stirrup spacing permitted by A23.3 Cl.11.2.8.2 can be determined from the equation

$$\frac{A_t}{s} \geq 0.03\sqrt{f_c'}\left(\frac{b_w}{f_y}\right) \qquad [7.19]$$

Consequently,

$$s_{max} \leq \frac{A_t}{0.03\sqrt{f_c'}\left(\frac{b_w}{f_y}\right)}$$

$$= \frac{100 \text{ mm}^2}{0.03 \times \sqrt{25 \text{ MPa}} \times \left(\frac{500 \text{ mm}}{400 \text{ MPa}}\right)} = 533 \text{ mm}$$

Hence, the maximum permitted spacing is the smallest of

- 400 mm
- 171 mm
- 533 mm

thus, $s_{max} = 171$ mm but for practical reasons the spacing will be rounded down to $s = 150$ mm

9. **Identify the region in the beam in which torsional reinforcement is required**
   a) Locate a region in the beam where torsional reinforcement is not required. Torsional reinforcement is not required in the region of the beam where $T_f \leq 0.25\, T_{cr}$, that is, where

   $$T_f \leq 12.5 \text{ kN} \cdot \text{m}$$

   Perform a linear interpolation to locate the distance from the left support (x), where

   $$T_{f@x} = 0.25\, T_{cr} = 12.5 \text{ kN} \cdot \text{m}$$

   $$\frac{T_{f@x}}{T_f} = \left(\frac{\frac{l_n}{2} - x}{\frac{l_n}{2}}\right)$$

   $$x = \frac{l}{2}\left(1 - \frac{T_{f@x}}{T_f}\right)$$

   $$x = \frac{5.6 \text{ m}}{2}\left(1 - \frac{12.5 \text{ kN} \cdot \text{m}}{43 \text{ kN} \cdot \text{m}}\right) \cong 2.0 \text{ m}$$

b) Develop the design torsional moment envelope diagram for the beam.

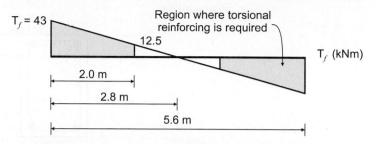

c) Sketch a beam elevation to illustrate the distribution of the transverse torsional reinforcement (stirrups).

Note that the first stirrup in the beam is located at a distance $s/2 = 75$ mm away from the face of the support (this is common practice related to the design of stirrup reinforcement).

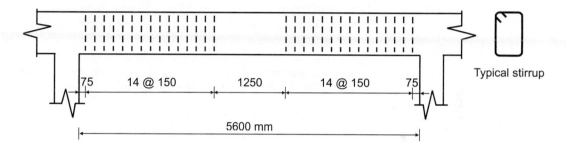

10. Calculate the area of longitudinal torsional reinforcement

a) The following equation can be used to obtain the required area of longitudinal tension reinforcement resisting torsion (A_{lt}):

$$T_r \geq F_{lt} = \frac{M_f}{d_v} + \left(\sqrt{(V_f - 0.5V_s)^2 + \left(\frac{0.45 p_h T_f}{2A_o} \right)^2} \right) \cot(\theta) \qquad [7.11]$$

Consider the section at the support, subjected to the maximum effects of shear and torsion and where $M_f \cong 0$ and $\theta = 35°$, thus

$$F_{lt} = \left(\sqrt{(57.4 \times 10^3)^2 + \left(\frac{0.45 \times 1840 \times (43 \times 10^6)}{2 \times 177\,735} \right)^2} \right) \cot(35°)$$

$$= 165.4 \times 10^3 \text{ N} = 165.4 \text{ kN}$$

Set $T_r = F_{lt} = 165.4$ kN, thus $A_{lt} = \dfrac{T_r}{\phi_s f_y} = \dfrac{165.4 \times 10^3}{0.85 \times 400} = 486 \text{ mm}^2$

b) Check whether the bar diameter is adequate (A23.3 Cl.11.2.7).
The code requires that the nominal bar diameter not be less than $s/16$:

$$\frac{s}{16} = \frac{150 \text{ mm}}{16} = 9 \text{ mm}$$

Furthermore, A23.3 Cl.11.2.7 requires that at least one longitudinal rebar be placed at each corner. Therefore, use four 25M bars (4-25M) (one at each corner). The area of tension steel is

$$A_{lt} = 2 \times 500 \text{ mm}^2 = 1000 \text{ mm}^2 > 484 \text{ mm}^2 : \text{okay}$$

It is recommended that the longitudinal bar spacing should not exceed 300 mm. Therefore, add four 10M bars (4-10M) in addition to the corner bars.

11. Provide a design summary showing the beam cross-section and the reinforcement arrangement

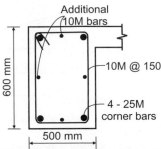

Torsion (torque) is a moment that tends to twist a member around its longitudinal axis. Reinforced concrete structures are often subjected to torsional moments, which cause shear stresses. Torsional effects in reinforced concrete members are significantly influenced by the type of structural system. In statically determinate structures, torsion develops as a result of the equilibrium between external and internal forces (equilibrium torsion). However, torsional moments in statically indeterminate structures develop due to the requirements of continuity (compatibility torsion). Torsion is often combined with shear, flexure, and/or axial loads.

Failure modes for reinforced concrete beams subjected to torsion

Failure in reinforced concrete beams subjected to torsion can occur by either the yielding of reinforcement (in lightly reinforced beams) or the crushing of concrete (in overreinforced beams). Reinforcement yielding failure is more desirable than concrete crushing failure.

The cracking pattern characteristic of the behaviour of beams subjected to torsion spirals around the beam. Therefore, both the longitudinal bars, distributed all around the beam cross-section, and the closed transverse reinforcement (stirrups) are essential for resisting torsional effects.

Torsional shear stresses are largest at the outer fibres of a beam cross-section. The outer skin of a beam cross-section is essential in providing torsional resistance, whereas the concrete core is not effective in resisting torsion due to the small shear stresses developed in that zone.

Key factors influencing the behaviour and resistance of concrete beams subjected to torsion

The thin-walled tube/space truss analogy method forms the basis of the CSA A23.3 simplified method for torsional design. The method includes the following two mechanics-based models for the behaviour of concrete beams subjected to torsion:

- Before cracking, the beam is idealized as a thin-walled hollow tube.
- After cracking, the beam is idealized as a hollow truss consisting of closed stirrups, longitudinal bars in the corners, and compression diagonals approximately centred on the stirrups.

The torsional resistance of concrete (T_{cr}) is defined as the torsional moment at the onset of cracking.

The ultimate factored torsion resistance of a concrete beam (T_r) is based on the resistance provided by the steel reinforcement only.

CSA A23.3 simplified method for torsional design

The *strength requirement* is the main CSA A23.3 requirement related to the simplified method for torsional design (A23.3 Cl.11.3.10.2).

The code also prescribes the procedure to determine

- *the factored torsional resistance of concrete* (T_{cr}) (Cl.11.2.9.1)
- *the factored torsional resistance of a beam section* (T_r) (Cl.11.3.10.3)

The following CSA A23.3 *limitations* are set related to torsion design:

- criteria related to the provision of torsional reinforcement (Cl.11.2.9.1)
- cross-sectional dimensions (Cl.11.3.10.4)
- minimum transverse reinforcement (Cl.11.2.8.2)

PROBLEMS

7.1. A reinforced concrete cantilever beam is shown in the figure here. The beam carries a uniform dead load (DL) of 20 kN/m (including beam self-weight) and a uniform live load (LL) of 50 kN/m. The load is applied eccentrically with regard to the vertical beam axis, as shown in section A-A. There is a 30 mm clear cover to the stirrups. A maximum aggregate size of 20 mm has been specified. The concrete and steel material properties are given on the next page.

Design the beam for the combined effects of shear and torsion according to the CSA A23.3 simplified method. Use 10M stirrups. Assume that transverse reinforcement is required throughout the beam length. Check whether the provided longitudinal reinforcement is adequate for the torsional resistance and make any required modifications. Provide a sketch summarizing the design.

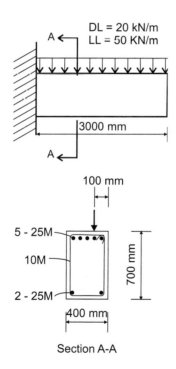

Section A-A

Given:

$f_c' = 30$ MPa   (normal-density concrete)
$f_y = 400$ MPa

7.2. Consider the one-way joist floor system shown in the figure below. The floor structure is supported by a 600 mm wide by 500 mm deep spandrel beam that is supported by 500 mm square columns. The beam is subjected to eccentric uniform factored load of 60 kN/m and a factored distributed moment of

40 kN·m/m due to joist reactions (see section A-A). In addition, the beam is subjected to a concentric uniform dead load of 13 kN/m (this load includes the beam self-weight). The beam is reinforced with 15M stirrups and 25M longitudinal rebars with a 30 mm clear cover to the stirrups. A maximum aggregate size of 20 mm has been specified. The concrete and steel material properties are given below.

Design the spandrel beam for the combined effects of shear and torsion according to the CSA A23.3 simplified method. Sketch a beam elevation and a typical cross-section showing the reinforcement arrangement.

Given:

$b = 600$ mm
$h = 500$ mm
$f_c' = 30$ MPa   (normal-density concrete)
$f_y = 400$ MPa

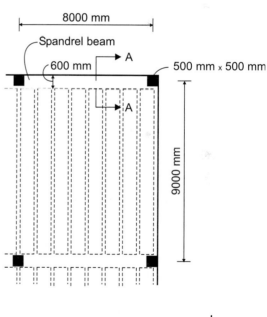

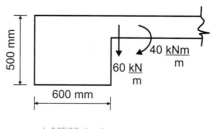

Section A - A

7.3. A 400 mm square highway signpost shown in the figure below is required to resist a specified wind load of 3.0 kPa. A maximum aggregate size of 20 mm has been specified. Use 10M stirrups and 25M flexural reinforcement in this design. There is a 40 mm clear cover to the stirrups. Ignore the effect of gravity load. The concrete and steel material properties are given below.

Design the signpost for the combined effects of shear and torsion according to the CSA A23.3 simplified method.

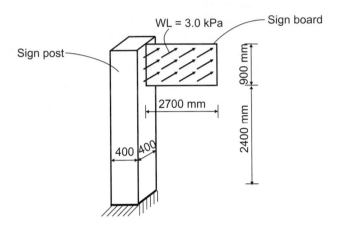

Given:

$f_c' = 30$ MPa   (normal-density concrete)
$f_y = 400$ MPa

7.4.   A cantilevered concrete wall with a 600 mm square beam at the top is required to resist a factored uniform lateral load of 15 kN/m applied at a height of 1 m above the top of the beam (see the figure below). A maximum aggregate size of 20 mm has been specified. Use 10M bars for stirrup reinforcement and 25M bars for flexural reinforcement. There is a 40 mm clear cover to the stirrups. The concrete and steel material properties are given on the next page.

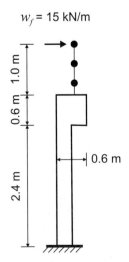

Design the beam atop the wall to resist the torsional effects. Perform the design according to the CSA A23.3 simplified method. Provide a sketch summarizing the design.

Given:

$f_c' = 30$ MPa   (normal-density concrete)
$f_y = 400$ MPa

7.5.   A concrete pedestrian girder supports cantilevered steel cross-beams. The factored load from pedestrians on one side of the girder is equal to 30 kN/m. The girder spans 20 m and the 900 mm wide supports are capable of resisting the torsional effects. A maximum aggregate size of 20 mm has been specified; 15M bars are used for the stirrup design and 35M bars for flexural reinforcement. There is a 40 mm clear cover to the stirrups. A typical girder cross-section is shown in the figure.

Design the girder to resist torsional effects. Perform the design according to the CSA A23.3 simplified method. Discontinue the torsional reinforcement where it is no longer required. Provide a sketch showing the stirrup arrangement along the girder length.

Given:

$f_c' = 30$ MPa   (normal-density concrete)
$f_y = 400$ MPa

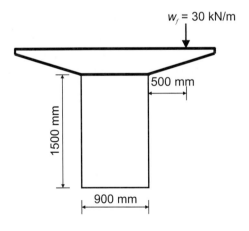

8 Columns

After reading this chapter, you should be able to

- identify the three main types of reinforced concrete columns
- outline the three main components of a reinforced concrete column and discuss the role of each component
- explain the three different types of column behaviour
- identify the five column design assumptions according to CSA A23.3
- outline the four key features of a column interaction diagram
- develop column interaction diagrams and use them to analyze and design concrete columns
- use the practical design guidelines related to the selection of column size and reinforcement
- describe the two key effects characteristic of the behaviour of slender columns and outline the CSA A23.3 slenderness criteria

8.1 INTRODUCTION

Columns are the most common vertical load-resisting elements in reinforced concrete structures. The primary role of a column in a typical building is to support floor structures (slabs, beams, and girders) and transmit the load to the lower levels and then to the foundations. Columns are mainly subjected to axial compression loads and are often called *compression members*. However, few reinforced concrete columns are subjected to purely axial compression loads. More often, bending moments are also present due to the eccentricity of applied loads, applied end moments, and/or lateral loading on the column. Such columns are subjected to *combined axial load and flexure.*

Columns have cross-sectional dimensions considerably less than their height. CSA A23.3 Cl.2.2 defines a column as a "member with a ratio of height to least lateral dimension of three or greater, used primarily to support axial compressive load" and refers to less slender members as pedestals.

In terms of their load-carrying capacity relative to material usage, columns are among the most efficient structural members. In many buildings, columns are the principal means of transmitting vertical loads to the foundation, and failure of a single column can potentially lead to progressive collapse of the entire structure. Given the potential for catastrophic failure and the relatively low additonal cost required to increase load-carrying capacity, it is recommended to design columns with some reserve capacity whenever possible.

This chapter discusses the behaviour and design of reinforced concrete columns. Different column types are outlined in Section 8.2, whereas the main components of a reinforced concrete column are discussed in Section 8.3. Column loads are discussed in Section 8.4. Section 8.5 is focused on the behaviour of columns subjected to combined axial load and flexure. The flexural resistance of columns subjected to combined axial load and flexure is discussed in Section 8.6. Column interaction diagrams, one of the most useful column design aids, are covered in Section 8.7. Section 8.8 discusses the CSA A23.3 column design requirements. Practical design guidelines for reinforced concrete columns are discussed in Section 8.9. A general design procedure for reinforced concrete columns is presented in Section 8.10. Structural drawings and details related to reinforced concrete columns are discussed in Section 8.11. Section 8.12 offers an overview of slenderness effects in reinforced concrete columns and the criteria and methods for the analysis of slender reinforced concrete columns in nonsway frames. (The design of slender columns in sway frames and the design of biaxially loaded columns are outside the scope of this book.) Finally, the calculation of column loads in multistorey buildings is discussed in Section 8.13.

DID YOU KNOW?

The 88-storey Petronas Twin Towers in Kuala Lumpur, Malaysia, shown in Figure 8.1, are the tallest concrete buildings in the world, with a height above ground of 452 m. Each building is supported by an outer ring of 16 reinforced concrete columns of 2400 mm diameter made of 70 MPa concrete to form a 46 m circle. At each level, the columns are connected with a haunched ring beam. Lateral loads are shared by the columns and the concrete core through the floor diaphragm. Each tower is supported by a 4.5 m thick foundation raft supported on 85 concrete piles, some of them as deep as 120 m! The towers were completed in 1999 at a cost of US$ 800 million (ENR 1999; Aga Khan Development Network, 2004).

Figure 8.1 Petronas Twin Towers in Kuala Lumpur, Malaysia.

(The Aga Khan Award for Architecture)

8.2 | TYPES OF REINFORCED CONCRETE COLUMNS

Reinforced concrete columns can be classified based on the type of transverse reinforcement, as illustrated in Figure 8.2:

- *Tied columns* are reinforced with longitudinal bars enclosed by lateral ties provided throughout the column length.
- *Spiral columns* are reinforced with longitudinal bars enclosed by closely spaced and continuously wound spiral reinforcement.
- *Composite columns* are reinforced longitudinally with a structural steel shape, either encased in or encasing the concrete, and with or without longitudinal bars (for example, a hollow steel pile filled with concrete and reinforced with steel bars).

The most common cross-sectional shapes for tied columns are square and rectangular, while spiral columns are usually of circular shape. This chapter is mainly focused on the design of tied reinforced concrete columns. The design of composite columns is governed by the steel design code CSA S16.1 and is outside the scope of this book.

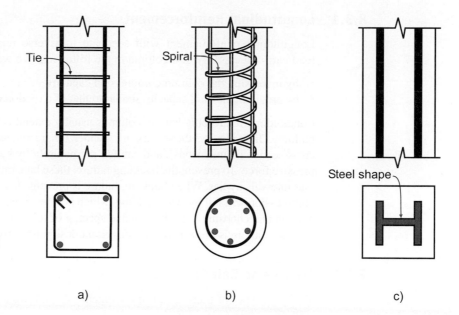

Figure 8.2 Types of reinforced concrete columns: a) tied columns; b) spiral columns; c) composite columns.

a)   b)   c)

Reinforced concrete columns can be classified based on their length into

* short columns
* long columns

Short columns are stocky and their behaviour is similar to that of reinforced concrete flexural members; this will be discussed in Section 8.5. Most of this chapter is focused on the design of short columns, which is very common for design practice.

Long (slender) columns are susceptible to buckling failure (similar to steel columns). The criteria related to slenderness effects in reinforced concrete columns and the design of slender columns are outlined in Section 8.12.

8.3 MAIN COMPONENTS OF A REINFORCED CONCRETE COLUMN

The main components of a reinforced concrete column are (see Figure 8.3)

* longitudinal reinforcement
* transverse reinforcement (ties or spirals)
* concrete core

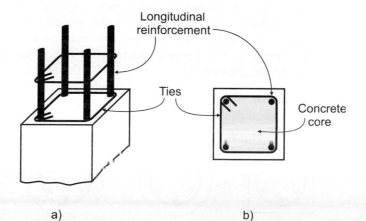

Figure 8.3 The anatomy of a reinforced concrete column: a) isometric view; b) cross-section.

a)   b)

8.3.1 Longitudinal Reinforcement

Longitudinal reinforcement with adequate transverse reinforcement contributes to the load-carrying capacity of a column in the following two ways:

- by increasing the column compression capacity
- by providing flexural capacity similar to that of a reinforced concrete beam

Longitudinal reinforcing bars in columns usually extend continuously between floors of a building. Individually, these bars are slender (characterized by a high ratio of length to cross-sectional dimensions), and can fail prematurely by buckling when subjected to compressive force. To prevent the buckling failure, these bars must be laterally supported at regular intervals by ties. When buckling failure is prevented, the longitudinal bars subjected to compression will fail by steel yielding, which is a desirable failure mode. Dimensioning the column and choosing both the size and spacing of the longitudinal bars and ties so that the buckling failure is prevented is an important design objective.

8.3.2 Transverse Reinforcement

The main types of transverse reinforcement (also known as hoops) are

- ties (rectangular, square, or circular)
- spirals

Tied columns are characterized by longitudinal bars restrained by ties, which are wired onto the bars. Though the ties are similar to closed stirrups in beams, their main purpose is not to resist shear forces but to prevent longitudinal bars from buckling outward under compressive loads.

Typical cross-sectional shapes of concrete columns reinforced with circular ties or spirals are shown in Figure 8.4. Most often, columns of circular cross-section are reinforced with spiral reinforcement, although in some cases rectangular or polygonal column sections are also used in design practice. A spiral is made up of either wire or bar and is formed in the shape of a helix.

Transverse reinforcement has a critical role in ensuring the effectiveness and structural integrity of reinforced concrete columns. The main functions of the transverse reinforcement are

- to provide lateral restraint to longitudinal bars under compression and prevent their premature buckling
- to increase the compressive strength and prevent the sudden bursting of the concrete core
- to hold the longitudinal reinforcement in place during construction
- to resist the shear and/or torsion when reinforcing for these effects is required

Transverse reinforcement in a column acts in a similar way to a trouser belt — it ties (or confines) the column and prevents it from falling apart when subjected to axial compression. Note that transverse reinforcement becomes effective only when the compressive stress in the concrete core reaches the ultimate value due to the increasing axial load in the column. At that stage, the transverse reinforcement is subjected to high strains (just like a belt stretched when a person eats too much). Experimental studies have shown that the confinement provided by transverse reinforcement can significantly enhance the strength and ductility of reinforced concrete compression members.

Due to their circular shape, *spirals* or *circular ties* work in tension and provide continuous confining pressure around the circumference of a column cross-section (this

Figure 8.4 Typical cross-sections of columns reinforced with spirals or circular ties.

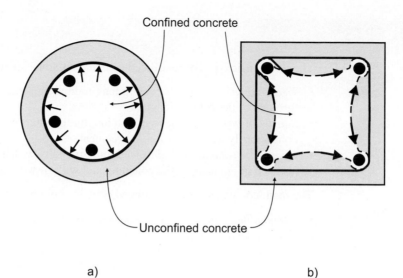

Figure 8.5 How ties work:
a) circular or spiral ties;
b) rectangular or square ties.

pressure is similar to that created by a fluid under pressure inside a pipe), as illustrated in Figure 8.5a. Note that *rectangular* or *square ties* are effective only near their corners because the concrete in the core pushes against the sides of the ties, tending to bend the sides outward. As a result, a considerable portion of the concrete cross-section may remain unconfined, as shown with a hatched pattern in Figure 8.5b.

The amount and distribution of transverse reinforcement are of critical importance for effective column design, as discussed in this chapter.

8.3.3 Concrete Core

Columns are compression members that make efficient use of concrete's compressive strength. However, concrete by itself is a brittle material, so steel reinforcement (longitudinal and transverse) must be provided in columns to prevent brittle failure and ensure ductile behaviour. It should be noted that closely spaced transverse reinforcement creates additional lateral pressure in concrete columns subjected to axial compression. This results in the compressive strength of the concrete core increasing beyond the specified compressive strength (f_c'), which is obtained by testing of unconfined concrete cylinders in uniaxial compression, that is, without any lateral pressure applied on a cylinder during the testing. The strength increase is proportional to the amount of lateral pressure created: the more closely spaced the transverse reinforcement, the larger the increase in compressive strength.

Experiments have shown that the concrete strength in confined concrete columns can be effectively enhanced in the following ways (Park and Paulay, 1975):

* by increasing the volume of transverse steel relative to the volume of concrete (the larger the volume of transverse reinforcement, the larger the confining pressure)
* by increasing the yield strength of the transverse steel (which gives an upper limit to the confining pressure)
* by decreasing the spacing of transverse steel (smaller spacing leads to more effective confinement)
* by increasing the diameter of the transverse reinforcement (resulting in larger flexural stiffness of rectangular or square ties)
* by providing an adequate amount and size of longitudinal reinforcement, since the longitudinal steel also confines the concrete
* by placing the longitudinal bars tightly against the transverse steel; otherwise, longitudinal bars need to move under the applied loads to bear onto the transverse reinforcement, and the effectiveness of the longitudinal bars is reduced.

The three basic types of reinforced concrete columns are as follows (see Figure 8.2):

- *Tied columns* are reinforced with longitudinal bars enclosed by lateral ties provided throughout the column length.
- *Spiral columns* are reinforced with longitudinal bars enclosed by closely spaced and continuously wound spiral reinforcement.
- *Composite columns* are reinforced longitudinally with a structural steel shape, either encased in or encasing the concrete, and with or without longitudinal bars.

The main components of a reinforced concrete column are as follows (see Figure 8.3):

- *Longitudinal reinforcement* increases the axial and flexural load resistance of a column.
- *Transverse reinforcement* (ties or spirals) provides lateral restraint or confinement to longitudinal bars under compression and prevents their premature buckling.
- The *concrete core* provides compression resistance to a column.

8.4 COLUMN LOADS: CONCENTRICALLY VERSUS ECCENTRICALLY LOADED COLUMNS

Based on the location of the axial load, columns can be classified into two categories:

- *Concentrically loaded columns* (also called compression members) are subjected to axial loads applied along the longitudinal column axis.
- *Eccentrically loaded columns* are subjected to axial load acting at an eccentricity with regard to the centroid of the column section; as a result, these columns are subjected to combined axial load and flexure.

In frame structures, bending moments are transferred from beams or slabs to the columns. In some cases, moments develop due to the eccentricity of the axial load itself or due to construction imperfections. Examples of eccentrically loaded columns include the following (see Figure 8.6):

- In a moment frame, due to the rigidity of the beam-to-column joints, the columns rotate along with the ends of the supported beams (Figure 8.6a).
- Interior columns supporting beams of equal spans receive unequal loads due to applied live load patterns (Figure 8.6b).
- In precast construction, a beam reaction is eccentrically applied on the column through the bracket (Figure 8.6c).

Consider a column loaded with an external axial load (P) acting at an eccentricity (e) with regard to the centroid of the cross-section (point CG), as illustrated in Figure 8.7a. When the two imaginary equal and opposite axial forces of magnitude P are added at the point CG (see Figure 8.7b), the upward force forms a couple with the external force P, with the lever arm e. The resulting bending moment (M) is

$$M = P \times e$$

Note that there is a remaining force (P) acting downward at the centroid of the cross-section (CG). Thus, the effect of an eccentrically applied axial force acting on a column cross-section is equivalent to the combined effect of a concentric axial force and a bending moment, as shown in Figure 8.7c.

In reality, the factored axial load (P_f) and the factored bending moment (M_f) acting on a column are often predetermined from analysis. The corresponding *design eccentricity* (e) is

$$e = \frac{M_f}{P_f}$$

[8.1]

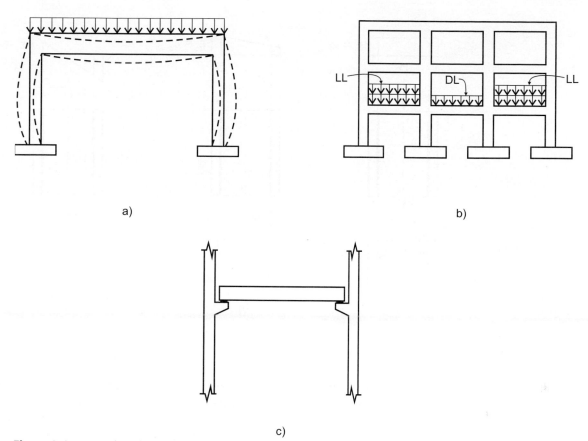

Figure 8.6 Examples of eccentrically loaded columns: a) moment frame; b) interior columns supporting unequally loaded beams; c) precast construction.

Figure 8.7 Axial load–bending moment eccentricity relationship: a) column loaded with the force P acting at an eccentricity e; b) imaginary forces (P) along the column axis; c) equivalent concentric load (P) and bending moment (M).

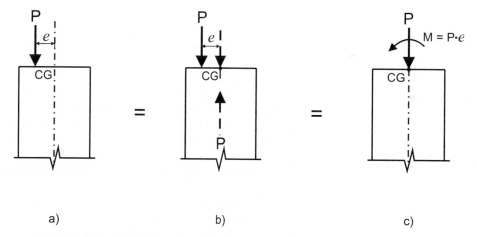

8.5 | BEHAVIOUR OF REINFORCED CONCRETE SHORT COLUMNS

The behaviour of reinforced concrete columns subjected to axial load depends primarily on their slenderness (short/long columns) and the magnitude of the load eccentricity. This section is focused on the behaviour of short columns. There are three distinct types of column behaviour, as follows (see Figure 8.8):

- concentrically loaded columns (zero eccentricity)
- eccentrically loaded columns (small eccentricity)
- eccentrically loaded columns (large eccentricity)

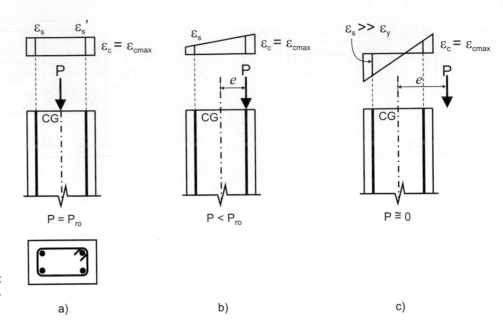

Figure 8.8 Variation in the strain distribution with the load eccentricity: a) concentric load ($P = P_{ro}$); b) small eccentricity; c) large eccentricity.

Concentrically loaded column (zero eccentricity)   Consider a reinforced concrete column subjected to a *concentric axial load* (*P*); that is, the load acts at the centroid of the column cross-section point (CG) and the load eccentricity is zero ($e = 0$). The load causes a uniformly distributed compressive strain in the cross-section. With increasing load, the strain in the longitudinal reinforcement reaches the yield strain (ε_y), which is approximately equal to 0.002 for Grade 400 steel. Under a further load increase, the inelastic strain in steel continues to increase until the strain in the concrete (ε_c) reaches the maximum value (ε_{cmax}) of 0.0035, as shown in Figure 8.8a. The maximum axial load capacity of a concentrically loaded column (P_{ro}) is the sum of the concrete contribution and the steel contribution. The failure of a concentrically loaded column occurs after the longitudinal reinforcement yields. The behaviour of a concentrically loaded column in the postyield phase depends on whether the transverse reinforcement consists of ties or spirals; this will be explained in Section 8.6.2. This failure will be referred to as *concrete-controlled* mode in what follows.

Eccentrically loaded column (small eccentricity)   When the load is applied at a small distance (*e*) from the centroid (CG), the strain distribution in the column is no longer uniform (see Figure 8.8b). The maximum load capacity of the column is reached when the strain at the compression face (ε_c) reaches the maximum value (ε_{cmax}) of 0.0035. In this case, the strains and stresses in the cross-section are less than those in the concentrically loaded column. Consequently, the maximum load the section can carry (*P*) is less than the axial load capacity of the concentrically loaded column; that is, $P < P_{ro}$. The extent of the decrease in the load capacity (*P*) depends on the magnitude of the eccentricity: the larger the eccentricity, the smaller the load capacity. Eccentrically loaded columns with small eccentricity fail in the *concrete-controlled* mode.

Eccentrically loaded column (large eccentricity)   When the load eccentricity (*e*) becomes large, the effect of the bending moment ($M = Pe$) becomes significant, and the column behaves like a beam subjected to flexure. In the extreme case, when the eccentricity becomes indefinitely large ($e = \infty$), the axial load becomes insignificant

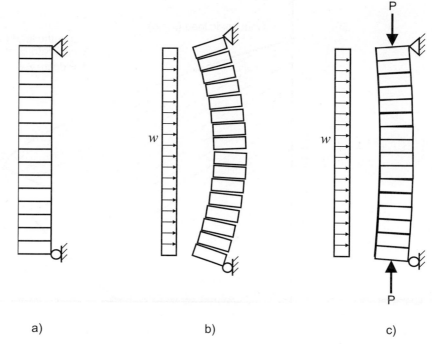

Figure 8.9 Effect of axial compression in eccentrically loaded columns with large eccentricity: a) a column represented as a stack of concrete blocks; b) the column lacks flexural resistance and falls apart; c) the column behaviour improves when axial compression is added.

a)   b)   c)

and can be taken as zero; that is, $P \cong 0$; this is denoted as a *pure bending* condition, as illustrated in Figure 8.8c. The column behaves like a properly reinforced beam failing in the steel-controlled mode, characterized by the yielding of steel reinforcement in tension before concrete crushing has taken place (for more details, see Section 3.4.3). Eccentrically loaded columns with large eccentricity fail in the *steel-controlled* mode.

In an axially loaded column where the load eccentricity is large, an increase in axial load leads to an increase in flexural (moment) resistance. While this behaviour may appear unusual, it can be explained by representing the column as a stack of unbonded concrete blocks laterally supported at the top and bottom, as shown in Figure 8.9a. When the stack of blocks is subjected to even a small uniformly distributed lateral load (w), it may exhibit flexural failure due to the absence of tensile resistance between the individual concrete blocks (see Figure 8.9b). However, if an axial compressive load (P) is introduced simultaneously with the lateral load (see Figure 8.9c), the compressive load can help compensate for the lack of tensile resistance and increase the magnitude of the moment resistance. If axial load is small, the magnitude of the moment resistance increases with an increase in the axial load level. This relationship continues until the axial load is sufficiently large such that the combined effects of axial load and flexure initiate a crushing failure at the column compression face. Further increases in axial load result only in a decrease in column moment resistance.

The *balanced condition* occurs when the strain in steel at the tension face reaches the yield strain ($\varepsilon_s = \varepsilon_y$) simultaneously with the concrete strain at the compression face reaching the maximum value ($\varepsilon_c = \varepsilon_{cmax}$). This condition represents the threshold between the small and large eccentricity load conditions discussed above. The load eccentricity corresponding to the balanced condition is called the *balanced eccentricity* (e_b).

The above discussed load conditions can be conveniently presented in diagram form, with bending moment on the horizontal axis and axial load on the vertical axis. Such a diagram is called a *column interaction diagram*, as shown in Figure 8.10. There are two distinct regions shown hatched on the diagram, corresponding to eccentrically loaded columns with small and large eccentricities, respectively. These two regions are divided by a line corresponding to the balanced condition. Features of column interaction diagrams and their use in column design will be discussed in Section 8.7.

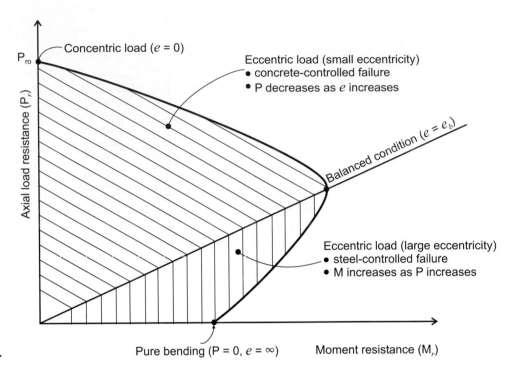

Figure 8.10 Characteristic types of behaviour summarized on a column interaction diagram.

KEY CONCEPTS

The behaviour of reinforced concrete columns subjected to axial load depends primarily on their slenderness and the magnitude of the load eccentricity. There are three distinct types of column behaviour, as follows:

Concentrically loaded column (zero eccentricity): The maximum axial load capacity of the column is reached when the compressive strain in concrete (ε_c) reaches the maximum value (ε_{cmax}) of 0.0035 ($\varepsilon_c = \varepsilon_{cmax}$). The maximum axial load capacity of a concentrically loaded column (P_{ro}) is the sum of the concrete contribution and the steel contribution. A concentrically loaded column fails after the longitudinal reinforcement yields. The behaviour of a concentrically loaded column in the postyield phase depends on whether the transverse reinforcement consists of ties or spirals. Concentrically loaded columns fail in the *concrete-controlled* mode.

Eccentrically loaded column (small eccentricity): The axial load resistance in eccentrically loaded columns with small eccentricity is reduced in comparison with concentrically loaded columns. The decrease in the load capacity depends on the magnitude of the eccentricity: the larger the eccentricity, the smaller the load capacity. Eccentrically loaded columns with small eccentricity fail in the *concrete-controlled* mode.

Eccentrically loaded column (large eccentricity): At a large load eccentricity (e), the effect of the bending moment ($M = P \times e$) becomes significant, and the column behaves like a beam subjected to flexure. In an eccentrically loaded column with large eccentricity, an increase in the axial load results in an increased flexural resistance. Eccentrically loaded columns with large eccentricity fail in the *steel-controlled* mode.

The balanced condition: When the strain in steel at the tension face reaches the yield strain ($\varepsilon_s = \varepsilon_y$) simultaneously with the strain at the concrete compression face reaching the maximum value ($\varepsilon_c = \varepsilon_{cmax}$), the balanced condition occurs. This condition represents the threshold between the small and large eccentricity load conditions.

8.6 AXIAL AND FLEXURAL LOAD RESISTANCE OF REINFORCED CONCRETE SHORT COLUMNS

8.6.1 Basic CSA A23.3 Column Design Assumptions

The design of reinforced concrete columns for axial load and flexure according to CSA A23.3 is based on the following design assumptions (note that the same assumptions apply to the flexural design of beams and slabs in Section 3.3.2):

1. *Plane sections remain plane:* Sections perpendicular to the axis of bending that are plane before bending remain plane after bending (Cl.10.1.2).
2. *Strain compatibility:* The strain in the reinforcement is equal to the strain in the concrete at the reinforcement location. This assumption implies perfect bond between the steel and the concrete.
3. *Stress-strain relationship:* Cl.10.1.7 proposes the use of an equivalent rectangular stress block instead of the actual stress distribution for concrete for limit states strength design.
4. *Concrete tensile strength:* The tensile strength of concrete is neglected in design (Cl.10.1.5).
5. *Maximum concrete compressive strain:* According to Cl.10.1.3, the maximum compressive strain in concrete (ε_{cmax}) is equal to 0.0035.

KEY CONCEPTS

The ultimate limit states design of concrete columns is based on the following basic assumptions outlined by CSA A23.3:

- Plane sections remain plane (linear strain distribution across the section) (Cl.10.1.2).
- Strain compatibility exists between concrete and steel.
- An equivalent rectangular stress block is used instead of the actual stress distribution for concrete (Cl.10.1.7).
- The concrete tensile strength is neglected in the design (Cl.10.1.5).
- The maximum concrete compressive strain (ε_{cmax}) is equal to 0.0035 (Cl.10.1.3).

8.6.2 Axial Load Resistance of a Concentrically Loaded Short Column

Consider the concentrically loaded reinforced concrete short column in Figure 8.11a. The total axial load resistance of a column is determined as the combined resistance provided by the concrete and the longitudinal reinforcement:

$$P_{ro} = P_{rco} + P_{rso} \qquad\qquad [8.2]$$

where

P_{ro} = the factored axial load resistance of a reinforced concrete column
P_{rco} = the factored axial load resistance provided by the concrete
P_{rso} = the factored axial load resistance provided by the longitudinal reinforcement.

The concrete resistance (P_{rco}) is determined as the product of the net concrete area ($A_g - A_{st}$), obtained from the gross cross-sectional area by deducting the steel area, and the corresponding stress ($\alpha_1 \phi_c f_c'$). The equivalent rectangular stress distribution (see the third design assumption in Section 8.6.1) and the material resistance factors according to CSA A23.3 are used, as follows:

$$P_{rco} = (\alpha_1 \phi_c f_c')(A_g - A_{st}) \qquad\qquad [8.3]$$

where

f_c' = the specified compressive strength of concrete
ϕ_c = the resistance factor for concrete

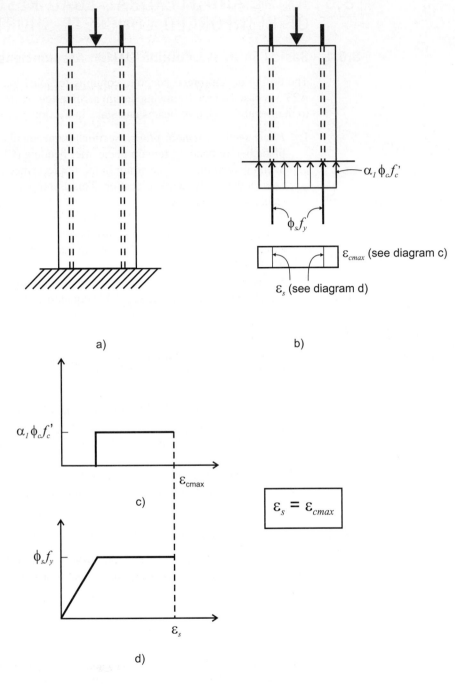

Figure 8.11 Axial load resistance of a concentrically loaded reinforced concrete column: a) elevation; b) stress and strain distribution; c) stress-strain diagram of concrete; d) stress-strain diagram of steel.

A_g = the gross area of a column cross-section
A_s = the total area of longitudinal reinforcement.

and

$$\alpha_1 \cong 0.8 \qquad\qquad\qquad\qquad [3.7]$$

The maximum force in the steel reinforcement (P_{rso}) can be calculated as

$$P_{rso} = \phi_s f_y A_{st} \qquad\qquad\qquad\qquad [8.4]$$

where

f_y = the specified yield strength of steel
ϕ_s = the resistance factor for steel.

According to A23.3 Cl.10.10.4, the factored axial load resistance of a reinforced concrete column is equal to the sum of the concrete resistance and the resistance provided by the yielded longitudinal reinforcement:

| A23.3 Eq. 10.10 | $$P_{ro} = \alpha_1 \phi_c f_c'(A_g - A_{st}) + \phi_s f_y A_{st}$$ | **[8.5]** |

Note that Eqn 8.5 is a simplified form of A23.3 equation 10.10; this simplified equation applies to nonprestressed columns and columns without internal or external structural steel shapes.

A23.3 Cl.10.10.4 accounts for the effect of unanticipated or accidental moments in concentrically loaded short columns by prescribing a reduction in the column axial load resistance.

For tied columns,

| A23.3 Eq. 10.9 | $$P_{rmax} = 0.80 P_{ro}$$ | **[8.6]** |

or

$$P_{rmax} = 0.80\,[\alpha_1 \phi_c f_c'(A_g - A_{st}) + \phi_s f_y A_{st}]$$

For spiral columns,

| A23.3 Eq. 10.8 | $$P_{rmax} = 0.85 P_{ro}$$ | **[8.7]** |

or

$$P_{rmax} = 0.85\,[\alpha_1 \phi_c f_c'(A_g - A_{st}) + \phi_s f_y A_{st}]$$

In these equations, P_{rmax} is the maximum axial load resistance of a concentrically loaded nonprestressed concrete column prescribed by CSA A23.3.

In practical design applications, the gross area of concrete is used in Eqn 8.3 instead of the net area. This is justified by the fact that the reinforcement area (A_{st}) represents a small fraction of the gross cross-sectional area (A_g). It follows that

$$P_{ro} \cong \alpha_1 \phi_c f_c' A_g + \phi_s f_y A_{st}$$ **[8.8]**

Consequently, the maximum axial load resistance can be calculated from the following *simplified equations* (note that these equations are not prescribed by CSA A23.3).

For tied columns,

$$P_{rmax} = 0.80\,(\alpha_1 \phi_c f_c' A_g + \phi_s f_y A_{st})$$ **[8.9]**

For spiral columns,

$$P_{rmax} = 0.85\,(\alpha_1 \phi_c f_c' A_g + \phi_s f_y A_{st})$$ **[8.10]**

The behaviour of tied and spiral columns subjected to axial load strongly depends on the extent of confinement provided by the transverse reinforcement. The extent of confinement depends on the size and spacing of the reinforcement; it increases with a decrease in the spacing and an increase in the size of transverse reinforcement. The behaviour of tied and spiral columns subjected to concentric axial load is shown on the stress-strain diagram in Figure 8.12. Tied and spiral columns demonstrate similar behaviour until the longitudinal reinforcement has begun to yield. The tied column with widely spaced ties (moderate confinement) fails shortly thereafter, as shown with a dashed line in Figure 8.12. The failure is accompanied by the crushing of concrete and the buckling of longitudinal bars between the ties.

In the spiral column, the yielding of the longitudinal reinforcement is accompanied by cracking or complete spalling of the concrete shell outside the spirals. After the shell has spalled off, the spirals become effective in preventing the buckling of the longitudinal bars. Under increasing load, the longitudinal bars bear on the spirals, which in turn cause larger lateral pressure to develop in the concrete core. As discussed in Section 8.3.3, the lateral pressure caused by the spirals enhances the compression resistance of the concrete core (see the solid line in Figure 8.12). Ultimately, the column fails when the spirals yield and the effectiveness of the confinement has been exhausted. The

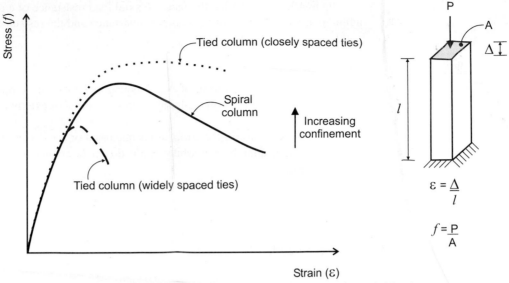

Figure 8.12   Behaviour of tied and spiral columns under increasing axial load.

behaviour of the spiral column will vary depending on the spiral size and pitch, which determine the extent of confinement.

A highly confined tied column with closely spaced ties behaves similarly to a spiral column and has a significant ability to deform before failure, as shown by the dotted line in Figure 8.12.

In conclusion, axially loaded columns with closely spaced ties or spirals demonstrate superior behaviour and are able to sustain larger deformations before failure than columns with limited confinement (widely spaced ties). For more details on the behaviour of confined concrete columns, the reader is referred to Sheikh and Uzumeri (1980) and Sheikh (2002).

KEY CONCEPTS

The factored axial load resistance of a reinforced concrete short column (P_{ro}) is determined as the combined resistance provided by the concrete (P_{rco}) and the longitudinal reinforcement (P_{rso}):

$$P_{ro} = P_{rco} + P_{rso}$$

A23.3 Cl.10.10.4 accounts for the effect of unanticipated or accidental moments in concentrically loaded short columns by prescribing a reduction in the column axial load resistance.

For tied columns,

| A23.3 Eq. 10.9 | $P_{rmax} = 0.80P_{ro}$ | **[8.6]** |

For spiral columns,

| A23.3 Eq. 10.8 | $P_{rmax} = 0.85P_{ro}$ | **[8.7]** |

The behaviour of axially loaded tied and spiral columns depends strongly on the extent of confinement provided by the transverse reinforcement. In general, *spiral columns* demonstrate superior behaviour to *tied columns* when subjected to a concentric axial load. Tied columns with widely spaced ties fail after the longitudinal reinforcement in the column yields. The failure is accompanied by concrete crushing and buckling of the longitudinal bars between the ties. Spiral columns fail when the spirals fracture after yielding at high axial deformations. Axially loaded confined columns with closely spaced ties also have an ability to sustain large deformations before failure.

Example 8.1

A typical cross-section of a concentrically loaded tied short column is shown in the figure below. The column is reinforced with 30M longitudinal bars and 10M ties. Use a 40 mm clear cover to the ties and a 20 mm maximum aggregate size. The concrete and steel material properties are given below.

Find the axial load resistance for this column.

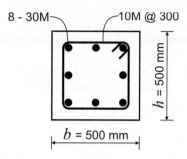

Given: $f_c' = 25$ MPa

$f_y = 400$ MPa

$\phi_c = 0.65$

$\phi_s = 0.85$

$b = 500$ mm

$h = 500$ mm

SOLUTION:

1. Calculate the areas of longitudinal reinforcement and concrete

Longitudinal reinforcement: 30M bars

From Table A.1,

Area $= 700$ mm^2

The total area of longitudinal steel (A_{st}) (eight bars in total) is

$A_{st} = 8(700 \text{ mm}^2) = 5600$ mm^2

Concrete

The gross cross-sectional area is

$A_g = b \times h = 500 \text{ mm} \times 500 \text{ mm} = 250\,000$ mm

| A23.3 Cl.10.10.4 |

2. Calculate the maximum axial load resistance (P_{rmax})

Determine the factored axial load resistance for a reinforced concrete column:

| A23.3 Eq. 10.10 |

$$P_{ro} = \alpha_1 \phi_c f_c'(A_g - A_{st}) + \phi_s f_y A_{st} \qquad [8.5]$$

All parameters in the above equation are known except for α_1:

$$\alpha_1 \cong 0.8 \qquad [3.7]$$

Calculate the P_{ro} value:

$P_{ro} = 0.8(0.65)(25 \text{ MPa})(250\,000 \text{ mm}^2 - 5600 \text{ mm}^2) + 0.85(400 \text{ MPa})(5600 \text{ mm}^2)$

$\cong 5081 \times 10^3 \text{ N} = 5081$ kN

The column under consideration is tied. Therefore, determine the maximum axial load resistance based on the following equation:

| A23.3 Eq. 10.9 |

$$P_{rmax} = 0.80 P_{ro} = 0.80(5081 \text{ kN}) = 4065 \text{ kN} \qquad [8.6]$$

Therefore, the maximum axial load this column can carry according to CSA A23.3 is 4065 kN.

For design purposes, calculate the P_{rmax} value using the simplified equation as

$$P_{rmax} \cong 0.8\,[\alpha_1\phi_c f_c' A_g + \phi_s f_y A_{st}] \qquad [8.9]$$
$$= 0.8\,[0.8 \times 0.65(25\text{ MPa})(250\,000\text{ mm}^2) + 0.85(400\text{ MPa})(5600\text{ mm}^2)]$$
$$= 4123 \times 10^3\text{ N} \cong 4120\text{ kN}$$

By comparison, the difference in the two values is less than 2%. Therefore, use the factored axial load resistance for this column obtained using the simplified equation:

$$P_{rmax} = 4120\text{ kN}$$

8.6.3   Axial and Flexural Resistance of Eccentrically Loaded Reinforced Concrete Short Columns

Consider a tied reinforced concrete short column of rectangular cross-section subjected to an axial load P_r acting at the eccentricity e; a vertical elevation of the column is illustrated in Figure 8.13a. The column is reinforced with the longitudinal reinforcement distributed symmetrically at the two column faces, as shown in the cross-section in Figure 8.13b. Due to the

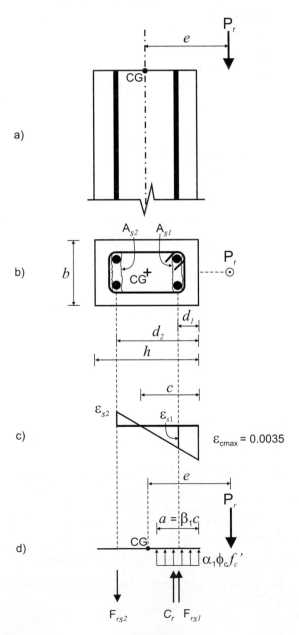

Figure 8.13   Eccentrically loaded column: a) vertical elevation; b) cross-section; c) strain distribution; d) internal forces.

eccentric location of the axial load, one column face is in the tension zone, whereas the remainder of the column section is in the compression zone.

The notation used in this section is summarized below:

b = column width,

h = overall depth,

A_{s1} = area of longitudinal reinforcement at a distance d_1 from the compression face of the column,

A_{s2} = area of longitudinal reinforcement at a distance d_2 from the compression face of the column,

a = depth of the equivalent rectangular stress block,

c = neutral axis depth (distance from the extreme compression fibre to the neutral axis).

Note that the symbols A_{s1} and A_{s2} are used to denote reinforcement located at different distances from the compression face of the column. This labelling is somewhat different from the previous chapters, and it is used in this section to enable the use of the same analysis procedure for columns with more than two layers of reinforcement (to be discussed in Section 8.7.2).

The strain distribution over the column section is shown in Figure 8.13c. The steel strain (ε_{s1}) in the reinforcement of area (A_{s1}) can be determined from similar triangles as

$$\frac{\varepsilon_{s1}}{c - d_1} = \frac{\varepsilon_{cmax}}{c}$$

The strain (ε_{s1}) can be calculated from the above equation as

$$\varepsilon_{s1} = \varepsilon_{cmax}\left(1 - \frac{d_1}{c}\right) \qquad [8.11]$$

where $\varepsilon_{cmax} = 0.0035$ is the maximum compression strain in concrete (see Section 8.6.1).

The strain (ε_{s2}) in the reinforcement (A_{s2}) can be obtained from similar triangles as

$$\varepsilon_{s2} = \varepsilon_{cmax}\left(1 - \frac{d_2}{c}\right) \qquad [8.12]$$

The steel reinforcement (areas A_{s1} and A_{s2}) may either yield or remain elastic, depending on several factors. The criteria related to the magnitude of the stress (f_{s1}) in the reinforcement of area (A_{s1}) corresponding to the strain (ε_{s1}) can be summarized as follows:

- If $\varepsilon_{s1} \geq \varepsilon_y$, then the steel yields and $f_{s1} = f_y$.
- If $\varepsilon_{s1} < \varepsilon_y$, then the steel remains elastic; that is, $f_{s1} = E_s\varepsilon_{s1} < f_y$.

The strain (ε_{s2}) in the reinforcement of area (A_{s2}) and the corresponding stress (f_{s2}) can be determined in an analogous way.

Note that the negative sign indicates tension in the yield strain; also, the yield strain (ε_y) can be determined from Hooke's law as $\varepsilon_y = f_y/E_s$ (see Section 2.6.2).

Let us consider the equilibrium of forces acting on the column cross-section in the vertical direction, as shown in Figure 8.13d. Based on the strain distribution, the compression is carried by the concrete (the corresponding stress resultant C_r) and the reinforcement A_{s1} corresponding to the resultant force F_{rs1}. However, the tension is carried by the reinforcement of area A_{s2} that is closest to the tension face of the column (the corresponding resultant force is denoted as F_{rs2}). Note that this concept is similar to flexure in reinforced concrete beams (see Section 3.5.1). These forces must be in equilibrium with the external force P_r as follows:

$$P_r = C_r + F_{rs1} + F_{rs2} \qquad [8.13]$$

where

$$C_r = \alpha_1\phi_c f_c' a\, b = \alpha_1\phi_c f_c'(c\beta_1)b \qquad [3.8]$$

is the compression stress resultant in the concrete,

$$F_{rs1} = \phi_s f_{s1} A_{s1} \qquad [8.14]$$

is the force in the reinforcement of area A_{s1},

$$F_{rs2} = \phi_s f_{s2} A_{s2}$$ [8.15]

is the force in the reinforcement of area A_{s2},

$$a = \beta_1 c$$ [3.4]

is the depth of the equivalent rectangular stress block, and

$$\alpha_1 \cong 0.8, \quad \beta_1 \cong 0.9$$ [3.7]

KEY EQUATIONS

Key equations related to axial and flexural resistance of eccentrically loaded columns are summarized below.

The force equilibrium equation is

$$P_r = C_r + F_{rs1} + F_{rs2}$$ [8.13]

The moment equilibrium equation is

$$P_r \times e = C_r\left(\frac{h}{2} - \frac{\beta_1 c}{2}\right) + F_{rs1}\left(\frac{h}{2} - d_1\right) + F_{rs2}\left(\frac{h}{2} - d_2\right)$$ [8.16]

The factored moment resistance is

$$M_r = P_r \times e$$ [8.17]

The strain in the reinforcement of area A_{s1} is

$$\varepsilon_{s1} = \varepsilon_{cmax}\left(1 - \frac{d_1}{c}\right)$$ [8.11]

If $\varepsilon_{s1} \geq \varepsilon_y$, then $f_{s1} = f_y$
If $\varepsilon_{s1} < \varepsilon_y$, then $f_{s1} = E_s \varepsilon_{s1} < f_y$

The force in the reinforcement of area A_{s1} is

$$F_{rs1} = \phi_s f_{s1} A_{s1}$$ [8.14]

The strain in the reinforcement of area A_{s2} is

$$\varepsilon_{s2} = \varepsilon_{cmax}\left(1 - \frac{d_2}{c}\right)$$ [8.12]

If $\varepsilon_{s2} \geq \varepsilon_y$, then $f_{s2} = f_y$
If $\varepsilon_{s2} < \varepsilon_y$, then $f_{s2} = E_s \varepsilon_{s2} < f_y$ (the negative sign indicates tension)

The force in the reinforcement of area A_{s2} is

$$F_{rs2} = \phi_s f_{s2} A_{s2}$$ [8.15]

The compression stress resultant in the concrete is

$$C_r = \alpha_1 \phi_c f_c' a b = \alpha_1 \phi_c f_c'(c\beta_1)b$$ [3.8]

where

$$a = \beta_1 c$$ [3.4]

is the depth of the equivalent rectangular stress block, and

$$\alpha_1 \cong 0.8 \quad \beta_1 \cong 0.9$$ [3.7]

Note that the neutral axis depth (c) has been used instead of (a) as a parameter in the above equations in order to decrease the number of unknown variables in the equations of equilibrium.

The second equation of equilibrium is obtained by equating the external moment to the sums of the moments of the factored stress resultants around the centroid of the column cross-section (point CG) as follows:

$$P_r \times e = C_r\left(\frac{h}{2} - \frac{\beta_1 c}{2}\right) + F_{rs1}\left(\frac{h}{2} - d_1\right) + F_{rs2}\left(\frac{h}{2} - d_2\right) \qquad \textbf{[8.16]}$$

Note that

$$M_r = P_r \times e \qquad \textbf{[8.17]}$$

where M_r denotes the factored moment resistance of a column section subjected to an axial force P_r acting at an eccentricity e.

Equations 8.13 and 8.16 can be used to calculate the axial force (P_r) and the corresponding moment resistance (M_r) for a column with given material properties, cross-sectional dimensions, and reinforcement.

8.6.4 Column Load Resistance Corresponding to the Balanced Condition

The balanced condition occurs when the strain in concrete is equal to the maximum value and the strain in steel is equal to the yield strain; that is,

$$\varepsilon_c = \varepsilon_{cmax} = 0.0035$$

and

$$\varepsilon_{s2} = \varepsilon_y = \frac{f_y}{E_s}$$

The following general steps can be followed to analyze a column section for the balanced condition.

1. A strain diagram for the section can be drawn and the neutral axis depth (c_b) calculated based on the strain distribution from similar triangles as

    $$c_b = \frac{\varepsilon_{cmax}}{\varepsilon_y + \varepsilon_{cmax}} d_2 \qquad \textbf{[8.18]}$$

    Note that, in this case, the neutral axis depth can be determined without using the equations of equilibrium.
2. The internal stresses and resultants can be determined based on Eqns 3.8, 8.14, and 8.15.
3. The axial force corresponding to the balanced condition (P_{rb}) can be determined from Eqn 8.13, and the balanced eccentricity (e_b) can be determined from Eqn 8.16. Subsequently, the corresponding moment resistance (M_{rb}) can be determined from Eqn 8.17.

Example 8.2

Consider the short reinforced concrete column with cross-sectional dimensions 300 mm by 300 mm in the figure that follows. Use a 40 mm clear cover to the ties. The material properties and the reinforcement are given.

Calculate the following column design parameters:

- *balanced axial load* (P_{rb})
- *balanced eccentricity* (e_b)
- *balanced moment resistance* (M_{rb})

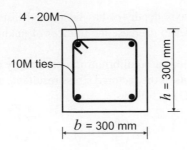

4 - 20M

10M ties

$h = 300$ mm

$b = 300$ mm

Given: $f_c' = 25$ MPa
$f_y = 400$ MPa
$\phi_c = 0.65$
$\phi_s = 0.85$
$E_s = 200\ 000$ MPa

SOLUTION: **1.   Calculate the neutral axis depth (c_b)**

a)   Calculate the reinforcement parameters (see Table A.1):

i)   Longitudinal reinforcement: 20M bars

Area = 300 mm^2

Diameter: $d_b \cong 20$ mm

ii)   Ties: 10M bars

Diameter: $d_{tie} \cong 10$ mm

iii)   The distances d_1 and d_2 (see the sketch below):

cover = 40 mm (given)

$$d_1 = \text{cover} + d_{tie} + \frac{d_b}{2} = 40\text{ mm} + 10\text{ mm} + \frac{20\text{ mm}}{2} = 60\text{ mm}$$

$$d_2 = h - \text{cover} - d_{tie} - \frac{d_b}{2}$$

$$= 300\text{ mm} - 40\text{ mm} - 10\text{ mm} - \frac{20\text{ mm}}{2} = 240\text{ mm}$$

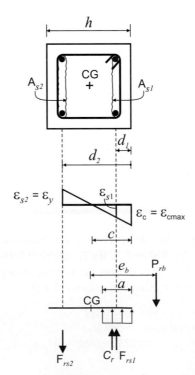

b) Determine the neutral axis depth (c_b):

$$c_b = \frac{\varepsilon_{cmax}}{\varepsilon_y + \varepsilon_{cmax}} d_2$$ [8.18]

where

$\varepsilon_c = \varepsilon_{cmax} = 0.0035$, and

$$\varepsilon_y = \frac{f_y}{E_s} = \frac{400 \text{ MPa}}{200\ 000 \text{ MPa}} = 0.002$$

Thus, $\varepsilon_{s2} = -0.002$ (negative sign indicates tension)

$$c_b = \frac{0.0035}{0.002 + 0.0035} (240 \text{ mm}) = 152.7 \text{ mm} \cong 153 \text{ mm}$$

2. **Calculate the forces in the concrete and steel**
   a) Calculate the concrete stress resultant using the following equations:

$$C_r = \alpha_1 \beta_1 \phi_c f_c' c\, b$$ [3.8]

where

$$\alpha_1 \cong 0.8, \quad \beta_1 \cong 0.9$$ [3.7]

Thus,

$$C_r = \alpha_1 \beta_1 \phi_c f_c' c\, b - 0.8 \times 0.9 \times 0.65\ (25 \text{ MPa})(153 \text{ mm})(300 \text{ mm})$$

$$= 537 \times 10^3 \text{ N} = 537 \text{ kN}$$

   b) Calculate the forces in the reinforcement.
      i) First, calculate the reinforcement area:

      Layer 1 (2-20M): $A_{s1} = 2\ (300 \text{ mm}^2) = 600 \text{ mm}^2$
      Layer 2 (2-20M): $A_{s2} = 2\ (300 \text{ mm}^2) = 600 \text{ mm}^2$

      ii) Next, check whether the reinforcement has yielded.
      Since $\varepsilon_{s2} = \varepsilon_y = -0.002$, the reinforcement A_{s2} has yielded.

$$\varepsilon_{s1} = \varepsilon_{cmax}\left(1 - \frac{d_1}{c}\right)$$ [8.11]

$$= 0.0035\left(1 - \frac{60 \text{ mm}}{153 \text{ mm}}\right) = 0.0021 > \varepsilon_y = 0.002$$

      so the reinforcement A_{s1} has yielded.
      Hence,

$$f_{s1} = f_y \text{ (compression)}$$
$$f_{s2} = -f_y \text{ (tension)}$$

      where

$$f_y = 400 \text{ MPa}$$

      iii) Finally, calculate the forces in the reinforcement as follows:

$$F_{rs1} \quad \phi_s f_{s1} A_{s1} - (0.85)(400 \text{ MPa})(600 \text{ mm}^2) - 204 \times 10^3 \text{ N} = 204 \text{ kN}$$ [8.14]

$$F_{rs2} = \phi_s f_{s2} A_{s2} = (0.85)(-400 \text{ MPa})(600 \text{ mm}^2) = -204 \text{ kN}$$ [8.15]

3. **Calculate the balanced axial force (P_{rb})**

The force equilibrium equation is

$$P_r = C_r + F_{rs1} + F_{rs2} \qquad\qquad\qquad\qquad \text{[8.13]}$$

Substitute the values for the component forces. Then, P_{rb} can be obtained as

$$P_{rb} = P_r = 537 \text{ kN} + 204 \text{ kN} - 204 \text{ kN} = 537 \text{ kN}$$

4. **Calculate the balanced eccentricity (e_b)**

   a) Compute the following:

$$C_r\left(\frac{h}{2} - \frac{\beta_1 c}{2}\right) = (537 \times 10^3 \text{ N})\left(\frac{300 \text{ mm}}{2} - \frac{0.9(153 \text{ mm})}{2}\right) = 43.6 \times 10^6 \text{ N·mm}$$

$$F_{rs1}\left(\frac{h}{2} - d_1\right) = (204 \times 10^3 \text{ N})\left(\frac{300 \text{ mm}}{2} - 60 \text{ mm}\right) = 18.4 \times 10^6 \text{ N·mm}$$

$$F_{rs2}\left(\frac{h}{2} - d_2\right) = (-204 \times 10^3 \text{ N})\left(\frac{300 \text{ mm}}{2} - 240 \text{ mm}\right) = 18.4 \times 10^6 \text{ N·mm}$$

   b) Determine the eccentricity from the moment equilibrium equation

$$P_r \times e = C_r\left(\frac{h}{2} - \frac{\beta_1 c}{2}\right) + F_{rs1}\left(\frac{h}{2} - d_1\right) + F_{rs2}\left(\frac{h}{2} - d_2\right) \qquad \text{[8.16]}$$

   or

$$(537 \times 10^3 \text{ N})e_b = 43.6 \times 10^6 \text{ N·mm} + 18.4 \times 10^6 \text{ N·mm} + 18.4 \times 10^6 \text{ N·mm}$$

$$e_b = 149.7 \text{ mm} \cong 150 \text{ mm}$$

5. **Calculate the balanced moment resistance (M_{rb})**

Calculate the factored moment resistance from the equation

$$M_r = P_r \times e \qquad\qquad\qquad\qquad\qquad\qquad\qquad\qquad \text{[8.17]}$$

$$M_{rb} = M_r = (537 \times 10^3 \text{ N})(150 \text{ mm}) = 80.6 \times 10^6 \text{ N·mm} \cong 81 \text{ kN·m}$$

8.7   COLUMN INTERACTION DIAGRAMS

8.7.1   Key Features

The axial and flexural resistance of reinforced concrete column sections can be expressed by a *column interaction diagram* (also called a P-M interaction diagram). As discussed in Section 8.5, the behaviour of an axially loaded column depends on the magnitude of the load eccentricity. Figure 8.14 shows a typical column interaction diagram; the axial load resistance is shown on the vertical axis on the diagram, whereas the moment resistance is shown on the horizontal axis. The regions of a column interaction diagram are associated with different types of column behaviour, as follows:

A *concentrically loaded column* is represented by point A on the diagram. This point corresponds to the zero eccentricity condition ($e = 0$).

An *eccentrically loaded column with small eccentricity* is represented by the region A-B on the diagram. This region is characterized by concrete-controlled failure initiated by the concrete crushing (this mode of failure is also characteristic of concentrically loaded columns).

An *eccentrically loaded column with large eccentricity* is represented by region B-C on the diagram. This region is characterized by a steel-controlled failure caused by the steel yielding before the concrete crushes. Note that the horizontal axis on the diagram represents an infinitely large eccentricity condition ($e = \infty$), corresponding to pure bending in the column.

Note that the threshold between these two types of behaviour is represented by the *balanced condition*, characterized by the balanced eccentricity (e_b), shown as line O-B on the diagram.

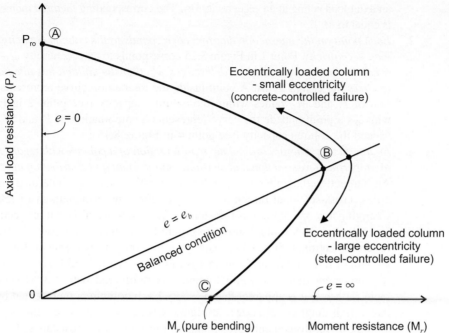

Figure 8.14   A column interaction diagram showing the types of column behaviour.

A column interaction diagram can also be used to predict the behaviour of an eccentrically loaded column subjected to an axial load (P_f) acting at the eccentricity (e), as follows:

1. If $e > e_b$ (the points along the portion B-C of the diagram), the column will experience steel-controlled failure.
2. If $e < e_b$ (the points along the portion A-B of the diagram), the column will experience concrete-controlled failure.

The key features of a column interaction diagram are explained below (see Figure 8.15):

1. *The points on the diagram represent the combinations of axial forces and bending moments corresponding to the resistance of a column cross-section.* Point 1 in Figure 8.15 corresponds to the *maximum factored axial load capacity* (P_{r1}) for a column loaded with

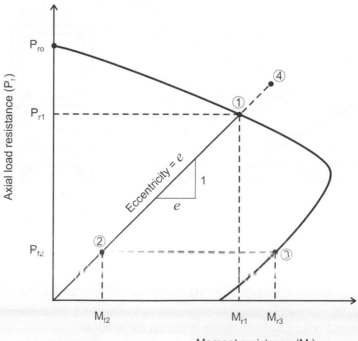

Figure 8.15   A typical column
interaction diagram.

an axial load acting at an eccentricity (e). The corresponding *factored moment resistance* is equal to M_{r1}.

2. *Each point on the interaction diagram corresponds to the column capacity at a specific load eccentricity.* Point 1 in Figure 8.15 corresponds to an eccentricity $e = M_{r1}/P_{r1}$.

3. *A column interaction diagram determines whether the column has adequate capacity to carry the design loads.* A point inside the interaction curve represents a combination of P_f and M_f values within the column capacity (see point 2 in Figure 8.15), whereas a point outside the curve represents a combination of P_f and M_f values that exceed the column capacity (see point 4 in Figure 8.15).

4. *An interaction diagram is a unique representation of a column with regard to its material properties, cross-sectional dimensions, and amount and distribution of reinforcement.* An interaction diagram corresponds to a specific column with defined material properties, cross-sectional dimensions, and amount and arrangement of reinforcement. Changing any one of these parameters would result in a different interaction diagram. Consider columns 1 and 2 shown in Figure 8.16. The columns are identical (in terms of cross-sectional dimensions and material properties), except for the different amount of reinforcement. Column 1 is reinforced with 4-20M bars (corresponding to reinforcement ratio ρ_1), while column 2 is reinforced with 4-30M bars (reinforcement ratio ρ_2); it is apparent that $\rho_1 < \rho_2$. Each column is represented by a different interaction diagram, as shown in Figure 8.16. Note that the column with the larger amount of reinforcement corresponds to the larger load resistance. For example, if columns 1 and 2 are subjected to the same axial load (P_f), the corresponding moment resistance values are M_{r1} and M_{r2}, respectively, where $M_{r1} < M_{r2}$.

A column interaction diagram must take into account the reduction in the axial load resistance due to the unanticipated or accidental moments in concentrically loaded columns, as discussed in Section 8.6.2. This is illustrated on the interaction diagram in Figure 8.17. The maximum axial load resistance of a concentrically loaded nonprestressed concrete column prescribed by A23.3 Cl.10.10.4 is equal to P_{rmax} (see Eqns 8.6 and 8.7). As a result, the horizontal line $P_r = P_{rmax}$ needs to be plotted on the diagram. This line intersects with the interaction diagram at point D in Figure 8.17. The points on the interaction diagram above the horizontal line (region A-D on the diagram) should be neglected in the design (shown with a hatched pattern in Figure 8.17). Note that point D defines the minimum accidental eccentricity (e_{min}) a column has to be designed for according to CSA A23.3 (this value used to be prescribed by the previous editions of CSA A23.3).

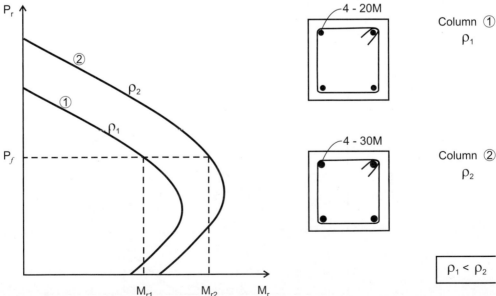

Figure 8.16 Interaction diagrams corresponding to columns 1 and 2 characterized by different reinforcement ratios (all other column properties are identical).

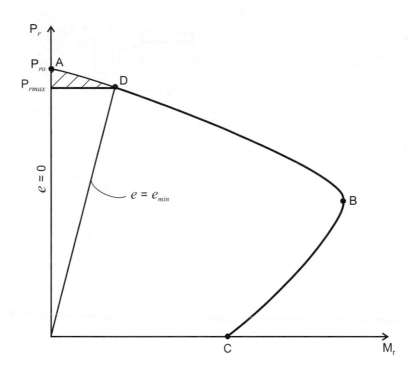

Figure 8.17 Minimum eccentricity in concentrically loaded columns.

The key features of a column interaction diagram are summarized below:

1. The points on the diagram represent the combinations of axial forces and bending moments corresponding to the resistance of a column cross-section.
2. Each point on the interaction diagram corresponds to the column capacity at a specific load eccentricity.
3. A column interaction diagram determines whether the column has adequate capacity to carry the design loads.
4. An interaction diagram is a unique representation of a column with regard to its material properties, cross-sectional dimensions, and amount and distribution of reinforcement.

Interaction diagrams can also be used to predict whether an eccentrically loaded column subjected to an axial load P_f acting at the eccentricity e is likely to fail in the concrete-controlled mode or the steel-controlled mode (see Figure 8.14), as follows:

1. If $e > e_b$ (the points along the portion B-C of the diagram), the column will experience steel-controlled failure.
2. If $e < e_b$ (the points along the portion A-B of the diagram), the column will experience concrete-controlled failure.

The threshold between these two types of behaviour is represented by the balanced condition, characterized by the *balanced eccentricity e_b*.

8.7.2 Development of a Column Interaction Diagram

A column interaction diagram can be developed by determining the P_r and M_r values corresponding to different strain distributions. These values correspond to points on the interaction diagram. The variation in strain distribution in different regions of a column interaction diagram is shown in Figure 8.18.

Points on the column interaction diagram can be determined using the procedure presented in Section 8.6.3, modified to account for several layers of reinforcement, as illustrated in Figure 8.19. A typical reinforcement layer is denoted by the subscript i,

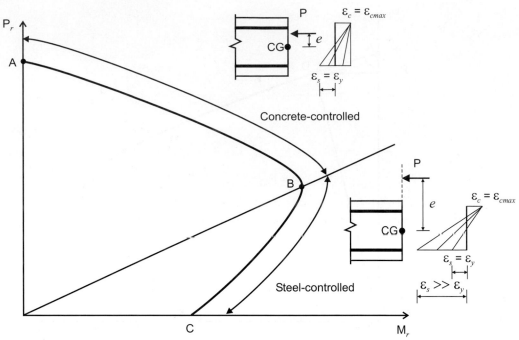

Figure 8.18 Strain distributions corresponding to the concrete-controlled and steel-controlled regions of the interaction diagram.

where i can vary from 1 to n (n denotes the total number of reinforcement layers in the column). The following notation is used in this procedure:

A_{si} = the area of the longitudinal reinforcement layer located at a distance (d_i) from the compression face of the column

ε_{si} = the strain in the reinforcement layer of area (A_{si}) (a positive sign is associated with compression strain)

f_{si} = the stress in the reinforcement layer of area (A_{si}); note that this stress can have either a positive or a negative value, depending on the sign of the strain (ε_{si}) (a positive sign corresponds to compression)

F_{rsi} = the force in the reinforcement layer of area (A_{si}).

The main steps in the general computational procedure used to determine a certain point on the column interaction diagram for a tied column of rectangular cross-section with more than two layers of reinforcement are as follows (see Figure 8.19):

1. Estimate the neutral axis depth (c).
2. Compute the strain in all reinforcement layers using the equation

$$\varepsilon_{si} = \varepsilon_{cmax}\left(1 - \frac{d_i}{c}\right), \quad i = 1 \text{ to } n \qquad \text{[8.19]}$$

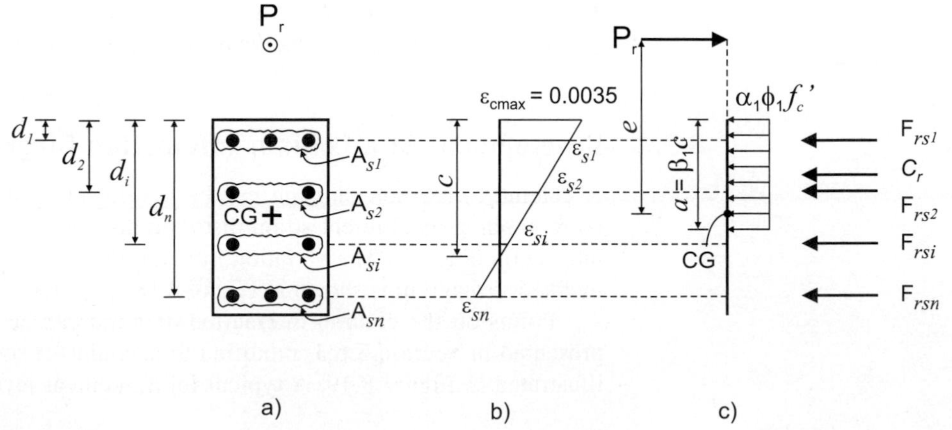

Figure 8.19 Reinforced concrete column with several reinforcement layers: a) cross-section; b) strain distribution; c) internal forces.

3. Compute the stresses in all reinforcement layers (1 to n).

$$\text{If } \varepsilon_{si} > \varepsilon_y, \text{ then } f_{si} = f_y \qquad \text{[8.20]}$$

$$\text{If } \varepsilon_{si} < \varepsilon_y, \text{ then } f_{si} = E_s \varepsilon_{si} < f_y \qquad \text{[8.21]}$$

4. Calculate the stress resultants in concrete and steel.

$$C_r = \alpha_1 \beta_1 \phi_c f_c' c\, b \qquad \text{[8.22]}$$

is the stress resultant in concrete, and

$$F_{rsi} = \phi_s f_{si} A_{si} \qquad \text{[8.23]}$$

is the force in the reinforcement layer of area A_{si}, $i = 1$ to n.

5. Calculate the force P_r based on the force equilibrium equation

$$P_r = C_r + \sum_{i=1}^{n} F_{rsi} \qquad \text{[8.24]}$$

6. Calculate the factored moment resistance (M_r) from the moment equilibrium equation around the point CG; assume that the centroid of the column cross-section is located at a distance $h/2$ from the column face:

$$M_r - M_{rc} + \sum_{i=1}^{n} M_{rsi} \qquad \text{[8.25]}$$

where

$$M_{rc} = C_r \left(\frac{h}{2} - \frac{\beta_1 c}{2} \right)$$

$$M_{rsi} = F_{rsi} \left(\frac{h}{2} - d_i \right)$$

The corresponding eccentricity (e) can be calculated as

$$e = \frac{M_r}{P_r}$$

The P_r and M_r values determined in Steps 1 to 6 above define a point on the column interaction diagram corresponding to the eccentricity e.

The following general steps are followed in the development of a column interaction diagram:

1. Calculate the axial load resistance (P_{ro}) corresponding to the zero eccentricity condition ($e = 0$).
2. Calculate the neutral axis depth corresponding to the balanced condition (c_b).
3. Determine the points on the column interaction diagram.

At least five points should be determined on the column interaction diagram following the above-outlined computational procedure (Steps 1 to 6):

- First, find the point corresponding to the balanced condition ($e = e_b$)
- Next, determine two points between the zero eccentricity condition and the balanced condition (corresponding to the condition $c > c_b$).
- Finally, determine two points corresponding to a large eccentricity and a small neutral axis depth ($c < c_b$).

4. Plot the interaction diagram.

    The points determined in Step 3 can be connected using straight lines. Note that the actual interaction diagram is obtained by performing the analysis considering several different strain distributions in the column cross-section.

| A23.3 Cl.10.10.4 |

5. Decrease the axial resistance of the column section to account for the accidental moments.

    The axial resistance of the column section (P_{ro}) calculated in Step 1 has to be decreased to the value P_{rmax} to account for the accidental moments, as prescribed by A23.3 Cl.10.10.4 (see Section 8.7.1 and Figure 8.17 for details).

8.7.3   Use of Column Interaction Diagrams in Design Applications

Interaction diagrams are widely used in design of reinforced concrete columns. To enable the use of interaction diagrams for columns of different cross-sectional dimensions and material properties, these diagrams are often shown in a nondimensional form for a column of a given shape (rectangular or circular), material properties, and reinforcement distribution. The values for $M_r/(A_g h f_c')$ are shown on the horizontal axis, whereas the $P_r/(A_g f_c')$ values are shown on the vertical axis. The reinforcement ratio values are varied so that several interaction diagrams are shown on the same figure, with each diagram representing a column with a certain reinforcement ratio. Such interaction diagrams are reproduced in several publications (including CAC, 2005) and can serve as an aid in performing the analysis or design of reinforced concrete columns. Also, the PCACOL software package developed by the Portland Cement Association (marketed in Canada by the Cement Association of Canada) can produce an interaction diagram for a column of an arbitrary shape and amount and arrangement of reinforcement according to the CSA A23.3 requirements.

The approach taken in this book is that the designer should understand the concepts behind the development of column interaction diagrams and then be able to develop a column interaction diagram. Developing an interaction diagram by performing hand calculations can be tedious, but it is also a very useful learning experience. In design practice, computer spreadsheets are often used for this purpose instead of hand calculations. The spreadsheets largely facilitate and expedite the column design process, since several columns of different dimensions and/or reinforcement are utilized in the design of a single building. Alternatively, the reader can use column interaction diagram software BPA COLUMN developed by Andreas Felber of EDI. The software is available for time-limited download on the companion web site. The software can generate P-M interaction diagrams for practical column reinforcement configurations corresponding to reinforcement ratios ranging from 1 to 4%. Column section can have rectangular, or hollow shape. Slenderness effects can be also considered by the software, by applying an appropriate moment magnifier to the flexural resistance for a short column. The software generates interaction diagrams for short and the corresponding slender columns on the same diagram.

The development of column interaction diagrams will be first illustrated in an example using hand calculations in tabular form, and then by using a computer spreadsheet.

KEY CONCEPTS

The following general steps are followed in the development of a column interaction diagram:

1. Calculate the axial load resistance (P_{ro}) corresponding to the zero eccentricity condition ($e = 0$).
2. Calculate the neutral axis depth corresponding to the balanced condition (c_b).
3. Determine the points on the column interaction diagram following the six-step procedure outlined in this section.
4. Plot the interaction diagram.
5. Decrease the axial resistance of the column section to account for accidental moments.

Example 8.3

A tied short column of square cross-section is shown in the figure below. The column is reinforced with 30M longitudinal bars and 10M ties. Use a 40 mm clear cover to the ties. The material properties are summarized below.

Develop the column interaction diagram.

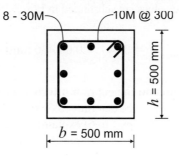

8 - 30M — 10M @ 300
$h = 500$ mm
$b = 500$ mm

Given: $f_c' = 25$ MPa
$f_y = 400$ MPa
$E_s = 200\,000$ MPa
$\phi_c = 0.65$
$\phi_s = 0.85$
$b = 500$ mm
$h = 500$ mm

SOLUTION: 1. **Calculate the axial load resistance corresponding to the zero eccentricity condition**
This step was performed in Example 8.1. The factored axial resistance P_{ro} is
$P_{ro} = 5081$ kN.
This value defines point 1 on the interaction diagram (note that $M_r = 0$ for this condition).

2. **Calculate the neutral axis depth corresponding to the balanced condition (c_b)**
The neutral axis depth (c_b) can be determined as

$$c_b = \frac{\varepsilon_{cmax}}{\varepsilon_y + \varepsilon_{cmax}} d \qquad [8.18]$$

where

$$\varepsilon_c = \varepsilon_{cmax} = 0.0035$$

$$\varepsilon_y = \frac{f_y}{E_s} = \frac{400 \text{ MPa}}{200\,000 \text{ MPa}} = 0.002$$

Calculate the column effective depth (d) (see the sketch below):

30M bars: $d_b \cong 30$ mm

10M ties: $d_{tie} \cong 10$ mm

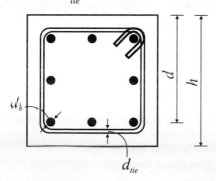

d_b
d
h
d_{tie}

$$d = h - \text{cover} - d_{tie} - \frac{d_b}{2}$$

$$= 500 \text{ mm} - 40 \text{ mm} - 10 \text{ mm} - \frac{30 \text{ mm}}{2} = 435 \text{ mm} \cong 430 \text{ mm}$$

After the above values are substituted in Eqn 8.19, the c_b value can be calculated as

$$c_b = \frac{0.0035}{0.002 + 0.0035} d = 0.64d = 0.64(430 \text{ mm}) \cong 275 \text{ mm}$$

3. **Determine the points on the column interaction diagram**
    At least five points need to be determined on the column interaction diagram using the generalized procedure (Steps 1 to 6) explained earlier in this section. These steps will be denoted as 3.1 to 3.6 in this example.

    3.1. Estimate the neutral axis depth (c).
    The neutral axis depth corresponding to the balanced condition was determined in the previous step; that is, $c_b = 275$ mm.

      The points 2 and 3 on the diagram need to be determined between the zero eccentricity condition and the balanced condition (corresponding to $c > c_b$). The choice of appropriate c values is a matter of the designer's judgment. The values should be in the range from $c = c_b = 275$ mm to $c = d = 430$ mm. Consider the following values:

    $c_2 = 400$ mm

    $c_3 = 325$ mm

      Point 4 on the interaction diagram corresponds to the balanced condition ($c_4 = c_b = 275$ mm).
      Finally, determine two more points (5 and 6) on the diagram corresponding to large eccentricity and small neutral axis depth; that is $c < c_b$. The values should be in the range from $c = 0$ to $c = c_b = 275$ mm. Consider the following values:

    $c_5 = 200$ mm

    $c_6 = 100$ mm

      The strain conditions and the corresponding neutral axis values corresponding to points 1 to 6 on the interaction diagram are illustrated in the sketch below.

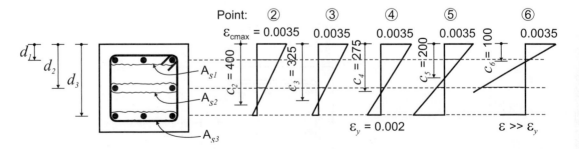

    3.2. Compute the strain in each reinforcement layer.
    First determine the reinforcement properties that will be used in this example. There are three reinforcement layers. Each layer is defined by the area A_{si} and the distance d_i from the compression face of the column to the centroid of the bar layer ($i = 1$ to 3). Note that 30M bars are used in this example; the area of

each bar is 700 mm^2 (see Table A.1). These properties are summarized in the table below.

Reinforcement layer	Specified reinforcing bars	A_{si} [mm^2]	d_i [mm]
1	3-30M	2100	70
2	2-30M	1400	250
3	3-30M	2100	430

Note that

$$d_1 = h - d = 500 \text{ mm} - 430 \text{ mm} = 70 \text{ mm}$$

$$d_2 = \frac{h}{2} = \frac{500 \text{ mm}}{2} = 250 \text{ mm}$$

$$d_3 = d = 430 \text{ mm}$$

Subsequently, determine the strain in each reinforcement layer using the equation

$$\varepsilon_{si} = \varepsilon_{cmax}\left(1 - \frac{d_i}{c}\right) \quad (i = 1 \text{ to } 3) \qquad [8.19]$$

Note that a negative sign corresponds to a tension strain.

Point	2	3	4	5	6
c [mm]	400	325	275	200	100
ε_{s1}	0.0029 (Y)	0.0027 (Y)	0.0026 (Y)	0.0023 (Y)	0.0011
ε_{s2}	0.0013	0.0008	0.0003	−0.0009	< (−0.002) (Y)
ε_{s3}	−0.0003	−0.0011	−0.002 (Y)	< (−0.002) (Y)	< (−0.002) (Y)

If $|\varepsilon_{si}| \geq \varepsilon_y = 0.002$, when the reinforcement in layer i yields (denoted with the letter Y in brackets).

3.3. *Compute the stresses* (f_{si}) *in each reinforcement layer* ($i = 1$ to 3).
   The following criteria apply to the stress calculations:

   If $\varepsilon_{si} \geq \varepsilon_y$, then $f_{si} = f_y$ [8.20]

   If $\varepsilon_{si} < \varepsilon_y$, then $f_{si} = E_s\varepsilon_{si} < f_y$ [8.21]

Point	2	3	4	5	6
c [mm]	400	325	275	200	100
f_{s1} [MPa]	400	400	400	400	220
f_{s2} [MPa]	260	160	60	−180	−400
f_{s3} [MPa]	−60	−220	−400	−400	−400

3.4. *Calculate the stress resultants in the concrete and steel.*
   Calculate the concrete stress resultant using the equation

$$C_r = \alpha_1\beta_1\phi_c f_c' c\, b \qquad [8.22]$$

where

$$\alpha_1 \cong 0.8, \beta_1 \cong 0.9 \qquad [3.7]$$

It follows that

$$C_r = \alpha_1\beta_1\phi_c f_c' c\, b = (0.8)(0.9)(0.65)(25 \text{ MPa})(c)(500 \text{ mm} \times 10^{-3}) \text{ [N]}$$
$$= 5.85c \text{ [kN]}$$

Calculate the forces in the reinforcement:

$$F_{rsi} = \phi_s f_{si} A_{si}$$ [8.23]

is the force in the reinforcement layer of area A_{si} ($i = 1$ to n), where

$$F_{rs1} = \phi_s f_{s1} A_{s1} = (0.85) f_{s1}(2100) = 1785 f_{s1}[N] = 1.785 f_{s1}[kN]$$

$$F_{rs2} = \phi_s f_{s2} A_{s2} = (0.85) f_{s2}(1400) = 1190 f_{s2}[N] = 1.19 f_{s2}[kN]$$

$$F_{rs3} = \phi_s f_{s3} A_{s3} = (0.85) f_{s1}(2100) = 1785 f_{s3}[N] = 1.785 f_{s3}[kN]$$

Point	2	3	4	5	6
c [mm]	400	325	275	200	100
$C_r = 5.85c$ [kN]	2340	1901.3	1609	1170	585
$F_{rs1} = 1.785 f_{s1}$ [kN]	714	714	714	714	392.7
$F_{rs2} = 1.19 f_{s2}$ [kN]	309.4	190.4	71.4	−214.2	−476
$F_{rs3} = 1.785 f_{s3}$ [kN]	−107.1	−392.7	−714	−714	−714

3.5. *Calculate the force P_r from the force equilibrium equation*

$$P_r = C_r + \sum_{i=1}^{n} F_{rsi}$$ [8.24]

Point	2	3	4	5	6
c [mm]	400	325	275	200	100
$P_r = C_r + \sum_{i=1}^{n} F_{rsi}$ [kN]	3256	2413	1680	956	−212

3.6. *Calculate the factored moment resistance (M_r) from the moment equilibrium equation around the point CG; assume that the centroid of the column cross-section is located at a distance $h/2$ from the column face:*

$$M_r = M_{rc} + \sum_{i=1}^{n} M_{rsi}$$ [8.25]

where

$$M_{rc} = C_r\left(\frac{h}{2} - \frac{\beta_1 c}{2}\right) = C_r\left(\frac{500}{2} - \frac{0.91c}{2}\right) \times 10^{-3}$$
$$= C_r(0.25 - 0.45 \times 10^{-3}c) \quad [kN \cdot m]$$

$$M_{rsi} = F_{rsi}\left(\frac{h}{2} - d_i\right) (i = 1 \text{ to } 3)$$

$$M_{rs1} = F_{rs1}\left(\frac{h}{2} - d_1\right) = F_{rs1}\left(\frac{500}{2} - 70\right) = 180 F_{rs1}[N] = 0.18 F_{rs1} \ [kN \cdot m]$$

$$M_{rs2} = F_{rs2}\left(\frac{h}{2} - d_2\right) = F_{rs2}\left(\frac{500}{2} - 250\right) = 0 \ [kN \cdot m]$$

$$M_{rs3} = F_{rs3}\left(\frac{h}{2} - d_3\right) = F_{rs3}\left(\frac{500}{2} - 430\right) = -180 F_{rs3}[N] = -0.18 F_{rs3} \ [kN \cdot m]$$

Point	2	3	4	5	6
c [mm]	400	325	275	200	100
$M_{rc} = C_r(0.25 - 0.45 \times 10^{-3}c)$ [kN·m]	163.8	197.3	203.1	187.2	120
$M_{rs1} = 0.18F_{rs1}$ [kN·m]	128.5	128.5	128.5	128.5	72.6
$M_{rs3} = (-0.18)F_{rs3}$ [kN·m]	19.3	70.7	128.5	128.5	128.5
$M_r = M_{rc} + \sum\limits_1^n M_{rsi}$ [kN·m]	312.0	397.0	460.0	444.0	321.0

4.   Plot the column interaction diagram

The column interaction diagram obtained as a result of the calculations performed in this example can be plotted by connecting points 1 to 6, as illustrated in the figure below.

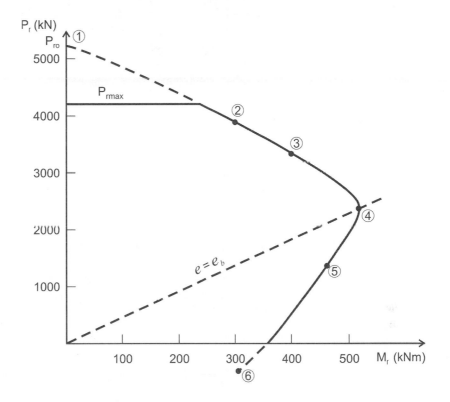

The load combinations corresponding to these points are summarized in the table below.

Point	1	2	3	4	5	6
P_r [kN]	5081	3256	2413	1680	956	−212
M_r [kN·m]	0	312	397	460	444	321

Cl.10.4.4

5.   Decrease the axial resistance of the column section

The axial resistance of the column section (P_{ro}) calculated in Step 1 has to be decreased to the value P_{rmax} to account for accidental moments. For tied columns,

$$P_{rmax} = 0.80P_{ro} = 0.80 \times 5081 \text{ kN} = 4065 \text{ kN} \qquad\qquad \textbf{[8.6]}$$

Plot the horizontal line $P_r = P_{rmax} = 4065$ kN on the interaction diagram and connect it with the portion of the diagram below this line. The points on the interaction diagram above the horizontal line should be neglected (this portion is shown on the diagram with a dashed line).

Example 8.4

Consider the short column with cross-sectional dimensions and reinforcement in the figure below. The column is subjected to an axial dead load of 500 kN and live load of 500 kN; both loads act at an eccentricity of 150 mm.

a)   *Generate an interaction diagram for the column using the BPA COLUMN software.*

b)   *Check whether the column is adequate to carry the design loads.*

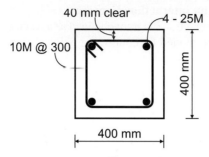

Given: $f_c' = 25$ MPa

$f_y = 400$ MPa

$\phi_c = 0.65$

$\phi_s = 0.85$

SOLUTION: **1.   Generate an interaction diagram for the column using the computer spreadsheet**

The printed output generated by the spreadsheet for this specific column is shown below. The reader should be able to generate the identical interaction diagram for the same input information.

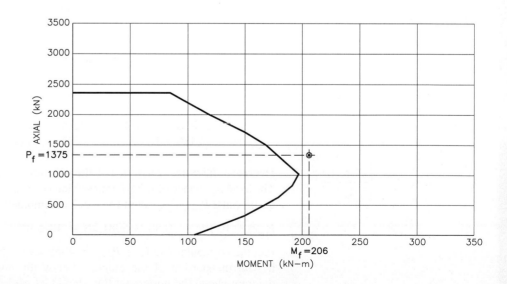

2. **Check whether the column is adequate to carry the design loads**
   a) Calculate the factored loads on the column.
      The factored loads on the column can be determined as

$$P_f = 1.25P_{DL} + 1.5P_{LL} = 1.25(500 \text{ kN}) + 1.5(500 \text{ kN}) = 1375 \text{ kN}$$

$$M_f = P_f \times e = 1375 \text{ kN} \times 0.15 \text{ m} = 206.25 \text{ kN} \cdot \text{m} \cong 206 \text{ kN} \cdot \text{m}$$

   b) Determine the moment and axial load resistances.
      Based on the column interaction diagram, for the axial load

$$P_f = P_r = 1375 \text{ kN}$$

   it follows that

$$M_r \cong 160 \text{ kN} \cdot \text{m}$$

   c) Check whether the column resistance is adequate for the given loads.
      Because

$$M_r = 160 \text{ kN} \cdot \text{m} < M_f = 206 \text{ kN} \cdot \text{m}$$

   a column reinforced with 4-25M vertical rebars does not have sufficient capacity to support the applied load.

Example 8.5

Consider the same column discussed in Example 8.4.
First, use the computer spreadsheet to generate column interaction diagrams for various practical configurations of vertical reinforcement (including 4, 8, and 12 bars); the reinforcement ratio should be in the range $0.01 \leq \rho \leq 0.04$.

Next, identify the reinforcement arrangement that is most adequate for the given design loads. The column is subjected to the same loads specified in Example 8.4.

Given: $f_c' = 25 \text{ MPa}$
$f_y = 400 \text{ MPa}$
$\phi_c = 0.65$
$\phi_s = 0.85$

SOLUTION: 1. **Determine the minimum and maximum amount of reinforcement**
The vertical reinforcement ratio is required to be within the range

$$0.01 \leq \rho \leq 0.04$$

For $\rho = 0.01$

$$A_{smin} = 0.01 \times (400 \text{ mm})^2 = 1600 \text{ mm}^2$$

For $\rho = 0.04$

$$A_{smax} = 0.04 \times (400 \text{ mm})^2 = 6400 \text{ mm}^2$$

2. **Identify the practical reinforcement configurations and find the corresponding ρ values**

    a) The practical configurations for this column are (see Table 8.1)

        i) 8-20M

       ii) 12-20M

      iii) 8-25M

      iv) 12-25M

    b) Calculate the reinforcement area and the corresponding ρ value for each configuration.

      The column cross-sectional dimension is

$$b = 400 \text{ mm}$$

     i) For 8-20M bars,

$$A_s = 8 \times 300 \text{ mm}^2 = 2400 \text{ mm}^2$$

$$\rho = \frac{A_s}{b^2} = \frac{2400 \text{ mm}^2}{(400 \text{ mm})^2} = 0.015$$

     ii) For 12-20M bars,

$$A_s = 12 \times 300 \text{ mm}^2 = 3600 \text{ mm}^2$$

$$\rho = \frac{3600 \text{ mm}^2}{(400 \text{ mm})^2} = 0.0225$$

    iii) For 8-25M bars,

$$A_s = 8 \times 500 \text{ mm}^2 = 4000 \text{ mm}^2$$

$$\rho = \frac{4000 \text{ mm}^2}{(400 \text{ mm})^2} = 0.025$$

    iv) For 12-25M bars,

$$A_s = 12 \times 500 \text{ mm}^2 = 6000 \text{ mm}^2$$

$$\rho = \frac{6000 \text{ mm}^2}{(400 \text{ mm})^2} = 0.0375$$

3. **Generate interaction diagrams for the selected reinforcement configurations**

The interaction diagrams for the four vertical reinforcement configurations are shown below. Note that these four reinforcement configurations represent all feasible configurations for a 400 mm by 400 mm square column.

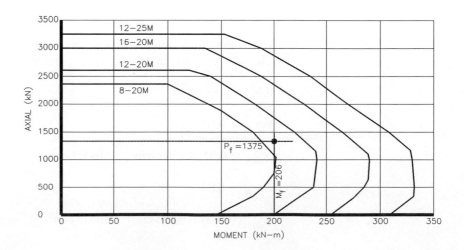

4. **Select the most suitable reinforcement configuration**
The point on the graph representing $P_f = 1375$ kN and $M_f \cong 206$ kN·m is located inside the interaction diagram for 12-20M bars. Therefore, 12-20M bars are suitable for this design; however, 12-25M bars would likely be chosen instead for a real design application. In this case, a minimal additional cost would be associated with the increase in the amount of reinforcement from 12-20M to 12-25M, but the corresponding increase in load-carrying capacity is as much as 30%.

In conclusion, it is recommended to design reinforced concrete columns with some reserve strength.

8.8 CSA A23.3 COLUMN DESIGN REQUIREMENTS

8.8.1 Reinforcement Requirements

A23.3 Cl.10.9

Longitudinal reinforcement ratio As in the case of flexural reinforcement in beams and slabs, the reinforcement ratio is used as an indicator of the relative amount of longitudinal reinforcement in the columns. The reinforcement ratio for longitudinal column reinforcement (ρ_t) can be defined as

$$\rho_t = \frac{A_{st}}{A_g}$$
[8.26]

where

ρ_t = the longitudinal reinforcement ratio
A_g = the gross area of a column section; for a rectangular column with width b and depth h, $A_g = b\,h$;
A_{st} = the *total area* of longitudinal reinforcement within a column cross-section.

Note a couple of differences between columns and beams with regard to the reinforcement ratio:

1) The reinforcement ratio in columns is determined based on the total longitudinal reinforcement area, whereas the reinforcement ratio for beams includes only the area of tension steel.
2) The column reinforcement ratio is determined based on the gross cross-sectional area (using overall dimensions), whereas the reinforcement ratio for beams uses the effective depth (d) instead of the overall depth (h).

In general, CSA A23.3 prescribes that the column reinforcement ratio (ρ_t) be in the range from 1% to 8%, as discussed below.

A23.3 Cl.10.9.1

Minimum longitudinal reinforcement The *lower limit* of $0.01A_g$ (corresponding to a ρ_t value of 1%) ensures that

- there is some flexural resistance even in members where calculations indicate no applied moment
- increased stresses in the steel reinforcement caused by creep and shrinkage do not lead to the eventual yielding of the reinforcement under service loads

According to A23.3 Cl.10.10.5, columns with a ρ_t value less than 1% but larger than 0.5% may be used, provided the factored axial and flexural resistance are multiplied by the ratio [illegible].

A23.3 Cl.10.9.2

Maximum longitudinal reinforcement The *upper limit* of $0.08A_g$ (corresponding to a ρ_t value of 8%) represents a practical maximum for longitudinal reinforcement based on economy and requirements for proper concrete placing. Since this limit also applies to regions containing lap splices, columns are usually designed with no more than 4% longitudinal reinforcement. In fact, the use of more than 4% reinforcement in regions outside of lap splices may result in difficulties in placing and compacting concrete. Special rules governing reinforcement limits at splice locations are provided in A23.3 Cl.12.17. (For more details on reinforcement splices, refer to Section 9.9.) The minimum spacing between the longitudinal bars may limit the practical amount of reinforcement that can be used.

| A23.3 Cl.10.9.3 |

Minimum number of longitudinal bars   The minimum number of longitudinal reinforcing bars shall be

- four for columns with rectangular or circular ties
- three for columns with triangular ties
- six for spiral columns

| A23.1 Cl.6.6.5.2 |

Minimum bar spacing   *Minimum bar spacing requirements* for columns are specified by the CSA A23.1 Standard, as discussed in Section 5.3.2. In general, the clear distance between parallel bars should not be less than the greatest of

- 1.4 times the bar diameter,
- 1.4 times the maximum size of the coarse aggregate, and
- 30 mm.

| A23.3 Cl.7.4.1.3 |

Maximum bar spacing   The clear distance between adjacent longitudinal bars should not exceed 500 mm.

| A23.3 Cl.7.6.5 |

Ties   Longitudinal reinforcement in the columns must be enclosed with *transverse reinforcement* consisting either of closed ties or spirals.

| A23.3 Cl.7.6.5.1 |

Tie size   Ties must have a diameter of at least 30% of that of the largest longitudinal bar for bar sizes up to 30M; 10M tie size or larger should be used for longitudinal bars over 30M and bundled bars.

| A23.3 Cl.7.6.5.2 |

Tie spacing   The tie spacing (s) must not exceed the least of

- 16 times the diameter of the smallest longitudinal bar,
- 48 tie diameters,
- the least lateral dimension of the compression member, and
- 300 mm when bundled bars are used.

Note that tie spacing is one of the critical parameters in column design. As discussed in Section 8.3.2, the role of ties is to provide lateral support to the longitudinal reinforcement. If the ties are spaced too far apart, the longitudinal bars between them will buckle. Figure 8.20 shows a reinforced concrete column with different tie spacing at the ends and in the middle portion. The column was severely damaged in the 1974 Peru earthquake. The failure occurred where the tie spacing was too large. It is generally recommended to vary tie spacing in columns. The use of closely spaced ties at the column ends is prescribed in regions of high seismic risk, and wider tie spacing is permitted in the middle portion. However, the tie spacing in the middle portion of the column shown on the photo is excessive.

Special care needs to be taken to ensure that the tie spacing specified on the design drawings is indeed implemented in the field. Figure 8.21 shows the failure of a column in the 1999 Chi Chi (Taiwan) earthquake due to widely spaced ties. Inadequate tie spacing due to improper design or construction practices could lead to the complete disintegration of concrete in reinforced concrete columns, as shown in this photo.

Ties are more labour intensive to install than longitudinal bars, and they often account for a large percentage of the total quantity of reinforcement in a column. In many cases, it

Figure 8.20 The effect of tie spacing on the earthquake performance of reinforced concrete columns. Note the closely spaced ties at the base and the widely spaced ties in the middle of the column damaged in the 1974 Peru earthquake.

(Stratta, 1987; reprinted by permission of Pearson Education, Inc.)

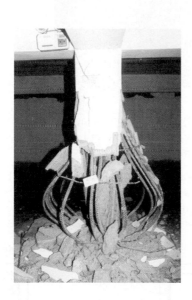

Figure 8.21 Complete disintegration of a column due to the absence of ties in the 1999 Chi Chi (Taiwan) earthquake.

(Earthquake Engineering Research Institute)

is more economical to use larger longitudinal bar sizes than necessary (observing the appropriate limits!) in order to increase the tie spacing.

A23.3 Cl.7.1.2

Tie anchorage Ties should be anchored by hooks with a bend of at least 135° unless the concrete cover surrounding the hook is restrained against spalling, in which case a 90° hook is permitted. Figure 8.22 illustrates 90° and 135° tie hooks. (Refer to Chapter 9 for more details related to the anchorage of reinforcement in concrete structures.) In general, it is recommended to use 135° tie hooks in all column applications, although A23.3 Cl.7.1.2 permits the use of 90° hooks for columns with $f_c' \leq 50$ MPa. The use of 135° tie hooks in the seismic zones of Canada is essential for the satisfactory performance of buildings in damaging earthquakes. Numerous experimental studies and past earthquakes have revealed the poor performance of columns with 90° hooks, as illustrated in Figure 8.23. For this reason, CSA A23.3 prescribes the use of 135° tie hooks for the seismic design of reinforced concrete columns in Canada.

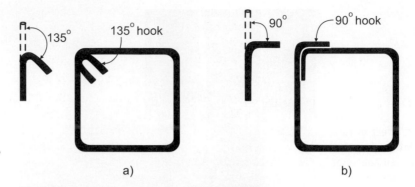

Figure 8.22   Tie hooks: a) 135° hooks; b) 90° hooks.

a)   b)

Figure 8.23   Poor performance of a column with 90° hooks in the 1999 Chi Chi (Taiwan) earthquake: Note that the hooks "opened up" due to the high lateral pressure.

(Earthquake Engineering Research Institute)

A23.3 Cl.7.6.5.5

Tie arrangement   Typical tie arrangements in reinforced concrete columns are illustrated in Figure 8.24. CSA A23.3 prescribes the following requirements for tie arrangement in concrete columns:

- The tie must enclose and laterally support every corner bar and every alternate longitudinal bar by the corner of a tie having an included angle of not more than 135°. (Note that every other bar is supported by 135° hooks in columns, as shown in Figure 8.24a and 8.24b.)

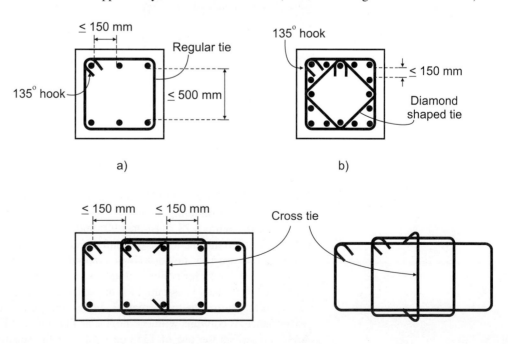

Figure 8.24   Tie arrangements: a) regular ties; b) diamond-shaped ties; c) ties for a large rectangular section.

a)   b)

c)

- A bar is adequately supported against lateral movement if it is located at a corner or if the clear distance to a laterally supported bar is less than 150 mm.
- The "free ends" of the ties must be bent into the confined core of the concrete, ensuring that the tie remains in place to support the longitudinal bars if the concrete cover spalls off.

Note that, besides the *regular ties* shown in Figure 8.24a, in some cases *diamond-shaped ties* are used to provide lateral support to longitudinal reinforcement in the columns (see Figure 8.24b). The centre of the column is kept open for easy concrete placement and vibration during construction. In some cases, cross-ties are used, as in the case of the rectangular column in Figure 8.24c. If used, cross-ties need to be anchored with a 135° hook at one end and at least a 90° hook at the other.

Spiral reinforcement   Spiral reinforcement is characterized by bar size d_b, outside diameter of the concrete core D_c, and pitch p, as illustrated in Figure 8.25.

A23.3 Cl.10.9.4   **Amount of spiral reinforcement**   The spiral reinforcement ratio (ρ_s) is equal to the ratio of the volume of spiral reinforcement to the volume of the concrete core. For the spiral column in Figure 8.25, the spiral reinforcement ratio for one spiral turn can be determined as

$$\rho_s = \frac{A_{sp} l_{sp}}{A_c p} \tag{8.27}$$

where

d_b = the spiral bar diameter

$A_{sp} = \dfrac{\pi d_b^2}{4}$ the volume of spiral reinforcement per unit length of column

$l_{sp} = \pi D_c$ the length of one spiral turn

D_c = the outside diameter of the concrete core

$A_c = \dfrac{\pi D_c^2}{4}$ the area of concrete core measured out to the spirals

p = the pitch (distance between successive spiral turns).

The spiral contribution to the compressive strength of a reinforced concrete column is approximately equal to $2f_{sy} A_{sp}$, where f_{sy} is the yield strength of spiral reinforcement and A_{sp} is the volume of spiral reinforcement per unit length of column. This additional strength develops only once the exterior column shell (concrete cover) spalls off. To ensure the effectiveness of spiral reinforcement, the spiral contribution to the column compressive strength should be at least equal to the lost concrete strength due to the spalling off of the exterior shell, as follows:

$$\alpha_1 f_c'(A_g - A_c) = 2f_{sy} A_{sp} \tag{8.28}$$

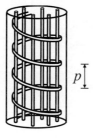

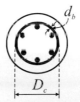

Figure 8.25   Spiral column.

If ρ_s is substituted from Eqn 8.27 into Eqn 8.28 and $\alpha_1 = 0.85$ is used, then the required spiral reinforcement ratio can be obtained from the equation

$$\rho_s = 0.425\left(\frac{A_g}{A_c} - 1\right)\frac{f_c'}{f_y}$$

A23.3 Cl.10.9.4 prescribes a minimum spiral reinforcement ratio of

A23.3 Eq. 10.7 $\qquad \rho_s = 0.45\left(\frac{A_g}{A_c} - 1\right)\frac{f_c'}{f_y}$ [8.29]

The above equation ensures that the spiral reinforcement is a bit stronger than the concrete shell. Adequately reinforced spiral columns demonstrate ductile behaviour and superior axial load resistance to tied concrete columns, as discussed in Section 8.6.2. The amount of spiral reinforcement in excess of the CSA A23.3 requirements (Eqn 8.29) causes a further increase in the compressive strength and larger deformations and cracking in the column. Therefore, it is not considered feasible to use a very large amount of spiral reinforcement in practical design applications.

A23.3 Cl.7.6.4.2 to 7.6.4.4

Spiral size and spacing

- The diameter of the reinforcing steel used for spirals (d_b) must not be less than 6 mm (Cl.7.6.4.2).
- The pitch (p) or distance between spiral turns must not exceed one sixth of the core diameter (Cl.7.6.4.3).
- The clear spacing between successive turns of a spiral must be between 25 mm and 75 mm (Cl.7.6.4.4).

A23.1 Cl.6.6.6.2.3

Concrete cover   Concrete cover must be provided to protect column reinforcement against corrosion and fire and to ensure bond between the reinforcement and the concrete.

KEY REQUIREMENTS

CSA A23.3 prescribes the following reinforcement requirements for reinforced concrete columns:

Longitudinal reinforcement

- minimum amount of reinforcement (Cl.10.9.1)
- maximum amount of reinforcement (Cl.10.9.2)
- minimum number of bars (Cl.10.9.3)
- minimum bar spacing (A23.1 Cl.6.6.5.2)
- maximum bar spacing (Cl.7.4.1.3)

Ties

- size (Cl.7.6.5.1)
- spacing (Cl.7.6.5.2)
- anchorage (Cl.7.1.2)
- arrangement (Cl.7.6.5.5)

Spirals

- amount (Cl.10.9.4)
- size and spacing (Cl.7.6.4.2 to 7.6.4.4)

Concrete cover (A23.1 Cl.6.6.6.2.3)

(For more details on concrete cover, see Section 5.3.1.) According to A23.1 Cl.6.6.6.2.3, minimum cover *to ties or spirals* should be provided as follows (see also Table A.2):

- 30 mm for interior, unexposed columns (Class N);
- 40 mm for columns exposed to freezing and thawing, or manure and/or sileage gasses (Classes F-1, F-2, S-1, S-2);
- 60 mm for columns exposed to chlorides (Classes C-XL, C-1, C-3, A-1, A-2, A-3).

8.9 PRACTICAL DESIGN GUIDELINES

The design of reinforced concrete columns for given external loads involves proportioning the steel and concrete areas and selecting properly sized and spaced ties or spirals. In general, there are many valid choices for the size of column that meets the design requirements. Inexperienced designers usually require some guidance in their choice of column dimensions, reinforcement sizes, and configurations. Broad guidelines related to these items are provided in this section.

Column cross-sectional dimensions   CSA A23.3 does not specify minimum dimensions for reinforced concrete columns. In general, the smallest dimension of a cast-in-place column should not be less than 200 mm, and preferably not less than 250 mm. The diameter of spiral columns should not be less than 300 mm. In choosing the overall cross-sectional dimensions, the designer should use multiples of 50 mm; that is, the dimensions should be rounded to the next higher 50 mm figure.

Based on the fire-resistance requirements, the smallest column dimension should be at least 220 mm for a 1-hour fire rating and 300 mm for a 2-hour or 3-hour fire rating.

In multistorey buildings, it is considered economical to keep the same cross-sectional dimensions to be able to reuse the forms from floor to floor. Column capacity requirements vary between floors, so the designer usually varies the amount of longitudinal reinforcement while preserving the same cross-sectional dimensions. It is common practice in the design of concrete high-rises to vary the column longitudinal reinforcement every three floors.

Size of transverse reinforcement (ties and spirals)   Typically, 10M bars are used for ties and spirals in reinforced concrete columns.

Size and configuration of longitudinal reinforcement   A minimum 20M longitudinal bar size should be used for most applications. Smaller bar sizes can buckle easily and require more accurately placed ties to ensure the effectiveness of the column design. Unless the least column dimension is 200 mm or less, longitudinal bar sizes of at least 20M should be used.

For columns of square (or almost square) cross-section, there are usually predetermined longitudinal bar configurations (in multiples of four bars), as illustrated in Figure 8.26.

Reinforcement areas corresponding to the bar configurations in Figure 8.26 are summarized in Table 8.1. Note that the framed boxes indicate configurations unlikely to be used. For example, the configurations in the bottom left corner require more ties. In general, it is more economical to use a smaller number of larger bars corresponding to the same area of steel. Also, the configurations in the top right corner of the table are unlikely to be used

4 bars   8 bars   12 bars   16 bars

Figure 8.26   Typical reinforcement configurations for tied square columns.

Table 8.1 Reinforcement areas for typical bar configurations (mm²)

	20M	25M	30M	35M
4 bars	1200	2 000	2 800	4 000
8 bars	2400	4 000	5 600	8 000
12 bars	3600	6 000	8 400	12 000
16 bars	4800	8 000	11 200	16 000
20 bars	7200	10 000	14 000	20 000
24 bars	9600	12 000	16 800	24 000

because larger bars require longer development lengths (to be discussed in Section 9.3). For example, if the required reinforcement area is $A_s = 4000$ mm², it is recommended to use 8-25M bars instead of 4-35M bars.

KEY CONCEPTS

The practical design guidelines related to the selection of column dimensions and reinforcement can be summarized as follows:

- The column cross-sectional dimensions should be a minimum of 200 mm (preferably 250 mm) for tied columns of rectangular or square shape and a minimum 300-mm diameter for spiral columns.
- The smallest size of ties and spirals should be 10M (a common size).
- The smallest size of longitudinal bars should be 20M.
- A few common reinforcement configurations for tied columns are shown in Figure 8.26 and also summarized in Table 8.1.

8.10 A GENERAL COLUMN DESIGN PROCEDURE

The main design objective is to determine the required column dimensions and reinforcement such that the column is able to carry the given design loads. In the case of a short column subjected to a combined factored axial load (P_f) and bending moment (M_f), the following strength requirements need to be satisfied according to A23.3 Cl.8.1.3:

$$P_f \leq P_r$$
$$M_f \leq M_r$$

Note that the design of slender columns will be discussed in Section 8.12. For now, the reader should consider that the column is classified as "short column" according to CSA A23.3 slenderness criteria.

It is common practice to design reinforced concrete columns using the interaction diagrams explained in Section 8.7. Interaction diagrams can be developed using computer spreadsheets (see Example 8.4) or generated using one of the commercially available software packages; alternatively, nondimensional interaction diagrams offered in some publications can be used. As mentioned in previous chapters, "cookbook" design solutions are generally not available. However, general practical guidelines for column design are outlined in Checklist 8.1, followed by a design example.

Note that the design can be refined by using the column reinforcement configurations from Table 8.1 corresponding to a reinforcement ratio within the allowable range

Checklist 8.1   Design of Reinforced Concrete Short Columns for Axial Load and Flexure

Step	Given: -Factored axial load (P_f) and bending moment (M_f) -Concrete and steel material properties (f_c' and f_y) Description	Code Clause
1	Estimate the column size. In some cases, architectural constraints limit the freedom of selecting column dimensions. If no constraints are present, a good rule-of-thumb estimate of the column gross cross-sectional area is $$A_g \geq \frac{P_f}{0.50 f_c'}$$ The above formula applies to columns of square cross-section, and it is based on the axial load resistance of the concrete portion only (the contribution of the reinforcement is neglected).	
2	Establish the desired value of the reinforcement ratio (ρ_t). Note that CSA A23.3 prescribes that $1\% \leq \rho_t \leq 4\%$ (considering the reinforcement laps). It is good practice to initially estimate the minimum reinforcement ratio as low as $\rho_t = 1\%$.	Cl.10.9.1 and 10.9.2
3	Determine the required area of longitudinal reinforcement (A_{st}) as $$A_{st} = \rho_t A_g$$ Select the number and size of longitudinal bars using the configurations in Table 8.1.	
4	Develop the column interaction diagram.	
5	Check whether the point corresponding to the design loads is inside the interaction diagram. • If the point corresponding to the factored axial load and bending moment is inside the diagram, the design is satisfactory. Proceed to the next step. • If the point is outside the diagram, take the maximum allowable reinforcement ratio ($\rho_t = 4\%$) and repeat Steps 3 and 4.	
6	Check whether the number and spacing of longitudinal bars are within the prescribed limits.	10.9.3 7.4.1.3 A23.1 6.6.5.2
7	Design the transverse reinforcement: • For ties, determine the size, spacing, and arrangement. • For spirals, determine the size, reinforcement ratio (ρ_s), and clear distance.	7.6.5.1, 7.6.5.2, 7.6.5.5 10.9.4
8	Summarize the design with a sketch (design summary)	

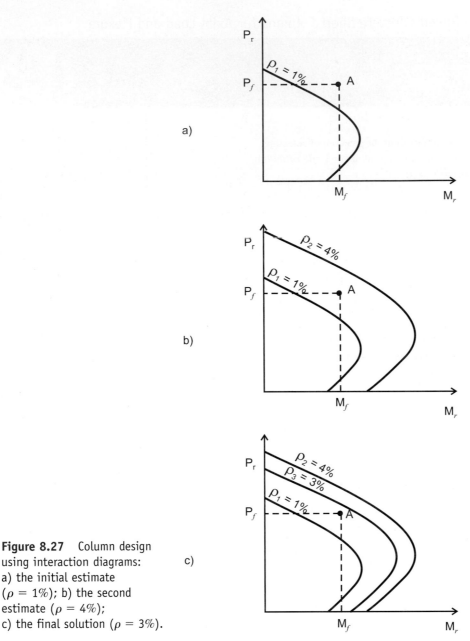

Figure 8.27 Column design using interaction diagrams:
a) the initial estimate ($\rho = 1\%$); b) the second estimate ($\rho = 4\%$);
c) the final solution ($\rho = 3\%$).

($1\% \leq \rho_t \leq 4\%$) and plotting the corresponding interaction diagrams until the point corresponding to the combined factored loads is close to the interaction curve (but still inside the diagram). This is an iterative process, as illustrated in Figure 8.27.

Example 8.6

A tied short column of square cross-section is subjected to a factored axial load of **1500 kN acting at an eccentricity of 100 mm. Use a 40 mm clear cover to the ties and a 20 mm maximum aggregate size. The column is reinforced with four longitudinal bars (one at each corner). The material properties are summarized below.**
Design the column, including cross-sectional dimensions and longitudinal and transverse reinforcement.

Given: $f_c' = 25$ MPa
$f_y = 400$ MPa
$\phi_c = 0.65$
$\phi_s = 0.85$

SOLUTION:

1. Estimate the column size

The column is of square cross-section, and the initial estimate can be made using the formula presented in Checklist 8.1:

$$A_g \geq \frac{P_f}{0.5f_c'} = \frac{1500 \times 10^3 \text{ N}}{0.5(25 \text{ MPa})} = 120\,000 \text{ mm}^2$$

For a square column,

$$A_g = b \times b$$

So,

$$b \geq \sqrt{120\,000 \text{ mm}^2} = 346 \text{ mm}$$

We could round the value to $b = 350$ mm; however, that is too close to the estimated lower bound value. Use the next larger size (use 50 mm increments), that is,

$$b = 400 \text{ mm}$$

A23.3 Cl.10.9.1 and 10.9.2

2. Establish the desired longitudinal reinforcement ratio

CSA A23.3 prescribes that $1\% \leq \rho_t \leq 4\%$ (considering the reinforcement laps). Use $\rho_t = 1\%$ as the starting point.

3. Determine the required area of longitudinal reinforcement

The gross cross-sectional area of the concrete is

$$A_g = b \times h = 400 \text{ mm} \times 400 \text{ mm} = 160\,000 \text{ mm}^2$$

So,

$$A_{st} = \rho_t A_g = 0.01(160\,000 \text{ mm}^2) = 1600 \text{ mm}^2$$

Refer to Table 8.1 to identify a suitable reinforcement configuration. The column is to be reinforced with four longitudinal bars (a design requirement). The most suitable reinforcement configuration has 4-25M longitudinal bars ($A_{st} = 2000 \text{ mm}^2$).

The longitudinal reinforcement ratio is

$$\rho_t = \frac{A_{st}}{A_g} = \frac{2000 \text{ mm}^2}{160\,000 \text{ mm}^2} = 0.012 = 1.2\%$$

Thus,

$$\rho_t = 1.2\% > 1\%$$

which is okay.

4. Develop the column interaction diagram

The interaction diagram for this column developed in Example 8.4 using a computer spreadsheet is shown on the next page.

5. Check whether the point corresponding to the design loads is inside the diagram

The design load eccentricity is

$$e = 100 \text{ mm (given)}$$

The given design loads are

$$P_f = 1500 \text{ kN}$$

$$M_f = P_f \times e$$

$$= (1500 \text{ kN})(0.1 \text{ m}) = 150 \text{ kN} \cdot \text{m}$$

The point with coordinates P_f and M_f is inside the interaction curve. Therefore, the design is satisfactory.

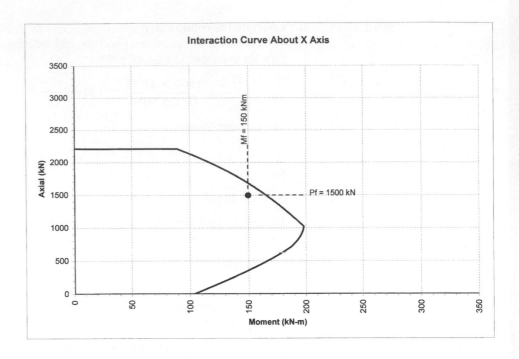

A23.3 Cl.10.9.3
7.4.1.3
A23.1 Cl.6.6.5.2

6. Check whether the number and spacing of longitudinal bars are within the prescribed limits

There are four longitudinal bars — this is the minimum number of bars for tied rectangular columns according to A23.3 Cl.10.9.3, which is okay.

According to A23.1 Cl.6.6.5.2, the clear spacing of longitudinal bars should be at least the largest of

- $1.4 \times d_b = 1.4 \times 25$ mm = 35 mm (For 25M bars, bar diameter is $d_b = 25$ mm (see Table A.1).)
- $1.4 \times$ (maximum size of the coarse aggregate) = 1.4×20 mm = 28 mm, and
- 30 mm.

The largest value is 35 mm. Use 10M ties (common size). The tie diameter is (see Table A.1)

$$d_{tie} \cong 10 \text{ mm}$$

The clear cover to the ties is

$$\text{cover} = 40 \text{ mm (given)}$$

Calculate the actual bar spacing:

$$x = b - (2 \times \text{cover} + 2 \times d_{tie} + 2 \times d_b)$$
$$= 400 \text{ mm} - (2 \times 40 \text{ mm} + 2 \times 10 \text{ mm} + 2 \times 25 \text{ mm}) = 250 \text{ mm}$$

Thus,

$$x = 250 \text{ mm} > 35 \text{ mm}$$

which is okay.

According to A23.3 Cl.7.4.1.3, the maximum permitted bar spacing is 500 mm:

$$x = 250 \text{ mm} < 500 \text{ mm}$$

So, a bar spacing (x) of 250 mm is okay.

| A23.3 Cl.7.6.5.1 |
| Cl.7.6.5.2 |
| Cl.7.6.5.5 |
| Cl.10.9.4 |

7. Check the transverse reinforcement requirements

a) Check the tie size.

Since 10M ties are used, it follows that

$$d_{tie} = 10 \text{ mm} > 0.3 d_b = 0.3 \times 25 \text{ mm} = 7.5 \text{ mm}$$

So, the size is okay.

| A23.3 Cl.7.6.5.2 |

b) Design the tie spacing.

The tie spacing must not exceed the least of

- $16 \times d_b = 16 \times 25 \text{ mm} = 400 \text{ mm}$,
- $48 \times d_{tie} = 48 \times 10 \text{ mm} = 480 \text{ mm}$, and
- 400 mm (the least lateral dimension of the compression member).

The smallest value is 400 mm. Use 10M ties @ 400 mm o.c.

| A23.3 Cl.7.6.5.5 |

c) Check the tie arrangement.

The corner bars are supported by ties with 135° hooks. The spacing of supported longitudinal bars was checked in Step 6.

8. Provide a design summary

The results of the design are summarized on a sketch showing the column cross-section and the reinforcement.

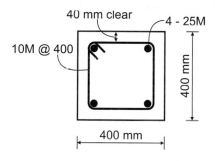

8.11 **STRUCTURAL DRAWINGS AND DETAILS FOR REINFORCED CONCRETE COLUMNS**

Columns are shown on structural drawings presenting foundation, floor, and roof plans. These drawings show the column locations, shape, and orientation. Columns are usually designated by the letter C and a number (for example, C1, C2, etc.). These designations are used to identify columns on the floor grid and to provide a tie with the column schedule. A column schedule is used to present design requirements for various columns in the building. An example of a column schedule is shown in Figure 8.28. The schedule shows the elevations of the top of the footing and each floor; column cross-sectional dimensions; number, bar size, and location; and location and details of splices. In a typical building, some columns are discontinuous near the top, while others do not extend down to the foundation; this information can be visually presented on the column schedule.

Column ties are critical components in each column, so it is essential to present all relevant tie information on the structural drawings. In a larger design project with several column types and corresponding tie arrangements, this information can be presented in tabular form, as shown in Figure 8.29.

In addition to the floor plans, structural drawings should contain all relevant details that may be of assistance to the contractor in his/her effort to completely understand the design requirements. Some typical details showing column reinforcement at the column-to-footing and column-to-floor joints are presented in Figure 8.30.

COLUMN SCHEDULE

FLOOR LEVEL / COLUMN TYPE	CONCRETE STRENGTH AT 28 DAYS	C8	C7	C6	C5	C4	C3	C2	C1
LEVEL: ROOF						18"x18" R/W 8-25M V 10M TIE @8 oc	18"x18" R/W 10-25M V 10M TIE @8 oc	18"x18" R/W 8-25M V 10M TIE @8 oc	18"x18" R/W 8-25M V 10M TIE @8 oc
LEVEL: B5									
LEVEL: B4						18"x18" R/W 16-25M V 10M TIE @8 oc	18"x18" R/W 18-25M V 10M TIE @8 oc	18"x18" R/W 16-25M V 10M TIE @8 oc	18"x18" R/W 12-25M V 10M TIE @8 oc
LEVEL: B3									
LEVEL: B2						24"x24" R/W 16-30M V 10M TIE @8 oc	24"x24" R/W 18-30M V 10M TIE @8 oc	24"x24" R/W 16-30M V 10M TIE @8 oc	24"x24" R/W 14-25M V 10M TIE @8 oc
LEVEL: P5 / B1			14"x30" R/W 10-25M V 10M TIE @8 oc	14"x30" R/W 10-25M V 10M TIE @8 oc	14"x30" R/W 10-25M V 10M TIE @8 oc				
LEVEL: P4		18"x48" R/W 10-30M V 10M TIE @8 oc	14"x30" R/W 12-25M V 10M TIE @8 oc	14"x30" R/W 12-25M V 10M TIE @8 oc	14"x30" R/W 12-25M V 10M TIE @8 oc		36"Ø R/W 20-35M 10M SPIRAL TIE @3"oc PITCH	36"Ø R/W 20-30M 10M SPIRAL TIE @3"oc PITCH	36"Ø R/W 14-30M 10M SPIRAL TIE @3"oc PITCH
LEVEL: P3		18"x48" R/W 12-30M V 10M TIE @8 oc	14"x30" R/W 12-30M V 10M TIE @8 oc		14"x30" R/W 12-30M V 10M TIE @8 oc				
LEVEL: P2				14"x30" R/W 12-30M V 10M TIE @8 oc					
LEVEL: P1		16"x48" R/W 14-30M V 10M TIE @8 oc	16"x48" R/W 14-30M V 10M TIE @8 oc	14"x72" R/W 18-30M V 10M TIE @8 oc	14"x30" R/W 14-30M V 10M TIE @8 oc	44"Ø R/W 24-30M 10M SPIRAL TIE @3"oc PITCH	36"Ø R/W 24-35M V 10M SPIRAL TIE @3"oc PITCH	36"Ø R/W 24-30M V 10M SPIRAL TIE @3"oc PITCH	36"Ø R/W 18-30M V 10M SPIRAL TIE @3"oc PITCH
LEVEL: FOUNDATION									
DOWELS	35 MPa					ALL DOWELS FROM FOOTINGS TO MATCH COLUMN VERTS.			
FLOOR LEVEL / COLUMN TYPE	CONCRETE STRENGTH AT 28 DAYS	C8	C7	C6	C5	NOT USED C4	C3	C2	C1

Figure 8.28   Column schedule.

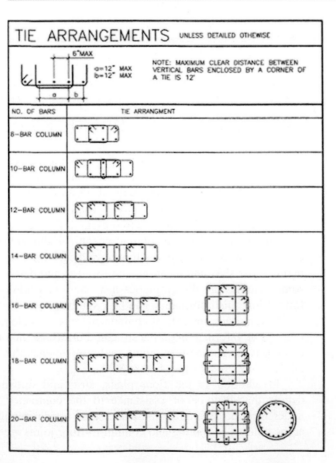

Figure 8.29   Tie arrangement table.

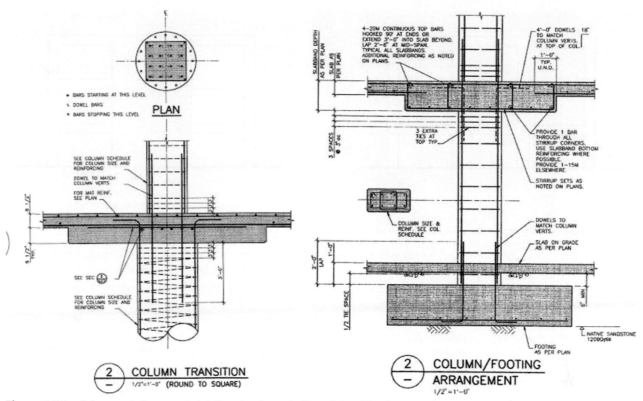

Figure 8.30 Column reinforcement details: a) column-to-floor joint; b) column-to-footing joint.

8.12 INTRODUCTION TO SLENDER COLUMNS

8.12.1 Slenderness of Concrete Columns

Columns can be categorized in terms of their length as short or slender columns. Previous sections in this chapter were focused on the discussion of short columns. The term *short column* is used to denote a column that has a strength determined based on the equations of equilibrium of forces developed in a column cross-section. Short columns fail due to *material failure* (steel-controlled, concrete-controlled, or balanced failure modes), as discussed in Section 8.5.

A column is considered to be *slender* if its least cross-sectional dimension is quite small in comparison with its length. Slender columns have smaller axial load-carrying capacity than short columns of otherwise similar characteristics and are susceptible to instability failures due to *buckling* and *second-order effects* prior to reaching their axial load resistance, as discussed in the next section.

The slenderness of a column is expressed in terms of its *slenderness ratio;* that is,

$$\frac{kl_u}{r}$$

where

l_u = the unsupported length of a compression member, taken as the clear distance between floor slabs, beams, or other members capable of providing lateral support in the direction being considered (A23.3 Cl.10.14.3.1);

r = the radius of gyration of the column cross section; r reflects the effect of cross-sectional size and shape on slenderness; for design purposes, r may be taken as $0.3h$, where h is the overall depth of a rectangular column, or $0.25D$, where D is the diameter of a circular column (A23.3 Cl.10.14.2);

k = the effective length factor for compression members.

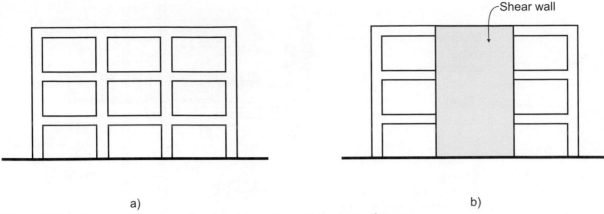

a)   b)

Figure 8.31   Sway versus nonsway frames: a) sway frame; b) nonsway frame.

The value of the k factor depends on the column support conditions and also on the existence of sidesway. Sidesway can be described as lateral deformations whereby one end of the column moves relative to the other. Based on the presence or lack of sidesway, concrete frames can be classified as *sway* or *nonsway frames*. Frames consisting of beams and columns only that are free to sway in the lateral direction are called *sway frames* (see Figure 8.31a). In reality, most concrete frames are braced against sidesway by means of stiff vertical structural members such as shear walls, elevator shafts, and stairwells; such frames are called *nonsway* (or braced) *frames* (see Figure 8.31b). Lateral loads due to wind and earthquakes are carried by frame members (beams and columns) in sway frames, whereas in nonsway frames these loads are carried by shear walls rather than the frame itself. A23.3 Cl.10.14.4 prescribes the stability index (Q) as a means of evaluating whether a particular storey in a frame can be classified as sway or nonsway.

This section is focused mainly on nonsway frames. Sway frames must be analyzed by means of a second-order analysis for multiple load combinations, and the design of such frames is beyond the scope of this book.

The value of the effective length factor (k) in nonsway frames varies from 0.5 to 1.0, depending on the column support conditions (pin or fixed), as illustrated in Figure 8.32. However, A23.3 Cl.10.15.1 prescribes that an effective length factor ($k = 1.0$) should be used unless a different value can be justified. This value corresponds to a pin–pin end support condition, and it represents a simple and conservative approximation appropriate for most nonsway frames. However, in reality, end conditions are never truly pinned or truly fixed. The designer can account for partial fixity of beam-to-column joints in nonsway frames by taking into account the stiffnesses of the other members framing at the same joint.

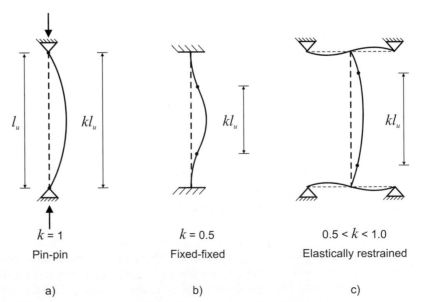

Figure 8.32   Effective length factor k for columns in nonsway frames: a) pin–pin support; b) fixed–fixed support; c) elastically restrained support.

$k = 1$
Pin-pin
a)

$k = 0.5$
Fixed-fixed
b)

$0.5 < k < 1.0$
Elastically restrained
c)

8.12.2 Behaviour of Slender Columns — Instability Failures

In general, slender columns are susceptible to *buckling* and *second-order effects* that might lead to an instability failure, as discussed in this section.

Buckling   Buckling is characterized by excessive bowing of columns between the supports due to axial load, ultimately leading to failure. The potential for buckling limits the axial load-carrying capacity of long columns. Buckling might also be caused by imperfections due to variations in material properties, construction tolerances, and load characteristics. It is expected that the reader has a background in buckling from a course on mechanics of materials. The buckling failure of a reinforced concrete column in the 2003 Boumerdes (Algeria) earthquake is shown in Figure 8.33.

The basic concept of the buckling failure of concentrically loaded slender columns was originally developed by L. Euler in 1759. He established that the critical axial load (P_c), also known as the Euler buckling load, is given by

$\boxed{\text{A23.3 Eq. 10.17}}$
$$P_c = \frac{\pi^2 E I}{(kl_u)^2} \qquad\qquad \text{[8.30]}$$

where E is the modulus of elasticity and I is the moment of inertia of the cross-section.

The product ($E I$) denotes the flexural stiffness of the member cross-section. The larger the flexural stiffness, the larger is the critical load.

The term kl_u denotes the effective length of a column. The larger the effective length, that is, the more slender the column, the smaller is the critical load. This indicates that long and slender columns buckle at lower load levels than shorter columns do.

Second-order ($P - \Delta$) effects   Second-order effects can be illustrated on an example of a simple reinforced concrete nonsway frame, as shown in Figure 8.34a. Assume that two vertical point loads (P) of increasing intensity are acting on the frame. The frame starts to sway due to the increasing load, causing lateral displacements, denoted as Δ, as illustrated in Figure 8.34b. Let us consider a free-body diagram of the column, as shown in Figure 8.34c. Internal moments (M) develop at the column ends to counterbalance the effect of the force couple (P) with the lever arm (Δ). These moments are equal to $M = P \times \Delta$. The magnitude of the end moment increases with an increase in the lateral deflection (Δ). These moments are also referred to as *second-order moments* since they develop in addition to any primary (first-order) end moments that result from applied loads on the frame with no consideration of geometry change. This behaviour is often referred to as the "$P - \Delta$" effect.

Figure 8.33   Buckling failure of a reinforced concrete column in the 2003 Boumerdes (Algeria) earthquake.

(Svetlana Brzev)

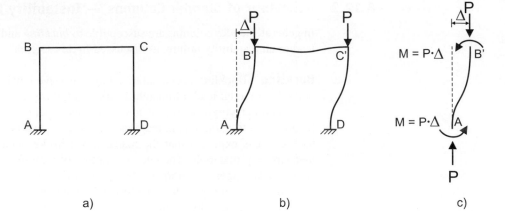

Figure 8.34 $P - \Delta$ effect:
a) unloaded sway frame;
b) swaying due to load P;
c) free-body diagram of
column AB.

The consequences of the $P - \Delta$ effect are often noticed after major earthquakes, in which gravity loads cause additional moments in the columns that sway in a lateral direction due to earthquake ground shaking. The greater the gravity loads, the more pronounced is the $P - \Delta$ effect (provided that lateral displacements are significant). In general, ground floor columns in buildings with discontinued shear walls are most severely affected.

The design of reinforced concrete columns must account for slenderness effects and the reduced axial load capacity caused by these effects.

8.12.3   When Slenderness Effects Should Be Considered

In reinforced concrete short columns, the strength corresponding to material failure (due to axial load and/or flexure) will be exhausted before a load large enough to cause buckling will occur. However, slender columns are susceptible to buckling failure. A23.3 Cl.10.15.2 prescribes that slenderness effects in *nonsway frames* can be ignored provided that the following condition is satisfied:

A23.3 Eq. 10.15

$$\frac{kl_u}{r} \leq \frac{25 - 10\left(\dfrac{M_1}{M_2}\right)}{\sqrt{\dfrac{P_f}{f_c' A_g}}}$$

[8.31]

where

$$\frac{M_1}{M_2} \geq -0.5$$

M_1 = the smaller factored end moment
M_2 = the larger factored end moment.

The term M_1/M_2 is positive if a column is bent in a single curvature, and negative in the case of double-curvature bending. Single- and double-curvature bending are illustrated in Figure 8.35.

Note that kl_u/r denotes the slenderness ratio of the axis of bending under consideration. In general, the values of the parameters k, l_u, and r can vary for different bending axes. The critical axis is the one that results in the larger slenderness ratio.

In general, concrete columns are sufficiently short and stocky and so the slenderness does not have a significant effect on the column design. In fact, for ordinary beam and column sizes and typical storey heights of concrete framing systems, slenderness effects may be neglected in more than 90% columns in nonsway (braced) frames and in about 40% columns in sway frames (PCA, 2002).

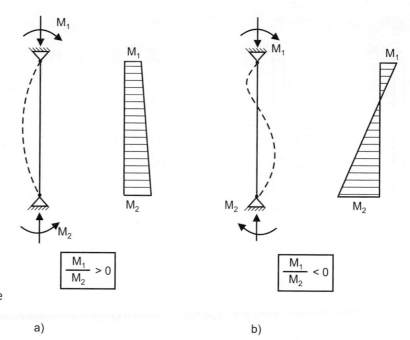

Figure 8.35 Bending in single and double curvature: a) single curvature; b) double curvature.

a)   b)

Columns are categorized in terms of their length as follows:

Short columns: the strength can be determined based on the equations of equilibrium of forces developed in a column cross-section. Short columns fail due to *material failure*.

Slender columns: the least cross-sectional dimension is small in comparison with their length. Slender columns have smaller axial load-carrying capacity than short columns of otherwise similar characteristics.

Slender columns are susceptible to the following two effects that might lead to an instability failure:

Buckling is excessive bowing of columns between the supports due to axial load, ultimately leading to failure.

Second-order (P − Δ) effects result in an increase in the bending moments developed in a column caused by lateral deflections (sidesway).

The slenderness of a column is expressed in terms of its *slenderness ratio* (kl_u/r). A23.3 Cl.10.15.2 prescribes that slenderness effects in *nonsway frames* can be ignored provided that the following condition is satisfied:

A23.3 Eq. 10.15

$$\frac{kl_u}{r} \leq \frac{25 - 10\left(\dfrac{M_1}{M_2}\right)}{\sqrt{\dfrac{P_f}{f_c' A_g}}}$$

[8.31]

8.12.4 Analysis of Slender Columns in Nonsway Frames

The design of slender reinforced concrete columns is one of the more complex aspects of reinforced concrete design. There are several methods available to account for slenderness effects in concrete columns, some of which require the use of second-order computer analysis. The easiest hand-calculation technique, called the *moment magnifier method*, will be presented in this section.

The purpose of the moment magnifier method can be explained using an example of an eccentrically loaded slender column, as shown in Figure 8.36a. The top and bottom of the column are pin supported and swaying is prevented. The column is subjected to a factored axial

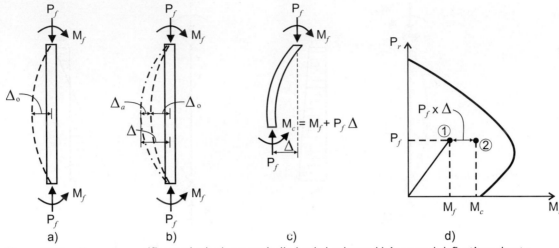

Figure 8.36 Moment magnifier method: a) eccentrically loaded column; b) increased deflections due to second-order effects; c) free-body diagram of the column; d) column interaction diagram.

load (P_f) and the bending moments (M_f) at the column ends. As a result, a lateral deflection develops at the column midheight, denoted as (Δ_o), as shown in Figure 8.36b. Due to the axial load (P_f), additional deflection develops in the column (denoted as Δ_a). The deflection component Δ_o is also called the *first-order* deflection and it is the result of the loads acting on the column, whereas the deflection component Δ_a is called the *second-order* deflection. The total deflection (Δ) causes the development of an additional (second-order) moment in the column equal to $P_f \times \Delta$. This is illustrated on the free-body diagram of the upper portion of the column in Figure 8.36c. As a result, the total bending moment at any point along the column height is

$$M_c = M_f + P_f \times \Delta$$

It is obvious that the moment M_c represents the magnified factored moment M_f increased due to the second-order effects. As discussed in the previous sections, second-order effects in slender columns may be significant, so $M_c > M_f$.

The same information can be presented on the column interaction diagram in Figure 8.36d. When the second-order deflection (Δ_a) is very small and can be neglected, then the load acting on the column can be represented by point 1 on the diagram corresponding to the axial load P_f and bending moment M_f. However, if the second-order deflections are significant due to the large column slenderness ratio, the column load is represented by point 2 on the diagram, corresponding to the moment M_c. In slender columns, lateral deflections cause an increase in the bending moments, which in turn cause a further increase in the lateral deflections, thus resulting in a nonlinear relation between axial forces and bending moments, as shown by the curve connecting points 0 and 2 on the diagram.

A23.3 Cl.10.15.3 prescribes the use of the moment magnifier method for the design of slender columns in nonsway frames. The method is applicable to eccentrically loaded slender columns. In the general case, a column is subjected to an axial load P_f acting at different eccentricities at the top and the bottom of the column, as illustrated in Figure 8.37. The eccentricity at the top of the column is denoted as e_1, whereas the eccentricity at the base of the column is denoted as e_2. Let us consider that $e_2 > e_1$; then $M_2 > M_1$.

The magnified design moment (M_c) to be used in combination with the factored axial load (P_f) is given by the equation

| A23.3 Eq. 10.16 | $M_c = \delta \times M_f$ | **[8.32]** |

where δ is the moment magnifier which can be determined from the following equation

$$\delta = \frac{C_m}{1 - \dfrac{P_f}{\phi_m \times P_c}}$$ **[8.33]**

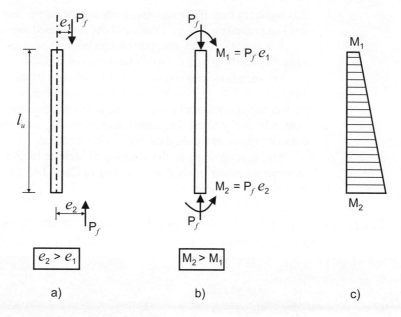

Figure 8.37 A slender column subjected to different end moments: a) column loaded with the force P_f acting at a different eccentricity at the top and bottom; b) equivalent axial force and end moments; c) bending moment diagram.

a)   b)   c)

where

C_m = the factor relating the actual moment diagram to an equivalent uniform moment diagram, determined based on the end moments (see below)

ϕ_m = 0.75 is the resistance factor for the moment magnifier

P_c = the critical axial load (Eqn 8.30).

The flexural stiffness ($E\,I$) depends on whether the section is cracked or uncracked. If the section is uncracked, the gross (uncracked) moment of inertia (I_g) should be used for the flexural stiffness calculations. For design purposes, concrete columns are considered to be cracked with a variable extent of cracking along the column length. As a result, a fraction of the gross moment of inertia is typically used for the stiffness calculations. According to A23.3 Cl.10.15.3.1, the flexural stiffness can be taken as

A23.3 Eq. 10.19
$$E\,I = \frac{0.4E_c I_g}{1 + \beta_d} \qquad [8.34]$$

where

β_d = the ratio of the maximum factored sustained axial load to the maximum factored axial load associated with the same load combination

E_c = the modulus of elasticity of concrete.

Note that M_2 is a minimum moment that accounts for an accidental eccentricity given by the following condition

$$M_2 \geq P_f(15 + 0.03h) \qquad [8.35]$$

where h is the column dimension in the direction being considered. Note that the above condition needs to be checked separately for each bending axis.

The term C_m in Eqn 8.33 is related to the influence of unequal end moments on the column. If the end moments are unequal, the maximum lateral displacement of the column will not occur at the centre. The magnification effect must take this into account. Alternatively, the end moments may be in opposite directions, putting the column into reverse curvature, which must be included in the analysis.

If there are no transverse loads between supports, C_m may be calculated as follows:

A23.3 Eq. 10.20
$$C_m = 0.6 + 0.4\frac{M_1}{M_2} \geq 0.4 \qquad [8.36]$$

It is apparent that, if the end moments are the same, that is, $C_m = 1$, the column will be designed based on a magnification of the larger end moment (M_2).

If the moments act in opposite directions, the value of the expression M_1/M_2 will be negative, resulting in a C_m value between 0.4 and 0.6.

For members with transverse loads between supports, $C_m = 1$ (A23.3 Cl.10.15.3.3). Note that, if $P_f > \phi_m P_c$ in Eqn 8.33, then the denominator is a negative number. This represents the condition when the column has become unstable and will buckle. In such a case, it is invalid to use this equation to obtain M_c since the column will certainly fail. The column properties would then need to be modified.

The general steps in the analysis of slender reinforced concrete columns using the moment magnifier method are outlined in Checklist 8.2, and the analysis will be demonstrated in an example.

Checklist 8.2 **Design of Slender Reinforced Concrete Columns in Nonsway Frames According to the Moment Magnifier Method (A23.3 Cl.10.15.3)**

Given:
- Eccentricity of the axial load (e)
- Factored axial load (P_f)
- Column cross-sectional dimensions (b and h)
- Amount and distribution of longitudinal reinforcement
- Concrete and steel material properties (f_c' and f_y)

Step	Description
1	Determine the following parameters for each direction: • the unbraced column length (l_u) • the effective length factor (k) • the radius of gyration (r)
2	Verify whether the slenderness effects should be considered for each direction: A23.3 Eq. 10.15 $\dfrac{kl_u}{r} \le \dfrac{25 - 10\left(\dfrac{M_1}{M_2}\right)}{\sqrt{\dfrac{P_f}{f_c'A_g}}}$ [8.31]
3	Determine the magnified moment for each direction: A23.3 Eq. 10.16 $M_c = \delta \times M_2 \ge M_2$ [8.32] Ensure that $M_2 \ge P_f(15 + 0.003h)$ [8.35]
4	Use the column interaction diagram to check whether the magnified bending moment (M_c) combined with the axial load (P_f) gives a point inside the diagram or on the diagram, as shown in Figure 8.38. In that case, $M_c < M_r$ and the column design is acceptable. If the point is outside the interaction diagram ($M_c > M_r$), the column is not adequate to carry the design loads. Either the cross-sectional dimensions or the reinforcement need to be modified.
5	Determine the size and spacing of the ties according to the CSA A23.3 requirements (see Section 8.8.1).

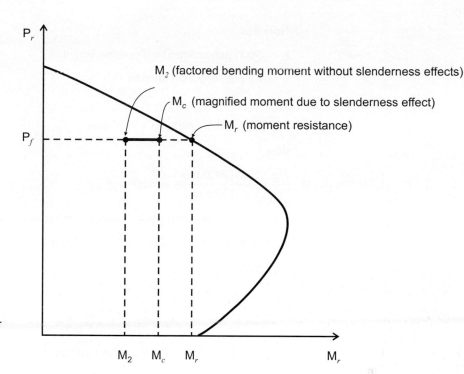

Figure 8.38 Column interaction diagram used in the slender column design according to the moment magnifier method.

Example 8.7

Consider the same 400 mm by 400 mm column discussed in Example 8.4. The column height is 5 m. The specified axial dead load and live load are equal to 500 kN. The factored bending moments (Mf) at the column ends are equal to 206 kNm. The column is bent in single curvature and it is pinned at both ends.

Design the column reinforcement using the BPA COLUMN software to generate interaction diagram. Consider slenderness effects in the design.

Given: $f_c' = 25$ MPa
$f_y = 400$ MPa

SOLUTION: 1. **Compute the design loads.**

$$P_f = 1.25P_{DL} + 1.5P_{LL} = 1.25 \times 500 + 1.5 \times 500 = 1375 \text{ kN}$$

$M_f = 206$ kNm

Since the moments are equal at the ends, it follows that

$M_1 = M_2 = M_f = 206$ kNm

Since the column is bent in the single curvature, it follows that

$$\frac{M_1}{M_2} = +1$$

2. **Check whether slenderness effects need to be considered (A23.3 Cl. 10.15.2).**
Slenderness effects do not need to be considered when

$$\frac{kl_u}{r} \leq \frac{25 - 10\left(\frac{M_1}{M_2}\right)}{\sqrt{\frac{P_f}{f_c' \cdot A_g}}}$$ [8.31]

Note that

$l_u = 5000$ mm unsupported column height

$k = 1.0$ pin-supported column

and

$r = 0.3h = 0.3 \cdot 400$ mm $= 120$ mm

thus

$$\frac{kl_u}{r} = \frac{1.0\,(5000\text{ mm})}{120\text{ mm}} = 41.7$$

Since this is a 400 mm square column, the gross cross-sectional area is equal to

$A_g = 400$ mm $\times 400$ mm $= 1.6 \times 10^5$ mm^2

Therefore,

$$\frac{25 - 10\left(\dfrac{M_1}{M_2}\right)}{\sqrt{\dfrac{P_f}{f_c' \cdot A_g}}} = \frac{25 - 10(1.0)}{\sqrt{\dfrac{1375 \cdot 10^3 \text{ N}}{25 \text{ MPa} \cdot (1.6 \cdot 10^5 \text{ mm}^2}}} = 25.6$$

Since $41.7 > 25.6$, it follows that slenderness effects need to be considered.

3. **Find the magnified moment (M_c).**
   a) Determine the moment magnifier δ.

   The moment magnifier can be determined from the following equation

   $$\delta = \frac{C_m}{1 - \dfrac{P_f}{\phi_m \times P_c}}$$ [8.33]

   where

   $$C_m = 0.6 + 0.4\left(\frac{M_1}{M_2}\right) = 0.6 + 0.4(1) = 1.0$$

   and

   $\phi_m = 0.75$

   The critical axial load needs to be determined from the following equation

   $$P_c = \frac{\pi^2 EI}{(kl_u)^2}$$ [8.30]

   Concrete modulus of elasticity:

   $$E_c = 4500\sqrt{f_c'} = 4500\sqrt{25 \text{ MPa}} = 22500 \text{ MPa}$$

   Gross moment of inertia:

   $$I_g = \frac{b \times h^3}{12} = \frac{400 \text{ mm} \times (400 \text{ mm})^3}{12} = 2.13 \times 10^9 \text{ mm}^4$$

   Note that β_d is the ratio of maximum factored axial dead load to the max factored axial load, that is,

   $$\beta_d = \frac{P_{DLf}}{P_f} = \frac{1.25 \times 500 \text{ kN}}{1375 \text{ kN}} = 0.45$$

Therefore,

$$EI = \frac{0.4E_c I_g}{1 + \beta_d}$$ [8.34]

$$= \frac{0.4(22500 \text{ MPa})(2.13 \times 10^9 \text{ mm}^4)}{1 + 0.45} = 13.2 \times 10^{12} \text{ Nmm}$$

Hence,

$$P_c = \frac{\pi^2 \times (13.2 \times 10^{12} \text{ Nmm})}{(1.0 \times 5000 \text{ mm})^2} = 5210 \text{ kN}$$

$$\delta = \frac{C_m}{1 - \dfrac{P_f}{\phi_m \times P_c}} = \frac{1.0}{1 - \dfrac{1375 \text{ kN}}{0.75 \times 5210 \text{ kN}}} = 1.55$$ [8.33]

b) Find the magnified moment (M_c).

$$M_c = \delta \times M_f = 1.55 \times 206 \text{ kN} = 320 \text{ kNm}$$ [8.32]

4. **Determine the required reinforcement.**

   Column interaction diagram can be developed using the software BPA COLUMN available on the web site. Several alternative reinforcement solutions are possible, as presented on the diagram below. It can be seen that a layout with 12-25M bars meets the design requirements.

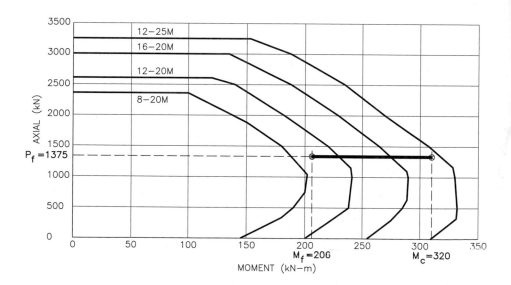

5. **Discussion**

   Recall from Examples 8.4 and 8.5 that the required column reinforcement was 12-20M (3600 mm²) when slenderness effects were disregarded and the column was treated as a short column. The amount of reinforcement has increased to 12-25M when slenderness effects were considered (the corresponding reinforcement area is 6000 mm²). This is a significant increase in the amount of steel–on the order of 67%.

8.13 COLUMN LOADS IN MULTISTOREY BUILDINGS

In design practice, the calculation of column loads in multistorey buildings is known as *column load take down*. This section presents a discussion of the tributary area used to determine axial loads in columns due to gravity loads. The main gravity loads considered in column design are dead load (DL), occupancy live load (LL), and snow load (SL). NBC 2010 permits a live load reduction in columns supporting large floor areas, as discussed below.

8.13.1 Tributary Area

In practice, columns and/or foundations are often designed before the detailed calculations of the horizontal elements have been completed. The designer must therefore approximate the gravity loads placed onto the columns; this is usually accomplished by approximating tributary areas that are supported by each column. The concept of a tributary area was introduced in Section 1.3.3. The tributary area for a column can be determined by carefully drawing lines of zero shear on a plan view of the structure. Each column is surrounded by an area enclosed within a set of these lines, and all load falling within that area is supported by the corresponding column.

For regular framing patterns in a one-way spanning system of relatively equal spans, the zero shear locations can be approximated at halfway between the supporting members. This is realistic for intermediate spans, but less so for end spans.

In general, it is advisable to increase the column load capacity by a minimum of 10% due to the uncertainty in maximum column loads determined in the load take down process.

8.13.2 Live Load Reductions

The loads in columns supporting large floor areas and multiple floors need to be determined to account for all applied loads and their effects. Due to a low probability that all floors will experience their maximum occupancy load simultaneously, NBC 2010 (Cl.4.1.5.9) allows for live load reduction in a structural member supporting a large floor area. The amount of reduction depends on the type of occupancy (for example, assembly occupancy, residential occupancy, etc.) and the load magnitude. The details of this provision are as follows:

1. For assembly occupancies designed for live load less than 4.8 kPa, no reduction is allowed.
2. For those occupancies, where a member supports a tributary area greater than 80 m^2 (including assembly occupancies designed for a live load of 4.8 kPa or more, storage, manufacturing, retail stores, garages etc.), the total specified live load due to use and occupancy, excluding snow, may be multiplied by a reduction factor (RF), determined as follows:

$$RF = 0.5 + \sqrt{\frac{20}{A}}$$

where A is the tributary area, in square metres.

3. For occupancies other than those indicated above, where a member supports a tributary area greater than 20 m^2, the total specified live load due to use and occupancy, excluding snow, may be multiplied by an RF factor given by

$$RF = 0.3 + \sqrt{\frac{9.8}{B}}$$

where B is the tributary area, in square metres. The column load take down will be illustrated in the following example.

Example 8.8

A partial floor plan and elevation of a reinforced concrete apartment building (not an assembly area) are shown in the figure below. The building has eight floors plus the roof. The floors and the roof carry a specified dead load (DL) of 7.2 kPa (including the self-weight) and a specified live load (LL) of 1.9 kPa. The roof is designed to carry a snow load (SL) of 2.5 kPa.

Determine the total factored axial load in column C1 at each level.

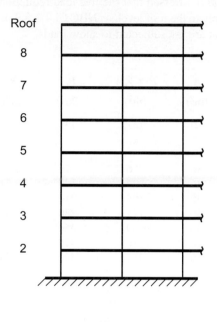

Elevation

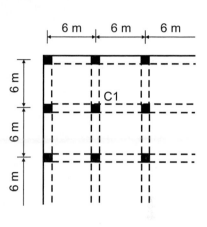

Floor plan

SOLUTION: To determine the column loads, the following needs to be calculated :

1. **Tributary area**

   $A = 6\,\text{m} \times 6\,\text{m} = 36\,\text{m}^2$

2. **Column axial dead load per floor**

   $P_{DL} = A \times \text{DL} = 36\,\text{m}^2 \times 7.2\,\text{kPa} = 259\,\text{kN} \cong 260\,\text{kN}$

3. **Column axial live load per floor**

   $P_{LL} = A \times \text{LL} = 36\,\text{m}^2 \times 1.9\,\text{kPa} = 68.4\,\text{kN} \cong 70\,\text{kN}$

4. **Column axial snow load per floor**

   $P_{SL} = A \times \text{SL} = 36\,\text{m}^2 \times 2.5\,\text{kPa} = 90\,\text{kN}$

5. **Live load reduction factor (RF)**
   This is a residential occupancy with a tributary area larger than 20 m², so

   $$RF = 0.3 + \sqrt{\frac{9.8}{B}}$$

   The tributary area B will be determined for each floor level separately. The values are summarized in the table below. Note that B is determined in a cumulative manner; that is, the lower floors have larger cumulative tributary area than the floors above. Consequently, the RF factor changes between floors.

6. **Column axial forces**

The column axial forces can be summarized in tabular form, as shown below. Axial loads due to dead load (P_{DL}), occupancy live load (P_{LL}), and snow load (P_{SL}) are considered in the table. Note that reduced axial live loads can be determined based on the equation

$$P_{LL}^R = P_{LL} \times \text{RF}$$

where the P_{LL} value from column (4) is multiplied by the RF value from column (5).

It should be stressed that the live load reduction applies to all floors; however, it does not apply to the roof level, as NBC Cl.4.1.5.9 permits the live load reduction only for areas that are not subjected to snow loads.

Level	A	$B = \Sigma A$	P_{LL}	RF	P_{LL}^R	P_{DL}	P_f
	(m²)	(m²)	(kN)	(kN)	(kN)	(kN)	(kN)
(1)	(2)	(3)	(4)	(5)	(6)	(7)	(8)
Roof (8)						260	460
8	36	36	70	0.82	57	520	870
7	36	72	140	0.67	94	780	1251
6	36	108	210	0.60	126	1040	1624
5	36	144	280	0.56	157	1300	1995
4	36	180	350	0.53	186	1560	2364
3	36	216	420	0.51	214	1820	2731
2	36	252	490	0.50	245	2080	3102

Note that the snow load P_{SL} = 90 kN (not shown in the table) is included in the cumulative load calculations, as explained below.

Column (8) shows the total cumulative factored load P_f at each floor:

$$P_f = 1.5P_{LL}^R + 1.5P_{SL} + 1.25P_{DL}$$

For example, the total cumulative factored load on the column from the fifth floor and above is

$$P_{f5} = 1.5(157 \text{ kN}) + 1.5(90 \text{ kN}) + 1.25(1300 \text{ kN}) = 1995 \text{ kN}$$

Similarly, the total cumulative factored load on the column from the second floor and above is

$$P_{f2} = 1.5(245 \text{ kN}) + 1.5(90 \text{ kN}) + 1.25(2080 \text{ kN}) = 3102 \text{ kN}$$

The designer can efficiently design columns by using the factored loads obtained from the column load table presented above. Subsequently, column interaction diagrams for practical reinforcement configurations can be developed to select the longitudinal reinforcement, as shown in Example 8.5.

To perform column load take down calculations, the reader can use a spreadsheet developed in Microsoft® Excel, which is posted on the book Web site.

SUMMARY AND REVIEW — COLUMNS

Columns are the most common vertical load-resisting elements in reinforced concrete structures. In general, columns are subjected to axial compression loads and are often called *compression members*. Some columns in concrete structures are subjected to purely axial compressive forces. However, more often, bending moments are also present due to the eccentricity of applied loads, applied end moments, and/or lateral loading on the column. Such columns are subjected to combined axial load and flexure. Columns are characterized by cross-sectional dimensions considerably less than their height.

Types of reinforced concrete columns

The three basic types of reinforced concrete columns are as follows:

- *Tied columns* are reinforced with longitudinal bars enclosed by lateral ties provided throughout the column length.
- *Spiral columns* are reinforced with longitudinal bars enclosed by closely spaced and continuously wound spiral reinforcement.
- *Composite columns* are reinforced longitudinally with a structural steel shape either encased in or encasing the concrete and with or without longitudinal bars.

Main components of a reinforced concrete column

The main components of a reinforced concrete column are as follows:

- Longitudinal reinforcement increases the axial and flexural load resistance of a column.
- Transverse reinforcement (ties or spirals) provides lateral restraint to longitudinal bars under compression and prevents their premature buckling.
- The concrete core provides compression resistance to a column.

Types of column behaviour

The behaviour of reinforced concrete columns subjected to axial load depends primarily on their slenderness and the magnitude of the load eccentricity. There are three distinct types of column behaviour.

Concentrically loaded column (zero eccentricity): The maximum axial load capacity of the column is reached when the compressive strain in concrete (ε_c) reaches a maximum value (ε_{cmax}) of 0.0035 ($\varepsilon_c = \varepsilon_{cmax}$). The maximum axial load capacity of a concentrically loaded column (P_{ro}) is the sum of the concrete contribution and the steel contribution. The failure of a concentrically loaded column occurs after the longitudinal reinforcement yields. The behaviour of a concentrically loaded column in the postyield phase depends on whether the transverse reinforcement consists of ties or spirals. These columns fail in the *concrete-controlled* mode.

Eccentrically loaded column (small eccentricity): The axial load resistance in eccentrically loaded columns with small eccentricities is less than in concentrically loaded columns. The decrease in the load capacity depends on the magnitude of the eccentricity: the larger the eccentricity, the smaller the load capacity. Eccentrically loaded columns with small eccentricity fail in the concrete-controlled mode.

Eccentrically loaded column (large eccentricity): At large load eccentricities (e), the effect of the bending moment ($M = P \times e$) becomes significant, and the column behaves like a beam subjected to flexure. In an eccentrically loaded column with large eccentricity, an increase in the axial load results in an increased flexural resistance. Eccentrically loaded columns with large eccentricity fail in the steel-controlled mode.

The *balanced condition:* When the strain in steel at the tension face reaches the yield strain ($\varepsilon_s = \varepsilon_y$) simultaneously with the strain at the concrete compression face reaching the maximum value ($\varepsilon_c = \varepsilon_{cmax}$), the balanced condition occurs. This condition represents the threshold between the small and large eccentricity load conditions.

CSA A23.3 column design assumptions

The ultimate limit states design of concrete columns is based on the following basic assumptions outlined by CSA A23.3:

- Plane sections remain plane (linear strain distribution across the section) (Cl.10.1.2).
- Strain compatibility exists between concrete and steel.
- An equivalent rectangular stress block is used instead of the actual stress distribution for concrete (Cl.10.1.7).
- The concrete tensile strength is neglected in the design (Cl.10.1.5).
- The maximum concrete compressive strain (ε_{cmax}) is equal to 0.0035 (Cl.10.1.3).

Key features of a column interaction diagram

The key features of a column interaction diagram are as follows:

1. The points on the diagram represent the combinations of axial forces and bending moments corresponding to the resistance of a column cross-section.
2. Each point on the interaction diagram corresponds to the column capacity at a specific load eccentricity.
3. A column interaction diagram determines whether the column has adequate capacity to carry the design loads.
4. An interaction diagram is a unique representation of a column with regard to its material properties, cross-sectional dimensions, and amount and distribution of reinforcement. Interaction diagrams can also be used to predict whether an eccentrically loaded column subjected to an axial load P_f acting at an eccentricity e is likely to fail in the concrete-controlled mode or the steel-controlled mode, as follows:
   a) If $e > e_b$, the column will experience steel-controlled failure.
   b) If $e < e_b$, the column will experience concrete-controlled failure.
   The threshold between these two types of behaviour is represented by the balanced condition, characterized by the *balanced eccentricity* (e_b).

Development of a column interaction diagram

The following general steps are followed in the development of a column interaction diagram:

1. Calculate the axial load resistance (P_{ro}) corresponding to the zero eccentricity condition ($e = 0$).
2. Calculate the neutral axis depth corresponding to the balanced condition (c_b).
3. Determine the points on the column interaction diagram following the six-step procedure outlined in Section 8.7.2.
4. Plot the interaction diagram.
5. Decrease the axial resistance of the column section to account for the accidental moments.

CSA A23.3 column design requirements

CSA A23.3 prescribes the following reinforcement requirements for reinforced concrete columns:

Longitudinal reinforcement

- minimum amount of reinforcement (Cl.10.9.1)
- maximum amount of reinforcement (Cl.10.9.2)
- minimum number of bars (Cl.10.9.3)
- minimum bar spacing (A23.1 Cl.6.6.5.2)
- maximum bar spacing (Cl.7.4.1.3)

Ties

- size (Cl.7.6.5.1)
- spacing (Cl.7.6.5.2)
- anchorage (Cl.7.1.2)
- arrangement (Cl.7.6.5.5)

Spirals

- amount (Cl.10.9.4)
- size and spacing (Cl.7.6.4.2 to 7.6.4.4)

Concrete cover (A23.1 Cl.6.6.6.2.3)

Practical guidelines for the selection of column size and reinforcement

Practical design guidelines for the selection of column dimensions and reinforcement can be summarized as follows:

- The column cross-sectional dimensions should be a minimum of 200 mm (preferably 250 mm) for tied columns of rectangular or square shape and a minimum 300 mm diameter for spiral columns.
- The smallest size of ties and spirals should be 10M (a common size).
- The smallest size of longitudinal bars should be 20M.
- A few common reinforcement configurations for tied columns are shown in Figure 8.26 and also summarized in Table 8.1.

Behaviour of slender columns and the CSA A23.3 slenderness criteria

Columns are categorized in terms of their length as follows:

Short columns: the strength can be determined based on the equations of equilibrium of forces developed in a column cross-section. Short columns fail due to *material failure*.

Slender columns: the least cross-sectional dimension is small in comparison with their length. Slender columns have smaller axial load-carrying capacity than short columns with otherwise similar characteristics.

Slender columns are susceptible to the following two effects that might lead to instability failure:

Buckling is the excessive bowing of columns between the supports due to axial load, ultimately leading to failure.

Second-order $(P - \Delta)$ *effects* result in an increase in the bending moments developed in a column caused by lateral deflections (sidesway).

The slenderness of a column is expressed in terms of its *slenderness ratio* (kl_u/r). A23.3 Cl.10.15.2 prescribes that the slenderness effects in nonsway frames can be ignored provided that the following condition is satisfied:

A23.3 Eq. 10.15

$$\frac{kl_u}{r} \leq \frac{25 - 10\left(\dfrac{M_1}{M_2}\right)}{\sqrt{\dfrac{P_f}{f_c' A_g}}}$$

[8.31]

PROBLEMS

8.1. Consider a reinforced concrete column subjected to an eccentrically applied axial load of constant magnitude.

Describe the column behaviour and mode of failure under gradually increasing load for the following three conditions:

a) zero eccentricity
b) small eccentricity
c) infinitely large eccentricity

8.2.
a) When does the balanced condition occur in reinforced concrete columns?
b) How is balanced eccentricity used in the design of reinforced concrete columns? Explain.

8.3.
a) What are the two main failure modes characteristic of eccentrically loaded reinforced concrete columns?
b) What are the key factors influencing the development of these failure modes?

8.4. Recall the discussion of failure modes in reinforced concrete beams presented in Section 3.4.3. Compare the failure modes for beams and columns and the factors influencing the development of a particular failure mode.

What is the main difference in the factors governing the failure modes in reinforced concrete beams and eccentrically loaded columns?

8.5. A tied short column of square cross-section is shown in the figure below. The column is reinforced with 4-20M longitudinal bars and 10M ties. Use a 40 mm clear cover to the ties. The material properties are given.

a) Determine the factored moment resistance (M_r) and the axial load resistance (P_r) at the balanced condition.

b) Derive the column interaction diagram.

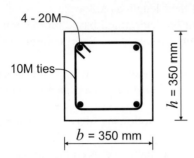

4 - 20M
10M ties
b = 350 mm
h = 350 mm

Given:

$f_c' = 30$ MPa
$f_y = 400$ MPa
$E_s = 200\,000$ MPa
$\phi_c = 0.65$
$\phi_s = 0.85$

8.6. Refer to the column discussed in Problem 8.5.

a) Determine the factored axial load resistance (P_{ro}).

b) If a factored bending moment $M_f = 40$ kN·m is applied to the column, determine the corresponding axial load resistance (P_r).

c) If the factored bending moment is gradually increased to $M_f = 80$ kN·m and is applied to the column, determine the corresponding axial load resistance (P_r).

d) Comment on the effect of the increase in the bending moment magnitude on the column axial load resistance.

8.7. Refer to the column discussed in Problem 8.5.

a) Determine the factored moment capacity (M_r) for the pure bending condition ($P = 0$).

b) If a factored axial load $P_f = 200$ kN is applied to the column, determine the factored moment resistance (M_r).

c) If the factored axial load is increased to $P_f = 400$ kN, determine the factored moment resistance (M_r).

d) Comment on the effect of the increase in the magnitude of axial load on the moment resistance.

8.8. A tied short column of square cross-section is shown in the figure below. The column is reinforced with 8-25M longitudinal bars and 10M ties. Use a 40 mm clear cover to the ties. The material properties are given.

Develop an interaction diagram for this column using the BPA COLUMN software.

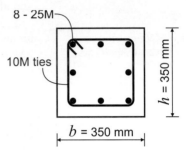

8 - 25M
10M ties
b = 350 mm
h = 350 mm

Given:

$f_c' = 30$ MPa
$f_y = 400$ MPa
$E_s = 200\,000$ MPa
$\phi_c = 0.65$
$\phi_s = 0.85$

8.9. A tied short column of square cross-section is shown in the figure below. The column is reinforced with 8-20M longitudinal bars and 10M ties. Use a 40 mm clear cover to the ties. The material properties are summarized below.

Develop an interaction diagram for this column using the BPA COLUMN software.

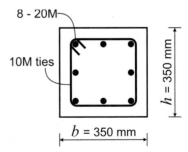

8 - 20M
10M ties
b = 350 mm
h = 350 mm

Given:

$f_c' = 30$ MPa
$f_y = 400$ MPa
$E_s = 200\,000$ MPa
$\phi_c = 0.65$
$\phi_s = 0.85$

8.10. A tied short column of square cross-section is shown in the figure below. The column is reinforced with

12-20M longitudinal bars and 10M ties. Use a 40 mm clear cover to the ties. The material properties are given. Develop an interaction diagram for this column.

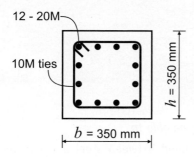

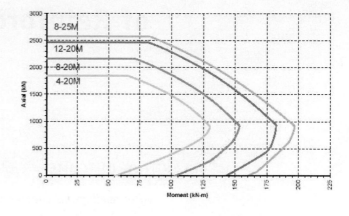

Given:

$f_c' = 30$ MPa

$f_y = 400$ MPa

$E_s = 200\,000$ MPa

$\phi_c = 0.65$

$\phi_s = 0.85$

8.11. The interaction diagrams for the 350 mm by 350 mm column section with reinforcement arrangements used in Problems 8.5 to 8.10 are shown below.

Design the required amount and distribution of longitudinal reinforcement for the following combinations of applied factored loads:

a) $P_f = 1500$ kN and $M_f = 30$ kN·m

b) $P_f = 1500$ kN and $M_f = 80$ kN·m

c) $P_f = 1500$ kN and $M_f = 100$ kN·m

d) $P_f = 400$ kN and $M_f = 150$ kN·m

e) $P_f = 2200$ kN and $M_f = 40$ kN·m

f) $P_f = 1800$ kN and $M_f = 40$ kN·m

g) $P_f = 0$ kN and $M_f = 150$ kN·m

Given:

$f_c' = 30$ MPa

$f_y = 400$ MPa

$E_s = 200\,000$ MPa

$\phi_c = 0.65$

$\phi_s = 0.85$

8.12. Consider the column section from Problem 8.11. For each load combination (a to g) determine the required reinforcement when slenderness effects are considered. Assume that the column is pin-supported, and that the bending moments are the same at both ends. The column is bent in single curvature. Consider an unsupported column height of 4 m.

8.13. Design the same column for load combinations discussed in Problem 8.12 assuming an unsupported height of 5 m. Compare the results for Problems 8.12 and 8.13. Identify cases where larger column dimensions are required to support the given load.

9 Bond and Anchorage of Reinforcement

LEARNING OUTCOMES

After reading this chapter, you should be able to

- describe the basic mechanism of bond between steel and concrete and the two characteristic modes of bond failure in reinforced concrete flexural members
- determine the development lengths for straight bars in tension and compression
- design hooked anchorages for flexural reinforcing bars
- apply the CSA A23.3 anchorage requirements to calculate bar cutoffs in simply supported reinforced concrete beams and slabs
- design lap splices for bars in tension and compression

9.1 INTRODUCTION

Steel reinforcing bars are embedded in concrete to increase the structural capacity of reinforced concrete members in flexure, compression, shear, and torsion. The bars act in either tension or compression — in order for the reinforcement to develop the required tension or compression forces, a *perfect bond* is required with the surrounding concrete. In this context, a perfect bond is one that permits no slippage at the steel-to-concrete interface.

To illustrate how essential a perfect bond is, it may be useful to visualize what would occur if the reinforcing bars were extremely smooth and there was no bond at the steel-to-concrete interface. In this case, the concrete and steel would move independently of each other — the bars would be free to slide inside the concrete. Consequently, it would not be possible for the force in steel reinforcement to form a couple with the concrete compression force and resist the effects of flexure or other load effects (as discussed in Section 3.5). In conclusion, the steel lacking a proper bond with concrete would serve little, if any, useful purpose.

To achieve adequate bond strength, the bars must be embedded or anchored in concrete over a certain length called the *development length*. Reinforcing bars must have a minimum development length in order to satisfy the basic strain compatibility requirement between steel and concrete, discussed in Section 3.3.2.

It is very important to understand that a proper bond between the steel and the concrete is critical for the satisfactory performance of reinforced concrete structures both at the service and ultimate load levels. At service loads, bond has an important role in controlling the magnitude of cracks and deflections. At ultimate loads, a proper bond is required to develop the yield strength in steel reinforcement.

The main objective of this chapter is to explain the concepts of bond, development length, and anchorage requirements for reinforced concrete members. The mechanism of the bond between the steel and the concrete and the characteristic modes of bond failure are explained in Section 9.2. The development length of straight bars in tension and compression is discussed in Section 9.3, whereas the development length of hooked bars is presented in Section 9.4. Detailing of hooks for stirrups and ties is discussed in Section 9.5. The concept of bar cutoffs in reinforced concrete flexural members is introduced in Section 9.6. The CSA A23.3 provisions related to anchorage design are outlined in Section 9.7. The procedure for calculating bar cutoffs is summarized in Section 9.8, along with an illustrative design example. Finally, detailing of bar splices for tension and compression reinforcement is outlined in Section 9.9.

9.2 BOND IN REINFORCED CONCRETE FLEXURAL MEMBERS

The bond between the steel reinforcement and the surrounding concrete will be explained using an example of a *pullout test* of a steel bar of diameter d_b embedded in a concrete block over a length l_d (see Figure 9.1a). When the bar is pulled by a tension force (T) of increasing magnitude, the bond stress (u) develops at the outside bar surface; u is defined as a local shearing stress transferred from the concrete to the bar interface. The distribution of bond stress along the embedded bar length is not uniform, as illustrated in Figure 9.1b. At the beginning of the test, the bond stress reaches a maximum near the free end of the embedded bar and zero at the embedded end, as shown with a dashed line in Figure 9.1b. However, at the end of the test, the distribution becomes more uniform and the bond stress is equal to zero at both bar ends (see the solid line in Figure 9.1b). In order to satisfy the equilibrium requirements, the resultant of the bond stress must be equal to the force developed in the steel reinforcement; that is,

$$u \times (l_d \times \pi \times d_b) = T \qquad\qquad [9.1]$$

where

$u =$ the average bond stress
$l_d \times \pi \times d_b$ is the outside bar surface area
$T = f_s \times A_b$ is the tension force in the steel
$A_b = \dfrac{\pi \times d_b^2}{4}$ is the bar cross-sectional area.

The tensile stress (f_s) in the bar reaches a maximum value at its free end and drops to zero at the embedded end. The variation of the steel stress is shown in Figure 9.1c. Note that a linear stress variation is shown in the diagram; that is, the stress builds up linearly from zero at the embedded end to the maximum at the free end of the bar. Research has shown that this assumption is conservative.

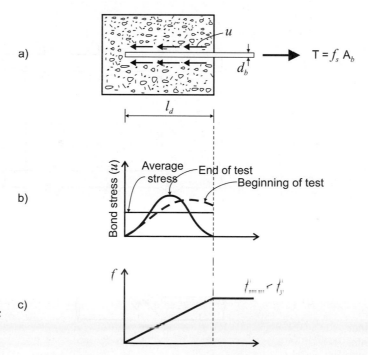

Figure 9.1 Pullout test on a steel bar embedded in concrete: a) test setup and stress distribution; b) bond stress distribution along the bar length; c) tensile stress distribution along the bar length.

It is expected that the bar will be pulled out of the concrete when the tension force (T) has reached a certain magnitude. Tests have shown that the pullout force depends on the embedment length (l_d): the longer the embedment length, the larger the force required to pull the bar out of the concrete. This relationship continues until the bar has an embedment length beyond which it cannot be pulled out; instead, the bar will yield when the stress in steel reaches the yield stress ($f_s = f_y$). At this point, the bar has developed its full tensile strength. The corresponding embedment length is called the *development length* and is denoted as l_d by CSA A23.3.

The actual mechanism of force transfer between the steel and the concrete is quite complex, especially in the case of deformed reinforcement, which is the most common type of reinforcement used in reinforced concrete design. Deformed bars have deformations (also called lugs) rolled into the bar surface during manufacture, as discussed in Section 2.6.3.

A deformed reinforcing bar embedded in concrete is shown in Figure 9.2. Initially, the bond between the steel and the concrete is achieved by chemical adhesion and friction (see Figure 9.2a). However, once the bar slips under an increasing load, most of the force is transferred by bearing on the bar deformations. These bearing forces (shown in Figure 9.2b) are balanced by the forces of the same magnitude and opposite direction acting on the concrete (see Figure 9.2c), which can be divided into a longitudinal component parallel to the bar and a radial component perpendicular to the bar, as shown in

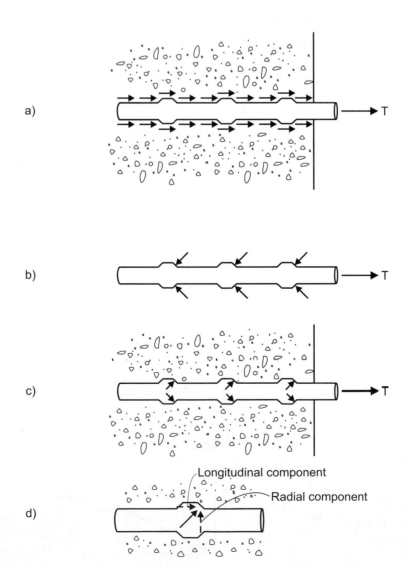

a)

b)

c)

d)

Longitudinal component

Radial component

Figure 9.2 Force transfer between the steel and the concrete: a) adhesion and friction forces; b) bearing forces acting on the bar; c) bearing forces acting on the concrete; d) longitudinal and radial force components.

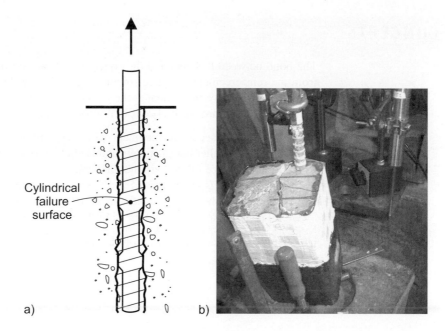

Figure 9.3 Pullout failure:
a) theoretical failure surface;
b) a laboratory test.

(S. J. Pantazopoulou)

Figure 9.2d. An increase in the tension force in the bar will result in further bond deterioration and cracking of the concrete, resulting in one of two failure modes:

- *Pullout failure* occurs in members with sufficiently large concrete cover and bar spacing and smaller bar diameters. Once the adhesive bond and friction resistance have been reached, the member fails by shearing along a cylindrical failure surface formed around the bars (see Figure 9.3).
- *Splitting failure* is characterized by the concrete splitting along the bar length. This failure occurs when cover, confinement, or bar spacing are insufficient to resist the lateral concrete tension resulting from the wedging effect of the bar deformations. Splitting cracks tend to develop along the shortest distance between the bar and the exterior face of the member or between two adjacent bars. These splitting cracks run parallel to the bar length in the horizontal and/or vertical direction. Splitting is the most visible sign of approaching bond failure in reinforced concrete beams. An example of splitting failure is shown in Figure 9.4. Splitting failure is also related to diagonal shear failure, discussed in Section 6.3.1.

Of these two modes, splitting failure is more common in reinforced concrete flexural members.

Note that deformed bars have significantly better bond resistance than smooth bars. In smooth bars, bond develops by adhesion between the bar and the concrete, plus a small amount of friction. Such adhesion deteriorates quite quickly with increasing load. For this reason, smooth bars are no longer used for flexural reinforcement in Canada.

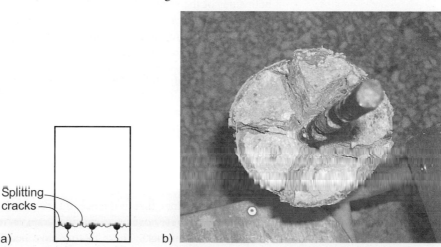

Figure 9.4 Splitting failure:
a) a beam section showing splitting cracks; b) a laboratory test.

(S. J. Pantazopoulou)

KEY CONCEPTS

The bond between the steel and the concrete allows the transfer of longitudinal forces from the reinforcement to the surrounding concrete. As long as the bond exists, there should be no relative movement at the interface between the steel and the concrete.

Two bond failure modes occur in reinforced concrete flexural members:

- *Pullout failure* occurs in members with sufficiently large concrete cover and bar spacing and smaller bar diameters and is characterized by shearing along a cylindrical failure surface formed around the bars.
- *Splitting failure* is characterized by the concrete splitting along the bar length. This failure occurs when cover, confinement, or bar spacing are insufficient to resist the lateral concrete tension resulting from the wedging effect of the bar deformations.

Of these two modes, splitting failure is more common in reinforced concrete flexural members.

9.3 | DEVELOPMENT LENGTH OF STRAIGHT BARS

9.3.1 Background

The concept of development length can be explained by a real-life example. Take a piece of chalk of a certain length. Wrap your fingers around the chalk at one end and have a partner grab the other end. Both of you should try to pull the chalk from each end simultaneously. The person that grabs a longer length will force the chalk to slip from the other person's fingers. This is due to the fact that the longer grip has a larger contact area with the chalk than the shorter grip. If this were a competition between the two people, each person would try to grab a longer length. The competition continues until both sides have grabbed a sufficient length of chalk to develop sufficient pull that the chalk breaks. This example illustrates the following important points related to the concept of development length:

1. *It takes a certain minimum length to develop a sufficient grip and break the chalk.* In reinforced concrete design, this minimum length of grip is known as the *development length,* that is, the minimum length required to develop the full tensile strength of the reinforcing bar.
2. *It takes two people, each with sufficient grip, to break the chalk.* To develop the full strength of a reinforcing bar, the required development length must be provided on both sides of the section at which full strength development is required.
3. *The thicker the chalk, the longer the grip required to break it.* This can be easily demonstrated by repeating the same competition with a larger diameter chalk. Larger diameter bars require longer lengths to develop their full strength.

The length of the bar from the point of maximum stress to its nearby free end must be at least equal to the development length (l_d). This is illustrated using an example of a simply supported beam subjected to uniform load, as shown in Figure 9.5a. The point of maximum stress is at the midspan, where the bending moment is maximum. The minimum development length for the reinforcement in Figure 9.5a must be provided at either side of the section. If that is the case, reinforcing bars should be capable of developing the yield stress (f_y) without experiencing bond failure. This is the basic requirement of anchorage design in reinforced concrete structures.

Some major factors that influence the required development length are concrete tensile strength, concrete cover, bar spacing, transverse reinforcement (stirrups), and bar location (top or bottom bar). These factors have been accounted for by the CSA A23.3 provisions for the development length.

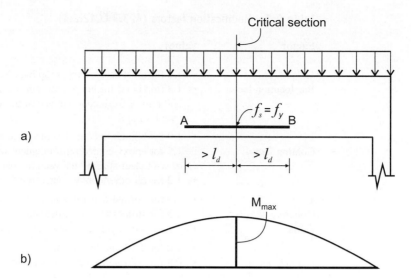

Figure 9.5 Development length: a) beam diagram showing flexural reinforcement; b) bending moment diagram.

This section discusses the development length for straight deformed bars. The development length for other types of reinforcement, such as welded wire fabric, is not discussed here (see A23.3 Cl.12.8 for more details).

9.3.2 Development Length of Straight Reinforcing Bars in Tension

The CSA A23.3 provisions for development length are based on the statistical analysis of a large amount of test data performed by Orangun and others in 1977. The basic equation for the development length, per A23.3 Cl.12.2.2, will not be discussed here. Instead, a simplified form of this equation, presented in A23.3 Cl.12.2.3, is considered to be more suitable for design applications and will be used in this text. The development length depends on the specific design application (Case 1 or Case 2) and it can be determined based on Eqns 9.2 and 9.3 below. These equations can be used when the clear cover is at least equal to d_b and the bar spacing is at least equal to $1.4d_b$.

For Case 1 applications (members containing minimum stirrups or ties within development length (l_d), or slabs, walls, shells, or folded plates with clear spacing between bars being developed not less than $2d_b$),

$$l_d = 0.45\, k_1 k_2 k_3 k_4 \frac{f_y}{\sqrt{f_c'}}\, d_b \qquad\qquad [9.2]$$

This is the basic equation and is appropriate for most design applications.

For Case 2 applications (all other members not meeting the requirements of Case 1),

$$l_d = 0.6\, k_1 k_2 k_3 k_4 \frac{f_y}{\sqrt{f_c'}}\, d_b \qquad\qquad [9.3]$$

where

l_d = the development length (mm)
k_1 = the bar location factor
k_2 = the coating factor
k_3 = the concrete density factor
k_4 = the bar size factor
d_b = the bar diameter (mm)
f_c' = the specified compressive strength of the concrete (MPa)
f_y = the specified yield strength of the steel reinforcement (MPa).

Note that the modification factors k_1, k_2, k_3, and k_4 account for various effects influencing the bar development length. The values of these factors as prescribed by A23.3

Table 9.1 Modification factors (A23.3 Cl.12.2.4)

Factor	Values
k_1 Bar location factor	• 1.0 for most cases, except top bars (see below) • 1.3 (top bars) for horizontal reinforcement so placed that more than 300 mm of fresh concrete is cast in the member below the development length or splice
k_2 Coating factor	• 1.0 for regular uncoated reinforcement • 1.5 for epoxy-coated reinforcement with clear cover less than $3d_b$ or with clear spacing between the bars being developed less than $6d_b$ • 1.2 for all other epoxy-coated reinforcement
k_3 Concrete density factor	• 1.0 for normal-density concrete • 1.3 for structural low-density concrete • 1.2 for structural semi-low-density concrete
k_4 Bar size factor	• 1.0 for 25M and larger bars • 0.8 for 20M and smaller bars and deformed wires

(*Source:* CSA A23.3-04 Cl.12.2.4, reproduced with the permission of the Canadian Standards Association)

Cl.12.2.4 are summarized in Table 9.1. According to the same clause, the product $k_1 \times k_2$ need not be taken greater than 1.7.

Note from Table 9.1 that the basic value for each modification factor is 1.0; however, the values can be different under certain circumstances. In general, typical l_d values for bottom bars are on the order of $30d_b$ for 10M, 15M, and 20M bars and approximately $40d_b$ for 25M and larger bars. Tension development lengths for various bar sizes are given in Table A.5. The background for these "modified" values is provided below.

Bar location factor (k_1) When bars are placed in the forms during construction so that a substantial amount of concrete is placed beneath them, there is a tendency for excess water and entrapped air to rise to the top of the fresh concrete during vibration. The water and air tend to accumulate on the underside of the bar, thus forming a small void pocket immediately beneath it. This is particularly common for top bars, that is, bars with more than 300 mm of fresh concrete beneath them at the time of construction. Tests have shown a significant loss in bond strength for top bars, so the development length must be increased accordingly. To account for the negative effects of top-bar casting position, CSA A23.3 prescribes an increased value for k_1 of 1.3 for top bars in beams with depths greater than 300 mm and horizontal steel in walls cast in lifts greater than 300 mm. (Note that this requirement does not apply to vertical reinforcement in walls.)

Bar coating factor (k_2) Epoxy-coated bars are sometimes used in projects where a structure is subjected to a corrosive environment or deicing chemicals — mostly bridges and parking garages. Tests have shown that the bond strength of epoxy-coated bars is significantly less than that of regular deformed bars. This is mainly due to the smooth epoxy coating, which significantly reduces the friction between the concrete and the bar deformations. Therefore, CSA A23.3 prescribes an increased k_2 value of 1.5 for epoxy-coated bars where the splitting failure is possible due to reduced cover and bar spacing and 1.2 where pullout failure is expected and the effect of epoxy coating is less pronounced.

Concrete density factor (k_3) Low-density concrete typically uses lighter-weight aggregate that will split at a lower tensile load. Also, the concentrated forces imposed by the bar deformations may cause localized crushing, which may allow the bar to slip.

Bar size factor (k_4) Tests have shown that bars of smaller diameters require a smaller development length.

A few other factors that are explicitly or implicitly accounted for in Eqns 9.2 and 9.3 are discussed below.

Concrete tensile strength The factor $\sqrt{f_c'}$ reflects the influence of the concrete tensile strength, which is an important factor, particularly as related to splitting failure. According to A23.3 Cl.12.1.2, $\sqrt{f_c'}$ is limited as follows:

$$\sqrt{f_c'} \le 8 \text{ MPa}$$

Steel yield strength Equations 9.2 and 9.3 indicate that the bar development length is directly proportional to the yield strength. The larger the yield strength, the larger is the required development length. This can be explained by referring to Eqn 9.1. The resultant of the bond stress along the bar length is equal to the tensile force developed in the bar.

Bar spacing Bar spacing requirements per CSA A23.1 were discussed in Section 5.3.2. In reinforced concrete beams, bars are usually spaced from d_b to $2d_b$ apart, whereas larger bar spacing is used in slabs, footings, and walls. Larger bar spacing results in increased bond strength as compared to closely spaced bars. Consequently, bars with larger spacing require smaller development lengths. Close bar spacing increases the possibility of the face shell spalling or the concrete splitting.

Concrete cover Concrete cover requirements per CSA A23.1 were discussed in Section 5.3.1. The effect of concrete cover on bond strength is similar to the effect of bar spacing. When the horizontal or vertical cover is increased, more concrete is available to resist the tension resulting from the wedging effect of the deformed bars; consequently, the resistance to splitting failure is improved.

Confinement (effect of transverse reinforcement) Transverse reinforcement (ties and stirrups) improves the resistance of tension reinforcement to splitting failure because the tensile force in the transverse reinforcement tends to prevent cracks from opening. Even if the face shell of the concrete member spalls off, transverse reinforcement can still hold the bars to the concrete core and maintain some bond resistance.

The effect of concrete cover, bar spacing, and confinement on bar development length is reflected in Eqns 9.2 and 9.3. For reinforced concrete members that satisfy the Case 1 requirements, a multiplier of 0.45 is used in Eqn 9.2. However, Case 2 members are generally characterized by smaller bar spacing and less confinement, so a multiplier of 0.60 is used, resulting in a 33% increase in the development length (provided that all other parameters are the same).

A23.3 Cl.12.2.5

Excess reinforcement The development length may be reduced when the area of tension reinforcement provided in the section ($A_{sprovided}$) is in excess of the required area ($A_{srequired}$). This can be accomplished by multiplying the development length by the ratio of the required and provided reinforcement areas:

$$l_d \times \frac{A_{srequired}}{A_{sprovided}}$$

For example, when the available geometry restricts bar length to provide only 50% of the normal development length, the designer is permitted to double the amount of required reinforcement. In terms of development, this is equivalent to having 100% of the reinforcement developed by the full length.

This provision should not be used where bars must be developed to ensure yielding at the maximum stress location (for example, development of positive moment reinforcement at supports). Whenever possible, the designer should avoid the use of this provision and ensure that the steel reinforcement is fully anchored in case the bars are fully stressed due to a sudden load increase.

The calculation of development length for straight deformed bars is illustrated in Examples 9.1 and 9.2.

Example 9.1

A simply supported reinforced concrete beam subjected to a uniform load is shown in the figure below. The beam is reinforced with shear reinforcement (stirrups) in excess of the minimum amount prescribed by CSA A23.3. The designer has determined that 10-30M bars 10 m long placed in two layers are required to resist the maximum bending moment at the midspan. Regular uncoated reinforcing bars have been used.

Determine the minimum development length required to develop the full design capacity of the reinforcing bars. Also, check whether the bar length is adequate for this design.

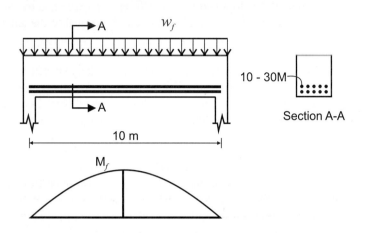

Given: $f_y = 400$ MPa

$f_c' = 25$ MPa (normal-density concrete)

SOLUTION: Determine the bar development length (l_d) according to A23.3 Cl.12.2.3. Because the beam is reinforced with the minimum shear reinforcement, this application can be classified as Case 1; consequently, use Eqn 9.2 to determine the development length.

1. **Establish the values of the modification factors**
   Determine the modification factors according to A23.3 Cl.12.2.4 (see Table 9.1):
   a) $k_1 = 1.0$ (bottom bars)
   b) $k_2 = 1.0$ (regular uncoated reinforcing bars)
   c) $k_3 = 1.0$ (normal-density concrete)
   d) $k_4 = 1.0$ (30M bars)

2. **Determine the development length (l_d)**
   Determine l_d for 30M bars (bar diameter $d_b = 30$ mm) as follows:

   $$l_d = 0.45\, k_1 k_2 k_3 k_4 \frac{f_y}{\sqrt{f_c'}}\, d_b \qquad\qquad \text{[9.2]}$$

   $$= 0.45(1.0)(1.0)(1.0)(1.0) \times \frac{400 \text{ MPa}}{\sqrt{25 \text{ MPa}}} \times 30 \text{ mm} = 1080 \text{ mm}$$

3. **Check whether the bar length is adequate**
   To resist the maximum bending moment, all rebars need to be extended by a minimum development length (l_d) on either side of the midspan. Since the beam spans 10 m, assume that the rebars are also 10 m long; consequently, the bar length from the midspan to the support is

   $$\frac{l}{2} = \frac{10\,000 \text{ mm}}{2} = 5000 \text{ mm}$$

Hence, the available length for 30M reinforcing bars at either side of the point of maximum stress is equal to 5000 mm. The calculated development length (l_d) is equal to 1080 mm (see the sketch below). So,

$$l_d = 1080 \text{ mm} < 5000 \text{ mm}$$

and the available bar length is adequate to develop its full design strength.

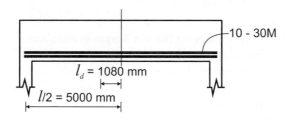

Example 9.2

A simply supported reinforced concrete slab with a thickness of 300 mm subjected to a uniform load is shown in the figure below. The designer has determined that 20M bars at 300 mm spacing on centre are required at the bottom face to resist the maximum bending moment at the midspan. The bar length is equal to 6 m and the clear cover to the bottom bars is equal to 25 mm. Due to the aggressive environmental conditions, epoxy-coated rebars are required for this design.

Determine the minimum development length required to develop the full design capacity of the reinforcing bars. Also, check whether the bar length is adequate for this design.

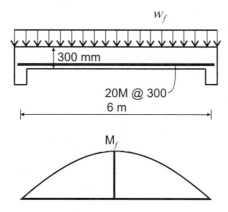

Given: $f_y = 400$ MPa

$f_c' = 30$ MPa (normal-density concrete)

SOLUTION: Determine the bar development length (l_d) according to A23.3 Cl.12.2.3. Because the slab is reinforced with 20M bars ($d_b = 20$ mm) at 300 mm spacing, that is,

$$2d_b = 40 \text{ mm} < 300 \text{ mm}$$

classify this application as Case 1, and use Eqn 9.2 to determine the development length.

1. **Establish the values of the modification factors**
    Determine the modification factors according to A23.3 Cl.12.2.4 (see Table 9.1).
    a) $k_1 = 1.0$ (bottom bars)
    b) $k_2 = 1.5$ (epoxy-coated bars with a 25 mm cover, that is, $3d_b = 60$ mm > 25 mm)
    c) $k_3 = 1.0$ (normal-density concrete)
    d) $k_4 = 0.8$ (20M bars)

2. **Determine the development length (l_d)**

    Determine l_d for 20M bars (bar diameter d_b = 20 mm) as follows:

    $$l_d = 0.45\, k_1 k_2 k_3 k_4 \frac{f_y}{\sqrt{f_c'}}\, d_b \qquad\qquad \textbf{[9.2]}$$

    $$= 0.45(1.0)(1.5)(1.0)(0.8) \times \frac{400\ \text{MPa}}{\sqrt{30\ \text{MPa}}} \times 20\ \text{mm} = 789\ \text{mm}$$

3. **Check whether the bar length is adequate**

    To resist the maximum bending moment, all rebars need to be extended by a minimum development length (l_d) on either side of the midspan. Because the slab spans 6 m, assume that the rebars are also 6 m long; consequently, the bar length from the midspan to the support is

    $$\frac{l}{2} = \frac{6\ \text{m}}{2} = 3\ \text{m}$$

    Hence, the available length for 20M reinforcing bars at either side of the point of maximum stress is equal to 3 m. The calculated development length (l_d) is equal to 789 mm (see the sketch below). So,

    $$l_d = 789\ \text{mm} < 3000\ \text{mm}$$

    and the available bar length is adequate to develop its full design strength.

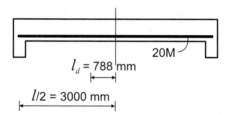

9.3.3 Development Length of Straight Bars in Compression

Compression development lengths are considerably shorter than tension development lengths because there are no tension cracks in the concrete anchorage region; also, the force transfer from the steel to the concrete is partially achieved by the bearing at the end of the bar.

A23.3 Cl.12.3.2 defines the basic compression development length as follows:

$$l_{db} = 0.24\, \frac{f_y}{\sqrt{f_c'}}\, d_b \le 0.044\, f_y d_b \qquad\qquad \textbf{[9.4]}$$

In order to obtain the hook development length, the basic development length (l_{db}) needs to be multiplied by modification factors with a cumulative value of not less than 0.6. The modification factors are as follows:

* $\dfrac{A_{srequired}}{A_{sprovided}}$ for reinforcement in excess of that required by analysis

*   0.75 for reinforcement enclosed with spiral reinforcement or with 10M ties in conformance with A23.3 Cl.7.6.5 and spaced not more than 100 mm on centre

The compression development length for a straight reinforcing bar should not be less than 200 mm (A23.3 Cl.12.3.1).

Development lengths in compression for different bar sizes are given in Table A.6.

In order to achieve adequate bond strength, reinforcing bars must be embedded or anchored in the concrete over a certain length called the *development length* (l_d). The development length depends on whether the bar is subjected to tension or compression and also on whether it is straight or hooked.

Development length of straight bars in tension (A23.3 Cl.12.2.3) *For Case 1 applications* (members containing minimum stirrups or ties within development length l_d, or slabs, walls, shells, or folded plates with clear spacing between bars being developed not less than $2d_b$),

$$l_d = 0.45\, k_1 k_2 k_3 k_4 \frac{f_y}{\sqrt{f_c'}}\, d_b \quad \text{(basic equation)}$$ **[9.2]**

For Case 2 applications (all other members not meeting the Case 1 requirements),

$$l_d = 0.6\, k_1 k_2 k_3 k_4 \frac{f_y}{\sqrt{f_c'}}\, d_b$$ **[9.3]**

The modification factors k_1, k_2, k_3, and k_4 account for various effects that influence the bar development length, as follows (refer to Table 9.1 for specific values):

- k_1 — the bar location factor
- k_2 — the coating factor
- k_3 — the concrete density factor
- k_4 — the bar size factor

According to A23.3 Cl.12.2.5, the development length may be reduced when the area of tension reinforcement provided in the section is in excess of the required area.

Development length of straight bars in compression (A23.3 Cl.12.3) The compression development length is considerably shorter than the tension development length and can be determined according to the equation

$$l_{db} = 0.24 \frac{f_y}{\sqrt{f_c'}}\, d_b \leq 0.044 f_y d_b$$ **[9.4]**

In some cases, the basic development length (l_{db}) needs to be multiplied by the modification factors that account for the excess reinforcement and the effect of confinement by spirals or ties.

9.4 STANDARD HOOKS IN TENSION

Special anchorage in the form of a hook is used when the available bar length is insufficient for full development (for example, end spans of beams or slabs). The use of hooked ends of reinforcing bars can provide sufficient anchorage over a shorter length than required for the tension development of straight bars. A hook consists of a straight bar which is curved at its end. Both the 90° and 180° hooks shown in Figure 9.6 are commonly used to provide anchorage for flexural reinforcement.

Consider a simply supported reinforced concrete beam subjected to a point load, as shown in Figure 9.7a. The bending moment diagram for the beam is shown in Figure 9.7b. The bending moment at section A-A is equal to M_A (see Figure 9.7a). The distance

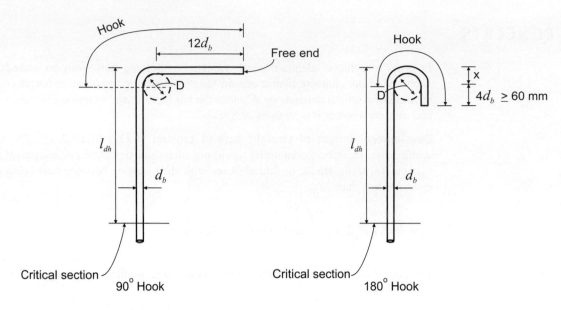

Figure 9.6   Standard hooks for flexural reinforcement.

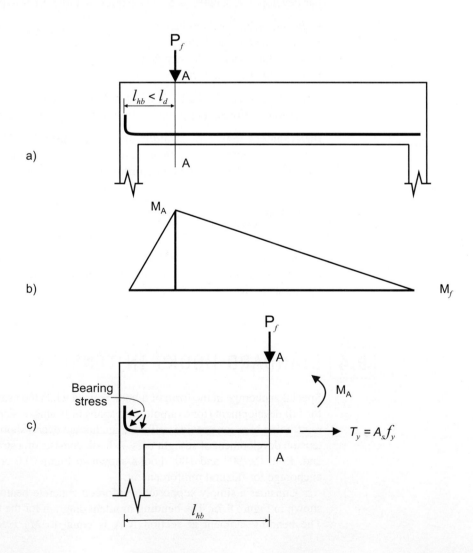

a)

b)

c)

Figure 9.7   Hook behaviour:
a) reinforced concrete beam
subjected to a point load;
b) bending moment diagram;
c) hooked anchorage for reinforcement fully developed
at section A-A.

between section A-A and the left beam support is less than the required development length (l_d) for the straight reinforcing bars used in this beam. Consequently, hooked anchorage is required, as shown in Figure 9.7c.

A hook helps develop the tensile capacity of the bar by the combined effect of the bond on the bar surface area and the bearing at the inside face of the curve (see Figure 9.7c). The hook has a tendency to straighten under tension. Tests have shown that the concrete splitting in the plane of the hook is the main cause of failure for hooked bars in tension. Splitting failure occurs due to the very high stresses in the concrete on the inner side of the hook. The size of the vertical and horizontal concrete cover to the hooked bar strongly influences the splitting resistance: the larger the cover, the larger the splitting resistance.

Note that hooks act in tension only and are not considered to help anchor bars in compression.

Hook dimensions and bend diameters have been standardized by CSA A23.1 Cl.6.6.2.2, which defines standard hooks for reinforced concrete structures. The basic hook development length (l_{hb}) is measured from the critical section to the farthest point on the hooked bar, as shown in Figure 9.7c. According to A23.3 Cl.12.5.2, the basic development length for a hooked bar with an f_y of 400 MPa is

$$l_{hb} = 100 \frac{d_b}{\sqrt{f_c'}}$$ [9.5]

In prescribing l_{hb} values, CSA A23.3 accounts for the combined contribution of bond along the straight bar ending with the hook and the hooked anchorage.

According to A23.3 Cl.12.5.1, the development length for hooked bars in tension (l_{dh}) should be obtained by multiplying the basic hook development length (l_{hb}) by appropriate modification factors (MF); that is,

$$l_{dh} = l_{hb} \times MF = 100 \frac{d_b}{\sqrt{f_c'}} \times MF$$ [9.6]

However, l_{dh} should not be less than $8d_b$ or 150 mm, whichever is greater.

The modification factors for hooked anchorages are summarized in Table 9.2 and illustrated in Figure 9.8. Hook development lengths for different bar sizes are given in Table A.7.

Table 9.2 Modification factors (MF) for hooked anchorages (A23.3 Cl.12.5.3)

Case	Description	Modification factor (MF)
1	For bars with f_y other than 400 MPa	$\frac{f_y}{400 \text{ MPa}}$
2	For 35M or smaller bars, where the side cover (normal to the plane of the hook) is not less than 60 mm (Figure 9.8a), and for 90° hooks, where the cover on the bar extension beyond the hook is not less than 50 mm (Figure 9.8b)	0.7
3	For 35M or smaller bars, where the hook is enclosed vertically or horizontally within at least three ties or stirrup ties spaced along a length at least equal to the inside diameter of the hook, at a spacing not greater than $3d_b$, where d_b is the diameter of the hooked bar (Figure 9.8c)	0.8
4	For reinforcement in excess of that required by analysis where the full development of a bar to achieve the yield strength f_y is not specifically required	$\frac{A_{s required}}{A_{s provided}}$
5	For structural low-density concrete	1.3
6	For epoxy-coated reinforcement	1.2
7	For all other applications	1.0

(*Source:* CSA A23.3-04 Cl.12.5.3, reproduced with the permission of the Canadian Standards Association)

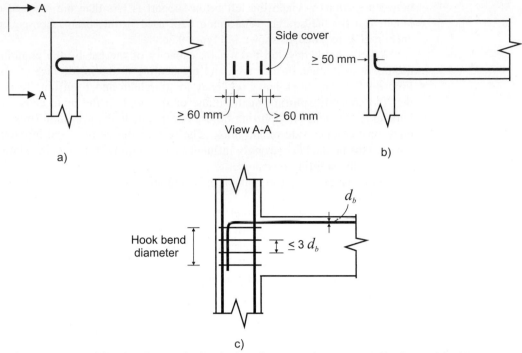

Figure 9.8 Modification factors for hooked anchorages: a) 35M or smaller bars with side cover of 60 mm or larger; b) 90° hooks where the cover on the bar extension is greater than or equal to 50 mm; c) vertically or horizontally enclosed hooks.

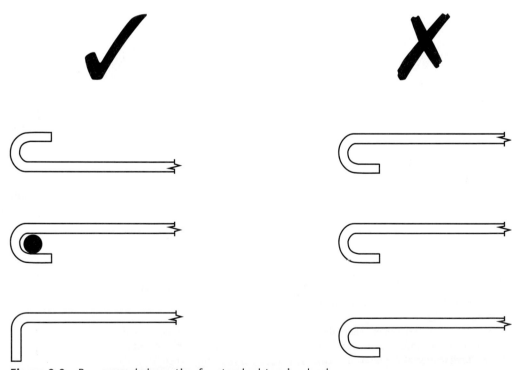

Figure 9.9 Recommended practice for standard tension hooks.

Some practical recommendations related to tension hooks are illustrated in Figure 9.9.

In some cases, there is not enough room for hooks or the necessary confinement steel, so special mechanical devices are required. These devices are in the form of welded plates, T-headed bars, or specially manufactured devices. The CSA A23.3 provisions related to anchorage by means of mechanical devices are provided in A23.3 Cl.12.6.

The design of a hooked anchorage will be illustrated by Example 9.3.

Example 9.3

A simply supported reinforced concrete beam subjected to the point load P_f is shown in the figure below. The designer has determined that 6-30M bars are required to resist the maximum bending moment (M_f) at the section 900 mm away from the centreline of the left support. Regular uncoated reinforcement has been used. A minimum 40 mm side cover at the supports should be provided.

Determine the required development length for the reinforcement at the section with the maximum bending moment. If it is not possible to provide straight bar anchorage into the left support, design the hooked anchorage such that the rebars develop full tension strength.

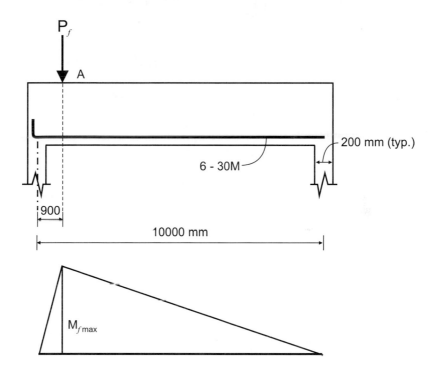

Given: f_y = 400 MPa

f_c' = 25 MPa (normal-density concrete)

SOLUTION:

1. **Determine the development length (l_d) for straight bars**

The development length for a 30M bar of the same characteristics and same modification factors was determined from Example 9.1 as

$$l_d = 1080 \text{ mm}$$

There is ample development length for the 30M bars on the right-hand side of the point load. However, the available length from the left side of the point load to the exterior face of the left support is

$$900 \text{ mm} + \frac{200 \text{ mm}}{2} = 1000 \text{ mm}$$

When an allowance for the side cover (40 mm) has been made, the available development length for a 30M bar is

$$1000 \text{ mm} - 40 \text{ mm} = 960 \text{ mm}$$

This is less than the minimum development length of 1080 mm required to develop the full bar strength, as shown on the sketch below. Therefore, hooked anchorage is required.

2. **Determine the development length (l_{dh}) for hooked anchorage**
   Determine the l_{dh} value according to A23.3 Cl.12.5.1 and Cl.12.5.2.

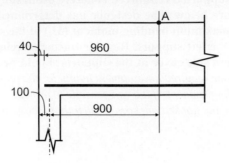

a) Find the bar diameter (d_b):
    30M bars have been used, so d_b = 30 mm (see Table A.1).
b) Determine the modification factor (MF):
    Since the cover is equal to 40 mm, MF = 1.0 (see Table 9.2).
c) Determine l_{dh}:

$$l_{dh} = 100 \frac{d_b}{\sqrt{f_c'}} \times MF \qquad [9.6]$$

$$= 100 \times \frac{30 \text{ mm}}{\sqrt{25 \text{ MPa}}} \times 1.0 = 600 \text{ mm}$$

3. **Design the hooked anchorage**
   The available length of the hooked bar from the point of maximum moment to the left end is 950 mm, which is greater than l_{dh} = 600 mm. Therefore, a hooked 30M bar can develop the full design strength at the point of the maximum moment. The hooked anchorage is shown on the sketch below. Note that it is not common in design practice to provide hooked anchorage for bottom reinforcement.

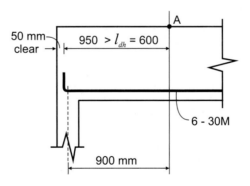

9.5 | HOOKS FOR STIRRUPS AND TIES

Hooks are used to provide anchorage for stirrups and ties in reinforced concrete beams and columns. According to A23.1 Cl.6.6.2.2(c), either 90° or 135° bends can be used for stirrup and tie hooks, as shown in Figure 9.10. However, A23.3 Cl.7.1.2 prescribes that a *standard stirrup hook* be characterized by a 135° (or larger) bend, unless the concrete cover surrounding the hook is restrained against spalling, in which case 90° bends can be used.

In some cases, it is necessary to provide closed stirrups in reinforced concrete members — this is often the case with members subjected to torsional effects (as discussed in Section 7.7). Closed stirrups should be designed to completely enclose a cross-section of the member, as shown in Figure 9.11a. However, closed stirrups can be obtained if cross ties are used in

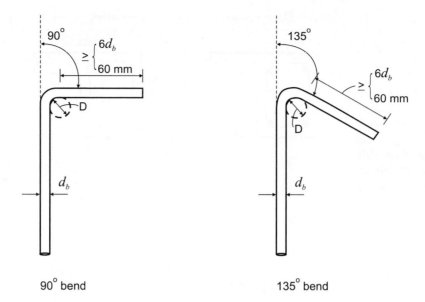

Figure 9.10 Standard hooks for stirrups and ties.

90° bend

135° bend

combination with standard stirrups, as shown in Figure 9.11b. A *cross tie* usually has a 135° hook at one end and a 90° hook at the other end (A23.3 Cl.7.1.3). Note that a more common alternative is shown in Figure 9.11c, where the top steel at the edge of the beam is hooked into the beam by a 90° hook and used in combination with an open stirrup.

Anchorage requirements for shear reinforcement are outlined in A23.3 Cl.12.13. The detailing of shear reinforcement in reinforced concrete beams was covered in Section 6.9, whereas the detailing of column ties was covered in Section 8.8.

KEY CONCEPTS

A hook consists of a straight bar that is curved at its end. Hooked bars are used under the following circumstances:

- to provide anchorage for flexural reinforcement when the available bar length of a straight bar is insufficient for its full development (for example, end spans of beams or slabs)
- to provide anchorage for stirrups and ties in reinforced concrete beams and columns

The use of hooked reinforcing bars can provide anchorage over a shorter length than required for the tension development of straight bars. CSA A23.1 Cl.6.6.2.2 defines standard hooks for reinforced concrete structures. Note that hooks act in tension only and are not considered to help anchor the bars in compression.

Hooked anchorage for straight bars in tension (A23.3 Cl.12.5.1 and 12.5.2)
Both 90° and 180° hooks are commonly used to provide anchorage for flexural reinforcement. The basic hook development length (l_{hb}) is measured from the critical section to the farthest point on the hooked bar. The development length for hooked bars in tension (l_{dh}) should be obtained by multiplying the basic hook development length (l_{hb}) by appropriate modification factors (MF):

$$l_{dh} = l_{hb} \times \mathrm{MF} = 100 \frac{d_b}{\sqrt{f_c'}} \times \mathrm{MF} \qquad \text{[9.6]}$$

Modification factors account for the effects of steel yield strength, confinement, and excess reinforcement (refer to Table 9.2 for specific values).

Hooked anchorage for stirrups and ties (A23.3 Cl.12.13) Hooks are used to provide anchorage for stirrups and ties in reinforced concrete beams and columns. A *standard stirrup hook* is characterized by a 135° (or larger) bend, unless the concrete cover surrounding the hook is restrained against spalling, in which case 90° bends can be used.

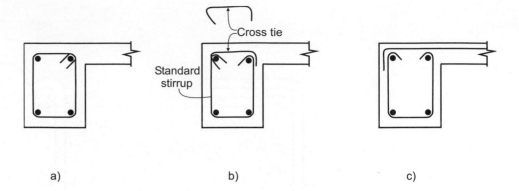

Figure 9.11 Closed stirrups: a) one-piece closed stirrup; b) combination of an open stirrup and a cross tie; c) top steel at the edge of the slab used in combination with an open stirrup (most common).

9.6 | BAR CUTOFFS IN SIMPLY SUPPORTED FLEXURAL MEMBERS

9.6.1 Background

In practice, the designer's aim is to produce cost-effective designs characterized by an optimal use of building materials. One way of achieving this objective is to terminate the rebars in reinforced concrete flexural members where they are no longer needed to resist bending moments. In design practice, bars that are terminated within a flexural member are called *cutoff bars,* and the corresponding points where these bars are terminated are called *cutoff points*. This section discusses the bar cutoff requirements for flexural reinforcement in reinforced concrete beams and slabs according to CSA A23.3. Note that both simply supported and continuous members will be discussed herein; however, the calculation of bar cutoffs in continuous members will be discussed in more detail in Section 11.4.

9.6.2 Theoretical Point of Cutoff

The concept of bar cutoffs will be explained using an example of a simply supported beam subjected to a uniform load, as shown in Figure 9.12a. The beam was designed based on the maximum factored bending moment (M_f) at the midspan. It was determined that six reinforcing bars of equal size are required at the midspan; the bars are labelled 1 to 6, as shown on the beam cross-section. The factored bending moment diagram is shown in Figure 9.12b. First, let us establish a correlation between the required amount of tension reinforcement and the corresponding factored moment resistance. Recall from Section 3.5.1 that the moment resistance for a flexural member can be determined as a product of the tension force in steel (T_r) and the lever arm ($d - a/2$) formed by the tensile force in the steel and the compression force in the concrete. This can be expressed by the following equation:

$$M_r = T_r \left(d - \frac{a}{2} \right)$$ [3.13]

where

$$T_r = \phi_s f_y A_s$$ [3.9]

For lightly reinforced flexural members, the lever arm ($d - a/2$) can be approximated as a constant with a value approximately equal to $0.9d$. At the stage of maximum flexural capacity, where the reinforcing steel has yielded, the tension force (T_r) is proportional to the area of tension steel (A_s). Consequently, A_s can be determined from Eqn 3.13 as follows:

$$A_s = \frac{M_r}{\phi_s f_y \left(d - \dfrac{a}{2} \right)} \cong \frac{M_r}{\phi_s f_y (0.9d)}$$

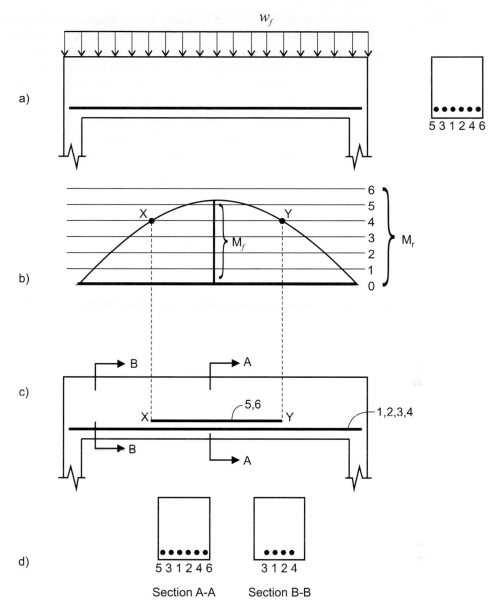

Figure 9.12 Theoretical bar cutoff points: a) simply supported beam and the reinforcement arrangement; b) bending moment diagram; c) beam diagram showing bar cutoffs; d) sections A-A and B-B.

Hence, the area of tension steel (A_s) can be approximated as directly proportional to the factored moment resistance (M_r).

Next, let us establish a correlation between the bending moment and the corresponding required amount of reinforcement at a particular section along the beam length. Refer again to the beam in Figure 9.12a. Since the beam is reinforced with six bars of equal size, the moment resistance can be divided into six segments, as illustrated by the horizontal lines in Figure 9.12b. In this manner, it is possible to determine the amount of moment resistance supplied by a single bar. Bars are labelled by the numbers 1 to 6 (see the beam cross-section in Figure 9.12a).

The designer is allowed to minimize the use of steel by terminating (cutting off) each bar at the point where it is no longer required. In practice, rebars are cut off in multiples of two; usually, 50% of the total number of rebars are cut off at the specified locations, although it is not uncommon to cut off 33% of the rebars. Suppose the designer intends to cut off 33% of the total number of rebars (in this case, six) required at the midspan where they are no longer required, so the bars with numbers 5 and 6 would be cut. The locations at which these bars are no longer required can be graphically determined as points X and Y at the intersection of the horizontal dashed line indicating the moment resistance for the remaining four bars (1 to 4) and the factored bending moment diagram (see Figure 9.12b). This information is usually presented on a scaled beam diagram showing the reinforcing bars and their respective lengths, as illustrated in Figure 9.12c. Notice from the beam diagram that bars 5 and 6 have been terminated, whereas the remaining bars (1 to 4) remain continuous throughout the beam to the supports. This is also illustrated on cross-sections A-A and B-B in Figure 9.12d.

   The reader should also note the following:

- The process of terminating the bars can be further continued by cutting off bars 3 and 4 or 1 and 2; however, some bars need to remain continuous over the supports (this will be discussed in more detail in Section 9.7).
- In the above example, the locations of bar cutoff points have been determined graphically; however, the same locations can be determined by calculations.
- The cutoff points for bars 5 and 6 are the points of zero stress in these bars ($f_s = 0$). These points are called *theoretical cutoff points*. However, due to shear and other effects that might cause cracking at the point at which the bars are to be terminated, the bars need to be extended beyond the theoretical cutoff point (to be discussed in Section 9.6.3).

In practice, an optimal bar cutoff solution strikes a balance between saving materials and increasing labour cost. Rebar placement can become an extremely time-consuming task when the worker in the field has to measure the position of every rebar. Consequently, the additional labour cost can outweigh savings in the material cost. More importantly, the designer must recognize that, in the interest of saving materials, the rebar cutoff design should not be too confusing for the rebar placer in the field; otherwise, incorrect rebar placement might lead to structural problems. The designer should attempt to specify a limited number of cutoff points along the member. Also, the cutoff lengths should always be designed with some reserve to allow for possible field errors during rebar placement.

   In short, rebar cutoffs should be designed in a simple and clear fashion so that the workers in the field are able to place the rebars without confusion or a significant increase in the labour cost.

   The concept of the theoretical cutoff point will be illustrated by Example 9.4.

Example 9.4

Consider a simply supported beam subjected to a specified uniform dead load (DL) of 60 kN/m and a specified uniform live load (LL) of 30 kN/m, as shown in the figure below. The beam is reinforced with 10-30M bars in two layers, but the designer would like to cut off 50% of the bars where they are no longer required by the design. Note that the same beam was discussed in Example 9.1.

Determine the theoretical cutoff points at which 50% of the bars can be terminated.

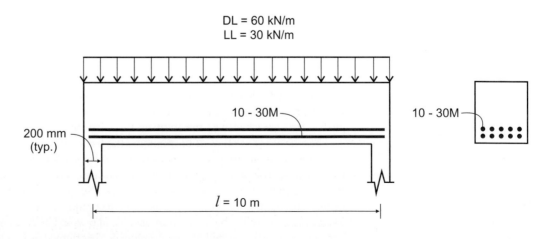

Given: $f_y = 400$ MPa
         $f_c' = 25$ MPa

SOLUTION: **1. Determine the required moment resistance at the cutoff location**
   a) Factored load:
       The factored load (w_f) is determined according to NBC 2010 Cl.4.1.3.2:

$$w_f = 1.25 \text{ DL} + 1.5 \text{ LL} = 1.25 \, (60 \text{ kN/m}) + 1.5(30 \text{ kN/m}) = 120 \text{ kN/m}$$

   b) The factored shear force at the support is

$$V_f = \frac{w_f l}{2} = \frac{120 \text{ kN/m} \times 10 \text{ m}}{2} = 600 \text{ kN}$$

   c) The maximum factored bending moment at the midspan (corresponding to 10-30M bars) is

$$M_{fmax} = \frac{w_f l^2}{8} = \frac{(120 \text{ kN/m}) \times (10 \text{ m})^2}{8} = 1500 \text{ kN} \cdot \text{m}$$

       It can be taken that $M_{r10} = M_{fmax} = 1500 \text{ kN} \cdot \text{m}$ (M_{r10} denotes the moment resistance corresponding to 10-30M bars).
   d) Determine the required moment resistance for 5-30M bars (M_{r5}).
       Determine the moment resistance as proportional to the number of bars, so (see the sketch below)

$$M_{r5} = \frac{M_{r10}}{2} = \frac{1500 \text{ kN} \cdot \text{m}}{2} = 750 \text{ kN} \cdot \text{m}$$

       In reality, the moment resistance for a section with five bars (M_{r5}) is slightly greater than 50% of M_{r10}, so this assumption is conservative.

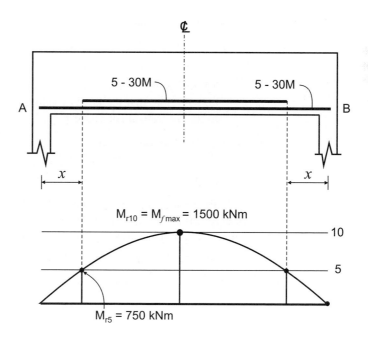

2. Determine the locations of the theoretical bar cutoff points
   Find the location(s) (x) along the beam length where the factored bending moment is

$$M_f = M_{r5} = 750 \text{ kN} \cdot \text{m}$$

a) Develop the general equation for the bending moment along the beam; that is,

$$M_x = V_f \times x - w_f \times \frac{x^2}{2}$$

$$= (600 \text{ kN}) \times x - (120 \text{ kN/m}) \times \frac{x^2}{2}$$

b) Find the location (x) where $M_x = 750 \text{ kN} \cdot \text{m}$.

This location can be determined from the free-body diagram of the beam in the sketch below.

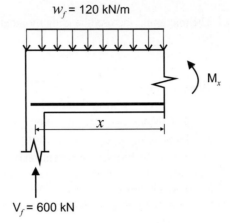

$w_f = 120 \text{ kN/m}$

M_x

x

$V_f = 600 \text{ kN}$

$$M_x = (600 \text{ kN}) \times x - (120 \text{ kN/m}) \times \frac{x^2}{2} = 750 \text{ kN} \cdot \text{m}$$

The above equation can be rearranged as a quadratic equation:

$$60\, x^2 - 600\, x + 750 = 0$$

The above equation can be solved for x as follows:

$$x = \frac{600 \pm \sqrt{600^2 - 4(750)(60)}}{2(60)} = 1.464 \text{ m}$$

Therefore, the theoretical bar cutoff point is 1.46 m from the centre-line of support A. Note that the larger x value, also a solution to the equation, corresponds to the theoretical bar cutoff point located at 1.46 m from support B.

3. **Determine the bar length**

The theoretical length of the 5-30M bars that are being terminated is

$$10 \text{ m} - 2 \times 1.46 \text{ m} = 7.08 \text{ m}$$

The final bar arrangement is presented in the sketch below. Note that this is only the theoretical bar cutoff length. The actual bar cutoff length prescribed by CSA A23.3 is longer and will be discussed in Section 9.7.3.

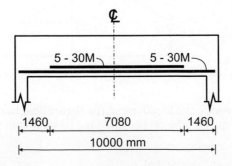

℄

5 - 30M   5 - 30M

1460   7080   1460

10000 mm

9.6.3 Bar Extensions

The theoretical cutoff points discussed in the previous section are based on the flexural requirements; that is, the bars are cut off where they are no longer needed to resist flexure. However, CSA A23.3 requires the bars to be extended beyond the theoretical cutoff point for the following reasons:

1. *Shear effects:* In reinforced concrete members subjected to shear, the actual tensile forces developed in the flexural reinforcement are larger than the forces determined based on the flexural requirements only. As a result, the required amount of flexural reinforcement at a beam section with diagonal tension cracks is larger than that determined based on the bending moment at that section. This additional reinforcement requirement can be satisfied by extending the available reinforcement beyond the theoretical cutoff point. According to A23.3 Cl.11.3.9.1, the additional bar extension needs to be at least equal to $d_v \cot \theta$.

2. *Stress concentrations:* When a bar is cut off, a significant stress transfer between the terminated bar and the continuing bars takes place. The continuing bars must suddenly resist the entire tensile force, and stress concentration at the bar cutoff locations may occur, thereby causing flexural cracks. When a bar is extended beyond the point of theoretical cutoff, the stress concentration moves toward the location of smaller stress and the consequences of bar cutoffs are less pronounced.

3. *Variations in bending moment values:* The actual bending moment may differ from the value determined by the analysis due to one of the following reasons: uncertainties in the actual load application, approximations in the structural models used in the analysis, unaccounted-for bending moments at the supports, lateral load effects, etc.

The concept of bar extension can be illustrated using the same beam example as shown in Figure 9.12 and discussed in Section 9.6.2. Points X and Y on the bending moment diagram (Figure 9.12b) denote the locations at which cutoff bars 5 and 6 are no longer required based on flexural requirements; that is, these are the theoretical cutoff points for bars 5 and 6. The same bending moment diagram is shown in Figure 9.13a. However, based on the above discussion, these bars need to be extended beyond the points X and Y by a length approximately equal to the member depth (h). These bar extensions are shown graphically on the bending moment diagram in Figure 9.13a: point X moves toward the left support along the horizontal line by a distance h to the point X', and point Y moves toward the right support by the same distance up to the point Y'. As a result, the length of bars 5 and 6 is extended from XY (based on the theoretical cutoff points) to X'Y' (due to

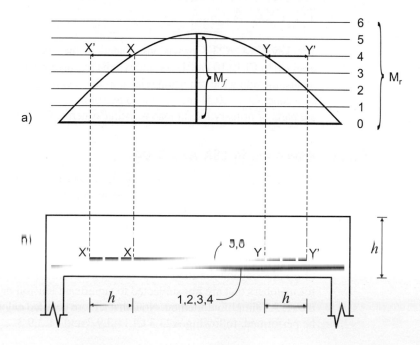

Figure 9.13 Bar extensions:
a) bending moment diagram;
b) beam elevation showing
bar extensions.

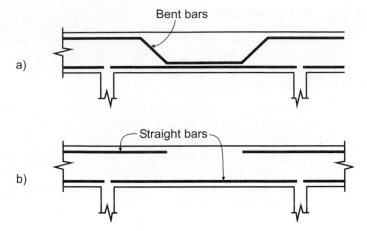

Figure 9.14 Flexural reinforcement in a continuous reinforced concrete beam: a) bent bars; b) straight bars.

the bar extension requirements), as shown in Figure 9.13b. Points X′ and Y′ are called *actual points of cutoff;* this will be discussed in more detail in Section 9.7.

9.6.4   Bent Bars

As an alternative to cutting off the flexural reinforcement, bottom bars may be anchored by bending them across the web and making them continuous with the top bars — such bars are called *bent bars*. Bent bars in a continuous reinforced concrete beam are shown in Figure 9.14a. (Note that Figure 9.14b shows the same beam with a straight bar arrangement.) It is more complex to detail and place bent bars than straight bars. However, note that the use of bent bars has an inherent benefit: bent bars are effective in resisting the spread of diagonal tension cracks; in fact, these bars act as shear reinforcement at the locations of the bends. The construction cost of members with bent bars is larger than that of those with straight bars. In Canada, the use of bent bars in concrete construction was abandoned in the 1960s; however, A23.3 Cl.12.10.1 permits their use in reinforced concrete flexural members.

9.7   ANCHORAGE DESIGN FOR FLEXURAL REINFORCEMENT ACCORDING TO CSA A23.3

The key CSA A23.3 provisions related to the anchorage of flexural reinforcement are outlined in Cl.12.10 and 12.11 and will be discussed in this section. Note that some requirements, including the provision related to the anchorage of flexural reinforcement in the tension zone (Cl.12.10.5) and the anchorage of positive moment reinforcement in members that are parts of primary lateral load resisting systems (Cl.12.11.2), are not covered in this book.

9.7.1   Revisions in CSA A23.3-04

CSA A23.3.04 does not contain any major revisions of the previous (1994) edition related to anchorage design. However, the provision related to bar extensions at the theoretical cutoff points (Cl.11.3.9.1 in the 2004 code) has been revised, which may affect the detailing of flexural reinforcement as compared to the 1994 version of the code.

A23.3 Cl.11.3.9.1   **Actual point of cutoff**   Flexural tension reinforcement should be extended a distance of $d_v \cot \theta$ beyond the theoretical point of cutoff. This detailing provision should be used for members that are not subjected to significant tension or torsion, designed according to the A23.3 simplified method. Note that a more detailed calculation of the bar extension can be performed, following A23.3 Cl.11.3.9.2 and 11.3.9.3.

However, according to Cl.11.3.8.1 of the previous edition (CSA A23.3-94), cutoff bars must be extended beyond the theoretical point of cutoff for a distance of d or $12d_b$, whichever is greater.

9.7.2 General Anchorage Requirement

A23.3 Cl.12.1.1

This is the basic provision governing the development of flexural reinforcement. *The design must ensure that each bar is properly embedded on both sides of the critical section within a reinforced concrete flexural member* (recall the chalk example in Section 9.3.1). If that is the case, the bar will yield at the ultimate condition; that is, the yield stress (f_y) is reached before the bond fails. In the case of straight reinforcing bars, the minimum required embedment length is equal to the development length (l_d) discussed in Section 9.3. Note that anchorage can also be achieved by means of hooks (discussed in Section 9.4) or mechanical devices.

9.7.3 Actual Point of Cutoff

A23.3 Cl.12.10.3
Cl.11.3.9.1

According to Cl.11.3.9.1, *flexural tension reinforcement should be extended a distance of* 1.3d *or* h, *whichever is greater, beyond the theoretical point of cutoff* (where d is the effective depth of the member and h denotes its overall depth). The point where the reinforcing bars are terminated is called the *actual point of cutoff*. This requirement is based on the increased tension forces in the bars due to shear and other effects discussed in Section 9.6.3 and applies to members that are not subjected to significant tension or torsion. Note that the above values have been derived from the following expression given in Cl.11.3.9:

$$d_v \cot \theta$$

where

d_v = the effective shear depth, taken as the greater of 0.9d and 0.72h (see Section 6.5.4)
θ = 35° (A23.3 Cl.11.3.6.3) if $f_y \leq 400$ MPa and $f_c' \leq 60$ MPa.

Since

$$\cot \theta = \cot (35°) = 1.43$$

$$d_v \cot \theta = \begin{bmatrix} 0.9d \\ 0.72h \end{bmatrix} \times 1.43 = \begin{bmatrix} 1.3d \\ h \end{bmatrix}$$

Note that this provision should be used *only* in conjunction with the CSA A23.3 simplified method of shear design because that method does not directly account for additional tension forces in the member caused by shear effects. This provision cannot be used in conjunction with the CSA A23.3 general method of shear design.

To illustrate the concept of the actual point of cutoff, let us consider a segment of the reinforced concrete beam in Figure 9.15. There are two types of bars: bars A that are to be

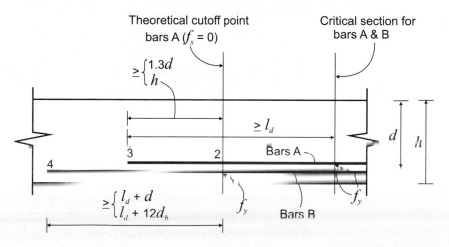

Figure 9.15 Actual point of cutoff.

terminated where no longer needed and continuing bars B (to be discussed in Section 9.7.4). Point 1 is located at the critical section characterized by the maximum bending moment. Point 2 is the point at which bars A are no longer needed based on the bending moment diagram, that is, the theoretical point of cutoff for bars A. Point 3 is the actual point of cutoff for bars A. The minimum required length for bars A beyond point 1, that is, the length of the segment 1-3, is equal to the greater of

- the development length (l_d), based on the general anchorage requirement prescribed by A23.3 Cl.12.1.1, and
- the sum of the lengths of segments 1-2 and 2-3, where segment 2-3 represents the bar extension beyond the theoretical cutoff point (based on Cl.11.3.9.1).

At this point, note that the *critical sections* for the development of reinforcement in flexural members are the points of maximum stress and the points where adjacent reinforcing bars may be terminated based on the flexural requirements (A23.3 Cl.12.10.2). In the example in Figure 9.15, point 1 is located at the critical section for bars A and B, whereas point 2 is located at the critical section for the continuing bars B.

9.7.4   Development of Continuing Reinforcement

A23.3 Cl.12.10.4

The continuing reinforcement should have an embedment length of not less than the development length (l_d) plus the longer of the effective depth of the member (d) and ($12d_b$) beyond the theoretical cutoff point for the tension reinforcement that is being terminated.

The concept of continuing reinforcement is illustrated in Figure 9.15. In this example, bars B are the continuing bars. A23.3 Cl.12.10.4 requires that bars B be extended by l_d plus the greater of d and $12d_b$ beyond point 2 (the theoretical cutoff point for bars A that are being terminated). Hence, the minimum length for bars B from the critical section (point 1) to the left end is equal to the length of segment 1-2 plus the greater of ($l_d + d$) and ($l_d + 12d_b$).

9.7.5   Development of Positive Moment Reinforcement at Supports

A23.3 Cl.12.11.1

At least one-third of the tension reinforcement in simply supported members and one quarter of the positive moment reinforcement in continuous members must extend into the support. In cast-in-place beams, this reinforcement shall extend into the support by at least 150 mm. This requirement is meant to ensure the ability of the structure to tolerate unanticipated actions such as load concentrations and pattern loadings in continuous members; however, it must also be used in the design of simply supported beams and slabs.

9.7.6   Anchorage of Negative Moment Reinforcement into Supporting Members

A23.3 Cl.12.12.1

This requirement is related to the negative moment reinforcement in continuous beams and slabs that are parts of building frames and the negative moment reinforcement in cantilevered members. *The negative moment reinforcement should be anchored in (or through) the supporting member by embedment length, hooks, or mechanical anchorage.*

Consider the typical interior span AB of a continuous beam that is part of a reinforced concrete frame, as shown in Figure 9.16a. Observe from the bending moment diagram that the maximum negative bending moment develops at the faces of the supports (points A and B) (see Figure 9.16b). Consequently, the top (negative) reinforcement that resists the negative bending moment needs to be extended beyond the face of the support into the adjacent member (beam), as shown in Figure 9.16c. When straight bars are used, this extension needs to be at least equal to the development length (l_d). However, in the case of the end span in the continuous beam in Figure 9.16d, the reinforcement is usually anchored by means of hooks; in that case, the corresponding hook development length is denoted as l_{dh}, as discussed in Section 9.4.

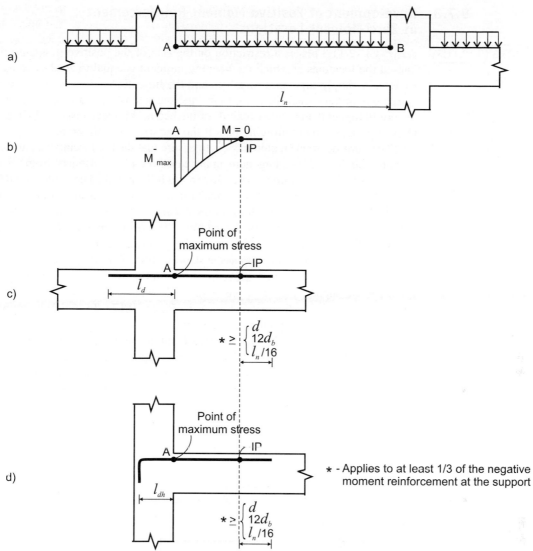

Figure 9.16 Development of negative moment reinforcement in continuous members: a) a typical interior span; b) distribution of negative bending moment in the support region; c) development of straight bars; d) hooked anchorage at the end span.

9.7.7 Development of Negative Moment Reinforcement at Inflection Points

A23.3 Cl.12.12.2

At least one-third of the total tension reinforcement that resists negative bending moments at the support should have an embedment length beyond the inflection point that is equal to or larger than the greatest of the following values:

- d (effective member depth),
- $12d_b$, and
- $l_n/16$, where l_n is the clear span of a beam or a slab (see Figure 9.16a)

This requirement is illustrated on the typical interior span AB of a continuous beam that is part of a reinforced concrete frame, as shown in Figure 9.16a. Observe from the bending moment diagram in Figure 9.16b that the bending moment reduces to zero within the beam span. The point of zero moment is called the *inflection point* (denoted by IP on the bending moment diagram). According to A23.3 Cl.12.12.2, at least one-third of the negative reinforcement at the support should be extended beyond IP, as denoted by asterisk (*) in Figure 9.16c. This requirement also applies to the end spans of continuous beams and slabs, as shown in Figure 9.16d.

9.7.8 Development of Positive Moment Reinforcement at Zero Moment Locations

A23.3 Cl.12.11.3 | A23.3 Cl.12.11.3 places a restriction on the size of the positive moment reinforcement bars at the locations at which the bending moment is equal to zero, such as the support regions in simply supported members and the inflection points in continuous members. In regions of very small bending moments, the rate of increase in positive moments is generally higher than in other regions of the beam; this can cause local slippage between the concrete and the reinforcing bars if the bar sizes are too large.

The purpose of this requirement is to ensure that the reinforcing bars are small enough to withstand the rapid bending moment increase over a short distance, from zero to the magnitude where these bars are required to develop full strength. The reader should recall from Section 9.3 that reinforcing bars of smaller diameter require a shorter development length.

A rapid increase in bending moment values can be explained by the correlation between bending moments and shear forces. The reader may recall from statics courses that the shear force at a certain section along the member length is equal to the slope on the bending moment diagram at the same section. This provision can be explained on an example of a simply supported beam shown in Figure 9.17a. The bending moment diagram for the support region of the beam is shown in Figure 9.17b, whereas the corresponding shear force diagram is shown in Figure 9.17c. In a simply supported beam, shear forces reach a maximum at the supports and then gradually decrease toward the midspan. Consequently, the slope on the bending moment diagram (equal to the shear force at the same section) is largest at the support region and decreases toward the midspan (the slope of the moment diagram is shown by the dashed line in Figure 9.17b).

In the case of simply supported members, this requirement can be stated as follows:

$$l_d \leq 1.3 \times \frac{M_r}{V_f} + l_a \qquad\qquad \textbf{[9.7]}$$

where

M_r = the factored moment resistance based on the bars continuing into the support
V_f = the factored shear force taken at the support
l_a = the embedment length beyond the centre of the support.

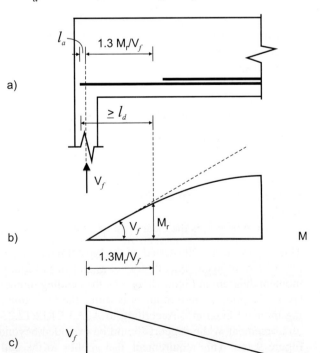

Figure 9.17 Bar development at the points of zero bending moment in simply supported members: a) support region; b) bending moment diagram; c) shear force diagram.

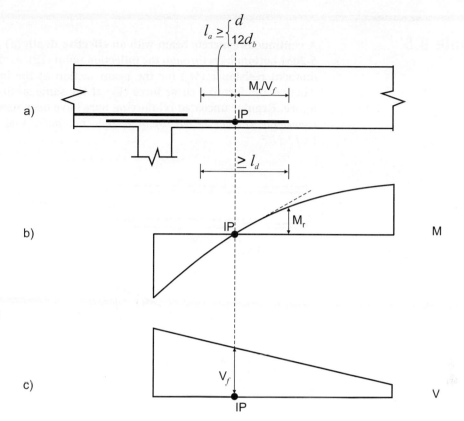

Figure 9.18 Bar development at the inflection point regions of a continuous member: a) beam elevation showing flexural reinforcement; b) bending moment diagram; c) shear force diagram.

The term M_r/V_f in Eqn 9.7 is approximately equal to the distance from the section under consideration to the location where the bars must develop their full strength, corresponding to the factored moment resistance (M_r), as shown in Figure 9.17b. In simply supported beams, this term is increased by 30% (hence the multiplier of 1.3). Hence, reinforcing bars of larger sizes may be used at the supports of simply supported beams with the same M_r/V_f ratio as at inflection points of continuous members. This is in part because the bar ends are confined by the compressive reaction at the support.

In the case of continuous beams and slabs, this requirement can be stated as follows (see Figure 9.18):

$\boxed{\text{A23.3 Eq. 12.6}}$ $\qquad l_d \leq \dfrac{M_r}{V_f} + l_a$ $\qquad\qquad\qquad\qquad\qquad\qquad$ **[9.8]**

where

$\qquad M_r$ = the factored moment resistance based on the reinforcement continuing through the inflection point (IP) (see Figure 9.18b)

$\qquad V_f$ = the factored shear force at the section through the inflection point (see Figure 9.18c)

$\qquad l_a$ = the embedment length beyond the inflection point, equal to the greater of d and $12d_b$ (see Figure 9.18a).

When this requirement is not satisfied, a smaller bar size needs to be used to decrease the l_d value; otherwise the area of tension steel needs to be increased in order to increase the M_r value to the required level.

This requirement is illustrated by Example 9.5.

Example 9.5

A continuous concrete beam with an effective depth (d) of 400 mm is reinforced with 6-20M bottom bars through the inflection point (IP), as shown in the figure below. The moment resistance (M_r) for the beam section at the inflection point is 270 kN · m, whereas the factored shear force (V_f) at the same section is 400 kN, as shown in the figure. Regular uncoated reinforcing bars have been used.

Determine whether the bar size at the inflection point is adequate according to CSA A23.3 Cl.12.11.3.

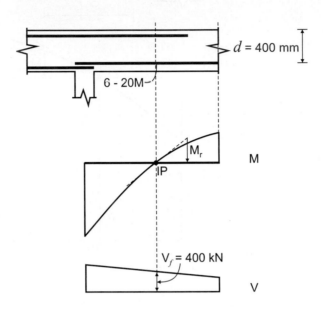

Given: $f_y = 400$ MPa
           $f_c' = 25$ MPa (normal-density concrete)

SOLUTION: **1. Determine the factored moment resistance**
           The factored moment resistance based on the bars continuing though the inflection point (6-20M) is

$$M_r = 270 \text{ kN} \cdot \text{m}$$

2. Determine the factored shear force
The factored shear force at the inflection point location is given in this case:

$$V_f = 400 \text{ kN}$$

3. Determine the embedment length (l_a) beyond the inflection point
Determine the embedment length (l_a) beyond the inflection point for 20M bars ($d_b = 20$ mm) as (see Figure 9.18a)

$$l_a = \begin{bmatrix} d \\ 12d_b \end{bmatrix} = \begin{bmatrix} 400 \text{ mm} \\ 12 \times 20 \text{ mm} = 240 \text{ mm} \end{bmatrix}$$

l_a is equal to the greater of these two values, so

$$l_a = 400 \text{ mm}$$

4. Determine the required development length (l_d)
The development length for 20M bars in this example (bottom bars, regular uncoated bars, normal-density concrete) can be determined from Table A.5 as

$$l_d = 575 \text{ mm}$$

5. **Find the maximum allowable development length according to A23.3 Cl.12.11.3**
    This provision (as applied to continuous beams) states that

A23.3 Eq. 12.6

$$l_d \leq \frac{M_r}{V_f} + l_a \qquad \qquad \text{[9.8]}$$

or

$$l_d \leq \frac{270 \text{ kN} \cdot \text{m}}{400 \text{ kN}} + 0.4 \text{ m} = 0.675 \text{ m} + 0.4 \text{ m} = 1.075 \text{ m} = 1075 \text{ mm}$$

This requirement is illustrated in the sketch below.

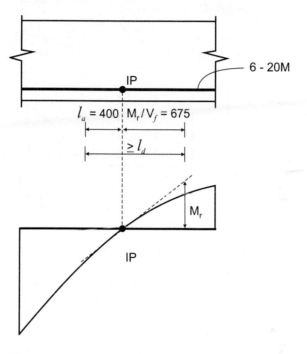

Since

$$l_d = 575 \text{ mm} < 1075 \text{ mm}$$

the requirement prescribed by A23.3 Cl.12.11.3 is satisfied.

9.7.9 Flexural Tension Side

A23.3 Cl.11.3.9.2

This requirement is related to the effect of shear reinforcement (stirrups) on the magnitude of tension force in flexural reinforcement. *The flexural reinforcement should be capable of resisting a tension force (T) given by*

A23.3 Eq. 11.14

$$T \geq F_{lt} = \frac{M_f}{d_v} + \left(V_f - 0.5 \, V_s\right) \cot \theta \qquad \qquad \text{[9.9]}$$

where

V_f = the factored shear force
V_s = the steel shear resistance provided by the stirrups (see Figure 9.19b)
M_f = factored bending moment
d_v = flexural lever arm corresponding to the factored moment resistance
θ = angle formed by diagonal cracks with regard to the longitudinal member axis (see Section 6.5.4).

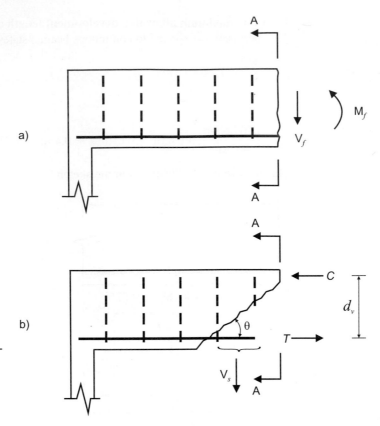

Figure 9.19 Flexural reinforcement: a) beam elevation showing flexural and shear reinforcement; b) internal force distribution.

It should be noted that the above equation is a simplified form of A23.3 Eq. 11.14 and it applies to nonprestressed members without any external axial load.

Consider a beam reinforced with flexural and shear reinforcement, as shown in Figure 9.19a. The tension force (T) in the flexural reinforcement in Figure 9.19b should be checked at section A-A according to A23.3 Cl.11.3.9.2. This requirement should be checked at every section containing flexural tension reinforcement.

KEY CONCEPTS

Flexural reinforcement can be terminated where no longer needed to resist bending moments in reinforced concrete flexural members. The bars that are terminated within a flexural member are called *cutoff bars*, and the corresponding points where these bars are terminated are called *cutoff powints*.

Theoretical cutoff points are the points where the rebars are no longer required to resist bending moments. However, due to shear and other effects that might cause cracking at the point where bars are to be terminated, the bars need to be extended beyond the theoretical cutoff point up to the *actual cutoff point*.

The key CSA A23.3 provision governing the development of flexural reinforcement is the general anchorage requirement. It states that *the design must ensure that each bar is properly embedded on both sides of the critical section within a reinforced concrete flexural member* (A23.3 Cl.12.1.1). The other CSA A23.3 requirements for bar cutoffs in reinforced concrete flexural members are related to

- actual point of cutoff (Cl.12.10.3)
- development of continuing reinforcement (Cl.12.10.4)
- development of positive moment reinforcement at supports (Cl.12.11.1)
- anchorage of negative moment reinforcement into supporting members (Cl.12.12.1)
- development of negative moment reinforcement at inflection points (Cl.12.12.2)
- development of positive moment reinforcement at zero moment locations (Cl.12.11.3)
- flexural reinforcement at supports (Cl.11.3.9.2)

9.8 CALCULATION OF BAR CUTOFF POINTS IN SIMPLY SUPPORTED FLEXURAL MEMBERS ACCORDING TO THE CSA A23.3 REQUIREMENTS

As discussed in the previous sections, the design of anchorage in reinforced concrete flexural members is a complex task that involves considering several requirements prescribed by CSA A23.3. The following general tasks need to be performed in order to calculate the bar cutoff points:

1. Determine the design bending moment diagram for the member under consideration.
2. Determine the points at which the bars are no longer required for flexure.
3. Extend the bars to satisfy the applicable CSA A23.3 anchorage requirements.

A summary of the anchorage requirements for flexural reinforcement in simply supported beams and slabs is provided in Figure 9.20. In addition, the key CSA A23.3 anchorage requirements for reinforcement in simply supported beams and slabs are outlined in Checklist 9.1 for the reader's reference.

In simply supported beams, applying the CSA A23.3 bar cutoff requirements generally leads to modest reductions in overall rebar lengths. Hence, it is not considered practical to cut off the flexural reinforcement in simply supported beams and slabs subjected to uniform loads because this practice results in insignificant savings. However, the bar cutoff procedure for simply supported beams will be illustrated by Example 9.6.

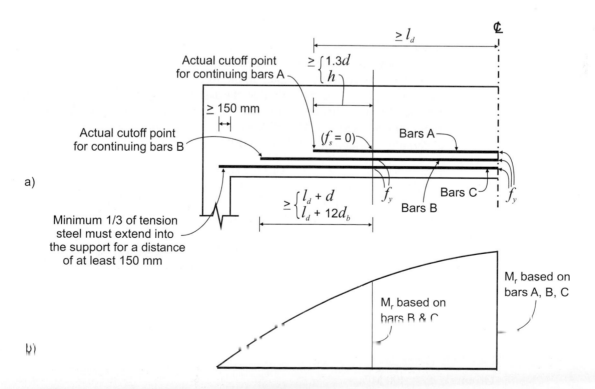

Figure 9.20 Cutoff requirements for simply supported beams.

Checklist 9.1 CSA A23.3 Bar Cutoff Requirements for Flexural Reinforcement in Simply Supported Beams and Slabs

Rule #	Description	Code Clause
1	Actual point of cutoff (Section 9.7.3) Cutoff bars must be extended beyond the theoretical point of cutoff for a a distance of $1.3d$ or h, whichever is greater.	12.10.3 11.3.9.1
2	General anchorage requirement (Section 9.7.2) Each bar must be embedded by a length greater than l_d from the point of maximum bar stress.	12.1.1
3	Continuing reinforcement (Section 9.7.4) The continuing reinforcement should have an embedment length not less than the greater of $(l_d + d)$ and $(l_d + 12d_b)$ beyond the theoretical point of cutoff for the bars being terminated.	12.10.4
4	Supports (Section 9.7.5) At least one third of the tension reinforcement should extend into the supports by at least 150 mm.	12.11.1
5	Zero moment locations (Section 9.7.8) The bar development length in the support region should be given by (see Figure 9.17) $$l_d \leq 1.3 \times \frac{M_r}{V_f} + l_a \qquad \textbf{[9.7]}$$ where l_a is the embedment length beyond the centre of the support.	12.11.3

Example 9.6

A simply supported reinforced concrete beam subjected to uniform load is shown in the figure below. The beam is 600 mm wide by 900 mm deep and is reinforced with 10-30M bars in two layers. The designer would like to cut off 3-30M bars from the top layer where they are no longer required by the design. Regular uncoated reinforcing bars have been used and the shear reinforcement (stirrups) is in excess of the CSA A23.3 minimum requirement. There is a 40 mm side cover to the reinforcement at the supports.

Determine the cutoff points for the 3-30M bars according to the CSA A23.3 requirements.

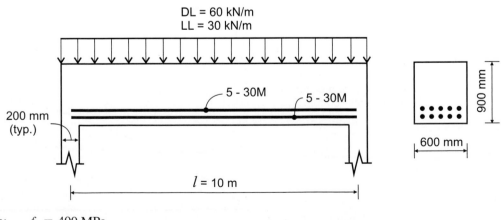

Given: $f_y = 400$ MPa
         $f_c' = 25$ MPa (normal-density concrete)

SOLUTION: **1.   Determine the required moment resistance at the cutoff location**

a)   Determine the factored load (w_f) according to NBC 2010 Cl.4.1.3.2:

$$w_f = 1.25DL + 1.5LL = 1.25(60 \text{ kN/m}) + 1.5(30 \text{ kN/m}) = 120 \text{ kN/m}$$

b)   Then, find the factored shear force at the support:

$$V_f = \frac{w_f l}{2} = \frac{120 \text{ kN/m} \times 10 \text{ m}}{2} = 600 \text{ kN}$$

c)   Next, determine the maximum factored bending moment at the midspan (corresponding to 10-30M bars):

$$M_{fmax} = \frac{w_f l^2}{8} = \frac{(120 \text{ kN/m}) \times (10 \text{ m})^2}{8} = 1500 \text{ kN} \cdot \text{m}$$

It can be taken that $M_{r10} = M_{fmax} = 1500 \text{ kN} \cdot \text{m}$, where M_{r10} denotes the factored moment resistance corresponding to 10-30M bars.

d)   Finally, find the required moment resistance for 7-30M bars (M_{r7}):
The moment resistance is approximately proportional to the number of bars, so (see the sketch that follows)

$$M_{r7} = \frac{7}{10} \times M_{r10} = \frac{7}{10} \times 1500 \text{ kN} \cdot \text{m} = 1050 \text{ kN} \cdot \text{m}$$

Note that, on the sketch, the terminating bars (3-30M) are denoted as bars A, while the continuing bars (7-30M) are denoted as bars B.

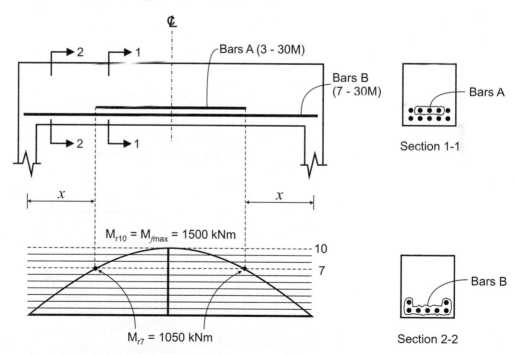

2.   Determine the locations of the theoretical bar cutoff points
Find the location(s) (x) along the beam length where the factored bending moment is

$$M_f = M_{r7} = 1050 \text{ kN} \cdot \text{m}$$

a)   Develop the general equation for the bending moment along the beam:

$$M_x = V_f \times x - w_f \times \frac{x \cdot L}{2}$$

$$= (600 \text{ kN}) \times x - (120 \text{ kN/m}) \times \frac{x^2}{2}$$

b) Find the location (x) where $M_x = 1050$ kN·m

Determine this location from the free-body diagram of the beam on the sketch below.

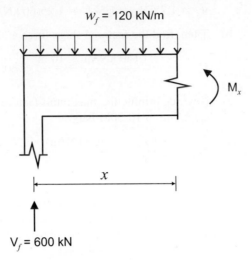

$w_f = 120$ kN/m

M_x

x

$V_f = 600$ kN

$$M_x = (600 \text{ kN}) \times x - (120 \text{ kN/m}) \times \frac{x^2}{2} = 1050 \text{ kN} \cdot \text{m}$$

Rearrange the above equation as a quadratic equation:

$$60x^2 - 600x + 1050 = 0$$

Solve the above equation for x as follows:

$$x = \frac{600 \pm \sqrt{600^2 - 4(1050)(60)}}{2(60)} = 2261 \text{ mm} \cong 2270 \text{ mm}$$

Note that the x value has been rounded up from 2261 mm to 2270 mm.

Therefore, the theoretical bar cutoff location is 2.27 m from the centre-line of the left support. Note that the larger x value, also a solution to the equation, corresponds to the theoretical bar cutoff point located 2.27 m from the right support. The reinforcement and theoretical cutoff points are shown in the beam diagram below.

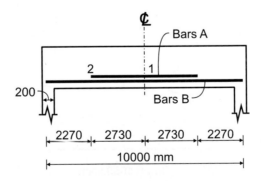

₵

Bars A

2   1

200

Bars B

2270   2730   2730   2270

10000 mm

3. **Determine the development length for the reinforcing bars**

Determine the bar development length (l_d) according to A23.3 Cl.12.2.3. Since the beam is reinforced with the minimum shear reinforcement, this application can be classified as Case 1; consequently, use Eqn 9.2 to determine the development length.

a) Establish the values of the modification factors according to A23.3 Cl.12.2.4 (see Table 9.1).

- $k_1 = 1.0$ (bottom bars)
- $k_2 = 1.0$ (regular uncoated reinforcing bars)
- $k_3 = 1.0$ (normal-density concrete)
- $k_4 = 1.0$ (30M bars)

b) Determine l_d for 30M bars (bar diameter $d_b = 30$ mm):

$$l_d = 0.45\, k_1\, k_2\, k_3\, k_4 \frac{f_y}{\sqrt{f_c'}}\, d_b \qquad\qquad \textbf{[9.2]}$$

$$= 0.45(1.0)(1.0)(1.0)(1.0) \times \frac{400 \text{ MPa}}{\sqrt{25 \text{ MPa}}} \times 30 \text{ mm} = 1080 \text{ mm}$$

4. **Determine the bar cutoffs according to the CSA A23.3 requirements**
As the next step, determine the cutoff points according to the CSA A23.3 requirements.
In this example, bars A are terminated, whereas bars B continue up to the supports. The five key bar cutoff rules (CSA A23.3 provisions) for a simply supported beam are summarized in Checklist 9.1 and will be referred to here.

a) For terminating bars A (3-30M), check the following rules.

i) Rule 1: Actual cutoff point (A23.3 Cl.11.3.9.1):
Bars A are no longer required for flexure at point 2, corresponding to a moment resistance of $M_f = M_{r7} = 1050$ kN · m, so point 2 is the theoretical cutoff point for these bars. However, according to Cl.11.3.9.1, these bars need to be extended beyond point 2 by the distance $1.3d$ or h, whichever is greater. In this case,

$h = 900$ mm

Because the beam has two layers of reinforcement, estimate d as (see Section 5.4.1)

$d = h - 110 \text{ mm} = 900 \text{ mm} - 110 \text{ mm} = 790 \text{ mm}$

So,

$1.3d = 1.3 \times 790 \text{ mm} = 1027 \text{ mm} \cong 1030 \text{ mm}$

Because 1030 mm > 900 mm, the bars need to be extended up to point 3, and the length of segment 2-3 is (see the sketch below)

$l_{2-3} = 1030$ mm

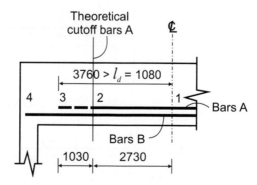

ii) Rule 2: General anchorage requirement (A23.3 Cl.12.1.1):
Based on Rule 1, the total distance from the point of maximum stress (point 1 on the above sketch) to the actual cutoff point (3) for bars A is

$l_{1-3} = l_{1-2} + l_{2-3} = 2730 \text{ mm} + 1030 \text{ mm} = 3760 \text{ mm}$

where

$$l_{1-2} = \frac{10\,000 \text{ m}}{2} - 2270 \text{ m} = 2730 \text{ m}$$

(see the sketch above).

Moreover, the development length (l_d) for the bars determined in Step 3 of this example is

$$l_d = 1080 \text{ mm}$$

Since $l_d = 1080 \text{ mm} < 3760 \text{ mm}$, Rule 2 is satisfied.

b) For continuing bars B (7-30M), check the following rules.

   i) Rule 3: Continuing reinforcement (A23.3 Cl.12.10.4):
   The total available length for continuing bars B, from point 2 to the exterior beam face, is

$$2270 \text{ mm} + \frac{200 \text{ mm}}{2} = 2370 \text{ mm} \text{ (where 200 mm is the support width).}$$

   When the side cover (40 mm minimum) is taken into account, the available length is

$$2370 \text{ mm} - 40 \text{ mm} = 2330 \text{ mm}$$

   According to A23.3 Cl.12.10.4, continuing bars B need to be extended beyond point 2 by the greater of

$$l_d + d = 1080 \text{ mm} + 790 \text{ mm} = 1870 \text{ mm}$$

   and

$$l_d + 12d_b = 1080 \text{ mm} + 12 \times 30 \text{ mm} = 1440 \text{ mm}$$

   The larger value (1870 mm) governs. Finally, check whether the required bar extension length is less than the available length:

$$1870 \text{ mm} < 2330 \text{ mm}$$

   Thus, Rule 3 is satisfied. The application of Rule 3 is illustrated in the sketch below.

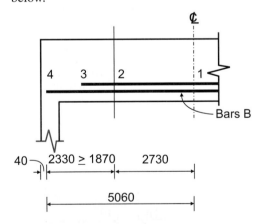

   ii) Rule 4: Supports (A23.3 Cl.12.11.1):
   At least one third of the tension reinforcement should extend into the supports by at least 150 mm.
       In this case, seven (out of ten) bars extend into the support. Since the support width is 200 mm and there is a 40 mm side cover, the extension length is

$$200 \text{ mm} - 40 \text{ mm} = 160 \text{ mm}$$

   Since

$$160 \text{ mm} > 150 \text{ mm}$$

   Rule 4 is satisfied.

iii) Rule 5: Zero moment locations (A23.3 Cl.12.11.3):
The factored moment resistance based on the bars continuing into the support (7-30M) is

$$M_{r7} = 1050 \text{ kN} \cdot \text{m}$$

The factored shear force at the support is

$$V_f = 600 \text{ kN}$$

The embedment length beyond the support centre-line is

$$l_a = \frac{200 \text{ mm}}{2} - 40 \text{ mm} = 60 \text{ mm} = 0.06 \text{ m}$$

This rule states that

$$l_d \le 1.3 \times \frac{M_r}{V_f} + l_a \qquad\qquad \textbf{[9.7]}$$

or

$$l_d \le 1.3 \times \frac{1050 \text{ kN} \cdot \text{m}}{600 \text{ kN}} + 0.06 \text{ m} = 2.335 \text{ m} = 2335 \text{ mm}$$

Since the required development length for straight bars is

$$l_d = 1080 \text{ mm} < 2335 \text{ mm}$$

Rule 5 is satisfied, as shown on the sketch that follows.

It should be noted that 7-30M bars extend into the supports, that is, 5-30M bars in the bottom layer and 2-30M bars in the top layer. Recall that Rule 4 requires continuing only one third of the bars into the supports.

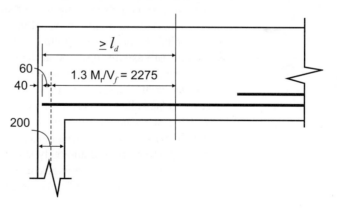

5. **Provide a sketch summarizing the reinforcement details**
Finally, summarize the results of the above calculations, as in the beam elevation sketch below. Note that the bar lengths have been specified from the beam centre-line (to make use of symmetry).

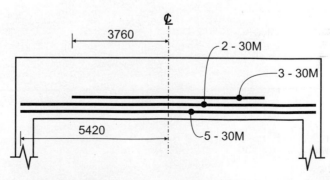

Learning from Examples

It can be concluded from this example that the calculation of bar cutoffs is a time-consuming process. In this case, the total savings associated with cutting off the bars is not more than 10 m of reinforcement, or less than 10% of the total length of flexural reinforcement. Consequently, the resulting cost savings may not be offset by the additional labour cost associated with time spent cutting and placing rebars of different lengths.

In general, cutting off rebars in simply supported beams is rarely justified from an economical perspective. Consequently, in practice it is not unusual to design simply supported beams and slabs without bar cutoffs. However, this is not the case with continuous beams and slabs. Bar cutoffs in continuous beams and slabs generally lead to significant savings in the overall construction cost, particularly on projects with a large number of similar spans (to be discussed in detail in Section 11.4).

9.9 SPLICES

9.9.1 Background

It is not possible to place all the concrete in a reinforced concrete structure at one time. Therefore, construction joints are required because of the stoppage in construction. (For more details on construction joints, refer to Sections 12.12.4 and 13.8.) Also, it is rarely possible to have continuous column rebars running from the foundations to the roof, or continuous floor rebars from one end of the building to the other — note that reinforcing bars are manufactured in standard lengths of 12 or 18 m! In addition to the above, it is not desirable to extend the rebars far beyond the construction joints; for example, in high-rise construction, column rebars are usually cut to a height of one storey. To ensure the continuity of reinforcement in concrete construction, it is necessary to splice the reinforcing bars; splices are usually installed at the joint locations.

Splicing can be defined as connecting shorter available rebar lengths in series. There are several ways by which splicing can be accomplished: welding of continuing bars, mechanical devices, and lap splices. Mechanically coupled splices use two basic splicing devices, namely couplers and end-bearing devices. *Couplers* are able to transfer both tension and compression forces, whereas *end-bearing devices* can transfer only compression forces. *Lap splices* are the simplest and most common type of splices and will be discussed in this section.

Splices are typically made at construction joints between footings and columns or walls, between columns and floors, or between walls and floors. Figure 9.21 illustrates

Figure 9.21 Splicing of wall reinforcement.

(John Pao)

the placing of wall reinforcement at the construction site. The workman is tying the wall reinforcing steel. Note the vertical bar splices at the bottom, with the footing dowels extended above the footing-to-wall construction joint. Also, note that the horizontal wall steel is also spliced; this is common practice for horizontal bars in walls, slabs, and beams.

Depending on whether the splicing is performed on reinforcing bars that act in tension or in compression, splices can be classified as *tension splices* or *compression splices*. Tension splices are used to splice tension steel in beams and slabs, whereas compression splices are used in columns and other compression members.

The location and type of splices should be shown on the construction drawings.

9.9.2 Tension Splices

Consider a lap splice consisting of two deformed rebars acting in tension, as shown in Figure 9.22. The splice is formed by overlapping two bars placed adjacent to each other over the lap length as shown in Figure 9.22a. The tension force first develops in one of the bars that forms the splice and is transferred to the concrete through inclined forces, as shown in Figure 9.22b. This force transfer is characteristic of the bond between the steel and the concrete, explained in Section 9.2 and Figure 9.2. As a result, the forces developed in the concrete are transferred to the other spliced bar, which in turn develops tension force as well. This force transfer is possible since the spliced bars are placed quite close to one another. Tension lap splices generally fail when the surrounding concrete splits. (Refer to Section 9.2 for more details on splitting bond failure.)

Consider the tension lap splice consisting of bars AB and CD in Figure 9.23. Bar CD has developed its full yield strength (f_y), and the corresponding tension force is equal to $A_b f_y$, where A_b denotes the bar area. At one end of the splice (point D), the stress in bar CD is equal to f_y ($f_s = f_y$), whereas the stress in bar AB is equal to 0 ($f_s = 0$). Due to the force transfer over the lap length, the stress in bar CD drops from f_y to 0, whereas the stress in bar AB increases from 0 to f_y, as shown in Figure 9.23. Therefore, this splice is effective in maintaining the tension force between bars CD and AB. Note, however, that this is a simplified explanation and that the stress distribution along the lap length is not ideally linear, as that shown in Figure 9.23. Also, the stress in spliced rebars might be less than f_y.

CSA A23.3 prescribes the required lap length (l_p) expressed in terms of the tension development length (l_d) for spliced bars. The key CSA A23.3 provisions related to tension lap splices are summarized below.

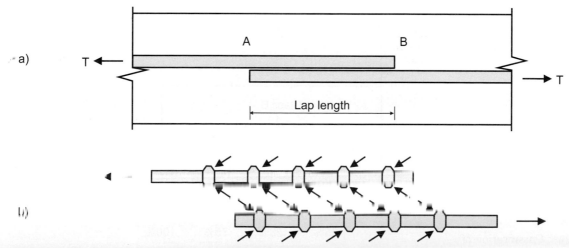

Figure 9.22 Tension lap splice: a) bar layout; b) force transfer between spliced bars.

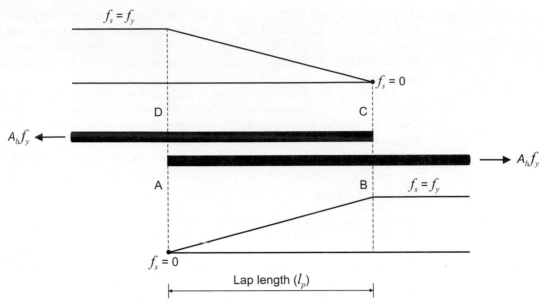

Figure 9.23   Stresses in a tension lap splice.

A23.3 Cl.12.15.1   **Classes of splices**   A23.3 Cl.12.15.1 identifies two splice classes, A and B, as follows:

- Class A corresponds to a lap length of l_d; that is, $l_p = l_d$
- Class B corresponds to a lap length of $1.3l_d$; that is, $l_p = 1.3l_d$

In any case, the lap length should not be less than 300 mm. In determining l_d, a modification factor of 1.0 should be used.

A23.3 Cl.12.15.2   **Classification of lap splices**   A23.3 Cl.12.15.2 prescribes the splice class (A or B) to be used for a particular design application, depending on the fraction of the bars being spliced within the required lap length, and the ratio of the tension steel area required by the design and the amount of reinforcement provided at the splice locations, that is,

$$\frac{A_{srequired}}{A_{sprovided}}$$

The latter requirement is related to the actual design stress in the tension steel as a fraction of the yield stress. This provision is illustrated in Figure 9.24.

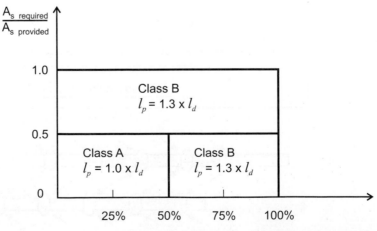

Figure 9.24   Classification of lap splices.

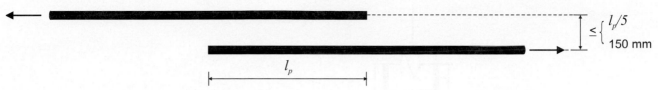

Figure 9.25   Spacing of bars in lap splices.

| A23.3 Cl.12.14.2.1 | **Bar sizes in spliced bars**   Bars of very large sizes (over 35M) should not be spliced using tension lap splices. |

| A23.3 Cl.12.14.2.3 | **Spacing of spliced bars**   Lapped bars in reinforced concrete flexural members cannot be spaced more than one fifth of the required lap splice length (l_p) or 150 mm, whichever is less, as shown in Figure 9.25. When the spliced bars are placed far apart, it is not possible to achieve the force transfer illustrated in Figure 9.22b. |

Except in slabs, tension lap splices should always be enclosed with stirrups, ties, or spirals. Special attention is required for tension splices in structures built in seismic areas.

9.9.3   Compression Splices

The strength of a compression lap splice depends considerably on the end bearing of the bars on the concrete; also, tension cracks are not characteristic of tension lap splices. Consequently, the required length of a compression lap splice is much smaller than that of a tension lap splice. According to A23.3 Cl.12.16.1, the length of the compression lap splice (l_p) should not be less than the greatest of

- $0.073 f_y d_b$ (when $f_y \leq 400$ MPa),
- $(0.133 f_y - 24) d_b$ (when $f_y > 400$ MPa), and
- 300 mm.

Compression lap splices are used in reinforced concrete columns, as shown in Figure 9.26.

9.9.4   Column Splices

Vertical reinforcing bars in reinforced concrete columns act primarily in compression, although in the case of combined flexure and axial load the reinforcement in one column face acts in tension. As a result, column rebars must be able to resist the compression from the very top of the column, where the loads are initially applied, all the way down to the base, where the column is supported by a footing. Column reinforcement is usually spliced at each floor level, as shown in Figure 9.26. The construction procedure usually requires the column pour to stop at the underside of the concrete floor. Subsequently, the concrete floor is placed, followed by the next storey column. In the construction of a concrete high-rise building, this procedure needs to be repeated several times (once for each floor).

When column bars start at the same level, the column reinforcement is preassembled at some convenient location at the construction site — this is called a *reinforcement cage*. An example of a column reinforcement cage with lap splices is shown in Figure 9.27. The column cage is hoisted and placed directly into the column form when required.

Hence, column splices are very common in reinforced concrete building construction. For columns with the same vertical reinforcement between floors, the reinforcing bars at the floor below must be "cranked" (bent) in order to properly lap with the reinforcing bars above. Cranked bars are shown in Figure 9.26. The compression force in the reinforcing

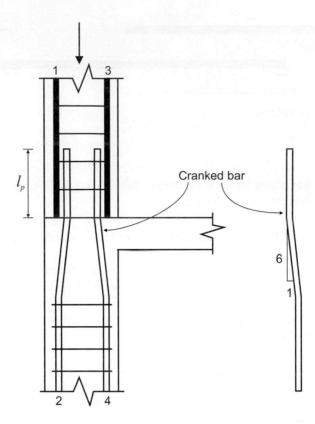

Figure 9.26 Compression lap splices in columns.

steel must be transferred from bars 1 and 3 above into the lapping steel, and then through the cranks to bars 2 and 4 in the floor below.

To ensure proper load transfer through the lap splices, it is important for the laps to be properly placed and for the cranks to stay within a certain limit off vertical. CSA A23.3 prescribes a maximum crank of 1:6 slope in vertical column reinforcement. This limit must be respected in order to avoid any significant horizontal forces resulting from vertical load transfer through the crank. Lap splice requirements for reinforced concrete columns are outlined in A23.3 Cl.12.17.3.

Figure 9.27   Column reinforcement cage with lap splices.

(John Pao)

KEY CONCEPTS

Splicing can be defined as connecting shorter available rebar lengths in series. Three ways by which splicing can be accomplished are:

- welding of continuing bars
- mechanical devices (couplers or end-bearing devices)
- lap splices

Lap splices are the simplest and most common type of splices. A lap splice is made by overlapping bars over the lap length (l_p). Lap splices can act in tension or compression. CSA A23.3 prescribes the required lap length for different design applications expressed in terms of development length. The splice length also depends on whether the spliced bars act in tension or compression.

Tension splices   According to A23.3 Cl.12.15.1, tension splices can be classified as follows:

- Class A: $l_p = l_d$
- Class B: $l_p = 1.3 l_d$

The other CSA A23.3 requirements related to tension lap splices include

- classification of lap splices (A23.3 Cl.12.15.2)
- bar sizes (A23.3 Cl.12.14.2.1)
- spacing of spliced bars (A23.3 Cl.12.14.2.3)

Compression splices (A23.3 Cl.12.16.1)   The required length of a compression lap splice is much smaller than that of a tension lap splice, since the strength of a compression lap splice depends considerably on the end bearing of the bars on the concrete; also, tension cracks are not characteristic of tension lap splices.

SUMMARY AND REVIEW — BOND AND ANCHORAGE OF REINFORCEMENT

The bond between the steel and the concrete allows the transfer of longitudinal forces from the reinforcement to the surrounding concrete. As long as the bond exists, there is no relative movement at the interface between the steel and the concrete. The bond is critical for the satisfactory performance of reinforced concrete structures at both the service and the ultimate load levels: at the service loads, bond has an important role in controlling cracks and deflections; at the ultimate loads, a proper bond is required to develop the yield strength in steel reinforcement.

Bond mechanism and failure modes

The following two bond failure modes occur in reinforced concrete flexural members:

- *Pullout failure* occurs in members with sufficiently large concrete cover and bar spacing and smaller bar diameters and is characterized by shearing along a cylindrical failure surface formed around the bars.
- *Splitting failure* is characterized by the splitting of concrete along the bar length. This failure occurs when cover, confinement, or bar spacing are insufficient to resist the lateral concrete tension resulting from the wedging effect of the bar deformations.

Of these two modes, splitting failure is more common in reinforced concrete flexural members.

Development length of straight bars

In order to achieve adequate bond strength, reinforcing bars must be embedded or anchored in the concrete over a certain length called the *development length* (l_d). The development length depends on whether the bar is subjected to tension or compression and also on whether it is straight or hooked.

Development length of straight bars in tension (A23.3 Cl.12.2.3) *For Case 1 applications* (members containing minimum stirrups or ties within development length (l_d), or slabs, walls, shells, or folded plates with clear spacing between bars being developed not less than $2d_b$),

$$l_d = 0.45\, k_1 k_2 k_3 k_4 \frac{f_y}{\sqrt{f_c'}}\, d_b \text{ (basic equation)}$$ [9.2]

For Case 2 applications (all other members not meeting the Case 1 requirements),

$$l_d = 0.6\, k_1 k_2 k_3 k_4 \frac{f_y}{\sqrt{f_c'}}\, d_b$$ [9.3]

The modification factors k_1, k_2, k_3, and k_4 account for various effects that influence the bar development length (refer to Table 9.1 for specific values):

- k_1 — the bar location factor
- k_2 — the coating factor
- k_3 — the concrete density factor
- k_4 — the bar size factor

According to A23.3 Cl.12.2.5, the development length may be reduced when the area of tension reinforcement provided in the section is in excess of the required area.

Development length of straight bars in compression (A23.3 Cl.12.3.2) The compression development length is considerably shorter than the tension development length and can be determined according to the equation

$$l_{db} = 0.24\frac{f_y}{\sqrt{f_c'}}\, d_b \le 0.044\, f_y\, d_b$$ [9.4]

In some cases, the basic development length (l_{db}) needs to be multiplied by modification factors that account for the excess reinforcement and the effect of confinement by spirals or ties.

Development length of hooked bars

A hook consists of a straight bar that is curved at its end. Hooked bars are used under the following circumstances:

- to provide anchorage for flexural reinforcement when the available bar length of a straight bar is insufficient for its full development (for example, end spans of beams or slabs)
- to provide anchorage for stirrups and ties in reinforced concrete beams and columns

The use of hooked reinforcing bars can supply anchorage over a shorter length than required for the tension development of straight bars. CSA A23.1 Cl.12.2.2 defines standard hooks for reinforced concrete structures. It should be noted that hooks act in tension and are *not* considered to help anchor the bars in compression.

Hooked anchorage for straight bars in tension (A23.3 Cl.12.5.1 and 12.5.2) Both 90° and 180° hooks are commonly used to provide anchorage for flexural reinforcement. The basic hook development length (l_{hb}) is measured from the critical section to the farthest point on the hooked bar. The development length for hooked bars in tension (l_{db}) should be obtained by multiplying the basic hook development length (l_{hb}) by appropriate modification factors (MF); that is,

$$l_{dh} = l_{hb} \times \text{MF} = 100\frac{d_b}{\sqrt{f_c'}} \times \text{MF}$$ [9.6]

Modification factors account for the effects of steel yield strength, confinement, and excess reinforcement (refer to Table 9.2 for specific values).

Hooked anchorage for stirrups and ties (A23.3 Cl.12.13)   Hooks are used to provide anchorage for stirrups and ties in reinforced concrete beams and columns. A *standard stirrup hook* is characterized by a 135° (or larger) bend, unless the concrete cover surrounding the hook is restrained against spalling, in which case 90° bends can be used.

Bar cutoffs for flexural reinforcement according to CSA A23.3

Flexural reinforcement can be terminated where no longer needed to resist bending moments in reinforced concrete flexural members. The bars that are terminated within a flexural member are called *cutoff bars,* and the corresponding points at which these bars are terminated are called *cutoff points.*

Theoretical cutoff points are the points at which the rebars are no longer required to resist the bending moments. However, due to shear and other effects that might cause cracking at the point at which bars are terminated, the bars need to be extended beyond the theoretical cutoff point to the *actual cutoff point.*

The general anchorage requirement is the key CSA A23.3 provision governing the development of flexural reinforcement. It states that *the design must ensure that each bar is properly embedded on both sides of the critical section within a reinforced concrete flexural member* (A23.3 Cl.12.1.1). The other CSA A23.3 requirements for bar cutoffs in reinforced concrete flexural members are related to

- actual point of cutoff (A23.3 Cl.12.10.3)
- development of continuing reinforcement (A23.3 Cl.12.10.4)
- development of positive moment reinforcement at supports (A23.3 Cl.12.11.1)
- anchorage of negative moment reinforcement into supporting members (A23.3 Cl.12.12.1)
- development of negative moment reinforcement at inflection points (A23.3 Cl.12.12.2)
- development of positive moment reinforcement at the zero moment locations (A23.3 Cl.12.11.3)
- flexural reinforcement at supports (Cl.11.3.9.2)

Bar splices

Splicing can be defined as connecting shorter available rebar lengths in series. Three ways by which splicing can be accomplished are

- welding of continuing bars
- mechanical devices (couplers or end-bearing devices)
- lap splices

Lap splices are the simplest and most common type of splices. A lap splice is made by overlapping bars over the lap length (l_p). Lap splices can act in tension or compression. CSA A23.3 prescribes the required lap length for different design applications expressed in terms of the development length. The splice length also depends on whether the spliced bars act in tension or compression.

Tension splices   According to A23.3 Cl.12.15.1, the tension splices can be classified as follows:

- Class A: $l_p = l_d$
- Class B: $l_p = 1.3 l_d$

The other CSA A23.3 requirements for tension lap splices include

- classification of lap splices (A23.3 Cl.12.15.2)
- bar sizes (A23.3 Cl.12.14.2.1)
- spacing of spliced bars (A23.3 Cl.12.14.2.3)

Compression splices (A23.3 Cl.12.16.1)   The required length of a compression lap splice is much smaller than that of a tension lap splice because the strength of a compression lap splice depends considerably on the end bearing of the bars on the concrete; also, tension cracks are not characteristic of tension lap splices.

PROBLEMS

9.1. Determine the required tension development lengths to be used in the following beam designs:
a) 15M top bars, regular uncoated reinforcement
b) 20M top bars, epoxy-coated reinforcement (cover and bar spacing not specified)
c) 30M bottom bars, regular uncoated reinforcement
First, perform the manual calculations according to CSA A23.3, and then compare the results using the appropriate design aids included in Appendix A.
  Note that the beams contain minimum stirrups according to the CSA A23.3 requirements.

Given:

$f_c' = 30$ MPa (normal-density concrete)
$f_y = 400$ MPa

9.2. A typical cross-section of a 200 mm thick cantilevered reinforced concrete floor slab supported by a reinforced concrete wall is shown in the figure below. The slab is subjected to uniform load and is reinforced with 15M top rebars at 300 mm spacing (15M@300). Regular uncoated reinforcement has been used in this design.
a) Determine the point of maximum stress in the reinforcement under the applied load.
b) Determine the minimum development length for straight bars required by CSA A23.3 to ensure that the reinforcement is properly anchored into the wall.
c) Design a hooked bar anchorage as an alternative solution. Summarize the design results on a sketch.

Given:

$f_c' = 25$ MPa (normal-density concrete)
$f_y = 400$ MPa

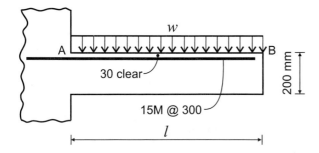

9.3. Consider a 600 mm deep reinforced concrete cantilever beam supported by a 600 mm square column, as shown in the figure that follows. The beam reinforcement is anchored into the column, as shown in the figure. The minimum stirrups are provided in the anchorage zone according to CSA A23.3. Regular uncoated bars have been used in this design.

a) Determine the largest beam rebar size that can be used without a hook in this design according to the CSA A23.3 requirements.
b) Do you believe that it is feasible to use straight bar anchorage for this design? Explain.

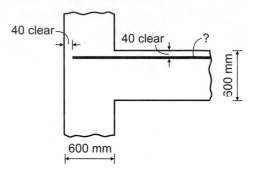

Given:

$f_c' = 35$ MPa (normal-density concrete)
$f_y = 400$ MPa

9.4. A 200 mm thick reinforced concrete wall is supported by a 1200 mm wide strip footing concentrically placed beneath the wall, as shown in the figure below. To resist the flexural effects due to soil pressure beneath the footing, 20M flexural reinforcement is provided at 400 mm spacing (20M@400). Point A can be considered as the point of maximum stress due to flexural effects.

Determine whether the rebar length AB is adequate to develop their full strength according to the CSA A23.3 requirements. If the answer is negative, calculate the maximum tensile stress that can safely develop in the bars.

Given:

$f_c' = 25$ MPa (normal-density concrete)
$f_y = 400$ MPa

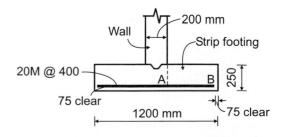

9.5. Consider the reinforced concrete beam ABC shown in the figure that follows. The beam is subjected to a factored uniform load (w_f) of 200 kN/m (including the beam self-weight). The beam is supported at points A and B, and the support width is

400 mm. The beam is located at the interior of a building. Regular uncoated deformed reinforcing bars have been used in this design. The maximum aggregate size is 20 mm. A 40 mm side cover is provided to the reinforcement. The beam is reinforced with the minimum shear reinforcement according to CSA A23.3.

a) Develop the factored shear force and bending moment diagrams for this beam.

b) Design the flexural reinforcement, including top and bottom steel, as required. Use 30M bars and 10M stirrups. Use two layers of bottom reinforcement and one layer of top reinforcement.

c) Suppose that the design requires a 50% cutoff for the top reinforcement where no longer required. Determine the minimum required bar lengths on each side of the support B centre-line according to the CSA A23.3 requirements. Provide a sketch showing the beam elevation with the top bar arrangement and lengths.

d) Is it necessary to provide hooked anchorage for the top reinforcement continuing up to point C of the beam according to CSA A23.3? Explain.

e) If the design requires a 50% cutoff for the bottom reinforcement near support B only, determine the minimum required lengths for cutoff bars according to the CSA A23.3 requirements. Provide a sketch showing the beam elevation and the bottom bar arrangement and lengths.

f) Is bottom reinforcement required in span BC of the beam according to the flexural requirements and/or CSA A23.3 provisions? Explain.

Given:

$f_c' = 30$ MPa (normal-density concrete)
$f_y = 400$ MPa

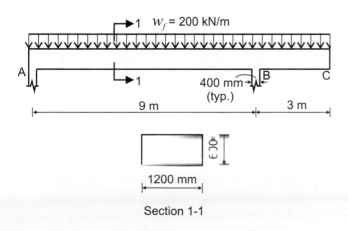

Section 1-1

9.6. A simply supported reinforced concrete beam subjected to a factored point load (P_f) of 500 kN is shown in the figure below. It has been determined that 6-25M bars are adequate for the section with maximum bending moments. The design requires a 50% cutoff for the flexural reinforcement where no longer required. Assume that the beam is reinforced with the minimum shear reinforcement according to CSA A23.3. Regular uncoated bars have been used. A 40 mm side cover is provided to the reinforcement.

a) Develop the factored bending moment diagram for the beam (ignore the self-weight).

b) Design the bar cutoffs. Use hooked anchorages where required.

c) Provide a sketch showing the beam elevation with the bar arrangement and lengths.

Given:

$f_c' = 25$ MPa (normal-density concrete)
$f_y = 400$ MPa

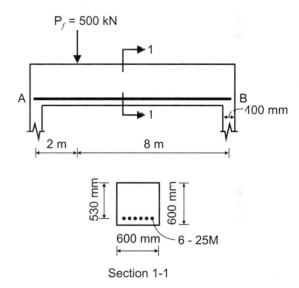

Section 1-1

9.7. A simply supported reinforced concrete beam subjected to a factored uniform load (w_f) of 80 kN/m (including the beam self-weight) is shown in the figure below. The bottom of the beam is flat, while the top is curved in the shape of a parabola (as requested by the architect). The designer has determined that 10-20M flexural bars are required at the beam midspan.

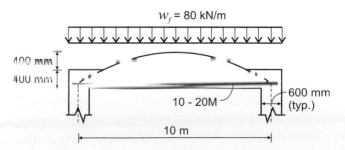

a) Determine the cutoff locations for 50% of the bottom reinforcement based on the CSA A23.3 requirements. Hint: The answer should be conceptual only; it is not necessary to perform detailed calculations.

b) Are hooks required to provide anchorage for the bottom reinforcement at the supports, assuming that the support width is 600 mm?

Given:

$f_c' = 25$ MPa (normal-density concrete)
$f_y = 400$ MPa

9.8. Calculate the required lap splice lengths for 20M top bars used in the following design applications (assume regular uncoated bars):

a) tension lap splice, 50% reinforcement to be spliced, $\dfrac{A_{srequired}}{A_{sprovided}} = 0.6$

b) compression lap splice

Given:

$f_c' = 25$ MPa (normal-density concrete)
$f_y = 400$ MPa

9.9. A 500 mm square reinforced concrete column subjected to compressive load is supported by a spread footing, as shown in the figure that follows.

The column is reinforced with 4-30M vertical bars. The 4-30M dowels are extended from the footing to provide force transfer at the footing-to-column interface and to ensure the continuity of reinforcement at the construction joint. The required amount of column longitudinal reinforcement (A_{sreq}) is equal to 2500 mm^2.

a) Determine the required lap length (l_p) for the dowels according to the CSA A23.3 requirements.

b) Check whether the footing thickness is adequate to develop full strength for the hooked dowels.

Given:

$f_c' = 25$ MPa (normal-density concrete)
$f_y = 400$ MPa

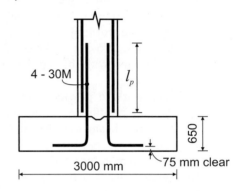

10 Behaviour and Analysis of Continuous Beams and Slabs

LEARNING OUTCOMES

After reading this chapter, you should be able to

- describe the key features of continuous reinforced concrete structures
- use load patterns prescribed by CSA A23.3 to develop moment envelopes for continuous beams and slabs
- compare the main approaches for the analysis of continuous reinforced concrete structures
- use the CSA A23.3 approximate frame method to determine bending moments and shear forces in continuous beams and slabs
- apply the CSA A23.3 moment redistribution provision that accounts for the behaviour of reinforced concrete continuous structures in the postcracking stage

10.1 INTRODUCTION

A significant advantage of reinforced concrete over other building materials, such as steel and timber, is the ease with which cast-in-place reinforced concrete beams and slabs can take advantage of the structural efficiency of continuous construction. In steel and timber construction, adjacent simple span beams are typically joined at the supports by means of fasteners in the form of bolts, welds, or nails. Unless specifically designed, such connections generally do not have the capacity to transfer any quantifiable amount of bending moment between the adjacent members. In comparison, cast-in-place reinforced concrete beams and slabs generally span continuously across several supports to form floor systems. These beams and slabs can be designed and constructed to behave as continuous structures and transfer bending moments across adjacent spans by strategic placement of tension reinforcement in the top and bottom regions. A properly designed continuous reinforced concrete beam is characterized by a more efficient use of reinforcement than a series of simply supported beams of similar geometric properties, with the benefit of smaller deflections and crack widths. The most common applications of continuous concrete construction are found in buildings and bridges.

In cast-in-place reinforced concrete structures, entire floors and roofs are typically cast together to form a monolithic unit. Beams are cast integrally with slabs over several spans, and beam-to-slab joints are generally sufficiently reinforced to limit excessive deformations and cracking in the support regions. Floor structures also act in unison with supporting columns. In this chapter, the effect of supporting columns will be ignored to simplify the concepts being presented. In reality, the effects of columns may or may not need to be considered in the flexural design of continuous reinforced concrete floor systems. This will depend on the actual structure under consideration and the desired accuracy in structural modelling.

This chapter discusses the critical aspects of the behaviour and structural analysis of continuous reinforced concrete beams and slabs subjected to flexure. Section 10.2 provides an overview of the fundamental concepts of continuous structures. The concept of load patterns and the construction of moment envelopes are presented in Section 10.3. Simplifications that are permissible in the structural analysis of continuous structures are outlined in Section 10.4, whereas different approaches to the analysis are discussed in Section 10.5. The application of CSA A23.3 approximate frame analysis is demonstrated in Section 10.6. A brief overview of the behaviour of reinforced concrete members in the postcracking stage is presented in Section 10.7, whereas the analysis of these structures is discussed in Section 10.8. Finally, the moment redistribution according to CSA A23.3 is discussed in Section 10.9.

Once the designer gains a good grasp of the fundamental concepts explained in this chapter, (s)he will be better prepared to exercise proper judgment in the efficient design of continuous reinforced concrete structures.

DID YOU KNOW?

The 13 km long Confederation Bridge (Figure 10.1), spanning the Northumberland Strait between Prince Edward Island and New Brunswick, was one of the largest public works projects in the history of Canada. The main bridge is a precast, prestressed concrete structure consisting of 44 spans, each 250 m long and weighing over 800 metric tons. The main segments consist of alternating continuous and simply supported drop-in spans. The majority of the bridge components were fabricated on land at precasting yards on either side of the bridge. The components were then moved into precise positions by a satellite-guided floating crane. The bridge, completed in 1997, is designed to last for 100 years in its aggressive marine environment, subjected to severe wind, wave, and ice loading (ACI, 2004).

Figure 10.1   Confederation Bridge.
(Straight Crossing Bridge Limited)

10.2 FUNDAMENTAL CONCEPTS OF CONTINUOUS REINFORCED CONCRETE STRUCTURES

10.2.1 Simple Versus Continuous Structures

The differences in the structural response of simple and continuous structures will be illustrated by the following example. Consider the simple (simply supported) reinforced concrete beam in Figure 10.2. The beam has a span (l) of 10 m and is subjected to a uniformly distributed load (w) of 1 kN/m. The maximum bending moment, which occurs at the midspan, is

$$M_{max} = \frac{wl^2}{8} = \frac{1 \text{ kN/m} \times (10 \text{ m})^2}{8} = 12.5 \text{ kN} \cdot \text{m}$$

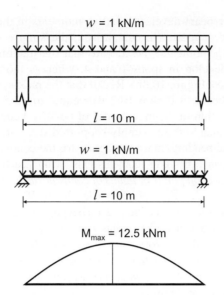

Figure 10.2 Bending moment distribution in a simple beam.

Next, consider the same beam as part of the four-span continuous structure in Figure 10.3a. The continuous beam is subjected to the same uniformly distributed load of 1 kN/m as the simply supported beam. All four spans, denoted by numbers 1 to 4, are 10 m in length, that is, equal to the span of the simply supported beam. The resulting bending moment diagram is shown in Figure 10.3b. Note from the bending moment

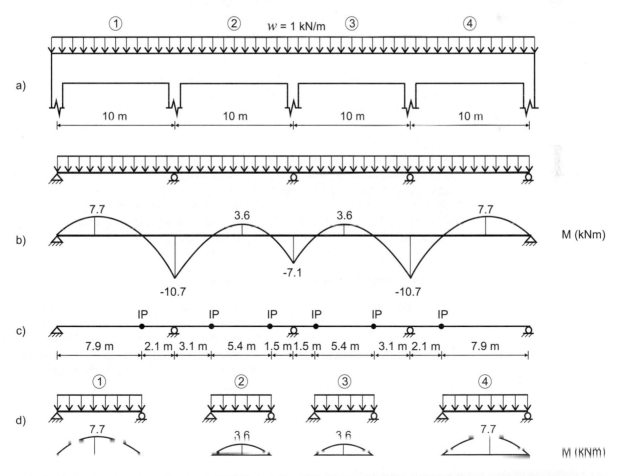

Figure 10.3 Four-span continuous reinforced concrete beam: a) beam elevation and loading; b) corresponding bending moment diagram; c) inflection point locations; d) equivalent simple beams and corresponding bending moment diagrams.

diagram that the beam develops positive moments in the midspan regions and negative moments at the supports (in comparison, the simply supported beam develops only positive bending moments). Next, note that the maximum positive bending moments of 7.7 kN·m develop in spans 1 and 4, whereas moments of 3.6 kN·m develop in spans 2 and 3 (see Figure 10.3b). Recall that the maximum bending moment in the simply supported beam in Figure 10.2 was equal to 12.5 kN·m. This example shows a significant reduction in the magnitudes of positive bending moments in the continuous structure as compared to the simply supported structure.

Consider the bending moment diagram for the continuous beam in Figure 10.3b. Both positive and negative bending moments exist within each span. As a result, there are two points (called *inflection points* or *points of contraflexure*) within each span where the bending moment is equal to zero; these points are denoted by IP on Figure 10.3c. (For more details on inflection points, refer to Section 3.2.) The portions of the beam with positive bending moments between the inflection points can be considered as equivalent simply supported beams (see Figure 10.3d). Note that the spans of these equivalent beams are significantly smaller than the 10 m span for the simply supported beam. For example, the equivalent beam in span 1 has a length of 7.9 m, whereas the beam in span 2 has a length of 5.4 m.

The above example highlights the following important features of continuous structures:

- Both positive and negative bending moments develop in continuous members subjected to gravity loads.
- Positive bending moments in the midspan regions of a continuous beam are less than the corresponding values for a simply supported structure. The extent of decrease depends on the load distribution and the beam properties.
- The decrease in positive bending moment values in continuous beams can be interpreted as a reduction in span lengths for equivalent simple beams spanning between the inflection points.

Continuous structures generally require smaller member thicknesses than simple structures with similar properties, as discussed below.

Table A.3, based on the CSA A23.3 indirect approach to deflection control, prescribes minimum beam and slab thicknesses for different end support conditions (see Section 4.5.2). For example, the minimum thickness required for a simply supported solid slab is equal to $l_n/20$, whereas for a slab with both ends continuous this value is equal to $l_n/28$. By comparing these two values, it can be seen that there is a nearly 30% reduction in the minimum slab thickness for continuous slabs. This significant reduction in slab thickness may, in turn, contribute to a substantial reduction in the overall building weight. Moreover, the use of slabs of smaller thickness will lead to a significant decrease in the capacity requirements for the supporting beams, columns, and footings.

10.2.2 Stiffness Distribution in Continuous Structures

The distribution of bending moments and shear forces in flexural members depends on load patterns, span lengths, and support conditions. In addition to these factors, the relative stiffness between adjacent spans also influences the distribution of internal forces in flexural members with continuous spans. As discussed in Section 4.3.1, the flexural stiffness of a member is directly proportional to its moment of inertia (I), that is, the larger the moment of inertia, the larger the flexural stiffness. Moreover, the flexural stiffness is inversely proportional to the span length (l) of the member, that is, the smaller the span length, the larger the flexural stiffness. Therefore, span length and moment of inertia are the two key parameters influencing the flexural stiffness of a member.

In a continuous structure with several spans of approximately equal lengths, spans characterized by larger cross-sectional dimensions have larger flexural stiffness. When subjected to flexure, members with stiffer adjoining members attract larger bending moments at the supports than members with adjoining members of smaller stiffness. The correlation between the stiffness distribution and bending moment magnitudes will be illustrated by a few examples.

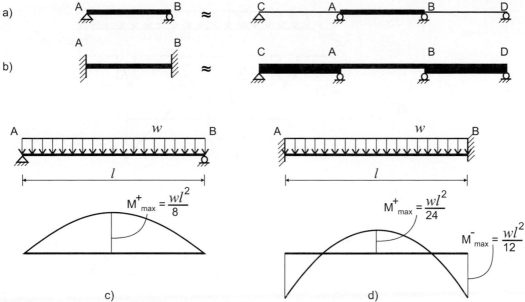

Figure 10.4   Simply supported versus fixed-end beam: a) simply supported beam and the equivalent continuous beam; b) fixed-end beam and the equivalent continuous beam; c) bending moment diagram for the simply supported beam; d) bending moment diagram for the fixed-end beam.

First, let us consider two beams characterized by the same span length (l) and subjected to the same uniform load (w) (see Figure 10.4). The beam in Figure 10.4a is simply supported, while the beam in Figure 10.4b is fixed at the ends. A simply supported beam can be modelled as a three-span continuous beam with extremely small stiffnesses in the end spans. Conversely, a beam with fixed ends can be modelled as a three-span continuous beam with extremely large stiffnesses in the end spans. The simply supported beam develops positive bending moments with a maximum value of $wl^2/8$ at the midspan (see Figure 10.4c), whereas the beam fixed at the ends develops both positive and negative moments, as shown in Figure 10.4d. Note that the positive bending moment developed at the midspan is equal to $wl^2/24$, that is, one-third of the corresponding value for the simply supported beam in Figure 10.4c. The negative bending moments at the supports are equal to $wl^2/12$, that is, twice the positive moment at the midspan. Hence, this beam requires a larger amount of top reinforcement at the supports than bottom reinforcement at the midspan. It can be concluded from this example that the presence of negative bending moments at the supports results in reduced positive bending moments at the midspan. This is true for any span within a continuous structure.

In reality, most reinforced concrete members are neither entirely pinned nor completely fixed at their ends. The above example illustrates the effect of extremely stiff and extremely flexible adjacent spans in a continuous structure. Hence, for a beam subjected to a uniformly distributed load, the bending moments at the supports vary between 0 and $-wl^2/12$, depending on the stiffness of the adjacent spans.

Next, let us consider a typical span in a continuous beam subjected to a uniformly distributed load, as shown in Figure 10.5a. It can be shown that the sum of the absolute values of the positive bending moment at the midspan (M_1) and the average of the negative bending moments at the supports (M_2) is always equal to $wl^2/8$, as shown on the bending moment diagrams in Figure 10.5 (this value is sometimes termed the *moment gradient*).

It is also important to discuss the location of the maximum bending moment in a continuous member. When the bending moments at the supports are of equal magnitude, the maximum positive bending moment acts exactly at the midspan (see Figure 10.5a); otherwise it is located off the midspan (see Figure 10.5b). In practice, the maximum bending moment is usually considered to act at the midspan, except when there are large differences in the magnitude of the bending moments at the supports.

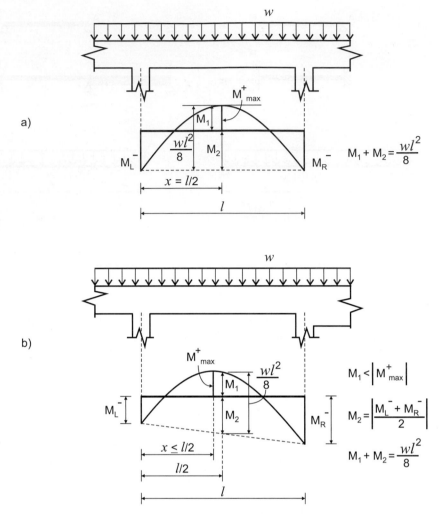

Figure 10.5 Moment distribution in a continuous beam: a) equal bending moments at the supports; b) unequal bending moments at the supports.

Let us examine the design situation in which there are large differences in end moments. Consider the three-span continuous beam ABCD subjected to a uniform load (w) in the centre span BC in Figure 10.6a. The beam is characterized by extremely different endspan stiffnesses: span AB is extremely stiff while span CD is very flexible. As a result, the centre span BC behaves similarly to a beam fixed at the left end and pinned at the right end (see Figure 10.6b). The bending moment diagram for this beam is shown in Figure 10.6c. Observe from the diagram that the maximum positive bending moment (approximately equal to $0.07\,wl^2$) acts at a distance $5l/8$ from the left end, whereas the maximum negative bending moment, equal to $wl^2/8$, acts at the left support. The magnitude of the positive bending moment at the midspan is equal to $wl^2/16$, that is, $0.063\,wl^2$ (see Figure 10.6d). The difference between the positive bending moments in Figure 10.6c and 10.6d ($0.07wl^2$ versus $0.063\,wl^2$) is about 11%. Therefore, when the maximum positive bending moment at the midspan is used in the design (instead of the bending moment at its exact location at a distance of $5l/8$ from the left end), the moment magnitude will be underestimated by 11%; also, the maximum moment location will be misplaced by $l/8$ (the difference between $5l/8$ and $l/2$). This example demonstrates an extreme case with the greatest possible moment differential between the supports. In most cases, the difference in the magnitudes of negative bending moments at the supports are not as significant.

Based on the above discussion, the maximum positive bending moments in a continuous beam or slab can be approximated in accordance with Table 10.1. This table can be used as a design tool, as a means to perform an independent check of analytical results, or, alternatively, when design decisions are based on a quick approximate analysis made at a construction site.

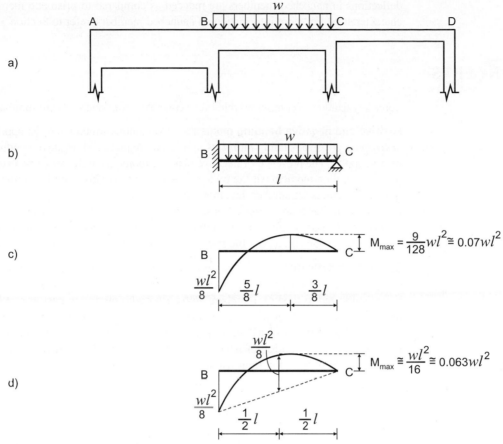

Figure 10.6 Maximum positive bending moment in a continuous beam: a) beam elevation and loading; b) analysis model for span BC; c) actual bending moment distribution; d) bending moment distribution with the maximum positive bending moment approximated at the midspan.

Table 10.1 Positive bending moments in continuous structures with different end span stiffnesses

Bending moments at the supports	Maximum positive bending moment (M_{max})	Location of maximum positive bending moment
$M_L^- \cong M_R^-$	$\dfrac{M_L^- + M_R^-}{2} + \dfrac{wl^2}{8}$	midspan
$M_L^- \cong 2M_R^-$	$\left(\dfrac{M_L^- + M_R^-}{2} + \dfrac{wl^2}{8}\right) \times 1.05$	midspan $+ \dfrac{l}{16}$ toward the right support (the smaller bending moment)
$M_L^- \gg M_R^-$	$\left(\dfrac{M_L^- + M_R^-}{2} + \dfrac{wl^2}{8}\right) \times 1.1$	midspan $+ \dfrac{l}{8}$ toward the right support (the smaller bending moment)

Note that the members considered in the above discussion are prismatic, that is, characterized by constant cross sectional dimensions within a particular span. However, nonprismatic sections (such as haunched beams) are often used in the design of longer span structures. Beams with haunches in the support regions tend to attract larger negative bending moments at the supports, resulting in a reduction in positive moments at the midspans. Consequently, deflections in haunched members are reduced as compared to prismatic members of similar characteristics. For more information on haunched members, refer to Section 3.2.

KEY CONCEPTS

Some key features of continuous reinforced concrete beams and slabs are summarized below.

Positive and negative bending moments   Continuous structures resist applied loads by developing positive bending moments at midspan regions and negative bending moments over supports. The presence of negative bending moments at the supports causes reduced positive bending moments at the midspan. As a result, positive bending moments in continuous structures are smaller than those developed in a series of simple structures characterized by similar span lengths. Consequently, cross-sectional dimensions in continuous beams and slabs are smaller than in simply supported members. This in turn leads to a more economical design and a possible significant reduction in the overall building weight.

Stiffness distribution   In a continuous structure with several spans of approximately equal lengths, spans with larger cross-sectional dimensions have larger flexural stiffness. When subjected to flexure, members with stiffer adjoining members attract larger negative bending moments at the supports than members with adjoining members of smaller stiffness. The stiffness distribution in continuous structures has a major impact on the magnitude and distribution of bending moments, and the designer must develop a good understanding of this important feature.

10.3 | LOAD PATTERNS

In general, structural members in a building structure should be designed for the worst combination of loads expected to occur during its service life. As discussed in Section 1.7, the loads that act on the structure consist of permanent loads such as dead load, and different transient loads, including live loads due to occupancy, snow loads, wind loads, and earthquake loads. For example, live loads due to human occupancy can be placed in the building in different ways, thus causing one part of the building to be more heavily loaded than the others. For example, imagine a floor in a college building with several classrooms in a row, wherein every other classroom is occupied at a certain time, while the others are vacant.

In a continuous structure, load acting on one span influences the bending moments and shear forces in other spans. Consequently, the structure must be designed to resist different live load distributions (also called *load patterns*) that can result in the worst possible effects.

It is difficult to establish general rules on the effect of load placement upon the maximum bending moments within a continuous structure. This task is particularly complex in continuous structures characterized by variable cross-sectional properties, amount of reinforcement, and stiffnesses across the spans (as discussed in Section 10.2.2).

In general, when alternate spans in a continuous structure are loaded, the maximum positive bending moments will develop in the loaded spans. On the other hand, maximum negative bending moments will develop when two adjacent spans are loaded at either side of the support. This is also reflected in the CSA A23.3 requirements. A23.3 Cl.9.2.3.1 prescribes the following load patterns to be considered in the design of continuous beams and slabs:

a)   factored dead load and factored permanent superimposed *dead load on all spans,* with factored partition load and *factored live load on two adjacent spans;* this will result in the maximum negative bending moment at the support between the two loaded spans;

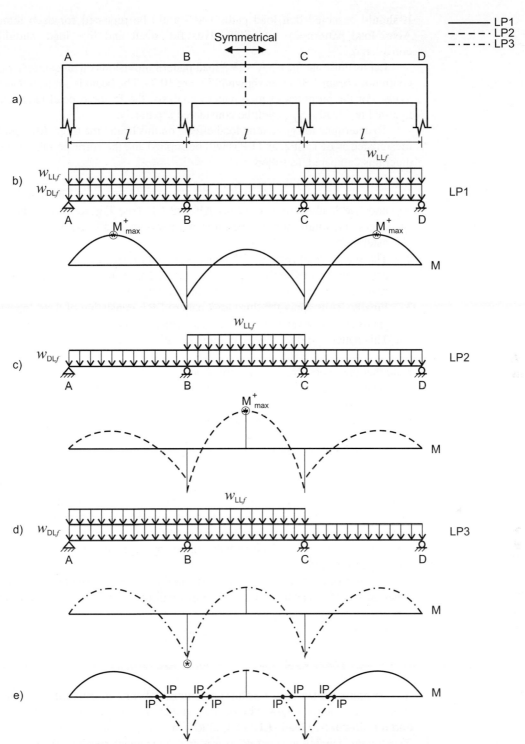

Figure 10.7 Bending moment diagrams for a three-span continuous beam: a) beam elevation; b) load pattern LP1; c) load pattern LP2; d) load pattern LP3; e) moment envelope.

b) factored dead load and factored permanent superimposed *dead load on all spans*, with factored partition load and *factored live load on alternate spans;* this will result in maximum positive moments at the midspan of the loaded spans and minimum positive moments at the middle of unloaded spans, and also the maximum negative moment at the exterior support;

c) *factored dead load and factored live load on all spans.*

It should be noted that load pattern a) should be repeated for each interior support. Also, load pattern c), which involves full dead and live load, should always be considered.

Let us illustrate the CSA A23.3 load pattern provisions on an example of a three-span continuous beam ABCD, as shown in Figure 10.7a. The beam is subjected to dead and live loads. For the purposes of load pattern analysis, the factored dead load ($w_{DL,f}$) and the factored live load ($w_{LL,f}$) will be considered separately.

To determine the governing load effects on this beam, the three different load patterns (designated as LP1, LP2, and LP3) will be applied and the corresponding bending moment diagrams compared, as follows:

1. First, let us apply the load pattern LP1 consisting of dead load acting on all spans and live loads acting on spans AB and CD (see Figure 10.7b). This pattern produces maximum positive bending moments in spans AB and CD (denoted by an asterisk *).
2. The second load pattern LP2 consists of dead load acting on all spans and live load acting on span BC only (see Figure 10.7c). This pattern results in the maximum positive bending moment in span BC (denoted by *).
3. Finally, let us apply the third load pattern LP3, consisting of dead load applied on all spans and live load applied on the two adjacent spans AB and BC (see Figure 10.7d). This pattern produces the maximum negative bending moment at support B.

Note that one more live load pattern should be considered, where spans BC and CD only are subjected to live load; this pattern gives the maximum negative bending moment at support C. However, this load pattern has been omitted in this case since the beam is symmetric with regard to the vertical axis. In addition, the load case involving the full dead and live load on all spans should be considered in the design.

Finally, let us superimpose the bending moment diagrams corresponding to all of the load patterns, as shown in Figure 10.7e. The composite diagram obtained in this manner represents the maximum bending moment at any point along the member and is called the *moment envelope*. It can be observed from the moment envelope diagram that the inflection points (denoted as IP) do not coincide for the positive and negative bending moments within the same span (note that these moments are caused by different load patterns). A shear force envelope can be obtained in a similar manner by superimposing shear force diagrams for different load patterns.

The use of load patterns in the structural analysis of continuous structures and the development of moment envelope diagrams will be illustrated by Example 10.1.

Example 10.1

A four-span continuous reinforced concrete beam of rectangular cross-section is shown in the figure below. The beam supports a uniform dead load (DL) of 16 kN/m and a uniform live load (LL) of 13.3 kN/m.

Develop the bending moment diagrams and the moment envelope for this beam considering the different load patterns prescribed by CSA A23.3.

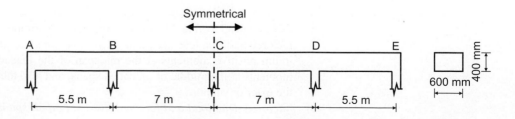

SOLUTION: **1. Determine the factored loads**
First, determine the factored loads for this analysis. In this example, the factored dead and live loads will need to be considered separately.
    The factored dead load is

$$w_{DL,f} = 1.25 \times DL = 1.25 \times 16 \text{ kN/m} = 20 \text{ kN/m}$$

while the factored live load is

$$w_{LL,f} = 1.5 \times LL = 1.5 \times 13.3 \text{ kN/m} = 20 \text{ kN/m}$$

2. Identify all possible load patterns
According to CSA A23.3 Cl.9.2.3.1, the following load patterns should be considered in the design of the four-span beam under consideration:

1. full dead load on all spans plus live load on spans AB and CD (LP1)
2. full dead load on all spans plus live load on spans AB and BC (LP2)
3. full dead load on all spans plus live load on spans BC and CD (LP3)
4. full dead load on all spans plus live load on spans BC and DE (LP4)
5. full dead load on all spans plus live load on spans CD and DE (LP5)
6. full dead load and live load on all spans (LP6)

Of all possible load patterns, only patterns LP1, LP2, and LP3 will be considered in this example. Since the beam is symmetric with regard to the vertical axis, load pattern LP4 represents a mirror image of pattern LP1 and will give the same results, while load pattern LP5 is a mirror image of load pattern LP2. Load pattern LP6 will be considered in the discussion on moment envelope later on in this example (see Step 4).

3. Develop the bending moment diagrams corresponding to the selected load patterns
In this example, the bending moment diagrams have been obtained from linear elastic analysis performed using a computer structural analysis program. Load patterns LP1, LP2, and LP3 and the corresponding bending moment diagrams are shown in Figure 10.8.

4. Develop the moment envelope diagram
Although the structural analysis of this beam has been performed three times considering different load patterns, it is hard to identify most critical bending moments by means of a visual comparison. Instead, a moment envelope diagram is developed by superimposing the bending moment diagrams for load patterns LP1, LP2, and LP3, as shown in Figure 10.9a. The moment envelope diagram can be constructed from portions of the bending moment diagrams corresponding to different load patterns (see Figure 10.9b). For example, the positive bending moments in span AB are due to pattern LP1, the maximum negative bending moments at support B are due to pattern LP2, and the maximum negative bending moments at support C are due to pattern LP3. Due to symmetry, the positive bending moments in spans CD and DE represent mirror images of the bending moments in spans AB and BC, while the portion of the negative bending moment diagram at support D is a mirror image of the diagram at support B.

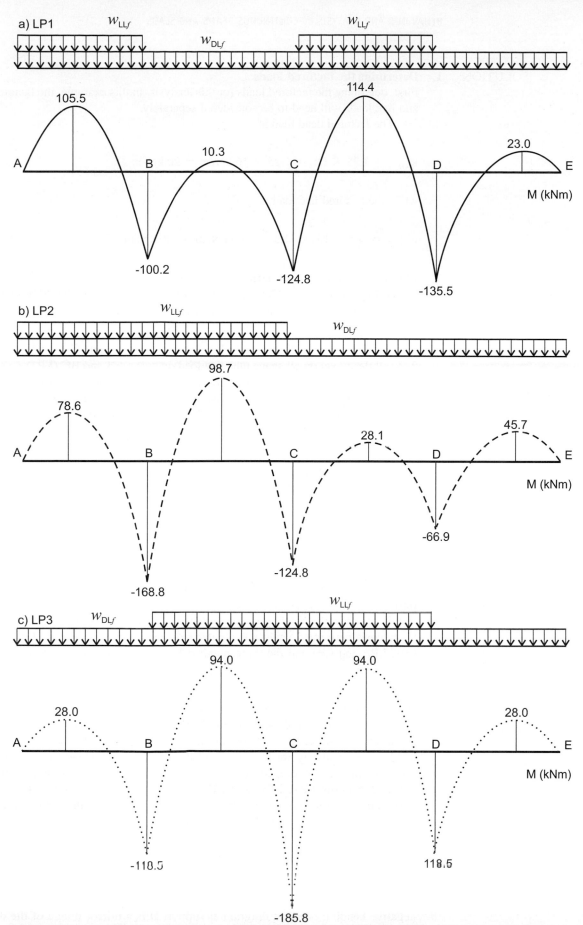

Figure 10.8 Bending moment diagrams: a) load pattern LP1; b) load pattern LP2; c) load pattern LP3.

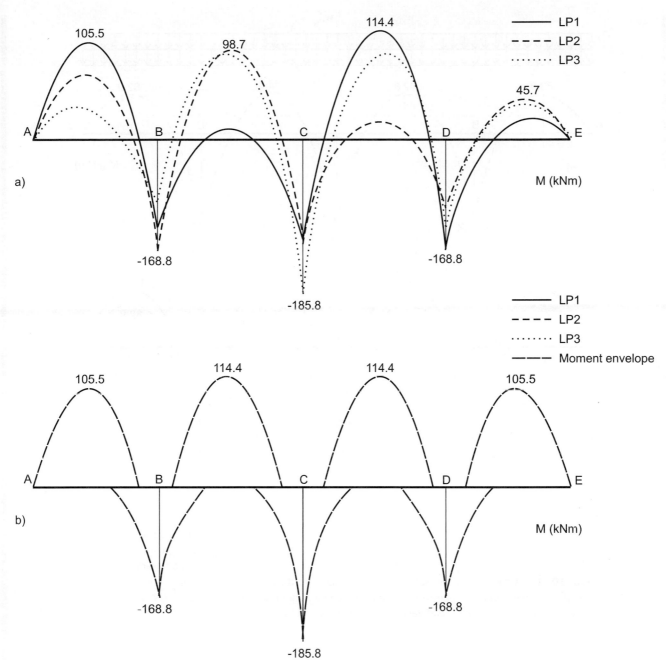

Figure 10.9 Development of a moment envelope diagram: a) bending moment diagrams for load patterns LP1, LP2, and LP3; b) corresponding moment envelope diagram.

Learning from Examples

The load acting on one span in a continuous structure influences the bending moments and shear forces in other spans. Consequently, the structure must be designed to resist a few different live load distributions (also called *load patterns*) that result in the worst possible effects. Load patterns with live load applied to alternate spans cause maximum positive bending moments in the loaded spans, while maximum negative bending moments will develop when the two adjacent spans are loaded at either side of the support. A23.3 Cl.9.2.3.1 prescribes three different load patterns to be considered in the design of continuous beams and slabs. Bending moment diagrams corresponding to different load patterns are superimposed to obtain the *moment envelope*, which represents the maximum bending moment values at any point along the member length.

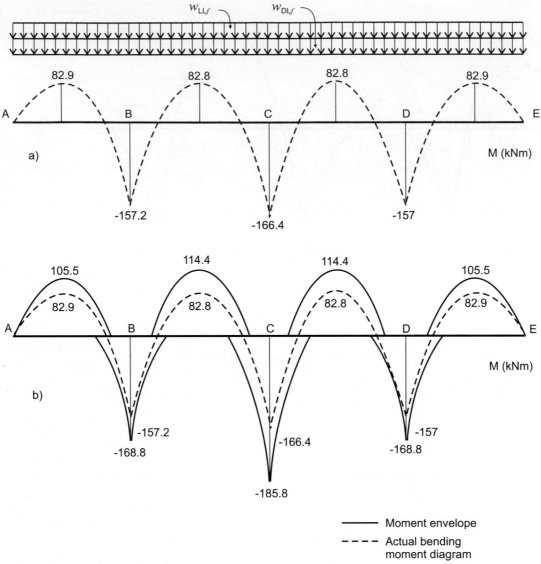

Figure 10.10 A comparison between the bending moment diagram and the moment envelope: a) bending moment diagram due to total factored load acting on all spans; b) bending moment diagram due to total factored load versus the moment envelope.

The purpose of Example 10.1 was to illustrate the development of the moment envelope diagram, which requires a repetitive analysis of the same structure by applying different load patterns. It is worth exploring whether there are significant differences between the design bending moments derived from the moment envelope and those obtained from a load case where the full uniform dead and live load are applied on every span.

The bending moment diagram due to the total factored load is shown in Figure 10.10a; this diagram (dashed line) is then superimposed on the moment envelope diagram (solid line), as shown in Figure 10.10b. The following observations can be made by comparing the bending moment values from these two diagrams:

• In the region of positive bending for span BC, the moment envelope gives a value of 114.4 kN·m, whereas the bending moment due to the total factored load on all spans is 82.8 kN·m; the difference between these two values is 38%.

• In the negative moment region (at support C), the moment envelope gives a bending moment of 185.8 kN·m, whereas the bending moment due to the total factored load on all spans is 166.4 kN·m; in this case, the difference between the two values is 12%.

It is clear that the moment envelope produces the governing bending moment values at each section along the beam length. However, in the case where the structure is loaded primarily with dead load, the effects of load patterns are less significant and may be neglected in the design. CSA A23.3 permits the design of two-way slabs without consideration of load patterns when the dead load exceeds 75% of the total load.

10.4 | SIMPLIFICATIONS IN THE ANALYSIS OF REINFORCED CONCRETE FRAME STRUCTURES

10.4.1 Actual Structure Versus Idealized Structural Model

In order to perform a structural analysis of a reinforced concrete structure, the designer needs to develop an appropriate structural model. In most cases, a three-dimensional frame structure is modelled as a series of plane (two-dimensional) frames.

Accurate modelling of reinforced concrete structures can be very complex. The complexity is due to many factors, including stiffness distribution between adjacent members, variable member properties (cracked/uncracked), and prismatic versus nonprismatic members. As an example, the supports for the simple frame in Figure 10.11 may be modelled as pinned, fixed, or semirigid, depending on the design application.

In general, three-dimensional beam and column members are represented by straight lines that usually coincide with their centroidal axes. This simplification is adequate for members with reasonably small cross-sectional dimensions, but it may not be appropriate for walls or large beams or columns. Therefore, good engineering judgment and experience are required to develop a suitable structural model that is a reasonable representation of the actual structure.

10.4.2 Frame Model

For gravity load analysis, a large multistorey reinforced concrete structure comprising multiple bays of beams/slabs and columns, as shown in Figure 10.12a, can be simplified by dividing the frame into simpler subframes. Each subframe consists of one continuous beam plus the top and bottom columns framing into it, as shown in Figure 10.12b. The far ends of the columns are considered to be fixed, except for the ground floor or basement columns with pinned bottom ends due to soil conditions. This approach is also stipulated by CSA A23.3 (Cl.9.3.2).

In a typical subframe, the bending moment and shear force distribution are obtained by applying various load patterns, as discussed in Section 10.3. The results (bending moments and shear forces) obtained by analyzing one or more typical subframes can be used to design

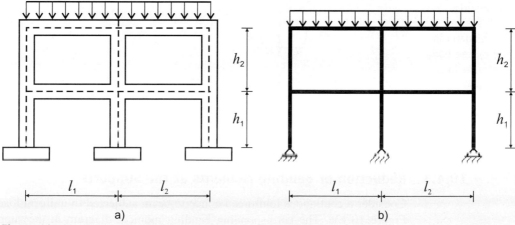

Figure 10.11 A simple reinforced concrete frame: a) actual structure; b) structural model.

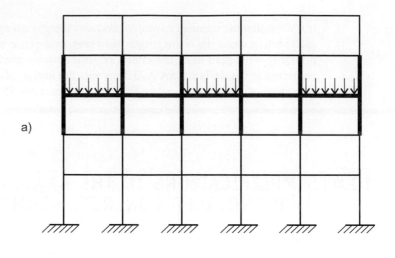

Figure 10.12 Frame model:
a) a complete frame; b) a sub-frame used for structural analysis.

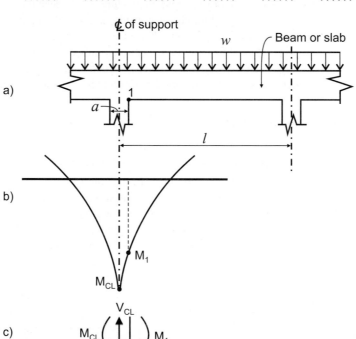

Figure 10.13 Bending moment at the face of the support:
a) typical beam/slab span;
b) bending moment diagram in the support region; c) free-body diagram showing internal force distribution in the support region.

beams and columns in the actual reinforced concrete frame structure. Analysis methods for continuous reinforced concrete frame structures will be discussed in Section 10.5.

10.4.3   Reduction of Bending Moments at the Supports

Consider a continuous reinforced concrete beam subjected to uniform load as shown in Figure 10.13a. The corresponding bending moment diagram in the support region is shown in Figure 10.13b. In general, the distribution of bending moments in reinforced concrete structures is determined based on the assumption that members span centre-to-centre across the supports, as discussed in Section 10.4.1. Therefore, beam and slab supports are concentrated at a single point. However, in reality reinforced concrete slabs and beams are supported by structural members (such as columns) of finite dimensions. Consequently, the peak bending moment (M_{CL}) at the column centre-line may overestimate the actual bending moment at the support (see Figure 10.13b). Studies have shown that the bending moment at the face of the support (M_1) is more representative of the actual bending moment distribution at the support region and should be used in the design.

To find the bending moment value at the face of the support, it is necessary to first determine the bending moment at the centre-line of the support (M_{CL}); this can be done using one of the analysis methods that will be discussed in Section 10.5. An accepted approach to determine the bending moment at the face of the support (M_1) is to consider a free-body diagram of the member in the support region, as shown in Figure 10.13c. The bending moment at the face of the support can be obtained by taking the sum of the bending moments at point 1. The moment at the face of the support can be determined as follows:

$$M_1 \cong M_{CL} - V_{CL} \times \frac{a}{2}$$ [10.1]

where

a = the column width
V_{CL} = the shear force at the support.

Note that the effect of the uniform load (w) in the support region has been ignored in the above equation.

In this calculation, the V_{CL} value is estimated by treating the span as a simply supported beam, as follows:

$$V_{CL} \cong \frac{w \times l}{2}$$

Note that the V_{CL} value obtained by the above approach is approximate; in reality, the actual shear force at the support may be higher. For design purposes, the moment at the face of the support can be used in lieu of the moment at the centre-line of the support. The above-outlined approach will be illustrated by Example 10.2.

KEY CONCEPTS

The following simplifications are often used in the structural analysis of continuous reinforced concrete structures:

- *Actual structure versus structural model:* A structural model for a reinforced concrete frame structure is developed by representing three-dimensional beam and column members by one-dimensional members represented by their centroidal axes.
- *Frame model:* For gravity load analysis, a complete reinforced concrete structure with multiple bays of beams/slabs and columns can be divided into simpler subframes; each subframe consists of one continuous beam plus the top and bottom columns framing into it.
- *Reduction of bending moments at the supports.* Reduced bending moments at the face of the support are used in the analysis instead of the peak bending moment value at the centre-line of the support.

Example 10.2

A typical span BC from the four-span reinforced concrete continuous beam from Example 10.1 and the corresponding moment envelope are shown in the figure below. The complete moment envelope for this beam was developed in Example 10.1. The beam is supported by 400 mm square reinforced concrete columns.

Determine the bending moment at the face of each support.

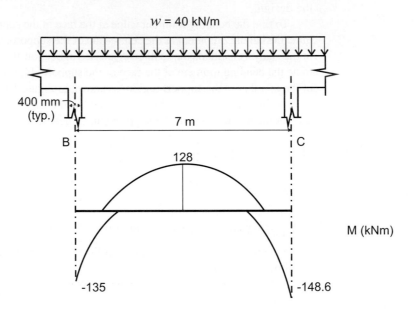

SOLUTION: The column width is

$a = 400$ mm

Estimate the shear force at the support as

$$V_{CL} \cong \frac{w \times l}{2} = \frac{40 \text{ kN/m} \times 7 \text{ m}}{2} = 140 \text{ kN}$$

Determine the bending moment at the face of the left support (M_1) as (see the sketch below)

$$M_1 \cong M_{CL} - V_{CL} \times \frac{a}{2} \qquad\qquad \textbf{[10.1]}$$

$$= 135 \text{ kN} \cdot \text{m} - 140 \text{ kN} \times \frac{0.4 \text{ m}}{2} = 107 \text{ kN} \cdot \text{m}$$

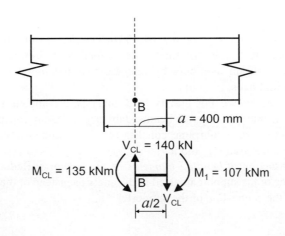

The bending moment at the face of the right support (M_2) is

$$M_2 = 148.6 \text{ kN} \cdot \text{m} - 140 \text{ kN} \times \frac{0.4 \text{ m}}{2} = 120.6 \text{ kN} \cdot \text{m} \cong 120 \text{ kN} \cdot \text{m}$$

The revised moment envelope diagram for span BC, taking into account the reduced bending moments at the support faces, is shown below.

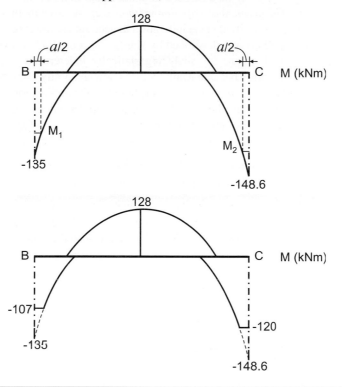

Learning from Examples

Based on this example, the bending moment values at the centre-line of the support can be compared with the corresponding values at the face of the support. At support B, the bending moment at the centre-line is equal to -135 kN·m, whereas the corresponding moment at the face of the support is equal to -107 kN·m. Similarly, the bending moment at the centre-line of support C is equal to -148.6 kN·m, whereas the corresponding moment at the face of the support is equal to -120 kN·m.

In both cases, the bending moments at the face of the support are reduced by approximately 20% as compared to the corresponding moments at the centre-lines. This is a significant difference, and it is recommended to use the reduced bending moments at the face of the supports in the design.

10.5 ANALYSIS METHODS FOR CONTINUOUS REINFORCED CONCRETE STRUCTURES

The two main approaches to the analysis of continuous reinforced concrete structures are

- linear elastic analysis
- inelastic analysis

Linear elastic analysis is commonly used to determine the internal forces and bending moments in continuous reinforced concrete beams and slabs. This approach takes into account the linear elastic behaviour of reinforced concrete members. Linear elastic analysis of reinforced concrete structures is usually performed using one of the following methods:

- *Approximate frame analysis* provides a means for a rapid estimate of internal forces and bending moments. This analysis can be performed using hand calculations, and it is adequate for simple and regular structures that meet the criteria prescribed by CSA A23.3 (to be discussed in detail in Section 10.6).
- *The moment distribution method* is an iterative method in which the fixed-end moments are modified in each iteration to account for the rotation and translation at the joint. The method was developed by Hardy Cross in 1932 and was used for many years as a standard means for analyzing statically indeterminate structures before computer-based methods became available. It is now generally considered obsolete.
- *Matrix structural analysis* is based on the application of matrix theory to structural analysis. This method enables a rapid computer-aided analysis of large statically indeterminate structures. Several computer software packages for structural analysis are commercially available; however, some consulting firms have their own in-house software.

Inelastic analysis simulates the response of reinforced concrete structures in the postcracking stage. Several approaches to inelastic analysis will be discussed in Section 10.8.

The designer should choose a suitable structural analysis method depending on the particular design application. Keeping in mind the complexity associated with modelling reinforced concrete structures, using sophisticated computer analysis is often not necessary for simple design applications. However, regardless of the analysis method used, the designer needs to ensure that the design assumptions and parameters are reasonable representations of the actual structure.

10.6 | APPROXIMATE FRAME ANALYSIS

The analysis of continuous reinforced concrete structures must account for the effects of loads applied within one span upon the bending moment distribution within adjacent spans, as discussed in Section 10.3. The detailed linear elastic analysis illustrated by Example 10.1 is a time-consuming effort that includes the consideration of several load patterns and the development of a moment envelope, so the solution given was computer generated.

CSA A23.3 permits the use of an approximate method for the analysis of continuous reinforced concrete beams and one-way slabs (Cl.9.3.3). This method, known as the *approximate frame analysis method* or the *coefficient method,* is suitable for the analysis of simple and regular structures. The method gives approximate expressions for bending moments and shear forces in continuous beams and slabs that can be used in the design.

The factored bending moment (M_f) can be determined based on the equation

$$M_f = C_m \times w_f \times l_n^2 \qquad \text{[10.2]}$$

where

C_m = the moment coefficient
w_f = the factored uniformly distributed load
l_n = the clear span from face to face of supports for positive bending moments, and an average of the adjacent clear span values for negative bending moments.

The factored shear force (V_f) can be determined based on the equation

$$V_f = C_v \times \frac{w_f \times l_n}{2} \qquad \text{[10.3]}$$

where C_v is the shear coefficient.

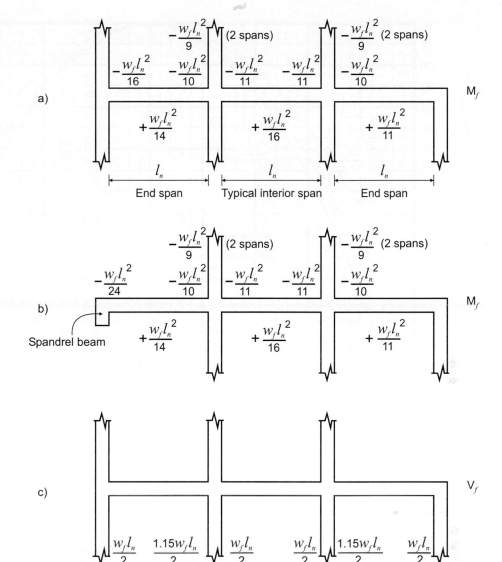

Figure 10.14 Moment and shear coefficients according to CSA A23.3 Cl.9.3.3:
a) M_f values — end support provided by columns;
b) M_f values — end support provided by spandrel beam or girder; c) V_f values — all cases.

The C_m and C_v values for frames with different end support conditions are prescribed by A23.3 Cl.9.3.3, as shown in Figure 10.14.

For simply supported beams and slabs,

$$C_m = \frac{1}{8}$$

and

$$C_v = 1.0$$

The approximate frame analysis per A23.3 Cl.9.3.3 can be used in the design provided that the following criteria have been satisfied:

1. There are at least two continuous spans.
2. The spans are approximately equal; that is, the span length of each span is within 20% of that of the adjacent spans.
3. The loads are uniformly distributed.
4. The factored live load does not exceed twice the factored dead load.
5. The members are prismatic, that is, of uniform cross-sectional dimensions throughout the span.

Note that the C_m and C_v coefficients have been determined from a linear elastic analysis taking into consideration the load patterns and moment envelopes discussed in Section 10.3.

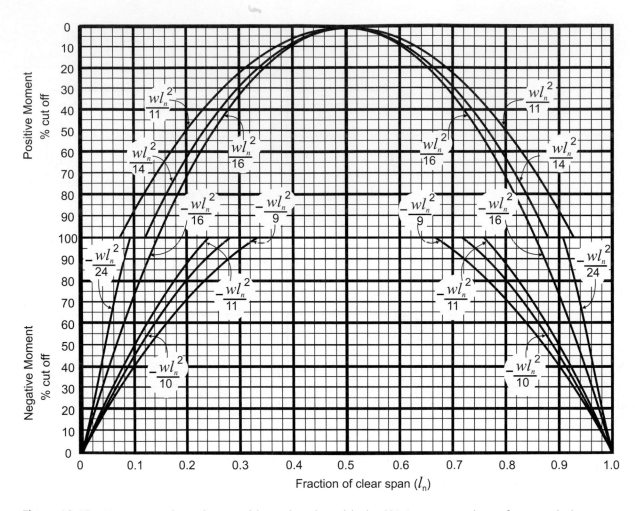

Figure 10.15   Moment envelope chart used in conjunction with the CSA A23.3 approximate frame analysis.

Therefore, it can be expected that the bending moment and shear force values obtained from the coefficient method are greater than the corresponding values obtained from any particular load case used in the analysis.

The development of a moment envelope, as presented in Example 10.1, is not required when the coefficient method is used. For design purposes, the generic moment envelope chart in Figure 10.15 can be used in conjunction with this method. The designer needs to identify the bending moment expression from Figure 10.14 depending on the support conditions. Subsequently, the corresponding bending moment diagram from Figure 10.15 may be used to determine the bar cutoffs for flexural reinforcement (to be discussed in Section 11.4).

The coefficient method will be used in Example 10.3 to determine the bending moment and shear force distribution in a continuous slab.

Example 10.3

A continuous five-span reinforced concrete one-way floor slab is shown in the figure that follows. The slab is subjected to a superimposed dead load (DL$_s$) of 1 kPa and a live load (LL) of 4.8 kPa in addition to its self-weight. The support beams are 300 mm wide and 600 mm deep.

First, determine the slab thickness according to the CSA A23.3 indirect approach to deflection control.

Then, determine the factored bending moments and shear forces according to the CSA A23.3 approximate method for frame analysis.

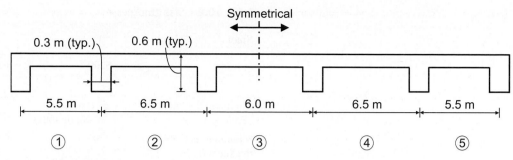

Given: $f_c' = 30$ MPa

$f_y = 400$ MPa

SOLUTION: This slab is symmetric with regard to the vertical axis shown in the figure above. In general, the designer should take advantage of symmetry wherever possible. In this case, all relevant design parameters (slab thickness, bending moments, and shear forces) need to be determined for spans 1, 2, and 3 only. Span 4 is a mirror image of span 2, while span 5 is a mirror image of span 1. Also, note that this slab behaves as one-way slab and that the design is based on a unit strip of width 1 m.

1. **Check whether the criteria for approximate frame analysis are satisfied**

   The criteria for approximate frame analysis according to CSA A23.3 Cl.9.3.3 are as follows:

   - There are more than two continuous spans.
   - The length of each span is within 20% of that of the adjacent spans.
   - The loads are uniformly distributed.
   - The factored live load does not exceed twice the factored dead load (including the slab self-weight) — this will be checked in Step 3.
   - The slab is prismatic (all sections are of uniform thickness).

   Hence, approximate frame analysis according to A23.3 Cl.9.3.3 can be used in this design.

2. **Determine the slab thickness**
   a) Determine the clear spans (l_n).

   Determine the values of the clear span (l_n) (clear distance between supports) for Spans 1 to 3, as shown on the sketch below (note that $b_s = 0.3$ m is the beam width).

Span 1

$$l_{n1} = l_1 - b_s = 5.5 \text{ m} - 0.3 \text{ m} = 5.2 \text{ m}$$

Span 2

$$l_{n2} = l_2 - b_s = 6.5 \text{ m} - 0.3 \text{ m} = 6.2 \text{ m}$$

Span 3

$$l_{n3} = l_3 - b_s = 6.0 \text{ m} - 0.3 \text{ m} = 5.7 \text{ m}$$

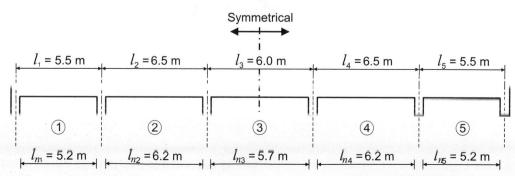

Table 10.2 Slab thicknesses

Span	1	2	3
l_n (mm)	5200	6200	5700
Minimum thickness (Table A.3)	$\dfrac{l_n}{24}$ (one end continuous)	$\dfrac{l_n}{28}$ (two ends continuous)	$\dfrac{l_n}{28}$
Minimum required thickness (mm)	220	220	200

b) To avoid detailed deflection calculations, determine the minimum slab thickness in accordance with the CSA A23.3 indirect approach to deflection control prescribed by C1.9.8.2.1 (see Section 4.5.2 and Table A.3). The calculations are summarized in Table 10.2.

It can be observed from Table 10.2 that the minimum required slab thickness is 220 mm for each span except for Span 3, where a 200 mm thickness is required. Given the small difference in the required thickness values, it may be practical to use a slab thickness of 220 mm for all spans. Therefore, the same thickness will be used throughout the slab length, as follows:

$$h = 220 \text{ mm}$$

The designer should always keep in mind that uniformity in member dimensions reduces the chances of construction complexities.

3. **Determine the factored loads**
   a) Calculate the dead load acting on the slab.
       First, calculate the slab self-weight:

$$\text{DL}_w = h \times \gamma_w = 0.22 \text{ m} \times 24 \text{ kN/m}^3 = 5.28 \text{ kPa}$$

where

$$\gamma_w = 24 \text{ kN/m}^3$$

is the unit weight of concrete.
The superimposed dead load is given as

$$\text{DL}_s = 1.0 \text{ kPa}$$

Finally, the total factored dead load is

$$w_{DL,f} = 1.25 \, (\text{DL}_w + \text{DL}_s)$$
$$= 1.25 \, (5.28 \text{ kPa} + 1 \text{ kPa}) = 7.85 \text{ kPa}$$

   b) Calculate the factored live load:

$$w_{LL,f} = 1.5 \times \text{LL}$$
$$= 1.5 \times 4.8 \text{ kPa} = 7.2 \text{ kPa}$$

   c) The total factored load is

$$w_f' = w_{DL,f} + w_{LL,f}$$
$$= 7.85 \text{ kPa} + 7.2 \text{ kPa} \cong 15 \text{ kPa}$$

This is an one-way slab and the design is based on a unit strip (1 m wide), so

$$w_f' = 15\,\text{kN/m} \times 1\,\text{m} = 15\,\text{kN/m}$$

Note that the factored live load $w_{LL,f} = 7.2$ kPa is less than the factored dead load, that is, $w_{DL,f} = 7.85$ kPa. Thus, the factored live load does not exceed twice the factored dead load, so the condition for the application of the CSA A23.3 coefficient method has been met.

4. **Determine the factored bending moments (M_f) and shear forces (V_f)**
   Next, determine the factored bending moments at critical sections using the moment coefficients prescribed by A23.3 Cl.9.3.3. In this case, the width of the spandrel beam is insufficient to ensure proper development of the top slab reinforcing for negative moment resistance at the support, hence the discontinuous end is considered to be unrestrained, as shown in Figure 10.14b. The results are summarized in Table 10.3.

   Finally, the moment envelope diagram and the shear force diagram can be plotted based on the values calculated in Table 10.3, as shown below.

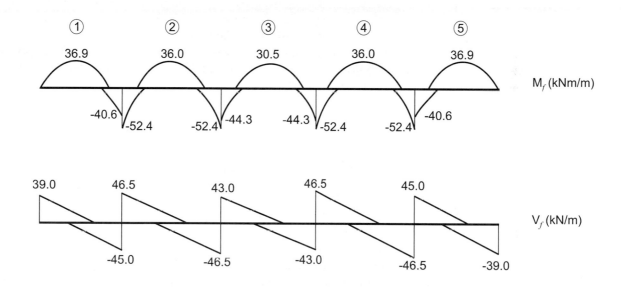

Table 10.3 Factored bending moments and shear forces (according to A23.3 Cl.9.3.3)

Span	1			2			3		
l_n (m)		5.2			6.2			5.7	
Location	Top	Bottom	Top	Top	Bottom	Top	Top	Bottom	Top
M_f (see Figure 10.14b)	0	$\dfrac{w_f l_n^2}{11}$	$-\dfrac{w_f l_n^2}{10}$	$-\dfrac{w_f l_n^2}{11}$	$\dfrac{w_f l_n^2}{16}$	$-\dfrac{w_f l_n^2}{11}$	$-\dfrac{w_f l_n^2}{11}$	$\dfrac{w_f l_n^2}{16}$	$-\dfrac{w_f l_n^2}{11}$
M_f (kN·m/m)	0 36.9	-40.6	-52.4 36.0	-52.4	-44.3	30.5	-44.3		
V_f (see Figure 10.14a)	$\dfrac{w_f l_n}{2}$		$1.15\dfrac{w_f l_n}{2}$	$\dfrac{w_f l_n}{2}$		$\dfrac{w_f l_n}{2}$	$\dfrac{w_f l_n}{2}$	—	$\dfrac{w_f l_n}{2}$
V_f (kN/m)	39.0	—	-45	46.5	—	-46.5	43	—	-43

The two main approaches to the analysis of continuous reinforced concrete structures are as follows:

- *Linear elastic analysis* takes into account the linear elastic behaviour of reinforced concrete members; however, it can indirectly account for the effects of cracking by using transformed section properties.
- *Inelastic analysis* simulates the response of reinforced concrete structures in the postcracking stage.

The linear elastic analysis of reinforced concrete structures is performed by one of the following methods:

- *Approximate frame analysis* (also known as the *coefficient method*) is based on approximate expressions for bending moments and shear forces in continuous beams and slabs. This analysis can be performed using hand calculations and is adequate for simple and regular structures that meet the criteria prescribed by CSA A23.3 Cl.9.3.3.
- *The moment distribution method* is an iterative method developed by Hardy Cross in 1932 in which the fixed-end moments are modified in each iteration to account for the rotation and translation at the joint.
- *Matrix structural analysis* is based on the application of matrix theory to structural analysis. This method enables a rapid computer-aided analysis of large statically indeterminate structures.

10.7 | BEHAVIOUR OF CRACKED CONTINUOUS REINFORCED CONCRETE FLEXURAL MEMBERS

10.7.1 Background

Linear elastic analysis is routinely used to determine internal forces and bending moments in reinforced concrete structures, as discussed in Section 10.5. The analysis considers the *linear elastic* behaviour of reinforced concrete members. However, once a reinforced concrete member cracks and enters the postcracking stage, it starts to behave in an *inelastic* manner.

This section is focused on explaining the basic concepts of the behaviour of continuous reinforced concrete flexural members in the postcracking stage (Section 10.7.2) and the main parameters influencing that behaviour, such as the amount and distribution of reinforcement (Section 10.7.3).

10.7.2 Elastic Versus Inelastic Behaviour

Consider a typical span of a continuous reinforced concrete beam subjected to a uniformly distributed load (w) of increasing magnitude, as shown in Figure 10.16a and in the corresponding bending moment diagram of Figure 10.16b. The beam is properly reinforced and is expected to demonstrate steel-controlled failure. A correlation between the bending moment at the midspan (section A-A) and the corresponding deflection (Δ)is shown in Figure 10.16c. At a low load level, the beam demonstrates essentially linear elastic behaviour, wherein the internal bending moments and deflections are directly proportional, as shown by the straight line on the moment deflection diagram in Figure 10.16c. Under increasing loads, cracks develop locally in the regions of the member where the bending moment (M) exceeds the cracking moment (M_{cr}); that is, $M > M_{cr}$. As a result, the cross-sectional properties in the cracked regions of the member change; these changes include a

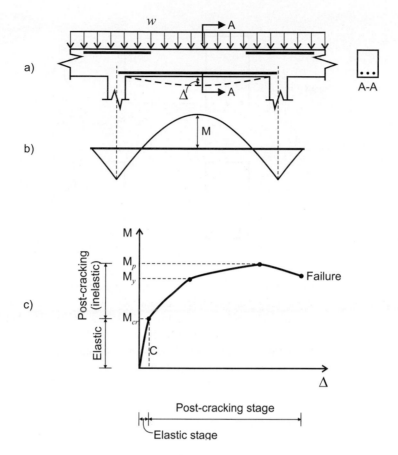

Figure 10.16 Inelastic behaviour of a continuous reinforced concrete flexural member:
a) typical span and loading;
b) bending moment diagram;
c) moment-deflection diagram for section A-A at the midspan.

decrease in the moment of inertia and flexural stiffness values. At that stage, the bending moments and the corresponding deflections are no longer proportional, and the beam is considered to show *inelastic behaviour*. A further significant change in the beam behaviour occurs when the tension reinforcement yields, that is, the internal bending moment in the section reaches the *yield moment* (M_y), as shown in Figure 10.16c.

Once the steel has begun to yield, the strain in the steel increases without a further stress increase; this behaviour is accompanied by significant deformations in the steel and by the crushing of concrete. With further load increase, the internal bending moment in the section will reach the *plastic moment* (M_p), that is, the maximum internal bending moment that the reinforced concrete section can resist (see Figure 10.16c). A localized region where M_p has been reached is called a *plastic hinge*. The formation of a plastic hinge precedes the fracture of the beam section, as shown in Figure 10.16c.

The yield moment (M_y) and the plastic moment (M_p) for a beam section are presented in Figure 10.17. Note that M_y corresponds to the yield stress (f_y) in the reinforcement at the strain ($\varepsilon_s = \varepsilon_y$), whereas M_p corresponds to the ultimate tensile strength (f_u) of steel, which is approximately equal to $1.25 f_y$ at a very large strain ($\varepsilon_s >> \varepsilon_y$) (see Section 2.6.2).

Practically, inelastic behaviour is characteristic of all reinforced concrete flexural members. This behaviour is particularly significant for continuous reinforced concrete structures because it has an effect on the distribution of bending moments in these structures. Once a continuous reinforced concrete structure begins to crack, the distribution of internal bending moments changes as compared to the one obtained by the linear elastic analysis of an uncracked structure. CSA A23.3 permits redistribution of bending moments obtained by linear elastic analysis (to be discussed in more detail in Section 10.9).

10.7.3   The Effect of Reinforcement Distribution in the Postcracking Stage

Once a continuous reinforced concrete flexural member cracks, the magnitude of the bending moment resisted by the cracked section is affected by the amount and distribution of

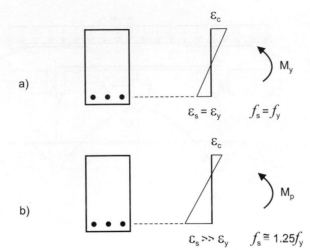

Figure 10.17 Internal bending moments in a beam section: a) yield moment (M_y); b) plastic moment (M_p).

reinforcement in the section. This unique feature of reinforced concrete structures will be explained by an example. Consider the two four-span continuous reinforced concrete beams supported by columns in Figures 10.18 and 10.19. The beam in Figure 10.18 is reinforced with bottom reinforcement only, whereas the beam in Figure 10.19 is reinforced with top reinforcement only. Both beams have an adequate amount of reinforcement to resist the applied moments for which they are designed.

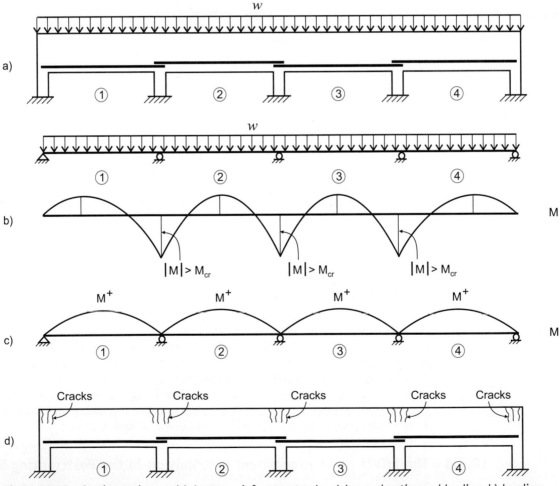

Figure 10.18 Continuous beam with bottom reinforcement only: a) beam elevation and loading; b) bending moment diagram; c) series of simple beams and the corresponding bending moment diagram; d) cracking pattern.

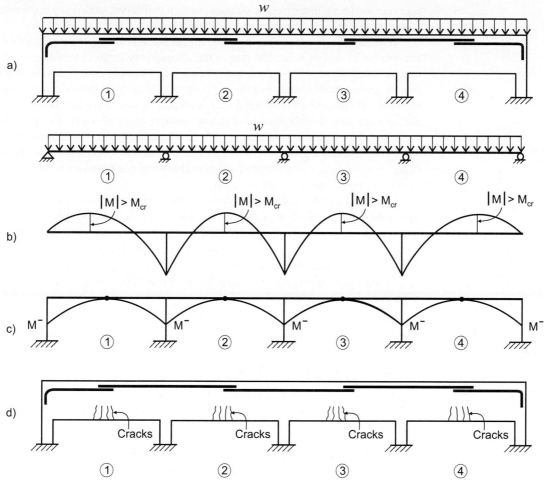

Figure 10.19   Continuous beam with top reinforcement only: a) beam elevation and loading; b) bending moment diagram; c) series of columns with projecting cantilevers and the corresponding bending moment diagram; d) cracking pattern.

At moderate loads, the beam with bottom reinforcement only will develop positive bending moments at the midspan and negative moments at the supports (see Figure 10.18b). Because the beam does not contain any top reinforcement, the negative moment capacity at the support is limited to the cracking moment (M_{cr}). Once the load is increased to the extent that causes the bending moments at the supports to exceed M_{cr}, that is, $M > M_{cr}$, the beam cracks and the bending moments cannot be resisted in the support regions. As a result, the four-span continuous beam will start to act like a series of four simply supported beams (see Figure 10.18c), in which the bending moment values reach a maximum at the midspan and drop to zero at the supports. The cracking pattern for this beam is shown in Figure 10.18d.

The beam with top reinforcement only in Figure 10.19a will behave in a manner similar to the other beam under moderate loads (before cracking has begun). However, since the beam does not contain bottom reinforcement, its positive moment capacity in the midspan regions is limited to M_{cr}. Once the load is increased to such an extent that cracking in the midspan regions takes place, the beam is no longer capable of resisting the positive bending moments (see Figure 10.19b). As a result, pinned joints will form in the cracked regions around the midspans, and the beam will behave as a series of columns with projecting cantilevers, as shown in Figure 10.19c. In this case, the bending moment values reach a maximum at the supports and drop to zero at the midspan. The cracking pattern for this beam is shown in Figure 10.19d.

It can be concluded from the above example that, due to different placement of flexural reinforcement, the behaviour of these two beams in the postcracking stage is very different. This example shows how the designer can control the behaviour of continuous reinforced concrete members, in terms of both their strength and serviceability, by appropriate distribution of flexural reinforcement.

It should be stressed that these beams represent extreme cases and are not really acceptable as design solutions due to the excessively large cracks that may develop in unreinforced areas. In general, most continuous members are designed with the top reinforcement continuous over the supports and bottom reinforcement at the midspans. However, it should be recognized that several solutions may be acceptable in terms of the amount and distribution of top and bottom reinforcement for the same design. In fact, any design solution that results in adequate flexural and shear capacity while satisfying the serviceability criteria can be considered acceptable. In practice, a good designer would derive an optimal solution that balances between being practical to construct and economical use of material.

Novice designers may find it difficult to accept the idea of having several alternative solutions to a design problem. The challenge is compounded by the uncertainty associated with predicting redistributed moments after cracking takes place in a continuous structure (to be discussed in the next section). However, an experienced designer can take advantage of flexibility associated with the design of continuous reinforced concrete structures and skilfully design a structure that is simple to construct with maximum repetition.

10.8 | ANALYSIS OF CRACKED CONTINUOUS REINFORCED CONCRETE FLEXURAL MEMBERS

10.8.1 Background

The effects of the cracking and inelastic behaviour of reinforced concrete flexural members can be simulated by a number of sophisticated structural analysis methods. The main objectives of such an analysis are: to provide a realistic estimate of the internal force distribution (including bending moments and shear forces) and to estimate the amount of moment redistribution in the postcracking stage of continuously reinforced concrete members. In many instances, the aim of such an analysis is to obtain a better estimate of deflection magnitudes than is possible by linear elastic analysis (as discussed in Section 4.6).

The following approaches can be used for this purpose:

- computer-aided iterative analysis
- plastic analysis (or limit analysis)
- finite element method

Computer-aided iterative analysis, which takes into account the effects of cracking and changes in member stiffness in the postcracking stage at the service load level, will be discussed in Section 10.8.2.

Plastic analysis (limit analysis) simulates the inelastic behaviour of continuous reinforced concrete structures at the ultimate load level. In theory, a continuous reinforced concrete structure is not expected to fail when the moment resistance at one critical section has been reached. Instead, the localized region within a structure that is subjected to the largest possible bending moment (called the plastic hinge) will experience large permanent deformations. (Examples of permanent deformations are cracking and crushing of concrete and yielding of steel reinforcement.) Once a plastic hinge forms at one location within a structure, a moment redistribution takes place that could lead to new plastic hinges forming at other locations subjected to maximum bending moments. Ultimately, collapse mechanism forms and the structure becomes statically determinate and fails due to a loss of overall stability. However, the sequence in which the hinges form in a particular structure and the number of hinges depend on the properties of the structure and the loads it is subjected to. Detailed coverage of this method is beyond the scope of this book; for more details, the reader is referred to Ferguson, Breen, and Jirsa (1988). CSA A23.3 Cl.9.7 discusses plastic analysis for reinforced concrete structures.

The *finite element method* (FEM) is an extension of the matrix structural analysis method discussed in Section 10.5. According to the FEM, the body or structure to be analyzed is modelled as an assembly of finite elements interconnected at specified nodal points. Depending on the material stress-strain relationship used, both linear elastic and inelastic structural analyses can be performed by means of the FEM. Several finite element software packages are commercially available; however, inelastic analysis using the FEM is rather complex and time-consuming and is used mainly for research studies or special design applications. CSA A23.3 Cl.9.5 prescribes the criteria under which the FEM can be used for the analysis of reinforced concrete structures.

As an alternative to a comprehensive inelastic analysis, CSA A23.3 prescribes a simplified *moment redistribution* procedure to account for the inelastic behaviour of continuous reinforced concrete structures (to be discussed in Section 10.9).

10.8.2 Computer-Aided Iterative Analysis of Cracked Continuous Reinforced Concrete Structures

Computer-aided iterative analysis can be used to determine the distribution of internal forces and deflections in cracked continuous reinforced concrete structures at service load level as an alternative to linear elastic analysis. In essence, the iterative analysis involves several repetitions of linear elastic analysis by varying section properties. The steps involved in this analysis are as follows:

1. Each span in a continuous structure is divided into several segments. Note that the number of segments per span depends on the desired accuracy. (A good guideline is to set the length of each segment in the range from one-half to full member depth.)
2. The key cross-sectional properties need to be determined; this includes the moment of inertia of the gross section (I_g) and the cracking moment (M_{cr}).
3. Initially, the analysis is performed using the uncracked (gross) cross-sectional properties for each segment. The bending moment diagram is obtained as an output of this analysis.
4. The amount and distribution of reinforcement are determined from the bending moment diagram. Subsequently, the moment of inertia of the cracked section (I_{cr}) is calculated taking into account the amount of reinforcement.
5. The structure is then divided into uncracked and cracked regions; these regions are defined based on the magnitude of the bending moments obtained in Step 3. Cracked segments characterized by $M > M_{cr}$ are identified in the positive and negative moment regions. Consequently, the gross moment of inertia (I_g) used in Step 3 is substituted by the cracked moment of inertia (I_{cr}) along the cracked segments. The remaining segments are located within the uncracked portions of the member; hence the gross moment of inertia (I_g) is used.
6. The analysis is repeated using the cross-sectional properties revised in the previous step. As a result, the peak bending moments decrease across cracked sections and increase in some of the remaining uncracked segments. The decrease in peak moment values leads to a decrease in the required amount of reinforcement in these regions. The user may choose to revise the reinforcement design and recalculate I_{cr}. The cross-sectional properties in the newly cracked segments (where $M > M_{cr}$) should be changed so that I_{cr} is used instead of I_g.
7. The analysis is repeated until the bending moments in all uncracked sections remain less than M_{cr}.
8. The bending moment diagram is obtained as an output from the final analysis.

Note that the iterative analysis procedure as presented in this section has certain limitations. In particular, it cannot realistically predict the magnitudes of internal bending moments and deflections in the postyield stage. After the steel reinforcement yields, significant deformations occur in the plastic hinge regions, accompanied by only a moderate increase in bending moment values. This procedure is therefore applicable when the internal bending moment at a section is below the yield moment (M_y) value, as shown in Figure 10.16c.

Computer-aided iterative analysis can be performed using one of the commercially available computer analysis programs and will be illustrated with Example 10.4.

Example 10.4

A four-span continuous reinforced concrete beam is shown in the figure below. The beam supports a uniform load (w) of 20 kN/m. The beam is reinforced with 4-20M bars top and bottom. The top reinforcement is continuous over the supports, while the bottom reinforcement is continuous in the midspan region, as shown in the figure. Consider an effective depth of 340 mm. The beam dimensions and material properties are given below.

First, develop the bending moment diagram for this beam using linear elastic analysis. Then, develop the bending moment diagram for the beam using computer-aided iterative analysis.

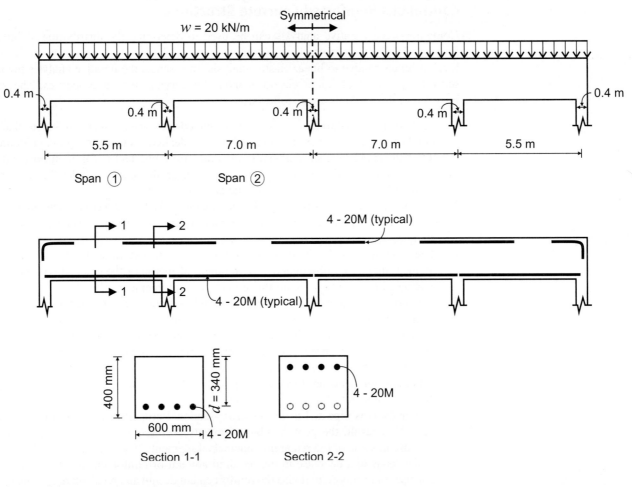

Given:　$f_c' = 25$ MPa

　　　　$f_y = 400$ MPa

SOLUTION:　This beam has been designed by means of linear elastic analysis, but the detailed design calculations are not included. Note also that this beam is symmetric with regard to the vertical axis and is subjected to a uniform load throughout its length. Therefore, in this example, it is justified to determine the design parameters for spans 1 and 2 only.

1. **Compute the gross moment of inertia (I_g)**

$$I_g = \frac{bh^3}{12} \hspace{4cm} [4.2]$$

$$= \frac{600 \text{ mm} \times (400 \text{ mm})^3}{12} = 3200 \times 10^6 \text{ mm}^4$$

2. **Compute the moment of inertia of the cracked section (I_{cr})**

   a) Compute the modular ratio (n).

   $$E_s = 200\,000 \text{ MPa} \hspace{3cm} [2.2]$$

   A23.3 Eq. 8.2
   $$E_c = 4500\sqrt{f_c'}$$
   $$= 4500\sqrt{25 \text{ MPa}} = 22\,500 \text{ MPa}$$

   $$n = \frac{E_s}{E_c} \hspace{4cm} [4.7]$$

   $$= \frac{200\,000 \text{ MPa}}{22\,500 \text{ MPa}} = 8.89 \cong 9$$

   b) Compute the neutral axis depth ($\bar{y}$).
   The area of a 20M bar is 300 mm² (refer to Table A.1), so

   $$A_s = 4 \times 300 \text{ mm}^2 = 1200 \text{ mm}^2$$

   Next, calculate the reinforcement ratio (ρ):

   $$\rho = \frac{A_s}{b\,d} \hspace{4cm} [3.1]$$

   $$= \frac{1200 \text{ mm}^2}{600 \text{ mm} \times 340 \text{ mm}} = 0.0059$$

   and

   $$n\rho = 9 \times 0.0059 = 0.053$$

   Finally, calculate the neutral axis depth ($\bar{y}$):

   $$\bar{y} = d\left(\sqrt{(n\rho)^2 + 2n\rho} - n\rho\right) \hspace{2cm} [4.9]$$

   $$= 340 \text{ mm} \times \left(\sqrt{0.053^2 + 2 \times 0.053} - 0.053\right) = 94 \text{ mm}$$

   The transformed cracked section is shown on the sketch below.

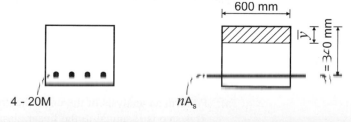

4 - 20M $\qquad$ nA_s

Actual cross section $\qquad$ Transformed cracked section

c) Compute the I_{cr} value:

$$I_{cr} = \frac{b\bar{y}^3}{3} + nA_s(d-\bar{y})^2 \qquad \textbf{[4.10]}$$

$$= \frac{600 \text{ mm}(94 \text{ mm})^3}{3} + 9(1200 \text{ mm}^2)(340 \text{ mm} - 94 \text{ mm})^2$$

$$= 820 \times 10^6 \text{ mm}^4$$

In this example, the same I_{cr} value can be used for the positive and negative moment regions. This is because the beam is reinforced with the same top and bottom reinforcement (4-20M). However, this might not be true in a general case when different amounts of top and bottom reinforcement are used in the design.

3. **Calculate the cracking moment (M_{cr})**
    a) Determine f_r:

A23.3 Eq. 8.3

$$f_r = 0.6\lambda\sqrt{f_c'} \qquad \textbf{[2.1]}$$

$$= 0.6(1)\sqrt{25 \text{ MPa}} = 3.0 \text{ MPa}$$

b) Determine y_t:

$$y_t = \frac{h}{2} = \frac{400 \text{ mm}}{2} = 200 \text{ mm}$$

c) Calculate M_{cr}:

$$M_{cr} = \frac{f_r I_g}{y_t} \qquad \textbf{[4.1]}$$

$$= \frac{(3.0 \text{ MPa}) \times (3200 \times 10^6 \text{ mm}^4)}{200 \text{ mm}} = 48 \text{ kN} \cdot \text{m}$$

4. **Perform a linear elastic analysis**
    The beam under consideration is shown in Figures 10.20a and 10.20b. The bending moment distribution for this beam was obtained as a result of the linear elastic analysis. The analysis can be performed using any commercially available computer program for structural analysis. Alternatively, the designer can use an appropriate beam diagram and the corresponding formulas in Appendix A (provided that the span lengths are equal).
    The following assumptions are taken in the analysis:

    1. Each span has been divided into 24 segments with the following properties:

    $$I_g = 3200 \times 10^6 \text{ mm}^4 \text{ and } E_c = 22\,500 \text{ MPa}$$

    2. The supports are considered to be pinned and do not influence the moment distribution in the beam.
    3. The beam shows linear elastic behaviour; that is, the load-deflection relationship is linear and proportional to the modulus of elasticity for concrete (E_c).

    The bending moment diagram obtained as a result of the linear elastic analysis is shown in Figure 10.20c.

5. **Perform a computer-aided iterative analysis**
    As the next step, perform a computer-aided iterative analysis that accounts for the effects of cracking and the related variation in flexural stiffnesses in the beam. The steps in the iterative analysis are as follows:
    a) Perform an analysis of the uncracked beam.
        This step is identical to the linear elastic analysis performed in Step 4 using the gross cross-sectional properties, so it does not need to be repeated here. Observe

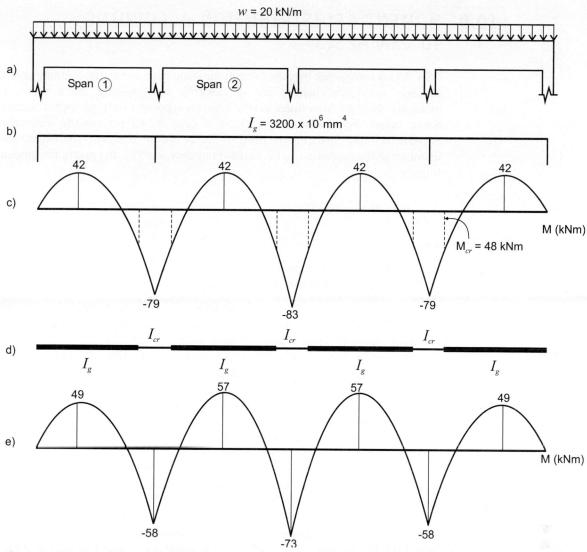

Figure 10.20 Computer-aided iterative analysis of a four-span continuous beam: a) beam elevation; b) structural model; c) bending moment diagram from the linear elastic analysis; d) moment of inertia distribution in cracked and uncracked regions; e) bending moment diagram from the computer-aided iterative analysis.

from the bending moment diagram obtained in Step 4 that the bending moment at each of the three interior supports has exceeded the cracking moment (M_{cr}) value of 48 kN·m (see Step 3); this indicates that the beam has cracked within the support regions.

b) Identify the cracked and uncracked regions and modify the cross-sectional properties accordingly.

Next, divide the beam into uncracked and cracked regions. Identify the cracked segments with $M > M_{cr}$ in the regions of positive and negative bending moments. Consequently, substitute the gross moment of inertia (I_g) used in the previous analysis by the cracked moment of inertia (I_{cr}) along the cracked segments. The remaining segments are located in the uncracked regions of the beam, so use the gross moment of inertia (I_g) (as in the previous analysis); see Figure 10.20d.

c) Repeat the analysis using the modified cross-sectional properties. The resulting bending moment diagram is shown in Figure 10.20e.

10.9 MOMENT REDISTRIBUTION ACCORDING TO CSA A23.3

CSA A23.3 recognizes that the bending moment distribution in continuous reinforced concrete structures changes due to cracking and yielding. As a result, bending moments obtained from linear elastic analysis represent only an approximation of the actual values. For that reason, Cl.9.2.4 of CSA A23.3 permits the *redistribution* of negative bending moments obtained by linear elastic analysis. The negative bending moments at the supports can be increased or decreased by the maximum amount Δ, as follows:

$$\Delta = 30 - 50 \times \frac{c}{d} \leq 20\%$$

where

c = the neutral axis depth
d = the effective depth of the cross-section.

Note that CSA A23.3 prescribes an upper limit for Δ of 20%; this corresponds to $c/d \leq 0.2$, which is characteristic of members expected to experience steel-controlled failure accompanied by large strains in reinforcement.

The application of the CSA A23.3 moment redistribution provision prescribed by Cl.9.2.4 is at the designer's discretion. Furthermore, the designer can choose to apply any amount of moment redistribution within the permitted limits; this leads to several acceptable moment redistribution schemes for the same design. It is important to note that the use of the CSA A23.3 moment redistribution provision is not permitted when the approximate frame analysis has been performed according to A23.3 Cl.9.3.3.

The CSA A23.3 moment redistribution provision will be illustrated on an example of a typical span of a continuous reinforced concrete structure shown in Figure 10.21a. The member is subjected to a uniformly distributed load (w); the corresponding bending moment diagram from the linear elastic analysis is depicted with a solid line in Figure 10.21b. According to CSA A23.3 Cl.9.2.4, the absolute value of the negative bending moment at the support can be increased by the amount $\Delta \times M$, where Δ should not exceed 0.2 (or 20%). As a result of this redistribution, the magnitude of the positive bending moment at the midspan decreases, as shown in Figure 10.21b.

An example of moment redistribution as applied to the moment envelope is shown in Figure 10.22. The envelope chart shown with a solid line depicts the envelope for the beam considered in Example 10.1. When the CSA A23.3 moment

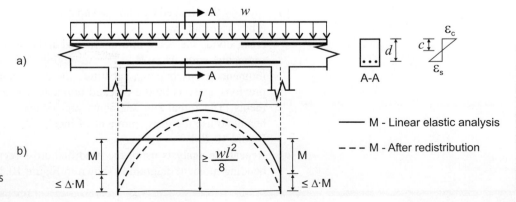

Figure 10.21 Moment redistribution in a continuous structure: a) a typical span; b) bending moment diagrams based on linear elastic analysis and after the redistribution.

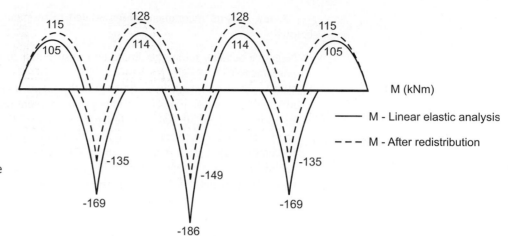

Figure 10.22 Moment envelope diagram for a continuous beam based on the linear elastic analysis and after the moment redistribution.

redistribution provision has been applied to this beam, the bending moments at the supports can decrease by 20%, while the bending moments at the midspan can increase accordingly. The resulting moment envelope is shown with a dashed line on the same diagram.

Because CSA A23.3 allows for a somewhat arbitrary redistribution of negative bending moments, designers are often concerned with whether the resulting bending moment distribution is realistic; this can be verified when the bending moment gradient within a span is determined. The moment gradient is always equal to $wl^2/8$, as explained in Section 10.2.2. Therefore, when the moment gradient within a span is greater than $wl^2/8$ (as shown in Figure 10.21b) and the reinforcement is designed accordingly, the member is considered to possess sufficient moment resistance. One of the primary objectives of sound design is to ensure that a continuous beam has sufficient combined moment resistance for positive and negative bending.

The benefits of moment redistribution will be illustrated in an example of a continuous reinforced concrete beam shown in Figure 10.23. Suppose that the reinforcement diagram in Figure 10.23a is based on the bending moment distribution obtained from the linear elastic analysis. The amount of reinforcement varies between the spans, but the variation is not significant. However, the designer may wish to use an equal amount of reinforcement in different spans in order to minimize the chances of field errors. This can be accomplished by developing an alternative design in accordance with the CSA A23.3 moment redistribution provision. The latter design, shown in Figure 10.23b, results in a repetitive reinforcement arrangement that is more suitable for field implementation in terms of both construction and inspection.

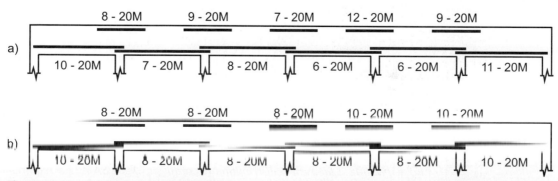

Figure 10.23 Detailing of flexural reinforcement in a continuous beam: a) based on linear elastic analysis; b) based on the CSA A23.3 moment redistribution provision.

A few general recommendations related to moment redistribution are summarized below:

- Negative bending moments should be redistributed so that the peak values at the supports decrease.
- Where possible (except for short spans), the members should be designed without significant differences in positive and negative moment resistances.
- In spans with small positive midspan moments, the required amount of positive reinforcement in the beam section may be governed by the CSA A23.3 minimum reinforcement requirement (rather than by the design bending moment). As a result, the section may have reserve strength in positive bending. The designer can take advantage of this by decreasing the negative moments at the supports so that the corresponding increase in the positive bending moment is less than or equal to the actual moment resistance.

The CSA A23.3 moment redistribution method will be illustrated by Example 10.5.

KEY CONCEPTS

The effects of cracking and inelastic behaviour of reinforced concrete flexural members can be simulated by performing a comprehensive structural analysis. The main objectives of such an analysis are to provide a realistic estimate of the internal force distribution (including bending moments and shear forces) and to estimate the amount of moment redistribution in the postcracking stage of continuous reinforced concrete structures. The following approaches can be used for this purpose:

- *Computer-aided iterative analysis* takes into account the effects of cracking and changes in member stiffnesses in the postcracking stage.
- *Plastic analysis* (*limit analysis*) simulates the inelastic behaviour of continuous reinforced concrete structures under progressively increasing loads.
- The *finite element method* (*FEM*) models the body or structure to be analyzed as an assembly of finite elements interconnected at specified nodal points. Depending on the material stress-strain relationship used, both linear elastic and inelastic structural analyses can be performed by means of the FEM.

As an alternative to a comprehensive inelastic analysis, Cl.9.2.4 of CSA A23.3 permits the redistribution of negative bending moments in continuous reinforced concrete structures obtained by linear elastic analysis. As a result, the negative bending moments at the supports can be increased or decreased by up to 20%.

Example 10.5

A four-span continuous reinforced concrete beam is shown in the figure that follows. The beam supports a factored uniform load (w_f) of 40 kN/m. The beam is reinforced with 4-20M bars top and bottom. The top reinforcement is continuous over the supports, while the bottom reinforcement is continuous in the midspan regions. Assume an effective depth of 340 mm. The beam dimensions and material properties are given.

First, perform a moment redistribution according to the CSA A23.3 requirements. Then, compare the actual moment resistance values with the corresponding redistributed moments.

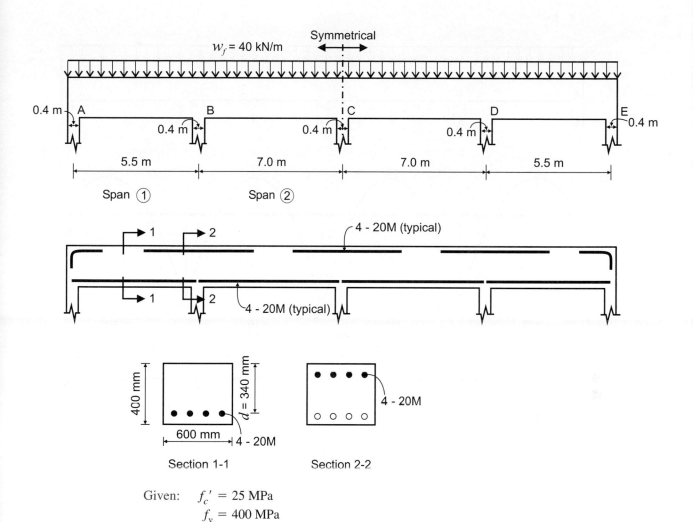

Given: $f_c' = 25$ MPa
         $f_y = 400$ MPa

SOLUTION: The beam elevation is shown in Figure 10.24a.

1. **Perform a linear elastic analysis**

   A linear elastic analysis has been performed using a computer program for structural analysis. The factored moments obtained as result of the linear elastic analysis are shown in Figure 10.24b.

2. **Apply the CSA A23.3 moment redistribution provision**

   Observe from Figure 10.24b that the bending moments at supports B, C, and D exceed the cracking moment (M_{cr}) value of 48 kN·m. Hence, it is likely that the negative bending moments obtained from the linear elastic analysis have been overestimated, whereas the positive moments have been underestimated. Perform the moment redistribution according to A23.3 Cl.9.2.4. In lieu of performing detailed calculations, a maximum (20%) reduction in the bending moments at the supports has been applied. The resulting bending moment diagram is shown in Figure 10.24c.

3. **Determine the actual moment resistance for the critical sections**

   Determine the moment resistance provided by the 600 mm wide by 400 mm deep beam section with 4-20M tension reinforcement.
   The effective depth is

   $d = 340$ mm

   The area of a 20M bar is 300 mm² (refer to Table A.1); therefore,

   $A_s = 4 \times 300$ mm² $= 1200$ mm²

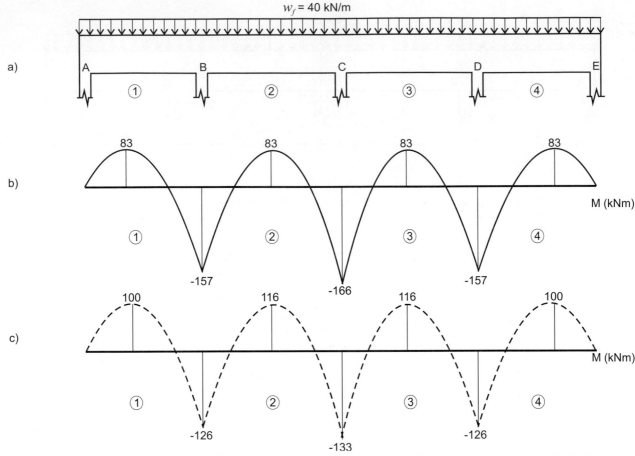

Figure 10.24   CSA A23.3 moment redistribution: a) beam elevation and load; b) bending moment diagram obtained by linear elastic analysis; c) bending moment diagram after the moment redistribution.

Then,

$$T_r = \phi_s f_y A_s \qquad \qquad \qquad \qquad \text{[3.9]}$$
$$= 0.85 \times 400 \text{ MPa} \times 1200 \text{ mm}^2 = 408 \times 10^3 \text{ N} = 408 \text{ kN}$$

Use the equation of equilibrium

$$C_r = T_r \qquad \qquad \qquad \qquad \text{[3.10]}$$

The compression force in concrete (C_r) is

$$C_r = \alpha_1 \phi_c f_c' \, a \, b \qquad \qquad \qquad \qquad \text{[3.8]}$$

where

$$\alpha_1 \cong 0.8 \qquad \qquad \qquad \qquad \text{[3.7]}$$

Calculate the depth of the rectangular stress block (a) from Eqn 3.10 as

$$a = \frac{T_r}{\alpha_1 \phi_c f_c' \, b} = \frac{408 \times 10^3 \text{ N}}{0.8 \times 0.65 \times 25 \text{ MPa} \times 600 \text{ mm}} = 52.3 \text{ mm} \cong 50 \text{ mm}$$

Calculate the moment resistance from the equation

$$M_r = T_r \left(d - \frac{a}{2} \right) \qquad \qquad \qquad \qquad \text{[3.13]}$$

$$= (408 \times 10^3 \text{ N}) \left(340 \text{ mm} - \frac{50 \text{ mm}}{2} \right) = 129 \text{ kN} \cdot \text{m}$$

A typical beam cross-section showing the internal force distribution is shown below.

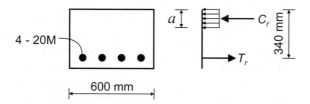

The beam is reinforced with the same tension and compression reinforcement (4-20M bars), so the moment resistance values at the midspan and the supports are equal.

4. **Compare the actual moment resistance values with the redistributed moments**
Consider the bending moment diagram for span 2 with the moment redistribution taken into account, as shown in Figure 10.25a.

The bending moment diagram (dashed line) can be superimposed with the moment resistance diagram (solid line) (see Figure 10.25b).

Observe from the bending moment diagram that the moment gradient is

$$\frac{wl^2}{8} = 245 \text{ kN} \cdot \text{m}$$

However, the sum of the positive moment resistance (equal to 129 kN·m) and the average negative moment resistances at the supports (also equal to 129 kN·m) is 258 kN·m; this is larger than the moment gradient of 245 kN·m. Hence, the amount and distribution of reinforcement in the beam seem to be appropriate.

However, observe that the moment resistance of 129 kN·m at the right support is slightly less than the redistributed moment of 133 kN·m; this difference is on the order of 3%. This amount of undercapacity is not significant, so the design can be considered acceptable.

a)

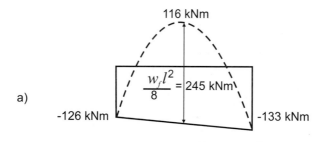

b)

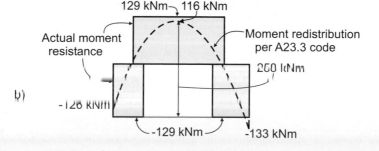

Figure 10.25 Bending moments in span 2; a) after the moment redistribution; b) actual moment resistance versus the bending moment diagram after the redistribution.

SUMMARY AND REVIEW — BEHAVIOUR AND ANALYSIS OF CONTINUOUS BEAMS AND SLABS

Cast-in-place reinforced concrete beams and slabs typically span continuously over several supports to form floor systems. These beams and slabs can be designed and constructed to behave as continuous structures by properly placing tension reinforcement in the top and bottom regions. Fundamental concepts related to the structural analysis of continuous reinforced concrete beams and slabs subjected to flexure of critical importance for the efficient design of these structures were discussed in this chapter.

Key features of continuous reinforced concrete structures

The key features characteristic of continuous reinforced concrete beams and slabs are summarized below.

Positive and negative bending moments   Continuous structures resist applied loads by developing positive bending moments at midspan regions and negative bending moments over supports. The presence of negative bending moments at the supports causes reduced positive bending moments at the midspan. As a result, positive bending moments in continuous structures are smaller than those developed in a series of simple structures characterized by similar span lengths. Consequently, the cross-sectional dimensions in continuous beams and slabs are smaller than in simply supported members. This in turn leads to a more economical design and a possible significant reduction in the overall building weight.

Stiffness distribution   In a continuous structure with several spans of approximately equal length, spans with larger cross-sectional dimensions will have larger flexural stiffness. When subjected to flexure, members with stiffer adjoining members will attract larger negative bending moments at the supports than members with adjoining members of smaller stiffness. The stiffness distribution in continuous structures has a major impact on the magnitude and distribution of bending moments.

Load patterns and moment envelope

Load acting on one span in a continuous structure influences bending moments and shear forces in other spans. Consequently, the structure must be designed to resist a few different live load distributions (also called *load patterns*) that result in the worst possible effects.

In general, when alternate spans in a continuous structure are loaded, maximum positive bending moments will develop in the loaded spans. On the other hand, maximum negative bending moments will develop when two adjacent spans are loaded at either side of the support. A23.3 Cl.9.2.3.1 prescribes three different load patterns to be considered in the design of continuous beams and slabs.

Bending moment diagrams corresponding to different load patterns are superimposed to obtain the *moment envelope,* which represents the maximum bending moment values at any point along the member length.

Approaches to the analysis of continuous structures

The two main approaches to the analysis of continuous reinforced concrete structures are as follows:

- *Linear elastic analysis* takes into account the linear elastic behaviour of reinforced concrete members; however, it can indirectly account for the effects of cracking by using transformed section properties.
- *Inelastic analysis* simulates the response of reinforced concrete structures in the post-cracking stage.

CSA A23.3 approximate frame analysis

Approximate frame analysis (also known as the *coefficient method*) is based on approximate expressions for bending moments and shear forces in continuous beams and slabs.

This analysis can be performed using hand calculations and is adequate for simple and regular structures that meet the criteria prescribed by CSA A23.3 Cl.9.3.3.

Moment redistribution according to CSA A23.3

The effects of cracking and inelastic behaviour of reinforced concrete flexural members can be simulated by performing a comprehensive structural analysis. The main objective of this analysis is to provide a realistic estimate of the internal force distribution in continuous reinforced concrete structures in the postcracking stage of behaviour. The following approaches can be used for this purpose:

- *Computer-aided iterative analysis* takes into account the effects of cracking and changes in member stiffness in the postcracking stage.
- *Plastic analysis* (limit analysis) simulates the inelastic behaviour of continuous reinforced concrete structures under progressively increasing loads.
- *The finite element method* (FEM) models the body or structure to be analyzed as an assembly of finite elements interconnected at specified nodal points. Depending on the material stress-strain relationship used, both linear elastic and inelastic structural analyses can be performed by means of the FEM.

As an alternative to a comprehensive inelastic analysis, Cl.9.2.4 of CSA A23.3 permits the redistribution of negative bending moments in continuous reinforced concrete structures obtained by linear elastic analysis. As a result, the negative bending moments at the supports can be increased or decreased by up to 20%.

PROBLEMS

10.1. Consider a simply supported beam subjected to a distributed triangular load, as shown in the figure below.
 a) Determine the maximum bending moment and its location relative to the left support.
 b) The designer has decided to approximate the actual loading by a uniformly distributed load of 10 kN/m. Determine the maximum bending moment for the beam under uniform load; also, determine the location of maximum bending moment relative to the left support.
 c) Do you believe that the approach taken in part b) gives a reasonable approximation of the maximum bending moment determined in part a)? Explain.

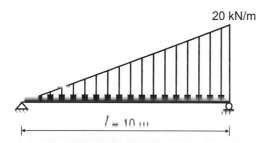

10.2. Consider a simply supported beam subjected to point loads (100 kN in total) in three different load configurations, as shown in the figure that follows.

 a) Draw the bending moment and shear force diagrams for each load configuration.
 b) Draw the bending moment and shear force diagram for the same beam subjected to a uniformly distributed load of 10 kN/m. Note that the resultant of the uniform load is 100 kN.
 c) Compare the maximum bending moment and shear force values for the three load configurations obtained in part a) with the corresponding values obtained in part b).
 d) Based on the comparison, do you believe that it is acceptable to approximate the load configurations from part a) by the uniformly distributed load from part b) in order to determine the maximum bending moments and shear forces?
 Hint: A 5% error is generally acceptable in design practice.
 e) Suppose that, in an actual design project, you have decided to use a 10 kN/m uniform load instead of the load configurations from part a). For each load configuration, determine the region along the beam (to the nearest 100 mm) where this approximation would lead to at least a 10% underdesign in the bending moment and shear force values.

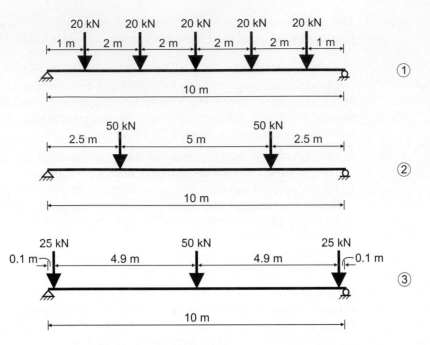

10.3. A typical span of a continuous reinforced concrete slab is shown in the figure below. The bending moment distribution has been obtained as a result of a linear elastic analysis. The slab is supported by 400 mm wide and 700 mm deep reinforced concrete beams.

Determine the bending moment at the face of each support.

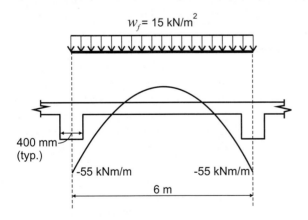

10.4. A four-span continuous reinforced concrete beam is shown in the figure below. The beam is subjected to a specified uniform dead load (DL) of 15 kN/m (including self-weight) and a specified uniform live load (LL) of 20 kN/m.

a) Determine the factored bending moment and shear force distribution using the CSA A23.3 approximate frame analysis method.

b) Determine the factored bending moment and shear force distribution using the load diagrams from Appendix A. Consider the total factored dead and live load on all spans.

c) Compare the maximum bending moments and shear forces obtained in parts a) and b). Based on the comparison, which approach would you consider more appropriate for this design? Explain.

10.5. Consider the four-span continuous reinforced concrete one-way slab shown in the figure on the next page. The slab is subjected to a specified uniform dead load (DL) of 6 kPa (including self-weight) and a specified uniform live load (LL) of 5 kPa.

a) Determine the factored bending moment and shear force distribution using the CSA A23.3 approximate frame analysis method.

b) Determine the factored bending moment and shear force distribution using the load diagrams from Appendix A. Consider the total factored dead and live load on all spans.

c) Compare the maximum bending moments and shear forces obtained from parts a) and b). Based

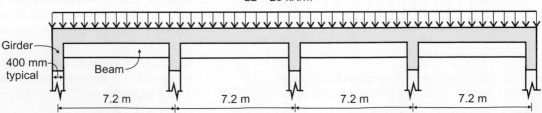

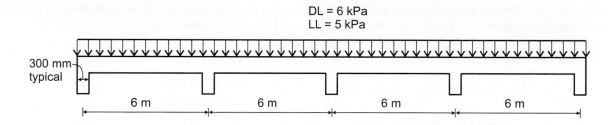

on the comparison, which approach would you consider more appropriate for this design? Explain.

10.6. A five-span continuous reinforced concrete beam is shown in the figure below. The beam is subjected to a factored uniform dead load (DL_f) of 50 kN/m (including self-weight) and a factored uniform live load (LL_f) of 50 kN/m. Ignore the flexural effect of the supports.

a) Develop the factored moment envelope diagram for this beam considering all relevant load

patterns according to the CSA A23.3 requirements. Perform a linear elastic analysis using a computer structural analysis program.

b) Compare the maximum factored bending moment values obtained from the envelope with the corresponding values for the load case of full dead and live load on all spans.

c) Based on the comparison, which approach would you use to determine the design bending moments for this design? Explain.

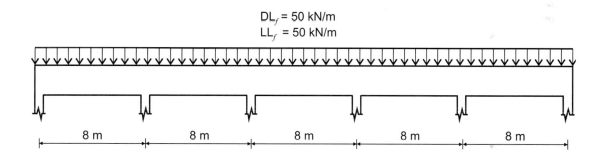

10.7. Consider the beam from Problem 10.6. The beam is subjected to a factored uniform dead load (DL_f) of 80 kN/m (including self-weight) and a factored uniform live load (LL_f) of 20 kN/m (see the figure below).

a) Develop the factored moment envelope diagram for this beam considering all relevant load patterns according to the CSA A23.3 requirements. Perform a linear elastic analysis using a computer structural analysis program.

b) Compare the maximum factored bending moments obtained from the moment envelope with the corresponding values for the load case that includes full dead and live load on all spans.

c) Based on the comparison you have made in Problems 10.6 and 10.7, discuss the effectiveness of the moment envelope approach versus the use of full dead and live load on all spans when the design bending moments in a continuous beam need to be determined.

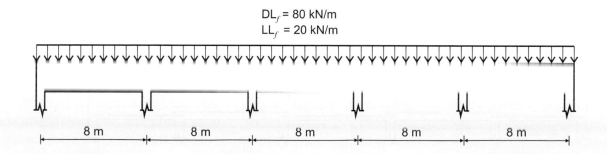

10.8. Consider the five-span continuous reinforced concrete beam shown in the figure below. The beam is subjected to a total factored load (w_f) of 120 kN/m. Ignore the effects of the support width.

a) Develop the factored bending moment diagram for this beam by performing a linear elastic analysis using a computer structural analysis program. Consider the load pattern with total factored load applied on all spans.

b) Determine the required amount of tension reinforcement corresponding to the positive and negative bending moments determined in part a).

c) Consider the three alternative reinforcement arrangements for this beam shown in the figure below. Which solution is most suitable for this design based on your design calculations in part b)?

d) Are the other two reinforcement arrangements safe, considering the required amount of reinforcement obtained in part b)?

e) If the answer to part d) is negative, redistribute the bending moments in the beam in accordance with the CSA A23.3 provisions by using the bending moment diagram from part a). Develop the diagrams for the redistributed bending moments corresponding to these two reinforcement arrangements. Is it possible to redistribute the bending moments so that these two reinforcement arrangements become acceptable from a strength (moment resistance) perspective?

Given:

$$f_c' = 30 \text{ MPa}$$
$$f_y = 400 \text{ MPa}$$
$$\phi_c = 0.65$$
$$\phi_s = 0.85$$

10.9. Consider the continuous reinforced concrete beam discussed in Problem 10.8.

a) Determine the factored bending moment diagram corresponding to each reinforcement arrangement (1, 2, and 3) by performing a computer-aided iterative analysis. Use 500 mm long beam segments to develop the model.

b) Based on the analysis performed in part a), which reinforcement arrangement best reflects the effect of cracking in the beam?

c) What conclusions can you draw on the acceptability of the two reinforcement arrangements not chosen in part b) from the strength and serviceability perspectives?

Use the following assumptions in the design:

1. The columns are dimensionless and do not resist bending moments.
2. The top reinforcement extends beyond the inflection points.

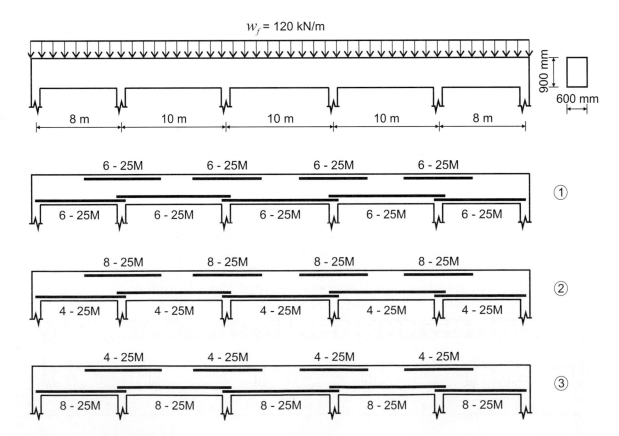

11

Design of Continuous Beams, Slabs, and Floor Systems

LEARNING OUTCOMES

After reading this chapter, you should be able to

- identify common floor systems in cast-in-place concrete construction and explain their key features
- design continuous reinforced concrete beams and slabs in accordance with the CSA A23.3 strength and serviceability requirements
- apply the CSA A23.3 anchorage requirements and calculate bar cutoffs for flexural reinforcement in continuous reinforced concrete beams and slabs

11.1 INTRODUCTION

In reinforced concrete buildings, continuous beams and slabs are typically used in cast-in-place floor construction. An example of a cast-in-place continuous concrete floor under construction is shown in Figure 11.1. In one-way slab and beam construction, beam reinforcement is usually placed first. The top reinforcement and any hooked rebars are then added in place. Subsequently, the bottom slab reinforcement is placed. After the beam reinforcement is completed, the top slab reinforcement is placed on top of the beam reinforcement at each span, along with any embedded mechanical and electrical conduits. (Refer to Section 5.4.2 for more details on the placement of reinforcement in beams and slabs.)

This chapter presents an approach to the comprehensive design of continuous beams and slabs, which includes the following major aspects:

- analysis of continuous beams and slabs (discussed in Chapter 10)
- design of one-way slabs and T-beams for flexure (discussed in Chapters 3 and 5)
- detailing of flexural reinforcement (to be discussed in Section 11.4)
- design for shear (discussed in Chapter 6)
- design for serviceability, including cracking and deflections (discussed in Chapter 4)

An overview of common floor systems is presented in Section 11.2, while a comprehensive case study of floor design is performed in Section 11.3. The detailing of flexural reinforcement and bar cutoffs in continuous beams and slabs is discussed in Section 11.4; this section builds on the material related to the detailing and anchorage of flexural reinforcement in simple beams and slabs presented in Chapter 9. Finally, structural drawings related to reinforced concrete floor systems, as well as general specifications, are discussed in Section 11.5. Note that the focus of this chapter is on the design of one-way floor systems.

513

Figure 11.1 Continuous concrete floor construction.

(Nebojsa Ojdrovic)

DID YOU KNOW ?

The speed of construction in reinforced concrete high-rises can be as high as 3 days per floor. In hot real estate markets, where residential units are sold very quickly, building owners often require an accelerated construction schedule in order to satisfy market demands. Due to the use of modern formwork and falsework equipment such as the "flyform table," the formwork can be stripped in 1 day or 2 days following the concrete construction on a particular floor. The formwork then "flies" to the next floor, ready for rebar placement; rough-ins by mechanical, HVAC, and electrical subtrades; and concrete placement on the next floor in only 3 days (see Figure 11.2).

Figure 11.2 A concrete high-rise under construction uses a flyform table system.

(John Pao)

11.2 FLOOR SYSTEMS IN CAST-IN-PLACE CONCRETE CONSTRUCTION

11.2.1 Background

The *floor system* (also called *floor framing*) is a key horizontal component in a building. Its primary role is to transfer gravity and lateral forces to the columns and/or walls, and needs to meet the following requirements:

- adequate strength to safely resist applied loads
- sufficient stiffness to limit deflections and vibrations
- required fire resistance
- adequate acoustical resistance

The floor system for a particular design should be selected based on economical and functional considerations.

Reinforced concrete floor systems can be classified as follows:

- In one-way systems, the primary flexural reinforcement in each structural element runs in one direction only.
- In two-way systems, the primary flexural reinforcement in at least some elements runs in two orthogonal directions.

Common cast-in-place one-way floor systems are:

- one-way joist floor systems
- slab-beam-and-girder floor systems
- slab band floor systems

Two-way floor systems include flat slab, flat plate, slab with beams, two-way joist floor, and waffle slab. Floor systems were briefly explained in Section 1.3.1. The design of two-way floor systems is covered in Chapter 12. For more details on various floor systems, refer to PCA (2000) and Domel and Ghosh (1990).

A brief overview of one-way floor systems is provided in this section. The cost of a floor system often constitutes the major part of the structural cost in the building, and cost considerations will be discussed in Section 11.2.5.

11.2.2 One-Way Joist Floor System

A one-way joist floor system consists of regularly spaced concrete joists (ribs) spanning in one direction. A thin reinforced concrete slab with a thickness ranging from 75 mm to 130 mm is cast integrally with the joists. The joists are supported by the wide beams spanning along the column lines. This system is also known as a ribbed slab or a panjoist floor. In general, the joists are 150 mm wide and 400 mm deep and they are spaced at 900 mm on centre; note that CSA A23.3 Cl.10.4 limits the joist dimensions. The joists act as continuous T-beams spanning between the wide beams. A vertical section through the joist floor showing the joists and the CSA A23.3 recommended dimensions is shown in Figure 11.3. The joists are formed using reusable panforms made of steel or fibreglass.

The beams that support the joists are of the same depth as the joists (usually 400 mm), but are generally much wider (on the order of 2400 mm). A typical floor plan illustrating the joist floor system is shown in Figure 11.4.

The one-way joist floor system can be used for spans ranging from 4.5 m to 12 m, but it is not economical for short spans. Its advantages include a lightweight, attractive ceiling, which also makes it possible to place electrical fixtures between joists. A drawback is that the formwork costs are higher than for the other systems and the construction is labour intensive.

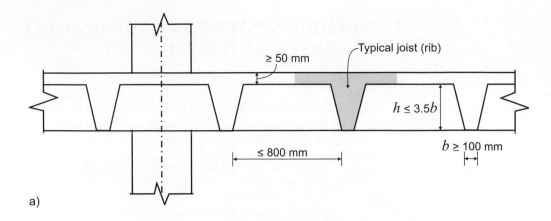

a)

Figure 11.3 One-way joist
floor system: a) vertical
section; b) example of
an actual structure.

(Svetlana Brzev)

b)

This floor construction was practised throughout Canada from the 1950s to the 1970s. It was abandoned in the 1970s due to the fire-rating requirements prescribed by the National Building Code of Canada, where a minimum 130 mm thick slab is required to provide a 2 hour fire-resistance rating. This requirement effectively eliminated the light-weight advantage of one-way joist floor systems.

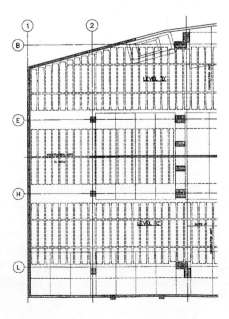

Figure 11.4 One-way joist
floor system.

(Read Jones Christoffersen Ltd.)

11.2.3 Slab-Beam-and-Girder Floor System

A slab-beam-and-girder floor consists of a series of parallel beams supported by girders, which are in turn supported by columns placed at regular intervals over the entire floor area. The beam and girder grid supports a one-way reinforced concrete slab. The beams are usually spaced at the midpoints, one-third points, or one-quarter points of the girders, as shown in Figure 11.5.

A plan of a typical slab-beam-and-girder floor is shown in Figure 11.6. Note that the building has four bays in the east-west (E-W) direction and three bays in the north-south (N-S) direction. Gridlines A to E are laid in the N-S direction, whereas gridlines 1 to 4 are laid in the E-W direction. The beams shown on the floor plan are labelled by the letter B (for example, B1, B2, B3, B4), whereas the girders are labelled by the letter G (for example, G1, G2, and G3).

In this floor system, the slabs are supported by the beams, which are in turn supported by the girders; finally, the girders are supported by the columns. The load transfer between these structural elements is explained below.

First, consider a slab strip of unit width (equal to 1 m), shown in the floor plan. (The concept of a slab strip was introduced in Section 3.6.) The slab spans in the N-S direction as indicated by arrow lines in Figure 11.6 (note a slab strip shown hatched in Figure 11.6). This is a one-way slab because the length-to-width aspect ratio for a typical slab panel between the two adjacent beams (shown shaded in Figure 11.6) is equal to

$$\frac{l_5}{l_3} = \frac{6.0 \text{ m}}{2.1 \text{ m}} = 2.9$$

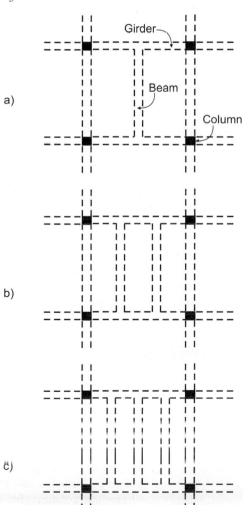

Figure 11.5 Possible beam layouts in slab-beam-and-girder floors: a) beams at midpoints; b) beams at third points; c) beams at quarter points.

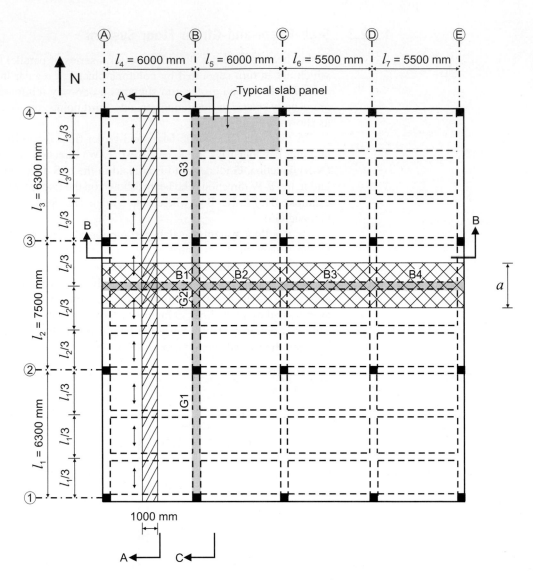

Figure 11.6 A floor plan of a one-way slab-beam-and-girder floor system.

that is, greater than 2. The slab strip is designed as a continuous structure, as shown on section A-A in Figure 11.7a.

Next, let us consider a typical beam B1-B2-B3-B4 spanning in the E-W direction (shown shaded in Figure 11.6). The beam elevation is shown on section B-B in Figure 11.7b. The beam carries the load transferred from the slabs over a certain tributary width. A typical interior beam supports the load from approximately one half the slab panel width on either side of the beam (north and south). The tributary area for the beam is shown cross-hatched and the tributary width is denoted as a in Figure 11.6, where

$$a = \frac{1}{2}\left(\frac{l_2}{3}\right) + \frac{1}{2}\left(\frac{l_2}{3}\right) = \frac{l_2}{3}$$

Note that the tributary width for the edge beams (for example, beams spanning along grid-lines 1 and 4) is equal to one half the tributary width for interior beams (provided that the slab panels are of the same width).

Since slabs, beams, and girders are built monolithically, beams and girders are usually designed as T-beams, thereby taking advantage of continuity. (The concept of T-beams was introduced in Section 3.7.) Beams are designed as continuous structures, as shown in Figure 11.7b. Also note that some beams frame into girders, like the beam under consideration; however, other beams frame into columns (for example, the beam spanning along gridline 3).

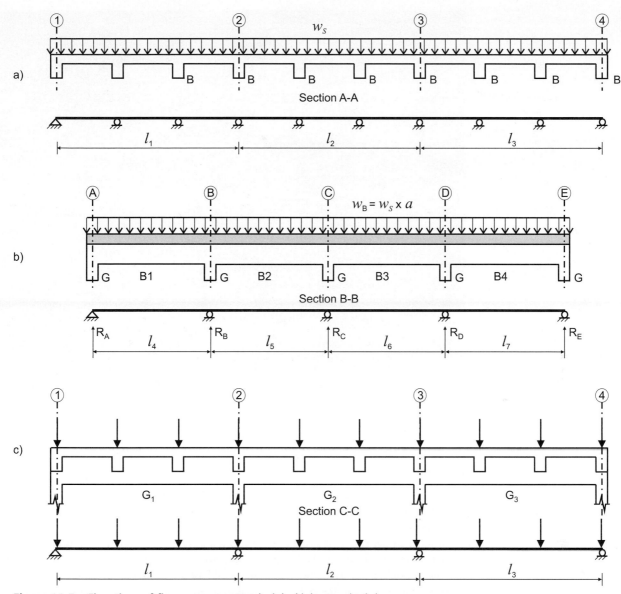

Figure 11.7 Elevations of floor components: a) slab; b) beam; c) girder.

Finally, let us consider the girder G1-G2-G3 spanning along bayline B (shown shaded in Figure 11.6). The girder is loaded by the beam reactions in the form of point loads and is in turn supported by the columns. Girders are usually designed as continuous T-beams. A typical girder elevation is shown on section C-C in Figure 11.7c. Because the girder forms a frame with the supporting columns, a frame analysis is usually performed to determine the load distribution between the girders and the columns (as discussed in Section 10.4.2).

Note that beams and girders always carry their own weight (self-weight) in addition to the load transferred to them from other elements.

The beam arrangement and column spacing in a slab-beam-and-girder floor system is determined based on economy, ease of construction, functional requirements of the floor space, and magnitude of applied loads. In general, beam slab and girder floors are suitable for live loads ranging from 2 kPa to 20 kPa. The typical span between columns ranges from 8 m to 18 m. Typical beam and girder depths range from $l/22$ to $l/12$, whereas the slab thickness ranges from $l/28$ to $l/20$. The beam depth-to-width aspect ratio is usually in the range from 1.5 to 3.0. An example of a slab-beam-and-girder system in an actual building is shown in Figure 11.8.

Figure 11.8 An example of a slab-beam-and-girder floor system in a parking garage.

(Svetlana Brzev)

In choosing the dimensions of floor components, the designer should consider the following:

- It is preferred to have floors of equal depth throughout the building so that the forms can be reused. Formwork is a major cost item in floor construction (to be discussed in Section 11.2.5).
- In most cases, it is more cost effective to increase the concrete strength or the amount of reinforcement than to vary the size of a structural member.
- Beam should be designed with the same width or wider than the column whenever possible. This results in the maximum cost efficiency because beam framing can proceed along a continuous line (see Figure 11.9).

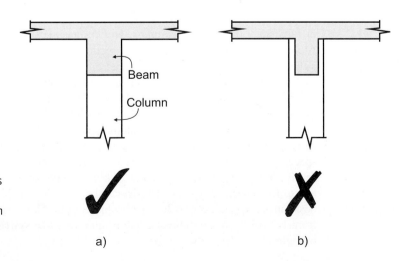

Figure 11.9 Beam/column intersection: a) the use of the same beam and column width is recommended; b) the use of a narrow beam and a wide column results in higher formwork costs.

11.2.4 Slab Band Floor System

In some cases, beams in lightly loaded floors can be omitted in one direction such that the one-way slab is carried directly by wide and shallow beams spanning across the columns. This floor system is called banded slab construction or the *slab band* system. These wide beams (called slab bands) are supported directly by the columns. CSA A23.3 Cl.2.2 defines a slab band as "a continuous extension of a drop panel between supports or between a support and another slab band." A partial floor plan of a building with a slab band floor system is shown in Figure 11.10a. This floor construction is common in

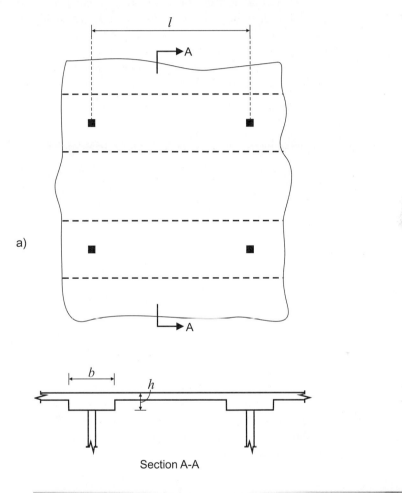

Section A-A

Figure 11.10 Slab band floor system: a) typical plan; b) example of an actual structure.

(Svetlana Brzev)

Western Canada and is typically used for parking garages, residential and office buildings; an example of a parking garage constructed using the slab band system is shown in Figure 11.10b.

The slab band is essentially a one-way floor system. Slab bands are designed as continuous beams for flexure and deflections, while the shear design is performed based on both the one-way shear requirements related to beams and the two-way shear requirements related to slabs. Typically, the width of the slab band (b) is in the range from 2200 mm to 2400 mm and the depth (h) is in the range from 400 mm to 600 mm (see Figure 11.10a). In some cases, shear reinforcement (stirrups) in slab bands is not required. Due to relatively shallow depths, flexural steel in the slab band is increased as compared to deeper beams and girders. However, wider slab bands result in smaller slab span and amount of reinforcement as compared to other one-way slabs of the same span. A typical reinforcement arrangement in a slab band floor is shown in Figure 11.11.

The most common application of the slab band floor system is in parking garages, for the following reasons:

- *Simple shoring and formwork.* Parking garages generally require a minimum structural clearance of 2.3 m, which requires a simple shoring system for the formwork. The formwork system is simple. The underside of the slab is formed with plywood sheets laid side by side. Typical slab band widths of 2400 mm (or 8 ft in Imperial units) can be easily formed with full plywood sheets (4 ft by 8 ft). The difference in the thickness between the slab and the slab band is usually formed with dimensional lumber with dimensions that range from 2 in by 4 in to 2 in by 12 in.
- *Flexible configurations.* Over the last few decades, parking stall configurations have become increasingly complex. There are several different types of parking stalls, such as small car stalls, regular car stalls, handicapped stalls, and tandem parking stalls. Furthermore, the parking stall requirements may vary between jurisdictions. The slab band system can easily accommodate irregular parking stall configurations without increasing the complexity of the formwork system and the reinforcing steel layout.

In multistorey wood frame apartment buildings with underground parking, the concrete slab on the main floor is required to carry loads from the entire wood frame structure above and transfer them to the columns in the underground parking structure. (These columns are typically not aligned with the load-bearing stud walls above.) A slab band system with wide bands is very suitable for this application. However, the slab bands are generally thicker than in regular garage floors due to significantly heavier loads. Also, stirrups are often required in column regions (which is not the case in regular garage floors).

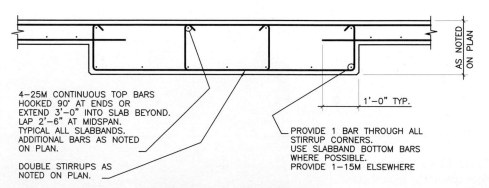

4−25M CONTINUOUS TOP BARS HOOKED 90° AT ENDS OR EXTEND 3'−0" INTO SLAB BEYOND. LAP 2'−6" AT MIDSPAN. TYPICAL ALL SLABBANDS. ADDITIONAL BARS AS NOTED ON PLAN.

DOUBLE STIRRUPS AS NOTED ON PLAN.

1'−0" TYP.

PROVIDE 1 BAR THROUGH ALL STIRRUP CORNERS. USE SLABBAND BOTTOM BARS WHERE POSSIBLE. PROVIDE 1−15M ELSEWHERE

AS NOTED ON PLAN

Figure 11.11 Typical reinforcement arrangement in a slab band.

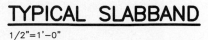

TYPICAL SLABBAND

1/2"=1'−0"

A typical slab span is on the order of 8 m, while slab bands typically span a maximum of 10 m. Depending on the applied loads, the slab band thickness ranges from $l/28$ to $l/24$ for interior spans and $l/24$ to $l/20$ for end spans (where l denotes the span between the adjacent columns supporting the slab bands).

11.2.5 Cost Considerations

Construction cost for elevated reinforced concrete floor systems is associated with the following items:

- shoring and scaffolding
- formwork
- concrete
- reinforcement

Depending on the height of the elevated floor, the cost of shoring, scaffolding, and formwork can account for 50% of the overall cost of a floor system. In order to design an easy-to-form floor system, the designer needs to be familiar with commonly available shoring equipment and formwork systems and materials. An effective approach to increased cost efficiency is characterized by consistent slab and beam dimensions and repetition in concrete member dimensions, and reinforcement throughout the building; this will result in the reuse of formwork and shoring equipment.

The cost of each of the above-listed items has the following components:

- material cost
- labour cost
- interest cost (affected by the speed of construction)

All of the above components may have a significant effect on the overall construction cost. The designer should realize that minimizing the material and labour costs is of primary importance for achieving cost-efficient design. In recent years, it has also become very important to produce designs that enable construction to proceed in a timely and efficient manner. Considering the high costs of prime properties in Canadian cities, contractors and building owners can achieve significant financial gains from designs which ensure that the construction schedule can be met.

Overall, a cost-effective design strikes an optimal balance between costs and the availability of materials, labour, and equipment.

KEY CONCEPTS

The *floor system* (also called *floor framing*) has a key role in transferring gravity and lateral forces to the columns and/or walls, and it needs to meet the following requirements:

- adequate strength to safely resist applied loads
- sufficient stiffness to limit deflections and vibrations
- required fire resistance
- adequate acoustical resistance

Reinforced concrete floor systems can be classified as follows:

- In one-way systems, the primary flexural reinforcement in each structural element runs in one direction only.
- In two-way systems, the primary flexural reinforcement in at least some elements runs in two orthogonal directions.

Common cast-in-place one-way floor systems include the following:

- *One-way joist floors* (also known as ribbed slab or panjoist floors) consist of regularly spaced concrete joists (ribs) spanning in one direction and a thin reinforced concrete slab cast integrally with the joists. The joists are supported by wide beams spanning along the column lines. Their advantages include a lightweight, attractive ceiling, which also makes it possible to place electrical fixtures between joists. A drawback is that the formwork costs are higher than for the other systems, and the construction is labour intensive. This system can be used for spans ranging from 4.5 m to 12 m.
- *Slab-beam-and-girder floors* consist of a series of parallel beams supported by girders, which are in turn supported by columns placed at regular intervals over the entire floor area. The beam and girder grid supports a one-way reinforced concrete slab. The typical span between the columns ranges from 8 m to 18 m.
- In *slab band floors*, the one-way slab is carried directly by wide and shallow beams spanning between the columns. These wide beams (called slab bands) are supported directly by the columns. Typically, the nominal width of the slab band (*b*) is 2400 mm and the depth (*h*) is in the range from 400 mm to 600 mm. Slab band floor construction is common in Western Canada and is typically used in parking garages, residential and office buildings. A typical slab span is on the order of 8 m, while slab bands typically span a maximum of 10 m.

11.3 A DESIGN CASE STUDY OF A SLAB-AND-BEAM FLOOR SYSTEM

The purpose of this section is to demonstrate the design of a typical one-way reinforced concrete floor system in the manner in which it is performed in design practice. A typical plan of a second floor in a reinforced concrete building is shown in Figure 11.12. The floor is to be constructed using a reinforced concrete one-way slab spanning in the east-west (E-W) direction supported by the beams. The beams are supported by 400 mm square reinforced concrete columns. Due to architectural constraints, the beam width is

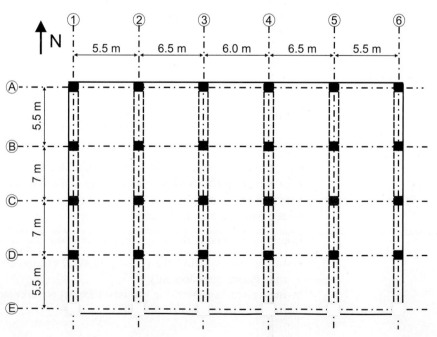

Figure 11.12 Plan view of the floor system.

limited to 300 mm and the depth is 700 mm. The building is scheduled for use as a retail occupancy.

Table 4.1.5.3 of NBC 2010 prescribes a minimum specified live load (LL) of 4.8 kPa for retail areas. It is expected that the floor area will be subdivided with partition walls, so a minimum partition dead load of 1 kPa needs to be considered in the design (Cl.4.1.4.1.3 of NBC 2010). In this design, the partition load will be considered as a superimposed dead load (DL_s).

As a part of the design criteria set up at the beginning of the project, the designer has decided to use concrete with a specified compressive strength (f_c') of 30 MPa for the floor design and deformed steel reinforcement of Grade 400W (f_y = 400 MPa).

Before we proceed with the detailed design, it is very important to understand the load transfer between the structural members.

First, consider the slab strip of unit width (equal to 1000 mm) shown hatched on the floor plan in Figure 11.13a. The slab spans in the E-W direction between the beams. This is clearly a one-way slab since beams exist only in the N-S direction (note the arrows showing the direction of load transfer in the slab). The slab strip will be designed as a five-span continuous structure with a rectangular cross-section of 1000 mm width, as shown in Figure 11.13b. The design of a typical slab strip will be demonstrated in Example 11.1.

Next, let us consider a typical beam B1-B2-B3-B4 spanning along gridline 3, as shown in Figure 11.14a. The beam elevation is shown in Figure 11.14b. The beam carries the load transferred from the slab over a tributary width a, shown hatched in Figure 11.14b, where

$$a = \frac{l_2}{2} + \frac{l_3}{2}$$

The beam design will be demonstrated in Example 11.2.

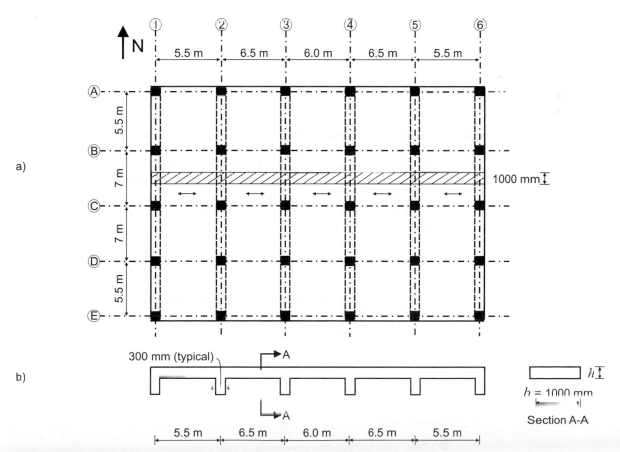

Figure 11.13 A five-span one-way continuous slab: a) plan view; b) elevation.

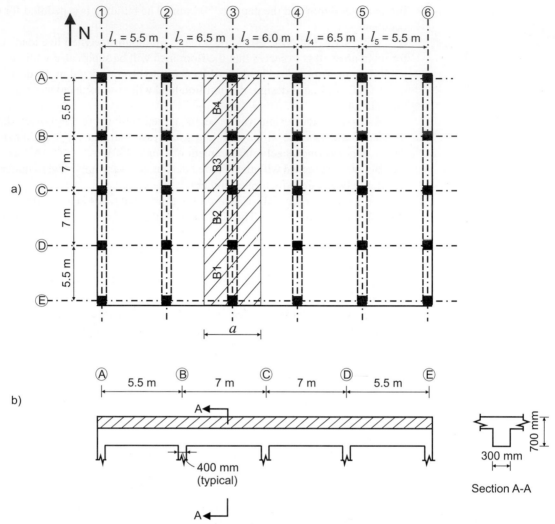

Figure 11.14 A four-span continuous reinforced concrete beam: a) plan view; b) elevation.

Example 11.1

Consider a one-way slab that is part of the floor system in Figure 11.13. A typical slab elevation is shown in the figure below. The slab is subjected to a specified superimposed dead load (DL$_s$) of 1 kPa and a specified live load (LL) of 4.8 kPa (in addition to its self-weight). Use 15M bars for flexural reinforcement. The supporting beams are 300 mm wide by 700 mm deep.

Design the slab, taking into account the CSA A23.3 strength and serviceability requirements.

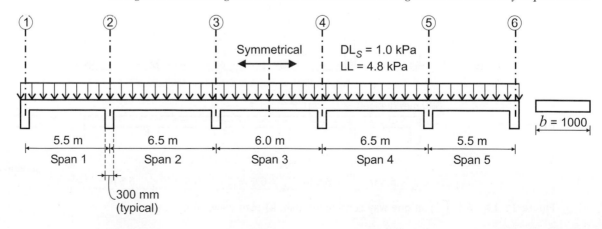

Given: $f_c' = 30$ MPa

$f_y = 400$ MPa

$\phi_c = 0.65$

$\phi_s = 0.85$

SOLUTION: This slab is symmetric with regard to its vertical axis, as shown on the figure above. In general, the designer should take advantage of symmetry wherever possible. In this case, all relevant design parameters (slab thickness, bending moments, and shear forces) need to be determined for spans 1, 2, and 3 only. Span 4 is a mirror image of span 2, while span 5 is a mirror image of span 1.

As discussed earlier in this section, this is a one-way slab and the design is based on a unit strip of width $b = 1000$ mm (as discussed in Section 3.6).

1. **Check whether the criteria for the CSA A23.3 approximate frame analysis are satisfied**

   The criteria for the approximate frame analysis, according to CSA A23.3 Cl.9.3.3, are as follows:

   - There are more than two continuous spans.
   - The length of each span is within 20% of that of the adjacent spans.
   - The loads are uniformly distributed.
   - The factored live load does not exceed twice the factored dead load (including the slab self-weight); this will be checked in Step 3.
   - The slab is prismatic (all sections are of uniform thickness).

   Hence, the approximate frame analysis according to CSA A23.3 Cl.9.3.3 can be used in this design. Note that the bending moments and shear forces for this slab were also determined in Example 10.3.

2. **Determine the slab thickness**

   Determine the values of the clear span (l_n) (clear distance between supports) for spans 1 to 3, as shown on the sketch below.

Span 1

$$l_{n1} = l_1 - b_s = 5.5 \text{ m} - 0.3 \text{ m} = 5.2 \text{ m}$$

Span 2

$$l_{n2} = l_2 - b_s = 6.5 \text{ m} - 0.3 \text{ m} = 6.2 \text{ m}$$

Span 3

$$l_{n3} = l_3 - b_s = 6.0 \text{ m} - 0.3 \text{ m} = 5.7 \text{ m}$$

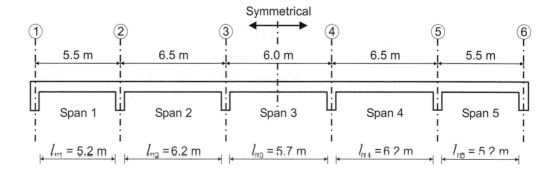

To avoid detailed deflection calculations, determine the minimum slab thickness in accordance with the CSA A23.3 indirect approach to deflection control prescribed by Cl.9.8.2.1 (see Section 4.5.2 and Table A.3). The calculations are summarized in Table 11.1.

Table 11.1 Slab thicknesses

Span	1	2	3
l_n (mm)	5200	6200	5700
Minimum thickness (Table A.3)	$\dfrac{l_n}{24}$ (one end continuous)	$\dfrac{l_n}{28}$ (both ends continuous)	$\dfrac{l_n}{28}$
Minimum required thickness (mm)	220	220	200

The minimum required slab thicknesses is 220 mm for each span except for span 3, where a 200 mm thickness is required (see Table 11.1). The use of a uniform thickness is recommended because it simplifies the construction; therefore, the same thickness will be used throughout the slab length:

$$h = 220 \text{ mm}$$

3. **Determine the factored loads**
   a) Calculate the dead load acting on the slab.
      First, calculate the slab self-weight:

      $$DL_w = h \times \gamma_w = 0.22 \text{ m} \times 24 \text{ kN/m}^3 = 5.28 \text{ kPa}$$

      where

      $$\gamma_w = 24 \text{ kN/m}^3$$

      is the unit weight of concrete.
      The superimposed dead load is given as

      $$DL_s = 1.0 \text{ kPa}$$

      Finally, the total factored dead load is

      $$w_{DL,f} = 1.25 \, (DL_w + DL_s)$$
      &

   b) Calculate the factored live load:

      $$w_{LL,f} = 1.5 \times LL$$
      $$= 1.5 \times 4.8 \text{ kPa} = 7.2 \text{ kPa}$$

   c) The total factored load is

      $$w_f = w_{DL,f} + w_{LL,f}$$
      $$= 7.85 \text{ kPa} + 7.2 \text{ kPa} \cong 15 \text{ kPa}$$

      This is a one-way slab and the design is based on the unit strip (1 m wide), so

      $$w_f = 15 \text{ kN} \cdot \text{m} \times 1 \text{ m} = 15 \text{ kN} \cdot \text{m/m}$$

      Note that the factored live load $w_{LL,f} = 7.2$ kPa is less than the factored dead load of $w_{DL,f} = 7.85$ kPa. Thus, the factored live load does not exceed twice the factored dead load, so the condition for the application of the CSA A23.3 coefficient method has been met.

4. **Determine the factored bending moments (M_f)**
   Next, determine the factored bending moments at critical sections using the approximate frame analysis according to A23.3 Cl.9.3.3. In this case, the width of the spandrel beam is insufficient to ensure proper development of the top slab reinforcing for negative moment resistance at the support, hence the discontinuous end is considered to be unrestrained, as shown in Figure 10.14b. The results are summarized in Table 11.2. The moment envelope diagram can be plotted based on the M_f values calculated in the table,

as shown in the figure below. The inflection points are determined based on the generic chart presented in Figure 10.15.

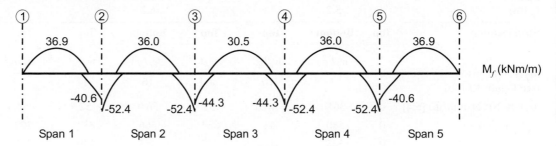

5. **Determine the area for the main flexural reinforcement**

Now determine the required area of tension reinforcement for each span.

a) Determine the effective depth (d).

Because the slab is of interior exposure, the cover can be determined from Table A.2 as

cover = 20 mm

Because 15M bars are to be used in this design, it follows that

$$d_b \cong 15 \text{ mm}$$

(see Table A.1). Therefore, determine the effective depth as

$$d = h - \text{cover} - \frac{d_b}{2}$$

$$= 220 \text{ mm} - 20 \text{ mm} - \frac{15 \text{ mm}}{2} = 192.5 \text{ mm} \cong 190 \text{ mm}$$

Use the same effective depth (d) for the positive and negative moment areas.

b) Determine the required area of tension reinforcement using the direct procedure according to the equation

$$A_s = 0.0015 f_c' b \left(d - \sqrt{d^2 - \frac{3.85 M_f}{f_c' b}} \right) \tag{5.4}$$

Note that b = 1000 mm (unit width).

c) Confirm that the minimum reinforcement requirement for the slab is satisfied:

$$A_g = h \times 1000 \text{ mm} = 220 \text{ mm} \times 1000 \text{ mm} = 220 \times 10^3 \text{ mm}^2$$

$$A_{smin} = 0.002 A_g$$

$$= 0.002(220 \times 10^3 \text{ mm}^2) = 440 \text{ mm}^2/\text{m}$$

d) Determine the required bar spacing (s).

The spacing is determined assuming 15M rebars with a bar area of

$$A_b = 200 \text{ mm}^2$$

(see Table A.1). Therefore,

$$s \leq A_b \frac{1000}{A_s} \tag{3.29}$$

e) Determine the provided area of reinforcement ($A_{sprovided}$) per metre of slab width.

$$A_{sprovided} = A_b \frac{1000}{s} \tag{3.28}$$

f) Check the reinforcement ratio (ρ):

$$\rho = \frac{A_s}{b d} \tag{3.1}$$

Table 11.2 Factored bending moments and the corresponding flexural reinforcement

Span l_n (m)		1 5.2			2 6.2			3 5.7	
Steel location	Top	Bottom	Top	Top	Bottom	Top	Top	Bottom	Top
M_f (kN·m/m) (see Figure 10.14b)	0	$\dfrac{w_f l_n^2}{11}$	$-\dfrac{w_f l_n^2}{10}$	$-\dfrac{w_f l_n^2}{11}$	$\dfrac{w_f l_n^2}{16}$	$-\dfrac{w_f l_n^2}{11}$	$-\dfrac{w_f l_n^2}{11}$	$\dfrac{w_f l_n^2}{16}$	$-\dfrac{w_f l_n^2}{11}$
M_f ($\times 10^6$ N·mm/m) [5.4]	0	36.9	−40.6	−52.4	36.0	−52.4	−44.3	30.5	−44.3
A_s (mm²/m)	0	580	641	837	566	837	702	477	702
s (mm)	0	345	312	239	353	239	285	420	285
Design reinforcement	0	15M@300	15M@225	15M@225	15M@350	15M@225	15M@225	15M@400	15M@225
$A_{sprovided}$ (mm²/m) [3.28]	0	667	890	890	571	890	890	500	890
ρ (%)	0	0.35	0.47	0.47	0.30	0.47	0.47	0.26	0.47
ρ_b (%)	2.7	2.7	2.7	2.7	2.7	2.7	2.7	2.7	2.7
A_{smin} (mm²/m)	440	440	440	440	440	440	440	440	440

g) Check whether the actual reinforcement ratio (ρ) is less than or equal to the balanced reinforcement ratio (ρ_b):

$$\rho_b = 0.027 \text{ for } f_c' = 30 \text{ MPa}$$

(see Table A.4).

    The design calculations for all parameters discussed under a) to f) are summarized in Table 11.2. The three critical sections in each span have been identified, of which two sections are at the supports and the third one is at the midspan, corresponding to the top and bottom steel, respectively.

    Notice from Table 11.2 that $A_s > A_{smin}$ for all sections, so the minimum reinforcement requirement has been satisfied. Also, because $\rho < \rho_b = 0.027$ for all sections, the slab is properly reinforced and will fail in a desirable, steel-controlled mode.

h) Confirm that the maximum bar spacing requirement is satisfied (CSA A23.3 Cl.7.4.1.2).

    First, calculate the maximum bar spacing (s_{max}) as the lesser of

$$3 \times h = 3 \times 220 \text{ mm} = 660 \text{ mm}$$

and

500 mm

The smaller value governs, so

$$s_{max} = 500 \text{ mm}$$

Next, compare the actual bar spacing with the maximum bar spacing. The bar spacing in this example varies from 225 mm to 400 mm (see Table 11.2), so

$$s = 400 \text{ mm} < 500 \text{ mm}$$

and the maximum bar spacing requirement is satisfied.

6. **Design the shrinkage and temperature reinforcement (A23.3 Cl.7.8.1 and 7.8.3)**

a) Determine the minimum area of shrinkage and temperature reinforcement:

$$A_{smin} = 0.002 \, A_g$$
$$= 0.002 (220 \times 10^3 \text{ mm}^2) = 440 \text{ mm}^2/\text{m}$$

b) Determine the maximum bar spacing (s_{max}).

s_{max} is the lesser of

$$5 \times h = 5 \times 220 \text{ mm} = 1100 \text{ mm}$$

and

500 mm

The smaller value governs, so

$s_{max} = 500$ mm

c) Determine the required bar spacing.
     The area of one 15M bar is 200 mm^2 (see Table A.1); that is,

$A_b = 200$ mm^2

Determine the required spacing as

$$s \leq A_b \frac{1000}{A_s} \qquad \text{[3.29]}$$

$$= (200 \text{ mm}^2) \frac{1000}{440 \text{ mm}^2/\text{m}} = 455 \text{ mm} \cong 450 \text{ mm}$$

d) Check whether the provided area of reinforcement is greater than or equal to the required amount of reinforcement:

$$A_s = A_b \frac{1000}{s} \qquad \text{[3.28]}$$

$$= (200 \text{ mm}^2) \frac{1000}{450 \text{ mm}} = 444 \text{ mm}^2/\text{m} > 440 \text{ mm}^2/\text{m}$$

so the provided area of reinforcement is slightly larger than the required area, which is okay. Therefore, the selected shrinkage and temperature reinforcement consists of 15M bars at 450 mm spacing, that is, 15M@450.

7. **Check the crack control parameter (z) (A23.3 Cl.10.6.1)**
    a) Compute the effective tension area per bar (A).

       i) Calculate the distance (d_s) from the centroid of the tension reinforcement to the tensile face of the concrete section:

       $d_s = h - d$

       $= 220 \text{ mm} - 190 \text{ mm} = 30 \text{ mm}$

       ii) Determine the effective tension area per bar (A).
            In this case, the A value can be determined directly for one bar with spacing s as follows (see the sketch below):

       $A = s(2d_s)$

       $= (400 \text{ mm})(2 \times 30 \text{ mm})$

       $= 24\ 000 \text{ mm}^2$

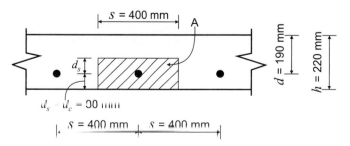

Note that the bar spacing in the slab varies from 225 mm to 400 mm. It is sufficient to perform the check for the largest spacing since it gives the most conservative results.

b) Determine the stress in the steel reinforcement (f_s) under the service load level. An approximate estimate for f_s permitted by A23.3 Cl.10.6.1 will be used in lieu of detailed calculations:

$$f_s = 0.6 f_y \qquad \text{[4.21]}$$
$$= 0.6 \times 400 \text{ MPa} = 240 \text{ MPa}$$

c) Determine the value of z:

$$d_c = d_s = 30 \text{ mm (one reinforcement layer)}$$

| A23.3 Eq. 10.6 |

$$z = f_s \sqrt[3]{d_c A} \qquad \text{[4.20]}$$
$$= 240 \text{ MPa} \sqrt[3]{(30 \text{ mm})(24\,000 \text{ mm}^2)} = 21\,510 \text{ N/mm}$$

Because

$$z = 21\,510 \text{ N/mm} < 30\,000 \text{ N/mm}$$

the CSA A23.3 cracking control requirements for interior exposure are satisfied.

8. **Check the shear resistance**

Check the shear resistance of the slab according to the CSA A23.3 simplified method for shear design. In general, reinforced concrete slabs are designed so that shear reinforcement is not required.

a) Determine the factored shear forces.

Determine the shear forces according to the CSA A23.3 approximate frame analysis (Cl.9.3.3). The results are summarized in Table 11.3.

b) Determine the concrete shear resistance (V_c) according to A23.3 Cl.11.3.4.

i) Find the effective shear depth (d_v), which is taken as the greater of

$$d_v = \begin{bmatrix} 0.9d \\ 0.72h \end{bmatrix} = \begin{bmatrix} 0.9 \times 190 \text{ mm} \\ 0.72 \times 220 \text{ mm} \end{bmatrix} = \begin{bmatrix} 171 \text{ mm} \\ 158 \text{ mm} \end{bmatrix}$$

Hence, $d_v \cong 170$ mm.

ii) Determine the β value:

$$\beta = 0.21$$

because

$$h = 220 \text{ mm} < 350 \text{ mm}$$

According to A23.3 Cl.11.3.6.2, $\beta = 0.21$ can be used for slabs with an overall thickness not greater than 350 mm.

iii) Finally, determine the V_c value as

| A23.3 Eq. 11.6 |

$$V_c = \phi_c \lambda \beta \sqrt{f_c'}\, b_w d_v \qquad \text{[6.12]}$$
$$= 0.65 \times 1.0 \times 0.21 \times \sqrt{30 \text{ MPa}} \times 1000 \text{ mm} \times 170 \text{ mm}$$
$$= 127 \times 10^3 \text{ N} = 127 \text{ kN/m}$$

Table 11.3 Factored shear forces in the slab

Span	1		2		3	
l_n (m)	5.2		6.2		5.7	
Location	Left	Right	Left	Right	Left	Right
V_f (kN/m) (A23.3 Cl.9.3.3)	$\dfrac{w_f l_n}{2}$	$-\dfrac{1.15\, w_f l_n}{2}$	$\dfrac{w_f l_n}{2}$	$-\dfrac{w_f l_n}{2}$	$\dfrac{w_f l_n}{2}$	$-\dfrac{w_f l_n}{2}$
V_f (kN/m)	39.0	−45.0	46.5	−46.5	43.0	−43.0

c) Determine whether shear reinforcement is required (A23.3 Cl.11.2.8.1).
   If $V_f < V_c$, then shear reinforcement is not required. In this example, the concrete shear resistance greatly exceeds the factored shear force in each span presented in Table 11.3; that is,

$$V_f < V_c = 127 \text{ kN/m}$$

so shear reinforcement does not need to be provided in the slab.

9. **Check for deflections in the slab**
   In this example, the slab thickness was determined in accordance with the CSA A23.3 indirect approach to deflection control prescribed by Cl.9.8.2.1. As a result, it is not required to perform detailed deflection calculations.

10. **Provide a design summary**
    Finally, a design summary showing the slab cross-section and the reinforcement is presented below.

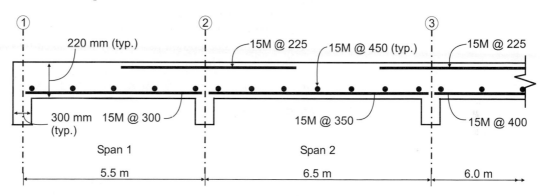

Example 11.2

Consider the beam in gridline 3 that is part of the floor system shown in Figure 11.14. The floor is subjected to a factored dead load of $w_{DL,f} = 7.85$ kPa and a factored live load of $w_{LL,f} = 7.2$ kPa (note that the beam self-weight is not included). The beam is 300 mm wide by 700 mm deep, as shown in the figure below. Use 25M bars for flexural reinforcement and 10M stirrups.

Design the beam by taking into account the CSA A23.3 strength and serviceability requirements.

Given: $f_c' = 30$ MPa
        $f_y = 400$ MPa
        $\phi_c = 0.65$
        $\phi_s = 0.85$

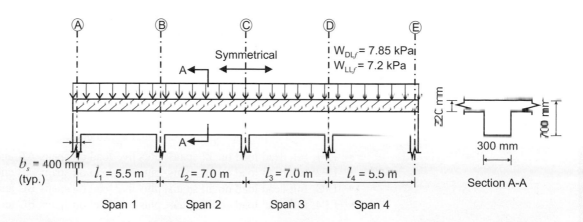

SOLUTION: **1. Check whether the criteria for the CSA A23.3 approximate frame analysis are satisfied**

The criteria for the approximate frame analysis according to CSA A23.3 Cl.9.3.3 are as follows:

- There are more than two continuous spans.
- The length of each span is within 20% of that of the adjacent spans.

The ratio between span lengths for spans 2 and 1 is equal to

$$\frac{l_2}{l_1} = \frac{7 \text{ m}}{5.5 \text{ m}} = 1.27$$

This indicates that the length of span 2 exceeds the length of span 1 by 27%, so the approximate frame analysis according to A23.3 Cl.9.3.3 cannot be used in this design. Therefore, a linear elastic analysis will be performed by using a computer analysis program.

2. Determine the factored loads

a) Determine the beam self-weight.

Only the portion of the beam below the slab will be considered for the self-weight calculation (the slab portion has already been taken into account), as shown on the sketch below. Hence,

$$DL_w = b \times h_{web} \times \gamma_w = 0.3 \text{ m} \times 0.48 \text{ m} \times 24 \text{ kN/m}^3 = 3.5 \text{ kN/m}$$

where

$$\gamma_w = 24 \text{ kN/m}^3$$

is the unit weight of concrete.

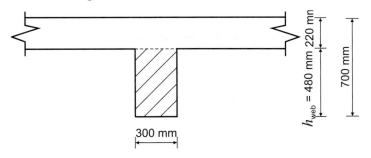

Section A-A

b) Determine the tributary width (a).

The tributary width for the beam is (see Figure 11.14a)

$$a = \frac{l_2}{2} + \frac{l_3}{2} = \frac{6.5 \text{ m}}{2} + \frac{6.0 \text{ m}}{2} = 6.25 \text{ m}$$

c) Find the factored dead load for the beam:

$$DL_f = w_{DL,f} \times a + 1.25 \times DL_w$$
$$= 7.85 \text{ kPa} \times 6.25 \text{ m} + 1.25 \times 3.5 \text{ kN/m} = 53.4 \text{ kN/m}$$

d) Find the factored live load for the beam:

$$LL_f = w_{LL,f} \times a = 7.2 \text{ kPa} \times 6.25 \text{ m} = 45 \text{ kN/m}$$

3. Identify all of the possible load patterns

According to CSA A23.3 Cl.9.2.3.1, the following load patterns should be considered in the design of the four-span beam under consideration:

- LP1: full dead load on all spans, plus live load on spans AB and CD
- LP2: full dead load on all spans, plus live load on spans AB and BC
- LP3: full dead load on all spans, plus live load on spans BC and CD
- LP4: full dead load on all spans, plus live load on spans BC and DE

- LP5: full dead load on all spans, plus live load on spans CD and DE
- LP6: full dead load and live load on all spans

Of all of the possible load patterns, only patterns LP1, LP2, LP3, and LP6 will be considered in this example. Because the beam is symmetric with regard to the vertical axis, load pattern LP4 represents a mirror image of load pattern LP1 and will give the same results, while load pattern LP5 is a mirror image of load pattern LP2.

4. **Develop the moment envelope diagram and determine the design bending moments**
   The work done in this step is similar to that in Example 10.1. First, moment diagrams need to be developed for each load pattern. Then, bending moment diagrams need to be superimposed on the same chart, as shown in the diagram below.

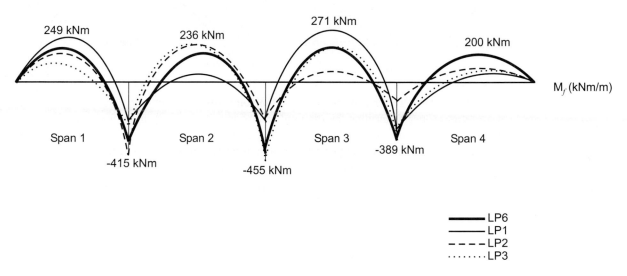

Subsequently, the moment envelope diagram is constructed from portions of the bending moment diagrams corresponding to different load patterns (see Figure 11.15a). Finally, the peak bending moments at the supports are reduced, as explained in Section 10.4.3 (see Figure 11.15b).

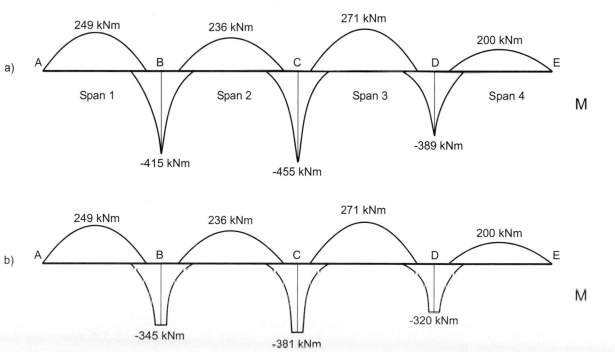

Figure 11.15 The moment envelope diagram for the beam: a) actual diagram; b) moment envelope with reduced moments at the supports.

Observe from the moment envelope diagram that the most critical positive bending moments occur in spans 1 and 3, while the most critical negative bending moments occur at supports B and C. These values will be used for the design, and a moment envelope diagram showing the critical values for the two spans is shown on the sketch below.

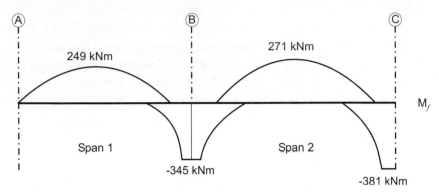

5. **Determine the area of tension reinforcement**

    Determine the required area of tension reinforcement for each span.

    a) Determine the effective depth (d).

        It is expected that the reinforcement will be placed in one layer, so estimate the d value as (see Section 5.4.1)

        $d = h - 70 \text{ mm} = 700 \text{ mm} - 70 \text{ mm} = 630 \text{ mm}$

        The same effective depth (d) will be used for the positive and negative moment areas.

    b) Determine the section width.

        Although this beam behaves as a T-beam, the slab contribution will be neglected for the sake of simplicity. Hence, the beam will be considered as a rectangular beam with width

        $b = 300 \text{ mm}$

        for both the negative and the positive moment areas.

    c) Determine the required area of tension reinforcement using the direct procedure according to the equation

        $$A_s = 0.0015 f_c' \, b \left(d - \sqrt{d^2 - \frac{3.85 \, M_f}{f_c' \, b}} \right) \qquad \text{[5.4]}$$

        Note that the provided reinforcement area ($A_{sprovided}$) is determined as the area of a 25M bar (500 mm^2) times the number of bars.

    d) Check the reinforcement ratio (ρ):

        $$\rho = \frac{A_s}{bd} \qquad \text{[3.1]}$$

    e) Check whether the actual reinforcement ratio (ρ) is less than or equal to the balanced reinforcement ratio (ρ_b):

        $\rho_b = 0.027$ for $f_c' = 30 \text{ MPa}$ (see Table A.4).

    f) Confirm that the minimum reinforcement requirement is satisfied (A23.3 Cl.10.5.1.2). Determine the minimum required reinforcement area (A_{smin}) as

        A23.3 Eq. 10.4

        $$A_{smin} = \frac{0.2\sqrt{f_c'}}{f_y} b_t h \qquad \text{[5.7]}$$

        $$= \frac{0.2 \left(\sqrt{30 \text{ MPa}} \right)}{400 \text{ MPa}} \times 300 \text{ mm} \times 700 \text{ mm} = 575 \text{ mm}^2$$

        where $b_t = b = 300 \text{ mm}$ is the width of the tension zone.

Design calculations for all parameters discussed under a) to f) are summarized in Table 11.4. The three critical sections in each span have been identified, of which two sections are at the supports and the third is at the midspan, corresponding to the top and bottom steel respectively. The bending moments have been determined from the moment envelope diagram presented at the end of Step 4.

Notice from Table 11.4 that $A_s > A_{smin}$ for all sections, so the minimum reinforcement requirement has been satisfied. Also, since $\rho < \rho_b = 0.027$ for all sections, the beam is properly reinforced and will fail in a desirable, steel-controlled mode.

g) Check whether the reinforcement can fit in the section.
  i) Determine the concrete cover:

  cover = 30 mm (Table A.2, Class N).

  ii) Calculate the minimum required bar spacing (s_{min}):
  - 25M tension reinforcement (d_b = 25 mm) (see Table A.1)
  - 10M stirrups (d_s = 10 mm) (see Table A.1)
  - maximum aggregate size of 20 mm (a_{max} = 20 mm)

  According to A23.1 Cl.6.6.5.2, the minimum spacing (s_{min}) between the reinforcement bars should be the largest of
  - $1.4 \times d_b = 1.4 \times 25$ mm = 35 mm
  - $1.4 \times a_{max} = 1.4 \times 20$ mm = 28 mm
  - 30 mm

  Hence, the minimum spacing between reinforcement bars is s_{min} = 35 mm.
  iii) Determine the number of bars that can fit in one layer.
  Check whether 3-25M bottom bars can fit in one layer considering a beam width of 300 mm. The minimum required beam width (b_{min}) is

  $$b_{min} = 3\,d_b + 2s + 2d_s + 2 \times \text{cover}$$
  $$= 3 \times 25 \text{ mm} + 2 \times 35 \text{ mm} + 2 \times 10 \text{ mm} + 2 \times 30 \text{ mm}$$
  $$= 225 \text{ mm}$$

  Because

  $$b_{min} = 225 \text{ mm} < 300 \text{ mm}$$

  3-25M bars can fit in one layer. In the negative moment areas, 4-25M bars need to fit in the top layer, but the spacing is less critical because some reinforcement can be placed in the slab.

Table 11.4 Factored bending moments and the corresponding flexural reinforcement

Span		1			2	
l (m)		5.5			7	
Steel location	**Top**	**Bottom**	**Top**	**Top**	**Bottom**	**Top**
M_f ($\times 10^6$ N·mm)	0	249	−345	−345	271	−381
A_u (mm²) [5,4]	0	1232	1763	1763	1350	1976
Design reinforcement	0	3-25M	4-25M	4-25M	3-25M	4-25M
$A_{provided}$ (mm²)	0	1500	2000	2000	1500	2000
ρ (%)	0	0.8	1.06	1.06	0.8	1.06
ρ_b (%)	2.7	2.7	2.7	2.7	2.7	2.7
A_{smin} (mm²)	575	575	575	575	575	575

iv) Determine the actual effective depth (d) as

$$d = h - \text{cover} - d_s - \frac{d_b}{2}$$

$$= 700 \text{ mm} - 30 \text{ mm} - 10 \text{ mm} - \frac{25 \text{ mm}}{2} = 647 \text{ mm} \cong 645 \text{ mm}$$

Because the actual d value exceeds the original estimate of 630 mm, there is no need to repeat the design calculations. (The estimated d value results in a more conservative design.)

6. **Check the crack control parameter (z) (A23.3 Cl.10.6.1)**
    a) Compute the effective tension area per bar (A).
        i) Calculate the distance between the centroid of the tensile reinforcement and the tensile face of the concrete section (d_s) (see the sketch below):

$$d_s = h - d$$

$$= 700 \text{ mm} - 645 \text{ mm} = 55 \text{ mm}$$

        ii) Then, determine the total effective tension area for all bars (A_e):

$$A_e = b(2d_s)$$

$$= (300 \text{ mm}) (2 \times 55 \text{ mm}) = 33\,000 \text{ mm}^2$$

        iii) Finally, calculate the effective tension area per bar (A) for the bottom reinforcement (3-25M).
               For 3 bars in total, $N = 3$. Hence,

$$A = \frac{A_e}{N}$$

$$= \frac{33\,000 \text{ mm}^2}{3} = 11\,000 \text{ mm}^2$$

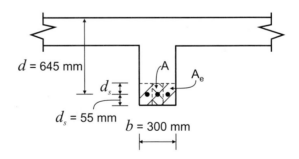

    b) Determine the stress in the steel reinforcement (f_s) under the service load level.
        Compute the stress in the steel reinforcement based on an approximate estimate permitted by CSA A23.3 Cl.10.6.1 in lieu of detailed calculations, as follows:

$$f_s = 0.6 f_y \qquad\qquad \textbf{[4.21]}$$

$$= 0.6 \times 400 \text{ MPa}$$

$$= 240 \text{ MPa}$$

    c) Determine the value of z:

$$d_c = d_s = 55 \text{ mm (one reinforcement layer);}$$

A23.3 Eq. 10.6

$$z = f_s \sqrt[3]{d_c A} \qquad\qquad \textbf{[4.20]}$$

$$= 240 \text{ MPa} \sqrt[3]{(55 \text{ mm})(11\,000 \text{ mm}^2)} = 20\,298 \text{ N/mm}$$

Because

$$z = 20\ 298\ \text{N/mm} < 30\ 000\ \text{N/mm}$$

the CSA A23.3 cracking control requirements for interior exposure are satisfied.

7. **Provide a design summary for the flexural reinforcement**
   Finally, a design summary showing the slab cross-section and the reinforcement is presented below.

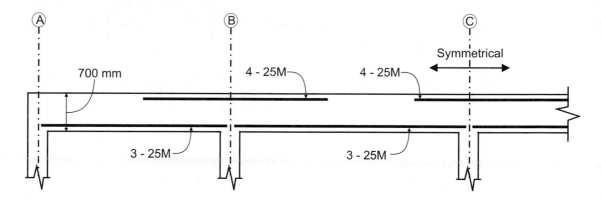

8. **Develop the shear envelope and determine the design shear forces**
   The shear envelope diagram is constructed in a similar manner to the moment envelope diagram discussed in Step 4. The same load patterns (LP1, LP2, LP3, and LP6) have been used and the resulting diagram is shown below. The maximum shear forces obtained from all load cases will be used in the design. The shear design of the beam will be performed according to the CSA A23.3 simplified method for shear design.

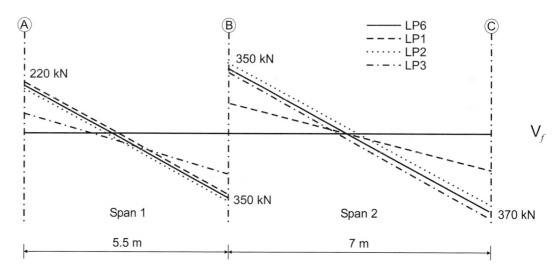

9. **Design the shear reinforcement for span 1**
   The following steps need to be followed to design the shear reinforcement for span 1:
   a) Calculate the design shear force ($V_{f@dv}$) at a distance d_v from the face of the support (A23.3 Cl.11.3.2).
      i) Calculate the effective shear depth (d_v), which can be taken as the greater of

$$d_v = \begin{bmatrix} 0.9d \\ 0.72h \end{bmatrix} = \begin{bmatrix} 0.9 \times 645\ \text{mm} \\ 0.72 \times 700\ \text{mm} \end{bmatrix} = \begin{bmatrix} 580\ \text{mm} \\ 504\ \text{mm} \end{bmatrix}$$

   Hence, $d_v = 580$ mm.
      ii) Then, determine the centre-to-centre span:

$$l = 5.5\ \text{m}$$

iii) Next, find the maximum shear force at the midspan ($V_{fmidspan}$):

$$V_{fmidspan} = \frac{LL_f\left(\frac{l}{2}\right)}{4} = \frac{(45 \text{ kN/m}) \times \left(\frac{5.5 \text{ m}}{2}\right)}{4} = 31 \text{ kN}$$

iv) Then, determine the factored shear force ($V_{f@dv}$) at a distance d_v from support A as

$$V_{f@dv} = V_{fmidspan} + \left(\frac{\frac{l}{2} - \left(d_v + \frac{b_s}{2}\right)}{\frac{l}{2}}\right)(V_f - V_{fmidspan})$$

$$= 31 \text{ kN} + \left(\frac{\frac{5.5 \text{ m}}{2} - \left(0.58 \text{ m} + \frac{0.4 \text{ m}}{2}\right)}{\frac{5.5 \text{ m}}{2}}\right)(220 \text{ kN} - 31 \text{ kN})$$

$$= 166.4 \text{ kN} \cong 166 \text{ kN}$$

The maximum shear force envelope shows shear forces at the centre-line of the supports. In order to determine $V_{f@dv}$, find the shear force at a distance $d_v + b_s/2$ from the support centre-line, where $b_s = 400$ mm is the width of the supporting column.

v) Determine the factored shear force ($V_{f@dv}$) at a distance d_v from support B as

$$V_{f@dv} = 31 \text{ kN} + \left(\frac{\frac{5.5 \text{ m}}{2} - \left(0.58 \text{ m} + \frac{0.4 \text{ m}}{2}\right)}{\frac{5.5 \text{ m}}{2}}\right)(350 \text{ kN} - 31 \text{ kN})$$

$$= 259.5 \text{ kN} \cong 260 \text{ kN}$$

vi) Sketch the shear envelope diagram for span 1.

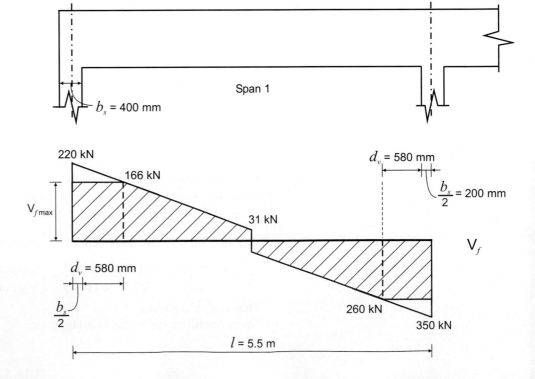

b) Determine the concrete shear resistance (V_c) (A23.3 Cl.11.3.4).
Consider that a larger than minimum shear reinforcement will be provided (as per A23.3 Cl.11.2.8.2), so

$$\beta = 0.18 \text{ (A23.3 Cl.11.3.6.3)}$$

and

A23.3 Eq. 11.6

$$V_c = \phi_c \lambda \beta \sqrt{f_c'} \, b_w d_v \qquad\qquad [6.12]$$
$$= 0.65(1)(0.18)\sqrt{30 \text{ MPa}} \, (300 \text{ mm}) (580 \text{ mm}) = 111 \text{ kN}$$

c) Determine whether shear reinforcement is required and identify the region(s) in which the reinforcement is required (A23.3 Cl.11.2.8.1).
If $V_f < V_c$, then shear reinforcement is not required.
i) At support A,

$$V_{f@dv} = 166 \text{ kN} > 111 \text{ kN}$$

ii) At support B,

$$V_{f@dv} = 260 \text{ kN} > 111 \text{ kN}$$

Therefore, shear reinforcement is required at both supports.

d) Calculate the stirrup spacing according to the CSA A23.3 requirements (A23.3 Cl.11.2.8.2 and Cl.11.3.8.1).
i) Use two-legged 10M stirrups (the area of one 10M stirrup is $A_b = 100 \text{ mm}^2$, see Table A.1):

$$A_v = (\text{\# of legs})(A_b)$$
$$= 2 \times 100 \text{ mm}^2 = 200 \text{ mm}^2$$

ii) Determine the maximum spacing of shear reinforcement based on the following equation (A23.3 Cl.11.2.8.2):

$$s = \frac{A_v f_y}{0.06\sqrt{f_c'} \, b_w} \qquad\qquad [6.19]$$
$$= \frac{200 \text{ mm}^2 \times 400 \text{ MPa}}{0.06\sqrt{30 \text{ MPa}} \, (300 \text{ mm})} = 811 \text{ mm}$$

iii) Check whether the spacing is within the maximum limits prescribed by A23.3 Cl.11.3.8.1:

$$s_{max} \leq \begin{bmatrix} 600 \text{ mm} \\ 0.7 d_v \end{bmatrix} = \begin{bmatrix} 600 \text{ mm} \\ 0.7 \times 580 \text{ mm} \end{bmatrix} = \begin{bmatrix} 600 \text{ mm} \\ 406 \text{ mm} \end{bmatrix}$$

Hence, the maximum permitted spacing s_{max} is the smallest of

- 811 mm
- 600 mm
- 406 mm

So, $s = 406$ mm $= 400$ mm

e) Check whether the strength requirement is satisfied (A23.3 Cl.11.3.1).
i) Find the shear capacity (V_s) of the shear reinforcement (A23.3 Cl.11.3.5.1).
Because this design has been performed according to the CSA A23.3 simplified method, Cl.11.3.6.3 prescribes that

$$\theta = 35° \text{ and } \cot\theta = 1.43$$

Thus,

A23.3 Eq. 11.7

$$V_s = \frac{\phi_s A_v f_y d_v \cot \theta}{s}$$ [6.9]

$$= \frac{0.85 \times 200 \text{ mm}^2 \times 400 \text{ MPa} \times 580 \text{ mm} \times 1.43}{400 \text{ mm}}$$

$$= 141 \times 10^3 \text{ N} = 141 \text{ kN}$$

ii) Find the factored shear resistance (V_r) (A23.3 Cl.11.3.3):

A23.3 Eq. 11.4

$$V_r = V_c + V_s + V_p$$ [6.11]

$$= 111 \text{ kN} + 141 \text{ kN} + 0 = 252 \text{ kN}$$

iii) Check whether the CSA A23.3 strength requirement is satisfied:

A23.3 Eq. 11.3

$$V_r \geq V_f$$ [11.1]

At support A,

$$V_r = 252 \text{ kN} > V_{f@dv} = 166 \text{ kN}$$

At support B,

$$V_r = 252 \text{ kN} < V_{f@dv} = 260 \text{ kN}$$

Thus, the strength requirement is satisfied because a 5% undercapacity at support B can be neglected. Therefore, the shear reinforcement is adequate.

f) Check the maximum factored shear resistance (V_r) (A23.3 Cl.11.3.3).
The factored shear resistance for the beam section is equal to

$$V_r = 252 \text{ kN}$$

According to A23.3 Cl.11.3.3, the maximum factored shear resistance ($maxV_r$) is

A23.3 Eq. 11.5

$$maxV_r = 0.25\phi_c f_c' b_w d_v$$ [6.17]

$$= 0.25 \times 0.65 \times 30 \text{ MPa} \times 300 \text{ mm} \times 580 \text{ mm}$$

$$= 848.2 \times 10^3 \text{ N} \cong 848 \text{ kN}$$

Because

$$V_r = 252 \text{ kN} < 848 \text{ kN}$$

the factored shear resistance is acceptable.

g) Sketch a beam elevation showing the distribution of shear reinforcement.

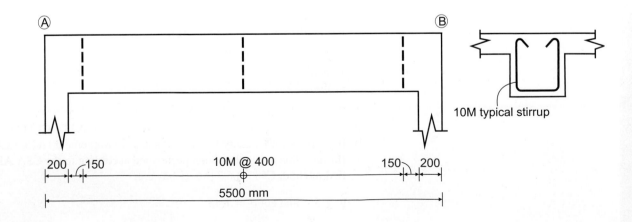

10. Design the shear reinforcement for span 2

    a)   Calculate the design shear force ($V_{f@dv}$) at a distance d_v from the face of the support (A23.3 Cl.11.3.2).

        i) Calculate the effective shear depth (d_v):

$$d_v = 580 \text{ mm (from Step 9)}$$

        ii) Then, determine the centre-to-centre span:

$$l = 7.0 \text{ m}$$

        iii) Next, find the maximum shear force at the midspan ($V_{fmidspan}$):

$$V_{fmidspan} = \frac{LL_f\left(\dfrac{l}{2}\right)}{4} = \frac{(45 \text{ kN/m}) \times \left(\dfrac{7.0 \text{ m}}{2}\right)}{4} = 39.4 \text{ kN} \cong 40 \text{ kN}$$

        iv) Then, determine the factored shear force ($V_{f@dv}$) at a distance d_v from support B as

$$V_{f@dv} = V_{fmidspan} + \left(\frac{\dfrac{l}{2} - \left(d_v + \dfrac{b_s}{2}\right)}{\dfrac{l}{2}}\right)(V_f - V_{fmidspan})$$

$$= 40 \text{ kN} + \left(\frac{\dfrac{7.0 \text{ m}}{2} - \left(0.58 \text{ m} + \dfrac{0.4 \text{ m}}{2}\right)}{\dfrac{7.0 \text{ m}}{2}}\right)(350 \text{ kN} - 40 \text{ kN})$$

$$= 281 \text{ kN}$$

        v) Next, determine the factored shear force ($V_{f@dv}$) at a distance d_v from support C as

$$V_{f@dv} = 40 \text{ kN} + \left(\frac{\dfrac{7.0 \text{ m}}{2} - \left(0.58 \text{ m} + \dfrac{0.4 \text{ m}}{2}\right)}{\dfrac{7.0 \text{ m}}{2}}\right)(370 \text{ kN} - 40 \text{ kN})$$

$$= 296.5 \text{ kN} \cong 300 \text{ kN}$$

        vi) Sketch the shear envelope diagram for span 2.

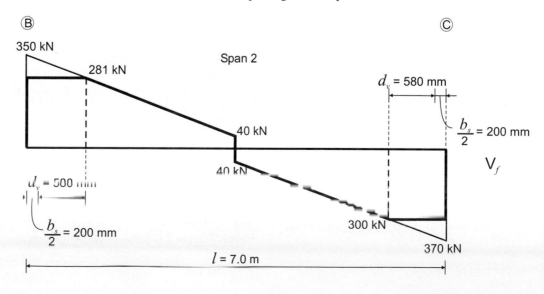

b) Determine the concrete shear resistance (V_c) (A23.3 Cl.11.3.4).

   i) In the regions where less than minimum shear reinforcement is provided, according to A23.3 Cl.11.3.6.3,

| A23.3 Eq. 11.9 |

$$\beta = \frac{230}{1000 + d_v}$$ **[6.13]**

$$= \frac{230}{1000 + 580} = 0.14 \text{ (see Table 6.1)}$$

Thus,

| A23.3 Eq. 11.6 |

$$V_c = \phi_c \lambda \beta \sqrt{f_c'}\, b_w d_v$$ **[6.12]**

$$= 0.65 \times 1.0 \times 0.14 \times \sqrt{30 \text{ MPa}} \times 300 \text{ mm} \times 580 \text{ mm}$$

$$= 86.7 \times 10^3 \text{ N} \cong 87 \text{ kN}$$

   ii) In the regions where more than minimum shear reinforcement will be provided (as per A23.3 Cl.11.2.8.2),

$$V_c = 111 \text{ kN (from Step 9)}$$

c) Determine the region(s) where shear reinforcement is required (A23.3 Cl.11.2.8.1). The following regions can be identified depending on the magnitude of the shear force:

- *Region 0:* Shear reinforcement is not required.
  When $V_f < V_c$, shear reinforcement is not required; that is,

  $$V_f < V_c = 87 \text{ kN}$$

  However, because the V_c value is small, the length of Region 0 will be very small, so it is not practical to consider this region in this design.

- *Region 1:* Minimum shear reinforcement is required.
  Region 1 is defined by the shear resistance corresponding to the maximum permitted shear reinforcement spacing prescribed by CSA A23.3. This was determined in Step 9, as follows:

     i) Steel shear resistance:

  $$V_s = 141 \text{ kN}$$

     ii) Concrete shear resistance V_c:
  It is anticipated that a larger than minimum shear reinforcement will be provided, so we can use the larger V_c value, that is,

  $$V_c = 111 \text{ kN}$$

     iii) Factored shear resistance V_r:

| A23.3 Eq. 11.4 |

$$V_{rmin} = V_c + V_s + V_p$$ **[6.11]**

$$= 111 \text{ kN} + 141 \text{ kN} + 0 = 252 \text{ kN}$$

Hence, Region 1 is defined within the range

$$V_f < 252 \text{ kN}$$

- *Region 2:* Full shear reinforcement is required.

  Region 2 requires the provision of shear reinforcement based on the steel shear resistance corresponding to the design shear force ($V_{f@dv}$); that is,

  $$252 \text{ kN} < V_f < V_{f@dv} = 281 \text{ kN} \qquad \text{(support B)}$$

  and

  $$252 \text{ kN} < V_f < V_{f@dv} = 300 \text{ kN} \qquad \text{(support C)}$$

d) Sketch a design shear envelope diagram showing the following regions in the beam:

  - *Region 1,* where the minimum shear reinforcement is required:

    $V_f < 252 \text{ kN}$ (shown shaded on the diagram);

  - *Region 2,* where the full shear reinforcement is required:

    $252 \text{ kN} < V_f < V_{f@dv} = 300 \text{ kN}$ (shown hatched on the diagram)

e) Locate Regions 1 and 2 on the shear envelope diagram.

  This will be done using linear interpolation on the shear envelope diagram that follows.

  i) Find the distance (x_1) from support B, where

    $$V_{f@x_1} = 252 \text{ kN}$$

    We have

    $$V_{f@x_1} = V_{fmidspan} + \left(\frac{\frac{l}{2} - x_1}{\frac{l}{2}} \right) (V_f - V_{fmidspan})$$

    $$252 \text{ kN} = 40 \text{ kN} + \left(\frac{\frac{7.0 \text{ m}}{2} - x_1}{\frac{7.0 \text{ m}}{2}} \right) (350 \text{ kN} - 40 \text{ kN})$$

    $$\frac{252 \text{ kN} - 40 \text{ kN}}{310 \text{ kN}} = \frac{3.5 \text{ m} - x_1}{3.5 \text{ m}}$$

    $$x_1 = 1.1 \text{ m}$$

  ii) Find the distance (x_2) from support C, where

    $$V_{f@x_2} = 252 \text{ kN}$$

    $$V_{f@x_2} = V_{fmidspan} + \left(\frac{\frac{l}{2} - x_2}{\frac{l}{2}} \right) (V_f - V_{fmidspan})$$

    $$252 \text{ kN} = 40 \text{ kN} + \left(\frac{\frac{7.0 \text{ m}}{2} - x_2}{\frac{7.0 \text{ m}}{2}} \right) (370 \text{ kN} - 40 \text{ kN})$$

    $$\frac{252 \text{ kN} - 40 \text{ kN}}{330 \text{ kN}} = \frac{3.5 \text{ m} - x_2}{3.5 \text{ m}}$$

    $$x_2 = 1.25 \text{ m}$$

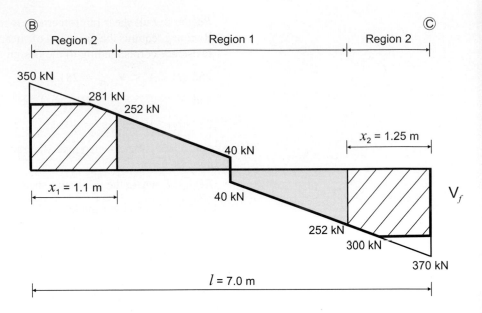

f)   Determine the required spacing of shear reinforcement in Regions 1 and 2 (A23.3 Cl.11.3.5.1).

- Region 1:

    $s_1 = 400$ mm (from Step 9).

- Region 2:

    i)   Determine the required V_s value at support C:

    $$V_s \geq V_f - V_c$$
    $$= 300 \text{ kN} - 111 \text{ kN} = 189 \text{ kN}$$
    $$V_s = 189 \text{ kN}$$

    ii)   Then, find the required stirrup spacing (s_2) from Eqn 6.16:

    $$s_2 = \frac{\phi_s A_v f_y d_v \cot \theta}{V_s}$$
    $$= \frac{0.85 \times 200 \text{ mm}^2 \times 400 \text{ MPa} \times 580 \text{ mm} \times 1.43}{189\,000 \text{ N}}$$
    $$= 298.4 \text{ mm} \cong 300 \text{ mm}$$

g)   Calculate the maximum stirrup spacing according to the CSA A23.3 requirements (A23.3 Cl.11.2.8.2 and Cl.11.3.8.1).
     This check was performed in Step 9. Because $s_{max} = 406$ mm > 300 mm, the spacing corresponding to the required V_s value for Region 2 governs. Therefore, use

     $s_2 = 300$ mm

h)   Check the maximum factored shear resistance (V_r) (A23.3 Cl.11.3.3).
     The factored shear resistance for the beam section at support C is

     $V_r = V_f = 300$ kN

     According to A23.3 Cl.11.3.3, the maximum factored shear resistance ($max V_r$) is

     | A23.3 Eq. 11.5 |

     $$max V_r = 0.25 \phi_c f_c' b_w d_v \qquad \textbf{[6.17]}$$
     $$= 0.25 \times 0.65 \times 30 \text{ MPa} \times 300 \text{ mm} \times 580 \text{ mm}$$
     $$= 848.2 \times 10^3 \text{ N} \cong 848 \text{ kN}$$

Because

$$V_r = 300 \text{ kN} < 848 \text{ kN}$$

the factored shear resistance is acceptable.

i) Sketch a beam elevation showing the distribution of the shear reinforcement. The shear reinforcement design for span 2 is summarized on the sketch below.

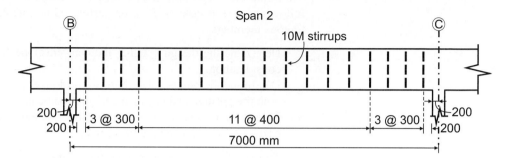

Note that the shear reinforcement arrangement is symmetric, although the design shear forces at supports B and C are different. The shear reinforcement in Region 2 was distributed over the length $x_2 = 1.25$ m.

11. **Check the beam deflections**

First, check the criteria for the CSA A23.3 indirect approach to deflection control prescribed by Cl.9.8.2.1 and summarized in Table A.3.

For span 1 (a beam with one end continuous),

$$\frac{l_n}{18} = \frac{5500 \text{ mm} - 400 \text{ mm}}{18} = 283 \text{ mm}$$

For span 2 (a beam with two continuous ends),

$$\frac{l_n}{21} = \frac{7000 \text{ mm} - 400 \text{ mm}}{21} = 314 \text{ mm}$$

Since the overall beam depth

$$h = 700 \text{ mm}$$

exceeds the minimum required depths determined according to A23.3 Cl.9.8.2.1, detailed deflection calculations are not required.

11.4 | DETAILING OF FLEXURAL REINFORCEMENT IN CONTINUOUS BEAMS AND SLABS

11.4.1 Background

The approach to the flexural design of continuous reinforced concrete members is similar to that for simply supported members outlined in Chapter 9. The amount of flexural reinforcement required at any section along the member is a function of the design bending moment. As the bending moment magnitude varies along the span, the amount of steel reinforcement can be modified or reduced from that required at the location of maximum moment. Flexural reinforcing bars in continuous members can be terminated where no longer required to resist bending moments. The CSA A23.3 provisions related to bar cutoffs are presented in Section 11.4.2, whereas practical recommendations related to the detailing of flexural reinforcement are offered in Section 11.4.3.

11.4.2 Bar Cutoffs in Continuous Members According to the CSA A23.3 Requirements

The calculation of bar cutoff points in continuous reinforced concrete beams and slabs is largely based on the CSA A23.3 provisions presented in Section 9.7 as related to simply supported members. In continuous concrete construction, the designer needs to determine the cutoff points for both positive and negative moment reinforcement. The following general tasks need to be performed to calculate the bar cutoff points in continuous members:

1. Determine the design bending moment envelope for the continuous member under consideration.
2. Determine the points where bars are no longer required to resist flexure considering both the positive and negative bending moments.
3. Extend the bars in positive and negative moment regions according to the applicable CSA A23.3 anchorage requirements.

The key CSA A23.3 anchorage requirements for *positive moment reinforcement* in continuous beams and slabs are outlined in Checklist 11.1 and illustrated in Figure 11.16.

The key CSA A23.3 anchorage requirements for *negative moment reinforcement* in continuous beams and slabs are outlined in Checklist 11.2 and illustrated in Figure 11.17.

The calculation of bar cutoffs for flexural reinforcement in continuous reinforced concrete members will be illustrated with Example 11.3.

Checklist 11.1 CSA A23.3 Bar Cutoff Requirements for Positive Moment Reinforcement in Continuous Beams and Slabs

Rule	Description	Code Clause
P-1	Actual point of cutoff (Section 9.7.3) Cutoff bars must be extended beyond the theoretical point of cutoff for a distance of $1.3d$ or h, whichever is greater.	12.10.3 11.3.9.1
P-2	General anchorage requirement (Section 9.7.2) Each bar must be embedded by a length greater than l_d from the point of maximum bar stress.	12.1.1
P-3	Continuing reinforcement (Section 9.7.4) The continuing reinforcement should have an embedment length not less than the greater of $(l_d + d)$ and $(l_d + 12d_b)$ beyond the theoretical point of cutoff for the bars being terminated.	12.10.4
P-4	Supports (Section 9.7.5) At least one quarter of the positive moment reinforcement should extend into the supports by at least 150 mm.	12.11.1
P-5	Inflection points (Section 9.7.8) The bar development length at the inflection points should be limited as follows (see Figure 9.18): $\boxed{\text{A23.3 Eq. 12.6}}\qquad l_d \le \dfrac{M_r}{V_f} + l_a \qquad\qquad \textbf{[9.8]}$ where l_a is the embedment length beyond the inflection point, limited to d or $12d_b$, whichever is greater (see Figure 9.18a in Chapter 9).	12.11.3

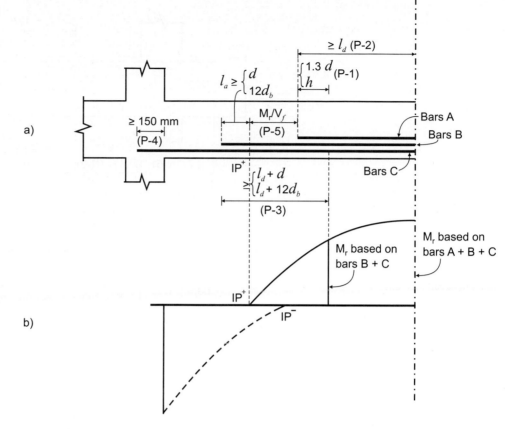

Figure 11.16 Bar cutoff requirements for positive moment reinforcement in continuous reinforced concrete flexural members: a) beam elevation and reinforcement arrangement; b) moment envelope.

Checklist 11.2 CSA A23.3 Bar Cutoff Requirements for Negative Moment Reinforcement in Continuous Beams and Slabs

Rule	Description	Code Clause
N-1	Actual point of cutoff (Section 9.7.3) Cutoff bars must be extended beyond the theoretical point of cutoff for a distance of $1.3d$ or h, whichever is greater.	12.10.3 11.3.9.1
N-2	General anchorage requirement (Section 9.7.2) Each bar must be embedded by a length greater than l_d from the point of maximum bar stress.	12.1.1
N-3	Continuing reinforcement (Section 9.7.4) The continuing reinforcement should have an embedment length not less than the greater of $(l_d + d)$ and $(l_d + 12d_b)$ beyond the theoretical point of cutoff for the bars being terminated.	12.10.4
N-4	Inflection points (Section 9.7.7) At least one third of the total negative moment reinforcement must be extended past the negative inflection point (IP $^-$) by the greatest of • d • $12\,d_b$ and • $l_n/16$	12.12.2
N-5	Anchorage into supporting members (Section 9.7.6) The negative moment reinforcement should be anchored in, or through, the supporting member by an embedment length l_d (for straight bars).	12.12.1

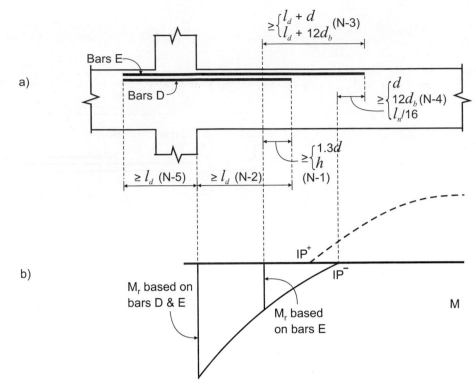

Figure 11.17 Bar cutoff requirements for negative moment reinforcement in continuous reinforced concrete flexural members: a) beam elevation and reinforcement arrangement; b) moment envelope.

Example 11.3

A four-span continuous reinforced concrete beam is shown in the figure that follows. The beam is loaded by a factored uniform dead load of $w_{DLf} = 20$ kN/m and a factored uniform live load of $w_{LLf} = 20$ kN/m. The moment envelope and the reinforcement for this beam are also shown in the figure. Regular uncoated reinforcing bars have been used, and the shear reinforcement (stirrups) is in excess of the CSA A23.3 minimum requirement. A 40 mm side cover to the reinforcement is provided at the end supports.

Determine the bar cutoffs for flexural reinforcement in this beam according to the CSA A23.3 requirements. Apply 50% bar cutoffs wherever feasible. Present the results on a sketch showing the bar layout.

Given: $f_c' = 25$ MPa
$f_y = 400$ MPa

SOLUTION: The objective of this example is to determine the lengths and cutoffs for the flexural reinforcement in the beam. The CSA A23.3 anchorage requirements from Checklists 11.1 and 11.2 will be referred to as required.

This beam is symmetric with regard to its vertical axis, as shown in the figure that follows. In this case, the detailing needs to be performed for spans 1 and 2 only. Span 3 is a mirror image of span 2, while span 4 is a mirror image of span 1.

First, examine the moment envelope for span 2 in the sketch that follows. The three sections along the span are characterized by the peak bending moments: the sections at the face of each support and the section at the midspan. The flexural reinforcement for this beam (4-20M bars top and bottom) was designed based on the bending moments at these sections. The bending moments at supports B and C will be used to detail the top reinforcement in the support regions.

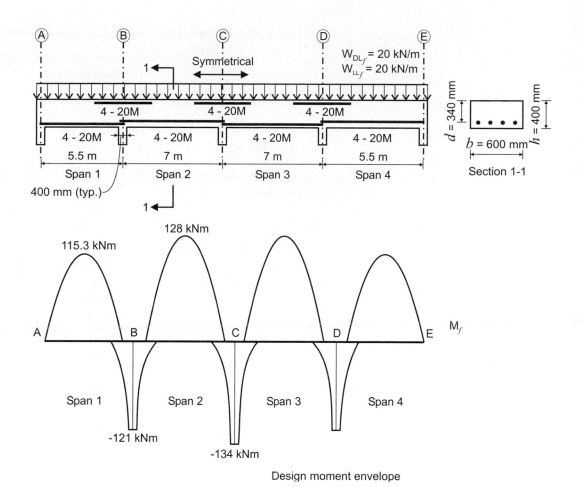

Design moment envelope

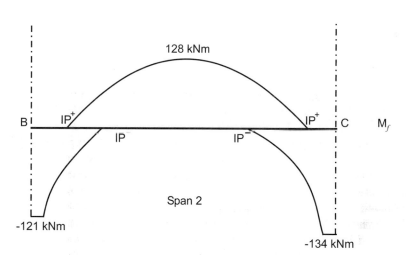

1. **Determine the bar development lengths according to the CSA A23.3 requirements**
   Both the positive and the negative moment reinforcement will be cut off in this example, so the corresponding development lengths need to be determined.

   a) Determine the development length (l_d) for the negative moment reinforcement (top bars) according to A23.3 Cl.12.2.3. Because the shear reinforcement exceeds the minimum amount required per CSA A23.3, this design can be classified as a Case 1 application. The l_d value can be determined using a design aid, that is, Table A.5 in Appendix A.

i)   Establish the values of the modification factors according to A23.3 Cl.12.2.4:

$k_1 = 1.3$ (top bars with more than 300 mm of concrete beneath)
$k_2 = 1.0$ (regular uncoated reinforcing bars)
$k_3 = 1.0$ (normal-density concrete)
$k_4 = 0.8$ (20M bars)

ii) Determine l_d from Table A.5 as

$l_d = 750$ mm

b) Determine the development length (l_d) for the positive moment reinforcement (bottom bars).

i)   Establish the values of the modification factors according to A23.3 Cl.12.2.4:

$k_1 = 1.0$ (bottom bars)
$k_2 = 1.0$ (regular uncoated reinforcing bars)
$k_3 = 1.0$ (normal-density concrete)
$k_4 = 0.8$ (20M bars)

ii) Determine l_d from Table A.5 as

$l_d = 575$ mm

2. **Determine the bar cutoffs for the negative moment reinforcement at support B (span 2)**

a) Apply Rule N-4 (Checklist 11.2): inflection points (A23.3 Cl.12.12.2).
The top reinforcement is theoretically not required beyond the negative moment inflection point. The inflection point for the negative bending moment at support B is located at $x_1 = 1.8$ m, as shown in Figure 11.18a.

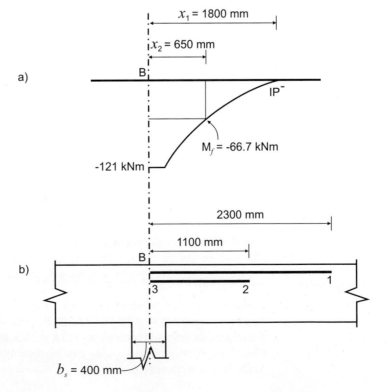

Figure 11.18 Negative moment reinforcement at support B (span 2): a) bending moment diagram; b) reinforcement arrangement.

Note that

$l_2 = 7000$ mm

is the span length for span 2 and

$b_s = 400$ mm

is the support width (given).

Find the clear span:

$$l_{n2} = l_2 - \left(\frac{b_s}{2} + \frac{b_s}{2}\right) = 7000 \text{ mm} - \left(\frac{400 \text{ mm}}{2} + \frac{400 \text{ mm}}{2}\right) = 6600 \text{ mm}$$

The rebar must extend beyond the inflection point by the greater of

- $d = 340$ mm (given),
- $12\, d_b = 12 \times 20$ mm $= 240$ mm (where $d_b = 20$ mm is the bar diameter for 20M bars), and
- $l_n/16 = 6600$ mm$/16 = 413$ mm

In this case, $l_n/16 = 413$ mm governs and will be used for further calculations.

Hence, the length of the negative moment reinforcement extending beyond point B is (see Figure 11.18b)

$$l_{1\text{-}3} = x_1 + \frac{l_n}{16} = 1800 \text{ mm} + 413 \text{ mm} = 2213 \text{ mm} \cong 2300 \text{ mm}$$

According to CSA A23.3 Cl.12.12.2, one-third of the total reinforcement is required to extend past the inflection point; in this case, there are 4-20M bars in total, so 2-20M bars can be terminated. It is common practice to terminate 50% of the reinforcing bars where no longer needed and extend the remaining rebars past the inflection point.

b) Determine the bar cutoffs for the negative moment reinforcement.

i) Determine the moment resistance for 2-20M bars (M_{r2}).
The area of a 20M bar is equal to 300 mm^2 (refer to Table A.1), so

$A_s = 2 \times 300$ mm$^2 = 600$ mm^2

Therefore,

$$T_r = \phi_s f_y A_s \hspace{2cm} \text{[3.9]}$$
$$= 0.85 \times 400 \text{ MPa} \times 600 \text{ mm}^2 = 204 \times 10^3 \text{ N} = 204 \text{ kN}$$

Use the equation of equilibrium

$$C_r = T_r \hspace{2cm} \text{[3.10]}$$

where

$$C_r = \alpha_1 \phi_c f_c' \, a \, b \hspace{2cm} \text{[3.8]}$$

is the compression force in concrete (C_r) and

$$\alpha_1 \cong 0.8 \hspace{2cm} \text{[3.7]}$$

Calculate the depth of the rectangular stress block (a) from Eqn (3.10)

$$a = \frac{T_r}{\alpha_1 \phi_c f_c' \, b} = \frac{204 \times 10^3 \text{ N}}{0.8 \times 0.65 \times 25 \text{ MPa} \times 600 \text{ mm}} \cong 26 \text{ mm}$$

Calculate the moment resistance as

$$M_{r2} = T_r \left(d - \frac{a}{2}\right) \hspace{2cm} \text{[3.13]}$$

$$= (204 \times 10^3 \text{ N})\left(340 \text{ mm} - \frac{26 \text{ mm}}{2}\right) - 66.7 \times 10^6 \text{ N} \cdot \text{mm}$$

Therefore, the factored moment resistance for a section with 2-20M bars is

$$M_{r2} = 66.7 \text{ kN} \cdot \text{m}$$

ii) Confirm that the CSA A23.3 minimum reinforcement requirement is satisfied. Determine the minimum required reinforcement area (A_{smin}) based on A23.3 Cl.10.5.1.2 as

A23.3 Eq. 10.4

$$A_{smin} = \frac{0.2\sqrt{f_c'}}{f_y} b_t h \qquad [5.7]$$

$$= \frac{0.2\left(\sqrt{25 \text{ MPa}}\right)}{400 \text{ MPa}} \times 600 \text{ mm} \times 400 \text{ mm} = 600 \text{ mm}^2$$

where $b_t = b = 600$ mm is the width of the tension zone.

Next confirm that $A_s \geq A_{smin}$. Since

$$A_s = A_{smin} = 600 \text{ mm}^2$$

the section with 2-20M bars satisfies the minimum reinforcement requirements.

iii) Determine the point on the moment envelope diagram where 50% of the bars can be cut off (the theoretical point of cutoff).

From the bending moment diagram in Figure 11.18a, determine the distance from point B where $M_f = M_{r2} = 66.7$ kN·m as

$$x_2 = 0.65 \text{ m} = 650 \text{ mm}$$

iv) Continuing bars (1-3) (Figure 11.18b):

Apply Rule N-3 (Checklist 11.2): continuing reinforcement (A23.3 Cl.12.10.4).

The development length for the negative reinforcement was determined in Step 1 as

$$l_d = 750 \text{ mm}$$

According to A23.3 Cl.12.10.4, the continuing bars need to be extended beyond the theoretical cutoff point by the greater of

- $l_d + d = 750 \text{ mm} + 340 \text{ mm} = 1090 \text{ mm}$
- $l_d + 12d_b = 750 \text{ mm} + 12 \times 20 \text{ mm} = 990 \text{ mm}$

The larger value (1090 mm) governs. Finally, check whether the required bar extension length is less than the available length of 2300 mm, as follows:

$$l_{1-3} = x_2 + (l_d + d) = 650 + 1090 = 1740 \text{ mm} < 2300 \text{ mm}$$

Hence, the available length of 2300 mm is satisfactory.

v) Terminating bars (2-3):

Apply Rule N-1 (Checklist 11.2): actual cutoff point (A23.3 Cl.12.10.3 and Cl.11.3.9.1).

According to A23.3 Cl.11.3.9.1, the terminating bars need to be extended beyond point 2 by the distance $1.3d$ or h, whichever is greater. In this case,

- $h = 400$ mm and
- $1.3d = 1.3 \times 340 \text{ mm} = 442 \text{ mm}$ (governs)

Based on this rule, the total length of the terminating bars beyond point B needs to be

$$l_{2-3} = x_2 + 1.3d = 650 \text{ mm} + 442 \text{ mm} = 1092 \text{ mm} \cong 1100 \text{ mm}$$

However, according to Rule N-2 of Checklist 11.2 (general anchorage requirement, A23.3 Cl.12.1.1), the bar length from the point of maximum stress (point B) to the actual cutoff point needs to be at least equal to the development length (l_d). Because

$$l_{2-3} = 1100 \text{ mm} > 750 \text{ mm}$$

the terminating bars (2-3) need to be extended by 1100 mm beyond point B.

The diagram showing the final length for the negative moment reinforcement at support B is presented in Figure 11.18b.

3. **Determine the bar cutoffs for the negative moment reinforcement at support C (span 2)**

The procedure that will be followed in this step is identical to that performed in Step 2.

a) Apply Rule N-4 (Checklist 11.2): inflection points (A23.3 Cl.12.12.2).
The top reinforcement is theoretically not required beyond the negative moment inflection point. The inflection point for the negative bending moment at support C is located at

$$x_1 = 2.0 \text{ m}$$

(as shown in Figure 11.19a).
The clear span is

$$l_{n2} = 6600 \text{ mm}$$

(from Step 2).
The rebar must extend beyond the inflection point by the greater of

- d
- $12 \, d_b$, and
- $l_n/16$

In this case, $l_n/16 = 413$ mm governs and will be used for further calculations (same as Step 2).
Hence, the length of the negative moment reinforcement extending beyond support C is (see Figure 11.19b)

$$l_{1-3} = x_1 + \frac{l_n}{16} = 2000 \text{ mm} + 413 \text{ mm} = 2413 \text{ mm} \cong 2400 \text{ mm}$$

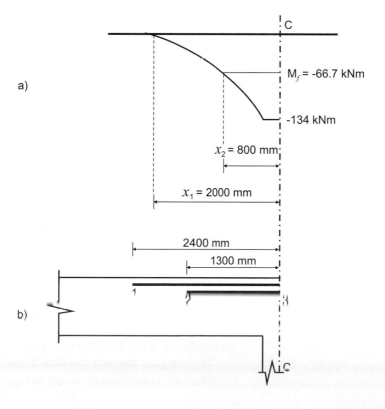

Figure 11.19 Negative moment reinforcement at support C (span 2): a) bending moment diagram; b) reinforcement arrangement.

b) Determine the bar cutoffs for the negative moment reinforcement (apply 50% cutoffs if possible).

In this case,

$$A_s = 2 \times 300 \text{ mm}^2 = 600 \text{ mm}^2$$

and

$$M_{r2} = 66.7 \text{ kN} \cdot \text{m}$$

(same as for support B).

i) Determine the point on the moment envelope diagram where 50% of the bars can be cut off (the theoretical point of cutoff).

From the bending moment diagram in Figure 11.19a, determine the distance from support C where

$$M_{r2} = M_f = 66.7 \text{ kN} \cdot \text{m}$$

that is,

$$x_2 = 0.8 \text{ m} = 800 \text{ mm}$$

ii) Continuing bars (1-3) (Figure 11.19b):

Apply Rule N-3 (Checklist 11.2): continuing reinforcement (A23.3 Cl.12.10.4).

The development length for the negative reinforcement was determined in Step 1 as

$$l_d = 750 \text{ mm}$$

According to A23.3 Cl.12.10.4, the continuing bars need to be extended beyond the theoretical cutoff point by the greater of

- $l_d + d = 750 \text{ mm} + 340 \text{ mm} = 1090 \text{ mm}$ and
- $l_d + 12d_b = 750 \text{ mm} + 12 \times 20 \text{ mm} = 990 \text{ mm}$

The larger value (1090 mm) governs. Finally, check whether the required bar extension length is less than the available length of 2400 mm, as follows:

$$l_{1-3} = x_2 + (l_d + d) = 800 \text{ mm} + 1090 \text{ mm} = 1890 \text{ mm} < 2400 \text{ mm}$$

iii) Terminating bars (2-3):

Apply Rule N-1 (Checklist 11.2): actual cutoff point (A23.3 Cl.12.10.3 and Cl.11.3.9.1).

According to A23.3 Cl.11.3.9.1, the terminating bars need to be extended beyond point 2 by the distance $1.3d$ or h, whichever is greater. In this case,

- $h = 400 \text{ mm}$ and
- $1.3d = 1.3 \times 340 \text{ mm} = 442 \text{ mm}$

Based on this rule, the total length of the terminating bars beyond support C is

$$l_{2-3} = x_2 + 1.3d = 800 \text{ mm} + 442 \text{ mm} = 1242 \text{ mm} \cong 1300 \text{ mm}$$

However, according to Rule N-2 of Checklist 11.2 (general anchorage requirement, A23.3 Cl.12.1.1), the bar length from the point of maximum stress (support C) to the actual cutoff point needs to be at least equal to the development length (l_d). Because

$$l_{2-3} = 1300 \text{ mm} > 750 \text{ mm}$$

the terminating bars (2-3) need to be extended by 1300 mm beyond support C.

The diagram showing the final length for the negative moment reinforcement at support C is presented in Figure 11.19b.

4. **Determine the bar cutoffs for the positive moment reinforcement in span 2**
   The key bar cutoff rules (CSA A23.3 provisions) for positive moment reinforcement are summarized in Checklist 11.1 and will be referred to here.
   First, detail the rebars on the left-hand side of the midspan.

   a) Terminating bars (1-2):

   i) Check Rule P-1 (Checklist 11.1): actual cutoff point (A23.3 Cl.11.3.9.1).
   Bars 1-2 are no longer required to resist flexure at the point where the moment resistance is $M_f = M_{r2} = 66.7$ kN·m (the theoretical point of cutoff).
   From the bending moment diagram in Figure 11.20a, determine the distance from support B where

   $$M_{r2} = M_f = 66.7 \text{ kN·m}$$

   that is,

   $$x_1 = 1.9 \text{ m} = 1900 \text{ mm}$$

   Because

   $$l_2 = 7000 \text{ mm (span length)}$$

   the distance from the midspan to the theoretical cutoff point is

   $$\frac{l_2}{2} - x_1 = \frac{7000 \text{ mm}}{2} - 1900 \text{ mm} = 1600 \text{ mm}$$

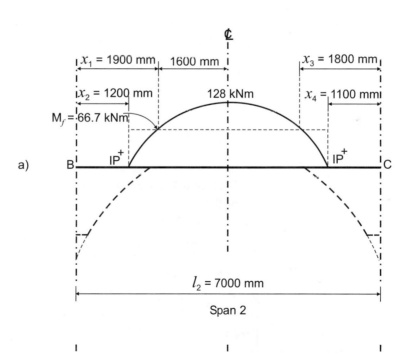

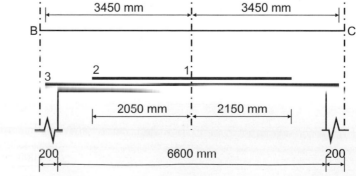

Figure 11.20 Positive moment reinforcement in span 2:
a) bending moment diagram;
b) reinforcement arrangement.

However, according to A23.3 Cl.11.3.9.1, the bars need to be extended beyond the theoretical point of cutoff by the distance $1.3d$ or h, whichever is greater. In this case,

- $h = 400$ mm   and
- $1.3d = 1.3 \times 340$ mm $= 442$ mm

Hence, the larger value (442 mm) governs and the actual cutoff length is (see Figure 11.20b)

$$l_{1\text{-}2} = 1600 \text{ mm} + 442 \text{ mm} = 2042 \text{ mm} \cong 2050 \text{ mm}$$

ii) Check Rule P-2 (Checklist 11.1): general anchorage requirement (A23.3 Cl.12.1.1). According to the general anchorage requirement of A23.3 Cl.12.1.1, the bar length from the point of maximum stress (midspan) to the actual cutoff point needs to be at least equal to the development length (l_d). Because

$$l_{1\text{-}2} = 2050 \text{ mm} > 575 \text{ mm}$$

the terminating bars (1-2) can be extended by 2050 mm beyond the midspan.

b) Continuing bars (1-3):

i) Check Rule P-3 (Checklist 11.1): continuing reinforcement (A23.3 Cl.12.10.4). According to Cl.12.10.4, the continuing bars need to be extended beyond the theoretical cutoff point by the greater of

- $l_d + d = 575$ mm $+ 340$ mm $= 915$ mm
- $l_d + 12d_b = 575$ mm $+ 12 \times 20$ mm $= 815$ mm

The larger value (915 mm) governs. Because the continuing bars will extend into the support (as shown in Figure 11.20b), this requirement is satisfied by default.

ii) Check Rule P-4 (Checklist 11.1): supports (A23.3 Cl.12.11.1).
At least one quarter of the tension reinforcement should extend into the supports by at least 150 mm.

    In this case, two (out of four) bars extend into the support and so this requirement is satisfied.

iii) Check Rule P-5 (Checklist 11.1): inflection point locations (A23.3 Cl.12.11.3). The factored moment resistance based on the 2-20M bars continuing into the support is

$$M_{r2} = 66.7 \text{ kN} \cdot \text{m}$$

The corresponding shear envelope diagram for span 2 is shown on the sketch below. The factored shear forces at the positive inflection points are shown on the diagram.

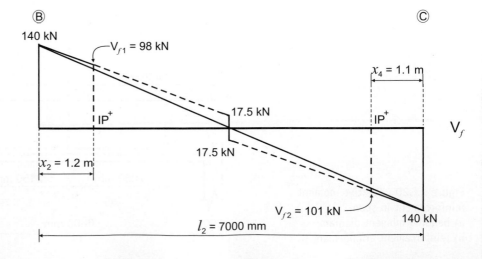

Note that the factored shear forces at the inflection points are

$$V_{f1} = 98 \text{ kN}$$

and

$$V_{f2} = 101 \text{ kN}$$

Hence, the factored shear force can be taken as

$$V_f \cong 100 \text{ kN}$$

at both inflection points.

Next, find l_a, that is, the embedment length beyond the inflection point, which is equal to the greater of

- $d = 340$ mm and
- $12d_b = 12 \times 20$ mm $= 240$ mm

Hence,

$$l_a = 340 \text{ mm}$$

Finally, check the following requirement:

| A23.3 Eq. 12.6 |

$$l_d \le \frac{M_r}{V_f} + l_a \qquad\qquad \textbf{[9.8]}$$

$$= \frac{66.7 \times 10^3 \text{ kN} \cdot \text{mm}}{100 \text{ kN}} + 340 \text{ mm} = 1007 \text{ mm}$$

Since

$$l_d = 575 \text{ mm} < 1007 \text{ mm}$$

this requirement is satisfied.

Next, detail the rebars on the right-hand side of the midspan. The procedure is identical to that used to detail the reinforcement on the left-hand side of the midspan performed above, so most checks will be omitted. The reader may perform these checks as a homework assignment.

c) Terminating bars (1-2):

Check Rule P-1 (Checklist 11.1): actual cutoff point (A23.3 Cl.11.3.9.1). Terminating bars (2-20M) are no longer required to resist flexure at the point where the moment resistance is $M_f = M_{r2} = 66.7 \text{ kN} \cdot \text{m}$ (the theoretical point of cutoff).

From the bending moment diagram in Figure 11.20a, determine the distance from support C where

$$M_{r2} = M_f = 66.7 \text{ kNm}$$

that is,

$$x_3 = 1.8 \text{ m} = 1800 \text{ mm}$$

Because

$$l_2 = 7000 \text{ mm (span length)}$$

determine the distance from the midspan to the theoretical cutoff point as

$$\frac{l_2}{2} - x_3 = \frac{7000 \text{ mm}}{2} - 1800 \text{ mm} = 1700 \text{ mm}$$

However, according to A23.3 Cl.11.3.9.1, the bars need to be extended beyond the theoretical point of cutoff by the distance 1.3d or h, whichever is greater. In this case,

- $h = 400$ mm, and
- $1.3d = 1.3 \times 340$ mm $= 442$ mm

Hence, the larger value (442 mm) governs and the actual cutoff length is (see Figure 11.20b)

$l_{1\text{-}2} = 1700$ mm $+ 442$ mm $= 2142$ mm $\cong 2150$ mm

Also, check Rule P-2 (Checklist 11.1): general anchorage requirement (A23.3 Cl.12.1.1).

d) Continuing bars (1-3):
Check the following rules:
  i) Rule P-3: continuing reinforcement (A23.3 Cl.12.10.4)
  ii) Rule P-4: supports (A23.3 Cl.12.11.1)
  iii) Rule P-5: inflection point locations (A23.3 Cl.12.11.3)

The diagram showing the final lengths for the positive moment reinforcement in span 2 is presented in Figure 11.20b.

5. Determine the bar cutoffs for the negative moment reinforcement at support B (span 1)
The procedure that will be followed in this step is identical to that performed in Step 2.

a) Apply Rule N-4 (Checklist 11.2): inflection points (A23.3 Cl.12.12.2).
The inflection point for the negative bending moment at the left of support B is located at $x_1 = 2.0$ m, as shown in Figure 11.21a.
Since

$l_1 = 5500$ mm

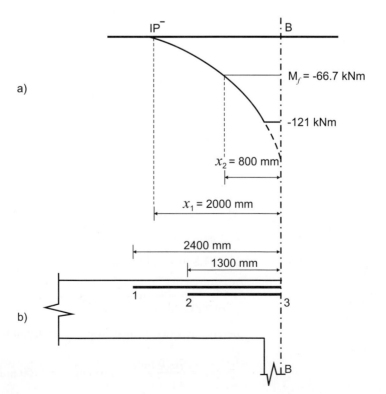

Figure 11.21 Negative moment reinforcement at support B (span 1): a) bending moment diagram; b) reinforcement arrangement.

is the span length for span 1, and

$$b_s = 400 \text{ mm}$$

is the support width (given), the clear span is

$$l_{n1} = l_1 - \left(\frac{b_s}{2} + \frac{b_s}{2} \right) = 5500 \text{ mm} - \left(\frac{400 \text{ mm}}{2} + \frac{400 \text{ mm}}{2} \right) = 5100 \text{ mm}$$

The rebar must extend beyond the inflection point by the greater of

- $d = 340 \text{ mm}$
- $12d_b = 12 \times 20 \text{ mm} = 240 \text{ mm}$

(where $d_b = 20 \text{ mm}$ is the bar diameter for 20M bars)

- $l_n/16 = 5100 \text{ mm}/16 = 319 \text{ mm}$

In this case, $d = 340 \text{ mm}$ governs and will be used in further calculations.
Hence, the length of the negative moment reinforcement extending beyond support B is (see Figure 11.21b)

$$l_{1\text{-}3} = x_1 + d = 2000 \text{ mm} + 340 \text{ mm} = 2340 \text{ mm} \cong 2400 \text{ mm}$$

b) Determine the bar cutoffs for the negative moment reinforcement.
In this case, the 50% bar cutoffs will be applied as well. For 2-20M bars,

$$A_s = 2 \times 300 \text{ mm}^2 = 600 \text{ mm}^2$$

and the corresponding factored moment resistance is

$$M_{r2} = 66.7 \text{ kN} \cdot \text{m}$$

   i) Determine the point on the moment envelope diagram where 50% of the bars can be cut off (the theoretical point of cutoff).
From the bending moment diagram in Figure 11.21a, determine the distance from support B where

$$M_{r2} = M_f = 66.7 \text{ kN} \cdot \text{m}$$

as

$$x_2 = 0.8 \text{ m} = 800 \text{ mm}$$

  ii) Continuing bars (1-3) (Figure 11.21b):
Apply Rule N-3 (Checklist 11.2): continuing reinforcement (A23.3 Cl.12.10.4).
The development length for the negative reinforcement was determined in Step 1 as

$$l_d = 750 \text{ mm}$$

According to A23.3 Cl.12.10.4, the continuing bars need to be extended beyond the theoretical cutoff point by the greater of

- $l_d + d = 750 \text{ mm} + 340 \text{ mm} = 1090 \text{ mm}$ and
- $l_d + 12d_b = 750 \text{ mm} + 12 \times 20 \text{ mm} = 990 \text{ mm}$

The higher value (1090 mm) governs. Finally, check whether the required bar extension length is less than the available length of 2400 mm:

$$l_{1\text{-}3} = x_2 + (l_d + d) = 800 \text{ mm} + 1090 \text{ mm}$$
$$= 1890 \text{ mm} < 2400 \text{ mm}$$

iii) Terminating bars (2-3):

Apply Rule N-1 (Checklist 11.2): actual cutoff point (A23.3 Cl.12.10.3 and Cl.11.3.9.1).

According to A23.3 Cl.11.3.9.1, the terminating bars need to be extended beyond point 2 by the distance $1.3d$ or h, whichever is greater. In this case,

- $h = 400$ mm, and
- $1.3d = 1.3 \times 340$ mm $= 442$ mm

Hence, 442 mm governs. Based on this rule, the total length of the terminating bars beyond support B is

$$l_{2\text{-}3} = x_2 + 1.3d = 800 \text{ mm} + 442 \text{ mm} = 1242 \text{ mm} \cong 1300 \text{ mm}$$

However, according to Rule N-2 of Checklist 11.2 (general anchorage requirement, A23.3 Cl.12.1.1), the bar length from the point of maximum stress (support B) to the actual cutoff point needs to be at least equal to the development length (l_d). Since

$$l_{2\text{-}3} = 1300 \text{ mm} > 750 \text{ mm}$$

the terminating bars (2-3) can be extended by 1300 mm beyond support B. The diagram showing the length of the negative moment reinforcement at support B is presented in Figure 11.21b.

6. **Determine the bar cutoffs for the positive moment reinforcement in span 1**

There is no need to cut off the positive moment reinforcement on the left-hand side of the midspan. In general, only a marginal savings can be made when the reinforcement at the end spans is terminated, because the bending moment is either equal to zero or is very small. Hence, it is proposed to extend the entire reinforcement into support A.

As a result, most detailing checks are not required in this case, except for Rule P-5: inflection point locations (A23.3 Cl.12.11.3). However, based on the check performed for span 2 (see Step 4), it can be concluded that the factored shear forces are small and hence this requirement is satisfied.

Next, detail the rebars on the right-hand side of the midspan. The procedure is identical to that used to detail the reinforcement performed in Step 4, so most checks will be omitted. The reader may perform these checks as a homework assignment.

a) Terminating bars (1-2):

Check Rule P-1 (Checklist 11.1): actual cutoff point (A23.3 Cl.11.3.9.1).

Terminating bars (2-20M) are no longer required to resist flexure at the point where the moment resistance is $M_f = M_{r2} = 66.7$ kN·m (the theoretical point of cutoff).

From the bending moment diagram in Figure 11.22a, determine the distance from support C where

$$M_{r2} = M_f = 66.7 \text{ kN·m}$$

that is,

$$x_2 = 2.1 \text{ m} = 2100 \text{ mm}$$

Because

$$l_1 = 5500 \text{ mm (span length)}$$

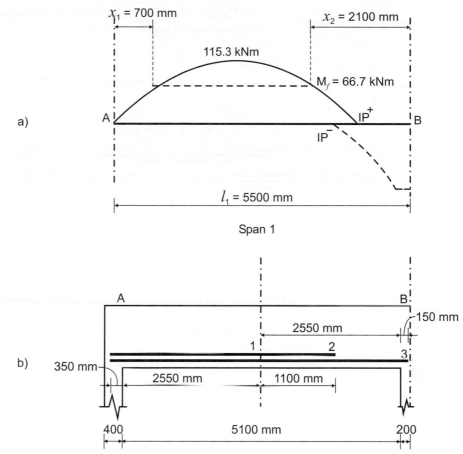

Figure 11.22 Positive moment reinforcement in span 1:
a) bending moment diagram;
b) reinforcement arrangement.

the distance from the midspan to the theoretical cutoff point is

$$\frac{l_2}{2} - x_2 = \frac{5500 \text{ mm}}{2} - 2100 \text{ mm} = 650 \text{ mm}$$

However, according to A23.3 Cl.11.3.9.1, the bars need to be extended beyond the theoretical point of cutoff by the distance $1.3d$ or h, whichever is greater. In this case,

- $h = 400$ mm and
- $1.3d = 1.3 \times 340 \text{ mm} = 442 \text{ mm}$

Hence, the larger value (442 mm) governs and the actual cutoff length is (see Figure 11.22b)

$$l_{1\text{-}2} = 650 \text{ mm} + 442 \text{ mm} = 1092 \text{ mm} \cong 1100 \text{ mm}$$

In addition to this, Check Rule P-2 (Checklist 11.1): general anchorage requirement (A23.3 Cl.12.1.1).

b) Continuing bars (1-3):
Check the following rules:

   i) Rule P-3: continuing reinforcement (A23.3 Cl.12.10.4)

   ii) Rule P-4: supports (A23.3 Cl.12.11.1)

   iii) Rule P-5: inflection point locations (A23.3 Cl.12.11.3)

The diagram showing the final length for the positive moment reinforcement in span 1 is presented in Figure 11.22b.

The reinforcement arrangement for the entire beam based on the above discussion is shown below.

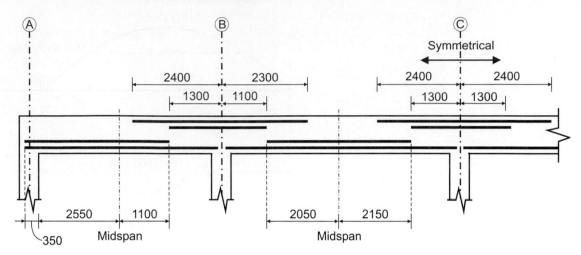

To simplify bar placement in the field, the top reinforcement should be extended over support B to a total length of 3.7 m; this will result in equal lengths for the top reinforcing bars at all supports. The bars will be placed in a *staggered* fashion (or STAG in the abbreviated form), in which alternating bars are extended by a longer distance. A plan view of a typical staggered arrangement for the negative moment reinforcement at the supports is shown below.

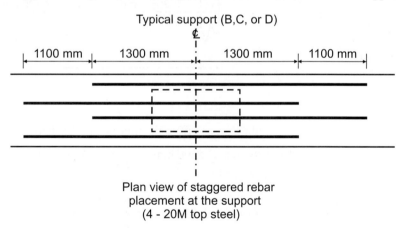

Plan view of staggered rebar
placement at the support
(4 - 20M top steel)

The positive moment reinforcement (bottom bars) can also be arranged in a staggered fashion. A plan view of the bottom reinforcement in span 1 is shown below.

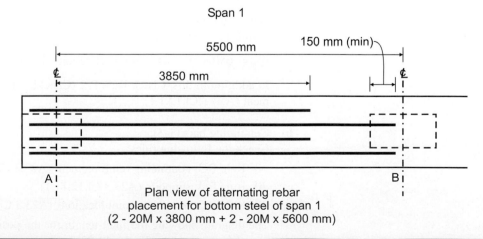

Plan view of alternating rebar
placement for bottom steel of span 1
(2 - 20M x 3800 mm + 2 - 20M x 5600 mm)

Note that CSA A23.3 (Cl.12.11.1) requires the bottom reinforcement to extend into the support by a minimum of 150 mm. In practice, the designer may wish to extend the bars into the far end of the support.

Finally, the reinforcement arrangement for the entire beam in the staggered fashion is shown in Figure 11.23. Note that the top reinforcing over columns B and D is longer than required by design; this is done to produce a staggered rebar arrangement symmetrical about each column that matches the top reinforcement over column C.

As another example of this approach, the reinforcing scheme for the slab designed in Example 11.1 in the staggered fashion is presented in Figure 11.24, while the reinforcing scheme for the beam designed in Example 11.2 in the staggered fashion is presented in Figure 11.25.

11.4.3 Practical Detailing Considerations for Flexural Reinforcement in Continuous Beams and Slabs

In design practice, the optimal usage of reinforcement is an important criterion in the design of reinforced concrete structures, but it is equally important for the design to take into consideration the ease and speed of construction. In general, an optimal design solution should take into consideration the entire construction process.

Experience has shown that the optimal approach to the detailing of flexural reinforcement in commercial, residential, and light industrial building construction involves 50% bar cutoffs in beams and slabs. Although other rebar cutoff ratios may prove to use less steel, the additional labour required to deal with a larger number of rebar arrangements and ensure

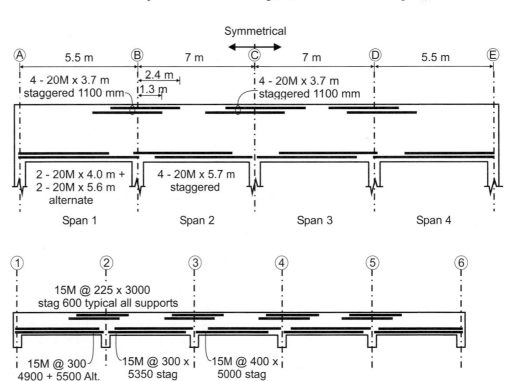

Figure 11.23 Staggered reinforcing arrangement for the beam designed in Example 11.3.

Figure 11.24 Staggered reinforcing arrangement for the slab designed in Example 11.1.

Figure 11.25 Staggered reinforcing arrangement for the beam designed in Example 11.2.

that every bar is placed to the designer's requirements, may be more expensive. Rebars are usually staggered in order to accomplish 50% cutoffs, as explained in Example 11.3.

In general, the simpler the design solution, the easier the construction process, with the benefits of faster construction and lower overall construction cost.

The following considerations may increase construction productivity and optimize the use of construction materials.

Repetition   The designer should try to prescribe reinforcement of similar size, amount, and lengths for different spans in a continuous beam or slab. This approach leads to less complexity in the field by facilitating the task of a construction worker who prefers a specification with a repetitive reinforcement arrangement. The additional benefits of this approach include reduced inspection time for the engineer and faster delivery of the completed structure to the owner due to increased overall efficiency.

Rounded rebar lengths   In Canada, most reinforcing bar sizes are produced in 18 m (60 feet) stock lengths. To suit the designer's requirements, the rebar supplier procures rebars in 18 m lengths and cuts them to the required lengths. The designer should take into consideration that most rebar lengths are cut from 18 m stock lengths and should try to reduce unnecessary scrap.

For example, consider a design that requires a large number of 5.8 m long rebars for various spans. In that case, the supplier would take the 18 m stock length and cut the rebar into three 5.8 m long segments; this will result in 0.6 m of leftover length that cannot be used for construction. In reality, the project has paid for three 6 m rebar lengths from an 18 m stock length. The designer should take this into consideration and prescribe the use of 6.0 m long rebars for this design.

Based on 18 m stock lengths, common rebar lengths that should be considered for the design are 18 m, 9 m, 7.2 m, 6 m, 4.5 m, 3.6 m, 3 m, 2.5 m, 2.25 m, 2 m, and so on. A stock length strategy should be employed by using long and short bars placed alternately in a span. For example, alternate bar pairs of 8 m and 10 m, 4 m and 5 m, 6 m and 7.2 m, and so on, can be used.

It should be noted that 12 m (40 feet) stock lengths are also available for some bar sizes, corresponding to bar pair lengths of 5 m and 7 m or 5.5 m and 6.5 m, or sets of three 4 m long bars.

Staggered bars   The designer should try to use staggered bars with a consistent amount of stagger throughout a project. For example, a consistent stagger of 600 mm for slabs and 1200 mm for beams should be maintained throughout a project. When different staggers must be used, the designer should try to vary the stagger in 300 mm increments.

When staggering the top reinforcement, the bars should be placed symmetrically with regard to the column centroidal axis. If unequal rebar lengths must be used on each side of the column, the designer should ensure that the rebar positions are clearly detailed on the drawings so that the potential of confusing the rebar placer at the construction site is minimized.

The approach to prescribing rebar lengths so that the use of stock lengths is maximized makes effective use of the reinforcement. CSA A23.3 prescribes minimum rebar lengths and bar cutoff requirements, but the designer can always choose to prescribe larger rebar lengths. Designers need to understand that larger rebar lengths can be supplied free of charge since short leftover lengths would otherwise be scrapped. This has an additional benefit associated with reducing bar-cutting effort (one less cut per bar), which has an ultimate effect on improving the overall construction efficiency.

As an example, let us rework the bar arrangement for Example 11.3 using the stock length approach shown in Figure 11.26.

First, let us consider the top reinforcement at the supports. According to the original bar arrangement in Figure 11.23, 3.7 m bar lengths are prescribed for bars at supports B and C. This length can be used with some wastage. However, these rebars are placed in staggered fashion, with a 1200 mm stagger. In this case, it may be better to use 4.0 m long

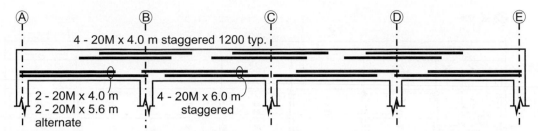

Figure 11.26 The reinforcement arrangement for the beam from Example 11.3 based on the stock length approach.

rebars with a 1200 mm stagger symmetrically over all supports. This arrangement has a number of practical advantages, as follows:

- The same top reinforcement is used at all supports.
- The use of 4 m rebar lengths allows for some field errors associated with rebar placement. Another advantage is the absence of additional material cost associated with using longer rebars in this case.
- The labour required to place 4 m long rebars with a 1200 mm stagger at all supports would certainly be less than that required to place the bars cut according to the actual design.

Span 1 uses 5.6 m bar lengths; 5.6 m does not divide into 18 m by itself. The designer may instead choose to use a 6.0 m bar length for this segment. The use of 6 m long bottom reinforcement will permit a further 400 mm bar extension into column B. This is a good practice — it is recommended to extend the bottom reinforcing into the support beyond the code-prescribed minimum length whenever feasible considering various constraints.

Obviously, there is more than one practical solution for the detailing of flexural reinforcement. In this case, any solution that, besides meeting other design requirements, reduces potential rebar wastage can be considered satisfactory.

11.5 STRUCTURAL DRAWINGS AND SPECIFICATIONS

11.5.1 Background

In a typical design project, a complete set of structural drawings includes several drawings related to floor structures. The key drawing showing elements of a floor system (beams, girders, slabs, etc.) is called a *floor plan*. In general, a floor plan is presented for each floor in a building. In the case of buildings with regular floor elements, a typical floor plan is included instead. The roof system is also shown on a separate drawing called the *roof plan*. In addition to the floor plans, there are other drawings that present information related to the floor design, such as the beam schedule and the beam elevations. CSA A23.3 describes the requirements for structural drawings related to reinforced concrete structures.

General specifications are an important component of the set of structural drawings. These specifications are usually summarized on the first sheet and called *General Notes*.

The objective of this section is to briefly describe the role of structural drawings in floor design.

Floor plans Structural floor plans generally show slabs and beams and any stairs and openings at a particular floor in a building. A typical floor plan for a slab-beam-and-girder system is shown in Figure 11.27. Notice from the plan that the beams are labelled with *beam marks,* such as B1, B2, and B3 (where B stands for beam), while the girders are denoted as G1, G2, G3, etc. The slab panels are labelled on the floor plan as S1, S2, etc. Note that different beams, girders, and slabs that are essentially the same are given the same mark.

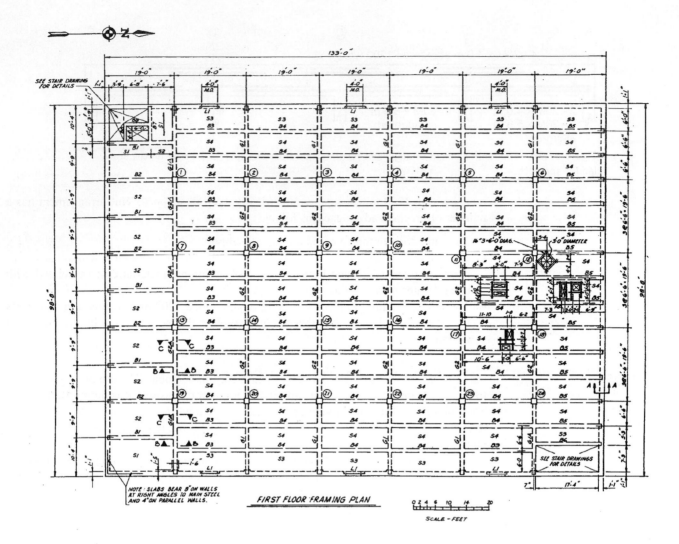

Figure 11.27   Floor plan showing beam and girder schedule.

(*Source:* ACI Committee 315, 1994, reproduced by permission of the American Concrete Institute)

The concrete dimensions and reinforcement can be shown directly on the floor plan with special designations and symbols to represent reinforcing lengths, stirrup spacing, number of bars and locations, to identify any hooks required, and so on. Alternatively, the beam reinforcing may be shown by either a reinforcing schedule or a beam elevation (to be discussed below).

Slab reinforcement is usually shown on the floor plan. In some cases, the slab thickness is indicated on the floor plan.

Beam schedule   When a large number of beams span in both directions, it may be advantageous to prepare a *beam schedule*. The beam schedule contains the beam design information in tabular form; this information includes beam marks, dimensions, top reinforcement, bottom reinforcement, and stirrups. Similar schedules can be prepared for girders and slabs. An example of a beam schedule is presented in Figure 11.27.

The use of beam schedules is appropriate when the building layout is regular and reinforcing patterns do not have large variations. A disadvantage of this approach is that it tends to be more prone to errors because it lacks the visual relationship between a particular beam and the corresponding amount of reinforcement.

CONCRETE

CONCRETE DESIGN CONFORMS TO CSA CAN3–A23.3.
PROVIDE CONCRETE AND PERFORM WORK TO CAN/CSA–A23.1. CONCRETE STRENGTHS TO BE VERIFIED BY INDEPENDENT TESTS TO CAN/CSA–A23.2 AT THE EXPENSE OF THE OWNER. FOR EACH 100 CUBIC METRES OR EACH DAY'S POUR, WHICHEVER IS LESS, A MINIMUM OF 3 TEST CYLINDERS SHALL BE CAST. COPIES OF THE TEST RESULTS SHALL BE SUBMITTED TO THE ENGINEER.

SUBMIT MIX DESIGNS FOR RECORD TO TESTING AGENCY AND ENGINEER.

CONCRETE MIX REQUIREMENTS:

LOCATION	MINIMUM SPECIFIED 28 DAY COMPRESSIVE STRENGTH, MPa	EXPOSURE CLASSIFICATION
1 1/2" CONCRETE TOPPING (150 LBS./CU.FT.) FOR WOOD FRAME CONSTRUCTION	20	–
FOOTINGS, WALLS BELOW GRADE, INTERIOR WALLS AND TOPPING ON STEEL DECK	25	–
EXTERIOR WALLS AND RETAINING WALLS	25	F2
TILT–UP PANELS AND SLABS ON GRADE	30	F2
INTERIOR SUSPENDED SLABS, BEAMS AND COLUMNS	30	–
EXTERIOR SUSPENDED SLABS, BEAMS AND COLUMNS	30	F2
SLAB ON GRADE AND PARKING DECK TOPPING EXPOSED TO DEICING CHEMICALS	32	C2
SUSPENDED PARKING SLABS AND ALL REINFORCED CONCRETE MEMBERS EXPOSED TO DEICING SALTS OR SEA WATER	35	C1

GROUT (FOR UNDER STEEL BASE PLATES AND WHERE SHOWN): NON–SHRINK, NON–FERROUS, PREMIXED GROUT DEVELOPING A MINIMUM COMPRESSIVE STRENGTH OF 50 MPa (7250 psi) AT 28 DAYS.

THE CONTRACTOR SHALL BE RESPONSIBLE FOR DESIGN OF ALL FORMWORK. USE MECHANICAL VIBRATORS THROUGHOUT TO COMPACT CONCRETE. USE KEYED CONSTRUCTION JOINTS AS DIRECTED BY THE ENGINEER, UNLESS DETAILED ON DRAWINGS. CHECK ALL APPLICABLE DRAWINGS FOR LOCATIONS OF BLOCKOUTS, ANCHORS, AND EMBEDDED INSERTS BEFORE CONCRETE IS PLACED. NO CONDUIT, BOXES OR OTHER INSERTS ARE PERMITTED IN COLUMNS WITHOUT THE PRIOR WRITTEN APPROVAL OF THE ENGINEER. DO NOT CUT SLAB REINFORCEMENT AT OPENINGS, SPREAD REINFORCEMENT AROUND OPENINGS. PROVIDE ALL NECESSARY CARRY BARS REQUIRED TO MAINTAIN REINFORCEMENT IN POSITION. REINFORCING STEEL MUST BE INSPECTED BY THE ENGINEER BEFORE CONCRETE IS PLACED.

DO NOT REMOVE FORMS FOR FOOTINGS AND WALLS UNTIL A MINIMUM OF 24 HOURS AFTER CONCRETE IS PLACED AND THE CONCRETE HAS ATTAINED A STRENGTH OF AT LEAST 10 MPa. NO EXCEPTIONS WILL BE PERMITTED. FORMS FOR MILD STEEL REINFORCED SUSPENDED SLABS MAY BE REMOVED AND RESHORING INSTALLED AFTER THE CONCRETE HAS ATTAINED AT LEAST TWO–THIRDS OF THE SPECIFIED STRENGTH. RESHORING TO REMAIN IN PLACE UNTIL CONCRETE HAS DEVELOPED FULL STRENGTH. STRENGTH OF CONCRETE AT TIME OF STRIPPING FORMS TO BE DETERMINED BY TESTING FIELD CURED CONCRETE CYLINDERS, COSTS FOR WHICH TO BE BORNE BY THE CONTRACTOR.

PROVIDE 1/8" CAMBER PER 8'–0" OF SPAN TO ALL SLAB BANDS.
RECESS WALLS WHERE REQUIRED TO SUPPORT BEAMS.
PROVIDE 3/4" CHAMFER ON ALL EXPOSED COLUMN AND SLABBAND CORNERS.

ALL COLD WEATHER AND HOT WEATHER CONCRETE WORK TO BE CARRIED OUT IN ACCORDANCE WITH CAN/CSA–A23.1.
WHEN TEMPERATURE IS EXPECTED TO FALL BELOW 5 DEGREES FAHRENHEIT WITHIN 3 DAYS OF PLACING CONCRETE, THE CONTRACTOR SHALL NOTIFY THE ENGINEER OF THE FOLLOWING:

A)   PROVISIONS FOR HEATING FRESH CONCRETE
B)   PROVISIONS FOR HEATING CONCRETE IN FORMS
C)   ALTERATIONS IN MIX DESIGN
D)   PROVISIONS FOR CURING

CONCRETE SHALL BE PROTECTED FROM ALL HARMFUL EFFECTS DURING CONSTRUCTION. PROVISIONS FOR COLD WEATHER CONCRETE SUBJECT TO THE APPROVAL OF THE ENGINEER. CONCRETE SHALL BE CURED BY APPROVED MEANS FOR AT LEAST 5 DAYS SUBSEQUENT TO POUR.

Figure 11.28   Concrete specification.

Beam elevations When beams are narrow and deep, it may be difficult to show beam reinforcing on the floor plan. In some cases, it is difficult to show beam information on a floor plan. In that case, the solution is to label all beams on the plan and to present corresponding beam elevations on separate drawing(s). Beam elevations show the precise position of the top and bottom reinforcing, stirrups, hooked bars, and reinforcing at the interface between the beams and the supporting columns and walls.

General specifications In addition to drawings, the designer must include specifications to clearly identify the required performance for various materials and structural components. General specifications can include information on the loading, including dead and live loads, concrete and steel strengths, concrete cover, bar anchorage lengths, splices, and any special coating requirements.

An excerpt of a typical general specification for concrete is shown in Figure 11.28. A typical reinforcing steel general specification is shown in Figure 11.29.

11.5.2 Sample Drawings for the Floor Design Case Study

In this section, a few structural drawings of the beam and slab design performed in the case study (Section 11.3) will be presented and discussed.

The floor plan is shown in Figure 11.30. The plan shows the slab and beams and the supporting columns at the second-floor level. The top slab reinforcement is shown by solid lines while the dashed lines represent the bottom reinforcing. The slab reinforcement is specified by the size, length, and spacing, for example 15M 5350@300 (15M bars, 5350 mm length, at 300 mm on centre spacing). Some bars are staggered; this is denoted on the drawing by the abbreviation STAG. Note that this slab was designed in Example 11.1 and that the staggered reinforcement arrangement was shown in Figure 11.24.

The top hooked bars are evenly distributed around the perimeter of the floor slab in order to reinforce the edge of the slab and mitigate any cracking problems at the edge beam (although the analysis did not account for any moment transfer between the slab and the edge beam and so the top bars are not required based on the design). This is common practice: bars have 90° hooks where the slab is supported by beams or walls below and

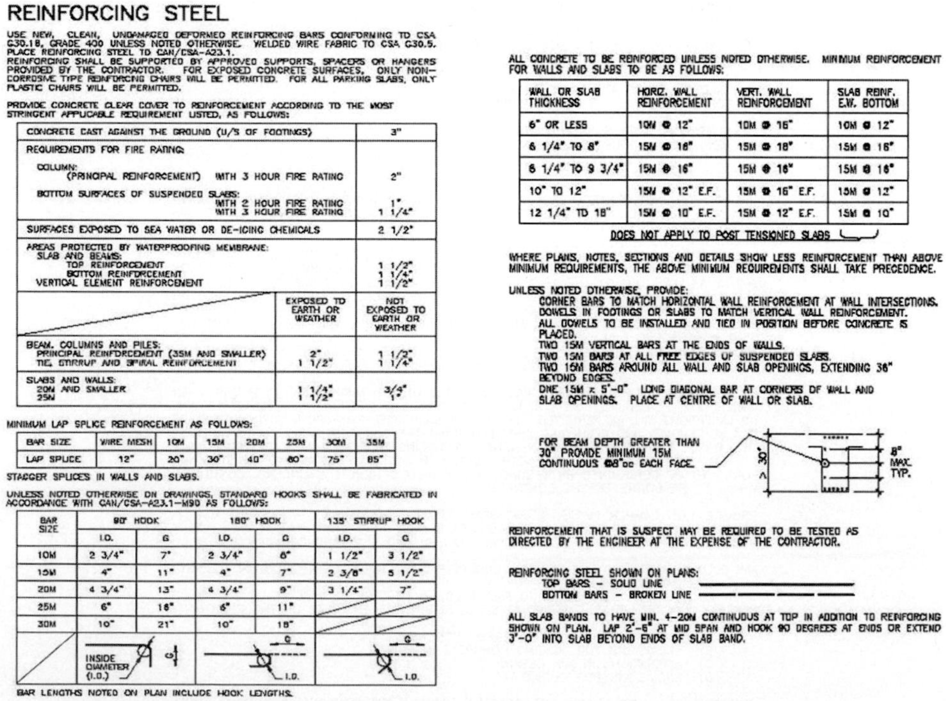
REINFORCING STEEL

USE NEW, CLEAN, UNDAMAGED DEFORMED REINFORCING BARS CONFORMING TO CSA G30.18, GRADE 400 UNLESS NOTED OTHERWISE. WELDED WIRE FABRIC TO CSA G30.5. PLACE REINFORCING STEEL TO CAN/CSA-A23.1.
REINFORCING SHALL BE SUPPORTED BY APPROVED SUPPORTS, SPACERS OR HANGERS PROVIDED BY THE CONTRACTOR. FOR EXPOSED CONCRETE SURFACES, ONLY NON-CORROSIVE TYPE REINFORCING CHAIRS WILL BE PERMITTED. FOR ALL PARKING SLABS, ONLY PLASTIC CHAIRS WILL BE PERMITTED.

PROVIDE CONCRETE CLEAR COVER TO REINFORCEMENT ACCORDING TO THE MOST STRINGENT APPLICABLE REQUIREMENT LISTED, AS FOLLOWS:

CONCRETE CAST AGAINST THE GROUND (U/S OF FOOTINGS)		3"
REQUIREMENTS FOR FIRE RATING: COLUMN: (PRINCIPAL REINFORCEMENT) WITH 3 HOUR FIRE RATING BOTTOM SURFACES OF SUSPENDED SLABS: WITH 2 HOUR FIRE RATING WITH 3 HOUR FIRE RATING		2" 1" 1 1/4"
SURFACES EXPOSED TO SEA WATER OR DE-ICING CHEMICALS		2 1/2"
AREAS PROTECTED BY WATERPROOFING MEMBRANE: SLAB AND BEAMS: TOP REINFORCEMENT BOTTOM REINFORCEMENT VERTICAL ELEMENT REINFORCEMENT		1 1/2" 1 1/4" 1 1/2"
	EXPOSED TO EARTH OR WEATHER	NOT EXPOSED TO EARTH OR WEATHER
BEAM. COLUMNS AND PILES: PRINCIPAL REINFORCEMENT (35M AND SMALLER) TIE, STIRRUP AND SPIRAL REINFORCEMENT	2" 1 1/2"	1 1/2" 1 1/4"
SLABS AND WALLS: 20M AND SMALLER 25M	1 1/4" 1 1/2"	3/4" 1"

MINIMUM LAP SPLICE REINFORCEMENT AS FOLLOWS:

BAR SIZE	WIRE MESH	10M	15M	20M	25M	30M	35M
LAP SPLICE	12"	20"	30"	40"	60"	75"	85"

STAGGER SPLICES IN WALLS AND SLABS.

UNLESS NOTED OTHERWISE ON DRAWINGS, STANDARD HOOKS SHALL BE FABRICATED IN ACCORDANCE WITH CAN/CSA-A23.1-M90 AS FOLLOWS:

BAR SIZE	90° HOOK		180° HOOK		135° STIRRUP HOOK	
	I.D.	G	I.D.	G	I.D.	G
10M	2 3/4"	7"	2 3/4"	6"	1 1/2"	3 1/2"
15M	4"	11"	4"	7"	2 3/8"	5 1/2"
20M	4 3/4"	13"	4 3/4"	9"	3 1/4"	7"
25M	6"	18"	6"	11"		
30M	10"	21"	10"	18"		

BAR LENGTHS NOTED ON PLAN INCLUDE HOOK LENGTHS.

ALL CONCRETE TO BE REINFORCED UNLESS NOTED OTHERWISE. MINIMUM REINFORCEMENT FOR WALLS AND SLABS TO BE AS FOLLOWS:

WALL OR SLAB THICKNESS	HORIZ. WALL REINFORCEMENT	VERT. WALL REINFORCEMENT	SLAB REINF. E.W. BOTTOM
6" OR LESS	10M @ 12"	10M @ 16"	10M @ 12"
6 1/4" TO 8"	15M @ 18"	15M @ 18"	15M @ 16"
8 1/4" TO 9 3/4"	15M @ 16"	15M @ 16"	15M @ 16"
10" TO 12"	15M @ 12" E.F.	15M @ 16" E.F.	15M @ 12"
12 1/4" TO 18"	15M @ 10" E.F.	15M @ 12" E.F.	15M @ 10"

DOES NOT APPLY TO POST TENSIONED SLABS

WHERE PLANS, NOTES, SECTIONS AND DETAILS SHOW LESS REINFORCEMENT THAN ABOVE MINIMUM REQUIREMENTS, THE ABOVE MINIMUM REQUIREMENTS SHALL TAKE PRECEDENCE.

UNLESS NOTED OTHERWISE, PROVIDE:
CORNER BARS TO MATCH HORIZONTAL WALL REINFORCEMENT AT WALL INTERSECTIONS.
DOWELS IN FOOTINGS OR SLABS TO MATCH VERTICAL WALL REINFORCEMENT.
ALL DOWELS TO BE INSTALLED AND TIED IN POSITION BEFORE CONCRETE IS PLACED.
TWO 15M VERTICAL BARS AT THE ENDS OF WALLS.
TWO 15M BARS AT ALL FREE EDGES OF SUSPENDED SLABS.
TWO 15M BARS AROUND ALL WALL AND SLAB OPENINGS, EXTENDING 36" BEYOND EDGES.
ONE 15M x 5'-0" LONG DIAGONAL BAR AT CORNERS OF WALL AND SLAB OPENINGS. PLACE AT CENTRE OF WALL OR SLAB.

FOR BEAM DEPTH GREATER THAN 30" PROVIDE MINIMUM 15M CONTINUOUS @8"oc EACH FACE.

REINFORCEMENT THAT IS SUSPECT MAY BE REQUIRED TO BE TESTED AS DIRECTED BY THE ENGINEER AT THE EXPENSE OF THE CONTRACTOR.

REINFORCING STEEL SHOWN ON PLANS:
TOP BARS - SOLID LINE
BOTTOM BARS - BROKEN LINE

ALL SLAB BANDS TO HAVE MIN. 4-20M CONTINUOUS AT TOP IN ADDITION TO REINFORCING SHOWN ON PLAN. LAP 2'-6" AT MID SPAN AND HOOK 90 DEGREES AT ENDS OR EXTEND 3'-0" INTO SLAB BEYOND ENDS OF SLAB BAND.

Figure 11.29 Reinforcing steel specification.

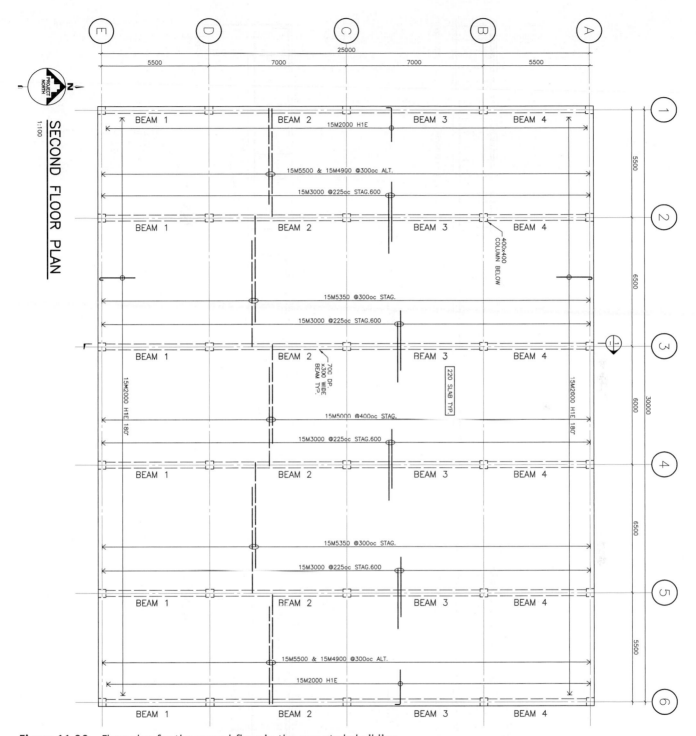

Figure 11.30   Floor plan for the second floor in the case study building.

180° hooks at free edges where the slab is too thin to permit a 90° hook. Note that hooked bars are denoted on the drawing by H1E (hooked one end) or H2E (hooked two ends).

The rebar lengths are drawn to scale to allow a visual check of the rebar lengths in each span. An important part of the design process is to confirm that the rebar quantities and lengths on the drawing comply with the design.

The elevation of a typical beam in gridline 3 is shown in Figure 11.31. Note that the same beam was designed in Example 11.2, while the staggered reinforcement arrangement was shown in Figure 11.25. The beam elevation shows the distribution of flexural and shear reinforcement and provides precise information related to bar lengths and stirrup spacings. In addition, a typical beam cross-section is often provided on the drawings, as shown in Figure 11.32.

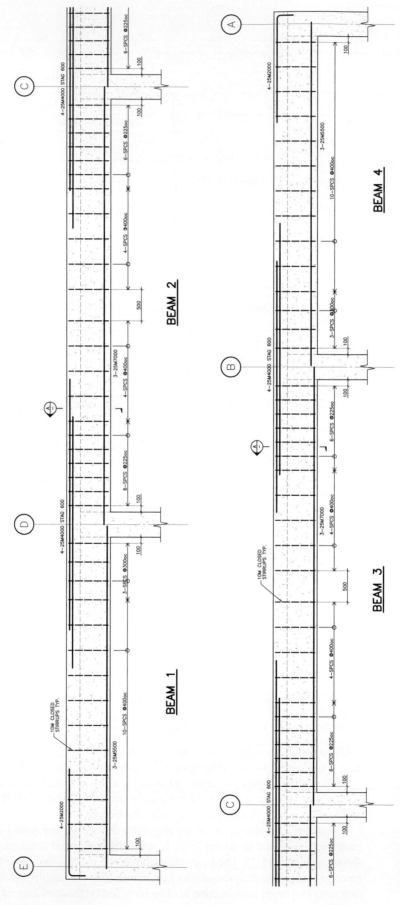

Figure 11.31   Elevation of the beam at gridline 3.

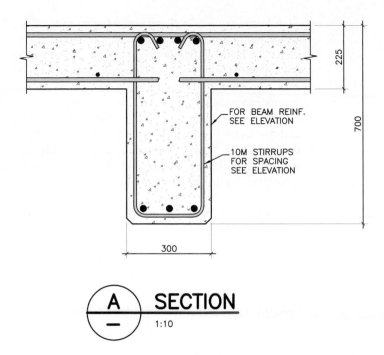

Figure 11.32 A typical beam cross-section.

In reinforced concrete buildings, continuous beams and slabs are typically used in cast-in-place floor construction. The floor system (also called floor framing) has a key role in transferring gravity and lateral forces to the columns and/or walls, and it needs to meet the following requirements:

- adequate strength to safely resist the applied loads
- sufficient stiffness to limit deflections and vibrations
- required fire resistance
- adequate acoustical resistance

Cast-in-place reinforced concrete floor systems

Reinforced concrete floor systems can be classified as follows:

- In one-way systems, the primary flexural reinforcement in each structural element runs in one direction only.
- In two-way systems, the primary flexural reinforcement in at least some elements runs in two orthogonal directions.

Common cast-in-place one-way floor systems are as follows:

- *One-way joist floors* (also known as ribbed slab or panjoist floors) consist of regularly spaced concrete joists (ribs) spanning in one direction and a thin reinforced concrete slab cast integrally with the joists. The joists are supported by wide beams spanning along the column lines. Their advantages include a lightweight, attractive ceiling, which also makes it possible to place electrical fixtures between joists. A drawback is that the formwork costs are higher than for the other systems and the construction is labour intensive. This system can be used for spans ranging from 4.5 m to 12 m.

- *Slab-beam-and-girder floors* consist of a series of parallel beams supported by girders, which are in turn supported by columns placed at regular intervals over the entire floor area. The beam and girder grid supports a one-way reinforced concrete slab. The typical span between the columns ranges from 8 m to 18 m.
- In a *slab band floor system*, a one-way slab is carried directly by wide and shallow beams spanning the columns. These wide beams (called slab bands) are supported directly by the columns. Typically, the nominal width of the slab band (b) is 2400 mm and the depth (h) is in the range from 400 mm to 600 mm. Slab band floor construction is common in Western Canada and is typically used in parking garages, residential and office buildings. The typical slab span is on the order of 8 m while slab bands typically span a maximum of 10 m.

Design of continuous reinforced concrete beams and slabs

The comprehensive design of continuous reinforced concrete beams and slabs includes the following major aspects:

- analysis of continuous beams and slabs (discussed in Chapter 10)
- design of one-way slabs and T-beams for flexure (discussed in Chapters 3 and 5)
- detailing of flexural reinforcement (discussed in Section 11.4)
- design for shear (discussed in Chapter 6)
- design for serviceability, including cracking and deflections (discussed in Chapter 4)

Bar cutoffs for flexural reinforcement in continuous beams and slabs according to CSA A23.3

The approach to detailing reinforcement in continuous reinforced concrete members is similar to that used for simply supported members: flexural reinforcing bars can be terminated where no longer required to resist bending moments. The following general tasks need to be performed to calculate the bar cutoff points in continuous members:

1. Determine the design bending moment envelope for the continuous member under consideration.
2. Determine the points where bars are no longer required to resist flexure, considering both the positive and negative bending moments.
3. Extend the bars in the positive and negative moment regions according to the applicable CSA A23.3 anchorage requirements.

The CSA A23.3 anchorage requirements related to bar cutoffs in *positive moment reinforcement* (Rules P-1 to P-5) are summarized in Checklist 11.1, while the rules for bar cutoffs in *negative moment reinforcement* (rules N-1 to N-5) are summarized in Checklist 11.2.

PROBLEMS

11.1. Consider the beam along gridline 1 that is part of the floor plan in Figure 11.12. The floor is subjected to a factored dead load of $w_{DL,f} = 7.85$ kPa (including the slab self-weight) and a factored live load of $w_{LL,f} = 7.2$ kPa (note that the beam self-weight is not included). The beam is 300 mm wide and 700 mm deep. Use 25M bars for flexural reinforce-

ment and 10M stirrups. Regular uncoated reinforcing bars have been used, and the shear reinforcement (stirrups) is in excess of the CSA A23.3 minimum requirement.

a) Develop the factored moment envelope diagram for this beam considering all relevant load patterns according to the CSA A23.3

requirements. Perform a linear elastic analysis using a computer structural analysis program.

b) Design the flexural reinforcement according to the CSA A23.3 requirements. Consider the rectangular section only. Show the reinforcement arrangement for the beam on an elevation sketch and a few typical sections.

c) Design the shear reinforcement for the beam according to the CSA A23.3 simplified method for shear design. Factored shear forces should be determined based on the full dead and live load pattern. Torsional effects should be neglected. Present the result of the design on a sketch showing the distribution of shear reinforcement.

d) Perform a serviceability check, including deflections and cracking, according to the CSA A23.3 requirements.

Use the design of the beam in gridline 3 performed in Example 11.2 as a reference.

Given:

$$f_c' = 30 \text{ MPa}$$
$$f_y = 400 \text{ MPa}$$
$$\phi_c = 0.65$$
$$\phi_s = 0.85$$

11.2. Consider a 125 mm thick one-way slab that is part of the floor system in Figure 11.6. The slab is subjected to a specified superimposed dead load (DL_s) of 1 kPa and a specified live load (LL) of 5 kPa (in addition to its self-weight). The supporting beams are 400 mm wide. Use 10M bars for flexural reinforcement.

a) Can this slab be analyzed using the CSA A23.3 approximate frame analysis method? Explain.

b) Determine the factored bending moment and shear force distribution for the slab. Use the CSA A23.3 approximate frame analysis method if permitted; otherwise, use linear elastic analysis.

c) Design the flexural reinforcement for the slab according to the CSA A23.3 requirements. Show the reinforcement arrangement on a slab elevation sketch.

d) Check whether shear reinforcement in the slab is required according to the CSA A23.3 simplified method for shear design.

e) Perform a serviceability check, including deflections and cracking, according to the CSA A23.3 requirements.

Given:

$$f_c' = 25 \text{ MPa}$$
$$f_y = 400 \text{ MPa}$$
$$\phi_c = 0.65$$
$$\phi_s = 0.85$$

11.3. Consider the four-span continuous reinforced concrete beam supported by 400 mm wide columns in the figure below. The beam is 500 mm wide and 600 mm deep, is located in the interior of a building, and is subjected to a specified uniform dead load (DL) of 15 kN/m (including self-weight) and a specified uniform live load (LL) of 30 kN/m. The beam is reinforced with 20M flexural reinforcement bars and 10M stirrups. Regular uncoated reinforcing bars have been used, and the shear reinforcement (stirrups) is in excess of the CSA A23.3 minimum requirement.

a) Determine the factored bending moment and shear force distribution using the CSA A23.3 approximate frame analysis method.

b) Design the flexural reinforcement in the beam according to the CSA A23.3 requirements.

c) Detail the flexural reinforcement in the beam. Apply 50% bar cutoffs where permitted. Show the reinforcement arrangement and lengths on a beam elevation sketch.

Given:

$$f_c' = 25 \text{ MPa}$$
$$f_y = 400 \text{ MPa}$$
$$\phi_c = 0.65$$
$$\phi_s = 0.85$$

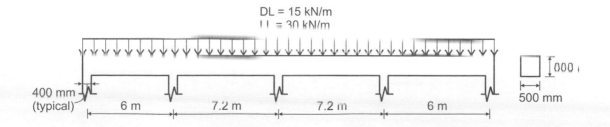

DL = 15 kN/m
LL = 30 kN/m

400 mm (typical)   6 m   7.2 m   7.2 m   6 m

000

500 mm

11.4. Consider the continuous reinforced concrete beam from Problem 11.3.

    a) Apply a linear elastic analysis to determine the moment envelope for this beam.

    b) Compare the design bending moments obtained using the CSA A23.3 approximate frame analysis method in Problem 11.3 with the elastic analysis performed in this problem. Discuss the results.

    c) Apply the CSA A23.3 moment redistribution provision. The redistributed bending moments at the supports should match the values obtained from the CSA A23.3 approximate frame analysis as closely as possible.

11.5. A two-span continuous reinforced concrete beam located in the interior of a building is shown in the figure below. The beam is subjected to a factored uniform load (w_f) of 200 kN/m. Use 20M flexural reinforcement and 10M stirrups. Regular uncoated reinforcing bars have been used, and the shear reinforcement (stirrups) is in excess of the CSA A23.3 minimum requirement. Assume that the magnitude

of the live load is very low compared to that of the dead load.

    a) Determine the factored bending moment and shear force distribution using the CSA A23.3 approximate frame analysis method.

    b) Design the top and bottom reinforcement for this beam according to the CSA A23.3 requirements. Detail the reinforcement by applying 50% bar cutoffs where permitted. Show the reinforcement arrangement and lengths on a beam elevation.

    c) The designer has decided to stagger the top reinforcement in the beam, as shown in the figure. Determine the length of each staggered bar. Show the reinforcement arrangement and lengths on a beam elevation sketch.

Given:

$$f_c' = 25 \text{ MPa}$$
$$f_y = 400 \text{ MPa}$$
$$\phi_c = 0.65$$
$$\phi_s = 0.85$$

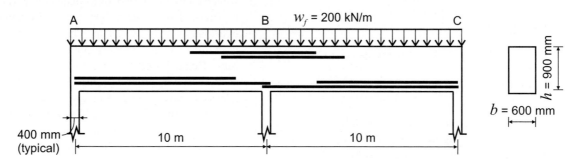

11.6. A typical span of an interior 200 mm thick reinforced concrete continuous slab is shown in the figure below. The bending moments at the supports have been obtained as a result of a linear elastic analysis. Suppose

that the reduced moment at the face of the support is equal to -48 kN·m/m and the inflection points are 1.7 m from each support, as shown in the figure. The slab is reinforced with 15M bars (regular, uncoated).

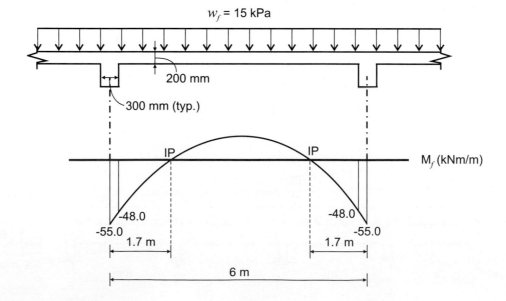

a) Design the top and bottom reinforcement for the slab according to the CSA A23.3 requirements.

b) Detail the flexural reinforcement by applying 50% bar cutoffs where permitted.

Given:

$$f_c' = 25 \text{ MPa}$$
$$f_y = 400 \text{ MPa}$$
$$\phi_c = 0.65$$
$$\phi_s = 0.85$$

11.7. Consider the slab from Problem 11.6. A maximum moment redistribution of 20% as permitted by CSA A23.3 was applied and the resulting bending moment diagram is shown below. The bending moment at the face of support is equal to $-39 \text{ kN} \cdot \text{m/m}$ and the inflection points are 1.3 m from each end.

a) Redesign the positive and negative reinforcement for the slab using 50% bar cutoffs.

b) What is the difference between the amount of positive moment reinforcement determined in Problems 11.6 and 11.7? Explain.

c) Which bending moment diagram would you use if you were asked to perform the design using the least amount of reinforcement while satisfying the CSA A23.3 requirements? Explain.

Given:

$$f_c' = 25 \text{ MPa}$$
$$f_y = 400 \text{ MPa}$$
$$\phi_c = 0.65$$
$$\phi_s = 0.85$$

11.8. Consider the continuous reinforced concrete slab designed in Example 11.1.

Detail the flexural reinforcement in the slab. Apply 50% bar cutoffs where permitted. Show the reinforcement arrangement and lengths on a slab elevation sketch.

11.9. Consider the continuous reinforced concrete beam designed in Example 11.2.

Detail the flexural reinforcement in the beam. Apply 50% bar cutoffs where permitted. Show the reinforcement arrangement and lengths on a beam elevation sketch.

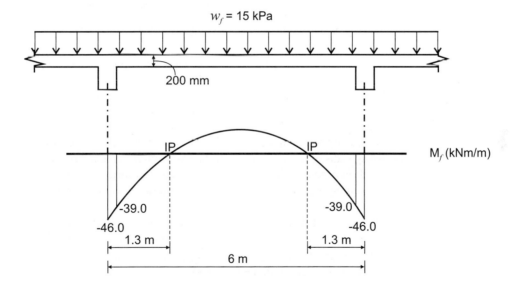

Design of Two-Way Slabs

LEARNING OUTCOMES

After reading this chapter, you should be able to

- describe the key features of reinforced concrete two-way slabs
- design two-way slabs for flexure in accordance with the CSA A23.3 design procedures
- evaluate shear resistance of two-way slabs considering one-way and two-way shear effects
- estimate immediate and long-term deflections in two-way slabs according to CSA A23.3 requirements

12.1 INTRODUCTION

This chapter builds on the fundamentals presented in Chapters 10 and 11, which explained how multiple span slabs and beams can be designed as continuous structures by the strategic placement of reinforcing steel in the top and bottom regions. Continuous reinforced concrete structures that span over several supports are characterized by greater flexural stiffness than single span structures with the same overall length. This could result in reduced member dimensions and gives reinforced concrete a significant economic advantage over other materials. The same advantage can apply to two-way slabs, which are continuous structures spanning in two directions. Two-way slabs are unique to reinforced concrete and are considered to be an advanced design topic for students and inexperienced engineers. Two-way slabs offer an economical design solution, characterized by minimal slab thickness and reduced building height, which is easy to form and fast to construct. As a result, two-way slabs are the most popular floor system for multi-storey building construction in Canada.

The intent of this chapter is to explain behaviour of two-way slab systems and assist readers in deriving simple, practical, and economical design solutions.

The design of two-way slabs encompasses the following four key steps (note the section numbers referenced in the brackets):

1. Selecting the slab thickness (Section 12.5.2)
2. Designing the slab for flexure according to one of following three procedures:
   a) Direct Design Method (Section 12.6)
   b) Equivalent Frame Method (Section 12.7.2) or
   c) Three-Dimensional Elastic Analysis (Section 12.7.3)
3. Designing the slab for shear (Section 12.9), and
4. Performing a deflection check, when required by CSA A23.3 (Section 12.10).

This chapter is divided into eleven sections. Section 12.2 provides an overview of various two-way slab systems and describes their main features. The key factors influencing the structural behaviour of two-way slabs subjected to flexure are discussed in Section 12.3.

Common definitions and terms associated with two-way slabs are explained in Section 12.4. General CSA A23.3 design provisions for two-way slabs, including minimum thickness requirements, are presented in Section 12.5. Section 12.6 explains the CSA A23.3 Direct Design Method, a statics-based procedure which can be used for flexural design of regular slab systems subjected to gravity loads. Section 12.7 discusses elastic analysis, which includes a two-dimensional Equivalent Frame Method and three-dimensional Finite Element Analysis. Section 12.8 outlines key concepts of the Yield Line Method. The application of these design methods is illustrated through a few examples. The same design problem is solved using different methods, and the results are compared to identify the advantages and disadvantages of each method. It is expected that the reader will be able to make an informed choice of the most appropriate method for a specific design application. Section 12.9 describes concepts and procedures related to design of two-way slabs for shear. Section 12.10 discusses deflection control and the relevant calculation procedures for two-way slabs. Finally, structural drawings and detailing of two-way slabs are discussed in Section 12.11.

12.2 TYPES OF TWO-WAY SLABS

12.2.1 Background

Most two-way slab systems in reinforced concrete buildings are floor and roof slabs supported by columns and/or walls. Various floor and roof systems typical in reinforced concrete construction were introduced in Section 1.3.1. These systems included both one-way and two-way slabs. One-way slabs transfer gravity loads in one direction, while two-way slabs are characterized by significant bending in two directions. The following two-way slab systems will be discussed in this chapter:

- Flat plates
- Flat slabs
- Waffle slabs
- Slabs with beams

The first two-way slab in reinforced concrete was a flat slab constructed in 1906 by C.A.P. Turner for the C.A. Bovey Building in Minneapolis, Minnesota in 1906 (Sozen and Siess, 1963). A significant amount of controversy and arguments accompanied the initial applications of the flat slabs. This was mostly due to their unique load path and structural design, which appeared strange to engineers who were used to designing timber and steel structures with one-way load paths. In spite of these challenges, flat slabs with column capitals and drop panels were widely used in the first half of the 20th century for industrial buildings with heavy loads and relatively long spans (drop panels and capitals are illustrated in Figure 12.2). After World War II, many buildings were built for lighter loading, such as residential occupancy, thus drop panels and capitals were not needed. The resulting system is called flat plate to distinguish it from flat slab.

Design provisions for two-way slabs were introduced in North American codes in 1941, through ACI 318-41 in the USA. Flat plate and flat slab systems are currently the most popular floor systems used for construction of multi-storey residential and office buildings in Canada. Waffle slabs and slabs with beams can also be used as economical light-weight solution for large spans.

Choice of the most appropriate slab system for a particular application depends on several factors, including span lengths (distance between the supports) and live loads. Economic considerations, in particular construction costs, are another important factor. In North America, formwork cost represents approximately 50% of the overall cost for two-way slabs floor system. Concrete material, and its placing and finishing, account for 30% of the cost, while the remaining 20% of the cost is associated with the material and placing cost of the reinforcing steel. Hence, labour and material costs are key considerations for cost-efficient design of two-way slabs. For a detailed discussion on economic aspects of two-way slabs the reader is referred to PCA (2000).

In most residential and commercial building designs, the slab self-weight is a significant proportion of the total load. Hence, a potential increase in the overall construction cost due to a live load increase from 2.5 to 5.0 kPa may be only 5 to 10%.

12.2.2   Flat Plates

A *flat plate* is a floor system that consists of solid slabs reinforced in two directions and supported directly by the columns or walls, as shown in Figure 12.1a. This system is economical for short and medium spans ranging from 6 to 9 m (20 to 30 ft). The slab thickness usually ranges from 160 to 250 mm (6.5 to 10 inches). Flat plates are suitable for light loads (live loads up to 5.0 kPa), and the slab thickness is usually controlled by long-term deflections.

A slab layout is defined by column locations. Parallel column lines (gridlines) in perpendicular direction form a rectangular slab *panel* (or a bay), as shown in Figure 12.1b. A slab panel is defined by the span lengths l_1 and l_2, and the ratio of span lengths (l_1/l_2) is called the *aspect ratio*. The simplest and most optimal design uses a column layout on a square grid (corresponding to the aspect ratio of 1.0).

In some cases, both flat plates and flat slabs can have shallow beams along the perimeter of the slab. These beams are called *spandrel beams* (or edge beams). These spandrel beams are effective in increasing the slab stiffness at the edges and thereby reducing slab deflections. A typical spandrel beam section is shown in Figure 12.1c.

A few important advantages of a flat plate system are outlined below:

1. Due to relatively thin structure, a flat plate system permits the construction of the maximum number of floors for a given building height. An additional benefit associated with the reduced floor height is an overall reduction in the building envelope (exterior walls) and partitions, utility shafts, and a significant reduction in seismic and wind loads.
2. Flat plate systems can be designed to satisfy most of the fire resistance requirements of the National Building Code of Canada without the need for additional fire protection. This is a significant advantage compared to floor systems in other materials, e.g. steel and wood.
3. Flat plate systems can be adapted to non-uniform column layout, usually without a significant cost premium. This is a significant advantage as it can efficiently accommodate flexibility in architectural design. Flat plates require simple formwork, resulting in minimal construction effort in terms of time, labour, and cost. Flat plates in high-rise buildings are usually constructed using prefabricated formwork systems (flyforms) that can be transported and lifted for reuse at the next floor level. In a high-rise building with flat plate design, a construction crew can form, reinforce, and place concrete for each floor in very short (three- to four-day) cycles.

In conclusion, a flat plate system is very efficient and economical, and for that reason it is the most widely used floor system for high-rise building construction in Canada. Typical applications include residential buildings, offices, hotels, hospitals, parking garages, etc. Two typical applications, a parking garage and a residential building, are shown in Figure 12.1d and e, respectively.

Figure 12.1 Flat plate system:
a) an isometric view; b) plan
view; c) vertical section;
d) a parking garage, and
e) a residential building under
construction
(Svetlana Brzev).

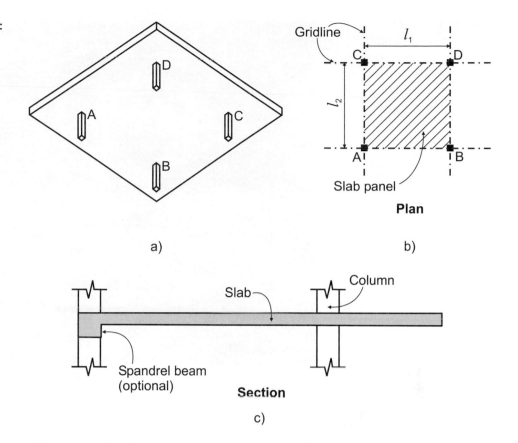

a)

b)

c)

d)

Figure 12.1   (*cont.*)

e)

12.2.3   Flat Slabs

The span-to-thickness ratio for a flat plate is usually limited by permitted long-term deflections. Thin slabs could also have punching shear issues in column regions. The punching shear capacity can be increased by thickening the slab locally around the columns, by way of *drop panels*. A flat plate system with drop panels is called a *flat slab*, as shown in Figure 12.2a. Drop panels cause an increase in the slab stiffness in regions subjected to negative bending moments, and therefore help to control deflections in midspan regions. A smaller amount of top reinforcement is required in slabs with drop panels than in flat plates.

Drop panels are usually square in plan, and plan dimensions are required to be one-third of the panel span length in each direction. Thickness of a drop panel is often governed by the formwork dimensions (e.g. 89 mm thickness can be specified when 2" x 4" plank is used on the edge). In some flat slab designs, column capitals are provided with or without drop panels. A *capital* is an upper portion of the column, usually of conical shape and with larger cross-sectional dimensions than the remaining portion of the column, as shown in Figure 12.2b. Rectangular-shaped capitals (also known as shear caps) are used in some designs as they are easier to form. The designer may use capitals, drop panels, or both to address punching shear and deflection issues in floor slabs. Both drop panels and capitals are effective in providing additional punching shear capacity of the slab. Typical building applications are shown in Figure 12.2c and d.

12.2.4   Waffle Slabs

A *waffle slab* is a floor system that meets requirements for larger spans and loads at a reduced slab weight, while also providing an aesthetically pleasing ceiling. Waffle slab systems consist of evenly spaced concrete joists spanning in two directions — this system is also known as a two-way joist system. The joists are commonly formed by using standard pans or domes installed in the forms to produce a coffered soffit in the slab. The perpendicular orientation of the joists results in evenly spaced square voids on the underside of the slab — this is the reason why the system is referred to as waffle slab. A plan view and a vertical section of waffle slab are shown in Figure 12.3a.

Figure 12.2 Flat slab system:
a) an isometric view and a vertical
section; b) column capital; c) a flat
slab with capitals, and d) a flat
slab with column capitals and drop
panels
(Svetlana Brzev).

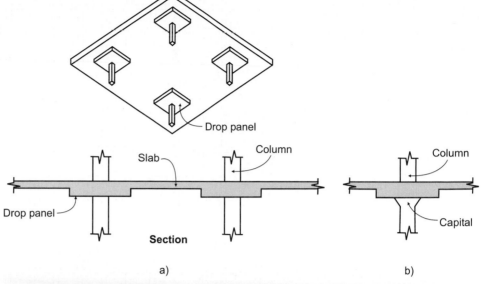

a)

b)

c)

d)

Figure 12.3 Waffle slab system:
a) plan and a vertical section,
and b) en example of a building
application.

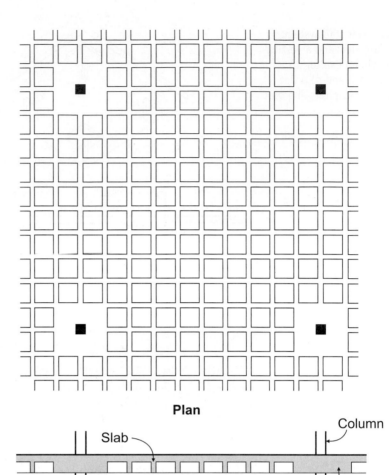

Plan

Section

a)

b)

Figure 12.4 Slab with beams:
a) typical plans and vertical
sections, and b) an example of
building application
(Svetlana Brzev).

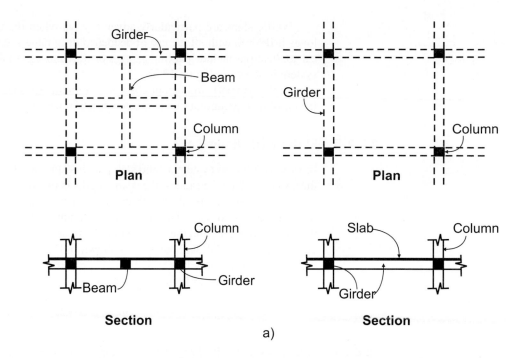

a)

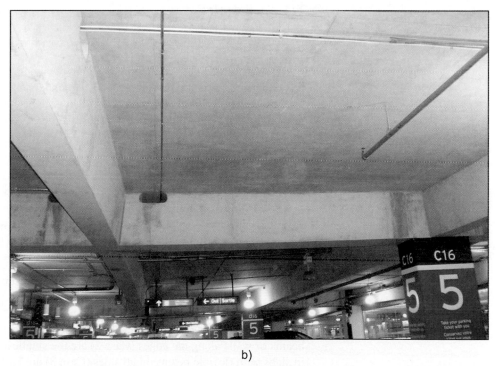

b)

Waffle slabs could offer economical design solution for spans in the range from 10 to 12 m, depending on the availability of forming pans and relevant construction skills. Slab thickness varies from 75 to 130 mm based on either fire resistance requirements or structural considerations. Note that a 130 mm slab is needed to meet 2-hour fire rating requirements. Joist width is 150 or 200 mm, and the depth ranges from 200 to 600 mm underside the slab. The spacing of joists is governed by the dome dimensions and code requirements. Maximum joist spacing is 750 mm. Solid slab portions (heads) are provided for increased shear strength at the column locations. For design purposes, waffle slabs are considered as flat slabs with the solid heads acting as drop panels. Waffle slab construction allows a considerable dead load reduction compared to conventional flat slab construction since the volume of concrete used can be minimized due to the short span between the joists.

Waffle slabs are particularly advantageous when the use of long span and/or heavy loads is desired without the use of deepened drop panels, capitals or support beams. Typical applications include parking garages or warehouses. A parking garage with a waffle slab system is shown in Figure 12.3b.

High formwork and labour costs are the main reasons for limited applications of waffle slabs in contemporary concrete construction practice.

12.2.5   Slabs with Beams

In *slab with beams* system a solid slab panel is supported by beams on all four sides. With this system, when the ratio of the span lengths is two or more, load is predominately transferred by bending in the short direction and the panel essentially acts as a one-way slab. As the panel approaches a square shape, a significant load is transferred by bending in both orthogonal directions, and the panel should be treated as a two-way slab. Plan views and vertical sections for slab with beams are shown in Figure 12.4a. The presence of beams may require a greater storey height. A typical application of the slab with beams system (parking garage) is shown in Figure 12.4b.

12.3   BEHAVIOUR OF TWO-WAY SLABS SUBJECTED TO FLEXURE

12.3.1 Background

Before we proceed with explaining different design approaches for two-way slabs later in this chapter, it is important for the reader to understand how these systems behave under gravity loads. Flexural behaviour of two-way slabs is significantly influenced by the support conditions. The following three cases will be considered:

1) Flat plate/slab panels supported only by columns at the corners; note the supports shown with arrows (see Figure 12.5a).
2) Slab panels supported either by stiff beams or reinforced concrete walls (slabs on stiff supports) have continuous supports along the edges, as shown in Figure 12.5b; note fixed supports on all sides.
3) Slab with beams on all sides - an intermediate case between 1) and 2), where a slab is supported by columns (shown with arrows) and beams (shown as spring supports), as illustrated in Figure 12.5c.

It is important to discuss the influence of the type of support conditions upon the magnitude of midspan displacements in two-way slabs. Flat plates and flat slabs (Case 1) are characterized by largest midspan displacements due to the slab flexibility in the edge regions (along column lines). Midspan displacements in slab panels supported by continuous stiff beams or walls along the edges (Case 2) are the smallest of all three cases. Displacements in slabs with flexible beams on all sides (Case 3) are larger compared to the Case 2, since beam displacements contribute to total displacements at the midspan. A diagram showing conceptual deflected shapes for all cases is presented in Figure 12.5d.

Consider two slabs characterized by the same thickness (180 mm) and span lengths (5 m square) that are subjected to the same uniform gravity load of 9.5 kPa. The only difference is in the support conditions: one of the slabs is supported by columns at the corners only (flat slab), while another is continuously supported at the edges by walls or other stiff supports. In the first case, displacements are restrained by the columns (which act like sticks with pinned connections at the top, similar to simple supports in beams), while rotations are possible throughout the slab. In the latter case, walls provide restraints both for displacements and rotations and act like fixed supports. These are two extreme support conditions: the most flexible and the most stiff. The displacement contours (regions with equal displacements)

Figure 12.5 Two-way slabs with different support conditions: a) flat slab/plate; b) slab on stiff supports; c) slab with beams on all sides, and d) deflected shapes for two-way slabs.

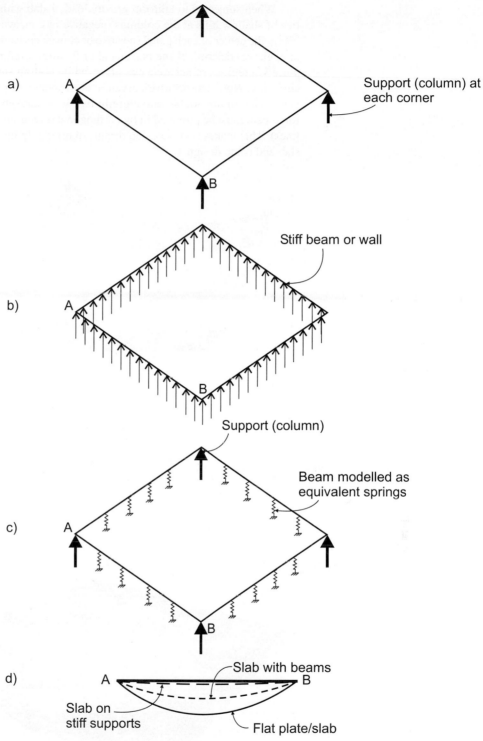

are shown in Figure 12.6. The maximum displacements for the flat slab supported only by the columns and the slab on fixed supports are 15 mm and 1 mm, respectively. The values can be seen from the legend corresponding to displacement contours. (Note that the approaches for deflection calculations in two-way slabs will be discussed in detail in Section 12.10).

It can be seen from the above examples that two-way slabs bend in double curvature like a plate. The extent and the shape of double curvature within each span depends on the support conditions, the applied loads, and the properties of the adjacent spans in each direction.

When subjected to uniform gravity load, a slab with several panels deforms into a series of shallow hills at the columns (negative or concave curvature), and bowl-shaped valleys at the center of each panel (positive or convex curvature). The sign of curvature (positive or negative) determines the placement of flexural reinforcement. For example, top steel is placed in regions of negative curvature, while bottom steel is placed in regions of positive curvature. Note that positive curvature corresponds to positive bending moment and vice versa. Due to the double curvature in two-way slabs, both top and bottom flexural reinforcement must be provided in two orthogonal directions (usually in line with column gridlines). This makes two-way slab design significantly more complex compared to one-way slab and beam design.

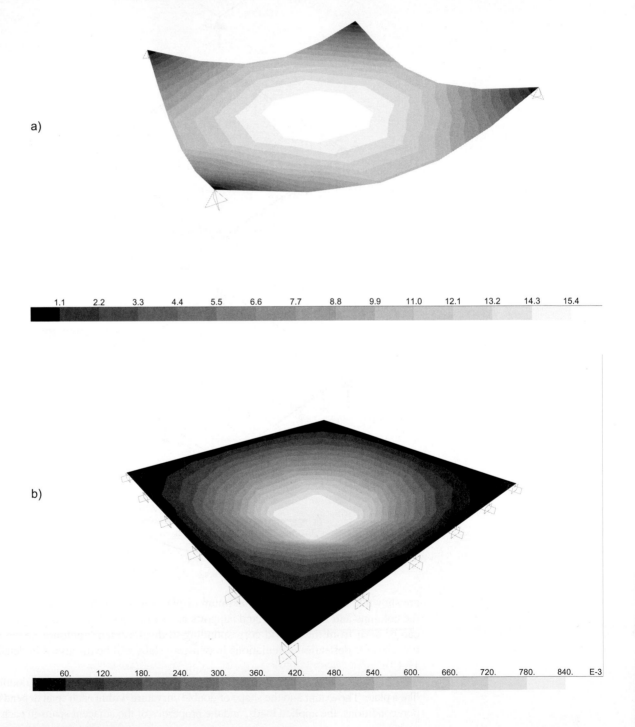

Figure 12.6 Displacements in two-way slabs: a) a flat plate supported on columns, and b) a slab on continuous supports along the edges (units: mm).

12.3.2 Gravity Load Paths in Two-Way Slabs

Before proceeding with the design of two-way slabs, the designer must understand the load path for this structural system. The concept of load path was first introduced in Section 1.3.3. Gravity load path in this context denotes the manner (or "the path") in which the gravity load is transferred from the slab to the supports. In reinforced concrete slabs, gravity load can be transferred to supports through either one-way or two-way load paths. A one-way load path is characteristic of one-way reinforced concrete slabs. Load distribution in one-way slabs is explained in Section 3.6. The concept of a one-way load path can be explained through example of a timber floor system consisting of parallel planks supported by the beams shown in Figure 12.7. When gravity load is applied on the floor, a plank (1) carries the load along its span and transfers it to the beams (2). These beams in turn transfer the plank reactions onto the supporting columns (3). In this system both planks and beams act like simply supported beams, thus the load transfer is achieved through one-way load path.

Figure 12.7 One-way slab system comprised of parallel timber planks supported by the beams on two sides.

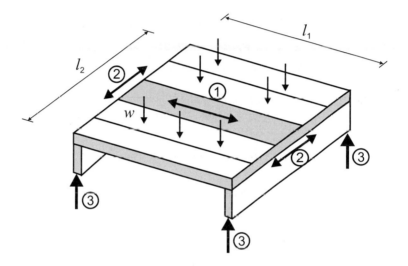

Reinforced concrete two-way slabs are characterized by load paths in two orthogonal directions. These load paths simultaneously transfer load from the slab to the supports. The proportion of load carried in each direction will depend on the relative flexural stiffness of the slab in each direction. To explain this concept, an example of a timber floor system with one-way load path will be expanded to idealize load path in two directions.

First, let us idealize the slab shown in Figure 12.8a as a system of timber planks laid in two orthogonal directions and simply supported at the ends. Consider load distribution for a system of two planks (A and B) which are laid on top of one another and connected at the intersection (see Figure 12.8b). The point load P is applied at the intersection of these two planks. Both planks jointly support the load, that is,

$$P = P_A + P_B$$

where P_A and P_B are the loads resisted by planks A and B, respectively. The planks have the same cross-sectional dimensions: 250 mm width by 50 mm depth, however their spans are different: plank A has a shorter span (l_a) while plank B has a longer span (l_n), say twice the span of plank A ($l_B = 2l_A$).

The proportion of load supported by each plank can be determined from the deflection compatibility at the point of intersection, that is, the deflections of the two planks at the point of intersection must be equal. The deflection for plank A at the intersection, δ_A, can be determined by treating the plank as a simply supported beam subjected to point load P_A, that is,

$$\delta_A = \frac{P_A \cdot l_A^3}{48 \cdot E \cdot I_A}$$

where E is the modulus of elasticity for timber, and I_A is the moment of inertia for plank A (refer to Table A.16 for deflection equation for a beam subjected to point load).

The maximum deflection for plank B at the intersection, δ_B, can be determined from the same equation as plank A, that is,

$$\delta_B = \frac{P_B \cdot l_B^3}{48 \cdot E \cdot I_B}$$

These deflections need to be equal based on compatibility requirements, thus

$$\delta_A = \delta_B$$

and

$$\frac{P_A \cdot l_A^3}{48 \cdot E \cdot I_A} = \frac{P_B \cdot l_B^3}{48 \cdot E \cdot I_B}$$

Since the planks have the same modulus of elasticity, E, and the same moment of inertia ($I_A = I_B$), the above equation can be simplified as follows

$$\frac{P_A}{P_B} = \frac{l_B^3}{l_A^3}$$

Since $l_B = 2l_A$, it follows that

$$\frac{P_A}{P_B} = 8$$

Since

$$P = P_A + P_B = \frac{9}{8} P_A$$

It can be concluded that 8/9 (approximately 89%) of the total load P is supported by plank A, that is,

$$P_A = \frac{8}{9} P$$

while the remaining 11% of the load P is supported by plank B.

This can be explained by a difference in flexural stiffness between planks A and B. Flexural stiffness is directly proportional to the moment of inertia (I), however it is inversely proportional to the third power of the span length (l^3). In this example, both planks have the same moment of inertia, however plank A has one-half of the span length of plank B. As a result, plank A has eight times larger stiffness compared to plank B, and thus carries a significant portion (about 90 %) of the total load P.

Also, note that that the resulting load on each plank support is quite different. Each plank A support carries approximately 45 % of the total load P, while the supports for plank B carry only 5% each.

Next, let us consider a modified version of the same example. Let us rotate plank B such that the longer cross-sectional dimension is placed in vertical direction, while plank A remains

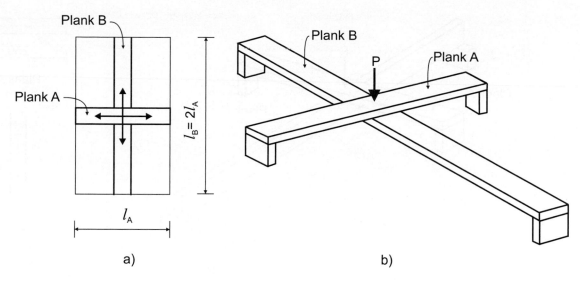

Figure 12.8 Two-way load path example: timber planks with the same stiffness: a) plan view, and b) isometric view.

the same as in the first example (see Figure 12.9). Planks A and B are now going to have different moments of inertia for bending about the horizontal axis (x-x), as follows

$$I_A = \frac{250 \cdot 50^3}{12} = 2.6 \cdot 10^6 \ \text{mm}^4$$

and

$$I_B = \frac{250^3 \cdot 50}{12} = 65.0 \cdot 10^6 \ \text{mm}^4$$

that is, $I_B = 25 I_A$.

Let us estimate a fraction of the total load P resisted by each plank by following the same approach as in the previous example. As shown earlier, deflections at the point of intersection have to be equal due to the compatibility requirement, that is,

$$\delta_A = \delta_B$$

In this case, the planks have the same modulus of elasticity, E, however their moments of inertia and the spans are different, hence

$$\frac{P_A}{P_B} = \frac{l_B^3 \cdot I_A}{l_A^3 \cdot I_B}$$

Since $l_B = 2 l_A$ and $I_B = 25 I_A$, it follows that

$$\frac{P_A}{P_B} = 0.32$$

Therefore

$$P = P_A + P_B = 1.32 P_B$$

Hence,

$$P_A = 0.24 P$$

and

$$P_B = 0.76 P$$

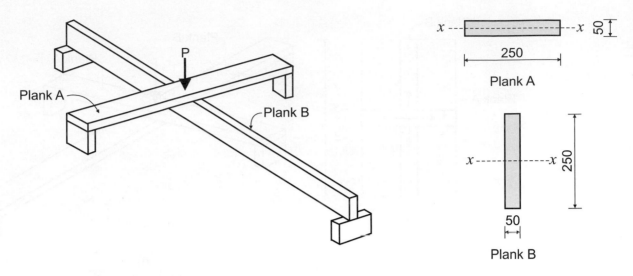

Figure 12.9 Two-way load path example: timber planks with different stiffness.

It can be concluded that only 24% of the total load P is supported by plank A, while the major portion of the load (76%) is supported by plank B. Plank B has 3 times higher flexural stiffness compared to plank A. As a result, plank B carries a major portion of the total load. Plank A now only supports 24% of the total load; this is a large decrease from 89% (see the previous example). The changes in support reactions are also significant, since 38% of the total load is transferred to each plank B support, and only 12% is transferred to each plank A support.

The above two examples show that, although planks A and B have equal spans, the support reactions in the first and the second example were significantly different due to a change in relative stiffness for these planks . We have also learned from the above examples that in structural systems with multiple load paths like two-way slabs, the load path associated with a larger flexural stiffness will always support a larger fraction of the total load. Since slabs have the same thickness in both directions, the stiffness usually depends on relative span lengths. In general, longer spans are more flexible and attract a smaller fraction of the total load, as discussed in the first example.

12.3.3   Distribution of Bending Moments in Two-Way Slabs

Statics-Based Approach for Estimating Bending Moments in Two-Way Slabs
When two-way slabs were first introduced in concrete construction practice at the beginning of the 20th century, initial design solutions were based solely on the statics-based approach which was originally proposed by J.R. Nichols in 1914 (Sozen and Siess, 1963). That approach may be useful to provide an insight into key parameters which influence moment distribution in two-way slabs, and it will be used in the following two examples.

Let us consider a two-way flat slab panel supported at the corners shown in Figure 12.10a. The panel has rectangular shape (span l_1 is longer than span l_2), and it is subjected to uniformly distributed load w. The four support reactions (R) are the same, that is,

$$R = \frac{w \cdot l_1 \cdot l_2}{4}$$

Next, let us idealize the slab as a beam with the span l_1. The beam is subjected to uniform load ($w \cdot l_2$) which corresponds to the tributary width l_2. Bending moment on Section 1-1 at midspan can be determined from a free-body diagram of the slab shown in Figure 12.10b, as follows

$$M_1 = \frac{\left(w \cdot l_2\right) l_1^2}{8}$$

Figure 12.10 Bending moments in a flat slab panel: a) a typical panel; b) a free-body diagram at Section 1-1, and c) a free-body diagram at Section 2-2.

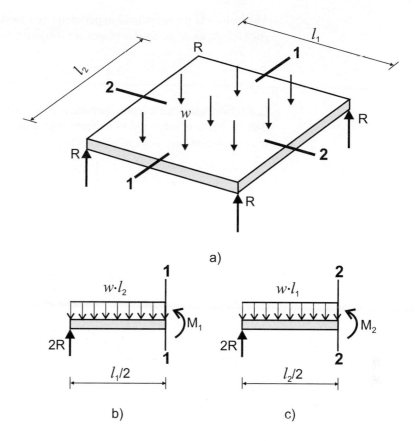

a)

b)   c)

Note that M_1 is equal to the maximum bending moment for a simply supported beam with the span l_1 subjected to uniform load $w \cdot l_2$. It is important to note that M_1 is a total bending moment which corresponds to the width l_2.

The same exercise can be performed to determine bending moment at Section 2-2 in other direction (see Figure 12.10c), as follows

$$M_2 = \frac{(w \cdot l_1)l_2^2}{8}$$

where M_2 is equal to the maximum bending moment in a simply supported beam with span l_2 and subjected to load $w \cdot l_1$.

The ratio of moments M_1 and M_2 is equal to

$$\frac{M_1}{M_2} = \frac{l_1}{l_2}$$

The above equation shows that bending moments at the midspan of a slab panel are proportional to the ratio of span lengths.

Statics-based approach will be also used to illustrate distribution of bending moments in different slab regions. Consider a square slab panel ABDE with span l supported by columns shown in Figure 12.11a. The structural action of the slab with a uniform load w is simulated by a grid of simply supported beams carrying the same uniform load, as shown in Figure 12.11b. The interior beams FG and JH simulate the midspan regions of the slab, while the perimeter beams simulate regions close to column gridlines. Because of symmetry the interior beams deflect the same amount and do not interact.

First, let us consider moment distribution in an interior beam FG shown in Figure 12.11c. The maximum bending moment for a simply supported beam subjected to uniform load is equal to

$$M_{FG} = \frac{w \cdot l^2}{8}$$

The beam FG transfers load to perimeter beams AB and DE through support reactions R. As a result, these perimeter beams are subjected to the point load R, that is,

$$R = \frac{w \cdot l}{2}$$

Load distribution for a typical perimeter beam AB is shown in Figure 12.11d. It can be seen that the beam is subjected to uniform load w in addition to the point load R. The resulting maximum bending moment is equal to

$$M_{AB} = \frac{w \cdot l^2}{8} + \frac{R \cdot l}{4} = 2\left(\frac{w \cdot l^2}{8}\right)$$

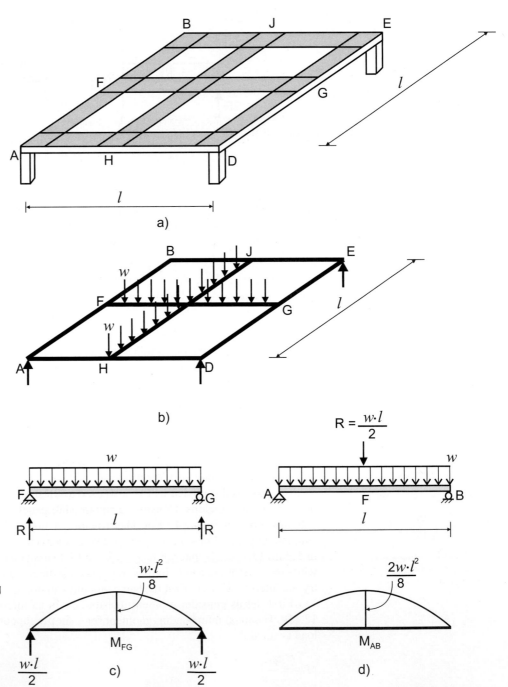

Figure 12.11 Flat plate/slab modelled as a beam grid:
a) an actual slab panel showing idealized beam strips;
b) an idealized beam grid;
c) moment distribution in interior beam FG, and
d) moment distribution in perimeter beam AB.

The ratio of maximum bending moments for a perimeter and an interior beam is equal to 2.0, that is,

$$\frac{M_{AB}}{M_{FG}} = 2.0$$

This example shows that bending moments in perimeter beams are significantly higher than corresponding moments in interior beams.

The key points from the above discussion are summarized below:

1. Bending moments at midspan of a column-supported slab are directly proportional to the ratio of their spans: larger bending moments occur in the longer span.
2. Bending moments along column gridlines are always higher than bending moments at midspan regions of the slab.

Bending Moments in Slabs with Beams   Relative beam-to-slab stiffness ratio is one of the key factors influencing moment distribution in two-way slabs with beams. The case of slab with beams is an intermediate case, and the moment distribution is in between the two extreme cases. In the first case, beam stiffness is equal to zero — behaviour is the same as for flat slabs where slabs carry the entire load and all moments are resisted by the slab. In the second case, the beam stiffness is extremely large — all the behaviour is similar to slabs on stiff supports.

A parametric study on the distribution of bending moments between beams and slab for slabs with beams was performed by Park and Gamble (2000). Charts showing a bending moment distribution in slabs with beams depending on the relative beam-to-slab stiffness ratio (expressed through the α factor) and the span length ratio l_1/l_2 are presented in Figure 12.12 (the α factor will be discussed in Section 12.4.5). Two different span length ratios (l_1/l_2), that is, 1.0 and 2.0, were considered in the study. It can be seen from Figure 12.12a that the negative bending moment at the support is distributed between the slab and the beam. When beam stiffness exceeds 3.0, a major portion of the moment is resisted by the beams (more than 50%). Also, there is a more significant sharing of bending moments between the beams and the slab for square column grid ($l_1/l_2 = 1.0$) compared to rectangular grid ($l_1/l_2 = 2.0$). The same tendency was observed as related to the distribution of positive moments at midspan, as illustrated in Figure 12.12b.

12.3.4   Moment Redistribution in Cracked Two-Way Slabs

Similar to other reinforced concrete flexural members, two-way slabs are expected to experience cracking under service loads. The effect of cracking and reinforcement distribution on the behaviour of continuous reinforced concrete members is discussed in Section 10.7. This section discusses the effect of cracking on the behaviour of two-way flat slabs, which are also continuous systems, but their behaviour is more complex than beams and one-way slabs.

Before the cracking, a two-way slab shows linear elastic behaviour, which is characterized by linear stress-strain relationship in concrete and steel. Cracking in the slab occurs when bending moments reach the cracking moment (M_{cr}) (see Section 4.2 for more details on the cracking moment). In flat slabs, cracking initially takes place in the vicinity of columns, since these areas are typically characterized by the largest negative bending moments. Once the cracking has been initiated, flexural stiffness of the slab decreases since cracked regions have significantly lower flexural stiffness than uncracked regions. (Recall that flexural stiffness is proportional to the moment of inertia, which decreases due to cracking, as discussed in Section 4.3.1.) Since sections with larger stiffness "attract" larger moments, uncracked slab regions will be subjected to larger bending moments than cracked regions. For example, positive bending moments in midspan regions will increase after the

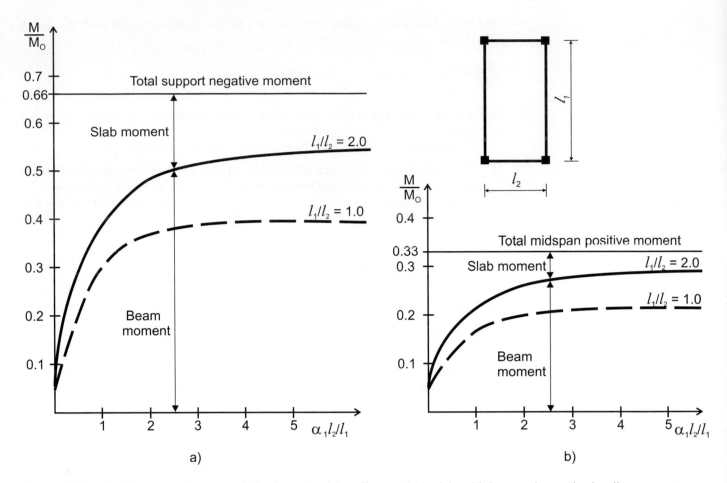

Figure 12.12 Bending moments versus relative beam-to-slab stiffness ratio in slabs with beams: a) negative bending moments at the supports, and b) positive bending moments at midspan.

cracking has taken place in the column regions. The resulting increase in moments could cause cracking in midspan regions; this will be accompanied by a drop in bending moments and flexural stiffness. This phenomenon is known as *moment redistribution*: bending moments redistribute (shift) between cracked and uncracked regions. The concept of moment redistribution was first introduced in Section 10.9 in relation to continuous beams and one-way slabs. With the further increase in gravity loads, moment redistribution will continue due to yielding of steel reinforcement. The ultimate load-bearing capacity will be reached when the reinforcement yields and experiences plastic deformations in localized regions called yield lines, which will be explained in Section 12.8.

A significant moment redistribution in two-way slabs subjected to gravity loads was reported by Rangan and Hall (1984). They tested a few half-scale models of the end panel of a flat plate system, as illustrated in Figure 12.13a. The models were subjected to progressively increasing gravity loads, and the behaviour was monitored at three load levels: service load (5.4 kPa), factored design load (9.0 kPa), and ultimate load/failure (22.5 kPa). Transverse moment distribution in the vicinity of column A (Section 1-1) is shown in Figure 12.13b. The ultimate moment at the column (M_{ULT}) was approximately 2.6 times higher than the design moment (M_{DES}). This indicates a significant reserve in bending moment capacity beyond the design load level.

It can be also seen that the moment magnitudes are highest at the column face, and that there is a significant drop in moment values just beyond. Moment redistribution was also observed in longitudinal direction, along Section 2-2 parallel with column line AB (see Figure 12.13c). A significant increase was observed in the bending moment resisted by the columns A and B relative to the total moment (moment gradient) M_o. For example, the bending

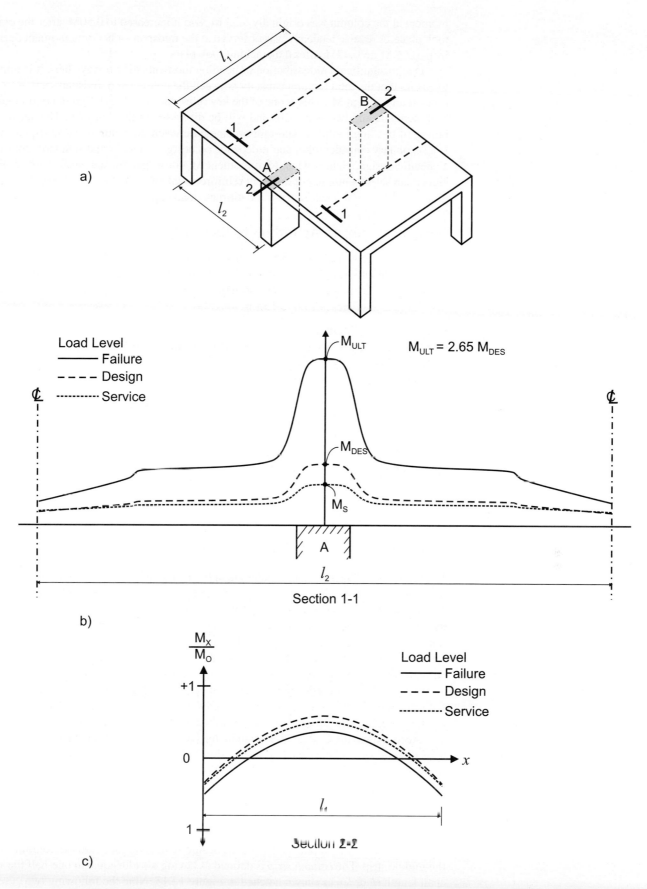

Figure 12.13   Moment redistribution in flat slabs: a) model layout; b) transverse moment distribution — Section 1-1 at the column A, and c) moment distribution in longitudinal direction — Section 2-2
(adapted from Rangan and Hall, 1984 with the permission of the American Concrete Institute).

moment at the column was originally 0.33 M_o and it increased to 0.50M_o after the cracking took place. A reverse tendency was observed at the midspan — bending moment decreased from 0.66 M_o to 0.47M_o after the cracking took place.

Due to significant redistribution of bending moments in two-way slabs, it is important to ensure that the total flexural capacity between the column and midspan sections is equal to the total moment M_o. This is one of the key rules related to the design of continuous flexural members and two-way slabs and will be discussed in Section 12.6. The actual distribution of bending moments and reinforcement between the column and midspan sections is a secondary consideration, and it depends on design practices and standards followed in a specific design. Hence, when total moment M_o for a specific slab span is given, the designer can select a number of different reinforcement amounts for midspan and column regions, which would result in the same ultimate load capacity for the span.

12.3.5 Design for Flexure According to CSA A23.3

According to CSA A23.3 Cl.13.5.1, a two-way slab system can be designed using any procedure which satisfies conditions of equilibrium and compatibility, provided that the strength and serviceability requirements have been met. The following design procedures are outlined in CSA A 23.3 (Clauses 13.6 to 13.9):

- Direct Design Method
- Elastic Frame Method (also known as the Equivalent Frame Method)
- Elastic Plate Theory, and
- Theorems of Plasticity (e.g. the Yield Line Method).

The design of two-way slab systems must be performed considering both gravity and lateral loads. Some of the CSA A23.3 design procedures, like the Direct Design Method, can be used for gravity load analysis only, while others can be used both for gravity and lateral load analysis. Common design procedures will be explained in detail and illustrated by design examples later in this chapter.

12.4 DEFINITIONS

12.4.1 Design Strip

A23.3 Cl.13.1.2

A *design strip* is the portion of a slab system bound laterally by the centrelines of the panels on each side of the column (CSA A23.3 Cl.13.9.2.1). This concept is illustrated in Figure 12.14, which shows a partial plan of a two-way floor slab without beams (flat plate). A typical slab panel is enclosed by four columns (e.g. panel ABEF shown in Figure 12.14a). A portion of the slab at each floor level may be modelled as an equivalent beam, and its width is referred to as a design strip. Equivalent beams at each floor level along with the supporting columns constitute a frame. In a general design scenario, several frames in both directions, east-west (E-W) and north-south (N-S), need to be considered. A typical frame ABC in the E-W direction is shown in Figure 12.14a, and the frame DBF in the N-S direction is shown in Figure 12.14b. The design strips in each direction are shown hatched in the figure.

According to the CSA A23.3 notation, the longitudinal direction of the frame is referred to as direction 1, and the span in this direction is denoted as l_1 (see Figure 12.14). The other (transverse) direction is referred to as direction 2 and the corresponding span is denoted as l_2. Both l_1 and l_2 are measured centre to centre of the support (column). The width of the design strip is denoted as l_{2a}. In many instances, slab span lengths are variable. In that case, the width of design strip l_{2a} needs to be determined by considering average l_2 value for the two slab spans adjacent to the gridline under consideration.

In flat slab and flat plate systems, the design strip is divided into the column strip and the middle strip. The *column strip* is defined as having a width equal to one-half the smaller span length (l_1 or l_2), as shown hatched in Figure 12.15. Note the following two cases:

1) For $l_1 < l_2$, an interior column strip width is equal to $l_1 / 2$ (see Figure 12.15a), and
2.) For $l_1 \geq l_2$, an interior column strip width is equal to $l_2 / 2$ (see Figure 12.15b).

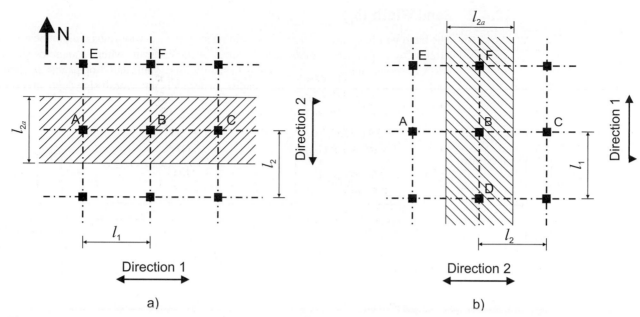

Figure 12.14 Design strip: a) frame ABC in the E-W direction, and b) frame DBF in the N-S direction.

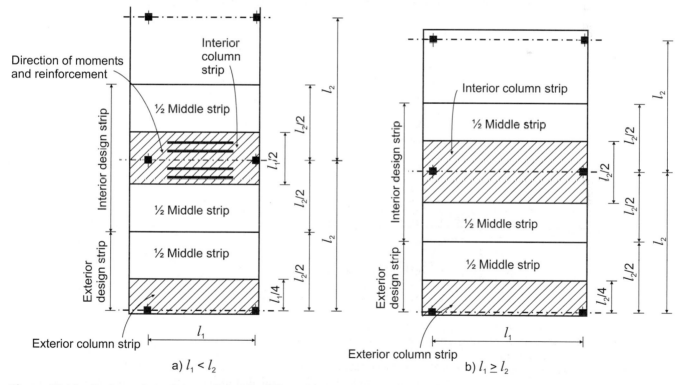

Figure 12.15 Design strip, column strip, and middle strip: a) $l_1 < l_2$, and b) $l_1 \geq l_2$ (courtesy of the Portland Cement Association).

The width of a column strip is determined using the smaller of l_1 or l_2, to account for the tendency for bending moment to concentrate about the column gridline when the span length of the design strip is smaller than its width.

The *middle strip* is a portion of the design strip outside the column strip. Each middle strip is bounded by two column strips.

Note that there are two types of design strips, depending on the frame location within a building plan: *interior* and *exterior* design strips, as shown in Figure 12.15. An interior design strip consists of a column strip and two half-middle strips, while an exterior strip consists of a column strip and one half-middle strip.

12.4.2 Band Width (b$_b$)

A23.3 Cl.2.3 │ The band width, b_b, is a portion of the column strip in two-way slabs without beams, which extends by a distance of $1.5h_s$ past the sides of the column (where h_s denotes slab thickness), as shown in Figures 12.16a and b. In slabs with capitals and drop panels, b_b should extend a distance $1.5h$ past the column capital, where h is the overall slab depth at drop panel location (see Figure 12.16c).

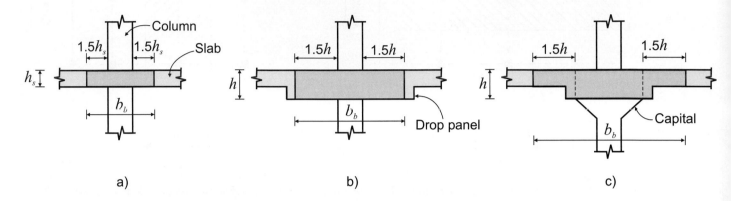

Figure 12.16 Band width b_b: a) flat plate; b) flat slab with drop panels, and c) flat slab with capitals and drop panels.

12.4.3 Clear Span

A23.3 Cl.13.9.2.3 │ *Span* denotes centre-to-centre distance between supports (e.g. l_1 span in Figure 12.17), while *clear span* denotes clear distance between supports (e.g. l_n span in Figure 12.17). CSA A23.3 Cl.13.9.2.3 requires that $l_n \geq 0.65l_1$. In some cases, slabs are supported by circular or polygonal-shaped columns. In this case, it is recommended that the supports are treated as square sections with equivalent cross-sectional area.

Figure 12.17 Clear span l_n for slabs supported by circular and polygonal columns
(adapted from CAC, 2005 with the permission of the Cement Association of Canada).

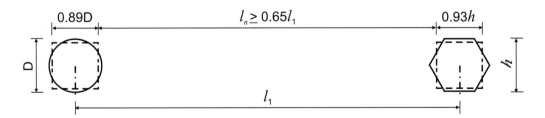

12.4.4 Effective Beam Section

A23.3 Cl.13.8.2.7 │ In two-way slabs with beams, a portion of the slab acts together with the beam as an L- or a T-section, as shown in Figure 12.18 (note that the concept of T-beams was introduced in Section 3.7). L-sections are found in end spans (edge beam shown in Figure 12.18a), while T-sections are characteristic for typical interior spans (interior beam shown in Figure 12.18b). The effective flange width for two-way slabs (noted as b_s in the figure) is different from the effective flange width for T-beams in one-way slabs discussed in Section 5.8.

The effective beam section for the two-way slab design is shown shaded in Figure 12.18, where

h_s = slab thickness
h = overall beam depth
h_w = beam web depth (below the slab soffit)
b_s = effective flange width
b_w = beam web width

Figure 12.18 Effective beam section: a) edge beam, and b) interior beam.

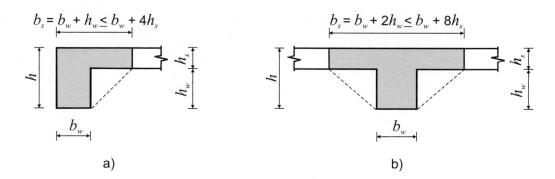

$$b_s = b_w + h_w \le b_w + 4h_s$$

$$b_s = b_w + 2h_w \le b_w + 8h_s$$

a)

b)

12.4.5 Beam-to-Slab Stiffness Ratio (α)

Consider an elevation of a two-way slab with beams as shown in Figure 12.19a. Beam cross-sectional dimensions, slab thickness, and spacing of adjacent beams will influence the relative stiffness distribution for beams and the slab. This is relevant for the design of two-way slabs with beams. The effect of beam stiffness on deflections and moment distribution in the slab can be taken into account through the beam-to-slab stiffness ratio, α. This section explains a procedure for finding α value for the design of two-way slabs with beams.

First, let us identify equivalent T- and L-beam sections, consisting of a beam web and a portion of the slab (note the sections shown shaded in the figure). These beams span between the column centers (in the direction perpendicular to the drawing). Flexural stiffness for a beam (k) can be determined from the following equation

$$k = \frac{4EI}{L} \tag{12.1}$$

where L denotes the beam span, E denotes the modulus of elasticity of concrete, and I denotes the moment of inertia for the beam section.

Next, let us divide the slab into sections, where section width is defined by adjacent panel centrelines and its depth is equal to the slab thickness. Slab sections for an end span and a typical interior span are shown shaded in Figure 12.19b. Flexural stiffness for a slab section can be determined from Eqn 12.1, by computing the moment of inertia for a rectangular section as shown in Figure 12.19c. In cast-in-place concrete construction, beams and slabs are placed monolithically, thus their E and L values are equal (provided that the same concrete mix was used for the beam and slab pour, otherwise the E value could be different). As a result, Eqn 12.1 can be simplified as follows

$$\alpha = \frac{I_b}{I_s} \tag{12.2}$$

where I_b denotes the beam moment of inertia and I_s denotes the slab moment of inertia. The dimensions of an effective beam section can be determined as shown in Figure 12.18. Note that an L-section is used to determine the moment of inertia for the edge beam, I_{be}, while a T-section is used for a typical interior beam (moment of inertia I_{bi}), as shown in Figure 12.19a.

The moment of inertia for a slab section can be determined as follows (see Figure 12.19c)

$$I_s = \frac{b_s \times h_s^3}{12}$$

where

$b_s = l_{be}$ section width for an end span (corresponding to the moment of inertia I_{se})
$b_s = l_{bi}$ section width for an interior span (corresponding to the moment of inertia I_{si})

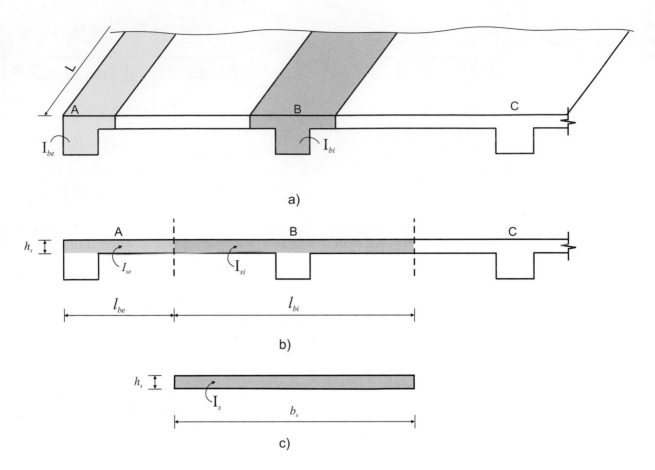

Figure 12.19   Beam and slab properties for an end span and an interior span: a) beam sections; b) slab properties, and c) a typical slab cross-section.

The beam moment of inertia, I_b, can be determined from the first principles, or an approximate value can be determined from the following simplified equation (CSA A23.3 Cl.13.2.5)

A23.3 Eq. 13.4 $I_b = \dfrac{b_s h^3}{12}\left[2.5\left(1-\dfrac{h_s}{h}\right)\right]$

The corresponding α value can be determined by substituting the I_b expression into Eqn 12.1 as follows

$$\alpha = \dfrac{2.5 b_w}{b_s}\left(\dfrac{h}{h_s}\right)^3\left(1-\dfrac{h}{h_s}\right) \qquad\qquad \textbf{[12.3]}$$

Alternatively, the beam moment of inertia can be determined from charts included in the Concrete Design Handbook (CAC, 2005).

12.5 │ GENERAL CSA A23.3 DESIGN PROVISIONS

12.5.1 Regular Two-Way Slab Systems

A23.3 Cl.2.2   Some of the CSA A23.3 design methods, namely the Equivalent Frame Method and the Direct Design Method, can be applied only to regular two-way slab systems. The reason is that most provisions related to these design methods are based on research studies performed on regular slab systems. A *regular* two-way slab system consists of approximately rectangular panels and supports primarily uniform gravity loading.

According to CSA A23.3, a regular slab system must meet the following geometric limitations illustrated in Figure 12.20:

(#a) within a panel, the ratio of longer to shorter span, centre-to-centre of supports, is not greater than 2.0, that is,

$$l_1 / l_2 \leq 2.0$$

(#b) for slab systems with beams between supports, the relative effective stiffness of beams in the two directions should be restricted as follows:

$$0.2 \leq \frac{\alpha_1 l_2^2}{\alpha_2 l_1^2} \leq 5.0$$

where α_1 and α_2 denote the beam-to-slab stiffness ratio for beams in directions 1 and 2 respectively (refer to Section 12.4.5 for an explanation of the beam-to-slab stiffness ratio)

(#c) column offsets are not greater than 20% of the span (in the direction of the offset) from either axis between the centrelines of successive columns; and

(#d) flexural reinforcement is placed on an orthogonal grid.

Note that the requirements #a, #c, and #d apply to flat plates and flat slabs, while an additional requirement (#b) applies to slabs with beams.

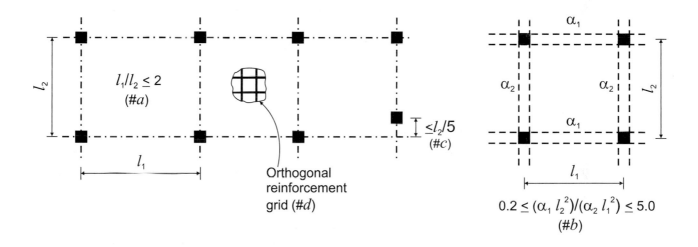

Figure 12.20   A summary of the CSA A23.3 requirements for regular two-way slab systems (#a to #d).

12.5.2   Minimum CSA A23.3 Slab Thickness Requirements for Deflection Control

A23.3 Cl.13.2   **Minimum Slab Thickness**   The thickness of two-way slabs under normal loading conditions is usually determined by deflection considerations. CSA A23.3 Cl.13.2 prescribes the minimum slab thickness, h_s, for the different types of two-way slabs (with and without beams), which will be discussed in this section. These minimum slab thickness/span ratios enable the designer to avoid detailed deflection calculations in routine designs; this is similar to the indirect approach for deflection control in flexural members discussed in Section 4.5.2. Note that the CSA A23.3 minimum thickness values are independent of design loading and concrete compressive strength (f_c'), and may lead to conservative design solutions in some cases. For example, the designer may be able to reduce thickness for slabs subjected to light loading (e.g. residential occupancies) by performing detailed deflection calculations. CSA A23.3 approaches for detailed deflection calculations are outlined in Section 12.10.

CSA A23.3 Cl.13.2.1 states that the minimum slab thickness, h_s, shall be based on serviceability requirements but should not be less than 120 mm. NBCC fire resistance requirements also limit the minimum slab thickness depending on the fire rating (see Section 1.8.4). For example, a minimum 130 mm thickness is required for a two-hour fire rating (Appendix D of NBCC 2010).

A23.3 Cl.13.2.3

Flat Plates   The minimum thickness, h_s, for the slab without drop panels depends on the slab span and the steel yield strength, f_y. The minimum thickness can be determined from the following equation:

A23.3 Eq. 13.1

$$h_s \geq \frac{l_n(0.6 + f_y/1000)}{30} \qquad \text{[12.4]}$$

where

f_y = steel yield strength (MPa)

l_n = longer clear span

Two-way slabs usually have same or similar spans in two orthogonal directions. However, in some cases these spans are different, as illustrated in Figure 12.21. The clear span, l_n, for slab thickness calculations (longer clear span) can be determined as follows

$$l_n = \max(l_{n1}, l_{n2})$$

For example, the clear span can be determined for the slab panel shown in Figure 12.21
Since

$$l_{n1} > l_{n2}$$

it follows that

$$l_n = l_{n1}$$

When Grade 400 steel is used, which is standard in Canada, f_y = 400 MPa, and the above equation can be simplified as follows:

$$h_s \geq \frac{l_n}{30} \qquad \text{[12.5]}$$

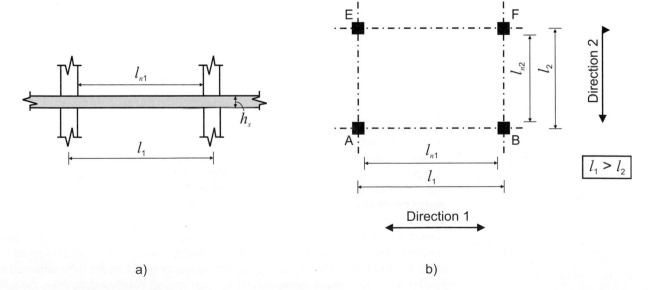

Figure 12.21   A typical two-way slab panel: a) elevation, and b) plan.

CSA A23.3 prescribes an additional requirement for end spans (discontinuous edges). In order to use the same equation, an edge beam should be provided with the stiffness ratio $\alpha \geq 0.80$. Alternatively, the minimum slab thickness should be increased by 10 % for panels with discontinuous edge(s), that is, the slab thickness should be at least $1.1h_s$, where h_s is determined from Eqn 12.4. These requirements are illustrated in Figure 12.22.

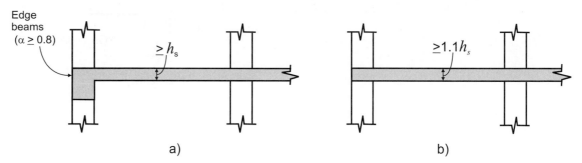

Edge beams ($\alpha \geq 0.8$)

$\geq h_s$

$\geq 1.1h_s$

a)   b)

Figure 12.22 Additional requirements for end spans: a) a slab with edge beam, and b) a slab without edge beam.

| A23.3 Cl.13.2.4 |

Flat Slabs (Slabs With Drop Panels) Flat slabs have drop panels, which are formed by thickening the bottom of the slab around the columns, as shown in Figure 12.23. Drop panels are effective in increasing slab stiffness in the areas around the columns; this results in smaller deflections compared to flat plates. For that reason, the minimum slab thickness, h_s, is somewhat reduced and it can be determined from the following equation

| A23.3 Eq. 13.2 |

$$h_s \geq \frac{l_n(0.6 + f_y/1000)}{30} - \left(\frac{2x_d}{l_n}\right)\Delta_h \qquad\qquad [12.6]$$

where

Δ_h = additional thickness of the drop panel underneath the slab, and

x_d = drop panel overhang (dimension from the face of the column to the edge of drop panel).

The smaller of the values determined in the two directions should be used for the slab thickness calculation, that is, $x_d = \min(x_{d1}, x_{d2})$ (see Figure 12.23).
The following dimensional limits should be met

$\Delta_h \leq h_s$

and

$x_d \leq l_n/4$

Note that the maximum x_d value results in a decrease in slab depth by approximately 45 mm.

Additional requirement for slab thickness at discontinuous edges is the same as that related to flat plates (see Figure 12.22).

| A23.3 Cl.13.2.5 |

Slabs with Beams Between All Supports The minimum thickness for slab panels with beams on all sides depends on the panel aspect ratio and the relative stiffness of beams in two directions. The minimum thickness, h_s, for slabs with beams between all supports is equal to

| A23.3 Eq. 13.3 |

$$h_s \geq \frac{l_n(0.6 + f_y/1000)}{30 + 4\beta\alpha_m} \qquad\qquad [12.7]$$

where

β = ratio of clear spans in long and short directions; for example, $\beta = l_1/l_2$ for a panel shown in Figure 12.21, and

Figure 12.23   Drop panel dimensions.

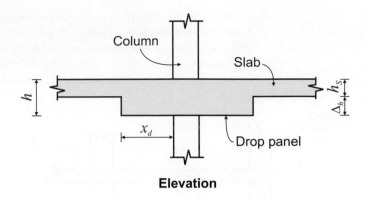

Elevation

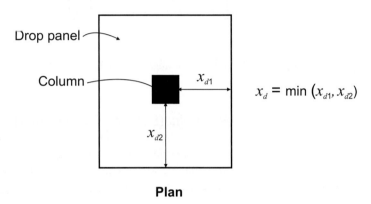

$$x_d = \min\,(x_{d1}, x_{d2})$$

Plan

α_m = average beam-to-slab stiffness ratio.

Note that

$1.0 \le \beta \le 2.0$, otherwise the slab should be treated as a one-way slab.

The beam-to-slab stiffness ratio, α_m, is an average value obtained considering all beams along the panel edges, and it can be determined using the procedures outlined in Section 12.4.5. For the given h_s value, the required α_m value can be determined from the following equation

$$\alpha_m = \frac{1}{4\beta}\left[\frac{l_n(0.6 + f_y/1000)}{h_s} - 30\right]$$

Relative stiffness of beams in the two orthogonal directions is an important parameter influencing the deflections in two-way slabs. For example, when beams in one direction are significantly stiffer, the slab tends to act as a one-way slab spanning between the stiffer beams (even if columns are located on an essentially square grid). Note that CSA A23.3 gives an upper bound to the α_m value as follows

$$\alpha_m \le 2.0$$

Minimum slab thicknesses for various slab types and different values of the key parameters β and α_m are summarized in Table 12.1.

Table 12.1   Minimum Thickness for Two-Way Slab Systems (Grade 400 Reinforcement)

Two-way Slab System	α_m	β	Minimum h_s
Flat plate with edge beams	-	≤ 2.0	$l_n/30$
Flat plate without edge beams	-	≤ 2.0	$l_n/27$
Slab with beams between all supports	1.0	1.0	$l_n/34$
		2.0	$l_n/38$
	≤ 2.0	1.0	$l_n/38$
		2.0	$l_n/46$

12.6 DESIGN FOR FLEXURE ACCORDING TO THE DIRECT DESIGN METHOD

A23.3 Cl.13.9

This section describes the underlying concepts and design provisions for the Direct Design Method (DDM), a statics-based method for the design of two-way slabs adopted by CSA A23.3 and other codes. The application of this method will be demonstrated through two design examples.

12.6.1 Limitations

A23.3 Cl.3.9.1

CSA A23.3 prescribes that the DDM can be used when the following requirements have been met (see Figure 12.24):

#1 A slab must be regular (refer to Section 12.5.1 for more details on regular two-way slabs).

#2 There must be at least three continuous spans in each direction.

#3 The successive span lengths (centre-to-centre of supports) in each direction must not differ by more than one-third of the longer span.

#4 The DDM can be used only for gravity load analysis; gravity loads must be uniformly distributed over the entire slab panel.

#5 The factored live load must not exceed two times the factored dead load.

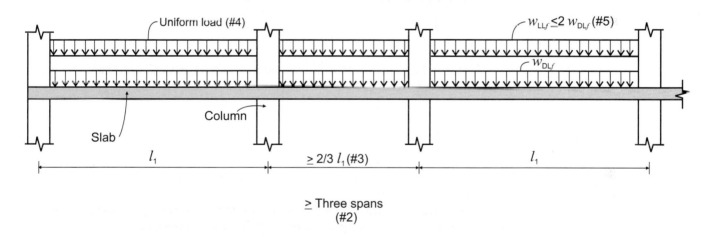

Figure 12.24   Limitations of the Direct Design Method.

Requirement #4 states that the DDM can be used only for gravity load analysis. The design considers a partial frame that consists of the design strip for the slab at a specific floor level and columns above and below that level, as shown in Figure 12.25 (the concept of design strip was introduced in Section 12.4.1). A similar model can be used to design slab at the top floor level. Note that lateral swaying of the frame is prevented by a roller support at the far end of the slab. Refer to Section 12.7.2 for discussion on frame models for gravity load analysis.

12.6.2 The Concept

A23.3 Cl.13.9.2

The distribution of bending moments in a two-way slab according to the DDM will be explained by an example. Let us first explain moment distribution in the longitudinal direction. Consider the span AB of a flat plate system shown in Figure 12.26a. The DDM is based on plane frame model and treats the slab as a wide beam with a width equal to the design strip. The slab is subjected to uniform area load w, which needs to be transformed into linear load $w \cdot l_{2a}$ (based on the design strip width l_{2a}), as shown in Figure 12.26b. A

Figure 12.25 A gravity frame model for the DDM.

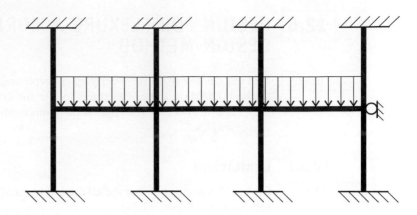

conceptual diagram illustrating moment distribution for span AB can be seen in Figure 12.26c. End moments M_A and M_B are negative due to the restraints (columns) at points A and B, while the midspan moment at point C (M_C) is positive. Moment distribution is similar to that for continuous beams and slabs discussed in Section 10.2.2. Magnitudes of bending moments at points A, B, and C depend on several factors, including the end support conditions and the type of slab system (slab on beams or flat plate/slab). However, the magnitude of moment gradient M_o is always equal to the sum of average value for bending moments at the supports A and B and the moment at the midspan C, as shown below

$$M_o = \frac{M_A + M_B}{2} + M_C$$

However, M_o is also equal to the maximum moment of an equivalent simply supported beam AB with the span l_n, subjected to uniform load $w \times l_{2a}$, that is,

$$M_o = \frac{(w \times l_{2a}) \times l_n^2}{8}$$

Note that bending moments are determined based on the clear span l_n (instead of the centre-to-centre span l_l). This is similar to the design approach for continuous beams and slabs presented in Chapter 10.

The above statement can be proven by considering a free-body diagram shown in Figure 12.26d. The beam support reaction at point A, R_A, is equal to

$$R_A = (w \times l_{2a}) \times l_n / 2$$

and the bending moment at point C is equal to

$$M_c = R_A \times l_n / 2 - (w \times l_{2a})(l_n / 2)(l_n / 4) = \frac{(w \times l_{2a}) \times l_n^2}{8}$$

which is equal to the moment gradient M_o.

Note that the DDM provisions refer to moment gradient M_o as the *total factored static moment* (CSA A23.3 Cl.13.9.2). M_o can be determined from the following equation:

| A23.3 Eq. 13.23 | $M_o = \dfrac{w_f \times l_{2a} \times l_n^2}{8}$ | **[12.8]** |

where

w_f = factored load per unit area of the slab

l_{2a} = width of the design strip

l_n = clear span, that is, length of span measured face-to-face of supports (columns, capitals, brackets, or walls).

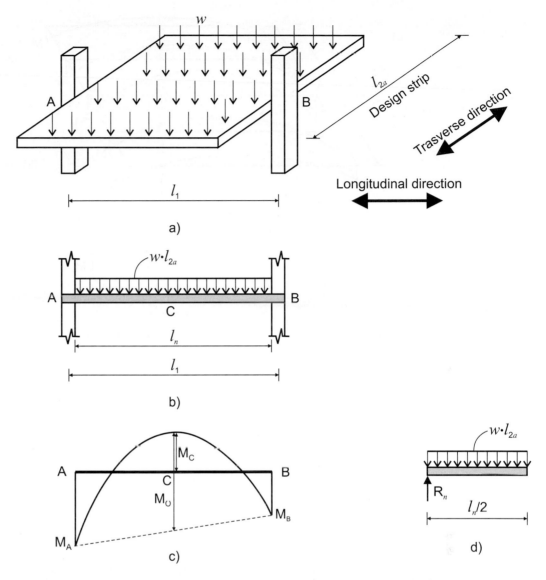

Figure 12.26 Bending moment distribution within a slab span: a) an isometric (3-D) model of the slab span; b) a linear (2-D) model; c) bending moment distribution, and d) a free-body diagram.

Next, let us discuss transverse distribution of bending moments. Figure 12.27a shows variation of bending moments in transverse direction. Note that points A and B denote the column locations, however columns have been omitted from the drawing for clarity. It can be seen that bending moments at support B are largest at the column location (e.g. moment M_3), and that the values drop towards the ends, that is, moment M_1 is the smallest. It should be noted that the distribution is symmetrical with regards to the column gridline. Moment variation at support B in transverse direction is shown in Figure 12.27b. It can be seen that the moments vary in a nonlinear manner; however, average bending moments are used for design. Two different bending moment values are assigned: larger value is assigned to a region close to the column lines (moment M_{ga}), and smaller value is assigned to a portion of the slab close to the panel centreline (moment M_{MB}).

In flat plate and flat slab systems these regions are called "column strip" and "middle strip", as shown in Figure 12.27c. Note that, in slab systems with beams, these regions are called "beam strip" (instead of "column strip") and "slab strip" (instead of "middle strip"). CSA A23.3 provisions for bending moment distribution in two-way slabs according to the DDM are outlined in the following sections.

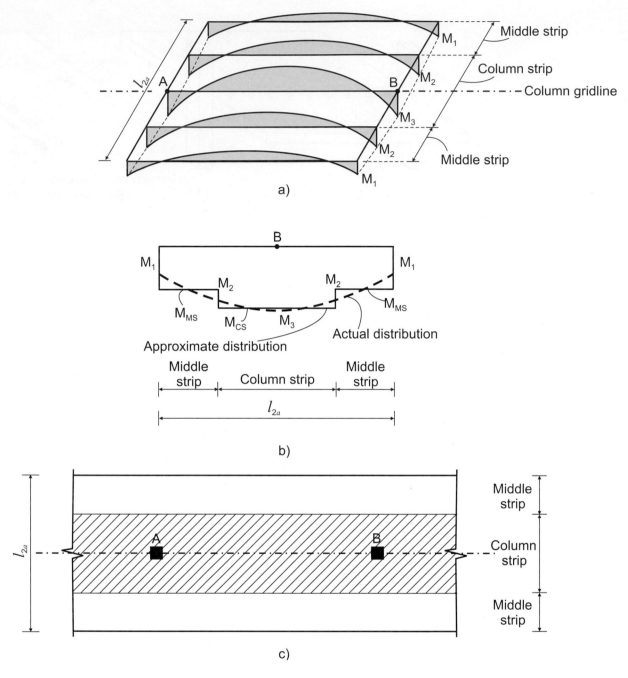

Figure 12.27 Transverse distribution of bending moments in a flat plate: a) an isometric view showing variation of bending moments; b) transverse distribution at support B, and c) column strip and middle strip.

12.6.3 Bending Moment Distribution in Flat Plates and Flat Slabs

A23.3 Cl.13.9.3 and 13.11

This section discusses distribution of the total factored moment, M_o, within a specific span of a two-way slab, which is performed in the following two steps:

1) Distribute M_o between critical locations (supports and midspan) - this is referred to as *distribution in longitudinal direction*, and

2) For each critical location, distribute the moment obtained in the previous step to column strip and middle strip — this is referred to as *transverse distribution*.

The distribution of bending moments in two-way slabs depends on the location of a slab span within a frame. For example, bending moment values in end spans are different compared to interior spans. Note that an exterior column in the end span is called an "exterior support", while all other columns are referred to as "interior supports" (see Figure 12.28).

Figure 12.28 Exterior and interior supports.

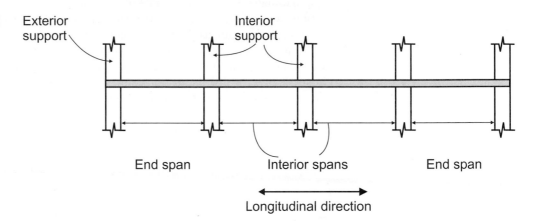

Bending moment distribution within a specific span of a flat slab depends on several factors, including the type of end supports (restrained/unrestrained), the type of slab (flat slab or flat plate), location of the span within a building (end span or interior span), and the location within a span (support or midspan). The sign of a bending moment depends on the location within a span — bending moments at the supports are negative, while the moments at the midspan are positive. Bending moment values are summarized in Tables 12.2 to 12.4. Note that most values refer to flat plates, while the values for flat slabs (where they are different) are shown in the notes beneath each table.

Table 12.2 shows the most typical case: a flat plate supported by columns at all points of support (including the exterior supports). Note the labelling for moments at critical sections, e.g. M_1 denotes the bending moment at the exterior support etc. The same labelling scheme is used in all tables.

Note that the moment distribution in the end span depends on the type of exterior support. The slab may be cast monolithically with a continuously reinforced concrete wall — this is referred to as a "fully restrained exterior edge" (see Table 12.3). Alternatively, a slab end span may be supported by a support which enables rotation - this is referred to as an "unrestrained exterior edge" (see Table 12.4).

These tables show the bending moments values at critical sections within a span (supports and midspan), that is, in longitudinal direction. Moment values are expressed as a fraction of M_o (according to CSA A23.3 Cl.13.9.3). Subsequently, each of these moments needs to be distributed transversely to the column strip and the middle strip; the corresponding moment values are also included in the tables. CSA A23.3 Cl.13.11 prescribes a range of moment values for column strip and middle strips as a fraction of the bending moment at a critical section (support/midspan).

Bending moments at the support may be different for two adjacent slab spans, but the designer should design slab sections for the larger of the two moments (Cl.13.9.3.4). For example, a negative bending moment at the first interior support (Section 3e) is equal to 0.70 M_o while the moment at the same support corresponding to an interior span (Section 3i) is -0.65 M_o. The designer should use the moment with the larger absolute value (-0.70 M_o) for the design at this location; an alternative would be to adjust spans so that the end span is shorter.

Note that Cl.9.3.3 permits an increase or decrease in negative or positive factored moments by 15%, provided that the total static moment M_o for a span in the direction under consideration is not less than that required by Eqn 12.8. This is known as "moment balancing" and it will be illustrated through Example 12.1. Note that the sum of positive and negative moments within a span must remain equal to M_o.

According to Cl.13.11.2, a major portion of the transverse bending moment at a critical section (support/midspan) is assigned to the column strip. The remaining portion is distributed to slab sections on both sides of the column strip, also known as middle strips (note that a middle strip is divided into two halves). The column strip and the middle strip constitute the design strip, as shown in Figure 12.15.

Transverse distribution of bending moments according to the DDM will be illustrated by the following example. Consider a typical interior span of a two-way flat plate shown in Figure 12.29a. The distribution of bending moments at the critical sections in the longitudinal direction for an interior span is illustrated in Figure 12.29b. The positive bending

Table 12.2   Flat slab or flat plate supported directly by columns (partially restrained exterior edge)

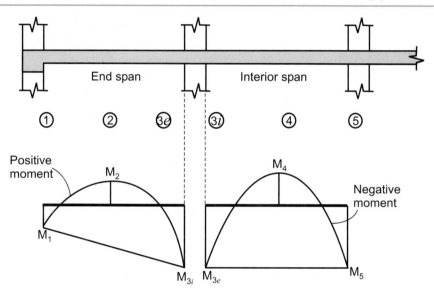

Type of span		\multicolumn{3}{c}{End Span}			\multicolumn{3}{c}{Interior Span}
	Section Location	(1) Exterior	(2) Midspan	(3e) First Interior Support	(3i) Interior Support	(4) Midspan Support	(5) Interior Support
		M_1	M_2	M_{3e}	M_{3i}	M_4	M_5
	Sign	Negative	Positive	Negative	Negative	Positive	Negative
Longitudinal Direction	Total Moment	$-0.26\,M_o$	$+0.52\,M_o$	$-0.70\,M_o$	$-0.65\,M_o$	$+0.35\,M_o$	$-0.65\,M_o$
Transverse	Column Strip Moment	$-0.26\,M_o$	$+(0.29 \text{ to } 0.34)\,M_o$	$-(0.49 \text{ to } 0.63)\,M_o$	$-(0.46 \text{ to } 0.59)\,M_o$	$+(0.19 \text{ to } 0.23)\,M_o$	$-(0.46 \text{ to } 0.59)\,M_o*$
Distribution	Middle Strip Moment	0	\multicolumn{5}{c}{= Total Moment - Column Strip Moment}

Note
$* = -(0.49 \text{ to } 0.59)\,M_o$ for flat slabs (with drop panels)

Table 12.3 Flat slab or flat plate with a fully restrained exterior edge

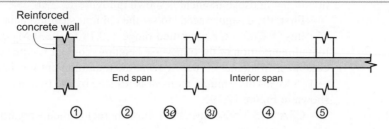

Type of span		End Span			Interior Span		
Section Location		(1) Exterior Support	(2) Midspan	(3e) First Interior Support	(3i) Interior Support	(4) Midspan	(5) Interior Support
		M_1	M_2	M_{3e}	M_{3i}	M_4	M_5
	Sign	Negative	Positive	Negative	Negative	Positive	Negative
Longitudinal Direction	Total Moment	$-0.65\,M_o$	$+0.35\,M_o$	$-0.65\,M_o$	$-0.65\,M_o$	$+0.35\,M_o$	$-0.65\,M_o$
Transverse	Column Strip Moment	$-0.65\,M_o$	$+(0.19 \text{ to } 0.23)\,M_o$	$-(0.46 \text{ to } 0.59)\,M_o$	$-(0.46 \text{ to } 0.59)\,M_o{}^*$	$+(0.19 \text{ to } 0.23)\,M_o$	$-(0.46 \text{ to } 0.59)\,M_o$
Distribution	Middle Strip Moment	0		= Total Moment - Column Strip Moment			

Note
$* = -(0.49 \text{ to } 0.59)\,M_o$ for flat slabs (with drop panels)

Table 12.4 Flat slab or flat plate with an unrestrained exterior edge

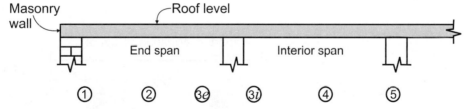

Type of span		End Span			Interior Span		
Section Location		(1) Exterior Support	(2) Midspan	(3e) First Interior Support	(3i) Interior Support	(4) Midspan	(5) Interior Support
		M_1	M_2	M_{3e}	M_{3i}	M_4	M_5
	Sign	Negative	Positive	Negative	Negative	Positive	Negative
Longitudinal Direction	Total Moment	0	$+0.66\,M_o$	$-0.75\,M_o$	$-0.65\,M_o$	$+0.35\,M_o$	$-0.65\,M_o$
Transverse	Column Strip Moment	0	$+(0.36 \text{ to } 0.43)\,M_o$	$-(0.53 \text{ to } 0.68)\,M_o$	$-(0.46 \text{ to } 0.59)\,M_o{}^*$	$+(0.19 \text{ to } 0.23)\,M_o$	$-(0.46 \text{ to } 0.59)\,M_o{}^*$
Distribution	Middle Strip Moment	0		= Total Moment - Column Strip Moment			

Note
$* = -(0.49 \text{ to } 0.59)\,M_o$ for flat slabs (with drop panels)

moment at the midspan (column 4 in Table 12.2) is equal to $+0.35\,M_o$. Transverse distribution of the midspan moment is shown in Figure 12.29d.

First, the designer needs to set the column strip moment. Let us set the value to $+0.2\,M_o$; this is within the permitted range $+(0.19$ to $0.23)\,M_o$ according to Table 12.2. The remaining portion of the transverse bending moment is equal to the difference between the column strip moment $(+0.2\,M_o)$ and the total moment for that section $(+0.35\,M_o)$, that is, $(+0.15\,M_o)$. This bending moment is resisted by the two mid-halves of the middle strip, as shown in Figure 12.29c.

CSA A23.3 prescribes the following requirements regarding the bending moments at column locations:

a) Interior columns: according to Cl.13.11.2.7, slab band b_b should be designed to resist at least one-third of the total factored negative moment at the column section equal to $-0.65\,M_o$ (note that the band b_b was introduced in Section 12.4.2). This is illustrated in Figure 12.30a. The total moment for an interior column section (M_{3i}) is equal to

Figure 12.29 Transverse distribution of bending moments in a flat plate: a) a typical interior span; b) the moment distribution at critical sections (supports and midspan); c) column and middle strips, and d) the transverse moment distribution at midspan - an isometric view.

a)

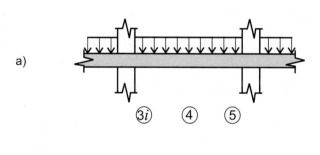

b)

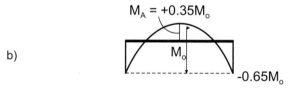

c)

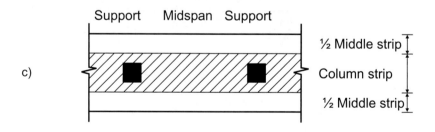

d)

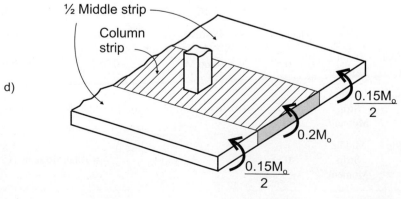

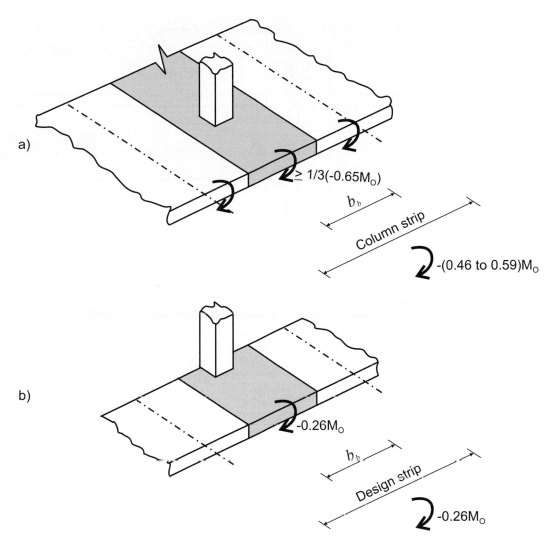

Figure 12.30 Moment distribution at columns: a) an interior column, and b) an exterior column.

$(-0.65M_o)$ (see Table 12.2). One-third of that moment is assigned to band b_b centered at the column. The moment for the column strip is equal to $-(0.46$ to $0.59)M_o$. The remaining bending moment, equal to the difference between the column strip moment and the moment at band b_b, should be resisted by the remaining portion of the column strip outside band b_b.

b)   Exterior columns: according to Cl.13.10.3, the total factored negative moment at an exterior column (M_1) equal to $-0.26 M_o$ should be resisted by the band b_b. This is illustrated in Figure 12.30b.

12.6.4   Bending Moment Distribution in Slabs With Beams Between all Supports

A23.3 Cl.13.9.3 and 13.12 Transverse distribution of bending moments in slabs with beams between all of their supports is different than that in flat slabs and flat plates. The design strip is divided into a beam strip and a slab strip. The beam strip is located in the proximity of column lines, similar to the column strips in flat plates and flat slabs. The width of the beam strip is equal to the effective flange width, b_v, shown in Figure 12.18. The remaining portion of the design strip is called the slab strip, and it is divided into two half-strips, as shown in Figure 12.31a.

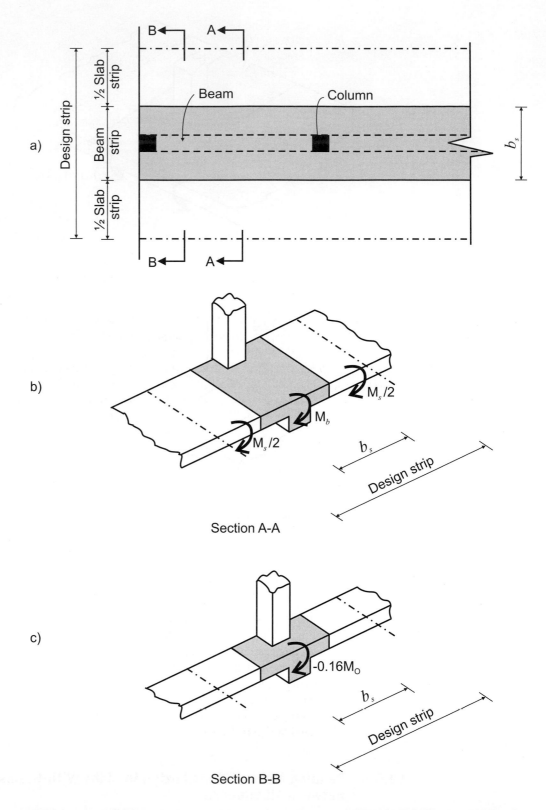

Figure 12.31 Design strip in a two-way slabs with beams: a) a plan view; b) an isometric view showing beam and slab strips and the corresponding bending moments, and c) moment distribution at the exterior column.

First, the bending moment (M) is determined at the critical section for the longitudinal direction (at the supports or the midspan), and the value is expressed as a fraction of the total factored moment, M_o. The transverse bending moments are resisted by the beam and slab strips (see Figure 12.31b), that is,

$$M = M_b + M_s$$

where M_s is the moment for the slab strip, and M_b is the moment for the beam strip M_b can be determined from the following equation (Cl.13.12.2.1):

$$M_b = \left[\frac{\alpha_1}{0.3 + \alpha_1} \left(1 - \frac{l_2}{3l_1} \right) \right] \times M \qquad [12.9]$$

where

l_1 and l_2 = slab spans in the direction 1 (longitudinal direction in the plane of the frame) and 2 (transverse direction) respectively, and

α_1 = the beam-to-slab stiffness ratio in direction 1 (corresponding to l_1), as discussed in Section 12.4.5.

The above equations apply to all slab locations (supports and midspan), except for the exterior column, where 100% of the negative bending moment (M_1) is assigned to the beam strip, as shown in Figure 12.31c (Cl.13.12.2.2). Note that the beam should be designed to resist its self-weight and 100% of the concentrated or distributed loads applied directly to the beam (e.g. load due to a partition wall) (Cl.13.12.2.3).

Table 12.5 summarizes the bending moment values at critical sections for slabs with beams.

Table 12.5 A slab with beams between all supports

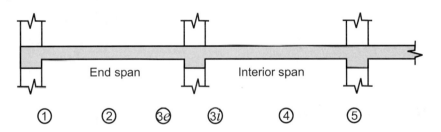

Type of span		End Span				Interior Span	
Section Location		(1) Exterior Support	(2) Midspan	(3e) First Interior Support	(3i) Interior Support	(4) Midspan	(5) Interior Support
		M_1	M_2	M_{3e}	M_{3i}	M_4	M_5
	Sign	Negative	Positive	Negative	Negative	Positive	Negative
Longitudinal Direction	Total Moment M	$-0.16\,M_o$	$+0.59\,M_o$	$-0.70\,M_o$	$-0.65\,M_o$	$+0.35\,M_o$	$-0.65\,M_o$
	Beam Strip Moment M_b	$-1.16M$		$M_b = \left[\dfrac{\alpha_1}{0.3 + \alpha_1}\left(1 - \dfrac{l_2}{2l_1}\right)\right] \times M$ [12.9]			
Transverse Distribution	Slab Strip Moment M_s	0		$M_s = M - M_b$			

Bending moment distribution for the selected values of α_1 (0.5 and 1.0) and l_2 / l_1 ratio (0.5, 1.0, and 2.0) has been summarized in the following tables. It can be seen from Table 12.6 that the bending moment in the beam strip significantly decreases with an increase in the l_2 / l_1 ratio; a similar trend can be observed in Table 12.7. This is illustrated in Figure 12.32.

Table 12.6   Transverse bending moments in slabs with beams: $\alpha_1 = 0.5$

l_2 / l_1	Beam strip (M_b)	Slab strip (M_b)
0.5	$0.52 \times M$	$0.48 \times M$
1.0	$0.42 \times M$	$0.58 \times M$
2.0	$0.20 \times M$	$0.80 \times M$

Table 12.7   Transverse bending moments in slabs with beams: $\alpha_1 = 1.0$

l_2 / l_1	Beam strip (M_b)	Slab strip (M_s)
0.5	$0.64 \times M$	$0.36 \times M$
1.0	$0.51 \times M$	$0.49 \times M$
2.0	$0.26 \times M$	$0.74 \times M$

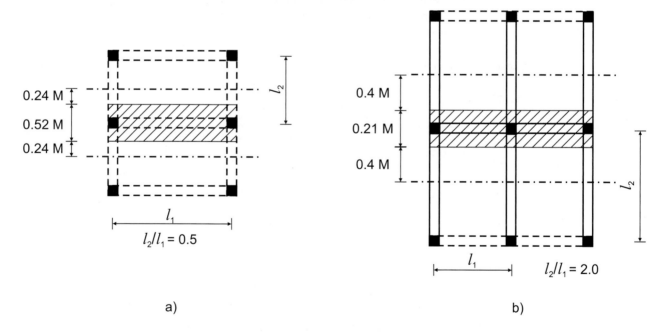

a)   b)

Figure 12.32   A variation in the bending moment distribution between the beam strip and the slab strip for $\alpha_1 = 0.5$: a) $l_2/l_1 = 0.5$, and b) $l_2/l_1 = 2.0$.

12.6.5 Unbalanced Moments

A23.3 Cl.13.9.4

One of the key issues associated with two-way slab systems is safety of slab-column connections. All loads supported by the slab converge on the column. This section explains the calculation procedure for unbalanced bending moments which occur due to an uneven distribution of live loads in adjacent slab spans.

Unbalanced moments in the slab are caused by an uneven distribution of live load in adjacent spans. This concept is illustrated in Figure 12.33. Slab span AB shown in Figure 12.33a is subjected to both dead and live load, while the adjacent span BC is subjected to dead load only. Bending moments at the interior column B due to loading in spans AB and BC (M_{BL} and M_{BR}) are shown in Figure 12.33b. The unbalanced bending moment M_{fu} shown in Figure 12.33c is equal to the difference between the slab bending moments for spans AB and BC, that is,

$$M_{fu} = M_{BL} - M_{BR}$$

The unbalanced moment is resisted by the connection and the column or wall above and below the slab, as follows (see Figure 12.33c)

$$M_{fu} = M_{B_1} + M_{B_2}$$

where M_{B_1} and M_{B_2} are bending moments in the column above and below the connection due to the unbalanced moment. Note M_{fu} that is distributed to the column in proportion to the flexural stiffness ($4EI/h$), where E is modulus of elasticity, I is moment of inertia, and h is column height (centre-to-centre distance between the floor slabs). The column segments shown in Figure 12.33c have different heights (h_{B_1} and h_{B_2}) and the moments of inertia (I_1 qnd I_2); this will result in different flexural stiffness and bending moments. When the column segments above and below the connection have the same height and corresponding stiffness, each segment will resist one-half of the unbalanced moment M_{fu}. Transfer of unbalanced moments through the connection will be discussed in Section 12.9.3.

The unbalanced moments are intended to account for uneven live load in adjacent spans when the design is performed according to the DDM. This is not required for the EFM, because it is possible to apply pattern loading which considers the effect of uneven live load (see Section 12.7.2 for a discussion on pattern loading).

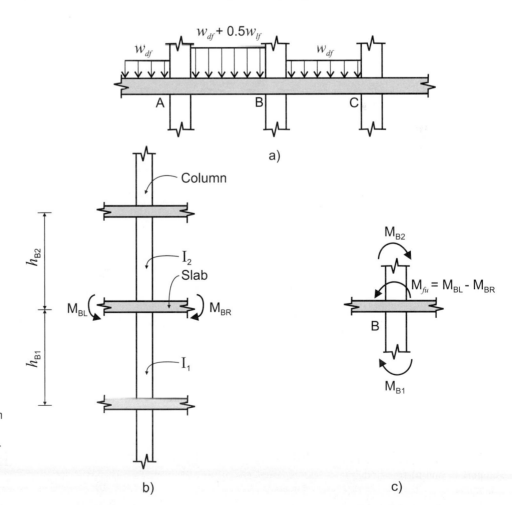

Figure 12.33 Unbalanced bending moments: a) an uneven load on adjacent slab spans, b) bending moments at connection B, and c) unbalanced bending moment transferred to the column.

Interior Columns

The unbalanced moment, M_{fu}, for an interior column is computed assuming that the longer span adjacent to the column (l_n) is subjected to the factored dead load and half the factored live load, while the shorter span (l'_n) is subjected to the factored dead load only. The factored unbalanced moment at an interior column is equal to the difference between the bending moments at adjacent spans with different lengths, that is,

A23.3 Eq. 13.24 $M_{fu} = 0.07\left[(w_{df} + 0.5w_{lf})l_{2a}l_n^2 - w'_{df}l'_{2a}(l'_n)^2 \right]$ **[12.10]**

where

w_{df} = factored dead load per unit area for the <u>longer span</u> corresponding to clear span l_n in longitudinal direction and design strip l_{2a}

w'_{df} = factored dead load per unit area for the <u>shorter span</u> corresponding to clear span l'_n in longitudinal direction and design strip l'_{2a}

w_{lf} = factored live load per unit area (longer span only)

The coefficient 0.07 in Eqn 12.10 is approximately equal to 0.65 times 1/8 (note that 0.65 is the multiplier used in the DDM to obtain the bending moment at the interior support location, and 1/8 is multiplier in Eqn 12.8).

The notation used in the above equation is presented in Figures 12.34a and b. Bending moments at the support are shown in Figure 12.34c. Note that

$$M_{BL} = 0.07\left[(w_{df} + 0.5w_{lf})l_{2a}l_n^2 \right]$$

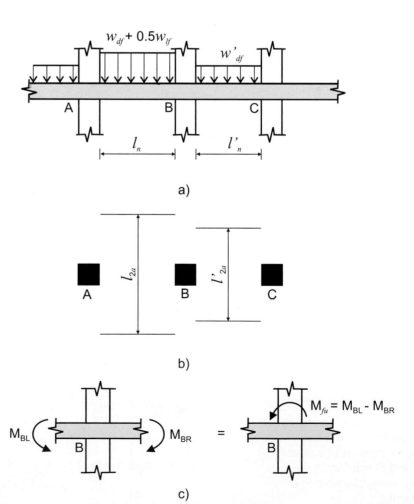

Figure 12.34 Unbalanced moment according to CSA A23.3: a) loading pattern causing an unbalanced moment at interior support B, and b) free-body diagrams.

and

$$M_{BR} = 0.07 \left[w'_{df} l'_{2a} (l'_n)^2 \right]$$

When adjacent span lengths are equal and subjected to live load of the same intensity, Eqn 12.10 can be simplified as follows

$$M_{fu} = 0.07 \left[(0.5 w_{lf}) l_{2a} l_n^2 \right]$$

Exterior Columns

The moment transferred at an exterior column is equal to the negative factored moment at the exterior support, thus the entire bending moment can be considered as unbalanced. In general, the slab capacity to transfer bending moments to the exterior columns is limited due to limited ability of the slab top reinforcement to develop flexural capacity at the edge. It is therefore recommended to reduce the design moments at the exterior columns to a minimum.

12.6.6 CSA A23.3 Reinforcement Requirements for Two-Way Slabs

It should be noted that the CSA A23.3 reinforcement requirements presented in this section apply to two-way slabs designed according to all design procedures.

Design of Flexural Reinforcement The amount of top and bottom reinforcement is determined considering the slab section with the width (b) and depth (d). The width depends on the location: it could be a column width, a slab width, or a band (b_b).

The design bending moment (M_f) corresponds to the section under consideration, and it is determined according to the procedures discussed earlier in this section. The required reinforcement area (A_s) can be found by applying the Direct Procedure discussed in Section 5.5.1 and Eqn 5.4 as follows

$$A_s = 0.0015 f_c' b \left(d - \sqrt{d^2 - \frac{3.85 M_f}{f_c' b}} \right) \tag{5.4}$$

The required bar spacing (s) can be determined from the following equation

$$s \le A_b \frac{b}{A_s} \tag{12.11}$$

where A_b denotes the bar area.

A23.3 Cl.13.10

Reinforcement Requirements for Flat Slabs and Flat Plates

Minimum reinforcement area (Cl.13.10.1)

CSA A23.3 prescribes the same minimum reinforcement area as for one-way slabs (see Section 5.7.1), that is (Cl.7.8.1),

$$A_{s\,min} = 0.002 A_g \tag{5.16}$$

and

$$A_g = b \cdot h_s$$

where A_g is the gross cross-sectional area of a slab section, h_s is the slab thickness, and b is the width of the slab strip (e.g. column strip or middle strip in flat plates). The minimum amount of reinforcement is intended to ensure shrinkage and temperature control in slabs.

Maximum amount of reinforcement (Cl.10.5.2)

This check is the same as for other reinforced concrete flexural members (see Section 5.6.1). In order to achieve the steel-controlled failure, reinforcement ratio (ρ) should not exceed the balanced reinforcement ratio (ρ_b) presented in Table A.4, that is,

$$\rho \le \rho_b$$

where

$$\rho = \frac{A_s}{b \cdot d}$$

[3.1]

Note that the flexural reinforcement in two-way slabs is usually closer to the minimum amount prescribed by CSA A23.3.

Reinforcement spacing

Maximum permitted reinforcement spacing (s_{max}) depends on the slab thickness, location within the slab (support or midspan), and the sign of bending moment (positive or negative). A summary of the key CSA A23.3 reinforcement spacing requirements for two-way slabs is presented in Table 12.8.

Table 12.8 CSA A23.3 Reinforcement Spacing Limits for Two-Way Slabs

Location	Symbol	Maximum spacing* (s_{max})	Code Clause
General requirement - all locations	s	$\le 5h_s$ ≤ 500 mm	7.8.1
Negative flexural reinforcement - band b_b	s_b	$\le 1.5h_s$ ≤ 250 mm	13.10.4
Negative flexural reinforcement - outside the band b_b	s^-	$\le 3h_s$ ≤ 500 mm	13.10.4
Positive flexural reinforcement	s^+	$\le 3h_s$ ≤ 500 mm	13.10.4

Note
* = lesser of the alternative values

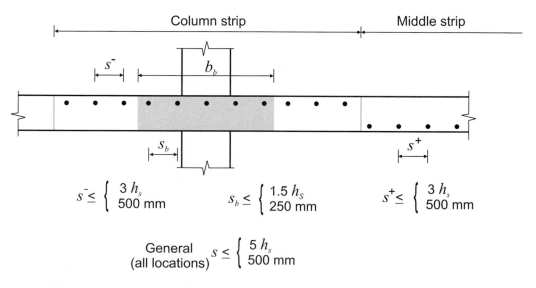

$$s^- \le \begin{cases} 3\,h_s \\ 500 \text{ mm} \end{cases} \qquad s_b \le \begin{cases} 1.5\,h_s \\ 250 \text{ mm} \end{cases} \qquad s^+ \le \begin{cases} 3\,h_s \\ 500 \text{ mm} \end{cases}$$

$$\text{General} \atop \text{(all locations)} \quad s \le \begin{cases} 5\,h_s \\ 500 \text{ mm} \end{cases}$$

Figure 12.35   Reinforcement distribution at column locations.

Reinforcement anchorage and curtailment (Cl.13.10.5 and 13.10.8)

The requirements regarding the reinforcement anchorage are included in Cl.13.10.5. It is a normal practice for the negative moment reinforcement (top reinforcement) to be hooked at edge locations, while the positive moment reinforcement (bottom reinforcement) consists of straight bars. Negative moment reinforcement at interior columns is usually continuous.

Hooked dowels that match the column reinforcement connect the column steel and the slab steel. The reinforcement must be adequately anchored into edge beams, columns, or walls. The slab reinforcement at discontinuous edges must extend up to the edge of the slab. Generic reinforcement lengths and arrangements are illustrated in Cl.13.10.8. Note that these recommendations apply only to regular slabs.

Structural integrity reinforcement (Cl.13.10.6)

In a flat slab structure, progressive collapse is usually initiated by shear failure around a column, which progresses to adjacent columns and ultimately leads to collapse of the entire slab; this is illustrated through a real-life case study presented in Section 12.9. Progressive collapse can be prevented by providing continuous bottom slab reinforcement through the column. These bars, known as *hanger bars* or *integrity reinforcement*, can support the slab in direct tension after the shear failure has taken place (provided that they are adequately anchored and that the area of reinforcement is sufficient). Note that the top reinforcement is ineffective because the bars tend to spall off concrete cover and lose their anchorage in the slab. The mechanism of shear failure and conceptual layout of hanger reinforcement are illustrated in Figure 12.36.

Figure 12.36 Shear failure of a flat plate showing integrity reinforcement (hanger bars).

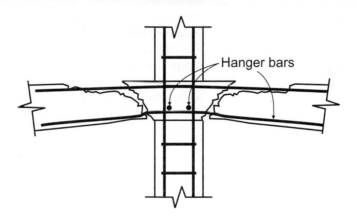

According to Cl.13.10.6.1, the total area of integrity reinforcement ($\sum A_{sb}$) connecting the slab or drop panel to the column or column capital can be calculated from the following equation

| A23.3 Eq. 13.26 | $$\sum A_{sb} = \frac{2V_{se}}{f_y}$$ | **[12.12]** |

where V_{se} denotes the shear force transmitted to the column or column capital due to specified loads (see Figure 12.37), which can be determined as follows

$$V_{se} = w \cdot A$$

and

$$w = DL + LL$$
$$w \geq 2 \cdot DL$$

where A is the tributary slab area for two-way shear calculations (shown shaded in the figure), and DL and LL are specified dead and live load, respectively.

According to Cl.13.10.6.2, integrity reinforcement should consist of at least two bottom bars or tendons that extend through the column core or column capital region in each span direction. It is essential for this reinforcement to be continuous through the column and to ensure an adequate anchorage. The following alternative arrangements are prescribed by Cl.13.10.6.3 (see Figure 12.38):

Figure 12.37 Tributary area A for the design of integrity reinforcement.

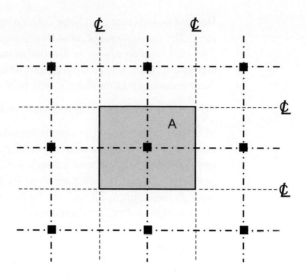

a) bottom reinforcement can be extended through the column, and Class A tension lap splice discussed in Section 9.9 can be used (Cl.13.10.6.3a);

b) additional bottom reinforcement can be placed over a column or column capital, such that an overlap of $2l_d$ is provided with the bottom reinforcement in adjacent spans, where l_d is bar development length discussed in Section 9.3 (Cl.13.10.6.3b), and

c) at discontinuous edges (end spans in the slab), bottom reinforcement needs to be extended and bent, hooked, or otherwise anchored over the supports such that the yield stress can develop at the face of column or column capital (Cl.13.10.6.3c).

Figure 12.38 CSA A23.3 provisions for integrity reinforcement:
a) bottom reinforcement lap spliced through the column;
b) additional bottom reinforcement placed over a column, and
c) bottom reinforcement at a discontinuous edge
(adapted from CAC, 2005 with the permission of the Cement Association of Canada).

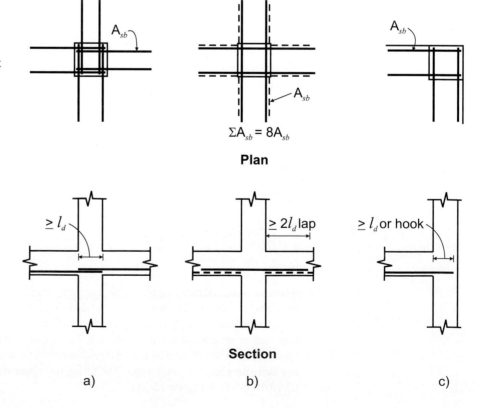

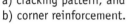 **Reinforcement Requirements for Slabs with Beams** Reinforcement in slabs with beams must comply with the same requirements as discussed in previous section regarding flat slabs and flat plates. In general, a slab strip resists a fraction of the bending moment and the reinforcement should be uniformly distributed over the slab width.

When beams are relatively stiff, slabs will develop torsional moments at exterior corners at 45° angle to the edges. These moments will cause tension both in the top and the bottom of the slab. The resulting cracking pattern in the slab is shown in Figure 12.39a. For further details on torsional moments in two-way slabs the reader is referred to Park and Gamble (2000).

Special *corner reinforcement* needs to be provided to resist these twisting moments in slabs with beams (Cl.13.12.5). The reinforcement should be designed to resist the maximum positive bending moment per unit width of the slab panel. The reinforcement should be provided within a band parallel to the diagonal in the top of the slab and a band perpendicular to the diagonal in the bottom of the slab (see Figure 12.39b). Alternatively, the reinforcement may be placed in two layers parallel to the edges of the slab in both the top and bottom of the slab. The reinforcement must extend at least one-fifth of the shorter span in each direction from the corner.

Figure 12.39 Corner reinforcement in slabs with beams: a) cracking pattern, and b) corner reinforcement.

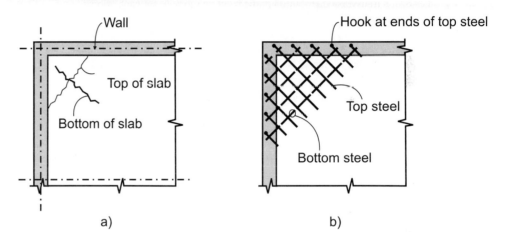

a)　　　　　　　　　b)

12.6.7 Design Applications of the Direct Design Method

The design of two-way slabs for flexure according to the DDM was discussed in detail in this section. General design steps are outlined in Checklist 12.1. Although the steps have been presented in specific sequence, it is not necessary to follow the same sequence in all design situations. Two design examples illustrating the design of a flat plate and a slab with beams according to the DDM are presented next.

Checklist 12.1 Design of Two-Way Slabs for Flexure According to the Direct Design Method

Step	Description	Code Clause
1	Check whether the Direct Design Method can be used (see Section 12.6.1).	13.9.1
2	Select slab thickness — use the CSA A23.3 minimum thickness requirements outlined in Section 12.5.2 as a reference.	13.2
3	Identify the design strip for frame under consideration as outlined in Section 12.4.1.	
3a	Flat plates and flat slabs. determine the widths for the column and middle strips	13.1.2 Commentary

(Continued)

Checklist 12.1 Continued

Step	Description	Code Clause
3b	Slabs with beams: determine the widths for the beam and slab strips. The width of the beam strip is equal to the flange width of the effective beam section (see Section 12.4.4).	13.8.2.7
4	Compute the total factored static moment M_o for the span. Determine the factored load for the span by treating the slab as a wide beam with the width equal to the design strip (see Section 12.6.2).	13.9.2
5	Distribute M_o in the longitudinal direction between critical sections (supports and midspan) within the span.	13.9.3
5a	Flat plates and flat slabs: refer to Tables 12.2 to 12.4 (see Section 12.6.3).	
5b	Slabs with beams: refer to Table 12.5 (see Section 12.6.4).	
6	Perform a transverse distribution of the bending moment for each critical section obtained in Step 5.	
6a	Flat plates and flat slabs: find bending moments for the column and middle strips (see Tables 12.2 to 12.4).	13.11
6b	Slabs with beams: find bending moments for the beam strip and the slab strip (see Table 12.5).	13.12
7	Design and detail flexural reinforcement for the slab. Determine the area of top and bottom reinforcement for various slab sections. Refer to the design procedures for rectangular beam sections outlined in Section 5.5.	
7a	Flat plates and flat slabs: distribute the reinforcement according to the CSA A23.3 reinforcement requirements summarized in Section 12.6.6 (in particular Table 12.8).	13.10
7b	Slabs with beams: refer to additional reinforcement requirements summarized in Section 12.6.6.	13.12
8	Slabs with beams and flat plates/slabs with edge beams: design the beams according to the CSA A23.3 design provisions for flexural members explained in Chapter 5.	

Example 12.1

Two-Way Flat Plate - Direct Design Method

Consider a floor plan of a two-way slab system without beams (flat plate) shown in the following figure. The plan shows an intermediate floor level, and a typical storey height is 3.0 m. Column dimensions are 300 mm by 600 mm, except for the corner columns (300 mm by 300 mm), as shown in the figure. Edge (spandrel) beams will be provided. The slab is subjected to specified live load (LL) of 3.6 kPa, and superimposed dead load (DL_s) of 1.44 kPa, in addition to its self-weight. Consider only the effect of gravity loads for this design - lateral loads are to be resisted by shear walls, which are omitted from the drawing. Use 15M bars for slab reinforcement.

Use the CSA A23.3 Direct Design Method to determine design bending moments and the amount and distribution of reinforcement for an interior frame along gridline 2.

Given: $f_c' = 30$ MPa
$f_y = 400$ MPa

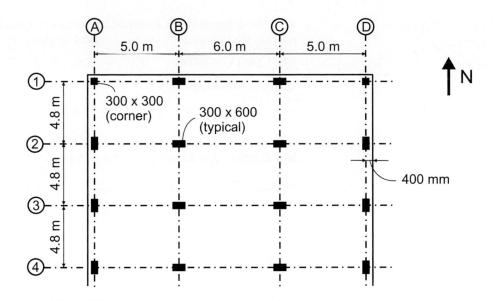

SOLUTION:
1. **Check whether the criteria for the CSA A23.3 Direct Design Method are satisfied (see Section 12.6.1):**

   CSA A23.3 Cl. 13.9.1 prescribes that the DDM can be used when the following requirements have been met:

   #1 A slab is regular (see Section 12.5.1).

   #2 There are three continuous spans in each direction.

   #3 The successive span lengths, centre-to-centre of supports, in each direction must not differ more than one-third of the longer span.

   E-W direction: Span 2/Span 1 = 6.0 m/5.0 m = 1.2 < 1.33
   N-S direction: all span lengths are equal (4.8 m)

   #4 The slab is subjected to uniformly distributed gravity loads.

   #5 The factored live load does not exceed two times the factored dead load (this will be confirmed in Step 3).

2. **Determine the required slab thickness based on deflection control requirements.**

   For two-way slab systems without beams, the minimum overall thickness (h_s) can be determined according to CSA A23.3 Cl.13.2.3 as follows (when Grade 400 reinforcement is used)

   $$h_s \geq \frac{l_n}{30}$$

   We need to determine the clear span for each span under consideration. Since l_n denotes the longer clear span, two clear span values need to be considered for each slab panel along gridline 2:

   Span 1 (end span AB)

   E-W direction: $l_n = 5.0 - (\frac{0.3}{2} + \frac{0.6}{2}) = 4.55$ m

   N-S direction: $l_n = 4.8 - (\frac{0.3}{2} + \frac{0.6}{2}) = 4.35$ m

   Therefore, the longer span (E-W direction) governs, that is,

   l_{n1} = 4.55 m

   The required slab thickness is

   $$h_s \geq \frac{l_{n1}}{30} = \frac{4550}{30} = 152 \text{ mm}$$

Since the design does not anticipate the provision of edge beams, CSA A23.3 Cl.13.2.3 requires the slab thickness to be increased by 10 %, that is,

$$h_s > 1.1 \times 152 = 167 \text{ mm}$$

Span 2 (interior span BC)

E-W direction: $l_n = 6.0 - (\frac{0.6}{2} + \frac{0.6}{2}) = 5.4$ m

N-S direction: $l_n = 4.8 - (\frac{0.3}{2} + \frac{0.3}{2}) = 4.5$ m

Therefore, the span in E-W direction governs, that is,

$$l_{n2} = 5.4 \text{ m}$$

The required slab thickness is

$$h_s \geq \frac{l_{n2}}{30} = \frac{5400}{30} = 180 \text{ mm}$$

In this case, the required thickness is larger for Span 2 (180 mm) than for Span 1 (152 mm). In practice the slab thickness should be uniform and the higher value should be used, that is,

$$h_s = 180 \text{ mm}$$

3. **Find the factored design loads.**

a)   Calculate the dead load acting on the slab.
First, calculate the slab's self-weight:

$$DL_w = h \times \gamma_w = 0.18 \text{ m} \times 24 \text{ kN/m}^3 = 4.32 \text{ kPa}$$

where $\gamma_w = 24$ kN/m^3 is the unit weight for normal-density concrete.
The superimposed dead load was given, and the value is as follows

$$DL_s = 1.44 \text{ kPa}$$

Finally, the total factored dead load is equal to

$$w_{DL,f} = 1.25(DL_w + DL_s) = 1.25(4.32 + 1.44) = 7.2 \text{ kPa}$$

b) Calculate the factored live load:

$$w_{LL,f} = 1.5 \times LL_w = 1.5 \times 3.6 \text{ kPa} = 5.4 \text{ kPa}$$

c) The total factored load is

$$w_f = w_{DL,f} + w_{LL,f} = 7.2 + 5.4 = 12.6 \text{ kPa}$$

Note that the factored live load $w_{LL,f} = 5.4$ kPa is less than twice the factored dead load $2 \times w_{DL,f} = 2 \times 7.2 \text{ kPa} = 14.4 \text{ kPa}$. Therefore, the requirement #5 for the application of DDM has been met.

4. **Determine the widths for design strip, column strip, and middle strip.**

a)   Design strip
The frame under consideration is laid along gridline 2; this is referred to as Direction 1, while the transverse direction is referred to as Direction 2. The corresponding spans are illustrated on the following sketch, that is,

$l_1 = 6$ m (let us consider the larger span)

and

$l_2 = 4.8$ m

The design strip is denoted as l_{2a} (Direction 2). The width of the design strip is to be determined by taking an average value for the two spans adjacent to the gridline under consideration. In this case, the spans in Direction 2 are equal, thus

$l_{2a} = l_2 = 4.8$ m

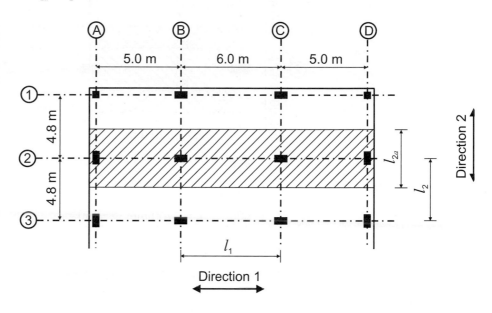

b)   Column strip and middle strip
The width of the column strip can now be determined, following the guidelines provided in Section 12.4.1. First, we need to compare the spans in both directions (1 and 2). For span BC, $l_2 = 4.8$ m and $l_1 = 6$ m, hence

$l_2 < l_1$

According to CSA A23.3, the shorter span (l_2) is used to find the width of the column strip. Note that the same conclusion would apply to span AB in E-W direction ($l_1 = 5$ m and $l_2 = 4.8$ m; hence $l_2 < l_1$).

As a result, the width of the column strip (l_c) is equal to

$l_c = l_2/2 = 4.8/2 = 2.4$ m

Note that the smaller of two spans is considered for the column strip width.
The middle strip (l_m) is a portion of the design strip outside the column strip, that is,

$l_m = l_{2a} - l_c = 4.8 - 2.4 = 2.4$ m

The column and middle strips are illustrated on the sketch below. Note that the middle strip is divided into two half-strips, shown cross-hatched on the following sketch.

5. **Find the factored bending moments in the slab.**
    a)   Find the total factored static moment M_o.
    The factored static moment can be determined from Eqn 12.8 as follows (see Section 12.6.2)

A23.3 Eq. 13.23

$$M_o = \frac{w_f \times l_{2a} \times l_n^2}{8}$$ [12.8]

where
$w_f = 12.6$ kPa is the total factored load
$l_{2a} = 4.8$ m is the width of the design strip

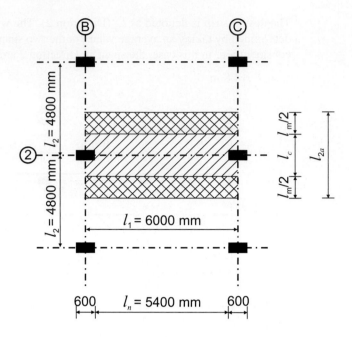

l_n is the clear span in the longitudinal direction (along gridline 2). Since the spans are different, this calculation needs to be performed for each span.

Span 1 (end span AB):

$$l_n = 5.0 - \left(\frac{0.3}{2} + \frac{0.6}{2}\right) = 4.55 \text{ m}$$

$$M_o = \frac{w_f \times l_{2a} \times l_n^2}{8}$$

$$= \frac{12.6 \text{ kPa} \times 4.8 \text{ m} \times (4.55 \text{ m})^2}{8} = 156 \text{ kNm}$$

Span 2 (interior span BC):

$$l_n = 6.0 - \left(\frac{0.6}{2} + \frac{0.6}{2}\right) = 5.4 \text{ m}$$

$$M_o = \frac{w_f \times l_{2a} \times l_n^2}{8}$$

$$= \frac{12.6 \text{ kPa} \times 4.8 \text{ m} \times (5.4 \text{ m})^2}{8} = 220 \text{ kNm}$$

b) Distribute the total factored static moment to critical locations in longitudinal direction, and subsequently distribute the moment at each critical location transversely to column strip and middle strip.

Distribution of bending moments in longitudinal direction will be performed according to Cl.13.9.3, while the transverse distribution of bending moments is performed according to Cl.13.11; these requirements are summarized in Table 12.2. Bending moments at critical locations in longitudinal direction (supports and midspan) are expressed in terms of the total factored moment M_o, as shown on the following sketch. The calculations are summarized in the following tables.

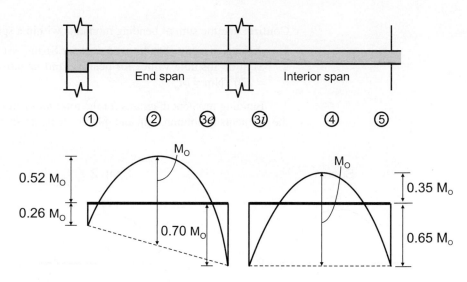

Table 12.9 Factored bending moments for Span 1 (End Span AB)

		A Negative moment (kNm)	Midspan Positive moment (kNm)	B Negative moment (kNm)
Longitudinal direction	Bending moments at critical sections	$-0.26M_o$ $=(-0.26) \times 156$ $=-41$	$+0.52M_o$ $=(+0.52) \times 156$ $=+81$	$-0.70M_o$ $=(-0.70) \times 156$ $=-109$
Transverse distribution - column strip	CSA A23.3 Provisions	$-0.26M_o$	$+(0.29 \text{ to } 0.34)M_o$	$(0.49 \text{ to } 0.63)M_o$
	Proposed value Design moment	$-0.26M_o$ -41	$+0.29M_o$ $+0.29 \times 156 = +45$	$-0.63M_o$ $-0.63 \times (156) = -98$
Transverse distribution - Middle strip	Design moment	0	$=81-45 = +36$	$=-109-(-98) = -11$

Confirm that the sum of bending moments within a span is equal to M_o:

1. Average negative bending moment $= (-41-109)/2 = -75$ kNm
2. Sum of absolute values for positive and negative bending moments $= 81+75 = 156$ kNm $= M_o$

Table 12.10 Factored bending moments for Span 2 (Interior Span BC)

		B Negative moment (kNm)	Midspan Positive moment (kNm)	C Negative moment (kNm)
Longitudinal direction	Bending moments at critical sections	$-0.65M_o$ $=(-0.65) \times 220$ 143	$+0.35M_o$ $=(+0.35) \times 220$ $=+77$	$-0.65M_o$ $=(-0.65) \times 220$ $=-143$
Transverse distribution - column strip	CSA A23.4 Provisions	$-(0.46 \text{ to } 0.59)M_o$	$+(0.19 \text{ to } 0.23)M_o$	$-(0.46 \text{ to } 0.59)M_o$
	Proposed value Design moment	$-0.59M_o$ $-0.59 \times (220) = -130$	$+0.19M_o$ $+0.19 \times (220) = +42$	$-0.59M_o$ $-0.59 \times (220) = -130$
Transverse distribution - Middle strip	Design moment	$=-143-(-130) = -13$	$=77-42 = +35$	$=-143-(-130) = -13$

Confirm that the sum of bending moments within a span is equal to M_o:

1. Average negative bending moment for Span 2 = $(-143-143)/2 = -143$ kNm
2. Sum of absolute values for positive and negative bending moments = $77+143$ = 220 kNm = M_o

Bending moment diagrams for the total moment in Span 1 and Span 2, as well as the moments in column strip and middle strip, are shown below.

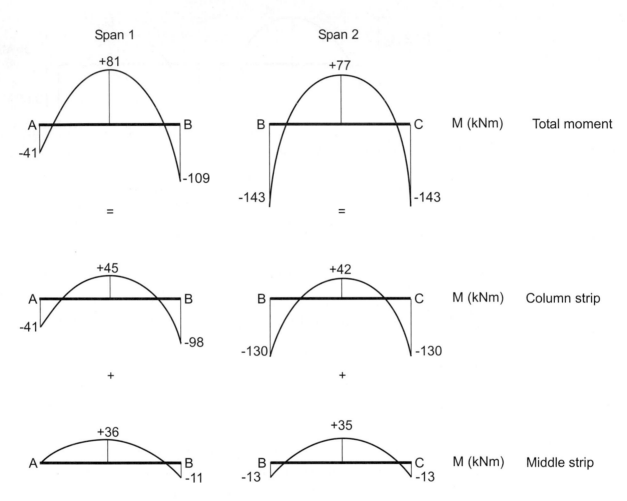

It can be seen that the negative bending moment at support B has different values for Span 1 (−109 kNm) and Span 2 (−143 kNm). In practice, the top reinforcement at support B would need to be designed for the greater of the two moments from adjacent spans. The solution will proceed by following that approach. However, a more effective solution can be obtained if these two moments are made equal by increasing the moment for Span 1 and/or decreasing the moment for Span 2; this procedure is often referred to as "balancing" of bending moments. Since the balancing approach is somewhat more complex, it will be discussed in Step 10 (at the end of the example).

6. **Design the flexural reinforcement.**

    The dimensions of the column strip and middle strip were found in Step 4. In this design, the column strip and the middle strip have the same width (2.4 m). The effective slab depth is

    $d = 180 - 25 - 15 = 140$ mm

    Note that this value has been determined assuming a 25 mm average cover (corresponding to the distance to the centre of the mat) and 15M bar size.

The flexural reinforcement calculation can be presented in a tabular form. The key equations are summarized below.

a) The required reinforcement area can be found using the Direct Procedure (see Section 5.5.1) and Eqn 5.4 as follows

$$A_s = 0.0015 f_c'b \left(d - \sqrt{d^2 - \frac{3.85 M_f}{f_c'b}} \right) \qquad [5.4]$$

b) The required spacing can be determined from the following equation

$$s \leq A_b \frac{b}{A_s}$$

Note that $A_b = 200$ mm^2 for 15M bars.

c) Maximum reinforcement spacing (s_b) - band width b_b

The reinforcement spacing requirements for negative reinforcement in column strips are outlined in Section 12.6.6. It follows that, within band width b_b, reinforcement spacing (s_b) is limited to the lesser of (Cl.13.10.4)

$$s_b \leq 1.5 h_s = 1.5 \times 180 = 270 \text{ mm}$$

or

$$s_b \leq 250 \text{ mm}$$

In this case,

$$s_b \leq 250 \text{ mm governs.}$$

d) Maximum reinforcement spacing - outside the band width b_b

Spacing for the reinforcement resisting negative bending moments in the column strip outside the band b_b is limited to the lesser of (Cl.13.10.4)

$$s^- \leq 3 h_s = 3 \times 180 = 540 \text{ mm}$$

or

$$s^- \leq 500 \text{ mm}$$

In this case, $s^- \leq 500$ mm governs.

Maximum spacing for the reinforcement resisting positive bending moments in the slab is the same as for the negative bending moments (Cl.13.10.4), that is, $s^+ \leq 500$ mm.

e) The minimum reinforcement requirement (Cl.7.8.1)

$$A_{s\,min} = 0.002 A_g$$

Note that the area A_g refers to the gross cross-sectional area for the section under consideration.

Column Strip Calculations

CSA A23.3 Cl.13.10.3 requires that the bending moment in the column strip of an exterior column be entirely resisted by the strip b_b centered at the column (see Figure 12.16). For an interior column, the band b_b should be designed to resist at least one-third of the total factored negative bending moment for the entire design strip (Cl.13.11.2.7). This is illustrated in Table 12.11. For exterior column A, design bending moment for the band b_b is equal to the total moment (−41 kNm). However, for interior columns (B and C), reinforcement within the band b_b is designed using one third of the total bending moment (−143 kNm), that is, the design moment is −48 kNm.

Find the width for band b_b.

i) Exterior column (A):

$$b_b = 600 + 3 \times 180 = 1140 \text{ mm} \cong 1200 \text{ mm}$$

where 600 mm is the column cross-sectional dimension in transverse direction, and 180 mm is the slab thickness.

ii) Interior columns (B and C):

$$b_b = 300 + 3 \times 180 = 840 \text{ mm} \cong 900 \text{ mm}$$

where 300 mm is the column cross-sectional dimension in transverse direction, and 180 mm is the slab thickness.

Note that the reinforcement calculations have been performed considering two different slab sections:

a) The section with the width b_b and the overall depth h_s (for the column locations only), and

b) The section with the remaining width (column strip minus the band b_b) and the overall depth h_s.

Table 12.11 Column strip - factored bending moments and flexural reinforcement calculations

		Exterior column (A)	Midspan	Interior columns (B and C)
	Steel location	**Top**	**Bottom**	**Top**
	Total bending moment (kNm)	−41	+81*	−143**
	Bending moment - column strip (kNm)	−41	+45	−130
	Column strip width (mm)	2400	2400	2400
Within b_b	Band width b_b	1200		900
	Design moment M_f (kNm)	−41		=−143/3=−48
	Required reinforcement area A_s (mm²)	900		1088
	Required spacing s (mm)	267		165
	Max spacing (mm)	250		250
	Design reinforcement (area in mm²)	6−15M@250 (1000)		6−15M@150 (1200)
	Min reinforcement area $A_{s\,min}$ (mm²)	432		324
Outside the band	Strip width (mm)	0	2400	=2400−900=1500
	Design moment M_f (kNm)	0	+45	=−130−(−48)=−82
	Required reinforcement area A_s (mm²)	0	958	1880
	Required spacing s (mm)		500	158
	Max spacing (mm)		500	250
	Design reinforcement (area in mm²)	0	6−15M@400 (1200)	11−15M@150 (2200)
	Min reinforcement area $A_{s\,min}$ (mm²)	432	864	540
	Design reinforcement - summary	6−15M@250 (centered over column within 1200 mm)	15M@4000	15M@150 (uniform spacing for the entire column strip)

Notes:

* - Larger midspan moment selected (+81 kNm for Span 1 versus +77 kNm for Span 2)

** - Larger negative moment for support B selected (−143 kNm for Span 2 versus −109 kNm for Span 1)

Note that the maximum reinforcement requirement check has been omitted from the table, since it does not govern due to the small amount of flexural steel. For example, top reinforcement at exterior column A within the band b_b (1200 mm width) is equal to 5-15M. The corresponding reinforcement ratio is

$$\rho = \frac{A_s}{b \cdot d} = \frac{5 \cdot 200}{1200 \cdot 140} = 0.006 \qquad [3.1]$$

This is significantly less than the balanced reinforcement ratio (ρ_b) of 0.027 for $f'_c = 30$ MPa presented in Table A.4, thus $\rho < \rho_b$.

Middle Strip Calculations

Note that there is no need to provide reinforcement in the middle strip at an exterior column, according to CSA A23.3 Cl.13.10.3. The calculations are summarized in Table 12.12.

Table 12.12   Middle strip - factored bending moments and flexural reinforcement calculations

	Exterior column (A)	Midspan	Interior columns (B and C)
Steel location	**Top**	**Bottom**	**Top**
Design bending moment - middle strip M_f (kNm)	0	+36*	−13**
Middle strip width (mm)	2400	2400	2400
Required reinforcement area A_s (mm²)	0	762	271
Required spacing s (mm)		630	1774
Max spacing (mm)	500	500	500
Min reinforcement area $A_{s\,min}$	864	864	864
Design reinforcement (area in mm²)	0	15M@400 (1400)	15M@400 (1400)

Notes:

* - Larger midspan moment selected (+36 kNm for Span 1 versus +35 kNm for Span 2)

** - Larger negative moment for support B selected (−13 kNm for Span 2 versus −11 kNm for Span 1)

7. Find the factored moments for the columns.

a)   Interior column (B)

The purpose of this calculation is to find unbalanced moments, which were explained in Section 12.6.5. Unbalanced moment for an interior column can be determined from the following equation

| A23.3 Eq. 13.24 |

$$M_{fu}= 0.07\left[(w_{df} +0.5w_{lf})l_{2a}l_n^2 - w'_{df}l'_{2a}(l'_n)^2 \right]$$ **[12.10]**

Since both Span 1 and Span 2 need to be considered, it is necessary to identify the longer and the shorter span in the plane of the frame. Based on the calculation from Step 2, it follows that Span 2 is longer, that is,

$l_n = 5.4$ m (longer span) and $l_{2a} = 4.8$ m (corresponding transverse span)

whereas

$l'_n = 4.55$ m (shorter span) and $l'_{2a} = 4.8$ m (corresponding transverse span)

In this case, factored dead load is equal for both spans, that is,

$w_{df} = w'_{df} = 7.2$ kPa

and the factored live load is

$w_{lf} = 5.4$ kPa

thus

$$M_{fu}= 0.07\left[(7.2 \text{ kPa} + 0.5(5.4 \text{ kPa}))\left(4.8 \text{ m}\right)\times\left(5.4 \text{ m}\right)^2 - \right.$$

$$\left. \left(7.2 \text{ kPa}\right)\times\left(4.8 \text{ m}\right)\times (4.55 \text{ m})^2\right]= 46.9 \text{ kNm}$$

Note that this unbalanced moment is transferred to the column segments above and below the slab proportional to their stiffness. If column segments have the same geometric properties (cross-sectional dimension and storey height), one-half of the moment is transferred to each segment.

b)   Exterior column (A)

CSA A23.3 Cl.13.10.3 requires that the entire exterior factored moment be transferred from the slab directly to the columns. Bending moment at the face of the support is equal to

$M_f = -41$ kNm

8. **Determine the slab integrity reinforcement.**

The purpose of integrity reinforcement is discussed in Section 12.6.6. According to CSA A23.3 Cl.13.10.6.1, the total area of bottom reinforcement (A_{sb}) connecting the slab, drop panel, or slab band to the column or column capital on all faces of the periphery of the column or column capital shall be at least equal to

A23.3 Eq. 13.26

$$\sum A_{sb} = \frac{2V_{se}}{f_y} \qquad\qquad\qquad [12.12]$$

where V_{se} is the shear force transmitted to the column or column capital due to specified loads, but it is not less than the shear corresponding to twice the self-weight of the slab. In this case, the tributary area for the shear design is determined for interior column B (see the sketch below)

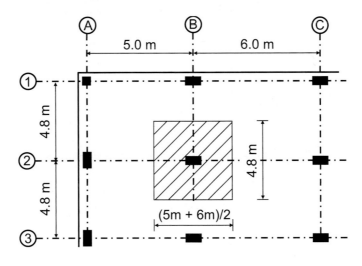

$$A = (4.8 \text{ m}) \times \left(\frac{5 \text{ m} + 6 \text{ m}}{2} \right) = 26.4 \text{ m}^2$$

Consider the following loads:

Total specified load $w_1 = 4.32$ kPa $+ 1.44$ kPa $+ 3.6$ kPa $= 9.36$ kPa

Twice the self-weight $w = 2 \times 4.32$ kPa $= 8.64$ kPa

In this case, the total specified load is larger and it governs. Next, the shear force can be determined as follows

$$V_{se} = w \times A = 9.36 \text{ kPa} \times 26.4 \text{ m}^2 = 247 \text{ kN}$$

Finally, the required area of integrity reinforcement can be determined as follows

$$\sum A_{sb} = \frac{2 \times 247 \times 10^3 \text{ N}}{400 \text{ MPa}} = 1235 \text{ mm}^2$$

It is required to provide 8-15M bars (total area 1600 mm²), that is, 4-15M bars in each direction.

9. **Present a design summary.**

A drawing summarizing the design solution is presented below. Note that the reinforcement should be laid out such that it is easy to construct. Rebar spacing should be specified using simple rounded numbers and it should preferably be repetitive. The same spacing should be used in both orthogonal directions to avoid confusion at the construction site. A good judgement is required to minimize potential construction errors and strike balance between labour, material usage, and cost.

    In this design, the calculations show 15M@400 mm o.c. bottom steel for column strips and 15M@500 mm o.c. for middle strips. It may be simpler to place 15M@400 mm in each direction throughout. Although this solution adds a few extra rebars in the middle strips, it will result in labour savings since simple specification will result in a more efficient placement.

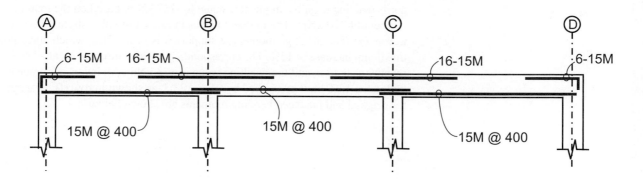

10. An alternative solution: balancing of bending moments at supports.

Balancing is achieved when bending moments at the support between two adjacent spans are made equal by increasing the moment for one span and/or decreasing the moment for other span. Balancing of bending moments is permitted by CSA A23.3 Cl.13.9.3.3 (see Section 12.6.3). To demonstrate the balancing process, let us increase the negative moment at support B in Span 1 by 15%, and decrease the negative moment at the same location in Span 2 by 15%. The balancing process is illustrated in the following table. Note that CSA A23.3 requires that the sum of positive and negative bending moments within a span must remain equal to M_o.

Table 12.13 Balancing of bending moments for Spans 1 and 2

		Span 1 = End Span AB $M_o = 156$ kNm			Span 2 = Interior Span BC $M_o = 220$ kNm		
		Support A	Midspan	Support B	Support B	Midpsan	Support C
		Negative moment (kNm) (1)	Positive moment (kNm) (2)	Negative moment (kNm) (3)	Negative moment (kNm) (4)	Positive moment (kNm) (5)	Negative moment (kNm) (6)
Original values	(a)	−41	+81	−109	−143	+77	−143
Moments varied by 15%	(b)			−109 × 1.15 = −125	−143/1.15 = −124		−143/1.15 = −124
Balanced moments	(c)	−41	156 − (41 + 125)/2 = +73	−125	−124	220 − (124 + 124)/2 = +96	−124
Balanced/original (d)/(a)	(d)	(−41)/ (−41) = 1.0	(+73)/(+81) = 0.90	(+125)/(+109) = 1.15	(−124)/(−143) = 0.87	(+96)/(+77) = 1.25	(−124)/(−143) = 0.87
(% difference)		(0%)	(−10%)	(+15%)	(−13%)	(+25% > 15%)	
Revised balanced moments	(e)	−41	156 − (41 + 125)/2 = 173	125	−(220−88) = −132	+77(1.15) = +88	−132
Balanced/original (e)/(a)	(f)	(−41)/ (−41) = 1.0	(+73)/(+81) = 0.90	(+125)/(+109) = 1.15	(−132)/(−143) = 0.92	(+88)/(+77) = 1.14	(−132)/(−143) = 0.92
(% difference)		(0%)	(−10%)	(+15%)	(−8%)	(+14%)	(−8%)

It can be seen that the positive bending moment in Span 2 (column 5) exceeds the 15% limit, since the balanced value is 95 kNm compared to the original value of 77 kNm. Therefore, the moments need to be balanced once again. Let us set the positive moment in Span 2 to the upper limit, that is, a 15% increase (88 kNm), as shown in row (e). The corresponding negative moment is equal to -132 kNm, based on the total factored static moment of 220 kNm. The moment values in Span 1 are also slightly revised, as shown in row (e). The negative moment at support B is -125 kNm, which corresponds to the maximum increase of 15%. The corresponding positive moment is calculated in the same manner as before, and the resulting value is +73 kNm. All balanced moment values are within the 15% limit prescribed by CSA A23.3, as shown in row (f). Diagrams showing the original and balanced bending moments are shown below.

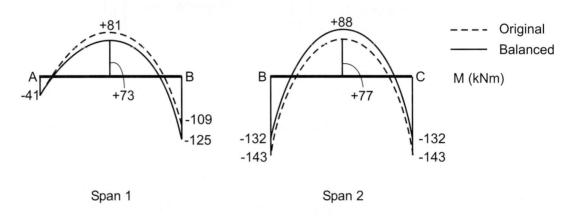

Example 12.2

Two-Way Slab with Beams - Direct Design Method

Consider a floor plan of a two-way slab system with beams shown in the following figure. Typical beam dimensions are 400 mm width by 600 mm overall depth. The plan shows an intermediate floor level, and a typical storey height is 3.0 m. Column dimensions are 400 mm by 400 mm. The slab is subjected to a specified live load (LL) of 3.6 kPa, and superimposed dead load (DL_s) of 1.44 kPa, in addition to its self-weight. Consider only the effect of gravity loads for this design - lateral loads are to be resisted by shear walls (not shown on the drawing).

Use the CSA A23.3 Direct Design Method to determine design bending moments for an interior frame along gridline 2.

Given: $f'_c = 30$ MPa

$f_y = 400$ MPa

SOLUTION: **1. Check whether the criteria for the CSA A23.3 Direct Design Method are satisfied.**

CSA A23.3 Cl. 13.9.1 prescribes that the DDM can be used when the following requirements have been met (see Section 12.6.1):

#1 A slab is regular (see Section 12.5.1).

Requirement b): for slab systems with beams between supports, the relative effective stiffness of beams in the directions 1 and 2 should be restricted as follows

$$0.2 \le \frac{\alpha_1 l_2^2}{\alpha_2 l_1^2} \le 5.0$$

where α_1 and α_2 denote the beam-to-slab stiffness ratio for beams in the directions 1 and 2, respectively. This requirement will be checked in Step 2.

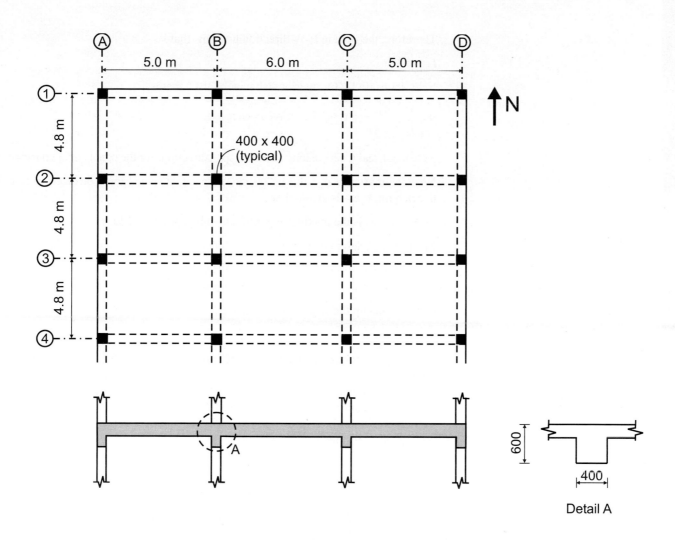

Detail A

#2 There are three continuous spans in each direction.

#3 The successive span lengths, centre-to-centre of supports, in each direction must not differ more than one-third of the longer span.

E-W direction: Span 2/Span 1 = 6.0 m/5.0 m = 1.2 < 1.33
N-S direction: all span lengths are equal (4.8 m)

#4 The slab is subjected to uniformly distributed gravity loads.

#5 The factored live load does not exceed two times the factored dead load (this will be confirmed in Step 3).

2. **Determine the required slab thickness based on deflection control requirements.**
For two-way slab systems with beams, the minimum overall thickness (h_s) can be determined according to CSA A23.3 Cl.13.2.5 as follows

A23.3 Eq. 13.3

$$h_s \geq \frac{l_n(0.6 + f_y/1000)}{30 + 4\beta\alpha_m}$$

[12.7]

We need to determine clear span l_n. Since the longer clear span governs, let us consider only interior span BC:

E-W direction: $l_n = 6.0 - (\frac{0.4}{2} + \frac{0.4}{2}) = 5.6$ m

N-S direction: $l_n = 4.8 - (\frac{0.4}{2} + \frac{0.4}{2}) = 4.4$ m

Note that the clear span was determined considering 400 mm wide beams.

Therefore, the span in E-W direction governs, that is,

$l_{n2} = 5.6$ m

Next, let us find β for interior slab panel 23BC, that is,

$$\beta = \frac{l_1}{l_2} = \frac{6.0 \text{ m}}{4.8 \text{ m}} = 1.25$$

Note that l_1 and l_2 denote long and short span directions for the panel under consideration.

Next, let us assume α_m value of 2.0, since the actual value cannot be determined unless the slab thickness is given, thus

$\alpha_m = 2.0$ (this is the maximum permitted value per Cl.13.2.5)

The required slab thickness is

$$h_s \geq \frac{l_n(0.6 + f_y/1000)}{30 + 4\beta\alpha_m} = \frac{5600(0.6 + 400/1000)}{30 + 4 \cdot (1.25) \cdot (2.0)} = 140 \text{ mm}$$

Let us round up the slab thickness, that is,

$h_s = 160$ mm

Note that the smaller thickness (say 150 mm) could have been used based on the slab thickness requirements.

Next, find the α_m value for the panel 23BC using the selected slab thickness. The simplified equation presented in Section 12.4.5 will be used for this purpose. We need to find α value for both directions (1 and 2). Different slab section widths (l_{b1} and l_{b2}) will be used for directions 1 and 2 respectively.

For Direction 1: $l_{b1} = \dfrac{5000 + 6000}{2} = 5500$ mm (an average value for spans AB and BC)

thus

$$\alpha_1 = \frac{2.5b_w}{l_{b1}}\left(\frac{h}{h_s}\right)^3\left(1 - \frac{h}{h_s}\right) = \frac{2.5 \cdot 400}{5500}\left(\frac{600}{160}\right)^3\left(1 - \frac{600}{160}\right) = 7.0 \qquad \textbf{[12.3]}$$

For Direction 2: $l_{b1} = 4800$ mm
thus

$$\alpha_1 = \frac{2.5b_w}{l_{b1}}\left(\frac{h}{h_s}\right)^3\left(1 - \frac{h}{h_s}\right) = \frac{2.5 \cdot 400}{4800}\left(\frac{600}{160}\right)^3\left(1 - \frac{600}{160}\right) = 8.0 \qquad \textbf{[12.3]}$$

The beam and slab dimensions used in the above equation are shown on the sketch below.

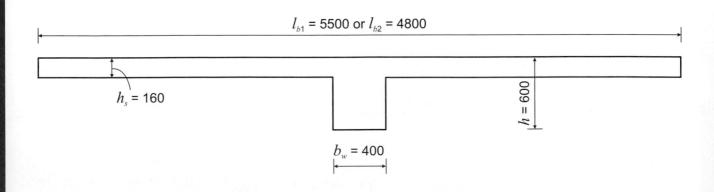

l_{b1} = 5500 or l_{b2} = 4800

h_s = 160

$h = 600$

b_w = 400

An average beam-to-slab stiffness ratio for the entire panel can be calculated as follows:

$$\alpha_m = \frac{2(7.0+8.0)}{4} = 7.5$$

Since

$$\alpha_m = 7.5 > 2.0$$

We are going to use $\alpha_m = 2.0$ for the remaining calculations.

Finally, let us confirm that the CSA A23.3 requirement #1 b) for regular slabs with beams has been met. We need to find the effective beam stiffness in two directions, as follows

$$\frac{\alpha_1 l_2^2}{\alpha_2 l_1^2} = \frac{2.0 \cdot (4.8)^2}{2.0 \cdot (6.0)^2} = 0.64$$

Since

$$0.2 < \frac{\alpha_1 l_2^2}{\alpha_2 l_1^2} = 0.64 < 5.0$$

It follows that the slab is regular according to CSA A23.3 Cl.2.2.

3. **Find the factored design loads.**

   a) Calculate the dead load acting on the slab.

   First, calculate the slab self-weight, based on the 160 mm slab thickness and 400 mm × 600 mm beams at 6 m spacing:

   $$DL_w = \left(0.16\,\text{m} + \frac{0.4\,\text{m} \cdot (0.6\,\text{m} - 0.16\,\text{m})}{6.0\,\text{m}}\right) \times 24\,\text{kN} / \text{m}^3 = 4.54\,\text{kPa}$$

   where $\gamma_w = 24\,\text{kN} / \text{m}^3$ is the unit weight of concrete.

   The superimposed dead load was given, and the value is as follows

   $$DL_s = 1.44\,\text{kPa}$$

   Finally, the total factored dead load is equal to

   $$w_{DL,f} = 1.25(DL_w + DL_s) = 1.25(4.54 + 1.44) = 7.5\,\text{kPa}$$

   b) Calculate the factored live load, as follows

   $$w_{LL,f} = 1.5 \times LL_w = 1.5 \times 3.6\,\text{kPa} = 5.4\,\text{kPa}$$

   c) The total factored load is

   $$w_f = w_{DL,f} + w_{LL,f} = 7.5 + 5.4 = 12.9\,\text{kPa}$$

   Note that the factored live load $w_{LL,f} = 5.4\,\text{kPa}$ is less than twice the factored dead load, that is, $2 \times w_{DL,f} = 2 \times 7.5\,\text{kPa} = 15.0\,\text{kPa}$. Therefore, the requirement #5 has been met.

4. **Determine the widths for design strip, beam strip, and slab strip.**

   a) Design strip

   The frame under consideration is laid along gridline 2; this is referred to as Direction 1, while transverse direction is referred to as Direction 2. The corresponding spans are illustrated on the following sketch, that is,

   $$l_1 = 6\,\text{m (let us consider the larger span)}$$

   and

   $$l_2 = 4.8\,\text{m}$$

   The design strip is denoted as l_{2a} (Direction 2). The design strip width is determined by taking an average value for the two spans adjacent to the gridline under consideration. In this case, the spans in Direction 2 are equal, thus

   $$l_{2a} = l_2 = 4.8\,\text{m}$$

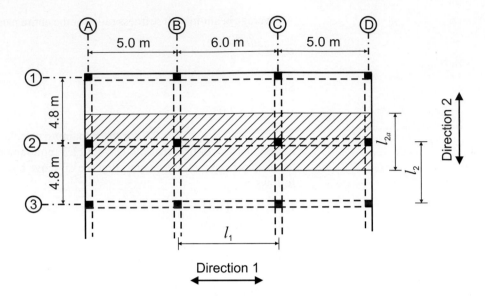

b)   Beam strip and slab strip

The width of the beam strip is determined based on the effective flange width b_s (see Section 12.4.3), that is,

$$b_s = b_w + 2h_w = 400 + 2 \cdot 440 = 1280 \text{ mm}$$

but

$$b_s \leq b_w + 8h_s = 400 + 8 \cdot 160 = 1680 \text{ mm}$$

hence

$$b_s = 1280 \text{ mm} \cong 1.3 \text{ m}$$

Dimensions of the equivalent beam section are shown on the following sketch.

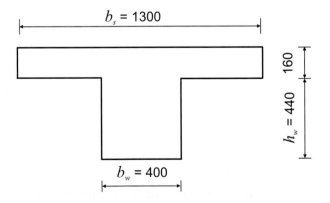

The width of the slab strip is equal to

$$l_{slab} = l_{2a} - l_{beam} = 4.8 - 1.3 = 3.5 \text{ m}$$

Beam and slab strips are illustrated on the following sketch. Note that the slab strip is divided into two half-strips, shown cross-hatched on the sketch.

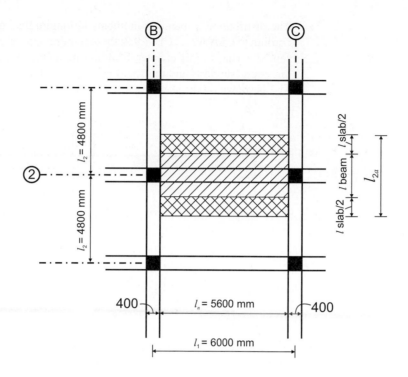

5. Find the factored bending moments in the slab.

a) Find the total factored static moment M_o.
The factored static moment can be determined from the following equation (see Section 12.6.2)

| A23.3 Eq. 13.23 |

$$M_o = \frac{w_f \times l_{2a} \times l_n^2}{8}$$ [12.8]

where
$w_f = 12.9$ kPa total factored load
$l_{2a} = 4.8$ m width of the design strip

l_n is the clear span in the longitudinal direction (along gridline 2). Since the spans are different, this calculation needs to be performed for each span.

Span 1 (end span AB)

$$l_n = 5.0 - (\frac{0.4}{2} + \frac{0.4}{2}) = 4.6 \text{ m}$$

$$M_o = \frac{w_f \times l_{2a} \times l_n^2}{8}$$

$$= \frac{12.9 \text{ kPa} \times 4.8 \text{ m} \times (4.6 \text{ m})^2}{8} = 164 \text{ kNm}$$

Span 2 (interior span BC)

$$l_n = 6.0 - (\frac{0.4}{2} + \frac{0.4}{2}) = 5.6 \text{ m}$$

$$M_o = \frac{w_f \times l_{2a} \times l_n^2}{8}$$

$$= \frac{12.9 \text{ kPa} \times 4.8 \text{ m} \times (5.6 \text{ m})^2}{8} = 243 \text{ kNm}$$

b) Distribute the total factored static moment at the critical locations in longitudinal direction.

The distribution of bending moments in longitudinal direction will be performed according to CSA A23.3 Cl.13.9.3; these requirements are summarized in Table 12.5 presented earlier in this chapter. Bending moments at critical locations in longitudinal direction (supports and midspan) are expressed in terms of the total factored moment M_o, as shown on the sketch below.

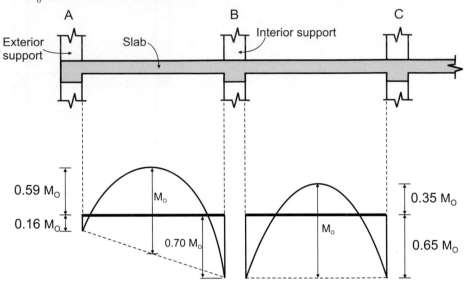

Distribution of bending moments in transverse direction is performed according to Cl.13.12 (see Table 12.5). The beam strip moment can be calculated from the following equation:

$$M_b = \left[\frac{\alpha_1}{0.3 + \alpha_1 \left(1 - \frac{l_2}{3l_1} \right)} \right] \times M \qquad [12.9]$$

For this design (considering interior span BC):

$\alpha_1 = 2.0$

$l_1 = 6.0$ m

$l_2 = 4.8$ m

Therefore,

$M_b = 0.64 \times M$

The slab strip moment is obtained as a difference between the total design moment for a particular location and the beam strip moment. The calculations are summarized in Tables 12.14 and 12.15.

Table 12.14 Factored bending moments for Span 1 (End Span AB)

		$M_o = 164$ **kNm**		
	Location	A	Midspan	C
Longitudinal direction	Bending moments at critical sections M	Negative moment (kNm)	Positive moment (kNm)	Negative moment (kNm)
		$-0.16M_o$ $= (-0.16) \times 164$ $= -26$	$+0.59M_o$ $= (+0.59) \times 164$ $= +97$	$-0.70M_o$ $= (-0.70) \times 164$ $= -115$
Transverse distribution	Beam strip moment $M_b = 0.64 \times M$	$-0.16M_o$ $= -26$	$= 0.64 \times (+97)$ $= +62$	$= 0.64 \times (-115)$ $= -74$
	Slab strip moment (total for the two halves)	0	$= 97 - 62 = +35$	$= -115 - (-74) = -41$

Confirm that the sum of bending moments within a span is equal to M_o:

1. Average negative bending moment $= (-26-115)/2 = -70$ kNm
2. Sum of absolute values for positive and negative bending moments $= 97+70 = 167$ kNm $\cong M_o$

Table 12.15 Factored bending moments for Span 2 (Interior Span BC)

		$M_o = 243$ kNm		
		B Negative moment (kNm)	Midspan Positive moment (kNm)	C Negative moment (kNm)
Longitudinal direction	Bending moments at critical sections M	$-0.65M_o$ $= (-0.65) \times 243$ $= -158$	$+0.35M_o$ $= (+0.35) \times 243$ $= +85$	$-0.65M_o$ $= (-0.65) \times 243$ $= -158$
Transverse distribution	Beam strip moment $M_b = 0.64 \times M$	$= 0.64 \times (-158)$ $= -101$	$= 0.64 \times (+85)$ $= +54$	$= 0.64 \times (-158)$ $= -101$
	Slab strip moment (total for the two halves)	$= -158-(-101) = -57$	$= 85-54 = +31$	$= -158-(-101) = -57$

Confirm that the sum of bending moments within a span is equal to M_o:

1. Average negative bending moment $= (-58-158)/2 = -158$ kNm
2. Sum of absolute values for positive and negative bending moments $= 85+158 = 243$ kNm $\cong M_o$

Bending moment diagrams for the total moment in Span 1 and Span 2, as well as the moments for beam and slab strips, are shown next.

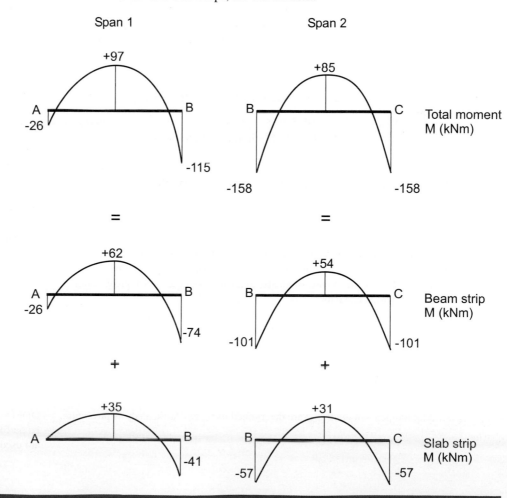

12.7 ELASTIC ANALYSIS

12.7.1 Background

This section presents design approaches which are based on elastic behaviour of reinforced concrete structures, hence the name Elastic Analysis. CSA A23.3 permits the following two elastic analysis approaches for design of reinforced concrete two-way slab systems:

- Equivalent Frame Method (referred to as "slab systems as elastic frames" by Cl.13.8), and
- Three-Dimensional Elastic Analysis (Cl.13.6).

According to the Equivalent Frame Method (EFM), an actual three-dimensional building structure consisting of slabs (with/without beams) and columns is divided into a series of parallel frames in longitudinal and transverse directions of the building. Each frame is modelled as a two-dimensional (2-D) structure called an *equivalent frame*. The design of two-way slab systems is based on the analysis of these equivalent frames in each principal direction, and the results are combined to create a design solution for an entire slab at the floor level.

According to the Three-Dimensional (3-D) Elastic Analysis, a floor system is idealized as a 3-D model, which takes into account properties of horizontal components (slabs and slab-column connections) and vertical components (columns and walls). This is a computer-based analysis procedure, and it is readily available due to the affordability of computer hardware and specialized software packages. In design practice, the use of 3-D elastic analysis for the design of two-way slabs is gaining popularity, especially for complex slabs with irregular (non-rectangular) plan shapes, or slabs characterized by large column and wall offsets relative to a rectangular grid.

It should be noted that the Elastic Analysis approaches presented in this section are based on the assumption of linear elastic behaviour of reinforced concrete structures, and that reinforced concrete is treated as a homogeneous material. Results of such analyses do not accurately simulate the behaviour of cracked slabs at service and ultimate load levels.

The Elastic Analysis approaches will be explained and illustrated by a few design examples.

12.7.2 Equivalent Frame Method

A23.3 Cl.13.8   **Features**   A comparison of key features for the Equivalent Frame Method and the Direct Design Method is outlined in Table 12.16.

Table 12.16   Key Features of the Equivalent Frame Method and the Direct Design Method

	Equivalent Frame Method (EFM)	Direct Design Method (DDM)
1.	Reinforced concrete is treated as a homogeneous material with linear elastic stress-strain relationship. As discussed in Section 12.3.4, reinforced concrete structures show inelastic (nonlinear) behaviour once the cracking has been initiated. Therefore, the results of an elastic analysis represent an approximation of actual structural behavior.	Same as the EFM.
2.	The slab must meet the requirements for regular two-way slabs outlined in CSA A23.3 Cl. 2.2 (see Section 12.5.1)	Same as the EFM.
3.	There are no restrictions with regards to the span dimensions between adjacent slab spans.	DDM contains constraints related to adjacent span lengths.
4.	Variations in slab thickness (moment of inertia) along the span have to be considered.	DDM considers only constant slab thickness.
5.	Bending moments in the frame are determined using an elastic analysis procedure.	DDM does not require analysis to be performed; empirical equations are used to find bending moments.

(continued)

Table 12.16   (Continued)

	Equivalent Frame Method (EFM)	Direct Design Method (DDM)
6.	There are no restrictions regarding the load magnitude, however pattern loadings have to be considered when a live-to-dead load ratio is large.	DDM does not directly consider pattern loadings; however there is a provision for unbalanced moments which takes into account the effect of live load variation in adjacent spans.
7.	The sum of average negative bending moments and the maximum positive moment for a typical span should not exceed the total factored moment M_o.	Same as the EFM.
8.	The EFM can be used to perform lateral load analysis.	DDM cannot be used for the lateral load analysis.

A23.3 Cl.13.8.1

Equivalent Frames: Concept   According to CSA 23.3 Cl. 13.8, a two-way slab system can be idealized as a series of parallel 2-D or "plane" frames in longitudinal and transverse directions of a building. CSA A23.3 refers to these frames as "elastic frames". The authors believe that the term "equivalent frame" is more appropriate and it will be used in this text (the same term is used in ACI 318 concrete design standard in the U.S.).

CSA A23.3 Cl.13.8.1 outlines provisions related to the frame geometry. Each equivalent frame is composed of line members intersecting at column and slab centrelines. The slab is modelled as a wide beam (referred to as "slab-beam"), and its width corresponds to the tributary portion of the slab extending on each side of the column midway between the adjacent column planes.

One of the first challenges a designer is faced with when implementing the EFM is how to isolate a 2-D frame from a 3-D building structure. This will be illustrated by an example. Figure 12.40a shows a floor plan of a building with a two-way flat slab system. An isometric view of the equivalent frame on gridline 2 is shown in Figure 12.40b. The frame is defined by columns laid along gridline 2. The slab is modelled as a wide beam (called slab-beam), which comprises a portion of the slab defined by vertical cutting planes located midway between the adjacent column gridlines 1-2 and 2-3. Gravity load acts over the slab-beam. A 2-D view of the frame is shown in Figure 12.40c. A notation related to frame span lengths was introduced in Section 12.4.1. For example, span lengths in the plane of the frame are labelled as l_1 (see span BC in Figure 12.40b), and the slab-beam width in the transverse direction is referred to as l_2; this is the same as slab design strip l_{2a} for the DDM, as discussed in Section 12.4.1.

A basic assumption of the equivalent frame model is that the width of the slab-beam is bounded by imaginary vertical cutting planes located midway between adjacent column gridlines on each side of the frame. In slab systems with uniform spans, the locations of these cutting planes generally correspond to the locations of zero shear forces and torsional moments. However, this assumption carries some errors, especially for end spans, where the location of zero shear is closer to the end column gridline. Each equivalent frame in transverse direction comprises a column strip bounded by two half-middle strips.

Structural analysis is performed to determine bending moment distribution in the frame. These frames are statically indeterminate systems, and internal forces can be obtained using established analysis procedures such as the moment distribution method or the Direct Stiffness Method.

Gross cross-sectional properties of slab-beams and columns are used for the frame model. For slabs with drop panels or slab bands, slab-beam properties need to reflect the gross cross-sectional dimensions of these members, including any variations along the span.

The results of elastic analysis are used to find bending moments and shear forces in longitudinal direction of the frame. However, transverse moment distribution is performed by applying empirical coefficients to find moments in the column strip and the middle strip (similar to the DDM).

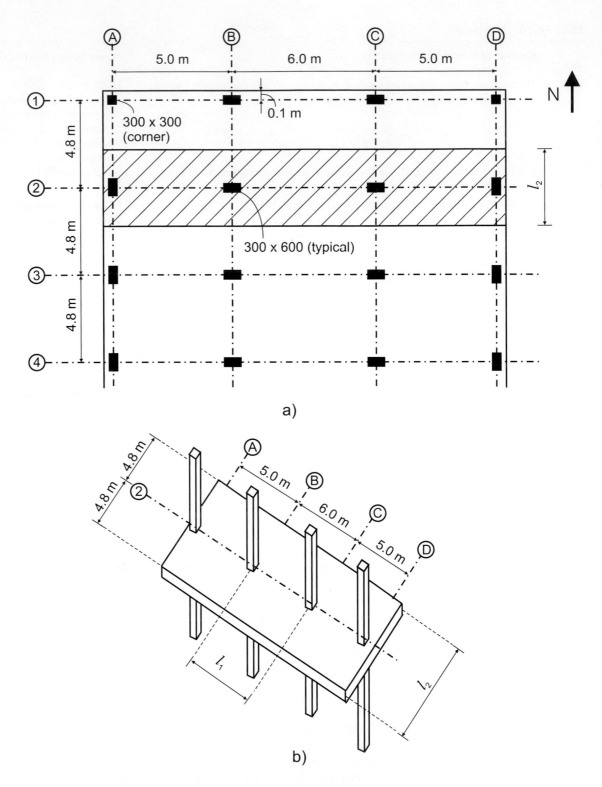

Figure 12.40 Defining an equivalent frame: a) a partial floor plan; b) isometric view of an equivalent frame on Gridline 2, and c) a 2-D frame for gravity load analysis.

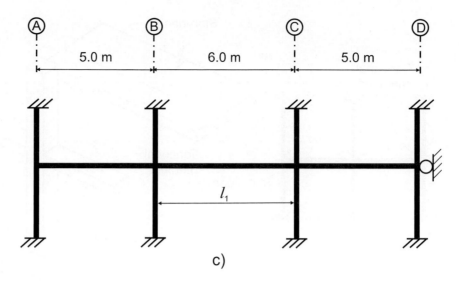

Figure 12.40 (*cont.*)

Each frame is analyzed separately, and the results for individual frames are superimposed to create the final design solution for building under consideration.

For gravity load analysis, each floor and roof slab can be analyzed separately. Far ends of the columns must be restrained by providing a roller support which prevents side sway of the frame (CSA A23.3 Cl.13.8.1.2). A frame model suitable for gravity load analysis is shown in Figure 12.40c.

| A23.3 Cl.13.8.2 and 13.8.3 | **Modelling of Equivalent Frame Members**   An equivalent 2-D frame consists of slab-beam and column members. The EFM assumes rigid beam-column connections, that is, all members joined at a connection undergo the same rotation. A partial view of an equivalent 2-D frame is shown in Figure 12.41a. It can be seen from the figure that slab-beams look like beam elements in a regular beam and column frame, however these beams are very wide (their width is equal to the tributary slab width l_2). |

Column members are more complex due to the presence of *attached torsional members*, that is, imaginary linear members which extend from the column in transverse direction, as shown in Figure 12.41b. The purpose of these members is to take into account reduced flexural stiffness of the columns at the locations of beam-column connections. This concept will be discussed in more detail later in this section.

Note that Direction 1 is along the plane of the frame, while Direction 2 denotes transverse direction (see Figure 12.41b).

An initial step in the frame analysis is to determine flexural stiffness for equivalent frame members. An appropriate stiffness must be estimated to simulate the behaviour of actual slab system (CSA A23.3 Cl.13.8.1.5). The flexural stiffness is proportional to *EI*, which is a product of the modulus of elasticity (*E*) and the moment of inertia (*I*).

Geometric properties (cross-sectional dimensions) of frame members can be determined using either non-prismatic or prismatic approach according to CSA A23.3 (see Section 3.2 for a definition of prismatic and non-prismatic flexural members). Simply put, *prismatic* refers to constant cross-sectional properties along the member length, while *non-prismatic* refers to variable cross-sectional properties. Both approaches use gross cross-sectional properties of frame members (the effect of cracking is not taken into account).

A *non-prismatic approach* is used in conjunction with the moment distribution method, which has been used since the EFM was first introduced in North American design codes in the 1960s. This method was suitable for the analysis of simple equivalent frames by digital calculators. Each non-prismatic slab and column member with variable cross-sectional properties along the span can be modelled as a single member (CSA A23.3 Cl.13.8.2). Cor-

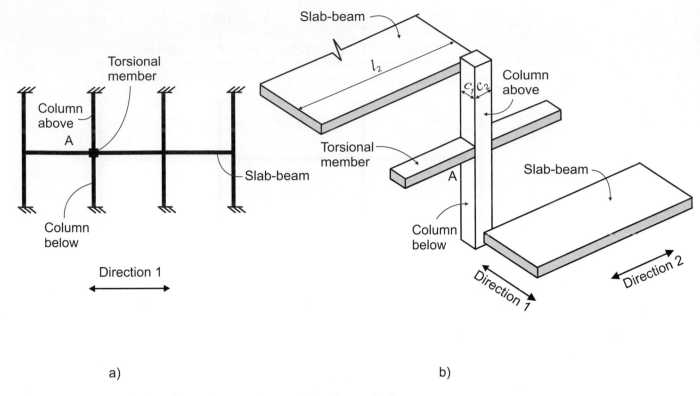

Figure 12.41   An equivalent frame: a) a 2-D view, and b) an isometric view.

rection factors are applied to modify member stiffness, carry-over factors, and fixed end bending moments for each frame member. Once the adjusted properties for all frame members have been determined, the moment distribution method can be used to derive bending moments at each joint of the equivalent frame. A detailed discussion on moment distribution method is beyond the scope of this book, however the reader is referred to MacGregor and Bartlett (2000) for more information.

A *prismatic approach* considers members with constant cross-sectional properties (CSA A23.3 Cl.13.8.3). This approach is used in conjunction with the Direct Stiffness Method and computer-based frame analysis. Since the prismatic approach is currently more widely used in design practice, it will be explained in the next section.

Let us illustrate modelling of slab-beam members for the EFM applications. Consider a typical span of a flat slab shown in Figure 12.42a. For modelling purpose, span length (l_1) is equal to distance between the column centrelines (AB). Cross-sectional dimensions of a typical slab section (1-1) are shown in Figure 12.42b. The slab-beam section has thickness h_s, and its width l_2 is equal to design strip width shown in Figure 12.40b. The corresponding moment of inertia is equal to I_1.

A slab-beam could be modelled as a single prismatic member provided that column and slab-beam dimensions are relatively small. However, member dimensions in reinforced concrete structures are significant and should be considered in the structural model. Therefore, moment of inertia of the slab-beam member between the column face and the column centreline must be modified to account for the column and slab-beam dimensions. This is reflected by the moment of inertia I_2 which corresponds to slab section 2-2 shown in Figure 12.42b. Note that I_2 is obtained by modifying I_1 value by the multiplier prescribed by CSA A23.3 Cl.13.8.2.3. A variation in the EI value along the slab span is shown in Figure 12.42c. For the non-prismatic approach, a slab span is modelled as a single member shown in Figure 12.42d, and EI variation along the slab span is taken into account by design aids (tables) used in conjunction with the moment distribution method. Alternatively, a slab-beam can be modelled as a series of connected prismatic segments shown in Figure 12.42e.

Each portion of the slab-beam with variable cross-sectional properties is treated as a separate segment; in this case, there are three segments (1, 2 and 3).

The above described technique can be applied to flat slabs with drop panels and/or columns with capitals; however, the model will become more complex because it needs to take into account geometric properties of drop panels and column capitals.

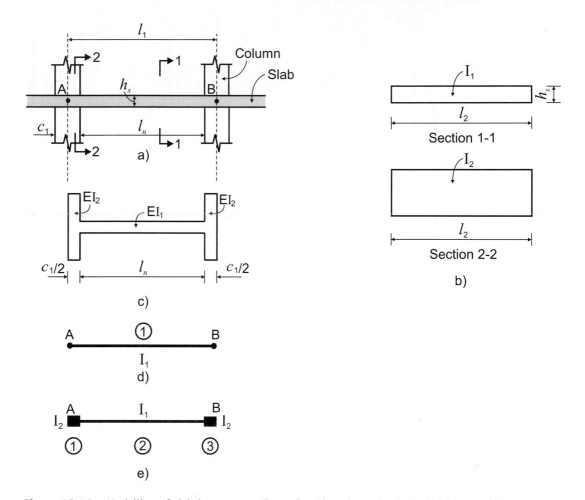

Figure 12.42   Modelling of slab-beam properties - elevation view: a) a typical slab span; b) cross-sections; c) variation of stiffness along the span; d) non-prismatic model, and e) prismatic model.

Figure 12.43 illustrates modelling of column properties for the EFM analysis. Figure 12.43a shows a typical column spanning between flat slabs at two adjacent floor levels. Column stiffness is determined using the column height l_c measured between the slab centrelines, as shown in the figure. Moment of inertia of the column outside the slab-column connection is based on the gross cross-sectional area of concrete (I_c) shown in Figure 12.43b. Note that the column moment of inertia is considered to be infinite in the slab-column connection region, which extends from the face of the slab-beam to the slab centreline at each end. A variation of EI along the column height is shown in Figure 12.43c. According to the non-prismatic approach, a column is modelled as a single member, as shown in Figure 12.43d, and the variation of column properties is accounted for by design aids used in conjunction with the moment distribution method. Alternatively, a prismatic approach can be used in conjunction with the Direct Stiffness Method, as shown in Figure 12.43e. It can be seen from the figure that there are three column segments (1, 2 and 3). The end segments (1 and 3) are assigned significantly larger (e.g. by an order of magnitude) moment of inertia values compared to I_c.

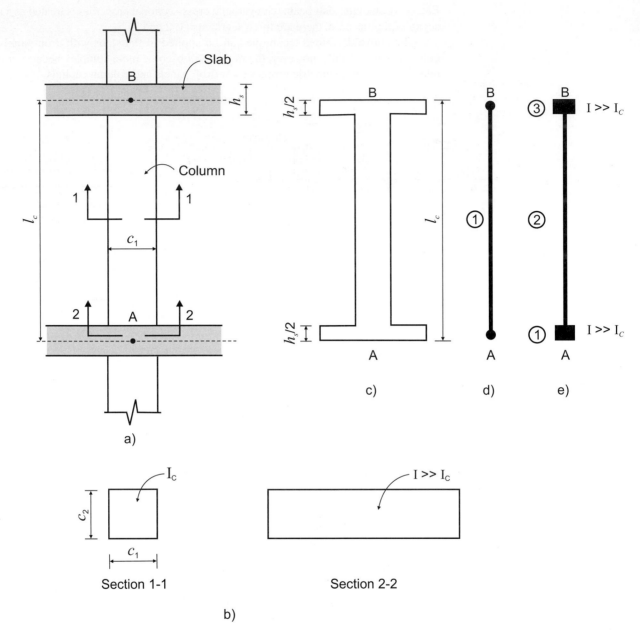

Figure 12.43 Modelling of column properties - an elevation view: a) a typical column; b) cross-sections; c) variation of stiffness along the column length; d) non-prismatic model, and e) prismatic model.

The equivalent frame models for slabs without beams are characterized by slab-beam members with significantly larger width compared to the columns; this has implications on the load transfer from the slab to the column.

Figure 12.44a shows gravity load path in the vicinity of a slab-to-column connection. The load in the vicinity of the column is transferred from the slab to the column directly through bending. However, load further away from the column centreline is transferred into the column by twisting of a slab strip on each side of the column. This has the effect of reducing flexural stiffness of the column relative to flexural stiffness of the slab. For analysis purpose, slab strip ABC (shown shaded in the figure) is considered as an attached torsional member, which accounts for a reduction in the column flexural stiffness. Figure 12.44b shows an isometric view of the attached torsional member ABC which has the length equal to the slab-beam width l_2. The rotation at the support (column) at point B is labelled as θ_B. It can be seen that the rotations at points A and C located at far ends of the slab section (θ_A

and θ_C respectively) are significantly larger than rotation θ_B at column face. This is due to the fact that the column has a minimal effect on slab rotations at far ends of the slab section.

Figure 12.44 Loads and deformations in the slab-column connection region: a) load transfer from the slab to the column, and b) rotation and twisting of the attached torsional member.

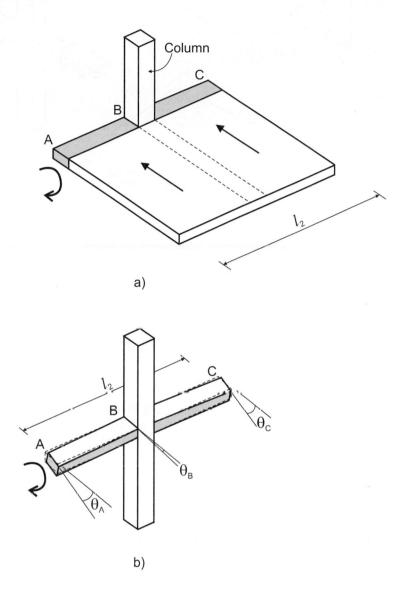

The above discussion is useful for understanding transfer of bending moments from the slab to the column. In a slab-and-column frame, bending moments from the slab are transferred directly to the column only over a rather narrow strip approximately equal to the column width, as shown in Figure 12.45a. The remaining bending moments in the slab must be transferred to the column through torsion of the attached torsional member, as shown in Figure 12.45b.

The rotational stiffness of the slab-to-column connection is a function of the torsional stiffness of the attached torsional member and the flexural stiffness of the columns framing into the connection from above and below. For analysis purpose, a column and the attached torsional members can be considered as an *equivalent column*, as shown in Figure 12.44b (CSA A23.3 Cl.13.8.2.5). The stiffness of an equivalent column (K_{ec}) can be determined from the following equation (CSA A23.3 Cl.13.8.2.5).

A23.3 Eq. 13.18

$$\frac{1}{K_{ec}} = \frac{1}{\sum K_c} + \frac{1}{K_t}$$

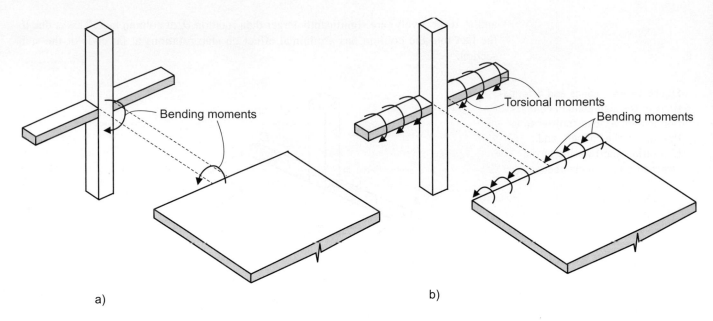

Figure 12.45   Transfer of moments from the slab to the column: a) bending moments at the column location, and b) torsional moments acting on the attached torsional member.

where $\sum K_c$ is the sum of flexural stiffness values for columns above and below the slab, and K_t is the stiffness of a torsional member. CSA A23.3 contains provisions for estimating the stiffness of attached torsional members (Cl.13.8.2.8 to 13.8.2.10).

A23.3 Cl.13.8.3 | **Prismatic Approach for Modelling Frame Sections**   A prismatic approach considers members with constant cross-sectional properties. Variation in slab properties within a span can be accounted for by considering prismatic segments with different gross cross-sectional properties. This approach is used in conjunction with the Direct Stiffness Method and it is suitable for computer applications.

A slab span is divided into several (usually 10 to 20) slab-beam segments which are joined together. Each segment has constant cross-sectional properties, and it simulates the stiffness of a specific slab section. This approach can take into account the effect of drop panels by considering segments with appropriate cross-sectional properties at drop panel locations.

A column is usually modelled as a single member. Column capital can be modelled as an additional segment both in the slab and the column with appropriate stiffness properties. CSA 23.3 Cl. 13.8.3.3 accounts for a reduction in column stiffness due to the attached torsional member through the column stiffness modification factor (ψ), which can be obtained from the expressions presented in Table 12.17.

Table 12.17   Column Stiffness Modification Factor ψ (CSA A23.3 Cl.13.8.3.3)

Span Ratio	Slabs without beams ($\alpha_1 = 0$)	Slabs with beams
$l_2/l_1 \le 1.0$	$\psi = 0.3$	$\psi = 0.3 + 0.7 \dfrac{\alpha_1 l_2}{l_1}$
$l_2/l_1 > 1.0$	$\psi = 0.6\left(\dfrac{l_2}{l_1} - 0.5\right)$	$\psi = 0.6\left(\dfrac{l_2}{l_1} - 0.5\right) + \left(1.3 - 0.6\dfrac{l_2}{l_1}\right)\dfrac{\alpha_1 l_2}{l_1}$

Notes:
l_1 and l_2 - slab spans in the plane of the frame (Direction 1) and transverse direction (Direction 2), respectively
α_1 - beam-to-slab stiffness ratio in Direction 1.

CSA A23.3 Cl.13.8.3.3 sets a range for the stiffness modification factor values, as follows

$$0.3 \le \psi \le 1.0$$

and

$$\frac{\alpha_1 l_2}{l_1} \le 1.0$$

For flat slabs and flat plates, ψ values vary between 0.3 (when $l_2/l_1 \le 1.0$) and 0.9 (when $l_2/l_1 = 2.0$). The ψ factor increases when the slab span in transverse direction is significantly larger than the span in longitudinal direction, because the column flexural stiffness is more significant and it attracts larger negative moments over the column strips.

To account for the effect of column flexural stiffness, the designer should use product $\psi \times I_c$ (instead of I_c) for columns in an equivalent frame model. For example, consider the equivalent frame discussed earlier in this section (see Figure 12.40). Gross cross-sectional properties, that is, column moment of inertia (I_c) and slab-beam moment of inertia (I_s), are presented in Figure 12.46a, while modified frame properties are shown in Figure 12.46b. Each slab-beam span is divided into several segments with the same properties (moment of inertia I_s), but column properties need to be modified by the ψ factor.

Figure 12.46 Equivalent frame with prismatic member properties: a) actual frame properties, and b) modified frame properties used for the design.

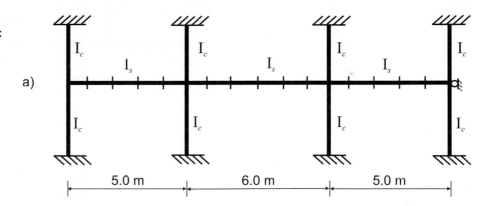

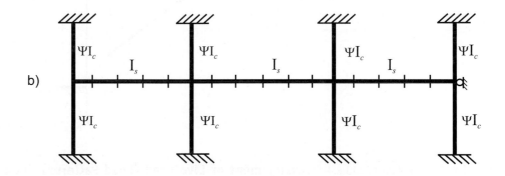

A23.3 Cl.13.8.5 **Design Bending Moments at Supports**   The members of an equivalent frame span centre-to-centre between supports. Therefore, bending moments and shear forces obtained from the EFM analysis are given at member centerlines. However, in reality supports (columns) in a reinforced concrete structure have finite dimensions, hence bending moments derived at centerlines may not be realistic. There may be a significant difference between the negative bending moments at the column centreline and the face of the column. CSA A23.3 Cl.13.8.5.1 permits the use of reduced bending moments at the face of support for design purpose. The locations where bending moments are reduced are referred to as *critical sections* by CSA A23.3.

An example illustrating the reduction of bending moments at the face of an interior column is shown in Figure 12.47. Critical sections at the face of a column are shown in Figure 12.47a. Note that a critical section should be taken at the column face, but the distance should not exceed $0.175\, l_1$ from the column centreline (this limit is set for columns with rectangular cross-sections where one dimension is significant compared to the slab span). Actual bending moment values for the slab at the support (M_L and M_R) are shown in Figure 12.47b; note that these values are different because the balance of these bending moments is transferred to the column. Reduced bending moments at the critical sections (M_{RL} and M_{RR}) should be used to design the flexural reinforcement. The reduced moments at the face of the column can be calculated from the equation presented in Section 10.4.3.

For exterior columns with column capitals Cl. 13.8.5.2 permits the critical section to be located at a distance equal to one-half of the capital projection from the face of the column, as shown in Figure 12.48.

Figure 12.47 Bending moment diagram at the critical section at the face of an interior column: a) critical sections at the column face, and b) bending moment diagram.

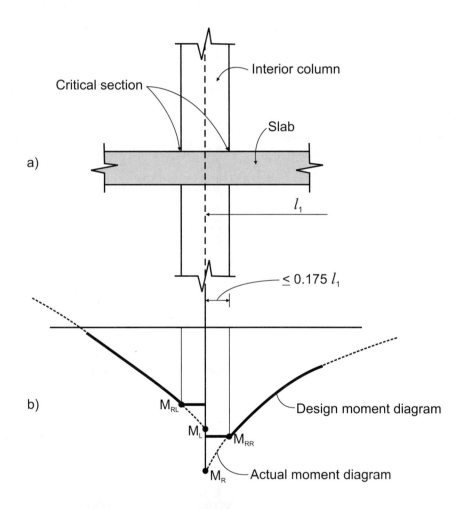

A23.3 Cl.13.8.4   **Arrangement of Live Load (Load Patterns)**   Live loads are transient loads, and a variation in their magnitude and arrangement may significantly influence bending moments and shear forces in reinforced concrete structures. Therefore, bending moments and shear envelopes obtained by considering variations in live load patterns must be taken into account by design (refer to Section 10.3 for more details on load patterns and moment envelopes). However, two-way reinforced concrete slabs usually have a significant self-weight, thus dead load often accounts for a major portion of the total factored load. In these cases, variations in the live load arrangement may not have a significant effect on the maximum bending moments and shear forces in the slab.

CSA A23.3 provisions related to the live load arrangement for design of two-way slabs according to the EFM consider the following three cases:

Figure 12.48 Bending moment diagram at the critical section at the face of an exterior column with the capital: a) critical section, and b) bending moment diagram.

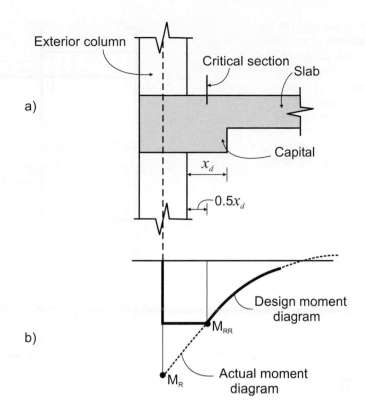

1) When the loading pattern is given, the frame should be analyzed considering that pattern (Cl.13.8.4.1). For example, in a warehouse with moving crane loads on the slab, the crane load should be considered with or without the occupancy live loads to determine the maximum bending moments and shear forces.

2) When the specified live load is uniformly distributed and does not exceed three-quarters (75%) of the specified dead load, the designer only needs to consider the full factored load on all spans (Cl. 13.8.4.2). This should apply to most two-way slab building applications with residential, office, and/or retail occupancy.

3) When the specified live-to-dead load ratio in the slab exceeds 75%, several live load patterns need to be considered to determine the maximum factored bending moments (Cl.13.8.4.3). These load patterns (LP1 to LP4) take into account partial factored live loads, as shown on an example of a two-way slab in Figures 12.49a to d. However, the factored moments should not be taken as less than those developed due to full factored live loads on all panels (Cl.13.8.4.4); this is illustrated by load pattern LP5 in Figure 12.49e.

Note that, when the specified live-to-dead load ratio exceeds 0.75, the designer may wish to increase the design dead load to bring the ratio down to below 0.75. In this manner, the designer avoids the need to consider pattern loading in the design. The resulting design solution uses a higher than required load and it is slightly more conservative, which may be a good practical approach.

| A23.3 Cl.13.5.2 and 13.5.3 |

Buildings with Two-Way Slab Systems: Gravity and Lateral Load Analysis A structural analysis of a reinforced concrete building with a two-way floor system can be performed by modelling a 3-D structure as a series of idealized parallel 2-D frames in each direction, as discussed earlier in this section. Key considerations related to the analysis of 2-D frames for gravity and lateral loads will be discussed in this section. Note that it is required to perform the gravity and lateral load analyses separately, and the results should be combined (Cl.13.5.3).

Figure 12.49 Load patterns for gravity load analysis of equivalent frames: a) positive design moment in span AB; b) positive design moment in span BC; c) negative design moment at support A; d) negative design moment at support B, and e) full factored load in all spans when the specified live load does not exceed 75% of the specified dead load.

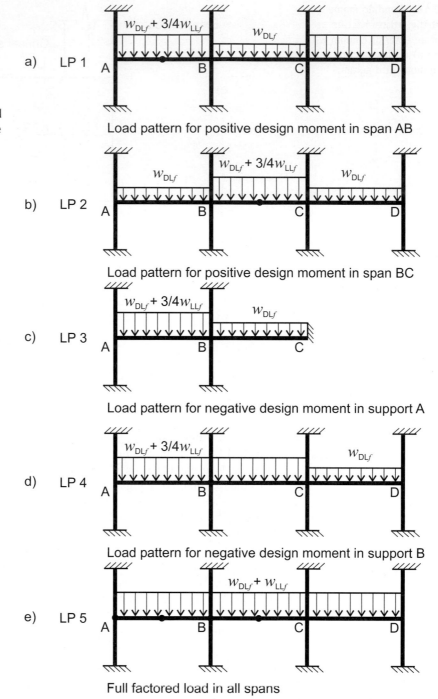

For the gravity load analysis, the designer needs to consider only a portion of the frame, which consists of a slab at the floor level under consideration and the column segments above and below that level. Consider a 2-D frame along gridline 2 (as shown in Figure 12.50a). It is not required to perform analysis of the entire frame for gravity loads - a partial frame can be used instead. An example of a partial frame for the slab at level 3 and the adjoining columns is shown in Figure 12.50b. Note that a lateral restraint in the form of a roller support need to be provided at the far end of the frame at the slab level to prevent lateral movements. Also, far ends of the columns are restrained by fixed supports. Gravity load analysis can be performed using either the Direct Design Method or the Equivalent Frame Method.

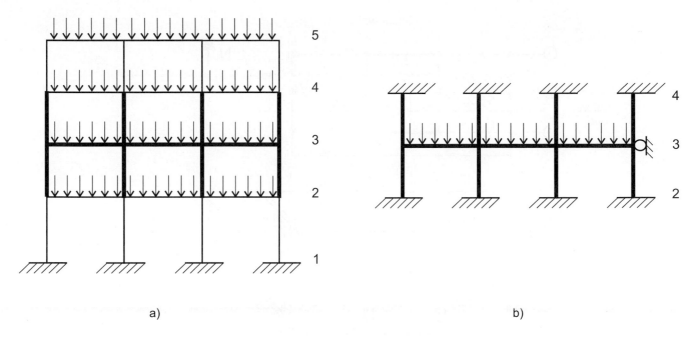

Figure 12.50 Frame model for gravity load analysis: a) a complete frame, and b) a partial frame used for gravity load analysis.

Lateral load model will be explained on an example of a four-storey building with a flat plate floor system shown in Figure 12.51a. The building can be modelled as a series of four frames in N-S direction and three frames in the E-W direction. These frames are defined by passing vertical cutting-planes through the structure midway between the column planes. Next, consider a typical frame along gridline 2 in E-W direction (which is shown shaded in Figure 12.51a). An isometric view of the same frame is shown in Figure 12.51b. The frame geometry is defined by column and slab properties. A 2-D view of the frame with the actual column and slab thicknesses is shown in Figure 12.51c; note the column and slab centrelines shown dashed in the figure. Finally, a 2-D model of the frame used for the analysis is shown in Figure 12.51d.

When frames constitute lateral load-resisting system, lateral load needs to be distributed between frames aligned in the same direction. For example, consider three frames in E-W direction (along gridlines 1, 2, and 3). The two exterior frames are supported by the walls along gridlines 1 and 3, while the interior frame along gridline 2 is supported by the columns. As a result, frames 1 and 3 will resist most of the lateral loads since their lateral stiffness will be significantly larger. Distribution of lateral loads in proportion to frame stiffness is appropriate when floors and the roof are relatively rigid compared to the vertical elements (columns or walls). In that case, a linked frame model can be used for lateral load analysis. An example of a linked frame model for E-W direction of the building is shown in Figure 12.51e. The model consists of three 2-D frames (gridlines 1, 2, and 3) which are connected by rigid pin-ended links. The entire lateral load is applied to the linked frame. Distribution of loads to individual frames is usually performed using structural analysis software.

The EFM can be used to perform lateral load analysis for buildings with two-way slab systems, however the DDM cannot be used for that purpose. Cl.13.5.2 states that the lateral load analysis should take into account the effects of cracking and reinforcement on stiffness of frame members.

In a building, flat plates or flat slabs supported by columns are rarely designed as the primary lateral load-resisting system in a building, and their application is not permitted in regions of high seismicity according to NBCC 2010. In most cases, reinforced concrete shear walls are used as principal components of the seismic force-resisting system in buildings with flat plate or flat slab floor systems. However, lateral swaying of floor and roof

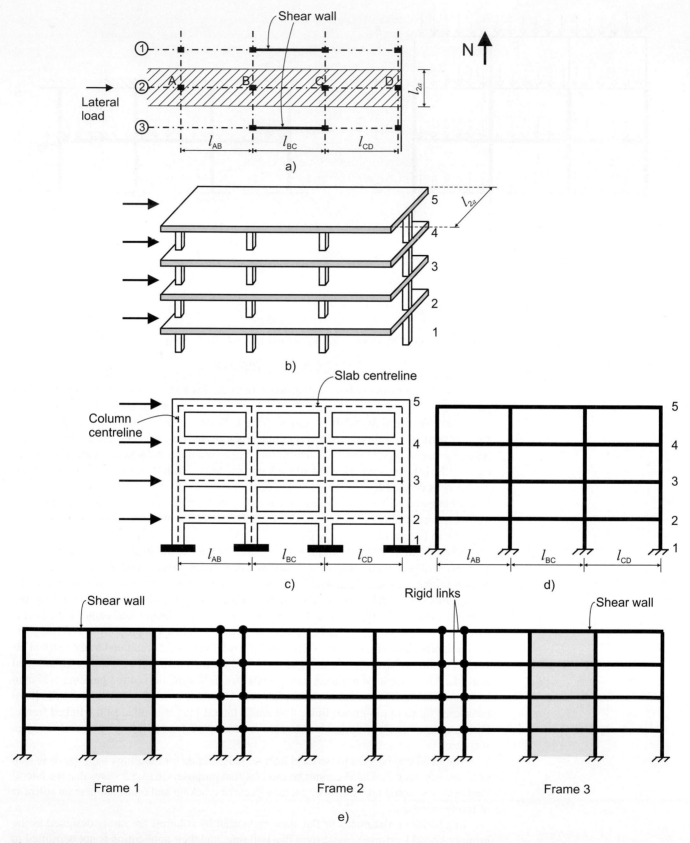

Figure 12.51 Frame models for lateral load analysis of two-way slabs: a) a typical building plan; b) an isometric view of a 2-D frame model; c) a 2-D frame showing actual column and slab dimensions; d) a 2-D model used for the lateral load analysis, and e) a linked frame model for lateral load analysis.

systems due to seismic or wind loading will have an effect upon the gravity load-bearing capacity of a slab system. When a building is subjected to lateral loading, lateral swaying will cause additional stresses in floor and roof slabs. Differential displacements between shear walls in a building might cause rotations in the floor slabs. The designer must carefully consider the effects of such lateral movements upon the punching shear capacity of flat plates and flat slabs.

The difference in lateral displacements between adjacent floors is called "inter-storey drift". Inter-storey drift for a frame is shown in Figure 12.52a. Note that inter-storey drift for floor levels i and j is denoted as Δ_{ij}, and the drift for floor levels j and k is denoted as Δ_{jk}. The total lateral drift for a building is equal to the lateral displacement at the roof level relative to the base of the building and it is denoted as Δ; this is relevant for seismic design. A magnitude of inter-storey drift at each floor level needs to be obtained from the seismic analysis of the primary lateral resisting structure. The designer needs to evaluate the effect of inter-storey drift on the floor and roof slabs by performing a frame analysis. A partial frame can be subjected to differential displacements equal to inter-storey drift values at adjacent floor levels, as shown in Figure 12.52b.

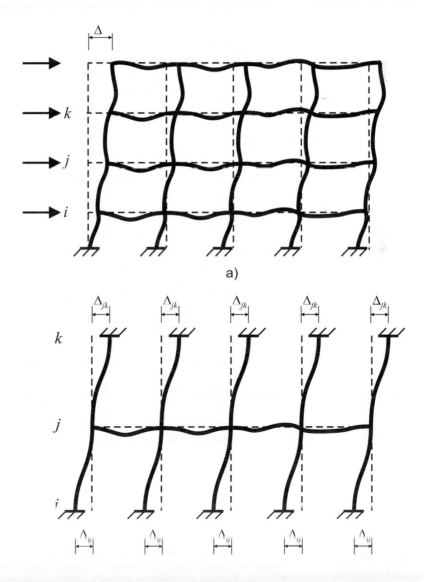

Figure 12.52 Lateral load analysis: a) inter-storey drift in the frame determined from lateral load analysis, and b) a partial frame subjected to lateral displacements at the supports.

The designer needs to combine internal forces due to inter-storey drift with those due to gravity loading. Analysis of reinforced concrete buildings for lateral loads is outside the scope of this book.

Guidelines for Effective Modelling of Equivalent Frames   Designers need to understand that an equivalent frame model is an approximation of the actual structural behavior. The EFM is an elastic analysis approach and it does not account for cracking and inelastic behaviour of reinforced concrete structures, hence the use of an overly accurate equivalent frame model may not be justified. Designers may keep in mind the following guidelines related to effective modelling of equivalent frames:

1. Bending moments are typically negative in support regions and positive in midspan regions of the slab. For a slab span subjected to uniformly distributed load, the sum of absolute values for positive bending moment at the midspan and average negative bending moment at the supports is equal to the moment gradient M_o (as discussed in Section 12.6.2).
2. It is important to estimate a reasonable range for maximum negative moments at the supports and maximum positive moments in midspan regions of the slab, and provide an adequate amount of reinforcement in these regions. A balance between the negative (top) reinforcement and the positive (bottom) reinforcement is required to minimize flexural cracking, a key design consideration for continuously reinforced concrete slab structures.
3. Continuously reinforced concrete slabs represent a cost-effective structural system, as discussed in Chapter 10. It is important to ensure an appropriate placement of reinforcement in order to take advantage of structural efficiency for a continuous span structure. The effective design of continuous slabs is also critical to minimize deflections. Section 10.2 discusses general concepts of continuously reinforced concrete structures; these concepts should be considered when optimizing a two-way slab system which has continuous spans in two directions.

The designer should keep in mind that two-way slabs are continuous structures, that is, statically indeterminate and redundant systems. Consequently, there may be several acceptable design solutions. For that reason, the designer may wish to perform a sensitivity analysis by varying positive and negative bending moments in each span to identify the most appropriate solution.

Furthermore, the designer can influence bending moment distribution in the slab either by increasing or decreasing column stiffness for modelling purposes. Two extreme solutions will be discussed below.

The first solution involves a model in which column stiffness is disregarded (equal to zero).The frame is modelled as a continuous beam system where the slab is supported by pin or roller supports instead of the columns. The resulting bending moments in the slab and the amount of reinforcement are expected to be larger compared to the design where column stiffness is taken into account; this is due to the fact that a model with columns characterized by larger stiffness results in reduced bending moments in the slab. In conclusion, the solution where column stiffness is ignored may be conservative but is considered acceptable and it is used in design practice.

The second solution involves the use of a very high column stiffness in the EFM. This could result in an increase in negative bending moments (and top reinforcing) at the supports, and a decrease in positive moments at the midspan. Note that increased column stiffness can lead to an increase in bending moments which are transferred from the slab into the columns. This may be particularly critical for columns that support two adjacent slab spans with significantly different lengths. In a flat slab, the capacity to transfer moments from slab into the column is limited by its punching shear capacity. Punching shear is a brittle failure mechanism and must be avoided. A suggestion for designers is to develop a frame model by using a reduced column stiffness, in order to restrict the magnitude of bending moment to be transferred from the slab to the column.

Design Applications of the Equivalent Frame Method The purpose of this section is to demonstrate design of two-way slabs without beams according to the EFM by three design examples.

Example 12.3

Two-Way Flat Plate (Slab without Beams) - Equivalent Frame Method

Consider a floor plan of a two-way slab system without beams designed in Example 12.1. The floor height is 3 m.

Use the CSA A23.3 Equivalent Frame Method to determine the design bending moments and reinforcement for an interior frame along gridline 2.

Given: $f'_c = 30$ MPa

$f_y = 400$ MPa

SOLUTION:

1. **Check whether the criteria for the CSA A23.3 Equivalent Frame Method (EFM) are satisfied.**

   The EFM can be applied when a slab is regular, that is, when the following CSA A23.3 Cl. 2.2 provisions have been met (see Section 12.5.1 for discussion on regular slabs):

   (#a) Within a panel, the ratio of longer to shorter span, centre-to-centre of supports, is not greater than 2.0. *In this case,* $l_1/l_1 = 6.0$ m$/4.8$ m $= 1.25 < 2.0.$

   (#b) For slab systems with beams between supports, the relative effective stiffness of beams in the two directions is not less than 0.2 or greater than 5.0. *This is not applicable, since this is a slab without beams.*

   (#c) Column offsets are not greater than 20% of the span (in the direction of the offset) from either axis between centerlines of successive columns. *There are no column offsets in this case.*

   (#d) The reinforcement is placed in an orthogonal grid. *The reinforcement will be placed in an orthogonal grid.*

   This is a regular slab according to CSA A23.3 Cl.2.2, therefore the EFM can be used for this design.

2. **Determine the required slab thickness.**

   The slab thickness is the same as in Example 12.1, that is,

   $h_s = 180$ mm

3. **Calculate the factored design loads.**

   a) Calculate the dead load acting on the slab.
   First, calculate the slab's self-weight:

   $DL_w = h_s \times \gamma_w = 0.18$ m$\times 24$ kN$/$m$^3 = 4.32$ kPa

   where $\gamma_w = 24$ kN$/$m^3 is the unit weight of concrete.
   The superimposed dead load is given, that is,

   $DL_s = 1.44$ kPa

   Finally, the total factored dead load is

   $w_{DL,f} = 1.25(DL_w + DL_s) = 1.25(4.32 + 1.44) = 7.2$ kPa

   b) Calculate the factored live load:

   $w_{LL,f} = 1.5 \times LL_w = 1.5 \times 3.6$ kPa $= 5.4$ kPa

   c) The total factored load is

   $w_f^* = w_{DL,f} + w_{LL,f} = 7.2 + 5.4 = 12.6$ kPa

d) Check whether the pattern loading needs to be considered according to CSA A23.3 Cl.13.8.4, that is, check the ratio of specified live load and dead load.
The total specified dead load is

$$w_{DL} = DL_w + DL_s = 4.32 + 1.44 = 5.76 \text{ kPa}$$

The specified live load is

$$w_{LL} = 3.6 \text{ kPa}$$

The ratio of the specified live and dead load:

$$\frac{w_{LL}}{w_{DL}} = \frac{3.6 \text{ kPa}}{5.76 \text{ kPa}} = 0.63 < 0.75$$

Since the ratio is less than 0.75, pattern loading does not need to be considered. The frame needs to be designed for the effects of total factored dead and live load on all spans. Hence, the design load is

$$w_f^* = 12.6 \text{ kPa}$$

This is a uniformly distributed area load which could be used for 3-D analysis. However, we need to find the design load for 2-D frame analysis, which can be obtained when a tributary slab width ($l_2 = 4.8$ m) is taken into account, as follows

$$w_f = w_f^* \times l_2 = 12.6 \text{ kPa} \times 4.8 \text{ m} = 60.5 \text{ kN/m}$$

4. **Develop a frame model.**

a) Determine the frame geometry.
In order to model an actual 3-D building as a series of parallel 2-D frames, the designer first needs to find the frame width. In this case, the width (l_2) is 4.8 m, as discussed above. This is shown on the sketch, for a frame along gridline 2.

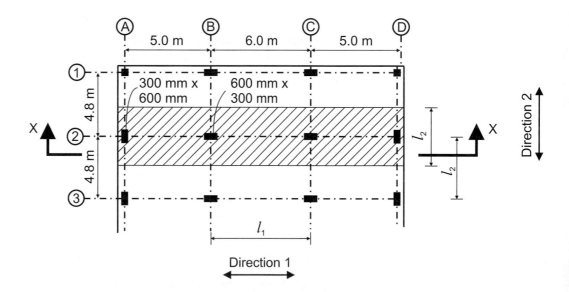

Next, it is necessary to sketch the frame and identify beam spans in horizontal direction and column heights in vertical direction. Note that the frame geometry is defined by slab and column centrelines. The following sketch shows Section X-X of the frame (note that actual dimensions of frame members are shown on the sketch).

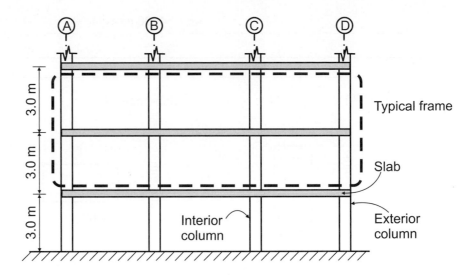

Section X-X

Finally, the designer needs to prepare a linear (wire) drawing of the frame. For gravity load analysis, CSA A23.3 permits the designer to consider one floor level plus adjacent columns above and below the floor. The sketch below shows frame geometry (beam and column spans). The frame axes correspond to the centrelines of slabs and columns. The column ends are shown fixed (no rotation). Note that the sketch also shows uniform design load (w_f) on the slab.

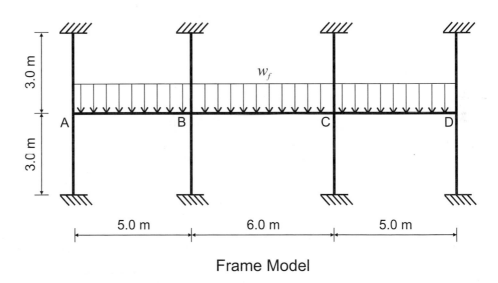

Frame Model

b)   Determine cross-sectional properties for the frame members.

Typical cross-sections for frame members (slab-beams and columns) are illustrated on the following sketch.

i)   Slab-beams

In the frame model, the slab is treated as a wide beam with the following gross cross-sectional properties: 4.8 m width (as discussed above) and 180 mm depth (equal to the slab thickness). Moment of inertia for the slab about axis x-x (see the sketch) is equal to

$$I_s = \frac{4800 \text{ mm} \times (180 \text{ mm})^3}{12} = 2.3 \times 10^9 \text{ mm}^4$$

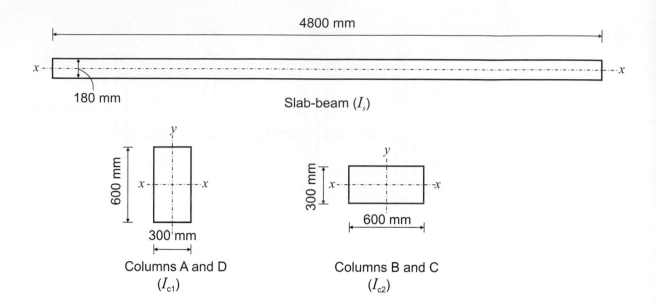

For prismatic modelling, each span can be divided into ten or more segments, where each segment has the same moment of inertia (I_s).

ii) Columns

Column gross cross-sectional dimensions are the same (300 mm by 600 mm). However, note that the layout is different for exterior columns (A and D) and interior columns (B and C). For the frame along gridline B bending in the columns occurs about the axis y-y (as shown on the sketch), therefore moment of inertia values for the columns are different.

Columns A and D (moment of inertia I_1):

$$I_1 = \frac{600 \text{ mm} \times (300 \text{ mm})^3}{12} = 1.35 \times 10^9 \text{ mm}^4$$

Columns B and C (moment of inertia I_2):

$$I_2 = \frac{300 \text{ mm} \times (600 \text{ mm})^3}{12} = 5.4 \times 10^9 \text{ mm}^4$$

For prismatic modelling, Cl.13.8.3.3 requires that the column moment of inertia be modified by the factor ψ. Since this is a slab without beams, $\alpha_1 = 0$ and $l_2/l_1 = 4.8 \text{ m}/6 \text{ m} = 0.8 < 1.0$, hence

$$\psi = 0.3$$

Therefore, the column moment of inertia values are $0.3 \cdot I_1$ and $0.3 \cdot I_2$.

c)   Sketch the final frame geometry and member properties.
A sketch showing the final frame geometry and section properties is shown next. Note that the frame has 11 members. Slab-beam members are labelled as 3, 6, and 9, and the remaining members are columns. For analysis purposes, the designer can choose to divide the slab-beam into 10 to 20 segments, depending on the desired level of accuracy. Keep in mind that the EFM is an approximate method, hence it is not necessary to divide slab-beam members into too many segments.

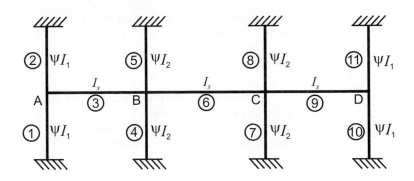

5. **Determine the factored bending moments (M_f).**
    This analysis was performed using a commercially available software package. The resulting bending moment and shear diagrams are shown next — note that the diagrams show only bending moments and shear forces in the slab-beam members.

    Negative bending moments transferred from the slab into the columns are higher at end supports A and D (38 kNm) than at interior supports B and C. Negative moment at supports B and C is equal to 7 kNm, that is, the difference in bending moments between adjacent spans (−181 kNm and −174 kNm). In this design, the spans seem to be reasonably balanced and one would expect that unbalanced moments at the interior supports (shared equally between the columns above and below the slab) are small.

6. **Calculate the reduced negative bending moments at the supports (CSA A23.3 Cl.13.8.5.1).**
    First, let us calculate a reduced moment at the face of the support B for span AB. The procedure outlined in Section 10.4.3 will be followed (see Figure 10.13).

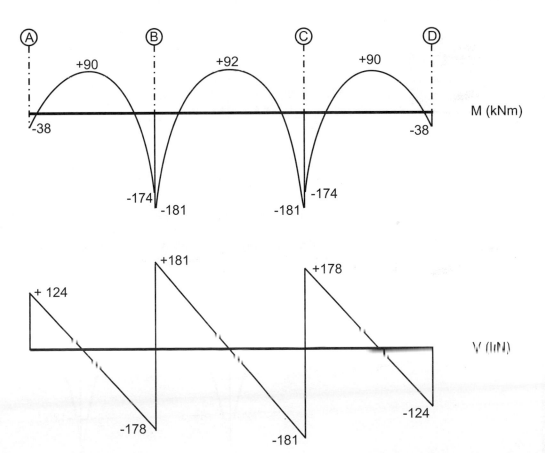

Reduced moment at the face of the support (M_1) will be calculated from the following equation

$$M_1 = M_{CL} - V_{CL} \times \frac{a}{2}$$

a)   Moment at support B for span AB (see the sketch below):
Support width: $a = 600$ mm
Moment at point B: $M_{CL} = -174$ kNm
Shear force at point B: $V_{CL} = -178$ kN
The reduced moment is equal to

$$M_1 = M_{CL} - V_{CL} \times \frac{a}{2} = -178 - (-178 \times \frac{0.6}{2}) = -120 \text{ kNm}$$

b)   Moment at support B for span BC (see the sketch below):
Support width: $a = 600$ mm
Moment at point B: $M_{CL} = -181$ kNm
Shear force at point B: $V_{CL} = -181$ kN
The reduced moment is equal to

$$M_1 = M_{CL} - V_{CL} \times \frac{a}{2} = -181 - (-181 \times \frac{0.6}{2}) = -127 \text{ kNm}$$

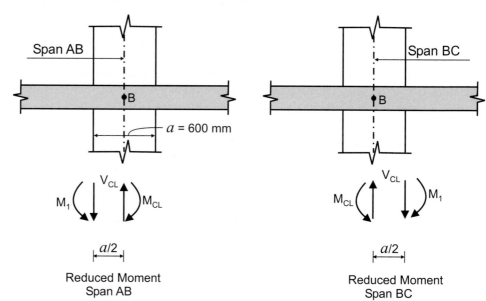

Reduced Moment
Span AB

Reduced Moment
Span BC

Below is a revised bending moment diagram with the reduced bending moments at the supports. Alternatively, bending moments at the critical sections could be obtained directly from the analysis software by placing nodes at the critical sections.

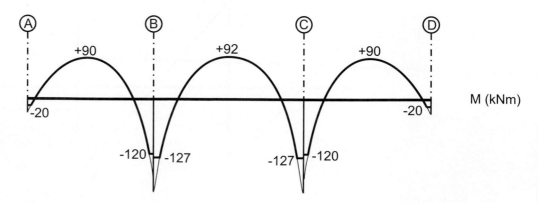

7. **Design the slab flexural reinforcement.**

   a)   Determine the effective depth (d).

   The effective depth will be determined based on the following parameters:

   Slab thickness: 180 mm
   Concrete cover: 20 mm (see Table A.2)
   Bar diameter: 15 mm (assume 15M bars)

   For a flat slab, reinforcement is provided in two directions at the top and at the bottom. An average effective depth can be estimated as follows

   $$d = 180 - 20 - 15 = 145 \text{ mm}$$

   However, note that a 25 mm concrete cover is often required for the bottom slab reinforcement due to the minimum two-hour fire rating requirement for the slab. For that reason, let us use a slightly reduced average effective depth for both top and bottom reinforcement, that is,

   $$d = 140 \text{ mm}$$

   b)   The required area of reinforcement can be found according to the Direct Procedure (see Section 5.5.1) by using the following equation

   $$A_s = 0.0015 f'_c b \left(d - \sqrt{d^2 - \frac{3.85 M_f}{f'_c b}} \right) \qquad \qquad \textbf{[5.4]}$$

   For this case,

   $$b = l_2 = 4800 \text{ mm}$$

   c)   The minimum area of reinforcement is calculated from the following equation (CSA A23.3 Cl.7.8.1)

   $$A_{s \text{ min}} = 0.002 A_g$$

   d)   The check for the maximum reinforcement ratio has been omitted from this design because the minimum reinforcement governs in most cases. If the check was to be performed, it would need to confirm that the reinforcement ratio for the slab-beam section under consideration is less than the balanced reinforcement ratio (see Table A.4), that is,

   $$\rho \le \rho_b$$

   e)   The spacing of bottom reinforcement is limited to the lesser of (CSA A23.3 Cl.13.10.4) (see Table 12.8)

   $$s \le 3h_s = 3 \times 180 = 540 \text{ mm}$$

   or

   $$s \le 500 \text{ mm}$$

   In this case, $s = 500$ mm governs.

   The results of these design calculations are summarized in Table 12.18. The slab is symmetrical, hence it is sufficient to perform calculations for spans AB and BC.

   The top reinforcing steel at interior column locations needs to be designed to resist greater of the two slab moments on either side of the column. For example, the top steel segment for support B should be designed for the bending moment of −127 kNm (span BC), which is larger than the moment of −120 kNm for span AD.

   The designer may choose to distribute the top reinforcement over columns by following the same procedure outlined in Section 12.6 in relation to the Direct Design

Table 12.18 Factored bending moments and the flexural reinforcement

Span	AB			BC		
Location	Top	Bottom	Top	Top	Bottom	Top
M_f (kNm)	−20	+90	−120	−127	+92	−127
A_s (mm²) [5.4]	415	1917	2744	2744	1961	2744
Top reinforcement	3–15M		14–15M			14–15M
Bottom reinforcement (required)		15M @ 500			15M @ 490	
Bottom reinforcement (design)		15M@400			15M@400	

Method. However, in practice the top reinforcement in middle strips has shown to be required only for slabs when the span ratio exceeds 1.0. Otherwise, the top reinforcement is more effective when concentrated in column strip regions.

8. **Provide a design summary.**

The design summary is presented below. Note that the spacing of bottom reinforcement is 400 mm, although 500 mm spacing is adequate according to the design calculations. In this case, it is deemed appropriate to use 400 mm spacing for bottom reinforcement to satisfy the spacing requirements for both spans (AB and BC), and provide a reserve strength to allow for construction errors. For that reason, it is recommended to place reinforcing steel spaced in increments of 100 mm.

Note that this example does not include the calculation of cut-off points for the reinforcement, which is an important part of the design. Refer to Chapter 11 for examples related to detailing of flexural reinforcement in continuous slabs.

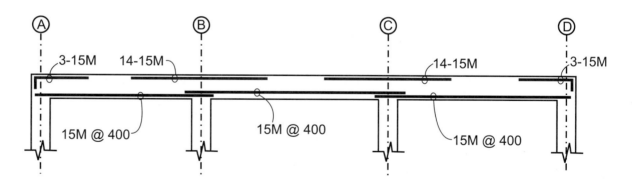

Example 12.4

Two-Way Flat Plate - A Simplified Solution Ignoring Column Stiffness

Consider the same slab discussed in Example 12.3, but ignore the column stiffness, that is, treat the slab as a continuous beam.

Determine bending moments and design flexural reinforcement for the frame along gridline 2.

SOLUTION: In this case, we are going to treat the slab as a continuous beam, as shown on the following sketch. This system is easier to analyze than the frame system. The beam is going to be subjected to the same factored load as the frame in Example 12.3, that is,

$$w_f = 60.5 \text{ kN/m}$$

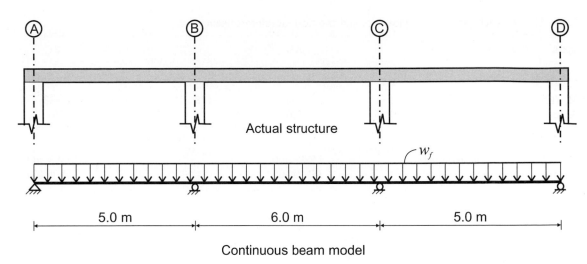

Continuous beam model

The bending moment and shear force diagrams obtained from the structural analysis are shown below. Reduced moments at the supports are also shown on the diagram.

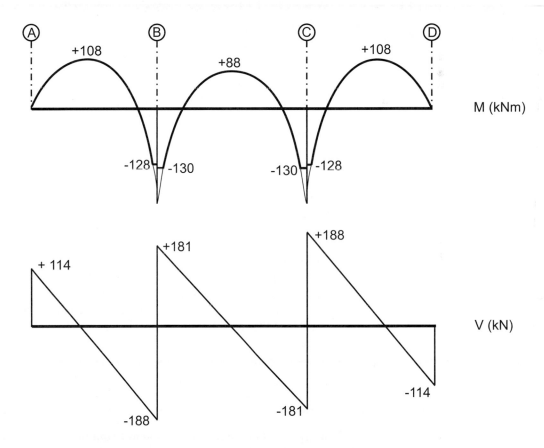

The required area and amount of reinforcement are summarized in Table 12.19. Note that the same procedure was used to design the reinforcement as in Example 12.3.

A design summary showing the reinforcement arrangement is shown on the next page.

A comparison of the bending moments obtained in this example with the values obtained in Example 12.3 both are s that the positive bending moment in the end span increased from 90 to 108 kNm (by approximately 20%). However, the positive bending moment at the midspan has decreased by 4%. Note that there are no negative moments at the end supports due to the initial analysis assumption (zero column stiffness). There

Table 12.19 Factored bending moments and the flexural reinforcement

Span	AB			BC		
Location	Top	Bottom	Top	Top	Bottom	Top
M_f (kNm)	0	+108	−128	−130	+88	−130
A_s (mm²) [5.4]	0	2316	2812		1873	2812
Top reinforcement			16−15M			16−15M
Bottom reinforcement (required)		15M@400			15M@500	
Bottom reinforcement (design)		15M@400			15M@400	

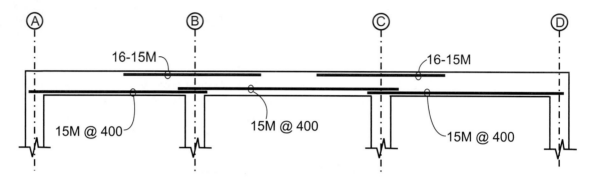

is a minimal increase in bending moment values at the interior columns. These observations are in line with the discussion presented earlier in this section: when column stiffness is ignored, bending moments in the slab will be higher than those obtained from the EFM analysis which takes into account column properties.

It can be seen that the amount of reinforcement is very similar to that obtained from the EFM analysis presented in Example 12.3. In both examples, the required bottom reinforcement is 15M@500. However, 15M@400 specification has been used for the design solution to achieve uniformity and avoid chances of construction errors. Therefore, the only difference between these examples is in the amount of top reinforcement: 16-15M for this example versus 14-15M for Example 12.3.

Both designs could be used in practice. This discussion is intended to illustrate a variety of available reinforcement arrangements for continuous slabs. Several solutions for top and bottom reinforcement are possible, and all of them should result in the same total factored moment M_o.

It can be concluded that the designs which take into account column stiffness result in higher negative moments at the supports and lower positive moments and deflections in midspan regions.

Example 12.5

Two-Way Flat Slab with Drop Panels - Equivalent Frame Method

Consider the same slab as discussed in Example 12.3, but consider drop panels at the interior column locations, as shown on the following floor plan sketch. Use drop panels with square plan dimensions (2 m × 2 m), and 150 mm thickness.

Use the CSA A23.3 Equivalent Frame Method to determine the design bending moments and size and spacing of reinforcement for an interior frame along gridline 2.

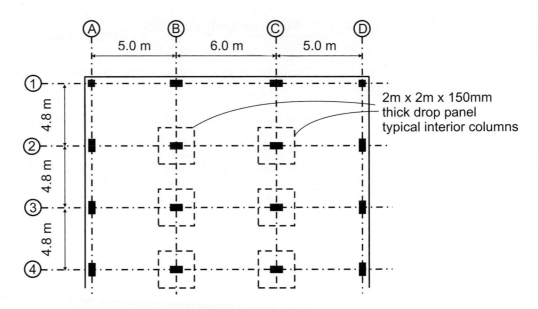

SOLUTION: **1. Determine the slab thickness.**

The dimensions of a typical drop panel are shown on the sketch below.

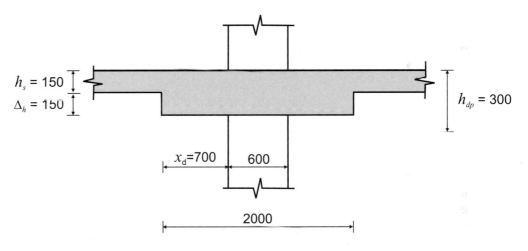

Span BC is characterized by the largest span

$l_n = 5400$ mm

and it governs. Drop panel plan dimensions are 2 m square. Let us confirm that the CSA A23.3 Cl.13.2.4 requirement has been met

$$x_d \leq \frac{l_n}{4} = \frac{5400}{4} = 1350 \text{ mn}$$

Since

$x_d = 700$ mm < 1350 mm OK

The proposed drop panel thickness is 150 mm, therefore

$\Lambda_h = 150$ mm

Note that CSA A23.3 Cl.13.2.4 requires that

$\Delta_h \leq h_s$

According to Cl.13.2.4, the minimum thickness for a slab with drop panels is equal to

$$h_s \geq \frac{l_n(0.6 + f_y/1000)}{30} - \left(\frac{2x_d}{l_n}\right)\Delta_h \qquad \text{(A23.3 Eq.13.2)}$$

$$= \frac{5400(0.6 + 400/1000)}{30} - \left(\frac{2 \cdot 700}{5400}\right) \cdot 150 = 141 \text{ mm}$$

Therefore, the slab thickness will be selected as follows

$h_s = 150$ mm

Note that the total slab thickness at drop panel locations is

$h_{dp} = h_s + \Delta_h = 150 + 150 = 300$ mm

Note that the slab thickness (150 mm) is less than the thickness used for flat plates without drop panels (180 mm) — see Example 12.3. This is a usual practice for slab designs where drop panels are provided at the columns.

2. **Calculate the factored design loads.**

    a) Calculate the dead load acting on the slab.
    First, calculate the slab self-weight:

    $$DL_w = h_s \times \gamma_w = 0.15 \text{ m} \times 24 \text{ kN/m}^3 = 3.6 \text{ kPa}$$

    where $\gamma_w = 24$ kN/m^3 is the unit weight of concrete.

    Next, let us calculate the additional dead load due to drop panels. The self-weight for a 150 mm thick drop panel is as follows:

    $$W_{dp} = (\Delta_h \times \gamma_w)(2 \text{ m} \times 2 \text{ m}) = (0.15 \text{ m} \times 24 \text{ kN/m}^3)(2 \text{ m} \times 2 \text{ m}) = 14.4 \text{ kN}$$

    The load due to drop panels can be distributed over a frame area (conservative). For example, the equivalent frame along the gridline 2 has 16 m length and 4.8 m width, and there are two drop panels in total. Therefore, dead load due to drop panels can be calculated as follows

    $$DL_{dp} = \frac{2W_{dp}}{16 \text{ m} \times 4.8 \text{ m}} = 0.38 \text{ kPa}$$

    The superimposed dead load is given, that is,

    $DL_s = 1.44$ kPa

    Finally, the total factored dead load is

    $$w_{DL,f} = 1.25(DL_w + DL_{dp} + DL_s) = 1.25(3.6 + 0.38 + 1.44) = 6.8 \text{ kPa}$$

    b) Calculate the factored live load:

    $$w_{LL,f} = 1.5 \times LL_w = 1.5 \times 3.6 \text{ kPa} = 5.4 \text{ kPa}$$

    c) The total factored area load is

    $$w_f{}^* = w_{DL,f} + w_{LL,f} = 6.8 + 5.4 = 12.2 \text{ kPa}$$

    The total specified dead load is equal to

    $w_{DL} = 3.6 + 0.38 + 1.44 = 5.42$ kPa

    The specified live load is

    $w_{LL} = 3.6$ kPa

    The ratio between specified live and dead load is

    $$\frac{w_{LL}}{w_{DL}} = \frac{3.6 \text{ kPa}}{5.42 \text{ kPa}} = 0.66 < 0.75$$

    Since the ratio is less than 0.75, pattern loading does not need to be considered, and the frame needs to be designed considering only total factored dead and live load on all spans.

d)   The total factored load on the frame is

$w_f^* = 12.2$ kPa

This is a uniform area load which could be used for 3-D analysis. However, we need to find the design load for 2-D frame analysis, which can be obtained when a tributary slab width ($l_2 = 4.8$ m) is taken into account, as follows

$w_f = w_f^* \times l_2 = 12.2$ kPa $\times 4.8$ m $= 58.6$ kN/m

3.   **Develop a frame model.**

a)   Determine the frame geometry.
A prismatic model is going to be used for the frame analysis. In this case, each span member is divided into 10 segments. For example, span BC contains 4 segments modelling drop panels and 6 segments for the slab-beam. Note that different lengths can be used for slab and drop panel segments.

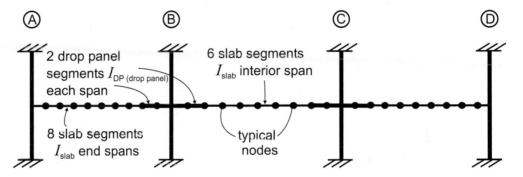

b)   Determine cross-sectional properties for frame members
Typical cross-sections for frame members (slab-beams and columns) are illustrated on the sketch below.

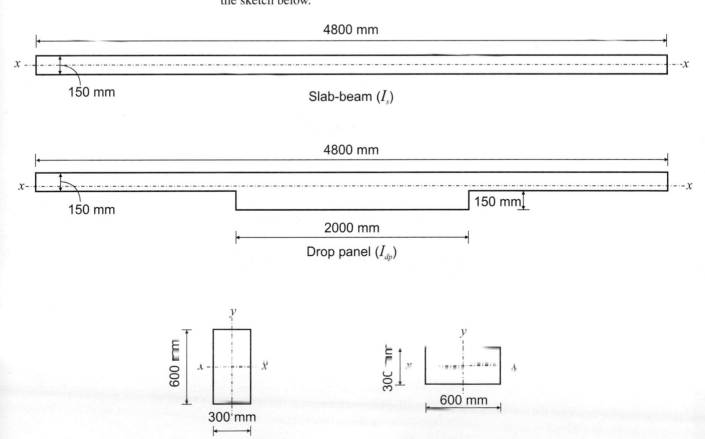

4800 mm

150 mm

Slab-beam (I_s)

4800 mm

150 mm

150 mm

2000 mm

Drop panel (I_{dp})

600 mm

300 mm

Columns A and D (I_1)

300 mm

600 mm

Columns B and C (I_2)

i)   Slab-beams

Slab is treated as a wide beam with the following gross cross-sectional properties: 4.8 m width and 150 mm depth. Moment of inertia for the slab about axis x-x (see the sketch) is equal to

$$I_s = \frac{4800 \text{ mm} \times (150 \text{ mm})^3}{12} = 1.35 \times 10^9 \text{ mm}^4$$

ii)   Drop panels

A drop panel is modelled as a T-section. The centroid can be determined as followed:

$$\bar{y} = \frac{4.8 \cdot 0.15 \cdot \dfrac{0.15}{2} + 2.0 \cdot 0.15 \cdot \left(0.15 + \dfrac{0.15}{2}\right)}{4.8 \cdot 0.15 + 2.0 \cdot 0.15} = 0.12 \text{ m} = 120 \text{ mm}$$

and the moment of inertia about axis x-x is equal to

$$I_{dp} = \frac{4.8 \cdot (0.15)^3}{12} + \frac{2.0 \cdot (0.15)^3}{12} + (4.8 \cdot 0.15)\left(\frac{0.15}{2} - 0.12\right)^2 +$$

$$(2.0 \cdot 0.15)\left(\frac{3 \cdot 0.15}{2} - 0.12\right)^2 = 6.7 \cdot 10^{-3} \text{ m}^4 = 6.7 \cdot 10^9 \text{ mm}^4$$

ii)   Columns

Column properties are identical to those in Example 12.3.
Columns A and D (moment of inertia I_1):

$$I_1 = \frac{600 \text{ mm} \times (300 \text{ mm})^3}{12} = 1.35 \times 10^9 \text{ mm}^4$$

Columns B and C (moment of inertia I_2):

$$I_2 = \frac{300 \text{ mm} \times (600 \text{ mm})^3}{12} = 5.4 \times 10^9 \text{ mm}^4$$

CSA A23.3 Cl.13.8.3.3 requires that a column's moment of inertia should be modified by the factor, ψ, which was determined in Example 12.3, as follows,

$$\psi = 0.3$$

Therefore, the column moment of inertia values are $0.3 \cdot I_1$ and $0.3 \cdot I_2$.

4. **Determine the factored bending moments (M_f).**
    Bending moment and shear force diagrams for gridline 2 are shown on the following sketch.
        The results show that the positive moments are significantly lower in both spans, however the negative moments in the slab and the columns are higher compared to Example 12.3.

5. **Design flexural reinforcement for the slab.**
    From the bending moment diagrams it is possible to determine the required amount of reinforcing steel for the slab. The calculation procedure is outlined below.

    a)   Determine the effective depth (d).
    The effective depth will be determined based on the following parameters:
    Slab thickness: 150 mm
    Concrete cover: 20 mm (see Table A.2)
    Bar diameter: 15 mm (assume 15M bars)
    The average effective depth can be estimated as follows:
    $d = 150 - 20 - 15 = 115$ mm
    The final effective depth for the slab (rounded down):
    $d = 110$ mm

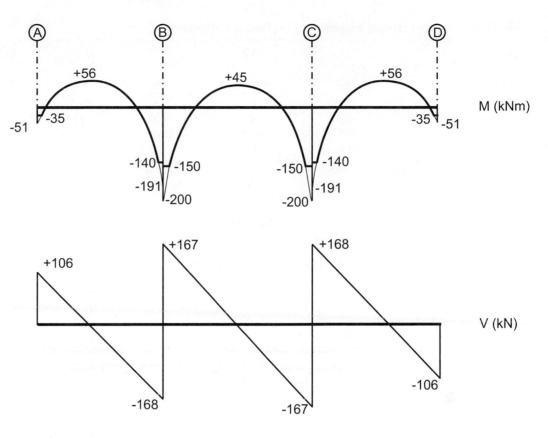

Note that the larger effective thickness needs to be considered at drop panel locations, since the overall thickness is 300 mm (slab plus drop panel). Therefore,

$d_{dp} = 300 - 20 - 15 = 265$ mm

The final rounded effective depth for drop panel locations:

$d_{dp} = 260$ mm

b) The required area of reinforcement can be found from the Direct Procedure using Eqn 5.4 as follows

$$A_s = 0.0015 f'_c b \left(d - \sqrt{d^2 - \frac{3.85 M_f}{f'_c b}} \right)$$ [5.4]

For this case,

$b = l_2 = 4800$ mm

c) The minimum area of reinforcement was calculated from the following equation (CSA A23.3 Cl.7.8.1)

$A_{s\,min} = 0.002\,A_g$

d) The spacing of bottom reinforcement is limited to the lesser of (CSA A23.3 Cl.13.10.4) (see Table 12.8)

or $3h_s = 3 \times 150 = 450$ mm

or

$s \leq 500$ mm

In this case, $s = 450$ mm governs.

Table 12.20 Factored bending moments and the flexural reinforcement

Span	AB			BC		
Location	Top	Bottom	Top	Top	Bottom	Top
d (mm)	115	115	265	265	115	265
M_f (kNm)	−35	56	−140	−150	45	−150
A_s (mm²) [5.4]	895	1448	1578	1695	1157	1695
A_{smin} (mm²)	1440	1440			1440	
Top reinforcement (required)	5–15M		8–15M	9–15M		9–15M
Top reinforcement (design)	6–15M		10–15M	10–15M		10–15M
Bottom reinforcement (required)		15M@450			15M@450	
Bottom reinforcement (design)		15M@400			15M@400	

6. Provide a design summary.

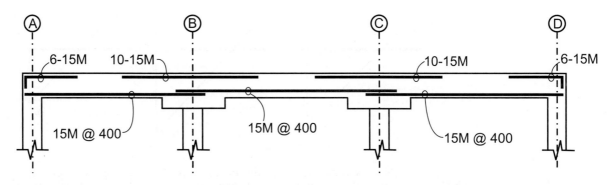

Learning from Examples

Previous examples were useful to evaluate the effect of column stiffness (Examples 12.3 and 12.4) and drop panels (Examples 12.3 and 12.5) on design of two-way slabs without beams according to the EFM using the prismatic modelling approach.

The effect of column stiffness is discussed below (based on Examples 12.3 and 12.4):

1. For slabs with relatively balanced spans, where interior spans are very similar and end spans are shorter than interior spans (70 to 90% of interior spans), bending moment values are usually not sensitive to column stiffness.
2. As a consequence of disregarding the column stiffness, positive moment in the end span would increase. In this case, end columns would attract negative moments along the slab edge(s); this would help to reduce the positive bending moments at the midspan. The designer needs to check the moment transfer capacity at exterior column locations to ensure that the slab is capable of transferring the bending moments derived from the analysis.
3. In general, an analysis where the effect of column stiffness is ignored generally leads to a more conservative design solution for the slab.

The effect of drop panels is discussed below (based on Examples 12.3 and 12.5):

1. There is a general decrease in reinforcing steel required in flat slab design with drop panels (Example 12.5) in comparison with flat plate design (Example 12.3). The amount of bottom reinforcement is governed by the maximum bar spacing requirements (CSA A23.3 Cl.13.10.4) - see Section 12.6.6 for more details.
2. The slab design with drop panels leads to an overall reduction in concrete volume by almost 20% compared to the flat plate design. Although some of the top reinforcement is reduced by 40%, an overall reduction in the amount of reinforcement

is approximately 10%. These potential cost savings are offset by additional labour and material required to form drop panels around each interior column. Flat plate construction is faster to form, thus it cuts down the construction time. The choice of a more efficient solution largely depends on the relative labour versus material costs for a particular geographic region. Often, the choice of design solution (flat plate versus drop panels) is also driven by contractor's expertise, experience, and preference, as well as extra construction costs versus savings due to chances for a reduced construction schedule.

12.7.3 Three-Dimensional Elastic Analysis

A23.3 Cl.13.6

The Concept   CSA A23.3 restricts applications of the DDM and 2-D EFM to regular two-way slabs which are in compliance with the requirements of Cl.2.2 (see Section 12.5.1). Neither of these methods can be used to design flat slabs with irregular shapes and non-rectangular column and/or wall grids, due to potentially significant errors associated with estimating internal forces and deflections in these structures. Irregular slab systems can be analyzed using the Finite Element Method (FEM), a numerical structural analysis method which enables a realistic determination of internal forces and deflections in complex 3-D structures. The FEM is a matrix method which idealizes the structure by modelling slabs and columns as a finite element mesh. In a typical FEM model of a building with two-way slabs, columns are modelled as linear (1-D) finite elements, while slabs are modelled as 2-D finite elements (called plate elements). Displacements within each element are expressed in terms of one or more degrees of freedom (displacements, slopes, etc.) specified at element nodal points. The element stiffness matrix is formed based on a displacement function and given stress-strain relationship for concrete and/or steel. The stiffness matrix of an entire slab is then assembled. The analysis is performed using standard matrix techniques for solving equilibrium equations. The results are in the form of displacements and internal forces (bending moments and shear forces), which can be used to proportion and detail the reinforcement. The FEM is based on the Elastic Plate Theory, and relevant provisions are outlined in CSA A23.3 Cl.13.6. Detailed coverage of the FEM is beyond the scope of this book, however the reader is referred to Zienkiewicz, Taylor and Zhu (2005); Bathe (1995); and Cook et al. (2001) for more information.

An analysis performed using the Finite Element Method will be referred to as the Finite Element Analysis (FEA) in this section. The basic terms associated with the FEA will be explained by an example of a flat plate/slab system consisting of a single panel supported by four columns at the corners and subjected to an uniform load w, as shown in Figure 12.53a. For analysis purposes, the panel is subdivided into a mesh of finite elements, as shown in Figure 12.53b. Note that gridlines in the mesh are parallel with x and y axes which are set in two orthogonal directions of the slab.

Internal forces in the slab can be examined on a small rectangular element cut from the slab by planes parallel with column centrelines shown in Figure 12.54. Note that internal shear forces (N_{xy} and N_{yx}) and axial forces (P_x and P_y) in the plane of the slab are known as membrane forces (see Figure 12.54a). The effect of membrane forces is disregarded in the flexural analysis of two-way slabs. Therefore, in this discussion we are going to focus on bending moments m_x and m_y shown in Figure 12.54b. Besides bending moments, torsional moments m_{xy} and m_{yx} are also present. A plan view of the same element in x-y plane is shown in Figure 12.54c, along with the bending and torsional moments. It is convenient to use moments in x-y plane for the design of typical orthogonal reinforcement mesh. However, it may be of interest to consider the maximum bending moments in the slab, called *principal moments* (denoted as m_1 and m_2), as shown in Figure 12.54d.

FEA is suitable for computer implementations and it has become popular due to rapid advancements in computer hardware and software in the last few decades. FEA software packages, both general-purpose and specialized for slab analysis, are commercially available. Most of these packages are based on linear elastic analysis, however a few packages are able to simulate inelastic (or nonlinear) behaviour of reinforced concrete flat slabs in the post-cracking stage, e.g. ADAPT (2010) and SAFE (CSI).

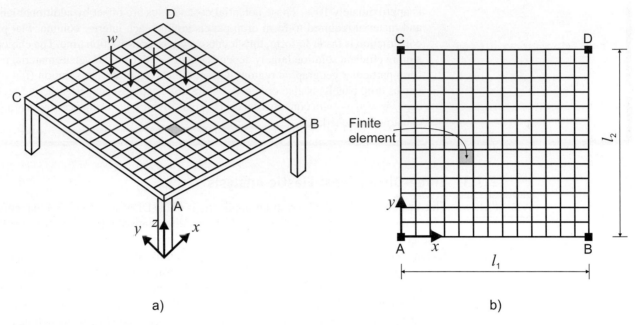

Figure 12.53   Flat plate/slab system: a) an isometric view, and b) a plan view
(courtesy of Gelacio Juárez Luna).

Figure 12.54   Internal forces and moments on a typical slab element: a) internal shear and axial forces (membrane forces); b) internal bending and torsional moments - an isometric view; c) a plan view in x-y plane, and d) principal moments.

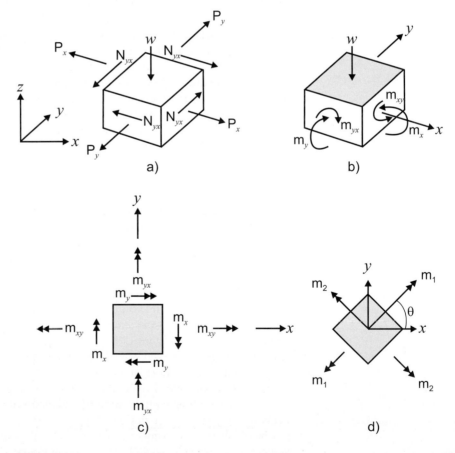

Design of two-way slabs according to the 3-D Elastic Analysis must consider the effect of bending moments and torsional effects. When reinforcement is placed as an orthogonal mat in the x- and y-direction, factored design bending moments (m_x) and (m_y) must be adjusted to account for the effect of torsional moment (m_{xy}) (CSA A23.3 Cl.13.6.4). For further details on torsional moments in two-way slabs the reader is referred to Park and Gamble (2000).

Example 12.6 illustrates an application of a 3-D FEA to a regular slab system and compares the results with the 2-D EFM.

Example 12.6

Two-Way Slab without Beams: 3-D Elastic Analysis

Consider a floor plan of a two-way slab system without beams designed earlier in this chapter using both the Direct Design Method (Example 12.1) and the 2-D Equivalent Frame Method (Example 12.3). The slab thickness is 180 mm. The slab is subjected to superimposed dead load of 1.44 kPa and live load of 3.6 kPa, in addition to its self-weight.

Use 3-D Elastic Analysis (FEA) to determine the design bending moments and the reinforcement layout along gridline 2.

SOLUTION: First, the slab system is modelled as a finite element mesh shown in Figure 12.55. This example uses rectangular-shaped finite elements with the length and width approximately equal to 1 m, and their thickness is equal to slab thickness (180 mm). Note that the mesh density, which is related to the number and size of finite elements, can affect the accuracy of numerical results for the model under consideration. For slabs with more complex geometry, a finer mesh with a larger number of smaller finite elements may improve the accuracy of calculated bending moments and shear forces in critical regions.

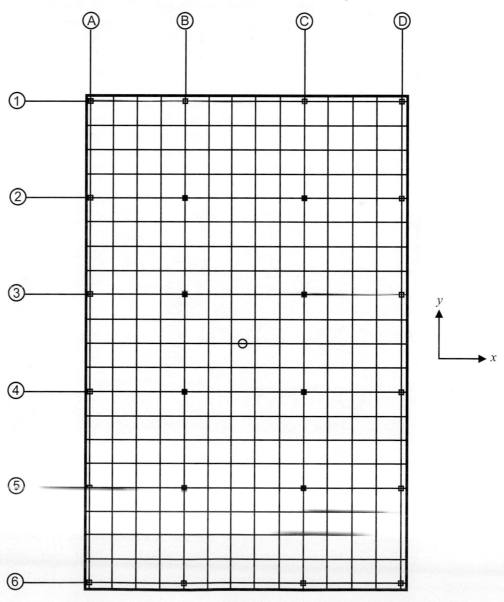

Figure 12.55 Finite element model of a two-way slab.

This FEA was carried out using a commercial software package widely used in design practice. The resulting bending moment distribution in each orthogonal direction is shown in Figure 12.56. The diagrams represent bending moment contours, which show distribution and magnitude of bending moments in the slab. Bending moment contours are indicated by different colours, and their magnitudes are indicated on the legend bar. Note that the m_x denotes bending moments about the y-axis, which are used to design reinforcement in y-direction. For example, reinforcement along gridline B must be proportioned using m_x values shown in Figure 12.56a. Cumulative (positive/negative) bending moments along gridline B shown on the diagram are equal to the sum of moments across the tributary slab width between centrelines of spans AB and BC. Similarly, moments m_y about y-axis shown in Figure 12.56b should be used to design reinforcement in x-direction, for example along gridline 2.

It can be seen from the diagrams that negative moments are concentrated in the column regions. Two-dimensional moment gradient profile is illustrated by a color contour diagram. It can be seen that the region of negative bending moments extends to approximately one-third of the span from the column centroid in each direction. It can be also seen that midspan regions along the column lines are characterized by very low or non-existent negative bending moment values. Note that the distribution of positive bending moments (marked by yellow-coloured contours) is uniform.

Cumulative bending moment values shown in Figure 12.56 can be used to perform a comparison with other design methods, such as the 2-D EFM discussed in Section 12.7.2. Consider the moment distribution along gridline 2 shown in Figure 12.56b. The largest negative bending moment is 181 kNm (column 2-B); this is equal to the value obtained from the EFM analysis (see Example 12.3). The maximum positive moment value for span AB is equal to 91.8 kNm (this is very close to the value of 90 kNm obtained from the EFM). Finally, the maximum positive moment for span BC is 91.1 kNm, which is very similar to the value of 92 kNm obtained from the EFM. Since both methods are based on an elastic analysis, the total bending moment values across a span width obtained from these two methods should be equal, subject only to minor deviations due to a slight difference in the models. This shows that the EFM can be considered as an acceptable design method for regular two-way slabs.

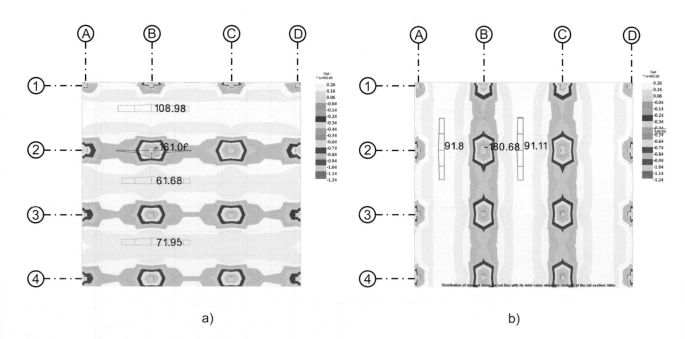

a)   b)

Figure 12.56   Bending moment contours for a regular slab: a) moment m_x, and b) moment m_y.

Moment contour diagrams obtained from the 3-D FEA show a continuous variation of bending moment values in horizontal plane. A reinforcement layout can be developed based on these moments. As with the EFM, the bending moments can be calculated at face of the supports (as described in Section 12.7.2). The aggregate values of design moments should be the same as that derived from the EFM. It is clear from these diagrams that the top reinforcement needs to be concentrated around the columns to resist negative bending moments, and that the distribution of the bottom reinforcement in midspan regions is almost uniform.

However, it is not realistic to place the reinforcing steel to match continuouously variable moments across the slab supports. In practice, the designer only needs to consider the total (cummulative) negative moment over each column region in each direction, and to account for concentration of bending moments in the most critical zone within the column region. Figure 12.57 illustrates a practical solution for the top reinforcement layout based on the 3-D FEA. This type of reinforcement layout is known as "mat reinforcement". The reinforcement is concentrated in the column region, corresponding to the negative moment distribution on each side of the column, and its spacing is uniform in each direction. Reinforcement in the approximate middle half of the mat, corresponding to the critical column region, is more closely spaced, that is, rebar spacing is reduced by one-half compared to the remainder of the mat. Usually, the number of reinforcing bars is rounded up to the next even number. The total moment capacity across the mat in each direction should not be less than the total moment across the negative moment region obtained by the 3-D FEA.

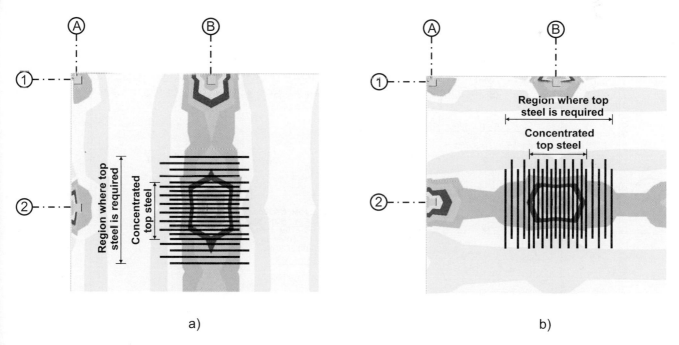

a)   b)

Figure 12.57   Moment and reinforcement distribution for column 2-B: a) gridline 2, and b) gridline B.

It is important to recognize that negative bending moments in midspan regions along each column line are either very low or nonexistent, thus it is acceptable to omit the top reinforcement in those regions (as shown in Figure 12.57). CSA A23.3 Cl.13.11.2.2 prescribes that a maximum 90% of the total negative bending moment at an interior column can be resisted by the column strip. To satisfy this requirement, the width of a top mat is always larger than the column strip, and a small fraction of the mat reinforcement is placed within the middle strip.

The design based on the FEA requires a provision of larger amount of top steel at the columns to suit higher negative moments, while the amount of reinforcement can be reduced or eliminated altogether in midspan regions due to low or nonexistent bending moment values. The bottom slab reinforcement can be uniformly distributed. Some designers prefer to provide additional bottom reinforcement at column centrelines to satisfy the CSA A23.3 requirement which states that 55% of the positive reinforcement should be provided within the column strip (Cl.13.11.2.2). Since the required amount of reinforcement is small, either the maximum bar spacing or the minimum area of reinforcement governs. As a result, it is possible to specify uniform spacing for the bottom reinforcement, which is the case in this example. The final reinforcement layout for this slab is shown in Figure 12.58.

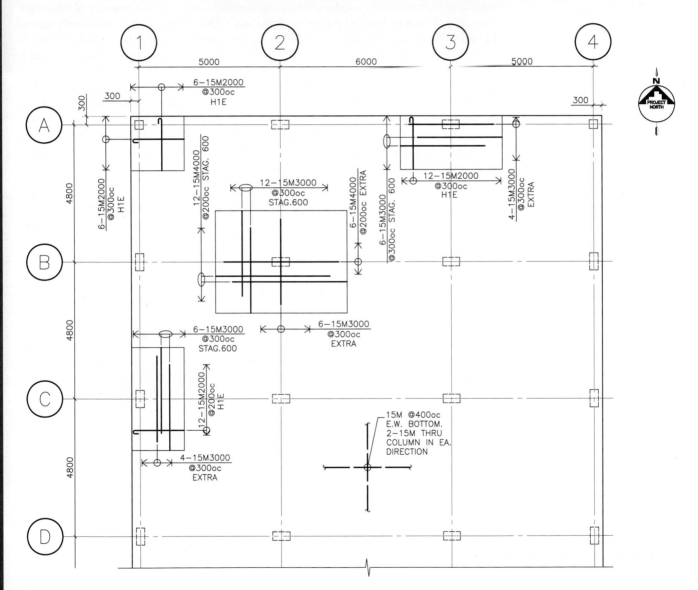

Figure 12.58   Reinforcement layout for the slab example based on the FEA.

As illustrated in Example 12.6, the total bending moments across each span obtained from the FEA should be comparable to other methods (EFM and DDM). The main difference is that a 3-D FEA model provides a moment variation in the slab in two directions, while a 2-D EFM model gives a variation in bending moment values in the plane of the frame; however, the EFM assumes that the transverse moment distribution is uniform across the slab width. This assumption is not applicable to thin flat plates and could lead to significant design errors.

In conclusion, different design solutions will be obtained depending on the method of analysis used to determine bending moments. A design solution based on the 3-D FEA will likely result in the top steel concentrated only over the columns, while both the DDM and EFM will require for the top reinforcement to be provided throughout the span.

3-D Finite Element Analysis of Irregular Two-Way Slabs

The behaviour of irregular two-way slabs is expected to be significantly affected by three-dimensional effects. For that reason, 3-D FEA should be used as a design tool to ensure a realistic prediction of actual moments in the slab. This analysis should produce a design solution that results in an appropriate amount and distribution of reinforcement. Application of 3-D FEA for design of irregular slabs will be discussed in this section.

Consider an office building with a complex floor plan and an irregular column grid shown in Figure 12.59. A 250 mm thick flat plate is used as the floor system. The CSA A23.3 requirements for regular two-way slab systems have not been met (Cl. 2.2). In spite of that, let us try to apply the EFM and check if it is workable for this design. First, we need to divide the slab into equivalent frames in each direction, but it is a challenging and a subjective exercise due to irregular slab shape and an irregular column and wall grid. Besides that, there is an additional column around the large opening on gridline 9 which makes the task of defining equivalent frames even more challenging.

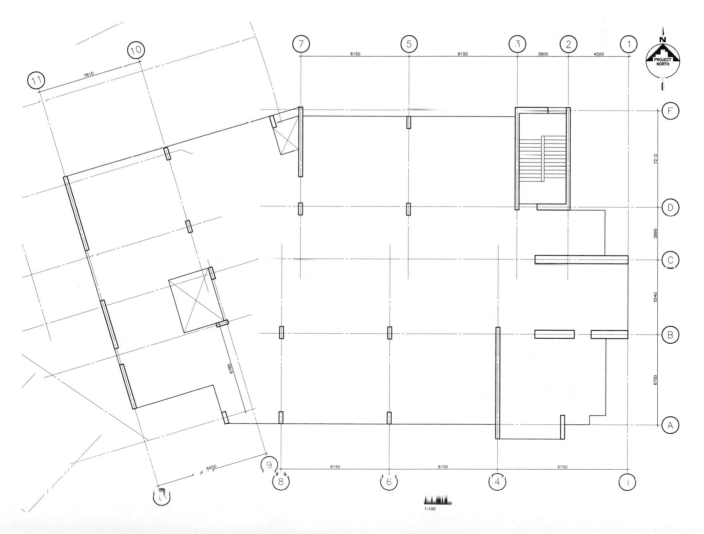

Figure 12.59 Floor plan of a building with an irregular column grid.

Figure 12.60 shows a possible layout of equivalent frames. These frames are defined by the corresponding strips which are shown hatched on the figure. Note that the strip along gridline 5 appears to be the most regular. The frame along gridline 7 is non-rectangular, particularly at the north end of the slab where an opening is provided for a mechanical shaft. The strips along gridlines B and D are skewed and characterized by variable widths. The strip along gridline B is particularly complex, due to a large opening in the slab near column B-10 which cuts away a significant portion of the strip.

It is clear that a 2-D EFM analysis could lead to uncertain results in terms of the accuracy of design solution. In this example, 2-D equivalent frame models would not be able to represent a 3-D structure with sufficient accuracy, and the resulting bending moment values would not be representative of actual moments in the slab. CSA A23.3 does not permit an application of the 2-D EFM in this case. Therefore, 3-D FEA appears to be the most appropriate analysis method for irregular slabs like the one discussed in this example.

Behaviour of a complex flat plate system subjected to gravity loads can be understood on a conceptual level by studying a deflection contour diagram shown in Figure 12.61a. The contours shown on the diagram denote regions characterized by equal deflections. It can be seen that the deflections are smallest (or nonexistent) at column and wall locations, and increasingly larger towards the midspan. Note that deflection values will depend on the span size and relative stiffness ratio between the supports and the slab.

The designer can get a sense for the deflection pattern in a complex flat plate slab by drawing an analogy with the deflection pattern of a deformed tent structure; this will be

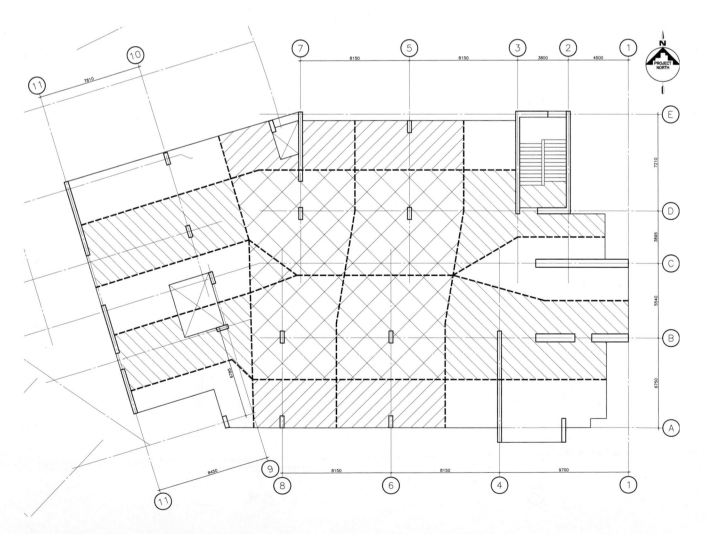

Figure 12.60   Floor plan of an irregular flat slab showing equivalent frames with dashed lines.

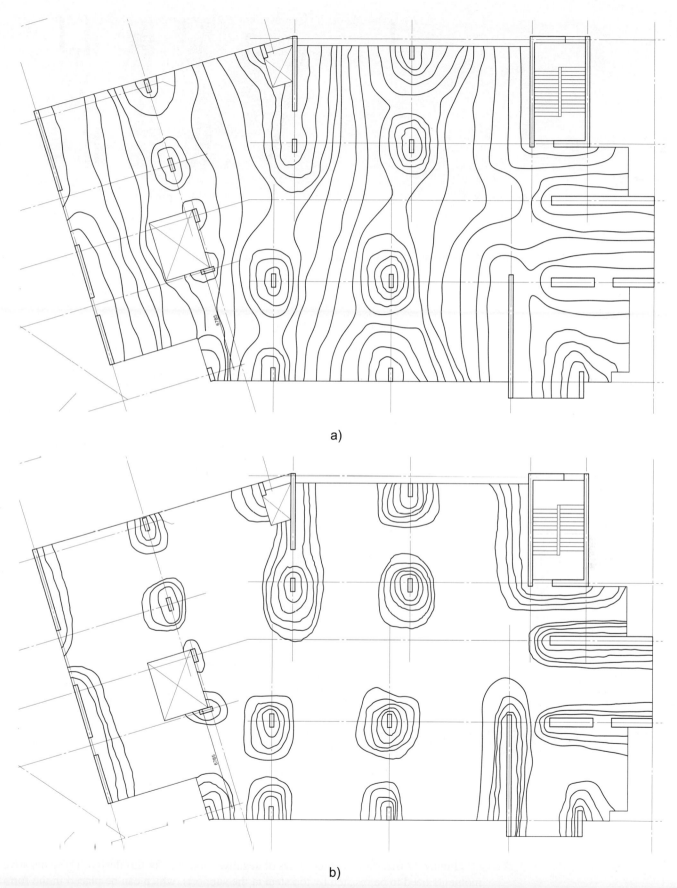

a)

b)

Figure 12.61 Tent analogy for a complex slab system: a) deflection contours, and b) bending moment contours.

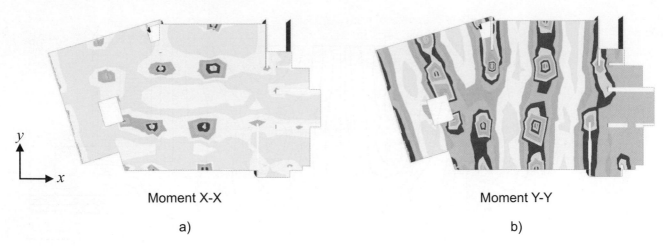

Moment X-X

a)

Moment Y-Y

b)

Figure 12.62 Bending moment contours for an irregular slab: a) moment m_x, and b) moment m_y.

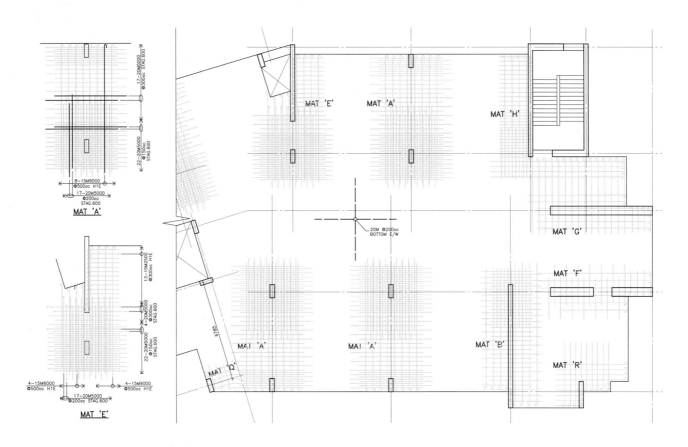

Figure 12.63 Reinforcement layout for the irregular flat slab example.

referred to as the Tent Analogy. Imagine that the floor structure is very flexible, like a large tent placed on top of the columns and walls. The tent material is resilient to tension and it is not going to tear apart under heavy load. The tent is expected to deform due to its self-weight in the most efficient shape, that is, in the form of two-way catenaries between the supports and volcano-like deflection contours around the supports. The Tent Analogy can be used to obtain qualitative deflection patterns for two-way slabs with a complex geometry.

Figure 12.61b shows contour lines of negative moments for this design. These negative moments need to be resisted by top steel at the supports, which can be placed in the form of reinforcement mats.

Bending moment diagrams obtained from the FEA are shown in Figure 12.62. Note that the negative moment regions tend to extend in the y-direction (see Figure 12.62b). The corresponding top steel extends through adjacent spans, as seen from the reinforcement layout shown in Figure 12.63.

12.8 YIELD LINE METHOD

A23.3 Cl.13.7 ### 12.8.1 Background

The Theorems of Plasticity include Hillerborg's strip method and the Yield Line Method (YLM). These methods are able to predict the ultimate load capacity of a slab: the YLM and the Hillerborg's strip method give an upper- and a lower-bound estimate respectively. The YLM was developed by a Danish engineer K.W. Johansen in 1943 and it will be discussed in this section.

The YLM is able to estimate a reserve in load capacity for a reinforced concrete slab beyond the onset of yielding in reinforcement. This is the main advantage of the YLM over alternative design methods (like the DDM and the Elastic Analysis methods), which are based on the Ultimate Limit States approach for flexural design of reinforced concrete members. The Ultimate Limit States approach considers the load corresponding to the onset of yielding in steel as the ultimate load for a structure under consideration, that is, it underestimates the ultimate load capacity of a member.

The intent of this section is to expose the reader to underlying concepts of the YLM and demonstrate its application through a design example. A detailed coverage of the YLM is beyond the scope of this book, however the reader is referred to other resources, such as Kennedy and Goodchild (2003) and Park and Gamble (2000).

12.8.2 The Concept

The YLM can predict the ultimate load (load capacity) for the slab panel under consideration. This method can be applied to slabs with a flexure-controlled behaviour subjected to increasing gravity loading. The yielding of reinforcement initially occurs at a region with the highest bending moments. The region of the slab where yielding has taken place reaches the maximum moment capacity and it is known as a *yield line*. With a further load increase, the moments due to increasing external loads are redistributed to other regions; this causes the reinforcement to yield in those regions and form additional yield lines. The slab's ultimate capacity is reached when enough yield lines exist to form a plastic mechanism, called the *yield pattern*, in which the slab can experience plastic rotations along yield lines without an increase in the applied load. A yield pattern for a two-way slab supported on four sides is shown in Figure 12.64; note yield lines that separate regions A, B, C, and D.

A yield pattern can be understood as a failure scenario for a particular slab. In general, there may be several yield patterns for a particular configuration of slab and loading, but the pattern that gives the least load at the ultimate stage (failure) governs and it is known as the *yield line solution*.

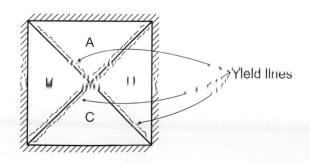

Figure 12.64 Yield pattern for a two-way slab supported on four sides.

The ultimate slab capacity is determined using the Virtual Work Method. This method is based on the underlying principle that the work done externally and internally must balance. At failure, the energy exerted by external loads is equal to the internal energy dissipated by rotations about the yield lines. The method states that the internal work (*IW*) and the external work (*EW*) are equal, as follows

$$IW = EW \tag{12.13}$$

Internal work (*IW*) is the work dissipated by internal moments on rotations along the yield lines, that is,

$$IW = \Sigma \, (m \cdot l \cdot \theta) \tag{12.14}$$

while external work (*EW*) is induced by applied external loads on the slab, as follows

$$EW = \Sigma \, (P \cdot \delta) \tag{12.15}$$

where

P = load(s) acting within a particular region
δ = the vertical displacement of the load(s) P on each region expressed as a fraction of unity
m = internal moment per unit length of yield line
l = length of the yield line
θ = the rotation of a region about its axis of rotation

It is important to note that the summation sign in the above equations denotes that both external and internal work are calculated for all regions of the slab under consideration.

The Virtual Work Method is illustrated by an example of a slab panel subjected to uniformly distributed load.

Figure 12.65a shows the slab which has developed a mechanism where the yielding of reinforcement and plastic rotations have occurred along the yield lines both at the supports and at the midspan.

Figure 12.65b shows a slab model which will be used for the design according to the YLM. Note that P_1 and P_2 are resultants of the uniform load w used for the external work calculations, and x_1 and x_2 denote the distances of these resultants from adjacent supports.

It should be noted that the governing yield pattern for a slab under consideration requires the least amount of internal work prior to failure. The design according to the YLM consists of identifying one or more valid yield patterns and performing calculations to determine the governing yield pattern. An application of the YLM will be illustrated by the following example.

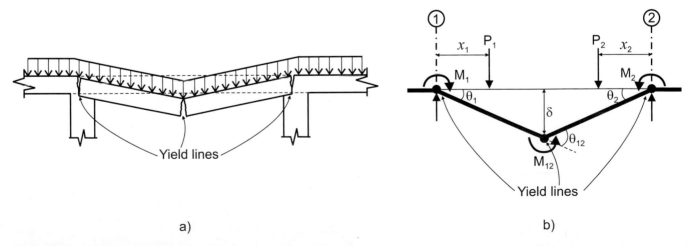

a) b)

Figure 12.65 Plastic mechanism in a two-way slab: a) actual slab span, showing locations where the reinforcement has yielded and the plastic rotations have occurred, and b) a slab model.

Example 12.7

Two-Way Flat Plate - Yield Line Method

Consider a floor plan of a flat plate slab system designed in Example 12.3. The factored design load (w_f) is 12.6 kPa.

Determine the ultimate slab capacity by the YLM: a) for span strip between gridlines 1 and 2, and b) for slab strip between gridlines 2 and 3.

SOLUTION:

1. **Determine possible yield patterns.**

    Several yield patterns are postulated to form and should be considered for design. Note that the positive bending yield lines at the end spans are not expected to form exactly at midspan due to lack of negative moment capacity at the slab edges. These yield lines tend to form closer towards the slab edge, away from the midspan.

    Let us consider the following three yield patterns:

    Yield Pattern 1 (YP1) – Positive bending yield lines are formed within each span, while negative bending yield lines are formed along interior gridlines; note that four patterns are shown on the diagram below, and they need to be checked individually.

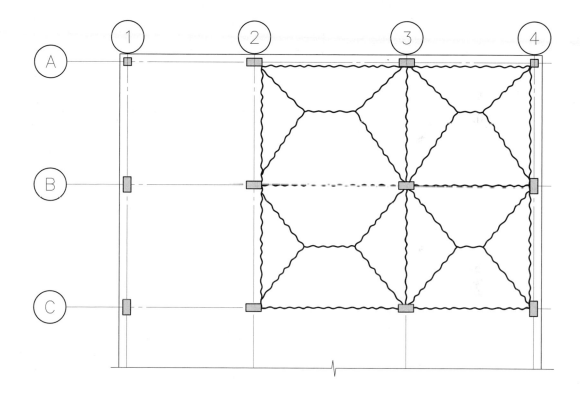

Yield Pattern 2 (YP2) – Parallel straight yield lines are formed in both the midspan regions and along the gridlines, as shown on the diagram below; note that similar patterns could be identified in the perpendicular direction (two patterns are shown on the diagram, and each needs to be checked individually).

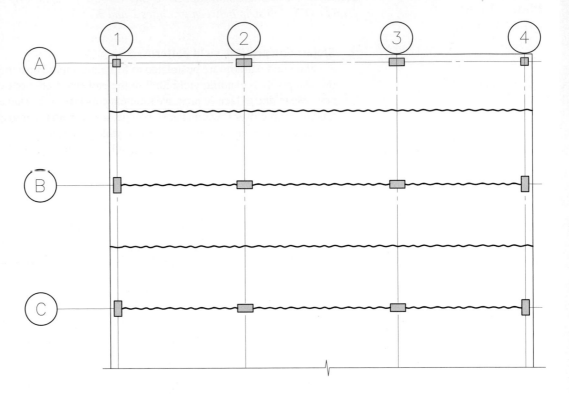

Yield Pattern 3 (YP3) – The largest rectangular yield pattern is formed within a set of interior columns, as shown on the diagram below.

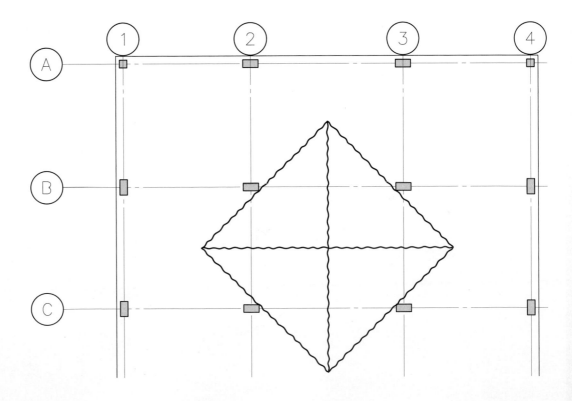

The yield pattern that requires the largest moment resistance along the yield lines governs. To assess the reserve capacity for the previously (*IWA*) designed slab areas, we are going to introduce the ratio of the Available Internal Work to the applied External Work (*IWA/EW*). This ratio is greater than 1.0 for yield patterns that contain more reinforcement than required to satisfy the equal work principle.

In general, all possible yield patterns should be considered to determine the *IWA/EW* ratio. In this example, the calculations have revealed that the governing yield pattern is pattern YP2 discussed above. Although this example shows only yield patterns for slab strips between gridlines 1 and 2 and 2 and 3, similar yield patterns in the perpendicular direction will also need to be considered.

Based on the above discussion, we are going to proceed using the pattern YP2. In many cases it is not obvious which yield pattern produces the least *IWA/EW* ratio, hence this process may need to be repeated for each possible yield pattern.

2. **Find the ultimate load capacity for the slab strip between gridlines 1 and 2.**

a) Determine relevant dimensions for the yield pattern.

For the end span 1-2, it is assumed that the positive steel yield line occurs at 0.4L from the slab edge (gridline 1), as shown on the diagram below. However, it can be shown that if the yield line is taken at midspan, the error is only about 3%, which is insignificant.

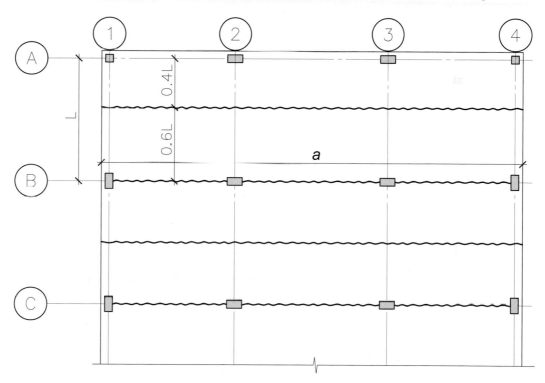

A vertical section of yield pattern (YP2) between gridlines 1 and 2 is shown below.

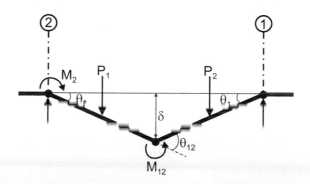

b)   Compute the External Work (*EW*).

Let us first calculate rotations (θ) along the yield lines, as shown on the sketch above. Note that the displacement (δ) is assigned a unit value for ease of calculation. The rotations can be found from the above sketch, as follows:

$$\theta_1 = \frac{\delta}{0.4L} = \frac{1}{0.4L}$$

$$\theta_2 = \frac{\delta}{0.6L} = \frac{1}{0.6L}$$

$$\theta_{12} = \theta_1 + \theta_2 = \frac{1}{0.24L}$$

The design load is equal to

$w_f = 12.6$ kPa

The load resultants for regions 1 and 2 can be determined as a product of the slab area and load (w_f), that is,

$$P_1 = (0.4L \cdot a) \cdot w_f$$

$$P_1 = (0.6L \cdot a) \cdot w_f$$

Finally, the external work can be calculated as follows:

$$EW = \Sigma(P \cdot \delta) \qquad\qquad\qquad [12.15]$$

$$= P_1 \frac{\delta}{2} + P_2$$

$$= \frac{1}{2}(0.4 + 0.6)L \cdot a \cdot w_f = \frac{L \cdot a \cdot w_f}{2}$$

$$= \frac{4.8\text{ m} \cdot (16\text{ m}) \cdot (12.6\text{ kPa})}{2} = 484\text{ kNm}$$

c)   Compute the Available Internal Work (*IWA*).

In order to calculate the *IWA*, it is required to find the moment resistance along the yield lines.

i)   Find the moment M_2.

The section considered for this calculation uses the total slab width, that is,

b = 16000 mm

and the effective depth based on average depth of two 15M bar layers, as follows

d = 180 mm − 20 mm − 15 mm = 145 mm

The reinforcements design (omitted from this example) requires four mats across the entire slab width between gridlines A and D: two interior mats with 18-15M bars, and two edge mats with 10-15M bars. Therefore, the top reinforcement consists of 56-15M bars. Hence,

$A_s = 56 \times 200\text{ mm}^2 = 11200\text{ mm}^2$

The factored moment resistance for a rectangular slab section will be determined from the procedure presented in Section 3.5, as follows

$T_r = \phi_s A_s F_y = 0.85(11200\text{ mm}^2)(400\text{ MPa}) = 3808\text{ kN}$

$C_r = \phi_c \cdot f_c' \cdot a \cdot b = 0.65(30\text{ MPa})(16000\text{ mm})(a)$

From the following equation of equilibrium

$T_r = C_r$

it can be found that

a = 12.22 mm

thus

$$M_2 = T_r\left(d - \frac{a}{2}\right) = 3808\,\text{kN}\left(145\,\text{mm} - \frac{12.2\,\text{mm}}{2}\right) = 529\,\text{kNm}$$

ii) Find the moment M_{12}.

This moment is calculated considering the bottom reinforcement 15M @ 400 mm across 16 m slab length. Hence,

$$A_s = \frac{16000\,\text{mm}}{400\,\text{mm}} \times 200\,\text{mm}^2 = 8000\,\text{mm}^2$$

$$T_r = \phi_s A_s F_y = 0.85\,(8000\,\text{mm}^2)(400\,\text{MPa}) = 2720\,\text{kN}$$

$$C_r = \phi_c \cdot f_c' \cdot a \cdot b = 0.65\,(30\,\text{MPa})\,(16000\,\text{mm})\,(a)$$

thus

$$a = 9\,\text{mm}$$

and

$$M_{12} = 2720\,\text{kN}\left(145\,\text{mm} - \frac{9\,\text{mm}}{2}\right) = 382\,\text{kNm}$$

iii) Finally, available internal work (IWA) at the supports and the midspan can be calculated as follows

$$IWA = IW = \Sigma(m \cdot l \cdot \theta)$$ [12.14]

$$= M_1\theta_1 + M_2\theta_2 + M_{12}\theta_{12}$$

where

$$M_1 = 0 \text{ at the slab edge}$$

and

$$l = 1 \text{ (assume unit length for the yield line)}$$

thus

$$IWA = M_2\theta_2 + M_{12}\theta_{12}$$

$$= M_2\left(\frac{1}{0.6L}\right) + M_{12}\left(\frac{1}{0.24L}\right)$$

$$= \frac{529\,\text{kNm}}{0.6\,(4.8\,\text{m})} + \frac{382\,\text{kNm}}{0.24\,(4.8\,\text{m})} = 515\,\text{kNm}$$

d) Find the *IWA/EW* ratio.

$$\frac{IWA}{EW} = \frac{515\,\text{kNm}}{484\,\text{kNm}} = 1.06$$

Therefore, for end span between gridlines 1 & 2, there is a 6% reserve capacity assuming that there is a zero negative moment capacity at the four edge columns. This reserve capacity translates to the actual ultimate load capacity, as follows

$$w_u = 1.06 \cdot w_f = 1.06 \cdot 12.6 = 13.4\,\text{kPa}$$

3. **Find the ultimate load capacity for the slab strip between gridlines 2 and 3.**

a) Find the relevant dimensions for the yield pattern.

The yield pattern is the same as above (YP2), and a vertical section is shown on next page.

b) Compute the External Work (*EW*).

In this case, the maximum deflection δ occurs at the midspan, hence

$$\theta_2 = \theta_3 = \frac{\delta}{0.5L} = \frac{1}{0.5L}$$

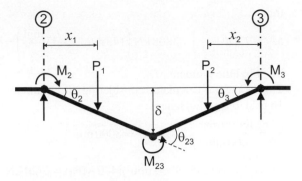

and

$$\theta_{23} = \theta_2 + \theta_3 = \frac{1}{0.25L}$$

The load resultants can be determined in the same manner as above, that is,

$$P_1 = P_2 = (0.5L \cdot a) \cdot w_f$$

$$EW = \Sigma(P \cdot \delta) \tag{12.15}$$

$$= P_1 \frac{\delta}{2} + P_2 \frac{\delta}{2} \quad = \frac{1}{2}(0.5 + 0.5)L \cdot a \cdot w_f = 484 \, \text{kNm}$$

c) Compute the Available Internal Work (IWA).
The same top and bottom reinforcement is used here as for the previous slab section, hence

$$M_2 = M_3 = 529 \, \text{kNm}$$

$$M_{23} = M_{12} = 382 \, \text{kNm}$$

and

$l = 1$ (assume unit length for the yield line)

$$IWA = IW = \Sigma(m \cdot l \cdot \theta) \tag{12.14}$$

$$= M_2 \theta_2 + M_3 \theta_3 + M_{23} \theta_{23} = 2 \, M_2 \theta_2 + M_{23} \theta_{23}$$

$$= 2 \cdot M_2 \left(\frac{1}{0.5L}\right) + M_{12} \left(\frac{1}{0.25L}\right)$$

$$= \frac{2(529)}{0.5(4.8)} + \frac{382}{0.25(4.8)} = 759 \, \text{kNm}$$

d) Find the *IWA/EW* ratio.

$$\frac{IWA}{EW} = \frac{759 \, \text{kNm}}{484 \, \text{kNm}} = 1.57$$

The ultimate load capacity is equal to

$$w_u = 1.57 \cdot w_f = 1.57 \cdot 12.6 = 19.8 \, \text{kPa}$$

The above calculation shows a significant reserve in the load capacity of 57% for interior span 2-3. This is due to the flexural capacity provided by the top steel at both support lines, compared to only one support line for end span 1-2, thereby resulting in a significantly smaller reserve capacity of 6%. This example illustrates that interior spans in two-way slabs designed in accordance with an elastic analysis procedure have a significant reserve in the load capacity.

12.8.3 Practical Design According to the Yield Line Method

Although the YLM is usually applied for evaluating the ultimate load capacity of existing slabs, it is also becoming more popular as a primary tool for design of new slabs. Some of the advantages of the YLM compared to elastic analysis methods are summarized below:

1. The YLM offers a more economical design solution, since it takes into consideration a reserve capacity beyond the onset of yielding in steel reinforcement.
2. The reinforcement solution is simpler, and the placement of steel tends to be more uniform and regularly arranged.
3. The YLM is a very appropriate design tool for slabs with complex configurations, and it gives the designer a better understanding of the overall structural capacity and the failure mechanisms.

The YLM can be used for design of both regular and irregular slabs. The design is usually performed by hand calculations and the designer is required to derive a solution from the basic principles.

In general, application of the YLM is a trial and error approach. The designer usually needs to carry out a few iterations using different yield patterns to identify the one which results in the minimum and the maximum load capacity for a particular slab panel. The complexity of yield patterns may initially present a challenge, and novice designers may not be confident in the resulting design solution. However, experienced designers may be able to intuitively eliminate the yield patterns that do not govern and consider only a few patterns that are likely to govern.

12.9 DESIGN FOR SHEAR

12.9.1 Background

Shear stresses in two-way slabs occur due to gravity loads and bending moments. These bending moments are caused by gravity loads (unbalanced moments) and/or lateral loads, and need to be transferred from the slab to the columns through slab-column connections. Slab area in the vicinity of a slab-column connection is subjected to the highest shear stresses. Shear failure in two-way slabs is sudden, and it must be carefully considered in the design to avoid potentially catastrophic consequences of shear failure, particularly in flat slabs and flat plates. The slab shear resistance in two-way slabs with beams is usually not critical, but the beams must still be designed for shear. This section builds on the background knowledge from Chapter 6 related to the shear design of beams and one-way slabs, and it also provides new information specific to the shear design of two-way slabs.

The main objective of shear design in two-way slabs is to check the concrete shear resistance for one-way and two-way shear. When the factored shear stress exceeds the concrete shear resistance, it is required to either provide shear reinforcement or modify the design (e.g. provide drop panels).

12.9.2 Shear Design for Two-Way Slabs without Beams

A23.3 Cl.13.3 | **CSA A23.3 Shear Design Requirements** The main CSA A23.3 shear design requirement for two-way slabs is the strength requirement. The maximum factored shear stress, v_f, should not exceed the factored shear stress resistance, v_r (Cl.13.3.1), as follows:

$$v_f \leq v_r \tag{12.16}$$

and

$$v_r = v_c + v_s$$

where

v_c is the factored concrete shear stress, and

A few cast-in-place reinforced concrete buildings with flat plate floor systems collapsed during construction in the last 50 years. The causes of failure were usually complex and could be attributed to several factors, but it appears that these collapses were primarily triggered by punching shear failures of flat plate floor slabs. For example, a 16-storey reinforced concrete building at 2000 Commonwealth Avenue in Boston, USA collapsed during its construction in January of 1971, killing 4 and injuring 30 construction workers (King and Delatte, 2004). Floor and roof systems consisted of cast-in-place flat plates supported by columns, and the slab thickness ranged from 190 to 230 mm. First, there was a punching shear failure in the roof slab at one of the column locations. The workers felt a drop in the roof slab of about 100 mm within a few seconds, which was followed by the complete collapse of the slab. This was followed by the total progressive collapse of the east side of the building, which left the floor plates stacked in the basement of the structure (see Figure 12.66). The most significant construction deficiencies were the lack of shoring under the roof slab, and low concrete strength. At the time of collapse, the concrete strength was reported to be in the range from 11 to 13 MPa (compared to 20 MPa design strength). No testing was done to confirm the strength before the shoring was removed. Also, actual loads on the roof were approximately 6.2 kPa, while structural plans specified construction loading of only 1.4 kPa (note that increased loads were due to construction equipment and boilers stored on the roof). One of the main lessons from the collapse is that redundancy within structural design is essential to prevent progressive collapse. Slabs in the collapsed building did not have any shear reinforcement or continuous integrity reinforcement through the columns, which is prescribed by CSA A23.3 and other codes to minimize chances of progressive collapse in these systems.

Figure 12.66 A view of the collapsed building with flat plate floor system in Boston, USA in 1971.

(from Boston Globe, January 26, 1971; republished with permission, courtesy of Getting Images/Boston Globe).

v_s is the factored shear stress in shear reinforcement (design of shear reinforcements is discussed in Section 12.9.4).

Note that the strength requirement is presented in the stress form, that is, it is required to compare stresses rather than forces. For example, the factored shear stress resistance, v_r, is expressed in stress units (e.g. MPa) while the factored shear resistance, V_r, is expressed in force units (e.g. kN). The stress approach is suitable for checking the two-way shear resistance, whereas the force approach can be used to check the one-way shear resistance; this is similar to the shear checks for beams and one-way slabs discussed in Chapter 6. The shear resistances, v_r and V_r, will be determined from CSA A23.3 equations and discussed later in this section.

According to A23.3 Cl.13.3.2, two different shear mechanisms must be considered in two-way slabs without beams:

- One-way shear (Cl.13.3.6) and
- Two-way shear (Cl.13.3.3 to 13.3.5).

A23.3 Cl.13.3.6 **One-Way Shear (Beam Shear)**   The one-way shear resistance for two-way slabs (often referred to as beam shear resistance) is determined in the same manner as for beams, one-way slabs, and footings specified in CSA A23.3 Cl.11.1 to 11.3 (see Section 6.5.4). Note that the one-way shear resistance usually does not govern in the design of flat plates or flat slabs, however it should still be checked.

When the slab has sufficient thickness, shear reinforcement is not required (Cl.11.2.8.1), that is,

$$V_f \le V_c$$

The factored shear force, V_f, is determined by considering the slab spanning as a wide beam, with the width corresponding to design strip b_w and span l_n, as shown in Figure 12.67a. The critical section is located at a distance d_v from the column face, as shown in Figure 12.67b. The design shear envelope is shown in Figure 12.67c. Note that both V_f and V_c can be determined based on a unit slab width (equal to 1 m), instead of the total design strip width, b_w. The unit shear forces are denoted as V_f' and V_c', and the final conclusion should be the

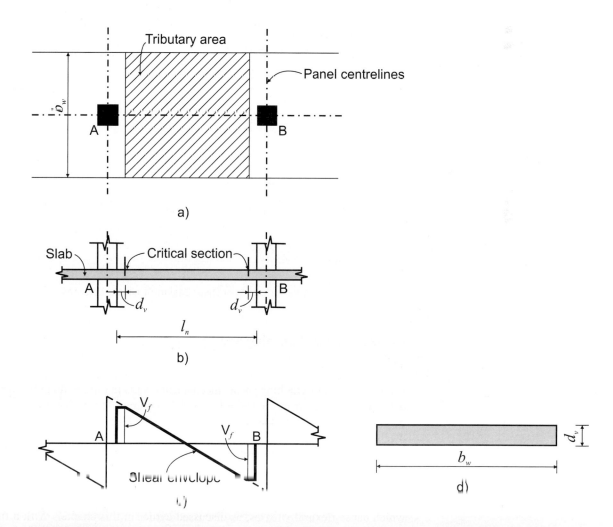

Figure 12.67   One-way shear design of two-way slabs: a) a plan view showing the tributary area; b) slab elevation showing the critical section; c) shear force envelope, and d) a typical slab cross-section for one-way shear design.

same, irrespective of whether the forces are calculated based on the unit width or the entire design strip width.

In two-way slabs with beams, the one-way shear resistance is provided by the beams (see Section 12.9.5 for more details).

The factored concrete shear resistance, V_c, can be determined from the following equation

| A23.3 Eq. 11.6 | $V_c = \phi_c \cdot \lambda \cdot \beta \cdot \sqrt{f_c'} \cdot b_w \cdot d_v$ | [6.12] |

where $\beta = 0.21$ (for slabs where $h_s \leq 350$ mm, A23.3 Cl.11.3.6.2)

A typical cross-section is shown in Figure 12.67d. Note that b_w is the width of the design strip, and d_v is the effective shear depth taken as the greater $0.9d$ of and $0.72h_s$.

When $h_s > 350$ mm, it is required to take into account the effect of depth on allowable shear strength, thus β should be determined from A23.3 Eq.11.9 (refer to Table 6.1).

For corner columns, V_c is determined from Eqn 6.12 considering

$$b_w = b_o$$

where b_o denotes perimeter of the critical section. It can be seen from Figure 12.68 that the critical section for a corner column may be extended into the slab overhang (cantilevered portion beyond the column) for a length not exceeding d (Cl.13.3.6.2).

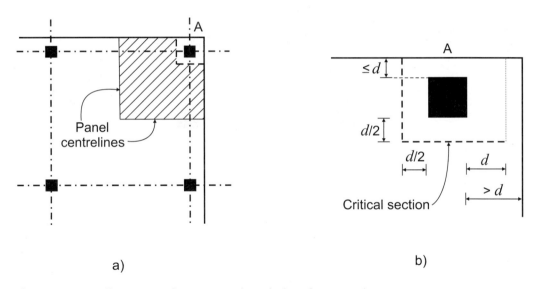

a)   b)

Figure 12.68   Tributary area for one-way shear design of corner columns.

Two-Way Shear (Punching Shear)

| A23.3 Cl.13.3.3 to 13.3.5 | **The mechanism** |

A two-way shear (punching shear) mechanism results in failure along the surface of a truncated pyramid (or a truncated cone) around the column. This failure mechanism takes place due to excessive gravity loads transmitted from the slab into the supporting columns. The mechanism develops when shear stresses on the area in the vicinity of column perimeter exceed the concrete shear strength.

The punching shear mechanism is illustrated in Figure 12.69a. Initially, circular and radial cracks develop on the top slab surface — these cracks are due to negative bending moments; the slab area in the vicinity of the column is subjected to significant bending moments which cause flexural stresses, as discussed earlier in this chapter. With a further load increase, diagonal tension cracks develop near the mid-depth of the slab and later propagate to the surface. It should be noted that these cracks first form at about one-half of the load corresponding to the punching shear failure, at a distance of approximately one-half of the

Figure 12.69 Punching shear failure: a) a vertical slab section showing the cracking pattern (adapted from Ghali and Hammill, 1992 with the permission of the American Concrete Institute), and b) an isometric view of the slab-column connection showing cracks at the top of the slab observed in an experimental study (courtesy of Min-Yuan Cheng, Gustavo Parra-Montesinos, and Carol K. Shield).

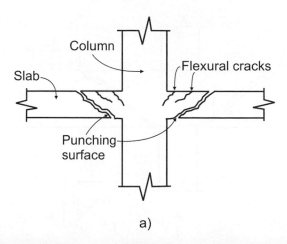

a)

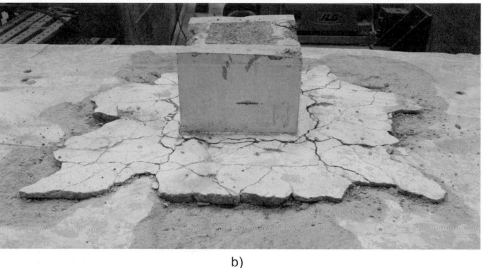

b)

slab depth from the column perimeter. Cracking pattern on the top slab surface characteristic of the punching shear mechanism is shown on an experimental specimen in Figure 12.69b.

The two-way shear mechanism is illustrated in Figure 12.70. Consider a two-way flat plate subjected to uniformly distributed gravity load, w. The slab tends to move uniformly downward due to the load, while the column (or other type of support) resists this movement. Shear (diagonal tension) stresses along the inclined planes are shown on slab element ABC in Figure 12.70a. Tensile stress, f_t, acts perpendicular to the inclined surface AC, while the shear stress, v, acts parallel with surfaces AB and BC. The cracking takes place when tensile stress in the slab reaches the concrete tensile resistance. The failed shape is in the form of a truncated pyramid, as shown on the isometric diagram in Figure 12.70a. For design purposes, CSA A23.3 permits the use of a simplified failure shape, geometrically similar to the column, as shown in Figure 12.70b. The critical section is located at a distance $d/2$ from the face of the column (note that d denotes the effective slab depth). The stress distribution is shown on slab element ABCD. Shear stress, v, acts downward, and it is transferred across the design shear surface with area $b_o \times d$, where b_o is the perimeter of the critical section (see isometric diagram in Figure 12.70b).

A23.3 Cl.13.3.3 **The critical section for two-way shear**

The critical section for two-way shear at an interior column is defined by four vertical faces, as shown in Figure 12.70b. The section should be taken at distance $d/2$ from the perimeter of the concentrated load or column (Cl.13.3.3.1), as shown on an example of a flat plate in Figure 12.71a. The critical section is perpendicular to the plane of the slab and located so that its perimeter b_o is at a minimum. Note that the critical section is shown with dashed

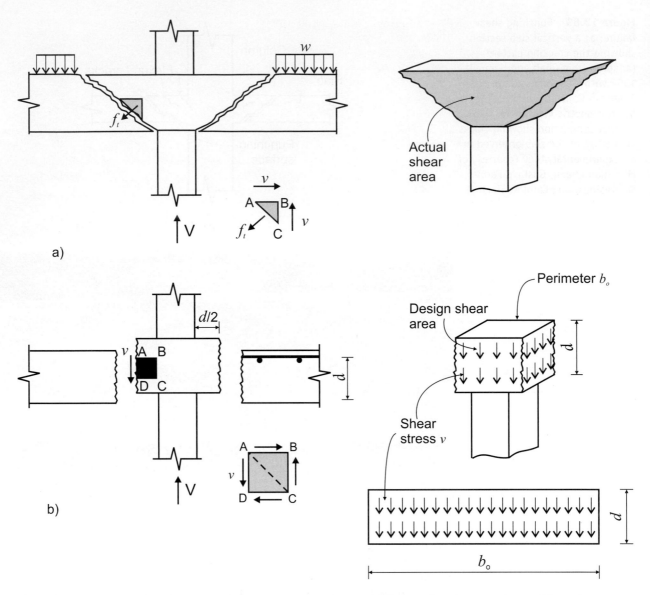

Figure 12.70 Two-way punching shear failure mechanism: a) actual failure surface (truncated pyramid), and b) an approximate failure surface consisting of the four vertical sides.

lines, and the tributary slab area for the factored shear force calculation is shown hatched on the drawing.

When a slab has variable thickness (for example, a flat slab with drop panels), shear failure can occur either through the thickened portion of the slab near the column, or through the slab portion outside the drop panel. As a result, there are two critical sections (Cl.13.3.3.2), as illustrated on an example shown in Figure 12.71b. Critical section 1 is located within the thicker portion (drop panel), at a distance $d_1/2$ from the face of the column (as shown with dashed lines in the figure). Critical section 2 is located in the slab at a distance $d_2/2$ from the end of drop panel (as shown with dashed and dotted lines in the figure); note that d_1 and d_2 denote the effective depth of thickened slab with drop panel and slab without drop panel, respectively. Figure 12.71c shows the notation related to the critical section shown in Figure 12.71a. Note that c_1 is column dimension in the direction of the span for which moments are determined, and c_2 is column dimension in the direction perpendicular to c_1. The perimeter of the critical section, b_o, can be determined as follows

$$b_o = 2(c_1 + d) + 2(c_2 + d)$$ [12.17]

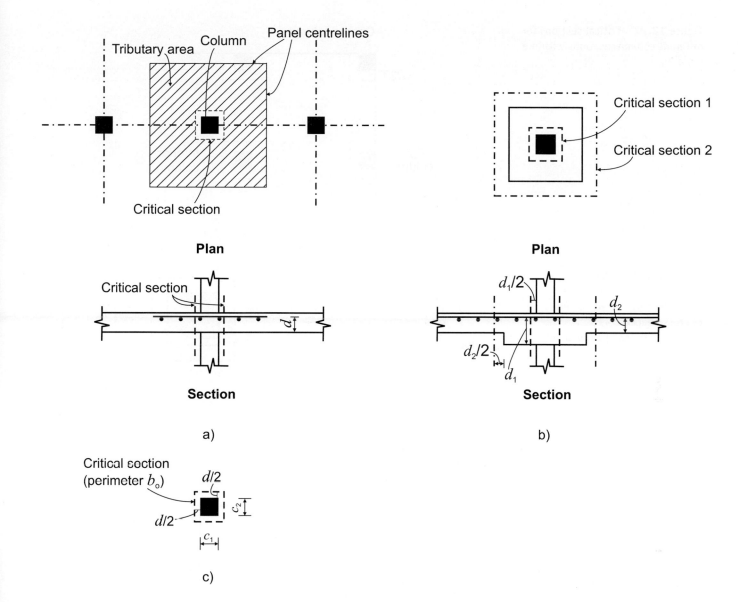

Figure 12.71 Critical section for two-way shear: a) a flat plate; b) a flat slab, and c) dimension notation.

The shape and size of the critical section depend on the location of the column within a building (Cl. 13.3.3.3). For a typical interior column of rectangular shape, critical section may be assumed to have four vertical sides, while for corner and edge columns this section may be assumed to have two or three vertical sides, respectively (see Figure 12.72).

Shape of a critical section also depends on the cross-sectional shape of the supporting column. Circular column sections may be found in many buildings. In theory, it would be possible for the critical section to have a circular shape, as shown in Figure 12.73a. However, the intent of Cl. 13.3.3.3 is that critical section has straight surfaces. Experimental studies have shown that punching shear strength for circular columns exceeds the strengths for square columns with the same cross-sectional area. Therefore, it is conservative and analytically simpler to idealize circular columns as square columns with the same cross-sectional area (ACI, 1990), this is shown in Figure 12.73a. Critical sections for some irregular column shapes are shown in Figure 12.73b and c.

Figure 12.72 Critical sections for different column locations within a building.

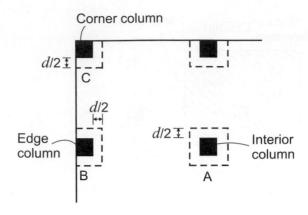

Figure 12.73 Critical section for different column shapes:
a) circular; b) L-shaped, and
c) cross-shaped
(courtesy of the American Concrete Institute).

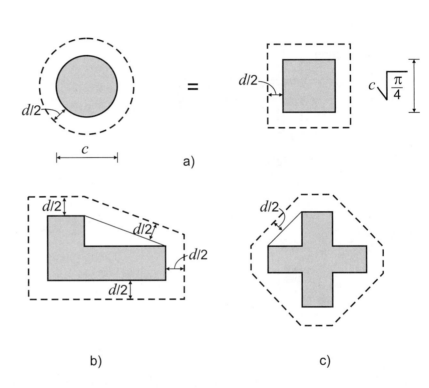

At the edge and corner columns, where the slab cantilevers beyond the exterior face of the support, critical section may be assumed to have three straight sides and extend into the cantilevered portion of the slab (overhang), as shown in Figure 12.74. The critical section can be assumed to extend into the cantilevered portion of the slab for a distance not exceeding d (Cl.13.3.3.3). The objective of this provision is to keep the perimeter of the critical section, b_o, at minimum.

Designers often need to deal with openings or slab edges located close to the columns (or other supports) in two-way slabs. One of the concerns regarding the openings is that shear flow in the slab is interrupted, hence the resulting shear capacity of the connection is reduced. This must be taken into account in the design.

Design provisions related to the critical perimeter at slab openings are outlined in Cl.13.3.3.4. Mechanical and electrical systems are often required to penetrate through the slab. Openings or holes placed in the vicinity of columns are often required due to architectural constraints. This presents an additional challenge to structural engineers when considering punching shear in critical regions around the columns. Vertical openings (holes) passing through the slab reduce shear strength when located within the intersecting column strips or at a distance closer than $10h_s$ to the column (where h_s denotes slab thickness). Horizontal openings (ducts) are often needed to provide space for electrical conduits. The effect of these openings can be disregarded, when their distance to the column does not exceed

Figure 12.74 Critical section
at the slab edge depending on
the overhang length and the
column shape
(adapted from CAC, 2005 with the
permission of the Cement Association
of Canada).

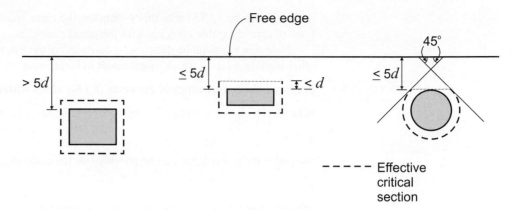

2h_s. Also, the size should be limited: the width should not exceed h_s, and the depth should
not exceed $h_s/3$. Holes located adjacent to the columns on all four sides must not be permitted. Note that CSA A23.3 does not include any provisions related to slab openings.

Factored shear stress (v_f)

The factored shear stress, v_f, is determined by dividing the factored shear force, V_f, by the
area over which the shear stresses are being transferred ($b_o \times d$), that is,

$$v_f = \frac{V_f}{b_o \times d} \qquad\qquad\qquad\qquad\qquad \textbf{[12.18]}$$

Note that $b_o \times d$ denotes the total area of vertical sides forming the critical section, as shown
in Figure 12.70b.

The factored shear force, V_f, is the resultant of the factored gravity load, w_f, acting on
the tributary column area, A, that is,

$$V_f = w_f \cdot A$$

The tributary area is shown in Figure 12.75. Note that the area for an interior column is
shown with a hatched pattern and the area for an edge column is shown with a cross-hatched
pattern. Note that the area enclosed by the perimeter of the critical section is not included
in the tributary area, because load acting over that area is transferred directly to the column.

Note that the tributary area is bound by lines of zero shear. For interior panels, these
lines are assumed to pass through the centre of the panel, while for the edge panels in flat
plates, the lines of zero shear correspond to a $0.45l_n$ distance from an exterior column, as

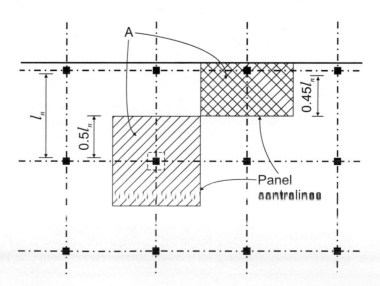

Figure 12.75 Tributary area for
the two-way shear design for an
interior and an edge column.

shown in Figure 12.75 (note that l_n denotes the clear span). Most designers consider the lines of zero shear that coincide with the panel centrelines.

Note that v_f should be determined considering the full load on all spans, as well as any other loading patterns which might result in larger stresses (Cl.13.3.3.1).

| A23.3 Cl.13.3.4 |

Factored shear resistance of concrete (v_c) for slabs without shear reinforcement

When a flat slab or flat plate is subjected to moderate gravity loads such that

$$v_c \le v_c$$

the entire shear resistance can be provided by the concrete, that is,

$$v_r = v_c$$

According to Cl.13.3.4.1, the factored shear resistance of concrete (v_c) should be taken as the <u>smallest</u> value obtained from the three equations outlined below.

i)   The effect of column shape

| A23.3 Eq. 13.5 |

$$v_c = \left(1 + \frac{2}{\beta_c}\right)0.19\lambda\phi_c\sqrt{f'_c} \ \text{(MPa)}$$ **[12.19]**

where

$\beta_c = h/b$ is the ratio of the long-side dimension (h) to the short-side dimension (b) of the column, the concentrated load, or reaction area (see Figure 12.76a). The A23.3 Eq. 13.5 takes into account the reduced shear strength in columns with elongated cross-sectional shape (where $\beta_c \ge 2$). As the column becomes more elongated, shear is mostly resisted along the short side, while the ultimate shear stress on the long side approaches the strength limit for beams or one-way slabs, that is, $0.19\lambda\phi_c\sqrt{f'_c}$. Figure 12.76b provides guidance on how to determine β_c for columns with irregular shapes.

Figure 12.76 The effect of column shape: a) regular column, and b) irregular column (courtesy of the Cement Association of Canada).

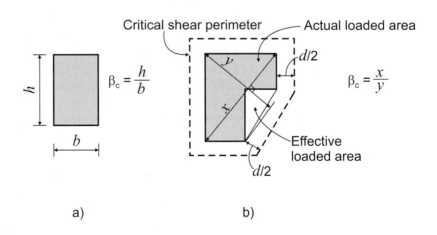

ii)   The effect of column location within a building

| A23.3 Eq. 13.6 |

$$v_c = \left(\frac{\alpha_s d}{b_o} + 0.19\right)\lambda\phi_c\sqrt{f'_c} \ \text{(MPa)}$$ **[12.20]**

where

$\alpha_s = 4$ for an interior column (see column A in Figure 12.72)
   $= 3$ for an edge column (see column B in Figure 12.72)
   $= 2$ for a corner column (see column C in Figure 12.72)

The A23.3 Eq. 13.6 may also govern when columns or capitals become very large. For example, this equation governs for an interior column when $b_o/d > 20$.

iii) Shear strength of plain concrete

| A23.3 Eq. 13.7 | $v_c = 0.38\lambda\phi_c\sqrt{f_c'}$ (MPa) | **[12.21]** |

Note that the A23.3 Eq. 13.7 is related to the shear strength of plain concrete. It gives a conservative estimate for v_c for interior and square-shaped columns. Note that, for a normal density concrete ($\lambda = 1.0$), this equation can be simplified as follows

$$v_c = 0.25\sqrt{f_c'} \ (MPa) \qquad\qquad \textbf{[12.22]}$$

The following two additional CSA A23.3 requirements, concrete compressive strength and size effect, need to be checked at this stage, as described below.

a) Concrete compressive strength (f_c') limit

It can be seen from above equations that v_c depends on $\sqrt{f_c'}$. This is due to the fact that the tensile strength of concrete (modulus of rupture) is proportional to $\sqrt{f_c'}$, and shear failure is primarily controlled by the concrete tensile strength, as illustrated in Figure 12.70a. Note that CSA A23.3 Cl.13.3.4.2 sets a limit for f_c', that is,

$$\sqrt{f_c'} \leq 8 \ (MPa)$$

b) Size effect

When $d > 300$ mm, Cl.13.3.4.3 requires that the v_c value needs to be modified to account for the size effect, as follows

$$v_c \times \frac{1300}{1000 + d}$$

12.9.3 Combined Moment and Shear Transfer at Slab-Column Connections

| A23.3 Cl.13.10.2 and 13.3.5 |

Unbalanced moments need to be transferred from the slab to the column through the slab-column connection (see Section 12.6.5 for discussion on unbalanced moments). The mechanism of transfer is somewhat complex, and involves both flexure and shear.

Consider a portion of a two-way slab in the vicinity of an interior column shown in Figure 12.77a. The slab-column connection needs to transfer the factored moment M_{fu} and the factored shear force V_f acting along the column axis. The transfer takes place through the following two mechanisms:

1. Flexure: a fraction of the unbalanced moment ($\gamma_f \times M_{fu}$) is transferred through flexure in the slab along the strip b_b, as shown in Figure 12.77b, and
2. Shear: the remaining fraction of the unbalanced moment ($\gamma_v \times M_{fu}$) is transferred by vertical shear stress (see Figure 12.77b). This stress is combined with punching shear stress caused by the shear force V_f, shown in Figure 12.77c.

The multiplier γ_v is required to find a fraction of the unbalanced moment transferred by shear ($\gamma_v \times M_{fu}$), and it can be determined from the following equation (Cl.13.3.5.3)

| A23.3 Eq. 13.8 | $\gamma_V = 1 - \dfrac{1}{1 + \dfrac{2}{3}\sqrt{\dfrac{b_1}{b_2}}}$ | **[12.23]** |

where

$b_1 = c_1 + d$

$b_2 = c_2 + d$

Note that b_1 denotes length of the critical section and c_1 is the dimension of the column along the span for which moments are determined (plane of the frame), and b_2 and c_2 are dimensions in the perpendicular direction (see Figure 12.77a).

The multiplier γ_f determines the remaining portion of the unbalanced moment ($\gamma_f \times M_{fu}$) which is to be transferred through flexure, that is,

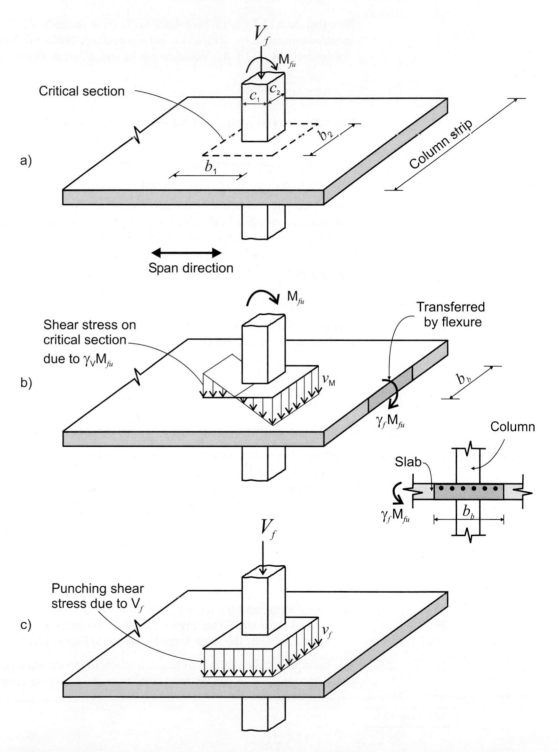

Figure 12.77 Shear and moment transfer: a) a portion of the slab at the slab-column connection showing the critical section; b) flexure and shear due to the moment M_{fu}, and c) punching shear stresses due to the shear force V_f.

and

A23.3 Eq. 13.25 $\gamma_f = 1 - \gamma_v$ **[12.24]**

or

$$\gamma_f = \frac{1}{1+\dfrac{2}{3}\sqrt{\dfrac{b_1}{b_2}}}$$ **[12.25]**

An increased amount of reinforcement within the strip b_b is prescribed to resist unbalanced moments at column locations (Clauses 13.3.5.6 and 13.10.2); refer to Section 12.4.2 for an explanation of strip b_b.

It can be seen from the above equations that for square column shapes, 60 % of the unbalanced moment M_{fu} is transmitted by conventional bending stresses at the column face ($\gamma_f \times M_{fu}$), while the remaining 40 % is transmitted by vertical shear stresses ($\gamma_v \times M_{fu}$); this ratio may vary depending on the column and slab geometry.

Shear stresses at the slab-column connection due to gravity load and the unbalanced moment are calculated according to Cl.13.3.5.5. The total factored shear stress at the perimeter of a critical slab section, v_{ftotal}, can be determined as follows (see Figure 12.78a)

$$v_{ftotal} = v_f + v_M$$ **[12.26]**

where v_f is the two-way (punching) shear stress due to V_f, and v_M is the shear stress due to unbalanced bending moment M_{fu} about axis y-y. It should be noted that, in a general case, bending moments may simultaneously occur in two directions (about axes x-x and y-y). In this case, the total shear stress needs to be expanded to account for an additional term, as follows

$$v_{ftotal} = v_f + (v_M)_x + (v_M)_y$$ **[12.27]**

where $(v_M)_x$ and $(v_M)_y$ refer to vertical shear stresses due to bending moments about axes x-x and y-y respectively. Note that the designer needs to consider multiple moment directions for edge and corner columns.

Distribution of shear stresses at an interior and an exterior column is illustrated in Figure 12.78.

Uniformly distributed punching shear stress (v_f) at the centroid of critical section can be calculated from Eqn. 12.18.

Shear stress, v_M, due to the bending moment ($\gamma_v \times M_{fu}$) transferred by the eccentricity of shear varies linearly about the centroid of critical section (see Figure 12.78), which can be determined from the following equation (Cl.13.3.5.4):

$$v_M = \frac{(\gamma_V \times M_{fu}) \times e}{J}$$ **[12.28]**

where

e = distance of the centroidal axis of the critical section perimeter to the point where shear stresses are being computed, and

J = property of the critical shear section analogous to the polar moment of inertia, equal to the sum of moments of inertia for the faces perpendicular to the centroidal axis C-C, plus the area of parallel faces times the square of the distance from those faces to axis C-C. Note that each face has its principal axes x-x and y-y; this is illustrated on an example of the face AA'BB' (see Figure 12.79a).

J value for an <u>interior column</u> can be determined from the following equation (using the notation shown in Figure 12.79a)

$$J = \frac{2(b_1 d^3)}{12} + \frac{2(d b_1^3)}{12} + 2(b_2 d)\left(\frac{b_1}{2}\right)^2$$ **[12.29]**

Note that the first and second term denote moments of inertia for the faces AA'BB' and EE'DD' respectively. The third term is equal to the product of area and the squared distance from axis C-C for the faces AA'DD' and BB'EE'. The J value calculated in this manner is

Figure 12.78 Shear stresses at the critical section: a) an interior column, and b) an exterior column.

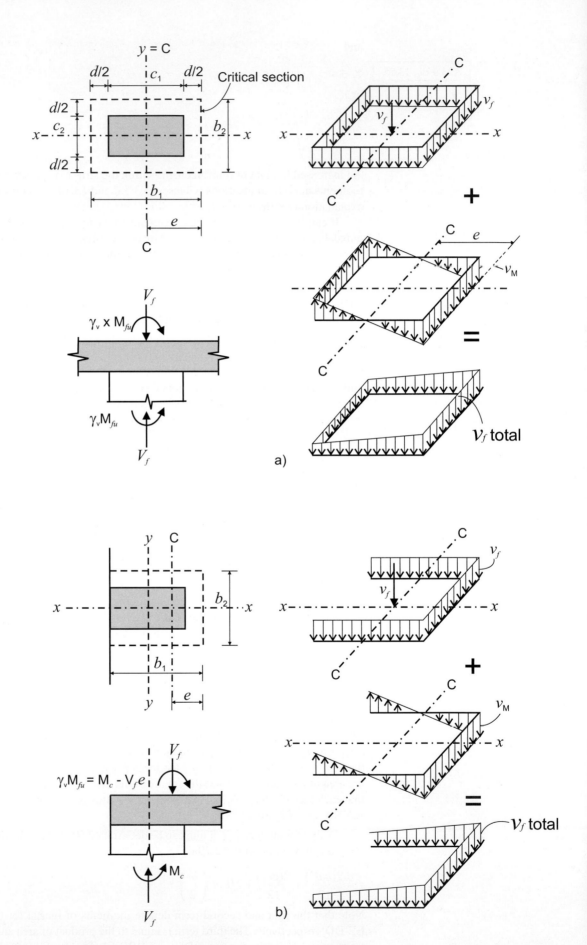

valid for bending about axis C-C. A similar equation could be derived for the bending about the perpendicular axis.

The J value for an exterior/edge column can be determined in a similar manner as for the interior column (see Figure 12.79b), that is,

$$J = \frac{2(b_1 d^3)}{12} + \frac{2(d b_1^3)}{12} + 2(b_1 d)\left(\frac{b_1}{2} - e\right)^2 + (b_2 d)e^2 \qquad \text{[12.30]}$$

The first and the second term in Eqn 12.30 are the same as for the interior column. The third term denotes the product of the area for the faces AA'BB' and EE'DD' and the squared distance $(b_1/2 - e)$ between their respective centroids and axis C-C. The last term is equal to the product of the area of the face BB'EE' and squared distance (e) from that face to axis C-C.

The axis C-C runs through the shear centre of the column perimeter. In case of an interior column shown in Figure 12.79a, shear center coincides with the geometric centre of the area ABCD. As a result, the distance (e) from axis C-C to the faces AA'DD' and BB'EE' is equal to $(b_1/2)$.

The location of the shear center for an exterior/edge column and its distance from the face BB'CC' can be found as centroid of the areas for faces of the critical column perimeter, with the reference axis running along BC. Distance (e) from the face BB'EE' to axis C-C can be determined as follows (see Figure 12.79b):

$$e = \frac{2(b_1 d)(b_1/2)}{2(b_1 d) + (b_2 d)}$$

The derivation of a J equation for a corner column is beyond the scope of this book (refer to MacGregor and Bartlett, 2000 for more details).

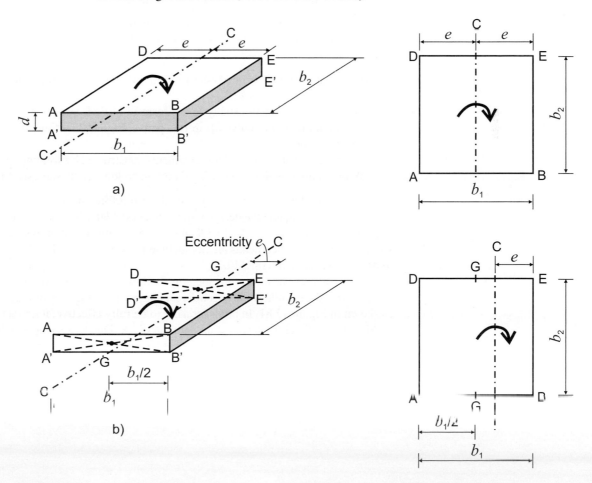

Figure 12.79 Calculation of the section property J: a) an interior column, and b) an exterior column.

12.9.4   Shear Reinforcement

| A23.3 Cl.13.3.7 | **CSA A23.3 Reinforcement Requirements**   Shear reinforcement is required when the total factored shear stress, v_{ftotal}, exceeds the shear resistance of concrete, v_c, that is,

$$v_{ftotal} > v_c$$

Before proceeding with the design of shear reinforcement, the designer may wish to explore alternative approaches for increasing the concrete shear resistance, v_c, such as: i) increasing the slab thickness, ii) providing drop panels, iii) increasing the column size (this will result in an increase in the perimeter of the critical section), and/or iv) increasing the concrete compressive strength f_c'.

When shear reinforcement is provided, the slab shear resistance is determined as follows (Cl.13.3.7.3)

$$v_r = v_c + v_s$$

where v_s is the factored shear stress in shear reinforcement. Note that CSA A23.3 requires that a reduced concrete shear resistance be used for slabs with shear reinforcement.

The shear reinforcement needs to resist the shear stress beyond the concrete shear resistance, that is,

$$v_s \geq v_{ftotal} - v_c$$

Note that CSA A23.3 sets the upper stress limit for shear resistance of two-way slabs with shear reinforcement, that is,

$$v_r \leq v_{max}$$

where v_{max} is the maximum allowed factored shear stress.

Vertical shear reinforcement can effectively increase the shear strength of slab-column connections in two-way slabs. CSA A23.3 Cl.13.3.7.1 permits the use of following three types of shear reinforcement for slabs:

1. Headed shear reinforcement (shear studs), in the form of large headed studs welded to steel strips (see Figure 12.80a).
2. Stirrup reinforcement, usually in the form of closed vertical stirrups enclosing horizontal bars radiating outwards in two perpendicular directions from the support (see Figure 12.80b).
3. Shearheads, that is, cross-shaped elements constructed by welding rolled steel sections (W or channel sections) into a rigid unit embedded in the slab (see Figure 12.80c).

Shear studs are the most common type of shear reinforcement used for two-way slab construction in Canada, and the design will be discussed later in this section. Design of stirrup reinforcement is covered in Cl.13.3.9 and the procedure is similar to shear stud design. Design provisions for shearheads are not included in CSA A23.3, but the designer is referred to ACI 318M-02/ACI 318RM-02.

Anchorage of shear reinforcement in shallow slabs is critical for effective prevention of punching shear failure. Shear reinforcement controls the size of diagonal shear cracks, as shown in Figure 12.81. In order for a bar to be fully effective, it needs to develop its full yield strength, f_{yv}, at the intersection with the crack. Due to a shallow slab depth, the reinforcement needs to be effectively hooked at the ends. Anchorage provisions are outlined in Clauses 7.1.2 and 12.13.

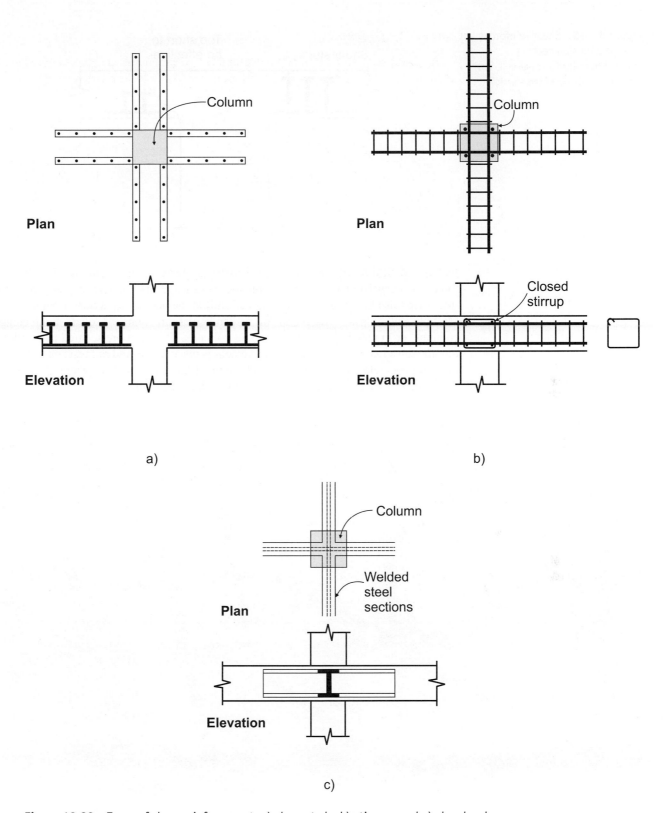

Figure 12.80 Types of shear reinforcement: a) shear studs; b) stirrups, and c) shearheads.

Figure 12.81 Shear reinforcement
in a cracked slab section
(adapted from Ghali and Hammill, 1992
with the permission of the American
Concrete Institute).

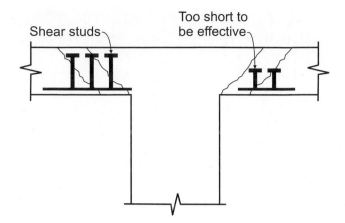

Shear studs

Too short to
be effective

A23.3 Cl.13.3.8

Design of Slabs with Shear Stud Reinforcement   Shear studs (headed shear reinforcement) consist of vertical bars with anchor heads at the top, and they are welded to a steel strip (also known as stud strip or stud rail) at the bottom, as shown in Figure 12.82a. Multiple stud rails are arranged in two perpendicular directions for square and rectangular columns, or in radial direction for circular columns. The studs are secured in position before the top and bottom flexural reinforcement is placed. The stud rail rests on bar chairs to maintain concrete cover below the steel. Figure 12.82b shows a two-way slab under construction with shear stud reinforcement arranged at a column location.

a)

Figure 12.82   Shear stud reinforcement: a) stud rails, and b) shear studs installed in a flat plate under construction
(Svetlana Brzev).

b)

Figure 12.82 (*cont.*)

| A23.3 Cl.13.3.8.2 | **Maximum factored shear stress** |

Shear studs can be used when the maximum factored shear stress in two-way slabs is less than the following limit

$$v_{ftotal} \le v_{max} = 0.75\lambda\phi_c\sqrt{f'_c} \tag{12.31}$$

| A23.3 Cl.13.3.8.3 | **Concrete shear resistance** |

The concrete shear resistance, v_c, in the zone reinforced with shear stud reinforcement is less than that for slabs without shear reinforcement, and it should be determined from the following equation:

$$v_c = 0.28\lambda\phi_c\sqrt{f'_c} \tag{12.32}$$

| A23.3 Cl.13.3.8.5 | **Steel shear resistance** |

The factored steel shear resistance for stud reinforcement, v_s, is determined in the similar manner as that for beams, as follows

| A23.3 Eq. 13.11 | $v_s = \dfrac{\phi_s f_{yv} A_{vs}}{b_o s}$ | [12.33] |

where

s = spacing of studs

f_{yv} = 345 MPa specified yield strength of studs (a typical value used in practice); must be ≤ 400 MPa

A_{vs} = the sum of areas of all studs along the perimeter of a critical section (b_o); it can be calculated as a product of the total number of stud rails around a column and the cross-sectional area of one stud.

Eqn 12.33 can be presented in an alternative form:

$$v_s = \frac{V_s}{b_o \cdot d}$$

where

$$V_s = \frac{\phi_s f_{yv} A_{vs} d}{s}$$

is the resultant for steel shear resistance, corresponding to the reinforcement with area A_{vs} and spacing s (similar to Eqn. 6.9 when $\theta = 45°$).

| A23.3 Cl.13.3.7.2 | **Critical sections** |

The shear stress needs to be checked both inside and outside the reinforced zone, thus there are the following two critical sections:

- Critical section 1 inside the reinforced zone at distance $d/2$ from the column face (same as for slabs without shear reinforcement), as per Cl.13.3.3.1, and
- Critical section 2 outside the shear-reinforced zone at distance $d/2$ from the outermost shear reinforcement.

| A23.3 Cl.13.3.7.4 | **Length of reinforced zone** |

The reinforcement needs to be extended at least by a distance $2d$ from the face of the column. At the section where reinforcement is discontinued, it is required that v_{ftotal} is less than the following limit:

$$v_{ftotal} \le 0.19\lambda\phi_c\sqrt{f'_c} \tag{12.33}$$

| A23.3 Cl.13.3.8.6 | **Spacing requirements** |

The following stud spacing requirements need to be followed:

a) For $v_f \le 0.56\lambda\phi_c\sqrt{f'_c}$

$$s_o \le 0.4d$$

and

$$s \le 0.75d$$

where s_o denotes distance of the first stud from the column face.

b) For $v_f > 0.56\lambda\phi_c\sqrt{f'_c}$

$$s_o \le 0.4d$$

and

$$s \le 0.5d$$

c) The distance between adjacent stud rails in the direction parallel to the column face: $g \le 2d$. This is not a CSA A23.3 requirement, but it is recommended by ACI (2008).

Stud layout and spacing requirements are illustrated in Figure 12.83. Note that the stud rails are placed at the column corners, and additional rails need to be provided depending on column dimensions. When more than two stud rails are placed along a column face, they should be evenly spaced. The minimum number of stud rails is 8, 6, and 4 for interior, edge, and corner columns, respectively (DECON, 2009).

Detailing of shear studs is critical for their effectiveness in providing the punching shear resistance in two-way slabs. Detailed recommendations for shear studs are summarized below (see Figure 12.84):

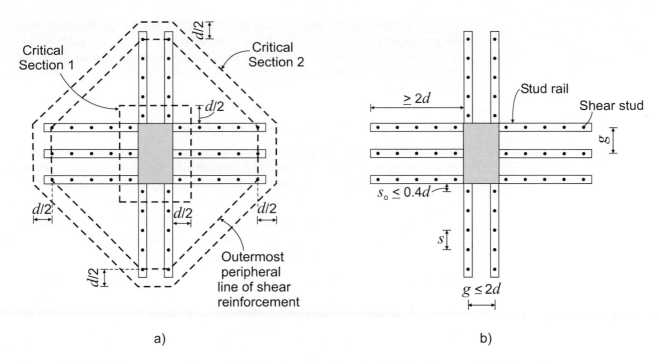

Figure 12.83 Stud arrangement for an interior column location: a) critical sections, and b) stud spacing.

1. Shear stud reinforcement should be located along concentric lines parallel to the perimeter of the column cross-section (Cl.13.3.8.4). Bottom stud rails should be aligned with the column faces in square or rectangular columns.
2. Shear studs must be mechanically anchored at each end by a plate or a head bearing against the concrete to develop bar yield strength (Cl.13.3.8.1). An effective anchorage can be achieved when the area of the top plate or the head is at least ten times the cross-sectional area of the bar.
3. The minimum concrete cover over the stud heads should be the same as the minimum cover for the flexural reinforcement (Cl.13.3.8.7). The concrete cover should not exceed the minimum cover plus one-half the bar diameter of the flexural reinforcement.

Stud rails with different specifications are commercially available (in terms of the number of studs per rail and the stud size). Overall stud height depends on the slab thickness. Stud

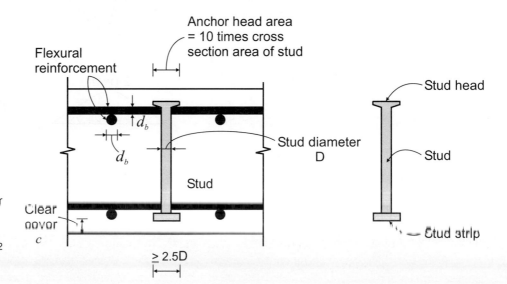

Figure 12.84 Location of shear stud reinforcement relative to flexural reinforcement (adapted from Ghali and Hammill, 1992 with the permission of the American Concrete Institute).

diameter ranges from 9.5 mm (3/8") to 19.1 mm (3/4"), and the rail thickness ranges from 4.8 mm (3/16") to 9.5 mm (3/8"). It has been found that 3/8" or 1/2" studs are most economical solutions for standard slabs. It is usually a good idea to keep the same stud diameter throughout the project, unless there is a wide range of slab thicknesses (DECON, 2009).

12.9.5 Shear Design of Two-Way Slabs with Beams

A23.3 Cl.13.4

Both beams and slab participate in transferring shear from slab to the columns in two-way slabs with beams. A fraction of load to be transferred by the beams depends i) on the beam-to-slab stiffness ratio for the beam under consideration, α_1, and ii) on the ratio of slab panel lengths, l_2/l_1. The following three scenarios need to be considered:

1. When, $\alpha_1 l_2/l_1 \geq 1.0$ beams are assumed to transfer the entire vertical load from the slab into the columns, that is, shear is resisted solely by the beams (Cl.13.4.1). It is not required to evaluate two-way shear resistance for the slab.

The beam shear resistance needs to be checked for all loads applied directly to the beam, plus the slab area enclosed by lines extending at 45° angle outward from the corners of the panel and the centrelines of adjacent panels on each side of the beam, as shown in Figures 12.85a and b. The beam shear resistance needs to be checked according to the design provisions for flexural members (beams and one-way slabs) prescribed by CSA A23.3 Cl.11 (see Section 6.5 for more details).

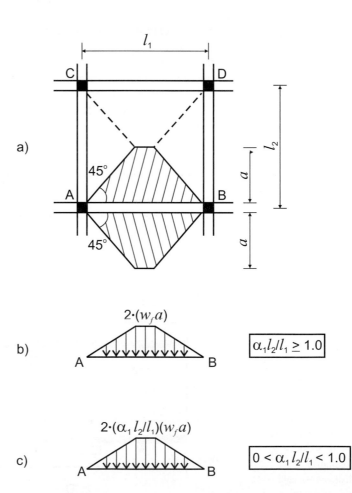

Figure 12.85 Tributary areas and loading in a two-way slab with beams: a) tributary area for an interior beam; b) beam loading when $\alpha_1 l_2/l_1 \geq 1.0$, and c) beam loading when $0 < \alpha_1 l_2/l_1 < 1.0$.

2. When $0 < \alpha_1 l_2/l_1 < 1.0$, the load distribution between the beams and the slab is determined by linear interpolation (Cl.13.4.2). A typical interior beam needs to be designed for a fraction of the total load, $2(w_f \cdot a)$, expressed as a multiplier of $\alpha_1 l_2/l_1$ (see Figure 12.85c), while the slab needs to be designed for two-way shear considering the fraction of the reduced factored shear force, V_f^*, obtained by using the multiplier $(1 - \alpha_1 l_2/l_1)$, that is,

$$V_f^* = V_f(1 - \alpha_1 l_2/l_1)$$ **[12.34]**

where V_f denotes the shear force corresponding to the two-way shear in the slab, as discussed in Section 12.9.2. Two-way shear resistance for a slab should be checked at the critical section with perimeter b_o, which needs to be reduced to account for the intersecting beams, as shown in Figure 12.86.

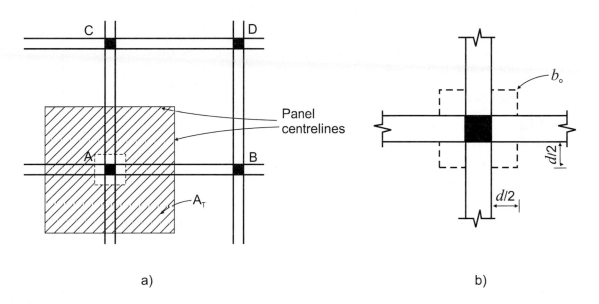

a)   b)

Figure 12.86 Distribution of load in a two-way slab when $0 < \alpha_1 l_2/l_1 < 1.0$: a) beam load, and b) critical section for two-way shear.

3. When $\alpha_1 l_2/l_1 = 0$, there are no beams, and the slab should be designed for shear as a two-way slab without beams discussed in Section 12.9.2.

One-way shear resistance for two-way slabs with beams should be checked in the same manner as for flat slabs and flat plates discussed in Section 12.9.2.

12.9.6 Design of Two-Way Slabs for Shear According to CSA A23.3: Summary and Design Examples

Key concepts related to the design of two-way slabs for shear have been presented in this section, and the key steps are outlined in Checklist 12.2. Although the steps are presented in a certain sequence, it is not necessary to follow the same sequence in all design situations. Three design examples will be presented to illustrate the application of the CSA A23.3 shear design provisions.

Checklist 12.2　Design of Two-Way Slabs for Shear

Step	Description	Code Clause
1	Check the one-way shear resistance (Section 12.9.2.2).	13.3.6
1a	Determine the factored shear force (V_f) by treating the slab as a wide beam (see Figure 12.67).	
1b	Determine the factored concrete shear resistance: $$\boxed{\text{A23.3 Eq.11.6}} \qquad V_c = \phi_c \cdot \lambda \cdot \beta \cdot \sqrt{f'_c} \cdot b_w \cdot d_v \qquad \textbf{[6.12]}$$	11.3.4
1c	Shear reinforcement is not required provided that $V_f \leq V_c$.	11.2.8.1
2	Check the two-way (punching) shear resistance (see Section 12.9.2).	13.3
2a	Determine the location and properties of critical section with perimeter b_o. The section should be taken at distance $d/2$ from the perimeter of the concentrated load or support (see Figure 12.71).	13.3.3
2b	Find the factored shear stress: $$v_f = \frac{V_f}{b_o \times d} \qquad \textbf{[12.18]}$$	
2c	Find the concrete shear resistance based on the following three criteria (the smallest value governs): i)　The effect of column shape $$\boxed{\text{A23.3 Eq.13.5}} \qquad v_c = \left(1 + \frac{2}{\beta_c}\right) 0.19\lambda\phi_c\sqrt{f'_c} \ \text{(MPa)} \qquad \textbf{[12.19]}$$ ii)　The effect of column location within a building $$\boxed{\text{A23.3 Eq.13.6}} \qquad v_c = \left(\frac{\alpha_s d}{b_o} + 0.19\right)\lambda\phi_c\sqrt{f'_c} \ \text{(MPa)} \qquad \textbf{[12.20]}$$ iii)　Shear strength of plain concrete $$\boxed{\text{A23.3 Eq.13.7}} \qquad v_c = 0.38\lambda\phi_c\sqrt{f'_c} \ \text{(MPa)} \qquad \textbf{[12.21]}$$	13.3.4
2d	Determine the shear stress, v_M, due to the a fraction of unbalanced bending moment ($\gamma_v \times M_{fu}$) transferred through the slab-column connection (see Section 12.9.3) $$v_M = \frac{\left(\gamma_V \times M_{fu}\right) \times e}{J} \qquad \textbf{[12.28]}$$	13.10.2 13.3.5
2e	Find the total factored shear stress: $$v_{ftotal} = v_f + v_M \qquad \textbf{[12.26]}$$	
3	Design the slab shear reinforcement. The shear reinforcement is required when $$v_{ftotal} > v_c$$	13.3.7
3a	Design the stud reinforcement (when required) - see Section 12.9.4. The shear studs can be used when the total factored shear stress is limited to $$v_{ftotal} \leq v_{max} = 0.75\lambda\phi_c\sqrt{f'_c} \qquad \textbf{[12.31]}$$	13.3.8

(Continued)

Checklist 12.2   Continued

Step	Description	Code Clause
3b	Find the concrete shear resistance for slabs with shear stud reinforcement: $$v_c = 0.28\lambda\phi_c\sqrt{f_c'} \qquad \textbf{[12.32]}$$	
3c	Determine the required steel shear resistance (v_s): $$v_s \geq v_{ftotal} - v_c$$	
3d	Find the total required area of stud reinforcement (A_{vs}) from the following equation $$\boxed{\text{A23.3 Eq.13.11}} \qquad v_s = \dfrac{\phi_s f_{yv} A_{vs}}{b_o s} \qquad \textbf{[12.33]}$$	
3e	Check the following spacing requirements: • stud spacing (s) • distance of the first stud from the column face (s_o) • stud rail spacing (g)	
4	Check the two-way shear resistance for slabs with beams (see Section 12.9.5).	13.4
4a	When $\alpha_1 l_2/l_1 \geq 1.0$, beams provide the entire shear resistance for the floor system (refer to Section 6.7 for shear design of beams).	
4b	When $0 < \alpha_1 l_2/l_1 < 0$, shear is resisted both by the beams and the slab (use linear interpolation to find the design shear forces and stresses).	

Example 12.8

Two-Way Flat Plate - Shear Design

Consider a plan view of a two-way flat plate floor system designed in Example 12.1, shown in the sketch below. Use the slab thickness of 180 mm and the effective depth of 140 mm. The factored area load is $w_f = 12.6$ kPa.

Design the slab for shear according to the CSA A23.3 requirements. Consider only an interior column at the intersection of gridlines 2 and B. Disregard the effect of unbalanced moments.

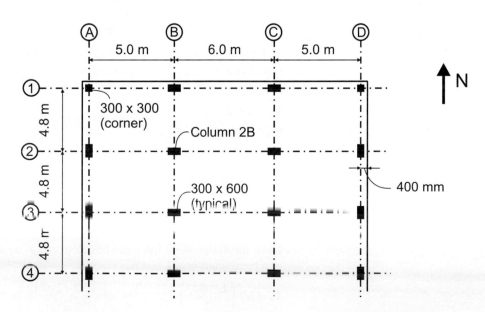

Given: $f'_c = 30$ MPa

$f_y = 400$ MPa

$\lambda = 1.0$ (normal-density concrete)

$\phi_c = 0.65$

$\phi_s = 0.85$

SOLUTION: **1. Check the one-way shear resistance (A23.3 Cl.13.3.6).**

a)   First, locate critical sections for the one-way shear design. The following two sections will be considered: 1-1 and 2-2 (one section in each horizontal direction of the building). Note that the critical sections are located at a distance d_v from the face of the column, as shown on the sketch below. Since $d = 140$ mm (given), let us determine d_v, which was previously defined as the effective shear depth taken as the greater of $0.9d$ and $0.72h_s$, that is,

$d_v = 0.9d = 0.9 \times 140$ mm $= 126$ mm

or

$d_v = 0.72h = 0.72 \times 180$ mm $= 130$ mm

The larger value governs, that is,

$d_v = 130$ mm

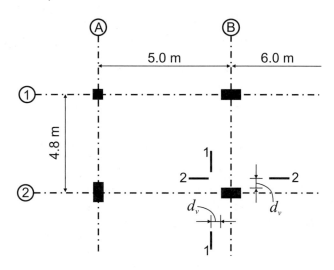

b)   Find the factored shear force V_f.

V_f is a design shear force at the critical section located at a distance d_v from the column face. Let us consider Section 1-1 for the slab span AB along gridline 2, as shown on the following sketch. The span is modelled as a wide beam with the width

$b_w = 4.8$ m

The clear span for span AB is

$l_n = 5.0$ m $- 0.6$ m$/2 - 0.3$ m$/2 = 4.55$ m

The factored area load $w_f = 12.6$ kPa needs to be transformed into the linear load w'_f acting on the wide beam, that is,

$w'_f = w_f \cdot b_w = 12.6$ kPa $\cdot 4.55$ m $= 57.3$ kN/m

Next, the design shear force can be calculated as an internal shear force at the support of a continuous beam, that is,

$$V_f = w'_f \left(\frac{l_n - 2d_v}{2} \right) = 57.3 \text{ kN/m} \left(\frac{4.55 - 2 \cdot 0.13}{2} \right) \cong 123 \text{ kN}$$

Since Section 1-1 corresponds to the tributary width of 4.8 m, we can find the shear force per 1 m slab width, that is,

$$V_f' = 123 \text{ kN}/4.8 \text{ m} = 26.0 \text{ kN/m}$$

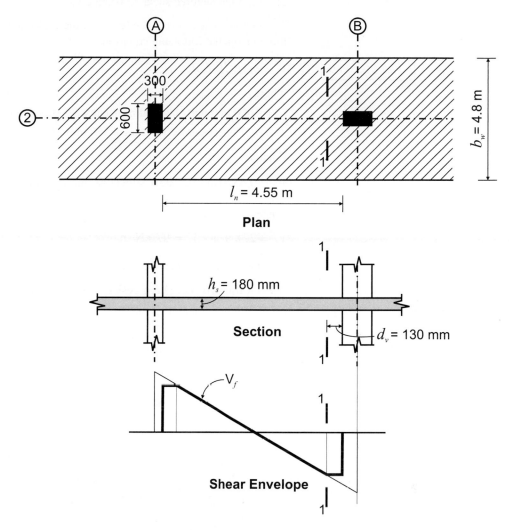

Plan

Section

$h_s = 180 \text{ mm}$

$d_v = 130 \text{ mm}$

$-V_f$

Shear Envelope

Calculations can be performed in a similar manner for Section 2-2. In this case, we are going to consider span 1-2 along gridline B. The clear span is

$$l_n = 4.8 \text{ m} - 2(0.3 \text{ m}/2) = 4.5 \text{ m}$$

and

$$b_w = \frac{5.0 \text{ m} + 6.0 \text{ m}}{2} = 5.5 \text{ m}$$

hence

$$w_f' = w_f \cdot b_w = 12.6 \text{ kPa} \cdot 5.5 \text{ m} = 69.3 \text{ kN/m}$$

Next, the design shear force can be calculated as an internal shear force at the support of a continuous beam, that is,

$$V_f = w_f' \left(\frac{l_n - 2d_v}{2} \right) = 69.3 \text{ kN/m} \left(\frac{4.5 - 2 \cdot 0.13}{2} \right) \cong 147 \text{ kN}$$

and the corresponding unit shear force is (based on the tributary width of 5.5 m)

$$V_f' = 147 \text{ kN}/5.5 \text{ m} = 27 \text{ kN/m}$$

The larger value governs, that is,

$V_f' = 27$ kN/m

c)   Find the factored shear resistance V_c.

The factored shear resistance (equal to the concrete shear resistance) should be determined from the following equation

A23.3 Eq. 11.6 $V_c = \phi_c \lambda \beta \sqrt{f_c'}\, b_w d_v$ [6.12]

where

$b_w = 1000$ mm unit slab width (because V_f' was determined based on the same width)
$\beta = 0.21$ because $h_s = 180$ mm ≤ 350 mm (Cl.11.3.6.2)

Finally,

$V_c' = 0.65 \times 1.0 \times 0.21 \sqrt{30 \text{ MPa}}\,(1000 \text{ mm})(130 \text{ mm}) = 97$ kN/m

Since

$V_f' = 27$ kN/m $< V_c' = 97$ kN/m

it can be concluded that the one-way shear resistance is satisfactory.

2.   Check the two-way shear resistance (CSA A23.3 Cl.13.3.4).

a)   Find the critical section.

The critical section is located at a distance $d/2 = 70$ mm from the face of the column, as shown on the sketch below. The perimeter of the critical section, b_o, is equal to:

$b_o = 2(300 \text{ mm} + 140 \text{ mm}) + 2(600 \text{ mm} + 140 \text{ mm}) = 2360$ mm

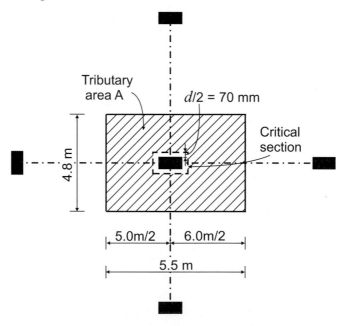

b)   Find the factored shear force (V_f) and the factored shear stress (v_f).

V_f is determined as a product of the factored load, w_f, and the tributary area, A, shown hatched on the sketch, that is,

$$V_f = w_f \cdot A = 12.6 \text{ kPa} \times \left[\left(\frac{5.0 \text{ m}+6.0 \text{ m}}{2}\right) \times (4.8 \text{ m}) - \right.$$

$$\left. (0.3 \text{ m}+0.14 \text{ m})(0.6 \text{ m}+0.14 \text{ m})\right] = 329 \text{ kN}$$

Next, find the factored shear stress, v_f, as follows

$$v_f = \frac{V_f}{b_o \times d} = \frac{329 \times 10^3 \, \text{N}}{2360 \, \text{mm} \times 140 \, \text{mm}} = 0.96 \, \text{MPa}$$ [12.18]

c) Find the concrete shear resistance, v_c.

Confirm that f_c' satisfies the requirements of CSA A23.3 Cl.13.3.4.2:

$$\sqrt{f_c'} \le 8 \, (\text{MPa})$$

Since

$$\sqrt{30} = 5.5 \, \text{MPa} \le 8 \, \text{MPa}$$

Use the following three criteria:

i) The effect of column shape:

| A23.3 Eq. 13.5 |

$$v_c = \left(1 + \frac{2}{\beta_c}\right) 0.19 \lambda \phi_c \sqrt{f_c'}$$ [12.19]

where

$$\beta_c = \frac{b_2}{b_1} = \frac{600 \, \text{mm}}{300 \, \text{mm}} = 2.0 \text{ is the ratio of longer and shorter column cross-sectional}$$

dimension
Therefore,

$$v_c = \left(1 + \frac{2}{2}\right) 0.19 \cdot 1.0 \cdot 0.65 \cdot \sqrt{30} = 1.35 \, \text{MPa}$$

ii) The effect of column location within a building:

| A23.3 Eq. 13.6 |

$$v_c = \left(\frac{\alpha_s d}{b_o} + 0.19\right) \lambda \phi_c \sqrt{f_c'}$$ [12.20]

where $\alpha_s = 4$ (interior column)

$$v_c = \left(\frac{4 \cdot 140}{2360} + 0.19\right) \cdot 1.0 \cdot 0.65 \cdot \sqrt{30} = 1.52 \, \text{MPa}$$

iii) Shear strength of plain concrete:

Since normal-density concrete is used, let us use Eqn 12.22, that is,

$$v_c = 0.25\sqrt{f_c'} = 0.25\sqrt{30} = 1.37 \, \text{MPa}$$ [12.22]

The smallest v_c value governs, hence

$$v_c = 1.35 \, \text{MPa}$$

Since the effect of unbalanced moments is disregarded in this example, it follows that

$$v_{ftotal} = v_f$$

and

$$v_{ftotal} = 0.96 \, \text{MPa} < v_c = 1.3 \, \text{MPa}$$

It can be concluded that the two-way shear resistance is adequate based on the punching shear requirement. However, it is required to check the total shear stresses due to punching shear and the transfer of unbalanced moments. This will be performed in the next example.

If the two-way shear resistance is not satisfied, the designer can find the effective depth, d, which satisfies the two-way shear resistance requirements using the procedure outlined in Section 14.4.1 related to the two-way shear design of spread footings. Alternatively, the designer can assume a higher d value and repeat the calculations until all requirements have been satisfied.

Example 12.9

Two-Way Flat Plate - Shear and Moment Transfer

Consider the same slab-column connection as discussed in Example 12.8.
Design the slab for two-way shear according to the CSA A23.3 requirements, but consider the effect of unbalanced moments in the design.

SOLUTION:

The objective of this example is to find the total shear stress due to the combined effect of punching shear and the unbalanced bending moment at the column (CSA A23.3 Cl.13.3.5.5).

The total shear stress at the perimeter of a critical slab section, v_{total}, can be determined as follows

$$v_{ftotal} = v_f + v_M$$

where $v_f = 0.96$ MPa is the punching shear stress determined in the previous step, and v_M is the shear stress due to unbalanced bending moment, M_{fu}. An unbalanced moment for the interior column under consideration was determined in Example 12.1 (Step 7), as follows

$$M_{fu} = 46.9 \text{ kNm}$$

The procedure for finding the v_M value is outlined below.

1. **Find the portion of unbalanced moment transferred from the slab by flexure ($\gamma_f \times M_{fu}$).**

   CSA A23.3 Cl.13.3.5 requires that a fraction of the unbalanced moment be transferred from the slab to the column by flexure. The corresponding moment is equal to

   $$\gamma_f \times M_{fu}$$

   Note that γ_f depends on the dimensions of the critical section at the specific column location, that is,

   $$\gamma_f = \frac{1}{1+\dfrac{2}{3}\sqrt{\dfrac{b_1}{b_2}}} = \frac{1}{1+\dfrac{2}{3}\sqrt{\dfrac{740}{440}}} = 0.54 \qquad \text{[12.25]}$$

   where (see the sketch below)

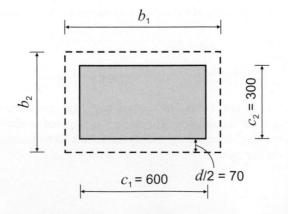

$$b_1 = c_1 + d = 600 + 140 = 740 \text{ mm}$$
$$b_2 = c_2 + d = 300 + 140 = 440 \text{ mm}$$

Therefore,

$$\gamma_f \times M_{fu} = 0.54 \times 46.9 = 25.3 \text{ kNm}$$

Find the required reinforcement corresponding to the bending moment above. For each 15M bar, it follows that

$$M_r \cong \phi_s A_b f_y (0.9d) = 0.85 \times 200 \text{ mm}^2 \times 400 \text{ MPa} \times (0.9 \times 140) = 8.6 \text{ kNm}$$

The required number of 15M bars is

$$\frac{25.3 \text{ kNm}}{8.6 \text{ kNm}} \cong 3.0$$

Use 3-15M bars. This is flexural reinforcement and it should be provided over the column within the band width b_b. However, the required flexural reinforcement provided at that location was previously determined as 6-15M@150 (see Example 12.1, Table 12.11), which is larger than the 3-15M bars determined from this calculation. In conclusion, there is no need to provide additional flexural reinforcement bars to ensure moment transfer from the slab to the columns. Note that this calculation should be performed as a part of the flexural design.

2. **Find the portion of unbalanced moment transferred from the slab by shear ($\gamma_v \times M_{fu}$).**

The remaining portion of unbalanced moment is transferred from the slab by shear. Since

$$\gamma_v = 1 - \gamma_f = 1 - 0.54 = 0.46$$

Therefore,

$$\gamma_v \times M_{fu} = 0.46 \times 46.9 = 21.6 \text{ kNm}$$

3. **Find the section property, J, for the interior column.**

The underlying equations were presented in Section 12.9.3, and the terms are illustrated on the sketch below.

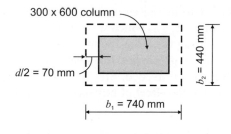

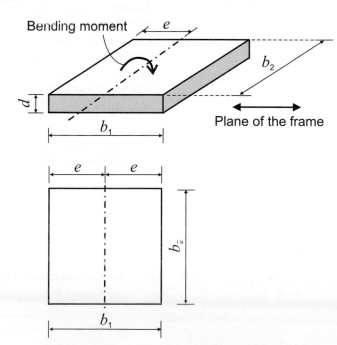

$$J = \frac{2(b_1 d^3)}{12} + \frac{2(db_1^3)}{12} + 2(b_2 d)\left(\frac{b_1}{2}\right)^2 \qquad \text{[12.29]}$$

where

$b_1 = c_1 + d = 600 + 2 \cdot 70 = 740$ mm

$b_2 = c_2 + d = 300 + 2 \cdot 70 = 440$ mm

Therefore,

$$J = \frac{2(740 \cdot 140^3)}{12} + \frac{2(140 \cdot 740^3)}{12} + 2(440 \cdot 140)\left(\frac{740}{2}\right)^2 = 2.67 \cdot 10^{10} \text{ mm}^4$$

iii) Find the shear stress v_M due to unbalanced moment $\gamma_v \times M_{fu}$.

Since this is a symmetrical section, the eccentricity is equal to

$$e = \frac{b_1}{2} = \frac{740}{2} = 370 \text{ mm}$$

Therefore,

$$v_M = \frac{(\gamma_v \times M_{fu}) \times e}{J} = \frac{(21.6 \cdot 10^6 \text{ Nmm})(370 \text{ mm})}{2.67 \cdot 10^{10} \text{ mm}^4} = 0.3 \text{ MPa} \qquad \text{[12.28]}$$

iv) Finally, find the total shear stress, v_{total}, that is,

$$v_{ftotal} = v_f + v_M = 0.96 + 0.3 = 1.26 \text{ MPa} \qquad \text{[12.26]}$$

Since

$$v_{ftotal} = 1.26 MPa < v_c = 1.35 \text{ MPa}$$

it follows that the two-way shear stress requirement has been satisfied.

Example 12.10

Two-Way Flat Plate - Design of Shear Stud Reinforcement

Consider a scenario where after the slab design discussed in Example 12.9 was completed, the total factored load had to be increased by 25%. Assume a proportional increase in the total factored shear stress (v_{ftotal}), which was originally equal to 1.26 MPa. Perform shear design calculations using the same 180 mm slab thickness, an effective depth of 140 mm, and the same material properties. Use steel stud reinforcement to satisfy the shear resistance requirements if needed. Steel yield strength for stud reinforcement is $f_{yv} = 345$ MPa. Disregard the effect of unbalanced moments.

SOLUTION:

Note that the one-way shear will not be considered in this design, since it was checked in Example 12.9 and it did not govern for the shear design.

1. **Find the critical sections for two-way shear design (CSA A23.3 Cl.13.3.7).**

In two-way slabs with shear reinforcement, there are two critical sections (see Figure 12.83):

i)   Critical section 1 inside the reinforced zone at the distance $d/2$ from the column face (same as for slabs without shear reinforcement), and

ii) Critical section 2 outside the shear-reinforced zone at a distance $d/2$ from the outermost shear reinforcement.

At this point, critical section 1 has been defined in Example 12.9 and the critical perimeter is equal to

$b_o = 2360$ mm

Properties of critical section 2 will be discussed later in this example.

2. **Find the factored shear stress (v_f) at critical section 1.**

Assume a 25% increase in the total factored shear stress from Example 12.9 (1.26 MPa), that is,

$v_{ftotal} = 1.25(1.26$ MPa$) = 1.56$ MPa

and

$V_f = 1.25(329$ kN$) = 411$ kN

3. **Check whether shear reinforcement is required for this design.**

We need to check whether the slab shear resistance is still adequate, considering a 25% stress increase. The concrete shear resistance was determined in Example 12.9, as follows:

$v_c = 1.35$ MPa

Since

$v_f = 1.56$ MPa $> v_c$

It follows that the shear reinforcement is required. Let us try to use stud reinforcement.

4. **Confirm that the stud reinforcement can be used (CSA A23.3 Cl.13.3.8.2).**

The stud reinforcement can be used when the maximum total shear stress at critical section 1 is below the limit prescribed by CSA A23.3, that is,

$$v_f \leq v_{max} = 0.75\lambda\phi_c\sqrt{f_c'} = 0.75\cdot1.0\cdot0.65\sqrt{30} = 2.7 \text{ MPa} \qquad [12.31]$$

Since

$v_f = 1.63$ MPa < 2.7 MPa

it follows that stud reinforcement can be used for this design.

5. **Find the reduced concrete shear resistance for slabs with stud reinforcement according to CSA A23.3 Cl.13.3.8.3.**

$$v_c = 0.28\lambda\phi_c\sqrt{f_c'} = 0.28\cdot1.0\cdot0.65\sqrt{30} = 1.0 \text{ MPa} \qquad [12.23]$$

Note that this v_c value should be used in the next steps. The v_c value of 1.35 MPa discussed in Step 3 can only be used for slabs without shear reinforcement.

6. **Design the stud reinforcement.**

a) Find the required steel shear resistance:

$v_s = v_f - v_c = 1.56 - 1.0 = 0.56$ MPa

b) Find the required stud spacing (Cl.13.3.8.6):

First, let us check the level of design shear stress, as follows:

$v_f = v_{ftotal}$

$$v_f \leq 0.56\lambda\phi_c\sqrt{f_c'} = 0.56\cdot1.0\cdot0.65\sqrt{30} = 2.0 \text{ MPa}$$

thus

$v_f = 1.56$ MPa < 2.0 MPa

CSA A23.3 stud spacing requirements for this stress level are outlined below.

Distance of the first stud from the column face is equal to

$$s_o \leq 0.4d = 0.4 \cdot 140 = 56 \text{ mm}$$

Typical stud spacing is

$$s \leq 0.75d = 0.75 \cdot 140 = 105 \text{ mm}$$

Let us proceed with the following values:

$$s_o = 55 \text{ mm}$$

and

$$s = 105 \text{ mm}$$

c)   Find the required area of stud reinforcement.

Since

$$v_s = \frac{\phi_s f_{yv} A_{vs}}{b_o s} \qquad\qquad\qquad \textbf{[12.33]}$$

it follows that the required area of stud reinforcement is equal to

$$A_{vs} = \frac{v_s \cdot b_o \cdot s}{\phi_s f_{yv}} = \frac{0.56 \text{ MPa} \cdot 2360 \text{ mm} \cdot 105 \text{ mm}}{0.85 \cdot 345 \text{ MPa}} = 473 \text{ mm}^2$$

Let us assume 9.5 mm (3/8 inch) studs (the smallest size), and the corresponding area per stud is $A_{1s} = 71 \text{ mm}^2$. The required number of studs (n) along the perimeter of the critical section is

$$n = \frac{A_{vs}}{A_{1s}} = \frac{473}{71} = 6.6 \cong 7$$

7.   **Develop a preliminary layout of stud reinforcement which meets the area and spacing requirements.**

The layout will be developed based on the spacing determined in the previous step (see Figure 12.82). First, we need to determine the layout of stud rails, based on the recommended spacing between rails (g), as follows:

$$g \leq 2d = 2 \cdot 140 = 280 \text{ mm}$$

Since the column dimensions are 300 mm by 600 mm, let us use 3 stud rails perpendicular to the long side, and 2 rails perpendicular to the short side of the column. Therefore, we will have 10 stud rails in total, as shown on the sketch below.

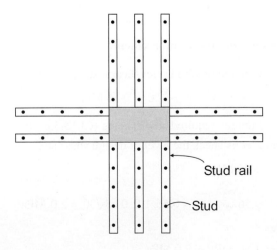

Stud rail

Stud

8. **Check whether the shear resistance is adequate.**

The selected stud parameters are as follows

$n = 10$ and $A_{1s} = 71$ mm^2

thus

$A_{vs} = n \cdot A_{1s} = 10 \cdot 71 = 710$ mm$^2 > 473$ mm^2

It can be concluded that the proposed stud layout is satisfactory. However, let us find the actual shear resistance, as follows

$v_r = v_c + v_s$

Since

| A23.3 Eq. 13.11 |

$$v_s = \frac{\phi_s f_{yv} A_{vs}}{b_o s} = \frac{0.85 \cdot 345 \text{ MPa} \cdot 710 \text{ mm}^2}{2360 \text{ mm} \cdot 105 \text{ mm}} = 0.84 \text{ MPa} \qquad [12.32]$$

and

$v_c = 1.0$ MPa

thus

$v_r = 1.0 + 0.84 = 1.84$ MPa

Let us perform the checks to confirm that v_r is adequate:

$v_r = 1.84$ MPa $> v_f = 1.56$ MPa

and

$v_r = 1.84$ MPa $< v_{max} = 2.7$ MPa

Therefore, it can be concluded that the design is adequate.

9. **Find the required length for stud rails (CSA A23.3 Cl.13.3.7.4).**

a) Estimate the required stud rail length.

CSA A23.3 prescribes that the reinforcement needs to be extended at least by a distance $2d$ from the face of the column. Therefore, the minimum length for a stud rail (l_s) is equal to

$l_s \geq 2d = 2 \cdot 140 = 280$ mm

Let us assume that a typical stud rail has 6 studs at 105 mm spacing, and 55 mm end spacing, that is,

$l_s = 5s + 2s_o = 5 \cdot 105 \text{ mm} + 2 \cdot 55 \text{ mm} = 635$ mm

A typical stud rail is shown on the sketch below.

b) Check the shear stress requirement.

We need to confirm that the proposed stud length is adequate. CSA A23.3 requires that the reinforcement should be extended such that the design shear stress is below the following limit:

$$v_f \leq 0.19 \lambda \phi_c \sqrt{f_c'} = 0.19 \cdot 1.0 \cdot 0.65 \sqrt{30} = 0.68 \text{ MPa}$$

This requirement will be confirmed by finding the total shear stress (due to gravity load) at critical section 2, as shown on the sketch below.

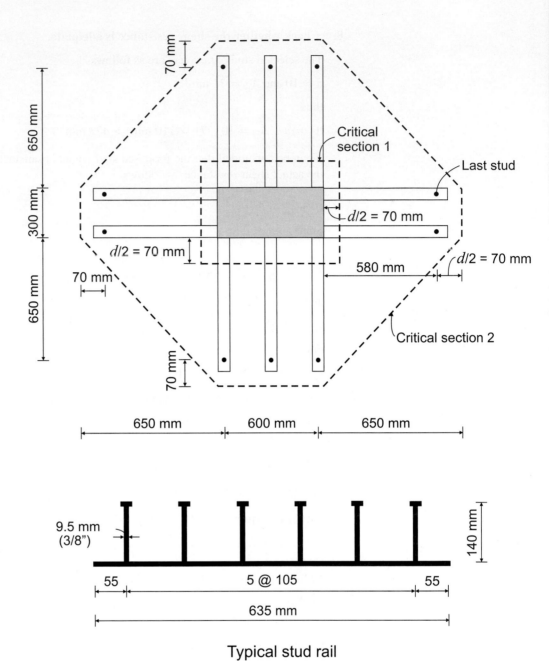

Typical stud rail

i) Find the properties for critical section 2.

First, let us find the dimensions of critical section. The critical section is located at a distance d/2 from the outermost shear reinforcement (the furthest stud within in a rail), as shown on the sketch above. The perimeter of critical section 2 is equal to

$$b_o = 2(600 \text{ mm} + 300 \text{ mm}) + 4 \cdot 920 \text{ mm} = 5480 \text{ mm}$$

ii) Find the factored shear stress v_f due to the shear force V_f:

$$v_f = \frac{V_f}{b_o \times d} = \frac{411 \times 10^3 \text{ N}}{5480 \text{ mm} \times 140 \text{ mm}} = 0.54 \text{ MPa}$$

iii) Confirm that the stress is within the permitted limits.

Since

$$v_f = 0.54 \text{ MPa} < 0.68 \text{ MPa}$$

It follows that the proposed stud length is adequate. However, if the effect of unbalanced moment was considered, the factored shear stress might exceed the shear resistance. In that case, it would be possible to increase the stud area by adding one more rail along the 600 mm column dimension, and the total number of stud rails would increase from 10 to 12. Alternatively, longer stud rails could be used; this would result in an increase in the critical shear perimeter.

10.   Provide a design summary.

Based on the design performed above, the following stud specification can be used:

Use 10 stud rails with overall height of 140 mm, 4.8 mm thickness, with 6-9.5 mm (3/8") diameter studs. The typical spacing is 105 mm on centre, and 55 mm end spacing.

12.10 │ DEFLECTIONS

│ A23.3 Cl.13.2.7 │ ### 12.10.1   Background

Serviceability limit states ensure that the intended use and occupancy of a building have been maintained, and include deflections, cracking, vibrations etc. Serviceability considerations for reinforced concrete flexural members are covered in Chapter 4. The discussion presented in this section is focused on deflections in two-way slabs. Deflections due to service loads must be limited to a tolerable amount. Excessive deflections can cause damage to non-structural elements such as partitions and glazing. Noticeable deflections appear unsafe and are not aesthetically pleasing.

The key factors influencing deflections in two-way slabs, such as concrete properties, creep and shrinkage, cracking, construction loads and procedures, and the placement of top reinforcement, will be discussed next. Other factors, such as slab geometry (span/thickness ratio), continuity, and support restraints, also influence deflections, and they are accounted for by deflection calculation procedures.

Concrete properties

The key concrete properties which influence deflections include the modulus of elasticity (E_c) and the modulus of rupture (f_r). Stiffness of an uncracked member increases in proportion to its modulus of elasticity, which is in turn directly proportional with the square root of the characteristic concrete compressive strength (f_c'), as discussed in Section 2.3.4. Key factors influencing the E_c value include aggregates, cement, silica fume, and admixtures. Lower water/cement ratio, a lower slump, and changes in concrete mix proportions can cause an increase in the modulus of elasticity (ACI 435R-95). Modulus of elasticity is a time-dependent property, and its value increases over time.

Modulus of rupture (f_r) denotes the concrete tensile strength, and it influences the onset of cracking since the cracking moment (M_{cr}) is directly proportional to f_r. The standard approach for determining the modulus of rupture reflects laboratory conditions, since the f_r value is determined from small specimens prepared and tested in a controlled environment. In practice, two-way slabs are constructed over large areas, under variable weather conditions and with a variation in the quality of workmanship. It is expected that the in-situ f_r values are significantly more variable compared to the laboratory conditions (Scanlon, 1999).

Creep and shrinkage

Time-dependent deflections in slabs are caused by creep and shrinkage. Shrinkage is a reduction in concrete volume over time due to the cement hydration, loss of moisture, and other factors, and it occurs independently of applied loading. Different types of shrinkage were discussed in Section 2.3.6. Shrinkage cracking in two-way slabs is mostly due to external and/or internal restraints. External supports, e.g. columns or walls, restrain the free horizontal movement in the slab. When shrinkage occurs, it can cause random cracking, thereby decreasing slab stiffness in cracked regions. Internal restraints are provided by flexural reinforcement. As the slab tries to shorten due to shrinkage, reinforcement placed on one side of the slab (top or bottom) shows a tendency to restrain the movement, thereby causing localized warping. This effect increases gradually and can cause progressive cracking in the slab over time. CSA A23.3-04 prescribes a reduced modulus of rupture for the cracking moment (M_{cr}) calculation in order to account for the effect of restraint shrinkage and other factors that cause cracking in two-way slabs at service load level (Cl.13.2.7).

Creep is demonstrated by an increase in concrete strains under sustained stresses, as discussed in Section 2.3.5. The effects of creep are more pronounced in reinforced concrete structures loaded at an early age, such as two-way flat slabs loaded at the construction stage. Creep-induced strains in concrete cause a significant increase in deflections due to sustained loads over time (by a factor of 2 or 3).

Cracking

Cracking is one of the key factors influencing deflections in two-way slabs. Slab deflections are sensitive to the extent of cracking, since two-way slabs are typically lightly reinforced (often requiring only minimum reinforcement for flexural strength). Scanlon (1999) concluded that slab sections with low flexural reinforcement ratios close to the minimum CSA A23.3 requirements are characterized by a significant difference between their cracked and gross stiffness. The cracked transformed moment of inertia, I_{cr}, is considerably lower than the gross moment of inertia, I_{gr}; often in the range of $I_{cr} = I_{gr} / 3$. Note that cracking is not uniform through the slab due to a variation in the bending moments. Regions with low bending moments remain uncracked and are characterized by significantly higher stiffness than the cracked regions.

This section outlines key methods for deflection calculations and their application is illustrated by a few examples.

12.10.2   CSA A23.3 Deflection Control Requirements

A23.3 Cl.13.2 — Deflections in two-way slabs are required to remain within acceptable limits. CSA A23.3 prescribes two approaches for deflection control: indirect approach and detailed deflection calculations.

According to the *indirect approach*, the designer is permitted to select the minimum slab thickness that results in a robust design, thus detailed deflection calculations are not required (Cl.13.2.2). The minimum CSA A23.3 slab thickness requirements are explained in Section 12.5.2. Note that the indirect approach can be applied only to regular two-way slabs discussed in Section 12.5.1.

Detailed deflection calculations must be performed for slabs with the span-to-thickness ratio below the CSA A23.3 limit. According to Cl.13.2.7, the deflections should be computed by taking into account the size and the shape of a slab panel, the support conditions, and the nature of restraints at the panel edges.

Both immediate and long-term deflections need to be considered in the design. *Immediate deflections* are initial deflections that occur as soon as the slab is constructed and the shoring is removed, while *long-term deflections* occur over time and may be caused by creep, shrinkage, and temperature strains. The CSA A23.3 procedures for estimating immediate and long-term deflections are discussed in Section 4.4.

The deflections in two-way slabs must be within the limits prescribed by CSA A23.3 Table 9.3. Note that CSA A23.3 prescribes the same deflection limits for one-way and two-way slabs.

Immediate slab deflections can be calculated using one of the following three methods:

1. *The Crossing Beam Method* is based on treating a two-way slab as an orthogonal one-way system, thus allowing the deflection calculations by beam analogy.
2. *The Equivalent Frame Method* uses a linear elastic analysis of 2-D frames. This method is discussed in detail in Section 12.7.2. An effective moment of inertia is used to account for the effect of cracking.
3. *The Finite Element Method* (FEM) is explained in Section 12.7.3, and it can be used to obtain both internal forces and deflections in two-way slabs. It is a computer-based method, and depending on the software capabilities it can be used either for linear elastic or nonlinear analysis which takes into account the effect of cracking.

For some types of slabs and support conditions, it is possible to obtain closed-form solutions for deflections based on the Elastic Plate Theory, which uses a partial differential equation for load-deflection response. The approach can be used to determine internal bending moments and shear forces. It is rarely used in practice due to the availability of computer-based methods such as the FEM.

Long-term deflections in two-way slabs can be estimated in the same manner as explained in Section 4.4.3. According to Cl.9.8.2.5, long-term deflections are obtained by multiplying the immediate deflection due to a sustained load by the factor ζ_s, given by

| A23.3 Eq. 9.5 | $\zeta_s = 1 + \dfrac{s}{1 + 50\rho'}$ | [4.15] |

Where s is the time-dependent factor for creep deflection under sustained load (see Table 4.1), and ρ' is the compression reinforcement ratio (usually taken as 0 for two-way slabs). For practical applications in slabs without compression steel, the ζ_s factor ranges from 1.0 (load sustained for 3 months) to 3.0 (load sustained for at least 5 years).

12.10.3 The Crossing Beam Method for Deflection Calculations

The Concept The Crossing Beam Method (also known as the Wide Beam Method) is an approximate method for calculating deflections in compliance with the CSA A23.3 requirements. The method will be explained on an example of a slab panel with spans l_x and l_y supported by the columns shown in Figure 12.87. The column strip in x-direction and the middle strip in y-direction are considered as continuous wide beams. The widths for column and middle strips are denoted as l_c and l_m, respectively (see Section 12.4.1 for an explanation of column and middle strips).

The maximum deflection for the column strip in x-direction is denoted as Δ_{cx}, as shown in Figure 12.88a. Similarly, the maximum deflection for middle strip in y-direction is denoted as Δ_{my}, as shown in Figure 12.88b. The slab deflection (Δ_{max}) is obtained by adding the maximum deflections for the column strip and the middle strip, as shown in Figure 12.88c. The same procedure can be applied by considering deflections for the column strip in y-direction (Δ_{cy}) and the middle strip in x-direction (Δ_{mx}).

For square panels (where $l_x/l_y = 1.0$), the maximum slab deflection can be determined from the following equation:

$$\Delta_{max} = \Delta_{cx} + \Delta_{my} = \Delta_{cy} + \Delta_{mx} \qquad [12.35]$$

where

Δ_{cx} and Δ_{cy} are the column strip deflections for x- and y-directions, respectively, and Δ_{mx} and Δ_{my} are the middle strip deflections for x- and y-directions, respectively.

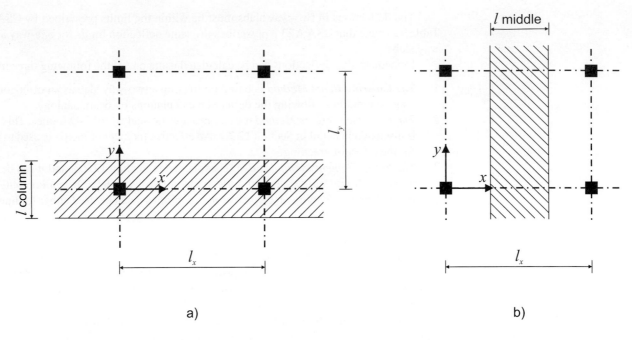

Figure 12.87   A slab panel for deflection calculations: a) column strip in x-direction, and b) middle strip in y-direction.

For rectangular panels, or for panels that have different properties in the two directions, an average deflection is calculated by considering the maximum deflections in both directions, as follows:

$$\Delta_{max} = \frac{\left(\Delta_{cx} + \Delta_{my}\right) + \left(\Delta_{cy} + \Delta_{mx}\right)}{2} \qquad [12.36]$$

The calculated deflections should be compared with the deflection limits specified in CSA A23.3 Table 9.3. Note that the deflection check for a slab panel should be performed by considering the span length (l) measured diagonally between columns, that is,

$$l = \sqrt{l_x^2 + l_y^2}$$

Deflection Calculations for Column Strip and Middle Strip   The column strip and the middle strip deflections can be determined according to the procedure for continuous beams explained in Section 4.6.3. A typical slab span considered for the deflection calculations is shown in Figure 12.89a. Note that the deflections are restrained at the supports; this is a simplifying assumption which does not reflect actual behaviour of a column-supported slab. The span is modelled as a continuous beam with span l_n (clear span), as shown in Figure 12.89b. The deflection (Δ) can be calculated from the following equation:

$$\Delta = k\left(\frac{5}{48}\right)\frac{M_m l_n^2}{E_c I_e} \qquad [4.12]$$

where the coefficient k can be determined as follows

$$k = 1.2 - 0.2\frac{M_o}{M_m} \qquad [4.17]$$

Deflection for a specific span depends on the end moments (M_1 and M_2) and the midspan moment (M_m), as shown in Figure 12.89c. Moment gradient (M_o) depends on bending moments at the supports and the midspan, that is,

$$M_o = M_m + (M_1 + M_2)/2 \qquad [12.37]$$

Figure 12.88 The Crossing Beam Method for deflection calculations: a) deflections in the column strip in x-direction; b) deflections in the middle strip in y-direction, and c) combined deflections (courtesy of the American Concrete Institute).

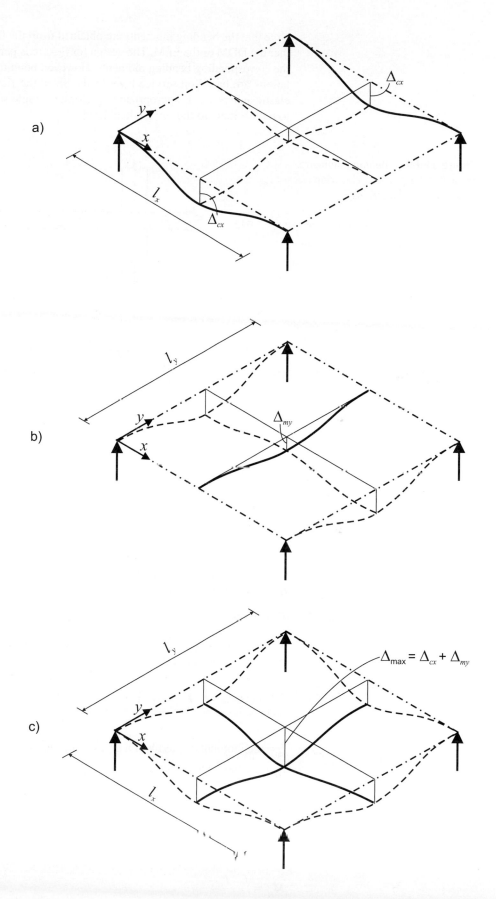

Note that the bending moments are obtained from the flexural design calculations using either the DDM or the EFM. The design for flexure is performed using the factored loads and the corresponding bending moments. However, bending moments for the deflection calculations are based on service level loads. Since the flexural design is performed using an elastic analysis, it is appropriate to perform a simple scaling of factored bending moments to reduce them to the service load level.

Figure 12.89   Bending moments for a slab strip: a) an elevation showing the deflections; b) cross-section for a column strip or middle strip section, and c) a bending moment distribution.

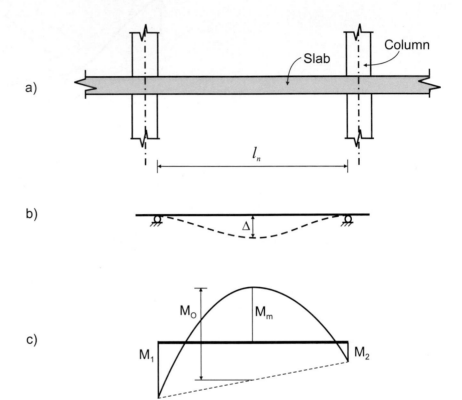

The modulus of elasticity for concrete (E_c) can be determined from the following equation (see Section 2.3.4):

A23.3 Eq. 8.2 $E_c = 4500\sqrt{f_c'}$ (MPa)   **[2.2]**

The effective moment of inertia (I_e) is determined in the same manner as for flexural members discussed in Section 4.3.5, that is,

A23.3 Eq. 9.1 $I_e = I_{cr} + \left(I_g - I_{cr}\right)\left(\dfrac{M_{cr}}{M_a}\right)^3 \leq I_g$ **[4.11]**

where

I_g = the moment of inertia of a gross concrete section

I_{cr} = the moment of inertia of the cracked section

Note that the moments of inertia I_g and I_{cr} are determined based on a wide beam section using either column strip or middle strip dimensions, as discussed above. The underlying equations are outlined in Sections 4.3.3 and 4.3.4.

The cracking moment (M_{cr}) is determined in the same manner as presented in Section 4.2, that is,

$$M_{cr} = \frac{f_r^* \cdot I_g}{y_t}$$ **[4.1]**

It should be noted that CSA A23.3-04 Cl.13.2.7 (revised in 2009) prescribes the use of a reduced modulus of rupture (f_r^*) equal to one-half of the f_r value for deflection calculations in two-way slabs, that is,

$$f_r^* = 0.5 \cdot f_r = 0.3\lambda\sqrt{f_c'} \text{ (MPa)} \tag{12.38}$$

where f_r denotes the modulus of rupture discussed in Section 2.3.2.

An average effective moment of inertia, $I_{e,avg}$, for continuous spans takes into account the moment of inertia at the supports and the midspan (Cl.9.8.2.4), as explained in Section 4.6.3. The following equations will be used for two-way slabs:

- For spans with two continuous ends (interior spans):

| A23.3 Eq. 9.3 |

$$I_{e,avg} = 0.7I_{e,m} + 0.15(I_{e1} + I_{e2}) \tag{4.18}$$

- For spans with one continuous end (end spans):

$$I_{e,avg} = 0.75I_{e,m} + 0.25I_{ec} \tag{12.39}$$

Note that the latter equation was proposed by CAC (2005) for column-supported end spans where partial fixity is provided by an exterior column; this equation is different from A23.3 Eq.9.4 which was used for deflection calculations in continuous reinforced concrete flexural members (see Eqn 4.19 in Chapter 4).

The designer is permitted to use other I_e values provided that the computed deflections are in reasonable agreement with the results of comprehensive tests (Cl.13.2.7).

The bending moment (M_a) is determined at the *service load level*, and its value depends on the load for which the deflection has been computed: dead load moment (M_D), live load moment (M_L), or dead plus live load moment (M_{D+L}). In the absence of detailed calculations, the bending moment due to the construction load (M_{const}) can be taken as twice the slab dead load for deflection calculations in multi-storey buildings, that is,

$$M_{const} = 2 \cdot M_D$$

Alternatively, the designer is referred to ACI 435R-95 (2003) for more details on construction load calculations, and also Scanlon and Supernant (2011) for a practical deflection calculation procedure which takes into account the construction loading history, time-dependent concrete properties, and the effect of cracking.

The effective moment of inertia (I_e) is a significant factor influencing the deflection magnitudes, therefore it is critical to use realistic I_e values for deflection calculations. The effective moment of inertia should be calculated based on the load level under consideration. For example, the dead load deflection (Δ_D) should be calculated using the I_e value based on the dead load moment (M_D), while the deflection due to combined dead plus live load (Δ_{D+L}) should be calculated using the I_e value based on the total load moment (M_{D+L}). As a result, two different I_e values need to be used for deflection calculations, as illustrated in Figure 12.90a; this approach will be referred to as the Standard Procedure in this section. However, construction loads usually govern over the combined dead and live load for deflection calculations in multi-storey flat slabs. It is expected that cracking takes place due to construction loads and its effect should be accounted for in deflection calculations. This can be accomplished by using the I_e value corresponding to the bending moment at the construction load level (M_{const}) for immediate deflection calculations, as shown in Figure 12.90b; this approach will be referred to as the Alternative Procedure. That procedure usually results in smaller live load deflections and larger immediate and sustained dead load deflections compared to the standard approach shown in Figure 12.90a. Note that this is not a CSA A23.3 requirement, but it was recommended by CAC (2005) and ACI 435R-95 (2003).

Immediate deflections for a typical span of a two-way slab can be computed according to the following procedure:

1. Calculate the dead load deflection (Δ_D) using the I_e value which corresponds to the dead load bending moment (M_D).

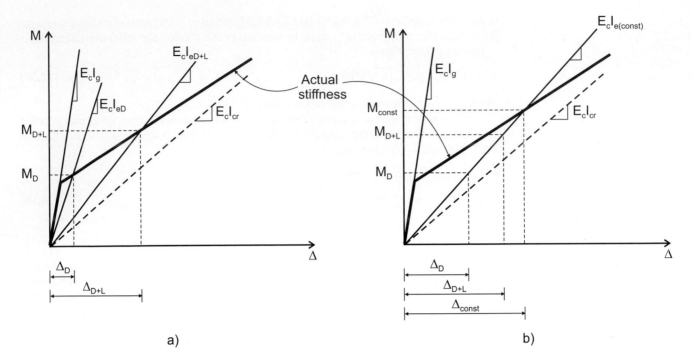

Figure 12.90   Effective moment of inertia for deflection calculations: a) Standard Procedure for flexural members, and b) Alternative Procedure for multi-storey flat slab construction
(adapted from ACI 435,1991 with the permission of the American Concrete Institute).

2.   Calculate the deflection (Δ_{D+L}) due to the combined dead and live load using the I_e value which corresponds to the total bending moment (M_{D+L}).

3.   The live load deflection (Δ_L) is calculated as follows:

$$\Delta_L = \Delta_{D+L} - \Delta_D$$

Alternatively, the procedure can take into account the effect of construction loading, as follows:

1.   Calculate the effective moment of inertia ($I_{e,const}$) corresponding to the bending moment at the construction load level (M_{const}).

2.   Calculate the dead load deflection (Δ_D) due to bending moment (M_D) using the I_e value calculated in Step 1.

3.   Calculate the deflection (Δ_{D+L}) due to combined dead and live loads using the bending moment (M_{D+L}) and the I_e value calculated in Step 1.

4.   The live load deflection (Δ_L) is calculated as follows:

$$\Delta_L = \Delta_{D+L} - \Delta_D$$

Note that both procedures use service (specified) bending moments, and not factored bending moments which are used for the strength design (e.g. moment and shear resistance).

Once the immediate deflections have been determined, a long-term deflection (Δ_t) can be determined from the following equation:

$$\Delta_t = \zeta_s \cdot \Delta_D + \Delta_L \qquad\qquad\qquad \textbf{[12.40]}$$

where the first term denotes deflection due to a sustained load which is magnified by a multiplier, ζ_s, and the second term denotes the deflection due to a live load; note that ζ_s should be calculated from Eqn. 4.15. This is the most basic case, where only the dead load is considered as a sustained load. In reality, a fraction of the live load can also be considered as a sustained load, and the long-term deflection can be determined according to the following procedure:

1. Find the deflection due to sustained load shortly after the construction is completed (e.g. installation of non-structural elements):

$$\Delta_{t1} = \zeta_{s1} \cdot (\Delta_D + \Delta_{LS})$$

2. Find the deflection due to the sustained load corresponding to the maximum creep effects:

$$\Delta_{t2} = \zeta_{s2} \cdot (\Delta_D + \Delta_{LS})$$

3. Finally, calculate the total long-term deflection due to the sustained load and the transient live load:

$$\Delta_t = (\Delta_{t2} - \Delta_{t1}) + (\Delta_L - \Delta_{LS}) \qquad \text{[12.41]}$$

where Δ_{LS} denotes deflection due to the sustained live load; this is usually expressed as a fraction of the live load, e.g. 20% of the live load can be sustained (this depends on the project requirements). See Section 4.4.3 for more details on long-term deflections.

The Crossing Beam Method for deflection calculations of two-way slabs is summarized in Checklist 12.3 and its application will be illustrated by two examples.

Checklist 12.3 Deflection Calculations for Two-Way Slabs According to the Crossing Beam Method

Step	Description	Code Clause
1	For a slab panel under consideration, select a column strip in one direction and a middle strip in the perpendicular direction (a flat plate/slab), or a beam strip in one direction and a slab strip in the perpendicular direction (a slab with beams). Perform the following calculations (steps 2 to 5) for each strip.	13.2.7
2	Find the moment of inertia values. Gross moment of inertia: $$I_g = \frac{b \cdot h_s^3}{12} \qquad \text{[4.2]}$$ Cracked moment of inertia: $$I_{cr} = \frac{b \cdot \overline{y}^3}{3} + n \cdot A_s \cdot (d - \overline{y})^2 \qquad \text{[4.10]}$$	
3	Determine the cracking moment: $$M_{cr} = \frac{f_r^* \cdot I_g}{y_t} \qquad \text{[4.1]}$$ where (f_r^*) is a reduced modulus of rupture: $$f_r^* = 0.5 \cdot f_r = 0.3\lambda\sqrt{f_c'} \text{ (MPa)} \qquad \text{[12.38]}$$	13.2.7
4	Compute the effective moment of inertia. A23.3 Eq.9.1 $\qquad I_e = I_{cr} + (I_g - I_{cr})\left(\dfrac{M_{cr}}{M_a}\right)^4 \le I_g \qquad \text{[4.11]}$	9.8.2.3

(Continued)

Checklist 12.3 Continued

Step	Description	Code Clause
5	Determine the immediate deflections due to service loads: $$\Delta = k\left(\frac{5}{48}\right)\frac{M_m l_n^2}{E_c I_e} \qquad \textbf{[4.12]}$$ where $$k = 1.2 - 0.2\frac{M_o}{M_m} \qquad \textbf{[4.17]}$$ and $$M_o = M_m + (M_1 + M_2)/2 \qquad \textbf{[12.37]}$$	
5a	Find the dead load deflection (Δ_D) using bending moments (M_m) and (M_o), and the effective moment of inertia I_e due to dead load (D).	
5b	Find the total (dead plus live) load deflection (Δ_{D+L}) using bending moments (M_m) and (M_o), and the effective moment of inertia I_e due to the total load ($D + L$).	
5c	Finally, the live load deflection (Δ_L) is calculated as follows: $$\Delta_L = \Delta_{D+L} - \Delta_D$$	
6	Calculate immediate deflections for the slab panel by combining the column strip and the middle strip live load deflections (see Figure 12.88): $$\Delta_{max} = \Delta_{cx} + \Delta_{my} = \Delta_{cy} + \Delta_{mx} \qquad \textbf{[12.35]}$$	
7	Calculate the maximum long-term deflections.	9.8.2.5
7a	Find the deflection due to sustained load shortly after the construction has been completed (e.g. installation of non-structural elements): $$\Delta_{t1} = \zeta_{s1} \cdot (\Delta_D + \Delta_{LS})$$	
7b	Find the deflection due to sustained load corresponding to the maximum creep effects: $$\Delta_{t2} = \zeta_{s2} \cdot (\Delta_D + \Delta_{LS})$$	
7c	Calculate the total long-term deflection due to sustained load and transient live load: $$\Delta_t = (\Delta_{t2} - \Delta_{t1}) + (\Delta_L + \Delta_{LS}) \qquad \textbf{[12.41]}$$	
8	Check whether deflections are within the limits prescribed by CSA A23.3.	9.8.5.3 (Table 9.3)

Example 12.11

Two-Way Flat Plate - Deflection Calculations According to the Crossing Beam Method Using the Standard Procedure

Consider a floor plan of a two-way slab system without beams (flat plate) which was designed for flexure in Example 12.1 (DDM) and Example 12.3 (EFM). Consider a design with 160 mm slab thickness (note that the previous examples used 180 mm slab thickness).

Use the Crossing Beam Method to find the immediate deflections for an end panel between column gridlines 1 and 2 in N-S direction, and gridlines B and C in E-W direction. Check whether immediate and long-term deflections are within the limits prescribed by CSA A23.3-04. Consider live load deflection limits for an occupancy where non-structural elements are not likely to be damaged by large deflections.

For long-term deflections, consider that non-structural elements have been installed after one month, and that 20% of the live load has been sustained.

Use the effective moment of inertia at the following two levels: i) the dead load, and ii) the total dead plus live load level.

Given: $f_c' = 30$ MPa

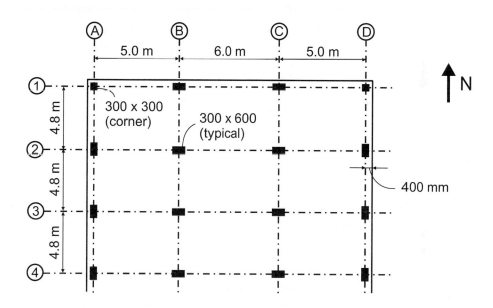

SOLUTION:

1. Check the slab thickness requirements (CSA A23.3 Cl.13.2.3).

This check was performed in Example 12.1, and it was concluded that the minimum 180 mm thickness is required in order to satisfy the indirect approach for deflection control, that is, deflection calculations are not required provided that the thickness exceeds 180 mm. However, the slab thickness is reduced to 160 mm in this example and detailed deflection calculations are required. Since the slab satisfies the CSA A23.3 requirements for regular slabs specified by CSA A23.3 Cl.2.2, it is possible to use the Crossing Beam Method.

2. Identify the design strips for deflection calculations.

It is necessary to calculate the deflections for an end panel between column gridlines 1 and 2 in N-S direction, and gridlines B and C in E-W direction. Let us consider column strip along gridline 2 (span BC), and a middle strip spanning between gridlines 1 and 2, as shown on the following sketch.

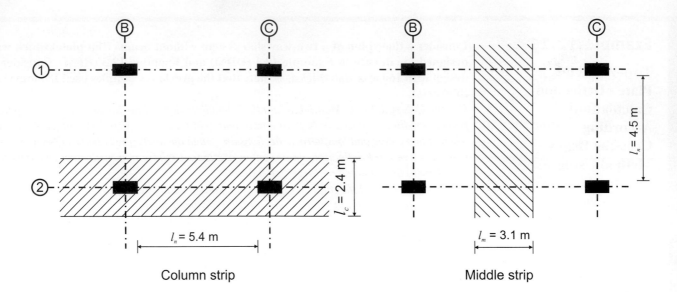

Column strip   Middle strip

The width of column strip for gridline 2 is equal to

$l_c = 2.4$ m (as discussed in Example 12.1).

The width of middle strip spanning between gridlines 1 and 2 can be determined as follows (see Figure 12.15):

$l_1 = 4.8$ m

$l_2 = (6.0 + 5.0)/2 = 5.5$ m design strip

Since $l_1 < l_2$, it follows that the column strip width is equal to

$l_c = l_1/2 = 4.8$ m$/2 = 2.4$ m

thus the width of middle strip is

$l_m = l_2 - l_c = 5.5 - 2.4 = 3.1$ m

3. **Determine the factored bending moments for the column strip and the middle strip.**

This step is based on Example 12.1, where the design was performed according to the DDM. However, the moments need to be recalculated due to different loads.

a) Perform the load analysis:

The slab's self-weight:

$DL_w = h \times \gamma_w = 0.16$ m $\times$ 24 kN/m^3 = 3.84 kPa

The superimposed dead load:

$DL_s = 1.44$ kPa

Live load:

$LL_w = 3.6$ kPa

The total factored load:

$w_f = 1.25(DL_w + DL_s) + 1.5 \times LL_w = 12.0$ kPa

Note that the factored load is smaller than that used in Example 12.1 (12.6 kPa) due to a reduced slab thickness (slab thickness in this example is 160 mm, which is less than the 180 mm thickness used in Example 12.1).

b) Calculate the factored bending moments for the column strip.

The moments are going to be calculated according to the DDM. Refer to Step 5 in Example 12.1.

Clear span:

$$l_n = 6.0 - (\frac{0.6}{2} + \frac{0.6}{2}) = 5.4 \text{ m}$$

Total factored static moment:

A23.3 Eq. 13.23

$$M_o = \frac{w_f \times l_{2a} \times l_n^2}{8} = \frac{12.0 \text{ kPa} \times 4.8 \text{ m} \times (5.4 \text{ m})^2}{8} = 210 \text{ kNm}$$ [12.8]

Note that the total factored moment in this example is slightly less than the moment obtained in Example 12.1 (220 kNm) - again, this is due to the smaller slab thickness.

Table 12.21 Factored bending moments for the column strip (Interior Span BC)

		Interior Span BC: $M_o = 210$ kNm		
		B Negative moment M_1 (kNm)	Midspan Positive moment M_m (kNm)	C Negative moment M_2 (kNm)
Longitudinal direction	Bending moments at critical sections	$-0.65M_o$ $= (-0.65) \times 210$ $= -137$	$+0.35M_o$ $= (+0.35) \times 210$ $= +73$	$-0.65M_o$ $= (-0.65) \times 210$ $= -137$
Transverse distribution - column strip	CSA A23.3 Provisions	$-(0.46 \text{ to } 0.59)M_o$	$+(0.19 \text{ to } 0.23)M_o$	$-(0.46 \text{ to } 0.59)M_o$
	Proposed value Design moment	$-0.59M_o$ $-0.59 \times (210) = -124$	$+0.23M_o$ $+0.23 \times 210 = +48$	$-0.59M_o$ $-0.59 \times (210) = -124$
Transverse distribution - middle strip	Design moment	$=-137-(-124) = -13$	$=73-48 = +25$	$=-137-(-124) = -13$

c) Calculate factored bending moments for the middle strip.

Clear span:

$$l_n = 4.8 - (\frac{0.3}{2} + \frac{0.3}{2}) = 4.5 \text{ m}$$

Total factored static moment:

$$M_o = \frac{w_f \times l_{2a} \times l_n^2}{8} = \frac{12.0 \text{ kPa} \times 5.5 \text{ m} \times (4.5 \text{ m})^2}{8} = 167 \text{ kNm}$$

Table 12.22 Factored bending moments for the middle strip (Span 1-2)

		End Span 1-2: $M_o = 167$ kNm		
		B Negative moment M_1 (kNm)	Midspan Positive moment M_m (kNm)	C Negative moment M_2 (kNm)
Longitudinal direction	Bending moments at critical sections	$-0.26M_o$ $= (-0.26) \times 167$ $= -43$	$+0.52M_o$ $= (+0.52) \times 167$ $= +87$	$-0.70M_o$ $= (-0.70) \times 167$ $= -117$
Transverse distribution - column strip	CSA A23.3 Provisions	$-0.26M_o$	$+(0.29 \text{ to } 0.34)M_o$	$-(0.49 \text{ to } 0.63)M_o$
	Proposed value Design moment	$-0.26M_o$ -43	$+0.3M_o$ $+0.3 \times 167 = +50$	$0.5M_o$ $-0.5 \times (167) = -84$
Transverse distribution - middle strip	Design moment	0	$=87-50 = +37$	$=-117-(-84) = -33$

4. **Perform deflection calculations for the column strip and the middle strip.**

Deflection calculation procedure as presented in this example can be presented in a tabular form. The key equations are summarized below.

a)   Material properties

The modulus of elasticity of concrete (E_c):

$$E_c = 4500\sqrt{f_c'} = 4500\sqrt{30} = 24650 \text{ MPa} \qquad \textbf{[2.2]}$$

Modular ratio:

$$n = \frac{E_s}{E_c} = \frac{200000}{24648} = 8.1 \qquad \textbf{[4.7]}$$

Reduced modulus of rupture (f_r^*):

$$f_r^* = 0.5 \cdot f_r = 0.3\lambda\sqrt{f_c'} = 0.3 \cdot 1.0 \cdot \sqrt{30} = 1.64 \text{ MPa} \qquad \textbf{[12.38]}$$

b)   Cross-sectional dimensions (see the sketch below)

Column strip:

b = 2400 mm

Middle strip:

b = 3100 mm

Slab thickness:

h_s = 160 mm

Effective depth (based on a 25 mm average cover and 15M rebar size):

d = 120 mm

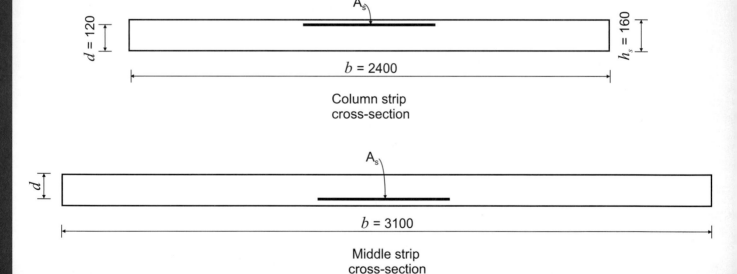

Column strip
cross-section

Middle strip
cross-section

b)   Section properties for deflection calculations (see Section 4.3)

• Gross moment of inertia for a rectangular section:

$$I_g = \frac{b \cdot h_s^3}{12} \qquad \textbf{[4.2]}$$

• Cracked section properties:
i)   Reinforcement ratio:

$$\rho = \frac{A_s}{b \cdot d} \qquad \textbf{[3.1]}$$

ii) Neutral axis depth for the cracked section:

$$\bar{y} = d\left(\sqrt{(n\rho)^2 + 2n\rho} - n\rho\right) \qquad [4.9]$$

iii) Moment of inertia for the cracked section:

$$I_{cr} = \frac{b \cdot \bar{y}^3}{3} + n \cdot A_s \cdot (d - \bar{y})^2 \qquad [4.10]$$

- The cracking moment (M_{cr}):

$$M_{cr} = \frac{f_r^* \cdot I_g}{y_t} \qquad [4.1]$$

where y_t is the distance from the centroid of the section to the extreme tension fibre. For a rectangular slab section it follows that

$$y_t = \frac{h_s}{2} = \frac{160}{2} = 80 \text{ mm}$$

- The effective moment of inertia (I_e):

A23.3 Eq. 9.1

$$I_e = I_{cr} + \left(I_g - I_{cr}\right)\left(\frac{M_{cr}}{M_a}\right)^3 \le I_g \qquad [4.11]$$

- Average effective moment of inertia
  i) For an interior span (e.g. column strip for span BC):

$$I_{e,avg} = 0.7I_{e,m} + 0.15(I_{e1} + I_{e2}) \qquad [4.18]$$

  ii) For an end span (e.g. middle strip for span 1–2) :

$$I_{e,avg} = 0.75I_{e,m} + 0.25I_{ec} \qquad [12.39]$$

c) Deflection calculation equations

The deflections need to be calculated for the dead load and the combined dead plus live load. The maximum deflection for a column strip or a middle strip can be determined from the following equation:

$$\Delta = k\left(\frac{5}{48}\right)\frac{M_m l_n^2}{E_c I_e} \qquad [4.12]$$

where

$$k = 1.2 - 0.2\frac{M_o}{M_m} \qquad [4.17]$$

and

$$M_o = M_m + (M_1 + M_2)/2 \qquad [12.37]$$

Note that the factored bending moments for column strips and middle strips were calculated in Step 3. Since the service load moments are required for deflection calculations, it is required to prorate the factored bending moment. In this example, two different levels are used: the dead load and the total dead plus live load. The following scaling factors are used to find service level bending moments:

Dead load deflection: $(DL_w + DL_s)/w_f = 5.28 \text{ kPa}/12.0 \text{ kPa} = 0.44$

Dead plus live load deflection: $[(DL_w + DL_s) + LL_w]/w_f = 8.88 \text{ kPa}/12.0 \text{ kPa} = 0.74$

The deflection calculation procedure is as follows.

1. Calculate the dead load deflection (Δ_D) due to the dead load bending moment (M_D).

2. Calculate the deflection (Δ_{D+L}) due to the combined dead and live load corresponding to the total bending moment (M_{D+L}).

3. Find the immediate live load deflection (Δ_L) :

$$\Delta_L = \Delta_{D+L} - \Delta_D$$

Deflection calculations are presented in the table below (see the sketch showing notation related to the bending moments). Note that the design of flexural reinforcement has been omitted from this example (refer to Example 12.1 for the reinforcement calculation).

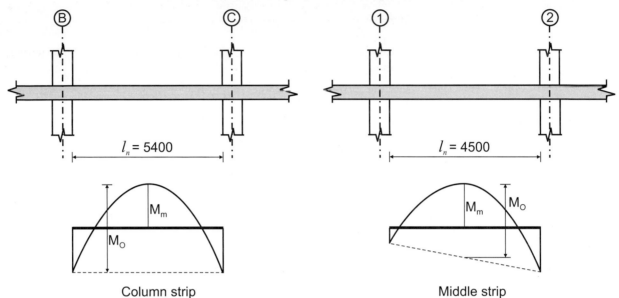

Column strip   Middle strip

Table 12.23 Immediate deflection calculations for the column strip and the middle strip

		Column Strip			**Middle Strip**		
		Support B	Midspan	Support C	Support 1	Midspan	Support 2
l_n	(mm)		5400			4500	
M_f	($\times 10^6$ Nmm)	124.0	48.0	124.0	0	37.0	33.0
A_s	(mm²)	3200	1200	3200	3200	1200	3200
ρ		0.011	0.004	0.011	0.009	0.003	0.009
$\bar{y}$	(mm)	41	27	41	37	24	37
I_{cr}	($\times 10^6$ mm⁴)	217	100	217	231	104	231
I_g	($\times 10^6$ mm⁴)	819	819	819	1060	1060	1060
M_{cr}	($\times 10^6$ Nmm)	16.8	16.8	16.8	21.7	21.7	21.7
M_D	($\times 10^6$ Nmm)	54.6	21.1	54.6	0	16.3	14.5
$M_{o,D}$	($\times 10^6$ Nmm)	75.7	75.7	75.7	23.5	23.5	23.5
k_D		0.48	0.48	0.48	0.91	0.91	0.91
I_{eD}	($\times 10^6$ mm⁴)	230	460	230	0	1060	1060
I_{eDavg}	($\times 10^6$ mm⁴)	390	390	390	1060	1060	1060
Δ_D	(mm)		3.2			1.2	
M_{D+L}	($\times 10^6$ Nmm)	92.0	36.0	92.0	0	27.0	24.0
$M_{o, D+L}$	($\times 10^6$ Nmm)	127.0	127.0	127.0	39.6	39.6	39.6
k_{D+L}		0.48	0.48	0.48	0.91	0.91	0.91
I_{eD+L}	($\times 10^6$ mm⁴)	220	180	220	0	580	810
$I_{eD+Lavg}$	($\times 10^6$ mm⁴)	190	190	190	640	640	640
Δ_{D+L}	(mm)		11.2			3.4	
Δ_L	(mm)		8.0			2.2	

5. **Check whether immediate deflections are within the CSA A23.3 limits.**

CSA A23.3 limits immediate live load deflections for an occupancy where the roof or floor construction is supporting or is attached to non-structural elements which are not likely to be damaged by large deflections (CSA A23.3 Table 9.3): $l_n/360$.

Live load deflections are determined by the following equation (see Step 4):

$$\Delta_L = \Delta_{D+L} - \Delta_D$$

The check is performed for the column strip, the middle strip, and the slab panel. Span length for the slab panel is determined as follows:

$$l_n = \sqrt{(5.4\ \text{m})^2 + (4.5\ \text{m})^2} = 7.0\ \text{m}$$

The deflection calculations are summarized below.

Table 12.24 Immediate deflections: summary calculations

		Column Strip	Middle Strip	Slab Panel
l_n	(mm)	5400	4500	7000
Δ_D	(mm)	3.2	1.2	=3.2+1.2=4.4
Δ_{D+L}	(mm)	11.2	3.4	=11.2+3.4=14.6
Δ_L	(mm)	8.0	2.2	=14.6−4.4=10.2
CSA A.23.3 deflection limit: $l_n/360$	(mm)	5400/360 = 15	4500/360 = 12	7000/360 = 19
Deflection check	(mm)	8.0<15.0 OK	2.2<12.0 OK	10.2<19.0 OK

6. **Check whether long-term deflections are within the CSA A23.3 limits.**

Long-term deflections will be calculated separately for the column strip and the middle strip. It is assumed that 20% of the live load is sustained, and the corresponding deflection is

$$\Delta_{LS} = 0.2\Delta_L$$

The procedure has the three steps, and it is presented in a tabular form. A sample calculation for the column strip is shown below.

a) Find the sustained deflections after 1 month.

Calculate $s = 0.5$ by interpolation from Table 4.1. Therefore,

$$\zeta_{s1} = 1 + \frac{s}{1+50\rho'} = 1 + \frac{0.5}{1+0} = 1.5 \qquad \text{[4.15]}$$

where $\rho' = 0$ since there is no compression reinforcement. Finally,

$$\Delta_{t1} = \zeta_{s1} \cdot (\Delta_D + \Delta_{LS}) = 1.5(3.2 + 0.2 \cdot 8.0) = 7.2\ \text{mm}$$

b) Find the sustained deflections after 5 years

Use $s = 2$ from Table 4.1, and calculate $\zeta_{s2} = 3$ from equation [4.15]. Finally,

$$\Delta_{t2} = \zeta_{s2} \cdot (\Delta_D + \Delta_{LS}) = 3.0(3.2 + 0.2 \cdot 8.0) = 14.4\ \text{mm}$$

c) Find the final long-term deflection.

$$\Delta_t = (\Delta_{t2} - \Delta_{t1}) + (\Delta_L - \Delta_{LS}) = (14.4 - 7.2) + (8.0 - 0.2 \cdot 8.0) = 13.6\ \text{mm} \qquad \text{[12.41]}$$

Once the column strip and middle strip deflections have been calculated, the long-term deflections for the slab panel can be calculated by summing up these two values.

Table 12.25 Long-term deflections: summary calculations

		Column Strip	Middle Strip	Slab Panel
l_n	(mm)	5400	4500	7000
Δ_D	(mm)	3.2	1.2	
Δ_L	(mm)	8.0	2.2	
Sustained deflections after 1 month $s = 0.5$ $\zeta_{s1} = 15$ $\Delta_{t1} = \zeta_{s1} \cdot (\Delta_D + \Delta_{LS})$	(mm)	7.2	2.5	
Sustained deflections after 5 years $s = 2$ $\zeta_{s2} = 3$ $\Delta_{t2} = \zeta_{s2} \cdot (\Delta_D + \Delta_{LS})$	(mm)	14.4	4.9	
$\Delta_t = (\Delta_{t2} - \Delta_{t1}) + (\Delta_L + \Delta_{LS})$	(mm)	13.6	4.2	=13.6+4.2=17.8
CSA A23.3 deflection limit: $l_n/240$	(mm)	5400/240 = 22	4500/240 = 19	7000/240 = 29
Deflection check	(mm)	13.6<22 OK	4.2<19 OK	17.8<29 OK

Note that the CSA A23.3 limits for immediate and long-term deflections may be different. In this example, the limit for immediate live load deflection is $l_n/360$, while the long-term deflection limit which includes the effect of sustained and transient loads is $l_n/240$. Both limits apply to roof or floor construction supporting or attached to non-structural elements not likely to be damaged by large deflections.

Example 12.12

Two-Way Flat Plate - Deflection Calculations According to the Crossing Beam Method and an Alternative Procedure to Account for the Effect of Construction Loads

Consider a floor plan of a two-way slab system without beams (flat plate) from Example 12.11.

Use the Crossing Beam Method to calculate immediate and long-term deflections for the slab panel, however use the effective moment of inertia at the construction load level for all deflection components. Assume that the construction load is equal to twice the dead load.

SOLUTION: The approach for solving this problem is similar to that taken in Example 12.11, with the following exceptions:

1) The effective moment of inertia for dead load, I_{eD}, was found assuming that the moment due to dead load, M_D, is twice the actual value. This calculation is illustrated below, for a column strip (bending moment at support B):

$$M_a = 2 \cdot M_D = 2 \cdot (54.6 \cdot 10^6) = 109.2 \cdot 10^6 \text{ Nmm}$$

$$I_{eD} = I_{cr} + \left(I_g - I_{cr}\right)\left(\frac{M_{cr}}{M_a}\right)^3 \qquad \text{[4.11]}$$

$$= 217\cdot10^6 + \left(819\cdot10^6 - 217\cdot10^6\right)\left(\frac{16.8\cdot10^6}{109.2\cdot10^6}\right)^3 \cong 220\cdot10^6 \ mm^4$$

Note that the deflections due to dead load are calculated using the actual M_D value (equal to 54.6×10^6 Nmm for the column strip at support B).

2) The effective moment of inertia for the total dead plus live load, I_{eD+L}, is the same I_{eD} as discussed above (note that two different values were used in Example 12.11).

The deflection calculation procedure was revised to take into account the effect of construction loads by considering the effective moment of inertia corresponding to the assumed cracked value, as shown in Figure 12.90.

Table 12.26   Immediate deflection calculations for the column strip and the middle strip considering the effect of construction loads

		Column Strip			Middle Strip		
		Support B	Midspan	Support C	Support 1	Midspan	Support 2
$\frac{n}{l}$	(mm)		5400			4500	
M_f	($\times10^6$ Nmm)	124.0	48.0	124.0	0	37.0	33.0
A_s	(mm²)	3200	1200	3200	3200	1200	3200
ρ		0.011	0.004	0.011	0.009	0.003	0.009
$\bar{y}$	(mm)	41	27	41	37	24	37
I_{cr}	($\times10^6$ mm⁴)	217	100	217	231	104	231
I_g	($\times10^6$ mm⁴)	819	819	819	1060	1060	1060
M_{cr}	($\times10^6$ Nmm)	16.8	16.8	16.8	21.7	21.7	21.7
M_D	($\times10^6$ Nmm)	54.6	21.1	54.6	0	16.3	14.5
$M_{o,D}$	($\times10^6$ Nmm)	75.7	75.7	75.7	23.5	23.5	23.5
k_D			0.48			0.91	
I_{eD}	($\times10^6$ mm⁴)	220	150	220	0	386	576
I_{eDavg}	($\times10^6$ mm⁴)	170	170	170	430	430	430
Δ_D	(mm)		7.5			2.9	
M_{D+L}	($\times10^6$ Nmm)	92.0	36.0	92.0	0	27.0	24.0
$M_{o,D+L}$	($\times10^6$ Nmm)	127.0	127.0	127.0	39.6	39.6	39.6
k_{D+L}			0.48			0.91	
I_{eD+L}	($\times10^6$ mm⁴)	220	150	220	0	386	576
$I_{eD+Lavg}$	($\times10^6$ mm⁴)	170	170	170	430	430	430
Δ_{D+L}	(mm)		12.6			4.9	
Δ_L	(mm)		5.1			2.0	

According to the approach presented in this example, the same effective moment of inertia (I_e) was used to calculate the dead load deflection (Δ_D) and the deflection due to total load (Δ_{D+L}). Therefore, we could have found (Δ_L) directly by finding the live load moment (M_L) and solving for (Δ_L). The deflection calculations are summarized in Table 12.27.

Table 12.27 Immediate deflections: summary calculations

		Column Strip	Middle Strip	Slab Panel
l_n	(mm)	5400	4500	7000
Δ_D	(mm)	7.5	2.9	=7.5+2.9=10.4
Δ_{D+L}	(mm)	12.6	4.9	=12.6+4.9=17.5
Δ_L	(mm)	5.1	2.0	=17.5−10.4=7.1
CSA A23.3 deflection limit: $l_n/360$	(mm)	5400/360 = 15	4500/360 = 12	7000/360 = 19
Deflection check	(mm)	5.1<15.0 OK	2.0<12.0 OK	7.1<19.0 OK

Table 12.28 Long-term deflections: summary calculations

		Column Strip	Middle Strip	Slab Panel
l_n	(mm)	5400	4500	7000
Δ_D	(mm)	7.5	2.9	
Δ_L	(mm)	5.1	2.0	
Sustained deflections after 1 month $s = 0.5$ $\zeta_{s1} = 1.5$ $\Delta_{t1} = \zeta_{s1} \cdot (\Delta_D + \Delta_{LS})$	(mm)	12.8	5.0	
Sustained deflections after 5 years $s = 2$ $\zeta_{s2} = 3$ $\Delta_{t2} = \zeta_{s2} \cdot (\Delta_D + \Delta_{LS})$	(mm)	25.6	10.0	
$\Delta_t = (\Delta_{t2} - \Delta_{t1}) + (\Delta_L + \Delta_{LS})$	(mm)	16.9	6.6	=16.9+6.6=23.5
CSA A23.3 deflection limit: $l_n/240$	(mm)	5400/240 = 22	4500/240 = 19	7000/240 = 29
Deflection check	(mm)	16.9<22 OK	6.6<19 OK	23.5<29 OK

12.10.4 Deflection Calculations Using the Computer-Aided Iterative Procedure and 2-D Equivalent Frames

The computer-aided iterative procedure uses structural analysis of 2-D equivalent frames to determine deflections in two-way slabs. Each member is divided into several segments characterized by different section properties. Cracked section properties are used where the bending moment exceeds the cracking moment, while gross section properties are used elsewhere along the span. The procedure is described in Section 4.6 as related to reinforced concrete flexural members. An application of this procedure to a two-way flat plate system is illustrated through the following example.

Example 12.13

Two-Way Flat Plate - Deflection Calculations Using The 2-D Computer- Aided Iterative Procedure

Consider the flat plate floor system discussed in Example 12.11.

Compute immediate deflections using the computer-aided iterative procedure.

Given: $f_c' = 30$ MPa

$E_c = 24650$ MPa

SOLUTION: In order to solve this problem, it is required to perform an analysis of two 2-D equivalent frame models: i) an equivalent frame along gridline 2 (between gridlines B and C), and ii) an equivalent frame along gridline B (between gridlines 1 and 2). The iterative analysis will be explained in detail for the equivalent frame along gridline 2.

1. **Find the cross-sectional properties required for the analysis.**

   a) Determine cross-sectional dimensions for the slab.

   The equivalent frame has the same properties as discussed in Example 12.3. The slab-beam has the width equal to the design strip, that is,

   $b = 4800$ mm

   and

   $h_s = 160$ mm

   b) Find the gross moment of inertia for the frame section.

   $$I_g = \frac{b \cdot h_s^3}{12} = \frac{4800 \cdot (160)^3}{12} = 16.4 \cdot 10^8 \text{ mm}^4 \qquad [4.2]$$

   c) Find the cracking moment (M_{cr}).

   $$M_{cr} = \frac{f_r^* \cdot I_g}{y_t} = \frac{1.64 \cdot (16.4 \cdot 10^8)}{80} = 33.5 \text{ kNm} \qquad [4.1]$$

   where the reduced modulus of rupture is $f_r^* = 0.5 \cdot f_r = 1.64$ MPa and $y_t = 80$ mm

   d) Find the cracked moment of inertia:

   $$I_{cr} = \frac{b \cdot \bar{y}^3}{3} + n \cdot A_s \cdot (d - \bar{y})^2 \qquad [4.10]$$

   The cracked moment of inertia is determined in the same manner as in Example 12.11. Note that I_{cr} values are different along the slab span, depending on the amount of top and bottom reinforcement.

2. **Perform an elastic analysis of the frame using gross cross-sectional properties.**

   The frame will be modelled using prismatic slab-beam elements. Each span can be divided into 10 or 20 equal rigidly joined segments. Initially, all segments are assigned a gross moment of inertia (I_g) and an elastic analysis is performed. The resulting bending moment diagram for equivalent frame along gridline 2 is shown on the following sketch. Note that the cracking moment value is superimposed on the diagram to identify cracked regions where bending moment exceeds the cracking moment, that is, $M > M_{cr}$

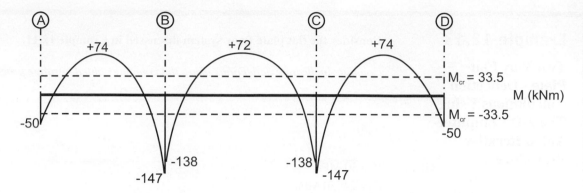

3. **Perform an iterative analysis using cracked and gross cross-sectional properties.**

The regions where bending moments exceed the cracking moment need to be identified, and the cracked moment of inertia needs to be assigned to those segments. The revised cross-sectional properties are used to perform an analysis and to find bending moments. A subsequent analysis may identify new segments where bending moments have exceeded the cracking moment. As a result, cracked moment of inertia should be assigned to those segments. The analysis continues until there is no change in the status (cracked/uncracked) for slab segments. Note that, once the cracked slab regions have been identified by the analysis, those segments must be assigned cracked sectional properties for all subsequent analyses.

In this example, three iterations were performed before the convergence has been reached. The final bending moment distribution (shown with dashed lines) superimposed on the original bending moment diagram is shown next.

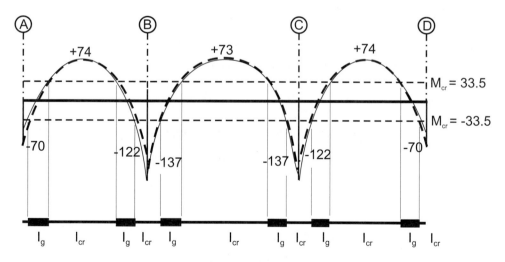

4. **Determine the slab deflections.**

Short-term deflections:

a)   The equivalent frame along gridline 2, between gridlines B and C:

Dead load: $\Delta_D = 6$ mm

Total dead+live load: $\Delta_{D+L} = 10$ mm

Live load: $\Delta_L = \Delta_{D+L} - \Delta_D = 10 - 6 = 4$ mm

b)   The equivalent frame along gridline B, between gridlines 1 and 2:

Dead load: $\Delta_D = 3$ mm

Total dead+live load: $\Delta_{D+L} = 5$ mm

Live load: $\Delta_L = \Delta_{D+L} - \Delta_D = 5 - 3 = 2$ mm

c)   Total deflections

Dead load: $\Delta_D = 6 + 3 = 9$ mm

Total dead+live load: $\Delta_{D+L} = 10 + 5 = 15$ mm

Live load: $\Delta_L = 4 + 2 = 6$ mm

12.10.5   Deflection Calculations Using the Computer-Aided Iterative Procedure and 3-D Finite Element Analysis

The underlying concept of this approach is similar to the 2-D iterative analysis, except that the slab system is modelled as a 3-D structure. The slab is modelled as a mesh of finite elements, as described in Section 12.7.3. The analysis is initially performed using gross cross-sectional properties, and the regions where cracking has taken place can be identified by the analysis software, depending on the cracking moment. A conservative estimate uses cracked section properties, however several iterations can be performed to reach convergence within a pre-defined tolerance. A few software packages are capable of performing an iterative 3-D analysis and deflection predictions for cracked two-way slabs. Before the solution is obtained, the reinforcement detailing of a slab system must be finalized. This method enables a more accurate prediction of long-term deflections compared to other methods. It is especially suitable for deflection predictions in slabs with a non-rectangular column grid, like the one discussed in 12.7.3, and the corresponding deflection contour diagram is shown in Figure 12.91.

Figure 12.91   Deflection contours for an irregular two-way slab.

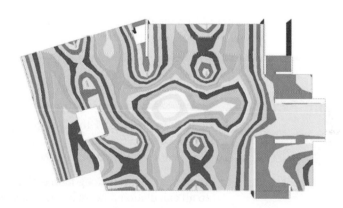

Example 12.14

Two-Way Flat Plate - Deflection Calculations Using The 3-D Finite Element Analysis

Consider the same flat plate floor system discussed in Example 12.11.

Calculate the immediate deflections using the 3-D Finite Element Analysis procedure.

SOLUTION:

The slab has been modelled as a mesh of finite elements and analyzed using a finite element software package which is able to consider the effect of cracking by performing an iterative computational analysis. The results of the analysis are presented in Figure 12.92.

It can be seen from the figure that the maximum dead load deflection for the slab panel between gridlines B and C and 1 and 2 is $\Delta_D = 6.7$ mm and the total dead plus live load deflection is $\Delta_{D+L} = 17.3$ mm. Finally, the immediate live load deflection is equal to:

$$\Delta_L = \Delta_{D+L} - \Delta_D = 17.3 - 6.7 = 10.6 \text{ mm}$$

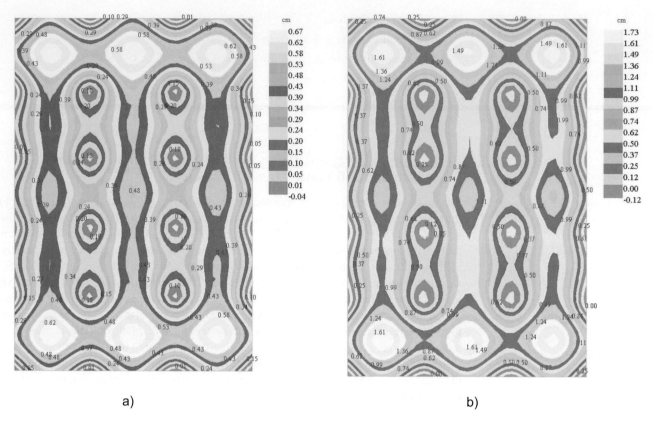

a)   b)

Figure 12.92 Deflection contours for a regular two-way flat slab: a) deflections due to dead load, and b) deflections due to the total load.

Learning from Examples

Previous examples (12.11 to 12.14) were useful to evaluate the results of different deflection calculation procedures. A summary of the results is presented in Table 12.29. A few conclusions are presented below:

1) Both immediate and long-term deflections are within the CSA A23.3 limits according to all calculation procedures, thus the general conclusion is the same irrespective of the method used.

2) Similar immediate live load deflection values are obtained for the Crossing Beam Method (Example 12.11) and the 3-D Finite Element Method (Example 12.14) (10.2 versus 10.6 mm). The value obtained from the Computer-Aided Iterative Procedure (Example 12.13) is the smallest of all (6.0 mm).

3) The results obtained from the Crossing Beam Method depend on whether the effect of construction load has been taken into account. For immediate live load deflections, the standard procedure used in Example 12.11 gives larger values (10.2 mm) compared to the alternative procedure which takes into account the effect of construction load (7.1 mm). However, note that the dead load deflection obtained by the standard procedure (4.4 mm) is significantly less than the value of 10.4 mm obtained from the alternative procedure. As a result, long-term deflections obtained from the alternative procedure (23.5 mm) are significantly higher (more conservative) that those obtained by the standard procedure (17.8 mm).

4) The 3-D FEA takes into account the effect of cracking and it is considered to be the most accurate of all methods. However, it is appropriate to use an approximate method such as the Crossing Beam Method to estimate deflections in regular slabs.

This comparison illustrates that the margin of error associated with deflection calculation could be large and it depends on the procedure used. The designer must have a good understanding of the governing principles behind various deflection calculation procedures.

Table 12.29   Deflection calculation procedures: a summary of the results for Examples 12.11 to 12.14

		Crossing Beam Method - Standard Procedure (Example 12.11)	Crossing Beam Method - Alternative Procedure (Example 12.12)	Computer - Aided Iterative Procedure (2-D) (Example 12.13)	3-D Finite Element Analysis (Example 12.14)	CSA A23.3 Deflection Limits
		Immediate deflections				
Δ_D	(mm)	4.4	10.4	9.0	6.7	
Δ_{D+L}	(mm)	14.6	17.5	15.0	17.3	$l_n/360 = 19$
Δ_L	(mm)	10.2	7.1	6.0	10.6	
		Long-term deflections				
Δ_t	(mm)	17.8	23.5	*	*	$l_n/240 = 29$

Note
* - The long-term deflections were not calculated for Examples 12.13 and 12.14.

12.10.6   Practical Guidelines for Deflection and Cracking Control

Deflection control is one of the key design considerations for two-way slabs. Several construction- or design-related options are available to ensure that deflections are within the CSA A23.3 allowable limits, and that the cracking is not excessive. The designer can ensure adequate deflection control for two-way slabs by means of the following construction procedures:

1. Specify additional construction procedures, such as delay strips and control joints, to allow the slab to shrink more freely during construction.
2. Specify a delayed formwork stripping and reshoring procedure to reduce the effect of creep.
3. Perform cambering of the slab by adjusting the formwork to have a camber or a "crown" at midspan regions. The purpose of cambering is to reduce the appearance of sag in the slab. The amount of camber is somewhat subjective, and can be as high as the total anticipated long-term deflection due to sustained loads. This strategy can be effective for reducing visual and functional effects of deflections in thin slabs, and particularly in slabs with a large dead/live load ratio.
4. Where long-term deflections are likely to induce significant strains in architectural building components, the problem can be mitigated by providing vertical slip joints in partitions and window walls. This measure does not restrict the deflections in the slab, but it prevents damage in non-structural elements.

Apart from design-related errors, some of the most common causes of excessive slab deflections are due to poor construction practices. Relevant construction considerations for two-way slabs are discussed in the next section.

12.11   CONSTRUCTION CONSIDERATIONS AND DRAWINGS

Efficiently designed two-way slabs are considered to be among the most cost-effective structural systems. The overall construction cost associated with two-way slabs is influenced by the amount of concrete and steel, and also labour and material costs associated with the formwork and shoring and reshoring. The labour costs are a significant component of two-way slab construction, thus a design that minimizes labour usage by taking into consideration simplicity and speed of the formwork erection and rebar placement often results in the most cost-efficient solution. An optimal design strikes the balance between the material usage and labour efficiency. To produce a safe and a cost-effective design solution, the designer needs to consider a few important constructability issues, such as i) simplify formwork requirements to speed up the construction process, ii) consider the shoring and reshoring in design, and ii) produce a simple reinforcement layout. These issues are discussed next.

Formwork When the design results in a labour-intensive formwork, the construction process is less efficient and the formwork cost is higher. Flat plates are easy to form, however forming the slabs with beams or drop panels/capitals is more difficult and time-consuming. When construction spced is important, the designer could attempt to simplify the formwork by designing a flat plate instead of a flat slab or a slab with beams. This may be done at the expense of a thicker slab and a larger amount of steel.

Shoring and reshoring After the concrete has been placed in the forms, formwork and shoring are required to support the slab before the concrete gains sufficient strength to be able to support the slab self-weight. Shoring must be adequately designed by considering the deflection and/or settlement tolerances in order to avoid additional stresses in the slab.

In multi-storey high-rise construction, contractors usually wish to pour one floor every few days, thus the formwork and the shoring need to be stripped once the concrete at a specific floor level has gained sufficient strength. The same process is repeated at each floor level. As the building construction progresses, a temporary construction load on each floor slab may exceed the permanent service (specified) design load. The problem is often compounded by the fact that the temporary construction load is placed on a partially cured slab.

The most vulnerable stage in the slab construction is after the shoring is removed and before the reshoring is installed, because young concrete hasn't reached sufficient design strength while being subjected to loads sometimes beyond the design service loads. Early age construction loading can lead to excessive immediate and long-term deflections in flat plates and flat slabs.

Construction sequencing must be considered in the design of flat slabs for multi-storey buildings. The designer may need to specify the maximum construction loads and work with the contractor to ensure that a proper amount of shoring and reshoring is in place in order to avoid overloading of slab during the construction. An example of a concrete building under construction is shown in Figure 12.93. The concrete is being placed onto the fly-form on the top floor, while partially cured slabs below have reshoring to help support the load from the young (wet) concrete.

Reinforcement placement and detailing Two-way slabs, especially slabs with complex geometry, are often designed using sophisticated computer-based analysis tools. Regardless of the analysis method used, the designer needs to exercise judgment while using the analysis results to prepare design drawings and specifications. The designer must use a good judgment by taking into account the underlying assumptions, and also recognize the approximations associated with the structural model. The designer can take advantage of the moment redistribution, discussed earlier in this chapter, to help even out negative bending moments in different spans. This could help produce similar bending moments and reinforcement for different spans. The resulting design solution and reinforcement details should be simple, repetitive, and uniform as much as possible. One of the key objectives is to ensure that the construction crew can easily understand the design drawings without confusion. For example, it is a good idea to specify reinforcement mats that are symmetrical over the columns. A repeated use of similar mats in both orthogonal directions is a good strategy for minimizing placement errors. In many cases, an optimal design solution includes an adequate consideration of practical construction constraints.

Figure 12.93 A concrete building under construction showing shoring and reshoring of flat slab floors (John Pao).

An example of a reinforcement plan for a flat plate designed in Examples 12.1 and 12.3 is presented in Figure 12.94.

It can be seen from the drawing that four different mats have been specified (A to D). Note that Mat A has been specified over all of the interior columns. The mat is symmetrical with regards to a column.

Additional relevant information concerning the reinforcement placement is provided on the notes and diagrams, such as the one shown in Figure 12.95. Note that the information related to anchorage is contained in the notes. For example, Note 6 specifies details of top reinforcement at end spans, which is usually hooked at one end (H1E).

A flat plate floor system under construction is shown in Figure 12.96.

It is important to ensure a reasonably accurate placement of reinforcement in compliance with the specified tolerances. Misplacement of reinforcement might have significant consequences upon the structural performance. For example, the placement of top reinforcement over columns at a lower level than specified by the design may lead to excessive cracking and rotation at the supports due to high bending moments, thereby resulting in increased midspan deflections.

The designer also needs to be familiar with the requirements related to the curing of concrete. Conformance to specified field curing as defined in the CSA A23.1 Tables 2 and 20 is essential to assure an adequate early strength gain. An insufficient early strength gain prior to stripping formwork might result in a significant reduction in the slab flexural stiffness and increased deflections.

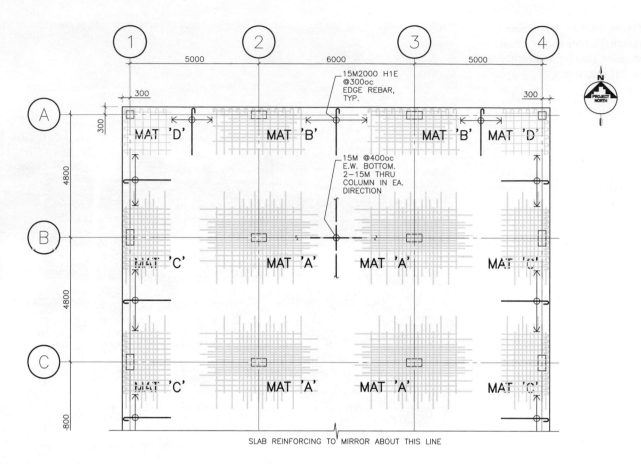

REINFORCEMENT PLAN

1:100

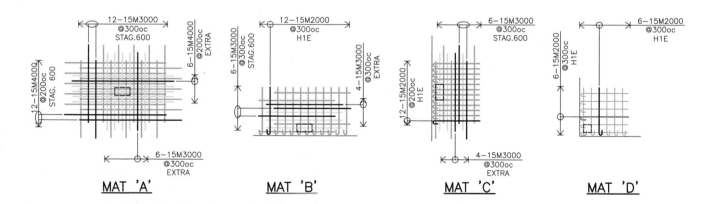

Figure 12.94 Reinforcement plan for a regular flat plate system.

NOTES:

1. THIS DRAWING TO BE READ IN CONJUNCTION WITH GENERAL NOTES ON DWG. S1.1 AND TYPICAL DETAILS ON DWG. S4.1, S4.2 AND S4.3. SEE ARCHITECTURAL DRAWINGS FOR ALL LAYOUT INFORMATION INCLUDING ELEVATIONS, COLUMN LOCATIONS, AND SLAB EDGE LOCATIONS.

2. REFER TO DWG. S3.1 FOR FOOTING, WALL AND COLUMN SCHEDULES.
REFER TO DWG. S3.2 FOR ZONE DETAILS.
REFER TO DWG. S3.3 FOR MAT DETAILS.

3. SLAB 7" THICK R/W 15M @12"oc E.W. BOTTOM U.N.O.

4. SPLICE BOTTOM REINF. WITH 3'-0" SPLICE AT COLUMN LINES ONLY. EXTEND BARS 18" PAST CENTRE OF SUPPORTING COLUMN.

5. BAR STAGGERS SHOWN ON PLAN ARE 24" U.N.O.

6. EDGE TOP HOOKED REINF. SHOWN ON PLAN 15M07'04 @12"oc H1E U.N.O.

7. DO NOT USE MAIN SLAB REINF. AS DROPPED CARRY BARS, ADD EXTRA REINF. FOR DROPPED CARRY BARS.

8. CRANK BARS SHOWN ON PLAN AS NEEDED TO SUIT SLAB SLOPES.

9. ⟨A⟩ DENOTES PUNCHING SHEAR REINFORCING SEE DWG. S3.3A

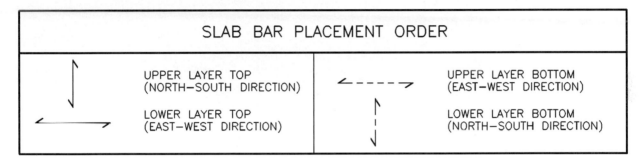

SLAB BAR PLACEMENT ORDER

UPPER LAYER TOP (NORTH-SOUTH DIRECTION)

LOWER LAYER TOP (EAST-WEST DIRECTION)

UPPER LAYER BOTTOM (EAST-WEST DIRECTION)

LOWER LAYER BOTTOM (NORTH-SOUTH DIRECTION)

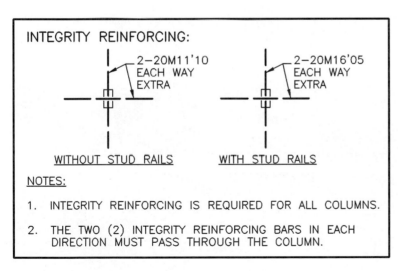

INTEGRITY REINFORCING:

2-20M11'10 EACH WAY EXTRA

2-20M16'05 EACH WAY EXTRA

WITHOUT STUD RAILS

WITH STUD RAILS

NOTES:

1. INTEGRITY REINFORCING IS REQUIRED FOR ALL COLUMNS.

2. THE TWO (2) INTEGRITY REINFORCING BARS IN EACH DIRECTION MUST PASS THROUGH THE COLUMN.

Figure 12.95 Notes related to reinforcement in two-way slabs.

Figure 12.96 A flat plate floor under construction: a) reinforcement mats (Ron Hopen), and b) hooked top reinforcement (Svetlana Brzev).

a)

b)

SUMMARY AND REVIEW — DESIGN OF TWO-WAY SLABS

Most two-way slab systems in reinforced concrete buildings are floor and roof slabs supported by columns and/or walls. These systems can be divided into one-way and two-way slabs depending on the manner in which they transfer gravity loads. One-way slabs transfer gravity loads in one direction, while two-way slabs are continuous structures spanning in two directions. Two-way slabs are easy to form and construct, and they are the most popular floor system for multi-storey building construction in Canada.

Types of two-way slabs

The four main types of two-way slab systems are as follows:

- *Flat plate* consists of solid slabs reinforced in two directions and supported directly by the columns or walls. This system is economical for short and medium spans. The slab thickness is usually controlled by long-term deflections.
- *Flat slab* is a column- or a wall-supported slab system with drop panels, that is, thickened regions of the slab centered at the columns. Drop panels are provided to increase the punching shear capacity of the slab and help to control deflections in midspan regions.
- *Waffle slab* consists of evenly spaced concrete joists spanning in two directions — this system is also known as a two-way joist system. The joists are formed by using standard pans or domes installed in the forms to produce a coffered soffit in the slab. Waffle slabs offer economical design solution for longer spans, and are particularly advantageous when the use of heavy loads is desired without the use of deepened drop panels, capitals, or support beams.
- *Slab with beams* consists of solid slab panels supported by beams on all four sides. When the ratio of the span lengths for a panel approaches 2.0, load is predominately transferred by bending in the short direction and the panel essentially acts as a one-way slab. As the panel approaches a square shape, a significant load is transferred by bending in both orthogonal directions, and the panel should be treated as a two-way slab.

Design of two-way slabs for flexure

The following four procedures can be used for flexural design of two-way slabs according to CSA A23.3:

1. Direct Design Method (DDM) is a statics-based method which can be used to design regular two-way slabs for flexure (Cl.13.9). The method treats a slab as a wide beam with a width equal to the tributary portion between column centrelines. Prescribed moment coefficient values can be used to determine bending moments at critical locations within a slab span (supports and midspan). The moment at each critical location is distributed transversely between column and middle strips in flat slabs and flat plates, or between beam and slab strips in slabs with beams.
2. Equivalent Frame Method (EFM) idealizes a 3-D building structure consisting of slabs and columns as a series of parallel 2-D equivalent frames in each principal direction of the building. Each equivalent frame is analyzed separately, and the results are combined to create a design solution for an entire slab at the floor level. CSA A23.3 Cl.13.8 refers to this method to as "slab systems as elastic frames".
3. Three-Dimensional Elastic Analysis idealizes a structure as a 3-D model, where slabs are modelled as 2-D (plate) finite elements, and columns are modelled as linear elements. This is a computer-based numerical analysis procedure based on the Elastic Plate Theory (Cl.13.6). This method is appropriate for design of complex slabs with irregular (nonrectangular) plan shapes, or slabs with regular plans characterized by large column and wall offsets relative to a rectangular grid.
4. The Theorems of Plasticity (Cl.13.7) are able to predict the ultimate load capacity of a slab: the Yield Line Method (YLM) and the Hillerborg Complemental give an upper and a lower-bound estimate respectively. The YLM can predict the ultimate load (load capacity) for a slab panel with flexure-controlled behaviour. The analysis is performed by applying the Virtual Work Method to compare possible yield patterns (failure scenarios) for a particular slab. The governing yield pattern is the one that gives the least load at the ultimate stage. When the ultimate load-bearing capacity in the slab has been reached,

plastic rotation occurs along yield lines, straight cracks across which the reinforcing bars have yielded, while regions between these yield lines act like rigid bodies.

Design of two-way slabs for shear

Shear failure in two-way slabs is sudden and must be carefully considered in the design. The two main CSA A23.3 shear design considerations are related to

- one-way shear (Cl.13.3.6) and
- two way shear (Cl.13.3.3 to 13.3.5).

The two-way shear (or punching shear) resistance is particularly critical for flat plates and flat slabs, due to chances of potentially catastrophic collapse associated with this failure mechanism. Shear stresses in two-way slabs are due to the combined effect of gravity loads and unbalanced bending moments transferred through slab-column connections. A fraction of bending moment at the slab-column connection is transferred through flexure and resisted by the flexural reinforcement, while the remainder is transferred through shear stresses at the critical perimeter. Shear resistance in two-way slabs is provided by concrete and reinforcement. Shear stresses in a slab with beams may be resisted jointly by the slab and the supporting beams.

Deflections in two-way slabs

Deflection control is one of the key design considerations for two-way slabs. Excessive deflections can cause damage to non-structural elements such as partitions and glazing, and noticeable deflections appear unsafe. Immediate and long-term deflections due to service loads must remain within the limits prescribed by CSA A23.3 Table 9.3.

CSA A23.3 prescribes the following two approaches for deflection control:

- Indirect approach permits the use of minimum slab thickness which results in a robust design, thus detailed deflection calculations are not required (Cl.13.2.2). This approach can be applied only to regular two-way slabs.
- Detailed deflection calculations must be performed for slabs with the span-to-thickness ratio below the CSA A23.3 limit.

Deflections in two-way slabs can be estimated by applying one the following three methods:

1. The Crossing Beam Method is based on treating a two-way slab as an orthogonal one-way system, thus allowing the deflection calculations by beam analogy.
2. The Equivalent Frame Method uses a linear elastic analysis of 2-D frames. An effective moment of inertia is calculated across the full width to account for cracking.
3. The Finite Element Method is a computer-based method which can be used to obtain both internal forces and deflections in two-way slabs. Depending on the software capabilities, it can be used either for linear elastic or nonlinear analysis which takes into account the effect of cracking.

PROBLEMS

Unless noted otherwise, the following material properties should be used: concrete $f_c' = 30$ MPa (normal density concrete) and steel $f_y = 400$ MPa.

12.1. Consider a 300 mm thick flat plate panel supported by walls on all sides shown in the figure. The slab is subjected to a 1000 kN point load at midspan (point P). Estimate the reactions at points A and C. Assume that imaginary 300 mm wide strips AB and CD carry the load simultaneously.

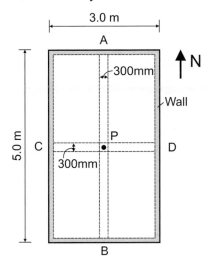

12.2. The slab of Problem 12.1 has been modified to include beam AB in the longitudinal direction, as shown in the figure. The slab is subjected to the same point load (1000 kN) as in Problem 12.1. Find the reactions at points A and C, by using the same strips AB and CD of Problem 12.1.

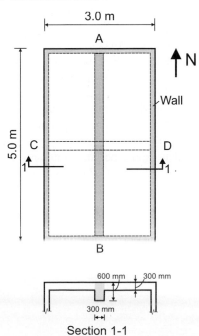

Section 1-1

12.3. Consider the flat plate system from Example 12.1. Use the Direct Design Method to determine design bending moments and reinforcement for an interior frame along gridline B. Use the same slab thickness (180 mm) as in Example 12.1. Consider unbalanced moments and moment transfer through slab-column connections in the design.

12.4. Redesign the slab of Problem 12.3 using the Equivalent Frame Method. Compare the bending moments and reinforcement obtained using the Equivalent Frame Method and the Direct Design Method.

12.5. A typical floor plan of a hospital building is shown in the figure. The floor system is a flat plate and it is subjected to specified live load (LL) of 3.6 kPa, and superimposed dead load (DL_s) of 1.5 kPa, in addition to its self-weight. Design a typical interior panel for flexure. Consider unbalanced moments and moment transfer through slab-column connections in the design. Select the slab thickness such that deflection check is not required according to CSA A23.3. Assume that edge beams are not provided.
a) Use the Direct Design Method.
b) Use the Equivalent Frame Method.
c) Compare the results (bending moments and reinforcement).

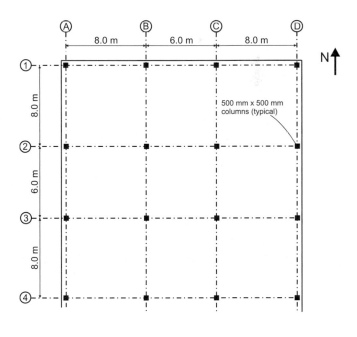

12.6. Redesign the slab from Problem 12.5 by considering the gravitational internal column locations only. Use minimum drop panel dimensions permitted by CSA A23.3. Revise the slab thickness based on the CSA A23.3 requirements for flat slabs. The slab plan dimensions

and loading are the same as in Problem 12.5. Design a typical interior panel for flexure.
a) Use the Direct Design Method.
b) Use the Equivalent Frame Method.
c) Compare the results (bending moments and reinforcement).

12.7. Design of the floor system from Problem 12.5 needs to be modified due to architectural constraints. Columns 2-A and 3-B need to be moved and are offset by 1 m relative to the gridlines, as shown in the figure. The slab dimensions and loading are the same as those in Problem 12.5. Design a typical interior panel for flexure.
a) Use the Direct Design Method.
b) Use the Equivalent Frame Method.
c) Compare the results (bending moments and reinforcement).

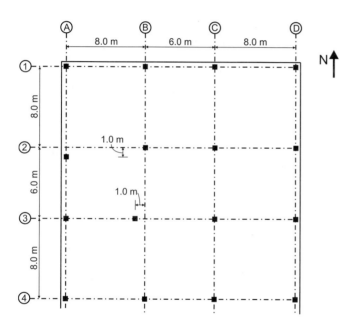

12.8. Consider the floor plan from Problem 12.5. The design needs to be modified by providing beams on all sides. Assume 400 mm square columns and beam webs matching the column width. The beam depth should be selected such that the reinforcement ratio is approximately equal to 40% of the balanced ratio (0.4 ρ_b). Revise slab thickness based on the CSA A23.3 requirements for slabs without beams. The plan di-

mensions and loading are the same as in Problem 12.5. Design a typical interior panel for flexure.
a) Use the Direct Design Method.
b) Use the Equivalent Frame Method.
c) Compare the results (bending moments and reinforcement).

12.9. Consider a two-way flat plate floor system designed in Problem 12.5. Use the slab thickness of 200 mm and an effective depth of 160 mm. The slab loading is the same as in Problem 12.5, resulting in the factored area load $w_f = 12.6 kPa$. Design the slab for shear according to the CSA A23.3 requirements. Consider only an interior column at the intersection of gridlines 2 and C.
a) Perform the design by disregarding the effect of unbalanced moments.
b) Consider the effect of unbalanced moments, and also shear and moment transfer at the slab-column connection. Use unbalanced moments calculated in Problem 12.5.

12.10. Consider the slab-column connection of Problem 12.9. Design shear stud reinforcement assuming that the total factored load had to be increased by 30%, but the slab thickness needs to remain unchanged.

12.11. Consider the flat plate system shown in Problem 12.5. A change in the building function took place after the design was performed, and the building is going to have a residential occupancy. The floor is subjected to specified live load (LL) of 1.9 kPa and superimposed dead load (DL_s) of 1.5 kPa, in addition to its self-weight. The owner requires 200 mm slab thickness for this design.
a) Calculate deflection for an interior slab panel according to the Crossing Beam Method. Use the reinforcement designed in Problem 12.5. Consider both immediate and long-term deflections. Consider 20% of the specified live load to be sustained for long-term deflection calculations.
b) Check whether deflections are within the CSA A23.3 limits. Note that, according to the design requirements, the underside of each concrete floor slab serves as the ceiling for the floor below. Consequently, non-structural elements are likely to be damaged by large deflections and that should be taken into account when considering CSA A23.3 deflection limits.

13 Walls

After reading this chapter, you should be able to

- identify the eight main types of reinforced concrete walls
- apply the CSA A23.3 requirements for detailing of wall reinforcement
- design bearing walls for gravity load effects
- design basement walls subjected to lateral earth pressure
- design shear walls for flexure and shear effects

13.1 INTRODUCTION

Walls are vertical structural members used to enclose or separate spaces. In addition, walls may be used to retain earth and liquids or resist wind pressures or to contain bulk materials in storage containers. Bearing walls can be designed like columns to support gravity loads or like beams to carry concentrated and uniformly distributed gravity loads and transfer them down to the foundations. Walls also have an important role in resisting lateral loads due to winds and earthquakes.

This chapter is focused on the conceptual design of bearing walls, basement walls, and shear walls. Different types of walls are outlined in Section 13.2. General CSA A23.3 design and detailing requirements are discussed in Section 13.3. The design of bearing walls is discussed in Section 13.4, whereas the design of basement walls is discussed in Section 13.5. Basic concepts related to the design of shear walls are discussed in Section 13.6. Structural drawings and details for reinforced concrete walls are discussed in Section 13.7. Brief coverage of joints in reinforced concrete walls is included in Section 13.8.

Advanced topics related to wall design, such as seismic design and detailing, are beyond the scope of this book. For more details on the seismic design of walls, the reader is referred to Paulay and Priestley (1992).

13.2 TYPES OF WALLS

Based on their function, reinforced concrete walls can be classified as follows:

- retaining walls
- basement walls
- grade beams
- bearing walls
- shear walls
- wall panels
- fire walls
- tilt-up walls

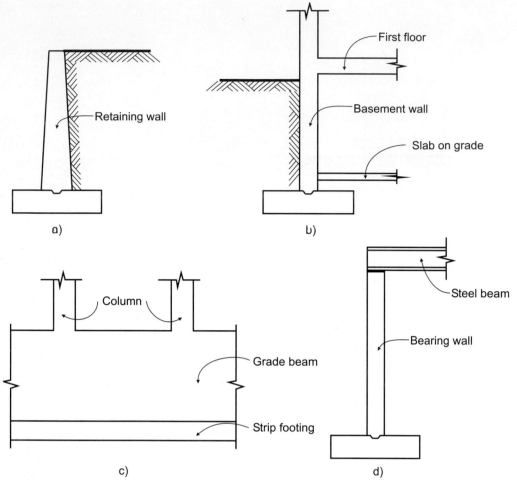

Figure 13.1   Types of walls: a) retaining wall; b) basement wall; c) grade beam; d) bearing wall.

Retaining walls retain earth or other materials, as shown in Figure 13.1a. When lateral support at the top of the wall is not provided, these walls are called cantilever retaining walls. Retaining walls are commonly used on projects where a sloped site requires to be reshaped to suit a particular development. Retaining walls are also used in the design of bridges, highways, and water reservoirs. The design of cantilever retaining walls is beyond the scope of this book.

Basement walls (also called foundation walls) are exterior walls, generally located below grade, that retain earth at the building exterior and prevent the entry of water into the building. These walls are usually supported laterally at both the top and the bottom (see Figure 13.1b).

Grade beams are basement walls that distribute the load from exterior columns above the wall to a strip footing beneath the wall (see Figure 13.1c). The top of the grade beam is located at the grade level. Grade beams are usually designed as deep beams and need to have longitudinal reinforcement at the top and the bottom of the wall. The design of grade beams is beyond the scope of this book.

Bearing walls provide enclosure as well as structural support for gravity loads, as shown in Figure 13.1d.

Shear walls resist lateral forces in the plane of the wall, mainly due to earthquakes and wind.

Wall panels are exterior walls that enclose the building above grade, resist wind forces acting perpendicular to the wall plane (out-of-plane forces), prevent the entry of water into the building, and provide a thermal barrier. Wall panels may be of precast construction.

Fire walls are usually nonbearing walls (partitions) that divide a building into sections and prevent the spread of fire from one section to another.

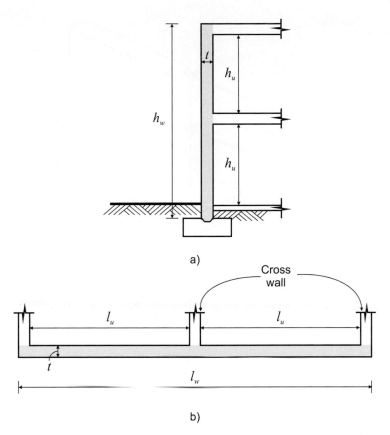

Figure 13.2 Wall terminology:
a) vertical elevation;
b) horizontal section.

Tilt-up walls are precast wall panels prefabricated at the building site. The walls are cast in a horizontal position, tilted to a vertical position, and then set in place. Tilt-up walls are used in the construction of warehouses, school buildings, and commercial buildings in Canada. The design of tilt-up walls is beyond the scope of this book. For more details on the subject, the reader is referred to PCA (1994).

CSA A23.3 Cl.2.2 defines a wall as "a vertical slab element, which may or may not be required to carry superimposed in-plane loads, in which the horizontal length (l_w)" is restricted as follows:

$$l_w \geq \begin{bmatrix} 6t \\ \dfrac{h_w}{3} \end{bmatrix}$$

The terminology used in this chapter is illustrated in Figure 13.2 and summarized below:

t = wall thickness,
l_w = horizontal length of a wall,
h_w = vertical length (height) of a wall,
l_u = unsupported wall length between vertical supports provided by cross walls,
h_u = unsupported wall height between the horizontal supports provided by floor and roof structures.

In general, walls can be subjected to the following types of loads:

• lateral loads (in-plane): shear walls (Figure 13.3a)
• lateral loads (out-of-plane): retaining walls, basement walls, wall panels (Figure 13.3b)
• gravity loads: bearing walls (Figure 13.3c)

At this point, reference is made to the concept of tributary area, covered in Section 1.5.5. The tributary area for walls subjected to gravity loads is determined in a similar manner as for columns. Consider the typical floor plan of a reinforced concrete building shown in Figure 13.4a. The tributary area for the interior bearing wall W2 is shown hatched in

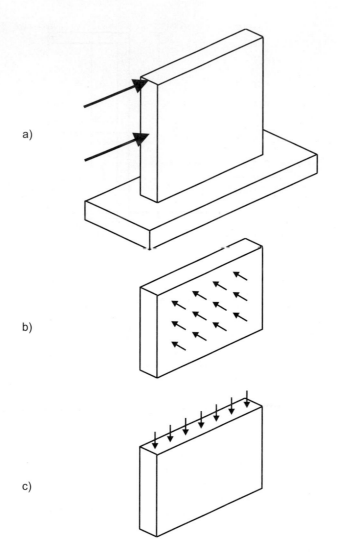

Figure 13.3 Types of loads acting on concrete walls:
a) lateral in-plane loads;
b) lateral out-of-plane loads;
c) gravity loads.

the figure. This area is bounded by the lines located halfway between the adjacent vertical supports (columns) and can be determined as follows:

$$A_2 = l_x \times \left(l_y + 2 \times \frac{l_y}{2} \right)$$

Similarly, the tributary area for the exterior bearing wall W3 shown cross-hatched in Figure 13.4a can be determined as

$$A_3 = \left(l_x + 2 \times \frac{l_x}{2} \right) \times \frac{l_y}{2}$$

Consider the building in Figure 13.4b in which the beams spanning in the N-S direction are supported by columns on one end and walls on the other end. In this case, the tributary area of wall W1, shown hatched in the figure, can be determined as

$$A_1 = (5l_x) \times \frac{l_y}{2}$$

When the factored floor or roof load (usually expressed in kilopascals) is multiplied by the tributary area (usually expressed in square metres), the resultant axial load acting on the wall is obtained in kilonewtons.

However, note that in the case of a wall loaded by a uniformly distributed gravity load along its length, it is common to use the *tributary width* instead of the tributary area. In the

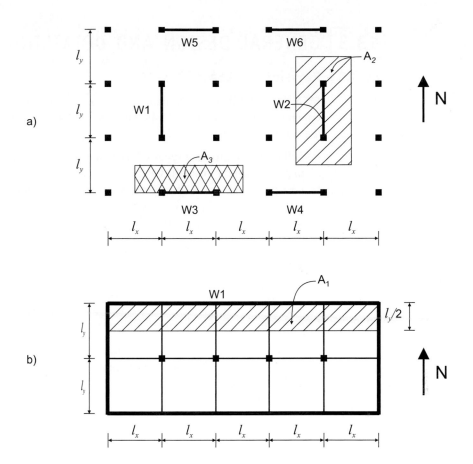

Figure 13.4 Wall tributary areas for gravity loads:
a) typical floor plan;
b) tributary width.

case of wall W1 in Figure 13.4b, the tributary width is equal to $l_y/2$. If the factored floor load (expressed in kilopascals) is denoted as w_f, the resulting uniformly distributed load along the wall length can be obtained as the product $w_f(l_y/2)$ (expressed in kilonewtons per metre).

KEY CONCEPTS

Walls are vertical structural members used to enclose or separate spaces. Based on their function, reinforced concrete walls can be classified into the following eight types:

- *Retaining walls* retain earth or other materials, including water.
- *Basement walls* retain earth at the building exterior and prevent the entry of water into the building.
- *Grade beams* are basement walls that distribute the load from exterior columns above the wall to a strip footing beneath the wall.
- *Bearing walls* provide support for gravity loads.
- *Shear walls* resist lateral forces in the plane of the wall.
- *Wall panels* are exterior walls that enclose the building above grade.
- *Fire walls* divide a building into sections and prevent the spread of fire from one section to another.
- *Tilt-up walls* are precast wall panels prefabricated at the building site.

Walls can be subjected to the following types of loads:

- lateral loads (in-plane)
- lateral loads (out-of-plane)
- gravity loads

13.3 GENERAL DESIGN AND DETAILING REQUIREMENTS

13.3.1 Revisions in CSA A23.3-04

CSA A23.3-04 does not contain any major revisions related to the design of reinforced concrete walls as compared to the previous (1994) edition. However, the revisions include the following items:

- definitions of different types of walls depending on the level of axial load and the primary loading on the wall (Cl.2.2);
- new provisions for wall reinforcement details, including concentrated and distributed reinforcement and ties for vertical reinforcement (Cl.14.1.8);
- a simplified design procedure for walls classified as bearing walls satisfying a number of restrictions (Cl.14.2);
- a new clause on shear wall design and detailing requirements for nonseismic conditions, including requirements for compression flanges for assemblies of interconnected walls (Cl.14.4).

13.3.2 CSA A23.3 Reinforcement Requirements

In order to comply with the strength and serviceability requirements, structural walls must be reinforced with an adequate amount of horizontal and vertical reinforcement. The CSA A23.3 wall reinforcement requirements are stated in Cl.14.1.8.

A23.3 Cl.14.1.8.1   **Distributed and concentrated wall reinforcement**   Walls should be reinforced with distributed vertical and horizontal reinforcement bars placed in one or two layers. The walls may also contain concentrated vertical reinforcement called "zone" reinforcing. Extra horizontal reinforcing may also be required to resist seismic shear forces. The CSA A23.3 wall reinforcement requirements are shown in Figure 13.5.

A23.3 Cl.14.1.8.5   **Minimum area of distributed vertical reinforcement**   The minimum required area of distributed vertical reinforcement is

$$A_{vmin} = 0.0015A_g \qquad\qquad [13.1]$$

where A_g is the gross cross-sectional area of a wall based on a strip of unit (1 m) length, as shown in Figure 13.6. The definition of gross area for bearing walls subjected to concentrated loads is different and will be discussed in Section 13.4.

A23.3 Cl.14.1.8.6   **Minimum area of distributed horizontal reinforcement**   The minimum required area of distributed horizontal reinforcement is

$$A_{hmin} = 0.002A_g \qquad\qquad [13.2]$$

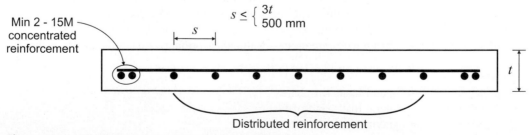

Figure 13.5 Distributed and concentrated wall reinforcement.

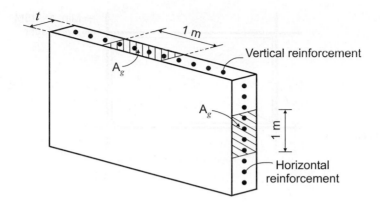

Figure 13.6 Gross cross-sectional area of a wall.

Note that horizontal wall reinforcement has an important role in controlling the size and distribution of cracks (mainly caused by shrinkage). Note also that the minimum required amount of horizontal reinforcement is larger than the minimum amount of vertical reinforcement due to the wall weight and other gravity loads that help prevent horizontal cracking in the walls.

A23.3 Cl.14.1.8.4 | **Maximum spacing of distributed reinforcement**   The maximum permitted bar spacing (s_{max}) for distributed wall reinforcement (horizontal and vertical) is equal to the lesser of $3t$ and 500 mm (same as for slabs and footings), where t denotes the wall thickness.

A23.3 Cl.14.1.8.2 | **Size of distributed reinforcement**   The nominal diameter of the bars used for the distributed wall reinforcement should not exceed $t/10$.

A23.3 Cl.14.1.8.3 | **Number of reinforcement layers and curtains**   The provision of both horizontal and vertical distributed reinforcement in the walls is mandatory. Horizontal and vertical reinforcement layers are usually tied together in an assemblage called a *curtain*. To facilitate the concrete placement, the walls should ideally be reinforced with only one curtain of reinforcement. This is common practice for basement and retaining walls in general, as well as other walls of thickness (t) less than 210 mm. If the wall thickness is greater than 210 mm, two curtains of reinforcement should be provided. Each reinforcement layer should be placed at a distance not greater than $t/3$ from the wall surface.

An example of the placement of wall reinforcement at the construction site is shown in Figure 13.7.

Figure 13.7 Placement of wall reinforcement in progress (note the formwork at the back and forming of the openings).

(Nebojsa Ojdrovic)

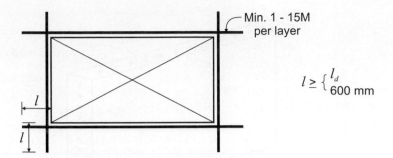

Figure 13.8 Reinforcement around openings.

| A23.3 Cl.14.1.8.8 |

Nominal concentrated vertical reinforcement In addition to the distributed reinforcement, concentrated vertical reinforcement consisting of a minimum of 2-15M bars at each wall end should also be provided (see Figure 13.5).

| A23.3 Cl.14.1.8.9 |

Reinforcement at openings Experience has shown that the concrete around openings (doors and windows) has a tendency to develop diagonal cracks starting at the re-entrant corners. In addition to the distributed and concentrated reinforcement requirements, a minimum of 2-15M bars or equivalent should be provided around all window and door openings, as shown in Figure 13.8.

Each end of 15M rebar should extend by at least 600 mm beyond each corner of the opening to develop the tensile strength of the bar.

| A23.1 Cl.6.6.6.2.3 |

Concrete cover The concrete cover requirements for reinforced concrete walls can be summarized as follows:

- 20 mm for interior walls (not exposed)
- 40 mm for exterior walls exposed to weather
- 75 mm for walls cast against and permanently exposed to earth (exterior face of cantilever retaining walls and basement retaining walls)

Refer to Table A.2 and Section 5.3.1 for more details on concrete cover requirements.

The minimum distributed wall reinforcement for walls of various thicknesses is summarized in Figure 13.9.

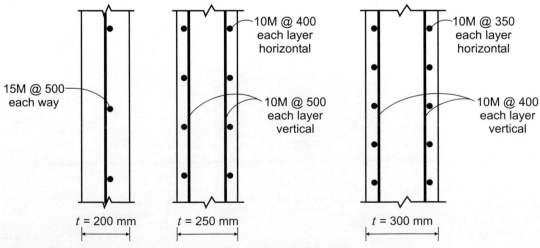

Figure 13.9 Minimum CSA A23.3 wall reinforcement requirements.

13.3.3 Wall Thickness

CSA A23.3 Cl.14.1.7.1 prescribes that the wall thickness should not be less than 1/25 of the height or length between the supports, whichever is less, but not less than 150 mm.

It is often difficult to ensure the proper placement and consolidation of concrete in thin walls. In addition, there is a very limited economic benefit associated with constructing thin walls. A significant cost associated with the construction of concrete walls is related to the formwork. Wall forms with two vertical faces are typically required in wall construction. Extensive labour and material resources are involved in building these forms. In the case of tall free-standing walls, significant additional expenses may be required to construct a temporary bracing system, which is required to provide lateral support for the forms to resist probable wind loads during construction. Therefore, constructing walls with reduced thickness decreases only the cost of concrete, which is a small fraction of the overall cost.

Unless thinner walls are required due to architectural constraints, difficulty in obtaining or handling concrete, or a need to limit the weight of the structure due to foundation problems, it is good practice to design and construct concrete walls with minimum 200 mm thickness for heights of about 2.5 m or less and minimum 300 mm thickness for wall height exceeding 4 m. For simplicity, wall thicknesses should vary in 50 mm increments (200 mm, 250 mm, 300 mm, etc.).

KEY REQUIREMENTS

CSA A23.3 prescribes the following requirements for the detailing of reinforcement in reinforced concrete walls:

- provision of distributed and concentrated wall reinforcement (Cl.14.1.8.1)
- minimum area of distributed vertical reinforcement (Cl.14.1.8.5)
- minimum area of distributed horizontal reinforcement (Cl.14.1.8.6)
- maximum spacing of distributed reinforcement (Cl.14.1.8.4)
- size of distributed reinforcement (Cl.14.1.8.2)
- number of reinforcement layers and curtains (Cl.14.1.8.3)
- nominal concentrated vertical reinforcement (Cl.14.1.8.8)
- provision of reinforcement at openings (Cl.14.1.8.9)
- concrete cover (A23.1 Cl.6.6.6.2.3)

The *wall thickness* should not be less than 200 mm for most design applications; A23.3 Cl.14.1.7.1 prescribes the minimum required thicknesses for different wall types.

13.4 BEARING WALLS

Bearing walls support vertical load in addition to their own weight. According to A23.3 Cl.2.2, bearing walls support the following loads:

a) factored in-plane vertical loads larger than $0.1 f_c' A_g$
b) bending moments around the weak axis (the horizontal axis parallel to the wall length)
c) the shear forces necessary to equilibrate those moments

Note that the definition of a *nonbearing wall* is very similar, except that a nonbearing wall supports in-plane vertical loads less than or equal to $0.1 f_c' A_g$ (A23.3 Cl.2.2).

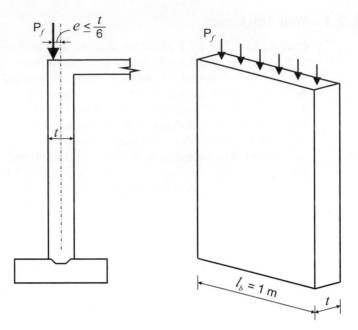

Figure 13.10 Bearing wall design according to the CSA A23.3 empirical method.

In general, bearing walls are relatively short walls spanning vertically and are subjected to gravity loads due to floor or roof structures. Such walls are usually designed using the *empirical method* stipulated in A23.3 Cl.14.2. The empirical method can be used provided that the following criteria have been met (see Figure 13.10):

a)   The wall has a solid rectangular cross-section that is constant along the wall height.
b)   The principal moments act about the weak axis (parallel to the plane of the wall).
c)   The resultant of all factored axial loads, including the effect of the principal moment, is located within the middle third of the overall wall thickness.
d)   The wall is supported against lateral displacement along at least the top and bottom edges.

According to the empirical method, the factored axial load resistance (P_r) of a bearing wall spanning vertically can be determined based on the equation

A23.3 Eq. 14.1 $$P_r = \frac{2}{3}\alpha_1\phi_c f_c' A_g \left[1 - \left(\frac{kh_u}{32t}\right)^2\right]$$ [13.3]

where

t = the wall thickness
h_u = the unsupported wall height between horizontal supports
k = the effective length factor, considering that the wall spans in the vertical direction.
$\alpha_1 \cong 0.8$ [3.7]

$$A_g = l_b \times t$$ [13.4]

is the gross area of a wall section; for walls subjected to uniform gravity load, A_g is determined based on the unit strip of length $l_b = 1000$ mm (see Figure 13.10).
According to A23.3 Cl.14.2, the effective length factor (k) shall use the following values (see Figure 13.11):

$k = 1.0$ for walls unrestrained against rotation at both ends (walls with pin supports)
$k = 0.8$ for walls restrained against rotation at one or both ends (top, bottom, or both)

In all design cases, the end support conditions in walls must be considered. If fixed end supports are considered in the design, the designer must ensure that the supports can develop end moments.

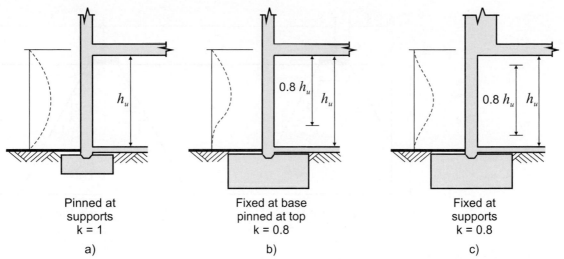

Figure 13.11   Effective length factor (*k*) for different wall support conditions: a) pinned at supports; b) fixed at the base and pinned at the top; c) fixed at supports.

According to A23.3 Cl.14.1.2, walls can be considered *laterally supported* if *both* of the following conditions are met:

a)   Walls or other vertical bracing elements are arranged in two directions to provide lateral stability to the structure as a whole.

b)   Connections between the wall and its lateral supports are designed to resist a horizontal force not less than 2% of the total factored vertical load which the wall is designed to carry at the level of the lateral support, but not less than 10 kN/m of the wall length.

When the conditions for the design according to the empirical method are not met, the wall should be designed as a column subjected to axial load and flexure, as discussed in Chapter 8.

In some cases, bearing walls are subjected to concentrated (point) loads (for example, loads transferred by columns or steel beams supported by the walls), as shown in Figure 13.12.

Figure 13.12   Bearing wall supporting steel trusses.

Nebojsa Ojdrovic)

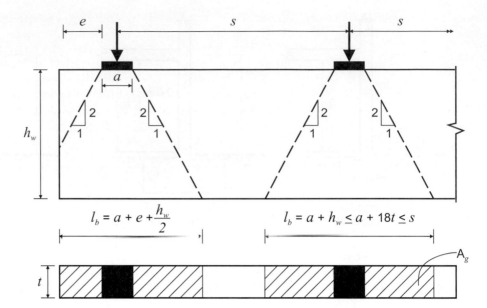

Figure 13.13 Horizontal wall length (l_b) for walls subjected to concentrated loads.

According to A23.3 Cl.14.1.3.1, each concentrated load should be considered as uniformly distributed over a horizontal wall length l_b determined based on the following three criteria:

i)   At any position below the concentrated load, the portion of l_b on each side of the centre of the concentrated load shall be half the width of the bearing plus the width enclosed by a line sloping downward by 2:1 (vertical:horizontal) on each side, limited by the intersection with the end of the wall.

ii) l_b should not exceed nine times the wall thickness on each side of the bearing area.

iii) For a wall subjected to more than one concentrated load, the design should consider the overlapping of uniformly distributed loads from each of the concentrated loads. Therefore, l_b cannot exceed the centre-to-centre spacing (s) of the point loads.

The definition of horizontal wall length (l_b) is illustrated in Figure 13.13. In the case of walls subjected to concentrated loads, the l_b value obtained based on the above criteria should be used in Eqn 13.4 to determine the gross area (A_g).

In the case of walls supporting concentrated loads, the designer also needs to check whether the bearing strength of the wall is adequate; that is, the design should ensure that the concrete is not crushed under excessively large concentrated loads. The concept of bearing strength is discussed in Section 14.10 as related to footings. The wall bearing resistance (B_r) can be determined based on the equation (A23.3 Cl.10.8.1)

$$B_r = 0.85\phi_c f_c' A_1 \qquad\qquad\qquad\qquad \text{[14.28]}$$

Note that the bearing resistance determined from Eqn 14.28 is the product of the maximum bearing stress developed in concrete ($0.85\phi_c f_c'$) and the loaded contact area (A_1). If the bearing walls support columns, A_1 is equal to the column area, whereas if the walls support steel members (beams or open web steel joists) resting on bearing plates, A_1 is equal to the area of a bearing plate.

The design of bearing walls under axial loads using the CSA A23.3 empirical method will be illustrated by the following example.

Example 13.1

A reinforced concrete bearing wall supports open web steel joists (OWSJs) spaced at 2 m on centre, as shown in the figure below (note that Figure 13.12 shows an example of a similar real-life design application). Each OWSJ rests on a 150 mm × 300 mm bearing plate. The wall is supported against lateral displacement at the top and bottom edges. Assume that the wall is simply supported at the top and bottom. The wall height is 4.1 m and its length is 10 m. The factored load (P_f) transferred by each OWSJ is 500 kN. Assume that the joist reaction is applied concentrically with regard to the vertical axis of the wall.
Design the wall for axial load effects.

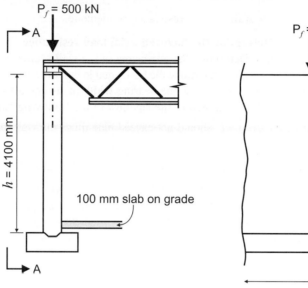

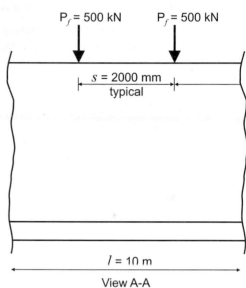

View A-A

Given: $f_c' = 25\ \text{MPa}$
$f_y = 400\ \text{MPa}$
$\phi_c = 0.65$
$\phi_s = 0.85$

SOLUTION: This wall meets the criteria for the CSA A23.3 empirical method stipulated in Cl.14.2 as follows:

a) The wall has a solid rectangular cross-section that is constant along the wall height.
b), c) Because the wall is subjected to concentric loads, these criteria do not apply.
d) The wall is supported against lateral displacement along at least the top and bottom edges.

1. **Determine the wall thickness**
    Determine the unsupported wall height as

    $h_u = 4100\ \text{mm} - 100\ \text{mm} = 4000\ \text{mm}$

    Note that the total wall height is 4100 mm and the thickness of the slab on grade is 100 mm.
    According to A23.3 Cl.14.1.7.1, the wall thickness should not be less than

    * $h_u/25 = 4000\ \text{mm}/25 = 160\ \text{mm}$ or
        160 mm (other cast in place walls)

        Since the wall length is 10 m, there is no need to check the required wall thickness based on the length criterion because it will not govern in this case.
        Therefore, the 160 mm value governs; however, for construction reasons use

    $t = 200\ \text{mm}$

2. **Check the bearing resistance (A23.3 Cl.10.8.1)**

   First, calculate the contact area (A_1):

   $$A_1 = 150\,\text{mm} \times 300\,\text{mm} = 45\,000\,\text{mm}^2$$

   Then, determine the bearing resistance as

   $$B_r = 0.85\phi_c f_c' A_1 \qquad\qquad\qquad\qquad \text{[12.28]}$$
   $$= 0.85 \times 0.65 \times 25\,\text{MPa} \times 45\,000\,\text{mm}^2 = 621.6 \times 10^3\,\text{N} = 622\,\text{kN}$$

   Because

   $$B_r = 622\,\text{kN} > P_f = 500\,\text{kN}:$$

   the wall bearing resistance is adequate.

3. **Determine the factored axial load resistance**

   a) Determine the gross area of the wall section (A_g).

   First, determine the horizontal length of the wall (l_b) according to A23.3 Cl.14.1.3.1.

   Because the bearing width (a) is equal to 300 mm in this case, l_b should be determined as the smallest of the following values:

   i) l_b should not exceed nine times the wall thickness on each side of the bearing area:

   $$l_b = a + 2 \times 9t = 300\,\text{mm} + 2 \times 9 \times 200\,\text{mm} = 3900\,\text{mm}$$

   ii) l_b should be determined by drawing lines sloping downward by 2:1 from each side of the bearing area. Note that l_b is equal to the width of the bearing plate plus two times the width enclosed by the sloping lines limited by the intersection with the lines corresponding to the adjacent point loads. For this design, the resulting l_b value is equal to 2000 mm, as shown in the sketch below.

   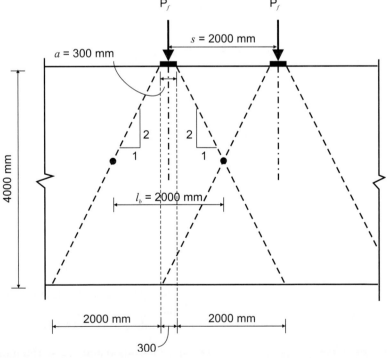

   iii) l_b should not exceed the centre-to-centre spacing between OWSJs; so, based on the above criterion

   $$l_b = 2000\,\text{mm}$$

   which governs as the smallest value.

   Determine the gross area:

   $$A_g = l_b \times t = 2000\,\text{mm} \times 200\,\text{mm} = 400 \times 10^3\,\text{mm}^2$$

b) Determine the factored axial load resistance:

| A23.3 Eq. 14.1 |

$$P_r = \frac{2}{3}\alpha_1 \phi_c f_c' A_g \left[1 - \left(\frac{kh_u}{32t}\right)^2 \right]$$ [13.13]

$$= \frac{2}{3} \times 0.8 \times 0.65 \times 25 \text{ MPa} (400 \times 10^3 \text{ mm}^2) \left[1 - \left(\frac{1.0 \times 4000 \text{ mm}}{32 \times 200 \text{ mm}}\right)^2 \right]$$

$$= 2112 \times 10^3 \text{ N} = 2112 \text{ KN}$$

where

$k = 1.0$ (simply supported wall),

$\alpha_1 \cong 0.8$ [3.7]

Since

$$P_r = 2112 \text{ kN} > P_f = 500 \text{ kN}$$

this wall is adequate for the given axial loads. Note that Eqn 13.13 does not take into account the effect of reinforcement. However, CSA A23.3 requires the provision of minimum distributed and concentrated reinforcement in bearing walls.

4. **Determine the distributed and concentrated wall reinforcement**
   a) First, consider the distributed horizontal reinforcement.
      i) Determine the area of reinforcement.
         First, determine the gross cross-sectional area of the wall as (1 m strip in the vertical direction; see Figure 13.6)

$$A_g = 1000 \text{ mm} \times t = 1000 \text{ mm} \times 200 \text{ mm} = 200 \times 10^3 \text{ mm}^2$$

According to A23.3 Cl.14.1.8.6,

$$A_{hmin} = 0.002 A_g$$ [13.2]

$$= 0.002 \times (200 \times 10^3 \text{ mm}^2) = 400 \text{ mm}^2/\text{m}$$

Hence,

$$A_s = A_{hmin} = 400 \text{ mm}^2/\text{m}$$

      ii) Determine the required bar spacing.
          For 15M bars,

$$A_b = 200 \text{ mm}^2 \text{ (see Table A.1).}$$

Determine the spacing of the reinforcement in the same way as for one-way slabs (see Section 3.6):

$$s \le A_b \frac{1000}{A_s}$$ [3.29]

$$= 200 \text{ mm}^2 \times \frac{1000 \text{ mm}}{400 \text{ mm}^2/\text{m}} = 500 \text{ mm}$$

Therefore, the required bar spacing is

$$s = 500 \text{ mm}$$

      iii) Check whether the spacing is within the limits prescribed by CSA A23.3. According to A23.3 Cl.14.1.8.4, the maximum permitted bar spacing (s_{max}) is equal to the lesser of $3t$ and 500 mm.
          Because

$$3t = 3(200 \text{ mm}) = 600 \text{ mm}$$

$$s_{max} = 500 \text{ mm} \text{ (governs)}$$

Since

$$s = s_{max} = 500 \text{ mm}$$

the spacing is within the prescribed limits. Use 15M bars at 500 mm spacing (15M@500).

iv) Check whether one layer of reinforcement is adequate.
According to A23.3 Cl.14.1.8.3, one layer of reinforcement is adequate because the wall thickness

$$t = 200 \text{ mm} < 210 \text{ mm}$$

b) Next, consider the distributed vertical reinforcement.

i) Determine the area of reinforcement.
First, determine the gross cross-sectional area of the wall (1 m strip in the vertical direction; see Figure 13.6). This is the same as in Step 4 a) i):

$$A_g = 200 \times 10^3 \text{ mm}^2$$

According to A23.3 Cl.14.1.8.5,

$$A_{vmin} = 0.0015 A_g \qquad \qquad \text{[13.1]}$$
$$= 0.0015 \times (200 \times 10^3 \text{ mm}^2) = 300 \text{ mm}^2/\text{m}$$

ii) Determine the required bar spacing.
For 15M bars,

$$A_b = 200 \text{ mm}^2 \text{ (Table A.1)}.$$

Determine the spacing of the reinforcement in the same way as for one-way slabs (see Section 3.6):

$$s \le A_b \frac{1000}{A_s} \qquad \qquad \text{[3.29]}$$
$$= 200 \text{ mm}^2 \times \frac{1000 \text{ mm}}{300 \text{ mm}^2/\text{m}} = 667 \text{ mm}$$

iii) Check whether the spacing is within the limits prescribed by CSA A23.3.
According to A23.3 Cl.14.1.8.4, the maximum permitted bar spacing (s_{max}) is equal to the lesser of $3t$ and 500 mm.
Because

$$3t = 3(200 \text{ mm}) = 600 \text{ mm}$$

$$s_{max} = 500 \text{ mm}$$

Since

$$s = 667 \text{ mm} > s_{max}$$

use $s_{max} = 500$ mm

Therefore, the distributed vertical reinforcement consists of 15M bars at 500 mm spacing (15M@500).

c) Finally, consider the concentrated reinforcement.
A23.3 Cl.14.1.8.8 also requires the provision of nominal concentrated reinforcement (2-15M bars) at the ends of the wall cross-section.

5. Provide a design summary

Finally, summarize the results of this design with a sketch showing the reinforcement distribution.

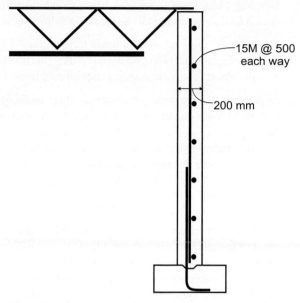

15M @ 500
each way

200 mm

Note that the same design could have been performed assuming the equivalent uniformly distributed load (w_f) instead of the concentrated loads (P_f). Because the point loads are spaced at $s = 2$ m, w_f can be determined by distributing the concentrated loads as follows (see the sketch below):

$$w_f = \frac{P_f}{s} = \frac{500 \text{ kN}}{2 \text{ m}} = 250 \text{ kN/m}$$

P_f P_f P_f

s s

$w_f = P_f/s$

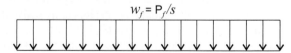

For a bearing wall subjected to a uniformly distributed load, the gross area is determined based on the unit strip; that is,

$l_b = 1000$ mm

and

$A_g = l_b \times t = 1000 \text{ mm} \times 200 \text{ mm} = 200 \times 10^3 \text{ mm}^2$

When this value is substituted in Eqn 13.3 (with all other parameters being the same as in the above example),

$P_r = 1056$ kN/m

since

$P_r = 1056 \text{ kN/m} > w_f = 250 \text{ kN/m}$

It can be concluded that the wall is adequate for the given loads.

This approximate method can be used when the wall is subjected to concentrated loads of the same weight spaced at relatively close and equal intervals.

Bearing walls support vertical loads in addition to their own weight. According to A23.3 Cl.2.2, bearing walls support the following loads:

a)   factored in-plane vertical loads larger than $0.1 f_c' A_g$
b)   bending moments around the weak axis (the horizontal axis parallel to the wall length)
c)   the shear forces necessary to equilibrate these moments

In general, bearing walls are relatively short walls spanning vertically and are subjected to vertical loads due to floor or roof structures. Such walls are usually designed using the *empirical method* stipulated in A23.3 Cl.14.2.

According to the empirical method, the factored axial load resistance of a bearing wall (P_r) can be determined based on the equation

$$\boxed{\text{A23.3 Eq. 14.1}} \quad P_r = \frac{2}{3}\alpha_1 \phi_c f_c' A_g \left[1 - \left(\frac{kh_u}{32t} \right)^2 \right] \qquad \text{[13.3]}$$

where k is the effective length factor, which depends on the end support conditions; in most cases, the end support conditions are considered as pinned, so $k = 1.0$, and A_g is the gross area of a wall section; for walls subjected to uniform load, A_g is determined based on the unit strip of length $l_b = 1000$ mm.

In some cases, bearing walls are subjected to concentrated (point) loads. Each concentrated load should be considered as uniformly distributed over the horizontal length of a wall (l_b), which needs to be determined based on the three different criteria outlined in A23.3 Cl.14.1.3.1.

13.5   BASEMENT WALLS

13.5.1   Background

Basement walls are exterior walls located below ground (grade) level in a building. Reinforced concrete basement walls are also used in various types of building construction, including masonry, timber frame, and framed steel construction.

Basement walls should be designed to resist the combined effects of lateral earth pressure and axial compressive loads transferred from the building superstructure above (see Figure 13.14). These walls are usually laterally supported at the bottom by the footings

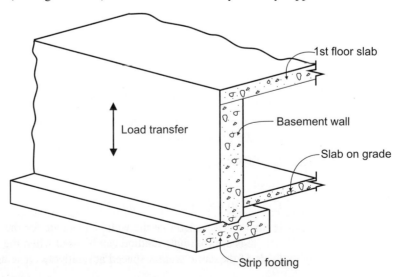

Figure 13.14   Components of a basement wall.

and slab on grade and at the top by the floor slab. The walls span like vertical slabs between these two supports. When the basement walls are long enough, that is, their length-to-height ratio is larger than two, they behave as one-way slabs. Short walls and corners of long walls usually span by two-way action. However, unless the bending moments developed in the wall are very large, it is common practice to design basement walls as one-way slabs without taking advantage of the reduction in bending moments due to two-way action.

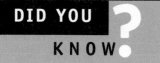

The One Wall Centre, a 48-storey tower and one of the tallest buildings in Vancouver, British Columbia, is shown in Figure 13.15. The building is 150 m tall and quite slender, with a 7:1 height-to-width aspect ratio. The building has an elliptical footprint with a width of around 21 m. To resist the effects of various gravity and lateral loads (including wind and seismic loads), the walls in the concrete core are up to 900 mm thick.

Figure 13.15 A view of the One Wall Centre building in Vancouver, British Columbia.

(John Pao)

13.5.2 Loads

The main loads acting on basement walls are

- gravity loads
- earth pressure
- surcharge
- seismic loads

Basement walls are designed for *gravity loads* in the same manner as bearing walls, which were discussed in Section 13.4.

Lateral *earth (soil) pressure* is due to the soil backfill outside the basement wall. For design purposes, lateral active earth pressure is treated as an equivalent fluid pressure. The magnitude of the pressure is considered as proportional to the depth below the ground surface and the surcharge soil weight in a manner similar to a fluid. This corresponds to a triangular load distribution with a zero value at the top and a maximum at the base of the wall. The soil pressure at the base of the wall (p_o) can be determined as follows (see Figure 13.16a):

$$p_o = h_w \gamma_o \text{ (kPa)}$$ [13.5]

where

h_w = the wall height (m)
$\gamma_o = K_o \gamma_s$ is the equivalent soil unit weight (kN/m³)
K_o = the coefficient of soil pressure at rest, used if the wall movement at the top is restricted (see soil mechanics textbooks for more details)
γ_s = the soil unit weight (kN/m³).

Note that the pressure has been determined over a unit wall strip (1 m length). The resultant soil pressure (H_o) is

$$H_o = \frac{1}{2} p_o h_w$$

The equivalent soil unit weight (γ_o) is generally in the range from 10 kN/m³ to 20 kN/m³. The lower value is assigned to loose granular soils and the higher value to well-compacted soils. The magnitude of the lateral soil pressure depends on the type of backfill soil, the depth of the groundwater table, and surcharge loads (note that the γ_o value will increase when there is a surcharge). A geotechnical engineer is generally responsible for recommending the design lateral soil pressure to be used on a project.

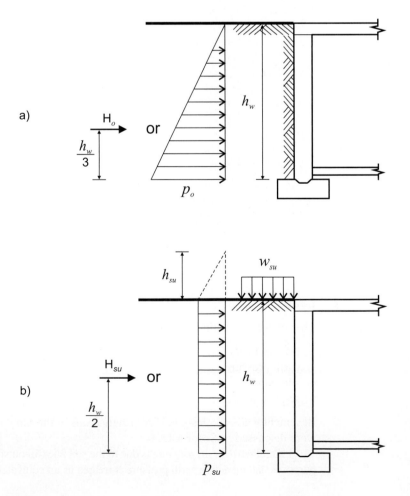

Figure 13.16 Loads acting on a basement wall: a) lateral soil pressure; b) surcharge.

The surcharge includes various dead and live loads imposed on a backfill surface behind the wall. These loads can be transformed into an equivalent soil height (h_{su}) as follows:

$$h_{su} = \frac{w_{su}}{\gamma_s}$$

where w_{su} is the soil surcharge load (kPa).

In effect, this adds a uniform pressure (p_{su}) behind the wall, that is,

$$p_{su} = h_{su} \times \gamma_o$$

with the resultant surcharge force (H_{su}) given by

$$H_{su} = p_{su} \times h_w$$

H_{su} is taken to act at the wall midheight, as shown in Figure 13.16b.

It is very important to note that lateral soil pressure represents a specified (unfactored) load and needs to be multiplied by an appropriate load factor to obtain the factored load used in the ultimate limit states design.

Seismic loads caused by earthquake ground motion must be considered in all regions of Canada, as discussed in Section 1.7.6. Magnitude of lateral seismic load acting on basement walls depend on the site seismicity, soil type, properties of the wall and the building above, and other factors. The load distribution over the wall height is of inverted triangular shape. However, seismic considerations are outside the scope of this book.

13.5.3 Construction Considerations

Contractors generally like to place backfill as soon as possible after wall construction. If the wall is designed to be supported laterally at the top, it is essential that the first floor slab be in place before the backfill is placed against the wall up to the top of the wall. Alternatively, the wall can be temporarily braced before backfilling until after the first floor concrete slab has been constructed. In any event, the concrete in the wall must gain sufficient strength before the backfill is placed.

Most walls are not designed to resist hydrostatic pressure. Therefore, it is very important to prevent the accumulation of water in the backfill. Proper drainage can be accomplished by placing granular backfill material with perforated drain pipes at the bottom to allow water to drain away from the wall.

13.5.4 Design

For design purposes, basement walls are treated as if they are composed of a series of vertical beams placed side by side, as illustrated in Figure 13.17a. Each beam is of rectangular cross-section with a 1 m width (b) and a depth (t) equal to the wall thickness (shown with a hatched pattern in Figure 13.17b). This concept is very similar to that of one-way slab design, explained in Section 3.6.

The key design considerations related to basement walls are

- axial load
- flexure
- shear
- sliding

Axial load   Refer to the bearing wall design in Section 13.4.

Flexure   In theory, basement walls should be designed for the combined effect of axial load and flexure. However, axial compression reduces the flexural tensile stresses in the wall. Consequently, the required amount of flexural reinforcement in the wall is smaller than in a wall subjected to flexural effects only. Therefore, it is common design practice to ignore the effect of axial loads in the flexural design of basement walls.

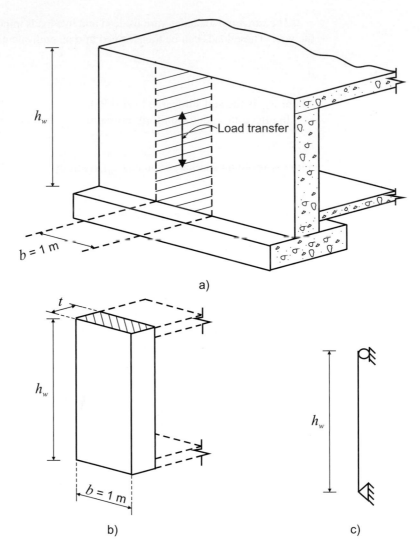

Figure 13.17 Basement wall design: a) an isometric view; b) a strip of unit width; c) a simply supported vertical beam.

As discussed earlier in this section, basement walls subjected to lateral earth pressure are considered as simply supported vertical beams (see Figure 13.17c). The top and bottom supports are modelled as pins, except when it is desired to use the partial backfill method. Multilevel basement walls span between the individual floors and should be designed as continuous vertical slabs.

In general, the flexural design of basement walls starts by developing the bending moment diagram and determining the required amount and spacing of reinforcement corresponding to the design bending moment (M_f). The wall thickness (t) is usually determined at the beginning of the design.

Flexural effects in basement walls are summarized in Figure 13.18. The bending moment diagram for a wall subjected to lateral earth pressure is shown in Figure 13.18a. Note that the flexural reinforcement for this loading condition is placed on the *inside* face of the wall, that is, the tension face. In the case of eccentrically loaded strip footings, discussed in Section 14.8 (for example, footing close to the property line), and where there will be backfill before the upper slab is present, the joint between the wall and the footing must be designed as a rigid joint. This joint must be able to transfer the footing moment due to the soil pressure. Consequently, the tension face at the base of the wall is located at the exterior. Depending on the magnitude of the lateral soil pressure in relation to the moment at the base of the wall, the tension face could shift to the inside face somewhere in the midheight region (see the bending moment diagram in Figure 13.18b).

In eccentric footings, where the applied load P acts with an eccentricity e with regard to the resultant of the foundation soil pressure (R), there is an applied moment equal to

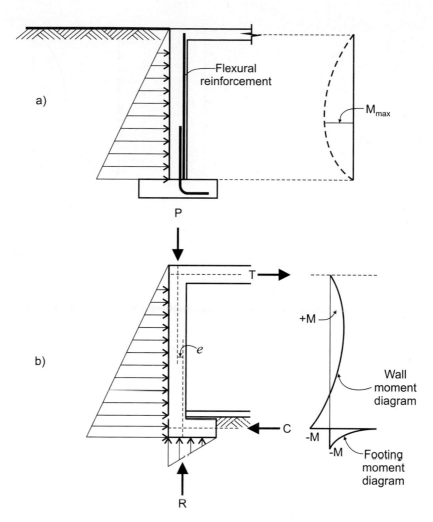

Figure 13.18 Flexural effects in basement walls: a) due to lateral earth pressure; b) due to the eccentrically loaded footing.

($P \times e$) that must be resisted to maintain equilibrium. In this case, the resisting couple is formed by a tension reaction (T) from the slab at the top of the wall and a compression reaction (C) from the slab on grade (see Figure 13.18b).

In addition to the lateral loads due to active soil pressure and surcharge, earthquake ground motion also causes lateral pressure on basement walls. The seismic load distribution over the wall height is of inverted triangular shape (maximum at the top of the wall and zero at the base), and it should be combined with other lateral loads acting on the wall in order to obtain design bending moments and shear forces.

In design practice, a triangular soil pressure distribution may be approximated by a uniform distribution, as shown in Figure 13.19a. The magnitude of the uniform load is equal to the average soil pressure acting at the wall midheight, that is, $p_o/2$. The maximum bending moments corresponding to these load distributions are approximately the same, so this approximation is acceptable for most design applications.

However, note that the design shear force should be determined based on the triangular load distribution because it gives a more accurate estimate than the uniform load does (see Figure 13.19b). The maximum shear force in a wall with a triangular load can be determined based on the equation

$$V_{max} = \frac{p_o \times h_w}{3}$$
[13.6]

The CSA A23.3 design requirements for wall reinforcement discussed in Section 13.3 need to be met. In many cases, minimum reinforcement will provide adequate flexural resistance against earth pressure, without consideration of seismic loads.

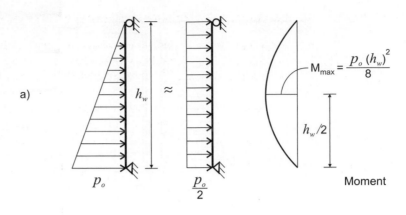

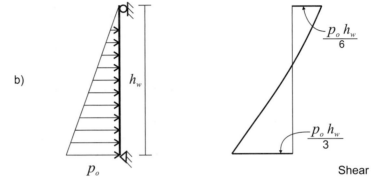

Figure 13.19 A pin-supported basement wall: a) equivalent uniform load and bending moment diagram; b) shear force diagram for the triangular load distribution.

Shear   Shear resistance rarely governs in basement wall design; however, it needs to be considered. The shear design of basement walls is performed in the same way as the shear design of slabs, discussed in Section 6.8. In general, basement retaining walls are designed such that shear reinforcement is not required. The design shear force (V_f) of a wall subjected to lateral earth pressure is determined based on the diagram in Figure 13.19. According to A23.3 Cl.11.2.8.1, shear reinforcement is *not required* if the wall has a large enough thickness such that

$$V_f \leq V_c$$

The designer may decide to increase the wall thickness in order to satisfy the above requirement.

The shear resistance of concrete (V_c) can be determined in the same way as for beams and slabs, discussed in Section 6.5.4 (A23.3 Cl.11.3.4). Therefore,

A23.3 Eq. 11.6 $$V_c = \phi_c \lambda \beta \sqrt{f_c'} \, b_w d_v$$ [6.12]

where

ϕ_c = the resistance factor for concrete
λ = the factor to account for concrete density ($\lambda = 1$ for normal-density concrete)
d_v = the effective shear depth taken as the greater of $0.9d$ and $0.72t$
f_c' = the specified compressive strength of concrete
b_w = 1000 mm is the width of a unit strip
β = the factor accounting for the shear resistance of the cracked concrete; according to A23.3 Cl.11.3.6.3b, "if the section contains no transverse reinforcement and the specified nominal maximum size of coarse aggregate is not less than 20 mm,"

A23.3 Eq. 11.9 $$\beta = \frac{230}{1000 + d_v}$$ [6.13]

Sliding   The sliding resistance of retaining walls is a stability consideration. Walls have a tendency to slide due to lateral forces, whereas gravity forces cause frictional resistance against sliding. In the case of basement walls, the resultant soil pressure (H_o) tends to move the wall in a horizontal direction; however, the friction force (F) developed at the base of the footing provides sliding resistance, as shown in Figure 13.20a. According to Coulomb's law, the friction force is

$$F = \Sigma W \times \mu$$

where

ΣW = the sum of the gravity loads at the base of the footing, including the loads transferred from the building superstructure above and the wall and footing self-weight
μ = the friction coefficient for the footing-to-soil interface.

The wall will not slide as long as the resultant soil pressure is less than the friction force; that is,

$$H_o \leq F$$

However, it is recommended that a factor of safety (FS) of 1.5 be provided to ensure an adequate safety against sliding:

$$FS = \frac{F}{H_o} \geq 1.5$$

When the basement slab on grade is cast against the wall, the slab may also resist the horizontal reaction by friction against the soil or by bearing against another wall with an

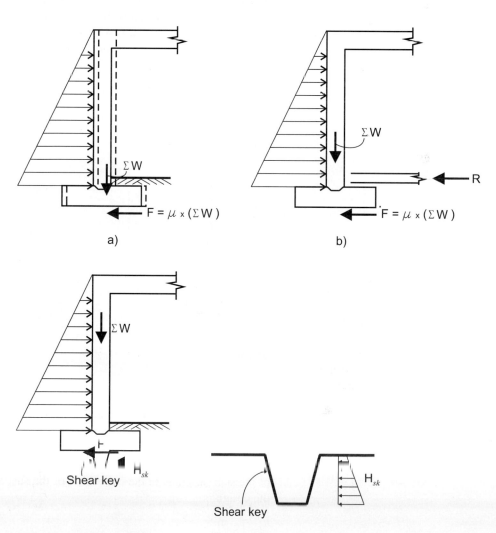

Figure 13.20 Sliding resistance of a basement wall: a) general case; b) wall with a slab on grade; c) shear key.

opposing horizontal reaction, as shown in Figure 13.20b. In that case, the force (R) developed in the slab adds to the friction resistance, so

$$H_o \leq F + R$$

In general, the sliding resistance of basement walls is adequate unless they support very light gravity loads and resist very large lateral soil pressure. If required, sliding resistance can be increased by providing a shear key beneath the footing, as shown in Figure 13.20c. The shear key is effective in increasing the sliding resistance by mobilizing the passive resistance of the soil behind it (with the resultant force H_{sk}). The dimensions of a shear key (especially its depth) are governed by the magnitude of the required additional sliding resistance. In the case of a footing with a shear key, the following relationship is valid:

$$H_o \leq F + H_{sk}$$

Usually, basement walls are placed all around the building perimeter. Consequently, these walls are not susceptible to sliding because the slab on grade acts in compression to resist horizontal forces from the base of the wall located on the opposite side of the building.

Example 13.2

A reinforced concrete basement wall is shown in the figure below. The wall supports a uniform dead load (DL) of 40 kN/m and a uniform live load (LL) of 20 kN/m. The wall is also subjected to lateral soil pressure. Based on the information provided by the geotechnical engineer, the backfill soil is characterized by a unit weight of $\gamma_s = 20$ kN/m³ and a coefficient of soil pressure at rest of $K_o = 0.5$. Use 15M bars for this design.
Design the wall for the given loads according to the CSA A23.3 requirements. Disregard seismic loads in this design.

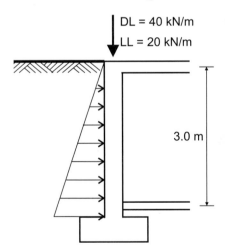

Given: $f_c' = 25$ MPa
         $f_y = 400$ MPa
         $\phi_c = 0.65$
         $\phi_s = 0.85$

SOLUTION: **1. Determine the design bending moments and shear forces**
     a)   Determine the wall height.
         Consider this wall as pin supported at the top (suspended floor slab) and the bottom (slab on grade). Therefore, the wall height (the distance from the centre-line of the slab on grade to the centre-line of the suspended floor slab) to be used in the design is

$$h_w = 3 \text{ m}$$

Note that the design is based on a unit wall strip (see Figure 13.17); that is,

$$b = 1000 \text{ mm}$$

b) Determine the magnitude of the lateral soil pressure:

$$\gamma_o = K_o \gamma_s$$
$$= 0.5 \times 20 \text{ kN/m}^3 = 10 \text{ kN/m}^3$$

The maximum soil pressure at the base of the wall is

$$p_o = h_w \gamma_o \qquad\qquad \textbf{[13.5]}$$
$$= 3 \text{ m} \times 10 \text{ kN/m}^3 = 30 \text{ kPa}$$

Based on load combination 2 from Table 4.1.3.2 of NBC 2010 (1.25DL + 1.5LL), the factored maximum soil pressure is

$$p_{of} = 1.5\, p_o = 1.5 \times 30 \text{ kPa} = 45 \text{ kPa}$$

c) Determine the factored bending moment.
As discussed in Section 13.5.4 (see Figure 13.19a), a triangular load distribution can be approximated by an equivalent uniform load distribution. Estimate the magnitude of the uniform load (w_f) as

$$w_f = \frac{p_{of}}{2} = \frac{45 \text{ kPa}}{2} = 22.5 \text{ kPa}$$

The maximum factored bending moment is the same as for a simply supported beam with a span h_w subjected to a uniform load w_f, that is,

$$M_f = \frac{w_f (h_w)^2}{8} = \frac{(22.5 \text{ kN/m}) \times (3 \text{ m})^2}{8} = 25.3 \text{ kN} \cdot \text{m/m}$$

d) Determine the factored shear force as a simply supported beam with a triangular load distribution (see Figure 13.19b):

$$V_f = \frac{p_{of} \times h_w}{3} \qquad\qquad \textbf{[13.6]}$$
$$= \frac{45 \text{ kN/m} \times 3 \text{ m}}{3} = 45 \text{ kN/m}$$

The bending moment and shear force diagrams are shown on the sketch below.

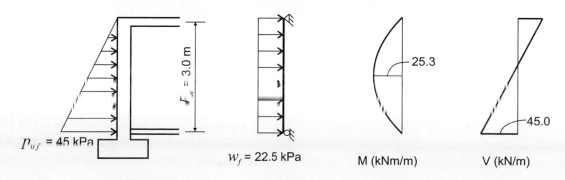

$p_{of} = 45 \text{ kPa}$ $h_w = 3.0 \text{ m}$ $w_f = 22.5 \text{ kPa}$ 25.3   45.0   M (kNm/m)   V (kN/m)

2. **Design the wall for combined effects of flexure and axial loads**
    a) Determine the wall thickness.
        According to A23.3 Cl.14.3.6.1, the wall thickness should not be less than

        - $h_u/25 = 3000$ mm$/25 = 120$ mm or
        - 190 mm (cast-in-place foundation walls).

        Therefore, the 190 mm value governs; however, a minimum thickness of 200 mm should be used (as discussed in Section 13.3.3); that is,

        $t = 200$ mm

    b) Determine the design loads.
        In addition to lateral soil pressure, the wall is subjected to gravity loads, including dead load and live load; that is,

        DL $= 40$ kN$/$m

        LL $= 20$ kN$/$m

        Therefore, the factored axial load is

        $P_f = 1.25$ DL $+ 1.5$ LL $= 1.25 \times 40$ kN$/$m $+ 1.5 \times 20$ kN$/$m $= 80$ kN$/$m

        The factored bending moment was determined in Step 1 as

        $M_f = 25.3$ kN$\cdot$m$/$m

        In theory, this wall should be designed for the combined effect of axial load and flexure. However, the effect of gravity loads will be neglected in this example. This is conservative because the axial compression reduces the flexural tensile stresses and the required amount of flexural reinforcement in the wall. This approach is generally followed in design practice.

    c) Calculate the required area of vertical tension reinforcement.

        i) First, determine the effective depth.
            Use 15M reinforcement bars ($d_b = 15$ mm; see Table A.1) and 20 mm cover (the wall's inside face is of interior exposure; see Table A.2):

            $$d = t - \text{cover} - \frac{d_b}{2}$$

            $$= 200 \text{ mm} - 20 \text{ mm} - \frac{15 \text{ mm}}{2} = 172.5 \text{ mm} \cong 170 \text{ mm}$$

        ii) Calculate the area of tension reinforcement using the direct procedure (Eqn 5.4):
             Set

             $M_r = M_f = 25.3$ kN$\cdot$m$/$m

             $b = 1000$ mm (unit strip)

             Then

             $$A_s = 0.0015 f_c' \, b \left(d - \sqrt{d^2 - \frac{3.85 \, M_r}{f_c' \, b}} \right) \qquad \textbf{[5.4]}$$

             $$= 0.0015 \times 25 \text{ MPa} \times 1000 \text{ mm}$$

             $$\times \left(170 \text{ mm} - \sqrt{(170 \text{ mm})^2 - \frac{3.85(25.3 \times 10^6 \text{ N} \cdot \text{mm})}{25 \text{ MPa} \times 1000 \text{ mm}}} \right)$$

             $$= 445 \text{ mm}^2/\text{m}$$

d) Select the amount of vertical reinforcement in terms of size and spacing.
Use 15M bars (the area of one 15M bar is 200 mm²; see Table A.1). Determine the required bar spacing as

$$s \leq A_b \frac{1000}{A_s}$$

$$= (200 \text{ mm}^2) \frac{1000}{445 \text{ mm}^2/\text{m}} = 449 \text{ mm} \cong 450 \text{ mm} \qquad \textbf{[3.29]}$$

Check whether the bar spacing is within the prescribed limits. According to A23.3 Cl.14.1.8.4, the maximum permitted bar spacing (s_{max}) is equal to the lesser of $3t$ and 500 mm.
Because

$$3t = 3(200 \text{ mm}) = 600 \text{ mm}$$

$$s_{max} = 500 \text{ mm (governs)}$$

Since

$$s = 450 \text{ mm} < s_{max}$$

use $s = 450$ mm. Therefore, the selected vertical reinforcement consists of 15M bars at 450 mm spacing, that is, 15M@450.

e) Confirm that the maximum tension reinforcement requirement is satisfied.
Check the reinforcement ratio:

$$\rho = \frac{A_s}{b\,d} \qquad \textbf{[3.1]}$$

$$= \frac{445 \text{ mm}^2}{1000 \text{ mm} \times 170 \text{ mm}} = 0.0026$$

Compare this ratio with the balanced reinforcement ratio (ρ_b) for $f_c' = 25$ MPa, given in Table A.4. Since

$$\rho = 0.0026 < \rho_b = 0.022$$

the maximum tension reinforcement requirement is satisfied.

f) Check the minimum reinforcement requirement.
First, determine the gross cross-sectional area of the wall (1 m strip in the vertical direction; see Figure 13.6):

$$A_g = 1000 \text{ mm} \times t = 1000 \text{ mm} \times 200 \text{ mm} = 200 \times 10^3 \text{ mm}^2$$

According to A23.3 Cl.14.1.8.5,

$$A_{vmin} = 0.0015A_g \qquad \textbf{[13.1]}$$

$$= 0.0015 \times (200 \times 10^3 \text{ mm}^2) = 300 \text{ mm}^2/\text{m}$$

Since

$$A_s = 445 \text{ mm}^2/\text{m} > 300 \text{ mm}^2/\text{m}$$

the vertical wall reinforcement is adequate.

3. **Design for shear**

   The design shear force at the base of the wall is

   $$V_f = 45 \text{ kN/m}$$

   a) Determine the concrete shear resistance (V_c).

      The effective shear depth (d_v) for the wall section is determined as the greater of

      $$\begin{bmatrix} 0.9d \\ 0.72t \end{bmatrix} = \begin{bmatrix} 0.9 \times 170 \text{ mm} \\ 0.72 \times 200 \text{ mm} \end{bmatrix} = \begin{bmatrix} 153 \text{ mm} \\ 144 \text{ mm} \end{bmatrix}$$

      Thus, $d_v = 153$ mm.
      Also

      $$b_w = 1000 \text{ mm}$$

      is the width of the unit strip.
      According to A23.3 Cl.11.3.6.3b, determine β as

      A23.3 Eq. 11.9
      $$\beta = \frac{230}{1000 + d_v} \qquad \qquad \textbf{[6.13]}$$

      $$= \frac{230}{1000 + 153} = 0.2$$

      Finally, determine V_c as

      A23.3 Eq. 11.6
      $$V_c = \phi_c \lambda \beta \sqrt{f_c'} \, b_w \, d_v \qquad \qquad \textbf{[6.12]}$$
      $$= 0.65(1)(0.2)\sqrt{25 \text{ MPa}} \, (1000 \text{ mm})(153 \text{ mm}) = 99.5 \text{ kN/m}$$

   b) According to A23.3 Cl.11.2.8.1, shear reinforcement is not required provided that

      $$V_f \leq V_c$$

      Because

      $$V_f = 45 \text{ kN/m} < V_c = 99.5 \text{ kN/m}$$

      shear reinforcement is not required.
      However, CSA A23.3 requires the provision of minimum horizontal reinforcement in the walls regardless of design calculations, as discussed below.

   c) Determine the minimum distributed horizontal reinforcement.

      i) Determine the area of reinforcement.

         First, determine the gross cross-sectional area of the wall for a 1 m strip in the vertical direction (see Figure 13.6) as in step 2 f):

         $$A_g = 200 \times 10^3 \text{ mm}^2$$

         According to A23.3 Cl.14.1.8.6,

         $$A_{hmin} = 0.002A_g \qquad \qquad \textbf{[13.2]}$$
         $$= 0.002 \times (200 \times 10^3 \text{ mm}^2) = 400 \text{ mm}^2/\text{m}$$

         Hence,

         $$A_s = A_{hmin} = 400 \text{ mm}^2/\text{m}$$

      ii) Determine the required bar spacing.
          For 15M bars,

          $$A_b = 200 \text{ mm}^2 \text{ (see Table A.1)}$$

The spacing of reinforcement can be determined in the same way as for one-way slabs (see Section 3.6):

$$s \leq A_b \frac{1000}{A_s}$$ [3.29]

$$= 200 \text{ mm}^2 \times \frac{1000 \text{ mm}}{400 \text{ mm}^2/\text{m}} = 500 \text{ mm}$$

Therefore, the required bar spacing is

$$s = 500 \text{ mm}$$

iii) Check whether the spacing is within the limits prescribed by CSA A23.3. According to A23.3 Cl.14.1.8.4, the maximum permitted bar spacing (s_{max}) is equal to the lesser of $3t$ and 500 mm.
Since

$$3t = 3(200 \text{ mm}) = 600 \text{ mm}$$

$$s_{max} = 500 \text{ mm (governs)}$$

Because

$$s = s_{max} = 500 \text{ mm}$$

the spacing is within the prescribed limits; therefore, use 15M bars at 500 mm spacing (15M@500).

iv) Check whether one layer of reinforcement is adequate.
According to A23.3 Cl.14.1.8.3, one layer of reinforcement is adequate when the wall thickness is less than 210 mm, which is true in this case:

$$t = 200 \text{ mm} < 210 \text{ mm}$$

4. **Provide a design summary**
Finally, summarize the results of the design of the basement wall with a sketch showing the reinforcement. The reinforcement of the slab and the footings is shown for completeness.

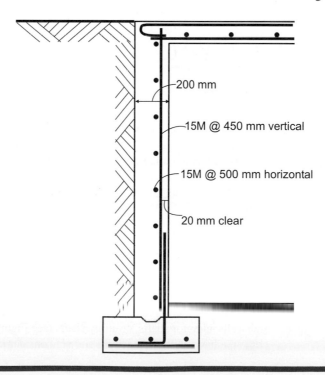

Basement walls are exterior walls located below ground (grade) level in a building. These walls should be designed to resist the combined effects of lateral earth pressure and axial compressive loads transferred from the building superstructure above.

Basement walls are usually laterally supported at the bottom by the footings and slab on grade and at the top by the suspended floor slab. The walls span like vertical slabs between these two supports.

The main loads acting on basement walls are

- gravity loads
- earth pressure
- surcharge
- seismic loads

Basement walls are designed for *gravity loads* in the same manner as bearing walls.

The magnitude of the *earth pressure* is considered to be proportional to the depth below the ground surface and the surcharge soil weight in a manner similar to a fluid. This corresponds to a triangular load distribution with a zero value at the top and a maximum value at the base of the wall.

The *surcharge* includes various dead and live loads imposed on a backfill surface behind the wall. For design purposes, the surcharge is modelled as a uniformly distributed load over the wall height.

The key design considerations related to basement walls are

- axial load
- flexure
- shear
- sliding

For design purposes, basement walls are treated as if they are composed of a series of vertical beams placed side by side (similar to the design of one-way slabs). Each beam has a rectangular cross-section with a 1 m width (b) and a depth (t) equal to the wall thickness.

13.6 SHEAR WALLS

13.6.1 Background

Reinforced concrete shear walls are vertical structural members used in medium and high-rise construction. The primary role of shear walls is to resist lateral loads due to earthquakes and wind. Shear walls provide resistance along the plane of their length (in-plane) and are weak when loaded perpendicular to that plane (out-of-plane). In-plane and out-of-plane wall loads are shown in Figure 13.3.

Based on their location and function in the building, shear walls can be classified as follows:

- *Bearing walls* are shear walls that also support a substantial fraction of gravity loads; these walls are also used as partition walls between adjacent apartments (see Figure 13.21a).
- *Frame walls* support mainly lateral loads, whereas gravity loads are carried by the reinforced concrete frame (called "gravity frame"); these walls are constructed between column lines (see Figure 13.21b).
- *Core walls* are incorporated into the central core area within a building, which may include stairwells, elevator shafts, or utility shafts (see Figure 13.21c). Walls located in the central core area have multiple functions and are considered to be an economical option.

Depending on the wall geometry, in particular the height/length (h_w/l_w) *aspect ratio,* shear walls are classified into two categories (A23.3 Cl.2.2):

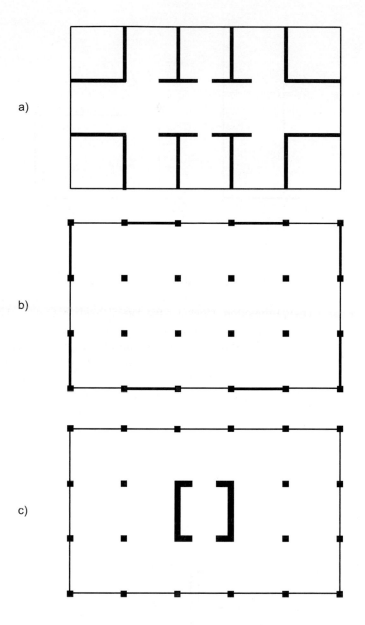

Figure 13.21 Functions of shear walls: a) bearing walls; b) frame walls; c) core walls.

- *Flexural shear walls* ($h_w/l_w > 2$) are tall and slender walls found in high-rise buildings (see Figure 13.22a).
- *Squat shear walls* ($h_w/l_w < 2$) are short and sturdy walls found in low-to-medium-rise buildings (see Figure 13.22b).

Flexural shear walls without openings (solid walls), like the one shown in Figure 13.21a, are also called *isolated walls,* whereas shear walls perforated with a regular pattern of openings (doors or windows) are called *coupled walls* (see Figure 13.22c).

13.6.2 Loads and Load Path

The lateral load path through a building was reviewed in Section 1.3.3 (see Figure 1.10). Lateral loads include wind and earthquake (seismic) loads. The procedures for determining the magnitude of wind and seismic loads are prescribed by NBC 2010 and are beyond the scope of this book. In general, wind loads act on the exterior wall panels that transfer the loads to the floor and roof structures. In the case of rigid reinforced concrete floors, the loads are transferred to the walls that serve as supports to the floors. Each wall carries a fraction of the total lateral load acting at a particular floor level, depending on its stiffness relative to other walls. (Refer to Section 4.3.1 for a discussion of flexural stiffness.)

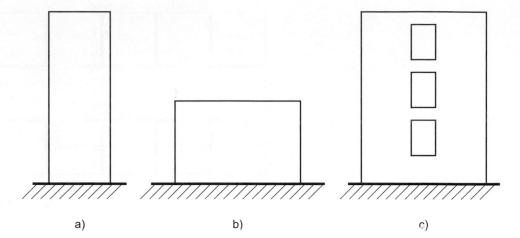

Figure 13.22 Types of shear walls: a) flexural shear walls; b) squat shear walls; c) coupled shear walls.

A plan of a typical four-storey shear wall building is shown in Figure 13.23a. Note that two shear walls each are provided in the north-south (N-S) direction and east-west (E-W) direction. Consider the lateral loads acting in the N-S direction. The loads are applied at the floor levels; for example, lateral force F_3 is resisted by the shear wall in bay-line 1 at the third floor level, as shown in Figure 13.23a. The distribution of lateral loads in bayline 1 over the building height is shown in Figure 13.23b. As mentioned earlier, the columns are expected to carry only gravity loads. Therefore, all lateral loads acting on the frame in bayline 1 are actually resisted by the shear wall, as shown in Figure 13.23c. To satisfy the equilibrium of horizontal forces, the sum of the lateral loads acting over the wall height must be resisted by the horizontal force acting at the base of the wall and transferred to the foundations and then to the ground, as shown in Figure 13.23c.

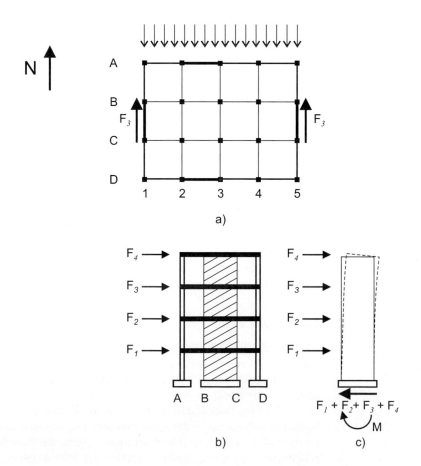

Figure 13.23 Shear wall building: a) plan of the third floor; b) vertical section through bayline 1 showing the distribution of lateral loads; c) force distribution along the wall height.

Earthquake (seismic) forces are lateral forces generated by ground shaking that lasts for a very short time (in general less than a minute!). Earthquake forces are proportional to the acceleration and the mass of a building and are called *inertia forces*. Most of the building mass is concentrated at the floor and roof levels, so the seismic forces are considered to act at these levels. The designs of shear walls for wind and earthquake loads are similar to a certain extent. However, the seismic design of shear walls and other structural members includes a number of special design and detailing requirements that are beyond the scope of this book.

Let us discuss the load distribution over the wall height. According to A23.3 Cl.2.2, shear walls are subjected to the following forces:

- vertical (axial) forces due to gravity loads
- moments about horizontal axes perpendicular to the plane of the wall
- shear forces due to lateral loads acting in the plane of the wall

In general, shear walls are subjected to lateral loads at the floor and roof levels, as shown in Figure 13.24a. The distribution of forces in shear walls is similar to that of vertical cantilevered beams fixed at their base. Figure 13.24b shows internal forces acting at section A-A at the base of the wall. Note that the wall section is subjected to a shear force (V) equal to the sum of the horizontal forces acting above the section under consideration, a bending moment (M) created by the horizontal forces acting above the section under consideration, and an axial force (N) equal to the sum of the axial loads acting on the wall. The shear force, bending moment, and axial force diagrams are shown in Figure 13.24c.

13.6.3   Behaviour and Failure Modes

Failure modes in *flexural shear walls* are strongly influenced by the amount of flexural reinforcement and the area of concrete under compression. These walls fail in a similar manner to beams subjected to flexure and shear, as follows:

- *Flexural failure.* Walls with a low flexural (vertical) reinforcement ratio fail in the flexural mode, with the vertical reinforcement yielding at the base of the wall. In general, only one

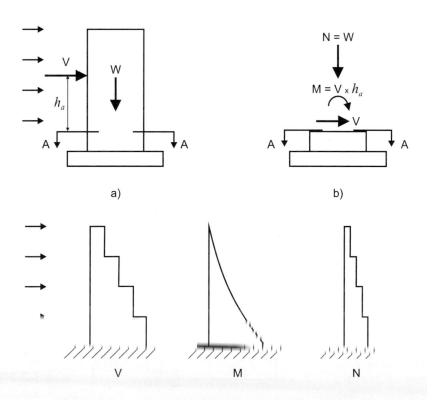

Figure 13.24 Load distribution in shear walls: a) wall elevation showing lateral and axial loads; b) section A-A at the base showing internal forces; c) shear force, bending moment, and axial load diagrams.

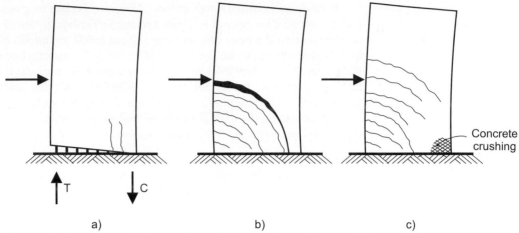

a)   b)   c)

Figure 13.25   Flexural shear walls — failure modes: a) flexural failure; b) flexural-shear failure; c) concrete crushing failure mode.

wide horizontal crack forms at the base of the wall before the reinforcement fractures (see Figure 13.25a).

- *Flexural-shear failure*. The walls fail in shear after a large number of flexural cracks form. The critical crack is a diagonal tension crack that propagates downward at 45°. At failure, some of the horizontal reinforcement across the critical crack may fracture and lead to the crack opening and crushing of the compression zone at the root of the crack (see Figure 13.25b).

- *Concrete crushing failure*. This is the most common failure mode, characterized by a considerable yielding of the flexural reinforcement and crushing of the concrete at the base of the wall (see Figure 13.25c).

In general, *squat shear walls* behave in a different manner than flexural shear walls. Due to a low aspect ratio, these walls fail in the following shear-controlled modes, as shown in Figure 13.26:

- *Diagonal tension failure* is characteristic of walls with a very low amount of horizontal shear reinforcement. As a result, an inclined corner-to-corner crack develops, as shown in Figure 13.26a, thus separating the wall into two segments. In longer walls, the failure occurs along a steeper failure plane (usually under an approximately 45° angle), as shown in Figure 13.26b.

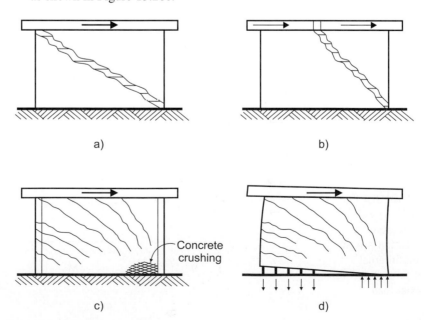

a)   b)

Figure 13.26   Failure modes for squat shear walls: a) and b) diagonal tension; c) diagonal compression; d) sliding shear.

c)   d)

- *Diagonal compression failure* occurs in walls with large flexural resistance (usually with boundary members provided) that are reinforced with an adequate amount of horizontal shear reinforcement. When subjected to high shear stresses, these walls fail by concrete crushing in the wall compression zone, as shown in Figure 13.26c. This failure mode leads to a significant reduction in the wall compression resistance, especially under seismic loading. In that case, the concrete crushing zone can spread over the entire wall length.
- *Sliding shear failure* occurs when the diagonal tension and diagonal compression failure modes are avoided by design and the wall fails in the flexural mode. When a large flexural crack develops at the base of the wall, the shear force has to be transferred across a limited contact area under compression, as shown in Figure 13.26d; this behaviour is associated with the shear friction mechanism explained in Section 6.10. A portion of the shear force is transferred by the *dowel mechanism,* which consists of the tension reinforcement kinking across the crack at the wall base combined with the *aggregate interlock* mechanism (these two mechanisms were discussed in Section 6.3.2). Sliding shear failure can occur in both squat and flexural walls; however, it is more common in squat walls due to the low gravity loads these walls are subjected to. In some cases, walls failing in sliding shear can experience large horizontal displacements.

Shear-controlled failure modes are undesirable and should be avoided by skillful design and detailing. This is especially important for structures built in high seismic zones. An example of the shear-controlled failure of a shear wall with an inadequate amount of horizontal reinforcement and poor quality concrete in the 2003 Boumerdes, Algeria earthquake is shown in Figure 13.27.

Based on the above discussion, it is apparent that the failure of squat walls under a distributed shear load (v) generally involves the development of inclined shear cracks, as shown in Figure 13.28. The areas of concrete between the cracks act as compression struts (shown hatched in Figure 13.28a). These compression struts are in equilibrium only if there is enough vertical reinforcement to develop a tensile force (t_v) opposite to the vertical component of the compression force in the strut (see Figure 13.28b). The tensile force developed in the horizontal reinforcement is denoted as t_h, as shown on the free-body diagram of the wall (Figure 13.28c). If the shear crack angle is equal to 45° and the effect of gravity loads is ignored, then the tensile forces developed in the horizontal and vertical reinforcement are equal; that is, $t_v = t_h$ (see Figure 13.28c). The design of squat shear walls is usually performed using the "strut and tie" approach outlined in A23.3 Cl.11.4.

Coupled walls act as a series of isolated walls that are mutually connected (coupled) by means of coupling beams or spandrel beams, as shown in Figure 13.29a. If the coupling beams are flexible, they act as links or "fuses" in case of extreme loads such as earthquake ground motion. Consequently, a coupled wall deforms as a series of cantilevered walls connected by

Figure 13.27 Shear failure of a reinforced concrete shear wall in the Mag 2003 Boumerdes, Algeria earthquake; note the fractured horizontal reinforcement.

(Svetlana Brzev)

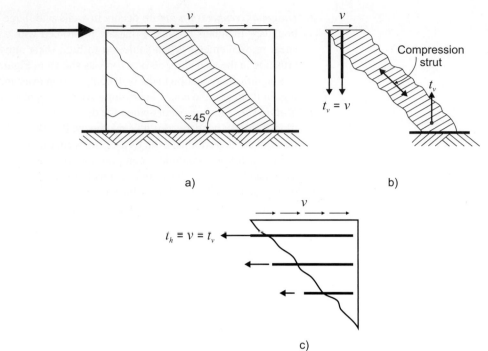

Figure 13.28 Behaviour of squat shear walls subjected to shear: a) crack pattern; b) typical compression strut; c) free-body diagram of a wall with horizontal reinforcement.

links, as shown in Figure 13.29b. The coupling beams are subjected to large shear forces and need to be designed accordingly. When the coupling beams are very stiff, a coupled wall acts as a unit. This is an undesirable design concept because the wall segments between the openings might fail in shear; note the diagonal tension cracks at the wall base in Figure 13.29c. The whole building could fail as a result of such a mechanism, as shown in Figure 13.29d.

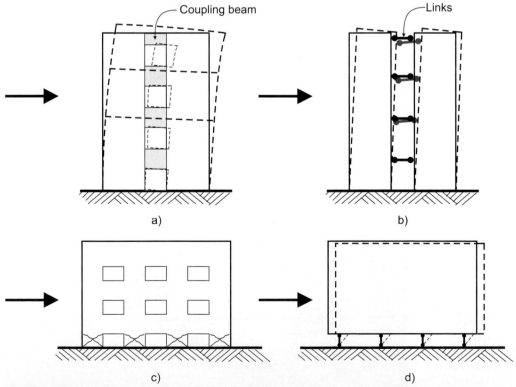

Figure 13.29 Failure modes in coupled shear walls: a) behaviour of a coupled shear wall with flexible coupling beams; b) a structural model showing coupling beams modelled as links; c) stiff coupling beams; d) a possible failure mechanism for a coupled wall with stiff beams.

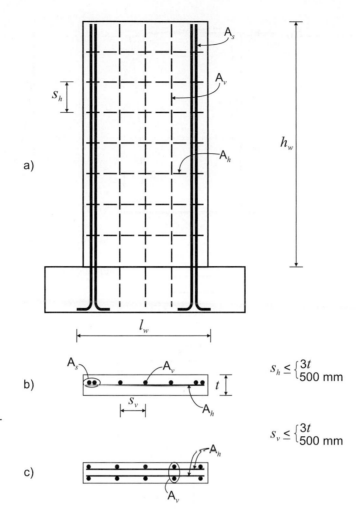

Figure 13.30 Shear wall reinforcement: a) wall elevation; b) wall cross-section showing one curtain of reinforcement; c) wall cross-section showing two curtains of reinforcement.

13.6.4 Design of Flexural Shear Walls

According to CSA A23.3 Cl.14.4.1, flexural shear walls need to be designed for the effects of flexure and axial load according to A23.3 Cl.10 (like columns) and for shear according to A23.3 Cl.11 (like beams). This section is focused on the design of flexural shear walls for flexure and shear effects. Flexural shear walls should be designed for axial loads like the bearing walls discussed in Section 13.4.

A typical reinforced concrete shear wall of height h_w, length l_w, and thickness t is shown in Figure 13.30. In general, a shear wall is reinforced with the following three types of reinforcement:

- distributed horizontal reinforcement (area A_h)
- distributed vertical reinforcement (area A_v)
- concentrated reinforcement (area A_s)

The main role of horizontal reinforcement is to resist shear effects, whereas both the distributed and the concentrated vertical reinforcement are responsible for resisting bending moments and axial loads. It should be noted that A_h and A_v refer to the area of one reinforcement bar only when there is one curtain of reinforcement (see Figure 13.30b) or two reinforcement bars when two curtains are provided, as shown in Figure 13.30c.

The CSA A23.3 reinforcement requirements for shear walls under nonseismic conditions are outlined in Section 13.3.2. Note that CSA A23.3 stipulates the provision of minimum distributed and concentrated wall reinforcement regardless of design calculations.

Design for flexure As discussed in the previous sections, flexural shear walls subjected to lateral load behave like vertical cantilevered beams. A typical flexural shear wall subjected

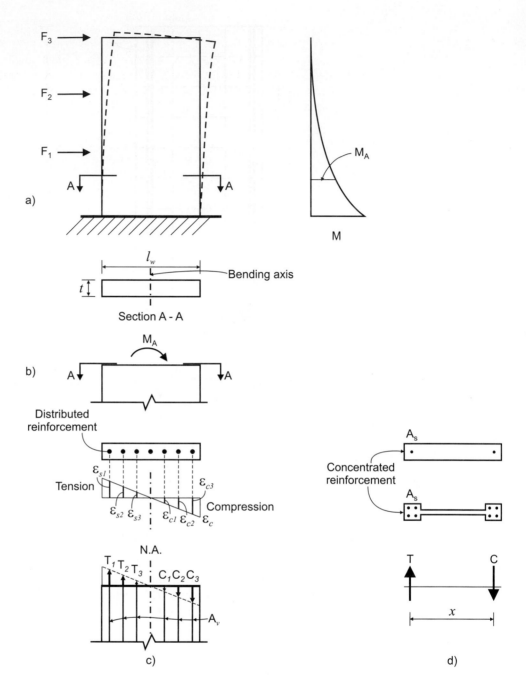

Figure 13.31 Shear wall design for flexure: a) distribution of lateral forces and bending moments; b) bending moment at section A-A; c) strains and internal forces in distributed reinforcement; d) internal forces in concentrated reinforcement.

to concentrated lateral loads is shown in Figure 13.31a. The wall is subjected to lateral load and the corresponding bending moment diagram looks like the bending moment diagram for a cantilever beam fixed at the base characterized by a zero value at the tip of the wall and a maximum at the base. Consider the wall section A-A at the base of the wall subjected to the bending moment M_A in Figure 13.31b; note that the bending moment at the base of the wall is also called the *overturning moment*. The bending moment causes tensile and compressive strains and corresponding flexural stresses. The wall is reinforced with uniformly distributed vertical reinforcement, as shown in Figure 13.31c. Depending on their location, some rebars act in tension (forces T_1, T_2, and T_3), while the remaining bars act in compression (forces C_1, C_2, and C_3). Due to the linear strain distribution, the bars farther away from the neutral axis develop larger forces than the bars located near the centre. Note that the concrete in the compression zone also develops an internal compressive force, similar to other flexural members such as beams and columns.

Alternatively, let us consider that the same wall has been reinforced with concentrated reinforcement of area A_s at the wall ends only, as shown in Figure 13.31d. In this case, the bending moment (M_A) will be resisted by the force couple comprising a tensile force in steel reinforcement (T) at one wall end and a concrete compressive force (C) at the other end. Note that the distance between the centroids of concentrated reinforcement regions is denoted as x. It follows that

$$M_A = T \times x$$

However, if we assume that the concentrated reinforcement of area A_s has yielded, then the tensile force (T) can be determined as

$$T = \phi_s f_y A_s$$

If the factored bending moment at a wall section is given, the required area of reinforcement can be determined as

$$A_s = \frac{M_A}{(\phi_s f_y)\, x}$$

This is a simple approximate method for determining the required amount of concentrated reinforcement in shear walls and can be used for preliminary design.

The end zones in shear walls with concentrated reinforcement behave like reinforced concrete columns. When a large amount of concentrated reinforcement is required, the wall section is increased at the ends to create a column-like element called a *boundary element,* as shown in Figure 13.31c.

Note that concentrated reinforcement has an important role in resisting flexure in tall flexural shear walls subjected to significant bending moments. However, in low-to-medium-rise concrete buildings, distributed vertical reinforcement can resist the effects of flexure and axial loads by itself. The moment resistance for walls with distributed vertical reinforcement such as the one in Figure 13.32a can be determined based on an approximate equation proposed

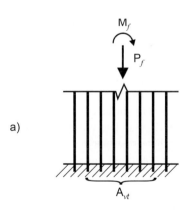

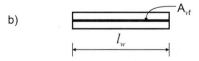

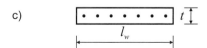

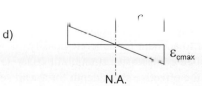

Figure 13.32 Shear wall with distributed vertical reinforcement: a) vertical elevation; b) equivalent cross-section; c) actual cross-section; d) strain distribution.

by Cardenas and Magura (1973). The equation was derived based on the assumption that the distributed wall reinforcement acts like a plate of length l_w and thickness such that the area A_{vt} (see Figure 13.32b) is the same as that provided by uniformly distributed reinforcement along the wall length (see Figure 13.32c). The moment resistance (M_r) can be determined as follows:

$$M_r = 0.5\phi_s f_y A_{vt} l_w \left(1 + \frac{P_f}{\phi_s f_y A_{vt}}\right)\left(1 - \frac{c}{l_w}\right) \qquad \text{[13.7]}$$

where

A_{vt} = the total area of distributed vertical reinforcement (see Figure 13.32a)
c = the neutral axis depth (see Figure 13.32d)

$$\frac{c}{l_w} = \frac{\omega + \alpha}{2\omega + \alpha_1 \beta_1} \qquad \text{[13.8]}$$

$$\omega = \frac{\phi_s f_y A_{vt}}{\phi_c f_c' l_w t} \qquad \text{[13.9]}$$

$$\alpha = \frac{P_f}{\phi_c f_c' l_w t} \qquad \text{[13.10]}$$

$$\alpha_1 \cong 0.8 \text{ and } \beta_1 \cong 0.9 \qquad \text{[3.7]}$$

In low-to-medium-rise buildings, the axial compressive load is generally quite low. Therefore, it is conservative to ignore axial loads and consider only the bending moments at the base of the wall; that is, $P_f = 0$. In that case, Eqn 13.7 can be further simplified as follows:

$$M_r = 0.5\phi_s f_y A_{vt} l_w \left(1 - \frac{c}{l_w}\right) \qquad \text{[13.11]}$$

In general, shear walls subjected to combined axial load and flexure can be designed in a similar manner to columns using interaction diagrams discussed in Section 8.7.

Design for shear   The shear resistance of reinforced concrete shear walls can be determined in a manner similar to beams, as discussed in Section 6.5. A shear wall subjected to lateral loads acts as a vertical cantilevered beam and the corresponding shear force diagram is shown in Figure 13.33a. The maximum shear force develops at the base of the wall. When the concrete shear resistance is exceeded, diagonal tension cracks develop at the base of the wall. From that point onward, horizontal shear reinforcement is critical in providing the wall shear resistance, as depicted by the free-body diagram in Figure 13.33b.

According to A23.3 Cl.11.3.3, the shear resistance of a shear wall section is supplied by the concrete and the horizontal reinforcement, as follows:

| A23.3 Eq. 11.4 | $V_r = V_c + V_s$ [6.1] |

where V_c is the concrete shear resistance and V_s is the steel shear resistance. The design of flexural shear walls for shear in nonseismic conditions can be performed using the CSA A23.3 simplified method (Cl.11.3.6.3).

The shear resistance of concrete (V_c) can be determined in the same way as for beams and slabs, as discussed in Section 6.5.4 (A23.3 Cl.11.3.4):

| A23.3 Eq. 11.6 | $V_c = \phi_c \lambda \beta \sqrt{f_c'} \, t \, d_v$ [6.12] |

where

ϕ_c = the resistance factor for concrete
λ = the factor to account for concrete density
f_c' = the specified compressive strength of concrete
$d_v = 0.8 \, l_w$ is the effective shear depth for a wall section

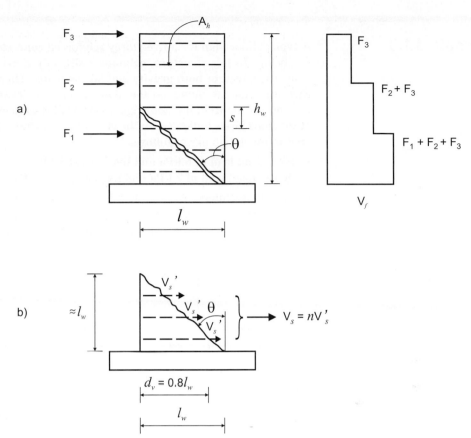

Figure 13.33 Behaviour of flexural shear walls subjected to shear loads: a) wall elevation and shear force diagram; b) free-body diagram of a cracked wall.

β = the factor accounting for the shear resistance of cracked concrete determined according to A23.3 Cl.11.3.6.3; $\beta = 0.18$ if the section contains minimum shear reinforcement

The shear resistance provided by the horizontal reinforcement (V_s) can be determined according to A23.3 Cl.11.3.5 (for more details refer to Section 6.4.4):

$$V_s = \frac{\phi_s A_h f_y d_v \cot \theta}{s}$$ [6.9]

where

s = the spacing of horizontal reinforcement (see Figure 13.33a)

A_h = the area of horizontal reinforcement crossing the cracks spaced at a distance s; note that A_h refers to the area of one reinforcement bar when there is one curtain of reinforcement and to the area of two reinforcement bars when two curtains are provided

f_y = the specified yield strength for horizontal reinforcement

ϕ_s = the resistance factor for steel

θ = 35° is the angle between a diagonal tension crack and the longitudinal wall axis (A23.3 Cl.11.3.6.3); see Figure 13.33a.

According to A23.3 Cl.11.2.8.1, shear reinforcement is not required if

$$V_f \leq V_n$$

However, a minimum amount of distributed horizontal reinforcement needs to be provided in any case (A23.3 Cl.14.1.8.6), as discussed in Section 13.3.2.

The design of flexural shear walls for the combined effects of flexure, axial load, and shear will be illustrated with an example.

Example 13.3

A typical floor plan for a five-storey reinforced concrete building is shown in the figure below. In this structure, columns resist only gravity loads, whereas shear walls resist the effects of both gravity and lateral loads. There are four shear walls, two in each direction, as shown on the plan. A vertical elevation of shear wall SW1 along bayline 2 is shown in the same figure. The wall is subjected to the specified wind loads at the floor levels indicated on the elevation drawing. The gravity loads to be considered in the design are as follows:

- **roof: dead load of 6 kPa and live load of 1 kPa**
- **floors: dead load of 7 kPa and live load of 4.8 kPa**

Use a wall thickness of 200 mm and 15M reinforcing bars in the design.
Design the critical section of shear wall SW1 for the effects of gravity and lateral loads. The design needs to be in compliance with CSA A23.3 and NBC 2010.

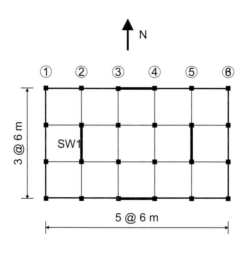

Typical floor plan

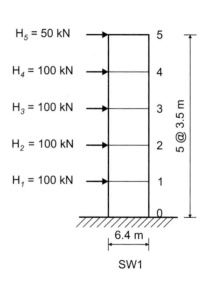

SW1

Given: $f_c' = 30$ MPa
          $f_y = 400$ MPa
          $\phi_c = 0.65$
          $\phi_s = 0.85$

SOLUTION: **1.   Determine the factored loads**

  a)   Identify the critical load combination(s).

  In this example, the combined effects of dead load (DL), live load (LL), and wind load (WL) need to be considered. According to Table 4.1.3.2 of NBC 2010, the following load combination needs to be considered in this design

  1.25 DL + 1.4 WL + 0.5 LL

  Note that a few other load combinations would need to be considered in the detailed design stage of a project.

  b)   Determine the factored wind load.

  In order to obtain the factored wind loads, multiply all the specified loads by 1.4; that is,

  $$H_{f5} = 1.4 \times 50 \text{ kN} = 70 \text{ kN}$$

  $$H_{f4} = H_{f3} = H_{f2} = H_{f1} = 1.4 \times 100 \text{ kN} = 140 \text{ kN}$$

c) Then, find the factored gravity loads.

Gravity loads, including dead and live loads, create an axial compressive load in wall SW1. The factored vertical loads at each level will be determined based on the above-specified NBC 2010 load combination. (Note that the wind does not represent a gravity load, so WL = 0.)

i) Roof level:

DL = 6 kPa and LL = 1 kPa

$$w_{froof} = 1.25\ DL + 0.5\ LL$$

$$= 1.25(6\ kPa) + 0.5(1.0\ kPa) = 8\ kPa$$

ii) Floor levels:

DL = 7 kPa and LL = 4.8 kPa

$$w_{ffloor} = 1.25\ DL + 0.5\ LL$$

$$= 1.25(7\ kPa) + 0.5(4.8\ kPa) = 11.2\ kPa$$

iii) Tributary area:

The tributary area for the gravity loads for wall SW1 is illustrated in the sketch below. This area is

$$A = 6\ m \times 12\ m = 72\ m^2$$

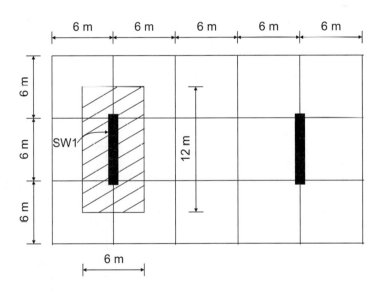

iv) Factored axial load:

Based on the above calculations, compute the axial force acting at a certain level by multiplying the factored load by the tributary area, as follows:

$$P_{f5} = w_{froof} \times A = (8\ kPa) \times (72\ m^2) = 576\ kN \cong 600\ kN$$

$$P_{f4} = P_{f3} = P_{f2} = P_{f1}$$

$$= w_{ffloor}\ A$$

$$= (11.2\ kPa) \times (72\ m^2) = 806\ kN \cong 800\ kN$$

2. **Determine the design forces**

a) Develop the load distribution diagrams.

Develop the bending moment diagrams based on fundamental static analysis. As shown in the load diagram below, the wall acts like a vertical cantilevered beam.

The lateral wind load (forces H_{f1} to H_{f5}) causes shear forces and bending moments; the corresponding diagrams are shown below.

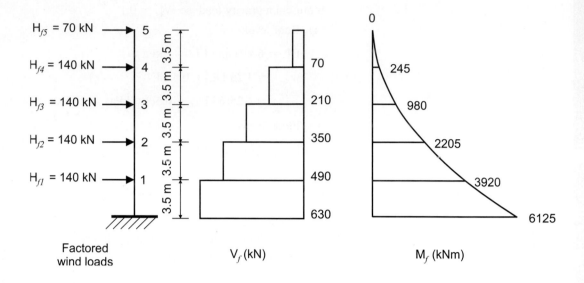

Factored wind loads

V_f (kN)

M_f (kNm)

Similarly, the axial forces (P_{f1} to P_{f5}) cause internal axial loads to develop in the wall; the corresponding axial load diagram is shown below.

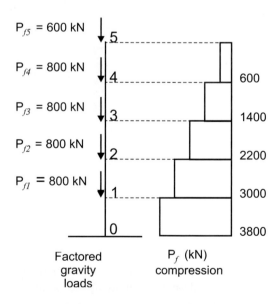

Factored gravity loads

P_f (kN) compression

b) Determine the design forces at the critical section.

Observe from the shear force, bending moment, and axial load diagrams that these forces increase from the top of the wall downward and that the largest values occur at the base of the wall. Use the following design forces (see the sketch that follows):

$M_f = 6100 \text{ kN} \cdot \text{m}$

$V_f = 630 \text{ kN}$

$P_f = 3800 \text{ kN}$

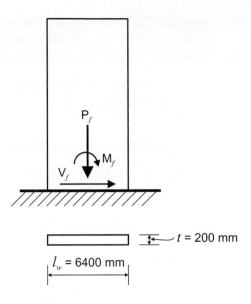

l_w = 6400 mm

t = 200 mm

3. **Design for shear**
    The overall wall length (l_w) is l_w = 6400 mm.
    The design shear force at the base of the wall is

    V_f = 630 kN

    a) Determine the concrete shear resistance (V_c).
        Determine the effective shear depth (d_v) for the wall section as

        $$d_v = 0.8l_w = 0.8(6400 \text{ mm}) = 5120 \text{ mm}$$

        According to A23.3 Cl.11.3.6.3, when the section contains minimum shear rein-
        forcement, then

        $$\beta = 0.18$$

        The wall thickness is

        $$t = 200 \text{ mm}$$

        Finally, determine V_c as

        | A23.3 Eq. 11.6 |

        $$V_c = \phi_c \lambda \beta \sqrt{f_c'} \, t \, d_v \qquad\qquad [6.12]$$
        $$= 0.65(1)(0.18)\sqrt{30 \text{ MPa}} \,(200 \text{ mm})(5120 \text{ mm})$$
        $$= 656.2 \times 10^3 \text{ N} = 656 \text{ kN}$$

    b) According to A23.3 Cl.11.2.8.1, shear reinforcement is not required provided that

        $$V_f \le V_c$$

        Because

        $$V_f = 630 \text{ kN} < V_c = 656 \text{ kN}$$

        shear reinforcement is not required.

However, CSA A23.3 requires the provision of minimum horizontal reinforcement in the walls regardless of design calculations.

c)   Determine the minimum distributed horizontal reinforcement.

i)   Determine the area of horizontal reinforcement.
First, determine the gross cross-sectional area of the wall for a 1 m strip in the vertical direction (see Figure 13.6) as

$$A_g = 1000 \text{ mm} \times t = 1000 \text{ mm} \times 200 \text{ mm} = 200 \times 10^3 \text{ mm}^2$$

According to A23.3 Cl.14.1.8.6,

$$A_{hmin} = 0.002A_g \qquad\qquad\qquad\qquad\qquad \textbf{[13.2]}$$
$$= 0.002 \times (200 \times 10^3 \text{ mm}^2) = 400 \text{ mm}^2/\text{m}$$

Hence,

$$A_s = A_{hmin} = 400 \text{ mm}^2/\text{m}$$

ii) Determine the required bar spacing.
For 15M bars,

$$A_b = 200 \text{ mm}^2 \text{ (see Table A.1)}.$$

Determine the required bar spacing in the same way as for one-way slabs (see Section 3.6):

$$s \le A_b \frac{1000}{A_s} \qquad\qquad\qquad\qquad\qquad \textbf{[3.29]}$$
$$= 200 \text{ mm}^2 \times \frac{1000 \text{ mm}}{400 \text{ mm}^2/\text{m}} = 500 \text{ mm}$$

Therefore, the required bar spacing is

$$s = 500 \text{ mm}$$

iii) Check whether the spacing is within the limits prescribed by CSA A23.3. According to A23.3 Cl.14.1.8.4, the maximum permitted bar spacing (s_{max}) is equal to the lesser of $3t$ and 500 mm. Since

$$3t = 3(200 \text{ mm}) = 600 \text{ mm}$$

then

$$s_{max} = 500 \text{ mm}$$

because

$$s = s_{max} = 500 \text{ mm}$$

the spacing is within the prescribed limits.
Use 15M bars at 500 mm spacing (15M@500).

iv) Check whether one layer of reinforcement is adequate.
According to A23.3 Cl.14.1.8.3, one layer of reinforcement is adequate if the wall thickness is less than 210 mm, which is true in this case:

$$t = 200 \text{ mm} < 210 \text{ mm}$$

4. **Design for flexure and axial load**
   The design axial load and bending moment at the base of the wall are

   $$P_f = 3800 \text{ kN}$$

   and

   $$M_f = 6125 \text{ kN} \cdot \text{m} \cong 6100 \text{ kN} \cdot \text{m}$$

   a) Estimate the vertical wall reinforcement.
      Assume that the wall contains the minimum amount of distributed reinforcement and check whether that amount is adequate.

      i) Determine the area of reinforcement.
         First, determine the gross cross-sectional area of the wall for a 1 m strip in the vertical direction (see Figure 13.6) from Step 3 c) i) as

         $$A_g = 200 \times 10^3 \text{ mm}^2$$

         According to A23.3 Cl.14.1.8.5,

         $$A_{vmin} = 0.0015 A_g \qquad \text{[13.1]}$$
         $$= 0.0015 \times (200 \times 10^3 \text{ mm}^2) = 300 \text{ mm}^2/\text{m}$$

         Hence,

         $$A_s = A_{vmin} = 300 \text{ mm}^2/\text{m}$$

      ii) Determine the required bar spacing.
          For 15M bars, $A_b = 200 \text{ mm}^2$ (see Table A.1).
          Determine the spacing of reinforcement in the same way as for horizontal reinforcement (see Step 3):

          $$s \leq A_b \frac{1000}{A_s} \qquad \text{[3.29]}$$

          $$= 200 \text{ mm}^2 \times \frac{1000 \text{ mm}}{300 \text{ mm}^2/\text{m}} = 667 \text{ mm}$$

          Therefore, the required bar spacing is

          $$s = 667 \text{ mm}$$

      iii) Check whether the spacing is within the limits prescribed by CSA A23.3.
           According to A23.3 Cl.14.1.8.4, the maximum permitted bar spacing (s_{max}) is equal to the lesser of $3t$ and 500 mm.
           Because

           $$3t = 3(200 \text{ mm}) = 600 \text{ mm}$$

           $$s_{max} = 500 \text{ mm}$$

           Since

           $$s = 666 \text{ mm} > s_{max}$$

           use $s_{max} = 500$ mm.

Therefore, the distributed vertical reinforcement consists of 15M bars at 500 mm spacing (15M@500), and the corresponding area is

$$A_v = \frac{A_b}{s} \times 1000 \text{ mm} = \frac{200 \text{ mm}^2}{500 \text{ mm}} \times 1000 \text{ mm} = 400 \text{ mm}^2/\text{m}$$

b) Determine the moment resistance (M_r) for the wall from Eqn 13.7 (see Figure 13.31).
    i) First, determine the total area of vertical reinforcement (A_{vt}) along the wall length:

$$A_{vt} = A_v \times l_w = (400 \text{ mm}^2/\text{m}) \times 6.4 \text{ m} = 2560 \text{ mm}^2$$

    ii) Then, calculate the parameters ω, α, and c/l_w:

$$\alpha_1 \cong 0.8 \text{ and } \beta_1 \cong 0.9 \qquad\qquad\qquad\qquad \text{[3.7]}$$

$$\omega = \frac{\phi_s f_y A_{vt}}{\phi_c f_c' l_w t} \qquad\qquad\qquad\qquad\qquad\qquad \text{[13.9]}$$

$$= \frac{0.85 \times 400 \text{ MPa} \times 2560 \text{ mm}^2}{0.65 \times 30 \text{ MPa} \times 6400 \text{ mm} \times 200 \text{ mm}} = 0.035$$

$$\alpha = \frac{P_f}{\phi_c f_c' l_w t} \qquad\qquad\qquad\qquad\qquad\qquad \text{[13.10]}$$

$$= \frac{3800 \times 10^3 \text{ N}}{0.65 \times 30 \text{ MPa} \times 6400 \text{ mm} \times 200 \text{ mm}} = 0.15$$

$$\frac{c}{l_w} = \frac{\omega + \alpha}{2\omega + \alpha_1 \beta_1} \qquad\qquad\qquad\qquad\qquad \text{[13.8]}$$

$$= \frac{0.035 + 0.15}{2 \times 0.035 + 0.8 \times 0.9} = 0.23$$

    iii) Finally, determine the moment resistance (M_r) as

$$M_r = 0.5\phi_s f_y A_{vt} l_w \left(1 + \frac{P_f}{\phi_s f_y A_{vt}}\right)\left(1 - \frac{c}{l_w}\right) \qquad \text{[13.7]}$$

$$= 0.5 \times 0.85 \times 400 \text{ MPa} \times 2560 \text{ mm}^2 \times 6400 \text{ mm}$$

$$\times \left(1 + \frac{3800 \times 10^3 \text{ N}}{0.85 \times 400 \text{ MPa} \times 2560 \text{ mm}^2}\right)(1 - 0.23)$$

$$= 11508 \times 10^6 \text{ N} \cdot \text{mm} \cong 11\,500 \text{ kN} \cdot \text{m}$$

Because

$$M_r = 11\,500 \text{ kN} \cdot \text{m} > M_f = 6100 \text{ kN} \cdot \text{m}$$

the minimum distributed reinforcement is sufficient to resist the effects of combined axial load and flexure in this wall. However, A23.3 Cl.14.1.8.8.1 also requires the provision of nominal concentrated reinforcement (2-15M bars) at the ends of the wall cross-section. The concentrated reinforcement increases the moment resistance of the wall. The effect of concentrated reinforcement can be approximately estimated as follows (see the sketch below).

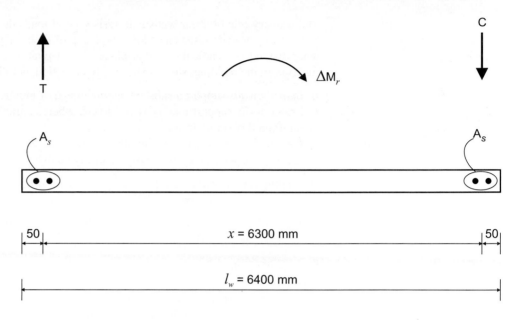

The total area of concentrated reinforcement at one end of the wall is

$$A_s = 2 \times 200 \text{ mm}^2 = 400 \text{ mm}^2$$

and the lever arm (distance between the centroids of the concentrated reinforcement) is

$$x = l_w - 2 \times 50 \text{ mm} = 6300 \text{ mm}$$

Finally, determine the moment resistance provided by the concentrated reinforcement only (assuming that the concentrated reinforcement has yielded) as

$$\Delta M_r = T \times x = (\phi_s f_y A_s)x = (0.85 \times 400 \text{ MPa} \times 400 \text{ mm}^2) \times 6300 \text{ mm}$$
$$= 857 \times 10^6 \text{ N} \cdot \text{mm} = 857 \text{ kN} \cdot \text{m}$$

ΔM_r can be added to the moment resistance (M_r) provided by the distributed reinforcement.

Note that the wall should also be designed for the effects of axial loads only, like a bearing wall, as shown in Example 13.1.

5. **Provide a design summary**

   Finally, summarize the results of this design with a sketch showing the wall cross-sectional dimensions and reinforcement distribution.

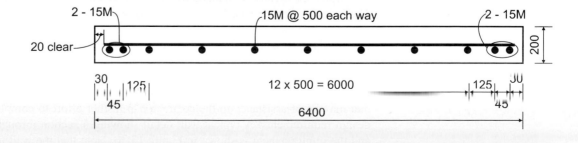

KEY CONCEPTS

The primary role of shear walls is to resist lateral loads due to earthquakes and wind. Shear walls provide resistance along the plane of their length (in-plane) and are weak when loaded perpendicular to that plane (out-of-plane). Based on their location and function in the building, shear walls can be classified as follows:

- *Bearing walls* support a substantial percentage of gravity loads.
- *Frame walls* support mainly lateral loads, whereas gravity loads are carried by the reinforced concrete frame.
- *Core walls* are incorporated into the central core area within a building, which may include stairwells, elevator shafts, or utility shafts.

Depending on the wall height/length (h_w/l_w aspect ratio, shear walls are classified into the following two categories (A23.3 Cl.2.2):

- *Flexural shear walls* ($h_w/l_w > 2$) are tall and slender walls in high-rise buildings.
- *Squat shear walls* ($h_w/l_w < 2$) are short walls in low-to-medium-rise buildings.

Shear walls perforated with a regular pattern of openings (doors or windows) are called *coupled walls*.

Flexural shear walls behave like vertical beams subjected to flexure and shear and are characterized by the following three failure modes:

- flexural failure
- flexural-shear failure
- concrete crushing failure

Due to a low aspect ratio, squat shear walls fail in the following shear-controlled modes:

- diagonal tension failure
- diagonal compression failure
- sliding shear failure

Flexural shear walls should be designed like columns for the combined effects of axial load and flexure and like beams for shear effects.

13.7 | STRUCTURAL DRAWINGS AND DETAILS FOR REINFORCED CONCRETE WALLS

In a typical reinforced concrete building, there are different types of reinforced concrete walls, such as bearing walls, shear walls, exterior walls, and basement walls. These walls have different design criteria and generally require specifically designed reinforcing steel.

Walls are usually designated by the letter W and a number (for example, W1, W2). These designations are used to identify walls on the floor plan and to provide a tie with the wall schedule. A wall schedule is used to present design requirements for various walls in the building and is a good way of conveying the design information to the contractor. An example of a wall schedule is shown in Figure 13.34, while an example of a zone schedule is shown in Figure 13.35.

In addition to the floor plans, structural drawings should contain all relevant details that may be of assistance to the contractor in his/her effort to completely understand the design requirements. A typical wall detail showing a section through the basement wall and the reinforcement is shown in Figure 13.36. Note that the wall footing and the wall-to-floor joint are also shown on the section.

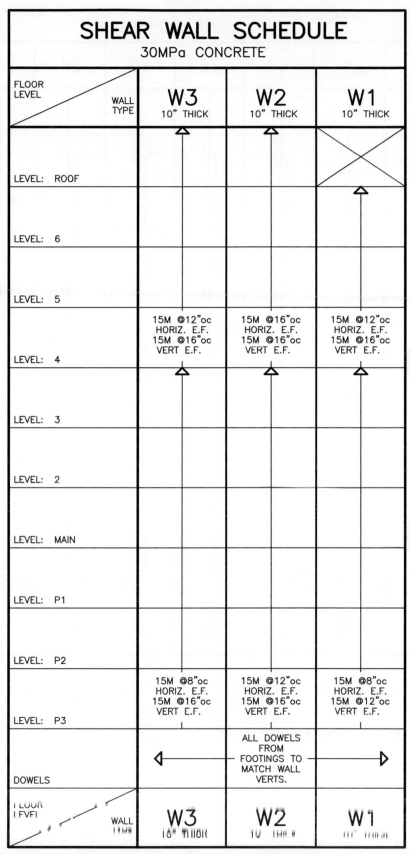

Figure 13.34 Wall schedule.

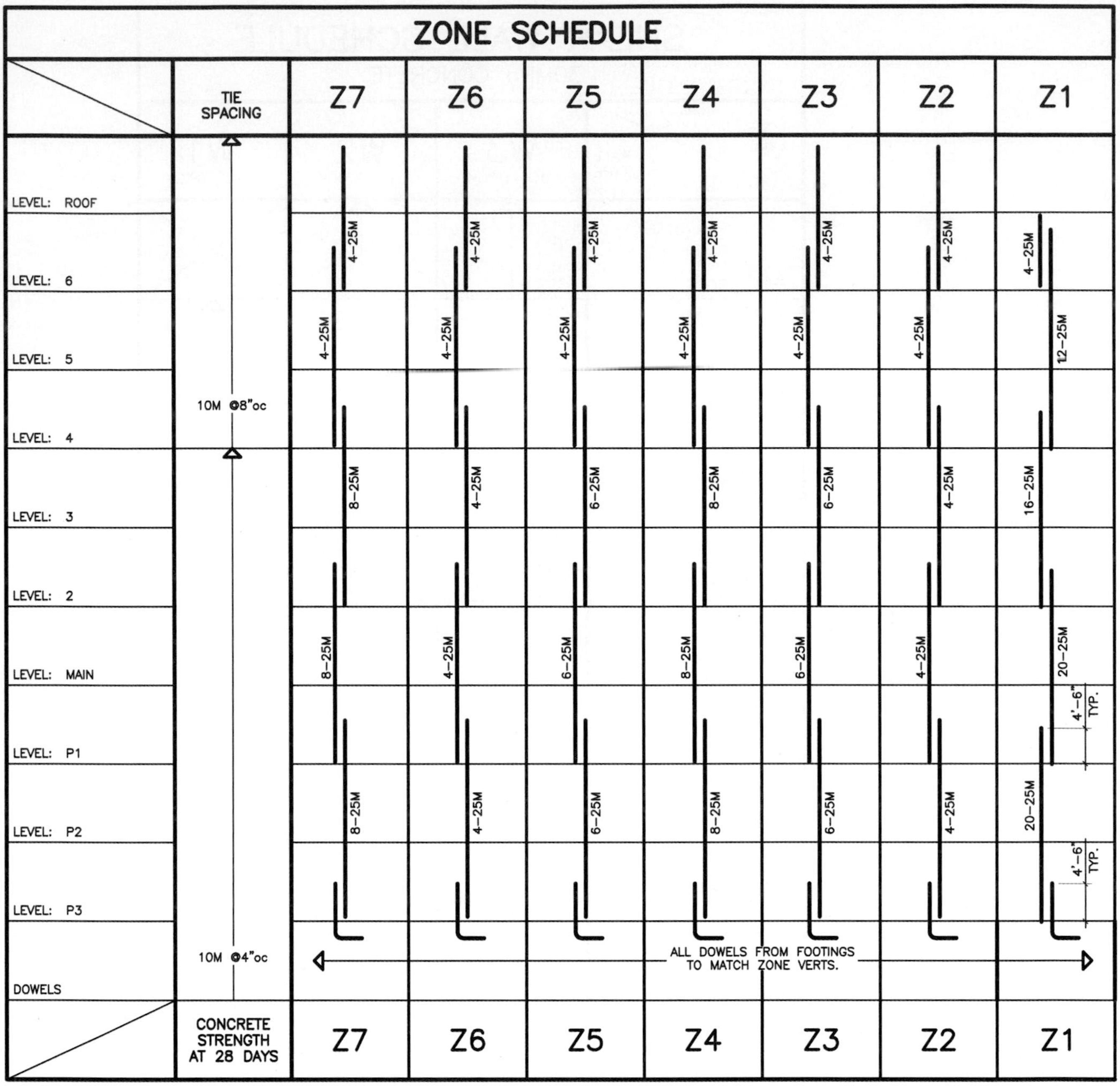

ZONE SCHEDULE								
	TIE SPACING	Z7	Z6	Z5	Z4	Z3	Z2	Z1
LEVEL: ROOF								
LEVEL: 6		4—25M	4—25M	4—25M	4—25M	4—25M	4—25M	4—25M
LEVEL: 5		4—25M	4—25M	4—25M	4—25M	4—25M	4—25M	12—25M
LEVEL: 4	10M ⌀8"oc							
LEVEL: 3		8—25M	4—25M	6—25M	8—25M	6—25M	4—25M	16—25M
LEVEL: 2								
LEVEL: MAIN		8—25M	4—25M	6—25M	8—25M	6—25M	4—25M	20—25M
LEVEL: P1								4'-6" TYP.
LEVEL: P2		8—25M	4—25M	6—25M	8—25M	6—25M	4—25M	20—25M
LEVEL: P3								4'-6" TYP.
DOWELS	10M ⌀4"oc	◄—	⌐	⌐	ALL DOWELS FROM FOOTINGS TO MATCH ZONE VERTS.			—►
	CONCRETE STRENGTH AT 28 DAYS	Z7	Z6	Z5	Z4	Z3	Z2	Z1

Figure 13.35 Zone schedule.

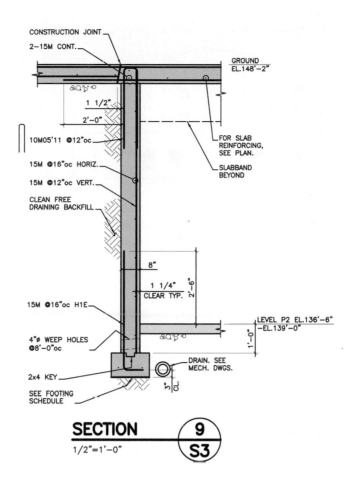

CONSTRUCTION JOINT
2–15M CONT.

GROUND
EL.148'–2"

1 1/2"

2'–0"

FOR SLAB
REINFORCING,
SEE PLAN.

SLABBAND
BEYOND

10M05'11 @12"oc

15M @16"oc HORIZ.

15M @12"oc VERT.

CLEAN FREE
DRAINING BACKFILL

8"

1 1/4"
CLEAR TYP.

2'–6"

15M @16"oc H1E

LEVEL P2 EL.136'–6"
–EL.139'–0"

1'–0"

4"ø WEEP HOLES
@8'–0"oc

2x4 KEY

DRAIN. SEE
MECH. DWGS.

3"
CL.

SEE FOOTING
SCHEDULE

SECTION   9
1/2"=1'–0"   S3

Figure 13.36 Basement wall reinforcement.

13.8 | JOINTS

Considering that reinforced concrete walls are structural elements with large surface areas, there are concerns that cracks might develop during their service life. As discussed in previous chapters, well distributed reinforcement prevents excessive cracking in reinforced concrete members. Causes of cracking and types of cracks were discussed in Section 4.7. In addition to the provision of minimum distributed reinforcement in the walls, *control joints* (also called contraction joints) need to be provided at proper intervals. The main role of control joints is to ensure that shrinkage cracking occurs at the predetermined locations. (Note that control joints in slabs on grade were discussed in Section 14.12.4.) These control joints are usually in the form of vertical grooves provided on the inside and outside of the wall. There are no exact rules for the location of control joints within a building. The joints must be placed such that the structural integrity is preserved. Broad recommendations related to the location of control joints are presented in Figure 13.37.

Another type of joint present in concrete construction is called a construction joint. *Construction joints* (also called cold joints) are stopping places for a day's work, as discussed in Section 12.12.4. It is common practice to limit concrete placement to a height of one storey. Construction joints in bearing walls are usually located at the underside of floor slabs, beams, and girders. The wall reinforcement intersecting the construction joint needs to have proper development on either side of the construction joint. Common locations for construction joints in reinforced concrete bearing walls are shown in Figure 13.38.

The locations of control and construction joints should be indicated on structural drawings. For further information on joints in reinforced concrete buildings, the reader is referred to PCA (1982).

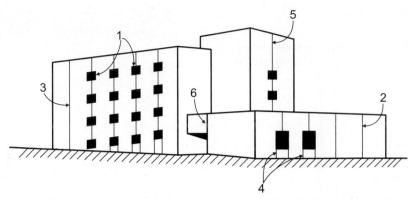

1 - 6 m apart in walls
    with frequent openings
2 - up to 6 m apart
    (walls without openings)
3 - within 3 to 5 m of
    a corner, if possible
4 - in line with each
    jamb at first storey level
5 - above first storey
    at centreline of opening
6 - jamb lines preferred

Figure 13.37   Location of control joints in a building. (Adapted from PCA, 1982 by permission of the Portland Cement Association)

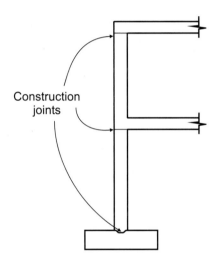

Construction joints

Figure 13.38   Location of construction joints in a reinforced concrete wall.

SUMMARY AND REVIEW — WALLS

Walls are vertical structural members that can be used to

- enclose or separate spaces
- retain earth and liquids
- support gravity loads
- resist lateral loads due to wind and earthquakes

Types of reinforced concrete walls

Based on their function, reinforced concrete walls can be classified into the following eight types:

- *Retaining walls* retain earth or other materials (used in nonbuilding applications).
- *Basement walls* retain earth at the building exterior and prevent the entry of water into the building.
- *Grade beams* are basement walls that distribute the load from exterior columns above the wall to a strip footing beneath the wall.
- *Bearing walls* support gravity loads.
- *Shear walls* resist lateral forces in the plane of the wall.

- *Wall panels* are exterior walls that enclose the building above grade.
- *Fire walls* are nonbearing walls that divide a building into sections and prevent the spread of fire from one section to another.
- *Tilt-up walls* are precast wall panels prefabricated at the building site.

Walls can be subjected to the following types of loads:

- lateral loads (in-plane)
- lateral loads (out-of-plane)
- gravity loads

CSA A23.3 wall detailing requirements

CSA A23.3 prescribes the following requirements related to the detailing of reinforcement in reinforced concrete walls:

- provision of distributed and concentrated wall reinforcement (Cl.14.1.8.1)
- minimum area of distributed vertical reinforcement (Cl.14.1.8.5)
- minimum area of distributed horizontal reinforcement (Cl.14.1.8.6)
- maximum spacing of distributed reinforcement (Cl.14.1.8.4)
- size of distributed reinforcement (Cl.14.1.8.2)
- number of reinforcement layers and curtains (Cl.14.1.8.3)
- nominal concentrated vertical reinforcement (Cl.14.1.8.8)
- provision of reinforcement at openings (Cl.14.1.8.9)
- concrete cover (A23.1 Cl.6.6.6.2.3)

The *wall thickness* should not be less than 200 mm for most design applications; A23.3 Cl.14.1.7.1 prescribes the minimum required thicknesses for different wall types.

Bearing walls

Bearing walls support vertical loads in addition to their own weight. According to A23.3 Cl.2.2, bearing walls should be able to sustain the following loads:

a) factored in-plane vertical loads larger than $0.1 f_c' A_g$
b) bending moments around the weak axis (the horizontal axis parallel to the wall length)
c) shear forces necessary to equilibrate these moments

In general, bearing walls are relatively short walls spanning vertically and are subjected to vertical loads due to floor or roof structures. Such walls are usually designed using the *empirical method* stipulated in A23.3 Cl.14.2.

According to the empirical method, the factored axial load resistance of a bearing wall (P_r) can be determined based on the equation

A23.3 Eq. 14.1

$$P_r = \frac{2}{3} \alpha_1 \phi_c f_c' A_g \left[1 - \left(\frac{kh_u}{32t} \right)^2 \right]$$ [13.3]

where k is the effective length factor, which depends on the end support conditions; in most cases, the end support conditions are considered as pinned, so $k = 1.0$; and A_g is the gross area of a wall section; for walls subjected to uniform load, A_g is determined based on a unit strip of length $l_b = 1000$ mm.

In some cases, bearing walls are subjected to concentrated (point) loads. Each concentrated load should be considered as uniformly distributed over the horizontal length (l_b) of a wall that needs to be determined based on the three different criteria outlined in A23.3 Cl.14.1.3.1.

Basement walls

Basement walls are exterior walls located below ground (grade) level in a building. These walls should be designed to resist the combined effects of lateral earth pressure and axial compressive loads transferred from the building superstructure above.

Basement walls are usually laterally supported at the bottom by the slab on grade and at the top by the suspended floor slab. The walls span like vertical slabs between these two supports.

The main loads acting on basement walls are

- gravity loads
- earth pressure
- surcharge

Basement walls are designed for *gravity loads* in the same manner as bearing walls.

The magnitude of the *earth pressure* is proportional to the depth below the ground surface and the surcharge soil weight in a manner similar to a fluid. This corresponds to a triangular load distribution with a zero value at the top and a maximum value at the base of the wall.

The *surcharge* includes various dead and live loads imposed on a backfill surface behind the wall. For design purposes, the surcharge is modelled as a uniformly distributed load over the wall height.

The key design considerations for basement walls are

- axial load
- flexure
- shear
- sliding

For design purposes, basement walls are treated as if they are composed of a series of vertical beams placed side by side (similar to the design of one-way slabs). Each beam is of rectangular cross-section with a 1 m width (*b*) and a depth (*t*) equal to the wall thickness.

Shear walls

The primary role of shear walls is to resist lateral loads due to earthquakes and wind. Shear walls provide resistance along the plane of their length (in-plane) and are weak when loaded perpendicular to that plane (out-of-plane). Based on their location and function in the building, shear walls can be classified as follows:

- *Bearing walls* support a substantial percentage of gravity loads.
- *Frame walls* support mainly lateral loads, whereas gravity loads are carried by the reinforced concrete frame.
- *Core walls* are incorporated into the central core area within a building, which may include stairwells, elevator shafts, or utility shafts.

Depending on the wall height/length (h_w/l_w) *aspect ratio,* shear walls are classified into two categories (A23.3 Cl.2.2):

- *Flexural shear walls* ($h_w/l_w > 2$) are tall and slender walls in high-rise buildings.
- *Squat shear walls* ($h_w/l_w < 2$) are short walls in low-to-medium-rise buildings.

Shear walls perforated with a regular pattern of openings (doors or windows) are called *coupled walls.*

Flexural shear walls behave like vertical beams subjected to flexure and shear and are characterized by the following three failure modes:

- flexural failure
- flexural-shear failure
- concrete crushing failure

Due to a low aspect ratio, squat shear walls fail in the following shear-controlled modes:

- diagonal tension failure
- diagonal compression failure
- sliding shear failure

Flexural shear walls should be designed like columns for the combined effects of axial load and flexure and like beams for shear effects.

PROBLEMS

13.1. Consider a 4.5 m high reinforced concrete bearing wall that supports precast concrete beams spaced at 2.5 m on centre. The beams are 200 mm wide and bear on the full wall thickness. The wall is pin supported at the top and bottom and restrained against lateral translation. The factored beam reactions (P_f) are equal to 300 kN. The reactions are applied concentrically with regard to the vertical wall axis. Use 15M bars for wall reinforcement.

Design the wall for the given loads according to CSA A23.3.

Given:

$f_c' = 25$ MPa
$f_y = 400$ MPa
$\phi_c = 0.65$
$\phi_s = 0.85$

13.2. Consider a 4.0-m-high reinforced concrete bearing wall subjected to a uniform dead load (DL) of 80 kN/m and uniform live load (LL) of 40 kN/m. The wall is pin supported at the top, fixed at the bottom, and restrained against lateral translation at both the top and the bottom. The load is applied concentrically with regard to the vertical wall axis. Use 15M bars for wall reinforcement.

Design the wall for the given loads according to CSA A23.3.

Given:

$f_c' = 25$ MPa
$f_y = 400$ MPa
$\phi_c = 0.65$
$\phi_s = 0.85$

13.3. A reinforced concrete basement wall is shown in the next figure. The wall is subjected to lateral earth pressure, as shown in the figure (note that the values refer to specified loads). Gravity loads do not need to be considered in this design. Maximum aggregate size is 20 mm. Use 15M or 20M bars for wall reinforcement.

Design the horizontal and vertical wall reinforcement according to the CSA A23.3 requirements.

Given:

$f_c' = 25$ MPa (normal-density concrete)
$f_y = 400$ MPa
$\phi_c = 0.65$
$\phi_s = 0.85$

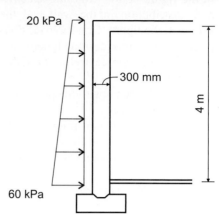

20 kPa

300 mm

4 m

60 kPa

13.4. A typical section of a 200 mm thick reinforced concrete basement wall is shown in the figure below. The wall is subjected to lateral earth pressure, as shown in the figure (note that the values refer to specified loads). Gravity loads do not need to be considered in this design. The designer has decided to place the reinforcement in the middle of the wall, as shown in the figure.
a) Design the wall for the given loads according to the CSA A23.3 requirements.
b) Comment on the effectiveness of placing the reinforcement in the middle of the wall in this design.
c) Does this reinforcement placement strategy represent the most cost-effective solution? Explain.

Given:

$f_c' = 25$ MPa (normal-density concrete)
$f_y = 400$ MPa
$\phi_c = 0.65$
$\phi_s = 0.85$

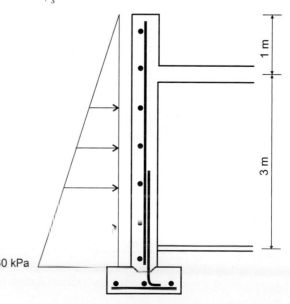

1 m

3 m

30 kPa

13.5. Consider an eccentrically loaded 300 mm thick footing supporting the 250 mm thick basement wall in the figure below. Note the specified earth pressure distribution in the wall and the footing, as shown in the figure. Gravity loads do not need to be considered in this design. Ignore the vertical distance between the centreline of slab on grade and the footing.

a) Draw the bending moment diagram for the wall and the footing.

b) Design the flexural reinforcing for the footing and the exterior wall face. The wall reinforcement needs to extend 600 mm past the inflection point.

c) Design the vertical and horizontal reinforcement at the interior wall face. The design should conform to the pertinent CSA A23.3 requirements.

Given:

$f_c' = 25$ MPa (normal-density concrete)
$f_y = 400$ MPa
$\phi_c = 0.65$
$\phi_s = 0.85$

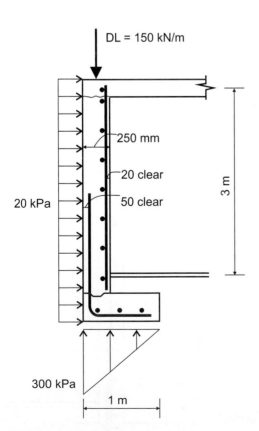

13.6. Consider a 300 mm thick basement wall supported by a strip footing shown in the next figure. The wall is subjected to two factored point loads (P_f) of 500 kN.

Design the vertical and horizontal wall reinforcement according to the CSA A23.3 requirements.

Given:

$f_c' = 25$ MPa (normal-density concrete)
$f_y = 400$ MPa
$\phi_c = 0.65$
$\phi_s = 0.85$

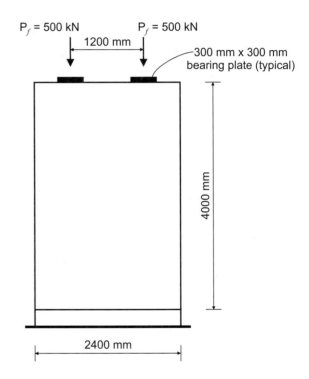

13.7. A two-storey reinforced concrete shear wall 200 mm thick is shown in the figure that follows. The wall is subjected to a specified uniform wind load of 150 kN/m and the gravity load as shown in the figure. Use 15M bars. Maximum aggregate size is 20 mm.

Design the wall for the effects of axial load, flexure and shear according to CSA A23.3.

Given:

$f_c' = 30$ MPa (normal-density concrete)
$f_y = 400$ MPa
$\phi_c = 0.65$
$\phi_s = 0.85$

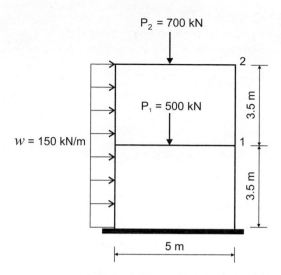

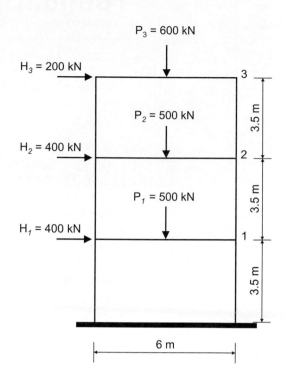

13.8. A three-storey reinforced concrete shear wall 250 mm thick is shown in the figure that follows. The wall is subjected to the specified wind and gravity loads as shown. Use 10M bars. Maximum aggregate size is 20 mm.

Design the wall for the effects of axial load, flexure and shear according to CSA A23.3.

Given:

$f_c' = 30$ MPa (normal-density concrete)
$f_y = 400$ MPa
$\phi_c = 0.65$
$\phi_s = 0.85$

14 Foundations

LEARNING OUTCOMES

After reading this chapter, you should be able to

- identify the six main types of shallow foundations
- differentiate between the allowable and factored soil pressure as used in foundation design
- apply the CSA A23.3 design requirements for strip and spread footings
- use the practical design guidelines for the selection of footing size and reinforcement
- design eccentrically loaded footings
- assess whether axial loads can be transferred from a column into the footing, and design dowel reinforcement if required

14.1 INTRODUCTION

Foundations carry loads from the building structure and transfer them to the soil. The *foundation* is the part of the structure below ground level (also called the substructure), while the *footing* is an individual member of the foundation supporting one or more columns, piers, or walls. The main function of the footing is to spread out the superimposed load so as not to exceed the safe load-bearing capacity of the underlying soil to which the load is delivered. Foundation design, including the selection of the foundation system and the footing dimensions, is strongly influenced by the type of soil and its properties. As a result, skilful foundation design blends knowledge of structural engineering and geotechnical engineering.

This chapter discusses the behaviour and design of reinforced concrete foundations. Different foundation types are outlined in Section 14.2, whereas the key geotechnical engineering considerations are discussed in Section 14.3. The design requirements of CSA A23.3 are discussed in Section 14.4. Section 14.5 offers practical guidelines for the design and construction of concrete foundations. The design of strip footings is explained in Section 14.6, whereas the design of spread footings is discussed in Section 14.7. The focus of this chapter is mainly on the design of concentrically loaded footings, except for Section 14.8, which offers an overview of the design of eccentrically loaded footings. Section 14.9 discusses the design of combined footings. The requirements related to load transfer from columns into footings are outlined in Section 14.10. Structural drawings for reinforced concrete footings are discussed in Section 14.11. Finally, basic concepts related to slabs on grade are discussed in Section 14.12.

The CN Tower in Toronto opened in 1976 as the world's tallest free-standing structure (Figure 14.1). This 553 m tall concrete structure was designed to be a television and radio antenna mast, but it has become a major Toronto tourist attraction and a national landmark. It has three observation levels, including the famous Glass Floor and the SkyPod, at 447 m the highest of the three. The tower has a hollow cross-section, and it is supported by 5.5 m deep concrete foundations. The construction joints in the tower are called "Friday Lines" because they mark where concrete pouring stopped for the weekend. (For more information on construction joints, see Section 13.8.)

Figure 14.1 A view of the CN Tower in Toronto, Ontario.

(Canada Lands Company CLC Limited)

14.2 TYPES OF FOUNDATIONS

Foundation systems can be classified into shallow and deep foundations.

Shallow foundations are used when favourable soil strata are located near the surface to support the building load. In shallow foundation systems, the footing provides an adequate bearing area below a wall or a column so that the bearing pressure developed beneath the footing does not exceed the soil bearing capacity.

Deep foundations are used when the soil near the ground level does not have an adequate bearing capacity. Piles and caissons are the most common types of deep foundations. Piles transfer the building loads from walls and columns deeper below ground level and allow the loads to be supported by more competent soils in the deeper strata. The design of deep foundations is beyond the scope of this book. For more details, the reader is referred to Coduto (2001).

This chapter is focused mainly on the design of shallow foundation systems. Shallow foundation systems include

- spread footings
- strip footings
- combined footings
- strap footings
- mat foundations
- raft foundations

Various types of shallow foundations are illustrated in Figure 14.2.

Spread footings, also called column footings or isolated footings, are used to support columns in buildings or piers in bridges (see Figure 14.2a). These footings are almost always placed concentrically beneath the columns. In most cases, spread footings have a square plan shape although footings of rectangular plan shape are not uncommon. Footings of other plan shapes, such as round, hexagonal, and trapezoidal, are not frequently used in design applications.

Strip footings, also called wall footings, are used to support walls along their length. This concept is similar to that of spread footings, except that the footing length is large compared to its width (see Figure 14.2b). Overhangs are provided in the transverse direction to ensure adequate footing bearing area. Strip footings are generally placed concentrically beneath the wall; however, a wall located adjacent to a property line could require a footing that is eccentric on one side.

Combined footings are used when two spread footings are located close together and need to be combined into a single footing. A combined footing is designed as a beam subjected to upward-acting soil pressure. Combined footings are used when a building is located close to the property line, as shown in Figure 14.2c.

Strap footings are similar to combined footings, except that the distance between the adjacent columns is larger than in combined footings. In this case, two separate spread footings are constructed, and a beam called a *strap* (or *grade beam*) is used to connect the two footings, as shown in Figure 14.2d.

Mat foundations are large continuous footings that cover a large plan area and support all column and wall bases within a building. Mat foundations are an economical solution if the soil bearing capacity is low, resulting in potentially large differential settlements between columns.

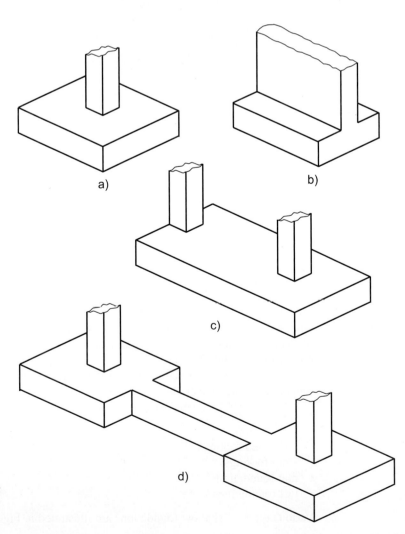

Figure 14.2 Types of shallow foundations: a) spread footings; b) strip footings; c) combined footings; d) strap footings.

Raft foundations are similar to mat foundations and are used where the soil has a limited bearing capacity. In this case, a mat is placed deep beneath ground level such that the weight of the excavated soil can compensate for the added building weight. This scheme is based on the approach that the soil bearing pressure is of similar magnitude to the overburden weight supported by the soil before construction.

This concept is often referred to as a *compensating foundation*. There are two types of compensating foundations:

- fully compensating foundations, where the amount of removed soil is exactly equal to the total building load;
- partially compensating foundations, where the weight of the removed soil amounts to only a portion of the building load.

The design of mat and raft foundations is beyond the scope of this book. For more details, the reader is referred to Coduto (2001).

KEY CONCEPTS

Note the difference between a foundation and a footing: the *foundation* is the part of the structure below ground level, while the *footing* is an individual member of the foundation supporting one or more columns, piers, or walls.

Foundations can be classified as follows:

- *Shallow foundations* are used when favourable soil strata are located near the surface to support the building load.
- *Deep foundations* are used when the soil near the ground level does not have an adequate bearing capacity.

The main six types of shallow foundations are (see Figure 14.2)

1. spread footings — footings that support individual columns
2. strip footings — footings that support walls
3. combined footings — a single footing supporting two or more columns
4. strap footings — two or more spread footings connected by a strap or grade beam
5. mat foundations — large continuous footings supporting all columns and walls within a building
6. raft foundations — continuous footings used when the soil has a limited bearing capacity

14.3 GEOTECHNICAL ENGINEERING CONSIDERATIONS

14.3.1 Soil Bearing Capacity

The magnitude of the soil bearing capacity is a very important foundation design parameter because it determines the footing plan dimensions. In most cases, a structural designer will be advised by a geotechnical engineer on site soil classification properties, including seismic site response, allowable soil pressures, load capacities, foundation depth, and other relevant considerations. A geotechnical engineer usually recommends the value (or a range of values) of *allowable soil bearing pressure* and lateral loads to be used for the foundation design. Gravity loads include the load imposed on the footing plus the total load of the column, pier, or wall, plus the footing weight. In some cases, a surcharge load is also included. Note that the allowable soil bearing pressure is always lower in value than the soil bearing capacity, which corresponds to the ultimate load condition. Lateral loads due to surcharge and seismic ground accelerations must also be considered.

The soil bearing capacity is usually determined by borings, test pits, or other types of soil investigations. In addition to the site-specific investigations, geotechnical engineers

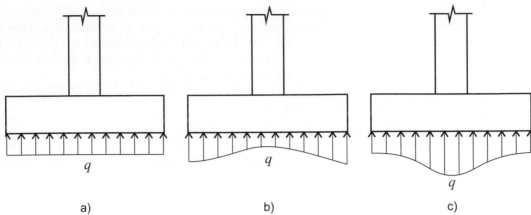

Figure 14.3 Soil pressure distribution: a) uniform pressure; b) actual pressure for cohesive soils; c) actual pressure for granular soils.

also consider the estimated size, shape, occupancy, and construction system for the proposed building, that is, the estimated column and wall loads, and the proposed foundation system in determining the soil bearing capacity.

Soil bearing pressure is normally considered to be uniform for a concentrically loaded footing. However, the actual pressure may be higher at the edges of the footing in the case of cohesive soil such as clay or at the centre for granular soil such as sandy soil, as shown in Figure 14.3.

The value of the allowable soil bearing pressure provided by the geotechnical engineer will ensure that excessive short- and long-term settlements and failures do not occur. The foundation designer needs to ensure that the pressures placed on the soil by the foundations are within the prescribed limits. Typical values for allowable soil bearing pressures (also called presumptive bearing pressures) are summarized in Table 14.1. Please note that these values are included for reference only and that the structural designer should use the advice of a geotechnical engineer related to a site-specific foundation design.

In most cases, the allowable soil bearing pressure is provided at the footing elevation. However, the exact footing elevation is often not known until the construction has started. Also, it may not be possible to predict in advance the exact soil conditions at each footing location, since only a limited number of drill tests are performed before the construction has started. Once the foundation excavation has been completed, a geotechnical engineer will perform a field review to verify the allowable soil bearing pressure at various footing locations. In some cases, the foundation design will need to be revised accordingly.

Table 14.1 **Allowable soil bearing pressures for various soil types**

Class	Material	Allowable soil pressure (kPa)
1	Crystalline bed rocks (granite, gneiss, trap rock, etc.)	600
2	Sedimentary rocks (hard shales, silt-stones, or sandstones) or foliated rocks (schist and slate)	200
3	Sandy gravel or gravel	150
4	Sand, silty sand, clayey sand, silty gravel, or clayey gravel	100
5	Clay, sandy clay, silty clay, or clayey silt	75

(*Source:* Table 1804.2, 2000 International Building Code. Copyright 2000. Falls Church, Virginia: International Code Council, Inc. Reproduced with permission. All rights reserved.)

Notes:

1. These bearing stresses apply to loading at the surface or where permanent lateral support for the bearing soil is not provided.

2. Fill material, organic material, and silt shall be deemed to be without presumptive bearing value except where, in the opinion of the enforcement officer, the bearing value is adequate for light frame structures. The bearing value of the material may be fixed on the basis of a test or other satisfactory evidence.

14.3.2 Foundation Depth

A minimum 150 mm depth from the top of the footing to the underside of basement or ground floor slab will often be necessary to ensure adequate soil compaction for supporting a slab on grade. A minimum depth to the bottom of the footing may be required for frost protection, and this may translate into a minimum depth of soil above the footing. In Canada, the frost protection depth varies from 0.45 m in coastal regions to as large as 1.6 m in cold climate regions.

Foundation depths are sometimes varied at the time of excavation to suit unforseen site conditions.

14.3.3 Allowable and Factored Soil Bearing Pressure

In general, the geotechnical engineer recommends the value of an allowable soil bearing pressure based on the working stress design method. According to this method, the maximum or ultimate stress (equal to the soil bearing capacity) is determined by taking into account the possible soil failure modes. The allowable soil bearing pressure is then obtained by dividing the soil bearing capacity by the factor of safety, in general on the order of 3.0.

In design practice, the actual soil pressure (q) is determined based on the total *specified (service) axial load* (P_s) placed on the footing by a column or a wall. For example, if the footing is subjected to specified dead and live load transferred from the column, the corresponding actual soil bearing pressure is

$$q = \frac{P_s}{A} = \frac{P_{DL} + P_{LL}}{A} \qquad [14.1]$$

where

P_{DL} = the axial load due to the specified dead load
P_{LL} = the axial load due to the specified live load
A = the footing plan area.

The actual soil pressure (q) due to the *specified loads* must be less than or equal to the *allowable soil bearing pressure* (q_{all}):

$$q \leq q_{all}$$

The above equations are often used to determine the required footing plan area as follows:

$$A \geq \frac{P_s}{q_{all}} \qquad [14.1a]$$

The structural design of the footing needs to be performed according to the limit states design method (see Section 1.8 for more details). Therefore, the footing thickness and reinforcement are selected based on the factored column or wall loads that correspond to the factored soil bearing pressure. Once the footing plan dimensions have been determined based on the allowable soil bearing pressure, the designer needs to calculate the *factored soil bearing pressure* (q_f) to be used in the footing design as follows:

$$q_f = \frac{P_f}{A} \qquad [14.2]$$

where P_f is the factored axial load acting on the footing. For example, if the footing is subjected to combined dead and live loads transferred from the column, the corresponding factored load is

$$P_f = 1.25\, P_{DL} + 1.5\, P_{LL}$$

Alternatively, P_f can be calculated based on the *average load factor* (α) as

$$P_f = \alpha\, P_s \qquad [14.3]$$

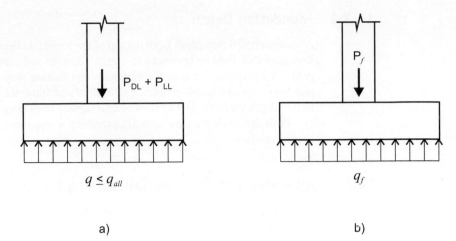

Figure 14.4 Soil pressure distribution: a) allowable soil pressure; b) factored soil pressure.

a)   b)

where

$$\alpha = \frac{P_f}{P_s} = \frac{1.25\,P_{DL} + 1.5\,P_{LL}}{P_{DL} + P_{LL}}$$ [14.4]

The concept of the allowable versus the factored soil pressure is illustrated in Figure 14.4. Note that geotechnical engineers generally do not provide the structural designer with the value of the *ultimate soil bearing capacity* (the soil pressure at which soil failure takes place).

It should be noted that lateral loads produce overturning load effects on some footings (usually called moment footings). These footings must be configured to distribute nonuniform (trapezoidal) soil pressures within the allowable values.

In some cases, the effect of surcharge is also considered in the footing design. *Surcharge* (or overburden) includes the superimposed uniform load due to soil, slab on grade, or other construction directly over the footing. The surcharge includes service loads acting on the floor (q_s), the soil overlay above the footing, and the footing thickness, as shown in Figure 14.5a.

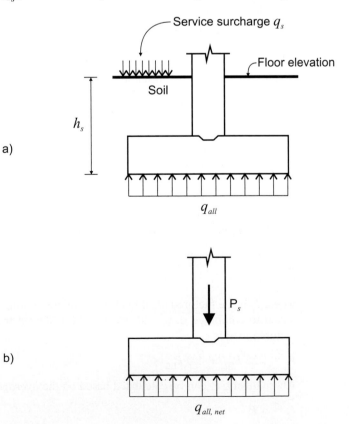

Figure 14.5 Surcharge effects: a) pressure distribution; b) net allowable pressure ($q_{all,\ net}$).

The net allowable soil pressure ($q_{all, net}$) (Figure 14.5b) can then be determined as

$$q_{all, net} = q_{all} - (q_s + \gamma_w \times h_s) \tag{14.5}$$

where

h_s = the foundation depth
γ_w = 24 kN/m^3 the unit weight density of concrete.

Note that the above equation uses the concrete weight density (γ_w) for the overall depth of the combined soil overlay and footing thickness. This is conservative because the concrete density is larger than the soil density.

When surcharge is considered in the footing design, $q_{all, net}$ should be used to determine the footing plan dimensions instead of q_{all}.

The concepts of allowable and factored soil pressure are illustrated with Example 14.1.

Example 14.1

Consider a spread footing supporting a reinforced concrete column, as shown in the figure below. The geotechnical engineer has stated that the allowable soil bearing pressure (q_{all}) at the bottom footing elevation is equal to 100 kPa. The column is subjected to a specified dead load (P_{DL}) of 50 kN and a specified live load (P_{LL}) of 130 kN. *Determine the required plan dimensions of the footing and the factored soil pressure to be used in the foundation design. The footing plan is square in shape.*

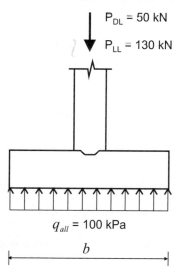

P_{DL} = 50 kN

P_{LL} = 130 kN

q_{all} = 100 kPa

b

SOLUTION: The allowable soil pressure is q_{all} = 100 kPa. Calculate the required footing plan area as

$$q = \frac{P_s}{A} \tag{14.1}$$

To determine footing plan area, set

$$q = q_{all}$$

Then, determine the footing plan area as

$$A \geq \frac{P_s}{q_{all}} = \frac{P_{DL} + P_{LL}}{q_{all}} = \frac{50 \text{ kN} + 130 \text{ kN}}{100 \text{ kPa}} = 1.8 \text{ m}^2 \tag{14.1a}$$

A square footing of length/width (b) is required for this design. Determine the minimum required footing size as

$$b \geq \sqrt{A} = \sqrt{1.8 \text{ m}^2} = 1.34 \text{ m} \cong 1.4 \text{ m}$$

Therefore, the actual footing plan area is

$$A = b^2 = (1.4 \text{ m})^2 = 1.96 \text{ m}^2 > 1.8 \text{ m}^2$$

Finally, calculate the factored soil bearing pressure as

$$q_f = \frac{P_f}{A} = \frac{1.25\,P_{DL} + 1.5\,P_{LL}}{A} \tag{14.2}$$

$$= \frac{1.25 \times 50\text{ kN} + 1.5 \times 130\text{ kN}}{1.96\text{ m}^2} = 131.4\text{ kPa}$$

KEY CONCEPTS

The *allowable soil bearing pressure* governs the footing plan dimensions (refer to Table 14.1 for typical values).

The soil pressure (q) is determined based on the total *specified (service) load* (P_s) placed on the footing with plan area A:

$$q = \frac{P_s}{A} = \frac{P_{DL} + P_{LL}}{A} \tag{14.1}$$

The actual soil pressure (q) due to the *specified loads* must be less than or equal to the *allowable soil bearing pressure* (q_{all}); that is,

$$q \le q_{all}$$

The required footing plan area can then be determined as

$$A \ge \frac{P_s}{q_{all}} \tag{14.1a}$$

The factored soil bearing pressure (q_f) is determined based on the *factored load* (P_f) as

$$q_f = \frac{P_f}{A} \tag{14.2}$$

It is very important to remember the following two concepts:

1. Footing plan dimensions are determined based on the allowable soil bearing pressure (q_{all}).
2. The factored soil bearing pressure (q_f) is used in footing design for flexure and shear.

14.4 | CSA A23.3 FOOTING DESIGN REQUIREMENTS

The general foundation design requirements are prescribed by A23.3 Cl.15. Other CSA A23.3 shear and flexure design requirements will also be referred to in this section as necessary.

14.4.1 Shear Design

The main design requirement for concrete footings subjected to shear as stated in A23.3 Cl.11.3.3 is the strength requirement:

A23.3 Eq. 11.3 $V_r \ge V_f$ [6.10]

The above requirement states that the factored shear resistance (V_r) must be greater than the factored shear force (V_f) acting at any section along a concrete member.

    The shear resistance of a footing section is provided by the concrete and shear reinforcement, as in the case of reinforced concrete beams and slabs (see Section 6.5.4); that is,

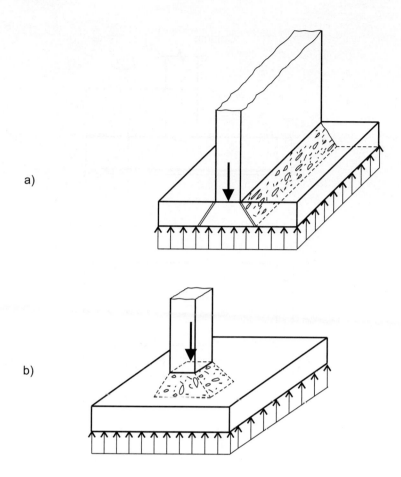

a)

b)

Figure 14.6 Shear effects in footings: a) one-way shear; b) two-way shear.

| A23.3 Eq. 11.4 | $V_r = V_c + V_s$ | [6.11] |

where V_r is the factored shear resistance of a footing section, V_c is the concrete shear resistance, and V_s is the steel shear resistance.

Shear reinforcement is *not required* if the footing has a large enough overall thickness that the Cl.11.2.8.1 requirement is satisfied, that is,

$$V_f \leq V_c$$

In that case,

$$V_s = 0$$

and

$$V_r = V_c$$

The concrete shear resistance depends on whether the footing is subjected to one-way shear or two-way shear. The concepts of one-way shear and two-way shear are related to two-way slabs and are mainly covered in Section 12.9 (see Figure 14.6).

One-way shear   One-way shear in a footing is determined in the same manner as for a slab or a beam, as prescribed in A23.3 Cl.11 and Cl.13. The critical section for one-way shear can be determined at a distance d from the face of the column (A23.3 Cl.13.3.6.1), as shown in Figure 14.7.

The one-way shear resistance of concrete (V_c) can be determined in the same way as for beams and slabs, as discussed in Section 6.5.4 (A23.3 Cl.11.3.4). Therefore,

| A23.3 Eq. 11.6 | $V_c = \phi_c \lambda \beta \sqrt{f_c'}\, b_w d_v$ | [6.12] |

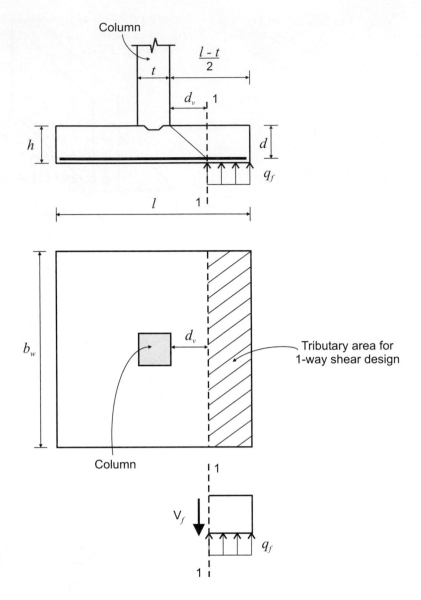

Figure 14.7 A critical section for one-way shear.

where

ϕ_c = the resistance factor for concrete

λ = the factor to account for concrete density ($\lambda = 1$ for normal-density concrete)

β = the factor accounting for the shear resistance of cracked concrete, determined according to A23.3 Cl.11.3.6

f_c' = the specified compressive strength of concrete

d_v = the effective shear depth, taken as the greater of $0.9d$ and $0.72h$

b_w = the minimum effective web width within depth d; this value is generally equal to the footing width (note that $b_w = 1$ m for strip footings).

The following values can be taken for the factor β:

1. $\beta = 0.21$ (A23.3 Cl.11.3.6.2) provided that

   a) the footing has an overall thickness not greater than 350 mm ($h \leq 350$ mm) or

   b) the distance from the point of zero shear to the face of the column, pedestal, or wall is less than three times the effective shear depth of the footing ($(l-t)/2 < 3\,d_v$).

2. For all other situations, β should be determined based on A23.3 Cl.11.3.6.3 (see Table 6.1 in Section 6.5.4).

The factored shear force (V_f) for one-way shear is determined from the equilibrium of the free-body diagram of an overhang on one side of a column or a wall cut at the critical section. The factored shear force (V_f) at the critical section used for the shear design is equal to the resultant of the factored soil pressure acting upward on the footing tributary area, shown hatched in Figure 14.7; that is,

$$V_f = q_f \times b_w \times \left(\frac{l - t}{2} - d \right)$$

Spread footings of rectangular plan should be checked for one-way shear in each direction.

Two-way shear (punching shear) Two-way shear behaviour is characterized by the punching of the column through the footing at the ultimate stage, as shown in Figure 14.6b. The reacting soil pressure will try to push the footing upward, except for the 45° pyramid-shaped area beneath the column.

For design purposes, the critical section for two-way shear can be taken at a distance $d/2$ from the face of the column (A23.3 Cl.13.3.3.1), as shown in Figure 14.8. The perimeter of the critical section (b_o) is called the *critical perimeter*. In the case of a footing supporting a square column of dimension t, the critical perimeter can be determined as (Figure 14.8b)

$$b_o = 4 \times \left(t + 2 \times \frac{d}{2} \right) = 4 (t + d) \qquad \text{[14.6]}$$

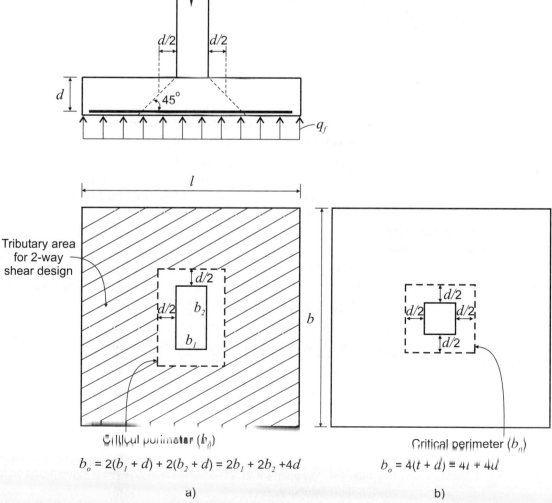

Critical perimeter (b_o)

$$b_o = 2(b_1 + d) + 2(b_2 + d) = 2b_1 + 2b_2 + 4d$$

Critical perimeter (b_o)

$$b_o = 4(t + d) = 4t + 4d$$

a)   b)

Figure 14.8 Critical shear perimeter: a) rectangular column; b) square column.

The two-way shear resistance of concrete (V_c) can be determined as

$$V_c = v_c \times b_o \times d \qquad \text{[14.7]}$$

where v_c is the factored shear stress resistance provided by the concrete.

According to A23.3 Cl.13.3.4.1, the factored shear stress resistance of the concrete (v_c) is equal to the smallest of the following three values:

| A23.3 Eq. 13.5 | 1. | $v_c = \left(1 + \dfrac{2}{\beta_c}\right) 0.19\lambda\phi_c\sqrt{f_c'}$ (MPa) | [14.8] |

where

$$\beta_c = \frac{b_2}{b_1}$$

which is the ratio of the long-side dimension (b_2) to the short-side dimension (b_1) of the column, concentrated load, or reaction area (see Figure 14.8a).

| A23.3 Eq. 13.6 | 2. | $v_c = \left(\dfrac{\alpha_s d}{b_o} + 0.19\right)\lambda\phi_c\sqrt{f_c'}$ (MPa) | [14.9] |

where

$$\alpha_s = 4 \text{ for interior columns}$$
$$= 3 \text{ for edge columns}$$
$$= 2 \text{ for corner columns}$$

| A23.3 Eq. 13.7 | 3. | $v_c = 0.38\,\lambda\,\phi_c\sqrt{f_c'}$ | [14.10] |

Note that $\sqrt{f_c'} \leq 8\,\text{MPa}$ (A23.3 Cl.13.3.4.2).

In performing design for two-way shear, it is common practice to consider that the v_c value obtained from Eqn 14.10 is smallest and governs the two-way shear resistance. When the effective depth of the footing needs to be determined based on the two-way shear, the following steps can be followed:

1. Determine the v_c value based on Eqn 14.10.
2. Substitute that value into Eqn 14.7 to express V_c in terms of b_o and d. Next, express b_o in terms of d. In the case of a square column, use Eqn 14.6 as follows:

$$\begin{aligned} V_c &= v_c \times b_o \times d \qquad \text{[14.7]}\\ &= v_c \times 4(t + d) \times d \end{aligned}$$

3. Determine the factored shear force (V_f) for two-way shear. V_f is equal to the stress resultant acting on a tributary area for two-way shear, as shown hatched in Figure 14.8a. However, structural designers often compute V_f as

$$V_f = q_f[A - (t + d)^2]$$

where A denotes the total plan area of the footing; note that the area within a distance d from the column does not contribute to the factored shear force.

4. To ensure that shear reinforcement according to A23.3 Cl.11.2.8.1 is not required, use the equation

$$V_f = V_c$$

5. Substitute V_c from Eqn 14.7 into the above equation and set up a quadratic equation in terms of d. The solution of this equation will give the d value based on the two-way shear requirements.
6. Use the b_o and d values obtained in the previous step to calculate the v_c values based on the other two criteria (equations 14.8 and 14.9) in order to confirm that these criteria do not govern in the two-way shear design.

14.4.2 Flexural Design

A key design requirement for concrete members subjected to flexure according to A23.3 Cl.8.1.3 is the ultimate limit states requirement or *strength requirement,* which states that

$$M_r \geq M_f \qquad \qquad \textbf{[14.11]}$$

The above requirement implies that the factored moment resistance (M_r) must be greater than or equal to the factored bending moment (M_f) acting at any section along a concrete flexural member.

Flexural design requirements govern the selection of flexural reinforcement (tension steel) in the footings. According to A23.3 Cl.15.4.3, the critical section for flexure exists at or close to the column face, as follows (see Figure 14.9):

a) For footings supporting a concrete column, pedestal, or wall, the critical section is at the face of the column, pedestal, or wall.

b) For footings supporting a masonry wall, the critical section is halfway between the middle and the edges of the wall.

c) For footings supporting a column with steel base plates, the critical section is determined by taking into account the dimensions and the stiffness of the base plate. In many cases, the critical section is taken halfway between the face of the column and the edge of the base plate.

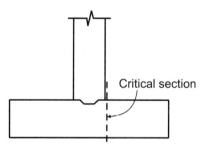

a) Supporting concrete column or wall

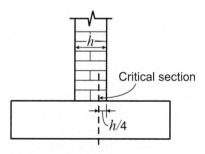

b) Supporting masonry wall

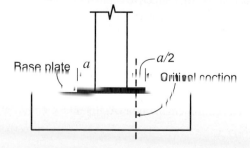

c) Supporting steel column

Figure 14.9 Critical sections for flexure in shallow footings: a) beneath a concrete column or wall; b) beneath a masonry wall; c) beneath a steel column.

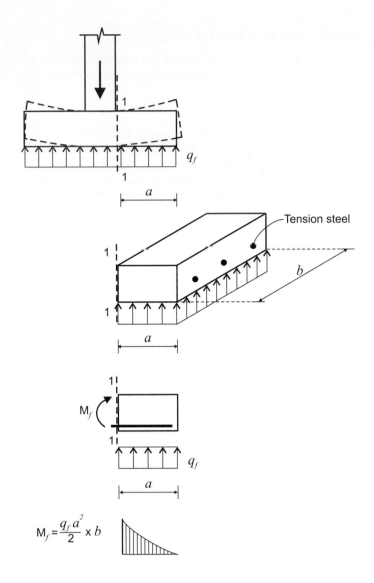

Figure 14.10   Flexural effects
in a spread footing.

$$M_f = \frac{q_f a^2}{2} \times b$$

According to A23.3 Cl.15.4.2, the factored bending moment (M_f) for the flexural design shall be determined by passing a vertical plane through the footing and computing the moment of the forces acting over the entire area of the footing on one side of that vertical plane, as shown in Figure 14.10.

In a strip footing, flexural reinforcement is placed perpendicular to the wall, whereas in a spread footing this reinforcement is placed in the two orthogonal directions, as shown in Figure 14.11.

In a strip footing, the main flexural reinforcement is placed in the bottom-most layer. If a similar amount of reinforcement is required in both directions for a spread footing, it is good practice to use the smaller effective depth for the design. If the design relies on having the reinforcement placed in a particular layering sequence, the correct values of the effective depth should be used, and the layering should be clearly indicated on the construction drawings.

| A23.3 Cl.7.8.1 | **Minimum footing reinforcement**   Minimum reinforcement in footings is intended to provide control of cracking due to shrinkage and temperature effects. Due to the large dimensions of concrete footings, this requirement often governs in the design. The minimum required reinforcement area (A_{smin}) is

$$A_{smin} = 0.002 A_g \qquad\qquad\qquad [14.12]$$

where A_g is the gross cross-sectional area of the footing, as shown in Figure 14.12.

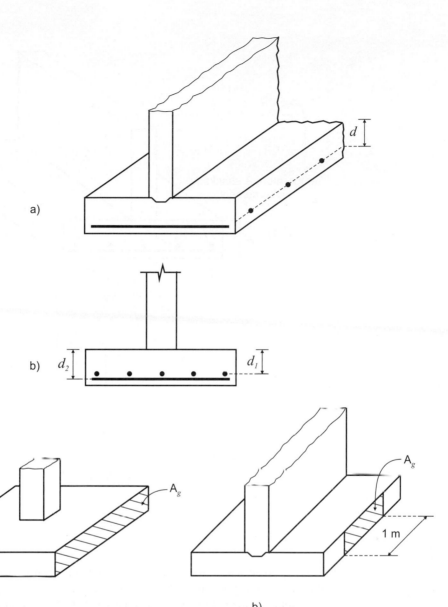

Figure 14.11 Main flexural reinforcement in shallow footings: a) strip footing; b) spread footing.

Figure 14.12 Gross cross-sectional area (A_g): a) spread footing; b) strip footing.

A23.3 Cl.7.4.1.2 and 13.10.4

Maximum spacing of flexural reinforcement The maximum permitted bar spacing (s_{max}) is equal to the lesser of $3h$ and 500 mm (same as slabs), where h denotes the footing thickness.

A23.3 Cl.7.8.1 and 7.8.3

Shrinkage and temperature reinforcement As discussed in Section 2.3.6, concrete shrinks during the curing process, thus causing cracks to develop in reinforced concrete members. If steel reinforcement is provided at close spacing, small hairline cracks develop in a desirable, uniformly distributed manner. In strip footings, this reinforcement is also of assistance in distributing concentrated loads over the wall length or in allowing the wall to span over localized soft spots. CSA A23.3 prescribes the minimum amount and spacing of reinforcement in each direction to ensure a uniform crack distribution due to shrinkage and temperature variations.

- The minimum amount of temperature reinforcement is the same as for flexural reinforcement ($A_{smin} = 0.002\,A_g$).
- The maximum permitted bar spacing (s_{max}) is equal to the lesser of $5h$ and 500 mm, where h denotes the footing thickness.

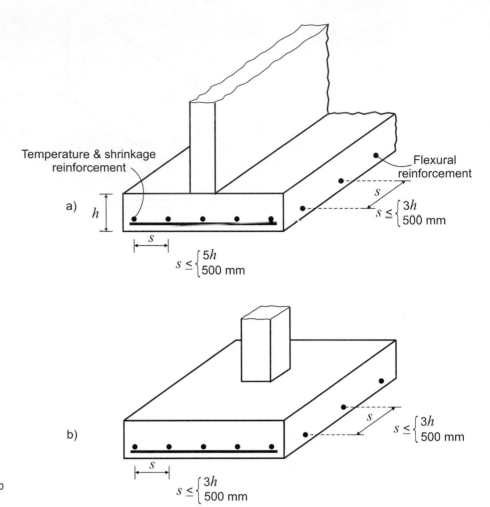

Figure 14.13 Footing reinforcement requirements: a) strip footing; b) spread footing.

There is no need to provide additional temperature reinforcement in spread footings because the reinforcement is provided in both directions.

Reinforcement spacing requirements for spread and strip footings, including flexural reinforcement and temperature and shrinkage reinforcement, are summarized in Figure 14.13.

A23.3 Cl.10.5.2

Maximum amount of flexural reinforcement   The requirements for the maximum amount of flexural reinforcement in footings are the same as described in Section 5.6.1 for reinforced concrete beams. It is unlikely that these requirements would govern in the footing design.

A23.3 Cl.15.4.4

Distribution of flexural reinforcement in rectangular footings   In rectangular footings, a portion of the total reinforcement in the *short direction* is placed in a band centred on the column with a width equal to that of the short side. The portion of the total required steel area within this band is equal to

$$\left(\frac{2}{\beta + 1}\right) A_s$$

where

$$\beta = \frac{b_1}{b_2}$$

where b_1 is the long-side dimension and b_2 is the short-side dimension. This is illustrated in Figure 14.14.

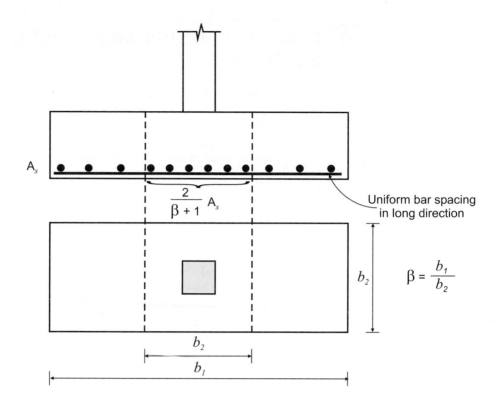

Figure 14.14 Distribution of flexural reinforcement in a rectangular spread footing.

Reinforcement in the *long direction* shall be distributed uniformly across the entire footing width.

A23.1 Cl.6.6.6.2.3

Concrete cover requirements An adequate concrete cover must be provided to protect the reinforcement against corrosion. Because the footing is in direct contact with the earth throughout its service life, A23.1 Cl.6.6.6.2.3 prescribes a 75 mm concrete cover for the footings, as summarized in Table A.2 (refer to Section 5.3.1 for more details on concrete cover requirements).

KEY REQUIREMENTS

The *strength requirement* is the key CSA A23.3 requirement for footing design for shear (Cl.11.3.3) and flexure (Cl.8.1.3).

Shear design The factored shear resistance (V_r) is determined according to A23.3 Cl.11.3.3.

The CSA A23.3 provisions for the concrete shear resistance (V_c) are related to

- one-way shear (Cl.11.3.4)
- two-way shear (Cl.13.3.3 and 13.3.4)

Shear reinforcement is not required provided that Cl.11.2.8.1 has been satisfied.

Flexural design The following key requirements are related to the detailing of flexural reinforcement:

- minimum area of reinforcement (A23.3 Cl.7.8.1)
- minimum spacing for reinforcement (A23.3 Cl.7.4.1.2 and 13.10.4)
- shrinkage and temperature reinforcement (A23.3 Cl.7.8.1 and 7.8.3)
- maximum amount of flexural reinforcement (A23.3 Cl.10.5.2)
- distribution of flexural reinforcement in rectangular footings (A23.3 Cl.15.4.4)
- concrete cover requirements (A23.1 Cl.6.6.6.2.3)

14.5 PRACTICAL DESIGN AND CONSTRUCTION GUIDELINES

In general, foundation construction is often characterized by a challenge as related to quality control, and it may be adversely affected by the soil and/or weather conditions at the construction site. Where unexpected conditions occur, the footing design is evaluated and revisions may be required.

Keeping in mind variable site conditions, the key practical advice is to refrain from "skimping" on the foundation dimensions; that is, foundations should be designed using larger than required dimensions wherever possible.

The consequences of inadequate footing design and construction can be significant. The designer should be aware of the complexities related to a possible footing upgrade after the construction is complete. To access an existing footing, it is necessary to remove the slab on grade and any interior or exterior construction supported by the slab, to remove the soil above the footing, and to deal with possible interference from underground utilities (electrical, plumbing, and water supply lines). As a result, the expenses related to the upgrade of an existing footing are often higher than those related to the construction of a new footing. Given the fact that the cost of a minor increase in footing concrete and reinforcement is usually insignificant relative to the overall project cost, it is recommended that footing dimensions be chosen slightly larger than required. This will allow for some construction errors and/or provide a somewhat higher load capacity than required by the design. Hence, it is suggested to always round up rather than round down when selecting footing plan dimensions, thickness, and reinforcement. A few other recommendations related to footing design and construction are outlined in this section.

Thickness   The thickness of a reinforced concrete footing is generally determined by the shear design requirements. For simple spread footings, it is economical to design the footing with sufficient thickness to allow the concrete to resist shear forces and avoid the use of shear reinforcement.

A23.3 Cl.15.7 prescribes a minimum 150 mm thickness of a footing above the bottom reinforcement. When a 75 mm clear cover is added to the above value, the resulting minimum footing thickness is on the order of 250 mm (see Figure 14.15).

In general, it is recommended to increase the footing thickness by an additional 25 mm to 50 mm beyond the actual design requirements whenever possible. The additional cost of concrete is minimal, so this is a cost-effective method of providing reserve strength in the footing. In general, 50 mm increments in footing thickness should be used to suit the sizes of the dimensional lumber used for the formwork.

Plan dimensions (length and width)   The allowable soil bearing pressure requires certain minimum plan dimensions to allow the soil mass to properly support loads. For example, an excessively narrow strip footing will apply pressure on the soil much like a knife edge cutting through a block of butter. Similarly, a spread footing that is too small will apply a load on the soil like the pointed end of a pencil and will punch though the soil at very low loads.

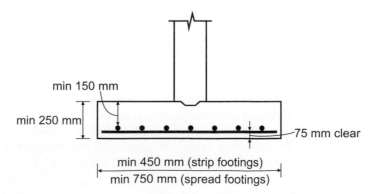

Figure 14.15 Minimum dimensions for spread and strip footings.

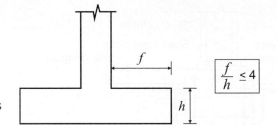

Figure 14.16 Recommended footing flange width-to-thickness ratio.

Typically, the minimum width of a strip footing should be on the order of 450 mm and the width of a spread footing should be at least 750 mm (see Figure 14.15). In general, the footing plan dimensions should be specified in increments of 150 mm.

Also, the footing dimensions should be chosen in such a way that they ensure that the footing is not too flexible. A flexible footing can result in a nonuniform soil pressure distribution beneath the footing (see Figure 14.3). To ensure that the footing is not too flexible, it is recommended to keep the footing *flange* (the portion of the footing outside the column or wall width) width-to-thickness aspect ratio (f/h) less than 4, and preferably even less than 3, as shown on Figure 14.16. It is interesting to note that a footing with a flange aspect ratio of 1 or less does not require reinforcement.

Reinforcement   In general, shallow footings should not require heavy reinforcement. If the above-recommended flange aspect ratio values have been followed, the footing reinforcement ratio (ρ) should be in the range of 0.33%. It is economical to use a smaller number of larger rebars to reduce labour requirements. For most applications, the use of 10M bars as flexural reinforcement should be avoided since these bars are too flexible and require more frequent chairing to avoid misplacement. In choosing the appropriate rebar size, it should be ensured that a rebar has ample development length to develop the yield strength at a critical section (as discussed in Chapter 9). Rebar sizes should be reduced if the footing flange length is close to the development length of a particular bar size.

It is often difficult to control the effective footing depth in the field because the tolerance in the reinforcement placement can be greater than expected. It is therefore recommended to increase the amount of reinforcement by 10% over that required by the design.

It is recommended to provide at least 50 mm side cover to bar ends in a formed footing. Also, rebars parallel to the footing edge should be placed with at least 75 mm cover to allow space for tying the intersecting bars, as shown in Figure 14.17.

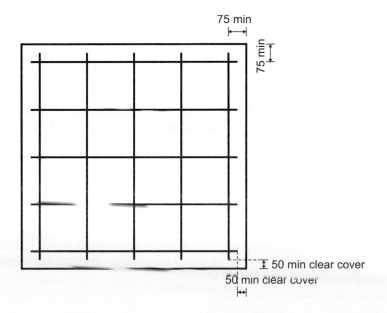

Figure 14.17 Plan of a spread footing showing the reinforcement arrangement.

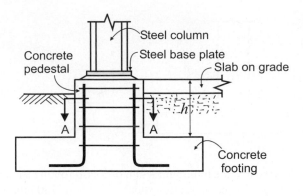

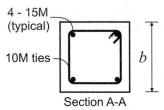

Section A-A

Figure 14.18 A reinforced concrete pedestal supporting a steel column.

Note: Steel column anchor bolts not shown

Pedestals   It is common practice to provide a concrete pedestal between the footing and steel columns, as shown in Figure 14.18. This applies to columns constructed using different materials, including steel and timber. The pedestal distributes the column load over a larger area of the footing, thereby contributing to a more economical footing design. Pedestals may be of either plain or reinforced concrete. According to A23.3 Cl.2.2, if the ratio of unsupported height (h) to the least lateral dimension (b) is less than 3.0, that is,

$$\frac{h}{b} < 3$$

then the member is categorized as a pedestal and theoretically may not require any reinforcement; otherwise, it must be designed as a reinforced concrete column.

The cross-sectional area of a pedestal is usually established by the bearing strength, the size of the steel base plate, or the desire to distribute the column load over a larger footing area.

It is common practice to design a pedestal in a manner similar to a column using a minimum of four corner bars (for a square or rectangular section) anchored into the footing and extending up through the pedestal (see Figure 14.18). Ties should be provided in pedestals according to the column design requirements.

KEY CONCEPTS

The practical design guidelines for the selection of footing dimensions and reinforcement can be summarized as follows:

- *Overall thickness:* The minimum thickness should be approximately 250 mm; the thickness should be increased by an additional 25 mm to 50 mm as compared to the actual design requirements whenever possible.
- *Plan dimensions:* The minimum width of a strip footing should be on the order of 450 mm and the width of a spread footing should be at least 750 mm.
- *Flange width-to-thickness ratio:* $f/h \leq 4$ (see Figure 14.16).
- *Reinforcement:* The footing reinforcement ratio (ρ) should be in the range of 0.33%; use a smaller number of larger rebars to reduce labour requirements; the minimum bar size should be 15M.

a)

Figure 14.19 Construction of a spread footing: a) soil excavation; b) a footing formed with dimensional lumber on the sides with top and bottom reinforcement and pedestal reinforcement in place.

(John Pao)

b)

Construction considerations The bottoms of the footings must always be cast against the soil on which they bear. The edges of footings may also be cast against the soil if a neat cut can be made in the soil that will remain vertical until the concrete is cast, as shown in Figure 14.19a. However, the edges of the footings must be formed if the soil is loose and could potentially collapse before the concrete construction. Usually, dimensional lumber of conventional Imperial sizes ($2'' \times 8''$, $2'' \times 10''$, or $2'' \times 12''$) is placed at the edges to form the sides of the footings. A footing formed with dimensional lumber is shown in Figure 14.19b. Note that larger tolerances in footing dimensions should be expected if the footing edges are not formed.

14.6 STRIP FOOTINGS

Strip footings, also called *wall footings,* are used to support walls along their length. A strip footing is usually placed concentrically beneath the wall. The width of a strip footing (l) is determined based on the allowable soil pressure, as follows (see Section 14.3.1):

$$A \geq \frac{P_s}{q_{all}}$$ [14.1a]

and

$$A = l \times b$$

where $b = 1.0$ m (unit strip).
Hence,

$$l \geq \frac{P_s}{b \times q_{all}}$$

Strip footing is usually designed such that its length exceeds the length of the wall it is supporting. A strip footing is subjected to flexure in the transverse direction (perpendicular to the wall length) and is generally designed, in a way similar to a one-way slab, by considering a typical unit (1 m wide) strip along the wall length, as illustrated in Figure 14.20.

The flexural design requirements discussed in Section 14.4.2 govern the amount and distribution of the reinforcement. The main flexural reinforcement is placed at the bottom of the footing in the transverse direction (short direction). In addition to this, it is necessary to provide reinforcement in the longitudinal direction (parallel to the wall) to help distribute concentrated loads beneath the wall and carry a part of the wall load over soft spots in the soil.

The required thickness of a strip footing is determined based on the one-way shear design requirements discussed in Section 14.4.1. It is usual to avoid the use of shear reinforcement by providing an adequate footing thickness.

Practical general guidelines for strip footing design are summarized in Checklist 14.1, followed by a design example.

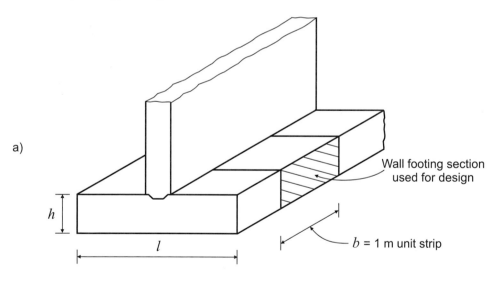

a)

h

l

Wall footing section used for design

$b = 1$ m unit strip

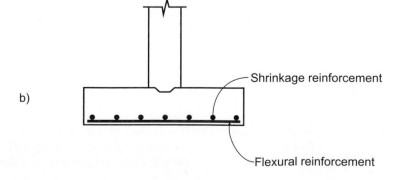

b)

Shrinkage reinforcement

Flexural reinforcement

Figure 14.20 Strip footing: a) isometric view; b) footing reinforcement.

Checklist 14.1 Design of Strip Footings According to CSA A23.3

	Given: - Specified axial load per metre of wall length (for example, P_{DL} and P_{LL}) - Allowable soil bearing pressure (q_{all}) - Concrete and steel material properties (f_c' and f_y)	
Step	**Description**	**Code Clause**
1	Determine the footing width. Determine the footing width based on the allowable soil pressure from (14.1a) as $$l \geq \frac{P_s}{b \times q_{all}} = \frac{P_{DL} + P_{LL}}{b \times q_{all}}$$ where $b = 1$ m is a strip of unit length. Note that the *specified (service) load* (P_s) is used to determine the footing width!	
2	Determine the factored soil pressure (q_f) to be used in the design for shear and flexure. $$q_f = \frac{P_f}{A} \qquad \text{[14.2]}$$ where $A = l \times b = l \times 1$ m and P_f is the factored axial load transferred from the wall to the footing.	
3	Determine the required footing thickness (h) based on the shear requirements (Section 14.4.1). $\boxed{\text{A23.3 Eq. 11.3}} \quad V_r \geq V_f \qquad \text{[6.10]}$ where $V_f \leq V_c$ and $V_s = 0$ (shear reinforcement not required). $\boxed{\text{A23.3 Eq. 11.6}} \quad V_c = \phi_c \lambda \beta \sqrt{f_c'}\, b_w d_v \qquad \text{[6.12]}$ Note that only one-way shear needs to be checked for strip footings. Determine V_f at the critical section (see Figure 14.7)	11.3.3 13.3.6.1 11.3.4
4	Determine the required flexural reinforcement (area A_s) based on the flexural design requirements (Section 14.4.2). $M_r \geq M_f \qquad \text{[14.11]}$ where M_r is the factored moment resistance for a rectangular section of unit width ($b = 1$ m) (see Figure 14.10), and M_f is the factored bending moment at a critical section for the flexural design.	15.4.2 15.4.3
5	Confirm that the minimum reinforcement requirement is satisfied $A_s > A_{smin} = 0.002 A_g \qquad \text{[14.12]}$ where A_g is the gross cross-sectional area of a footing for a unit strip ($A_g = h \times 1$ m).	7.8.1

(Continued)

Checklist 14.1   Continued

Given:
- Specified axial load per metre of wall length (for example, P_{DL} and P_{LL})
- Allowable soil bearing pressure (q_{all})
- Concrete and steel material properties (f_c' and f_y)

Step	Description	Code Clause
6	Confirm that the maximum flexural reinforcement requirement is satisfied (ensure a properly reinforced section), that is, $$\rho \leq \rho_b$$ where $$\rho = \frac{A_s}{bd} \qquad \text{[3.1]}$$ and ρ_b is the balanced reinforcement ratio (see Table A.4).	10.5.2
7	Determine the required bar spacing. The bar spacing can be determined once the reinforcing bar size (area A_b) has been estimated: $$s \leq A_b \frac{1000}{A_s} \qquad \text{[3.29]}$$ The maximum permitted bar spacing (s_{max}) is equal to the lesser of $3h$ and 500 mm.	7.4.1.2 13.10.4
8	Design the minimum reinforcement to be provided in the longitudinal direction. $$A_s > A_{smin} = 0.002 A_g \qquad \text{[14.12]}$$ where A_g is the gross cross-sectional area of a footing ($A_g = l \times h$). The maximum permitted spacing (s_{max}) is equal to the lesser of $5h$ and 500 mm.	7.8.1 7.8.3
9	Summarize the design with a sketch (design summary).	

Example 14.2

A typical cross-section of a strip footing beneath a 200 mm thick wall is shown in the figure that follows. The footing is subjected to a specified dead load (P_{DL}) of 60 kN/m and a specified live load (P_{LL}) of 60 kN/m. According to the information provided by the geotechnical engineer, the allowable soil bearing pressure is 100 kPa. Use 15M bars for flexural reinforcement and 20 mm maximum aggregate size.

Design the footing according to the CSA A23.3 design requirements.

Given: $f_c' = 25$ MPa (normal-density concrete)
$f_y = 400$ MPa
$\phi_c = 0.65$
$\phi_s = 0.85$

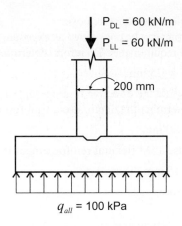

$P_{DL} = 60$ kN/m

$P_{LL} = 60$ kN/m

200 mm

$q_{all} = 100$ kPa

SOLUTION:
1. **Determine the footing width**

   The allowable soil pressure is $q_{all} = 100$ kPa. Calculate the required footing plan area from Eqn 14.1a by considering a strip of unit length $b = 1$ m:

   $$l \geq \frac{P_s}{b \times q_{all}} = \frac{P_{DL} + P_{LL}}{b \times q_{all}}$$

   $$= \frac{60 \text{ kN/m} + 60 \text{ kN/m}}{1 \text{ m} \times 100 \text{ kPa}} = 1.2 \text{ m}$$

   Therefore, use $l = 1.2$ m for the width.

2. **Determine the factored soil pressure (q_f)**

   First, calculate the footing plan area for a strip of unit width (see the sketch below):

   $$A = l \times b = 1.2 \text{ m} \times 1 \text{ m} = 1.2 \text{ m}^2$$

   Then, calculate the factored soil bearing pressure as

   $$q_f = \frac{P_f}{A} \qquad\qquad\qquad \textbf{[14.2]}$$

   $$= \frac{1.25 \, P_{DL} + 1.5 \, P_{LL}}{A} = \frac{1.25 \times 60 \text{ kN/m} + 1.5 \times 60 \text{ kN/m}}{1.2 \text{ m}^2}$$

   $$= 137.5 \text{ kPa}$$

   Use a rounded value for the further calculations:

   $$q_f = 137.5 \text{ kPa} \cong 140 \text{ kPa}$$

   Note that the factored soil pressure is used both for the flexural and shear design.

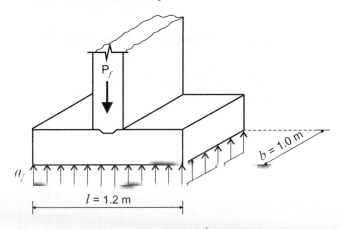

P_f

q_f

$b = 1.0$ m

$l = 1.2$ m

A23.3 Cl.11.3.3 3. **Determine the required footing thickness (h) based on the shear design requirements**
   a) First, determine the factored shear force (V_f) at the critical section at a distance d from the face of the wall.

i) Calculate the concrete cover.

A footing can be considered as exposure class F-1 or F-2 according to the CSA A23.1 requirements. Therefore, determine the concrete cover from Table A.2 as

cover = 75 mm

ii) Then, find the effective footing depth (d).

To determine the depth, use a trial footing thickness of

h = 250 mm

Because 15M flexural reinforcement is used,

$d_b \cong$ 15 mm (Table A.1).

Therefore,

$$d = h - \text{cover} - \frac{d_b}{2}$$

$$= 250 \text{ mm} - 75 \text{ mm} - \frac{15 \text{ mm}}{2} \cong 165 \text{ mm}$$

Finally, determine the factored shear force (V_f) at the critical section as (see the sketch below)

$$V_f = q_f \times b \times \left(\frac{l-t}{2} - d_v \right)$$

$$= 140 \text{ kPa} \times 1.0 \text{ m} \times \left(\frac{1.2 \text{ m} - 0.2 \text{ m}}{2} - 0.18 \text{ m} \right) = 45 \text{ kN/m}$$

b) Next, determine the concrete shear resistance (V_c) at the critical section based on Eqn 6.12 (see the sketch below).

i) Determine the effective shear depth (d_v) as the greater of

- $0.9d = 0.9 \times 165 \text{ mm} = 148.5 \text{ mm}$
- $0.72h = 0.72 \times 250 \text{ mm} = 180 \text{ mm}$

Hence,

d_v = 180 mm

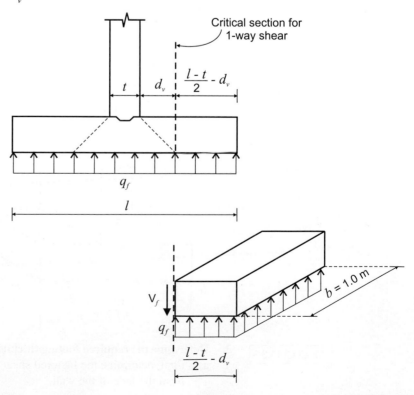

ii) Then, find the β value.
According to A23.3 Cl.11.3.6.2,

$\beta = 0.21$

because $h = 250$ mm < 350 mm (see Section 14.4.1 for more details).

iii) Finally, determine V_c for a unit width $b_w = b = 1000$ mm as

| A23.3 Eq. 11.6 |

$$V_c = \phi_c \lambda \beta \sqrt{f_c'}\, b_w\, d_v \qquad\qquad [6.12]$$
$$= 0.65(1)(0.21)\sqrt{25 \text{ MPa}}\,(1000 \text{ mm})(180 \text{ mm}) = 122.8 \text{ kN/m}$$

c) To avoid the use of shear reinforcement in the footing (A23.3 Cl.11.2.8.1), it is required that $V_f \le V_c$.
Because

$V_f = 45$ kN/m $< V_c = 122.8$ kN/m

this requirement is satisfied.

In practice, V_f can be determined at the face of the wall instead of at the critical section located a distance d from the face of the wall. The calculation would be simpler but more conservative and would result in a larger V_f value. Fortunately, shear design generally does not govern the thickness of strip footings, so such a simplification is appropriate.

4. **Determine the required flexural reinforcement based on the flexural design requirements**
   a) First, determine the factored moment resistance at the critical section (face of the wall) (see the sketch below):

$$M_f = q_f \left(\frac{l-t}{2}\right)\left(\frac{l-t}{4}\right) \times b$$
$$= 140 \text{ kPa} \times \left(\frac{1.2 \text{ m} - 0.2 \text{ m}}{2}\right)\left(\frac{1.2 \text{ m} - 0.2 \text{ m}}{4}\right) \times 1 \text{ m}$$
$$= 17.5 \text{ kN} \cdot \text{m/m} \cong 18 \text{ kN} \cdot \text{m/m}$$

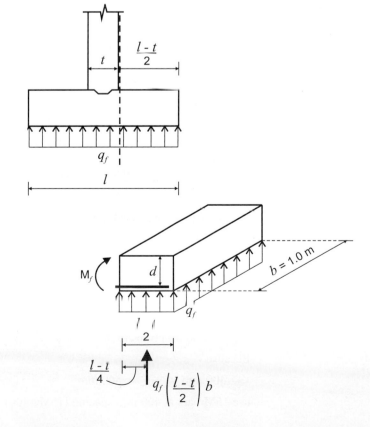

b) Determine the required area of reinforcement. The strength design requirement states that

$$M_r \geq M_f$$

Set $M_r = M_f = 18 \text{ kN} \cdot \text{m/m}$.

Then, calculate the required area of flexural reinforcement from Eqn 5.4 (direct procedure) as

$$A_s = 0.0015 f_c' b \left(d - \sqrt{d^2 - \frac{3.85 \, M_r}{f_c' b}} \right) \tag{5.4}$$

$$= 0.0015 \, (25 \text{ MPa}) \, (1000 \text{ mm})$$

$$\times \left(165 \text{ mm} - \sqrt{(165 \text{ mm})^2 - \frac{3.85 \, (18 \times 10^6 \text{ N} \cdot \text{mm/m})}{25 \text{ MPa} (1000 \text{ mm})}} \right)$$

$$= 323 \text{ mm}^2/\text{m}$$

| A23.3 Cl.7.8.1 |

5. Confirm that the minimum reinforcement requirement is satisfied

First, determine the gross cross-sectional area of the footing section:

$$A_g = h \times b = 250 \text{ mm} \times 1000 \text{ mm} = 250 \times 10^3 \text{ mm}^2$$

Then,

$$A_{smin} = 0.002 \, A_g \tag{14.12}$$

$$= 0.002 \times (250 \times 10^3 \text{ mm}^2) = 500 \text{ mm}^2/\text{m}$$

Because

$$A_s = 323 \text{ mm}^2/\text{m} < A_{smin} = 500 \text{ mm}^2/\text{m}$$

the minimum reinforcement criterion is satisfied, so

$$A_s = A_{smin} = 500 \text{ mm}^2/\text{m}$$

| A23.3 Cl.10.5.2 |

6. Confirm that the maximum reinforcement requirement is satisfied

In this case, the maximum reinforcement requirement is satisfied by default because the required amount of reinforcement is governed by the minimum reinforcement requirement.

7. Determine the required bar spacing for flexural reinforcement

For 15M bars,

$$A_b = 200 \text{ mm}^2 \quad \text{(see Table A.1)}$$

Therefore,

$$s \leq A_b \frac{1000}{A_s} \tag{3.29}$$

$$= 200 \text{ mm}^2 \times \frac{1000 \text{ mm}}{500 \text{ mm}^2/\text{m}} = 400 \text{ mm}$$

So, the required bar spacing is

$$s = 400 \text{ mm}$$

According to A23.3 Cl.7.4.1.2 and Cl.13.10.4, the maximum permitted bar spacing (s_{max}) is equal to the lesser of $3h$ and 500 mm.

Because

$$3h = 3 \, (250 \text{ mm}) = 750 \text{ mm}$$

$$s_{max} = 500 \text{ mm (governs)}$$

Because

$$s = 400 \text{ mm} < s_{max} = 500 \text{ mm}$$

use 15M bars at 400 mm spacing (15M@400)

A23.3 Cl.7.8.1 and 7.8.3

8. **Design the minimum reinforcement in the longitudinal direction**

First, determine the gross cross-sectional area of the footing:

$$A_g = h \times l = 250 \text{ mm} \times 1200 \text{ mm} = 300 \times 10^3 \text{ mm}^2$$

Then,

$$A_{smin} = 0.002 A_g \qquad \qquad [14.12]$$
$$= 0.002 \times (300 \times 10^3 \text{ mm}^2) = 600 \text{ mm}^2$$

Use 3-15M bars ($A_b = 200 \text{ mm}^2$):

$$A_s = 3 \times 200 \text{ mm}^2 = 600 \text{ mm}^2$$

The maximum permitted bar spacing (s_{max}) is equal to the lesser of $5h$ and 500 mm, so

$$s_{max} = 500 \text{ mm}$$

It is practical to place the edge bars 100 mm from the footing edge, thus resulting in 500 mm spacing of the longitudinal rebars.

9. **Provide a design summary**

Finally, summarize the results of this design with a sketch showing the footing cross-sectional dimensions and reinforcement.

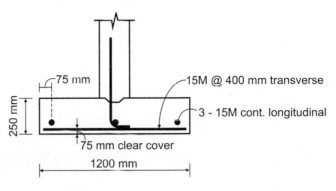

14.7 SPREAD FOOTINGS

14.7.1 Manual Design Procedure

Spread footings (also known as column footings or isolated footings) are used to support columns or piers and are the most common, simplest, and most economical of all footing types. These footings are almost always placed concentrically under the columns, as shown in Figure 14.21. In most cases, spread footings are of square plan shape, although a rectangular shape is also used in some cases.

A spread footing behaves like a two-way cantilevered slab with overhangs (flanges) beyond the column edge on all four sides. The plan dimensions of a spread footing are determined based on the allowable soil pressure q_{all} discussed in Section 14.3.1.

The amount and distribution of reinforcement are governed by the flexural design requirements discussed in Section 14.4.2. The footing is reinforced with two bar layers perpendicular to each other and parallel to the footing edges.

The required thickness of a thin footing is determined based on the shear design requirements discussed in Section 14.4.1. In this case, both the one-way and the two-way shear effects need to be considered. It is usual practice to avoid the use of shear reinforcement by providing an adequate footing thickness.

Broad guidelines related to spread footing design are summarized in Checklist 14.2, followed by a design example.

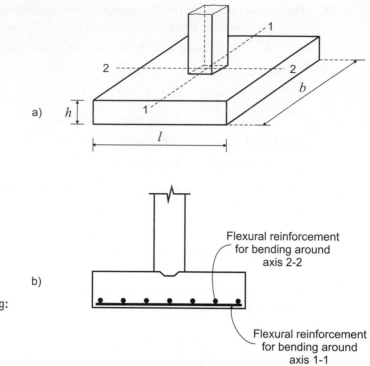

Figure 14.21 Spread footing:
a) isometric view; b) footing
reinforcement.

Checklist 14.2   Design of Spread Footings According to CSA A23.3

		Code Clause
	Given: - **Specified axial load (for example, P_{DL} and P_{LL})** - **Allowable soil bearing pressure (q_{all})** - **Concrete and steel material properties (f_c' and f_y)**	
Step	**Description**	**Code Clause**
1	Determine the footing plan dimensions. Determine the footing area based on the allowable soil pressure (Eqn 14.1a) as $$A \geq \frac{P_s}{q_{all}} \qquad \textbf{[14.1a]}$$ $$= \frac{P_{DL} + P_{LL}}{q_{all}}$$ It is common practice to use a square footing. Note that the *specified (service) load* (P_s) is used to determine the footing plan dimensions!	
2	Determine the factored soil pressure (q_f) to be used in the shear and flexural design of the footing. $$q_f = \frac{P_f}{A} \qquad \textbf{[14.2]}$$ where $A = l \times b$	
3	Determine the required footing thickness (h) based on the shear requirements (Section 14.4.1). A23.3 Eq. 11.3 $\qquad V_r \geq V_f \qquad \textbf{[6.10]}$ where $V_f \leq V_c$	11.3.3

(Continued)

v) Next, determine the overall footing depth (h). Because 25M rebars bars have been given, the bar diameter is

$d_b \cong 25$ mm (see Table A.1)

and the cover is

cover $= 75$ mm (see Table A.2, exposure class F-1 or F-2)

The overall footing thickness is

$$h = d + \frac{d_b}{2} + \text{cover}$$

$$= 540 \text{ mm} + \frac{25 \text{ mm}}{2} + 75 \text{ mm} = 628 \text{ mm}$$

However, in this case it is good practice to round up a footing thickness to 650 mm or even 700 mm to account for site conditions such as uneven ground. Thus, use

$h = 650$ mm

so

$$d = h - \frac{d_b}{2} - \text{cover} = 650 \text{ mm} - \frac{25 \text{ mm}}{2} - 75 \text{ mm} = 562 \text{ mm} \cong 550 \text{ mm}$$

b) Finally, check the one-way shear requirements (A23.3 Cl.11.3.4).

i) First, determine the factored shear force (V_f) at the critical section at a distance d away from the face of the column (see the sketch below):

$$V_f - q_f \times b \times \left(\frac{b-t}{2} - d_v \right)$$

$$- 305 \text{ kPa} \times 3.0 \text{ m} \times \left(\frac{3.0 \text{ m} - 0.5 \text{ m}}{2} - 0.5 \text{ m} \right) = 686 \text{ kN}$$

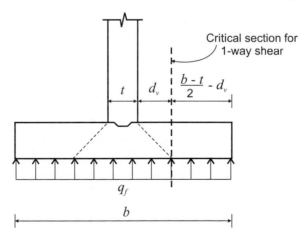

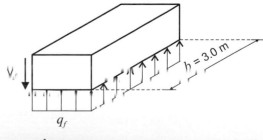

ii) The concrete shear resistance (V_c) at the critical section is based on Eqn 6.12. Determine the effective shear depth (d_v) as the greater of

- $0.9d = 0.9 \times 550$ mm $= 495$ mm
- $0.72h = 0.72 \times 650$ mm $= 468$ mm

Hence,

$d_v = 495$ mm $\cong 500$ mm

The β value can be determined based on A23.3 Cl.11.3.6.2 provided that there is no transverse reinforcement and the maximum aggregate size is not less than 20 mm, as follows (see Table 6.1):

| A23.3 Eq. 11.9 |

$$\beta = \frac{230}{1000 + d_v} \qquad \qquad [6.13]$$

$$= \frac{230}{1000 + 500} = 0.15$$

Finally, V_c can be determined for a section with width $b_w = b = 3000$ mm as

| A23.3 Eq. 11.6 |

$$V_c = \phi_c \lambda \beta \sqrt{f_c'}\, b_w\, d_v \qquad \qquad [6.12]$$
$$= 0.65\,(1)(0.15)\sqrt{25 \text{ MPa}}\,(3000 \text{ mm})(500 \text{ mm}) = 731.3 \text{ kN} \cong 730 \text{ kN}$$

iii) To avoid the use of shear reinforcement in the footing (A23.3 Cl.11.2.8.1),

$V_f \le V_c$

Since

$V_f = 686$ kN $< V_c = 730$ kN

shear reinforcement is not required.

4. **Determine the required flexural reinforcement based on the flexural design requirements**
   a) First, determine the factored bending moment resistance at the critical section (face of the column), as follows (see the sketch below):

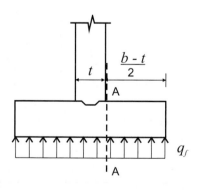

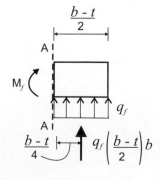

$$M_f = q_f \left(\frac{b-t}{2}\right)\left(\frac{b-t}{4}\right)b$$

$$= 305 \text{ kPa} \times \left(\frac{3.0 \text{ m} - 0.5 \text{ m}}{2}\right) \times \left(\frac{3.0 \text{ m} - 0.5 \text{ m}}{4}\right) \times 3.0 \text{ m}$$

$$= 714.8 \text{ kN} \cdot \text{m} \cong 715 \text{ kN} \cdot \text{m}$$

b) Determine the required area of reinforcement. The strength design requirement states that

$$M_r \geq M_f \qquad \qquad \textbf{[14.11]}$$

Set

$$M_r = M_f = 715 \text{ kN} \cdot \text{m}$$

Then, calculate the required area of flexural reinforcement from (5.4) (direct procedure) as

$$A_s = 0.0015 f_c' \, b \left(d - \sqrt{d^2 - \frac{3.85 \, M_r}{f_c' \, b}}\right) \qquad \qquad \textbf{[5.4]}$$

$$= 0.0015\,(25 \text{ MPa})\,(3000 \text{ mm})$$

$$\times \left(550 \text{ mm} - \sqrt{(550 \text{ mm})^2 - \frac{3.85\,(715 \times 10^6 \text{ N})}{25 \text{ MPa} \,(3000 \text{ mm})}}\right)$$

$$= 3875 \text{ mm}^2$$

| A23.3 Cl.7.8.1 |

5. Confirm that the minimum reinforcement requirement is satisfied
First, determine the gross cross-sectional area of the footing as

$$A_g = h \times b = 650 \text{ mm} \times 3000 \text{ mm} = 195 \times 10^4 \text{ mm}^2$$

Then,

$$A_{smin} = 0.002 A_g \qquad \qquad \textbf{[14.12]}$$

$$= 0.002 \times (195 \times 10^4 \text{ mm}^2) = 3900 \text{ mm}^2$$

Because

$$A_s = 3875 \text{ mm}^2 < A_{smin} = 3900 \text{ mm}^2$$

the minimum reinforcement requirement governs in this case. Use 8-25M bars:

$$A_s = 8 \times 500 \text{ mm}^2 = 4000 \text{ mm}^2$$

| A23.3 Cl.7.4.1.2 and 13.10.4 |

6. Confirm that the CSA A23.3 bar spacing requirements are satisfied
Determine the bar spacing assuming 100 mm end spacing:

$$s = \frac{3000 \text{ mm} - (2 \times 100 \text{ mm})}{7} = 400 \text{ mm}$$

According to A23.3 Cl.7.4.1.2 and Cl.13.10.4, the maximum permitted bar spacing (s_{max}) is equal to the lesser of $3h$ and 500 mm.
Because

$$3h = 3\,(650 \text{ mm}) = 1950 \text{ mm}$$

$$s_{max} = 500 \text{ mm (governs)}$$

$$s = 400 \text{ mm} < s_{max} = 500 \text{ mm}$$

| A23.3 Cl.10.5.2 |

7. Confirm that the maximum reinforcement requirement is satisfied
In this case, the maximum reinforcement requirement is satisfied by default because the amount of reinforcement is governed by the minimum reinforcement requirement.

A23.3 Cl.12.2.3 and 12.2.4

8. **Check the development length for the flexural reinforcement**

The required development length for 25M reinforcement (bottom bar) is

$$l_d = 0.45 k_1 k_2 k_3 k_4 \frac{f_y}{\sqrt{f_c'}} d_b \qquad [9.2]$$

$$= 0.45(1.0)(1.0)(1.0)(1.0) \frac{400 \text{ MPa}}{\sqrt{25 \text{ MPa}}} (25 \text{ mm}) = 900 \text{ mm}$$

where

$$k_1 = k_2 = k_3 = k_4 = 1.0 \text{ (see Table 9.1)}.$$

Consider a 100 mm cover at the edge of the footing; therefore, the available length of the reinforcing bar from the edge of the footing to the face of the column is

$$l - \frac{b-t}{2} \quad 100 \text{ mm}$$

$$= \frac{3000 \text{ mm} - 500 \text{ mm}}{2} - 100 \text{ mm} = 1150 \text{ mm}$$

Since

$$l = 1150 \text{ mm} > 900 \text{ mm}$$

the flexural reinforcement has sufficient length for proper development.

9. **Provide a design summary**

Finally, summarize the results of this design with a sketch showing the footing cross-sectional dimensions and reinforcement.

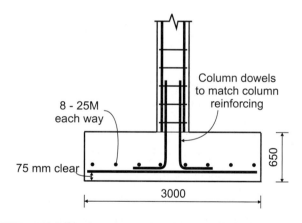

14.7.2 Design of Spread Footings Using Computer Spreadsheets

In general, each building is supported by a large number of footings. Most often, concentrically loaded spread footings are used, provided that the underlying soil is reasonably competent.

In design practice, it may be too cumbersome to design spread footings by considering the applied load in each column within a building. Once the designer gains sufficient background in design principles and procedures, the *capacity design approach* may be considered. According to this approach, the factored axial load resistances for various footing sizes are calculated by using the given allowable soil pressure, material properties, and column dimensions; the results are usually presented in tabular form. The designer can then select an adequate footing size to resist the applied column load from the table. The table can be generated by using a programmed computer spreadsheet (using, for example, Microsoft® Excel software).

An example of a computer spreadsheet developed for the design of concentrically loaded spread footings is shown in Figure 14.22. The input to the spreadsheet includes the following design parameters (see Figure 14.23):

Square Footing Table (ver 4.0)

Units (M or I) =	M		Example 11.3
Allowable soil pressure =	250	kPa	
f'c =	25	MPa	
Square pedestal/col width =	500	mm	smaller dimension of pedestal or column
Average load factor =	1.38		(1.25D + 1.5L)/(D + L)
Surcharge depth =	0	mm	top of soil/slab to top of footing
Surcharge density =	0	kN/m^3	110 (17.6) for compact soil. 150 (24) for concrete

Width mm	Thick mm	Factored Capacity kN	Allowable Capacity kN	Allowable Uplift kN	Top Rebar with uplift -	Bottom Rebar Area mm^2	Bottom Rebar -
1000	200	345	250	2	none	403.41	5-10M E.W.
1250	250	539	391	5	none	633.86	7-10M E.W.
1500	300	776	563	8	none	942.28	10-10M E.W.
1750	350	1057	766	13	none	1388.49	7-15M E.W.
2000	400	1380	1000	19	none	1917.56	7-20M E.W.
2250	450	1747	1266	27	none	2529.47	9-20M E.W.
2500	500	2156	1563	38	none	3224.20	11-20M E.W.
2750	600	2609	1891	54	none	3580.07	8-25M E.W.
3000	650	3105	2250	70	none	4394.85	9-25M E.W.
3250	700	3644	2641	89	none	5292.49	11-25M E.W.
3500	750	4226	3063	110	none	6272.98	9-30M E.W.
3750	800	4852	3516	135	none	7336.31	11-30M E.W.
4000	900	5520	4000	173	none	7922.35	12-30M E.W.
4250	950	6232	4516	206	none	9106.77	14-30M E.W.
4500	1000	6986	5063	243	none	10373.93	15-30M E.W.
4750	1050	7784	5641	284	none	11723.86	12-35M E.W.
5000	1150	8625	6250	345	none	12494.71	13-35M E.W.
5250	1200	9509	6891	397	none	13966.46	14-35M E.W.
5500	1300	10436	7563	472	none	14839.49	15-35M E.W.
5750	1400	11407	8266	555	none	16100.00	17-35M E.W.
6000	1450	12420	9000	626	none	17405.33	18-35M E.W.
6250	1550	13477	9766	727	none	19375.00	20-35M E.W.
6500	1650	14576	10563	837	none	21450.00	22-35M E.W.
6750	1750	15719	11391	957	none	23625.00	24-35M E.W.
7000	1800	16905	12250	1058	none	25200.00	26-35M E.W.
7250	1850	18134	13141	1167	none	26825.00	27-35M E.W.
7500	1900	19406	14063	1283	none	28500.00	29-35M E.W.
7750	2000	20722	15016	1442	none	31000.00	32-35M E.W.
8000	2050	22080	16000	1574	none	32800.00	33-35M E.W.
8250	2100	23482	17016	1715	none	34650.00	35-35M E.W.
8500	2200	24926	18063	1907	none	37400.00	38-35M E.W.
8750	2250	26414	19141	2067	none	39375.00	40-35M E.W.
9000	2300	27945	20250	2236	none	41400.00	42-35M E.W.

Figure 14.22 Microsoft® Excel spreadsheet for spread footing design. (Copyright: Garry Kirkham)

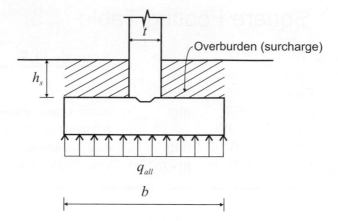

Figure 14.23 Parameters for design of spread footings using the Microsoft® Excel spreadsheet.

1. units: metric (M) or Imperial (I)
2. allowable soil pressure (q_{all}) (kPa)
3. concrete strength for the footing (f_c') (MPa)
4. width of a square-shaped pedestal or column (b) (mm)
5. average load factor (α); see Eqn 14.4
6. overburden (surcharge) depth (h_s) (mm)
7. overburden density (kN/m³)

Note that the spreadsheet gives an opportunity to take into account the effect of overburden (surcharge). When the overburden is not considered, the parameters 6 and 7 should be made equal to 0.

In the example in Figure 14.22, various footings supporting a 500 mm square reinforced concrete column or pedestal are generated by the spreadsheet. The allowable soil bearing pressure is equal to 250 kPa. Note that the column dimensions and the allowable soil pressure are the same as those used in Example 14.3. Also, the average load factor (α) was determined using Eqn 14.4 as

$$\alpha = \frac{1.25\,P_{DL} + 1.5\,P_{LL}}{P_{DL} + P_{LL}} \qquad \text{[14.4]}$$

$$= \frac{1.25 \times 1000\ \text{kN} + 1.5 \times 1000\ \text{kN}}{1000\ \text{kN} + 1000\ \text{kN}} = 1.38$$

The footing sizes generated in this example vary from a 1000 mm square to a 9000 mm square in 250 mm increments. Note that the initial two values for the footing dimensions need to be entered in the spreadsheet — in this case, 1000 mm and 1250 mm have been entered in the "Width" column.

The footing of plan dimensions 3000 mm by 3000 mm generated by the spreadsheet corresponds to the one designed in Example 14.3. Note that the spreadsheet has generated a thickness of 650 mm and reinforcement of 9-25M bars each way. The corresponding axial load capacity is 3105 kN, which is larger than the factored load (P_f) of 2750 kN. Therefore, the footing is designed to sustain a load in excess of the actual design load. This approach is in accordance with the previously given advice to refrain from skimping on foundation dimensions (see Section 14.5).

In order to use this spreadsheet effectively, the designer should compute the factored axial load for each column and then choose the footing size and reinforcement from the spreadsheet. This spreadsheet offers an opportunity to vary all input parameters and can be easily used to design spread footings for an entire building.

A copy of the footing design spreadsheet is posted on the book Web site as an example of a design aid routinely used in design practice. However, the reader is encouraged to develop their own design spreadsheets and avoid the "black box" syndrome. The benefits of developing one's own spreadsheets were discussed in Section 1.11.

In design applications with variable column sizes and allowable soil bearing pressure in localized areas, or where allowable soil pressure depends on the footing size, the designer is advised to generate several spreadsheets to allow for these variations.

14.8 ECCENTRICALLY LOADED FOOTINGS

Footings are often subjected to a concentric axial load (P) acting at the column base, as shown in Figure 14.24a. (Note that the vertical column axis coincides with the vertical footing axis.) As a result, the soil pressure (q) under the footing is uniform and inversely proportional to the footing plan area (A), as follows:

$$q = \frac{P}{A} \hspace{3cm} [14.1]$$

However, in some cases a bending moment (M) (due to lateral wind or seismic loads) needs to be transferred from the column to the footing in addition to the axial load (P), as shown in

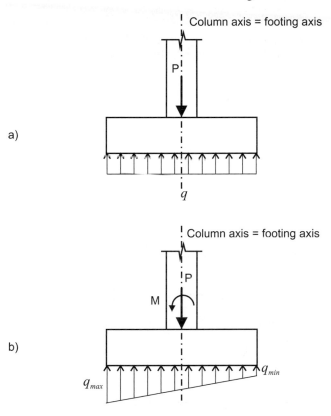

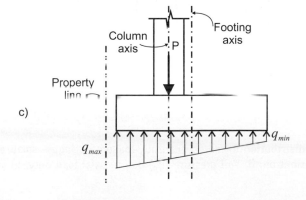

Figure 14.24 Soil pressure distribution: a) concentrically loaded footing — uniform pressure; b) column subjected to combined flexure and axial load — nonuniform pressure; c) eccentricity caused by property line — nonuniform pressure.

Figure 14.24b. The effect of the combined axial load and bending moment is equivalent to the effect of an axial load P acting at an eccentricity e (refer to Section 8.4 for more details), or

$$M = P \times e \tag{14.13}$$

External constraints such as a property line or underground utilities may require an eccentrically loaded footing, as illustrated in Figure 14.24c. In this case, the column vertical axis is eccentric with regard to the footing axis. The load eccentricity results in a nonuniform soil pressure distribution.

Let us examine the soil pressure distribution in the case of an eccentrically loaded footing with length l and width b. The following three cases can be considered based on the magnitude of the load eccentricity (e):

- small load eccentricity ($e < l/6$)
- boundary case ($e = l/6$)
- large load eccentricity ($e > l/6$)

Note that the case of an eccentrically loaded footing with large load eccentricity ($e > l/6$) is not recommended and should be avoided whenever possible.

Small load eccentricity ($e < l/6$)   First, consider a case where the load eccentricity is small; that is, $e < l/6$. The soil pressure has two components (see Figure 14.25), soil pressure due to axial load (q_P) and soil pressure due to bending moment (q_M), as follows:

$$q_P = \frac{P}{A}$$

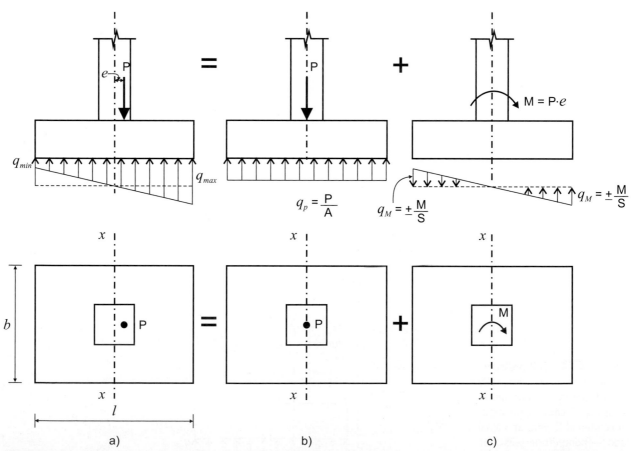

Figure 14.25 Soil pressure distribution for an eccentrically loaded footing — small eccentricity ($e < l/6$): a) resultant soil pressure distribution; b) soil pressure (q_P) due to axial load only; c) soil pressure (q_M) due to - bending moment only.

$$q_M = \frac{M}{S}$$

where

$$A = b \times l$$

is the footing plan area and

$$S = \frac{b \times l^2}{6} = \frac{A \times l}{6} \qquad [14.14]$$

is the elastic section modulus for flexure around the axis x-x (see Figure 14.25). The maximum and minimum soil pressures can be determined as

$$q_{min} = \frac{P}{A} - \frac{M}{S} \qquad [14.15]$$

and

$$q_{max} = \frac{P}{A} + \frac{M}{S} \qquad [14.16]$$

Note that the positive sign indicates compression. The above equations can be further transformed by substituting M from Eqn 14.13 and S from Eqn 14.14 into Eqns 14.15 and 14.16, as follows:

$$q_{min} = \frac{P}{A} - \frac{6P \times e}{A \times l} = \frac{P}{A}\left(1 - \frac{6e}{l}\right) \qquad [14.15a]$$

and

$$q_{max} = \frac{P}{A}\left(1 + \frac{6e}{l}\right) \qquad [14.16a]$$

The above equations can be used to determine the soil pressure distribution for the case of small eccentricity. However, when the eccentricity is large, that is, $e > l/6$, then

$$q_{min} < 0$$

In mathematical terms, this means that the portion of the soil beneath the footing is subjected to tension. However, the soil does not have any tensile resistance, so negative soil pressure beneath the footing (indicating tension) should be ignored. As a result, Eqns 14.15a and 14.16a are no longer valid.

Boundary case ($e = l/6$)   As the next step, let us consider the soil pressure distribution assuming the eccentricity $e = l/6$, resulting in a triangular soil pressure distribution, as illustrated in Figure 14.26. The soil pressure resultant (R) must act along the same line as the load P (this is based on the fundamental concepts of statics), and these forces must be in equilibrium, so

$$R = P \qquad [14.17]$$

The soil pressure resultant (R) is located at a distance $l/3$ from the face of the footing subjected to the maximum soil pressure (at the centroid of the triangular pressure diagram) and can be determined as

$$R = \frac{q_{max} \times l}{2} \times b$$

The maximum soil pressure (q_{max}) can be obtained from the above equation as

$$q_{max} = \frac{2 \times R}{b \times l}$$

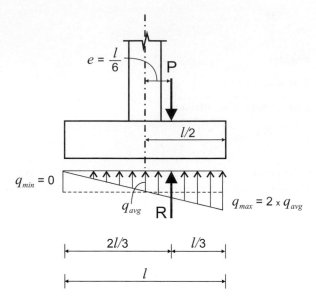

Figure 14.26 Soil pressure distribution for an eccentrically loaded footing — boundary case ($e = l/6$).

Based on Eqn 14.17,

$$q_{max} = \frac{2 \times P}{b \times l} \qquad [14.18]$$

and

$$q_{min} = 0$$

Note that the maximum soil pressure (q_{max}) can be expressed in terms of the *average soil pressure* (q_{avg}) as

$$q_{max} = 2 \times q_{avg} \qquad [14.19]$$

where

$$q_{avg} = \frac{P}{b \times l}$$

The average soil pressure can be defined as a uniform soil pressure acting over the portion of the footing under compression and at the centre of that area, that is, at a distance $l/2$ from the compressed face of the footing. In this case, the entire footing plan area (A) is under compression and therefore

$$q_{avg} = \frac{P}{A}$$

and

$$q_{max} = 2 \times \frac{P}{A} \qquad [14.20]$$

Equation 14.20 is often used in design because it is easy to memorize.

Large load eccentricity ($e > l/6$)   Finally, consider the case of an eccentrically loaded footing with a large eccentricity ($e > l/6$), as shown in Figure 14.27. In this case, the soil pressure distribution is triangular (similar to the case when $e = l/6$), except that the pressure acts over only a portion of the total footing area. The maximum soil pressure (q_{max}) can be obtained from Eqn 14.18 as

$$q_{max} = \frac{2 \times P}{b \times l'} \qquad [14.21]$$

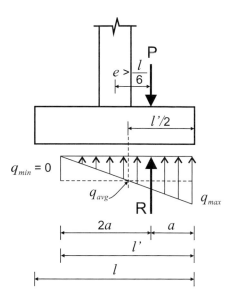

Figure 14.27 Soil pressure distribution for an eccentrically loaded footing with large eccentricity ($e > l/6$).

where

$$l' = 3 \times a = 3\left(\frac{l}{2} - e\right)$$ [14.22]

is the total length of footing under the soil pressure, and

$$a = \frac{l}{2} - e$$

a is the distance from the stress resultant (R) to the location of maximum soil pressure. When l' from Eqn 14.22 is substituted into Eqn 14.21, q_{max} can be determined as

$$q_{max} = \frac{2P}{3b\left(\frac{l}{2} - e\right)}$$ [14.23]

However, note that q_{max} can be expressed in terms of the average soil pressure (q_{avg}) by using Eqn 14.19, where

$$q_{avg} = \frac{P}{b \times l'}$$ [14.24]

In this case, the average soil pressure (q_{avg}) is an equivalent uniform pressure acting over an area of width b and length l' subjected to an axial load P, as shown in Figure 14.27; q_{avg} acts at a distance $l'/2$ from the face of the footing under compression. Therefore,

$$q_{max} = 2 \times \frac{P}{b \times l'}$$ [14.25]

A further increase in the magnitude of the eccentricity ($e \cong l/2$) could cause a significant increase in the maximum soil pressure. In any case, the maximum soil pressure beneath the footing should not exceed the allowable pressure (q_{all}); that is,

$$q_{max} \leq q_{all}$$ [14.26]

The above examples apply to eccentric footings where the stress resultant (R) is aligned with the applied load (P), but this may not always be the case. In a situation where a strip footing is located very close to a property line, the wall may have the beneficial effect of

3. *Large load eccentricity* ($e > l/6$):
The maximum soil pressure is

$$q_{max} = \frac{2P}{3b\left(\dfrac{l}{2} - e\right)}$$ [14.23]

or

$$q_{max} = 2 \times \frac{P}{b \times l'} \text{ (more convenient for design purposes)}$$ [14.25]

Eccentrically loaded footings with large load eccentricity ($e > l/6$) are not recommended and should be avoided whenever possible.

14.9 COMBINED FOOTINGS

Combined footings support more than one column or wall. The simplest case is the *two-column combined footing*, as shown in Figure 14.29a, which is used in the following situations:

- An exterior column is located close to the property line where the isolated spread footing is not able to carry the column loads.
- Two closely spaced columns cause their individual footings to be too closely spaced.

Alternatively, when two columns are spaced at a larger distance and one of them is close to the property line, strap footing may be a feasible solution. A *strap footing* consists of two individual column footings connected by a strap beam (grade beam), as shown in Figure 14.29b. The strap beam is designed to transfer the eccentric load applied on the footing adjacent to the property line.

The key considerations in the design of combined footings are outlined below (see Figure 14.30).

Plan shape and dimensions   Combined footings are usually of rectangular or trapezoidal plan shape. The footing shape is governed by the difference in column loads and by dimensional constraints. A rectangular shape is preferred whenever possible (see Figure 14.30a).

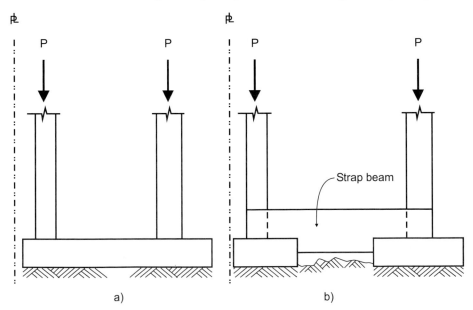

Figure 14.29 Combined footings: a) supporting two closely spaced columns; b) a strap footing consisting of two spread footings connected by a strap beam.

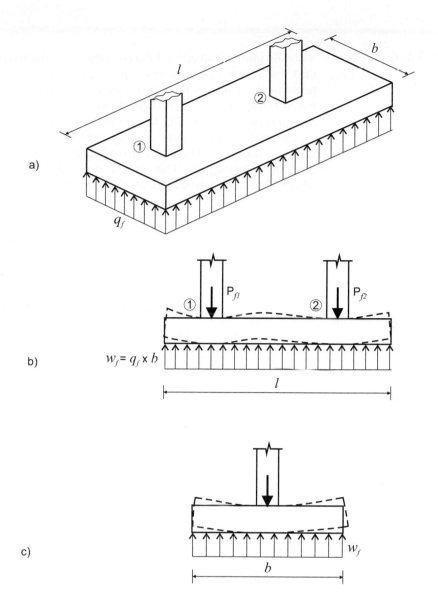

Figure 14.30 Behaviour of a combined footing: a) isometric view; b) longitudinal direction; c) transverse direction.

As in the case of other shallow footings, the plan dimensions are governed by the allowable soil pressure. In order to ensure uniform soil pressure distribution in combined footings, the centroid of the footing area should coincide with the line of action of the resultant of the column loads.

Design for flexure   A combined footing acts like a beam in the *longitudinal direction*, as shown in Figure 14.30b. Therefore, the flexural reinforcement is determined based on the bending moment distribution in the longitudinal direction. The main flexural steel is located at the top of the footing. However, bottom steel also needs to be provided (based on the bending moments developed in the overhangs beyond the column locations).

Flexural reinforcement needs to be provided in the *transverse direction* as well. The amount of this reinforcement is determined in the same way as for the spread footings discussed in Section 14.7 (see Figure 14.30c). The transverse reinforcement is uniformly ⬚⬚⬚⬚⬚⬚⬚⬚⬚⬚⬚⬚⬚⬚⬚⬚ column.

Design for shear   Both one-way and two-way shear effects need to be checked, as in the case of spread footings.

The design of combined footings will be illustrated by Example 14.6.

Example 14.6

Two 500 mm by 500 mm reinforced concrete columns (columns 1 and 2) are located at 3.9 m centre-to-centre spacing, as shown in the figure below. Due to its proximity to the property line, the column 2 footing can be extended by a maximum of 0.6 m beyond the column vertical axis on the left side. It is therefore required to design a combined footing to support both columns. The column loads are shown on the sketch below (note that the column 1 loads are the same as discussed in Example 14.3 in relation to a spread footing design). Use a 6 m long by 3 m wide combined footing for this design and 25M bars for flexural reinforcement. Maximum aggregate size is 20 mm. Ignore the effect of pattern loading in the design.

Design the combined footing according to the CSA A23.3 requirements.

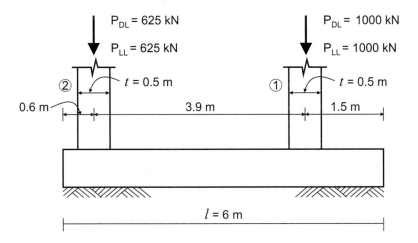

Given: $f_c' = 25$ MPa (normal-density concrete)
$f_y = 400$ MPa
$\phi_c = 0.65$
$\phi_s = 0.85$

SOLUTION: **1.   Determine the factored soil pressure (q_f)**

    a)   Determine the factored column loads:
       Column 1: $P_{f1} = 2750$ kN (see Example 14.3)
       Column 2:

$$P_{f2} = 1.25P_{DL} + 1.5P_{LL}$$
$$= 1.25 \times 625 \text{ kN} + 1.5 \times 625 \text{ kN} = 1719 \text{ kN}$$

    b)   Determine the location of the resultant (R) of these two loads (x). The resultant force is

$$R = P_{f1} + P_{f2} = 2750 \text{ kN} + 1719 \text{ kN} = 4469 \text{ kN}$$

The sum of the moments around point A can be determined as follows (see the sketch that follows):

$$\sum M_A = P_{f1} \times a_1 + P_{f2} \times a_2 = 2750 \text{ kN} \times 4.5 \text{ m} + 1719 \text{ kN} \times 0.6 \text{ m}$$
$$= 13\ 406 \text{ kN} \cdot \text{m}$$

or

$$\sum M_A = R \times x = (4469 \text{ kN})x$$

Based on the above two equations,

$$x = \frac{P_{f1} \times a_1 + P_{f2} \times a_2}{R} = \frac{13\,406 \text{ kN} \cdot \text{m}}{4469 \text{ kN}} = 3 \text{ m}$$

Because the resultant is located at the midpoint of the footing ($l = 6$ m), the footing is subjected to uniform soil pressure.

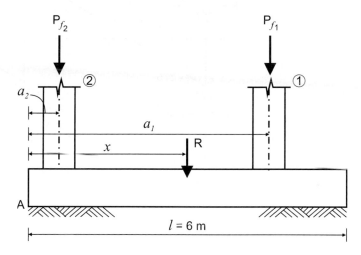

c) Next, calculate the footing plan area as

$$A = b \times l = 3 \text{ m} \times 6 \text{ m} = 18 \text{ m}^2$$

Then, calculate the factored soil bearing pressure as

$$q_f = \frac{P_f}{A} \qquad\qquad [14.2]$$

$$= \frac{4469 \text{ kN}}{18 \text{ m}^2} = 248 \text{ kPa}$$

2. **Determine the shear force and bending moment distribution in the longitudinal direction**

    This footing can be modelled as an inverted beam supported by the columns and subjected to an upward uniform load. The factored uniform load (w_f) acting on the footing is equal to the factored soil pressure determined in the previous step acting over the footing width (b) as

    $$w_f = q_f \times b$$
    $$= 248 \text{ kPa} \times 3 \text{ m} = 744 \text{ kN/m}$$

    The bending moment and shear force diagrams are shown in the figure that follows. Use these diagrams to check the one-way shear requirements in Step 3 and the flexural requirements in Step 4.

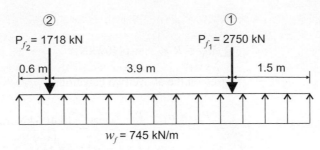

Load diagram

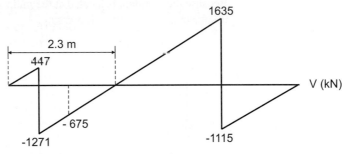

Shear diagram

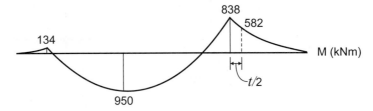

Moment diagram

<table>
<tr><td>A23.3 Cl.11.3.3
Cl.13.3.3
Cl.13.3.4</td></tr>
</table>

3. Check the shear design requirements

a) Take the same footing thickness as used in Example 14.3:

$$h = 650 \text{ mm}$$

and

$$d = 550 \text{ mm}$$

b) First, check the two-way shear requirements (Cl.13.3.3 and Cl.13.3.4). The footing underneath column 1 was previously designed in Example 14.3 for the same factored load

$$P_{f1} = 2750 \text{ kN}$$

Thus, the design resulted in a footing of the same overall thickness as chosen in this example. Therefore, perform the two-way shear design check for column 2 only.

i) Determine the factored load to be considered for two-way shear:

$$V_f = P_{f2} = 1719 \text{ kN}$$

ii) Determine the concrete shear resistance for two-way shear (V_c).
Find v_c as

A23.3 Eq. 13.7

$$v_c = 0.38 \lambda \phi_c \sqrt{f_c'}$$ **[14.10]**

$$= 0.38(1)(0.65)\sqrt{25 \text{ MPa}} = 1.24 \text{ MPa} = 1240 \text{ kPa}$$

At this point, proceed without checking the other two criteria (as discussed in Section 14.4.1); therefore,

$$v_c = 1.24 \text{ MPa}$$

iii) Determine the two-way shear resistance of concrete (V_c) as

$$V_c = v_c \times b_o \times d \qquad \text{[14.7]}$$

Determine the shear perimeter (b_o) for column 2 as (see the sketch below)

$$b_o = 3(t + d) = 3(0.5 \text{ m} + 0.55 \text{ m}) = 3.15 \text{ m}$$

where

$t = 0.5$ m (column width)

Thus,

$$V_c = 1240 \text{ kPa} \times 3.15 \text{ m} \times 0.55 \text{ m} \cong 2150 \text{ kN}$$

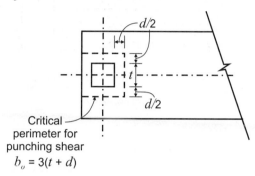

Critical
perimeter for
punching shear
$b_o = 3(t + d)$

iv) At this point, check the remaining two v_c criteria:

A23.3 Eq. 13.5

$$v_c = \left(1 + \frac{2}{\beta_c}\right) 0.19 \lambda \phi_c \sqrt{f_c'} \qquad \text{[14.8]}$$

$$= \left(1 + \frac{2}{1}\right) 0.19 \, (1)(0.65)\sqrt{25 \text{ MPa}} = 1.85 \text{ MPa}$$

where

$$\beta_c = \frac{t}{t} = 1$$

and

A23.3 Eq. 13.6

$$v_c = \left(\frac{\alpha_s d}{b_o} + 0.19\right) \lambda \phi_c \sqrt{f_c'} \qquad \text{[14.9]}$$

$$= \left(\frac{3 \times 0.55 \text{ m}}{3.15 \text{ m}} + 0.19\right) \times 1 \times 0.65 \sqrt{25 \text{ MPa}} = 2.32 \text{ MPa}$$

where $\alpha_s = 3$ (exterior column).
Therefore, the assumption that the third criterion ($v_c = 1.24$ MPa) governs was correct.
Because

$$V_c = 2150 \text{ kN} > V_f = 1719 \text{ kN}$$

shear reinforcement is not required.

c) Next, check the one-way shear requirements (A23.3 Cl.11.3.4).
Again, the one-way shear requirements for column 1 were checked in Example 14.3, so only column 2 needs to be checked in this example.

i) First, determine the factored shear force (V_f) at the critical section at a distance d from the right hand face of column 2, as follows (see the sketch that follows):

$$a = 0.6 \text{ m} + \frac{t}{2} + d = 0.6 \text{ m} + \frac{0.5 \text{ m}}{2} + 0.55 \text{ m} = 1.4 \text{ m}$$

$$V_f = P_{f2} - w_f \times a = 1719 \text{ kN} - 745 \text{ kN/m} \times 1.4 \text{ m} = 676 \text{ kN} \cong 675 \text{ kN}$$

Note that the same value may be obtained from the shear force diagram presented in Step 2.

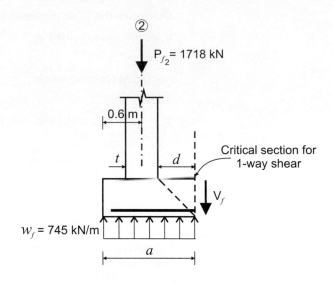

ii) Determine the one-way shear resistance of concrete (V_c) in the same way as for beams and slabs discussed in Section 6.5.4 (A23.3 Cl.11.3.4). Therefore, determine V_c for a section with width $b_w = b = 3000$ mm and $d_v = 500$ mm (from Example 14.3) as

A23.3 Eq. 11.6

$$V_c = \phi_c \lambda \beta \sqrt{f_c'}\, b_w\, d_v \qquad\qquad [6.12]$$

$$= 0.65(1)(0.15)\sqrt{25\text{ MPa}}\,(3000\text{ mm})(500\text{ mm})$$

$$\cong 730\text{ kN (same as Example 14.3)}$$

Note that $\beta = 0.15$, as in Example 14.3.

iii) To avoid the use of shear reinforcement in the footing (A23.3 Cl.11.2.8.1), it is required that

$$V_f \le V_c$$

Since

$$V_f = 675\text{ kN} < V_c = 730\text{ kN}$$

shear reinforcement is not required.

4. **Determine the required amount and distribution of flexural reinforcement in the longitudinal direction**

In this case, the bending moment distribution in the diagram presented in Step 2 shows that both positive and negative bending moments develop along the footing length. Determine the flexural reinforcement for both regions, as discussed below.

Positive moment region

a) Determine the factored bending moment M_f^+.

From the bending moment diagram, the maximum positive bending moment of 838 kN·m is developed at the column 1 centre-line. However, the critical section for the flexural design is located at the face of the column. Therefore, design the reinforcement based on the bending moment at the right face of Column 1, as illustrated in the sketch that follows.

The distance (a) from the end of the footing to the right face of column 1 is

$$a = 1.5\text{ m} - \frac{0.5\text{ m}}{2} = 1.25\text{ m}$$

Hence, determine the factored bending moment as

$$M_f^+ = w_f \times \frac{a^2}{2}$$

$$= 745 \text{ kN/m} \times \frac{(1.25 \text{ m})^2}{2} = 582 \text{ kN} \cdot \text{m}$$

Note that the same bending moment can be obtained from the bending moment diagram in Step 2.

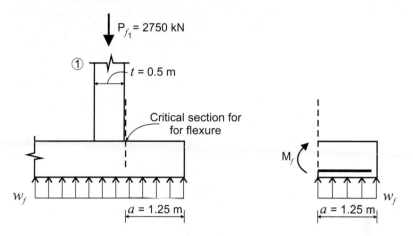

b) Determine the required area of reinforcement.
The strength design requirement states that

$$M_r \geq M_f^+ \qquad\qquad\qquad\qquad\qquad\qquad\text{[14.11]}$$

Set

$$M_r = M_f^+ = 582 \text{ kN} \cdot \text{m}$$

Then, calculate the required area of flexural reinforcement from Eqn 5.4 (direct procedure) as

$$A_s = 0.0015 f_c' \, b \left(d - \sqrt{d^2 - \frac{3.85 M_r}{f_c' \, b}} \right) \qquad\qquad\text{[5.4]}$$

$$= 0.0015 \, (25 \text{ MPa}) \, (3000 \text{ mm})$$

$$\times \left(550 \text{ mm} - \sqrt{(550^2 \text{ mm}) - \frac{3.85 \, (582 \times 10^6 \text{ N} \cdot \text{mm})}{(25 \text{ MPa}) \, (3000 \text{ mm})}} \right)$$

$$= 3135 \text{ mm}^2$$

Use 7-25M bars (bottom reinforcement):

$$A_s = 7 \times 500 \text{ mm}^2 = 3500 \text{ mm}^2$$

c) Determine the bar spacing.
Use 100 mm end spacing, so

$$s = \frac{3000 \text{ mm} - (2 \times 100 \text{ mm})}{6} = 466.7 \text{ mm} \cong 465 \text{ mm}$$

According to A23.3 Cl 7.4.1.2 and Cl.13.10.4, the maximum permitted bar spacing (s_{max}) is equal to the lesser of $3h$ and 500 mm.
Because

$$3h = 3(650 \text{ mm}) = 1950 \text{ mm}$$

$$s_{max} = 500 \text{ mm (governs)}$$

Negative moment region

Since

$$s = 465 \text{ mm} < s_{max} = 500 \text{ mm}$$

d) First, determine the factored bending moment M_f^-.

Based on the bending moment diagram developed in Step 2, the maximum negative moment that occurs between columns 1 and 2 is

$$M_f^- = 950 \text{ kN} \cdot \text{m}$$

e) Determine the required area of reinforcement.

The cover to the reinforcing steel at the top of the footing can be reduced to 50 mm because only the footing face cast against the earth requires a 75 mm cover. Therefore, determine the effective depth (d) for negative bending as

$$d = h - \text{cover} - \frac{d_b}{2} = 650 \text{ mm} - 50 \text{ mm} - \frac{25 \text{ ıııııı}}{2} = 587.5 \text{ mm} \cong 580 \text{ mm}$$

where $d_b = 25$ mm (25M bars).

The strength design requirement states that

$$M_r \geq M_f^- \tag{14.11}$$

Set

$$M_r = M_f^- = 950 \text{ kN} \cdot \text{m}$$

Then calculate the required area of flexural reinforcement from Eqn 5.4 (direct procedure) as

$$A_s = 0.0015 f_c' \, b \left(d - \sqrt{d^2 - \frac{3.85 \, M_r}{f_c' b}} \right) \tag{5.4}$$

$$= 0.0015(25 \text{ MPa})(3000 \text{ mm})$$

$$\times \left(580 \text{ mm} - \sqrt{(580 \text{ mm})^2 - \frac{3.85 \, (950 \times 10^6 \text{ N} \cdot \text{mm})}{(25 \text{ MPa})(3000 \text{ mm})}} \right)$$

$$= 4915 \text{ mm}^2$$

Use 10-25M bars (top reinforcement):

$$A_s = 10 \times 500 \text{ mm}^2 = 5000 \text{ mm}^2$$

A23.3 Cl.7.8.1   **5. Confirm that the minimum reinforcement requirement is satisfied**

First, determine the gross cross-sectional area of the footing as

$$A_g = h \times b = 650 \text{ mm} \times 3000 \text{ mm} = 195 \times 10^4 \text{ mm}^2$$

Then,

$$A_{smin} = 0.002 A_g \tag{14.12}$$

$$= 0.002 \, (195 \times 10^4 \text{ mm}^2) = 3900 \text{ mm}^2$$

In the positive moment region (7-25M bottom bars),

$$A_s = 3500 \text{ mm}^2 < A_{smin} = 3900 \text{ mm}^2$$

so the minimum reinforcement requirement is *not* satisfied. Therefore, it is necessary to use 8-25M bottom bars ($A_s = 4000 \text{ mm}^2$). In the negative moment region (10-25M top bars),

$$A_s = 5000 \text{ mm}^2 > A_{smin} = 3900 \text{ mm}^2$$

so the minimum reinforcement requirement *is* satisfied.

6. **Confirm that the maximum reinforcement requirement is satisfied**

   | A23.3 Cl.10.5.2 |

   In this case, the maximum reinforcement requirement is satisfied by default because the amount of reinforcement is close to the minimum reinforcement.

7. **Determine the required amount and distribution of flexural reinforcement in the transverse direction**

   The amount of reinforcement required in the transverse direction is the same as that determined for the spread footing designed in Example 14.3, that is, 8-25M bars. (See Steps 4 to 6 of Example 14.3 for detailed calculations.) Therefore, 8-25M bars need to be placed beneath each column in the combined footing; this will result in a total of 16-25M bars, as shown in the design summary below.

8. **Provide a design summary**

   A design summary of the combined footing showing the amount and distribution of reinforcement is presented below.

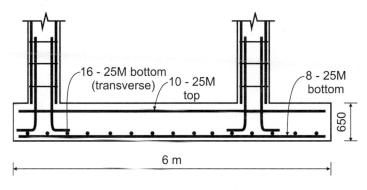

14.10 LOAD TRANSFER FROM COLUMN INTO FOOTING

14.10.1 Bearing Strength

Bearing strength should be considered in foundation design. The designer must ensure that the concrete in the column is not crushed at the interface between the footing and the column and that the footing is not crushed locally beneath the column, as shown in Figure 14.31. This is related to the *bearing strength* of concrete, that is, its ability to resist the compressive stresses between the two members in direct contact with one another. Bearing failure is usually localized around the contact area. The concept of bearing stress is covered in mechanics of materials textbooks as a form of normal stress developed at the interface between structural members subjected to axial compression.

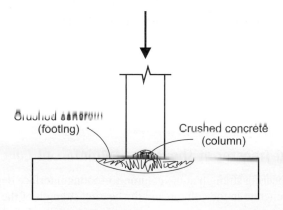

Figure 14.31 Bearing failure at the column-to-footing interface.

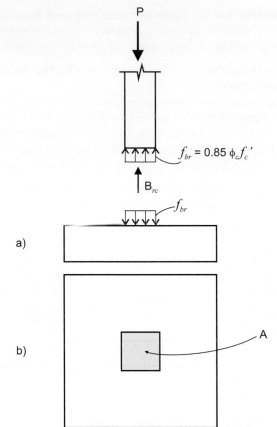

Figure 14.32 Bearing stress at the column-footing interface: a) free-body diagrams of the column and the footing; b) footing plan showing the contact area.

Consider a reinforced concrete column supported by a spread footing, as shown in Figure 14.32a. The column is subjected to a concentric axial load P. Consider a free-body diagram of the column and footing. The column is subjected to axial compression causing bearing stress at its base. When the load acting on the column increases, the bearing stress at the base reaches its maximum value (f_{br}). According to A23.3 Cl.10.8.1, f_{br} is given by

$$f_{br} = 0.85\phi_c f_c'$$

where f_c' is the specified compressive strength of the concrete. Considering that $\phi_c = 0.65$, it follows that

$$f_{br} \cong 0.55 f_c'$$

It is considered that the bearing stress acts over the column contact area (equal to the gross area of the column section). As a result, the *column bearing resistance* (B_{rc}) can be determined from the equation for normal stress as follows:

$$f_{br} = \frac{B_{rc}}{A}$$

where A is the column-footing contact area, shown hatched in Figure 14.32b.

Therefore, the column bearing resistance can be determined as the product of the concrete bearing resistance and the contact area as

$$B_{rc} = f_{br} \times A = (0.85\phi_c f_c') \times A \qquad \text{[14.27]}$$

Note that the concept of bearing strength applies to other structural members, for example steel beams and open web steel joists resting on reinforced concrete walls (see Section 13.4).

14.10.2   Load Transfer at the Base of Reinforced Concrete Columns

The bearing strength at the column-to-footing interface depends on the bearing strength of the concrete in the column and the bearing strength of the concrete in the footing. If these strengths are not sufficient to transfer the axial forces across the interface, additional steel reinforcement must be provided, as discussed in this section.

Based on Eqn 14.27, the *column bearing resistance* (B_{rc}) is (A23.3 Cl.10.8.1)

$$B_{rc} = 0.85\phi_c f_c' A_1 \qquad \text{[14.28]}$$

where

A_1 = the loaded contact area (equal to the column gross cross-sectional area)
$f_c' = f_{cc}'$ is the specified compressive strength for the concrete in the column.

According to A23.3 Cl.10.8.1, the bearing resistance (B_{rf}) of the concrete contact area of the *supporting member (footing)* is

$$B_{rf} = 0.85\phi_c f_{cf}' A_1 \sqrt{\frac{A_2}{A_1}} \qquad \text{[14.29]}$$

where

A_2 = the maximum area of the portion of the supporting surface that is geometrically similar and concentric to the loaded area (see Figure 14.33)
f_{cf}' = the specified compressive strength for the concrete in the footing.

When

$$\sqrt{\frac{A_2}{A_1}} > 2$$

then the footing bearing resistance is limited to

$$B_{rfmax} = 1.7\phi_c f_{cf}' A_1 \qquad \text{[14.30]}$$

If the factored axial load (P_f) acting at the base of the column exceeds the concrete bearing resistance, that is,

$$P_f \geq B_{rc}$$

then reinforcing bars, usually in the form of dowels, needs to be provided to resist the excess load (A23.3 Cl.15.9.1) so that

$$B_{rd} \geq P_f - B_{rc} \qquad \text{[14.31]}$$

where B_{rd} denotes the dowel contribution to the bearing resistance, determined with the assumption that dowels yield in compression as follows:

$$B_{rd} = A_d \phi_s f_y \qquad \text{[14.32]}$$

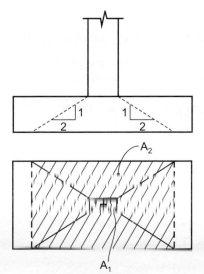

Figure 14.33 Load transfer at the column base — a reinforced concrete column.

where A_d is the total cross-sectional area of all dowels crossing the interface and f_y is the specified yield strength of the dowel reinforcement.

According to A23.3 Cl.15.9.2.1, the area of the dowel reinforcement crossing the column-footing interface (A_d) should be restricted as follows:

$$A_d \geq 0.005 A_1 \qquad\qquad [\textbf{14.33}]$$

When dowel reinforcement is provided, the overall bearing resistance (the combined contribution of the concrete and dowel reinforcement) of the column section at the interface with the footing (B_{rc}) can be determined according to A23.3 Cl.15.9.2.1 and Cl.10.8.1 as

$$B_{rc} = 0.85 \phi_c f'_{cc} A_1 + A_d \phi_s f_y \qquad\qquad [\textbf{14.34}]$$

while the footing bearing resistance (B_{rf}) is

$$B_{rf} = 0.85 \phi_c f'_{cf} A_1 \sqrt{\frac{A_2}{A_1}} + A_d \phi_s f_y \qquad\qquad [\textbf{14.35}]$$

It is good practice to use at least four dowels for this purpose. The dowels should be placed at the corners, adjacent to the column longitudinal bars. The dowels must be developed on both the footing and the column side; they must be projected at least by the development length in compression (l_{db}), according to A23.3 Cl.12.3; that is,

$$l_{db} = 0.24 \frac{f_y}{\sqrt{f'_c}} d_b \leq 0.044 f_y d_b \qquad\qquad [\textbf{9.4}]$$

In some cases, different dowel compression development length (l_{db}) may be required for the footing and the column due to the different concrete strengths used in these two members; refer to Section 9.3.3 for more information on compression development length. In practice, the specified concrete compressive strength in the column (f'_{cc}) may be higher than the specified compressive strength in the footing (f'_{cf}).

In general, dowels are provided with 90° hooks and are placed on the top of the bottom footing reinforcement mat. In this way, the dowels are tied in place and the chances of them being dislodged during construction are minimized. It should be noted that hooks are not effective in compression. For more details on hooked anchorage of reinforcement, refer to Section 9.4.

In general, it is not feasible to embed the full-length column reinforcement into the footing, due to its rather large unsupported height above the footing and a lack of proper column formwork at the time of the footing pour. Therefore, it is common practice to provide dowels projecting from the footing. The dowels usually have hooked anchorage into the footing and they extend into the column. These bars, called *starter bars,* should be overlapped with the column longitudinal reinforcement. Dowels and starter bars are shown in Figure 14.34.

14.10.3   Load Transfer at the Base of Steel Columns

Where structural steel columns and base plates are used, the total load is usually transferred entirely by bearing on the concrete contact area. The design bearing strength for the

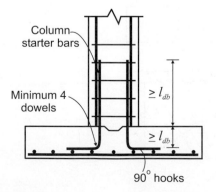

Figure 14.34 Dowels and column starter bars.

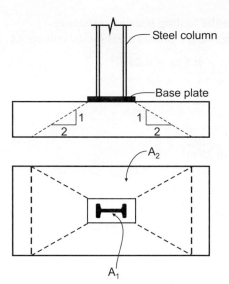

Figure 14.35 Load transfer at
the column base — steel column.

footing can be checked using Eqn 14.29. In this case, the area A_1 denotes the base plate
area in Figure 14.35.

When a column base plate is inadequate to transfer the total load, the following adjustments can be made:

- Increase the column base plate dimensions.
- Use a higher-strength concrete for the pedestal or footing.
- Increase the supporting area A_2 until the ratio A_2/A_1 reaches the maximum value.

Example 14.7

**Consider the spread footing supporting a 500 mm by 500 mm column designed in
Example 14.3. The footing plan dimensions are 3 m by 3 m and the thickness is 650 mm,
as shown in the figure below. The column carries a factored axial load (P_f) of 3750 kN
and is reinforced with four 25M vertical rebars.**

*Check whether the column-footing interface is adequate for the load transfer and design
the dowel reinforcement if required.*

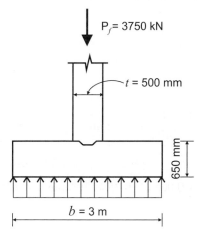

Given: $f_c' = 25$ MPa (both footing and column)
$f_y = 400$ MPa
$\psi_c = 0.65$
$\phi_s = 0.85$

SOLUTION: In this example, the concrete compressive strength in the column and the footing is the
same; that is,

$$f_c' = f_{cc}' = f_{cf}' = 25 \text{ MPa}$$

1. **Determine the footing bearing resistance**
   First, determine the contact areas for the column (A_1) and the footing (A_2) as follows:

   $A_1 = 0.5 \text{ m} \times 0.5 \text{ m} = 0.25 \text{ m}^2$

   $A_2 = 3 \text{ m} \times 3 \text{ m} = 9 \text{ m}^2$

   According to A23.3 Cl.10.8.1, since

   $$\sqrt{\frac{A_2}{A_1}} = \sqrt{\frac{9 \text{ m}^2}{0.25 \text{ m}^2}} = 6 > 2$$

   then

   $$B_{rf} = B_{rfmax} = 1.7\phi_c f_c' A_1 \qquad \text{[14.30]}$$
   $$= 1.7 \times 0.65 \times 25 \text{ MPa} \times (0.25 \times 10^6 \text{ mm}^2) - 6906 \text{ kN}$$

   Since

   $$B_{rf} = 6906 \text{ kN} > P_f = 3750 \text{ kN}$$

   the footing bearing resistance is satisfactory.

2. **Determine the column bearing resistance**
   Determine the column bearing resistance as

   $$B_{rc} = 0.85\phi_c f_c' A_1 \qquad \text{[14.28]}$$
   $$= 0.85 \times 0.65 \times 25 \text{ MPa} \times (0.25 \times 10^6 \text{ mm}^2) = 3453 \text{ kN}$$

   Because

   $$B_{rc} = 3453 \text{ kN} < P_f = 3750 \text{ kN}$$

   the column bearing resistance is not satisfactory. Dowels need to be provided.

3. **Design the dowel reinforcement**
   Determine the required dowel area as follows:

   $$B_{rd} \geq P_f - B_{rc} \qquad \text{[14.31]}$$
   $$= 3750 \text{ kN} - 3453 \text{ kN} = 297 \text{ kN}$$

   $$B_{rd} = A_d \phi_s f_y \qquad \text{[14.32]}$$

   Hence,

   $$A_d = \frac{B_{rd}}{\phi_s f_y}$$
   $$= \frac{297 \times 10^3 \text{ N}}{0.85 \times 400 \text{ MPa}} = 874 \text{ mm}^2$$

   According to A23.3 Cl.15.9.2.1, A_d should be restricted as follows:

   $$A_d \geq 0.005 A_1 = 0.005 \times (0.25 \times 10^6 \text{ mm}^2) = 1250 \text{ mm}^2 \qquad \text{[14.33]}$$

   In this case, the minimum reinforcement requirement governs. A minimum of four dowels needs to be provided. Use 4-25M bars to match the column reinforcement. The area of a 25M bar is 500 mm² (see Table A.1), so

   $$A_d = 4 \times 500 \text{ mm}^2 = 2000 \text{ mm}^2 > 1250 \text{ mm}^2$$

   so the minimum reinforcement requirement is satisfied.

   The column bearing resistance including dowels is (using the result for B_{rc} from Step 2)

   $$B_{rc} = 0.85\phi_c f_c' A_1 + A_d \phi_s f_y \qquad \text{[14.34]}$$
   $$= 3453 \times 10^3 \text{ N} + (2000 \text{ mm}^2 \times 0.85 \times 400 \text{ MPa}) = 4133 \text{ kN}$$

Since

$$B_{rc} = 4133 \text{ kN} > P_f = 3750 \text{ kN}$$

the resistance is satisfactory.

Finally, detail the dowels; that is, determine the required development length for straight bars in compression (l_{db}) according to A23.3 Cl.12.3 as

$$l_{db} = 0.24 \frac{f_y}{\sqrt{f_c'}} d_b \qquad\qquad \text{[9.4]}$$

$$= 0.24 \times \frac{400 \text{ MPa}}{\sqrt{25 \text{ MPa}}} \times 25 \text{ mm} = 480 \text{ mm}$$

where $d_b = 25$ mm for 25M bars.

However, according to A23.3 Cl.12.3, l_{db} is restricted as follows:

$$l_{db} \leq 0.044 \, d_b f_y = 0.044 \times 25 \text{ mm} \times 400 \text{ MPa} = 440 \text{ mm}$$

Therefore, use

$$l_{db} = 440 \text{ mm}$$

Check whether the available depth in the footing is adequate. Use the same values as in Example 14.3; that is,

$$h = 650 \text{ mm}$$

is the overall footing depth,

$$\text{cover} = 75 \text{ mm}$$

is the clear cover to the flexural reinforcement, and

$$d_b = 25 \text{ mm}$$

is the diameter for the footing reinforcement, so

$$l = h - \text{cover} - 2d_b = 650 \text{ mm} - 75 \text{ mm} - 2 \times 25 \text{ mm} = 525 \text{ mm}$$

Because

$$l = 525 \text{ mm} > 440 \text{ mm}$$

round the dowel length to

$$l_{db} = 500 \text{ mm}$$

It follows that the 25M dowels can be anchored in the footing. The dowels need to be extended by at least the same development length into the column. These dowels will be spliced with the column starter bars.

4. **Provide a design summary**

The dowel layout for this example is illustrated on the sketch below. Note that the footing reinforcement is omitted from the sketch. It is usual practice to provide dowel reinforcement with 90° hooked anchorage, as shown in the sketch below.

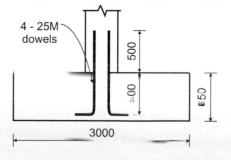

The bearing strength at the column-to-footing interface depends on the bearing strength of the concrete in the column and the bearing strength of the concrete in the footing. If these strengths are not sufficient to transfer the axial forces across the interface, additional steel reinforcement (called dowel reinforcement) must be provided.

The *column bearing resistance* is (A23.3 Cl.10.8.1)

$$B_{rc} = 0.85\phi_c f'_{cc} A_1 \qquad \qquad [14.28]$$

The bearing resistance of the *supporting member (footing)* is

$$B_{rf} = 0.85\phi_c f'_{cf} A_1 \sqrt{\frac{A_2}{A_1}} \qquad \qquad [14.29]$$

If

$$\sqrt{\frac{A_2}{A_1}} > 2$$

then the footing bearing resistance is limited to

$$B_{rfmax} = 1.7\phi_c f'_{cf} A_1 \qquad \qquad [14.30]$$

When the factored axial load (P_f) acting at the base of the column exceeds the concrete bearing resistance, dowel reinforcement needs to be provided to carry the excess load (A23.3 Cl.15.9.1) such that

$$B_{rd} \geq P_f - B_{rc} \qquad \qquad [14.31]$$

where

$$B_{rd} = A_d \phi_s f_y \qquad \qquad [14.32]$$

The required area (A_d) of dowel reinforcement crossing the column-footing interface is restricted as follows (A23.3 Cl.15.9.2.1):

$$A_d \geq 0.005 A_1 \qquad \qquad [14.33]$$

14.11 STRUCTURAL DRAWINGS AND DETAILS FOR REINFORCED CONCRETE FOOTINGS

In a typical design project, a complete set of structural drawings includes a foundation plan drawing. An example of a foundation plan is shown in Figure 14.36. The *foundation plan* shows the locations of all footings in the building. The footings are usually labelled as F1, F2, etc. (where F stands for footing), or in alphabetical order (A, B, C, etc.). Information related to footing plan dimensions and reinforcement (number and size of rebars) is often summarized in tabular form as a *footing schedule*. An example of a footing schedule is shown in Figure 14.37. The same drawing should contain all relevant details, showing column-to-footing connections and reinforcement arrangements. Column and wall details may be presented on the same drawing; alternatively, these details may be provided on separate drawings (this depends on the project size).

Typically, the footing reinforcement arrangement is shown on a vertical section through the footing. An example of a spread (column) footing section is shown in Figure 14.38; note that the footing reinforcement information is omitted because it is included in the footing schedule. An example of a section through the combined footing is shown in Figure 14.39.

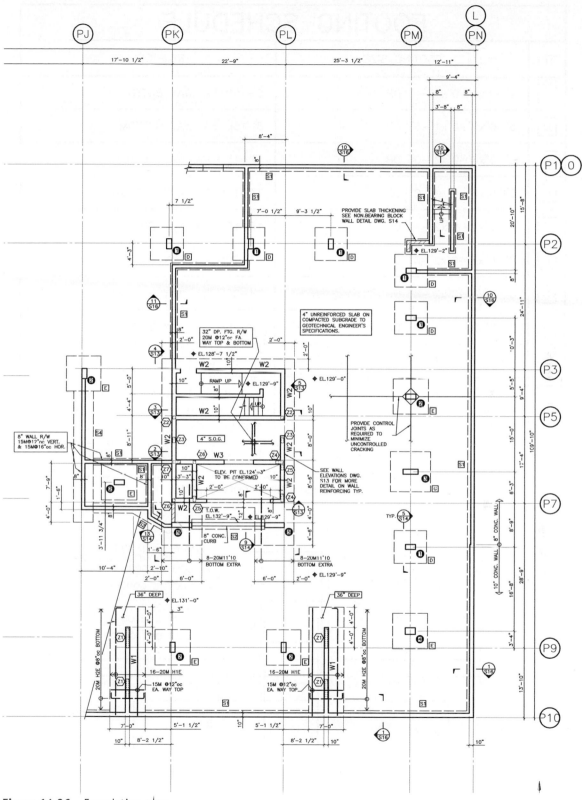

Figure 14-36 Foundation plan.

FOOTING SCHEDULE

TYPE	FOOTING SIZE	REINFORCING
A	4'-0"x4'-0"x1'-4" DP.	5-15M EA. WAY BOTTOM
B	5'-0"x5'-0"x1'-8" DP.	8-15M EA. WAY BOTTOM
C	6'-0"x6'-0"x2'-0" DP.	12-15M EA. WAY BOTTOM
D	6'-6"x6'-6"x2'-2" DP.	14-15M EA. WAY BOTTOM
E	7'-0"x7'-0"x2'-4" DP.	16-15M EA. WAY BOTTOM
F	7'-6"x7'-6"x2'-6" DP.	13-20M EA. WAY BOTTOM
G	8'-0"x8'-0"x2'-8" DP.	14-20M EA. WAY BOTTOM
H	9'-0"x9'-0"x3'-0" DP.	20-20M EA. WAY BOTTOM
S1	1'-4" WIDEx10" DP. STRIP FTG.	2-15M CONT. LONGIT. BOTTOM
S2	2'-0" WIDEx10" DP. STRIP FTG.	2-15M CONT. LONGIT. BOTTOM
S3	3'-0" WIDEx10" DP.	SEE SECTION
S4	4'-0" WIDEx16" DP. STRIP FTG.	5-15M CONT. LONGIT. BOTTOM & 15M @8"oc TRANSVERSE BOTTOM

ALL FOOTINGS TO BE CENTERED UNDER COLUMN/WALL TYP. U.N.O.

Figure 14.37 Footing schedule.

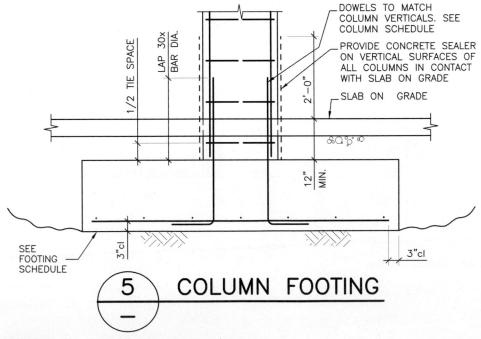

Figure 14.38 Spread footing cross-section.

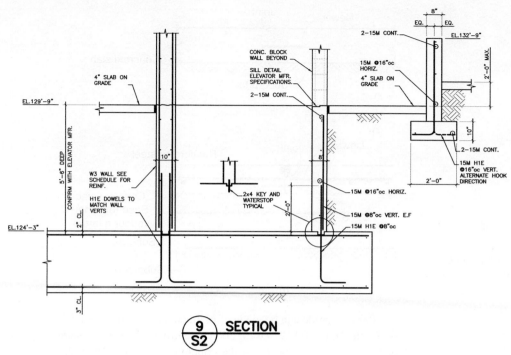

Figure 14.39 A combined footing section.

14.12 SLAB ON GRADE

14.12.1 Background

A slab on grade is a common element of concrete construction. According to ACI 360 R-92 (2004), a slab on grade is "a slab, continuously supported by the ground, whose total loading when uniformly distributed would impart a pressure to the grade or soil that is less than 50% of the allowable bearing capacity thereof." In some references, slabs on grade are called "slabs on ground" or "floor slabs." The design of slabs on grade is significantly different from the design of elevated floor slabs discussed in Chapter 11 of this book.

Slabs on grade are used in concrete structures of various functions, including industrial facilities, warehouses, garages, airports, and some types of water reservoirs. Concrete buildings often have a slab on grade, either at the basement level or at the ground floor level.

The main components of a slab on grade are the subgrade and the concrete slab, as shown in Figure 14.40a. Other (optional) components, based on site and environmental conditions, include the subbase course, the vapour retarder, and the base course (see Figure 14.40b).

The *subgrade* is the natural soil at the site on which the slab is built. The top 150 mm to 500 mm is usually compacted before the slab is placed. It is very important for the subgrade to provide uniform support to the slab. The floor strength is achieved most economically by building strength into the concrete slab with the optimum use of low-cost materials under the slab.

The *subbase* is a compactible, free draining granular fill with a typical thickness of 75 mm to 100 mm placed on top of the prepared subgrade.

The *vapour retarder* is usually a thin layer of a material such as polyethylene film that minimizes the transmission of moisture from the soil through the slab. Moisture protection is required for slabs on grade that are to be covered by tile, wood, or carpet, or in areas where the floor will be in contact with any moisture-sensitive equipment or product. In some cases, a vapour barrier in the form of a membrane is used to provide a higher level of moisture control.

The *base* is a compactible granular material placed on top of the vapour retarder and beneath the concrete slab (not always mandatory in slab on grade construction).

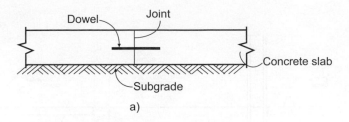

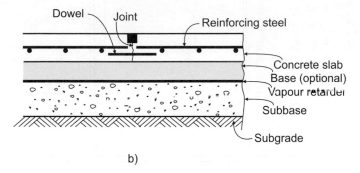

Figure 14.40 Components of a slab on grade: a) simple layout; b) complex layout.

Slabs on grade can be made of plain concrete, reinforced concrete, or prestressed concrete. The ACI 360 R.92 (ACI, 2004) identifies six types of slabs on grade, depending on the type of reinforcement and the concrete properties. Plain concrete slabs are the most economical type of construction for common building applications.

There are several factors influencing the performance of slabs on grade (PCA, 2001):

- uniformity of the subgrade and adequacy of its bearing capacity
- quality of concrete
- slab thickness
- type and spacing of joints
- type, magnitude, and position of loads and corresponding deformations

Slabs must be designed for the most critical combination of these loading conditions by considering the maximum load, its contact area, and its spacing.

A brief overview of these factors is presented in the following sections. For more details related to the design and construction of concrete slabs on grade, the reader is referred to PCA (2001), ACI 360R-92 (2004), ACI 209R-92 (2004), Ringo and Anderson (1996), and ACI (1998).

14.12.2   Loads

Loads on slabs on grade include

- vehicle wheel loads
- concentrated loads
- line and strip loads
- uniform loads
- construction loads
- environmental effects, including expansive soils
- effects of differential settlement

Loads on a slab on grade depend on the function of the structure in which it is located. In residential buildings, slabs on grade are often used as floors in underground parkades, in which case these slabs are subjected to vehicle wheel loads. In industrial buildings, slabs on grade may be subjected to a variety of loads, including vehicle loads and loads due to equipment (concentrated or line/strip loads).

The design approach for slabs on grade is strongly influenced by the type of load. Flexural stress is critical for the design of slabs subjected to loads acting over a smaller area (such as lift trucks with individual wheel contact areas). Compressive stresses in slabs

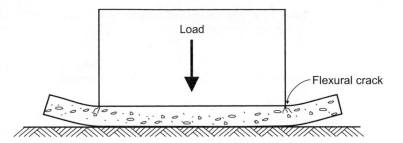

Figure 14.41 Flexural stresses in a slab on grade.

on grade are usually much smaller than the concrete compressive strength. However, tensile stresses may reach the concrete tensile strength (modulus of rupture f_r), which is usually small in magnitude. As a result, cracks develop in the zone subjected to the largest stresses, as shown in Figure 14.41.

The load contact area may be critical in some cases, especially for heavy loads acting over a small area, resulting in bearing or punching shear failure (for example, in the case of heavy loads transferred at the base of a post with inadequate base plate size). In the case of uniform loads acting over a larger area, negative moments developed in the slab cause tensile stresses at the top of the slab.

14.12.3 Design of Slabs on Grade

Design methods for slabs on grade are based on theories originally developed for airport and highway pavements. The first rational approach to slab on grade design was developed by Westergard in the 1920s. The approach was based on treating the slab as a homogeneous and isotropic structure resting on an elastic subgrade acting as a linear spring. The Portland Cement Association (PCA) developed design procedures for concrete slabs on grade subjected to various loads (PCA, 2001). The modern approach to slab on grade design involves the finite element method. Specialized software packages for slab on grade design, such as PCAMATS, developed by PCA, are commercially available.

The key design factors related to slabs on grade are

- subgrade properties
- concrete strength
- slab thickness
- slab reinforcement

Subgrade properties The modulus of subgrade reaction (k) is a soil parameter commonly used in design procedures for concrete pavements and slabs on grade. It is a spring constant that depends on the kind of soil, the degree of compaction, and the moisture content. The k value reflects the response of the subgrade under temporary (elastic) conditions and small deflections, usually 1.25 mm or less, and it is measured by plate-loading tests taken on top of the compacted subgrade.

In lieu of tests performed at the construction site, k values can be taken based on Table 14.2 (PCA, 2001). In general, materials with k value greater than 125 lb/in³ may be used as subgrade. Soils with k below this value, as well as low-compressibility organic material and high-compressibility silt, are to be avoided. It should be noted that the k value can vary largely, depending on the degree of compaction and moisture content of the soil.

Concrete strength Good quality concrete is essential to ensure satisfactory load-bearing capacity and wear resistance in slabs on grade. In general, the specified compressive strength (f_c') for slabs on grade used in office buildings and residential and commercial applications should be at least 20 MPa. For industrial floor applications, the strength should be at least 30 MPa, and even 35 MPa for industrial floors subjected to heavy traffic. For more guidance on the required concrete properties of slabs on grade, refer to Table 4.5 of PCA (2001).

Table 14.2 Recommended modulus of subgrade reaction (k) values

Type of soil	Subgrade strength	CBR[2] (%)	Design k value	
			lb/in^3	MPa/m
Silts and clays of high compressibility[1] at natural density	Low	2 or less	50	13.6
Silts and clays of high compressibility[1] at compacted density	Average	3	100	27.1
Silts and clays of low compressibility[1]				
Sandy silts and clays, gravelly silts and clays				
Poorly graded sands				
Gravelly soils, well-graded sands, and sand-gravel mixtures relatively free of plastic fines	High	10	200	54

(*Source:* PCA, 2001; reproduced with permission of the Portland Cement Association)
Notes:
1. High-compressibility soil characterized by a liquid limit equal to or greater than 50; otherwise low-compressibility soil
2. California bearing ratio

The wear resistance of the top layer of the slab on grade is also important, and it is strongly influenced by the finishing method. For more information on finishing methods, refer to Chapter 8 of PCA (2001).

Slab thickness   The design of slab thickness may be a complex task, especially in the case of industrial floors subjected to heavy traffic. The design may be performed following one of the established design procedures (PCA, 2001; Ringo and Anderson, 1996). In most cases of regular design applications of slabs subjected to moderate loads, the required slab thickness can be selected without performing detailed design calculations. In general, the thickness of unreinforced slabs depends on the subgrade properties and joint spacing and is on the order of 100 mm to 150 mm. Note that the increase in the slab flexural strength is proportional to the square of the slab thickness. For example, an increase in the slab thickness from 100 mm to 125 mm results in an approximately 50% increase in the flexural strength (assuming the same concrete properties).

Slab reinforcement   Reinforcement in concrete slabs on grade is provided either to increase the structural load-bearing capacity or to control cracking due to temperature and shrinkage effects. In the first case, the reinforcement should be designed using one of the detailed procedures (PCA, 2001; Ringo and Anderson, 1996). Shrinkage and temperature reinforcement needs to be provided in slabs with large spacing of control joints or when the joints are unacceptable for floor use. In general, if the joints are closely spaced, temperature reinforcement is not needed unless cracks must be tightly held together. There are several alternative methods for calculating the required amount of temperature and shrinkage reinforcement (see Ringo and Anderson, 1996). The *subgrade drag equation* is frequently used to determine the amount of shrinkage and temperature reinforcement by equating the force in the steel reinforcement and the subgrade friction, as shown in Figure 14.42 (assuming that the slab will shrink in such a manner that each end will move by an equal distance toward the centre). The amount of reinforcement is determined based on the equation

$$A_s f_s = \left(w \frac{L}{2}\right) F$$

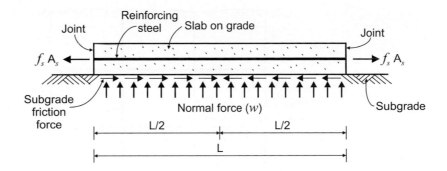

Figure 14.42 Free-body diagram of forces producing subgrade drag action.

where

A_s = the cross-sectional area of the reinforcing steel (in square millimetres per metre of slab width)

F = the coefficient of friction between the base and the slab on grade; $F = 1.5$ is a common value (Ringo and Anderson, 1996)

L = the slab length between joints (m)

w = the weight of the concrete slab (N/m^2)

f_s = the allowable steel stress (MPa); use $f_s = 0.67 f_y$, so $f_s = 270$ MPa for Grade 400 steel.

In general, slab reinforcement should be provided in each direction with a reinforcement ratio of at least 0.1% to prevent cracking, especially in colder climates. This corresponds to a minimum area of reinforcement A_{smin} of

$$A_{smin} = 0.001A_g$$

where A_g is the gross cross-sectional area of a slab. For a slab strip of width $b = 1000$ mm and depth h, $A_g = 1000h$.

The reinforcement can be in the form of deformed reinforcing bars (usually of 10M or 15M size) or welded wire mesh. It is recommended (ACI 360 R.92, 2004) that the reinforcement be located at or above the middepth of the slab. A common practice is to specify that the steel be 40 mm or 50 mm below the top concrete surface or at one third the slab depth below the surface.

14.12.4 Joints

Joints in concrete slabs on grade control cracking caused by volume changes in concrete, in particular drying shrinkage and cooling contraction. These volume changes cause tensile stresses in concrete because the slab is restrained by the supporting subgrade. If joints are not provided, the concrete will crack once its tensile strength is reached. Joints allow for slight movement of the slab and provide stress relief for concrete restrained by soil.

To create a joint, a slab is usually saw cut to a depth of one quarter of the slab thickness, creating a weakened plane for the cracks to form, as shown in Figure 14.40b. Load transfer across the closely spaced joints is provided by the interlocking action of the aggregate particles at the fractured faces of the crack that forms below saw cuts. Load transfer across wider cracks is developed by a combination of aggregate interlock and mechanical devices such as dowels. *Dowels* are provided at the joint locations to control joint opening, as is shown in Figure 14.40. One end of the dowel is usually bonded to the concrete on one side of the joint, and the other end is greased or fit with a sleeve so that it is free to move as the joint opens and closes. If provided, dowels must be carefully aligned horizontally and vertically to enable freedom of movement.

The following types of joints are characteristic of concrete slabs on grade:

- control joints
- isolation joints
- construction joints

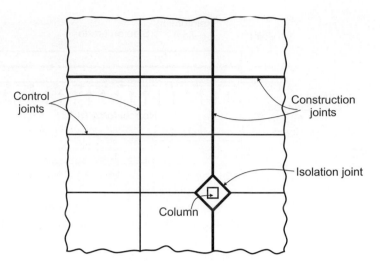

Figure 14.43 Joint layout for a typical slab on grade.

Control joints (contraction joints) are provided to control tensile stresses caused by shrinkage and restraint provided by the subgrade. The friction between the subgrade and the slab restrains the movement of the slab, creating the tensile stresses that cause cracking. Control joints are not intended to eliminate cracks — on the contrary, these joints are used to induce "planned" cracks at the joint locations.

Isolation joints (expansion joints) are used to separate the slab from any adjacent structure such as another slab, a wall, a column, or an adjacent building. These joints prevent bond and allow for independent movement between two elements.

Construction joints are stopping places for a day's work. The use of construction joints should be avoided when possible — ideally, these joints are detailed and built to function as and align with control joints or isolation joints. Any reinforcement in the slab should be continuous through a construction joint unless the joint functions as a control or isolation joint.

A joint layout (jointing plan) is a drawing showing a plan view of the slab on grade with joint spacing and locations of typical joints, as in Figure 14.43. The joint layout should be prepared by the floor designer for all slab on grade work.

Joints should be spaced closely to minimize the chances of random cracking in the slab. For unreinforced concrete slabs, the maximum recommended spacing between control joints depends on the slab thickness, the potential for drying shrinkage and contraction, and the curing conditions. Table 14.3 (based on PCA, 2001) summarizes the recommended spacing of control joints for slabs of different thicknesses. In general, joint spacing ranges from 24 to 36 times the slab thickness, depending on the aggregate size and the slump. Ideally, an aggregate size of 20 mm or larger should be used in slab on grade construction and the slump should be in the range of 75 mm to 100 mm.

Table 14.3 Spacing of control joints in metres*

Slab thickness (mm)	Maximum-size aggregate less than 19 mm	Maximum-size aggregate 19 mm and larger
125	3.0	3.75
150	3.75	4.5
175	4.25	5.25**
200	5.0**	6.0**
225	5.5**	6.75**
250	6.0**	7.5**

(*Source:* PCA, 2001; reproduced with permission of the Portland Cement Association)
Notes:
  *Spacings are appropriate for slump between 100 mm and 150 mm; for slump less than 100 mm, the joint spacing can be increased by 20%.
  **When the joint spacing exceeds 4.5 m, load transfer by aggregate interlock decreases significantly.

SUMMARY AND REVIEW — FOUNDATIONS

Foundations carry loads from the building structure and transfer them to the soil. The *foundation* is the part of the structure below ground level (also called the substructure), while the *footing* is an individual member of the foundation supporting one or more columns, piers, or walls. Foundation design, including the selection of the foundation system and the footing dimensions, is strongly influenced by the type of underlying soil and its properties. Foundations can be classified into two types:

- *Shallow foundations* are used when favourable soil strata are located near the surface to support the building load.
- *Deep foundations* are used when the soil near the ground level does not have an adequate bearing capacity.

Types of shallow foundations

The main six types of *shallow foundations* are (see Figure 14.2) as follows:

1. *Spread footings* support individual columns.
2. *Strip footings* support walls.
3. *Combined footings* support two or more columns.
4. *Strap footings* are two or more spread footings connected by a strap (grade) beam.
5. *Mat foundations* are large continuous footings supporting all the columns and walls within a building.
6. *Raft foundations* are continuous footings used when the soil has a limited bearing capacity.

Allowable and factored soil bearing pressure

The *allowable soil bearing pressure* governs the sizing of footing plan dimensions (refer to Table 14.1 for typical values).

The soil pressure (q) is determined based on the total *specified (service) load* (P_s) placed on the footing with plan area (A):

$$q = \frac{P_s}{A} = \frac{P_{DL} + P_{LL}}{A} \qquad \text{[14.1]}$$

The actual soil pressure (q) due to the *specified loads* must be less than or equal to the *allowable soil bearing pressure* (q_{all}):

$$q \leq q_{all}$$

The required footing plan area can then be determined as

$$A \geq \frac{P_s}{q_{all}} \qquad \text{[14.1a]}$$

The factored soil bearing pressure (q_f) is determined based on the *factored load* (P_f) as

$$q_f = \frac{P_f}{A} \qquad \text{[14.2]}$$

It is very important to remember the following two concepts:

1. The footing plan dimensions are determined based on the allowable soil bearing pressure (q_{all})
2. The factored soil bearing pressure (q_f) is used in the footing design for flexure and shear.

CSA A23.3 design requirements for strip and spread footings

The *strength requirement* is the key CSA A23.3 requirement related to the footing design for shear (Cl.11.3.3) and flexure (Cl.8.1.3).

Shear design   The factored shear resistance (V_r) is determined according to A23.3 Cl.11.3.3. The CSA A23.3 provisions for the concrete shear resistance (V_c) are related to

- one-way shear (Cl.11.3.4)
- two-way shear (Cl.13.3.3 and 13.3.4)

Shear reinforcement is not required provided that A23.3 Cl.11.2.8.1 has been satisfied.

Flexural design   The following key requirements are related to the detailing of flexural reinforcement:

- minimum footing reinforcement (A23.3 Cl.7.8.1)
- maximum spacing of flexural reinforcement (A23.3 Cl.7.4.1.2 and Cl.13.10.4)
- shrinkage and temperature reinforcement (A23.3 Cl.7.8.1 and Cl.7.8.3)
- maximum amount of flexural reinforcement (A23.3 Cl.10.5.2)
- distribution of flexural reinforcement in rectangular footings (A23.3 Cl.15.4.4)
- concrete cover requirements (A23.1 Cl.6.6.6.2.3)

Practical guidelines for the selection of footing size and reinforcement

Practical design guidelines for the selection of footing dimensions and reinforcement can be summarized as follows:

- *Overall thickness:* The minimum thickness should be approximately 250 mm. The thickness should be increased by an additional 25 mm to 50 mm as compared to the actual design requirements whenever possible.
- *Plan dimensions:* The minimum width of the strip footing should be on the order of 450 mm and the width of the spread footing should be at least 750 mm.
- *Flange width-to-thickness ratio:* $f/h \leq 4$ (see Figure 14.16).
- *Reinforcement:* The footing reinforcement ratio (ρ) should be in the range of 0.33%. Use a smaller number of larger rebars to reduce labour requirements. The minimum bar size should be 15M or larger.

Eccentrically loaded footings

Eccentrically loaded footings are subjected to an axial load P acting at an eccentricity e with regard to the footing vertical axis. The load eccentricity results in a nonuniform soil pressure distribution. The maximum and minimum soil pressures can be determined for the three different levels of load eccentricity (small, boundary, and large) as follows:

1. *Small load eccentricity ($e < l/6$)*
   The minimum soil pressure is

$$q_{min} = \frac{P}{A}\left(1 - \frac{6e}{l}\right)$$ **[14.15a]**

   The maximum soil pressure is

$$q_{max} = \frac{P}{A}\left(1 + \frac{6e}{l}\right)$$ **[14.16a]**

2. *Boundary case ($e = l/6$)*
   The minimum soil pressure is

$$q_{min} = 0$$

   The maximum soil pressure is

$$q_{max} = \frac{2P}{b \times l}$$ **[14.18]**

   or

$$q_{max} = 2 \times \frac{P}{A}$$ **[14.20]**

3. *Large load eccentricity ($e > l/6$)*
   The maximum soil pressure is

$$q_{max} = \frac{2P}{3\,b\left(\dfrac{l}{2} - e\right)}$$ **[14.23]**

or

$$q_{max} = 2 \times \frac{P}{b \times l'} \qquad \text{[14.25]}$$

Eccentrically loaded footings with a large load eccentricity ($e > l/6$) are not recommended and should be avoided whenever possible.

Load transfer from column into footing

The designer must ensure that the concrete in the column is not crushed at the junction between the footing and the column and that the footing is not crushed locally beneath the column. The bearing strength at the column-to-footing interface depends on the bearing strength of the concrete in the column and in the footing. If these strengths are not sufficient to transfer the axial forces across the interface, additional steel reinforcement (called dowel reinforcement) must be provided.

The *column bearing resistance* is (A23.3 Cl.10.8.1)

$$B_{rc} = 0.85\phi_c f_{cc}' A_1 \qquad \text{[14.28]}$$

The bearing resistance of the *supporting member (footing)* is

$$B_{rf} = 0.85\phi_c f_{cf}' A_1 \sqrt{\frac{A_2}{A_1}} \qquad \text{[14.29]}$$

If

$$\sqrt{\frac{A_2}{A_1}} > 2$$

then the footing bearing resistance is limited to

$$B_{rfmax} = 1.7\phi_c f_{cf}' A_1 \qquad \text{[14.30]}$$

When the factored axial load (P_f) acting at the base of the column exceeds the concrete bearing resistance, dowel reinforcement needs to be provided to carry the excess load (A23.3 Cl.15.9.1) such that

$$B_{rd} \geq P_f - B_{rc} \qquad \text{[14.31]}$$

where

$$B_{rd} = A_d \phi_s f_y \qquad \text{[14.32]}$$

The required area (A_d) of dowel reinforcement crossing the column-footing interface is limited as follows (A23.3 Cl.15.9.2.1):

$$A_d \geq 0.005 A_1 \qquad \text{[14.33]}$$

PROBLEMS

14.1. The purpose of footing design is to determine the following design parameters for the given design loads: the footing plan dimensions, the thickness, and the amount and distribution of reinforcement.
a) Which of the above design parameter(s) is governed by shear?
b) Which parameter is governed by flexure?
c) How does the allowable soil pressure affect the footing design?
  Briefly explain your answers.

14.2. Consider a typical section of a 250 mm thick spread footing supporting a reinforced concrete column, as shown in the figure that follows. The allowable soil bearing pressure is 200 kPa. The column is subjected to a specified dead load (P_{DL}) of 300 kN and a specified live load (P_{LL}) of 500 kN. The footing is square in plan.
a) Determine the required plan dimensions of the footing and the factored soil pressure.
b) Determine the required footing plan dimensions considering the effect of surcharge. The

surcharge includes the floor service load of 4.8 kPa and the soil overlay with a thickness of 300 mm. The unit weight density of concrete is 24.0 kN/m³.

c) Comment on the effect of the surcharge with regard to the footing plan dimensions and the factored soil pressure.

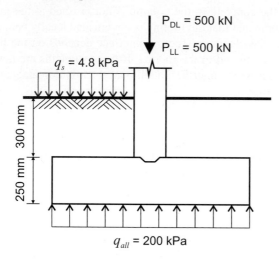

P_{DL} = 500 kN

P_{LL} = 500 kN

q_s = 4.8 kPa

300 mm

250 mm

q_{all} = 200 kPa

14.3. A typical cross-section of a strip footing beneath a 300 mm thick wall is shown in the figure below. The footing is 1.5 m wide and is subjected to a concentric specified dead load (P_{DL}) of 140 kN/m and a specified live load (P_{LL}) of 150 kN/m. The allowable soil bearing pressure is 200 kPa. Use 20M bars for flexural reinforcement. Maximum aggregate size is 20 mm.

Design the footing according to the CSA A23.3 design requirements.

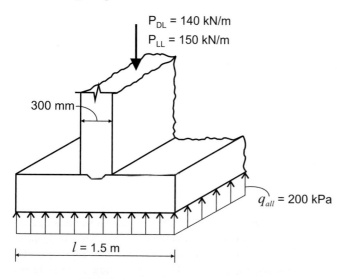

P_{DL} = 140 kN/m

P_{LL} = 150 kN/m

300 mm

q_{all} = 200 kPa

l = 1.5 m

Given:

f_c' = 25 MPa (normal-density concrete)
f_y = 400 MPa
ϕ_c = 0.65
ϕ_s = 0.85

14.4. A typical cross-section of a strip footing beneath a 200 mm thick wall is shown in the figure below. The footing is subjected to a concentric specified dead load (P_{DL}) of 100 kN/m and a specified live load (P_{LL}) of 100 kN/m. Use an allowable soil bearing pressure of 200 kPa and 15M bars for flexural reinforcement. Maximum aggregate size is 20 mm.

Design the footing according to the CSA A23.3 design requirements.

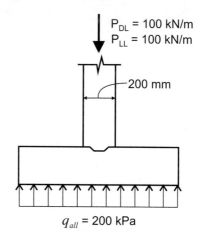

P_{DL} = 100 kN/m

P_{LL} = 100 kN/m

200 mm

q_{all} = 200 kPa

Given:

f_c' = 25 MPa (normal-density concrete)
f_y = 400 MPa
ϕ_c = 0.65
ϕ_s = 0.85

14.5. Consider that the wall supported by the footing designed in Problem 14.4 is located within 200 mm of the property line, as shown in the figure below.

a) If the allowable soil bearing pressure is 200 kPa, would the footing designed in Problem 14.4 be able to carry the design loads without violating the equilibrium of forces acting on it?

b) If the answer to the above question is negative, determine the maximum dead load (P_{DL}) and live load (P_{LL}) which this footing would be able to carry. Assume that the dead load and live load values are equal.

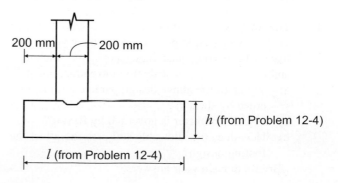

200 mm   200 mm

h (from Problem 12-4)

l (from Problem 12-4)

Given:

$f_c' = 25$ MPa (normal-density concrete)
$f_y = 400$ MPa
$\phi_c = 0.65$
$\phi_s = 0.85$

14.6. Consider that the wall supported by the strip footing discussed in Problem 14.5 is also supported by a reinforced concrete floor slab at the top and a slab on grade at the bottom, as shown in the figure below. Use the same allowable soil pressure and load values as in Problem 14.4.

a) Determine the required footing dimensions for the following design scenarios:

   i) triangular soil distribution beneath the footing
   ii) uniform soil distribution beneath the footing

b) Sketch the bending moment diagrams for the footing and the wall for both scenarios.

Given:

$f_c' = 25$ MPa
$f_y = 400$ MPa
$\lambda = 1$ (normal-density concrete), 20-mm maximum aggregate size
$\phi_c = 0.65$
$\phi_s = 0.85$

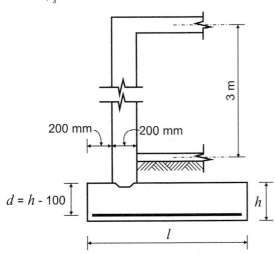

14.7. Based on the solution to Problem 14.6, comment on the effects of a triangular versus a uniform soil pressure distribution in an eccentrically loaded footing.

14.8. A typical cross-section of a square-shaped spread footing which supports a 400 mm square interior concrete column is shown in the next figure. The column carries a specified dead load (P_{DL}) of 1000 kN and a specified live load (P_{LL}) of 500 kN. The allowable soil bearing pressure is 400 kPa. Use 20M bars for flexural reinforcement. Maximum aggregate size is 20 mm.

Design the footing according to the CSA A23.3 design requirements.

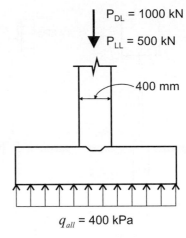

Given:

$f_c' = 30$ MPa (normal-density concrete)
$f_y = 400$ MPa
$\phi_c = 0.65$
$\phi_s = 0.85$

14.9. Refer to the spread footing designed in Problem 14.8.

a) Determine the footing capacity, expressed in terms of the axial load resistance P_r, for each of the following undesirable (but possible) construction scenarios:

   i) The footing was constructed off centre with an eccentricity of 100 mm in one direction.
   ii) Due to an excavation error, the footing was constructed 75 mm too shallow, thus resulting in a reduced overall depth (h).
   iii) The reinforcement was placed 50 mm too high, thus resulting in a reduced effective depth (d).

b) Check whether the footing capacity for the construction scenarios in part a) is adequate to withstand the design loads given in Problem 14.8.

14.10. Refer to the footing discussed in Problem 14.9.

Keeping in mind the possible inaccuracies in the footing construction discussed in Problem 14.9, which measures would you as a designer consider to minimize the impact of potential construction problems? Explain.

14.11. The free-standing sign shown in the figure that follows is designed to resist a service wind pressure of 2 kPa. The designer has decided to use a square spread footing buried deep in the ground to counteract the effects of the load eccentricity. The dimensions of the footing are 2 m by 2 m and the overall depth is 600 mm. The allowable soil bearing pressure is 150 kPa. Maximum aggregate size is 20 mm.

Determine the minimum required foundation depth (t) to ensure resistance against overturning with a factor of safety of 1.5.

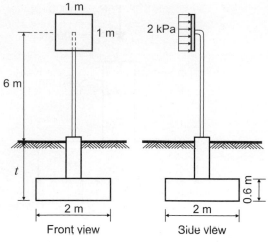

Front view   Side view

Given:

$f_c' = 25$ MPa (normal-density concrete)
$f_y = 400$ MPa
$\phi_c = 0.65$
$\phi_s = 0.85$
$\gamma_w = 24$ kN/m³
$\gamma_s = 17$ kN/m³

14.12. A spread footing of 2 m by 2 m plan dimensions and 400 mm depth shown in the figure below is built on soil of unknown allowable bearing pressure. The footing supports a 300 mm square reinforced concrete column. It is reinforced with 25M bars at 200 mm spacing in each direction with a 75 mm clear cover to the reinforcement. Maximum aggregate size is 20 mm.

Determine the maximum factored axial load that can be applied through the column.

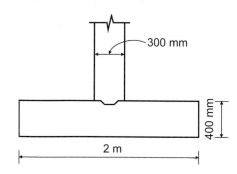

Given:

$f_c' = 25$ MPa (normal-density concrete)
$f_y = 400$ MPa
$\phi_c = 0.65$
$\phi_s = 0.85$

14.13. A column shown in the figure that follows is located over an existing sewer line that cannot be moved. The column carries a specified dead load (P_{DL}) of 700 kN and a specified live load (P_{LL}) of 500 kN. Use an allowable soil bearing pressure of 100 kPa and 25M bars

for the flexural reinforcement. Maximum aggregate size is 20 mm.

Design the footing according to the CSA A23.3 design requirements, considering that the footing width is limited to 4 m.

Given:

$f_c' = 25$ MPa (normal-density concrete)
$f_y = 400$ MPa
$\phi_c = 0.65$
$\phi_s = 0.85$

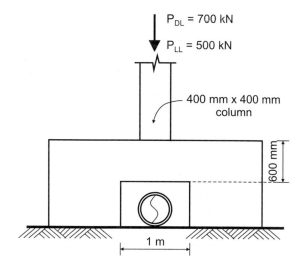

14.14. Two 500 mm square reinforced concrete columns (columns 1 and 2) are spaced at 4.0 m on centre, as shown in the figure that follows. It is required to design a combined footing to support these two columns. The footing width is restricted to 4.5 m, whereas the length is not limited. The total service loads (P_s) for columns 1 and 2 are equal to 2500 kN and 3500 kN, respectively, and the average load factor (α) is 1.4. Use an allowable soil bearing pressure of 200 kPa. Ignore the effect of the footing weight in the design. Maximum aggregate size is 20 mm.

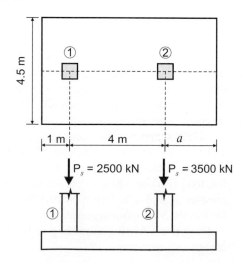

Design the combined footing according to the CSA A23.3 requirements.

Given:

$f_c' = 25$ MPa (normal-density concrete)
$f_y = 400$ MPa
$\phi_c = 0.65$
$\phi_s = 0.85$

14.15. A spread footing of plan dimensions 4 m by 4 m supports a 400 mm square column. The column carries a factored axial load (P_f) of 4000 kN (see the figure below). The column is reinforced with four 30M longitudinal bars (4-30M).

Check whether the column-footing interface is adequate for the load transfer and design the dowels to match the column reinforcement as required.

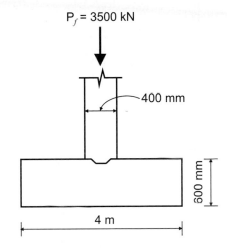

$P_f = 3500$ kN

400 mm

600 mm

4 m

Given:

$f_c' = 25$ MPa (footing)
$f_c' = 30$ MPa (column)
$f_y = 400$ MPa
$\phi_c = 0.65$
$\phi_s = 0.85$

14.16. Consider a concentrically loaded spread footing supporting a 400 mm square reinforced concrete column shown in the figure below.

a) Develop a computer spreadsheet to generate a table containing the following information:

- footing size (ranging from 1 m × 1 m to 5 m × 5 m in increments of 0.5 m)
- overall thickness (h)
- factored capacity (P_f)
- allowable capacity (P_s)
- flexural reinforcement (area and distribution)

Use an average load factor (α) of 1.4 and an allowable soil bearing pressure of 200 kPa. Maximum aggregate size is 20 mm.

b) Use the spreadsheet posted on the Web site and compare the results obtained by the two spreadsheets.

Given:

$f_c' = 25$ MPa (normal-density concrete)
$f_y = 400$ MPa
$\phi_c = 0.65$
$\phi_s = 0.85$

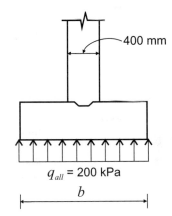

400 mm

$q_{all} = 200$ kPa

b

APPENDIX A
Design Aids

A.1 DESIGN AIDS

Table A.1 Properties of deformed reinforcing bars

Nominal bar size	Nominal dimensions			Mass per unit length (kg/m)
	Diameter (mm)	Area (mm^2)	Perimeter (mm)	
10M	11	100	36	0.8
15M	16	200	50	1.6
20M	20	300	61	2.4
25M	25	500	79	4.0
30M	30	700	94	5.5
35M	36	1000	112	8.0
45M	44	1500	137	12.0
55M	56	2500	177	20.0

Table A.2 Concrete cover requirements

Exposure condition (see Table 2.1)	Exposure class		
	N	F-1, F-2, S-1, S-2	C-XL, C-1, C-3, A-1, A-2, A-3
• Cast against and permanently exposed to earth	—	75 mm	75 mm
• Beams, girders, columns, and piles	30 mm	40 mm	60 mm
• Slabs, walls, joists, shells, and folded plates	20 mm	40 mm	60 mm
• Ratio of cover to nominal bar diameter	1	1.5	2
• Ratio of cover to nominal maximum aggregate size	1	1.5	2

(*Source:* CSA A23.1–04 Table 17, reproduced with permission by the Canadian Standards Association)

Table A.3 Thickness below which deflections must be computed for nonprestressed beams or one-way slabs not supporting or attached to partitions or other construction likely to be damaged by large deflections

	Minimum thickness, h			
	Simply supported	One end continuous	Both ends continuous	Cantilever
Solid one-way slabs	$\dfrac{l_n}{20}$	$\dfrac{l_n}{24}$	$\dfrac{l_n}{28}$	$\dfrac{l_n}{10}$
Beams or ribbed one-way slabs	$\dfrac{l_n}{16}$	$\dfrac{l_n}{18}$	$\dfrac{l_n}{21}$	$\dfrac{l_n}{8}$

Note: The values specified in this table shall be used directly for members with normal-density concrete where $y_c > 2150$ kg/m^3 and the reinforcement is Grade 400. For other conditions, the values shall be modified as follows:
a) for structural low-density concrete and structural semi-low-density concrete, the values shall be multiplied by $(1.65 - 0.0003y_c)$, but not less than 1.0, where y_c is the density in kilograms per cubic metre; and
b) for f_y other than 400 MPa, the values shall be multiplied by $(0.4 + f_y/670)$.
(*Source:* CSA A23.3–04 Table 9.2, reproduced with permission by the Canadian Standards Association)

Table A.4 Balanced reinforcement ratio (for grade 400 steel)

f_c'	25 MPa	30 MPa	35 MPa	40 MPa
ρ_b	0.022	0.027	0.030	0.034

Table A.5 Tension development lengths (mm) for straight deformed bars with $f_y = 400$ MPa and normal-density concrete (CSA A23.3 Cl.12.2.3 and Cl.12.2.4)

Modification factors (Cl.12.2.4)	f_c' (MPa)	Nominal bar size					
		10M	15M	20M	25M	30M	35M
$k_1 = 1.0$ "bottom" bar	20	375	525	650	1025	1225	1450
$k_2 = 1.0$ uncoated bar	25	325	475	575	925	1075	1300
$k_3 = 1.0$ normal-density concrete	30	300	425	525	850	1000	1175
	35	275	400	500	775	925	1100
	40	275	375	450	725	850	1025
	45	250	350	425	700	825	975
	50	250	325	400	650	775	925
$k_1 = 1.0$ "bottom" bar	20	550	775	950	1525	1825	2175
$k_2 = 1.5$ epoxy coated bar	25	500	700	850	1375	1625	1950
$k_3 = 1.0$ normal-density concrete	30	450	650	775	1250	1475	1775
	35	425	600	725	1175	1375	1650
	40	400	550	675	1100	1275	1525
	45	375	525	650	1025	1225	1450
	50	350	500	600	975	1150	1375
$k_1 = 1.3$ "top" bar	20	475	675	825	1325	1575	1875
$k_2 = 1.0$ uncoated bar	25	425	600	750	1200	1400	1675
$k_3 = 1.0$ normal-density concrete	30	400	550	675	1100	1275	1525
	35	375	525	625	1000	1200	1425
	40	350	475	600	950	1125	1325
	45	325	450	550	900	1050	1250
	50	300	425	525	850	1000	1200
$k_1 \times k_2 = 1.7$ top location and epoxy coated bar	20	625	875	1075	1750	2050	2450
$k_3 = 1.0$ normal-density concrete	25	575	800	975	1550	1850	2200
	30	525	725	875	1425	1675	2000
	35	475	675	825	1325	1550	1850
	40	450	625	775	1225	1450	1750
	45	425	600	725	1175	1375	1650
	50	400	575	700	1100	1300	1550

Note: Values in this table have been rounded to the next 25 mm.
(*Source:* CPCA, 1995; reproduced with permission by the Cement Association of Canada)

Table A.6 Compression development lengths (mm) for straight deformed bars with $f_y = 400$ MPa (CSA A23.3 Cl.12.3.1)

f_c' (MPa)	Nominal bar size					
	10M	15M	20M	25M	30M	35M
20	250	350	425	550	650	775
25	225	325	375	500	600	700
≥ 30	200	300	350	450	525	650

Note: Values in this table have been rounded to the next 25 mm.
(*Source:* CPCA, 1995; reproduced with permission by the Cement Association of Canada)

Table A.7 Tension development lengths (mm) for hooked deformed bars with f_y = 400 MPa
(CSA A23.3 Cl.12.5.1)

f_c' (MPa)	Nominal bar size					
	10M	**15M**	**20M**	**25M**	**30M**	**35M**
20	275	375	450	575	675	800
25	250	325	400	525	600	725
30	225	300	375	475	550	675
35	200	275	350	450	525	625
40	200	275	325	400	475	575
45	175	250	300	400	450	550
50	175	250	300	375	425	525

Note: Values in this table have been rounded to the next 25 mm.
(*Source:* CPCA, 1995; reproduced with permission by the Cement Association of Canada)

Table A.8 Practical reinforcing bar configurations and areas

Rebar configuration	Area (mm^2)	Area (mm^2) for components with even number of bars only
4-20M	1200	1200
5-20M or 3-25M	1500	
6-20M	1800	1800
4-25M	2000	2000
7-20M or 3-30M	2100	
8-20M	2400	2400
5-25M	2500	
9-20M	2700	
4-30M	2800	2800
10-20M or 6-25M	3000	3000
7-25M or 5-30M	3500	
8-25M or 4-35M	4000	4000
6-30M	4200	4200
9-25M	4500	
7-30M	4900	
10-25M or 5-35M	5000	5000
11-25M	5500	
8-30M	5600	5600
12-25M	6000	6000
9-30M	6300	
13-25M	6500	
14-25M or 10-30M	7000	7000

Table A.9 Dead loads for floors, ceilings, and roofs

	Load (kN/m^2)
Floorings	
Normal-density concrete topping, per 10 mm of thickness	0.25
Semi-low-density concrete (1900 kg/m^3) topping, per 10 mm	0.2
Low-density concrete (1500 kg/m^3) topping, per 10 mm	0.15
22 mm hardwood floor on sleepers, clipped to concrete without fill	0.25
40 mm terrazzo floor finish directly on slab	1.0
40 mm terrazzo floor finish on 25 mm mortar bed	1.5
25 mm terrazzo floor finish on 50 mm concrete bed	1.8
20 mm ceramic or quarry tile on 12 mm mortar bed	0.8
20 mm ceramic or quarry tile on 25 mm mortar bed	1.1
8 mm linoleum or asphalt tile directly on concrete	0.1
8 mm linoleum or asphalt tile on 25 mm mortar bed	0.6
20 mm mastic floor	0.5
Hardwood flooring, 22 mm thick	0.2
Subflooring (softwood), 20 mm thick	0.15
Asphaltic concrete, 40 mm thick	0.9
Ceilings	
12.7 mm gypsum board	0.1
15.9 mm gypsum board	0.15
19 mm gypsum board directly on concrete	0.25
20 mm plaster directly on concrete	0.3
20 mm plaster on metal lath furring	0.4
Suspended ceilings	0.1
Acoustical tile	0.1
Acoustical tile on wood furring strips	0.15
Mechanical duct allowance	0.2
Roofs	
Five-ply felt and gravel (or slag)	0.3
Three-ply felt and gravel (or slag)	0.3
Five-ply felt composition roof, no gravel	0.2
Three-ply felt composition roof, no gravel	0.15
Asphalt strips shingles	0.15
Slate, 8 mm thick	0.6
Gypsum, per 10 mm of thickness	0.1
Insulating concrete, per 10 mm	0.1

(*Source:* CPCA, 1995; reproduced with permission by the Cement Association of Canada)

Table A.10 Dead loads for walls

Walls	Load (kN/m²) Unplastered	One side plastered	Both sides plastered
Walls (brick, concrete block, or tile)			
100 mm brick wall	2.0	2.2	2.4
200 mm brick wall	3.8	4.0	4.3
300 mm brick wall	5.6	5.8	6.1
100 mm hollow normal-density concrete block	1.4	1.6	1.9
150 mm hollow normal-density concrete block	1.7	2.0	2.2
200 mm hollow normal-density concrete block	2.1	2.4	2.6
250 mm hollow normal-density concrete block	2.5	2.8	3.0
300 mm hollow normal-density concrete block	3.0	3.2	3.4
100 mm hollow low-density block or tile	1.1	1.4	1.6
150 mm hollow low-density block or tile	1.3	1.6	1.8
200 mm hollow low-density block or tile	1.6	1.8	2.1
250 mm hollow low-density block or tile	2.0	2.2	2.4
300 mm hollow low-density block or tile	2.3	2.6	2.8
100 mm brick, 100 mm hollow normal-density block backing	3.3	3.6	3.8
100 mm brick, 200 mm hollow normal-density block backing	4.0	4.2	4.5
100 mm brick, 300 mm hollow normal-density block backing	4.9	5.1	5.4
100 mm brick, 100 mm hollow low-density block backing	3.0	3.2	3.5
100 mm brick, 200 mm hollow low-density block backing	3.6	3.8	4.0
100 mm brick, 300 mm hollow low-density block backing	4.2	4.4	4.6

Walls (others)	Load (kN/m²)
Windows, glass, frame, and sash	0.4
100 mm stone	2.6
Steel or wood studs, lath, 20 mm plaster	1.0
Steel or wood studs, lath, 15.9 mm gypsum board each side	0.3
Steel or wood studs, 2 layers 12.7 mm gypsum board each side	0.5
Exterior stud walls with brick veneer	2.3

(*Source:* CPCA, 1995; reproduced with permission by the Cement Association of Canada)

Table A.11 Minimum design loads for materials

Material	Load (kN/m³)	Material	Load (kN/m³)
Bituminous products:		Ceramic tile	24
Asphaltum	13	Charcoal	2
Graphite	22	Cinder fill	9
Paraffin	9	Cinders, dry, in bulk	7
Petroleum, crude	9	Coal	
Petroleum, refined	8	Anthracite, piled	8
Petroleum, benzine	7	Bituminous, piled	8
Petroleum, gasoline	7	Lignite, piled	8
Pitch	11	Peat, dry, piled	4
Tar	12	Concrete, plain:	
Brass	83	Cinder	18
Bronze	87	Expanded-slag aggregate	16
Cast-stone masonry (cement, stone, sand)	23	Haydite (burned-clay aggregate)	14
Cement, portland, loose	15	Slag	21

(*Continued*)

Table A.11 Continued

Material	Load (kN/m³)	Material	Load (kN/m³)
Stone (including gravel)	23	Masonry, rubble mortar:	
Vermiculite and perlite aggregate,		Granite	24
non–load bearing	4–8	Limestone, crystalline	23
Other light aggregate, load-bearing	11–17	Limestone, oolitic	22
Concrete, reinforced:		Marble	25
Cinder	18	Sandstone	22
Slag	22	Mortar, hardened:	
Stone (including gravel)	24	Cement	21
Copper	88	Lime	18
Cork, compressed	2.5	Particleboard	7
Earth (not submerged):		Plywood	6
Clay, dry	10	Riprap (not submerged):	
Clay, damp	18	Limestone	13
Clay and gravel, dry	16	Sandstone	14
Silt, moist, loose	13	Sand:	
Silt, moist, packed	15	Clean and dry	14
Silt, flowing	17	River, dry	17
Sand and gravel, dry, loose	16	Slag:	
Sand and gravel, dry, packed	18	Bank	11
Sand and gravel, wet	19	Bank screenings	17
Earth (submerged):		Machine	15
Clay	13	Sand	8
Soil	11	Slate	27
River mud	14	Steel, cold-drawn	77
Sand or gravel	10	Stone, quarried, piled:	
Sand or gravel, and clay	10	Basalt, granite, gneiss	15
Gravel, dry	17	Limestone, marble, quartz	15
Gypsum, loose	11	Sandstone	13
Gypsum wallboard	8	Shale	15
Ice	9	Greenstone, hornblende	17
Iron:		Terra cotta, architectural:	
Cast	71	Voids filled	19
Wrought	75	Voids unfilled	12
Lead	112	Tin:	72
Lime:		Water:	
Hydrated, loose	5	Fresh	10
Hydrated, compacted	7	Sea	10
Masonry, ashlar:		Wood, seasoned:	
Granite	26	Ash, commercial white	7
Limestone, crystalline	26	Cypress, southern	6
Limestone, oolitic	22	Fir, Douglas, coast region	6
Marble	27	Hem fir	5
Sandstone	23	Oak, commercial reds and whites	8
Masonry, brick:		Pine, southern yellow	6
Hard (low absorption)	20	Redwood	5
Medium (medium absorption)	18	Spruce, red, white, and Sitka	5
Soft (high absorption)	16	Western Hemlock	5
		Zinc, rolled sheet	72

(Source: CPCA, 1995, reproduced with permission by the Cement Association of Canada)

A.2 UNITS

In Canada, structural design is mainly performed using the Système International (SI) units (also called "metric units"). SI consists of a limited number of base units that establish fundamental quantities and a large number of derived units describing other quantities.

The basic quantities in SI are (the corresponding units are included in brackets)

- length (m)
- mass (kg)
- time (s)
- angle (rad)

Some derived quantities and the corresponding SI units relevant to the design of concrete structures are listed in Table A.12.

Table A.12   Derived quantities and units in SI

Quantity	Unit
Acceleration	m/s^2
Area	m^2 or mm^2
Mass density	kg/m^3
Weight density	N/m^3 or kN/m^3
Force	$1\ N = 1\ kg \cdot m/s^2$
	$1\ kN = 1000\ N$
Stress	$1\ Pa = 1\ N/m^2$
	$1\ MPa = 1\ N/mm^2$
Volume	m^3 or mm^3
Moment of inertia	m^4 or mm^4
Bending moment	$1\ kN \cdot m = 10^6\ N \cdot mm$
Force per unit length	$1\ kN/m$
Force per unit area	$1\ kPa = 1\ kN/m^2$

Observe from Table A.12 that in several cases there are two units corresponding to the same quantity. In design practice, it is common to use the units summarized in Table A.13.

Table A.13   Common quantities and SI units used in reinforced concrete design

Quantity	Unit
Length (span, cross-sectional dimensions)	mm
Area (reinforcement, cross-section)	mm^2
Moment of inertia	mm^4
Weight density (steel, concrete)	kN/m^3
Modulus of elasticity	GPa ($1\ GPa = 10^3\ MPa$)
Force	kN
Stress/strength	MPa
Bending moment	$kN \cdot m$
Beam/girder loads	kN/m
Slab loads, soil bearing capacity	kPa

In many instances, the designer needs to perform unit conversion in design calculations. Also, in some cases the Imperial system of units is used. The conversion of units between SI and Imperial is summarized in Tables A.14 and A.15.

Table A.14 Metric units conversion factors

Metric unit	Imperial equivalent
Length	
1 millimetre (mm)	0.04 inch
1 metre (m)	39.4 inches
	3.28 feet
	1.1 yards
1 kilometre (km)	0.62 mile
Length/Time	
1 metre per second (m/s)	3.28 feet per second
1 kilometre per hour (km/h)	0.62 mile per hour
Area	
1 square millimetre (mm^2)	0.0016 square inch
1 square metre (m^2)	10.76 square feet
1 hectare (ha)	2.47 acres
1 square kilometre (km^2)	0.39 square mile
Volume	
1 cubic millimetre (mm^3)	0.000 061 cubic inch
1 cubic metre (m^3)	35.3 cubic feet
	1.3 cubic yards
1 millilitre (mL)	0.035 fluid ounce
1 litre (L)	0.22 gallon
Mass	
1 gram (g)	0.035 ounce
1 kilogram (kg)	2.2 pounds
1 tonne (t) (= 1000 kg) (also called	1.1 tons
metric ton)	2200 pounds
Mass/Volume	
1 kilogram per cubic metre (kg/m^3)	0.062 pound per cubic foot
Force	
1 newton (N)	0.225 pound-force
Stress	
1 megapascal (MPa) (= 1 N/mm^2)	145 pounds-force per square inch
Loading	
1 kilonewton per square metre (kN/m^2)	20.9 (~21) pounds-force per square foot
1 kilonewton per metre (kN/m) (= 1 N/mm)	68.5 pounds-force per foot
Moment	
1 kilonewton·metre (kN·m)	738 pound-force foot
Miscellaneous	
1 joule (J)	0.00095 British thermal unit (Btu)
	1 watt-second
1 watt (W)	0.0013 electric horsepower
1 degree Celsius (°C)	32 + 1.8(°C) degrees Fahrenheit (°F)

Table A.15 Imperial units conversion factors

Imperial Unit	Metric Equivalent
Length	
1 inch	25.4 mm
	0.0254 m
1 foot	0.305 m
1 yard	0.91 m
1 mile	1.6 km
Length/Time	
1 foot per second	0.305 m/s
1 mile per hour	1.6 km/h
Area	
1 square inch	645 mm^2
1 square foot	0.093 m^2
1 acre	0.4 ha
1 square mile	2.59 km^2
Volume	
1 cubic inch	16 387 (~16400) mm^3
1 cubic foot	0.028 m^3
1 cubic yard	0.765 m^3
1 fluid ounce	28.4 mL
1 gallon	4.55 L
Mass	
1 ounce	28.35 g
1 pound	0.45 kg
1 ton (2000 pounds)	0.91 t
1 pound	0.000 45 t
Mass/Volume	
1 pcf	16.1 kg/m^3
Force	
1 pound	4.45 N
Stress	
1 psi	0.007 MPa
Loading	
1 psf	0.048 kN/m^2
1 plf	0.0146 kN/m
Moment	
1 pound-force foot	0.001 36 kN·m
Miscellaneous	
1 British thermal unit (Btu)	1055 J
1 watt-second	1 J
1 horsepower	746 W
1 degree Fahrenheit (°F)	(°F − 32)/1.8 degrees Celsius (°C)

$$\boxed{\textbf{A.3}} \ \ \textsf{BEAM LOAD DIAGRAMS}$$

Table A.16 Beam diagrams

Simple beam - uniformly distributed load

$R = V$ .... $= \dfrac{wl}{2}$

V_x .... $= w\left(\dfrac{l}{2} - x\right)$

M_{max} (at centre) .... $= \dfrac{wl^2}{8}$

M_x .... $= \dfrac{wx}{2}(l - x)$

$\triangle_{max}$ (at centre) .... $= \dfrac{5wl^4}{384\ EI}$

$\triangle_x$ .... $= \dfrac{wx}{24\ EI}(l^3 - 2lx^2 + x^3)$

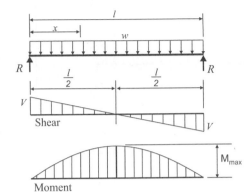

Simple beam - concentrated load at any point

$R_1 = V_1$ (max. when $a<b$) .... $= \dfrac{Pb}{l}$

$R_2 = V_2$ (max. when $a>b$) .... $= \dfrac{Pa}{l}$

M_{max} (at point of load) .... $= \dfrac{Pab}{l}$

M_x (when $x<a$) .... $= \dfrac{Pbx}{l}$

$\triangle_{max}\left(\text{at } x = \sqrt{\dfrac{a\,(a+2b)}{3}} \quad \text{when } a>b\right)$ .... $= \dfrac{Pab\,(a+2b)\sqrt{3a(a+2b)}}{27EI\,l}$

$\triangle_a$ (at point of load) .... $= \dfrac{Pa^2b^2}{3\ EI\ l}$

$\triangle_x$ (when $x<a$) .... $= \dfrac{Pbx}{6\ EI\ l}(l^2 - b^2 - x^2)$

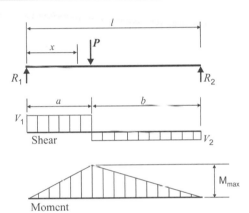

Cantilever beam - uniformly distributed load

$R = V$ .... $= wl$

V_x .... $= wx$

M_{max} (at fixed end) .... $= \dfrac{wl^2}{2}$

M_x .... $= \dfrac{wx^2}{2}$

$\triangle_{max}$ (at free end) .... $= \dfrac{wl^4}{8\ EI}$

$\triangle_x$ .... $= \dfrac{w}{24\ EI}(x^4 - 4l^3x + 3l^4)$

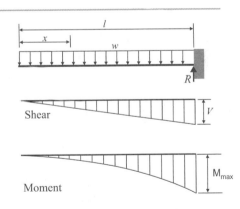

Cantilever beam - concentrated load at any point

$R = V$ .... $= P$

M_{max} (at fixed end) .... $= -Pb$

M_x (when $x>a$) .... $= P(x - a)$

$\triangle_{max}$ (at free end) .... $= \dfrac{Pb}{6\ EI}(3l - b)$

$\triangle_a$ (at point of load) .... $= \dfrac{Pb^3}{3\ EI}$

$\triangle_x$ (when $x<a$) .... $= \dfrac{Pb^2}{6\ EI}(3l - 3x - b)$

$\triangle_x$ (when $x>a$) .... $= \dfrac{P(l - x)^2}{6\ EI}(3b - l + x)$

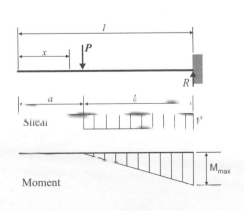

Table A.17   Bending moments and reactions for continuous beams under uniformly distributed loads

Moment = Coefficient · wl^2

Reaction = Coefficient · wl

w = uniform load per unit length

l = length of one span

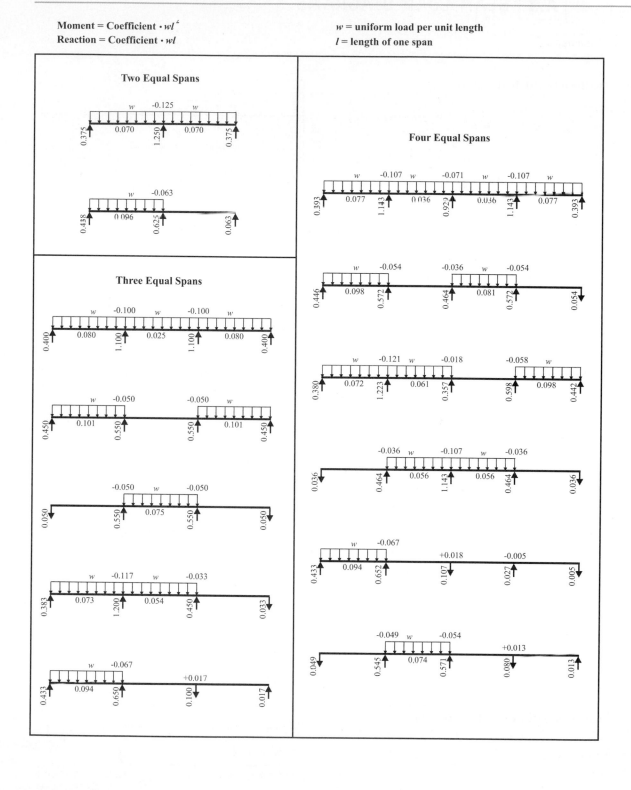

APPENDIX B
Notation

a = depth of equivalent rectangular stress block

A = effective tension area of concrete surrounding the flexural tension reinforcement (Chapter 4)

A = footing plan area (Chapter 12)

A_b = area of a reinforcing bar

A_c = gross cross-sectional area of a rectangular section (Chapter 7)

A_c = area of concrete core measured out-to-out of the spirals (Chapter 8)

A_{cs} = area of a concrete strip along an exposed face

A_d = total cross-sectional area of all dowels crossing the interface

A_e = total effective tension area

A_g = gross cross-sectional area

A_h = area of distributed horizontal reinforcement (Chapter 13)

A_{hmin} = minimum required area of distributed horizontal reinforcement

A_l = total area of longitudinal reinforcement resisting torsion (Chapter 7)

A_o = area enclosed by the centre-line of the walls in the equivalent tube section (Chapter 7)

A_{oh} = area enclosed by the centre-line of the transverse reinforcement (Chapter 7)

A_s = area of tension steel

$A_s{}'$ = area of compression steel

A_{sb} = balanced reinforcement area; integrity reinforcement (Chapter 12)

A_{sk} = area of skin reinforcement

A_{smin} = minimum required area of reinforcement

A_{sp} = volume of spiral reinforcement per unit length of column

$A_{sprovided}$ = provided area of reinforcement

$A_{srequired}$ = required area of reinforcement

A_{st} = total area of longitudinal column reinforcement

A_t = area of one leg of closed transverse reinforcement (Chapter 7)

A_{t+v} = area of transverse reinforcement to resist combined torsion and shear

A_{tr} = area of transformed uncracked section

A_v = area of shear reinforcement perpendicular to the axis of a member within a distance s

A_v = area of distributed vertical reinforcement (walls)

A_{vmin} = minimum required area of distributed vertical reinforcement

A_{vt} = total area of distributed vertical reinforcement

A_v' = area of shear reinforcement per leg (Chapter 7)

A_1 = loaded contact area

A_2 = maximum area of the portion of the supporting surface

b = width of a beam or column section

b_b = band width extending by a distance of $1.5h_s$ past the sides of the column in a two-way slab without beams

b = width of a footing

b_f = effective flange width

b_o = perimeter of the critical section

b_s = effective flange width

b_t = width of a beam tension zone (Chapter 5)

b_w = web width (T-beams) (Chapters 3 and 5)

b_w = minimum web width within depth d (Chapter 6)

b_1 = short-side dimension

b_2 = long-side dimension

B_r = wall bearing resistance

B_{rc} = column bearing resistance

B_{rd} = dowel bearing resistance

B_{rf} = footing bearing resistance

c = distance from extreme compression fibre to neutral axis (neutral axis depth)

C = compression force in concrete

C_m = factor relating actual moment diagram to an equivalent uniform moment diagram (columns)

C_r = resultant factored compression force in concrete

C_t = torsional property of a section

d = effective depth of the cross-section (distance from the extreme compression fibre to the centroid of the tension steel)

d_b = bar diameter

d_c = distance from the extreme tension fibre to the centre of the longitudinal bar located closest thereto

d_s = distance from the extreme tension fibre to the centroid of flexural reinforcement

d_v = effective shear depth

d' = distance from the top compression fibre to the centroid of the compression steel

D = dead load

D_c = outside diameter of the concrete core (columns)

e = eccentricity of axial load with regard to the centroid of the cross-section; distance of the centroidal axis of the critical section perimeter to the point where shear stresses are being computed

e_b = axial load eccentricity corresponding to the balanced condition (columns)

E_c = modulus of elasticity of concrete (CSA A23.3 Cl.8.6.2)

E_s = modulus of elasticity of steel (CSA A23.3 Cl.8.5.4.1)

f = flexural stress

f_b = maximum column bearing stress

f_c' = specified compressive strength of concrete

f_{cc}' = specified compressive strength for the column concrete

f_{cf}' = specified compressive strength for the footing concrete

f_r = modulus of rupture of concrete (CSA A23.3 Cl.8.6.4)

f_r^* = a reduced modulus of rupture, equal to one-half of the f_r value, to be used for deflection calculations in flexural members and two-way slabs (CSA A23.3 Clauses 9.8.2.3 and 13.2.7)

f_s = stress in tension steel

f_s' = stress in compression steel

f_{sy} – yield strength of spiral reinforcement (columns)

f_u = ultimate tensile strength of steel reinforcement

f_y = specified yield strength of steel reinforcement

F = coefficient of friction between base and slab on grade (Chapter 14)

F = friction force developed at the base of the footing (Chapter 13)

F_{rs} = resultant factored force in one reinforcement layer

G = modulus of rigidity

h = overall depth of a cross-section

h_f = flange (slab) thickness in T-beams

h_s = thickness of a two-way slab (Chapter 12); surcharge depth (Chapter 14)

h_{su} = equivalent soil height

h_u = unsupported wall height between horizontal supports

h_w = beam web depth below the slab soffit (Chapter 12); wall height (Chapter 13)

H_o = resultant soil pressure

H_{sk} = passive soil resistance provided by a shear key

H_{su} = surcharge resultant

I = moment of inertia of a cross-section

I_{cr} = moment of inertia of the cracked transformed section

I_e = effective moment of inertia

I_{eavg} – weighted average effective moment of inertia

I_{em} – effective moment of inertia at the midspan

I_{e1} = value of I_e at end one of a continuous beam span

I_{e2} = value of I_e at end two of a continuous beam span

I_{ec} = value of I_e at the continuous end

I_g = gross (uncracked) moment of inertia

$I_{g,tr}$ = moment of inertia for the gross transformed section

J = a property of the critical shear section analogous to the polar moment of inertia

k = effective length factor for compression members

k = coefficient that takes into account different end support conditions (Chapter 4)

k_1 = bar location factor

k_2 = coating factor

k_3 = concrete density factor

k_4 = bar size factor

K_o = coefficient of soil pressure at rest

K_t = torsion stiffness

l = span (centre-to-centre distance between the supports) of a beam or slab

l = footing length (Chapter 14)

l' = total length of footing under soil pressure

l_a = embedment length beyond the centre of the support (simply supported members) or beyond the inflection point (continuous members)

l_b = horizontal wall length

l_d = development length

l_{db} = basic compression development length

l_{dh} = development length for a hooked bar in tension

l_{hb} = basic development length for a hooked bar

l_n = clear span (clear distance between the supports)

l_p = lap length

l_{sp} = length of one spiral turn (columns)

l_u = unsupported column or wall length

l_w = horizontal length of a wall

l_1 = slab span in longitudinal direction

l_2 = slab span in transverse direction

l_{2a} = width of the design strip for a two-way slab

L = live load

m_t = distributed torque

M = bending moment

M_a = maximum bending moment in the member at the load stage at which the deflection is computed, or at any previous stage

M_c = magnified design moment in a slender column

M_{cr} = cracking moment

M_D = bending moment due to dead load only

M_{D+L} = bending moment due to combined dead and live load

M_f = factored bending moment

M_{fu} = unbalanced bending moment

M_L = bending moment due to live load only

M_m = net moment at the midspan of a continuous structure

M_r = factored moment resistance

M_o = maximum bending moment for a simply supported beam based on a clear span l_n; total factored moment (Chapter 12)

M_1 = smaller factored end moment (columns)

M_2 = larger factored end moment (columns)

MF = modification factor

n = modular ratio

n = number of stirrups crossing an inclined crack (Chapter 6)

N = axial tension force

N = number of bars (Chapter 4)

p = pitch (distance between successive spiral turns)

p_c = perimeter of a rectangular section (Chapter 7)

p_h = perimeter of the transverse reinforcement centre-line (Chapter 7)

p_o = perimeter of the centre-line of the equivalent tube section (Chapter 7)

p_o = soil pressure at the base of the wall (Chapter 13)

p_{su} = uniform soil pressure behind the wall

P = point load (beams, slabs)

P = axial load (columns)

P_c = critical axial load (columns)

P_{DL} = axial load due to specified dead load

P_f = factored axial load

P_{LL} = axial load due to specified live load

P_r = factored axial load resistance

P_{rb} = factored axial load resistance at the balanced condition

P_{ro} = factored axial load resistance of a reinforced concrete column

P_{rco} = factored axial load resistance provided by the concrete

P_{rmax} = maximum axial load resistance of a concentrically loaded nonprestressed concrete column

P_{rso} = factored axial load resistance provided by the longitudinal reinforcement

$P_{\parallel}$ = specified (service) axial load

q = shear flow (Chapter 7)

q = actual soil pressure (Chapter 14)

q_{all} = allowable soil bearing pressure

q_{allnet} = net allowable soil pressure

q_{avg} = average soil pressure

q_f = factored soil bearing pressure

q_M = soil pressure due to bending moment

q_{max} = maximum soil pressure

q_{min} = minimum soil pressure

q_P = soil pressure due to axial load

q_s = floor service load (Chapter 14)

r = radius of gyration for the column cross-section

R = soil pressure resultant

s = centre-to-centre spacing of reinforcing bars

s = clear distance between reinforcing bars (Chapter 5)

s = time-dependent factor for creep deflections (Chapter 4)

s_{max} = maximum permitted bar spacing

s_z = crack spacing parameter

s_{ze} = equivalent crack spacing parameter

S = elastic section modulus

S = snow load

t = wall thickness

T = external torque (Chapter 7)

T_{cr} = torsional cracking resistance

T_f = factored torsional moment

T_r = factored tension force in steel reinforcement

T_r = factored torsional resistance (Chapter 7)

u = average bond stress

v = shear stress

v_c = factored shear stress resistance provided by the concrete

v_M = shear stress due to unbalanced bending moment

v_r = factored shear stress resistance

v_s = factored steel shear stress

v_t = torsion shear stress

V = shear force

V_a = interface shear force

V_{ay} = vertical component of the interface shear force

V_c = concrete shear resistance

V_d = dowel-shear force

V_f = factored shear force

V_p = component of the effective prestressing force in the direction of applied shear

V_r = factored shear resistance

V_s = steel shear resistance

V_s' = tension force in a stirrup

w = uniformly distributed load

w_f = factored uniformly distributed load

w_s = total specified (service) uniform load

w_{su} = soil surcharge load

W = weight (gravity load)

W = wind load

x = distance from the side face to the centre of the skin reinforcement

x_d = drop panel overhang (dimension for the face of the column to the edge of drop panel)

$\bar{y}$ = neutral axis depth

y_t = distance from the centroid of the section to the extreme tension fibre

z = crack control factor

α = beam-to-slab stiffness ratio for a two-way slab with beams

α_f = angle between the shear friction reinforcement and the shear plane

α_s = parameter dependent on column location (Chapters 12 and 14)

α_t = coefficient of thermal expansion

α_1 = ratio of average stress in rectangular compression block to specified strength (CSA A23.3 Cl.10.1.7)

β = factor accounting for shear resistance of cracked concrete (Chapter 6); ratio of slab spans in long and short directions (Chapter 12)

β_c = ratio of long versus short dimension of the support (Chapters 12 and 14)

β_d = ratio of the maximum factored sustained axial load to the maximum factored axial load associated with the same load condition

β_1 = ratio of depth of rectangular compression block to depth of neutral axis (CSA A23.3 Cl.10.1.7)

γ_c = mass density of concrete

γ_f = a multiplier used to determine a fraction of the unbalanced moment transferred by flexure

γ_o = equivalent soil unit weight

γ_s = soil unit weight

γ_v = a multiplier used to determine a fraction of the unbalanaced moment transferred by shear

γ_w = unit weight of concrete

Δ = deflection

Δ_{all} = allowable deflection

Δ_D = dead load deflection

$\Delta_{D+L}=$ deflection due to combined dead and live loads

$\Delta_{DS}=$ deflection due to sustained dead load only

$\Delta_h=$ thickness of the drop panel underneath the slab

$\Delta_i=$ immediate deflection

$\Delta_L=$ live load deflection

$\Delta_{LS}=$ deflection due to the sustained portion of live load

$\Delta_t=$ total long-term deflection

$\Delta_{tD}=$ total deflection due to dead load only

$\varepsilon_c=$ strain in the extreme concrete compression fibre

$\varepsilon_{cmax}=$ maximum strain in the extreme concrete compression fibre at the ultimate

$\varepsilon_s=$ strain in the tension steel

$\varepsilon_s'=$ strain in the compression steel

$\varepsilon_{sh}=$ shrinkage strain

$\varepsilon_y=$ steel yield strain

$\zeta_s=$ long-term deflection factor (CSA A23.3 Cl.9.8.2.5)

$\zeta_{sD}=$ long-term deflection factor that depends on the dead load duration

$\zeta_{sL}=$ long-term deflection factor that depends on the live load duration

$\theta=$ angle formed by diagonal cracks with regard to the longitudinal axis of the member

$\theta_t=$ angle of twist

$\lambda=$ factor to account for concrete density (CSA A23.3 Cl.8.6.5)

$\mu=$ friction coefficient for the footing-to-soil interface

$\rho=$ reinforcement ratio (tension steel)

$\rho'=$ reinforcement ratio (compression steel)

$\rho_b=$ balanced reinforcement ratio

$\rho_s=$ spiral reinforcement ratio

$\rho_{sk}=$ skin reinforcement ratio

$\rho_t=$ longitudinal steel reinforcement ratio (columns)

$\rho_v=$ ratio of shear friction reinforcement

$\sigma=$ effective normal stress

$\sigma_1=$ maximum principal stress

$\sigma_2=$ minimum principal stress

$\phi_c=$ resistance factor for concrete (CSA A23.3 Cl.8.4.2)

$\phi_m=$ resistance factor for moment magnifier (columns)

$\phi_p=$ resistance factor for prestressing tendons (CSA A23.3 Cl.8.4.3)

$\phi_s=$ resistance factor for steel reinforcement and embedded steel anchors (CSA A23.3 Cl.8.4.3)

$\psi=$ column stiffness modification factor

References

ACI, *Practitioner's Guide to Slabs on Ground,* ACI Publication PP-4, American Concrete Institute, Detroit, 1998, 567 pp.

ACI 209R-92, Prediction of Creep, Shrinkage and Temperature Effects in Concrete Structures (Reapproved 1997), *ACI Manual of Concrete Practice,* American Concrete Institute, Detroit, 2012.

ACI 302.1R-04, Guide for Concrete Floor and Slab Construction, *ACI Manual of Concrete Practice,* American Concrete Institute, Detroit, 2012.

ACI 360R-92, Design of Slabs on Grade (Reapproved 1997), *ACI Manual of Concrete Practice,* American Concrete Institute, Detroit, 2012.

ACI 224.1R-93, Causes, Evaluation, and Repair of Cracks in Concrete Structures (Reapproved 1998), *ACI Manual of Concrete Practice,* American Concrete Institute, Detroit, 2012.

ACI 315, *ACI Detailing Manual — 2004,* ACI Publication SP-66 (04), American Concrete Institute, Detroit, 2004, 212 pp.

ACI 318-11, Building Code Requirements for Structural Concrete and Commentary, *American Concrete Institute,* 2011, 503 pp.

ACI 352.1R, Recommendations for Design of Slab-Column Connections in Monolithic Reinforced Concrete Structures, *ACI Structural Journal,* Vol. 85, No. 6, 1988, pp. 675–696.

ACI 421.1R-08, Guide to Shear Reinforcement for Slabs, Joint ASCE-ACI Committee 421, *American Concrete Institute,* 2008, 23 pp.

ACI 435, State-of-the-Art Report on Control of Two-Way Slab Deflections, *ACI Structural Journal,* Vol. 88, No. 4, 1991, pp. 501–514.

ACI, Essential Requirements for Reinforced Concrete Buildings, *American Concrete Institute,* 2002, 248 pp.

ACI 435, R-95, Control of Deflections in Concrete Structures (Reapproved 2000), *ACI Manual of Concrete Practice,* American Concrete Institute, Detroit, 2012.

ACI 438, Tentative Recommendations for the Design of Reinforced Concrete Members to Resist Torsion, *ACI Journal, Proceedings,* Vol. 66, No. 1, 1969, pp. 1–8.

ACI, *Concrete: A Pictorial Celebration,* American Concrete Institute, Detroit, 2004, 264 pp.

ACI/ASCC, *The Contractor's Guide to Quality Concrete Construction,* American Concrete Institute and American Society for Concrete Construction, Detroit, 1998, 107 pp.

ADAPT, ADAPT-Builder 2010 - Structural Concrete Design Suite, *ADAPT Corporation,* Redwood City, California, 2010.

Aga Khan Development Network, The Aga Khan Award for Architecture, 2004, Petronas Towers, Kuala Lumpur, Malaysia, www.akdn.org/agency/akaa/ninthcycle/imp_07m.htm, 2004.

Allen, E., *Fundamentals of Building Construction — Materials and Methods,* John Wiley & Sons, Inc., New York, 1999, 852 pp.

Anderson, B.G., Rigid Frame Failures, *Journal of the American Concrete Institute,* Vol. 28, No. 7, 1957, pp. 625–636.

ASCE-ACI Committee 426, The Shear Strength of Reinforced Concrete Members, *Journal of the Structural Division, Proceedings of the American Society of Civil Engineers,* Vol. 99, No. ST6, 1973, pp. 1091–1187.

ASCE-ACI Committee 445, Recent Approaches to Shear Design of Structural Concrete, *ASCE Journal of Structural Engineering,* Vol. 124, No. 12, 1998, pp. 1375–1417.

Bartlett, F.M., and MacGregor, J.G., Statistical Analysis of the Compressive Strength of Concrete in Structures, *ACI Materials Journal,* Vol. 96, No. 2, 1999, pp. 261–270.

Bathe, K.J., *Finite Element Procedures,* Seventh Edition, Prentice Hall, Upper Saddle River, New Jersey, 2003, 1037 pp.

Bedard, C., Composite Reinforcing Bars: Assessing Their Use in Construction, *Concrete International,* American Concrete Institute, Vol. 14, No. 1, 1992, pp. 55–59.

Branson, D.E., *Deformation of Concrete Structures,* McGraw-Hill, Inc., New York, 1977, 546 pp.

Bresler, B., and MacGregor, J.G., Review of Concrete Beams Failing in Shear, *Journal of the Structural Division, Proceedings of the American Society of Civil Engineers,* Vol. 93, No. ST1, 1967, pp. 343–372.

CAC, *Concrete Design Handbook,* Cement Association of Canada, Ottawa, 2005.

CAC, *Design and Control of Concrete Mixtures,* Seventh Canadian Edition, Cement Association of Canada, Ottawa, 2002, 355 pp.

Cardenas, A.E., and Magura, D.D., Strength of High-Rise Shear Walls — Rectangular Cross Section, *Response of Multistory Concrete Structures to Lateral Forces,* ACI Publication SP-36, American Concrete Institute, Detroit, 1973, pp. 119–150.

CCA, Canadian Construction Industry Forecast November 2003, Canadian Construction Association, http://www.cca-acc.com/factsheet/factsheet.html, 2003.

Coduto, D.P., *Foundation Design — Principles and Practices,* Second Edition, Prentice Hall, Upper Saddle River, New Jersey, 2001, 883 pp.

Coduto, D.P., *Geotechnical Engineering — Principles and Practices,* Prentice Hall, Upper Saddle River, New Jersey, 1999, 759 pp.

Collins, M.P., and Mitchell, D., *Prestressed Concrete Basics,* Canadian Prestressed Concrete Institute, Ottawa, 1987, 614 pp.

Cook, R.D., Malkus,D.S., Plesha,M.E., and Witt,R.J., *Concepts and Applications of Finite Element Analysis,* Fourth Edition, John Wiley & Sons, Inc., New York, 2001, 736 pp.

CPCA, *Concrete Design Handbook,* Canadian Portland Cement Association, Ottawa, 1995, 650 pp.

CPCI, *Design Manual — Precast and Prestressed Concrete,* Third Edition, Canadian Prestressed Concrete Institute, Ottawa, 1996, 400 pp.

CPCA, *Modern Multi-Storey Concrete Buildings,* Canadian Portland Cement Association, Ottawa, 1989, 31 pp.

CRSI, *Manual of Standard Practice,* 27th Edition, Concrete Reinforcing Steel Institute, Schaumburg, Illinois, 2001, 100 pp.

CRSI, *Placing Reinforcing Bars,* Seventh Edition, Concrete Reinforcing Steel Institute, Schaumburg, Illinois, 1997, 250 pp.

CSA A23.3-04, *Design of Concrete Structures,* including the 2005, 2007 and 2009 amendments, Canadian Standards Association, Rexdale, Ontario, 2004, 233 pp.

CSI, SAFE - Integrated Design of Flat Slabs, Foundation Mats and Spread Footings, *Computers and Structures, Inc.,* Berkeley, California.

DECON, *Decon Studrail Design Manual,* Decon Canada, Brampton, Ontario, 2009.

Domel, A.W., and Ghosh, S.K., *Concrete Floor Systems — Guide to Estimating and Economizing,* Portland Cement Association, Skokie, Illinois, 1990, 33 pp.

Drysdale, R.G., Hamid, A.A., and Baker, L.R., *Masonry Structures: Behaviour and Design,* Second Edition, The Masonry Society, Boulder, Colorado, 1999, 888 pp.

ENR, 1997 Malaysian Twin Towers Aspire to World's Tallest Ranking, *Engineering News Record,* McGraw-Hill, New York, Vol. 243, Iss. 22, 1999, p. 35.

Ferguson, P.M., Breen, J.E., and Jirsa, J.O., *Reinforced Concrete Fundamentals,* Fifth Edition, John Wiley & Sons, Inc., New York, 1988, 746 pp.

Fling, R.S., *Practical Design of Reinforced Concrete,* John Wiley & Sons, New York, 1987, 516 pp.

Gergely, P., and Lutz, L.A., Maximum Crack Width in Reinforced Concrete Flexural Members. Causes, Mechanism and Control of Cracking in Concrete, *ACI Publication SP-20,* American Concrete Institute, Detroit, 1973, pp. 87–117.

Ghali, A., and Hammill, N., Effectiveness of Shear Reinforcement in Slabs, *Concrete International,* American Concrete Institute, Vol. 14, No. 1,1992, pp. 60–65.

Graham, C.J., and Scanlon, A., Long-Time Multipliers for Estimating Two-Way Slab Deflections, *Journal of the American Concrete Institute,* Vol. 83, No. 6, 1986, pp. 899–908.

IBC, International Building Code 2000, International Code Council, Inc., Falls Church, Virginia, 2000.

ISIS, *Reinforcing Concrete Structures with Fibre Reinforced Polymers,* Design Manual No. 3, ISIS Canada, Winnipeg, 2001, 156 pp.

Kennedy, G. and Goodchild, C., *Practical Yield Line Design,* British Cement Association, Crowthorne, UK, 2003, 171 pp.

King, S. and Delatte, N.J., Collapse of 2000 Commonwealth Avenue: Punching Shear Case Study, *Journal of Performance of Constructed Facilities,* ASCE, Vol. 18, No. 1, 2004, pp. 54–61.

MacGregor, J.G., and Bartlett, F.M., *Reinforced Concrete: Mechanics and Design,* First Canadian Edition, Prentice-Hall Canada Inc., Scarborough, Ontario, 2000, 1042 pp.

Mitchell, D., and Collins, M.P., Detailing for Torsion, *ACI Journal, Proceedings,* Vol. 73, No. 9, 1976, pp. 506–511.

Nawy, E.G., *Reinforced Concrete: A Fundamental Approach,* Fifth Edition, Pearson Education, Inc., Upper Saddle River, New Jersey, 2003, 821 pp.

Neville, A.M., *Neville on Concrete — An Examination of Issues in Concrete Practice,* American Concrete Institute, Detroit, 2003, 450 pp.

Neville, A.M., *Properties of Concrete,* Fourth Edition, John Wiley & Sons, Inc., New York, 1977, 844 pp.

Neville, A.M., and Brooks, J.J., *Concrete Technology,* Educational Low-Priced Books Scheme, Longman Group UK Ltd., 1987, 438 pp.

Newman, A., *Structural Renovation of Buildings — Methods, Details, and Design Examples,* MacGraw-Hill, Inc., New York, 2001, 866 pp.

Nilson, A.H., Darwin, D., and Dolan, C.W., *Design of Concrete Structures,* Thirteenth Edition, McGraw-Hill Higher Education, New York, 2004, 779 pp.

NRC, *National Building Code of Canada 2010,* Canadian Commission for Building and Fire Codes, National Research Council, Ottawa, 2010, 571 pp.

Park, R. and Gamble, W., *Reinforced Concrete Slabs,* Second Edition, John Wiley and Sons Inc., New York, 2000, 736 p.

Park, R., and Paulay, T., *Reinforced Concrete Structures,* John Wiley & Sons, Inc., New York, 1975, 769 pp.

Paulay, T., and Priestley, M.J.N., *Seismic Design of Reinforced Concrete and Masonry Buildings,* Wiley Interscience, New York, 1992, 774 pp.

PCA, *Notes on ACI 318-08 Building Code Requirements for Structural Concrete with Design Applications,* Portland Cement Association, Skokie, Illinois, 2008, 950 pp.

PCA, *Concrete Floors on Ground,* Third Edition, Portland Cement Association, Skokie, Ilinois, EB075, 2001, 136 pp.

PCA, *Concrete Structural Floor Systems and More,* CD-ROM Publication, Portland Cement Association, Skokie, Illinois, 2000.

PCA, *Concrete Floor Systems—Guide to Estimating and Economizing,* Portland Cement Association, Skokie, Illinois, 2000, 45 pp.

PCA, *Tilt-Up Load-Bearing Walls,* Portland Cement Association, Skokie, Illinois, 1994, 28 pp.

PCA, *Building Movements and Joints,* Portland Cement Association, Skokie, Illinois, 1982, 64 pp.

Popov, E., *Engineering Mechanics of Solids,* Second Edition, Prentice Hall, Upper Saddle River, New Jersey, 1999, 864 pp.

Rangan, B.V., and Hall, A.S., Moment Redistribution in Flat Plate Floors, *Journal of the American Concrete Institute,* Vol. 81, No. 6, 1984, pp. 601–608.

Raphael, J.M., Tensile Strength of Concrete, *ACI Journal, Proceedings,* Vol. 81, No. 2, 1984, pp. 158–165.

RSIC, *Reinforcing Steel-Manual of Standard Practice,* Reinforcing Steel Institute of Canada, Richmond Hill, Ontario, 2004, 128 pp.

RSIO, *Scotia Plaza — Case History Report,* Reinforcing Steel Institute of Ontario, Willowdale, Ontario, Vol. 2, No. 3, 1988, 4 pp.

RSIO, *SkyDome — Case History Report,* Reinforcing Steel Institute of Ontario, Willowdale, Ontario, Vol. 3, No. 2, 1989, 4 pp.

Ringo, B.C., and Anderson, R.B., *Designing Floor Slabs on Grade,* Second Edition, The Aberdeen Group, Addison, Illinois, 1996, 266 pp.

Scanlon, A. and Suprenant, B.A., Estimating Two-Way Slab Deflections, *Concrete International,* American Concrete Institute, Vol. 33, No. 7, 2011, pp. 29–34.

Scanlon, A., Design and Construction of Two-Way Slabs for Deflection Control, The Design of Two-Way Slabs, *ACI Special Publication SP-183,* American Concrete Institute, Detroit, Michigan, 1999, pp.145–160.

Sheikh, S.A., and Uzumeri, S.M., Strength and Ductility of Tied Concrete Columns, *Journal of Structural Division,* ASCE, Vol. 106, ST5, 1980, pp. 1079–1102.

Sheikh, S.A., Design of Confining Reinforcement in Columns for Seismic Performance, S.M. Uzumeri Symposium — Behavior and Design of Concrete Structures for Seismic Performance, *ACI Special Publication SP-197,* American Concrete Institute, Detroit, Michigan, 2002, pp. 149–167.

Sozen, M.A., and Siess, C.P., Investigation of Multiple-Panel Reinforced Concrete Floor Slabs Design Methods - Their Evolution and Comparison, *Journal of the American Concrete Institute*, Vol. 60, No. 8, 1963, pp. 999–1028.

Stratta, J., *Manual of Seismic Design,* Prentice-Hall, Englewood Cliffs, New Jersey, 1987, 268 pp.

Van Bokhoven, M., Slab Bands, Lecture Notes for the Course E3 — Reinforced Concrete Design, Certificate Program in Structural Engineering, Vancouver Structural Engineers Group Society, Vancouver, 2003, 10 pp. (unpublished).

Zienkiewicz, O.C., Taylor, R.L., and Zhu, J.Z., *The Finite Element Method: Its Basis and Fundamentals*, Sixth Edition, Butterworth-Heinemann, United Kingdom, 2005, 752 pp.

Index